**Twelfth Edition**

# Wardlaw's Contemporary NUTRITION

WARDLAW'S CONTEMPORARY NUTRITION, TWELFTH EDITION

Published by McGraw Hill LLC, 1325 Avenue of the Americas, New York, NY 10121.
Copyright ©2022 by McGraw Hill LLC. All rights reserved. Printed in the United States of
America. Previous editions ©2019, 2016, 2013. No part of this publication may be reproduced
or distributed in any form or by any means, or stored in a database or retrieval system, without
the prior written consent of McGraw Hill LLC, including, but not limited to, in any network or
other electronic storage or transmission, or broadcast for distance learning.

Some ancillaries, including electronic and print components, may not be available to customers
outside the United States.

This book is printed on acid-free paper.

1 2 3 4 5 6 7 8 9 LWI 26 25 24 23 22 21

ISBN 978-1-260-69548-9 (bound edition)
MHID 1-260-69548-4 (bound edition)
ISBN 978-1-260-79004-7 (loose-leaf edition)
MHID 1-260-79004-5 (loose-leaf edition)

Portfolio Manager: *Lauren Vondra*
Product Developer: *Darlene Schueller*
Marketing Manager: *Tami Hodge*
Content Project Managers: *Ann Courtney, Tammy Juran*
Buyer: *Susan K. Culbertson*
Design: *David W. Hash*
Content Licensing Specialist: *Brianna Kirschbaum*
Cover Image: *Thawornmat/123RF*
Compositor: *MPS Limited*

All credits appearing on page are considered to be an extension of the copyright page.

### Library of Congress Cataloging-in-Publication Data

Smith, Anne M., 1955- author. | Collene, Angela, author. | Spees,
   Colleen K., author. | Wardlaw, Gordon M. Wardlaw's temporary nutrition.
   Wardlaw's contemporary nutrition/Anne Smith, Angela Collene,
   Colleen Spees.
   Contemporary nutrition
   Twelfth edition. | New York : McGraw-Hill LLC, [2022] |
   Includes index.
   LCCN 2020008067 (print) | LCCN 2020008068 (ebook) |
   ISBN 9781260695489 (hardcover) | ISBN 9781260790061 (ebook)
   LCSH: Nutrition—Textbooks.
   LCC QP141 .W378 2021 (print) | LCC QP141 (ebook) |
   DDC 612.3—dc23
   LC record available at https://lccn.loc.gov/2020008067
   LC ebook record available at https://lccn.loc.gov/2020008068

The Internet addresses, photo, or mention of a specific product listed in the text were accurate
at the time of publication. The inclusion of a website does not indicate an endorsement by the
authors or McGraw Hill LLC, and McGraw Hill LLC does not guarantee the accuracy of the
information presented at these sites.

mheducation.com/highered

# Brief Contents

Breakfast bowl: Alexis Joseph/McGraw-Hill Education; dietary fiber health food: Marilyn Barbone/Shutterstock; micronutrients/phytochemicals: Alexis Joseph/McGraw-Hill Education; swimmer: Erik Isakson/Blend Images LLC; portrait of a couple expecting a baby sitting on a park bench: Stockbyte/Getty Images

## Dear Students,

*Welcome to the fascinating world of nutrition! Because we all eat several times a day and the choices we make can have a dramatic influence on health, nutrition is our favorite area of science. At the same time, though, the science of nutrition can seem a bit confusing. One reason for all the confusion is that it seems like "good nutrition" is a moving target; different authorities have different ideas of how we should eat, and nutrition recommendations sometimes change! Second, there are so many choices. Did you know that the average supermarket carries about 40,000 food and beverage products? With so many food manufacturers vying for your attention, how can you identify a healthy product? Third, as a nation, we eat many of our meals and snacks away from home. When we eat foods someone else has prepared for us, we surrender control over what is in our food, where the food came from, and how much of it goes on our plates. Undoubtedly, you are interested in what you should be eating and how the food you eat affects you.*

*Wardlaw's Contemporary Nutrition is designed to accurately convey changing and seemingly conflicting messages to all kinds of students. Our students commonly have misconceptions about nutrition, and many have a limited background in biology or chemistry. We teach complex scientific concepts at a level that will enable you to apply the material to your own life.*

*This marks the twelfth edition of Wardlaw's Contemporary Nutrition. We have added several unique and timely features to this edition, most notable, information from the recently released Dietary Guidelines for Americans, 2020–2025. To help sort fact from fiction, we start each chapter with a new Fake or Fact feature to call out common misconceptions about foods and health, then we explain what science has to say. The Magnificent Microbiome feature explores the many ways the gut microbiota influences your health. Sustainability topics are highlighted in Sustainable Solutions found in every chapter. We have also redesigned our vitamin and mineral facts into "flash card" infographics that concisely summarize the food sources, main functions, recommendations, and other pertinent features of each vitamin and mineral. Finally, in the COVID Corner, we have addressed many ways the global outbreak both affects and is affected by food and nutrition.*

*We have written this book to help you make informed choices about the food you eat. We will take you through explanations of the nutrients in food and their relationship to health and make you aware of the multitude of other factors that drive food choices. To guide you, we refer to many reputable research studies, books, policies, and websites throughout the book. With this information at your fingertips, you will be well equipped to make your own informed choices about what and how much to eat. There is much to learn, so let's get started!*

*Anne Smith*
*Angela Collene*
*Colleen Spees*

## About the Cover

To expand upon our wildly successful *Farm to Fork* feature, we are excited to introduce the sustainable practice of hydroponics in this edition! As our cover image displays, this alternative agricultural practice involves growing plants in soilless and nutrient-rich root mediums in a variety of controlled environments, including gutters, pipes (as pictured growing vertically on the cover), and other space-saving and inexpensive containers. Learners will also enjoy our new *Sustainable Solutions* feature, created to promote environmentally friendly and applicable tips that students can use to improve their corner of the world.

Briana Zabala

The authors, Anne Smith, Colleen Spees, and Angela Collene, at the Hope Garden, a living research, teaching, and service-learning laboratory at The Ohio State University, academic home of the author team.

Twelfth Edition

# Wardlaw's Contemporary NUTRITION

**Anne M. Smith, PhD, RDN, LD**
Department of Human Sciences,
College of Education and Human Ecology
The Ohio State University

**Angela L. Collene, MS, RDN, LD**
Department of Human Sciences,
College of Education and Human Ecology
The Ohio State University

**Colleen K. Spees, PhD, MEd, RDN, LD, FAND**
Division of Medical Dietetics and Health Sciences,
College of Medicine
The Ohio State University

McGraw Hill

# About the Authors

Monty Soungpradith/Open Image Studio LLC

**ANNE M. SMITH, PhD, RDN, LD,** is an associate professor emeritus at The Ohio State University. She was the recipient of the Outstanding Teacher Award from the College of Human Ecology, the Outstanding Dietetic Educator Award from the Ohio Dietetic Association, the Outstanding Faculty Member Award from the Department of Human Nutrition, and the Distinguished Service Award from the College of Education and Human Ecology for her commitment to undergraduate education in nutrition. Dr. Smith's research in the area of vitamin and mineral metabolism has appeared in prominent nutrition journals, and she was awarded the Research Award from the Ohio Agricultural Research and Development Center. She is a member of the American Society for Nutrition and the Academy of Nutrition and Dietetics.

Tim Klontz, Klontz Photography

**ANGELA L. COLLENE, MS, RDN, LD,** began her career at her alma mater, The Ohio State University, as a research dietitian for studies related to diabetes and aging. Other professional experiences include community nutrition lecturing and counseling, owner of a personal chef business, and many diverse and rewarding science writing and editing projects. She is currently interested in the intersection between nutrition and mental health and—quite predictably for the mother of three little girls—maternal and child nutrition. Mrs. Collene currently teaches introductory nutrition and life cycle nutrition at The Ohio State University. She is a member of the Academy of Nutrition and Dietetics.

Ralphoto Studio

**COLLEEN K. SPEES, PhD, MEd, RDN, LD, FAND,** is an academic instructor and researcher at The Ohio State University College of Medicine. In addition to teaching Evidence-Based Practice and Nutritional Genomics, Dr. Spees's primary research focus involves conducting garden-based biobehavioral clinical interventions aimed at providing optimal nutrition for high-risk populations (see http://go.osu.edu/hope). In addition, Dr. Spees is the recipient of several national awards from the Academy of Nutrition and Dietetics, including the Distinguished Practice Award; Award for Excellence in Oncology Nutrition Research; Outstanding Dietetic Educator Award; Nutrition Informatics Video Challenge Teaching Award; and the Top Innovator in Education Teaching Award. In addition, she is the most recent recipient of the Early Professional Achievement Award from the Society for Nutriiton Education and Behavior and serves on the Scientific Panel for the American Cancer Society's Dietary and Physical Activity Guideline. Dr. Spees is also a recognized Fellow of the Academy of Nutrition and Dietetics.

# Acknowledgments

It is because of the tireless efforts of a cohesive team of talented professionals that we can bring you the twelfth edition of *Wardlaw's Contemporary Nutrition*. We consider ourselves massively blessed to work with the top-notch staff at McGraw Hill. We thank Lauren Vondra, Portfolio Manager, for her effective leadership of our team. We value the efforts of our Executive Marketing Manager, Tami Hodge, who consistently connects instructors with our work and brings us constructive feedback from the field. We are immensely grateful to our Product Developer, Darlene Schueller, who strategically led the day-to-day efforts of the entire editorial team. We especially appreciate her longevity over many editions of the *Contemporary Nutrition* products and value her keen eye for detail, strong work ethic, and organizational expertise. We are grateful to our Content Project Manager, Ann Courtney, and her staff for their patience and careful coordination of the numerous production efforts needed to create the very appealing and accurate twelfth edition. We thank Tammy Juran, our Assessment Content Project Manager, for her efforts and assistance. We appreciate the meticulous work of our copyeditor, Heath Lynn Silberfeld; proofreaders Kevin Campbell and David Heath; and our Content Licensing Specialist, Brianna Kirschbaum. We thank our Designer, David Hash, who ensured that every aspect of our work is visually appealing—not just on the printed page but also in a variety of digital formats. Finally, we are indebted to our colleagues, friends, and families for their constant encouragement, honest feedback, and shared passion for the science of nutrition.

## Reviewers

In the preparation of each edition, we have been guided by the collective wisdom of reviewers who are excellent teachers. They represent experience in community colleges, liberal arts colleges, institutions, and universities. We have followed their recommendations, while remaining true to our overriding goal of writing a readable, student-centered text.

Amy Bolinger
*Greenville Technical College*

Diane E. Carson
*California State University, Long Beach*

Susan Krug
*Butte College*

Sharon Lawless
*Allen Community College*

Jennifer L. Newton
*Greenville Technical College*

Susan Wakeman
*Greenville Technical College*

## *Ask the RDN* Contributors

We are grateful to our reputable and talented RDN colleagues who authored several new *Ask the RDN* features in this edition. It was exciting to share the spotlight and include their evidence-based expertise and applicable, down-to-earth recommendations. Many thanks to the following:

Leslie Bonci, MPH, RDN, CSSD, LDN

Zachari Breeding, MS, RDN, LDN, FAND

Karen K. Collins, MS, RDN, CDN, FAND

Alexis Joseph, MS, RDN, LD

Sally Kuzemchak, MS, RDN

Leah McGrath, RDN, LDN

Chris Vogliano, MS, RDN

## Student-Informed Reviews

We are very pleased to have been able to incorporate real student data points and input, derived from thousands of our SmartBook® users, to help guide our revision. SmartBook heat maps provided a quick visual snapshot of usage of portions of the text and the relative difficulty students experienced in mastering the content. With these data, we were able to hone not only our text content but also the SmartBook probes.

With the twelfth edition of *Wardlaw's Contemporary Nutrition*, we remember its founding author, Gordon M. Wardlaw. Dr. Wardlaw had a passion for the science of nutrition and the research that supports it and demonstrated an exceptional ability to translate scientific principles into practical knowledge. This skill is what made his book truly "contemporary." He was tireless when it came to staying current and relevant to a changing world. It has been a privilege for all of us to join Dr. Wardlaw as coauthors of this textbook. For Anne Smith, he was an extraordinary colleague, mentor, and friend. Angela Collene was blessed to have been one of his graduate students at The Ohio State University, where she first began to assist with revisions to his books. Colleen Spees was a student in Dr. Wardlaw's first nutrition class at The Ohio State University and now holds his previous tenured faculty position. Like so many other students, colleagues, and friends, we remember Dr. Wardlaw as a source of vast knowledge, good humor, and inspiration. The best way we know to honor our dear friend and mentor is to carry on his legacy of outstanding textbooks in introductory nutrition. *Wardlaw's Contemporary Nutrition* will continue to evolve and reflect current trends and breakthroughs in nutrition science, but Dr. Wardlaw's fingerprints will remain on every page.

# Instructors: Student Success Starts with You

## Tools to enhance your unique voice

Want to build your own course? No problem. Prefer to use our turnkey, prebuilt course? Easy. Want to make changes throughout the semester? Sure. And you'll save time with Connect's auto-grading too.

**65%**
**Less Time Grading**

Laptop: McGraw-Hill; Woman/dog: George Doyle/Getty Images

## Study made personal

Incorporate adaptive study resources like SmartBook® 2.0 into your course and help your students be better prepared in less time. Learn more about the powerful personalized learning experience available in SmartBook 2.0 at **www.mheducation.com/highered/connect/smartbook**

## Affordable solutions, added value

Make technology work for you with LMS integration for single sign-on access, mobile access to the digital textbook, and reports to quickly show you how each of your students is doing. And with our Inclusive Access program you can provide all these tools at a discount to your students. Ask your McGraw-Hill representative for more information.

Padlock: Jobalou/Getty Images

## Solutions for your challenges

A product isn't a solution. Real solutions are affordable, reliable, and come with training and ongoing support when you need it and how you want it. Visit **www.supportateverystep.com** for videos and resources both you and your students can use throughout the semester.

Checkmark: Jobalou/Getty Images

SUPPORT AT every step

# Students: Get Learning that Fits You

## Effective tools for efficient studying

Connect is designed to make you more productive with simple, flexible, intuitive tools that maximize your study time and meet your individual learning needs. Get learning that works for you with Connect.

## Study anytime, anywhere

Download the free ReadAnywhere app and access your online eBook or SmartBook 2.0 assignments when it's convenient, even if you're offline. And since the app automatically syncs with your eBook and SmartBook 2.0 assignments in Connect, all of your work is available every time you open it. Find out more at
**www.mheducation.com/readanywhere**

*"I really liked this app—it made it easy to study when you don't have your textbook in front of you."*

- Jordan Cunningham,
  Eastern Washington University

## Everything you need in one place

Your Connect course has everything you need—whether reading on your digital eBook or completing assignments for class, Connect makes it easy to get your work done.

## Learning for everyone

McGraw-Hill works directly with Accessibility Services Departments and faculty to meet the learning needs of all students. Please contact your Accessibility Services Office and ask them to email accessibility@mheducation.com, or visit
**www.mheducation.com/about/accessibility**
for more information.

# Connecting Teaching and Learning

vicushka/123RF

## Mc Graw Hill NutritionCalc Plus

NutritionCalc Plus is a **powerful dietary analysis tool** featuring more than 30,000 foods from the reliable and accurate ESHA Research nutrient database, which is comprised of data from the latest USDA Standard Reference database, manufacturer's data, restaurant data, and data from literature sources. NutritionCalc Plus allows users to track food and activities, and then analyze their choices with a robust selection of intuitive reports. The interface was updated to accommodate ADA requirements and modern mobile experience native to today's students.

## Virtual Labs

While the sciences are hands-on disciplines, instructors are now often being asked to deliver some of their lab components online, as full online replacements, supplements to prepare for in-person labs, or make-up labs.

These simulations help each student learn the practical and conceptual skills needed, then check for understanding and provide feedback. With adaptive pre-lab and post-lab assessment available, instructors can customize each assignment.

From the instructor's perspective, these simulations may be used in the lecture environment to help students visualize processes, such as digestion of starch and emulsification of lipids.

## *Dietary Analysis Case Studies* in Connect®

One of the challenges instructors face with teaching nutrition classes is having time to grade individual dietary analysis projects. To help overcome this challenge, assign auto-graded dietary analysis case studies. These tools require students to use NutritionCalc Plus to analyze dietary data, generate reports, and answer questions to apply their nutrition knowledge to real-world situations. These assignments were developed and reviewed by faculty who use such assignments in their own teaching. They are designed to be relevant, current, and interesting!

**Ava Ponce**
22 years old, Female, 5'6", 145 lbs
Weight gain/loss: 0 lbs/week
Activity level: Active

Pkchai/Shutterstock

McGraw Hill Create® is a self-service website that allows you to create customized course materials using McGraw Hill's comprehensive, cross-disciplinary content and digital products.

Tegrity in Connect is a tool that makes class time available 24/7 by automatically capturing every lecture. With a simple one-click start-and-stop process, you capture all computer screens and corresponding audio in a format that is easy to search, frame by frame. Students can replay any part of any class with easy-to-use, browser-based viewing on a PC, Mac, or other mobile device.

Educators know that the more students can see, hear, and experience class resources, the better they learn. Tegrity's unique search feature helps students efficiently find what they need, when they need it, across an entire semester of class recordings. Help turn your students' study time into learning moments immediately supported by your lecture.

# Assess My Diet

**Auto-graded personalized dietary analysis.** Students are using NutritionCalc Plus to analyze their own dietary patterns. But how can instructors integrate that information into a meaningful learning experience? With Assess My Diet, instructors can now assign auto-graded, personalized dietary analysis questions within Connect. These questions refresh their memory on the functions and food sources of each nutrient and prompt the students to evaluate their own eating behaviors. Students can compare their own nutrient intakes to current Dietary Reference Intakes and demonstrate their ability to perform calculations on their own data, such as percentage of calories from saturated fat. They can compare the nutrient density of their own food selections to see which of their food choices provides the most fiber or iron. A benefit of the Assess My Diet question bank is that it offers assignable content that is personalized to the students' data, yet it is still auto-graded. It **saves time** and keeps all assignments in one place.

## *Prep* for Nutrition

To help you **level-set your classroom**, we've created Prep for Nutrition. This question bank highlights a series of questions, including Basic Chemistry, Biology, Dietary Analysis, Mathematics, and Student Success, to give students a refresher on the skills needed to enter and be successful in their course! By having these foundational skills, you will feel more confident your students can begin class, ready to understand more complex concepts and topics. Prep for Nutrition is **course-wide for ALL nutrition titles** and can be found in the Question Bank dropdown within Connect.

## Campus

**McGraw Hill Campus** integrates all of your digital products from McGraw Hill with your school's Learning Management System for quick and easy access to best-in-class content and learning tools.

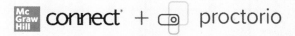 connect MASTER     connect + proctorio

### Writing Assignment

Available within McGraw Hill Connect® and McGraw Hill Connect® Master, the Writing Assignment tool delivers a learning experience to help students improve their written communication skills and conceptual understanding. As an instructor you can assign, monitor, grade, and provide feedback on writing more efficiently and effectively.

### Remote Proctoring & Browser-Locking Capabilities

New remote proctoring and browser-locking capabilities, hosted by Proctorio within Connect, provide control of the assessment environment by enabling security options and verifying the identity of the student.

Seamlessly integrated within Connect, these services allow instructors to control students' assessment experience by restricting browser activity, recording students' activity, and verifying students are doing their own work.

Instant and detailed reporting gives instructors an at-a-glance view of potential academic integrity concerns, thereby avoiding personal bias and supporting evidence-based claims.

# Connecting Students to Today's Nutrition

## Understanding Our Audience

We have written *Wardlaw's Contemporary Nutrition* while assuming that our students have a limited background in college-level biology, chemistry, or physiology. We have been careful to include the essential science foundation needed to adequately comprehend certain topics in nutrition, such as protein synthesis in Chapter 6. The science in this text has been presented in a simple, straightforward manner so that undergraduate students can master the material and apply it to their own lives. The Concept Maps and detailed, annotated figures bring complex topics into view for students from any major.

## Check Out the Functional Approach

An alternative presentation of the contents of this book is available as *Wardlaw's Contemporary Nutrition: A Functional Approach*. The difference, as shown in the side-by-side tables of contents below, is in Part Three. Instead of describing these nutrients in their traditional categories (e.g., water-soluble vitamins), we discuss them in groups based on their roles in fluid and electrolyte balance, body defenses, bone health, energy metabolism, blood health, and brain health. This format enables students to understand in more detail how these nutrients interact in food and in our bodies to support key functions that sustain our health.

# Featuring the Latest Guidelines and Research

Nutrition is a dynamic field. Ongoing research continually reshapes our knowledge of nutritional science. The twelfth edition has been carefully updated to reflect current scientific understanding, as well as the latest health and nutrition guidelines. For everyday dietary and activity planning, students will learn about the *Dietary Guidelines for Americans, MyPlate,* and *Physical Activity Guidelines for Americans*. In discussions about specific nutrition concerns, the most recent data and recommendations from the Academy of Nutrition and Dietetics, American Heart Association, American Diabetes Association, American Cancer Society, National Academy of Medicine, and American Psychological Association are included in this edition.

*Newsworthy Nutrition,* a feature in each chapter, highlights the use of the scientific method in recently published research studies that relate to the chapter topics. In addition, assignable questions in Connect take learning a step further by asking students to read primary literature and apply what they have learned.

## Newsworthy Nutrition

### Calcium supplements and cardiovascular disease risk

**INTRODUCTION:** Although several studies have shown a beneficial effect of calcium intake on cardiovascular effects, others have shown that calcium intake, especially from calcium supplements, is associated with increased mortality or the risk of heart attack and stroke. **OBJECTIVE:** The goal of this study was to explore the associations between calcium from dietary and supplemental intakes and cardiovascular disease (CVD) risks. **METHODS:** The study design was a systematic review and meta-analysis of 16 randomized controlled trials and 26 prospective cohort studies of dietary or supplemental intake of calcium, with or without vitamin D, and cardiovascular outcomes. Data was from PubMed, Cochrane Central, Scopus, and Web of Science, published up to March 2019. **RESULTS:** Results of cohort studies indicated that there were no associations between dietary calcium intakes and the risk of CVD, coronary heart disease (CHD), and stroke, for intakes ranging from 200 to 1500 mg/day. Results showed that calcium supplements, ranging from 1000 to 1400 mg/day, did not increase the risk of CVD and stroke; however, the risk of CHD increased by 20% and the risk of heart attack increased by 21% with the use of oral calcium supplements. **CONCLUSIONS:** Keeping in mind that very high calcium intakes are difficult if not impossible to achieve by dietary sources alone, the authors conclude that calcium intake from dietary sources does not increase the risk of CVD, and they suggest that adequate dietary calcium intakes are beneficial to cardiovascular protection. They conclude that calcium supplements might raise CHD risk, especially heart attack, and therefore the concerns regarding potential adverse cardiovascular risks are related to the use of calcium supplements.

Source: Yang C and others: The evidence and controversy between dietary calcium intake and calcium supplementation and the risk of cardiovascular disease: A systematic review and meta-analysis of cohort studies and randomized controlled trials. *Journal of the American College of Nutrition* 18:1, 2019. DOI:10.1080/07315724.2019.1649219.

In light of the COVID-19 pandemic, we have added the *COVID Corner* feature to highlight the many ways the global outbreak both affects and is affected by food and nutrition.

### COVID CORNER

In 2020, online shopping and curbside pick-up of groceries increased dramatically as a result of the COVID-19 pandemic. The pandemic also had devastating effects on the economy which severely impacted the food budget of many families.

*Ask the RDN* appears in every chapter to answer frequently asked questions we hear from our students and colleagues. For many topics, including plant-based eating, sustainability, and child nutrition, we have reached out to additional experts in their fields to answer questions. This feature will highlight the function and ability of the RDN to translate the latest scientific findings into easy-to-understand, practical, and applicable nutrition information.

## ASK THE RDN    Plant-Based Eating

*Dear RDN:* I am hearing more and more about the health benefits of a plant-based eating pattern. Can you give me some tips on replacing meat and dairy with high-quality plant proteins?

Regularly consuming foods high in plant proteins, such as legumes (including tofu and other soybean products), whole grains, nuts, and seeds can help prevent and reverse a slew of chronic conditions, including cancers, diabetes, and heart disease. Plant foods are packed with fiber and phytochemicals that support immunity, combat inflammation, and promote healthy bacteria in our gut. As an added bonus, plant proteins are far more affordable, sustainable, and lower in terms of environmental impact than animal proteins.

The good news is that you don't have to swear off meat forever to reap these benefits. Research suggests that following a flexitarian diet (increasing plant-based foods and reducing, but not eliminating, animal foods) yields similar health benefits, like reduced risk of heart disease and diabetes. Eating less meat doesn't mean you're going to suffer from protein deficiency any time soon, either. It is important to note that protein is found in *almost all* foods; and it is nearly impossible not to get enough protein if you're eating enough calories.

In order to transition to a more plant-centric dietary pattern, start small. Overturning your entire eating pattern in a day can be a bit overwhelming initially. Instead of jumping to extremes, pick two small changes to implement each week. First, it may be swapping cow's milk with unsweetened almond or coconut milk. The great thing about nondairy beverages is that they're lower in calories, and some pack more calcium and vitamin D than dairy milk. Make your morning oatmeal with almond milk and stir in a tablespoon of peanut butter and chia seeds for a protein boost. Chia seeds are a hydrating powerhouse, made up of 20% protein and 25% fiber while absorbing up to 30 times their weight in water.

Did you know ¼ cup of pumpkin seeds has 7 grams of protein? Or that hemp seeds are the highest-protein seed, with 3 grams of protein per tablespoon? Peanuts boast the most protein in the nut category, with 7 grams per serving. For a tasty chocolate-banana shake, blend together 1 large frozen overripe banana, 1 tablespoon peanut butter, 1 tablespoon hemp seeds, 1 tablespoon cocoa powder, a handful of spinach, and 1 cup unsweetened vanilla almond milk. Breakfast is served!

For lunches, try power bowls made with a base of wild rice or quinoa, which yield 6 and 8 grams of protein per 1 cup serving, respectively. Top with ½ cup beans, chickpeas, or baked tempeh, a handful of arugula, avocado slices, a drizzle of tahini and lemon juice, and a sprinkle of hemp seeds for a calcium boost. If you're craving a sandwich, stuff a sprouted wheat wrap with ½ cup of black beans, a sprinkle of corn, salsa, avocado, crunchy romaine, and hot sauce.

Consider pasta night. Instead of refined white pasta, try one of the many bean or lentil-based noodles on the market. You can find spaghetti, fusilli, and penne made from black beans, lentils, or chickpeas that all boast 13 grams of protein or more per one cup serving. Stick with 100% whole grain pasta, and you've still got 8 grams of protein and 25% of the Daily Value of fiber per one cup serving. On top of pasta, instead of parmesan, sprinkle nutritional yeast, a cheesy-tasting inactive yeast that's packed with protein and vitamin B-12. Drizzle a tablespoon of tahini and a tablespoon of hemp seeds on your green salad for another 6 grams of protein.

For stir-fry night, swap the chicken for high-protein edamame, which you can usually find in the freezer section of your grocery store. Soy is not only a complete plant protein, but it also has a high concentration of branched-chain amino acids, which are beneficial to athletic performance. Many stores sell marinated tofu (or try tempeh) that's delicious in stir-fry as well. For a tasty peanut sauce, whisk together ¼ cup natural peanut butter, ¼ cup almond milk, 4 teaspoons honey, and 4 teaspoons reduced sodium soy sauce.

When you're craving chili, swap the meat for a couple cans of kidney beans. Adding sautéed mushrooms to the mix will up the umami factor and add meatiness. Boost spices like oregano and chili powder for extra flavor. High-plant-protein dinner is served!

When it comes to baking, experiment with nut- and seed-based flours. Peruse your favorite food blogs for chocolate chip cookies or banana bread made with almond flour or coconut flour for a protein boost. These versions are lower in carbohydrates and super moist thanks to the healthy fat content.

There's no doubt about it—plant proteins are trending *for good*. Do your health and wallet a favor and hop on the bandwagon!

Enjoy your plant proteins,

*Alexis Joseph, MS, RD, LD*

Dietitian, Nutrition Consultant, Founder of Hummusapien: Co-Owner of Alchemy Brands

©Raul Velasco

The *Medicine Cabinet* feature presents information on common medications used to treat diseases that have a nutrition connection. These features highlight the ways medications can affect nutritional status, as well as ways food and nutrients can affect how medications work.

## Medicine Cabinet

**Proton pump inhibitors (PPIs)** are medications that inhibit the ability of gastric cells to secrete hydrogen ions and thus reduce acid production. Low doses of this class of medications may be available without a prescription. Because stomach acid is important for the absorption of vitamin B-12, prolonged use of PPIs could impair vitamin B-12 status.

Examples:

- Omeprazole (Prilosec®)
- Lansoprazole (Prevacid®)
- Rabeprazole (Aciphex®)
- Esomeprazole (Nexium®)

**H₂ blockers** impede the stimulating effect of histamine on acid-producing cells in the stomach.

Examples:

- Cimetidine (Tagamet®)
- Nizatidine (Axid®)
- Famotidine (Pepcid®)

# Connecting with a Personal Focus

## Valuing Our Food Supply

In this edition, we continue our *Farm to Fork* feature. Each chapter spotlights one or two fruits or vegetables and traces their path to our plates. Where does it grow? How do you select the most flavorful and nutritious foods? What are the best ways to store and prepare foods to maximize nutritional value?

**FARM to FORK**   **Carrots and Beets**

Alexis Joseph/McGraw-Hill Education

Carrots and beets are root crops with a wide range of nutrients, flavors, and health-promoting benefits. Red beet juice has also been touted to enhance athletic performance.

**Grow**
- The wild ancestor of carrots was purple, but most carrots grown in the United States are orange, a good indicator of the nutrients and other phytochemicals they contain, especially beta-carotene. Farmers are again producing purple carrot varieties, which are sweeter and higher in beta-carotene and the purple pigments, anthocyanins.
- Red beets are high in betalains, phytochemicals that may reduce the risk of cancer and other diseases.

## Applying Nutrition on a Personal Level

Throughout the twelfth edition, we reinforce the fact that each person responds differently to nutrients. To further convey the importance of applying nutrition to their personal lives, we include many examples of people and situations that resonate with college students. We also stress the importance of learning to intelligently sort through the seemingly endless range of nutrition messages to recognize reliable information and to sensibly apply it to their own lives. Our goal is to provide students the tools they need to eat healthfully and make informed nutrition decisions after they complete the class. Many of these features can be assigned and graded through Connect to help students learn and apply the information and engage with the text.

## Challenging Students to Think Critically

The pages of *Wardlaw's Contemporary Nutrition* contain numerous opportunities for students to learn more about themselves and their diet and to use their new knowledge of nutrition to improve their health. These pedagogical elements include *Ask the RDN, Case Studies, Nutrition and Your Health,* and *Newsworthy Nutrition.* Many of the thought-provoking topics highlighted in these features are expanded upon in the online resources found in Connect.

## Fake or Fact

To counter the nutrition misinformation we see from all directions, we open each chapter with *Fake or Fact*. This new feature highlights a relevant fake news topic followed by an explanation of the evidence that either supports or refutes the claim.

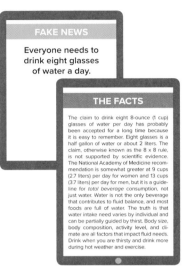

**FAKE NEWS**

Everyone needs to drink eight glasses of water a day.

**THE FACTS**

The claim to drink eight 8-ounce (1 cup) glasses of water per day has probably been accepted for a long time because it is easy to remember. Eight glasses is a half gallon of water or about 2 liters. The claim, otherwise known as the 8 x 8 rule, is not supported by scientific evidence. The National Academy of Medicine recommendation is somewhat greater at 9 cups (2.7 liters) per day for women and 13 cups (3.7 liters) per day for men, but it is a guideline for *total beverage* consumption, not just water. Water is not the only beverage that contributes to fluid balance, and most foods are full of water. The truth is that water intake need varies by individual and can be partially guided by thirst. Body size, body composition, activity level, and climate are all factors that impact fluid needs. Drink when you are thirsty and drink more during hot weather and exercise.

## Sustainable Solutions

In response to the drive toward sustainability—including environmental, health, social, and economic issues—we have introduced a new feature, *Sustainable Solutions*, in which we highlight sustainability topics related to each chapter's content.

More than half (54%) of the respondents to the *2019 Food and Health Survey* said that environmental sustainability is important when they purchase and consume foods. Being labeled as locally grown, sustainably sourced, non-GMO/not bioengineered, and organic are the primary characteristics used to identify sustainable foods, along with foods with recyclable or minimal packaging.

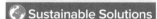

🌐 **Sustainable Solutions**

## Magnificent Microbiome

In light of the rapidly expanding body of research related to the gut microbiome and its impact on human health, we have added a new feature, *Magnificent Microbiome*, that explores relevant interactions between chapter topics and the gut microbiota.

magnificent

**Did you know that human breast milk is not sterile?** Along with human milk oligosaccharides, the unique collection of microorganisms from the mother's breast helps to colonize the infant's GI tract.

microbiome

# Connecting to Engaging Visuals

## Attractive, Accurate Artwork

Illustrations, photographs, infographics, and tables in the text were created to help students master complex scientific concepts.

- Many illustrations were redesigned or replaced to inspire student inquiry and comprehension and to promote interest and retention of information. Several new infographics have been created to present materials in a more attractive, contemporary style.

- In many figures, color-coding and directional arrows make it easier to follow events and reinforce interrelationships. Process descriptions appear in the body of the figures. This pairing of the action and an explanation walks students step-by-step through the process and increases teaching effectiveness.

- Throughout the chapters, every photo and caption has been chosen with the intention to spark critical thinking.

The final result is a striking visual program that holds readers' attention and supports comprehension and critical thinking. The attractive layout and design of this edition are clean, bright, and inviting. This creative presentation of the material is geared toward engaging today's visually oriented students.

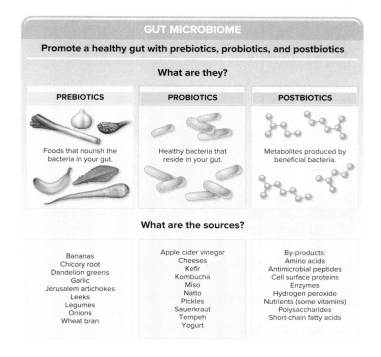

**FIGURE 9-23** ▶ Food sources, functions, and recommendations for calcium. The fill of the background color (none, 1/3, 2/3, or completely covered) within each food group on MyPlate indicates the average nutrient density for calcium in that group. The figure shows the calcium content of several foods in each food group. Overall, the richest sources of calcium are dairy foods (and dairy alternatives), legumes, green leafy vegetables, and fortified foods. mayonnaise: Iconotec/Alamy Stock Photo; orange juice: Sergei Vinogradov/seralexvi/123RF; broccoli: lynx/iconotec.com/Glow Images; two jars of yogurt, Foodcollection; slice of bread: Ingram Publishing/Age Fotostock; tofu: chengyuzheng/iStock/Getty Images; MyPlate: U.S. Department of Agriculture Sources: Office of Dietary Supplements, Dietary Supplements Fact Sheets, available from https://ods.od.nih.gov/factsheets/list-all; USDA FoodData Central, available from https://fdc.nal.usda.gov.

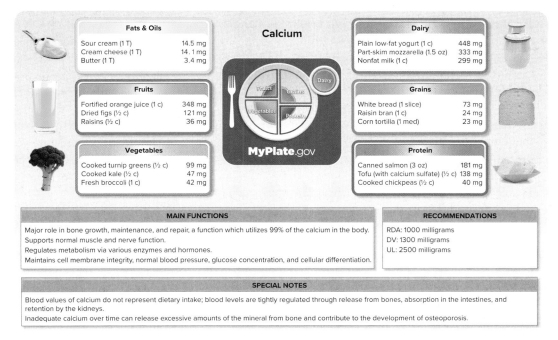

# Connecting with the Latest Updates

## Global Changes

- Chapters have been updated with information from the recently released *Dietary Guidelines for Americans, 2020–2025*.

- A new feature, *Sustainable Solutions*, highlights sustainability topics related to each chapter's content.

- Current research reveals many new ways in which the gut microbiome influences human health. *Magnificent Microbiome*, another new feature, explores relevant interactions between chapter topics and the gut microbiota.

- The *Newsworthy Nutrition* feature has been updated to include headings (Objectives, Methods, Results, Conclusions) within the research summary, and the type of study (e.g., systematic review, meta-analysis, cross-sectional studies, animal studies, and case reports) has been added in italics.

- In a new *COVID Corner* feature, we have addressed food and nutrition issues related to the novel coronavirus pandemic that dramatically impacted our lives beginning in 2020.

- Are you looking for *What Would You Choose, What the Dietitian Chose,* and *Rate Your Plate?* Although these features have been removed from the chapters, they still exist as assignable content within Connect.

- The *Ask the RDN* feature has been enhanced with new contributions from experts in the field as well as photos and short bios of the authors.

- Two new *Farm to Fork* features (about mushrooms and about pineapples) have been added.

- Several tables and figures have been transformed into eye-catching infographics that help students to map complex concepts.

- The terms "plant-focused" and "plant-forward" are now used along with "plant-based" to describe dietary patterns that emphasize plants.

- The content about alcohol has been moved from Chapter 16 to Chapter 1.

- The order of Chapters 12 and 13 have been switched so that topics in the *Protecting Our Food Supply* chapter (where our food comes from, how it is grown, and the safety of our food supply) are discussed before related issues in the *Global Nutrition* chapter.

## Chapter-by-Chapter Revisions

### Chapter 1: *Nutrition, Food Choices, and Health*

- A new *Fake or Fact* feature dispels myths about the effect of nutritional supplements on viral infections.

- Figures 1-2 and 1-3 have been updated with recent Loss-Adjusted Food Availability Data.

- The impact of the COVID-19 pandemic on food access and utilization and the ability to maintain a healthy diet is discussed in the new *COVID Corner* feature.

- The section *What Influences Your Food Choices* has been updated with results from the 2019 Food and Health Survey.

- The importance of environmental sustainability during the purchase and consumption of foods is discussed in the new *Sustainable Solutions* feature.

- Figure 1-4 has been updated with the most recent data on the 10 leading causes of death.

- Figure 1-5 is a new infographic illustrating the major functions of the various classes of nutrients.

- Figure 1-6 is a new infographic showing plant foods and their phytochemicals organized by their color families.

- Section 1.5 has been enhanced with information on the strength of scientific evidence and includes a new infographic (Figure 1-9: Hierarchy of Evidence) and discussion of additional types of studies (systematic review, meta-analysis, cross-sectional studies, and case reports).

- Figure 1-11 has been updated with the most recent statistics on the rates of obesity among adults.

- A new *Newsworthy Nutrition* feature, "Obesity-Related Cancers on the Rise in Young Adults," summarizes a 2019 publication from *The Lancet*.

- The *Nutrition and Your Health* topic has been changed to "Nutritional Implications of Alcohol Consumption" to facilitate the discussion of alcohol earlier in the course.

- Figure 1-12 has been expanded to include the approximate alcohol, carbohydrate, and calorie contents of alcoholic beverages along with the standard drink sizes that provide about 14 grams of alcohol.

- The effects of moderate versus heavy drinking are now illustrated in an infographic (Figure 1-13).

- *Further Readings* include 11 new references.

### Chapter 2: *Designing a Healthy Eating Pattern*

- Chapter 2 content has been streamlined and reduced to be more conversational and focused.

- A new Learning Objective (LO 2.9) has been added to reflect the addition of identifying foods and nutrition issues relevant to college students.

- The new *Fake or Fact* feature dispels the myth that healthy food is bland and boring.

- Figure 2-1 has been updated to clarify the comparisons of sugar-sweetened beverages with low-fat milk. Percentage contributions for protein, vitamin A, vitamin C, and riboflavin have been updated to reflect the current USDA nutrient values.

- Two new equation boxes were added to explain the concepts of nutrient and energy density.

- A new figure (2-2) has been added to show simple nutrient-dense swaps to common foods and beverages.

- The Healthy Eating Index (HEI) term has been defined in section 2-1 and added to the glossary.

- Energy density categories and examples of items in each group are now illustrated in a new infographic (Figure 2-3).

- A new infographic (Figure 2-4) highlights the foundational guidelines and recommendations of the *Dietary Guidelines*.

- Figure 2-5 has been added to show the percent of the U.S population (above age 1) who are at or below each dietary goal the *Dietary Guidelines*.

- A new *Newsworthy Nutrition*, "Impact of diet quality on health outcomes," presents a systemic literature review and meta-analysis published in the *Journal of the Academy of Nutrition and Dietetics*.

- A *Fake or Fact* feature is presented in Section 2.2 and highlights plant proteins.

- The *Farm to Fork* feature highlights citrus fruit, including oranges, tangelos, grapefruit, lemons, and limes.

- A new infographic (Figure 2-7) emphasizes Smart and Simple Food Swaps to improve overall dietary patterns.

- The new *Magnificent Microbiome* feature in Section 2.2 briefly describes the impact of dietary and physical activity patterns on gut microbes.

- Table 2-1 has been updated to reflect the most recent U.S. Department of Health and Human Services *Physical Activity Guidelines,* and the sections on Key Guidelines for Children and Adolescents and Older Adults have been moved to later chapters focusing on specific cohorts. The glossary terms "aerobic" and "anaerobic" now appear with other glossary terms in the margin.

- The image and calorie values for Figure 2-10 have been updated to illustrate that all dairy is not the same.

- A new infographic (Figure 2-11: "MyPlate Build-a-Meal Wizard") is designed to help students design meals aligning with the MyPlate food sources.

- A new critical thinking feature about solid fats has been added to the margin in Section 2.3.

- Figure 2-12 has been created to display simple swaps for breakfast, lunch, dinner, and snacking to promote nutrient-dense options that align with the *Dietary Guidelines*.

- A new glossary term, *hidden hunger,* has been added to Section 2.4 in the Undernutrition section.

- The *y*-axis for Figure 2-14 has been updated so the word *suboptimal* replaced *poor* when describing nutritional status and body functions.

- The title for Section 2.5 has been updated to "Measuring Nutritional Status." The content of Assessing Nutritional Status in this section has been streamlined to provide more

detail and clarification. Figure 2-15, the ABCDEs of Nutritional Assessment, is now presented in a colorful infographic.

- Section 2.6 has been shortened and renamed "Nutrient Recommendations" to align with USDA verbiage. The DRI Chronic Disease Risk Reduction Intake (CDRR) category has been added and defined in his section. Figure 2-16 has been updated to detail varying nutrient intake levels and their relationship to the DRIs.

- A new *Ask the RDN*, authored by dietitian Zach Breeding, discusses transgender nutrition-related issues and provides recommendations specific to transgender populations.

- Figure 2-17 has been added to Section 2.7 to help evaluate reliable nutrition information. In addition, the text has been updated to reflect tips for evaluating nutrition and health information.

- New information on nutrition and fitness apps has been added to Section 2.7 and includes a link to the Academy of Nutrition and Dietetics's science-based reviews of the most current and popular nutrition apps for smart phones and tablets.

- Table 2-3 has been shortened to emphasize key nutrient claims on food labels. The label terms *reduced* and *low cholesterol* have been added.

- Section 2.8 includes a new "Top 8 Food Allergens" in the margin box to reinforce common food-related allergens. Approved health claims have also been updated.

- Section 2.9 has been added to Chapter 2 (previously located in Chapter 1) and focuses on eating well as a student. The subheadings on food choices and weight control have been updated to reflect the current scientific evidence. This section also contains a section on alcohol and binge drinking. The *Case Study* presents a college student who is trying to improve his dietary patterns while maintaining a busy college schedule and tight budget.

- The new *Sustainable Solutions* feature reinforces the positive impact of a plant-focused dietary pattern.

- There are 24 new *Further Readings* in Chapter 2.

## Chapter 3: *The Human Body: A Nutrition Perspective*

- Two new *Fake or Fact* features highlight common misconceptions about peptic ulcers and diverticular disease.

- *Magnificent Microbiome* focuses on the hygiene hypothesis.

- Sections 3.1 and 3.2 have been reorganized to separate the discussion of the structure and function of the cell from the simple introduction to the concept of metabolism.

- Section 3.6 has been reworked to provide a more detailed introduction to the endocrine system, which comes up repeatedly throughout the text.

- Section 3.9 serves as an introduction to the concepts of the microbiota and microbiome, topics that are now woven throughout the text as part of the new *Magnificent Microbiome* features.

- Figure 3-3 has been updated to accentuate the organs of each organ system.

- Figures 3-1, 3-2, 3-4, 3-6, 3-7, 3-9, and 3-14 have been updated to an infographic style to more clearly explain key features of the body systems.

- *Newsworthy Nutrition* has been updated to present recent research on fecal transplant as a treatment for antibiotic-associated diarrhea.

- *Sustainable Solutions* presents the immune system as the human body's natural solution for sustainability!

- *COVID Corner* summarizes the evidence for and against probiotic supplements to boost immune function.

- The discussion of nutrition and genetics has been moved to Chapter 6, where it appears alongside pertinent information on gene expression.

- The discussion of diverticulosis and diverticulitis has been moved from Chapter 4 into the "Nutrition and Your Health Section" of Chapter 3.

- Chapter 3 includes 23 new *Further Readings.*

### Chapter 4: *Carbohydrates*

- The chapter opening image has been replaced with a colorful spread of carbohydrate-containing foods.

- The *Fake or Fact* feature dispels the myth that carbohydrates are fattening and should be avoided.

- Figure 4-1 has been updated to include the photosynthesis equation as a component of the figure.

- The Carbohydrate Concept Map is now presented in three separate figures that align with the text (Figure 4-2: Monosaccharides and Disaccharides; Figure 4-5: Polysaccharides; and Figure 4-7: Full Carbohydrate Concept Map). In addition, the chemical structures of each type of carbohydrate are now included in these figures to simplify these complex topics for improved comprehension.

- Section 4.2 contains a new Table 4-1 detailing the classifications of fiber (type, components, physiological effects, and major food sources).

- The *Magnificent Microbiome* feature in Section 4.3 details microbiota-accessible carbohydrates and their importance.

- The Food Sources of Carbohydrate bar graph has been replaced by a colorful flashcard-type infographic (Figure 4-8) that clearly displays the MyPlate food sources, common foods, and images of select foods that contain carbohydrates. All nutrient values have been updated to align with the updated USDA values.

- A new Smart Beverage Choices infographic (Figure 4-9) was added to Section 4.3 to help students recognize healthier beverage choices throughout the day, including while purchasing drinks at smoothie stands and coffee shops.

- Figure 4-10 shows the average intake of grain subgroups compared to the *Dietary Guidelines* for adults.

- Figure 4-11, the Whole Grains Council stamps for use on grain products has been updated from two to three stamps with details for each plus the minimum requirements for whole grains per serving. Table 4-3 has been updated with 13 additional whole and ancient grains, with gluten-free grains clearly identified.

- The new *Sustainable Solutions* feature discusses the systems-level collaborations needed to encourage improved dietary patterns.

- In Section 4.3, energy and sports drinks are added to emphasize these as growing sources of added sugars in the United States.

- Approved by the FDA in 2019, allulose is now included under alternative sweeteners.

- A *Fake or Fact* feature dispels the myth that artificial sweeteners cause cancer.

- Table 4-5 includes updates and additional information about artificial sweeteners, including the newest approved sweetener (allulose), brand names, acceptable daily intakes (ADI), and the amount needed to reach the ADI.

- Figure 4-12 has been updated as a colorful infographic to improve the visual appeal and highlight the main organs involved in carbohydrate digestion and absorption.

- A new infographic (Figure 4-14) is included in Section 4.6 to visually present the carbohydrate-specific recommendations from the *Dietary Guidelines.*

- A new *Newsworthy Nutrition*, "Americans are decreasing consumption of sugar-sweetened beverages," is a cross-sectional study based on research published in the *Journal of Obesity.*

- A new figure (4-16) was created to visually display the top sources and average intakes of added sugars in the population.

- Figure 4-15, has been added to help explain the concept of insulin resistance.

- Figure 4-18 has been updated to present the characteristics of metabolic syndrome as a simple and appealing infographic.

- There are 12 new *Further Readings* in Chapter 4.

### Chapter 5: *Lipids*

- A *Fake or Fact* feature weighs the pros and cons of eggs as part of a heart-healthy dietary pattern.

- Section 5.1 has been revised to improve students' understanding of the chemistry of lipids, which is often daunting for non-majors.

- Figure 5-6 and the section on food sources of lipids have been revised to present the food sources of lipids in a more visually appealing manner and give the students practical tips for choosing healthy fats.

- The discussion of *trans* fats has been revised in response to the ban on *trans* fats that went into effect in 2018. *Trans* fats have been removed from Figure 5-7.

- Food sources of omega-3 fatty acids are now presented as Table 5-1.

- *Sustainable Solutions* educates students about aquaculture as a means of providing a safe and sustainable supply of seafood to meet growing global demand.

- Figure 5-10 has been updated with an infographic style to more clearly present the complex topic of digestion and absorption of lipids.

- *COVID Corner* explores the evidence regarding omega-3 fatty acid supplements for COVID-19 prevention or treatment.

- Age-specific guidance on saturated fat intake has been updated to reflect the most recent edition of the *Dietary Guidelines for Americans*.

- *Newsworthy Nutrition* presents a randomized, controlled trial conducted with college students to explore the health benefits of nuts.

- The content of Section 5.7 has been simplified to focus less on medical treatments and more on dietary strategies to lower risk for cardiovascular disease. A new Figure 5-15 contrasts heart attack symptoms for men and women. In addition, a new Figure 5-17 summarizes risk factors for cardiovascular disease.

- *Magnificent Microbiome* explores the impact of postbiotics on heart health.

- Chapter 5 has been updated with 16 new *Further Readings.*

## Chapter 6: *Proteins*

- The chapter opens with a *Fake or Fact* feature dispelling the myth that it is impossible to eat too much protein.

- Figure 6-2 now depicts the metabolic reaction that is compromised in the genetic disorder phenylketonuria.

- In Section 6.2 the discussion of protein organization appears before protein synthesis.

- Figure 6-7 has been revised to present the food sources of proteins more visually.

- Figure 6-8 has been enhanced to include photos of possible plant group combinations in which proteins complement each other. Figure 6-9 is a new illustration showing that plant-based proteins are found in a variety of food sources.

- The following topics are discussed in the *COVID Corner* feature: (1) the impact of the pandemic on the livestock and meat industry; (2) the potential beneficial effect of the Mediterranean diet against infections such as COVID-19; and (3) protein recommendations for patients who are ill with infections such as COVID-19.

- A *Sustainable Solution* feature in Section 6.3 discusses the potential of the Mediterranean diet to support sustainable food production.

- In Section 6.4, Figure 6-12 has been redesigned for easier understanding of protein digestion and absorption, and the new *Magnificent Microbiome* feature discusses how different types of protein sources can impact gut microbiota and obesity in both positive and negative ways.

- In Section 6.5, Figure 6-15: Protein Concept Map has been updated to illustrate the regulatory functions of protein.

- Section 6.6 includes a new infographic, Figure 6-17, illustrating the protein-specific recommendations from the new *Dietary Guidelines for Americans, 2020–2025*, as well as a new *Ask the RDN* entitled "Active Eating Advice" written by Leslie Bonci, a sports dietitian for collegiate, professional, Olympic, and recreational athletes as well as performance artists.

- The discussion of nutrition and genetics now appears as Section 6.8, and the difference between nutrigenetics and nutrigenomics is now illustrated in an infographic, Figure 6-19.

- Different types of vegetarians and the protein sources they consume are now illustrated in an infographic, Figure 6-21.

- Table 6-3: Food Plan for Vegetarians is now color-coded based on MyPlate.

- Section 6.9 includes a new *Ask the RDN* on plant-based eating, written by Alexis Joseph, dietitian, nutrition consultant, founder of Hummusapien and co-owner of Alchemy Brands. She provides tips on replacing meat and dairy with high-quality plant proteins.

- Eight new articles are included in the *Further Readings*.

## Chapter 7: Energy Balance

- The chapter title has been updated to focus on energy balance versus weight control. This aligns with our updated text presenting a more weight-neutral approach and also including the evidence-based recommendations for those choosing to reduce their body weight.

- The chapter opening image has been updated to display the key components of health—nutrients including water, fruits, and vegetables, as well as hand weights to remind us of physical activity.

- The *Fake or Fact* feature dispels the myth that you should avoid all carbohydrates to lose weight.

- Figure 7-1 has been updated to reflect the most current data related to adult obesity trends in the U.S.

- Figure 7-2 has been updated to be more visually appealing.

- A new infographic (Figure 7-5) presents the components of energy output, including the contributions of basal metabolic rate, voluntary activity, and the thermic effect of food on energy output. In addition, the concepts of nonexercise-activity thermogenesis (NEAT) and exercise-induced thermogenesis (EAT) are introduced.

- The new *Sustainable Solutions* feature in Section 7.1 includes the definition of a sustainable dietary pattern from the Food and Agriculture Organization of the United Nations (FAO).

- Figure 7-6 has been revised to simplify the concepts. The indirect calorimetry image has also been updated in Section 7.2.

- Section 7.3 has been updated with more current images of the BodPod, skinfold and waist circumference measurements, and DXA. This section also now includes information about personal body fat scales with body composition measures that sync with mobile devices for easier tracking. There is also new information on utilizing NutritionCalc Plus in Connect to estimate energy needs.

- Table 7-1 has been updated with all WHO categories and subcategories of body mass indices. Table 7-2 has been replaced by a new infographic (Figure 7-11) visually displaying the health issues related to excess body fatness.

- New text has been added related to obesity posing a threat to national security, and *Further Readings* contains a reference for those with additional interest in this topic.

- The weight-loss triad, Figure 7-20, has been updated to emphasize the interrelated components of successful weight loss and regular physical activity.

- The *Magnificent Microbiome* feature in Section 7.6 describes the connection between gut microbes and appetite.
- A *Fake or Fact* feature dispels the myth that calories alone make you feel full.
- A new infographic, Figure 7-21, visually presents the concept of energy density with various food choices.
- Table 7-3 has been updated with the most recent USDA nutrient analysis food values. The images for Figure 7-22 have also been updated to promote reading of Nutrition Facts labels as part of a sustainable weight management plan.
- The new *Magnificent Microbiome* feature, in Section 7.7, details the impact of regular physical activity and protein intake on the gut microbiome.
- Table 7-4 has been condensed to include popular physical activities and estimated calorie costs associated with each activity, and a new Concept Check question has been added.
- The concept of intuitive eating is now included in this chapter and reinforces respect for one's body and eating without judgment. In addition, a new infographic (Figure 7-25) presents weight bias, its impact, and examples of people-first language.
- Table 7-6 presents the medications approved for obesity treatment, including information about the FDA approval of Plenity and discontinuation of Belviq XR (lorcaserin) given its potential link with cancer.
- Cryolipolysis, patented under the name CoolSculpting™, is described in Section 7.8.
- The *Ask the RDN* features the current evidence on intermittent fasting.
- Table 7-8 and related text have been updated to include new details of the strengths, limitations, and possible side effects of bariatric surgeries performed on youth.
- Section 7.10 includes new information related to specific tips for gaining healthy body weight that should be of interest to students struggling with underweight or those attempting to gain weight.
- Presentation of the five best overall diet plans has been updated in the text and presented in Table 7-8.
- *Further Readings* have been updated with 21 new references.

## Chapter 8: *Vitamins and Phytochemicals*

- This chapter title now includes both vitamins and phytochemicals, and the opening image is now a beautiful photograph of fresh produce rich in vitamins and phytochemicals. The learning objectives have been updated to streamline the vitamin content throughout the chapter.
- The new *Fake or Fact* feature dispels the myth that vitamin-fortified foods are a healthier option to include in your dietary pattern.
- Table 8-2: Food Sources of Some Phytochemicals Under Study has been moved from Chapter 1 to Chapter 8.
- Figure 8-2 has been revised to add more visual appeal and enhance learning.

- An updated section on phytochemicals has been included to reflect current research and updates on these health-promoting plant-based chemicals. A new image and text have been added on reusable produce bags that are environmentally friendly.
- Examples of zoochemicals and phytochemicals were added to provide further clarification of these compounds. Carotenoid information has been updated and expanded in Section 8.1.
- The previous food source figures have been replaced throughout the chapter by vitamin-specific flashcard infographics that enhance readability and concisely summarize key information on the food sources of each vitamin, their main functions, relevant recommendations, and other pertinent notes.
- The *Sustainable Solutions* feature discusses biofortification to increase food crop nutrient density.
- New information from the *Dietary Guidelines for Americans, 2020–25*, about vitamin D has been included.
- Factors that impair vitamin D status have been condensed and moved to a margin box. An additional *Farm to Fork* feature focuses on mushrooms, an excellent source of vitamins and minerals and a vitamin D precursor.
- Fig. 8-15 has been condensed to simply how the B vitamins are essential components of many coenzymes involved in energy metabolism.
- The *Magnificent Microbiome* feature discusses endogenous vitamin production that occurs within our GI tract.
- Figure 8-25: Vitamin B-12 Digestion and Absorption has been expanded and reformatted to improve comprehension.
- Figure 8-30 has been condensed to highlight the dietary supplement industry's projected sales through 2024.
- New *Ask the RDN* by contributor Karen Collins discusses dietary supplements and current evidence surrounding their use. The Top 5 Dietary Supplements has been updated in Section 8.17.
- *COVID Corner* presents evidence and recommendations to support a healthy immune system to fight disease and enviromental insults.
- The "Nutrition and Your Health" section on nutrition and cancer contains the 2020 American Cancer Society's evidence-based guideline for diet and physical activity.
- A *Fake or Fact* feature in Section 8.18 dispels the myth that sugar feeds cancer.
- New *Concept Checks* and other questions have been added throughout to stimulate critical thinking.
- *Further Readings* have been updated to include 10 new articles.

## Chapter 9: *Water and Minerals*

- There are two new *Fake or Fact* features in Chapter 9 that explain (1) the truth behind whether everyone needs to drink eight glasses of water a day and (2) if we should feed a fever.
- Food source figures for water and each mineral have been updated to be more attractive and include sources, as well as functions and recommendations, for water and minerals.

- Alternatives to bottled water use are discussed in the *Sustainable Solutions* feature.

- Section 9.2: The Water Balancing Act includes a new infographic (Figure 9-10) explaining the guidelines for the safe use of water bottles, and a new section entitled "Is Sparkling or Seltzer Water Harmful for Teeth?"

- This chapter presents the latest Dietary Reference Intakes for sodium and potassium released by the National Academies of Sciences in March 2019, including the new category Chronic Disease Risk Reduction Intake (CDRR).

- New information from the *Dietary Guidelines for Americans, 2020–25*, about sodium, potassium, and calcium has been included.

- New content on types of salt is included in the section "Table Salt, Kosher Salt, Sea Salt—Which One Is Best?"

- An additional *Farm to Fork* discusses bananas and ties into the section on "Getting Enough Potassium."

- The factors that enhance and inhibit calcium absorption are now shown in a new infographic (Figure 9-22) that includes a diagram of the gastrointestinal tract.

- There are two new *Newsworthy Nutrition* features summarizing research on (1) calcium supplements and cardiovascular disease risk and (2) the effect of dietary potassium and arterial stiffness.

- The interaction between iron and the intestinal microbiota is discussed in the *Magnificent Microbiome* feature.

- Section 9.17: Minerals and Hypertension highlights the new 2017 guidelines for the prevention, detection, evaluation, and management of high blood pressure in adults. Figure 9-33 has been updated with new effects of lifestyle changes on blood pressure. Also included are the goals and strategies from the recently released *The Surgeon General's Call to Action to Control Hypertension*.

- *COVID Corner* examines the following topics: (1) the suggestion to drink water frequently to help prevent COVID-19 infection; (2) the effect of nutrient supplements, including zinc, on the risk or severity of viral infections; (3) links between selenium status and COVID-19; and (4) the identification of hypertension as one of the strongest risk factors for suffering a severe case of COVID-19 infection.

- Fifteen new *Further Readings* are included.

### Chapter 10: *Nutrition: Fitness and Sports*

- The Chapter 10 opening image has been updated, and physical activity recommendations throughout the chapter align with the revised *Physical Activity Guidelines for Americans* published by the U.S. Department of Health and Human Services.

- The *Fake or Fact* feature dispels the myth that higher protein intake alone equates to greater muscle mass.

- Figure 10-1 has been updated to present the latest research related to the physical and mental health benefits of physical activity.

- A *Magnificent Microbiome* feature in Section 10.2 presents ongoing research linking physical activity to health-promoting bacteria in the gut.

- Tips for measuring your heart rate and calculating your target heart rate have been simplified, and details on using the "try to talk test" have been added to a margin note.

- Table 10-1 has been updated to align with the new physical activity guidelines.

- The Rating of Perceived Exertion (RPE) scale in Figure 10-3 has been updated to reflect the common Borg Scale of Perceived Exertion.

- A new *Newsworthy Nutrition* feature presents an animal study evaluating the impact of branched-chain amino acids on health and lifespan.

- The *Farm to Fork* feature presents carrots and beets, root crops often consumed by athletes before, during, and after competition.

- Figure 10-11 and a new text section detailing relative energy deficiency in sport (RED-S) have been added to Section 10.4.

- A *Fake or Fact* feature dispels the myth that thirst is a valid indicator of hydration status, and a *Sustainable Solutions* feature discusses the waste associated with water intake when individuals use single-use containers.

- Tables 10-4 and 10-8 have been updated with new energy drink, bar, gel, and chew products and their nutrient contents.

- Tables 10-7 and 10-9 have been edited to reflect the latest USDA nutrient composition values and emphasize more plant-based meal options for sports nutrition.

- A new *Ask the RDN* related to enhancing metabolism presents recommendations to modify lifestyle behaviors to promote weight maintenance. An emphasis on plant-based protein sources is included and emphasized throughout the chapter.

- Based upon the evidence, the title for Section 10.5 has been changed to Recommendations for Endurance, Strength, and Power Athletes.

- The lists of commonly used sports supplements and illegal substances have been expanded and updated in Tables 10-11 and 10-12.

- Seven new *Further Readings* have been added.

### Chapter 11: *Eating Disorders*

- A new *Fake or Fact* feature debunks some widely held myths about the origins of eating disorders.

- A new *Ask the RDN* by Alexis Joseph explains the connections among energy availability, regular menstrual function, and bone health.

- The *Magnificent Microbiome* feature explores the role of the gut microbiota in recovery from eating disorders.

- A new Figure 11-2 illustrates Russell's sign.

- *COVID Corner* examines the impact of stay-at-home orders (to prevent the spread of COVID-19) on the eating disorders community.

- Section 11.6 now includes content about the combined problems of disordered eating and binge drinking, a growing problem on college campuses.

- Twenty-two new articles have been incorporated into the list of *Further Readings*.

## Chapter 12: Protecting Our Food Supply

- The *Fake or Fact* feature debunks the myth that organic products have increased nutritional value compared to conventional products.

- The chapter now begins with the section on Food Production Choices to discuss where our food comes from and how it is produced. Topics include organic food production, biotechnology, and sustainable agriculture.

- A new *Ask the RDN* feature by guest contributor and supermarket dietitian Leah McGrath answers questions about whether the positive effects of organic and non-GMO foods justify their added expense.

- The *Sustainable Solutions* features explain the sustainable practice of hydroponics, and what it means to be a LOHAS (lifestyles of health and sustainability) consumer.

- A new *Magnificent Microbiome* feature highlights the potential impact of fermented foods on gut health and the effect of the gut microbiome on norovirus infection.

- Figure 12-3 is a map from the CDC that graphically depicts the reported foodborne disease outbreaks across the U.S., and Figure 12-5 shows the step-by-step process by which reported outbreaks prompt an investigation and, once confirmed, are shared with the public.

- All foodborne illness prevalence data and food import data have been updated throughout the chapter.

- Information on recent examples, onset, symptoms, and sources of foodborne illness outbreaks has been updated and moved to Tables 12-5, 12-6, and 12-7 on the bacterial, viral, and parasitic causes of foodborne illnesses.

- The *COVID Corner* feature reviews the inability to contract COVID-19 from food or packaging.

- Twenty-one new resources are cited in the chapter and included in the *Further Readings*.

## Chapter 13: Global Nutrition

- The chapter opening image has been updated to reinforce global dietary patterns.

- The new *Fake or Fact* feature dispels the myth that there is not enough food on the planet to meet our dietary needs.

- The most recent statistics on both domestic and global poverty and hunger from the Food and Agriculture Organization (FAO) of the United Nations and the U.S. Census Bureau are included.

- A new infographic, Figure 13-4, has been added to describe how food insecurity can contribute to wasting,

stunting, and micronutrient deficiencies. Figure 13-5 has been revised to a simpler format.

- A new margin note details the top 10 countries at highest risk for a humanitarian crisis.

- The new *Fake or Fact* feature in Section 13.1 dispels the myth that malnutrition equates to hunger.

- Updates of the characteristics of users and impact of the federally subsidized nutrition programs, including SNAP, WIC, and Senior Nutrition Services that supply food for people in the U.S., has been expanded and updated in Table 13-2.

- A new infographic, Figure 13-6, visually displays the impact of malnutrition on health.

- This chapter's *COVID Corner* about the impact of the COVID-19 pandemic on food security levels and malnutrition is presented in Section 13.2.

- The title of Section 13.3 has been updated to Malnutrition in the Developing World to emphasize both under- and overnutrition influencing global malnutrition.

- The *Sustainable Solutions* feature describes the history of agricultural practices dating back 1.5 million years.

- Figure 13-7 presents the complex factors that contribute to malnutrition in developing countries.

- Figure 13-8 presents the data pertaining to global access to safe water aligning with the Sustainable Development Goal for clean water and sanitation.

- The *Farm to Fork* in this section highlights pineapples, often seen as a global staple providing key nutrients and an economic advantage for developing countries.

- A new *Newsworthy Nutrition* presents a case-control study assessing maternal depression and acute malnutrition in children.

- A new Sustainable Development Goal (SDG) progress report includes SDG updates on key goals and metrics.

- The new *Magnificent Microbiome* feature details the microbiome of a vibrant living community—our soil!

- Figure 13-10 is a new infographic, adapted from the FAO, describing the many facets of sustainable intensification in agriculture.

- A new *Ask the RDN,* authored by Chris Vogliano, presents the issues related to global food waste and details the EPA's Food Recovery Hierarchy to counter food waste.

- *Further Readings* includes 28 new references.

## Chapter 14: Nutrition During Pregnancy and Breastfeeding

- This chapter includes three new *Fake or Fact* features. One examines the science behind pregnancy cravings. The second cautions against tight sodium restrictions during pregnancy. A third explains the lack of evidence to support maternal dietary restrictions for the prevention of food allergies.

- In Section 14.1, the latest evidence-based guidelines on treatment of polycystic ovary syndrome are discussed.

- In Section 14.4, the discussion of macronutrient needs of pregnant women has been separated into subsections to improve readability. New tables have been added to summarize the energy needs of pregnant and lactating women. Also, a separate subsection on fluid needs during pregnancy has been added.

- The graphics in Figure 14-8 have been updated to match the latest version of MyPlate.

- In Section 14.7, we have included information about the Baby-Friendly Hospital Initiative and defined a new term: *human milk oligosaccharides.* We have also added a new section on expressing and storing human milk.

- The *Magnificent Microbiome* feature highlights the probiotic and prebiotic properties of human milk.

- The *Sustainable Solutions* feature touts the role of breast-feeding in meeting several Sustainable Development Goals of the United Nations.

- *COVID Corner* reviews safety issues regarding breastfeeding during the pandemic.

- *Further Readings* includes 56 updated resources.

## Chapter 15: *Nutrition from Infancy Through Adolescence*

- A *Fake or Fact* feature reinforces the latest recommendations from the American Academy of Nutrition and Dietetics, American Academy of Pediatric Dentistry, American Academy of Pediatrics, and American Heart Association about healthy beverage choices. Information about selecting healthy beverages has been incorporated into sections 15.3, 15.4, 15.5, and 15.6.

- Figures 15-1 and 15-2 have been annotated to show students how to plot growth and BMI on growth charts.

- We have added several new margin notes throughout the chapter to summarize nutrient recommendations at various stages of childhood.

- In Section 15.2, information about iron supplementation has been updated to reflect the recommendations of the American Academy of Pediatrics.

- In Section 15.3, a new subsection has been added about baby-led introduction to solid foods and a new term has been defined: *responsive feeding.* Tables 15-5, 15-6, and 15-7 have been updated to match the newest *Dietary Guidelines'* healthy U.S.-style dietary patterns for various stages of childhood.

- Section 15.4 now includes the most recent recommendations from the *Physical Activity Guidelines for Americans.*

- A new *Ask the RDN* from child nutrition expert Sally Kuzemchak provides practical advice to help caregivers cope with picky eating. A new term, *neophobia*, is defined.

- The discussion about links between nutrition and autism spectrum disorder is updated.

- The *Magnificent Microbiome* feature reviews the role of the gut microbiota in the development and possible treatment of autism spectrum disorder.

- The *Sustainable Solutions* feature proposes partnerships between schools and local farmers as a way to promote child health and economic growth in communities.

- *COVID Corner* highlights the impact of the pandemic on nutrition security for families with children.

- An updated *Newsworthy Nutrition* feature describes recent research on the connections between dietary patterns and acne.

- Section 15.7 has been updated with new statistics about food allergies and intolerances and the latest recommendations from the *Dietary Guidelines* and the American Academy of Pediatrics for preventing food allergies.

- Fifty-nine new references have been added to the list of *Further Readings.*

## Chapter 16: *Nutrition During Adulthood*

- All demographic and prevalence data on aging have been updated in Section 16.1. In addition, the entire chapter has been revised to emphasize healthy aging and active living during this stage of the life cycle.

- A *Fake or Fact* feature shows that longevity is not just genetic, but largely results from a lifetime of healthy behaviors.

- A new figure from the World Health Organization (WHO) (Fig. 16-2) on aging and health emphasizes key influences of aging and recommendations for healthy aging. A new box also highlights healthy aging.

- A new *Newsworthy Nutrition* feature highlights research assessing the impact of dietary patterns on telomeres, a biomarker of aging.

- Figure 16-4 has been updated to match the most recent version of the Nutrition Screening Initiative's Nutrition Checklist for Older Adults, which utilizes the DETERMINE mnemonic.

- The dietary reference intakes for sodium have been updated in Section 16.2, including Figures 16-5 and 16-6.

- Figure 16-7 has been revised to provide a more comprehensive array of strategies to cope with the physiological changes of aging as adapted by the National Institute on Aging.

- *COVID Corner* reviews the connections between malnutrition and COVID-related morbidity and mortality among older adults.

- A new Table 16-2 summarizes the key physical activity guidelines for adults and older adults.

- Table 16-3 has been updated with the latest research on popular herbal remedies used by American adults.

- A new Table 16-4 has been added to illustrate the healthy U.S.-style dietary patterns for older adults from the latest edition of the *Dietary Guidelines*.

- A box feature introduces the rising popularity of home-delivered meal kits and the most popular services available.

- Section 16.7 summarizes the role of nutrition for brain health throughout the life span with an emphasis on prevention of neurodegenerative diseases. *Newsworthy Nutrition* touts the benefits of a Mediterranean diet for prevention of Alzheimer's disease.

- *Further Readings* have been updated with 43 new references.

# Contents

## Part One: Nutrition: A Key to Health

# Part Two: Energy Nutrients and Energy Balance

# Chapter 7  Energy Balance  236

# Part Three: **Vitamins, Minerals, and Water**

Micronutrients/phytochemicals: Alexis Joseph/McGraw-Hill Education; lemon water: Susan Rusnak

# Part Four: **Nutrition: Beyond the Nutrients**

   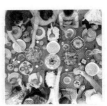

# Part Five: **Nutrition: A Focus on Life Stages**

Portrait of a couple expecting a baby sitting on a park bench: Stockbyte/Getty Images; Germany, Bavaria, Munich, mother and daughter preparing salad, son eating: Westend61/Getty Images; Ingrid Adams: Mary Jon Ludy/McGraw-Hill Education

# Nutrition, Food Choices, and Health

# Student Learning Outcomes

**Chapter 1 is designed to allow you to:**

**1.1** Describe how our food choices are affected by the flavor, texture, and appearance of food; eating habits and food availability; advertising; dining out; convenience; cost; sustainability; nutrition; and hunger and appetite.

**1.2** Identify dietary and lifestyle factors that contribute to the 10 leading causes of death in North America.

**1.3** Define *nutrition, carbohydrate, protein, lipid, alcohol, vitamin, mineral, water, phytochemical, kilocalorie,* and *fiber.*

**1.4** Determine the total calories (kcal) of a food or meal using the weight and calorie content of the energy-yielding nutrients, convert English to metric units, and calculate percentages, such as percent of calories from fat in a meal.

**1.5** Understand the scientific method as it is used in forming and testing hypotheses in the field of nutrition; determine the strength of scientific evidence related to nutrition.

**1.6** List the major characteristics of the North American dietary pattern, the food habits that often need improvement, and the aims of the "Nutrition and Healthy Eating" objectives of *Healthy People 2030.*

**1.7** Describe a basic plan for health promotion and disease prevention and what to expect from good nutrition and a healthy lifestyle.

**1.8** Compare benefits of moderate alcohol use to the risks of alcohol abuse.

## FAKE NEWS

### Some nutritional supplements protect you from viruses.

## THE FACTS

Although some nutrients such as vitamin C and zinc are important for immune health, no nutritional supplement will cure or prevent viral infections, including the novel COVID-19 coronavirus. Unproven claims to this effect can give a false sense of protection and may lead to toxicity. Until evidence supporting use is published, patients and healthcare providers should not rely on dietary supplements to prevent or cure viral infections including COVID-19.

Research has clearly shown that a lifestyle that includes a dietary pattern rich in fruits, vegetables, whole grains, and lean meat or plant protein, coupled with regular exercise, can enhance our current quality of life and keep us healthy for many years to come. Unfortunately, this healthy lifestyle is not always easy to follow. When it comes to "nutrition," it is clear that some of our eating patterns are out of balance with our metabolism, physiology, and physical activity level.

We begin this chapter with some questions. What influences your daily food choices? How important are factors such as taste, appearance, convenience, or cost? Is nutrition one of the factors you consider? Are your food choices influencing your quality of life and long-term health? By maintaining a healthy eating pattern, we can bring the goal of a long, healthy life within reach. This is the primary theme of this chapter and throughout this book.

The ultimate goal of this book is to help you find the best path to good nutrition. The information presented is based on emerging science that is translated into everyday actions that improve health. We begin each chapter with the Fake or Fact feature covering a relevant "Fake News" topic followed by the truth or "Facts" based on the state of reliable evidence. After completion of your nutrition course, you should understand the science behind the food choices you make and recommend to others. We call this achievement of making food choices that are healthy for you "nutrition literacy."

# 1.1 Why Do You Choose the Food You Eat?

In your lifetime, you will eat about 88,000 meals and 75 tons of food. Many factors—some internal, some external—influence our food choices. This chapter begins with a discussion of these factors and ends with a conversation about alcohol consumption and its relationship to our health. In between, we examine the powerful effect of eating patterns in determining overall health and take a close look at the general classes of nutrients—as well as the calories—supplied by the food we eat. We also discuss the major characteristics of North American eating patterns, the food habits that often need improvement, and where we stand on the "Nutrition and Weight Status" objectives of *Healthy People 2020*. A review of the scientific process behind nutrition recommendations is also included, along with an introduction to our *Farm to Fork, Newsworthy Nutrition, Magificent Microbiome, Sustainable Solutions,* and *Ask the RDN* features that appear in each chapter.

Understanding what drives us to eat and affects our food choices will help you understand the complexity of factors that influence eating, especially the effects of our routines and food advertising (Fig. 1-1). You can then appreciate why foods may have different meanings to different people and thus why others' food habits and preferences may differ from yours.

## WHAT INFLUENCES YOUR FOOD CHOICES?

Food means much more to us than nourishment—it reflects much of what we think about ourselves. The Bureau of Labor Statistics estimated that in 2018, Americans spent 72 minutes a day eating and drinking.[1] If we live to be 80 years old, that will add up to 4 years of eating and drinking. Overall, our daily food choices stem from a complicated mix of biological and social influences (see Fig. 1-1). The 2019 Food and Health Survey

**FIGURE 1-1 ▶** Food choices are affected by many factors. Which have the greatest impact on your food choices?
Florian Franke/Purestock/SuperStock

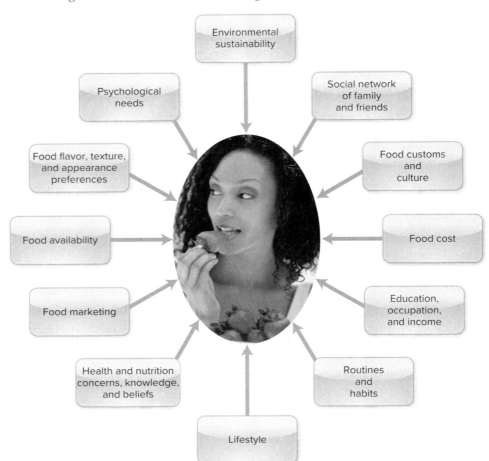

found that, from a list of six factors, 86% said that taste influenced their food purchases, followed by brand name (69%), price (68%), healthfulness (62%), convenience (57%), and environmental sustainability (27%).[2] Let's examine some of the key reasons why we choose what we eat and then ask your instructor about the *Rate Your Plate* activities: "Examine Your Eating Habits More Closely" and "Observe the Supermarket Explosion" in Connect.

*Flavor, texture, and appearance* are the most important factors determining our food choices. Creating more flavorful foods that are both healthy and profitable is a major focus of the food industry. The challenge is to combine the "taste" of the foods we prefer with the best nutrition and health characteristics. The good news is that chefs and "food bloggers" are dedicating themselves to creating nutritious food that is also delicious.

*Early influences* related to various people, places, and events have a continuing impact on our food choices. Many food customs, including ethnic eating patterns, begin as we are introduced to foods during childhood. Parents can lay a strong foundation knowing that exposure to food choices during early childhood is important in influencing later health behaviors. Developing healthy patterns during childhood will help ensure healthy preferences and choices when we are teenagers and adults.

*Eating habits, food availability, and convenience* strongly influence choices. Recent Food Availability and Consumption data from the United States Department of Agriculture (USDA) show that Americans consume more than the recommended amounts from the meat, eggs, and nuts group and the grains group. Potatoes and tomatoes are the most commonly consumed vegetables (Fig. 1-2), with French fries and pizza contributing to their popularity. Oranges and apples are the most commonly consumed fruits, but they are consumed mostly in juice form. Fluid milk and cheese, especially mozzarella cheese, comprise most of dairy consumption; fluid milk consumption has shown a big decline, while cheese consumption has doubled (Fig. 1-3).

*Marketing and advertising* are major tools for capturing the food interest of the consumer. Consumers have more food choices than ever, and the food industry in the United States spends billions on advertising. Some of this advertising is helpful, as it promotes the importance of healthy food components such as calcium and fiber. However, the food industry also advertises highly sweetened cereals, cookies, snacks, and soft drinks because they bring in the greatest profits. Studies have shown

▲ Exposing children to growing, preparing, and eating healthy food options, such as this veggie pasta, will lay a strong foundation for healthful choices throughout life. ©Jeff Laubert

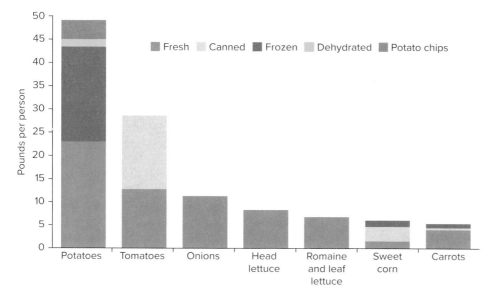

**Most Commonly Consumed Vegetables Among U.S. Consumers, 2017**

**FIGURE 1-2** ◄ According to food availability data, the favorite vegetables of Americans are potatoes and tomatoes. In 2017, Americans consumed 49.2 pounds per person of potatoes and 28.7 pounds of tomatoes, with 56% as canned tomatoes. French fries and pizza contribute to the high consumption of these two vegetables. Onions were the third highest consumed vegetable at 11.3 pounds per person. USDA, Economic Research Service, Loss Adjusted Food Availability Data, 2017

**FIGURE 1-3** ▶ According to food availability data from the USDA, Americans consumed a similar amount of dairy products (1.5 cup-equivalents of dairy products per person per day) in 1977 and in 2017. This is half the recommended amount for a 2000-calorie diet. Although the overall quantity is the same, the types of dairy products consumed have changed. Fluid milk consumption decreased from 0.9 to 0.5 cup per person per day, while cheese consumption has doubled.
USDA, Economic Research Service, Loss Adjusted Food Availability Data, 2017

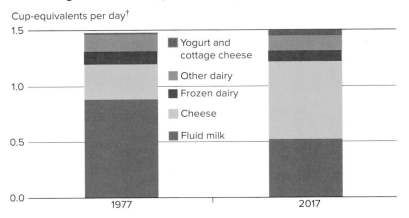

**Average U.S. Consumption of Dairy Products, 1977 and 2017**

Cup-equivalents per day[†]

Legend: Yogurt and cottage cheese; Other dairy; Frozen dairy; Cheese; Fluid milk

[†]Based on a 2000-calorie-per-day diet. One cup-equivalent for dairy is: 1 cup milk or yogurt; 1½ ounces natural cheese or 2 ounces of processed cheese or ½ cup shredded cheese; 1 cup frozen yogurt or 1½ cups ice cream; 2 cups cottage cheese. Loss-adjusted food availability data are proxies for consumption. "Other dairy" includes evaporated milk, condensed milk, dry milk products, half and half, and eggnog.

Jacob is majoring in nutrition and is well aware of the importance of a healthy dietary pattern. He has recently been analyzing his meals and is confused. He notices that he eats a great deal of high-fat foods, such as peanut butter, cheese, chips, ice cream, and chocolate, and few fruits, vegetables, and whole grains. He also has become hooked on his daily cappuccino with lots of whipped cream. **What three factors may be influencing Jacob's food choices? What advice would you give him on how to have his dietary pattern match his needs?** Ingram Publishing

an association between TV advertising of foods and drinks and childhood obesity in the United States. A recent study found that preschoolers who were not hungry and who watched a show embedded with food advertisements consumed more calories from snacks than those who saw nonfood advertisements.[3] These findings suggest that exposure to food advertisements may encourage eating behaviors that promote obesity in the very young. Concern for the negative effects of advertising and marketing on the dietary patterns and health of children led the Council of Better Business Bureaus to establish the Children's Food and Beverage Advertising Initiative. Participants are 18 of the largest food and beverage companies that represent about 80% of child-directed television food advertising. The initiative is designed to shift the mix of foods advertised to children to encourage healthier dietary choices and healthy lifestyles.[4] Research also indicates that mass media influences the onset of eating disorders through its depiction of extremely thin models as stereotypes of attractive bodies.

*Restaurant dining* plays a significant role in our food choices. Restaurant food is often calorie dense, in large portions, and of poorer nutritional quality compared to foods made at home. Fast-food and pizza restaurant menus typically emphasize meat, cheese, fried foods, and carbonated beverages. In response to recent consumer demands, restaurants have placed healthier options on their menus, and many are listing nutritional content there as well. Posting of the calorie content of restaurant items is now mandated by law. The law requires chain restaurants with 20 or more locations to post the calorie content of their offerings on menus or menu boards with other nutritional information available upon request. The intent of the law is to provide consumers with clear and consistent nutrition information so that they can make informed and healthful choices. While research is showing that using calorie menu labels is associated with purchasing fewer calories, a study of customers at McDonald's restaurants found that there are significant socioeconomic disparities among customers who notice and use calorie menu labels.[5] These findings suggest that targeted education campaigns would help improve the use of menu labeling across all sociodemographic groups.

*Time and convenience* have become significant influences affecting food choices. Current lifestyles limit the time available for food preparation. A recent study in Seattle found that working adults who placed a higher priority on convenience than on home-cooked meals spent the least amount of time cooking. They spent more money eating away from home, especially at fast-food restaurants, suggesting that time is a key ingredient in the development of healthier eating habits.[6] Restaurants, supermarkets, and meal delivery services have responded to our demanding work schedules and long hours away from home by supplying prepared meals, microwavable entrees, online grocery shopping and curbside pick-up, and quick-prep meals delivered to your door.

*Cost and economics* play a role in our food choices. The 2019 Food and Health Survey indicates that, after taste, cost is the number-two reason why people choose the food they do. While the average American now spends less on food than in the past, young adults and those with higher incomes spend the most on food. As income increases, so do meals eaten away from home, and as calorie intake increases, so does the food bill.

*Sustainability* is a relatively new factor that is affecting our food choices. With future generations in mind, many consumers are becoming more socially responsible to care for the environment. College students have become a big part of the movement to purchase local, seasonal, and sustainable food and to spread awareness that the way we produce and eat food can slow the rate of global warming, build strong communities, and improve our health. While the 2019 Food and Health Survey found that environmental sustainability was an important driver of food purchasing, 63% of consumers continue to find it hard to know whether the food choices they make are environmentally sustainable. These consumers agreed that environmental sustainability had a greater influence on food choices if it was easier to know which choices were in fact environmentally sustainable. When making environmentally sustainable animal protein purchases, consumers surveyed looked for labels such as "no added hormones," "grass-fed animals," and "locally raised."[7]

*Nutrition*—or what we think of as "healthy foods"—also directs our food purchases. North Americans who tend to make health-related food choices are often well-educated, middle-class professionals who are generally health oriented, have active lifestyles, and focus on weight control. The 2019 Food and Health Survey found that about 62% of consumers sometimes use healthfulness as a driver when purchasing food. The survey found that 23% of consumers say they actively seek out foods or follow a diet for health benefits such as weight loss, energy, digestive health, and heart health. Nutrition and health information on food package labels has also been shown to affect food choices. A recent study found that people are less likely to buy sugary drinks if they see warning labels that include graphic pictures of health consequences such as obesity, diabetes, and tooth decay.[8]

More than half (54%) of the respondents to the *2019 Food and Health Survey* said that environmental sustainability is important when they purchase and consume foods. Being labeled as locally grown, sustainably sourced, non-GMO/not bioengineered, and organic are the primary characteristics used to identify sustainable foods, along with foods with recyclable or minimal packaging.

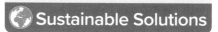 Sustainable Solutions

## WHY ARE YOU SO HUNGRY?

Two drives, **hunger** and **appetite,** influence our desire to eat. These drives differ dramatically. Hunger is primarily our physical, biological drive to eat and is controlled by internal body mechanisms. For example, as foods are digested and absorbed, the stomach and small intestine send signals to the liver and brain telling us to reduce further food intake.

Appetite, our primarily psychological drive to eat, is affected by many of the external factors we discussed in the preceding section, such as environmental and psychological factors and social cues and customs (see Fig. 1-1). Appetite can be triggered simply by seeing a tempting dessert or smelling popcorn at the movie theater. Fulfilling either or both drives by eating sufficient food normally brings a state of **satiety,** a feeling of satisfaction that temporarily halts our desire to continue eating.

The *feeding center* and the *satiety center* are in the **hypothalamus,** a region of the brain that helps regulate satiety. They work in opposite ways, like a tug-of-war, to promote adequate availability of nutrients at all times. When we haven't eaten for a while, stimulation of the feeding center signals us to eat. As we eat, the nutrient content in the blood rises, and the satiety center is stimulated. This is why we no longer have a strong desire to seek food after a meal. Admittedly, this concept of a tug-of-war between the feeding and satiety centers is an oversimplification of a complex process. The various feeding and satiety messages from body cells to the brain do not single-handedly determine what we eat. We often eat because food comforts us.[9] Almost everyone has encountered a mouthwatering dessert and devoured it,

**hunger** The primarily physiological (internal) drive to find and eat food.

**appetite** The primarily psychological (external) influences that encourage us to find and eat food, often in the absence of obvious hunger.

**satiety** A state in which there is no longer a desire to eat; a feeling of satisfaction.

**hypothalamus** A region of the forebrain that controls body temperature, thirst, and hunger.

**nutrients** Chemical substances in food that contribute to health, many of which are essential parts of a dietary pattern. Nutrients nourish us by providing calories to fulfill energy needs, materials for building body parts, and factors to regulate necessary chemical processes in the body.

**essential nutrient** In nutritional terms, a substance that, when left out of a dietary pattern, leads to signs of poor health. The body either cannot produce this nutrient or cannot produce enough of it to meet its needs. If added back to a dietary pattern before permanent damage occurs, the affected aspects of health are restored.

even on a full stomach. It smells, tastes, and looks good. We might eat because it is the right time of day, we are celebrating, or we are seeking emotional comfort to overcome the blues. After a meal, memories of pleasant tastes and feelings reinforce appetite. If stress or depression sends you to the refrigerator, you are mostly seeking comfort, not food calories. Appetite may not be a physical process, but it does influence food intake.

When food is abundant, appetite—not hunger—more frequently triggers eating. Satiety associated with consuming a meal may reside primarily in our psychological frame of mind. Also, because satiety regulation is not perfect, body weight can fluctuate. We become accustomed to a certain amount of food at a meal. Providing less than that amount leaves us wanting more. One way to use this observation for weight-loss purposes is to train your eye to expect less food by slowly decreasing serving sizes to more appropriate amounts. Your appetite then readjusts as you expect less food. You should now understand that daily food consumption is a complicated mix of biological and social influences. Keep track of what triggers your eating for a few days. Is it primarily hunger or appetite?

### ✅ CONCEPT CHECK 1.1

1. What are the factors that influence our food choices?
2. Which two vegetables are the most commonly consumed in the U.S. and why?
3. How do hunger and appetite differ in the way they influence our desire to eat?
4. What factors influence satiety?

## 1.2 How Is Nutrition Connected to Good Health?

Fortunately, the foods we eat can support good health in many ways, depending on their components. You just learned, however, that lifestyle habits and other factors may have a bigger impact on our food choices than the food components themselves. Unfortunately, many North Americans suffer from diseases that could have been prevented if they had known more about the foods and, more importantly, had applied this knowledge to plan meals and design their eating pattern. We will now look at the effect these choices are having on our health both today and in the future.

### WHAT IS NUTRITION?

Nutrition is the science that links foods to health. It includes the processes by which the human organism ingests, digests, absorbs, transports, uses, and excretes food substances.

### NUTRIENTS COME FROM FOOD

What is the difference between food and **nutrients**? Food provides the energy (in the form of calories) as well as the compounds needed to build and maintain all body cells. Nutrients are the substances obtained from food that are vital for growth and maintenance of a healthy body throughout life. For a substance to be considered an **essential nutrient,** three characteristics are needed:

- At least one specific biological function of the nutrient must be identified in the body.
- Omission of the nutrient from the dietary pattern must lead to a decline in certain biological functions, such as production of blood cells.
- Replacing the omitted nutrient in the dietary pattern before permanent damage occurs will restore those normal biological functions.

## WHY STUDY NUTRITION?

We all may feel like nutrition experts because we all eat several times a day. Nutrition knowledge can be confusing, however, and seem like a moving target. Recommendations may seem to differ depending on their source, and there are so many choices when shopping for food or eating out. You just learned that nutrition is only one of many factors that influence our eating habits. Studying nutrition will help you erase any misconceptions you have about food and nutrition and will assist you in making informed choices about the foods you eat and their relationship to health.

Nutrition is a lifestyle factor that is a key to developing and maintaining an optimal state of health. A poor dietary pattern and a sedentary lifestyle are known to be **risk factors** for life-threatening **chronic** diseases such as **cardiovascular (heart) disease, hypertension,** type 2 **diabetes,** and some forms of **cancer.** Together, these and related disorders account for two-thirds of all deaths in North America (Fig. 1-4).[11] Not meeting nutrient needs in our younger years makes us more likely to suffer health consequences, such as bone fractures from the disease **osteoporosis,** in later years. In 2020, COVID-19 was a major cause of death in the United States, with older adults and people with preexisting medical conditions such as heart or lung disease or diabetes at higher risk.

The combination of a poor eating pattern and too little physical activity may be the second-leading cause of death in the United States. In fact, U.S. government statistics indicate that a poor eating pattern combined with a lack of sufficient physical activity contributes to hundreds of thousands of fatal cases of cardiovascular disease, cancers, and diabetes each year. In addition, **obesity,** which the American Medical Association declared as a disease in 2013, is considered the second-leading cause of preventable death in North America (use of tobacco products is the first). Obesity and other chronic diseases are often preventable, and the cost of prevention, usually when we are children and young adults, is small compared to the cost of treating these diseases when we are older.

The good news is that an increased interest in health, fitness, and nutrition in Americans has been associated with long-term decreasing trends for heart disease, cancers, and **stroke** (three of the leading causes of death). Mortality from heart disease, the leading cause of death, has been declining steadily since 1980. As you gain understanding about your nutritional habits and increase your knowledge about optimal nutrition, you will have the opportunity to dramatically reduce your risk for many common health problems. Recent research has shown that those following the healthiest eating pattern overall had a 65% lower risk of dying from cancers, or any other cause, than those who had followed the worst eating pattern. A healthy eating pattern was defined as one with a high proportion of vegetables, fruits, whole grains, proteins, and dairy.[12]

**risk factors** A term used frequently when discussing the factors contributing to the development of a disease. A risk factor is an aspect of our lives, such as heredity, lifestyle choices (e.g., use of tobacco products), or nutritional habits.

**chronic** Long-standing, developing over time. When referring to disease, this term indicates that the disease process, once developed, is slow and lasting. A good example is cardiovascular disease.

**cardiovascular (heart) disease** A general term that refers to any disease of the heart and circulatory system. This disease is generally characterized by the deposition of fatty material in the blood vessels (hardening of the arteries), which in turn can lead to organ damage and death. Also termed *coronary heart disease (CHD)* or simply, *heart disease,* as the vessels of the heart are the primary sites of the disease.

**hypertension** A condition in which blood pressure remains persistently elevated. Obesity, inactivity, alcohol intake, excess salt intake, and genetics may each contribute to the problem.

**diabetes** A group of diseases characterized by high blood **glucose.** Type 1 diabetes involves insufficient or no release of the hormone insulin by the pancreas and therefore requires daily insulin therapy. Type 2 diabetes results from either insufficient release of insulin or general inability of insulin to act on certain body cells, such as muscle cells. Persons with type 2 diabetes may or may not require insulin therapy.

**glucose** A six-carbon sugar that exists in a ring form; found as such in blood, and in table sugar bound to fructose; also known as *dextrose,* it is one of the simple sugars.

**cancer** A condition characterized by uncontrolled growth of abnormal cells.

**osteoporosis** The presence of a stress-induced fracture or a T-score of −2.5 or lower. The bones are porous and fragile due to low mineral density.

**obesity** Disorder involving excessive body fat that increases the risk of health problems.

**stroke** A decrease or loss in blood flow to the brain that results from a blood clot or other change in arteries in the brain. This in turn causes the death of brain tissue. Also called a *cerebrovascular accident.*

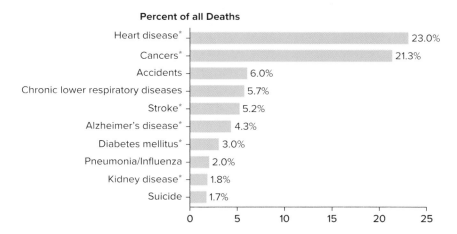

**Percent of all Deaths**

| Cause | Percent |
|---|---|
| Heart disease* | 23.0% |
| Cancers* | 21.3% |
| Accidents | 6.0% |
| Chronic lower respiratory diseases | 5.7% |
| Stroke* | 5.2% |
| Alzheimer's disease* | 4.3% |
| Diabetes mellitus* | 3.0% |
| Pneumonia/Influenza | 2.0% |
| Kidney disease* | 1.8% |
| Suicide | 1.7% |

**FIGURE 1-4** ▲ Ten leading causes of death in the United States.

Source: Centers for Disease Control and Prevention, *National Vital Statistics Report,* Deaths: Leading Causes for 2017.

* Causes of death in which diet plays a part.

1. How do we define *nutrition?*
2. What are the three leading causes of death in which the dietary pattern plays a part?

## 1.3 What Are the Classes and Sources of Nutrients?

To begin the study of nutrition, let's start with an overview of the six classes of nutrients. You are probably already familiar with the terms **carbohydrates, lipids** (fats and oils), **proteins, vitamins,** and **minerals.** These nutrients, plus **water,** make up the six classes of nutrients found in food.

Nutrients can be assigned to three functional categories: (1) those that primarily provide us with calories to meet energy needs (expressed in **kilocalories [kcal]**); (2) those important for growth, development, and maintenance; and (3) those that act to keep body functions running smoothly. Some overlap in function exists among these categories (Fig. 1-5). The energy-yielding nutrients (carbohydrates, lipids, and protein) along with water are needed in relatively large amounts, so they are called **macronutrients.** Vitamins and minerals are needed in such small amounts in the dietary pattern that they are called **micronutrients.**

### CARBOHYDRATES

Chemically, carbohydrates can exist in foods as simple sugars and complex carbohydrates. **Simple sugars,** frequently referred to as *sugars,* are relatively small molecules. These sugars are found naturally in fruits, vegetables, and dairy products. Table sugar, known as sucrose, is a simple sugar that is added to many foods we eat. Glucose, also known as blood sugar or dextrose, is a simple sugar in your blood. **Complex carbohydrates** are formed when many simple sugars are joined together. Plants store carbohydrates in the form of **starch,** a complex carbohydrate made up of hundreds of

**carbohydrate** A compound containing carbon, hydrogen, and oxygen atoms. Most are known as *sugars, starches,* and *fibers.*

**lipid** A compound containing much carbon and hydrogen, little oxygen, and sometimes other atoms. Lipids do not dissolve in water and include fats, oils, and cholesterol.

**protein** Food and body compounds made of more than 100 amino acids; proteins contain carbon, hydrogen, oxygen, nitrogen, and sometimes other atoms in a specific configuration. Proteins contain the form of nitrogen most easily used by the human body.

**vitamin** An essential organic (carbon-containing) compound needed in small amounts in the dietary pattern to help regulate and support chemical reactions and processes in the body.

**mineral** Element used in the body to promote chemical reactions and to form body structures.

**water** The universal solvent; chemically, $H_2O$. The body is composed of about 60% water. Water (fluid) needs are about 9 (women) or 13 (men) cups per day; needs are greater if one exercises heavily.

**kilocalorie (kcal)** Heat energy needed to raise the temperature of 1000 grams (1 L) of water 1 degree Celsius.

**macronutrient** A nutrient needed in gram quantities in a dietary pattern.

**micronutrient** A nutrient needed in milligram or microgram quantities in a dietary pattern.

**simple sugar** Monosaccharide or disaccharide in the diet.

**complex carbohydrate** Carbohydrate composed of many monosaccharide molecules. Examples include glycogen, starch, and fiber.

**starch** A carbohydrate made of multiple units of glucose attached together in a form the body can digest; also known as *complex carbohydrate.*

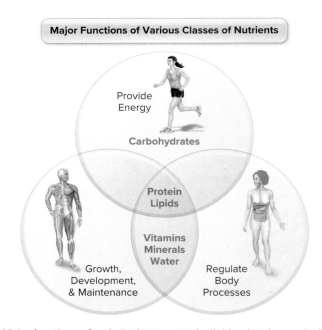

**FIGURE 1-5** ▲ Major functions of carbohydrates, protein, lipids, vitamins, and minerals. Notice that proteins and most lipids participate in all of the major function categories, whereas carbohydrates function primarily as sources of energy.

glucose units. Breads, cereals, grains, and starchy vegetables are the main sources of complex carbohydrates.

During digestion, complex carbohydrates are broken down into single sugar molecules (such as glucose) and absorbed into the bloodstream via **cells** lining the small intestine. However, the **bonds** between the sugar molecules in certain complex carbohydrates, called **fiber,** cannot be broken down by human digestive processes. Fiber passes through the small intestine undigested to provide bulk for the stool (feces) formed in the large intestine (colon).

Aside from enjoying their taste, we need sugars and other carbohydrates in our eating patterns primarily to help satisfy the calorie needs of our body cells. Carbohydrates provide a major source of calories for the body, on average 4 kcal per gram. Glucose, a simple sugar that the body can derive from most carbohydrates, is a major source of calories for most cells. When insufficient carbohydrate is consumed, the body is forced to make glucose from proteins—not a healthy alternative. Most foods high in carbohydrates, such as fruits, vegetables, and whole grains, are also excellent sources of vitamins, minerals, and phytochemicals.

## LIPIDS

Lipids (mostly fats and oils) in the foods we eat also provide energy. Lipids yield more calories per gram than do carbohydrates—on the average, 9 kcal per gram—because of differences in their chemical composition. They are also the main form for energy storage in the body.

In this book, the more familiar terms *fats* and *oils* will generally be used, rather than *lipids.* Lipids do not dissolve in water. Generally, fats are lipids that are solid at room temperature, and oils are lipids that are liquid at room temperature. We obtain fats and oils from animal and plant sources. Animal fats, such as butter or lard, are solid at room temperature. Plant oils, such as corn or olive oil, tend to be liquid at room temperature. To promote heart health, most people would benefit from using more plant oils in place of solid fats.

Certain fats are essential nutrients that must come from our dietary pattern. These key fats that the body cannot produce, called essential fatty acids, perform several important functions in the body: they help regulate blood pressure and play a role in the synthesis and repair of vital cell parts. However, we need only about 4 tablespoons of a common plant oil (such as olive or soybean oil) each day to supply these essential fatty acids. A serving of fatty fish, such as salmon or tuna, at least twice a week is another healthy source of fats. The unique fatty acids in these fish complement the healthy aspects of common plant oils.

## PROTEINS

Proteins are the main structural material in the body. For example, proteins constitute a major part of bone and muscle; they are also important components in blood, body cells, **enzymes,** and immune factors. Proteins can also provide calories for the body—on average, 4 kcal per gram. Typically, however, the body uses little protein for the purpose of meeting daily calorie needs. Proteins are formed when **amino acids** are bonded together. Some amino acids are essential nutrients.

Protein in our eating pattern comes from animal and plant sources. The animal products meat, poultry, fish, dairy, and eggs are significant sources of protein in most eating patterns. Beans, grains, nuts, seeds, and some vegetables are good plant protein sources and are important to include in **vegetarian** eating patterns. If protein consumption is greater than what is needed for body functions, the excess is used for calorie needs and carbohydrate production but ultimately can be converted to and stored as fat.

## VITAMINS

The main function of vitamins is to enable many **chemical reactions** to occur in the body. Some of these reactions help release the energy trapped in carbohydrates, lipids,

**cell** The structural basis of plant and animal organization. In animals it is bounded by a cell membrane. Cells have the ability to take up compounds from and excrete compounds into their surroundings.

**bond** A linkage between two atoms formed by the sharing of electrons, or attractions.

**fiber** Substances in plant foods not digested in the human stomach or small intestine. These add bulk to feces. Fiber naturally found in foods is also called *dietary fiber.*

▲ Salmon is a fatty fish that is a healthy source of essential fatty acids.
Olga Nayashkova/Shutterstock

**enzyme** A compound that speeds up the rate of a chemical reaction but is not altered by the reaction. Almost all enzymes are proteins (some are made of genetic material).

**amino acid** The building block for proteins containing a central carbon atom with nitrogen and other atoms attached.

**vegetarian** Referring to a dietary pattern that includes primarily foods of plant origin.

**chemical reaction** An interaction between two chemicals that changes both chemicals.

and proteins. Remember, however, that vitamins themselves contain no usable calories for the body.

The 13 vitamins are **organic compounds** divided into two groups: four are **fat-soluble** because they dissolve in fat (vitamins A, D, E, and K); nine are **water-soluble** because they dissolve in water (the B vitamins and vitamin C). The two groups of vitamins have different sources, functions, and characteristics. Water-soluble vitamins are found mainly in fruits and vegetables, whereas dairy products, nuts, seeds, oils, and fortified breakfast cereals are good sources of fat-soluble vitamins. Cooking destroys water-soluble vitamins much more readily than it does fat-soluble vitamins. Water-soluble vitamins are also excreted from the body much more readily than are fat-soluble vitamins. Thus, the fat-soluble vitamins, especially vitamin A, have the ability to accumulate in excessive amounts in the body, which then can lead to **toxicity.**

## MINERALS

Minerals are structurally simple, **inorganic** substances that do not contain carbon **atoms.** Minerals such as sodium and potassium typically function independently in the body, whereas minerals such as calcium and phosphorus function together in tissue, such as bone. Because of their simple structure, minerals are not destroyed during cooking, but they can still be lost if they dissolve in the water used for cooking and that water is then discarded. Minerals provide no calories for the body but are critical players in nervous system functioning, water balance, structural (e.g., skeletal) systems, and many other cellular processes.

The essential minerals required in the dietary pattern for good health are divided into two groups—**major minerals** and **trace minerals**—because dietary needs and concentrations in the body vary enormously. If daily needs are less than 100 milligrams, the mineral is classified as a trace mineral; otherwise, it is a major mineral. Minerals that function based on their electrical charge when dissolved in water are also called **electrolytes;** these include sodium, potassium, and chloride. Many major minerals are found naturally in dairy products and fruits, whereas many trace minerals are found in meats, poultry, fish, and nuts.

## WATER

Water makes up the sixth class of nutrients. Although sometimes overlooked as a nutrient, water (chemically, $H_2O$) has numerous vital functions in the body. It acts as a **solvent** and lubricant, as a vehicle for transporting nutrients and waste, and as a medium for temperature regulation and chemical processes. For these reasons, and because the human body is approximately 60% water, the average man should consume about 3 liters—about 13 cups—of water and/or other fluids every day; women need closer to 2.2 liters or about 9 cups per day. Fluid needs vary widely, however, based on differences in body mass and environmental conditions.

Water is obviously available from all beverages and is also the major component in some foods, such as many fruits and vegetables (e.g., lettuce, grapes, and melons). The body even makes some water as a by-product of **metabolism.**

## OTHER IMPORTANT COMPONENTS IN FOOD

Another group of compounds called **phytochemicals** are found in foods from plant sources, especially within the fruit and vegetable groups. Although these phytochemicals are not considered essential nutrients, they provide significant health benefits. Considerable research is focused on the ability of various phytochemicals to reduce the risk for certain diseases. For example, evidence from animal and laboratory studies indicates that compounds such as polyphenols in blueberries and strawberries prevent the growth of certain cancer cells. Research also suggests that

**organic compounds** In chemistry, any chemical compounds that contain carbon.

**fat-soluble** Soluble in fats, oils, or fat solvents.

**water-soluble** Capable of dissolving in water.

**toxicity** Capacity of a substance to produce injury or illness at some dosage.

**inorganic** Any substance lacking carbon atoms bonded to hydrogen atoms in the chemical structure.

**atom** Smallest combining unit of an element, such as iron or calcium. Atoms consist of protons, neutrons, and electrons.

**major mineral** Vital to health, a mineral that is required in the dietary pattern in amounts greater than 100 milligrams per day.

**trace mineral** Vital to health, a mineral that is required in the dietary pattern in amounts less than 100 milligrams per day.

**electrolyte** A mineral that separates into positively or negatively charged ions in water. Electrolytes are able to transmit an electrical current.

**solvent** A liquid substance in which other substances dissolve.

**metabolism** Chemical processes in the body by which energy is provided in useful forms and vital activities are sustained.

**phytochemical** A chemical found in plants. Some phytochemicals may contribute to a reduced risk of cancer or cardiovascular disease in people who consume them regularly.

| PHYTOCHEMICALS FOUND IN PLANT PIGMENT COLORS | FOUND IN | GOOD FOR |
|---|---|---|
| CAROTENOIDS & HESPERETIN | Carrots, sweet potatoes, oranges, lemons | Eye health, immune function, heart health |
| ANTHOCYANOSIDES & RESVERATROL | Blueberries, blackberries, grapes, beets, eggplants | **Antioxidants,** anti-inflammatory, heart health, cognition |
| SULFORAPHANE & LUTEIN | Leafy greens, broccoli, cabbages, green tea | Anti-carcinogenic, anti-inflammatory, liver function, eye health |
| ALLYL SULFIDES & FLAVONOIDS | Garlic, onions, leeks, apples, buckwheat | Cholesterol lowering, immune function, anti-allergy, estrogen metabolism |
| LYCOPENE & ELLAGIC ACID | Tomatoes, cherries, radishes, strawberries | Antioxidants, heart health, immune function |

**FIGURE 1-6** ◄ Plant foods are packed with colorful and powerful phytochemicals that support health. Eating plant foods from a variety of color families will provide a variety of health benefits. oranges: Lluis Real/AGE Fotostock; berries: Noppadon sakulsom/Janecocoa/123RF; broccoli: spafra/iStock/Getty Images; garlic: Maks Narodenko/Shutterstock; tomatoes: Tim UR/Shutterstock

**antioxidant** A substance that has the ability to prevent or repair the damage caused by oxidation.

the health benefits of phytochemicals are best obtained through the consumption of whole foods rather than dietary supplements. Food sources of phytochemical compounds under study include soybeans and other legumes, which provide isoflavones, and cruciferous vegetables, especially broccoli, which provide isothiocyanates and indoles.

Foods with high phytochemical content are sometimes called "superfoods" because of the health benefits they are thought to confer. There is no legal definition of the term *superfood,* however, and there is concern that it is being overused in marketing certain foods. Figure 1-6 lists some noteworthy phytochemicals that are under study according to their plant pigment or color family. Although there is not enough evidence to link individual phytochemicals with specific health benefits, there is enough proof to suggest that consuming phytochemical-rich foods and beverages may help prevent disease. Tomatoes are an important source of phytochemicals and are discussed in this chapter's *Farm to Fork* feature. (*Farm to Fork* appears in every chapter and presents practical information on how to grow, shop, store, and prepare various fruits and vegetables to obtain and preserve their flavor and nutrients.)

## SOURCES OF NUTRIENTS

Now that we know the six classes of nutrients, it is important to understand the quantities of the various nutrients that people consume. On a daily basis, we consume about 500 grams, or about 1 pound, of protein, fat, and carbohydrate combined. In contrast, the typical daily mineral intake totals about 20 grams (about 4 teaspoons), and the daily vitamin intake totals less than 300 milligrams (1/15 of a teaspoon). Although we require a gram or so of some minerals, such as calcium and phosphorus, we need only a few milligrams or less of other minerals, such as zinc, each day.

The nutrient content of the foods we eat also differs from the nutrient composition of the human body. This is because growth, development, and later maintenance of the human body are directed by the genetic material (DNA) inside body cells. This genetic blueprint determines how each cell uses the essential nutrients to perform body functions. These nutrients can come from a variety of sources. Cells are not concerned about whether available amino acids come from animal or plant sources. The

## FARM to FORK — Tomatoes

Adrian Burke/Getty Images

Tomatoes are very good sources of the antioxidants lycopene and vitamin C and are also rich in beta-carotene, manganese, and vitamin E.

### Grow
- Naturally ripened tomatoes are more nutritious and flavorful than the artificially ripened tomatoes sold in supermarkets. Look for local tomatoes, including heirloom varieties, at nearby farmers' markets.
- Consider growing your own tomatoes, even in containers, to enjoy nutritious varieties harvested at the peak of ripeness.

### Shop
- Choose tomatoes with the darkest red color for the most nutrients and highest amount of the phytochemical lycopene.
- Purchase smaller tomatoes for their sweetness and flavor, and the most lycopene and vitamin C.
- Buy processed tomato products, including jars or cans of paste and sauce, for their highly bioavailable lycopene.
- Choose tomato products in glass jars, aseptic-coated paper containers, or BPA-free cans. BPA, or bisphenol A, is a synthetic estrogen found in the coatings of some food cans and has possible health effects.

### Store
- To preserve the flavor of fresh tomatoes, store them stem side up at room temperature. Flavor and aroma quickly decrease when tomatoes are stored in the refrigerator.
- Grape tomatoes should be stored in plastic clamshells (i.e., original packaging) to prevent them from drying out.
- Tomatoes are ripe and ready to eat when they are a deep color but still firm. Eat ripe tomatoes within two or three days.

### Prep
- Use the whole tomato. The juice contains the flavor enhancer glutamate, and the skin and seeds provide vitamin C and lycopene.
- Snack on nutrition-packed grape tomatoes, and slice or chop them for salads, omelets, sandwiches, or tacos.
- Cooking tomatoes increases the bioavailability of nutrients and phytochemicals.
- Add tomato paste to recipes as a concentrated source of flavor, color, nutrients, and phytochemicals, with no added sugar or salt.

Source: Robinson J: Tomatoes: Bringing back their flavor and nutrients. In *Eating on the Wild Side*. New York: Little, Brown and Company, 2013.

Alfio Roberto Silvestro/123RF

carbohydrate glucose can come from sugars or starches. The food that you eat provides cells with basic materials to function according to the directions supplied by the genetic material (genes) housed in body cells.

### ✓ CONCEPT CHECK 1.3

1. What are the six classes of nutrients?
2. What are the three general functions of nutrients in the body?
3. What are phytochemicals?

## 1.4 What Math Concepts Will Aid Your Study of Nutrition?

### CALORIES

We obtain the energy we need for body functions and physical activity from various calorie sources: carbohydrates (4 kcal per gram), fats (9 kcal per gram), and proteins (4 kcal per gram). Foods generally provide more than one calorie source. Plant oils, such as soybean or olive oil, are one exception; these are 100% fat at 9 kcal per gram.

Alcohol is also a potential source of calories, supplying about 7 kcal per gram. It is not considered an essential nutrient, however, because it is not required for human function. Still, alcoholic beverages, such as beer—also rich in carbohydrate—are a contributor of calories to the eating patterns of many adults. The nutritional implications of alcohol consumption are discussed in Section 1.8.

The body releases energy (measured in calories) from the chemical bonds in carbohydrate, protein, and fat (and alcohol) in order to:

- Build new compounds.
- Perform muscular movements.
- Promote nerve transmission.
- Maintain electrolyte balance within cells.

The energy in food is often expressed using the term *calories* on food labels. A calorie is the amount of heat energy it takes to raise the temperature of 1 gram of water 1 degree Celsius (1°C, centigrade scale). A calorie is a tiny measure of heat relative to the amount of calories we eat and use. Food energy is more conveniently expressed in terms of the kilocalorie (kcal), which equals 1000 calories. (If the "c" in calories is capitalized, this also signifies kilocalories.) A kilocalorie is the amount of heat energy it takes to raise the temperature of 1000 grams (1 liter) of water 1°C. The abbreviation *kcal* is used throughout this book. On food labels, the word *calorie* (without a capital "C") is also used loosely to mean *kilocalorie.* Any values given on food labels in calories are actually in kilocalories (Fig. 1-7). A suggested intake of 2000 calories per day on a food label is technically 2000 kcal.

**WHOLE WHEAT BREAD**

Fullerene/iStock/Getty Images

# Nutrition Facts

19 servings per container

Serving size      **1 slice (36g)**

**Amount per serving**

# Calories    80

| | % Daily Value* |
|---|---|
| **Total Fat** 1g | **2%** |
| Saturated Fat 0g | **0%** |
| *Trans* Fat less than 1g | ** |
| **Cholesterol** 0mg | **0%** |
| **Sodium** 200mg | **8%** |

| | % Daily Value* |
|---|---|
| **Total Carbohydrate** 15g | **5%** |
| Dietary Fiber 2g | **8%** |
| Total Sugars 3g | |
| Includes 2g Added Sugars | **4%** |
| **Protein** 3g | |

Vitamin D 0mcg 0% • Calcium 30mg 3% • Iron 1mg 5% • Potassium 70mg 2%

\* The % Daily Value (DV) tells you how much a nutrient in a serving of food contributes to a daily diet. 2,000 calories a day is used for general nutrition advice.

**FIGURE 1-7** ▲ Use the nutrient values on the Nutrition Facts label to calculate calorie content of a food. Based on carbohydrate, fat, and protein content, a serving of this food (whole wheat bread) contains 81 kcal ([15 × 4] + [1 × 9] + [3 × 4] = 81). The label lists 80, suggesting that the calorie value was rounded down.

 Carbohydrate
4 kcal per gram

 Fat
9 kcal per gram

 Protein
4 kcal per gram

 Alcohol
7 kcal per gram

▲ Calorie content of energy nutrients and alcohol. The weights illustrate their relative energy potential per gram.

## CALCULATING CALORIES

The calorie estimates for carbohydrate, fat, protein, and **alcohol** (4-9-4-7) can be used to determine the calorie content of a food. Consider these foods:

**alcohol** Ethyl alcohol or ethanol ($CH_3CH_2OH$) is the compound in alcoholic beverages.

*1 Grilled Chicken Sandwich*

Burke/Triolo/Brand X Pictures

| | |
|---|---|
| Carbohydrate | 46 grams × 4 = 184 kcal |
| Fat | 14 grams × 9 = 126 kcal |
| Protein | 45 grams × 4 = 180 kcal |
| Alcohol | 0 gram × 7 =    0 kcal |
| **Total** | **490 kcal** |

*8-Ounce Piña Colada*

C Squared Studios/Getty Images

| | |
|---|---|
| Carbohydrate | 57 grams × 4 = 228 kcal |
| Fat | 5 grams × 9 =  45 kcal |
| Protein | 1 gram × 4 =   4 kcal |
| Alcohol | 23 grams × 7 = 161 kcal |
| **Total** | **438 kcal** |

You can also use the 4-9-4 estimates to determine what portion of total kilocalorie intake is contributed by the various calorie-yielding nutrients. Assume that one day you consume 290 grams of carbohydrates, 60 grams of fat, and 70 grams of protein. This adds up to a total of 1980 kcal ([290 × 4] + [60 × 9] + [70 × 4] = 1980). The percentage of your total kilocalorie intake derived from each nutrient can then be determined:

$$\% \text{ of kcal as carbohydrate} = (290 \times 4) \div 1980 = 0.59 \ (\times 100 = 59\%)$$
$$\% \text{ of kcal as fat} = (60 \times 9) \div 1980 = 0.27 \ (\times 100 = 27\%)$$
$$\% \text{ of kcal as protein} = (70 \times 4) \div 1980 = 0.14 \ (\times 100 = 14\%)$$

Check your calculations by adding the percentages together. Do they total 100%?

## PERCENTAGES

You will use a few mathematical concepts in studying nutrition. Besides performing addition, subtraction, multiplication, and division, you need to know how to calculate percentages and convert English units of measurement to metric units.

The term *percent* (%) refers to a part of the total when the total represents 100 parts. For example, if you earn 80% on your first nutrition examination, you will have answered the equivalent of 80 out of 100 questions correctly. This equivalent also could be 8 correct answers out of 10; 80% also describes 16 of 20 (16/20 = 0.80 or 80%). The decimal form of percents is based on 100% being equal to 1.00. Percentages are used frequently when referring to menus and nutrient composition as we saw in the previous calculation of the percentage of total calorie intake from each nutrient. The best way to master this concept is to calculate some percentages. Some examples follow:

| Question | Answer |
|---|---|
| What is 6% of 45? | 6% = 0.06, so 0.06 × 45 = 2.7 |
| What percent of 99 is 3? | 3/99 = 0.03 or 3% (0.03 × 100) |

Joe ate 15% of the adult Recommended Dietary Allowance for iron (RDA = 8 milligrams) at lunch. How many milligrams did he eat?

0.15 × 8 milligrams = 1.2 milligrams

## THE METRIC SYSTEM

The basic units of the metric system are the meter, which indicates length; the gram, which indicates weight; and the liter, which indicates volume. Appendix F in this textbook lists conversions from the metric system to the English system (feet, pounds, and cups) and vice versa. Here is a brief summary:

A centimeter is 1/100 of a meter, 2.54 centimeters equals 1 inch.

A gram (g) is about 1/30 of an ounce (an ounce weighs 28 grams).

   5 grams of sugar or salt is about 1 teaspoon.

A pound (lb) weighs 454 grams.

A kilogram (kg) is 1000 grams, equivalent to 2.2 pounds.

To convert weight in pounds to kilograms, divide it by 2.2.

   A 154-pound man weighs 70 kilograms (154/2.2 = 70).

A gram can be divided into 1000 milligrams (mg) or 1,000,000 micrograms (μg or mcg).

   10 milligrams of zinc (approximate adult need) would be a few grains of zinc.

Liters are divided into 1000 units called milliliters (ml); 100 milliliters is a deciliter (dl).

   One teaspoon equals about 5 milliliters (ml), 1 cup is about 240 milliliters, and 1 quart (4 cups) equals almost 1 liter (L) (0.946 liter to be exact).

Examples:

You see on the label that a 5.3-ounce (oz) container of Greek yogurt contains 15 grams of sugar. How many teaspoons of sugar does this equal?

Answer: 15 grams ÷ 5 grams/teaspoon = 3 teaspoons of sugar in the 5.3-oz yogurt.

You are trying to drink at least 8 cups of water each day. You know 8 cups equals 64 ounces or 2 quarts of water because there are 8 ounces in a cup. Your water bottle, however, holds 500 milliliters (ml). How many milliliters or liters should you drink to equal 8 cups?

Answer: 8 cups × 240 ml/cup = 1920 ml = 1.92 liters (almost four 500 ml bottles).

If you plan to work in any scientific field, you will need to learn the metric system. In the field of nutrition, it is important to remember that a kilogram equals 2.2 pounds, an ounce weighs 28 grams, 2.54 centimeters equals 1 inch, and a liter is almost the same as a quart. In addition, know the fractions that the following prefixes represent: micro (1/1,000,000), milli (1/1000), centi (1/100), and kilo (1000).

### ✓ CONCEPT CHECK 1.4

1. What are the energy (kilocalorie) values for each of the "energy nutrients"?
2. If you weigh 154 pounds, what is your weight in kilograms?

### CASE STUDY   Choosing a Quick but Healthy Breakfast

Harrison was awake last night until 2:30 A.M. finishing a class project. Unfortunately, his Psychology 101 class meets at 9:00 A.M. this morning. When his alarm goes off at 7:30 A.M., he decides to sleep those extra 20 minutes it would take to sit down and enjoy breakfast at the dining hall. When Harrison finally rolls out of bed he must decide what to do about a morning meal. He considers skipping breakfast altogether, grabbing a snack and coffee from a vending machine, eating cereal and yogurt in his room, or picking up a breakfast sandwich to eat during class. Answer the questions below to determine the healthiest yet time-saving breakfast option.

1. Harrison is considering skipping breakfast and consuming a few extra calories at lunch and dinner. How will this plan affect Harrison's morning energy levels? What do we know about eating breakfast and weight control?
2. Would a low-fat granola bar and iced coffee from the vending machines in his dorm be a good source of calories and nutrients? Would this choice satisfy his hunger for very long?
3. Harrison has the quickest breakfast choice in his own room: a quick bowl of whole grain cereal with a banana and low-fat milk along with a yogurt. Explain why this convenient choice might also be the most nutritious choice.
4. Harrison could also pick up a ham, egg, and cheese bagel to eat during class. How do the calorie, fat, and sodium contents of this fast-food breakfast sandwich compare to the other options? Would this be a healthy breakfast choice every day?

*Complete the Case Study. Responses to these questions can be provided by your instructor.*

▲ With a little bit of planning, breakfast can be both quick and healthy. Stockbyte/Getty Images

## 1.5 How Do We Know What We Know About Nutrition?

The knowledge we have about nutrient needs comes from research. Like other sciences, the research that sets the foundation for nutrition knowledge has developed using the *scientific method,* a testing procedure designed to detect and eliminate error.

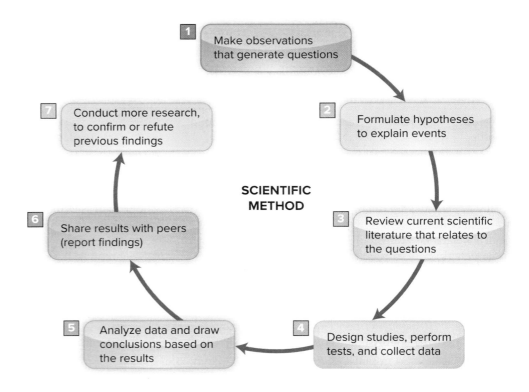

**FIGURE 1-8** ▶ The scientific method.
Scientists consistently follow these steps
when testing all types of hypotheses.
Scientists do not accept a nutrition
or other scientific hypothesis until it
has been thoroughly tested using the
scientific method.

**hypotheses** Tentative explanations by a
scientist to explain a phenomenon.

**scurvy** The vitamin C deficiency disease
characterized by weakness, fatigue, slow
wound healing, bone pain, fractures,
sore and bleeding gums, diarrhea, and
pinpoint hemorrhages on the skin.

**epidemiology** The study of how disease
rates vary among different population
groups.

**theory** An explanation for a phenomenon
that has numerous lines of evidence to
support it.

**peer review** Evaluation of work by
professionals of similar competence
(peers) to the producers of the work
to maintain standards of quality and
credibility. Scholarly peer review is
used to determine if a scientific study is
suitable for publication.

**systematic review** A thorough summary
of the results of available carefully
designed health care studies (controlled
trials) in a particular area.

## THE SCIENTIFIC METHOD

The first step of the scientific method is the observation of a natural phenomenon
(Fig. 1-8). Scientists then suggest possible explanations, called **hypotheses,** for the phenomenon. At times, historical events have provided clues to important relationships in
nutrition science, such as the link between the need for vitamin C and the development
of the disease **scurvy.** Another approach is for scientists to study dietary and disease patterns among various populations, a research method called **epidemiology.**

Thus, hypotheses about the role of the dietary pattern in various health problems
can be suggested by historical and epidemiological findings. *Proving* the role of particular dietary components, however, requires controlled experiments. The data gathered
from experiments may either support or refute each hypothesis. If the results of many
experiments support a hypothesis, scientists accept the hypothesis as a **theory.** Often,
the results from one experiment suggest a new set of questions. Figure 1-8 shows how
the scientific method is used to test a hypothesis.

Once an experiment is complete, scientists summarize the findings and seek to
publish the results in scientific journals. Generally, before articles are published in
scientific journals, they undergo a critical **peer review** by other scientists familiar with
the subject, which helps to ensure that only high-quality, objective research findings
are published. Peer review occurs between steps 5 and 6 in Figure 1-8.

Keep in mind that one experiment is never enough to prove a particular hypothesis
or provide a basis for nutritional recommendations. Rather, through follow-up studies,
the results obtained in one laboratory must be confirmed by similar experiments conducted in other laboratories and, possibly, under varying circumstances. Only then can
we really trust and use the results.

## STRENGTH OF SCIENTIFIC EVIDENCE

When answering questions about nutrition needs, we look for the best available evidence. The hierarchy of evidence (Fig. 1-9) is based on the rigor (strength and precision) of the research methods used and provides a framework to help us locate the best
evidence. The first step is to search for a recent well-conducted **systematic review.** A

**Trusting the Evidence**

Strongest

Meta-analyses & systematic reviews

Randomized controlled trials

Cohort studies

Case-control studies

Cross-sectional studies & surveys

Case reports & case studies

Animal & *in vitro* studies

Editorials, expert opinion papers

Quality of evidence

Weakest

**FIGURE 1-9** ◄ Hierarchy of scientific evidence with the strongest types of evidence at the top progressing to the weakest types of evidence at the bottom. *Adapted from thelogicofscience.com.*

systematic review is a thorough analysis of the results of all available studies in a particular area. If the studies have the same outcome measures, a systematic review may include a **meta-analysis** of the statistical results of the studies.

If a current systematic review is not available, we move down to the next level of evidence, the primary studies. The most rigorous type of controlled experiment, a **randomized controlled trial,** follows a study design that is **double-blind** and **placebo** controlled. In this type of study, a group of participants—the experimental group— follows a specific protocol (e.g., consuming a certain food or nutrient), and participants in a corresponding **control group** follow their normal habits or consume a placebo. People are randomly assigned to each group. Scientists then observe the experimental group over time to see if there is any effect not found in the control group.

**Cohort studies** are further down the hierarchy because they are observational studies that look at large groups of people and examine their exposure to certain risk factors for disease. If data are gathered going forward, they are called prospective studies; if data that are already collected are assessed, they are considered retrospective studies. The Nurses' Health Study is an example of a cohort study that has followed hundreds of thousands of women in North America and found many links between lifestyle choices and health.

A **case-control study** is typically retrospective and compares individuals who have a disease or condition, such as lung cancer, to individuals who do not have the condition. A **cross-sectional study** looks at data from a population group at one specific point in time. **Case reports** are descriptive studies based on uncontrolled observations of patients.

While all of these human experiments provide convincing evidence about relationships between nutrients and health, they are often not practical or ethical to conduct. Thus, much of what we know about human nutritional needs and functions has been gleaned from **animal experiments.** The use of animal experiments to study the role of nutrition in certain human diseases depends on the availability of an animal model in which a disease in laboratory animals closely mimics a particular human disease. Often, if no animal model is available and human experiments are ruled out, scientific knowledge cannot advance beyond what can be learned from epidemiological studies.

Finally, editorials and expert opinion papers provide an overview of a specific topic but do not qualify as adequate evidence to answer research questions. It is just as important to understand what is not scientific evidence. Personal anecdotes, YouTube videos,

**meta-analysis** A statistical examination of data from multiple scientific studies of the same subject in order to determine overall trends.

**randomized controlled trial** An experimental design that is double-blind and placebo controlled.

**double-blind** A study or trial in which any information which may influence the behavior of the tester or the subject is withheld until after the test.

**placebo** Generally, an inactive medicine or treatment used to disguise the treatments given to the participants in an experiment.

**control group** Participants in an experiment who are not given the treatment being tested.

**cohort studies** Observational studies that look at large groups of people, prospectively or retrospectively, studying their exposure to certain risk factors for disease.

**case-control study** A study in which individuals who have a disease or condition, such as lung cancer, are compared with individuals who do not have the condition.

**cross-sectional study** Type of observational study that analyzes data from a population group at one specific point in time and based on particular variables of interest.

**case reports** Descriptive studies based on uncontrolled observations of individual patients.

**animal experiments** Use of animals to study disease to understand more about human disease.

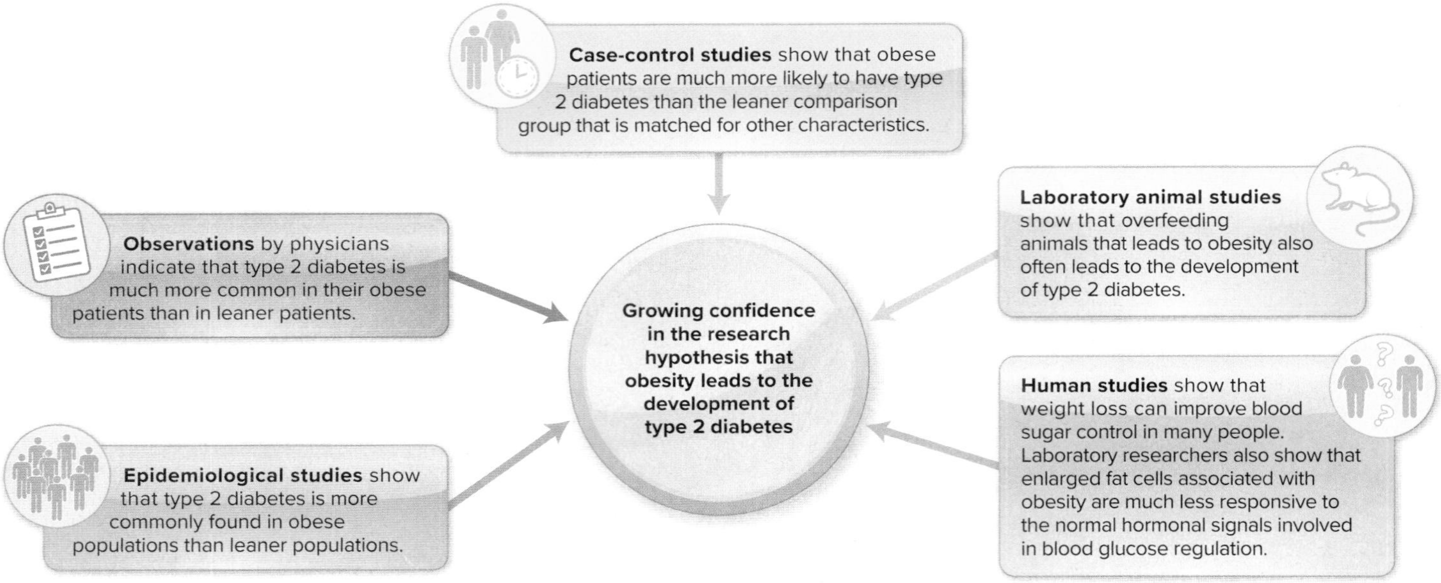

**FIGURE 1-10** ▲ Data from a variety of sources can come together to support a research hypothesis. This diagram shows how various types of research data support the hypothesis that obesity leads to the development of type 2 diabetes.

# Newsworthy Nutrition

Throughout this textbook, we have highlighted the use of the scientific method in research studies in the feature *Newsworthy Nutrition*. These are recently published studies that relate to chapter topics and that have made a significant impact on our nutrition knowledge. We have selected articles that have used a variety of study designs from the hierarchy of scientific evidence. You will find the first *Newsworthy Nutrition* study in Section 1.6 on the increased incidence of obesity-related cancer in young adults.

and websites like mercola.com, naturalnews.com, greenmedinfo.com, and whale.to are not credible sources of scientific information.

As shown in Figure 1-10, the more lines of evidence available to support an idea, the more likely it is to be true. Epidemiological studies may suggest hypotheses, but controlled experiments are needed to rigorously test hypotheses before nutrition recommendations can be made. For example, epidemiologists found that smokers who regularly consumed fruits and vegetables had a lower risk for lung cancer than smokers who ate very few fruits and vegetables. Scientists proposed that beta-carotene, a pigment present in many fruits and vegetables, may be responsible for reducing the damage caused by tobacco smoke in the lungs. They hypothesized that providing dietary supplements of beta-carotene would reduce the risk of lung cancer. However, in double-blind studies of heavy smokers, the risk of lung cancer was *higher* for those who took beta-carotene supplements than for those who did not (this is not true for the small amount of beta-carotene found naturally in foods). Soon after these results were reported, two other large federally funded studies using beta-carotene supplements were stopped on the basis that these supplements are ineffective in preventing both lung cancer and cardiovascular disease.

## ✅ CONCEPT CHECK 1.5

1. What are the seven steps used in the scientific method?
2. Name the various types of research studies that can be done to test a hypothesis.

# 1.6 What Is the Current State of North American Eating Patterns and Health?

## DOES OBESITY THREATEN OUR FUTURE?

There is no doubt that the obesity epidemic threatens the future health of Americans. It is estimated that 39.8% of adults were obese in 2015–2016, with *obesity* defined as having an excessive amount of body fat relative to lean tissue. Considered more broadly,

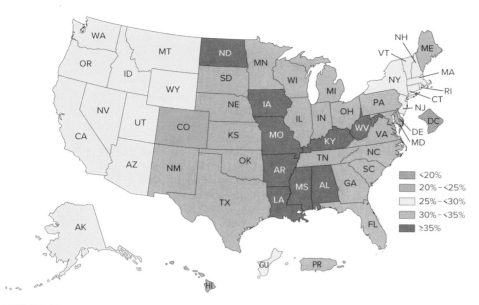

**FIGURE 1-11** ▲ Percents of adults who are obese,* by state, 2018.

Source: CDC, Prevalence of Self-Reported Obesity Among U.S. Adults by State and Territory, BRFSS, 2018.

*Body mass index (BMI) > 30, or about 30 pounds overweight for a 5′9″ person, based on self-reported weight and height.

two-thirds of adults and one-third of children are **overweight** or obese. According to the Centers for Disease Control and Prevention (CDC), in 2015–2016 the average American adult man weighed 197.9 pounds, and the average American adult woman weighed 170.6 pounds. Where you live is also a factor, with **obesity** rates varying by state. State by state, self-reported obesity data from the CDC (Fig. 1-11) indicate that in 2018, 22 states and Puerto Rico had an adult obesity rate of 30% to less than 35% and 9 states (Alabama, Arkansas, Iowa, Kentucky, Louisiana, Mississippi, Missouri, North Dakota, and West Virginia) had the highest adult obesity rates—over 35%. In 2018, most states with the lowest obesity rates were in the Northeast or the West, and the South and the Midwest had the highest prevalence of obesity.[13] In 2006, only one state was above 30%.

All the data from national surveys indicate that the adult obesity rate has been rising for decades.[14] Although obesity rates held at around 34% and 35% between 2005 and 2012, the most recent data show the rates approaching 40%. There is work being done, however, to help combat these obesity trends. The annual *State of Obesity* reports have documented how policies and programs at the local, state, and federal levels have helped Americans eat healthier.[15]

It is well documented that this extra weight will continue to have dangerous consequences. Obesity plays a role in chronic illness, including heart disease, stroke, high blood pressure, high cholesterol, diabetes, arthritis, and certain cancers. An increase in the incidence of obesity-related cancers has been found in young adults and is discussed in the *Newsworthy Nutrition* feature in this section.[16] It is estimated that obesity kills more than 200,000 Americans each year. Because of its role in so many chronic disorders, obesity is an expensive condition, with more than $190 billion spent annually on health care related to obesity. Because of numerous medical conditions, obese individuals are absent from work more often than those of healthy weight. Health economists estimate that obesity-related absenteeism costs employers as much as $8.65 billion a year, whereas loss of on-the-job productivity due to pain, shortness of breath, or other obstacles costs another $30 billion.[17] It has become obvious that the answers to the obesity crisis are not simple. From a nutrition perspective, however, the problem can be clearly stated. Most of us continue to eat too much, especially foods with a high number of calories and a low number of nutrients, and we do not engage in enough physical activity.

**overweight** A ratio of weight to height that is moderately higher than what is associated with optimal health. For adults, this is defined as BMI within the range of 25.0 up to 30. For children, this is defined as BMI-for-age from the 85th up to the 95th percentile.

**obesity** Ratio of weight to height that is significantly higher than what is associated with optimal health, usually due to excessive body fat. For adults, this is defined as BMI of 30 or higher. For children, this is defined as BMI-for-age at the 95th percentile or higher.

**COVID CORNER**

During 2020 obesity was linked to a higher risk of developing severe symptoms and complications of COVID-19, independent of other illnesses, such as cardiovascular disease.

# Newsworthy Nutrition

▲ From 1995 to 2014, the incidence of 12 obesity-related cancers increased in young adults. Maria Dryfhout/Cutcaster

## Obesity-related cancers on rise in young adults

**INTRODUCTION:** There is concern that the increased prevalence of overweight and obesity among young people could be increasing their risks of obesity-related cancers and reverse recent progress made in reducing cancer mortality. An increase in early onset colorectal cancer, which could partly reflect the obesity epidemic, was found in a previous study. **METHODS:** In this *cohort study,* 20 years of incidence data were examined for 30 common cancers, including 12 associated with obesity and excess body weight, among adults ages 25 to 84, in 25 states from the North American Association of Central Cancer Registries' Cancer in North America database. Data were separated into five-year age cohorts. **RESULTS:** Incidence of six of the 12 cancers related to obesity (colorectal, endometrial, multiple myeloma, gallbladder, kidney, and pancreatic cancers) increased significantly from 1995 to 2014 in adults between the ages of 25 and 49. Steeper rises were found in successively younger generations. For example, the risk of colorectal, endometrial, pancreas, and gallbladder cancers in millennials was found to be about double the rate baby boomers faced at the same age. Of the 18 additional cancers analyzed, the incidence of only two—non-cardia gastric cancer and leukemia—increased among young adults in the same time period. **CONCLUSIONS:** Although the results do not provide sufficient information to determine a causal relationship, the risk of developing an obesity-related cancer seems to be increasing in a stepwise manner in successively younger birth cohorts in the U.S. Further studies are needed to determine exposures responsible for these emerging trends, including excess bodyweight and other risk factors.

Source: Sung H and others: Emerging cancer trends among young adults in the USA: Analysis of a population-based cancer registry. *The Lancet* 4(3):PE137, 2019. DOI: 10.1016/S2468d-2667(18)30267-6.

## ASSESSING THE CURRENT NORTH AMERICAN EATING PATTERN

▲ Although positive changes in eating habits have begun, about half the carbohydrates that North Americans consume come from simple sugars; and the other half come from starches in foods such as pastas, breads, and potatoes. Two-thirds of protein consumption is from animal sources such as the burgers, cheese, and hot dogs shown in this pile of "junk" food. mphillips007/iStock/Getty Images

With the aim of finding out what North Americans eat, federal agencies conduct surveys to collect data about food and nutrient consumption and the connections between dietary patterns and health. In the United States, the U.S. Department of Health and Human Services monitors food consumption with the National Health and Nutrition Examination Survey (NHANES). In Canada, this information is gathered by Health Canada in conjunction with Agriculture and Agrifood Canada. Survey data from 2013–2014 indicate that North American adults consume about 16% of their calorie intake as proteins, 48% as carbohydrates, 34% as fats, and 3% as alcohol. These percentages fall within the ranges recommended by the Food and Nutrition Board (FNB) of the National Academy of Sciences. The FNB advocates that 10% to 35% of calories come from protein, 45% to 65% from carbohydrate, and 20% to 35% from fat. These standards apply to people in both the United States and Canada.

Food-consumption data also indicate that about two-thirds of protein intake is from animal sources for most North Americans, whereas plant sources supply only about one-third. In many other parts of the world, it is just the opposite: plant proteins—from rice, beans, corn, and other grains and vegetables—dominate protein intake. About half the carbohydrate in North American dietary patterns comes from simple sugars; the other half comes from starches (such as in pastas, breads, and potatoes). About 60% of dietary fat comes from animal sources and 40% from plant sources.

Evidence of positive changes in eating patterns have begun to appear. Results from the recent NHANES show that calories consumed daily by the typical U.S. adult are declining for the first time in over 40 years. One of the most significant declines has been in the amount of sugar-sweetened soda consumed. Keep in mind that while these changes in calories consumed are a step in the right direction and appear to stem from our growing awareness of the dangers of eating and drinking too much, we often do not choose the foods that will meet all our nutrient needs.

In the next section, we discuss recommendations to consume a variety of nutrient-dense foods within and across the food groups, especially whole grains, fruits, vegetables, low-fat or fat-free milk or milk products, and lean meats and other protein sources. These foods will provide nutrients that are often overlooked, including various vitamins, minerals, fiber, and many phytochemicals. Daily intake of a balanced multivitamin and mineral supplement is another strategy to help meet nutrient needs but does not make up for a poor eating pattern. Also keep in mind that use of nutrient supplements should be discussed with your primary care provider to avoid potentially harmful side effects.

Experts also recommend that we pay more attention to balancing calorie intake with needs. An excess intake of calories is usually tied to overindulgence in sugar, fat, and alcoholic beverages. Many North Americans would benefit from a healthier balance of food in their eating patterns. Moderation is the key for some foods that are high in sugar and fat calories. For other foods, such as fruits and vegetables, increased quantity and variety are warranted. Few adults currently meet the recommendation to "fill half your plate with fruits and vegetables" promoted by many health authorities.

## HEALTH OBJECTIVES FOR THE UNITED STATES

Health promotion and disease prevention have been public health strategies in North America for the past several decades. Every 10 years, the U.S. Department of Health and Human Services (HHS) issues a collection of health objectives for the nation. These objectives are developed by experts in federal agencies, target major public health concerns, and set goals for the coming decade.

In August 2020, the HHS's Office of Disease Prevention and Health Promotion released *Healthy People 2030,* the nation's 10-year plan for addressing our most critical public health priorities and challenges. *Healthy People 2030* is the fifth edition of *Healthy People,* and includes 355 core, measurable objectives with 10-year targets. Objectives are organized under five topics (1) health conditions; (2) health behaviors; (3) populations; (4) settings and systems; and, for the first time, (5) social determinants of health. *Healthy People 2030* continues to emphasize objectives from the past decade that prioritize health disparities, health equity, and health literacy. There are also new objectives related to opioid use disorder and youth e-cigarette use, and resources for adapting *Healthy People 2030* to emerging public health threats like COVID-19.

The overarching goals of *Healthy People 2030* are to:

- Attain healthy, thriving lives and well-being free of preventable disease, disability, injury, and premature death.
- Eliminate health disparities, achieve health equity, and attain health literacy to improve the health and well-being of all.
- Create social, physical, and economic environments that promote attaining the full potential for health and well-being for all.
- Promote healthy development, healthy behaviors, and well-being across all life stages.
- Engage leadership, key constituents, and the public across multiple sectors to take action and design policies that improve the health and well-being of all.

*Healthy People 2030* includes a specific nutrition topic area called Nutrition and Healthy Eating, and its overall goal is to improve health by promoting healthy eating and making nutritious foods available. The Nutrition and Healthy Eating objectives aim to help people get the recommended amounts of healthy foods—like fruits, vegetables, and whole grains—to reduce their risk for chronic

▲ An increase in the consumption of some foods, such as fruits and vegetables, can lead to a healthier balance of food in the North American eating pattern. xefstock/ Getty Images

diseases and improve their health and also focus on helping people get recommended amounts of key nutrients, like calcium and potassium.

*Healthy People 2030* includes 14 general Nutrition and Healthy Eating objectives which aim to encourage public health interventions to: (1) reduce household food insecurity and hunger; (2) eliminate very low food security in children; (3) reduce iron deficiency in children aged 1 to 2 years; and (4) increase the proportion of schools that don't sell less healthy foods and drinks. Additional objectives for people aged 2 years and over aim to increase consumption of fruits; vegetables (particularly dark green vegetables, red and orange vegetables, and beans and peas); whole grains; calcium; potassium; and vitamin D; and reduce consumption of added sugars; saturated fat; and sodium. In addition to these general nutrition objectives, there are others that relate to nutrition issues specific to different life stages and diseases, including obesity, as well as objectives that focus on physical activity as well as food safety.

Progress on the *Healthy People* objectives are evaluated throughout each respective decade. At the midcourse progress review of *Healthy People 2020* in 2016, some of the Nutrition and Weight Status objectives were met or improved, but there was little or no detectable change for several of them. Objectives that showed improvement were these: very low food security among children in the past 12 months; mean daily intake of whole grains; mean percent of total daily calorie intake from solid fats and **added sugars;** and mean total daily calcium intake. The only target that exceeded the objective was schools not offering calorically sweetened beverages.[18]

**added sugars** Nutritive sweeteners (e.g., sugars and syrups) that are not naturally present in foods, but are added during processing for the purpose of flavoring and/or preserving foods.

### ✔ CONCEPT CHECK 1.6

1. Surveys indicate that we could improve our eating patterns by increasing which types of food sources?

2. The consumption of which types of foods should be reduced to attain and maintain good health?

## 1.7 What Can You Expect from Good Nutrition and a Healthy Lifestyle?

The obesity epidemic and prevalence of chronic diseases in the United States show that something is not right with many of our eating patterns and/or lifestyles. The strong association between obesity and poor health is clear. The reverse is also well documented: when an overweight person loses just 5% to 10% of body weight, that person's risks of many chronic diseases are greatly reduced.

### HEALTHY WEIGHT

Because weight gain is one of the greatest lifelong nutrition challenges, we encourage you to seek a lifestyle that will make gaining weight more difficult and maintaining a healthy weight easier. Preventing obesity in the first place is the easiest approach. Unfortunately, many aspects of our society make it hard not to gain weight. The earlier (preferably in childhood) we develop lifestyle habits of good nutrition, regular physical activity, and the avoidance of addictions to salt, fat, sweets, high-calorie foods, and sedentary lifestyles, the better our chances for a long, healthy life. Aim to live in a city or town that has opportunities for physical activity such as bike paths, walking trails, and parks, as well as access to fresh fruits and vegetables through farmers' markets and community gardens. Seek out and join running or walking clubs. Shop at grocery stores that offer a good selection of fruits, vegetables, and other healthy foods. When dining out, choose restaurants that have tasty but healthy options on their menu.

▲ Access to fresh fruits and vegetables through farmers' markets and community gardens is important to a healthy lifestyle. Mary-Jon Ludy/McGraw-Hill Education

Fortunately, many eating habits have improved during the past decade. Today, we can choose from a wide variety of food products as a result of continual innovation by food manufacturers. Our cultural diversity, varied cuisines, and general lack of nutrient deficiencies should be points of pride for North Americans.

## LONGER, HEALTHIER LIVES

Today, North Americans live longer than ever and enjoy better general health, partly because of better medical care and dietary patterns. Affluence, however, has also led to sedentary lifestyles and high intakes of animal fat, salt, and alcohol. This lifestyle pattern has led to problems such as cardiovascular disease, hypertension, diabetes, and, of course, obesity. Greater efforts are needed by the general public to lower intake of animal fats and to improve variety in our dietary patterns, especially from fruits, vegetables, and whole grains. With better technology and greater choices, we can have a much healthier eating pattern today than ever before—if we know what choices to make!

## THE TOTAL DIETARY PATTERN

Nutrition experts generally agree that there are no "good" or "bad" foods, but some foods provide relatively few nutrients in comparison to calorie content. Health experts have prepared many reports and outlined numerous objectives to get us closer to being a "Healthy People." As you reexamine your nutritional habits, remember your health is largely your responsibility. Your body has a natural ability to heal itself. Offer it what it needs, and it will serve you well. Be aware that confusing and conflicting health messages hinder change in our eating patterns. We have addressed some of these conflicting health messages or myths in our *Fake or Fact* feature in every chapter.

Prevention of disease is an important investment of one's time, even during the college years. The following recommendations will help promote your health and prevent chronic diseases: (1) consume enough essential nutrients, including fiber, while moderating calories, solid fat, and added sugar; (2) engage in adequate, regular physical activity (at least 30 to 60 minutes on most or all days); (3) minimize alcohol intake (no more than two drinks per day for men and one drink for women); and (4) do not use tobacco products or e-cigarettes. In addition to these recommendations, you can optimize your health by getting adequate sleep (7 to 9 hours per night), consuming sufficient water (9 to 13 cups per day from foods and beverages), reducing stress, using medications prudently, and, of course, abstaining from use of illicit drugs. Because of the widespread use and abuse of alcohol, Section 1.8 provides more information on the nutritional implications of alcohol consumption.

▲ Regular physical activity complements a healthy dietary pattern. Whether it is all at once or in segments throughout the day, incorporate 30 to 60 minutes or more of such activity into your daily routine. Monkey Business Image/age fotostock

✓ **CONCEPT CHECK 1.7**

1. What are some eating patterns, physical activities, and lifestyle recommendations for health promotion and disease prevention?

## ASK THE RDN | Who's the Expert?

*Dear RDN: I am interested in making positive changes to my eating pattern to reach a healthy weight and feel better. How can I find a qualified nutrition expert who will give me personalized nutrition advice?*

You have already made a big step toward better nutrition by taking this nutrition course! The information in this textbook is written by authors who are all qualified nutrition experts, namely *registered dietitian nutritionists* (RDN). The textbook and your instructor will provide a solid foundation in nutrition, but be aware that some people call themselves "nutritionists" without qualified training in nutrition. The best approach to finding answers about your personal nutritional state is to consult your primary care provider, **registered dietitian (RD),** or **registered dietitian nutritionist (RDN).** The RD/RDN has been certified by the Commission on Dietetic Registration of the Academy of Nutrition and Dietetics (Academy) after completing rigorous classroom and clinical training in nutrition. The RD/RDN must also complete continuing education. The RD credential was recently updated to RDN to better reflect the scope of practice of dietitians. While both titles signify the same credential, we will use RDN when referring to dietitians in this book.

You can begin your search for a local RDN by asking your instructor, primary care provider, or health insurance company for a referral. You can also find an RDN by using the Academy national referral service, called *Find a Registered Dietitian Nutritionist.* This service links consumers with qualified nutrition practitioners who are members of the Academy and provide reliable, objective nutrition information. Visit the Academy's website, www.eatright.org, and click on "Find an Expert." (In Canada, visit the Dietitians of Canada website, www.dietitians.ca, and click on "Find a Dietitian.") Enter your zip code or state to display the providers in your area. Select additional specialties that may apply to your specific needs. The website will display a list of providers. A professional with the RD or RDN credential after his or her name is a qualified nutrition expert who is trained to help you separate facts from fads and optimize your health with better food choices. You can trust an RDN to translate the latest scientific findings into easy-to-understand nutrition information.

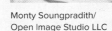

We will use this feature, *Ask the RDN,* in every chapter to answer questions about topics that may seem to have conflicting viewpoints.

Your nutrition expert,

**Anne M. Smith, PhD, RDN, LD**

Associate Professor Emeritus, The Ohio State University, Author of *Contemporary Nutrition*

Monty Soungpradith/
Open Image Studio LLC

**registered dietitian (RD)** A person who has completed a baccalaureate degree program approved by the Accreditation Council for Education in Nutrition and Dietetics (ACEND), performed at least 1200 hours of supervised professional practice, passed a registration examination, and complies with continuing education requirements.

**registered dietitian nutritionist (RDN)** The RDN is the updated credential formerly abbreviated RD. The credential was updated to better reflect the scope of practice of the dietitian and to align with the new name of the professional organization for dietitians, the Academy of Nutrition and Dietetics.

▶ An RD or RDN is a qualified nutrition expert trained to help you separate facts from fads and to optimize your health with better food choices. You can trust an RDN, like this supermarket dietitian, to translate the latest scientific findings into easy-to-understand nutrition information.
Hero Images/Getty Images

# 1.8 Nutrition and Your Health
## Nutrition Implications of Alcohol Consumption

Given the wide spectrum of alcohol use and abuse, knowledge of alcohol consumption and its relationship to overall health is essential to the study of nutrition. Alcoholic beverages contain the chemical form of alcohol known as **ethanol.** Although not a nutrient per se, alcohol is a source of calories (about 7 kcal per gram). Over half of American adults drink alcohol. On average, alcohol accounts for about 5% of total calories in the average North American dietary pattern.

The *Dietary Guidelines for Americans* defines an alcoholic drink equivalent as 14 grams of alcohol[19]. For beer or wine coolers, this equates to a 12-fluid-ounce serving. Most cans or bottles of beer are 12 fluid ounces, but some may contain as much as 40 fluid ounces. Malt liquor and most craft beers have a slightly higher alcohol content than regular beer, so the equivalent drink size is 8 fluid ounces. For wine, a 5-fluid-ounce glass is the equivalent. An equivalent drink of hard liquor, such as whiskey or rum, is the size of a shot glass—1.5 fluid ounces. The alcohol, carbohydrate, and calorie contents of standard drink sizes are depicted in Figure 1-12.

**Moderate drinking** is defined by the CDC as up to two drinks per day for men and up to one drink per day for women. **Heavy drinking** is usually defined as consuming 15 or more drinks per week for men and 8 or more drinks per week for women. **Binge drinking** is characterized by a pattern of drinking within a short period of time (usually within a few hours) causing blood alcohol concentration (BAC) to rise above the legal limit of 0.08%. It is defined as five or more drinks for men or four or more drinks for women in about two hours. Although binge drinking is certainly linked to negative effects on physical and emotional health, it is not necessarily an **alcohol use disorder,** which will be discussed next.

Moderate consumption of alcohol by most adults is viewed as an acceptable practice with mixed health benefits.[20] However, only about half of alcohol consumed is done so in moderation. One in six U.S. adults binge drinks, consuming about eight drinks per binge, about four times a month. Binge drinking is most common among younger adults ages 18 to 34 and is about twice as prevalent among men as among women.[21] Problem drinking that becomes severe is given the medical diagnosis of alcohol use disorder. Over 16 million people in the United States suffer from an alcohol use disorder (often referred to as *alcohol dependence* or *alcoholism*). By far, alcohol is the most commonly abused drug. In 2010, the cost of excessive alcohol consumption in the U.S. was almost $250 billion.[22]

## How Alcoholic Beverages Are Produced

The basis of alcohol production is fermentation, a process by which microorganisms break down simple sugars (e.g., glucose or maltose) to alcohol, carbon dioxide, and water in the absence of oxygen. High-carbohydrate foods encourage the growth of yeast, the microorganism responsible for alcohol production. Wine is formed by the fermentation of grape or other fruit juices. Beer is made from malted cereal grain. Distilled spirits (e.g., vodka, gin, and whiskey) are made from any number of fruits, vegetables, and grains. Production temperatures, the composition of the food used

Ingram Publishing

for fermentation, and aging techniques determine the characteristics of the product. Alcohol proof represents twice the volume of alcohol in percentage terms. Thus, 80 proof vodka contains 40% alcohol.

**ethanol** Chemical term for the form of alcohol found in alcoholic beverages.

**moderate drinking** For men, consuming no more than two drinks per day, and for women, consuming no more than one drink per day.

**heavy drinking** Any pattern of alcohol consumption defined as consuming 15 drinks or more per week for men and 8 drinks or more per week for women.

**binge drinking** Drinking sufficient alcohol within a 2-hour period to increase blood alcohol content to 0.08% or higher; for men, consuming 5 or more drinks in a row; for women, consuming 4 or more drinks in a row.

**alcohol use disorder** Problem drinking characterized by a compulsive pattern of alcohol use that leads to significant impairment or distress.

| | 12 fl oz of regular beer | 8–9 fl oz of malt liquor (shown in a 12 oz glass) | 5 fl oz of table wine | 3–4 fl oz of fortified wine (such as sherry or port; 3.5 oz shown) | 2–3 fl oz of cordial, liqueur, or aperitif (2.5 oz shown) | 1.5 fl oz of brandy or cognac (a single jigger or shot) | 1.5 fl oz shot of 80-proof spirits ("hard liquor"— whiskey, gin, rum, vodka, tequila, etc.) |
|---|---|---|---|---|---|---|---|

The percent of "pure" alcohol, expressed here as alcohol by volume (alc/vol), varies by beverage.

| | | | | | | | |
|---|---|---|---|---|---|---|---|
| Alcohol (%) | 5 | 7 | 12 | 17 | 24 | 40 | 40 |
| Alcohol (grams) | 14 | 12 | 15 | 12 | 23 | 15 | 14 |
| Carbohydrate (grams) | 13 | 8 | 4 | 20 | 18.5 | – | – |
| Calories (kcal) | 153 | 118 | 125 | 236 | 238 | 105 | 97 |

**FIGURE 1-12** ▲ The standard drink sizes shown provide about 14 grams of alcohol. Keep in mind that alcoholic beverages served in bars and restaurants can be 20% to 45% larger than a standard drink.

Source: NIAAA.

*magnificent*

Some fermented foods contain live microorganisms, called **probiotics**. During processing, live organisms and any microbiota benefits are removed from beer and wine. **Microbiota** refers to the entire population of microorganisms (bacteria, fungi, viruses, and parasites) living in a specific environment. **Microbiome** includes the microbiota, their genetic code, and their immediate environment. Look for interesting facts about the magnificent microbiome in every chapter.

*microbiome*

**Rethink Your Drinks** Have you ever wondered how much alcohol is in your drink? How many calories are in it? What is your cost per week, month, or year to drink? Visit the alcohol calculator at https://www.rethinkingdrinking.niaaa.nih.gov.

breath, the basis for the breathalyzer test. Most alcohol (90% to 98%), however, is metabolized. The liver is the primary site for alcohol metabolism, and some may also be metabolized by the cells lining the stomach. The main pathway of alcohol metabolism involves the enzymes **alcohol dehydrogenase** and **acetaldehyde dehydrogenase.**

As a person's alcohol consumption exceeds the body's capacity to metabolize it, blood alcohol concentration rises, the brain is exposed to alcohol, and symptoms of intoxication appear (Table 1-1). Absorption and metabolism of alcohol depend on

**probiotics** Live microorganisms that, when administered in adequate amounts, confer health benefits on the host.

**microbiota** Community of microorganisms living in a particular region; with regard to our discussion of probiotics, the community of microorganisms coexisting on and within the human body.

**microbiome** Entire collection of microorganisms, their genes, and their environment.

**alcohol dehydrogenase** An enzyme used in alcohol (ethanol) metabolism that converts alcohol into acetaldehyde.

**acetaldehyde dehydrogenase** An enzyme used in ethanol metabolism that eventually converts acetaldehyde into carbon dioxide and water.

## Absorption and Metabolism of Alcohol

Alcohol requires no digestion. It is absorbed rapidly from the GI tract by diffusion, making it the most efficiently absorbed of all calorie sources. Once absorbed, alcohol is freely distributed into all the fluid compartments within the body. About 1% to 3% of alcohol is excreted via urine, and about 1% to 5% evaporates via the

**TABLE 1-1 ■ Blood Alcohol Concentration and Symptoms**

| Concentration[a] | Sporadic Drinker | Chronic Drinker | Hours for Alcohol to Be Metabolized[b] |
|---|---|---|---|
| 50 (0.05%) | Congenial euphoria, decreased tension, and noticeable impairment in driving and coordination | No observable effect | 2–3 |
| 75 (0.075%) | Gregarious | Often no effect | 3–4 |
| 80–100 (0.08%–0.1%) | Uncoordinated, 0.08% is legal driving limit in U.S. and Canada | Minimal signs | 4–6 |
| 125–150 (0.125%–0.15%) | Unrestrained behavior, episodic uncontrolled behavior | Pleasurable euphoria or beginning of uncoordination | 6–10 |
| 200–250 (0.2%–0.25%) | Alertness lost, lethargic | Effort is required to maintain emotional and motor control | 10–24 |
| 300–350 (0.3%–0.35%) | Stupor to coma | Drowsy and slow | 10–24 |
| > 500 (> 0.5%) | Some deaths | Coma | > 24 |

[a]Milligrams of alcohol per 100 milliliters of blood.

[b]For a social drinker; alcohol metabolism is somewhat faster in chronic alcohol abusers.

Source: Modified from Goldman L, Schafer AI: *Goldman's Cecil Medicine,* 24th edition. Philadelphia: Elsevier Health Sciences, 2012.

numerous factors: genetics, sex, body size, physical condition, meal composition, gastric emptying rate, alcohol content of the beverage, use of certain drugs, chronic alcohol use, and how much sleep one has had. Women absorb and metabolize alcohol less efficiently than men. The amount of alcohol metabolized by the cells lining the stomach is greater in men than in women. Women also have less body water in which to dilute the alcohol than do men. Overall, women develop chronic alcohol-related ailments, such as cirrhosis of the liver, more rapidly than men do with the same alcohol-consumption habits.

## Moderate Alcohol Use

When used in moderation, alcohol is linked to several health benefits. Benefits of alcohol use are associated with specific intakes of no more than two drinks per day for men and no more than one drink per day for women. Socialization and relaxation are among

▲ Of all the alcohol sources, red wine in moderation is often singled out as the best choice because of the added bonus of the many studied phytochemicals present (e.g., resveratrol). These are leached out from the grape skins as the red wine is fermented. Dark beer is also a source of phytochemicals. Ingram Publishing/Alamy Stock Photo

the intangible benefits of moderate alcohol use by people of legal drinking age. In terms of physiological benefits, moderate drinkers experience lower risk of developing cardiovascular diseases and type 2 diabetes. However, the American Institute for Cancer Research (AICR) *Third Expert Report on Diet, Nutrition, Physical Activity and Cancer: A Global Perspective* states that just one drink per day can increase the risk for breast cancer. Previous consumers of alcohol no longer experience the benefits of alcohol when consumption ceases. See Figure 1-13 for the possible benefits of moderate alcohol consumption compared to the effects of heavy drinking.

## Heavy Drinking

An alcohol use disorder is a formal psychiatric diagnosis defined in the latest edition of the *Diagnostic and Statistical Manual of Mental Disorders (DSM-5)* as a problematic pattern of alcohol use leading to significant impairment or distress. This new definition integrates both alcohol abuse and alcohol dependence into a single disorder with mild, moderate, and severe subclassifications. According to the *DSM-5,* diagnosis depends on meeting two or more of the following criteria within the past year:

- Use of alcohol in larger amounts or over a longer period than intended
- Persistent desire or unsuccessful efforts at cutting down or controlling alcohol use
- Spending a great deal of time obtaining, using, and recovering from the effects of alcohol
- Experiencing cravings for alcohol
- Repeated use of alcohol that results in failure to fulfill major obligations at school, work, or home

| Effects of Moderate Drinking | Effects of Heavy Drinking |
|---|---|
| **Brain & Nervous System:** Enhanced brain function and decreased risk of dementia by increasing brain blood circulation. Some relaxation. Provides some benefit to socialization and leads to relaxation by increasing brain neurotransmitter activity. | Brain tissue damage and decreased memory. Loss of nerve sensation and nervous system control of muscles. Fragmented sleep patterns; worsens sleep apnea. Contributes to violent behavior and agitation. |
| **Cancer:** Decreased risk of colon, basal cell, ovarian, and prostate cancers; increased risk of breast cancer. | Increased risk of at least 11 cancers. |
| **Dietary Intake:** May supply some B vitamins, phytochemicals, and iron. | **Immune function:** Reduced function and increased infections. |
| **Cardiovascular System:** Decreased risk of death in those at high risk for coronary heart disease–related death primarily by increasing HDL cholesterol, decreasing blood clotting, and relaxing blood vessels. Mild decrease in blood pressure, lower rates of ischemic stroke in people with normal blood pressure. Decreased risk of peripheral vascular disease due to reduced blood clotting. | Heart rhythm disturbances, heart muscle damage, increased blood triglycerides and blood clotting. Increased blood pressure (hypertension), more ischemic and hemorrhagic stroke. Increased risk of diabetic arterial disease. |
| | **Muscle:** Skeletal muscle damage. |
| | **Liver:** Fatty infiltration and eventual liver cirrhosis (especially if a person is also infected with hepatitis C), iron toxicity. |
| **Pancreas:** Decreased risk of developing type 2 diabetes, decreased risk of death from cardiovascular disease among those diagnosed with diabetes. | Hypoglycemia, reduced insulin sensitivity, and damage to pancreas. |
| | Inflammation of the stomach and pancreas, absorptive cell damage leading to malabsorption of nutrients. |
| **GI Tract:** Decreased risk of certain bacterial infections in the stomach. | **Obesity:** Increased abdominal fat deposition; contributes to weight gain. |
| | Contributes to impotence and decreased libido in both men and women. Variety of toxic effects on the fetus when alcohol is consumed by pregnant women. |
| **Bone Health:** Some increase in bone mineral content in women, linked to estrogen output. | Loss of active bone-forming cells and eventual osteoporosis, increased risk of gout. |

**FIGURE 1-13** ▲ Possible benefits of moderate alcohol consumption are shown on the left. The risks of heavy drinking are more numerous and harmful and are shown on the right.

- Continued use of alcohol despite personal problems created by the effects of heavy drinking
- Avoiding social, occupational, or recreational activities due to use of alcohol
- Recurrent alcohol use in situations in which it is physically hazardous
- Continuing to use alcohol even after realizing one has a problem caused by heavy drinking
- Developing a **tolerance** to the effects of alcohol
- Experiencing symptoms of **withdrawal** in the absence of alcohol use

Alcohol use disorders affect about 17% of adult men and about 8% of adult women at some point in their lives. Studies suggest that about half of a person's risk for developing these disorders is genetic. Therefore, people with a family history of heavy drinking, particularly children of parents with alcohol use disorders, should be especially aware of their alcohol consumption.

Early diagnosis of alcohol use disorders can prevent multiple health problems and save millions in health care costs. Asking a person about the quantity and frequency of alcohol consumption is an important means of detecting problematic behaviors (see CAGE Questionnaire sidebar). Observable warning signs of an alcohol use disorder may include an alcohol odor on the breath, flushed face and reddened skin, nervous system disorders (such as tremors), unexplained work or school absences,

**tolerance** Needing more of a substance to achieve the desired effect (e.g., intoxication) or experiencing diminished effects of a given amount of a substance after repeated use.

**withdrawal** Physical symptoms related to cessation of substance use, such as sweating, rapid pulse, shakiness, insomnia, nausea and vomiting, anxiety, and even seizures.

## CAGE Questionnaire

The CAGE Questionnaire is used to identify alcohol use disorders. More than one positive response suggests an alcohol problem.

**C:** Have you ever felt you ought to _cut_ down on drinking?

**A:** Have people _annoyed_ you by criticizing your drinking?

**G:** Have you ever felt bad or _guilty_ about your drinking?

**E:** Have you ever had a drink first thing in the morning to steady your nerves or get rid of a hangover (an _eye-opener_)?

**FIGURE 1-14** ▲ Effects of alcohol on the liver. Alcohol is particularly damaging to this organ. Pictured are (a) a healthy liver and (b) a liver with cirrhosis. There is no cure for this disease except a liver transplant. Arthur Glauberman/Science Source

frequent accidents, and falls or injuries of vague origin. Laboratory evidence (e.g., impaired liver function, enlarged red blood cells, and elevated triglycerides) is also helpful for diagnosis of alcohol use disorders.

Despite the few benefits of regular, moderate use, the risks of heavy drinking are numerous and harmful. Although it is one of the most preventable health problems, excessive consumption of alcohol contributes to 4 of the 10 leading causes of death in North America: heart failure, certain cancers, motor vehicle and other accidents, and suicides (Figure 1-13 lists additional health risks). In the United States, about $249 billion is spent annually in terms of lost productivity, medical care, and property damage associated with alcohol use disorders. Overall, alcohol use disorders typically reduce a person's life expectancy by up to 30 years.

Alcohol is most damaging to the liver. **Cirrhosis** develops in up to 20% of cases of alcohol use disorders and is the second-leading reason for liver transplants, affecting about 2 million people in the U.S. This chronic and usually relentlessly progressive disease is characterized by fatty infiltration of the liver. Fatty liver occurs in response to increased synthesis of fat and decreased use of it for energy by the liver. Eventually, the enlarged fat deposits choke off the blood supply, depriving the liver cells of oxygen and nutrients. Liver cells eventually accumulate so much fat that they burst, die, and are replaced by connective (scar) tissue. At this stage, the liver is deemed cirrhotic (Fig. 1-14). Early stages of alcoholic liver injury are reversible, but advanced stages are not. Once a person has cirrhosis, there is a 50% chance of death within 4 years, a far worse prognosis than for many cancers. Although no specific level of alcohol consumption guarantees cirrhosis, some evidence suggests that damage is caused by a dose as low as 40 grams per day for men (3 beers) and 20 grams per day for women (1.5 beers).

Alcoholic beverages have little nutritional value, and thus nutrient deficiencies are a common result of alcohol use disorders. The protein and vitamin contents are extremely low, except in beer, where they are marginal at best. Iron content varies widely between drinks, with red wine ranking high in

**cirrhosis** A loss of functioning liver cells, which are replaced by nonfunctioning connective tissue. Any substance that poisons liver cells can lead to cirrhosis. The most common cause is chronic, excessive alcohol intake. Exposure to certain industrial chemicals also can lead to cirrhosis.

Ethnicity plays an important role in both the probability of becoming a heavy drinker and the negative health risks associated with heavy drinking. Native Americans suffer the highest rates of alcohol-related events (e.g., driving under the influence, domestic abuse, suicide, and accidents). African Americans with alcohol use disorders are at greater risk than other racial groups for tuberculosis, hepatitis C, HIV/AIDS, and other infectious diseases. Asian Americans have lower alcohol dehydrogenase production, which intensifies the unpleasant effects of alcohol at low intakes.

iron. Deficiencies arise mainly from poor nutrient intakes, but increased urinary losses and fat malabsorption (linked to poor pancreatic function) also contribute. Vitamins most susceptible to depletion from heavy drinking include vitamins A, D, E, and K; thiamin; niacin; folate; vitamins B-6 and B-12; and vitamin C. Mineral deficiencies of calcium, phosphorus, potassium, magnesium, zinc, and iron are also possible. Conversely, vitamin and mineral toxicity is also of concern with heavy drinking. Damage to the GI tract and liver, as well as high levels of some minerals in alcoholic beverages, may lead to toxicity of vitamin A, iron, lead, or cobalt. In nutritional treatment of alcohol use disorders, the immediate aim is eliminating alcohol intake, followed by restoration of nutrient stores.

Older adults are uniquely vulnerable to alcohol use disorders, perhaps due to an abundance of free time, social events involving drinking, interactions with medications, loneliness, or depression. Common symptoms of alcohol use disorders—trembling hands, slurred speech, sleep problems, memory loss, and unsteady gait—can be easily overlooked as signs of aging. Slower alcohol metabolism and decreased body water allow older adults to become intoxicated from a smaller amount of alcohol than their younger counterparts. Even moderate alcohol consumption can exacerbate some chronic health conditions, such as diabetes and osteoporosis. As well, small amounts of alcohol can react negatively with various medications used by older persons. The adverse health effects of drinking may be amplified in older adults, so the National Institute of Alcohol Abuse and Alcoholism recommends

For many people, drinking and smoking go hand in hand. **What health problems arise from the combination of these behaviors?** Ingram Publishing/Getty Images

people over the age of 65 limit alcohol consumption to no more than 7 drinks in one week.

Once a diagnosis of an alcohol use disorder is established, a primary care provider can arrange appropriate treatment and counseling for the person and his or her family. Treatment often includes the use of targeted medications, counseling, and social support. Total abstinence must be the ultimate objective. *Alcoholics Anonymous* or other reputable therapy programs can support those struggling with alcohol use disorders and their families as they recover from this devastating disease. Read more about Alcohol and Binge Drinking, including the warning signs and symptoms of alcohol poisoning, in Section 2.9, *Eating Well as a Student*.

## Guidance Regarding Alcohol Use

No government agencies recommend drinking alcohol. The *Dietary Guidelines for Americans* provide the following advice regarding use of alcoholic beverages:

- Adults of legal drinking age can choose not to drink, or to drink in moderation by limiting intake to 2 drinks or less in a day for men and 1 drink or less in a day for women, when alcohol is consumed. Drinking less is better for health than drinking more.
- Alcoholic beverages are not a component of the USDA Dietary Patterns. The amount of alcohol and calories in beverages varies and should be accounted for within the limits of healthy dietary patterns, so that calorie limits are not exceeded.
- The *Dietary Guidelines* do not recommend that individuals who do not drink alcohol start drinking for any reason.
- There are also some people who should not drink at all, such as if they are pregnant or might be pregnant; under the legal age for drinking; if they have certain medical conditions or are taking certain medications that can interact with alcohol; and if they are recovering from an alcohol use disorder or if they are unable to control the amount they drink.
- Individuals should not drink if they are driving, are planning to drive or operate machinery, or are participating in other activities requiring skill, coordination, and alertness.
- It is not safe for women to drink any type or amount of alcohol during pregnancy. Women who drink alcohol and become pregnant should stop drinking immediately and women who are trying to become pregnant should not drink at all. Alcohol can harm the baby at any time during pregnancy, even during the first or second month when a woman may not know she is pregnant.
- Not drinking alcohol also is the safest option for women who are lactating. Generally, moderate consumption of alcoholic beverages by a woman who is lactating (up to 1 standard drink in a day) is not known to be harmful to the infant, especially if the woman waits at least 2 hours after a single drink before nursing or expressing breast milk.

To learn more about alcohol use disorders, visit these websites:

- National Institute on Alcohol Abuse and Alcoholism: www.niaaa.nih.gov
- American Society of Addiction Medicine: www.asam.org
- Centers for Disease Control and Prevention: www.cdc.gov /alcohol

### ✓ CONCEPT CHECK 1.8

1. Describe two benefits and two risks of alcohol consumption.
2. What is the CAGE Questionnaire?
3. The *Dietary Guidelines* recommend that which groups refrain from alcohol consumption?

# Summary (Numbers refer to numbered sections in the chapter.)

**1.1** The flavor, texture, and appearance of foods primarily influence our food choices. Several other factors also help determine food habits and choices: food availability and convenience, early childhood experiences and ethnic customs, nutrition and health concerns, advertising, restaurants, environmental sustainability, and economics. A variety of external (appetite-related) forces affect satiety (feeling of satisfaction that halts our desire to continue eating). Hunger cues combine with appetite cues, such as easy availability of food, to promote food intake.

**1.2** Nutrition is a lifestyle factor that is a key to developing and maintaining an optimal state of health. A poor eating pattern and a sedentary lifestyle are known to be risk factors for life-threatening chronic diseases such as heart disease, hypertension, diabetes, and cancer. Not meeting nutrient needs in younger years makes us more likely to suffer poor health consequences in later years. Too much of a nutrient also can be harmful. Drinking too much alcohol is another problem associated with many health problems.

**1.3** Nutrition is the study of how the body uses food substances to promote and support growth, maintenance, and reproduction of cells. Essential nutrients in foods fall into six classes: (1) carbohydrates, (2) lipids (mostly fats and oils), (3) proteins, (4) vitamins, (5) minerals, and (6) water. The first three, along with alcohol, provide calories for the body to use. Phytochemicals are plant chemicals that may contribute to a reduced risk of disease in people who consume them.

**1.4** The body transforms the energy contained in carbohydrate, protein, and fat into other forms of energy that in turn allow the body to function. Fat provides, on average, 9 kcal per gram, whereas both protein and carbohydrate provide, on average, 4 kcal per gram. Alcohol also supplies about 7 kcal per gram. Calculating percentages and converting English units to metric units are important skills needed for the study of nutrition.

**1.5** The scientific method is the process for testing the validity of possible explanations of a phenomenon, called hypotheses. Experiments are conducted to either support or refute a specific hypothesis. Once we have enough experimental information to support a specific hypothesis, it then can be called a theory. All of us need to be skeptical of new ideas in the nutrition field, waiting until many lines of experimental evidence support a concept before adopting any suggested dietary practice. Systematic reviews, randomized controlled trials, and cohort studies provide the most reliable scientific evidence.

**1.6** The obesity problem has worsened, with 39.8% of people in the United States reported to be obese in 2015–2016. This increase is a result of eating too much, especially foods with a high number of calories and a low number of nutrients, and not engaging in enough physical activity. Results from large nutrition surveys in the United States and Canada suggest that some of us need to concentrate on consuming foods that supply more of certain vitamins, minerals, and fiber. *Healthy People 2030* is a national initiative that includes Nutrition and Healthy Eating objectives related to a healthful dietary pattern. The Nutrition and Healthy Eating objectives aim to help people get the recommended amounts of healthy foods—like fruits, vegetables, and whole grains—to reduce their risk for chronic diseases and improve their health and also focus on helping people get recommended amounts of key nutrients, like calcium and potassium.

**1.7** A basic plan for health promotion and disease prevention includes following a varied eating pattern, performing regular physical activity, not using tobacco products, not abusing nutrient supplements (if used), consuming adequate water and other fluids, getting enough sleep, limiting alcohol intake (if consumed), and limiting or appropriately coping with stress. The primary focus of nutrition planning should be on food, not on dietary supplements. The focus on foods to supply nutrient needs avoids the possibility of severe nutrient imbalances.

**1.8** Alcohol contributes approximately 7 kcal per gram, requires no digestion, and is metabolized primarily in the liver. The benefits of alcohol use are associated with low to moderate alcohol consumption. These benefits include the pleasurable and social aspects of alcohol use, a reduction in various forms of cardiovascular disease, increase in insulin sensitivity, and protection against some harmful stomach bacteria. Heavy drinking, however, contributes significantly to 5 of the 10 leading causes of death in North America. If alcohol is consumed, it should be consumed in moderation with meals. Women (and older adults, in general) are advised to drink no more than one drink per day; men should drink no more than two drinks per day.

# Check Your Knowledge (Answers are available at the end of this question set.)

1. Our primary psychological drive to eat that is affected by many external food-choice mechanisms is called
   a. hunger.   b. appetite.   c. satiety.   d. feeding.

2. Energy-yielding nutrients include
   a. vitamins, minerals, and water.
   b. carbohydrates, proteins, and fats.
   c. trace minerals and fat-soluble vitamins.
   d. iron, vitamin C, and potassium.

3. The *essential* nutrients
   a. must be consumed at every meal.
   b. are required for infants but not adults.
   c. can be made in the body when they are needed.
   d. cannot be made by the body and therefore must be consumed to maintain health.

4. Sugars, starches, and dietary fibers are examples of
   a. proteins.       c. carbohydrates.
   b. vitamins.       d. minerals.

5. Which nutrient classes are most important in the regulation of body processes?
   a. Vitamins        c. Minerals
   b. Carbohydrates   d. Both a and c

6. A food that contains 10 grams of fat would yield _____ kcal.
   a. 40       b. 70       c. 90       d. 120

7. A kcal is a
   a. measure of heat energy.
   b. measure of fat in food.
   c. heating device.
   d. term used to describe the amount of sugar and fat in foods.

8. If you consume 300 grams of carbohydrate in a day that you consume 2400 kcal, the carbohydrates will provide _____% of your total energy intake.
   a. 12.5     b. 30     c. 50     d. 60

9. Which of the following is true about North American eating patterns?
   a. Most of our protein comes from plant sources.
   b. About half of the carbohydrates come from simple sugars.
   c. Most of our fats come from plant sources.
   d. Most of our carbohydrates come from starches.

10. Alcohol is most damaging to the
    a. brain cells, because alcohol can be used as an energy source even before glucose.
    b. kidney cells, because this is where alcohol is excreted.
    c. gallbladder, because this is where alcohol is stored.
    d. liver cells, because this is where alcohol is metabolized.

Answer Key: 1. b (LO 1.1), 2. b (LO 1.3), 3. d (LO 1.3), 4. c (LO 1.3), 5. d (LO 1.3), 6. c (LO 1.4), 7. a (LO 1.4), 8. c (LO 1.4), 9. b (LO 1.5), 10. d (LO 1.8)

# Study Questions (Numbers refer to Learning Outcomes.)

1. What part of the brain controls hunger and satiety in the body? List other factors that influence our food choices. **(LO 1.1)**

2. Describe how your food preferences have been shaped by the following factors:
   a. Exposure to foods at an early age
   b. Advertising (What is the newest food you have tried?)
   c. Eating out
   d. Peer pressure
   e. Economic factors **(LO 1.1)**

3. What products in your supermarket reflect the consumer demand for healthier foods? For convenience? **(LO 1.1)**

4. Name one chronic disease associated with poor nutrition habits. Now list a few corresponding risk factors. **(LO 1.2)**

5. Describe two sources of fat, and explain why the differences are important in terms of overall health. **(LO 1.3)**

6. Identify three ways that water is used in the body. **(LO 1.3)**

7. Explain the concept of calories as it relates to foods. What are the values used to calculate kilocalories from grams of carbohydrate, fat, protein, and alcohol? **(LO 1.4)**

8. A bowl of broccoli cheddar soup contains 21 grams carbohydrate, 13 grams fat, and 12 grams protein. Calculate the percentage of kilocalories derived from fat. **(LO 1.4)**

9. List the steps of the scientific method used to test a hypothesis. **(LO 1.5)**

10. According to national nutrition surveys, which nutrients tend to be underconsumed by many North Americans? Why do you think this is the case? **(LO 1.6)**

11. List recommendations that can promote your health and prevent chronic diseases. **(LO 1.7)**

12. List two risks of heavy drinking. Should a nondrinker take up drinking for the health benefits? **(LO 1.8)**

# Further Readings

1. U.S. Bureau of Labor Statistics: *American Time Use Survey.* https://www.bls.gov/tus/home.htm#news. 2019.

2. International Food Information Council Foundation (IFIC): *2019 Food & Health Survey.* https://foodinsight.org/wp-content/uploads/2019/05/IFIC-Foundation-2019-Food-and-Health-Report-FINAL.pdf.

3. Emond JA and others: Randomized exposure to food advertisements and eating in the absence of hunger among preschoolers. *Pediatrics* 138:2361, 2016. DOI:10.1542/peds.2016-2361.

4. Enright M and Eskenazi L: *The Children's Food and Beverage Advertising Initiative in Action, A Report on Compliance and Progress During 2016.* Better Business Bureau, December 2016.

5. Green JE and others: Sociodemographic disparities among fast-food restaurant customers who notice and use calorie menu labels. *Journal of the Academy of Nutrition and Dietetics* 115:1093. 2015. DOI:10.1016/j.jand.2014.12.004.

6. Monsivais P and others: Time spent on home food preparation and indicators of healthy eating. *American Journal of Preventive Medicine* 47:796, 2014. DOI:10.1016/j.amepre.2014.07.033.

7. Sims T: *Where do Sustainable and Healthy Food Choices Intersect?* https://foodinsight.org/healthy-diets-environmental-sustainability. July 25, 2019.

8. Donnelly G and others: The effect of graphic warnings on sugary-drink purchasing. *Psychological Science* 8:1321, 2018. DOI:10.1177/0956797618766361.

9. Yanover T and Sacco WP: Eating beyond satiety and body mass index. *Eating Weight Disorders* 3:119, 2008. DOI:10.1007/BF03327612.

10. Naja F and Hamadeh R: Nutrition amid the COVID-19 pandemic: A multi-level framework for action. *European Journal of Clinical Nutrition* April 20:1–5, 2020. DOI:10.1038/s41430-020-0634-3.

11. Kochanek KD and others: Death: Final data for 2017. *National Vital Statistics Reports* 68(9), 2019.

12. Deshmukh AA and others: The association between dietary quality and overall and cancer-specific mortality among cancer survivors, *NHANES III JNCI Cancer Spectrum,* Vol 2, DOI:/10.1093/jncics/pky022. June 2018.

13. Centers for Disease Control and Prevention: *Overweight and Obesity 2019.* https://www.cdc.gov/obesity/data/prevalence-maps.html. Accessed November 10, 2018.

14. Flegal KM and others: Trends in obesity among adults in the United States, 2005 to 2014. *Journal of the American Medical Association* 315:2284, 2016. DOI:10.1001/jama.2016.6458.

15. Warren M and others: *The State of Obesity: Better Policies for a Healthier America 2019.* Trust for America's Health. Robert Wood Johnson Foundation. September 2019. tfah.org/stateofobesity2019.

16. Sung H and others: Emerging cancer trends among young adults in the USA: Analysis of a population-based cancer registry. *The Lancet* 4(3), PE137, 2019. DOI:10.1016/S2468-2667(18)30267-6.

17. Andreyeva T and others: State-level estimates of obesity-attributable costs of absenteeism. *Journal of Occupational and Environmental Medicine* 56(11), 1120, 2014. DOI:10.1097/JOM.0000000000000298.

18. National Center for Health Statistics. Chapter 29: Nutrition and Weight Status. Healthy People 2020 Midcourse Review. Hyattsville, MD. 2016.

19. U.S. Department of Agriculture and U.S. Department of Health and Human Services. *Dietary Guidelines for Americans, 2020–2025.* 9th Edition. December 2020. Available at: DietaryGuidelines.gov.

20. National Institute on Alcohol Abuse and Alcoholism: *Rethinking Drinking: Alcohol and Your Health.* December 2015. http://pubs.niaaa .nih.gov/publications/RethinkingDrinking/Rethinking_Drinking.pdf.

21. Dafna Kanny D and others: Binge drinking—United States, 2011. *Morbidity and Mortality Weekly Report (MMWR)* 62(03), 77, 2013.

22. Sacks JJ and others: 2010 National and state costs of excessive alcohol consumption. *American Journal of Preventive Medicine* 49, e73–e79, 2015. DOI:10.1016/j.amepre.2015.05.031.

# Chapter 2

# Designing a Healthy Eating Pattern

Makistock/Shutterstock

# Student Learning Outcomes

**Chapter 2 is designed to allow you to:**

**2.1** Outline the basic principles of a healthy dietary pattern using nutrient density and energy density.

**2.2** List the purpose and key recommendations of the *Dietary Guidelines* and the *Physical Activity Guidelines for Americans*.

**2.3** Design a meal that conforms to MyPlate recommendations.

**2.4** Describe the types of nutritional status and characteristics of each.

**2.5** Outline the **A**nthropometric, **B**iochemical, **C**linical, **D**ietary, and **E**nvironmental (ABCDE) measurements used in nutritional assessment.

**2.6** Describe the specific categories of nutrient recommendations within the Dietary Reference Intakes.

**2.7** Identify reliable sources of nutrition information.

**2.8** Describe the components of the Nutrition Facts label and the various health claims and label descriptors that are allowed.

**2.9** Identify food and nutrition issues relevant to college students.

## FAKE NEWS

## Healthy food is bland and boring.

## THE FACTS

Fruits and vegetables are abundant in flavor and versatile in terms of preparation. Spices and fresh herbs are all you need to jazz up a plate of healthy food. Experiment with making different sauces like homemade pesto or tahini. Homemade marinara sauce is delicious drizzled over a plate of roasted vegetables. The more herbs, spices, sauces, and dressings you can incorporate with your fruits and veggies, the greater your flavor journey will be!

How many times have you heard amazing claims about how miraculous certain foods are for you? Reading claims such as "Drink pomegranate juice to guard your body against free radicals" makes you think that food manufacturers have all the answers.

Advertising aside, nutrient intakes that are out of balance with our needs—such as excess calories, saturated fat, salt, alcohol, and **added sugars**—are linked to many leading causes of death in North America, including obesity, hypertension, cardiovascular disease, cancers, liver disease, and type 2 diabetes. Physical inactivity is also much too common. In this chapter, you will explore the components of healthy eating and lifestyle patterns—an approach that will minimize your risks of developing nutrition-related diseases. The goal is to provide you with a firm understanding of these concepts before you study the nutrients in detail. Just like the captivating photograph on the previous page, lifelong dietary patterns can be fostered from an early age and the joy of consuming nutritious foods and beverages can be passed through generations. This chapter is designed to provide you with the tools you need to craft a healthy and sustainable eating pattern to promote optimal health and wellness.

▲ A dietary pattern rich in fruits, vegetables, whole grains, nuts, seeds, and legumes will help ward off disease and control body weight. Ken Karp/McGraw-Hill Education

**nutrient density** The ratio derived by dividing a food's nutrient content by its calorie content. When the food's overall nutrient contribution exceeds its contribution to our calorie needs, the food is considered to have a favorable nutrient density.

# 2.1 A Food Philosophy That Works

You may be surprised that what you should eat to minimize the risk of developing the nutrition-related diseases seen in North America is exactly what you have heard many times before. Health professionals have recommended the same basic dietary principles for many years:

- Meet your nutritional needs primarily from foods and beverages.
- Choose a variety of options from each food group.
- Pay attention to portion size. Controlling portion size is one strategy to help achieve energy balance.

A healthy lifestyle does not have to mean deprivation and misery; it simply requires some basic nutrition know-how and planning. Many nutrition experts agree that there are no exclusively *good* or *bad* foods. Even so, many U.S. adults have eating patterns that miss the mark when it comes to the foundations of a healthy lifestyle. Eating patterns overloaded with fatty meats, fried foods, sugar-sweetened beverages, and highly processed foods can result in substantially increased risk for obesity and nutrition-related chronic diseases.

We can optimize health when we put those basic dietary principles into action. Let's now describe these principles in more detail and outline ways you can apply them in your own life. We also introduce two very important concepts that will help us to make healthy food choices: nutrient density and energy density.

## MEET NUTRITIONAL NEEDS PRIMARILY FROM FOODS AND BEVERAGES

You learned in Chapter 1 that foods and beverages provide the nutrients that serve as fuel, building blocks, and regulators for a healthy body. However, we cannot consume unlimited amounts of foods and beverages to meet our daily nutrient needs. Rather, we each have an individualized daily calorie requirement—the amount of energy we need to maintain vital functions and support physical activity.

Habitually exceeding your daily calorie "budget" can lead to undesirable changes in body weight and associated health risks. To meet our daily nutrient needs within our calorie limits, we must choose nutrient-dense foods and beverages. The **nutrient density** of a food or beverage is a comparison of its nutrient content with the amount of calories it provides. Foods and beverages are characterized as nutrient dense if they provide a large amount of a nutrient (or nutrients) relative to calories. To summarize the concept:

$$\text{Nutrient density} = \frac{\text{Amount of nutrient per serving}}{\text{Amount of calories per serving}}$$

Generally, nutrient density is evaluated with respect to individual nutrients. For example, let's compare a whole, fresh orange to a serving of orange juice. Like many fruits and vegetables, both of these options are sources of vitamin C. One medium orange provides 70 milligrams of vitamin C and 60 kcal. One cup of orange juice provides 85 milligrams of vitamin C and 120 kcal.

| **1 medium orange** | **1 cup of orange juice** |
|---|---|
| $\dfrac{70 \text{ milligrams of vitamin C}}{60 \text{ kcal}}$ = 1.2 milligrams/kcal | $\dfrac{85 \text{ milligrams of vitamin C}}{120 \text{ kcal}}$ = 0.7 milligrams/kcal |

This example shows that the whole, fresh fruit is a more nutrient-dense source of vitamin C than the fruit juice. You are more likely to meet your nutrient needs without exceeding your daily calorie needs if you choose whole fruits instead of fruit juice. This is just one example of simple swaps you can use to make more nutrient-dense choices

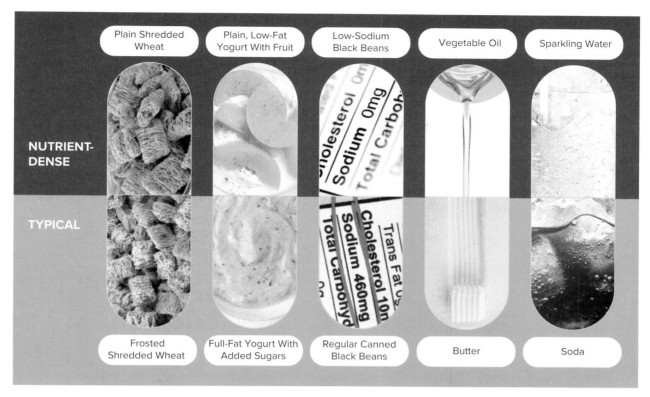

| | Plain Shredded Wheat | Plain, Low-Fat Yogurt With Fruit | Low-Sodium Black Beans | Vegetable Oil | Sparkling Water |

**NUTRIENT-DENSE**

**TYPICAL**

| | Frosted Shredded Wheat | Full-Fat Yogurt With Added Sugars | Regular Canned Black Beans | Butter | Soda |

**FIGURE 2-1** ▲ The *Dietary Guidelines* shows simple swaps to increase the nutrient density of your dietary pattern. Choose breakfast cereals with less added sugars. Choose fresh fruit or canned fruit packed in juice rather than sweetened liquids. Choose roasted or steamed vegetables instead of fried vegetables. Work to swap energy dense foods with nutrient dense foods to build a healthy dietary pattern.

Source: U.S. Department of Agriculture and U.S. Department of Health and Human Services. *Dietary Guidelines for Americans,* 2020–2025. 9th Edition. December 2020. Available at DietaryGuidelines.gov.

in your eating pattern. Figure 2-1 also shows other ways you can apply nutrient-dense swaps to improve your overall dietary patttern.

## CHOOSE A VARIETY OF OPTIONS FROM EACH FOOD GROUP

Variety in your dietary pattern means choosing foods from all the food groups and subgroups. A varied dietary pattern permits individuals to enjoy different foods and beverages tailored to meet both cultural and personal preferences. This is important because no single food will meet all your nutrient needs. For example, meat provides protein and iron but little calcium and no vitamin C. Eggs are a source of protein, but they provide little calcium because the calcium is mainly in the shell. Cow's milk contains calcium but very little iron. None of these foods contains fiber.

One way to balance your eating pattern as you consume a variety of foods is to select foods from each of these five major MyPlate food groups every day:

USDA

▲ Vegetables, such as sliced cucumbers, avocados, artichokes, and more can be added to salads, sandwiches, pizza, burritos, soups, and grain bowls to increase your phytochemical intake.

Jupiterimages/Polka Dot/Getty Images

MyPlate is a practical food guide discussed in Section 2.3, endorsed by the USDA, that provides a visual reminder and guidance to help you make smart choices from each of these food groups. MyPlate can be used in various settings and easily adapted to meet personal preferences, cultural foodways, traditions, and budgetary needs. For

James usually eats the same foods every day: corn flakes for breakfast, a ham and cheese sandwich for lunch, and meat with potatoes for dinner. **What are some practical tips James can use to increase his fruit and vegetable intake?** Image Source, all rights reserved.

**energy density** A comparison of the calorie (kcal) content of a food with the weight of the food. An energy-dense food is high in calories but weighs very little (e.g., potato chips), whereas a food low in energy density has few calories but weighs a lot (e.g., an orange).

Salsa is full of phytochemicals and a great, low-energy-density alternative to higher-calorie cheese-based chip dips. It helps to balance out the high-energy-density tortilla chips. **What is the difference between energy density and nutrient density?** Alexis Joseph/ McGraw-Hill Education

instance, a dinner consisting of a black bean burrito, a lettuce and tomato salad with oil-and-vinegar dressing, a glass of low-fat milk, and an apple covers all five food groups.

Carrots, a source of fiber that contains a pigment that forms vitamin A, may be your favorite vegetable. However, if you consume carrots every day as your only vegetable source, you may miss out on other vitamins, such as folate. Other vegetables, such as broccoli and asparagus, are rich sources of this nutrient. Hopefully, you're beginning to get a sense of how different foods and food groups vary in the nutrients they contain. For now, just recognize that you need a variety of foods in your eating pattern because the required nutrients are scattered among many foods.

An added bonus of dietary variety, especially within the fruit and vegetable groups, is the inclusion of a rich supply of phytochemicals.[1] Recall that many of these substances provide significant health benefits such as reducing the risk for cancers, diabetes, and cardiovascular disease. To obtain these beneficial plant compounds, it is important to consume a wide variety of whole, plant-based foods. Keep in mind that all forms of foods, including fresh, canned, dried, frozen, and 100% juices, in nutrient-dense forms, can be included in healthy dietary patterns. Many studies show reduced cancers and chronic disease risk among people who regularly consume a dietary pattern rich in fruits and vegetables.[2]

## PAY ATTENTION TO PORTION SIZE

The term *portion size* is used to describe the amount of a food or beverage consumed or served in one eating occasion. The term *serving size*, included on the Nutrition Facts label, can be used to help you chose appropriate portion sizes. Both of these terms are relevant to the concept of moderation. Eating in moderation requires paying attention to portion sizes and planning your daily eating pattern. It is especially important to choose foods that help you limit your intake of saturated fat, added sugars, salt, and alcohol. For example, if you plan to eat a bacon cheeseburger (relatively high in fat, salt, and calories) at lunch, you should select more nutrient-dense foods, such as fruits and vegetables, at other meals that day.[1] If you prefer whole milk to low-fat or fat-free milk, try to reduce the fat elsewhere throughout your day. Remember that dietary patterns, over time, largely predict many health outcomes. Occasional treats, eaten in moderation, are fine for most individuals. Plan for these occasions and balance them out with nutrient-dense foods throughout the rest of your week.

**Energy density** is a measurement that best describes the calorie content of a food. Energy density of a food is determined by comparing the calorie (kilocalorie) content with the weight of food. A food that is rich in calories but weighs relatively little is considered *energy dense*. Sources of energy-dense, empty calories include sugar-sweeetened beverages, cookies, fried foods, and even some fat-free snacks. Foods with low energy density include fruits, vegetables, and any food that incorporates lots of water during cooking, such as oatmeal (Fig. 2-2).

$$\text{Energy density} = \frac{\text{Amount of energy (kcal) per serving}}{\text{Weight or volume of serving}}$$

People consume fewer calories in a meal if most of the food choices are low in energy density, compared with foods high in energy density. An eating pattern low in energy density can aid in losing and managing weight.[3] Overall, foods with lots of water and fiber (i.e., low-energy-dense foods) contribute fewer calories even though they help one feel full. Alternatively, foods with high energy density must be eaten in greater amounts to promote fullness and a feeling of satiety. This is one more reason to fill your plate with a variety of fruits, vegetables, and whole grain breads and cereals—a dietary pattern that is typical of many traditional ethnic diets throughout the world.

Many foods, such as peanut butter, nuts, and seeds, are both energy and nutrient dense. Energy-dense foods can have a place in your dietary pattern, but you will have to plan for them. For example, chocolate is a very energy-dense food, but a small portion at the end of a meal can supply a satisfying finale. In addition, foods with high energy density can help individuals with poor appetites, such as some older adults, to maintain or gain weight.

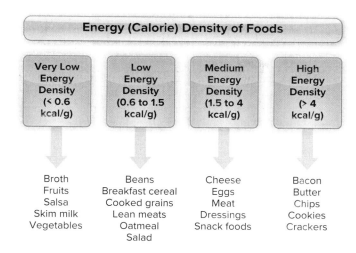

### Energy (Calorie) Density of Foods

| Very Low Energy Density (< 0.6 kcal/g) | Low Energy Density (0.6 to 1.5 kcal/g) | Medium Energy Density (1.5 to 4 kcal/g) | High Energy Density (> 4 kcal/g) |
|---|---|---|---|
| Broth Fruits Salsa Skim milk Vegetables | Beans Breakfast cereal Cooked grains Lean meats Oatmeal Salad | Cheese Eggs Meat Dressings Snack foods | Bacon Butter Chips Cookies Crackers |

**FIGURE 2-2** ◄ Energy (calorie) dense foods are those that are rich in calories relative to their nutrient content. Energy-dense foods tend to be heavily processed, high in added sugars and saturated fats, and low in fiber. Making smart swaps, like replacing a high energy-dense food with a low energy-dense food, will reduce calories and will often provide more nutrients.

Data adapted from Rolls B: *The Ultimate Volumetrics Diet.* New York: HarperCollins, 2012.

### ✓ CONCEPT CHECK 2.1

1. List three benefits of eating plenty of fruits and vegetables.
2. How would you express the concepts of nutrient density and energy density as equations?
3. Describe one change you could make to improve the nutrient density of your typical food choices.

## 2.2 Dietary and Physical Activity Guidelines

For well over 100 years, researchers have been translating the science of nutrition into practical dietary advice. Early food guidance systems aimed to reduce the risk for nutrient deficiencies, but severe nutrient deficiency diseases are no longer common in industrialized countries. Marginal deficiencies of calcium, iron, folate and other B vitamins, vitamin C, vitamin D, vitamin E, potassium, magnesium, and fiber remain a problem, but for many North Americans, major health problems more frequently stem from overconsumption of calories, added sugars, saturated fat, alcohol, and sodium.

The following sections of this chapter describe guidelines and tools for planning healthy lifestyles. You will notice how those core concepts of nutrient density, variety, and portion control keep showing up throughout our discussions of the *Dietary Guidelines,* the *Physical Activity Guidelines,* and MyPlate.

### DIETARY GUIDELINES—THE BASIS FOR MENU PLANNING

Since 1980, the U.S. Department of Agriculture (USDA) and Health and Human Services (HHS) have jointly published the *Dietary Guidelines for Americans.*[4] This newly released 2020–2025 edition of the *Dietary Guidelines* now provides recommendations for all stages of life—from birth through older adulthood, including pregnancy and breast feeding. The latest guidelines also includes a call to action—*Make Every Bite Count with the Dietary Guidelines.* Updated at least every 5 years, these science-based public health guidelines aim to improve dietary patterns to promote health, reduce risk of chronic disease, and meet nutrient needs. It is notable that the 2020–2025 *Dietary Guidelines* is the first edition to include recommendations for healthy dietary patterns for infants and toddlers.

The foods and beverages that people consume over a lifetime have a profound impact on their health status. The fundamental premise of the *Dietary Guidelines* is that all individuals, no matter their age, sex, race, ethnicity, economic circumstances, or health status, can benefit from shifting adopting healthy dietary patterns.[5] This is critical in promoting health and preventing disease. The federal government also uses these guidelines as the foundation of federal food, nutrition, and health policies and programs. In addition, the *Dietary Guidelines* are used to develop nutrition education materials for the public, such as the MyPlate resources (see http://myplate.gov).

Because the *Dietary Guidelines* have a public health orientation, these guidelines are not intended to serve as clinical recommendations for treating disease. As you learned in Chapter 1, chronic diseases result from genetic, biological, behavioral, socioeconomic, and environmental factors. Those with chronic conditions have complex health care needs that should be addressed by a primary care provider.

The *Dietary Guidelines* provides four guidelines that promote healthy dietary patterns at each stage of life (Fig. 2-3). These new guidelines also emphasize that a healthy

**Make Every Bite Count with the *Dietary Guidelines for Americans***

**Follow a healthy dietary pattern at every life stage.**
- For about the first 6 months of life, exclusively feed infants human milk.
- At about 6 months, introduce infants to nutrient-dense complementary foods.
- From 12 months through older adulthood, follow a healthy dietary pattern across the lifespan to meet nutrient needs, help achieve a healthy body weight, and reduce the risk of chronic disease.

**Customize and enjoy nutrient-dense food and beverage choices to reflect personal preferences, cultural traditions, and budgetary considerations.**
- A healthy dietary pattern can benefit all individuals regardless of age, race, or ethnicity, or current health status.
- The *Dietary Guidelines* provides a framework intended to be customized to individual needs and preferences, as well as the foodways of the diverse cultures in the United States.

**Focus on meeting food group needs with nutrient-dense foods and beverages, and stay within calorie limits. The core elements include:**
- Vegetables of all types—dark green; red and orange; beans, peas, and lentils; starchy; and other vegetables.
- Fruits, especially whole fruit.
- Grains, at least half of which are whole grain.
- Diary, including fat-free or low-fat milk, yogurt, and cheese, and/or lactose-free versions and fortified soy beverages and yogurt as alternatives.
- Protein foods, including lean meats, poultry, and eggs; seafood; beans, peas, and lentils; and nuts, seeds, and soy products.
- Oils, including vegetable oils and oils in food, such as seafood and nuts.

**Limit foods and beverages higher in added sugars, saturated fat, and sodium, and limit alcoholic beverages.**
- Added sugars—Less than 10% of calories per day starting at age 2.
- Saturated fat—Less than 10% of calories per day starting at age 2.
- Sodium—Less than 2,300 milligrams per day—and even less for children younger than age 14.
- Alcoholic beverages—Adults can choose not to drink, or to drink in moderation by limiting intake to 2 drinks or less in a day for men and 1 drink or less in a day for women.

**FIGURE 2-3** ▲ Key Recommendations from the *Dietary Guidelines for Americans*.

Source: U.S. Department of Agriculture and U.S. Department of Health and Human Services. *Dietary Guidelines for Americans, 2020 -2025.* 9th Edition. December 2020. Available at DietaryGuidelines.gov.

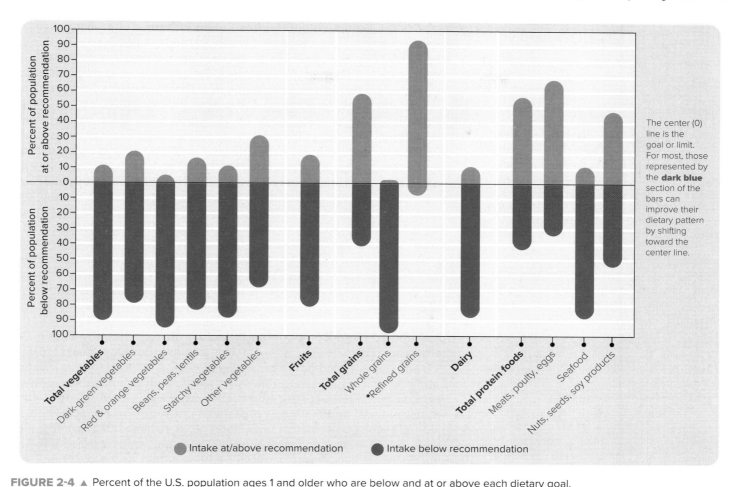

**FIGURE 2-4** ▲ Percent of the U.S. population ages 1 and older who are below and at or above each dietary goal.
Note: Recommended daily intake of whole grains is to be at least half of total grain consumption, and the limit for refined grains is to be no more than half of total grain consumption.
Source: Analysis of What We Eat in America, NHANES 2013–2016, ages 1 and older, 2 days dietary intake data, weighted. *Recommended Intake Ranges:* Healthy U.S.-Style Dietary Patterns

eating pattern is both flexible and customizable, so individuals can make tailored and affordable choices that meet their personal, cultural, budgetary, and traditional preferences.

**Healthy Eating Patterns.** The ultimate goal of a nutritious dietary pattern is to support a healthy body weight and help reduce the risk of chronic disease. The *Dietary Guidelines* use the Healthy U.S.-Style Dietary Pattern to exemplify the specific amounts of food groups and other components that make up healthy eating patterns. It is based on foods Americans typically consume, but in nutrient-dense forms and appropriate amounts. The **Healthy Eating Index (HEI)** is an objective measure of concordance with the *Dietary Guidelines*. The current HEI score in the U.S. is 59 (out of 100). Clearly, there is much room for improvement in the dietary patterns of Americans.[5]

The *Dietary Guidelines* also include the Healthy Mediterranean-Style Dietary Pattern and the Healthy Vegetarian Dietary Pattern as alternative dietary patterns. More information about healthy eating patterns will be presented later in this chapter.

As seen in Figure 2-4, the average American exceeds or nearly meets recommendations for grains and protein, whereas other food groups require significant improvement. Many Americans have dietary patterns that are low in vegetables, fruits, whole grains, and dairy and high in meat, refined grains, added sugars, saturated fats, and sodium.

**Healthy Eating Index (HEI)** A measure of diet quality that can be used to assess compliance with the *Dietary Guidelines*.

| | Calorie (kcal) Level | |
|---|---|---|
| Children | Sedentary ⟶ | Active |
| 2–3 years | 1000 ⟶ | 1400 |
| Females | | |
| 4–8 years | 1200 ⟶ | 1800 |
| 9–13 | 1400 ⟶ | 2200 |
| 14–18 | 1800 ⟶ | 2400 |
| 19–30 | 1800 ⟶ | 2400 |
| 31–50 | 1800 ⟶ | 2200 |
| 51+ | 1600 ⟶ | 2200 |
| Males | | |
| 4–8 years | 1200 ⟶ | 2000 |
| 9–13 | 1600 ⟶ | 2600 |
| 14–18 | 2000 ⟶ | 3200 |
| 19–30 | 2400 ⟶ | 3000 |
| 31–50 | 2200 ⟶ | 3000 |
| 51+ | 2000 ⟶ | 2800 |

**FIGURE 2-5** ▲ Estimates of calorie (kcal) needs based on activity levels.
Source: *Dietary Guidelines for Americans.*

**Balancing Calories Within Healthy Eating Patterns.** The balance between calories consumed (from foods and beverages) and calories expended (through physical activity and metabolic processes) determines our body weight. The total number of calories you need each day varies on a number of factors, including age, sex, height, weight, physical activity patterns, and pregnancy or lactation status. In addition, a desire to lose, maintain, or gain weight affects how many calories should be consumed. Consuming too many calories without increasing physical activity will inevitably lead to weight gain. Meeting nutrient needs within calorie limits could alleviate many chronic diseases, especially cardiovascular disease, type 2 diabetes, cancers, and osteoporosis. In light of the current epidemic of overweight and obesity, the *Dietary Guidelines* encourage all Americans to achieve and maintain a healthy body weight. Knowing how many calories you need each day is a good place to start (Fig. 2-5). You can also use an online calculator, such as NutritionCalc Plus in Connect, to calculate your estimated calorie needs. Next, try to become familiar with the calorie content of foods and beverages you commonly consume. Many free apps are available to help with this. Finally, monitoring your weight over time will allow you to see how your food and physical activity choices are balancing out.

**Customized Food Components to Include in Your Healthy Eating Pattern.** The *Dietary Guidelines* recognize our wonderfully diverse population and provide food-based recommendations to avoid being prescriptive. Eating should be an enjoyable experience, and this flexible approach encourages you to customize your meals and snacks with a variety of nutrient-dense foods and beverages that align with your own needs and preferences. Build your eating pattern on a foundation of vegetables, fruits, whole grains, seafood, eggs, beans, peas, and lentils, unsalted nuts and seeds, fat-free and low-fat dairy products, and lean meats and poultry. Use added sugars, solid fats, and salt sparingly. This type of dietary pattern will: (1) contribute to overall nutrient adequacy; (2) lower intake of added sugars, salt, and saturated fats; (3) improve gastrointestinal function; (4) aid in weight management; and (5) decrease risk for chronic diseases.[6]

A basic premise of the *Dietary Guidelines* is that nutritional needs should be met primarily from consuming whole foods. Each of the food groups provides an array of nutrients, so eating a variety of foods from each food group is important for overall health and wellness. For instance, eating a variety of vegetables from all five subgroups (dark green; red and orange; beans, peas, and lentils; starchy, and other) is recommended because each subgroup contributes different combinations of key nutrients. The *Farm to Fork* feature in this chapter highlights citrus fruit. Although citrus fruits are rich in carotenoids and key nutrients, you should strive to eat an abundance of colorful fruits to meet your needs over time. Additional fruits and vegetables are highlighted in the *Farm to Fork* features in other chapters.[7]

**Food Components to Limit in Your Healthy Eating Pattern.** Because typical North American dietary patterns contain too much added sugars, saturated fats, and sodium, the *Dietary Guidelines* emphasize the need to limit these components. Foods with added sugars and saturated fats are often energy dense (high in calories) and provide few essential nutrients. Overconsumption of these foods increases the risk for obesity, type 2 diabetes, hypertension, cardiovascular disease, and cancers.

Added sugars include sugars and other sweeteners that are added during food processing or cooking. The recommendation is to limit the intake of added sugars to less than 10% of total calories per day for those over age 2. Saturated fats are found mostly in animal sources, such as butter and beef fat, and tropical oils, such as palm oil and coconut oil. They may also be added during food processing and cooking. The recommendation is to limit intake of saturated fats to less than 10% of total

## Smart and Simple Food Swaps

| | |
|---|---|
| White bread | → Whole wheat bread |
| Sugar-sweetened cereal | → Low-sugar, high-fiber cereal with fresh fruit |
| Cheeseburger and French fries | → Hamburger and baked beans |
| Potato salad | → Three-bean salad |
| Doughnut | → Bran muffin or bagel with light cream cheese |
| Regular soft drinks | → Water or unsweetened beverages |
| Boiled vegetables | → Steamed vegetables |
| Canned vegetables (typically salted) | → Fresh or frozen vegetables |
| Fried meats | → Baked or broiled meats |
| Processed meats | → Lean beef, chicken, or fish |
| Whole milk | → Low-fat or fat-free milk |
| Mayonnaise or cream-based salad dressing | → Oil-and-vinegar dressings or light creamy dressings |

**FIGURE 2-6** ▲ Smart and simple food swaps, like those presented here, are quick and easy ways to improve overall dietary patterns by incorporating more nutrient-dense foods into your eating plan.

calories per day for age 2 and above. Sodium should be limited to less than 2300 mg per day for those 14 years and older.

Finally, the *Dietary Guidelines* encourage the use of multiple strategies across all segments of society to promote healthy eating and physical activity behaviors. Figure 2-6 provides examples of ways to incorporate the *Dietary Guidelines* into practice by making smart and simple food swaps. The *Newsworthy Nutrition* feature in this chapter also highlights the positive health impacts associated with high-quality dietary patterns.

## FARM to FORK · Citrus Fruit

Jason Patrick Ross/Shutterstock

Oranges, tangelos, grapefruit, lemons, and limes make up the glorious and sunny citrus fruit family. The most popular citrus is the navel orange, easily identified by its *belly button* on the blossom end of the fruit. Indeed, more than 11 million tons of oranges are grown in the United States each year. Yet few know that the most nutrient-dense part of citrus is the white inner membrane found just below the skin. Called the *pith*, this spongy substance contains the phytochemicals naringenin, hesperidin, and others that have known **antioxidant,** antibacterial, antiviral, anti-inflammatory, and anti-allergenic properties!

### Grow
- Most citrus is grown in California, Arizona, and Texas, while most orange juice and grapefruit are produced in Florida.
- To grow an indoor citrus tree, buy premixed potting soil formulated specifically for citrus trees. Most citrus trees require 8 to 12 hours of sunlight each day in a south-facing window.

### Shop
- The first crop of U.S. oranges typically hits the supermarket shelves in October each year.
- When purchasing citrus, look for the largest fruits with the deepest colors; these have had more time to ripen on the tree.
- Instead of purchasing bottled orange juice, select juice concentrate to boost your flavonoid intake by 45%.

### Store
- Citrus fruits do not continue to ripen after they have been picked, so it is best to enjoy them soon after purchase.
- If you can't enjoy citrus fruits within a few days of purchasing, refrigerate them. Do not place them in a plastic bag as this will increase moisture and mold.
- If you cannot eat your oranges within a few weeks, squeeze them and make orange juice!

### Prep
- Many recipes call for citrus peels or zest to add a splash of exotic flavor to favorite dishes.
- Sliced citrus makes a perfect addition to salads and adds an ample dose of phytochemicals to any meal.
- Citrus juice is also a component of many recipes and dressings.
- And let's not forget the simple pleasure of adding citrus slices to enhance the flavor of ice water.

Source: Robinson J. Citrus fruits: Beyond vitamin C, in *Eating on the Wild Side.* New York, NY: Little, Brown and Company, 2013.

lynx/iconotec.com/Glow Images

## PHYSICAL ACTIVITY GUIDELINES FOR AMERICANS

In line with its goal for all Americans to live healthier, more prosperous, and more productive lives, the U.S. Department of Health and Human Services updated and published the second *Physical Activity Guidelines for Americans* as a complement to the *Dietary Guidelines.*[8,9] The overarching idea continues to be that regular physical activity for people of all ages, races, ethnicities, and physical abilities produces immediate health benefits.

The key physical activity guidelines provide measurable physical activity standards for Americans ages 3 and older. Specific recommendations also apply to special population groups, including pregnant and postpartum women, adults with disabilities, and people with chronic medical conditions. The adult guidelines are presented in Table 2-1. You will learn more about specific physical activity guidelines for children, adolescents, and older adults in subsequent chapters. Overall, the main message is to *move more and sit less.* For adults, major health benefits occur with at least 150 to

**aerobic** Requiring oxygen; with reference to physical activity, all forms of activity that are intense enough and performed long enough to maintain or improve an individual's cardiorespiratory fitness.

**anaerobic** Not requiring oxygen; with reference to physical activity, high-intensity activity that exceeds the capacity of the cardiovascular system to provide oxygen to muscle cells for the usual oxygen-consuming metabolic pathways.

**TABLE 2-1 ■ Types of Physical Activity and Adult Recommendations**

| Type | Description |
|---|---|
| Aerobic | Includes forms of activity that are intense enough and performed long enough to maintain or improve an individual's cardiorespiratory fitness. |
| Anaerobic | Refers to high-intensity activity that exceeds the capacity of the cardiovascular system to provide oxygen to muscle cells for the usual oxygen-consuming metabolic pathways. |
| Muscle-strengthening | Activities that maintain or improve muscular strength, endurance, or power. |
| Bone-strengthening | Movements that create impact and muscle-loading forces on bone. |
| Balance training | Training activities and movements that safely challenge postural control. |
| Flexibility training | Also called stretching, these activities improve the range and ease of movement around a joint. |
| Mind-body | Typically combines muscle strengthening, balance training, light-intensity aerobic activity, and flexibility in one package. |

| Age | Description |
|---|---|
| Adults | Adults who sit less and do any amount of moderate to vigorous physical activity gain health benefits.<br>• For substantial health benefits, adults should do at least 150 to 300 minutes a week of moderate-intensity, or 75 to 150 minutes a week of vigorous-intensity, aerobic physical activity, or an equivalent combination of moderate and vigorous-intensity aerobic activity. Preferably, aerobic activity should be spread throughout the week.<br>• Additional health benefits are gained by engaging in physical activity beyond 300 minutes of moderate-intensity physical activity each week.<br>• Adults should also do muscle-strengthening activities of moderate or greater intensity that involve all major muscle groups on 2 or more days each week. |
| Guidelines for safe physical activity | To do physical activity safely and reduce risk of injuries and other adverse events, people should:<br>• Understand the risks, yet be confident that physical activity can be safe for almost everyone.<br>• Choose types of physical activity that are appropriate for their current fitness level and health goals, because some activities are safer than others.<br>• Increase physical activity gradually over time to meet key guidelines or health goals. Inactive people should "start low and go slow" by starting with lower-intensity activities and gradually increasing how often and for how long activities are done.<br>• Protect themselves by using appropriate gear and sports equipment, choosing safe environments, following rules and policies, and making sensible choices about when, where, and how to be active.<br>• Be under the care of a health care provider if they have chronic conditions or symptoms. People with chronic conditions and symptoms can consult a health care professional or physical activity specialist about the types and amounts of activity appropriate for them. |

Source: *Physical Activity Guidelines for Americans* (https://health.gov/paguidelines/second-edition).

# Newsworthy Nutrition

## Impact of diet quality on health outcomes

**INTRODUCTION:** Improving dietary patterns and quality is a modifiable behavior that has been shown to improve health outcomes in previous studies. **OBJECTIVE:** This *systematic literature review* and *meta-analysis* were conducted to determine if diet quality, measured in terms of the Healthy Eating Index, the Alternate Healthy Eating Index, and the Dietary Approaches to Stop Hypertension score, influenced health status for U.S. adults. The authors hypothesized that diets of the highest quality would be associated with lower risk of noncommunicable disease. **METHODS:** A literature search was performed to identify studies published from 2014 to 2017 using electronic databases PubMed, Scopus, and Embase. Summary risk ratios (RRs) and confidence intervals were analyzed for over 1,670,000 participants and stratified by high versus low adherence categories. **RESULTS:** Higher dietary pattern scores were associated with a significant reduction in the risk of all-cause mortality (22%), cardiovascular disease (22%), cancer (16%), type 2 diabetes (18%), and neurodegenerative disease (15%). High-quality diets were also associated with a significant reduction in the risk of overall mortality (12%) and cancer mortality (10%) among cancer survivors. **CONCLUSION:** Diets of the highest quality, as assessed by the HEI, AHEI, and DASH score, resulted in a significant risk reduction for disease and all-cause mortality.

Source: Schwingshackl L and others: Diet quality as assessed by the Healthy Eating Index, Alternate Healthy Eating Index, Dietary Approaches to Stop Hypertension Score, and health outcomes: An updated systematic review and meta-analysis of cohort studies. *J Acad Nutr Diet* 2018;118(1)74.

*magnificent*

300 minutes per week of moderate-intensity aerobic activity. Ideally, adults should also engage in muscle-strengthening activities at least 2 days each week. Children and adolescents should strive to include 60 minutes of physical activity per day. For optimum benefits, include both aerobic and muscle-strengthening activities. Overall, physical activity should be enjoyable and safe for each individual.

**As with nutritious dietary patterns, regular physical activity is linked to increased beneficial microbial species, greater microbial diversity, enhanced short-chain fatty acid synthesis, and improved metabolism of carbohydrates.**

### ✓ CONCEPT CHECK 2.2

1. What are four of the key guidelines of the *Dietary Guidelines for Americans?*
2. The 2020–2025 *Dietary Guidelines* includes recommendations for which life stages?
3. How many minutes of moderate-intensity physical activity are advised per week in the *Physical Activity Guidelines for Americans?*

*microbiome*

## 2.3 MyPlate—A Menu-Planning Tool

The titles, food groupings, and shapes of food guides have evolved since the first edition published by the USDA a century ago. The most recent food-guidance systems have provided a means for individualization of dietary advice via interactive tools availale online. MyPlate shapes the key recommendations from the *Dietary Guidelines* into an easily recognizable and extremely applicable visual: a meal place setting (Fig. 2-7). MyPlate also serves as a reminder of how to build a healthy plate at mealtimes.[10] It emphasizes important areas of the American diet that are in need of improvement. Recall from the discussion of the *Dietary Guidelines* that Americans need to increase the relative proportions of fruits, vegetables, whole grains, and fat-free or low-fat dairy products while simultaneously decreasing consumption of refined grains and high-fat meats.

The MyPlate icon (Fig. 2-7) includes five food groups which are detailed below:

1. Vegetables should include a variety of colors. This nutrient-dense food group includes all fresh, frozen. canned, dried, and cooked vegetables.
2. Fruits include all fresh, frozen, canned, dried, and 100% fruit juices. Fruits and vegetables should cover half of your plate.

**FIGURE 2-7** ▲ MyPlate is a visual representation of the advice contained in the *Dietary Guidelines for Americans*. Use MyPlate to assist with meal planning. **How many food groups are represented in this meal?**
Alexis Joseph/McGraw-Hill Education;
U.S. Department of Agriculture

Focus on nutrient-rich foods as you strive to meet your nutrient needs. The more colorful the food on your plate, the greater the content of nutrients and phytochemicals. **Can you name the MyPlate sources of phytochemicals?** ©Mary-Jon Ludy, Bowling Green State University, Garden of Hope images

3. Grains should cover just over 25% of your plate with a goal of eating at least 50% of all grains as whole grains.
4. Dairy and fortified soy alternatives choices should be fat-free or low-fat. Aim for 2 to 3 cups per day.
5. Protein foods include meats, poultry, eggs, seafood, nuts, seeds, and soy products. Try to focus on lean and plant-based proteins in your dietary pattern.

MyPlate does not display a separate group for fats and oils, as they are mostly incorporated into other foods. The MyPlate food guide recommends limiting solid fats and focusing instead on plant oils, which are sources of essential fatty acids and vitamin E.

## MAKE EVERY BITE COUNT WHEN BUILDING A HEALTHY EATING PATTERN

MyPlate can help you build your own dietary pattern that you can tailor and maintain over time. Remember a healthy dietary pattern is essential at every stage of life and has a profound impact on both physical and mental health. The MyPlate.gov website provides ideas and tips to help you create a healthier eating style that meets your individual needs and improves your health. If you prefer to use apps, try the *Start Simple with MyPlate App*.

Think about how you can put the following guidelines into action, over the course of your day or week, to create a healthy and sustainable eating routine.

### Choices Matter—Focus on Those that Work for You!

- Focus on making healthy food and beverage choices from all five food groups, including fruits, vegetables, grains, protein foods, and dairy, to get the nutrients you need.
- Eat the appropriate amount of calories for you based on your age, sex, height, weight, and physical activity level to achieve and maintain a healthy weight and reduce your risk of many chronic diseases.

### Choose an Eating Style Low in Saturated Fat, Sodium, and Added Sugars

- Use Nutrition Facts labels and ingredient lists to locate the amounts of saturated fat, sodium, and added sugars in the foods and beverages you choose.
- Incorporate food and beverage choices that are lower in saturated fat, sodium, and added sugars.

### Make Small Changes to Create a Healthier Eating Pattern

- Think of each positive change as a personal "win" on your path to living healthier. Create little victories that fit into your lifestyle and celebrate them!
- Start with a few of these small changes:
  ○ Make half your plate fruits and vegetables.
    - Focus on whole fruits.
    - Vary your veggies to consume a rainbow of colors.
  ○ Make half your grains whole grains.
  ○ Move to low-fat and fat-free dairy or alternatives.
  ○ Vary your lean protein routine.
  ○ Eat and drink the right amount to support a healthy weight.

## MyPLATE DAILY PLAN

On the myplate.gov website, you will find an interactive tool, *MyPlate Plan,* that estimates your calorie needs and suggests a food pattern based on your age, sex, height, and weight (Table 2-2). These daily food plans provide useful information for each food group, including recommended daily amounts in common household measures. Modified daily food plans are also available for preschoolers, pregnant or breastfeeding mothers, and those interested in losing weight. Be sure to visit the site to generate your own daily food plan.

**TABLE 2-2 ■ Healthy U.S.-Style Dietary Pattern for Adults Ages 19 Through 59 from MyPlate Food Groups**

| | Daily Amount of Food from Each Group Based on Calorie Level | | | | | |
|---|---|---|---|---|---|---|
| Calorie Level | 1200 | 1600 | 2000 | 2400 | 2800 | 3200 |
| Fruits | 1 cup | 1.5 cups | 2 cups | 2 cups | 2.5 cups | 2.5 cups |
| Vegetables | 1.5 cups | 2 cups | 2.5 cups | 3 cups | 3.5 cups | 4 cups |
| Grains | 4 oz-eq | 5 oz-eq | 6 oz-eq | 8 oz-eq | 10 oz-eq | 10 oz-eq |
| Protein | 3 oz-eq | 5 oz-eq | 5.5 oz-eq | 6.5 oz-eq | 7 oz-eq | 7 oz-eq |
| Dairy | 2 cups | 3 cups | 3 cups | 3 cups | 3 cups | 3 cups |

Note: oz-eq stands for ounce equivalent.

A "cup" of **Fruit** is equivalent to 1 cup of whole fruit, 1 cup of fruit juice, or ½ cup of dried fruit.

A "cup" of **Vegetables** is equivalent to 1 cup of raw or cooked vegetables, 1 cup of vegetable juice, or 2 cups of raw, leafy greens.

An "ounce-equivalent" of **Grains** is equal to 1 slice of bread, 1 cup of ready-to-eat breakfast cereal, or ½ cup of cooked pasta, rice, or cereal.

An "ounce-equivalent" of **Protein** refers to 1 ounce of meat, fish, or poultry, 1 egg, 1 tablespoon of nut butter, ¼ cup of cooked legumes, or ½ ounce of nuts or seeds. NOTE: Dry beans and peas can count either as vegetables or protein foods.

A "cup" of **Dairy** is equivalent to 1 cup of milk, soy milk, or yogurt, 1.5 ounces of natural cheese, or 2 ounces of processed cheese.

The recommended serving sizes are provided in cups for vegetables, fruits, and dairy foods. Grains and protein foods are listed in ounces. Figure 2-8 shows a convenient guide to estimate common serving-size measurements and a description of what counts as a MyPlate serving. Pay close attention to the stated serving size for each choice when following your daily food plan to help control your calorie intake. Common household units are also listed in Appendix F with their metric equivalents. Note that ounces and fluid ounces differ: ounces are a measure of weight (think solid foods), whereas fluid ounces are a measure of volume (think liquids).

*Discretionary calories* are excess calories to enjoy once your required nutrient needs are met. MyPlate sets limits for discretionary calories from saturated fats, alcohol, and added sugars. Other than calories, these sources contribute few nutrients. Saturated fats are solid at room temperature and include butter, beef fat, and shortening. Some saturated fats, such as the white marbling in a ribeye steak and the fat contained in milk, are naturally present in foods (Fig. 2-9). Others, such as the shortening used to make a flaky croissant, are added during food processing or preparation. Added sugars include sugars and syrups that are added to foods during processing or preparation. Examples of foods that are major contributors of discretionary calories in the American dietary pattern are cakes, cookies, pastries, soft drinks, energy drinks, cheese, pizza, ice cream, and processed meats. The MyPlate Plan also makes some allowance for discretionary calories throughout the day. About 85% of the calories you consume daily are needed to meet the nutrient-dense food group DGA recommendations. The remaining 15% of calories are discretionary calories that may be used for added sugars or saturated fat intake. For most Americans, this equates to 250 to 350 discretionary calories per day.

## MENU PLANNING WITH MYPLATE

Overall, MyPlate exemplifies the foundations of a healthy eating pattern you have already learned. To achieve optimal nutrition, remember the following points when using MyPlate to plan your daily menus:

- Use a nutrient-dense, whole-food philosophy to build your own plate.
- Variety is a key to successful implementation of MyPlate. There is no single, perfect food. Likewise, no food group is more important than another; each food group

## PORTION SIZES

### Fruits

1/2 cup sliced fruit
Small/medium fruit

= ½ to ⅔ cup

### Vegetables

Large apple or orange
1 cup green leafy vegetables
1 cup broccoli

= 1 cup

### Grains/ Carbohydrates

1/2 small bagel
1/2 cup pasta
1/2 baked potato

= 1 slice bread

### Protein

3 oz meat, poultry, or fish

= ½ to ¾ cup

### Fats/Dairy

2 tbsp salad dressing, peanut butter
1 oz hard cheese

= 2 tbsp

**FIGURE 2-8** ▲ A tennis ball, softball, hockey puck, deck of cards, and dice are standard-size objects that make convenient guides for judging serving sizes. C Squared Studios/Getty Images (tennis ball); shpakdm/Shutterstock (softball); C Squared Studios/ Getty Images (hockey puck); Photodisc Collection/ Getty Images (stack of playing cards); Ron Chapple Stock/FotoSearch/Glow Images (red dice)

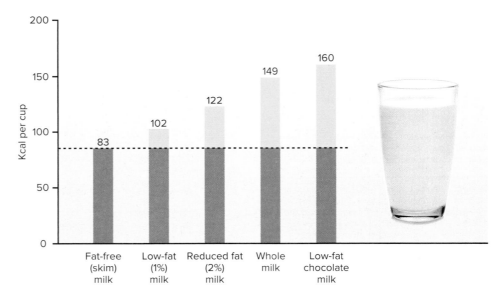

**FIGURE 2-9** ▲ Not all dairy is the same. This bar graph compares the difference in calories from various types of milk. The yellow bars represent the added calories from fat and sugar in various milks compared to fat-free milk. Note that milk also contains some natural sugars contributing to calories. (Milk) NIPAPORN PANYACHAROEN/Shutterstock

makes an important, distinctive contribution to nutritional intake. Choose a variety of foods from each food group. For a sample meal plan, see Fig. 2-10. The foods within a group may vary widely with respect to nutrients and calories. For example, the calorie content of 3 ounces of baked potato is 79 kcal, whereas 3 ounces of potato chips is 452 kcal.

- Choose primarily low-fat or fat-free items from the dairy group. By reducing calorie intake in this way, you can select more items from other food groups. If milk causes intestinal gas and bloating, consider fortified soy alternatives, yogurt, or cheese.
- Include plant foods that are good sources of protein, such as nuts, seeds, soy products, beans, peas, and lentils, at least several times a week because many are rich in vitamins (such as vitamin E), minerals (such as magnesium), and fiber.
- For vegetables and fruits, try to include a dark-green or orange vegetable for vitamin A, and a vitamin C-rich fruit, such as an orange, every day. Try not to focus primarily on starchy vegetables for your main vegetable choice. According to the Centers for Disease Control and Prevention (CDC), only 12% of American adults meet the guidelines for fruit, and 9% meet the standard for vegetables.[11] Increased consumption of these foods is important because they contribute vitamins, minerals, fiber, and phytochemicals.
- Choose whole grain varieties of breads, cereals, rice, and pasta because they contribute nutrients such as vitamin E and fiber. A daily serving of a whole grain, ready-to-eat breakfast cereal is an excellent choice because the vitamins (such as vitamin B-6) and minerals (such as zinc), along with fiber, help fill in common nutritional gaps.
- Include some unsaturated plant oils on a daily basis, such as those in salad dressing, and try to eat fish at least twice a week. This supplies you with health-promoting essential fatty acids.

**MyPlate Build-a-Meal Wizard**

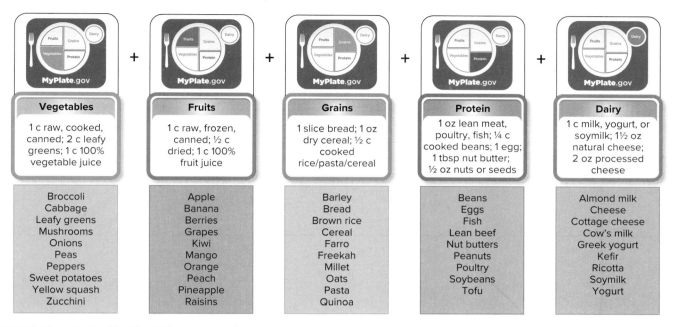

| Vegetables | Fruits | Grains | Protein | Dairy |
|---|---|---|---|---|
| 1 c raw, cooked, canned; 2 c leafy greens; 1 c 100% vegetable juice | 1 c raw, frozen, canned; ½ c dried; 1 c 100% fruit juice | 1 slice bread; 1 oz dry cereal; ½ c cooked rice/pasta/cereal | 1 oz lean meat, poultry, fish; ¼ c cooked beans; 1 egg; 1 tbsp nut butter; ½ oz nuts or seeds | 1 c milk, yogurt, or soymilk; 1½ oz natural cheese; 2 oz processed cheese |
| Broccoli<br>Cabbage<br>Leafy greens<br>Mushrooms<br>Onions<br>Peas<br>Peppers<br>Sweet potatoes<br>Yellow squash<br>Zucchini | Apple<br>Banana<br>Berries<br>Grapes<br>Kiwi<br>Mango<br>Orange<br>Peach<br>Pineapple<br>Raisins | Barley<br>Bread<br>Brown rice<br>Cereal<br>Farro<br>Freekah<br>Millet<br>Oats<br>Pasta<br>Quinoa | Beans<br>Eggs<br>Fish<br>Lean beef<br>Nut butters<br>Peanuts<br>Poultry<br>Soybeans<br>Tofu | Almond milk<br>Cheese<br>Cottage cheese<br>Cow's milk<br>Greek yogurt<br>Kefir<br>Ricotta<br>Soymilk<br>Yogurt |

**FIGURE 2-10** ▲ The Build-a-Meal Wizard can guide you through creating nutritious meals using the MyPlate food groups. Aim to incorporate 4 or 5 foods from each different food group for each meal. The serving sizes listed (in the white boxes above) will assist with portion control.
Adapted from USDA https://www.myplate.gov/eat-healthy/healthy-eating-budget/make-plan

## HOW DOES YOUR PLATE RATE?

The *Dietary Guidelines* emphasize the totality of your eating patterns over all life stages. In other words, a dietary pattern is more than the sum of its individual parts. Healthy eating patterns should not be rigid plans that are difficult to follow. Rather, dietary patterns should be flexible and adapted to include foods you enjoy that meet your personal preferences and fit within your lifestyle, traditions, culture, and budget.

Comparing your daily intake with your personalized food plan recommendations is a simple way to evaluate the quality of your overall eating pattern. Identify the nutrients that are suboptimal in your eating pattern based on the nutrients found in each food group. For example, if you do not consume enough from the dairy group, your calcium intake is most likely too low. Look for foods that you enjoy that supply calcium, such as calcium-fortified orange juice or Greek yogurt.

Food quality is just as important as food quantity when it comes to good nutrition. Review the simple meal and snack swaps leading to simple and more nutrient-dense dietary patterns (Fig. 2-11). For a more detailed analysis of your current eating pattern, use the NutritionCalc Plus function in Connect to compare your food choices to MyPlate. Also ask your instructor about the "Rate Your Plate: Does Your Diet Compare to MyPlate?" in Connect. With a detailed dietary analysis, you can compare your intakes of individual nutrients to recommendations and clearly see the areas that need improvement. Even small changes to your dietary and physical activity patterns can have positive results.

## MEDITERRANEAN DIET PYRAMID

The Mediterranean Diet Pyramid is a useful alternative to MyPlate. It is based on the dietary patterns of the Southern Mediterranean region, which has enjoyed low rates of chronic diseases and long life expectancies.[12,13] As shown on the pyramid in

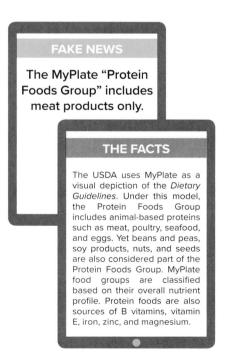

**FAKE NEWS**

The MyPlate "Protein Foods Group" includes meat products only.

**THE FACTS**

The USDA uses MyPlate as a visual depiction of the *Dietary Guidelines*. Under this model, the Protein Foods Group includes animal-based proteins such as meat, poultry, seafood, and eggs. Yet beans and peas, soy products, nuts, and seeds are also considered part of the Protein Foods Group. MyPlate food groups are classified based on their overall nutrient profile. Protein foods are also sources of B vitamins, vitamin E, iron, zinc, and magnesium.

# Smart MyPlate Meal Swaps

## Breakfast Swaps

**Egg, Sausage & Cheese Breakfast Burrito**

| Calories | Saturated Fat | Sodium | Added Sugars |
|---|---|---|---|
| 629 | 14g | 1178mg | 1g |

How this food fits into MyPlate:

**2 Tacos on Corn Tortillas with Egg, Black Beans, Cheese & Salsa**

| Calories | Saturated Fat | Sodium | Added Sugars |
|---|---|---|---|
| 356 | 4g ↓ | 921mg ↓ | 0g ↓ |

How this food fits into MyPlate:

## Lunch Swaps

**Peanut Butter & Jelly Sandwich & Small Bag of Potato Chips**

| Calories | Saturated Fat | Sodium | Added Sugars |
|---|---|---|---|
| 557 | 4g | 543mg | 12g |

How this food fits into MyPlate:

**Peanut Butter & Banana Sandwich with Baby Carrots**

| Calories | Saturated Fat | Sodium | Added Sugars |
|---|---|---|---|
| 313 | 2g ↓ | 379mg ↓ | 3g ↓ |

How this food fits into MyPlate:

## Dinner Swaps

**Fettuccine Alfredo with Italian Sausage**

| Calories | Saturated Fat | Sodium | Added Sugars |
|---|---|---|---|
| 964 | 26g | 1581mg | 1g |

How this food fits into MyPlate:

**Whole Wheat Spaghetti & Meatballs with a Garden Salad**

| Calories | Saturated Fat | Sodium | Added Sugars |
|---|---|---|---|
| 533 | 4g ↓ | 857mg ↓ | 3g |

How this food fits into MyPlate:

## Snack Swaps

**Carrots & Celery with Creamy Dip**

| Calories | Saturated Fat | Sodium | Added Sugars |
|---|---|---|---|
| 193 | 5g | 317mg | 0g |

How this food fits into MyPlate:

**Carrots & Celery with Hummus**

| Calories | Saturated Fat | Sodium | Added Sugars |
|---|---|---|---|
| 137 | 1g ↓ | 232mg ↓ | 0g |

How this food fits into MyPlate:

**FIGURE 2-11** ▲ These quick and simple meal swaps will result in a more nutrient-dense dietary pattern. Note the nutrient comparisons under each meal and the colored MyPlate icons depicting the food groups represented in each meal. Source: *Dietary Guidelines for Americans, 2020–2025.* 9th Edition. December 2020. Available at DietaryGuidelines.gov.

Solid fats contribute over 15% of total calories in the typical American's dietary pattern, but they have little to offer in terms of essential nutrients and dietary fiber. Instead of solid fats, healthier fats, like those pictured here, are recommended. **What types of fats are naturally found in salmon, avocado, olives, almonds, and plant-based oils?** Mary-Jon Ludy /McGraw-Hill Education

Figure 2-12, foods from plant sources form the foundation of this plan. Eat 5 to 10 servings of fruits and non-starchy vegetables a day. For most produce, each serving is ½ cup cooked or 1 cup raw. In this plan, olives, olive oil, and avocados are considered healthy sources of fat. You also can include a handful of nuts or seeds daily as good sources of healthy fats, fiber, and protein. Legumes (beans) are also plant sources of fiber and protein. Eat ½ cup of cooked legumes, including hummus, at least twice a week. Up to four (1-ounce) portions of whole grain bread, pasta, or quinoa are included each day. A serving of grains is 1 slice of bread or ½ cup of cooked pasta, rice, or cereal. Plenty of herbs and spices are also used in the Mediterranean dietary plan and have known antioxidant and anti-inflammatory properties.

As you make your way up the pyramid, eat a 4-ounce serving of fish two to three times a week. Moderate portions (3 or 4 ounces) of lean meat and poultry or 2 eggs are recommended once a week or every few days. Choose low-fat dairy (up to three 1-cup servings a day) from cultured sources such as yogurt or kefir. These are easier to digest and supply beneficial bacteria.

Water is always the beverage of choice. Very moderate drinking of wine (one to two 5-ounce glasses of red wine per day) has some health benefits - but proceed with caution. Finally, remember that regular physical activity that promotes a healthy weight is an important aspect of the Mediterranean lifestyle as is eating meals with family and friends.

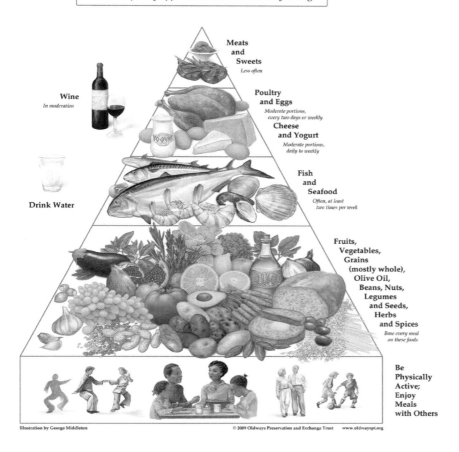

**Mediterranean Diet Pyramid**

*A contemporary approach to delicious, healthy eating*

Meats and Sweets
*Less often*

Wine
*In moderation*

Poultry and Eggs
*Moderate portions, every two days or weekly*

Cheese and Yogurt
*Moderate portions, daily to weekly*

Fish and Seafood
*Often, at least two times per week*

Drink Water

Fruits, Vegetables, Grains (mostly whole), Olive Oil, Beans, Nuts, Legumes and Seeds, Herbs and Spices
*Base every meal on these foods*

Be Physically Active; Enjoy Meals with Others

Illustration by George Middleton          © 2009 Oldways Preservation and Exchange Trust     www.oldwayspt.org

**FIGURE 2-12** ◄ The Mediterranean Diet Pyramid is based on dietary patterns from the Mediterranean region. Base every meal on fruits, vegetables, whole grains, olive oil, beans, nuts, legumes, and seeds; eat fish and seafood at least two times per week; eat poultry and eggs every two days or weekly; eat cheese and yogurt daily to weekly; eat meats and sweets less often; drink water; drink red wine in moderation; be physically active and enjoy meals with others.

2009 Oldways Preservation & Exchange Trust, oldwayspt.org

## ✓ CONCEPT CHECK 2.3

1. What is the website where you can find all of the tools associated with MyPlate?
2. What are the five major food groups represented on MyPlate?
3. List the main components of the Mediterranean Diet.

## **2.4** Nutritional Health

The ultimate intent of the sound nutrition advice found in the *Dietary Guidelines* and the MyPlate food guide is to promote optimal **nutritional status** for individuals. The amount of each nutrient needed to achieve this goal is the basis for published dietary intake recommendations. We have already discussed general dietary guidelines and will cover more specific nutrient recommendations later in this chapter. Adequate nutritional status is needed to ensure body tissues have enough of each nutrient to support normal metabolic functions and surplus stores that can be used in times of increased nutritional need. An optimal nutritional state can be achieved by obtaining essential nutrients from a variety of foods and adhering to the *Dietary Guidelines*.

**nutritional status** The nutritional health of a person as determined by anthropometric measurements (height, weight, circumferences, and so on), biochemical measurements of nutrients or their by-products in blood and urine, a clinical (physical) examination, a dietary analysis, and economic evaluation; also called *nutritional state*.

**malnutrition** Failing health that results from chronic dietary patterns that do not coincide with nutritional needs.

**overnutrition** A state in which nutritional intake greatly exceeds the body's needs.

**undernutrition** Failing health that results from a long-standing dietary intake that is suboptimal and does not meet nutritional needs.

**hidden hunger** A lack of vitamins and minerals that occurs when the quality of foods people eat does not meet their nutrient requirements.

**symptom** A change in health status noted by the person with the problem, such as stomach pain.

**subclinical** Stage of a disease or disorder not severe enough to produce symptoms that can be detected or diagnosed.

On the other end of the spectrum, **malnutrition** refers to both **overnutrition** or **undernutrition.** Neither state is conducive to good health. Furthermore, it is possible to be both overnourished (e.g., consume excess calories) and undernourished (e.g., consume too few essential vitamins and minerals) at the same time.

## UNDERNUTRITION

Undernutrition occurs when nutrient intake does not meet nutrient needs. At first, any surpluses are put to use; then, as stores are exhausted, nutritional status begins to decline. Many nutrients are in high demand due to constant cell loss and regeneration in the body, such as in the gastrointestinal (GI) tract. For this reason, the stores of certain nutrients, including many of the B vitamins, are exhausted rapidly and therefore must be replenished regularly. In addition, some women in North America do not consume sufficient iron to compensate for monthly menstrual losses and eventually deplete their iron stores (Fig. 2-13).

**Hidden hunger** describes a state of micronutrient deficiency, when the quality of food consumed does not meet the nutrient requirements for normal metabolic functions and maintenance. Once availability of a nutrient falls too low, the body's metabolic processes slow down or stop. At this state of nutrient deficiency, there are often no observable **symptoms;** thus, it is termed a **subclinical** deficiency. A subclinical deficiency can go on for some time before individuals suffer detectable nutrient deficiency symptoms. Eventually, clinical symptoms will develop. Clinical evidence of a nutritional deficiency—perhaps in the skin, hair, nails, tongue, or eyes—can occur within months, but overt symptoms may take years to develop. Often, clinicians do not detect a problem until a deficiency produces observable symptoms, such as excessive bruising from a vitamin C deficiency.

## OVERNUTRITION

Prolonged consumption of more nutrients than the body needs can lead to overnutrition. In the short term (e.g., 1 to 2 weeks), overnutrition may cause only a few symptoms, such as stomach distress from excess iron intake. If an excess intake continues, however, some nutrients may accumulate to toxic amounts, which can lead to serious consequences. For example, too much vitamin A during pregnancy can cause birth defects.

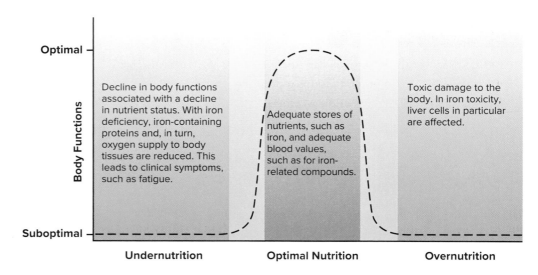

**FIGURE 2-13** ▲ The general scheme of nutritional status. Green reflects optimal nutritional status, yellow indicates a marginal nutritional status, and red reflects a poor nutritional status (undernutrition or overnutrition). This general concept can be applied to all nutrients. Iron was chosen as an example because iron deficiency is the most common nutrient deficiency worldwide.

The most common form of overnutrition in developed nations is an excess intake of calories that leads to overweight and obesity. In the long run, obesity leads to other serious diseases, such as type 2 diabetes and certain forms of cancer. The USDA's Food and Nutrition Information Service has many useful resources about weight control, and more useful information will be discussed later in the text.

### ✓ CONCEPT CHECK 2.4

1. What are the main differences between an optimal nutritional state and a state of malnutrition?
2. What are the two categories of malnutrition?
3. Describe the characteristics of hidden hunger.

## 2.5 Measuring Nutritional Status

To find out how nutritionally fit you are, a nutritional assessment needs to be performed. Generally, this is performed by a primary care provider, often with the aid of a registered dietitian nutritionist (RDN).

### ANALYZING BACKGROUND FACTORS

Because your health history plays an important role in determining nutritional and health status, it must be carefully recorded and critically analyzed as part of a nutritional assessment. Other related information includes a: (1) medical history, especially for any disease states or treatments that could decrease nutrient absorption or ultimate use; (2) list of medications; (3) social history (e.g., marital status and living conditions); (4) health literacy level; and (5) economic status to determine the ability to access and prepare food.

### ASSESSING NUTRITIONAL STATUS USING THE ABCDEs

In addition to background factors, five nutritional-assessment categories contribute to providing a comprehensive profile of nutritional status. **Anthropometric assessment** is used to assess the size, shape, and composition of the human body. These measurements often include body mass index (height, weight), circumference measures, bioelectrical impedance, and skinfold measures. Most measures of body composition are easy to obtain and are generally reliable. However, an in-depth examination of nutritional health is inadequate without the more expensive process of **biochemical assessment.** This involves the measurement of nutrients, by-products of nutrients; or factors known to affect the digestion, absorption, and/or metabolism of nutrients. A **clinical assessment** is often necessary to determine physical evidence (e.g., hypertension) of diet-related diseases or deficiencies. Then, a close look at the person's eating pattern (**dietary assessment**), including a record of previous dietary intake or food frequency would help to determine any possible problem areas. Finally, adding the **environmental assessment** (from the background information) provides further details about the living conditions, education level, and ability to access and prepare foods needed to maintain optimal health. Taken together, these five assessments form the ABCDEs of nutritional assessment (Fig. 2-14).

### LIMITATIONS OF NUTRITIONAL ASSESSMENT

A long time may elapse between the initial development of suboptimal nutritional health and the first clinical evidence of a problem. For instance, an eating pattern high in saturated fats often increases blood cholesterol without producing any clinical evidence for years. However, when the blood vessels become sufficiently blocked

**anthropometric assessment** Measurement of body weight and the lengths and proportions of parts of the body.

**biochemical assessment** Measurement of biochemical functions (e.g., concentrations of nutrient by-products or biologic activities in the blood, feces, or urine) related to a nutrient's function.

**clinical assessment** Examination of general appearance of skin, eyes, and tongue; sense of touch; ability to cough and walk; and evidence of rapid hair loss.

**dietary assessment** Estimation of typical food choices relying mostly on the recounting of one's usual intake or a record of one's previous days' intake.

**environmental assessment** Includes details about living conditions, education level, and the ability of the person to purchase, transport, and prepare food. The person's weekly budget for food purchases is also a key factor to consider.

**FIGURE 2-14** ▶ A complete nutritional assessment includes anthropometric, biochemical, clinical, dietary, and environmental information. This information comes from a combination of patient report and history, physical assessment, and evaluation of other objective indicators (e.g., lab values). Anthropometrics scale: DNY59/E+/Getty Images; Biochemical test tube: TippaPatt/Shutterstock; Clinical physical exam: Rick Brady/McGraw-Hill Education; Clinical physical exam: RobMattingley/E+/Getty Images; Clinical physical exam: Hannamariah/Shutterstock

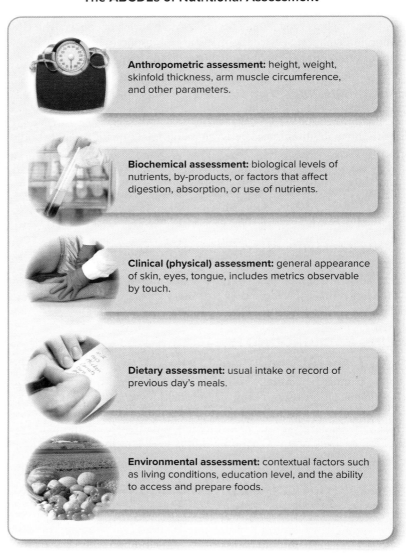

## The ABCDEs of Nutritional Assessment

**Anthropometric assessment:** height, weight, skinfold thickness, arm muscle circumference, and other parameters.

**Biochemical assessment:** biological levels of nutrients, by-products, or factors that affect digestion, absorption, or use of nutrients.

**Clinical (physical) assessment:** general appearance of skin, eyes, tongue, includes metrics observable by touch.

**Dietary assessment:** usual intake or record of previous day's meals.

**Environmental assessment:** contextual factors such as living conditions, education level, and the ability to access and prepare foods.

**heart attack** Rapid fall in heart function caused by reduced blood flow through the heart's blood vessels. Often part of the heart dies in the process. Technically called a *myocardial infarction*.

by cholesterol and other substances, chest pain or a **heart attack** may occur. Another example of a serious health condition with delayed symptoms is low bone density resulting from a calcium deficiency—a particularly relevant issue for adolescent and young adult females. Many young women do not consume the needed amount of calcium but suffer no obvious effects in their younger years; however, the bone structures of these women with low calcium intakes do not reach full potential during the years of growth, increasing the risk for osteoporosis later in life. Furthermore, clinical symptoms of some nutritional deficiencies (e.g., diarrhea, inability to walk normally, and facial sores) are not very specific. These may have causes other than poor nutrition.

## ✓ CONCEPT CHECK 2.5

1. What are the ABCDE categories used in assessing nutritional status?
2. What are the limitations of nutritional assessment?

# 2.6 Nutrient Recommendations

The overarching goal of any healthy eating plan is to meet nutrient needs. To begin, we must determine what amount of each essential nutrient is necessary to maintain health. Most of the terms that describe nutrient needs fall under one umbrella term: **Dietary Reference Intakes (DRIs).** The development of DRIs is an ongoing, collaborative effort between the National Academies of Sciences, Engineering, and Medicine in the United States and Health Canada. Included under the DRI umbrella are **Recommended Dietary Allowances (RDAs), Adequate Intakes (AIs), Estimated Energy Requirements (EERs), Tolerable Upper Intake Levels (Upper Levels or ULs)**, and **Chronic Disease Risk Reduction Intakes (CDRRs).**

## RECOMMENDED DIETARY ALLOWANCE

A Recommended Dietary Allowance (RDA) is the daily amount of a nutrient that will meet the needs of nearly all individuals (about 98%) of a particular age and sex. A person can compare his or her daily intake of specific nutrients to the RDA. Although slight deviations in nutrient intakes above or below the RDA for a particular nutrient are typically no reason for concern, a significant deviation below (about 70%) or above (about 300% for some nutrients) the RDA for an extended time can result in a deficiency or toxicity of that nutrient, respectively.

## ADEQUATE INTAKE

An RDA can be set for a nutrient only if there is sufficient information on the human needs for that particular nutrient. Today, there is not enough information on some nutrients, such as chromium, to set such a precise standard as an RDA. For these nutrients, the DRIs include a category called an Adequate Intake (AI). This standard is based on the dietary intakes of people who appear to be maintaining nutritional health. That amount of intake is assumed to be adequate, as no evidence of a nutritional deficiency is apparent.

## ESTIMATED ENERGY REQUIREMENT

For calorie needs, we use the Estimated Energy Requirement (EER) instead of an RDA or AI. In contrast to the RDAs, which are set somewhat higher than the average needs for nutrients, the EER is set for the average person. While a slight excess of vitamins and minerals is not harmful, a long-term excess of even a small amount of calories will lead to weight gain. Therefore, the calculation of EER needs to be more specific, taking into account age, biologic sex, height, weight, and physical activity (e.g., sedentary or moderately active). In some cases, the additional calorie needs for growth and lactation are also included. Note that the EER is based on the average person. Thus, it can only serve as a starting point for estimating calorie needs.

## TOLERABLE UPPER INTAKE LEVEL

A Tolerable Upper Intake Level (Upper Level or UL) has been set for some vitamins and minerals (Appendix G). The UL is the highest amount of a nutrient unlikely to cause adverse health effects in the long run. As intake exceeds the UL, the risk of ill effects increases. These amounts generally should not be exceeded day after day, as toxicity could develop. For people eating a variety of foods and/or using a balanced multivitamin and mineral supplement, exceeding the UL is unusual. Problems are more likely to arise with eating patterns that promote excessive intakes of a limited variety of foods, with the use of many fortified foods, or with excessive doses of individual vitamins or minerals.

## CHRONIC DISEASE RISK REDUCTION INTAKES

DRIs are reference values that provide recommendations for adequate and safe intakes in apparently healthy individuals. Yet after reviewing the body of scientific evidence,

**Dietary Reference Intakes (DRIs)** Term used to encompass nutrient recommendations made by the Food and Nutrition Board of the National Academies of Sciences, Engineering, and Medicine. These include RDAs, AIs, EERs, CDRRs, and ULs.

**Recommended Dietary Allowance (RDA)** Nutrient intake amount sufficient to meet the needs of 97% to 98% of the individuals in a specific life stage.

**Adequate Intake (AI)** Nutrient intake amount set for any nutrient for which insufficient research is available to establish an RDA. AIs are based on estimates of intakes that appear to maintain a defined nutritional state in a specific life stage.

**Estimated Energy Requirement (EER)** Estimate of the energy (kcal) intake needed to match the energy use of an average person in a specific life stage.

**Tolerable Upper Intake Level (UL)** Maximum chronic daily intake level of a nutrient that is unlikely to cause adverse health effects in almost all people in a specific life stage.

**Chronic Disease Risk Reduction Intake (CDRR)** Category of DRIs based upon chronic disease risk.

## Nutrition Facts

12 servings per container

| Serving size | 1 donut (about 52g) |
|---|---|

**Amount per serving**

| Calories | 200 |
|---|---|

| | % Daily Value* |
|---|---|
| **Total Fat** 12g | **18**% |
| Saturated Fat 3g | **15**% |
| *Trans* Fat 4g | |
| **Cholesterol** 5mg | **1**% |
| **Sodium** 95mg | **4**% |
| **Total Carbohydrate** 22g | **7**% |
| Dietary Fiber <1g | **1**% |
| Total Sugars 12g | |
| Includes 10g Added Sugars | **20**% |
| **Protein** 2g | |
| Vitamin D 0mcg | 0% |
| Calcium 60mg | 6% |
| Iron 1mg | 6% |
| Potassium 39mg | 1% |

\* The % Daily Value (DV) tells you how much a nutrient in a serving of food contributes to a daily diet. 2,000 calories a day is used for general nutrition advice.

The Daily Value is the nutrient standard used on the Nutrition Facts portion of the food label. **What calorie level is the percent Daily Value on a Nutrition Facts label based upon?**

the National Academies of Sciences, Engineering, and Medicine established the new Chronic Disease Risk Reduction Intakes (CDRR) DRI category. These are the first DRIs that are disease-specific and target risk reduction for chronic disease. The CDRR value for sodium was set after it was found to be linked to the risk of cardiovascular disease.

## DAILY VALUE

**Daily Value (DV)** Quantity (expressed in percentage) of a specific nutrient that corresponds to the total percentage of the daily requirements for a particular nutrient based on a 2000 kcal diet.

A nutrition standard more relevant to everyday life is the **Daily Value (DV).** This is a generic standard used on food labels. It is applicable to both genders from 4 years of age through adulthood and is based on consuming a 2000 kcal diet. DVs are mostly set at or close to the highest RDA value or related nutrient standard seen in the various age and gender categories for a specific nutrient and are listed in Appendix A. DVs have been set for vitamins, minerals, total fat and carbohydrate, and other dietary components. For fat and cholesterol, the DVs represent a maximum level, not a goal one should strive to reach. DVs allow consumers to compare their intake from a specific food to desirable (or maximum) intakes.

## APPLICATION OF NUTRIENT STANDARDS

As nutrient intake increases, the RDA for the nutrient, if set, is eventually met and a deficient state is no longer present (Fig. 2-15). An individual's needs most likely will be met since RDAs are set high to include almost all people. Related to the RDA concept of meeting an individual's needs are the standards of AI and the EER. These can be used to estimate an individual's needs for some nutrients and calories, respectively. Still, keep in mind that these standards do not share the same degree of accuracy as the RDA. For example, EER may have to be adjusted upward if the individual is very physically active. Finally, as nutrient intake increases above the UL, poor nutritional health is again likely. However, this poor health is due now to the toxic effects of a nutrient, rather than those of a deficiency.

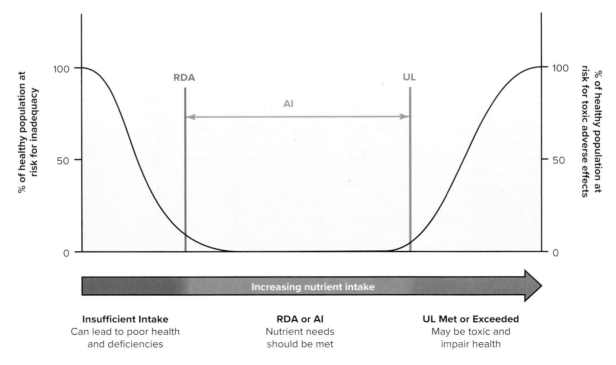

**FIGURE 2-15** ▲ This figure shows the relationship of the Dietary Reference Intakes (DRIs) to each other and the percentage of the population covered by each. At intakes between the RDA and the UL, the risk of either an inadequate diet or adverse effects from the nutrient in question is close to zero. The UL is then the highest level of nutrient intake likely to pose no risks of adverse health effects to almost all individuals in the general population. At intakes above the UL, the margin of safety to protect against adverse effects is reduced. The AI is set for some nutrients instead of an RDA. There is no established benefit for healthy individuals if they consume nutrient intakes above the RDA or AI.

The type of standard set for nutrients depends on the quality of available evidence. A nutrient recommendation backed by lots of experimental research will be expressed as an RDA. For a nutrient that still requires more research, only an AI is presented. We use the EER as a starting point for determining calorie needs. Some nutrients also have a UL if information on toxicity or adverse health effects is available. Periodically, new DRIs or categories of values become available as expert committees review and interpret the available research. As mentioned, a new DRI category, called Chronic Disease Risk Reduction Intake (CDRR) was established in 2019 for sodium based upon scientific evidence linking it to cardiovascular disease.

RDAs and related standards are intended mainly for diet planning. Specifically, an eating pattern should aim to meet the RDA or AI as appropriate and not to exceed the UL over the long term. Specific RDA, AI, EER, CDRR, and UL standards are found in Appendix G. To learn more about these nutrient standards, visit Dietary Guidance at the Food and Nutrition Information Center's website.[14]

## ASK THE RDN    Transgender Issues

*Dear RDN:* *The transgender issue seems to be more common currently, especially in teens and young adults. Should nutrition requirements for transgender individuals be based on birth-assigned sex or identified/expressed gender?*

To better understand potential differences in nutrient requirements for transgender individuals, let us start with defining important terms. The term *transgender* refers to someone's expression of gender. *Gender expression* (or *gender identity*) is unrelated to the physical attributes of a person (i.e., *sex*). For transgender individuals, the sex they were assigned at birth and their own gender identity do not match. On the other hand, *cisgender* individuals share the same gender as their birth-assigned *sex*. Because most people identify as cisgender, we use this term less often. Sexual orientation, or the gender to which one is attracted, is not related to gender identity.

In the medical field, understanding an individual's gender identity is sometimes complicated because an individual's birth-assigned sex is often listed as *gender* on medical documentation regardless of the person's gender identity. In fact, many medical institutions and insurance companies do not accept transgender identity as an option on medical charts.

Simply asking about and acknowledging a person's gender expression and preferred pronoun are important first steps to improving the overall health care experience for transgender individuals. According to a survey of over 6000 transgender individuals, almost 30% of participants reported postponing medical care due to perceived discrimination from their health care providers, while 19% reported being refused medical care completely.[15] By establishing a climate of respect and rapport, the provider will be better positioned to help an individual make changes to improve health.

In this chapter, you learned about many Dietary Reference Intakes, including Estimated Energy Requirements, which are different for males and females. Should we rely on the nutrient recommendations for the transgender individual's birth-assigned sex or the individual's gender identity?

Calorie, protein, and fluid requirements are typically no different between transgender and cisgender individuals. However, we don't have enough research data to know for sure how specific nutrient recommendations may vary for transgender individuals. Nutrition professionals who work with transgender clients have started relying on gender-neutral estimates for calorie, protein, and fluid needs, which are based on body weight. Slight differences between male and female recommendations can be easily adjusted by a registered dietitian nutritionist.

Nutritional requirements certainly do change as a result of physical (i.e., surgical) or hormonal interventions. To promote healing after transition surgery, protein and calorie needs will increase. Transgender individuals may also elect to utilize hormonal therapy as part of the transition process (with or without surgical interventions). Although the timing of these effects may vary, hormone therapy may alter a person's metabolic rate. For instance, transgender men (female to male) who use testosterone hormone therapy can experience an increase in muscle and bone mass. Changes in lean muscle mass will certainly increase calorie and protein needs long term. For transgender females (male to female) who use progesterone, weight gain is likely, which can also impact daily calorie needs.[16] As you will learn, current recommendations for some micronutrients also vary by sex. At this time, there are no specific DRIs for transgender individuals.

In order to ensure consistent and supportive care, close collaboration and honest communication between transgender individuals and the health care team are essential. As more research data are gathered, we will have greater insight into the specific nutrient requirements for this community.

*Cheers,*

**Zachari Breeding, MS, RDN, LDN, FAND**

Clinical dietitian, professional chef, and owner of Sage Nutritious Solutions

©Christopher Lake

1. How do the definitions of RDA and AI differ? Can a nutrient have both an RDA and an AI?
2. Which DRI category includes the highest amounts of a nutrient unlikely to cause adverse health effects?
3. Chronic Disease Risk Reduction Intakes (CDRR) have been established for what nutrient?

## 2.7 Evaluating Nutrition Information

In addition in Figure 2-16, the following should be considered when making logical nutrition decisions:

1. Apply the basic principles of nutrition along with the *Dietary Guidelines* to any nutrition claim, including those on websites. Do you note any inconsistencies? Do reliable references support the claims? Beware of the following:
   - Testimonials about personal experience
   - Nonreputable publications without **peer review**
   - Promises of dramatic and often rapid results
   - Lack of evidence from other scientific studies
2. Examine the background and scientific credentials of the individual, organization, or authors making the nutritional claim. Usually, a reputable author is one with a nationally recognized university or medical center that offers programs or courses in the field of nutrition, medicine, or other health-related specialty.
3. Be wary if the answer is *Yes* to any of the following questions about a nutrition claim:
   - Are only advantages discussed and possible disadvantages ignored?
   - Are claims made about *curing* disease? Do they sound too good to be true?
   - Is extreme bias against the medical community or traditional medical treatments evident?
   - Is the claim touted as a *new* or *secret* scientific breakthrough?
4. Note the size and duration of any study cited in support of a nutrition claim. The larger it is and the longer it went on, the more dependable its findings. Also consider the type of study. Investigate the group studies. Are they relevant to you? Also keep in mind that *contributes to, is linked to,* or *is associated with* do not mean *causes.*
5. Beware of news conferences and social media hype regarding the latest findings. Much of this will not survive more detailed scientific evaluation.
6. When you meet with a nutrition professional, you should expect that he or she will do the following:
   - Ask questions about your medical history, lifestyle, and current eating patterns.
   - Formulate a dietary pattern tailored to your specific needs.
   - Schedule follow-up visits to track your progress, answer any questions, and help keep you motivated.
   - Involve family members in the conversation when appropriate.
7. Avoid individuals who prescribe **megadoses** of vitamin and mineral supplements for everyone, especially those who profit from these products.
8. Examine product labels carefully. Be skeptical of any promotional information about a product that is not clearly stated on the label.

Remember from the *Ask the RDN* feature in Chapter 1 that the best approach to finding answers about your nutritional state is to consult your primary care provider or registered dietitian nutritionist first.[17] This chapter's *Ask the RDN* focuses on nutritional recommendations for transgender individuals. Appendix E also lists many reputable sources of nutrition advice for your use. Overall, nutrition is a rapidly advancing field, and there are always new findings. In addition, the USDA Fraud and Nutrition Misinformation website can help you evaluate ongoing nutrition and health claims.[18]

▲ Registered dietitian nutritionists (RDN) are a reliable source of nutrition advice. Stuart Jenner/Shutterstock

**megadose** Large intake of a nutrient well beyond estimates of needs or what would be found in a balanced diet; 2 to 10 times above human needs is typically a starting point.

**Evaluating Nutrition Information**

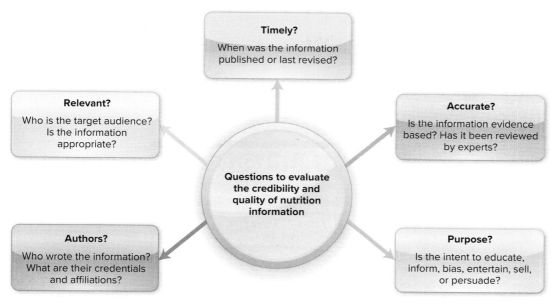

FIGURE 2-16 ▲ Adapted from the CRAAP Test, used to evaluate information and its quality.
Adapted from https://library.csuchico.edu/help/source-or-information-good.

## NUTRITION AND FITNESS APPS

Over the last decade, there has been a significant increase in mobile phone app development related to nutrition and physical activity. Numerous options have entered the marketplace, including those that provide nutrition education, enable tracking of dietary and physical activity patterns, or provide feedback and accountability. With thousands to choose from, navigating these apps can be an overwhelming experience for consumers. It is important to pick an app based on the underlying nutrition-related concern at hand, as research shows some have little basis in behavioral theory and may not assist in behavior change.[19] Working with nutrition professionals to gauge the most appropriate options is a good place to start, and as options continue to emerge they are regularly reviewed by the Academy of Nutrition and Dietetics.[20]

In recent years, use of diet, fitness, and other health apps has skyrocketed. **What benefits have you found from tracking your nutrition and fitness? Are there any potential drawbacks?** ipopba/123RF

### ✓ CONCEPT CHECK 2.7

1. What are some characteristics that suggest that references supporting a nutrition claim are unreliable?

2. What should your expectations be when meeting with a nutrition professional?

## 2.8 Food Labels and Dietary Pattern Planning

Today, nearly all foods sold in stores must be in a package that has a label containing the following information: the product name, name and address of the manufacturer, amount of product in the package, and ingredients listed in descending order by weight. This food and beverage labeling is monitored by government agencies such as the Food and Drug Administration (FDA) in the United States. The listing of certain food constituents is also required—specifically, on a Nutrition Facts label (Fig. 2-17). Consumers

**FIGURE 2-17** ▲ Food packages must list product name, name and address of the manufacturer, amount of product in the package, and ingredients. The Nutrition Facts label is required on virtually all packaged food products. The % Daily Value listed on the label is the percent of the amount of a nutrient needed daily that is provided by a single serving of the product. Canadian food labels use a slightly different group of health claims and label descriptors. U.S. Food and Drug Administration

can use the information in the Nutrition Facts label to learn more about what they eat. The following components must be listed:

- Total calories (kcal)
- Total fat
- Saturated fat
- *Trans* fat
- Cholesterol

- Sodium
- Total carbohydrate
- Fiber
- Total sugars
- Added sugars

- Protein
- Vitamin D
- Calcium
- Iron
- Potassium

In addition to these required components, manufacturers can choose to list polyunsaturated and monounsaturated fat, additional vitamins and minerals, and others. Listing an additional nutrient becomes *required* if the food is fortified with that nutrient or if a claim is made about the health benefits of the specific nutrient.

Remember that the Daily Value is a generic standard used on the food label. The percentage of the Daily Value (% Daily Value or % DV) is usually given for each nutrient per serving. These percentages are based on a 2000 kcal diet and must be adjusted for people who require considerably more or less than 2000 kcal per day with respect to fat and carbohydrate intake. DVs are mostly set at or close to the highest RDA value or related nutrient standard seen in the various age and sex categories for a specific nutrient.

Serving sizes on the Nutrition Facts label must be consistent among similar foods. This means that all brands of ice cream, for example, must use the same serving size on their label. These serving sizes may differ from those of MyPlate because those on food labels are based on more typical portion sizes. In addition, food claims made on packages must follow legal definitions. A long list of definitions for nutrient claims allowed on food labels is given in Table 2-3. For example, if a product claims to be "low sodium," it must have 140 milligrams of sodium or less per serving.

Many manufacturers list the DVs set for dietary components such as fat, cholesterol, and carbohydrate on the Nutrition Facts label. This can be useful as a reference point. As noted, they are based on 2000 kcal; if the label is large enough, amounts based on 2500 kcal are listed as well. As mentioned, DVs allow consumers to compare their intake from a specific food to desirable (or maximum) daily intakes.

## CHANGES TO NUTRITION LABELS

In 2016, the FDA finalized new rules for the Nutrition Facts label. The changes were designed to promote healthier eating and combat obesity, aiming to help consumers make informed decisions about the foods they eat. Several changes, such as increasing the type size for "Calories," "Servings per container," and the "Serving size" declaration, and bolding the number of calories and the "Serving size" declaration, were aimed at making it easier for Americans to know how many calories they are consuming. The Daily Value information for nutrients was updated and includes the actual amount, in addition to percent Daily Value of vitamin D, calcium, iron, and potassium. The Daily Value footnote was changed to better explain the meaning of percent Daily Value. Along with potassium and vitamin D, "Added Sugars" was added to the label, whereas calories from fat was eliminated. These changes to sugars and fat were based on scientific data showing that it is difficult to meet nutrient needs and stay within calorie limits if added sugars are more than 10% of total daily calories and that the type of fat is more important than the amount. The FDA has also provided guidance to provide clarity on the labeling of added sugars for honey, maple syrup, allulose, and other single ingredient sugars and syrups. It is not required to declare the number of grams of these added sugars in a serving of the product, but the label must still include the percent Daily Value for added sugars so consumers can discern how these products contribute to their total dietary patterns.

One of the most significant changes was that serving size information more accurately reflects how much is consumed in one sitting. For example, the serving sizes for a soda went from 8 ounces to 12 ounces, and ice cream servings increased from ½ cup to ⅔ cup. To eliminate confusion about the number of servings in a container and calories in a serving, larger packages, such as a pint of ice cream, have two columns on the labels, one for "per serving" and one for "per package." The majority of food packaging will have the new label implemented by early 2021. More information on the Nutrition Facts label is available at https://www.fda.gov/food/nutrition-education-resources-materials/new-nutrition-facts-label.[21]

## MENU PLANNING WITH LABELS

All of the tools discussed in this chapter greatly aid in menu planning. Menu planning can start with MyPlate. The totality of choices made within the groups can then be

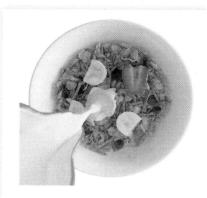

Breakfast is a great time to start the day off right.[22] Check the Nutrition Facts label to be sure your cereal meets these criteria:

1. Go for whole grains. First two ingredients are whole grain or bran.
2. Check the serving sizes, which range from 30 grams (1 oz) for light cereals to 55 grams (2 oz) for heavy cereals.
3. Choose cereals that do not have more than 1.2 teaspoons (7 grams) of total sugar for light cereals and 2.5 teaspoons (11 grams) for heavy cereals.
4. Get enough unprocessed fiber, such as wheat bran, whole grain wheat, and oats.
5. Look for cereals with less than 2.5 grams of saturated fat.

**List some examples of breakfast cereals that provide at least 3 grams of fiber per serving.**
Peter Cade/Photodisc/Getty Images

## TABLE 2-3 ■ Common Nutrient Claims Allowed on Food Labels

### Sugar

**Sugar free:** less than 0.5 gram (g) per serving

**No added sugars; without added sugars:** no sugar or sugar-containing ingredient added

**Reduced sugar:** at least 25% less sugar per serving than reference food

### Calories

**Calorie free:** less than 5 kcal per serving

**Low calorie:** 40 kcal or less per serving

### Fiber

**High fiber:** 5 grams or more per serving

**Good source of fiber:** 2.5 to 4.9 grams per serving

**More or added fiber:** at least 2.5 grams more per serving than reference food

### Fat

**Fat free:** less than 0.5 gram of fat per serving

**Low fat:** 3 grams or less per serving

### Sodium

**Sodium free:** less than 5 milligrams per serving

**Low sodium:** 140 milligrams or less per serving

### Other Terms

**Enriched:** replacing nutrients lost in processing

**Fortified:** adding nutrients not originally present in the food

**Healthy:** an individual food that is low fat and low saturated fat and has no more than 360 to 480 mg sodium or 60 mg cholesterol per serving and provides at least 10% of the Daily Value for vitamin A, vitamin C, protein, calcium, iron, or fiber

**Light or lite:** contains ⅓ fewer calories or ½ less fat of the reference food; also can describe characteristics such as texture and color

**Good source:** provides at least 10% to 19% of the Daily Value for a particular nutrient

**High:** provides 20% or more of the Daily Value for a particular nutrient

Source: U.S. Department of Agriculture

**Organic:** grown without the use of pesticides, synthetic fertilizers, sewage sludge, genetically modified organisms, or ionizing radiation; meat, poultry, eggs, and dairy products from animals free of antibiotics or growth hormones; at least 95% of ingredients (by weight) must meet guidelines to be labeled "organic" on the front of the package. If the front label instead says "made with organic ingredients," only 70% of the ingredients must be organic

**Natural:** free of food colors, synthetic flavors, or any other synthetic substance

### Meat and Poultry Products

**Extra lean:** less than 5 grams of fat, 2 grams of saturated fat, and 95 milligrams of cholesterol per serving (or 100 grams of an individual food)

**Lean:** less than 10 grams of fat, 4.5 grams of saturated fat, and 95 milligrams of cholesterol per serving (or 100 grams of an individual food)

**Reduced:** at least 25% less than the usual product

**Low cholesterol:** 20 milligrams or less and 2 grams or less of saturated fat per serving

Many definitions are from FDA's *Dictionary of Terms,* as established in conjunction with the 1990 Nutrition Labeling and Education Act (NLEA).
Source of USDA Organic seal: https://www.ams.usda.gov/rules-regulations/organic/organic-seal. U.S. Department of Agriculture

evaluated using the *Dietary Guidelines.* Individual foods that make up a dietary pattern can be examined more closely using the Daily Values listed on the Nutrition Facts label of the product. For the most part, these Daily Values are in line with the Recommended Dietary Allowances and related nutrient standards. The Nutrition Facts label is especially useful in identifying nutrient-dense foods (foods high in a specific nutrient, such as vitamin A, but low in the relative amount of calories provided) and energy-dense foods (foods that provide a lot of calories but don't typically fill you up). Research has shown that individuals who read the Nutrition Facts when shopping for food report healthier nutrient consumption compared to nonusers.[18] Calorie labeling, in isolation, will not likely impact the obesity epidemic, as research has documented that many consumers do not alter their calorie intake based upon this information. Yet, it is plausible that subsequent changes may follow, thus setting the stage for ultimate success. For instance, raising public awareness may improve consumer behaviors and understanding. Consumers may then demand lower calorie alternatives in both grocery outlets and in restaurants. Food manufacturers and industry may respond by reducing the calorie content of high calorie items.[23]

## EXCEPTIONS TO FOOD LABELING

Foods such as fresh fruits, vegetables, and fish currently are not required to have Nutrition Facts labels. However, many grocers have voluntarily chosen to provide their customers with information about these products on posters or pamphlets that may contain recipes that can assist you in your endeavor to improve your eating pattern.

The % Daily Value for protein is not mandatory on foods because protein deficiency is not a public health concern in the United States. If the % Daily Value for protein is given on a label, FDA requires that the product be analyzed for protein quality. This procedure is expensive and time-consuming, so many companies opt not to list a % Daily Value for protein. However, labels on food for infants and children under 4 years of age must include the % Daily Value for protein, as must the labels on any food carrying a claim about protein content.

## LABELING OF FOOD ALLERGENS

The Food Allergen Labeling and Consumer Protection Act (FALCPA) requires manufacturers to label food products that contain an ingredient that is or contains protein from a major food **allergen**. According to the FDA, there are eight allergens that have to be labeled: milk, eggs, fish, crustacean shellfish, tree nuts, peanuts, wheat, and soybeans. This information can be stated in one of two ways. The first option is to include the name of the food source in parentheses following the common or usual name of the major food allergen in the list of ingredients if the name of the food source of the major allergen does not appear elsewhere in the ingredients list. The second option is to put the word *Contains* followed by the name of the food source from which the major food allergen is derived immediately after or adjacent to the list of ingredients in type size that is no smaller that the ingredient type size (e.g., Contains Wheat, Milk, Eggs, and Soy).

**allergen** A foreign protein, or antigen, that induces excess production of certain immune system antibodies; subsequent exposure to the same protein leads to allergic symptoms. Whereas all allergens are antigens, not all antigens are allergens.

| ▶ Top 8 Food Allergens | |
| --- | --- |
| Eggs | Shellfish |
| Fish | Soybeans |
| Milk | Tree nuts |
| Peanuts | Wheat |

## HEALTH CLAIMS ON FOOD LABELS

As a marketing tool directed toward the health-conscious consumer, food manufacturers like to claim that their products have all sorts of health benefits. FDA has legal oversight over most food products and permits some health claims with certain restrictions. Overall, claims on foods fall into one of four categories:

- Health claims—closely regulated by FDA
- Preliminary health claims—regulated by FDA, but evidence may be scant for the claim
- Nutrient claims—closely regulated by FDA
- Structure/function claims—these are not FDA approved, or necessarily valid

▲ Use the Nutrition Facts label to learn more about the nutrient content of the foods you eat. Nutrient content is expressed as a percent of Daily Value. Mary-Jon Ludy/McGraw-Hill Education

Table 2-3 lists the definitions for nutrient claims on food labels. Currently, FDA limits the use of health messages to specific instances in which there is significant scientific agreement that a relationship exists between a nutrient, food, or food constituent and the disease. The claims allowed at this time may show a link ("may" or "might" qualifier must be used in the statement) between an eating pattern and the following:

- Enough calcium and vitamin D and a reduced risk of osteoporosis
- Low in total fat and a reduced risk of some cancers
- Low in saturated fat and cholesterol and a reduced risk of heart disease
- Rich in fiber—containing grain products, fruits, and vegetables—and a reduced risk of some cancers
- Low in sodium and a reduced risk of hypertension and stroke
- Rich in fruits and vegetables and a reduced risk of some cancers
- Adequate in the synthetic form of the vitamin folate (folic acid) and a reduced risk of neural tube defects
- Sugarless gum and a reduced risk of tooth decay
- Rich in fruits, vegetables, and grain products that contain fiber and a reduced risk of cardiovascular disease

▶ **APPROVED NUTRITION FACT LABEL HEALTH CLAIMS:**

- Calcium and vitamin D - osteoporosis
- Dietary lipids - cancer
- Folate - neural tube defects
- Saturated fat and cholesterol - coronary heart disease
- Sodium - hypertension
- Soluble fiber - coronary heart disease
- Soy - coronary heart disease
- Stanols/sterols - coronary heart disease
- Whole grains, fruits, and vegetables - cancer

- A diet rich in whole grain foods and other plant foods, as well as low in total fat, saturated fat, and cholesterol, and a reduced risk of cardiovascular disease and certain cancers
- Low in saturated fat and cholesterol that also includes 25 grams of soy protein and a reduced risk of cardiovascular disease (the statement "one serving of the (name of food) provides _____ grams of soy protein" must also appear as part of the health claim)
- Fatty acids from oils present in fish and a reduced risk of cardiovascular disease
- Margarines containing plant stanols and sterols and a reduced risk of cardiovascular disease

In addition, before a health claim can be made for a food product, it must meet two general requirements. First, the food must be a "good source" (before any fortification) of fiber, protein, vitamin A, vitamin C, calcium, or iron. (The legal definition of "good source" appears in Table 2-3.) Second, a single serving of the food product cannot contain more than 13 grams of fat, 4 grams of saturated fat, 60 milligrams of cholesterol, or 480 milligrams of sodium. If a food exceeds any one of these requirements, no health claim can be made for it, despite its other nutritional qualities. For example, even though whole milk is high in calcium, its label can't make the health claim about calcium and osteoporosis because whole milk contains 5 grams of saturated fat per serving. In another example, a health claim regarding fat and cancer can be made only if the product contains 3 grams or less of fat per serving, the standard for low-fat foods.

The FDA's Consumer Health Information for Better Nutrition Initiative also allows the use of qualified health claims when there is emerging evidence for a relationship between a food, food component, or dietary supplement and reduced risk of a disease or health-related condition. In this case, the evidence is not well enough established to meet the significant scientific agreement standard required for an FDA-authorized health claim. The following are examples of "Qualified Health Claims Subject to Enforcement Discretion."[24,25]

✓ **CONCEPT CHECK 2.8**

1. What calorie level is used to determine the percent Daily Value on a food label?
2. What are the two general requirements that must be met for a health claim on a food label?

# 2.9 Nutrition and Your Health
## Eating Well as a Student

Fancy Collection/SuperStock

Whether you are a traditional college student or a returning student balancing school, work, and family, studies show that the eating patterns of college students are not optimal. Typically, students fall short of nutrition recommendations for whole grains, vegetables, fruits, milk, and meat, opting instead to max out on fats, sweets, and alcohol. This information is disturbing because students are forming many health behaviors that will persist throughout life.

What is it about the student lifestyle that makes it so difficult to build healthy habits? In this section, we discuss several topics and provide possible solutions.

## Food Choices

For traditional college students, these years are often a time for independence and a chance to make personal lifestyle decisions. Yet these years also pose some challenges that impact behaviors. For example, when you are writing papers and cramming for exams, balanced meals are all too easily replaced by high-fat and high-calorie fast foods, convenience items, and sugary, caffeinated beverages. Physical activity is often sacrificed in favor of study time. In a recent study of college students living on and off campus, two-thirds of the students reported skipping meals, with "no time to prepare" the major reason for this behavior.[26]

Also consider that campuses have a wide variety of dining choices. Dining halls, food trucks, fast-food establishments, bars, and vending machines combine to offer food 24 hours per day. While it is certainly possible to make wise food choices at each of these outlets, the temptations of convenience, taste, and value (i.e., inexpensive, oversized portions) may persuade the college student to select unhealthy options.

Meals and snacks are also times to socialize. You may unintentionally eat a big lunch at noon without regard to hunger if your classmates are meeting in the dining hall or food court to catch up. While chatting, it is easy to lose track of portions and to overeat. In addition, food may be a source of familiarity and comfort in a new and stressful place.

## Weight Management

Studies show that many college students gain weight during their first year.[27,28] The *freshman 15* is a term used to describe the weight gained by students during their first year of college. Although it is becoming evident that most freshmen actually do not gain 15 pounds, it is common for students dietary patterns to vary widely and result in weight loss or gain. Two lifestyle factors that made a difference in weight gain among the students include heavy drinking and working during college.

There are several reasons to maintain a healthy weight. Research clearly demonstrates that setting several small, achievable goals can spur motivation. Body weight is a balancing act between calories in and calories out. Try keeping track of your calorie consumption for several days and comparing that to your energy needs, based on your age, sex, and activity level. You can use the NutritionCalc Plus application in Connect to estimate your energy needs.

For those electing to lose weight, a healthy rate of weight loss is 1 to 2 pounds per week. Greater rates of weight loss will not likely be sustained over time. Remember that the numbers on the scale are not as important as your body composition—the amount of fat

◄ Fancy coffee beverages, such as lattes and cappuccinos, can increase calorie consumption by 200 kcal or more per day. BananaStock/ PunchStock

# CASE STUDY College Student Eating Habits

Andy is like many other college students. He grew up on a quick bowl of cereal and milk for breakfast and a hamburger, French fries, and cola for lunch, either in the school cafeteria or at a local fast-food restaurant. At dinner, he generally avoided eating any of his vegetables, and by 9:00 P.M. he was deep into chips and cookies. Andy has taken most of these habits to college. He prefers coffee for breakfast and possibly a chocolate donut. Lunch is still mainly a hamburger, French fries, and cola, but pizza and tacos now alternate more frequently than when he was in high school. One thing Andy really likes about the restaurants on campus is that, for a few cents more, he can make his hamburger a double or get extra cheese and pepperoni on his pizza. This helps him stretch his food dollar; searching out large-portion value meals for lunch and dinner has become part of a typical day. Now that he is in college, some of Andy's calories also come from alcohol. He will often have a beer with dinner a couple nights a week and will binge on a six-pack or more while tailgating before Saturday football games.

▲ Now that he is a college student, Andy could use some advice on developing a healthy, adult eating pattern. Dinodia Photos/Alamy Stock Photo

Provide Andy some advice about his eating pattern. Start with his positive habits and then provide some constructive criticism, based on what you now know.

Answer the following questions, and as you make suggestions for Andy, think about your favorite food choices, why they are your favorites, and whether these are positive choices.

1. Recalling what you learned in Chapter 1, list the factors that are influencing Andy's current food choices.
2. Start with Andy's positive habits: What healthy choices are being made when Andy eats at local restaurants?
3. Now provide some constructive feedback:
   a. What are some of the less than ideal items available at fast-food restaurants?
   b. Why is ordering the "value meals" a potential problem over time?
   c. What nutritious substitutions could he make throughout the day?
   d. What concerns would you share with Andy about his weekly alcohol intake?

*Complete the Case Study. Responses to these questions can be provided by your instructor.*

---

▶ Simple Tips to Avoid Weight Gain

- **Eat a nutritious breakfast.** Rev up your metabolism with a lean protein source such as an egg or Greek yogurt, a serving of whole grains such as a fiber-rich breakfast cereal, and a fruit.
- **Limit liquid calories.** Drink water instead of high-calorie soft drinks, fruit juice, alcohol, or flavored coffee drinks. If you drink alcohol, limit it to no more than two drinks per day for men and one drink per day for women.
- **Move more and sit less.** Engage in regular physical activity whenever possible.
- **Plan ahead.** Eat a nutrient-dense meal or snack every few hours.
- **Stock the fridge and pantry.** Keep a stash of nutrient-dense snacks such as string cheese, popcorn, nuts, and fruit (fresh, canned, or dried).

in relation to lean mass. In order to lose weight, you must create an energy deficit, either by restricting energy intake below what you need to maintain your current weight or by increasing your physical activity. For an adult with excess weight, an energy deficit of approximately 500 kcal per day will result in weight loss of about 25 pounds over a year's time. As weight is lost, energy needs gradually decrease, such that further deficits will be required to lose additional weight.

Although it may be tempting to skip, consuming a nutritious breakfast sets the stage for a healthy eating pattern. Starting the day off with a serving of lean protein (e.g., egg, Greek yogurt, or protein shake), a fortified whole grain breakfast cereal, low-fat milk, and a serving of fruit puts you on the right path for meeting recommendations for fiber, calcium, and fruit. Studies also show that eating breakfast may prevent overeating later in the day.

One of the biggest contributors to weight gain for college students is consuming several hundred calories per day in the form of sugar-sweetened or alcoholic beverages. One 12-ounce can of regular cola contains about 140 kcal. A 12-ounce can of regular beer has 150 kcal. Consuming gourmet coffee beverages, such as lattes and cappuccinos, can increase average calorie consumption by 200 kcal or more per day. Even fruit juices have at least 100 kcal per 8-ounce glass. A convenient stash of water is the best way to quench your thirst.

Physical activity is very important to any weight loss and weight maintenance plan, but sticking with it is hard to do. When you find yourself short on time, physical fitness is often the first thing that goes. To ensure your success at boosting daily activity, choose activities you enjoy, such as working out with friends at the campus recreation center, participating in intramural sports, or taking an activity class like dancing. Don't forget to include brisk walking to and from classes.

## Alcohol and Binge Drinking

Excessive alcohol consumption is a big problem on college campuses and in general for many young adults. Many college students consider drinking alcohol, legally or not, to be a rite of passage into adulthood. On campuses, binge drinking—consuming five or more drinks in about two hours for men or four drinks or more for women—has become an epidemic. A new level of "extreme drinking" goes far beyond binge drinking. Drinking games contribute to extreme drinking during many parties and 21st birthday celebrations.[27]

The statistics on the impact of binge drinking on college campuses are sobering. It is estimated that about 40% of students on college campuses participate in binge drinking. Each year, over 1800 college students between the ages of 18 and 24 die from alcohol-related unintentional injuries, including motor vehicle crashes.[28] In addition to deaths and injuries, other problems stemming from binge drinking include unsafe sexual behavior, long-term health problems, suicides, academic issues, legal troubles, and alcohol abuse or dependence. Twenty percent of college students meet the criteria for alcohol use disorders.

In addition, alcohol consumption definitely contributes to weight gain—by virtue of its own calories and the increased food consumption at events where drinking occurs. If you choose to drink alcohol, do so in moderation—no more than two drinks per day for men and one drink per day for women. Be aware of the warning signs and dangers of alcohol poisoning shown here.

▶ **The warning signs and symptoms of alcohol poisoning:**
- Cold, clammy, pale, or bluish skin
- Semiconsciousness or unconsciousness
- Slow respiration of 8 or fewer breaths per minute or lapses between breaths of more than 8 seconds
- Strong odor of alcohol, which usually accompanies these symptoms

## Eating Disorders

As many as 30% of college students are at risk of developing an eating disorder. Disordered eating is a short-term change in eating patterns. Sometimes, disordered eating habits may lead to an eating disorder, such as anorexia nervosa, bulimia nervosa, or binge eating disorder. Advice on what to do if you suspect that someone you know is suffering from an eating disorder will be discussed later in the text.

Starving the body also starves the brain, which limits performance in academics and beyond. The negative consequences of disordered eating may last a lifetime. Frequently, what begins as a weight-loss fad diet spirals into a much larger problem. Eating disorders are not just diets gone bad: they require professional intervention. Left unchecked, eating disorders can lead to serious adverse effects, such as loss of menstrual periods, bone disorders, gastrointestinal problems, kidney issues, heart abnormalities, and even death.

## Choosing a Plant-Forward Lifestyle

Many young adults experiment with or adopt a vegetarian or **vegan** eating pattern. Plant-focused dietary patterns can meet nutrition needs and decrease risk of many chronic diseases, but they require appropriate planning at all life stages.

Plant-forward or plant-focused dietary patterns celebrate, but are not limited to, plant-based foods. Plant food sources include fruits and vegetables (produce); whole grains; beans, other legumes (pulses), and soy foods; nuts and seeds; plant oils; and herbs and spices. Dietary patterns rich in plant foods reflect evidence-based principles supporting personal and environmental health and sustainability.

### Sustainable Solutions

Protein is not typically deficient in vegetarian eating patterns, even with a vegan diet, which contains no animal products. However, vegetarians, and especially vegans, may be at risk for deficiencies of several vitamins and minerals. Consuming a fortified ready-to-eat breakfast cereal is an easy and inexpensive way to obtain these nutrients.

Restaurants and campus dining services have responded to the growing interest in plant-forward meals by offering a variety of vegetarian options. For optimal health benefits, choose foods that are baked, steamed, or stir-fried rather than deep-fried; select whole grains rather than refined carbohydrates; and consume foods fortified with vitamins and minerals. Even if you do not follow a plant-based eating pattern all the time, choosing several plant-focused meals each week can help with weight control and boost intake of fiber and beneficial phytochemicals. The MyPlate Plan recommends that the largest portion of your plate be filled with plant foods, including grains, fruits, and vegetables.

Many students adopt a vegetarian eating pattern during college. Guidelines for planning a nutritious vegetarian eating pattern with items such as this veggie-stuffed lasagna dish are presented in Chapter 6. **Which plant-based meals do you enjoy?**
Francesco83/Shutterstock

**vegan**  Referring to a dietary pattern that only includes foods of plant origin.

## Fuel for Competition: Student Athletes

Students who compete in sports, such as intramural and intercollegiate athletics, need to consume more calories and nutrients. Despite an emphasis on a lean physique, athletes at all levels must take care not to severely restrict calories, as this could negatively impact performance and health. Muscles require adequate carbohydrates for fuel and protein for growth and repair. Fat, as well, is an important source of stored energy for use during physical activity. In addition to the calories needed to fuel the body, fluids are essential for health and performance. Water is adequate to replenish losses for most activities.

Athletes also should take care not to be wooed by the supplement industry. Simply increasing food intake to meet the energy demands of athletic training should be sufficient to meet most vitamin and mineral needs. Individual vitamin, mineral, amino acid, or herbal supplements are rarely advised, in spite of the hype of supplement makers. More about sports nutrition will be discussed later in the text.

## Tips for Eating Well on a Student's Budget

Because higher education can be hard on the wallet, it is good to know that it is possible to eat well on campus on a budget. In fact, a recent review found that over 30% of college students experience some food insecurity, including disruptions in eating patterns and reduced food intake.[29] If you live on campus, try to participate in a prepaid campus meal plan. These plans are generally designed to offer great food value with a variety of nutritious foods. If you live off campus or have your own kitchen, try to plan ahead. Packing a lunch from home rather than grabbing lunch on the run will save you money and put you in control of healthy choices. Some campuses now have food pantries for students. This is an excellent resource when food access issues arise.

Avoid shopping on an empty stomach: everything will look good, and you'll likely buy more. Instead, stock your fridge and pantry with healthy foods so they are the first things on hand when you get hungry. Try to have a list in hand and stick to it, because impulse buys tend to drain your wallet. Try store-brand rather than name-brand items. Keep healthy snacks around and limit junk food. Eat more fruits and vegetables. Make use of canned and frozen fruits and vegetables; they are just as nutritious, particularly if you choose low-sodium and low-sugar options. Canned (fruits, tuna) and dry (oatmeal) foods can be nutritious and last a long time, so you can avoid waste. Drink water and limit sugary, alcoholic, and highly caffeinated beverages. Finally, avoid using food to combat stress and try working out instead.

**✓ CONCEPT CHECK 2.9**

1. Name three aspects of the student lifestyle that make it difficult to build nutritious eating habits.
2. What are the current statistics regarding weight gain during college?
3. Specify three tips for eating well on a student's budget.

# Summary (Numbers refer to numbered sections in the chapter.)

**2.1** A healthy dietary pattern includes a variety of nutrient-dense foods to meet nutrient needs within calorie limits. Such an eating pattern helps to minimize the risk of developing nutrition-related diseases.

Nutrient density reflects the nutrient content of a food in relation to its calorie content. Nutrient-dense foods are relatively rich in nutrients in comparison with calorie content.

The energy density of a food is determined by comparing calorie content with the weight of food. A food rich in calories but weighing relatively very little, such as cookies, fried foods in general, and most snack foods (including fat-free brands), is considered energy dense. Foods with low energy density include fruits, vegetables, and any food that incorporates lots of water during cooking, such as oatmeal.

**2.2** *Dietary Guidelines for Americans* have been issued to help improve the health of all Americans throughout every life stage from birth through older adulthood. The guidelines emphasize a healthy eating pattern that includes a variety of vegetables from all of the subgroups, whole fruits, whole grains, fat-free or low-fat dairy, a variety of lean and plant-based protein foods, and it limits saturated fats, added sugars, sodium, and alcohol.

**2.3** MyPlate and accompanying online tools are designed to translate nutrient recommendations into a food plan at every stage of life. It emphasizes eating a variety of fruits, vegetables, grains, dairy or fortified soy alternatives, and protein foods. When deciding what to eat or drink, make every bite count.

**2.4** A person's nutritional status can be categorized as optimal when the body has adequate nutrient stores for times of increased needs. Malnutrition encompasses both states of undernutrition and overnutrition: undernutrition, which may be present with or without clinical symptoms, and overnutrition, which can lead to vitamin and mineral toxicities and various obesity-related chronic diseases.

**2.5** Evaluation of nutritional status involves analyzing background factors, as well as anthropometric, biochemical, clinical, dietary, and environmental assessments. It is not always possible to detect nutritional inadequacies via nutritional assessment because symptoms of deficiencies are often nonspecific and may not appear for many years.

**2.6** Recommended Dietary Allowances (RDAs) are set for many nutrients. These amounts yield enough of each nutrient to meet the needs of healthy individuals within specific sex and age categories. Adequate Intake (AI) is the standard used when not enough information is available to set a more specific RDA. Estimated Energy Requirements (EERs) set calorie needs for both sexes at various ages and physical activity patterns. Tolerable Upper Intake Levels (Upper Levels or ULs) for nutrient intake have been set for some vitamins and minerals. Chronic Disease Risk Reduction Intakes (CDRRs) are set for sodium and are the only DRIs specific to disease risk. All of these dietary standards fall under the term Dietary Reference Intakes (DRIs).

Daily Values are used as a basis for expressing the nutrient content of foods on the Nutrition Facts label and are based for the most part on the RDAs.

**2.7** Apply the basic principles of nutrition to evaluate any nutrition claim. Several indicators of nutrition misinformation include insufficient scientific evidence to support a product claim, lack of credible sources, promises of unbelievable results, or distrust of the medical community. To sort nutrition fact from fiction, seek the advice of a registered dietitian nutritionist.

**2.8** Food labels, especially the Nutrition Facts labels, are a useful tool to track your nutrient intake and learn more about the nutritional characteristics of the foods you eat. Changes to the Nutrition Facts label have been approved and include increasing the type size for Calories, Servings per container, and Serving size. Any health claims listed must follow FDA-set criteria.

**2.9** Eating patterns and other health habits of college students often fail to align with the *Dietary Guidelines*. This information is disturbing from a public health standpoint, because young adulthood is the time when many health behaviors are formed and will likely persist throughout life. Issues of particular importance for students in college are weight control, food choices, alcohol consumption, food security, and eating disorders.

# Check Your Knowledge (Answers are available at the end of this question set.)

1. Anthropometric measurements include
   a. height, weight, skinfolds, and body circumferences.
   b. blood concentrations of nutrients.
   c. a diet history of the previous days' intake.
   d. blood levels of enzyme activities.

2. Foods with *high* nutrient density offer the _____ nutrients for the _____ calories.
   a. least, lowest
   b. least, most
   c. most, lowest
   d. most, most

3. A meal of a bean burrito, cucumber salad, and glass of milk represents foods from all MyPlate food groups except
   a. dairy.
   b. protein.
   c. vegetables.
   d. fruits.

4. The *Dietary Guidelines for Americans* provide advice for Americans
   a. from birth and older.
   b. from age 2 and older.
   c. from age 6 and older.
   d. from age 18 and older.

5. The *Dietary Guidelines* recommend that we increase which of the following foods?
   a. refined grains
   b. whole milk products
   c. seafood
   d. added sugars

6. How many minutes of moderate-intensity physical activity are recommended for adults in the *Physical Activity Guidelines for Americans?*
   a. 150 to 300 minutes per week
   b. 30 to 60 minutes every day
   c. 50 to 150 minutes every day
   d. 30 to 60 minutes three days a week

7. The term *Daily Value* is used on
   a. restaurant menus.
   b. food labels.
   c. medical charts.
   d. health claims.

8. The Tolerable Upper Intake Level, or UL, is used to
   a. estimate calorie needs of the average person.
   b. evaluate the highest amount of daily nutrient intake unlikely to cause adverse health effects.
   c. evaluate your current intake for a specific nutrient.
   d. compare the nutrient content of a food to approximate human needs.

9. The current food label must list
   a. a picture of the product.
   b. a uniform and realistic serving size.
   c. the RDA for each age group.
   d. ingredients alphabetically.

10. The most common type of undernutrition in industrialized nations, such as the United States, is
    a. anorexia.
    b. protein deficiency.
    c. obesity.
    d. iron deficiency.

11. A behavior that will decrease the risk of weight gain in college is to
    a. skip breakfast.
    b. drink more liquid calories.
    c. stock your fridge with nutritious snacks.
    d. move less and sit more.

Answer Key: 1. a (LO 2.5), 2. c (LO 2.1), 3. d (LO 2.3), 4. a (LO 2.2), 5. c (LO 2.2), 6. a (LO 2.2), 7. b (LO 2.8), 8. b (LO 2.6), 9. b (LO 2.8), 10. d (LO 2.4), 11. c (LO 2.9)

# Study Questions (Numbers refer to Learning Outcomes.)

1. How would you explain the concepts of nutrient density and energy density to a fourth-grade class? (**LO 2.1**)

2. Describe the intent of the *Dietary Guidelines for Americans*. Based on the discussion of the *Dietary Guidelines*, suggest two key dietary changes the typical North American adult should consider making. (**LO 2.2**)

3. What changes to your eating pattern would you need to make to comply with the healthy eating guidelines exemplified by MyPlate on a regular basis? (**LO 2.3**)

4. Describe how the nutritional status of a person might change with malabsorption issues. (**LO 2.4**)

5. What steps would you follow to evaluate the nutritional state of an undernourished person? (**LO 2.5**)

6. How do RDAs and AIs differ from Daily Values in intention and application? (**LO 2.6**)

7. What would you list as the top five sources of reliable nutrition information? What makes these sources reliable? (**LO 2.7**)

8. Dietitians encourage all people to read labels on food packages to learn more about what they eat. What four nutrients could easily be tracked in your diet if you read the Nutrition Facts labels regularly on food products? (**LO 2.8**)

9. Define the USDA definition for the term *organic*. (**LO 2.8**)

10. List some specific health claims that can be made on food labels. (**LO 2.8**)

11. List five strategies to avoid weight gain during college. (**LO 2.9**).

# Further Readings

1. Hingle MD, Kandiah J, and Maggi A: Practice Paper of the Academy of Nutrition and Dietetics: Selecting Nutrient-Dense Foods for Good Health, *J Acad Nutr Diet*;116:1473, 2016. DOI:https://doi.org/10.1016/j .jand.2016.06.375.

2. Taylor C and others: Fruits, vegetables, and health: A comprehensive narrative, umbrella review of the science and recommendations for enhanced public policy to improve intake, *Crit Rev Food Sci Nutr*; 3:1, 2019. DOI: 10.1080/10408398.2019.1632258.

3. Rolls B: Dietary energy density: Applying behavioural science to weight management. *Nutr Bull* 42:246, 2017. DOI:10.1111/nbu.12280.

4. U.S. Department of Agriculture and U.S. Department of Health and Human Services. Dietary Guidelines for Americans, 2020–2025. 9th Edition. December 2020. Available at: DietaryGuidelines.gov. DietaryGuidelines.gov.

5. Ivens BJ and others: Translating the *Dietary Guidelines* to promote behavior change: Perspectives from the Food and Nutrition Science Solutions Joint Task Force. *J Acad Nutr Diet* 2016;116:1697. DOI:10.1016/j. jand.2016.07.014.

6. Schwingshackl L and others: Diet quality as assessed by the Healthy Eating Index, Alternate Healthy Eating Index, Dietary Approaches to Stop Hypertension Score, and health outcomes: An updated systematic review and meta-analysis of cohort studies. *J Acad Nutr Diet* 2018;118(1)74.

7. Robinson J: *Eating on the Wild Side: The Missing Link to Optimum Health.* New York: Little, Brown and Company, 2013.

8. U.S. Department of Health and Human Services: *2018 Physical Activity Guidelines for Americans.* 2018. www.health.gov/PAGuidelines. Accessed January 12, 2020.

9. U.S. Department of Health and Human Services: *2018 Physical Activity Guidelines Advisory Committee Scientific Report.* 2018. https://health.gov /paguidelines/second-edition/report/pdf. PAG_Advisory_Committee _Report.pdf. Accessed January 24, 2020.

10. U.S. Department of Agriculture: USDA's MyPlate. July 2015. www.choosemyplate.gov. Accessed January 12, 2020.

11. Lee-Kwan SH and others: Disparities in state-specific adult fruit and vegetable consumption—United States. *Morb Mortal Wkly Rep*;66:1241, 2017. DOI:http://dx.doi.org/10.15585/mmwr.mm6645a1.

12. Oldways Preservation Trust: Mediterranean diet. https://oldwayspt.org /traditional-diets/mediterranean-diet. Accessed January 15, 2020.

13. American Heart Association: Mediterranean Diet. April 18, 2018. https:// www.heart.org/en/healthy-living/healthy-eating/eat-smart/nutrition -basics/mediterranean-diet. Accessed January 15, 2020.

14. USDA National Agricultural Library, Food and Nutrition Information Center: Dietary Reference Intakes. https://www.nal.usda.gov/fnic /dietary-reference-intakes. Accessed January 20, 2020.

15. Grant JM, Mottet LA, Tanis J, and others: Injustice at every turn: A report of the National Transgender Discrimination Survey. Washington, DC: National Center for Transgender Equality and National Gay and Lesbian Task Force 2011;1–228.

16. Weinand JD and Safer JD: Hormone therapy in transgender adults is safe with provider supervision: A review of hormone therapy sequelae for transgender individuals. *Journal of Clinical and Translational Endocrinology*; 2(2):55–60, 2015. DOI:10.1016/j.jcte.2015.02.003.

17. Academy of Nutrition and Dietetics: Practice Paper of the Academy of Nutrition and Dietetics: Communicating accurate food and nutrition information. *J Acad Nutr Diet* 112:759, 2012. DOI:10.1016/j. jand.2012.03.006.

18. USDA National Agricultural Library: Food and Nutrition Information Center. Fraud and nutrition misinformation. https://www.nal.usda.gov /fnic/fraud-and-nutrition-misinformation. Accessed January 20, 2020.

19. Fakih EK and others: The effects of dietary mobile apps on nutritional outcomes in adults with chronic diseases: A systematic review and meta-analysis. *J Acad Nutr Diet* 119(4), 2019. DOI: https://doi .org/10.1016/j.jand.2018.11.010.

20. Academy of Nutriton and Dietetics: *Food and Nutrition Magazine*. https://foodandnutrition.org/tag/apps. Accessed January 10, 2020.

21. U.S. Food and Drug Administration: Industry Resources on the Changes to the Nutrition Facts Label. https://www.fda.gov/food /food-labeling-nutrition/industry-resources-changes-nutrition-facts-label. Accessed January 25, 2020.

22. Center for Science in the Public Interest: Five Things to Check Before You Buy Breakfast Cereal. https://cspinet.org/tip/five-things-check-you -buy-breakfast-cereal. December 5, 2016. Accessed January 8, 2020.

23. Miller LM and others: Relationships among food label use, motivation, and dietary quality. *Nutrients* 7:1068, 2015.

24. U.S. Food and Drug Administration: Qualified Health Claims: Letters of Enforcement Discretion. https://www.fda.gov/food/food-labeling -nutrition/qualified-health-claims-letters-enforcement-discretion. Accessed January 25, 2020.

25. U.S. Food and Drug Administration: Authorized Health Claims That Meet the Significant Scientific Agreement (SSA) Standard. https://www .fda.gov/food/food-labeling-nutrition/authorized-health-claims-meet -significant-scientific-agreement-ssa-standard. Accessed January 25, 2020.

26. Choi S and Lee Y: Relationship of college students' residence to frequency of meal skipping and snacking pattern. *J Acad Nutr Diet* 112:A24, 2012. DOI: https://doi.org/10.1016/j.jand.2012.06.082.

27. Brister HA and others: 21st birthday drinking and associated physical consequences and behavioral risks. *Psychol Addict Behav* 25:573, 2011. DOI:10.1037/a0025209.

28. Alcohol Rehab Guide: College Alcoholism. https://www.alcoholrehabguide .org/resources/college-alcohol-abuse. Accessed January 25, 2020.

29. Bruening M and others: The struggle is real: A systematic review of food insecurity on postsecondary education campuses. *J Acad Nutr Diet* 117:1767, 2017. DOI:10.1016/j.jand.2017.05.022.

**Design element credits:** pills: Peter Dazeley/Photographer's Choice/Getty Images; germs and bacteria: Alena Ohneva/Shutterstock; globe: McGraw-Hill

# Chapter 3

# The Human Body: A Nutrition Perspective

# Student Learning Outcomes

**Chapter 3 is designed to allow you to:**

**3.1** Understand some basic roles of nutrients in human physiology.

**3.2** Outline the functions of cell components and how cells work.

**3.3** Define *metabolism* and differentiate between anabolic and catabolic reactions.

**3.4** Identify the roles of the cardiovascular and lymphatic systems in nutrition.

**3.5** List basic characteristics of the urinary system and its role in nutrition.

**3.6** List basic characteristics of the nervous system and its role in nutrition.

**3.7** List basic characteristics of the endocrine system, especially the pancreas, and its role in nutrition.

**3.8** List basic characteristics of the immune system and its role in nutrition.

**3.9** Describe the roles of the mouth, stomach, small intestine, large intestine, liver, gallbladder, and pancreas in digestion and absorption of nutrients.

**3.10** Discuss the importance of the microbiota for human health.

**3.11** Understand how nutrients are stored in the body.

**3.12** Identify the major nutrition-related gastrointestinal health problems and approaches to treatment.

## FAKE NEWS

## Eating spicy foods causes ulcers.

### THE FACTS

Medical experts once thought that ulcers were caused by stress or by eating spicy or acidic foods. However, in 1982, scientists discovered that most cases of peptic ulcers are caused by an infection with the acid-resistant bacterium *Helicobacter pylori*.[1] This microorganism disrupts the thick layer of mucus that lines the stomach, allowing acids and enzymes that normally digest foods to irritate and erode the stomach lining.

While spicy or acidic foods may irritate an existing ulcer, they do not cause ulcers. In fact, foods containing capsaicin, the spicy compound in chili peppers, may help to reduce inflammation, which is a good thing for gut health![2]

How does eating food nourish you?. Your body must first digest the food by breaking it down into usable forms of the essential nutrients that can be absorbed into the bloodstream. Once nutrients are taken up by the bloodstream, they can be distributed to and used by body cells.

We rarely think about digesting and absorbing foods. Except for a few voluntary responses—such as deciding what and when to eat, how well to chew food, and when to eliminate the remains—most digestion and absorption processes control themselves. We don't consciously decide when the pancreas will secrete enzymes into the small intestine or how quickly food will be propelled along the intestinal tract. Hormones and nerve signals control these functions. Your only awareness of these involuntary responses may be a hunger pang right before lunch or a "full" feeling after eating that last slice of pizza.

You've learned about cells, tissues, and organs before, but now let's look at the human body from a nutrition perspective. Refresh your memory of the basic anatomy (structure) and physiology (function) of the cardiovascular, lymphatic, urinary, nervous, endocrine, and immune systems. In particular, as you focus on the digestive system, you will gain an in-depth understanding of how the food you eat nourishes your body.

## 3.1 Cells, Tissues, and Organs

### CELLS

A cell is the basic structural and functional component of life—and your body has trillions of them! Each cell is a self-contained, living entity, specialized to perform particular functions. In the human body, all cells have a few common features: membranes, cytoplasm, and **organelles** that perform specialized functions (Fig. 3-1).

**organelles** Compartments, particles, or filaments that perform specialized functions within a cell.

**Cell (Plasma) Membrane.** There is an outside and an inside to every cell, separated by the cell (plasma) membrane. (Please note that cell membranes are not the same as cell walls, which are found in plant cells.) The cell membrane itself is not an organelle, but it holds the cellular contents (cytoplasm and organelles) together and regulates the flow of substances into and out of the cell. Cell-to-cell communication also occurs by way of the cell membrane.

The cell membrane, illustrated in Figure 3-2, is a lipid bilayer. It consists of a double layer of **phospholipids.** A phospholipid is a unique type of lipid that has a water-soluble head and a fat-soluble tail. In a lipid bilayer, the water-soluble heads of many phospholipids face the watery environments that exist both inside and outside the cell. The fat-soluble tails are tucked into the interior of the cell membrane. Molecules of **cholesterol,** another type of lipid, are also embedded within the lipid bilayer. Cholesterol adds some rigidity and stability to the cell membrane. You will learn more about phospholipids and cholesterol in Chapter 5.

**phospholipid** Any of a class of fat-related substances that contain phosphorus, fatty acids, and a nitrogen-containing component. Phospholipids are an essential part of every cell.

**cholesterol** A waxy lipid found in all body cells. It has a structure containing multiple chemical rings. Cholesterol is found only in food ingredients of animal origin.

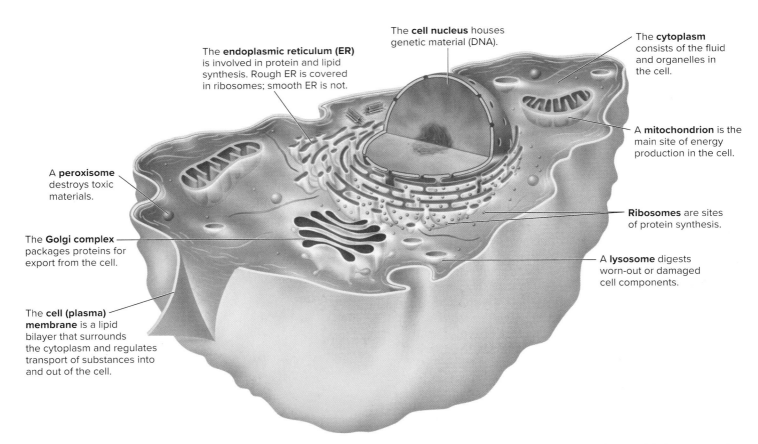

The **cell nucleus** houses genetic material (DNA).

The **endoplasmic reticulum (ER)** is involved in protein and lipid synthesis. Rough ER is covered in ribosomes; smooth ER is not.

The **cytoplasm** consists of the fluid and organelles in the cell.

A **mitochondrion** is the main site of energy production in the cell.

A **peroxisome** destroys toxic materials.

**Ribosomes** are sites of protein synthesis.

The **Golgi complex** packages proteins for export from the cell.

A **lysosome** digests worn-out or damaged cell components.

The **cell (plasma) membrane** is a lipid bilayer that surrounds the cytoplasm and regulates transport of substances into and out of the cell.

**FIGURE 3-1** ▲ An animal cell. Almost all human cells contain the organelles described above. Shown here, but not discussed in the text, are the nucleolus, nuclear envelope, and centrioles. The nucleolus participates in genetic-related functions. The nuclear envelope encloses the nucleus. The centrioles participate in cell division.

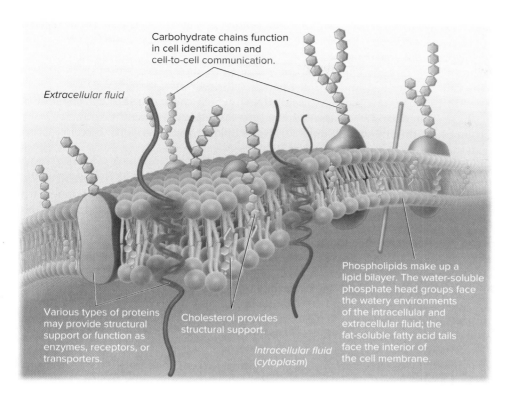

Extracellular fluid

Carbohydrate chains function in cell identification and cell-to-cell communication.

Various types of proteins may provide structural support or function as enzymes, receptors, or transporters.

Cholesterol provides structural support.

Intracellular fluid (cytoplasm)

Phospholipids make up a lipid bilayer. The water-soluble phosphate head groups face the watery environments of the intracellular and extracellular fluid; the fat-soluble fatty acid tails face the interior of the cell membrane.

**FIGURE 3-2** ◀ Cell membrane. The cell membrane is composed of a lipid bilayer. Carbohydrates and proteins serve important roles in the cell membrane, too.

A variety of proteins are part of the cell membrane. Some proteins provide structural support. Others function as **enzymes,** which regulate chemical reactions (see Section 3.8). Some proteins serve as gates or transporters to move substances across the cell membrane. Still other proteins on the outside surface of the cell membrane act as receptors, binding to essential substances that the cell needs and drawing them into the cell. Proteins are covered in more detail in Chapter 6.

Besides lipids and proteins, the membrane also contains carbohydrates that mark the exterior of the cell. These carbohydrates are combined with either protein or fat, and they help to send messages to the cell's organelles and serve as identification markers for the cell. In the immune response, these carbohydrate tags help the immune system to detect invaders and initiate defensive actions. You will learn more about carbohydrates in Chapter 4.

**Cytoplasm.** The **cytoplasm** (also known as *cytosol*) is the combination of fluid material and organelles within the cell, not including the nucleus. A small amount of the energy used by the cell can be produced by chemical processes that occur in the cytoplasm. At least 15 different organelles can be found within the cytoplasm; the nutritional relevance of just six of the organelles will be discussed in the next few subsections.

**Mitochondria.** Mitochondria (singular, mitochondrion) are sometimes called the "power plants" or the "powerhouse" of the cell. These organelles are largely responsible for converting the chemical energy in carbohydrates, lipids, proteins, and alcohol from foods and drinks into a form of energy that cells can use. Except for red blood cells, all cells contain mitochondria.

**Nucleus.** With the exception of red blood cells, all cells have one or more nuclei (singular, nucleus). The **nucleus** is surrounded by its own double membrane, similar to the cell membrane. The role of the nucleus is to store and protect the cell's "code book" of directions for making the substances (i.e., proteins) the cell needs. These directions exist in the form of **deoxyribonucleic acid (DNA),** which is a double strand of nitrogenous

**cytoplasm** The fluid and organelles (except the nucleus) in a cell; also called *cytosol.*

**mitochondria** (singular, mitochondrion) Organelles that are the main sites of energy-yielding chemical reactions in a cell.

**nucleus** Membrane-bound organelle that contains the genetic information (DNA) for protein synthesis and cell replication.

**deoxyribonucleic acid (DNA)** Double strand of nucleic acids that carries hereditary information in cells; DNA directs the synthesis of cell proteins.

**chromosome** A single, large DNA molecule and its associated proteins; contains many genes to store and transmit genetic information.

**gene** A specific segment on a chromosome. Genes provide the blueprint for the production of cell proteins.

**gene expression** Use of DNA information on a gene to produce a protein.

**ribonucleic acid (RNA)** The single-stranded nucleic acid involved in the transcription of genetic information and translation of that information into protein structure.

**ribosomes** Cytoplasmic particles that mediate the linking together of amino acids to form proteins; may exist freely in the cytoplasm or attached to endoplasmic reticulum.

**endoplasmic reticulum (ER)** An organelle composed of a network of canals running through the cytoplasm. Part of the endoplasmic reticulum contains ribosomes.

**Golgi complex** The cell organelle near the nucleus that packages proteins and lipids for secretion or distribution to other organelles.

**secretory vesicles** Membrane-bound vesicles produced by the Golgi complex; contain protein and other compounds to be secreted by the cell.

**lysosome** A cellular organelle that contains digestive enzymes for use inside the cell for turnover of cell parts.

**peroxisome** A cell organelle that destroys toxic products within the cell.

**catalase** Enzyme that catalyzes the decomposition of hydrogen peroxide into water and oxygen.

**tissues** Collections of cells adapted to perform a specific function.

**epithelial tissue** The surface cells that line the outside of the body and all external passages within it.

**connective tissue** Protein tissue that holds different structures in the body together. Some body structures are made up of connective tissue—notably, tendons and cartilage. Connective tissue also forms part of bone and the nonmuscular structures of arteries and veins.

**muscle tissue** A type of tissue adapted to contract to cause movement.

**nervous tissue** Tissue composed of highly branched, elongated cells that transport nerve impulses from one part of the body to another.

**organ** A group of tissues designed to perform a specific function—for example, the heart, which contains muscle tissue, nervous tissue, and so on.

**organ system** A collection of organs that work together to perform an overall function.

bases that are arranged in a very specific sequence. The DNA in cells is packaged as structures called **chromosomes.** A segment of DNA on a chromosome that codes for a specific protein is called a **gene.**

The process of **gene expression,** in which DNA directs the synthesis of proteins in the cell, is described in detail in Chapter 6. Briefly, a segment of a DNA strand is copied in the form of **ribonucleic acid (RNA),** which can move through pores in the nuclear membrane and travel through the cytoplasm to protein-synthesizing sites called **ribosomes.** At the ribosomes, amino acids are linked together to form proteins according to the genetic code transmitted by RNA.

**Endoplasmic Reticulum.** The outer membrane of the cell nucleus is continuous with a network of tubes called the **endoplasmic reticulum (ER).** Part of the endoplasmic reticulum is covered in ribosomes, where the RNA code is translated to synthesize new proteins. The sections of the endoplasmic reticulum that are covered in ribosomes are called rough (as opposed to smooth) ER. Other parts of the endoplasmic reticulum are involved in lipid synthesis, detoxification of harmful substances, and storage of calcium in the cell.

**Golgi Complex.** The **Golgi complex** (also known as the *Golgi apparatus* or *Golgi body*) is a packaging site for proteins and lipids produced in the cell. It consists of sacs within the cytoplasm, where proteins and lipids are packaged into **secretory vesicles** for transport within the cell or secretion from the cell.

**Lysosomes.** **Lysosomes** are the cell's digestive system. They are sacs that contain enzymes for the digestion of foreign material. Sometimes known as "suicide bags," they are responsible for digesting worn-out or damaged cell components. Certain cells associated with immune function contain many lysosomes.

**Peroxisomes.** **Peroxisomes** contain enzymes that detoxify harmful chemicals. Peroxisomes get their name from the fact that hydrogen peroxide ($H_2O_2$) is formed as a result of some detoxification reactions inside this organelle. To counter the damaging effects of hydrogen peroxide within the cell, peroxisomes also contain a protective enzyme called **catalase,** which breaks down hydrogen peroxide into water and oxygen. Peroxisomes also have a minor role in metabolizing alcohol.

## TISSUES

When groups of similar cells work together to accomplish a specialized task, the arrangement is referred to as a **tissue.** Humans are composed of four primary types of tissue: epithelial, connective, muscle, and nervous tissue.

- **Epithelial tissue** is composed of cells that cover surfaces both inside and outside the body. For example, epithelial cells make up the lining of the respiratory tract. Epithelial cells secrete important substances, absorb nutrients, and excrete waste.
- **Connective tissue** supports and protects the body, stores fat, and produces blood cells.
- **Muscle tissue** is designed for movement.
- **Nervous tissue,** which is found in the brain and spinal cord, is designed for communication.

## ORGANS

One, two, or more types of tissue combine to form more complex structures called **organs.** At a still higher level of organization, several organs that work together form an **organ system,** such as the digestive system. All organs contribute to nutritional health, and a person's overall nutritional state determines how well each organ functions. Examine Figure 3-3 to learn about the nutritional relevance of the organ systems.

## Cardiovascular System

**Major components**
heart, blood vessels, and blood

**Functions**
Carries blood and regulates blood supply

Transports nutrients, waste products, hormones, and gases (oxygen and carbon dioxide) throughout the body

Regulates blood pressure

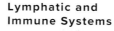

## Lymphatic and Immune Systems

**Major lymphatic components**
lymph, lymphocytes, lymphatic vessels, and lymph nodes

**Major immune components**
white blood cells, lymph vessels and nodes, spleen, thymus gland, and other lymph tissues

**Lymphatic functions**
Removes foreign substances from blood and lymph

Maintains tissue fluid balance

Aids fat absorption

**Immune functions**
Provides defense against pathogens

Formation of white blood cells

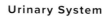

## Urinary System

**Major components**
kidneys, urinary bladder, and the ducts that carry urine

**Functions**
Removes waste products from the blood and forms urine

Regulates blood acid–base (pH) balance, overall chemical balance, and water balance

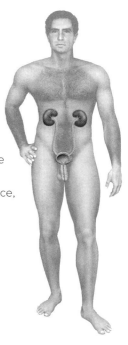

## Nervous System

**Major components**
brain, spinal cord, nerves, and sensory receptors

**Functions**
Detects and interprets sensation

Controls movements, physiological, and intellectual functions

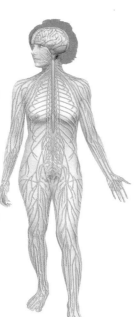

## Endocrine System

**Major components**
hypothalamus, pituitary gland, thyroid gland, thymus gland, pancreas, adrenal glands, and gonads (ovaries in females; testes in males)

**Functions**
Regulates metabolism, growth, reproduction, and many other functions by producing and releasing hormones

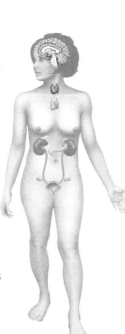

## Digestive System

**Major components**
mouth, esophagus, stomach, intestines, and accessory organs (liver, gallbladder, and pancreas)

**Functions**
Performs the mechanical and chemical processes of digestion of food, absorption of nutrients, and elimination of wastes

Assists the immune system by destroying some pathogens and forming a barrier against foreign materials

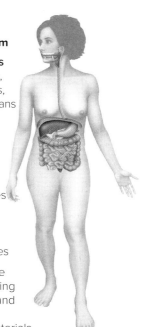

**FIGURE 3-3** ▲ Organ systems of the body.

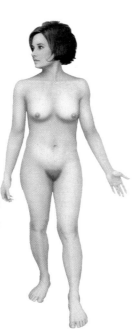

### Integumentary System

**Major components**
skin, hair, nails, and sweat glands

**Functions**
Protects the body

Regulates body temperature

Prevents water loss

Produces vitamin D

### Skeletal System

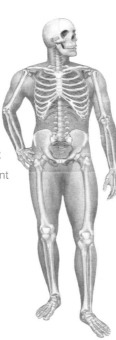

**Major components**
bones, cartilage, ligaments, and joints

**Functions**
Protects organs

Supports body weight

Allows body movement

Produces blood cells

Stores minerals

### Muscular System

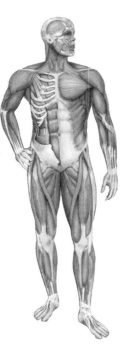

**Major components**
smooth, cardiac, and skeletal muscle

**Functions**
Produces body movement, heartbeat, and body heat

Propels food in the digestive tract

Maintains posture

### Respiratory System

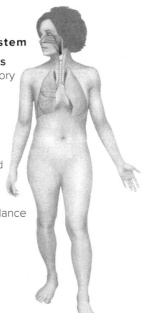

**Major components**
lungs and respiratory passages

**Functions**
Exchanges gases (oxygen and carbon dioxide) between the blood and the air

Regulates blood acid–base (pH) balance

### Reproductive System

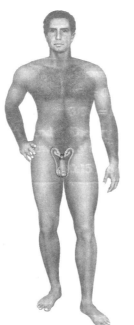
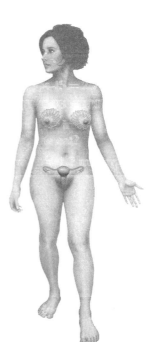

**Major components**
gonads (ovaries and testes), genitals, and breasts

**Functions**
Performs the processes of sexual maturation and reproduction

Influences sexual functions and behaviors

Produces human milk to nourish an infant

**FIGURE 3-3** ▲ Organ systems of the body (*continued*).

Of course, there are lots of interactions among the various organ systems. Sometimes organs within a system can serve another system. For example, the basic function of the digestive system is to convert the food we eat into absorbable nutrients. At the same time, the digestive system serves the immune system by preventing dangerous pathogens from invading and causing illness in the body. As you study nutrition, you will note the multiple roles played by many organs.

The main objective of this chapter is to understand the actions of nutrients as they affect different cells, tissues, organs, and organ systems. As we explore several key organ systems—cardiovascular, lymphatic, urinary, nervous, endocrine, immune, and digestive—look for the ways each system both *affects* and *is affected by* nutrition.

## ✓ CONCEPT CHECK 3.1

1. Choose three organelles and explain their relevance to human nutrition.
2. List the four types of tissues and give an example of where you could find each in the body.
3. Examine Figure 3-3. Provide three examples of ways the organs of one system support the functions of another system.

# 3.2 Metabolism Is the Chemistry of Life

Metabolism refers to the entire collection of chemical processes (reactions) involved in maintaining life. It encompasses all of the sequences of chemical reactions that occur in the body's cells. Some of these reactions take place in the cytoplasm and organelles we have just discussed. They enable us to release and use energy from foods, store sources of fuel for later use, convert toxic substances into less harmful products, and prepare waste products for excretion.

Metabolic reactions can be categorized as either anabolic or catabolic. In **anabolic** reactions, molecules are joined together to synthesize new, larger products. Anabolic reactions require energy. Other reactions are **catabolic,** in which larger materials are broken down into smaller molecules. Catabolic reactions release energy. The catabolism of carbohydrates, fats, and proteins yields energy. Energy metabolism begins in the cytoplasm with the initial breakdown of glucose. The remaining steps of energy metabolism take place in the mitochondria. Ultimately, these reactions harness the chemical energy in food to make the high-energy compound **adenosine triphosphate (ATP),** which our cells can use to do work.

Chemical reactions occur constantly in every living cell; the production of new substances (anabolism) is balanced by the breakdown of older ones (catabolism). An example is the constant formation and breakdown of bone. For bone turnover to occur, cells require a continuous supply of energy derived from dietary carbohydrates, lipids, and proteins. Cells also need water, building materials (e.g., protein and minerals), and chemical regulators (e.g., hormones, enzymes, vitamins, and minerals). Almost all cells also need a steady supply of oxygen from the lungs.

Are you beginning to see how important nutrition is to all the functions of the human body?

▲ **Where do you get the energy to work and play?** Your cells use the energy stored in the chemical bonds of carbohydrates, fats, and proteins to generate ATP. RubberBall Productions/ Photodisc/Getty Images

**anabolic** Relating to pathways that use small, simple compounds to build larger, more complex compounds.

**catabolic** Relating to pathways that break down large compounds into smaller compounds.

**adenosine triphosphate (ATP)** The main form of energy used by cells. ATP energy is used to promote ion pumping, enzyme activity, and muscular contraction.

## ✓ CONCEPT CHECK 3.2

1. What are the differences between anabolic and catabolic reactions?
2. What is ATP?
3. Describe three ways essential nutrients support cell functions.

# 3.3 Cardiovascular System and Lymphatic System

The body has two separate organ systems that circulate fluids in the body: the **cardiovascular system** and the **lymphatic system.** Some texts group these two systems together as the *circulatory system,* but each system has distinct components and functions. The cardiovascular system consists of the heart and blood vessels. The lymphatic system consists of lymphatic vessels and a number of lymph tissues. Blood flows through the cardiovascular system, while **lymph** flows through the lymphatic system.

**cardiovascular system** The body system consisting of the heart, blood vessels, and blood. This system transports nutrients, waste products, gases, and hormones throughout the body and plays an important role in immune responses and regulation of body temperature.

**lymphatic system** A system of vessels and lymph that accepts fluid surrounding cells and large particles, such as products of fat absorption. Lymph eventually passes into the bloodstream from the lymphatic system.

**lymph** A clear fluid that flows through lymph vessels; carries most forms of fat after their absorption by the small intestine.

## CARDIOVASCULAR SYSTEM

The heart is a muscular pump that normally contracts and relaxes 50 to 90 times per minute when the body is at rest. This continual pumping, measured by taking your pulse, keeps blood moving through the blood vessels. The blood that flows through the cardiovascular system is composed of **plasma, red blood cells, white blood cells, platelets,** and many other substances. It travels two basic routes. In the first route, blood circulates from the right side of the heart, through the lungs, and then back to the heart. In the lungs, blood picks up oxygen and releases carbon dioxide. After this exchange of gases has taken place, blood is *oxygenated* and returns to the left side of the heart. In the second route, the oxygenated blood circulates from the left side of the heart to all other body cells, eventually returning back to the right side of the heart (Fig. 3-4). After blood has circulated throughout the body, it is *deoxygenated.* (As you review anatomy diagrams in this book, recognize that *left* and *right* designations refer to the left and right sides of your body, not of the diagram in front of you.)

**plasma** The fluid, extracellular portion of blood.

**red blood cells** Cells that transport oxygen and carbon dioxide through the blood; also called *erythrocytes.*

**white blood cells** Variety of immune cells that circulate in the lymph and blood and work to neutralize, detoxify, and/or destroy pathogens and other foreign proteins; also called *leukocytes.*

**platelets** Protoplasmic discs in the blood that promote coagulation; also called *thrombocytes.*

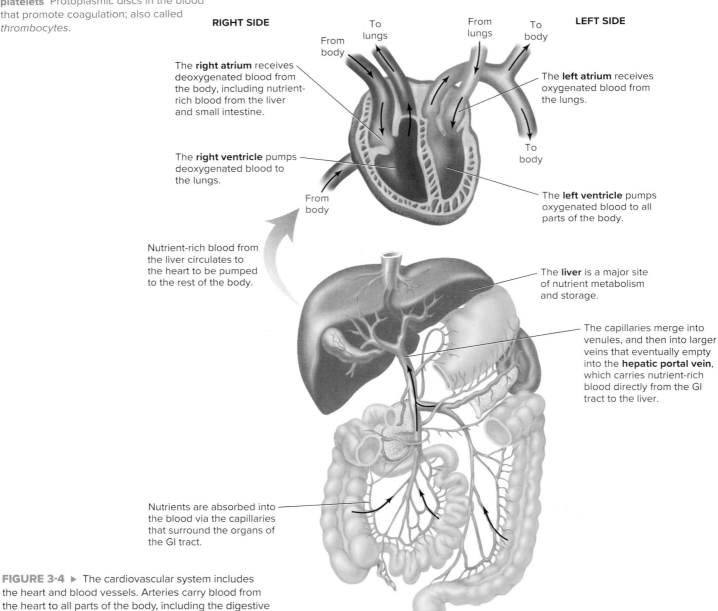

**RIGHT SIDE**

To lungs

From lungs

To body

**LEFT SIDE**

From body

The **right atrium** receives deoxygenated blood from the body, including nutrient-rich blood from the liver and small intestine.

The **left atrium** receives oxygenated blood from the lungs.

The **right ventricle** pumps deoxygenated blood to the lungs.

To body

From body

The **left ventricle** pumps oxygenated blood to all parts of the body.

Nutrient-rich blood from the liver circulates to the heart to be pumped to the rest of the body.

The **liver** is a major site of nutrient metabolism and storage.

The capillaries merge into venules, and then into larger veins that eventually empty into the **hepatic portal vein**, which carries nutrient-rich blood directly from the GI tract to the liver.

Nutrients are absorbed into the blood via the capillaries that surround the organs of the GI tract.

**FIGURE 3-4 ▶** The cardiovascular system includes the heart and blood vessels. Arteries carry blood from the heart to all parts of the body, including the digestive system. Blood that leaves the gastrointestinal (GI) tract is rich in nutrients.

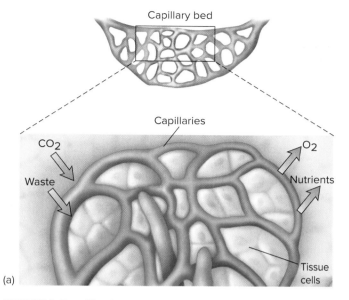

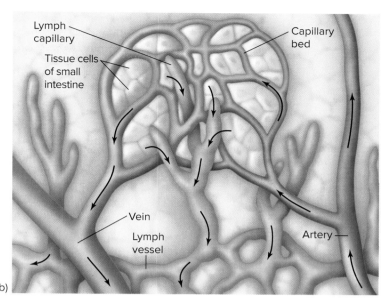

**FIGURE 3-5** ▲ Blood vessels and lymph vessels. (a) Exchange of oxygen ($O_2$) and nutrients for carbon dioxide ($CO_2$) and other waste products occurs between the capillaries and the surrounding tissues. (b) Lymph vessels are also present in capillary beds, such as in the small intestine. Lymph vessels in the small intestine are also called *lacteals*. The lymph vessels have closed ends and are important for fat absorption.

In the cardiovascular system, blood leaves the heart via **arteries,** which branch into **capillaries,** a network of tiny blood vessels that are just one cell layer thick. Exchange of nutrients, oxygen, and waste products between the blood and cells occurs through the tiny, weblike pores of the capillaries (Fig. 3-5). The blood then returns to the heart via the **veins.**

The cardiovascular system facilitates the exchange of oxygen, nutrients, and wastes between the body's internal and external environments. Other functions include delivery of hormones to their target cells, maintenance of a constant body temperature, and distribution of white blood cells throughout the body.

**Portal Circulation in the Gastrointestinal Tract.** Once absorbed through the stomach or intestinal wall, nutrients reach one of two destinations. Some nutrients are taken up by cells in the intestines and portions of the stomach to nourish those organs. Most of these water-soluble nutrients from recently eaten foods, however, are transferred into the **hepatic portal circulation.** (The term *hepatic* refers to the liver. There are other portal systems in physiology, but the simpler terms *portal circulation* or *portal vein* usually refer to hepatic portal circulation.) To enter portal circulation, the nutrients pass from the intestinal capillaries into veins that eventually merge into a very large vein called the **hepatic portal vein.** Unlike most veins in the body—which carry blood back to the heart—this portal vein leads directly to the liver. This enables the liver to process absorbed nutrients before they enter the general circulation of the bloodstream. Overall, hepatic portal circulation represents a special form of circulation in the cardiovascular system.

## LYMPHATIC SYSTEM

The lymphatic system consists of a network of lymphatic vessels and the fluid (lymph) that moves through them. The lymph vessels take up excess fluid that collects between cells and return it to the bloodstream. Lymph is similar to blood, consisting largely of plasma (fluid portion of the blood) that has found its way out of capillaries and into the

**artery** A blood vessel that carries blood away from the heart.

**capillary** A microscopic blood vessel that connects the smallest arteries and veins; site of nutrient, oxygen, and waste exchange between body cells and the blood.

**vein** A blood vessel that carries blood to the heart.

**hepatic portal circulation** The portion of the circulatory system that uses a large vein (portal vein) to carry nutrient-rich blood from capillaries in the intestines and portions of the stomach to the liver.

**hepatic portal vein** Large vein that carries absorbed nutrients from the gastrointestinal tract to the liver.

**lacteal** Lymphatic vessel that absorbs fats from the small intestine.

**urinary system** The body system consisting of the kidneys, urinary bladder, and the ducts that carry urine. This system removes waste products from the circulatory system and regulates blood acid–base balance, overall chemical balance, and water balance in the body.

**ureter** Tube that transports urine from the kidney to the urinary bladder.

**urethra** Tube that transports urine from the urinary bladder to the outside of the body.

**urea** Nitrogenous waste product of protein metabolism; major source of nitrogen in the urine.

**pH** A measure of relative acidity or alkalinity of a solution. The pH scale is 0 to 14. A pH of 7 is neutral; a pH below 7 is acidic; a pH above 7 is alkaline.

**nervous system** The body system consisting of the brain, spinal cord, nerves, and sensory receptors. This system detects sensations, directs movements, and controls physiological and intellectual functions.

**neuron** The structural and functional unit of the nervous system. Consists of a cell body, dendrites, and an axon.

spaces between cells. Lymph also contains white blood cells, which support immune function, as well as dietary fats that have been absorbed from the small intestine. However, neither red blood cells nor platelets are present. Lymph is collected in tiny lymph vessels all over the body and moves through even larger vessels until it eventually empties into the cardiovascular system through a duct near the heart. The lymphatic system does not have a pump (like the heart); its flow is driven by muscle contractions arising from normal body movements.

**Lymphatic Circulation in the Gastrointestinal Tract.** The lymphatic vessels that serve the gastrointestinal tract are specifically known as **lacteals.** Besides contributing to the defense of the body against invading pathogens, lacteals play an important role in nutrition. Most dietary fats are too large to enter the capillaries that surround the cardiovascular system. Instead, most dietary fats enter the lacteals and travel through the lymphatic system until lymph is emptied into the bloodstream by a duct near the heart.

**✓ CONCEPT CHECK 3.3**

1. Describe how nutrients, oxygen, and wastes are exchanged between the body's internal and external environments.
2. What is hepatic portal circulation?
3. Which nutrients are absorbed into the lymph? Why?

## **3.4** Urinary System

The **urinary system** is composed of two kidneys, one on each side of the spinal column. Each kidney is connected to the bladder by a **ureter.** The bladder is emptied by way of the **urethra** (Fig. 3-6). The main function of the kidneys is to remove waste from the body. The kidneys are constantly filtering blood to control its composition.

**FIGURE 3-6** ▶ Organs of the urinary system. The urinary system of the female is shown. The male's urinary system is the same, except that the urethra extends through the penis.

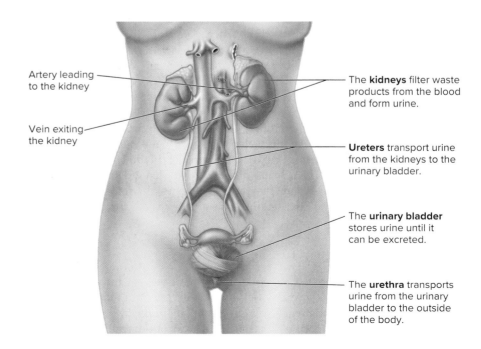

Artery leading to the kidney

Vein exiting the kidney

The **kidneys** filter waste products from the blood and form urine.

**Ureters** transport urine from the kidneys to the urinary bladder.

The **urinary bladder** stores urine until it can be excreted.

The **urethra** transports urine from the urinary bladder to the outside of the body.

This results in the formation of urine, which is composed of water, dissolved waste products of metabolism (e.g., **urea**), and excess or unneeded water-soluble vitamins and various minerals.

Together with the lungs, the kidneys also maintain the acid–base balance **(pH)** of the blood. The kidneys contribute to bone health because they convert a form of vitamin D into its active hormone form. The kidneys also produce a hormone that stimulates red blood cell synthesis. During times of fasting, the kidneys even produce glucose from certain amino acids. Thus, the kidneys perform many important functions related to nutrition.

The proper function of the kidneys is closely tied to the strength of the cardiovascular system, particularly its ability to maintain adequate blood pressure, and the consumption of sufficient fluid. Uncontrolled diabetes, hypertension, and drug abuse are harmful to the kidneys. See *Farm to Fork* in this section to learn how some phytochemicals in cranberries may protect the health of the urinary system.

### ✅ CONCEPT CHECK 3.4

1. List three functions of the kidneys.
2. Trace the path of waste products out of the body.

## 3.5 Nervous System

The **nervous system** is a regulatory system that centrally controls most body functions. The nervous system can detect changes occurring in various organs and the external environment and initiate corrective action when needed to maintain a constant internal body environment. The nervous system also regulates activities that change almost instantly, such as voluntary muscle contractions and the body's response to stress or danger. The body has many receptors that receive information about what is happening within the body and in the outside environment. For the most part, these receptors are found in our eyes, ears, skin, nose, and stomach. We act on information from these receptors via the nervous system.

The basic structural and functional unit of the nervous system is the **neuron.** Neurons are elongated, highly branched cells. The body contains about 100 billion neurons. Neurons respond to electrical and chemical signals, conduct electrical impulses, and release chemical regulators. Overall, neurons allow us to perceive what is occurring in our environment, engage in learning, store vital information in memory, and control the body's voluntary (and involuntary) actions.

The brain stores information, reacts to incoming information, solves problems, and generates thoughts. In addition, the brain plans a course of action based on the other sensory inputs. Responses to the stimuli are carried out mostly through the rest of the nervous system.

### FARM to FORK  Cranberries

F1 ONLINE/SuperStock

Phytochemicals in cranberries impede the ability of some bacteria to adhere to epithelial tissue. Thus, cranberries may be beneficial for reducing infections of the GI and urinary tracts.

**Grow**
- Cranberries grow in wet, mossy areas of land known as *bogs*. Growers sometimes flood the bogs to harvest cranberries, knocking them off their vines with special farm equipment, then quickly collecting the berries when they float to the surface.
- Cold growing temperatures actually increase the sugar content of the berries, making them less tart.
- Fresh cranberries are widely available during the winter months, but the frozen or dried varieties are quite nutritious and can be enjoyed throughout the year.

**Shop**
- Look for firm berries with the deepest red color. The red pigments (anthocyanins) are powerful cancer-fighting phytochemicals! Phytochemicals in cranberries may impede the ability of some bacteria to adhere to and grow on epithelial tissue (e.g., in the urinary tract).
- Food manufacturers add sugars to dried cranberries. Choose varieties that are made with less sugar.
- Although it doesn't pack quite the same disease-fighting punch as fresh cranberries, cranberry juice may be useful for fending off GI and urinary tract infections. Look for brands with less added sugar.

**Store**
- Fresh cranberries can be stored in the refrigerator for 1 week. They should be kept in the crisper drawer in a perforated bag (i.e., the original packaging) to maintain optimal water content and exposure to air.
- If you don't plan to eat fresh cranberries within 1 week of purchase, freezing the berries is the best option to preserve nutrients.

**Prep**
- Add dried cranberries to salads or trail mix to add a dose of antioxidants as well as flavor. In recipes, pairing cranberries with sweeter fruits, such as apples and pears, can strike a nice balance between sweet and tart.
- Fresh, frozen, or dried cranberries make a colorful addition to baked goods, but for maximum health benefits, enjoy cranberries raw. Cooking the berries greatly reduces their antioxidant content.

Source: Robinson J: Strawberries, cranberries, and raspberries: Three of our most nutritious fruits. In *Eating on the Wild Side.* New York: Little, Brown and Company, 2013.

Pixtal/AGE Fotostock

**synapse** The space between one neuron and another neuron (or cell).

**neurotransmitter** A compound made by a nerve cell that allows for communication between it and other cells.

**serotonin** A neurotransmitter involved in the regulation of mood, sleep, and appetite.

**norepinephrine** A neurotransmitter from nerve endings and a hormone from the adrenal gland. It is released in times of stress and is involved in hunger regulation, blood glucose regulation, and other body processes.

**epinephrine** A hormone also known as *adrenaline;* it is released by the adrenal glands at times of stress. It acts to increase glycogen breakdown in the liver, among other functions.

How are nutrients involved in the function of the nervous system? Transmission of a signal (nerve impulse) occurs by way of a change in the concentrations of two minerals, sodium and potassium, across the cell membrane of a neuron. As you will learn in Chapter 9, sodium and potassium are also called electrolytes because they can conduct an electrical current when they are dissolved in water. As a neuron responds to a stimulus, these electrolyte minerals facilitate the transmission of an electrical impulse along the neuron.

When the signal must bridge a gap **(synapse)** from one neuron to the next or between a neuron and its target tissue (e.g., muscle), the electrical message can be converted into a chemical signal called a **neurotransmitter** (Fig. 3-7). Most neurotransmitters are made from amino acids, derived from the protein in foods. For example, the amino acid tryptophan is used to make **serotonin,** a neurotransmitter involved in the regulation of mood, sleep, and many other body functions. The amino acid tyrosine is used to make the neurotransmitters **norepinephrine** and **epinephrine** (also called adrenaline), which are involved in our body's response to stress.

Other nutrients also play a role in the nervous system. Calcium is needed for the release of neurotransmitters from neurons. Vitamin B-12 plays a role in the formation of the

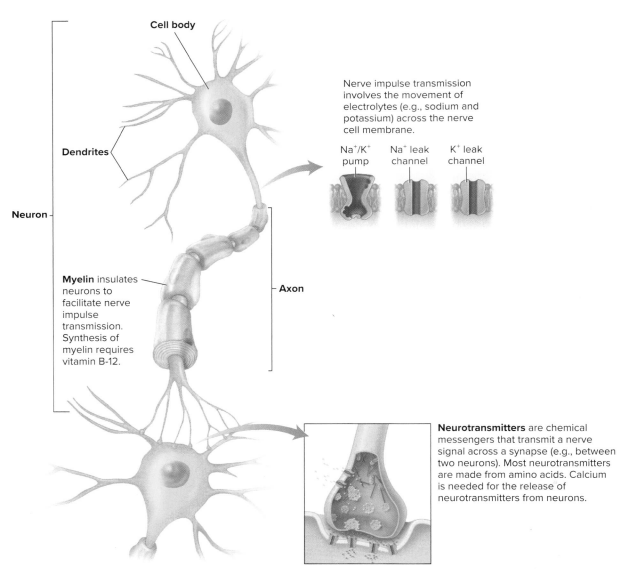

Cell body

Nerve impulse transmission involves the movement of electrolytes (e.g., sodium and potassium) across the nerve cell membrane.

Na⁺/K⁺ pump   Na⁺ leak channel   K⁺ leak channel

Dendrites

Neuron

Myelin insulates neurons to facilitate nerve impulse transmission. Synthesis of myelin requires vitamin B-12.

Axon

**Neurotransmitters** are chemical messengers that transmit a nerve signal across a synapse (e.g., between two neurons). Most neurotransmitters are made from amino acids. Calcium is needed for the release of neurotransmitters from neurons.

**FIGURE 3-7** ▲ A neuron. Transmission of nerve impulses relies on many essential nutrients.

**myelin** sheath, which provides insulation around specific parts of most neurons. Finally, a regular supply of carbohydrate in the form of glucose is important for supplying fuel for the brain. The brain can use other energy sources but generally relies on glucose.

**myelin** A combination of lipids and proteins that covers nerve fibers.

### ✓ CONCEPT CHECK 3.5

1. Why are sodium and potassium important for the work of the nervous system?
2. How are signals transmitted between one neuron and the next? Why are amino acids important in this process?
3. Which nutrient is the brain's preferred source of energy?

## 3.6 Endocrine System

The **endocrine system** plays a major role in the regulation of metabolism, reproduction, water balance, and many other functions through the action of hormones (Table 3-1). Think of hormones as the chemical messengers of the body. They are produced in the **endocrine glands**, released into the blood, and eventually cause changes in target tissues throughout the body. Hormones are not taken up by all cells in the body but only

**endocrine system** The body system consisting of the various glands and the hormones these glands secrete. This system has major regulatory functions in the body, such as reproduction and cell metabolism.

**endocrine gland** A hormone-producing gland.

**TABLE 3-1 ■ Some Hormones with Nutritional Significance**

| Hormone | Gland/Organ | Target | Effect | Role in Nutrition |
|---|---|---|---|---|
| Leptin | Adipose tissue | Hypothalamus | Decreases food intake | Helps to regulate energy intake and body weight |
| Ghrelin | Stomach cells | Hypothalamus | Increases food intake, promotes fat storage, stimulates release of growth hormone | Helps to regulate energy intake and body weight |
| Cholecystokinin | Small intestine cells, brain | Stomach, pancreas, gallbladder | Slows movement of food through the GI tract, stimulates release of bile from the gallbladder and pancreatic juice from the pancreas | Improves the efficiency of digestion |
| Insulin | Pancreas | Adipose, muscle, and liver cells | Decreased blood glucose | Uptake and storage of glucose, fat, and amino acids by cells |
| Glucagon | Pancreas | Liver | Increased blood glucose | Release of glucose from liver stores, synthesis of glucose from amino acids, release of fat from adipose tissue |
| 1,25 dihydroxy-vitamin $D_3$ (calcitriol) | Skin (and food sources); activated in liver and kidneys | Bone, small intestine, kidneys | Regulates blood calcium and phosphorus levels | Promotes bone mineralization |
| Epinephrine, Norepinephrine | Adrenal glands | Heart, blood vessels, brain, lungs | Increased body metabolism and blood glucose | Release of glucose and fat into the blood |
| Growth hormone | Pituitary gland | Most cells | Promotion of amino acid uptake by cells, increased blood glucose | Promotion of protein synthesis and growth, increased fat use for energy |
| Thyroid hormones | Thyroid gland | Most organs | Increased oxygen consumption, overall growth, development of the nervous system | Protein synthesis, increased metabolic rate |

**receptor** A site in a cell at which compounds (such as hormones) bind. Cells that contain receptors for a specific compound are partially controlled by that compound.

**leptin** A hormone made by adipose tissue in proportion to total fat stores in the body that influences long-term regulation of fat mass. Leptin also influences appetite and the release of insulin.

**ghrelin** A hormone produced by stomach cells and the brain that stimulates appetite.

**cholecystokinin** A hormone produced by the small intestinal cells that stimulates enzyme release from the pancreas and bile release from the gallbladder.

**insulin** A hormone produced by the pancreas. Insulin allows for the movement of glucose from the blood into body cells and signals the synthesis of glycogen.

**glucagon** A hormone made by the pancreas that stimulates the breakdown of glycogen in the liver into glucose; this ends up increasing blood glucose. Glucagon also increases the generation of glucose from noncarbohydrate substances.

by those with the correct **receptor** protein. These binding sites, which generally are found on the cell membrane, are highly specific for a certain hormone. Often, binding of a hormone to a receptor on the cell membrane activates additional compounds called second messengers within the cell to carry out the assigned task. A few hormones can penetrate the cell membrane and eventually bind to receptors on the DNA in the nucleus.

There are at least 50 different hormones at work in the body. Throughout this text, you will learn about the nutritional relevance of just a few.

- Several hormones assist in the regulation of food intake and digestion. **Leptin** and **ghrelin** are two hormones that regulate appetite. **Cholecystokinin** sends a signal to digestive organs to secrete bile and digestive enzymes. Other hormones regulate how quickly food moves along the gastrointestinal tract.[3]
- **Insulin** and **glucagon** are two hormones that are synthesized in and released from the pancreas to control the amount of glucose in the blood (Fig. 3-8). When blood glucose rises above normal (usually after a meal), insulin is released from the pancreas and travels through the bloodstream to the muscle, adipose, and liver cells of the body. Among its many functions, insulin allows cells to take up and store glucose. Glucagon has the opposite effect on blood glucose. When blood glucose

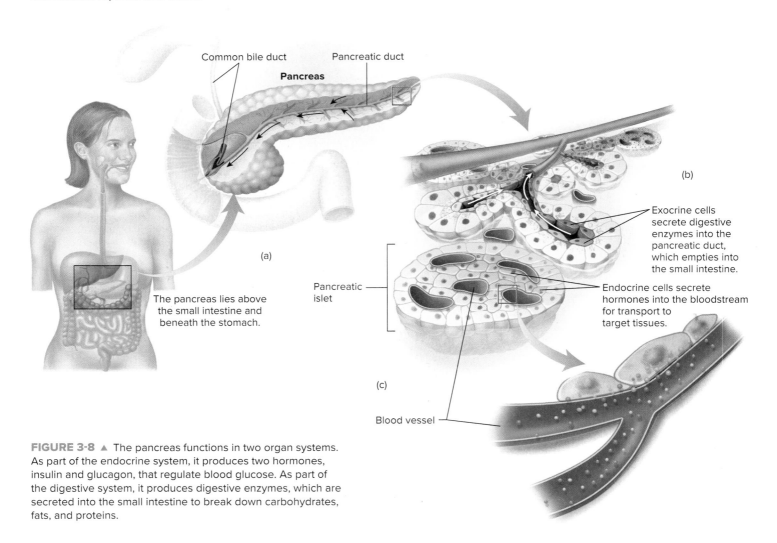

**FIGURE 3-8** ▲ The pancreas functions in two organ systems. As part of the endocrine system, it produces two hormones, insulin and glucagon, that regulate blood glucose. As part of the digestive system, it produces digestive enzymes, which are secreted into the small intestine to break down carbohydrates, fats, and proteins.

levels are lower than normal, glucagon triggers the release of stored glucose and the conversion of certain amino acids into glucose, which causes blood glucose to increase.

- Vitamin D is a *prohormone*, meaning that it can be converted into an active hormone in the body. When you consume foods that contain vitamin D, it is in an inactive form. Chemical reactions in the liver and kidneys convert vitamin D into its active form, **1,25 dihydroxyvitamin D₃,** also called *calcitriol*. The active vitamin D hormone regulates a variety of body processes, including maintenance of blood calcium levels, nerve and muscle development, and immune function.

- Iodine (an essential mineral) is important for thyroid hormone function. **Thyroid hormones,** synthesized in and released from the thyroid gland, help to control the body's metabolic rate—the rate at which you break down carbohydrates, fats, and proteins to make ATP.

### ✓ CONCEPT CHECK 3.6

1. Examine Figure 3-8. What are the *endocrine* roles of the pancreas? What are the *exocrine* roles of the pancreas?

2. What effect does insulin have on the storage of nutrients?

3. List at least three hormones that tend to increase blood sugar.

4. If a person has hypothyroidism, the thyroid gland produces low levels of thyroid hormone. Will a person with hypothyroidism tend to lose weight or gain weight? Explain your answer.

## 3.7 Immune System

Cells throughout the body—skin and intestinal cells—work in concert with the cells and tissues of the immune system to defend the body against infection. The immune system is a collection of diverse tissues that work together to prevent infection, break down aged and dying cells, and remove abnormal cells. **Lymphoid tissue** and white blood cells are specific to the immune system, but other body systems also support the immune system: the skin and GI tract provide physical and chemical barriers against invading **pathogens**. In addition, specialized **gut-associated lymphoid tissues (GALT)** are scattered throughout the intestinal tract. GALT assists the cells of the GI tract in keeping pathogens from entering the bloodstream. The lymphatic system is another major site of immune activity: the **lymph nodes** trap pathogens, and the lymph itself transports white blood cells through the body.

We are born with some aspects of immune function, such as physical and chemical barriers against infection, the inflammatory response, and the ability of some white blood cells to engulf microorganisms by **phagocytosis.** These are termed **nonspecific (innate) immunity** because they protect the body against invasion by any microorganism. The skin and the intestinal cells support the immune system by forming an important barrier against invading microorganisms. If the integrity of either one of these barriers is compromised, microorganisms can invade the body and cause illness. Substances secreted by the skin and intestinal cells can also destroy pathogens.

If the body's nonspecific immune defenses are unable to block a microorganism's entry into the bloodstream, cells and chemicals involved in **specific (adaptive) immunity** will identify and destroy the invading pathogen. Specific immunity involves the process by which some types of white blood cells produce **antibodies** (also called *immunoglobulins*) that target specific microorganisms or foreign proteins (known as **antigens**). After initial exposure to an antigen, a "memory" is created such that a second exposure to the substance will produce a more vigorous and rapid response.

**1,25-dihydroxyvitamin D₃** Biologically active form of vitamin D that regulates blood calcium levels; also called *calcitriol* or abbreviated *1,25(OH)D₃*.

**thyroid hormones** Hormones produced by the thyroid gland that regulate growth and metabolic rate.

**lymphoid tissue** Specialized cells that participate in the immune response; includes the thymus, spleen, lymph nodes, and white blood cells.

**pathogen** A microorganism that can cause disease.

**gut-associated lymphoid tissues (GALT)** Clusters of lymphoid cells located throughout the gastrointestinal tract that destroy pathogens.

**lymph nodes** Clusters of lymphoid tissue, situated along the lymph vessels, that trap and destroy pathogens.

**phagocytosis** A process in which a cell forms an indentation, and solid particles enter the indentation and are engulfed by the cell.

**nonspecific immunity** Defenses that stop the invasion of pathogens; requires no previous encounter with a pathogen; also called *innate immunity*.

**specific immunity** Function of white blood cells directed at specific antigens; also called *adaptive immunity*.

**antibody** Blood protein that binds foreign proteins found in the body; also called *immunoglobulin*.

**antigen** Any substance that induces a state of sensitivity and/or resistance to microorganisms or toxic substances after a lag period; a foreign substance that stimulates a specific aspect of the immune system.

An optimally functioning immune system is our body's way of renewing and sustaining itself. The immune system detects and destroys harmful or malfunctioning components (e.g., old and worn out cells, mutated cells, or microbial invaders) to maintain proper function. To do this work, the immune system requires adequate energy, protein, essential fatty acids, and a variety of micronutrients. In addition, a healthy gut microbiome stimulates proper immune system development and supports our body's natural defenses. Eating well to support the immune system is a step towards sustainability of the human body—healthy people for a healthy community!

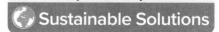 Sustainable Solutions

The immune system provides a very clear example of the interrelationship between nutrition status and organ system function. In developing nations, where food shortages are common, malnutrition increases susceptibility to infectious diseases, such as diarrheal disease. The turnover of many cells of the immune system is quite rapid—only a few hours or days. The constant synthesis of new cells requires steady nutrient intake. Nutrients that are important for the health of the immune system include essential fatty acids; protein; and vitamins A, C, D, and some B vitamins (Chapter 8); and the minerals iron, copper, and zinc (Chapter 9).

**✓ CONCEPT CHECK 3.7**

1. Contrast nonspecific (innate) and specific (adaptive) immunity.
2. What are the roles of antigens and antibodies in the immune response?
3. List three nutrients that support the immune system.

## 3.8 Digestive System

The foods and beverages we consume, for the most part, must undergo extensive alteration by the **digestive system** to provide us with usable nutrients. The digestive system is composed of six hollow organs that make up the **gastrointestinal (GI) tract** as well as three accessory organs that secrete important substances into the GI tract. The processes of **digestion** and **absorption** take place inside the GI tract (Fig. 3-9). The open space inside

**digestive system** System consisting of the gastrointestinal tract and accessory structures (liver, gallbladder, and pancreas). This system performs the mechanical and chemical processes of digestion, absorption of nutrients, and elimination of wastes.

**gastrointestinal (GI) tract** The main sites in the body used for digestion and absorption of nutrients. It consists of the mouth, esophagus, stomach, small intestine, large intestine, rectum, and anus. Also called the *digestive tract.*

**digestion** Process by which large ingested molecules are mechanically and chemically broken down to produce basic nutrients that can be absorbed across the wall of the GI tract.

**absorption** The process by which substances are taken up from the GI tract and enter the bloodstream or the lymph.

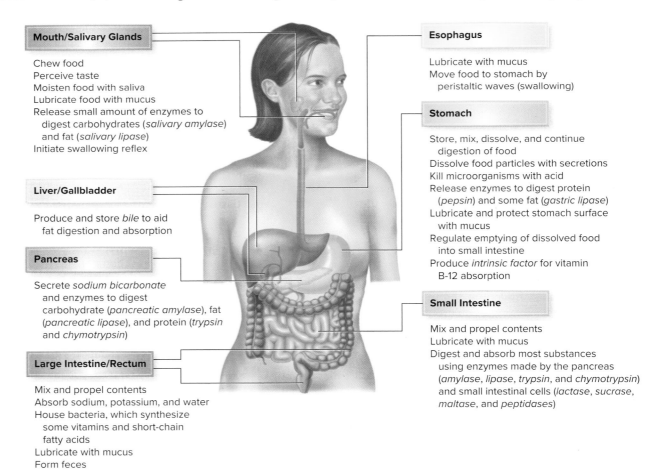

**FIGURE 3-9** ▲ Physiology of the GI tract. Many organs cooperate to facilitate the digestion and absorption of nutrients from foods. Partially digested food spends about 2 to 3 hours in the stomach (longer for large meals). Passage through the small intestine takes 3 to 10 hours, followed by up to 72 hours in the large intestine. On average, digestion and absorption of a meal take about 2 days. Food matter tends to pass more quickly through the GI tract of men than women.

the GI tract is called the **lumen.** Nutrients from the food we eat must pass through the walls of the GI tract—from the lumen through the cells lining the GI tract—to be absorbed into the bloodstream.

**lumen**  The hollow opening inside a tube, such as the GI tract.

There are two ways food is broken down in the GI tract: *mechanical* digestion and *chemical* digestion. Mechanical digestion takes place as soon as you begin chewing your food and continues as muscular contractions simultaneously mix and move food through the length of the GI tract. Chemical digestion refers to the chemical breakdown of foods by acid and enzymes secreted into the GI tract.

Enzymes are a key part of chemical digestion. Each enzyme is specific to one type of chemical process. For example, the enzyme that recognizes and digests table sugar (sucrose) ignores milk sugar (lactose). Besides working on only specific types of chemicals, enzymes are sensitive to acidic and alkaline conditions, temperature, and the types of vitamins and minerals they require to function. Digestive enzymes that work in the acidic environment of the stomach do not work well in the alkaline environment of the small intestine. The pancreas and small intestine produce most of the digestive enzymes; however, the mouth and the stomach also contribute their own enzymes to the process of digestion. The organs of the digestive system are able to fine-tune the production of each type of digestive enzyme in response to the nutritional makeup and amount of food consumed. Overall, the enzymes of the digestive system work together to hasten the breakdown of ingested food into absorbable nutrients (Fig. 3-10).

As food moves along the GI tract, nutrients are absorbed. The primary site of nutrient absorption is the small intestine. By the time the meal contents reach the large intestine, most of the usable nutrients have been absorbed. What remains is waste. The final role of the digestive system is elimination of wastes.

Most of the processes of digestion and absorption are under *autonomic* control; that is, they are involuntary. The functions involved in digestion and absorption are controlled by signals from the nervous system, hormones from the endocrine system, and hormone-like compounds. Many common ailments arise from problems with the digestive system. Several of these digestive problems are discussed in Section 3.11.

Take a moment to study Figure 3-9. In the next few subsections, we will examine the functions of each organ in detail. Can you label each organ on a diagram? Can you briefly describe the role of each organ in digestion and absorption?

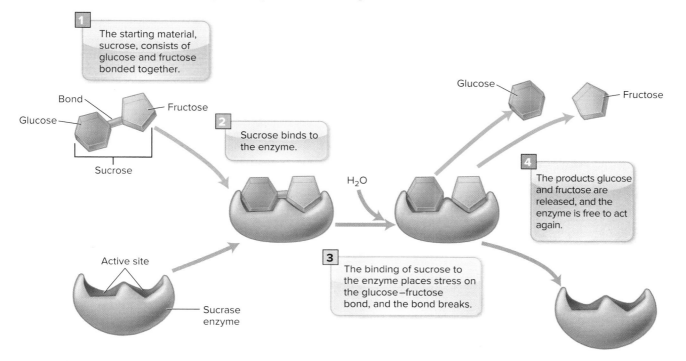

**1** The starting material, sucrose, consists of glucose and fructose bonded together.

Bond

Fructose

Glucose

Sucrose

**2** Sucrose binds to the enzyme.

Glucose

Fructose

$H_2O$

**4** The products glucose and fructose are released, and the enzyme is free to act again.

Active site

Sucrase enzyme

**3** The binding of sucrose to the enzyme places stress on the glucose–fructose bond, and the bond breaks.

**FIGURE 3-10** ▲ A model of enzyme action. The enzyme sucrase splits the sugar sucrose into two simpler sugars, glucose and fructose. Energy (in the form of ATP) is needed to make some reactions occur. Sometimes, enzyme activity depends on the presence of specific vitamin or mineral cofactors.

**umami** A brothy, meaty, savory flavor in some foods. Monosodium glutamate enhances this flavor when added to foods.

**oleogustus** A taste for fat. The presence of fatty acids in foods stimulates taste receptors in the mouth; this sensation is unpleasant.

**saliva** Watery fluid, produced by the salivary glands in the mouth, which contains lubricants, enzymes, and other substances.

**amylase** Starch-digesting enzyme produced by the salivary glands and the pancreas.

**lipase** Fat-digesting enzyme produced by the salivary glands, stomach, and pancreas.

**mucus** A thick fluid secreted by many cells throughout the body. It contains a compound that has both carbohydrate and protein parts. It acts as a lubricant and means of protection for cells.

**esophagus** A tube in the GI tract that connects the pharynx with the stomach.

**pharynx** A cavity located at the back of the oral and nasal cavities, commonly known as the throat. It is part of the digestive tract and the respiratory tract.

**epiglottis** The flap that folds down over the trachea during swallowing.

**bolus** A moistened mass of food swallowed from the oral cavity into the pharynx.

**trachea** The airway that extends from the throat, down the neck, to the lungs. Also called the *windpipe*.

## MOUTH

The mouth has the unique ability to sense the taste of the foods we consume. The tongue, through the use of its taste buds, identifies foods on the basis of their specific flavor(s). Sweet, sour, salty, bitter, **umami,** and **oleogustus** comprise the primary taste sensations we experience. Surprisingly, the nose and our sense of smell greatly contribute to our ability to sense the taste of food. When you chew a food, chemicals are released that stimulate the nasal passages. Thus, it makes perfect sense that when your nose is congested, even your favorite foods will not taste as good as they normally do. As the COVID-19 pandemic has evolved, alterations in the senses of taste and smell have emerged as early symptoms of viral infection.[4]

The taste of food, or the anticipation of it, signals the rest of the GI tract to prepare for the digestion of food, which begins in the mouth. The chewing action of the teeth contributes to the mechanical digestion of foods. Chemical digestion begins with **saliva,** produced by the salivary glands. Saliva functions as a solvent so that food particles can be further separated and tasted. In addition, saliva contains a starch-digesting enzyme, salivary **amylase,** and a fat-digesting enzyme, **lipase. Mucus,** another component of saliva, is a lubricant that makes it easier to swallow a mouthful of food. The food then travels to the esophagus. The important secretions and products of digestion are listed in Table 3-2.

## ESOPHAGUS

The **esophagus** is a long tube that connects the **pharynx** with the stomach. Near the pharynx is a flap of tissue (called the **epiglottis**) that prevents a **bolus** of swallowed food from entering the **trachea** (windpipe) (Fig. 3-11). During swallowing, food lands

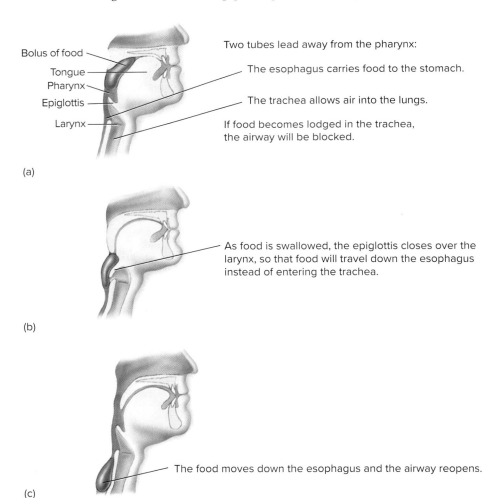

Bolus of food
Tongue
Pharynx
Epiglottis
Larynx

Two tubes lead away from the pharynx:

The esophagus carries food to the stomach.

The trachea allows air into the lungs.

If food becomes lodged in the trachea, the airway will be blocked.

(a)

As food is swallowed, the epiglottis closes over the larynx, so that food will travel down the esophagus instead of entering the trachea.

(b)

The food moves down the esophagus and the airway reopens.

(c)

**FIGURE 3-11** ▲ The process of swallowing.

**TABLE 3-2** ■ **Important Secretions of the Digestive Tract**

| Secretion | Site of Production | Purpose |
| --- | --- | --- |
| Saliva | Mouth | Contains enzymes that make a minor contribution to starch and fat digestion<br>Lubrication of food for swallowing |
| Mucus | Mouth, esophagus, stomach, small intestine, large intestine | Protects GI tract cells<br>Lubricates food as it travels through the GI tract |
| Enzymes | Mouth, stomach, small intestine, pancreas | Promote digestion of carbohydrates, fats, and proteins into forms small enough for absorption (examples: amylases, lipases, proteases) |
| Acid | Stomach | Promotes digestion of protein<br>Destroys pathogens<br>Solubilizes some minerals<br>Activates some enzymes |
| Bile | Liver (stored in gallbladder) | Aids fat digestion in the small intestine by suspending fat in water using **bile acids**, cholesterol, and phospholipids |
| Bicarbonate | Pancreas, small intestine | Neutralizes stomach acid when it reaches the small intestine |
| Hormones | Stomach, small intestine, pancreas | Stimulate production and/or release of acid, enzymes, bile, and bicarbonate<br>Help regulate movement of food matter through the GI tract |
| Intrinsic factor | Stomach | Facilitates absorption of vitamin B-12 in the small intestine |

**bile acid** A compound produced by the liver. Bile acids are the main component of bile, which aids in emulsification of fat during digestion in the small intestine.

**peristalsis** A coordinated muscular contraction used to propel food down the gastrointestinal tract.

**lower esophageal sphincter** A circular muscle that constricts the opening of the esophagus to the stomach. Also called the *gastroesophageal sphincter* or the *cardiac sphincter.*

on the epiglottis, folding it down to cover the opening of the trachea. Breathing also stops automatically. These responses ensure that swallowed food will only travel down the esophagus. If food becomes lodged in the trachea, choking will occur (the victim will not be able to speak, cough, or breathe). A group of techniques to treat a choking person is called the Heimlich maneuver (see www.heimlichinstitute.org for details).

At the top of the esophagus, nerve fibers release signals to tell the GI tract that food has been consumed. This results in an increase in GI muscle action, called **peristalsis.** These waves of muscular contractions force the food in one direction along the digestive tract from the mouth toward the anus (Fig. 3-12).

At the end of the esophagus is the **lower esophageal sphincter,** a ring of muscle that constricts (closes) after food enters the stomach. In general, the function of sphincters is to prevent the backflow of GI tract contents. Sphincters respond to various stimuli, such as signals from the nervous system, hormones, acidic conditions, and pressure that builds up around the sphincter. The primary function of the lower esophageal sphincter is to prevent the acidic contents of the stomach from flowing back up into the esophagus. Dysfunction of this sphincter can cause heartburn (Section 3.11).

No digestion or absorption occurs in the esophagus; it serves merely to transport food from the mouth to the stomach. The cells of the esophagus secrete mucus to lubricate the passage of food, but no digestive enzymes are produced.

## STOMACH

The stomach is a large sac that can hold up to 4 cups (or 1 quart) of food for several hours until all of the food has been moved into the small intestine. Stomach size varies individually and can be reduced surgically as a radical treatment for obesity (more on this in Section 7.9). While in the stomach, the food is mixed with gastric juice, which contains water, hydrochloric acid, and enzymes. (*Gastric* is a term pertaining to the stomach.) The acid in the gastric juice halts the biological activity of proteins,

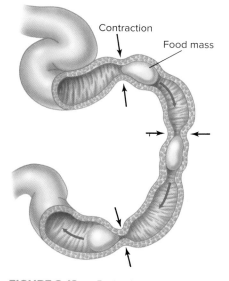

Contraction

Food mass

**FIGURE 3-12** ▲ Peristalsis. Peristalsis is a progressive type of movement, propelling material from point to point along the GI tract. To begin this, a ring of contraction occurs where the GI wall is stretched, passing the food mass forward. The moving food mass triggers a ring of contraction in the next region, which pushes the food mass even farther along. The result is a ring of contraction that moves like a wave along the GI tract, pushing the food mass down the tract.

converts inactive digestive enzymes into their active form, partially digests food protein, and makes dietary minerals soluble so that they can be absorbed. The mixing that takes place in the stomach produces a watery food mixture, called **chyme,** which slowly leaves the stomach a teaspoon (5 milliliters) at a time and enters the small intestine. Following a meal, the stomach contents are emptied into the small intestine over the course of 1 to 4 hours. The **pyloric sphincter,** located at the base of the stomach, controls the rate at which the chyme is released into the small intestine (Fig. 3-13). There is very little absorption of nutrients from the stomach, except for some water and alcohol.

**chyme** A mixture of stomach secretions and partially digested food.

**pyloric sphincter** Ring of smooth muscle between the stomach and the small intestine.

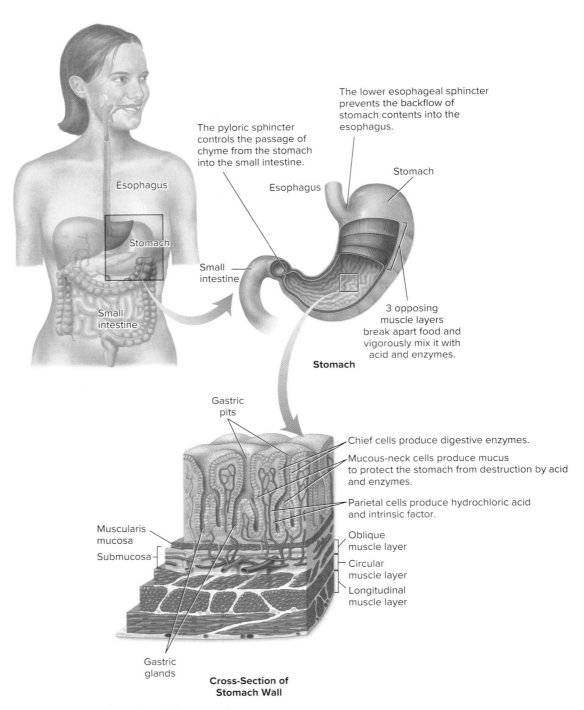

FIGURE 3-13 ▲ Physiology of the stomach.

If the acid and enzymes in the stomach are strong enough to break apart food proteins and kill pathogens, how does the stomach tissue itself withstand destruction by these harsh chemicals? First, the production of acid and enzymes in the stomach is regulated by hormones (e.g., gastrin) that are released when we are eating or thinking about eating. Between meals, the stomach cells do not produce much acid and enzymes. Second, the stomach produces a thick layer of mucus that forms a protective barrier over the stomach lining.

One other important function of the stomach is the production of a substance called **intrinsic factor.** This vital proteinlike compound is essential for the absorption of vitamin B-12. You will learn about the digestion and absorption of vitamin B-12 in Section 8.14.

## SMALL INTESTINE

The small intestine is considered "small" because its diameter is only 1 inch (2.5 centimeters). It is actually quite long—about 10 feet (3 meters), beginning at the stomach and extending to the large intestine (Fig. 3-14). The three parts of the small intestine are the **duodenum** (first 10 inches), the **jejunum** (second 4 feet), and the **ileum** (last 5 feet).

Most of the digestion and absorption of food occurs in the small intestine. As chyme moves from the stomach into the first part of the small intestine, it is still very acidic. You just learned that the stomach secretes a thick layer of mucus to protect itself from the strong acid. However, if the small intestine were coated with mucus, digestion and absorption would be very limited. Therefore, the pancreas and intestinal cells secrete **bicarbonate** to neutralize the acid. The neutral pH also optimizes the activity of the digestive enzymes that work in the small intestine. Muscular contractions move the chyme through the small intestine and thoroughly mix food particles with digestive juices (review Fig. 3-12). These juices contain enzymes that break down carbohydrates, protein, and fat into absorbable units.

The physical structure of the small intestine is very important to the body's ability to digest and absorb the nutrients it needs. The lining of the small intestine is called the mucosa and is folded many times; within these folds are fingerlike projections called **villi.** These "fingers" are constantly moving, which helps them trap food to enhance absorption. Each individual villus (singular) is made up of many **absorptive cells** (also called enterocytes), and the mucosal surface of each of these cells is folded even further into **microvilli.** The combined folds, villi, and microvilli in the small intestine increase its surface area 600 times beyond that of a simple tube (see Fig. 3-14).

The absorptive cells have a short life. New intestinal absorptive cells are constantly produced in the crypts of the lining of the small intestine (see Fig. 3-14) and appear daily along the surface of each villus. This is probably because absorptive cells are subjected to a harsh environment, so renewal of the intestinal cell lining is necessary. This rapid cell turnover leads to high nutrient needs for the small intestine. Fortunately, many of the old cells can be broken down and their parts can be reused. The health of the cells is further enhanced by various hormones and other substances that participate in or are produced as part of the digestive process.

The small intestine absorbs nutrients through the intestinal wall through various means and processes, as illustrated in Figure 3-15.

- **Passive diffusion:** When the nutrient concentration is higher in the lumen of the small intestine than in the absorptive cells, the difference in nutrient concentration drives the nutrient into the absorptive cells by diffusion. Fats, water, and some minerals are examples of nutrients that move down a concentration gradient to be absorbed by passive diffusion.
- **Facilitated diffusion:** Some compounds require a carrier protein to follow a concentration gradient into absorptive cells. This type of absorption is called facilitated diffusion. Fructose is one example of a compound that makes use of such a carrier to allow for facilitated diffusion.

**intrinsic factor** A proteinlike compound produced by the stomach that enhances vitamin B-12 absorption in the ileum.

**duodenum** First segment of the small intestine that receives chyme from the stomach and digestive juices from the pancreas and gallbladder. This is the site of most chemical digestion of nutrients; approximately 10 inches in length.

**jejunum** Middle segment of the small intestine; approximately 4 feet in length.

**ileum** Last segment of the small intestine; approximately 5 feet in length.

**bicarbonate** Alkaline compound produced as part of the body's buffer systems. For example, the pancreas secretes bicarbonate to neutralize the hydrochloric acid in chyme in the small intestine.

**villi** (singular, villus) The fingerlike protrusions into the small intestine that participate in digestion and absorption of food.

**absorptive cells** The intestinal cells that line the villi and participate in nutrient absorption; also known as *enterocytes*.

**microvilli** Extensive folds on the muscosal surface of the absorptive cells.

**passive diffusion** Movement of a substance across a semipermeable membrane from an area of higher solute concentration to an area of lower solute concentration. This type of transport does not require a carrier and does not require energy.

**facilitated diffusion** Movement of a substance across a semipermeable membrane from an area of higher solute concentration to an area of lower solute concentration. This type of transport does not require energy, but it does require a carrier.

**FIGURE 3-14** ▶

Organization of the small intestine. (a) The small intestine is divided into three segments: duodenum, jejunum, and ileum. (b) The walls of the small intestine have four layers: mucosa, submucosa, muscularis, and serosa. (c) The epithelial tissue on the inner surface of the mucosa has many villi. (d) The surface of each villus is covered in many absorptive cells with microvilli.

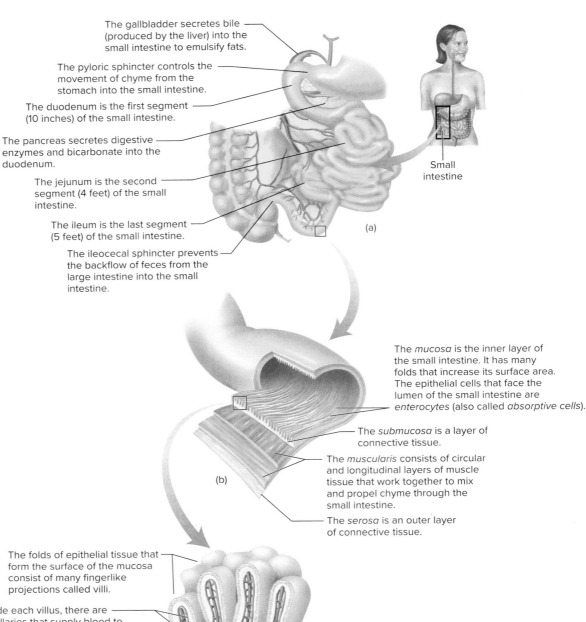

The gallbladder secretes bile (produced by the liver) into the small intestine to emulsify fats.

The pyloric sphincter controls the movement of chyme from the stomach into the small intestine.

The duodenum is the first segment (10 inches) of the small intestine.

The pancreas secretes digestive enzymes and bicarbonate into the duodenum.

The jejunum is the second segment (4 feet) of the small intestine.

The ileum is the last segment (5 feet) of the small intestine.

The ileocecal sphincter prevents the backflow of feces from the large intestine into the small intestine.

Small intestine

(a)

The *mucosa* is the inner layer of the small intestine. It has many folds that increase its surface area. The epithelial cells that face the lumen of the small intestine are *enterocytes* (also called *absorptive cells*).

The *submucosa* is a layer of connective tissue.

The *muscularis* consists of circular and longitudinal layers of muscle tissue that work together to mix and propel chyme through the small intestine.

The *serosa* is an outer layer of connective tissue.

(b)

The folds of epithelial tissue that form the surface of the mucosa consist of many fingerlike projections called villi.

Inside each villus, there are capillaries that supply blood to the cells of the small intestine and carry absorbed nutrients from the GI tract to the liver.

Each villus also contains a lacteal, a small vessel of the lymphatic system, which transports dietary lipids from the GI tract to the bloodstream.

New absorptive cells develop daily in the crypts between the villi.

(c)

Each absorptive cell on the inner surface of the mucosa is covered in microvilli, which further increase the surface area to maximize nutrient absorption.

(d)

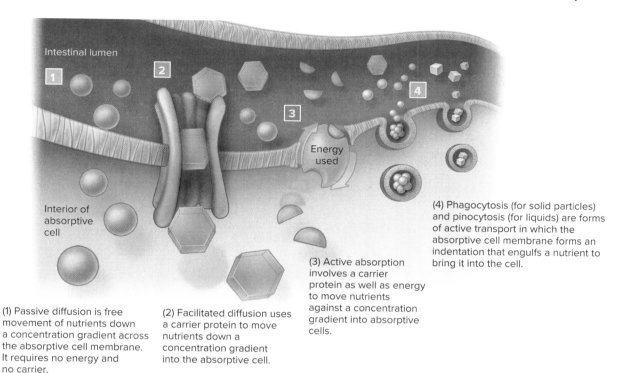

Intestinal lumen

1  2

3  4

Energy used

Interior of absorptive cell

(4) Phagocytosis (for solid particles) and pinocytosis (for liquids) are forms of active transport in which the absorptive cell membrane forms an indentation that engulfs a nutrient to bring it into the cell.

(3) Active absorption involves a carrier protein as well as energy to move nutrients against a concentration gradient into absorptive cells.

(1) Passive diffusion is free movement of nutrients down a concentration gradient across the absorptive cell membrane. It requires no energy and no carrier.

(2) Facilitated diffusion uses a carrier protein to move nutrients down a concentration gradient into the absorptive cell.

**FIGURE 3-15** ▲ Nutrient absorption relies on four major absorptive processes. Passive diffusion (1) and facilitated diffusion (2) move solutes *down a concentration gradient* (i.e., from an area of high nutrient concentration to an area of low nutrient concentration), so they do not require energy. Active absorption (3) and phagocytosis (4) move solutes *against a concentration gradient* (i.e., from an area of low nutrient concentration to an area of high nutrient concentration), so they do require energy.

- **Active absorption:** In addition to the need for a carrier protein, some nutrients also require energy input to move from the lumen of the small intestine into the absorptive cells. This mechanism makes it possible for cells to take up nutrients even when they are consumed in low concentrations (i.e., against a concentration gradient). Some sugars, such as glucose, are actively absorbed, as are amino acids.
- Phagocytosis and **pinocytosis:** In a further means of active absorption, absorptive cells literally engulf solid particles (phagocytosis) or liquids (pinocytosis). A cell membrane forms an indentation and when particles or fluids move into the indentation, the cell membrane surrounds and engulfs them. This process is used when an infant absorbs immune substances from human milk (see Section 14.7).

Once absorbed, water-soluble compounds such as glucose and amino acids are transported by the capillaries to the hepatic portal vein, which leads directly to the liver. Most fats are absorbed into the lymph vessels, which eventually empty into the bloodstream (review Figs. 3-4 and 3-5).

Undigested food cannot be absorbed into cells of the small intestine. Any undigested food that reaches the end of the small intestine must pass through the **ileocecal sphincter** on the way to the large intestine. This sphincter prevents the contents of the large intestine from reentering the small intestine.

**active absorption** Movement of a substance across a semipermeable membrane from an area of lower solute concentration to an area of higher solute concentration. This type of transport requires energy and a carrier.

**pinocytosis** A process in which a cell forms an indentation, and fluid enters the indentation and is engulfed by the cell.

**ileocecal sphincter** The ring of smooth muscle between the end of the small intestine and the beginning of the large intestine.

## LARGE INTESTINE

When the contents of the small intestine enter the large intestine through the ileocecal sphincter, the material that is left bears little resemblance to the food that was originally eaten. If the previous steps of digestion and absorption are working normally, only a minor amount (5%) of carbohydrate, protein, and fat escapes absorption to reach the large intestine.

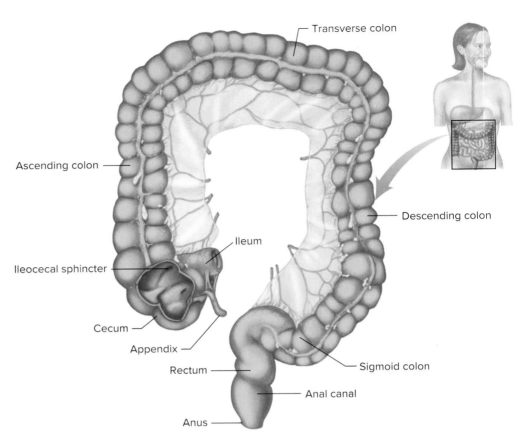

FIGURE 3-16 ▲ The parts of the large intestine include the cecum, ascending colon, transverse colon, descending colon, and sigmoid colon. Overall, the large intestine is about 3.5 feet (1.1 meters) long.

**cecum** A pouch at the first part of the large intestine that houses many bacteria.

**ascending colon** Segment of the large intestine that carries feces from the cecum, up the right side of the abdomen, to the transverse colon.

**transverse colon** Segment of the large intestine that carries feces from the ascending colon, from right to left across the top of the abdomen, to the descending colon.

**descending colon** Segment of the large intestine that carries feces from the transverse colon, down the left side of the abdomen, to the sigmoid colon.

**sigmoid colon** Last segment of the large intestine that carries feces from the descending colon to the rectum.

**feces** Mass of water, fiber, tough connective tissues, bacterial cells, and sloughed intestinal cells that passes through the large intestine and is excreted through the anus; also called *stool*.

The large intestine (sometimes called the *colon*) can be subdivided into five main segments: the **cecum, ascending colon, transverse colon, descending colon,** and **sigmoid colon** (see Fig. 3-16). Physiologically, the large intestine differs from the small intestine in that there are no villi or digestive enzymes. The absence of villi means that less absorption takes place in the large intestine in comparison to the small intestine. Nutrients absorbed from the large intestine include water, some vitamins, some fatty acids, and the minerals sodium and potassium (Table 3-3). Unlike the small intestine, the large intestine has a number of mucus-producing cells. The mucus secreted by these cells functions to hold the **feces** together and protect the large intestine from the bacterial activity within it.

The large intestine is home to a large population of bacteria (over 500 different species) which are collectively called the **microbiota**. Bacteria in the large intestine are able to break down some of the remaining food products that enter the large intestine, such as the milk sugar lactose (in lactose-intolerant people) and some components of fiber. Also, the bacteria that live in the large intestine produce some vitamins (vitamin K and biotin) that can be absorbed. Some of the products of bacterial metabolism in the large intestine, which include various fatty acids and gases, can be absorbed and exert health effects in other areas of the body. A growing body of research shows that intestinal bacteria play a significant role in the maintenance of health, not just for the colon, but throughout the body. See Section 3.9 for more information about the importance of the gut microbiota for human health.

Some water remains in the material that enters the large intestine because the small intestine absorbs only 70% to 90% of the fluid it receives, which includes large amounts of GI-tract secretions produced during digestion (review Table 3-2). The remnants of a meal also contain some minerals and some fiber. Because water is removed from the

**TABLE 3-3** ■ **A Summary of Absorption Along the GI Tract**

| Organ | Primary Nutrients Absorbed |
|---|---|
| Stomach | Alcohol (20% of total)<br>Water (minor amount) |
| Small intestine | Calcium, magnesium, iron, and other minerals<br>Glucose<br>Amino acids<br>Fats<br>Vitamins<br>Water (70% to 90% of total)<br>Alcohol (80% of total)<br>Bile acids |
| Large intestine | Sodium<br>Potassium<br>Some fatty acids<br>Gases<br>Water (10% to 30% of total) |

large intestine, its contents change from liquid to semisolid as they pass through the organ. By the time it is expelled from the body, what remains in the feces is undigested carbohydrates (i.e., fiber); tough connective tissues (from animal foods); bacteria from the large intestine; some body wastes (e.g., parts of dead intestinal cells); and a small amount of water.

## RECTUM

The feces (also known as *stool*) remains in the last portion of the large intestine, the **rectum,** until muscular movements push it into the **anus** to be eliminated. The presence of feces in the rectum stimulates elimination. The anus contains two **anal sphincters** (internal and external), one of which is under voluntary control (external sphincter). Relaxation of this sphincter allows for elimination.

## ACCESSORY ORGANS

The liver, **gallbladder,** and pancreas work with the GI tract and are considered accessory organs to the process of digestion (review Fig. 3-9). These accessory organs are not part of the GI tract (i.e., food never touches the accessory organs), but they play necessary roles in the process of digestion. These organs secrete digestive fluids into the GI tract and facilitate the digestion of food into absorbable nutrients.

**Liver and Gallbladder.** The liver produces a substance called **bile.** The bile is stored and concentrated in the gallbladder until the gallbladder receives a hormonal signal to release the bile. This signal is induced by the presence of fat in the small intestine. Bile is released and delivered to the duodenum via a tube called the bile duct (Fig. 3-17).

In action, bile is like soap. Components of the bile enable large portions of fat to break into smaller bits so that they can be suspended in water (Chapter 5 will cover this process in detail). Interestingly, some of the bile constituents can be "recycled" in a process known as **enterohepatic circulation:** components of bile are reabsorbed from the small intestine, returned to the liver via the portal vein, and reused.

The liver also releases some waste products (e.g., excess minerals, breakdown products from cell metabolism) into the bile. These wastes will eventually end up in the large intestine and will be excreted as part of the feces. The liver functions in this manner to remove unwanted substances from the blood. (Recall from Section 3.4, the kidneys also remove waste products from the blood and excrete these wastes as part of urine.)

**rectum** Terminal section of the large intestine, where feces are held prior to expulsion.

**anus** Last portion of the GI tract; serves as an outlet for the digestive system.

**anal sphincters** A group of two sphincters (inner and outer) that help control expulsion of feces from the body.

**gallbladder** An organ attached to the underside of the liver; site of bile storage, concentration, and eventual secretion.

**bile** A liver secretion stored in the gallbladder and released through the common bile duct into the first segment of the small intestine. It is essential for the digestion and absorption of fat.

**enterohepatic circulation** A continual recycling of compounds such as bile acids between the small intestine and the liver.

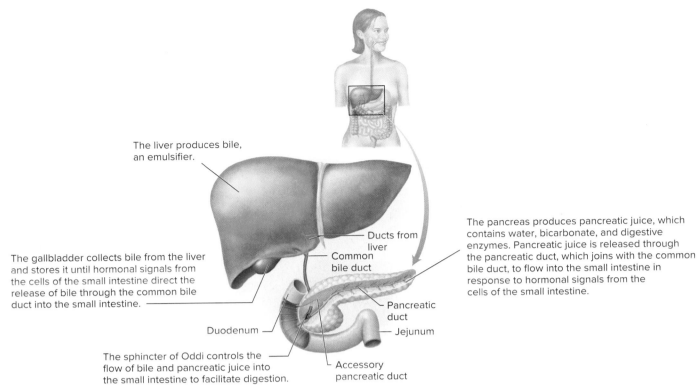

The liver produces bile, an emulsifier.

The gallbladder collects bile from the liver and stores it until hormonal signals from the cells of the small intestine direct the release of bile through the common bile duct into the small intestine.

The pancreas produces pancreatic juice, which contains water, bicarbonate, and digestive enzymes. Pancreatic juice is released through the pancreatic duct, which joins with the common bile duct, to flow into the small intestine in response to hormonal signals from the cells of the small intestine.

Ducts from liver

Common bile duct

Pancreatic duct

Jejunum

Duodenum

The sphincter of Oddi controls the flow of bile and pancreatic juice into the small intestine to facilitate digestion.

Accessory pancreatic duct

**FIGURE 3-17** ▲ Although food does not come into contact with these accessory organs, the liver, gallbladder, and pancreas are important for digestion.

**Pancreas.**   The pancreas has both endocrine and digestive functions. As a gland of the endocrine system, the pancreas manufactures hormones—insulin and glucagon—that are secreted into the blood to regulate blood glucose levels (review Fig. 3-8). As an organ of the digestive system, it produces "pancreatic juice," a mixture of water, bicarbonate, and a variety of digestive enzymes capable of breaking apart carbohydrates, proteins, and fats into small fragments. Bicarbonate neutralizes the acidic chyme as it moves from the stomach into the duodenum. As noted earlier, the small intestine does not have a protective layer of mucus because mucus would impede nutrient absorption. Instead, the neutralizing capacity of bicarbonate from the pancreas protects the walls of the small intestine from erosion by acid, which would otherwise lead to the formation of an ulcer (see Section 3.11).

### ✓ CONCEPT CHECK 3.8

1. Choose three secretions of the digestive system. Where is each secreted? What is the role of each in the process of digestion?

2. What are enzymes? Is bile an enzyme?

3. How do mucus and surface area affect absorption?

4. Which absorptive processes use energy? How does the *concentration gradient* factor into this?

5. What did you have for lunch today? Trace the path of your meal through the digestive system. Where is each nutrient broken down as it passes through the GI tract? Where is each absorbed?

## 3.9 The Human Microbiota

As soon as we are born (and perhaps before birth!), we are exposed to a vast array of microorganisms from the world around us. Some of these microorganisms take up residence in and on the human body—the skin, the respiratory tract, the genitourinary tract, and the gastrointestinal tract. Over the first few years of life, a fairly stable community of microorganisms colonizes the human host. This community of more than 100 trillion microorganisms is known as the human microbiota. Scientists have come to recognize the human microbiota as a human organ, given its high metabolic activity and impact on nearly every aspect of human health.

The precise composition of the microbiota varies from person to person. So far, more than 2300 different strains of microorganisms have been identified in the human microbiota, but each person is home to a unique community of only a few hundred species. The composition of bacteria in your GI tract depends on a variety of factors: genetics, age, environmental exposures, dietary pattern, and other lifestyle factors.[5] The way you came into the world (vaginal or Cesarean birth), the way you were fed as an infant (breast milk or formula), and the use of antibiotics to treat acute infections were a few of the choices that have directed the development of your gut microbiota.

Whatever the specific composition, there are at least as many microbial cells living in and on the human body as there are human cells. Also consider that all of these microbial cells contain their own genes. As a whole, the **microbiome** contains about 150 times as many genes as the human genome![6] Pathogenic (i.e., disease-causing) strains of microorganisms synthesize compounds that can make us sick. For example,

## Newsworthy Nutrition

### Fecal transplant reduces recurrence of diarrheal disease

**INTRODUCTION:** Antibiotics are helpful for treating infections. However, they may alter the normal balance of microbes that live in the GI tract, allowing pathogenic strains, such as *Clostridium difficile* (*C. diff*), to flourish. *C. diff* infections cause severe and recurrent diarrhea, leading to dehydration and possibly to death. Traditional medical care for antibiotic-associated diarrhea involves prescribing even stronger antibiotics to eradicate *C. diff*, but these drugs have side effects, are not always effective, and may lead to the development of antibiotic-resistant strains of bacteria. Rather than attempting to kill the pathogenic strains, some practitioners have become interested in ways to repopulate the GI tract with beneficial microbes, either with probiotic supplements or a sample of feces from a healthy donor. **OBJECTIVES:** In this experiment, the researchers wanted to determine if a fecal transplant from a healthy donor is a safe and effective way to treat *C. diff* infection. They hypothesized that a fecal transplant would reduce the rate of relapse among patients being treated for *C. diff* infection. **METHODS:** 46 patients with recurrent *C. diff* infections were randomized to receive fecal microbial transplants from either their own fecal output (autologous FMT) or fecal samples from healthy donors (heterologous FMT). Patients were followed for 8 weeks after FMT to look for recurrence of diarrhea, fever, or other adverse events and to assess their fecal composition. **RESULTS:** By the end of 8 weeks, 20 of 22 patients (91%) in the heterologous FMT group achieved clinical cure (no further recurrence of diarrhea, no fever), compared to 15 of 24 patients (63%) in the autologous FMT group. **CONCLUSION:** This study demonstrated with a randomized, controlled, double-blind trial that a fecal microbial transplant from healthy donors can assist with recovery among patients with *C. diff* infection.

Source: Kelly CR and others: Effect of fecal microbiota transplantation on recurrence in multiply recurrent *Clostridium difficile* infection: A randomized trial. *Annals of Internal Medicine* 165(9):609, 2016.

**FIGURE 3-18** ▶ Bacteria in the gastrointestinal tract

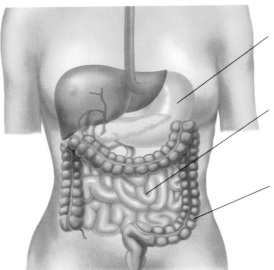

The stomach is sparsely populated by about 1000 colony-forming units (CFU) per gram of contents. The low pH limits growth of most microorganisms.

As the pH of the small intestine increases through the duodenum, jejunum, and ileum, microorganisms begin to flourish. There are at least 1 million CFU per gram of contents inside the ileum.

The large intestine houses more than 100,000,000,000 CFU per gram of contents.

there is evidence that certain patterns in the microbiome are linked to the occurrence of cardiovascular disease, type 2 diabetes, and chronic liver diseases. Some bacteria produce toxins that are linked to the development of colon cancer. However, many microbial genes code for proteins that affect human health in beneficial ways. Some affect the integrity of the epithelial tissue that lines the large intestine. Some may enhance the absorption of minerals, such as iron and calcium. Others may affect cholesterol synthesis. Many of these microbes, including lactic acid bacteria and *bifidobacteria*, enhance immune function.

By far, the community of microorganisms living in the gastrointestinal tract—the *gut microbiota*—has been the most studied area of the human microbiota. Microbes live all along the GI tract, but the large intestine is the organ most heavily colonized with bacteria (Fig. 3-18). As we continue to look at the human body from a nutrition perspective, let's see how the gut microbiota specifically influences human health.

### PROBIOTICS, PREBIOTICS, SYNBIOTICS, AND POSTBIOTICS

**Probiotics** are live microorganisms that have positive effects on human health if they are consumed in sufficient quantities. These live microorganisms can be ingested as part of foods (e.g., yogurt and kefir), or they can be administered as supplements (e.g., pills, powders, or suppositories). Common probiotic bacteria used in foods and supplements include species of *Lactobacillus, Lactococcus,* and *Streptococcus* (frequently grouped together as lactic acid bacteria because they produce lactic acid when they are used to ferment dairy foods), as well as some species of the *Bifidobacterium* genus. Notice that the definition of probiotics is not limited to bacteria; although many probiotic microorganisms are bacteria, *Saccharomyces boulardii* is a strain of yeast that can also promote human health.[7]

**prebiotic** Selectively fermented ingredient that results in specific changes in the composition and/or activity of the gastrointestinal microbiota, thus conferring benefits upon the host.

**Prebiotics,** on the other hand, are not living microorganisms. Prebiotics are ingredients (typically carbohydrates) that are not well digested by human enzymes but serve as fuel for beneficial bacteria in the gut. For example, fructooligosaccharides (FOS) are short chains of carbohydrates that can be fermented (i.e., broken down) and used for energy by probiotic bacteria. Some other examples of prebiotics are galactooligosaccharides (GOS), trans-galactooligosaccharides (TOS), and lactulose. Human milk, as you will read in Chapter 14, contains prebiotics that selectively promote the growth of beneficial bifidobacteria in the digestive tracts of infants.

**synbiotic** Combination of pro- and prebiotics taken to confer health benefits on the host.

**Synbiotics** are food products or dietary supplements that contain both probiotics and prebiotics. The two components may work in the same or different parts of the GI tract, but they both work to improve human health.

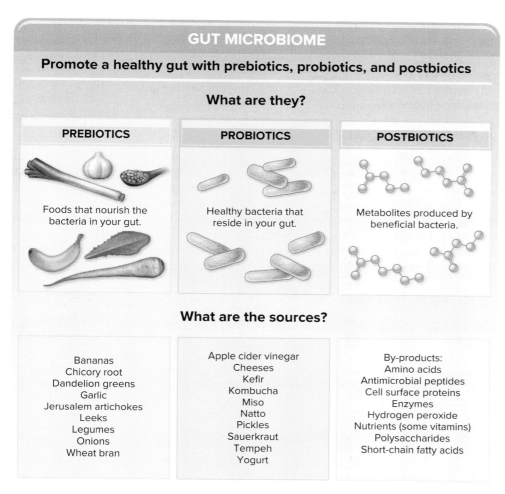

**FIGURE 3-19** ◄ Components of the gut microbiota, including prebiotic foods, probiotic bacteria, and postbiotic metabolites.

Now researchers are focusing on **postbiotics,** the metabolic products of probiotic microorganisms.[8] When probiotic microorganisms colonize the body, they produce compounds that can influence the health of their human hosts. For example, when bifidobacteria and lactobacilli ferment undigested carbohydrates in the colon, they produce short-chain fatty acids that can serve as an energy source for the cells of the GI tract. Furthermore, some postbiotics can be absorbed and transported through the blood to have diverse and far-reaching effects throughout the body. See Figure 3-19 for a summary of prebiotics, probiotics, and postbiotics.

**postbiotics** Metabolic by-products of the microorganisms that colonize the human body.

## GASTROINTESTINAL HEALTH

In healthy humans, the GI tract is home to some pathogenic microbes as well as beneficial microbes. The beneficial species keep the pathogenic species in check by competing with them for living space and food and by producing some antimicrobial compounds. The activity of microorganisms in the gut stimulates the maturation of the epithelial tissue that lines the gastrointestinal tract. Picture the thin layer of cells that line the intestine. There are tiny spaces between these cells that allow some materials to pass through but keep other materials in the lumen of the intestine and out of the blood supply. The junctions between these cells are normally stitched together tightly by proteins. Some compounds produced by the microbiota influence the synthesis of these tight junction proteins, which affects the integrity of the intestine's epithelial tissue. This is important for nutrient absorption and immune function.

**Dysbiosis** is a term that refers to an imbalance of "good" and "bad" microbes in the gut. Sometimes, such a disruption is short-lived, such as a viral infection that leads to diarrhea. The immune system works to destroy the pathogen, and beneficial microbes also play a role in restoring balance (see *Newsworthy Nutrition* in this

**dysbiosis** A disturbance in the balance of beneficial and pathogenic microorganisms in the microbiota.

section). However, research evidence shows us that gut dysbiosis is related to the development of many long-term gastrointestinal diseases, including irritable bowel syndrome and inflammatory bowel diseases (see Section 3.11).[9] At this time, it is unclear if gut dysbiosis is a cause or a consequence of these chronic GI diseases, but researchers are actively looking for ways to manipulate the gut microbiota as a way to prevent or treat these diseases.

## NUTRITIONAL STATUS

Some strains of bacteria contribute to human nutritional status as well. Although human enzymes cannot break down dietary fiber, bacterial enzymes can metabolize some of those carbohydrates. As they do so, they effectively harvest additional energy from these foods for us. The short-chain fatty acids produced as a by-product of microbial metabolism of fiber serve as fuel for the cells that line the intestine. In addition, bacteria in the GI tract synthesize several vitamins. Although the majority of vitamins are absorbed in the small intestine, some of the vitamin K and several B vitamins produced by microbes may contribute to human nutritional status.[6]

## IMMUNE FUNCTION

The gut microbiota profoundly influences immune function. In laboratory animals that are born and raised in a sterile environment, the immune system is underdeveloped and the organisms are more susceptible to autoimmune diseases and allergies. It is evident that a healthy gut microbiota is crucial for proper immune system development and maintenance (see *Sustainable Solutions* in this section).

During infancy and childhood, beneficial organisms support the proper development of the immune system. Exposure of GALT to microorganisms promotes the maturation of the immune cells. GALT makes up about 70% of all immune tissue in the body, and much of the GALT is situated in the large intestine, where the microbiota flourishes. Through early interactions with microorganisms in the gut, the cells of the immune system "learn" to recognize and differentiate between what is harmful and what is not. Production of certain antibodies by the mucosal cells is increased.

As mentioned, normal microbial activity in the gut stimulates the proper development of the epithelial tissue that lines the gastrointestinal tract, which serves as a barrier against pathogenic organisms. The cells that line the intestines are knitted together by tight junctions, which are a network of tiny strands of protein that regulate the absorption of nutrients and block the entry of large molecules and pathogens. If the integrity of these tight junctions is compromised, the intestinal lining becomes more permeable. Increased gut permeability might allow large molecules, including microorganisms, to move from the lumen of the GI tract into body tissues, leading to infections. Some researchers wonder if increased gut permeability is involved in the development of allergies and autoimmune diseases. Genetics alone cannot explain the increased rates of these disorders in recent years, leading researchers to suspect that environmental factors are at work. Blame may rest with changes in eating patterns, infectious disease rates, use of antibiotics, or the makeup of the microbiota.[10]

Beneficial strains of microorganisms may also directly compete with and reduce the activity of disease-causing organisms in the GI tract.[11,12] This occurs by several mechanisms. The metabolic activity of beneficial strains changes the acidity of the environment in the gut, which makes it less hospitable to the pathogenic strains. Increasing loads of beneficial bacteria will compete with pathogenic bacteria for binding sites and sources of food. Finally, some probiotic strains directly bind to pathogens or secrete substances that kill them.

Beyond these direct interactions, the gut microbiota also produces compounds that can be absorbed into the blood and regulate immune function. For example, the amino acids and short-chain fatty acids produced by the gut microbiota can influence the synthesis of inflammatory compounds, not just in the gut but throughout the body.[13]

**COVID CORNER**

Prior research linking the gut microbiota to several aspects of respiratory health has prompted interest in the use of probiotic foods or supplements to support immune function and hasten recovery from COVID-19. However, to date, there is not yet enough evidence to recommend the use of specific probiotic foods or supplements for protection against COVID-19.[14,15]

## MUCH MORE TO LEARN

In this section, we have introduced a few of the many health effects of the human microbiota. Researchers are interested in how the microbiota and their metabolites may influence a variety of other health outcomes, including weight control,[16] cardiovascular disease,[17] blood glucose,[6] and even mental health.[18] In later chapters, you will see additional information about the microbiota and human health in the *Magnificent Microbiome* features.

Continued study of the microbial genome will help researchers to understand how microorganisms exert their influences on human health. The Human Microbiome Project in the United States and the Metagenomics of the Human Intestinal Tract program in Europe are two large-scale efforts to map the microbial genome and understand the interactions between human and microbial genes. We have come a long way in our understanding of the ways probiotics and prebiotics can influence human health, especially gastrointestinal health and body defenses. Even so, there are many things left to uncover.

*magnificent*

According to the *hygiene hypothesis*, our tendency to sanitize every surface and treat every infection with antibiotics reduces exposure of the immune system to a diverse community of microorganisms. The understimulated immune system becomes sensitized to harmless proteins, such as food proteins, beneficial bacteria, or the body's own cells. Being *too* clean might make us more likely to develop allergies and autoimmune diseases.

*microbiome*

### ✓ CONCEPT CHECK 3.9

1. In simple terms, explain the differences among *probiotics, prebiotics,* and *postbiotics*.
2. Other than yogurt, list two food sources of probiotic microorganisms.
3. Describe two ways the gut microbiota may work to support human health.

## 3.10 Nutrient Storage Capabilities

The human body must maintain reserves of nutrients; otherwise, we would need to eat continuously. Storage capacity varies for each different nutrient. Most fat is stored in **adipose tissue.** Short-term storage of carbohydrate occurs in muscle and the liver in the form of glycogen. The blood maintains a small reserve of glucose and amino acids. Many vitamins and minerals are stored in the liver, whereas other nutrient stores are found in other sites in the body.

When people do not meet their needs for certain nutrients, blood levels of these nutrients can be maintained by breaking down body tissues. For example, calcium is taken from bone and protein is taken from muscle. In cases of long-term deficiency, these nutrient losses weaken and harm these tissues.

Many people believe that if too much of a nutrient is obtained—for example, from a vitamin or mineral supplement—only what is needed is stored and the rest will be excreted by the body. Though true for some nutrients, such as vitamin C, the large dosages of other nutrients frequently found in supplements, such as vitamin A and iron, can cause harmful side effects because they are not readily excreted. This is one reason why obtaining your nutrients primarily (or exclusively) from a balanced diet is the safest means to acquire the building blocks you need to maintain the good health of all organ systems.

**adipose tissue** Connective tissue made up of cells that store fat; also cushions and insulates the body.

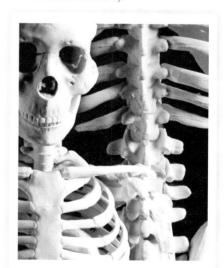

The skeletal system provides a reserve of calcium for day-to-day needs. **If your dietary intake of calcium is inadequate, what happens to your bones?** Jason Reed/ Ryan McVay/Getty Images

### ✓ CONCEPT CHECK 3.10

1. What is the body's most efficient form of energy storage?
2. Why is it important to consume nutrients daily?
3. When it comes to vitamins and minerals, is consuming more than the RDA or AI a good way to ensure optimal nutrition status? Why or why not?

# 3.11 Nutrition and Your Health
## Common Problems with Digestion

When suffering from persistent heartburn or GERD, see a doctor if you have

- difficulty swallowing or pain when swallowing.
- heartburn that has persisted for more than 10 years.
- initial onset of heartburn after age 50.
- heartburn that resists treatment with medications.
- sudden, unexplained weight loss.
- chest pain.
- blood loss or anemia.
- blood in stool or vomit.

but in individuals with GERD, it is relaxed at other times as well. Increased pressure against the lower esophageal sphincter (e.g., as a result of pregnancy or obesity) heightens risk for heartburn. The hormonal changes of pregnancy also tend to relax the lower esophageal sphincter. For some people, slow movement of gastric contents from the stomach to the small intestine complicates the problem.

**gastroesophageal reflux disease (GERD)** Disease that results from stomach acid backing up into the esophagus. The acid irritates the lining of the esophagus, causing pain.

Without fanfare, the digestive system does the important work of extracting nutrients from the food you eat to supply the needs of your body's trillions of cells. It is not until something goes awry that you notice digestion at all. In this section, you will learn about nutritional strategies to cope with heartburn, ulcers, constipation, diverticulosis and diverticulitis, hemorrhoids, diarrhea, irritable bowel syndrome (IBS), gallstones, and celiac disease.

## Heartburn

About half of North American adults experience occasional heartburn, also known as acid reflux (Fig. 3-20). This gnawing pain in the upper chest is caused by the movement of acid from the stomach into the esophagus. Unlike the stomach, the esophagus has very little mucus to protect it, so acid quickly erodes the lining of this organ. Acid reflux symptoms may include pain, nausea, gagging, cough, or hoarseness. The recurrent and therefore more serious form of the problem is called **gastroesophageal reflux disease (GERD).** GERD is diagnosed when symptoms occur two or more times per week.

Heartburn is caused by relaxation of the gastroesophageal sphincter. Typically, it should be relaxed only during swallowing,

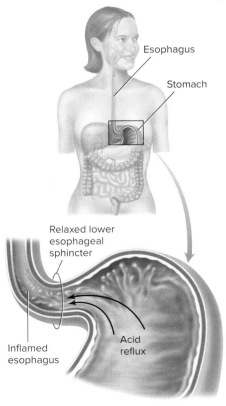

**FIGURE 3-20** ▲ Heartburn is a sign of reflux of stomach acid into the esophagus.

If left untreated, heartburn can damage the lining of the esophagus, leading to chronic esophageal inflammation and an increased risk of esophageal cancer. Heartburn sufferers should follow the general recommendations given in Table 3-4. For occasional heartburn, quick relief can be found with over-the-counter antacids. Taking antacids will reduce the acid in the stomach but will not stop the acid reflux. For more persistent (few days a week or every day) heartburn or GERD, $H_2$ blockers or **proton pump inhibitors (PPIs)** may be needed (see the *Medicine Cabinet* feature in this section). PPIs provide long-lasting relief by reducing stomach acid production and should be taken before the first meal of the day because they take longer to work. Medications that improve GI **motility** may also be useful. If the proper medications are not effective at controlling GERD, surgery may be needed to strengthen the weakened esophageal sphincter.[19]

## Peptic Ulcers

A **peptic ulcer** occurs when the lining of the esophagus, stomach, or small intestine is eroded by the acid secreted by stomach cells (Fig. 3-21). A disruption of the layer of mucus that usually protects the stomach allows acid and protein-digesting enzymes to damage the stomach lining. Acid can also erode the lining of the esophagus and the first part of the small intestine, the duodenum. This can cause pain, blood loss, and even **perforation.** At any given time, about 4.5 million people in the United States are affected by peptic ulcers. In young people, most ulcers occur in the small intestine, whereas in older people, they are most common in the stomach.

**proton pump inhibitor (PPI)**  A medication that inhibits the ability of gastric cells to produce acid.

**motility**  Generally, the ability to move spontaneously. In this context, it refers to movement of food through the GI tract.

**peptic ulcer**  Erosion of the tissue lining, usually in the stomach or the upper small intestine. As a group, these are generally referred to as peptic ulcers.

**perforation**  A hole made by boring or piercing. With reference to the gastrointestinal tract, a hole may form in the wall of the esophagus, stomach, intestine, rectum, or gallbladder. Complications include bleeding and infection.

**TABLE 3-4 ■ Nutrition and Lifestyle Recommendations for Care of Heartburn and Peptic Ulcers**

| | Heartburn | Peptic Ulcers |
|---|:---:|:---:|
| Avoid smoking. | ✓ | ✓ |
| Avoid large doses of aspirin, ibuprofen, and other NSAID compounds unless a physician advises otherwise.[a] | ✓ | ✓ |
| Achieve or maintain a healthy body weight. | ✓ | ✓ |
| Eat small, low-fat meals. | ✓ | ✓ |
| Limit alcohol consumption. | ✓ | ✓ |
| Limit consumption of caffeine (e.g., coffee, some soft drinks). | ✓ | ✓ |
| Consume a nutritionally complete diet with adequate fiber. | ✓ | ✓ |
| Avoid foods that worsen symptoms:[b] | | |
| • Acidic foods (e.g., orange juice, tomato products) | ✓ | ✓ |
| • Highly spiced foods (e.g., chili, cayenne, and black pepper) | ✓ | ✓ |
| • Foods that relax the lower esophageal sphincter (e.g., peppermint, spearmint, chocolate) | ✓ | |
| • Carbonated beverages | ✓ | ✓ |
| • Onions and garlic | ✓ | |
| Avoid tight-fitting clothing. | ✓ | |
| Elevate the head of the bed 6 to 8 inches. | ✓ | |
| Avoid eating at least 3 to 4 hours before lying down. | ✓ | |
| Wash hands often and follow food safety guidelines. | | ✓ |

[a] For people who must use these medications, FDA has approved an NSAID combined with a medication to reduce gastric damage. The medication reduces gastric acid production and enhances mucus secretion.

[b] These foods do not *cause* heartburn or ulcers, but some may irritate sites of existing damage in the esophagus or stomach.

## CASE STUDY Gastroesophageal Reflux Disease

Caitlin is a 20-year-old college sophomore. Over the last few months, she has been experiencing regular bouts of heartburn. This usually happens after a large lunch or dinner. Occasionally, she has even bent down after dinner to pick up something and had some stomach contents travel back up her esophagus and into her mouth. This especially frightened Caitlin, so she visited the University Health Center.

The nurse practitioner at the center told Caitlin it was good that she came in for a checkup because she suspects Caitlin has gastroesophageal reflux disease. She tells Caitlin that this can lead to serious problems, such as a rare form of cancer, if not controlled. She provides Caitlin with a pamphlet describing GERD and schedules an appointment with a physician for further evaluation.

▲ Caitlin was wise to see a health professional about her persistent heartburn. Rocketclips, Inc./ Shutterstock

1. What is the difference between heartburn and GERD?
2. What dietary and lifestyle habits may have contributed to Caitlin's symptoms of GERD?
3. What is the dietary and lifestyle management advice that will help Caitlin cope with this health problem?
4. What types of medications have been especially useful for treating this problem?
5. Why is management of GERD so important?

*Complete the Case Study. Responses to these questions can be provided by your instructor.*

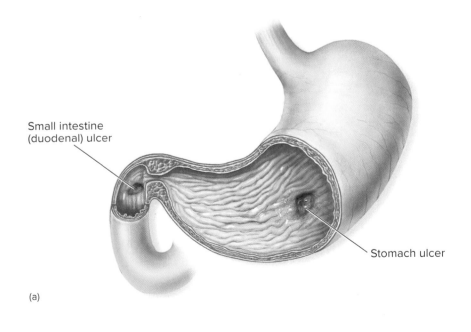

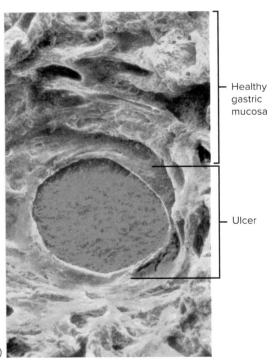

**FIGURE 3-21** ▲ (a) A peptic ulcer in the stomach or small intestine. *H. pylori* bacteria and NSAIDs (e.g., ibuprofen) cause ulcers by impairing mucosal defense, especially in the stomach. Smoking, genetics, and stress also can impair mucosal defense or cause an increase in the release of pepsin and stomach acid. (b) Close-up of a stomach ulcer. This needs to be treated or eventual perforation of the stomach is possible. J. James/Science Source

How do you know if you have a peptic ulcer? Some people experience no symptoms at all, but most notice stomach pain about 2 hours after eating. Stomach acid acting on a meal irritates the ulcer after most of the meal has moved from the site of the ulcer. Other symptoms may include weight loss, lack of appetite, nausea and vomiting, or bloating. Vomiting blood or what looks like coffee grounds and the appearance of black, tarry stools are signs of bleeding in the GI tract. Any evidence of GI bleeding warrants immediate medical attention.

The two chief culprits of peptic ulcer disease are infection of the stomach by the acid-resistant bacterium *Helicobacter pylori* (see *Fake News* at the beginning of this chapter) and heavy use of **nonsteroidal anti-inflammatory drugs (NSAIDS)**. Both *H. pylori* and NSAIDs disrupt the thick layer of mucus that normally protects the stomach from the action of acids and enzymes. Conditions that cause excessive stomach acid production also play a role. In addition, cigarette smoking is known to cause ulcers, increase ulcer complications such as bleeding, and lead to ulcer treatment failure.

NSAIDs are medications for painful inflammatory conditions such as arthritis. Aspirin, ibuprofen, and naproxen are the most commonly used types. As a side effect, NSAIDs reduce the mucus secreted by the stomach. Newer medications, called COX-2 inhibitors (e.g., celecoxib [Celebrex®]), have been used as a replacement for NSAIDs because they are less likely to cause stomach ulcers. They do offer some advantages over NSAIDs, but they may not be totally safe for some people, especially those with a history of cardiovascular disease or strokes.

The primary risk associated with an ulcer is the possibility that it will erode entirely through the stomach or intestinal wall. The GI contents could then spill into the body cavities, causing a massive infection. In addition, an ulcer may damage a blood vessel, leading to substantial blood loss. For these reasons, it is important to never ignore the early warning signs of ulcer development, which include a persistent gnawing or burning near the stomach that may occur immediately following a meal or awaken you at night.

Today, a combination approach is used for ulcer therapy.[20] People infected with *H. pylori* are given antibiotics and stomach acid–blocking medications (see the *Medicine Cabinet* feature in this section). There is a 90% cure rate for *H. pylori* infections in the first week of this treatment. Recurrence is unlikely if the infection is cured, but an incomplete cure almost certainly leads to repeated ulcer formation.

Are dietary changes effective for prevention or treatment of peptic ulcers? Many people think that eating spicy or acidic foods can cause ulcers. Contrary to popular belief, these foods do not cause ulcers. However, once an ulcer has developed, these foods may irritate damaged tissues. Thus, for some people, avoidance of spicy or acidic foods may help to relieve symptoms.

In the past, milk and cream were thought to help cure ulcers. Clinicians now know that milk and cream are two of the worst foods for a person with ulcers because the calcium in these foods stimulates acid secretion and actually inhibits ulcer healing.

Overall, medical treatment of *H. pylori* infection has so revolutionized ulcer therapy that dietary changes are of minor importance. People with ulcers should refrain from smoking and minimize the use of NSAIDs. Current dietary therapy approaches simply recommend avoidance of foods that tend to worsen ulcer symptoms (see Table 3-4).

## Medicine Cabinet

***Proton pump inhibitors (PPIs)*** are medications that inhibit the ability of gastric cells to secrete hydrogen ions and thus reduce acid production. Low doses of this class of medications may be available without a prescription. Because stomach acid is important for the absorption of vitamin B-12, prolonged use of PPIs could impair vitamin B-12 status.

Examples:

- Omeprazole (Prilosec®)
- Lansoprazole (Prevacid®)
- Rabeprazole (Aciphex®)
- Esomeprazole (Nexium®)

***H₂ blockers*** impede the stimulating effect of histamine on acid-producing cells in the stomach.

Examples:

- Cimetidine (Tagamet®)
- Nizatidine (Axid®)
- Famotidine (Pepcid®)

## Constipation

What does it mean to be "regular" when it comes to bowel function? Individuals vary, but in general, normal bowel frequency ranges from three times per day to three times per week. **Constipation**, a condition characterized by difficult or infrequent evacuation of the bowels, is commonly reported by adults. As you learned in Section 3.8, the primary role of the large intestine is to absorb fluid. If fecal material moves too slowly through the large intestine, so much fluid is absorbed that the feces become dry, hard, and difficult to pass.

There are many possible causes for constipation. The muscular movement of feces through the GI tract is regulated by neurological and hormonal signals, so disorders of the nervous system or muscular function could alter bowel motility. A physical obstruction in the GI tract could also be to blame. Constipation may be a side effect of certain medications (e.g., antacids) or dietary supplements (e.g., iron or calcium). More often, however, constipation arises due to inadequate dietary fiber and/or fluid intake or poor toileting habits.

Increasing dietary intakes of fiber and fluid are usually safe and effective strategies for treating mild cases of constipation.[21] Whole grain breads and cereals, beans, and dried fruits are excellent sources of fiber. Fiber stimulates peristalsis by drawing water into the large intestine and helping to form a bulky, soft fecal output. Additional fluid should be consumed to facilitate

**NSAIDs** Nonsteroidal anti-inflammatory drugs; include aspirin, ibuprofen (Advil®), and naproxen (Aleve®).

**constipation** A condition characterized by difficult and/or infrequent bowel movements (i.e., fewer than three bowel movements per week).

▲ Dried fruits are a natural source of fiber and can help prevent constipation when consumed with an adequate amount of fluid.

C Squared Studios/Getty Images

fiber's action in the large intestine. Also, people with constipation may need to develop more regular bowel habits. When people regularly ignore their normal bowel reflexes (e.g., because it is inconvenient to interrupt occupational or social activities), feces can become hard and dry. Allowing the same time each day for a bowel movement can help to train the large intestine to respond routinely. Regular physical activity can also stimulate the GI tract to function normally.[22]

In more severe cases, **laxatives** can alleviate constipation. Some laxatives work by irritating the intestinal nerve junctions to stimulate peristalsis, while others that contain fiber draw water into the intestine to enlarge fecal output. The larger output stretches the peristaltic muscles, making them rebound and then constrict. Regular use of laxatives, however, should be supervised by a primary care provider. Overall, if laxatives are necessary, the bulk-forming fiber laxatives, such as **psyllium** husk, are the safest to use.[22]

## Hemorrhoids

**Hemorrhoids,** also called *piles,* are swollen veins of the rectum and anus. The blood vessels in this area are subject to intense pressure, especially during bowel movements. Added stress to the

**laxative** A medication or other substance that stimulates evacuation of the intestinal tract.

**psyllium** Mostly soluble type of dietary fiber found in the seeds of the plantago plant; common ingredient in bulk-forming laxatives, such as Metamucil®.

**hemorrhoid** A swollen vein in the rectum or anus.

**FODMAPs** **F**ermentable **o**ligosaccharides, **d**isaccharides, **m**onosaccharides, **a**nd **p**olyols. These carbohydrates may be poorly digested and lead to GI symptoms such as bloating, gas, and diarrhea in some people.

vessels from pregnancy, obesity, prolonged sitting, violent coughing or sneezing, or straining during bowel movements (particularly with constipation) can lead to a hemorrhoid.

Hemorrhoids can develop unnoticed until a strained bowel movement precipitates symptoms. Itching (caused by moisture in the anal canal), swelling, and irritation are the most common symptoms. Pain, if present, is usually aching and steady. Bleeding may result from a hemorrhoid and appear in the toilet as a bright red streak in the feces. The sensation of a mass in the anal canal after a bowel movement is a symptom of an internal hemorrhoid that protrudes through the anus.

Anyone can develop a hemorrhoid, and about half of adults over age 50 do. Diet, lifestyle, and heredity may contribute to the problem. For example, a low-fiber diet can lead to hemorrhoids as a result of straining during bowel movements. If you think you have a hemorrhoid, you should consult your primary care provider. Rectal bleeding, although usually caused by hemorrhoids, may also indicate other problems, such as cancer.

A physician may suggest a variety of self-care measures for hemorrhoids. Pain can be lessened by applying warm, soft compresses or sitting in a tub of warm water for 15 to 20 minutes. Dietary recommendations are the same as those for treating constipation, emphasizing the need to consume adequate fiber and fluid. Over-the-counter remedies, such as Preparation H®, can also offer relief from symptoms.

## Diverticular Disease

**Diverticulosis** is a common GI problem that affects about half of adults over the age of 60 years. This occurs when parts of the inner layer (mucosa) of the intestinal wall (usually in the large intestine) protrude outward between the surrounding bands of muscle, forming small pouches called **diverticula** (Fig. 3-22). The presence of diverticula is called diverticulosis. The exact cause of diverticulosis

**diverticulosis** The condition of having many diverticula in the large intestine.

**diverticula** Pouches that protrude through the exterior wall of the large intestine.

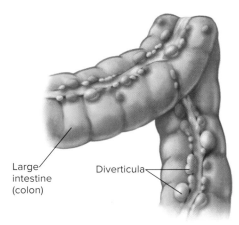

Large intestine (colon)

Diverticula

**FIGURE 3-22** ▲ Diverticulosis is a condition in which small pouches (diverticula) form in the wall of the intestine. When diverticula become inflamed or infected, the condition is called diverticulitis.

is unknown. Scientists used to think that a low-fiber diet led to diverticulosis, but recent studies demonstrate that dietary fiber intake is not consistently related to the development of the condition. Genetics play a role. Modifiable risk factors include obesity, inactivity, excessive use of alcohol, smoking, and use of certain medications. Alterations in the gut microbiota may be involved in development of diverticular disease.[23]

Although many adults have diverticulosis, it usually does not cause any noticeable symptoms. Among a small proportion of individuals with diverticulosis, the diverticula become inflamed or infected. This condition is known as **diverticulitis**. If diverticulitis develops, it is characterized by intense abdominal pain, sometimes accompanied by bowel irregularities (constipation is most common), nausea, vomiting, and fever. Rarely, inflamed diverticula may

**diverticulitis** Inflammation of the diverticula, which may be related to bacterial activity inside the diverticula.

## FAKE NEWS

# Individuals with diverticulosis should avoid eating nuts and seeds.

## THE FACTS

For many years, health care providers recommended that people with diverticulosis should avoid nuts, seeds, and other foods with tough, indigestible hulls (e.g., popcorn) based on the premise that these small bits of food matter could get stuck in the diverticula and lead to increased bacterial activity, inflammation, and infection. However, current research shows that consuming nuts, seeds, and foods with hulls is *not* associated with the occurrence of diverticulitis. Thus, it is not necessary for people with diverticulosis to restrict these nutrient-rich foods from their dietary patterns.[25]

rupture, leading to bleeding, formation of an abscess, or leakage of intestinal contents into the abdominal cavity.[24]

During flare-ups of diverticulitis, a low-fiber, liquid diet may be helpful to promote healing. Antibiotics may be prescribed. However, once the inflammation subsides, a high-fiber dietary pattern is generally recommended to maintain intestinal health. Advice to avoid certain foods, such as popcorn, nuts, or seeds, is outdated. Evidence shows no relationship between intake of popcorn, nuts, or seeds and occurrence of diverticulitis.[25]

## Irritable Bowel Syndrome

An estimated 10% to 15% of adults have irritable bowel syndrome (IBS), characterized by a combination of bloating, abdominal pain, and irregular bowel function (diarrhea, constipation, or alternating episodes of both). It is about twice as common in women as it is in men. The disease leads to about 3.5 million visits to primary care providers in the United States each year. Although it does not lead to cancer or other serious digestive problems, the physical discomfort and anxiety of IBS can significantly impact quality of life.

It is difficult to pinpoint an exact cause for IBS. Alterations in some of the hormones that regulate the movement of food matter through the GI tract may be to blame. Also, inflammatory responses in the GI tract could be involved for some people with IBS. Recent studies demonstrate alterations in the activity of gut microorganisms in people with IBS. Gut microorganisms produce compounds that affect many body systems. Perhaps IBS leads to changes in the microbiota, or perhaps imbalances in the gut microbiota trigger the range of problems that plague people with IBS.

The majority of people who suffer from IBS perceive that their symptoms are related to food, but there is little evidence of actual food allergies or intolerances. When it comes to specific foods, poorly digested carbohydrates are a prime suspect (see the discussion of **FODMAPs** in Chapter 4).[26,27] Fructose, sugar alcohols, and other carbohydrates may lead to diarrhea or excessive gas if they reach the large intestine undigested. Depression and stress are also associated with IBS; up to 50% of sufferers report a history of verbal or sexual abuse.

Given the diversity of symptoms and possible causes, therapy must be individualized. Medications that target nerves or alter the bacterial population in the GI tract may help people who suffer from frequent diarrhea. For IBS patients whose primary complaint is constipation, medications are available to stimulate peristalsis or block abdominal pain.

Medications may be expensive, and some have side effects. Therefore, dietary strategies to cope with IBS are of great interest. Some, but not all, patients with IBS experience improvements with a low-FODMAP dietary pattern. For several weeks, they eliminate (or greatly reduce) their intake of foods containing FODMAPs, including wheat, onions, legumes, and dairy products. Then, foods are gradually added back to the dietary pattern to determine which can be tolerated and which should be avoided long term.[27] Probiotics and peppermint oil have been shown to decrease symptoms of IBS and improve overall quality of life.[28] The patient should limit or eliminate caffeine-containing foods and beverages. Low-fat and more frequent, small meals may help because large meals can trigger contractions of the large intestine. Other strategies include a reduction in stress, psychological counseling, and antidepressant medications. Hypnosis has been shown to relieve symptoms in severe cases.

Following an eating pattern that eliminates certain foods or entire food groups can limit nutritional adequacy. Indeed, research indicates that intakes of some nutrients, including calcium and vitamin A, are inadequate among people with IBS. An experienced registered dietitian nutritionist is a valuable resource to help a person with IBS identify problem foods and plan a nutritionally adequate dietary pattern.

## Diarrhea

**Diarrhea** is defined as increased fluidity, frequency, or amount of bowel movements compared to a person's usual pattern. Most cases of diarrhea are of short duration and result from viral or bacterial infections. These microorganisms produce substances that cause the intestinal cells to secrete fluid rather than absorb fluid. Another form of diarrhea can be caused by consumption of substances that are not readily absorbed, such as sorbitol, a sugar alcohol found in sugarless gum (see Section 4.3) or large amounts of a high-fiber source such as bran. When consumed in large amounts, the unabsorbed substance draws water into the intestines, leading to diarrhea.

The goal of nutrition therapy for any form of diarrhea is to prevent dehydration. Increasing intake of water and electrolytes is the first line of defense against dehydration. Prompt treatment of dehydration—within 24 to 48 hours—is critical, especially for infants and older adults. Diarrhea that lasts more than 7 days in adults should be investigated by a primary care provider as it can be a sign of a more serious intestinal disease, especially if there is also blood in the stool.

For diarrhea caused by infection, dietary changes (besides increased fluid intake) are usually not necessary. Some health care providers recommend temporarily decreasing intake of caffeine, fat, fiber, and poorly absorbed carbohydrates, but other sources show that maintaining a regular diet speeds recovery. Foods containing probiotics may assist recovery. For diarrhea caused by a poorly absorbed substance, such as excess sugar alcohols or lactose, avoidance of the offending substance is the key to relief.[29]

## Gallstones

Gallstones are a major cause of illness and surgery, affecting 10% to 20% of U.S. adults. Gallstones are pieces of solid material that develop in the gallbladder when substances in the bile—primarily cholesterol (80% of gallstones)—form crystal-like particles. They may be as small as a grain of sand or as large as a golf ball (Fig. 3-23). These stones are caused by a combination of factors, with excess weight being the primary modifiable risk factor, especially among women. Other factors include genetic background (e.g., Native Americans), advanced age (>60 years), pregnancy, reduced activity of the gallbladder (contracts less than normal), altered bile composition (e.g., too much cholesterol or not enough bile salts), diabetes, and eating pattern (e.g., low-fiber diets). In addition, gallstones may develop during rapid weight loss or prolonged fasting (as the liver metabolizes more fat, it secretes more cholesterol into the bile).

Gallstones may cause intermittent pain in the upper right abdomen, gas and bloating, nausea or vomiting, or other health problems. Medications are available to dissolve gallstones, but

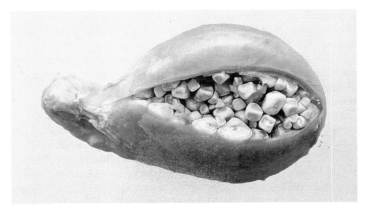

**FIGURE 3-23** ▲ Gallbladder and gallstones seen after surgical removal from the body. Size and composition of the stones vary from one case to another. The Sydney Morning Herald/Fairfax Media/ Getty Images

these take a long time to work, and the recurrence of gallstones after therapy is common. Therefore, surgical removal of the gallbladder is the most common method for treating gallstones (500,000 surgeries per year in the United States).

The best prevention strategy is to avoid becoming overweight, especially for women. Avoiding rapid weight loss (>3 pounds per week), limiting animal protein and focusing more on plant protein intake (especially nuts), and following a high-fiber diet can help as well. Regular physical activity is also recommended, as are moderate to no caffeine and alcohol intake.[30]

## Celiac Disease and Gluten Sensitivity

**Celiac disease** (sometimes called *celiac sprue*) affects about 1% of the U.S. population. Development of celiac disease depends on two factors: a genetic predisposition and dietary exposure to a protein called **gluten.** Gluten is a type of protein found in certain grains: wheat, rye, and barley. Protein-digesting enzymes in the GI tract break down some of the bonds in gluten, but digestion is incomplete. These partially digested proteins can be absorbed into the cells lining the small intestine. When people with a genetic predisposition for celiac disease are exposed to these small proteins from gluten, they experience an inflammatory reaction. Although many people think celiac disease is a food allergy, it is actually an *autoimmune* response: the immune system attacks and destroys its own cells. (You will learn more about food allergies in Section 6.3 and Section 15.7.)

The autoimmune response that occurs after exposure to gluten targets the cells of the small intestine, causing a flattening of the villi, which thereby reduces the absorptive surface (Fig. 3-24). The

**diarrhea** Increased fluidity, frequency, or amount of bowel movements (i.e., three or more loose stools per day).

**celiac disease** Chronic, immune-mediated disease precipitated by exposure to dietary gluten in genetically predisposed people.

**gluten** Poorly digested protein found in wheat, barley, and rye.

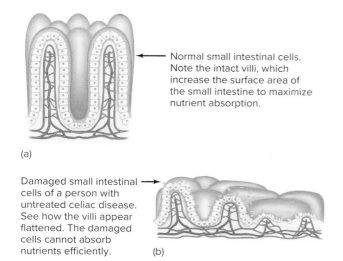

Normal small intestinal cells. Note the intact villi, which increase the surface area of the small intestine to maximize nutrient absorption.

(a)

Damaged small intestinal cells of a person with untreated celiac disease. See how the villi appear flattened. The damaged cells cannot absorb nutrients efficiently.

(b)

**FIGURE 3-24** ▲ Cross-section of the lining of (a) a normal small intestine and (b) the small intestine of an individual with untreated celiac disease.

production of some digestive enzymes is decreased, and the ability of the small intestine to absorb nutrients is impaired. Malabsorption leads to a variety of GI complaints: diarrhea, bloating, cramps, and flatulence. In fact, it is common for celiac disease to be misdiagnosed as IBS. However, the pathology underlying celiac disease has far worse consequences than IBS. Over time, malabsorption of nutrients can lead to fatigue, weight loss (or poor growth in children), anemia, infertility, and bone loss.[31]

If celiac disease is suspected, the first step in making a formal diagnosis is a blood test for the presence of antibodies to gluten. This may be followed by one or more biopsies of the small intestine to confirm the pathological defects. There is also a genetic test for celiac disease, but having the gene does not always predict development of the disease.

Strict dietary avoidance of food products containing wheat, rye, and barley is the only proven way to manage the disease.[31] On food labels, food manufacturers must identify the presence of wheat (one of eight major food allergens). However, rye and barley are not as easy to spot. Therefore, people following a gluten-free diet must learn to carefully interpret the list of ingredients to identify sources of gluten. Within the grains group, rice, potato flour, cornmeal, buckwheat, arrowroot, and soy are gluten free, but ingredients such as wheat, rye, barley, bran, graham flour, semolina, spelt, and malt are sources of gluten and must be avoided. Oats do not traditionally contain gluten, but contamination in the field or during food processing could introduce gluten into this grain as well.

People with celiac disease quickly learn that wheat, barley, and rye can be hidden ingredients in any food group. Wheat and its derivatives are used to thicken sauces and condiments, as flavoring agents in dairy products and many other processed foods, and in breading for deep-fried vegetables and meats. It is helpful that many food manufacturers now voluntarily disclose the presence or absence of gluten in their products. However, not all products clearly identify gluten. Dining out is yet another challenge: even a dusting of wheat flour can have adverse effects for a person with celiac disease.

After several weeks on a gluten-free diet, the small intestine lining regenerates, GI symptoms subside, and nutrient absorption improves. So far, the gluten-free diet is the only proven way to manage celiac disease, but research on other treatments is underway. Food scientists are working toward developing strains of wheat, barley, and rye that do not contain gluten. From a gastroenterological perspective, other approaches are to supply digestive enzymes that will break down the gluten proteins before they stimulate an autoimmune response and to use polymers that will bind to gluten in the GI tract and prevent it from being absorbed. From an immunological perspective, researchers are looking at medications that could block immune responses that damage the small intestine.

A related issue is **nonceliac wheat sensitivity (NCWS),** sometimes called *gluten sensitivity* or *gluten intolerance*. Some people experience symptoms of celiac disease after ingestion of gluten, but they do not have the small intestine pathology of celiac disease, nor do they express the antibodies typical of celiac disease. Some reports indicate that for each person who is diagnosed with celiac disease, as many as six others have NCWS. Aside from GI symptoms, patients with NCWS may also report fatigue, headache, muscle and joint pain, and/or sleep disorders. Symptoms subside with a gluten-free diet but reappear when gluten is reintroduced. The medical community recognizes NCWS as a verifiable condition, but the immunological mechanism that causes it is not well understood. There is no diagnostic test for the condition at this time—only the effectiveness of the gluten-free diet in alleviating symptoms. Many questions remain: Is NCWS a permanent condition? Is there a level of gluten intake that would not trigger symptoms? It seems that there are multiple immunological reactions to gluten that are predicted by different genetic traits.[32,33]

Overall, the prevalence and awareness of celiac disease and NCWS seem to be on the rise. A cause for the increased prevalence has not been pinpointed, but some scientists speculate that changes in wheat production or widespread use of wheat in the food supply may be to blame. Others suspect that an infection or exposure to some environmental toxin could lead to gluten sensitization.

## Summary

The conditions discussed here can be very serious, possibly leading to malnutrition, internal bleeding, and life-threatening infections. It is important to seek competent medical advice if you or someone you know suspects a GI disorder. However, you should feel empowered to know that you can control some risks and complement medical treatment with nutrition and other lifestyle changes. Overall, keeping body weight within a healthy range, meeting recommendations for fiber and fluid intake, and avoiding tobacco and overuse of NSAID medications are useful strategies that can help you cope with several common disorders of the GI tract.

**nonceliac wheat sensitivity (NCWS)** One or more of a variety of immune-related conditions with symptoms similar to celiac disease that are precipitated by the ingestion of gluten in people who do not have celiac disease.

## ASK THE RDN    Gluten-Free Diet

**Dear RDN:** *I have heard that a gluten-free diet is a healthier choice. Will a gluten-free diet help me lose weight?*

The media is brimming with popular advice to eliminate wheat and other grains from your eating pattern, promising everything from a clear mind to a trim waistline. For individuals with celiac disease and nonceliac wheat sensitivity, avoiding gluten is a health priority. It will likely correct gastrointestinal complications and a range of other symptoms such as fatigue and body pain. Weight loss, however, is not one of the benefits of a gluten-free diet.

The only reason a gluten-free diet might induce weight loss is because it can be restrictive. For the person who regularly overconsumes bread, pasta, pizza, and baked goods, eliminating all sources of gluten could limit these food choices, thus leading to weight loss. Important point: weight loss is the result of a calorie deficit, not the lack of gluten. Now, with the increased availability of gluten-free options in grocery stores and restaurants, you will soon realize that a gluten-free diet may not be that restrictive after all.

In fact, you may actually *gain* weight on a gluten-free diet. For a person with celiac disease, as the small intestine heals, nutrient absorption will increase and appetite will likely improve. In addition, many gluten-free products contain more calories than their wheat-based counterparts. To compensate for losses of taste and texture, some gluten-free products incorporate extra fat or sugar. For example, a typical slice of whole-grain wheat bread is about 80 kcal; some brands of gluten-free bread provide as many as 140 kcal per slice. Keep in mind that the "Gluten-Free" claim does not make a food healthy!

Lastly, grains are an important part of a balanced eating pattern. They provide calories, of course, but they also supply dietary fiber and essential vitamins and minerals. Wheat flour is fortified with thiamin, niacin, riboflavin, folic acid, and iron. So, if you are planning to try a gluten-free diet, work with a registered dietitian nutritionist to ensure that your eating pattern meets your nutrient needs. Choose whole gluten-free grains and unprocessed fruits and vegetables to replace wheat and balance calories to avoid weight gain.

With a grain of truth,

**Angela Collene, MS, RDN, LD**
Lecturer, The Ohio State University, Author of *Contemporary Nutrition*

Tim Klontz, Klontz Photography

## ✓ CONCEPT CHECK 3.11

1. What are the two leading causes of ulcers?

2. Describe the relationship between spicy foods and ulcers: Do spicy foods cause ulcers? Should a person with an ulcer eat spicy foods?

3. What are the two best dietary strategies to prevent or treat constipation?

4. What is the most important nutritional concern for a person experiencing diarrhea?

5. Based on what you have learned about digestion and absorption of nutrients, is taking laxatives an effective way to prevent fat gain from excess calorie intake? Why or why not?

# Summary (Numbers refer to numbered sections in the chapter.)

**3.1** Cells join together to make up tissues, tissues unite to form organs, and organs work together in organ systems. Cells are the basic structural units of the human body. Almost all cells contain the same organelles, but cell structure varies according to the type of job that cells must perform. Epithelial, connective, muscle, and nervous tissues are the four primary types of tissues in the human body. Each organ system both affects and is affected by nutrient intake.

**3.2** Metabolism refers to all the chemical reactions involved in maintaining life, including the synthesis (anabolism) of new compounds and the breakdown (catabolism) of carbohydrates, fats, and proteins to yield energy in the form of adenosine triphosphate (ATP). Some nutrients (e.g., vitamins and minerals) are important regulators of metabolic reactions.

**3.3** From the cells of the GI tract, water-soluble nutrients are absorbed into capillaries and fat-soluble nutrients are absorbed into lymph vessels, which eventually connect to the bloodstream. Blood delivers nutrients and oxygen to cells and picks up waste products as it circulates around the body.

**3.4** In the urinary system, the kidneys are responsible for filtering the blood, removing body waste, and maintaining the chemical composition of the blood.

**3.5** The nervous system allows for communication and regulation. Vitamin B-12 is part of the insulation that surrounds neurons. Transmission of nerve impulses relies on sodium and potassium. Neurotransmitters are made from amino acids.

**3.6** The endocrine system produces hormones—protein-based chemical messengers—to regulate metabolic reactions and the levels of nutrients in the blood.

**3.7** With assistance from the skin and the gastrointestinal tract, the immune system protects the body from pathogens. Optimal immune system function relies on protein; essential fatty acids; vitamins A, C, and D; some B vitamins; and the minerals iron, zinc, and copper.

**3.8** The GI tract consists of the mouth, esophagus, stomach, small intestine, large intestine (colon), rectum, and anus. The liver, gallbladder, and pancreas are accessory organs that participate in digestion and absorption.

Spaced along the GI tract are sphincters that regulate the flow of food matter. Peristalsis is the movement of food matter along the GI tract. Nerves, hormones, and other substances control the activity of sphincters and peristaltic muscles.

Digestive enzymes are secreted by the mouth, stomach, small intestine, and pancreas. Bile from the liver aids the digestion of fat. Some protein and fat are digested in the stomach, but most digestion occurs in the small intestine. In the large intestine, no further digestion by human enzymes takes place, but bacterial enzymes break down some dietary components.

Most absorption occurs through the cells of the villi, which line the small intestine. Absorptive processes include passive diffusion, facilitated diffusion, active transport, phagocytosis, and pinocytosis. The large intestine absorbs water, a few minerals, and some products of microbial fermentation. Any remaining undigested materials are eliminated in the feces.

**3.9** The community of microorganisms living in and on the human body is called the human microbiota. The microbiome includes the microbiota and its genome. Gut dysbiosis, or an imbalance between beneficial and pathogenic strains of microorganisms in the gut, may contribute to many common health conditions. Probiotics are beneficial bacteria that can be consumed as part of food or supplements, colonize the gut, and confer health benefits on the host. Prebiotics are food substances (usually carbohydrates) that serve as fuel for the probiotic microorganisms. Postbiotics are the by-products of microbial metabolism. Some postbiotics affect the GI tract locally and others can be absorbed. The gut microbiota is now recognized to influence human health, not only in the GI tract, but throughout the body.

**3.10** Limited stores of nutrients are present in the blood for immediate use. Some nutrients, such as minerals and fat-soluble vitamins, can be stored extensively in bone, adipose, and liver tissues. Excessive storage of nutrients can be toxic. Conversely, breakdown of vital tissues can supply nutrients in times of need, but continued breakdown eventually leads to ill health.

**3.11** Common GI tract diseases, such as heartburn, constipation, and irritable bowel syndrome, can be treated with a combination of dietary changes and medications.

# Check Your Knowledge (Answers are available at the end of this question set.)

1. Which of these nutrients is (are) important for the proper function of cell membranes?
   a. Lipids
   b. Proteins
   c. Carbohydrates
   d. All of these

2. The chemical reactions that break down carbohydrates, fats, and proteins to yield energy are _____ reactions.
   a. anabolic
   b. catabolic

3. The stomach is protected from digesting itself by producing
   a. bicarbonate.
   b. a thick layer of mucus.
   c. hydroxyl ions to neutralize acid.
   d. antipepsin that destroys enzymes.

4. The lower esophageal sphincter is located between the
   a. esophagus and stomach.
   b. stomach and duodenum.
   c. ileum and cecum.
   d. colon and anus.

5. A muscular contraction that propels food along the GI tract is called
   a. a sphincter.
   b. enterohepatic circulation.
   c. gravitational pull.
   d. peristalsis.

6. Bicarbonate ions ($HCO_3^-$) from the pancreas
   a. neutralize acid in the esophagus.
   b. are synthesized in the pyloric sphincter.
   c. neutralize bile in the duodenum.
   d. neutralize acid in the duodenum.

7. Most chemical digestion occurs in the
   a. mouth.
   b. stomach.
   c. small intestine.
   d. large intestine.
   e. liver.

8. Bile is formed in the _____ and stored in the _____.
   a. stomach, pancreas      c. liver, gallbladder
   b. duodenum, kidney      d. gallbladder, liver

9. Which of the following terms refers to the entire collection of microorganisms, their genes, and their environment?
   a. Microbiota      c. Probiotic
   b. Microbiome      d. Dysbiosis

10. Treatment of ulcers may include
    a. $H_2$ blockers.      c. antibiotics.
    b. proton pump inhibitors.      d. all of these.

Answer Key: 1. d (LO 3.1), 2. b (LO 3.3), 3. b (LO 3.9), 4. a (LO 3.9), 5. d (LO 3.9) 6. d (LO 3.9), 7. c (LO 3.9), 8. c (LO 3.9), 9. b (LO 3.10), 10. d (LO 3.12).

## Study Questions (Numbers refer to Learning Outcomes.)

1. Draw and label parts of the cell, and explain the function of each organelle as it relates to human nutrition. **(LO 3.2)**

2. Identify at least one nutrition-related function of the 12 organ systems. **(LO 3.4)**

3. How is blood routed to and from the small intestine? Which classes of nutrients enter the body via the blood? Via the lymph? **(LO 3.4)**

4. How are neurotransmitters and hormones different? How are they the same? Give one example of each. **(LO 3.6, LO 3.7)**

5. Explain why the small intestine is better suited than the other GI tract organs to absorb nutrients. **(LO 3.9)**

6. What is one role of acid in the process of digestion? Where is it secreted? **(LO 3.9)**

7. Contrast the processes of active absorption and passive diffusion of nutrients. **(LO 3.9)**

8. Identify two accessory organs that secrete digestive substances into the small intestine. How do the substances secreted by these organs contribute to the digestion of food? **(LO 3.9)**

9. In which organ systems would the following substances be found? chyme **(LO 3.9)**, plasma **(LO 3.4)**, lymph **(LO 3.4)**, urine **(LO 3.5)**

10. Choose one common digestive disorder and explain how dietary modifications can be used to prevent or treat the disorder. **(LO 3.12)**

## Further Readings

1. Hunt RH and others: The stomach in health and disease. *Gut* 64:1650, 2015. DOI:10.1136/gutjnl-2014-307595.

2. Satyanarayana MN: Capsaicin and gastric ulcers. *Critical Reviews in Food Science and Nutrition* 46:275, 2006. DOI: 10.1080/1040-830491379236.

3. McCulloch M: Appetite hormones. *Today's Dietitian* 17:26, 2015.

4. Kaye R and others: COVID-19 anosmia reporting tool: initial findings. *Otolaryngology Head and Neck Surgery* April 28, 2020 (Epub ahead of print; DOI:10.1177/0194599820922992).

5. Kelly D and Mulder IE: Microbiome and immunological factors. *Nutrition Reviews* 70:S18, 2012. DOI:10.1111/j.1753-4887.2012.00498.x.

6. Wang B and others: The human microbiota in health and disease. *Engineering* 3:71, 2017. DOI: 10.1016/J.ENG/2017.01.008.

7. International Scientific Association for Prebiotics and Probiotics: Effects of probiotics and prebiotics on our microbiota. 2017. Available at https://isappscience.org. Accessed November 28, 2019.

8. Aguilar-Toala JE and others: Postbiotics: An evolving term within the functional foods field. *Trends in Food Science and Technology* 75:105, 2018. DOI:10.1016/j.tifs.2018.03.009.

9. Guinane CM and Cotter PD: Role of the gut microbiota in health and chronic gastrointestinal disease: Understanding a hidden metabolic organ. *Therapeutic Advances in Gastroenterology* 6:295, 2013. DOI:10.1177/1756283X13482996.

10. de Oliveira GLV and others: Intestinal dysbiosis and probiotic applications in autoimmune diseases. *Immunology* 152:1, 2017. DOI:10.1111/imm.12765.

11. Liang D: Involvement of gut microbiome in human health and disease: Brief overview, knowledge gaps and research opportunities. *Gut Pathogens* 10:3, 2018. DOI:0.1186/s13099-018-0230-4.

12. Ximenez C and Torres J: Development of microbiota in infants and its role in maturation of gut mucosa and immune system. *Archives of Medical Research* 48:666, 2017. DOI:10.1016/j.arcmed.2017.11.007.

13. Danneskiold NB and others: Interplay between food and gut microbiota in health and disease. *Food Research International* 115:23, 2019. DOI:10.1016/j.foodres.2018.07.043.

14. Dhar D and Mohanty A: Gut microbiota and Covid-19—possible link and implications. *Virus Research* May 13, 2020 (Epub ahead of print; DOI:10.1016/j.virusres.2020.198018).

15. Mak JWY, Chan FKL, and Ng SC: Probiotics and COVID-19: One size does not fit all. *The Lancet Gastroenterology & Hepatology* April 24, 2020 (Epub ahead of print; DOI:10.1016/S2468-1253(20)30122-9).

16. Moran-Ramos S, Lopez-Contreras BE, and Canizales-Quinteros S: Gut microbiota in obesity and metabolic abnormalities: A matter of composition or functionality? *Archives of Medical Research* 48:735, 2017. DOI:10.1016/j.arcmed.2017.11.003.

17. Jie Z and others: The gut microbiome in atherosclerotic cardiovascular disease. *Nature Communications* 8:845, 2017. DOI:10.1038/s41467-017-00900-1.

18. Fung TC, Olson CA, and Hsiao EY: Interactions between the microbiota, immune and nervous systems in health and disease. *Nature Neuroscience* 20:145, 2017. DOI:10.1038/nn.4476.

19. National Digestive Diseases Information Clearinghouse: Acid reflux (GER and GERD) in adults. 2014. Available at https://www.niddk.nih.gov/health-information/digestive-diseases/acid-reflux-ger-gerd-adults. Accessed November 28, 2019.

20. National Digestive Diseases Information Clearinghouse: Peptic ulcers (stomach ulcers). 2014. Available at https://www.niddk.nih.gov/health-information/digestive-diseases/peptic-ulcers-stomach-ulcers. Accessed November 28, 2019.

21. National Digestive Diseases Information Clearinghouse: Constipation. 2018. Available at https://www.niddk.nih.gov/health-information/digestive-diseases/constipation. Accessed November 28, 2019.

22. Scarlata K: Get things moving: A dietitian's guide to relieving constipation. *Today's Dietitian* 18:10, 2016.

23. Feuerstein JD and Falchuk KR: Diverticulosis and diverticulitis. *Mayo Clinic Proceedings* 91:1094, 2016. DOI:10.1016/j.mayocp.2016.03.012.

24. National Digestive Diseases Information Clearinghouse: Diverticular disease. 2016. Available at https://www.niddk.nih.gov/health-information/digestive-diseases/diverticulosis-diverticulitis. Accessed November 28, 2019.

25. Stollman N and others: American Gastroenterological Association institute guideline on the management of acute diverticulitis. *Gastroenterology* 149:1944, 2015. DOI:10.7326/M15-2499.

26. Scarlata K: FODMAPs: Overview of the emerging science. *Today's Dietitian* 20:14, 2018.

27. Wolfram T: Why the Low-FODMAP Diet is a growing dietitian-led treatment for people with IBS. *Food & Nutrition Magazine* Sept/Oct 2016:18, 2016.

28. National Digestive Diseases Information Clearinghouse: Irritable bowel syndrome. 2017. Available at https://www.niddk.nih.gov/health-information/digestive-diseases/irritable-bowel-syndrome. Accessed November 28, 2019.

29. National Digestive Diseases Information Clearinghouse: Diarrhea. 2016. Available at https://www.niddk.nih.gov/health-information/digestive-diseases/diarrhea. Accessed November 28, 2019.

30. National Digestive Diseases Information Clearinghouse: Gallstones. 2017. Available at https://www.niddk.nih.gov/health-information/digestive-diseases/gallstones. Accessed November 28, 2019.

31. Lebwohl B and others: Celiac disease and non-celiac gluten sensitivity. *BMJ* 351:h4347, 2015. DOI:10.1136/bmj.h4347.

32. Zelman K: Is gluten sensitivity real? New research causes new thinking. *Food and Nutrition Magazine* Sept/Oct 2015:16, 2015.

33. Uhde M and others: Intestinal cell damage and systemic immune activation in individuals reporting sensitivity to wheat in the absence of coeliac disease. *Gut* 65:1930, 2016. DOI:10.1136/gutjnl-2016-311964.

# Chapter
## 4

# Carbohydrates

# Student Learning Outcomes

**Chapter 4 is designed to allow you to:**

**4.1** Explain how carbohydrates are created and their role in a healthy dietary pattern.

**4.2** Identify the basic structures of the major carbohydrates: monosaccharides, disaccharides, and polysaccharides.

**4.3** Describe food sources of carbohydrates and list some alternative sweeteners.

**4.4** Explain how carbohydrates are taken in and used by the body, including the processes of digestion, absorption, metabolism, and glucose regulation.

**4.5** List the functions of carbohydrates in the body, the problems that result from not eating enough carbohydrates, and the beneficial effects of fiber on the body.

**4.6** State the RDA for carbohydrate and guidelines for carbohydrate intake.

**4.7** Identify the consequences of diabetes, and explain appropriate lifestyle behaviors that will reduce the adverse effects of this chronic disease.

## FAKE NEWS

## All carbohydrates are fattening and should be avoided.

## THE FACTS

All carbs are *not* created equal. Many people think carbohydrate-rich foods cause weight gain, but pound for pound, carbohydrates supply fewer calories than fats and oils. Furthermore, high-carb foods, especially fiber-rich foods such as fruits, vegetables, whole grains, and legumes, provide significant health benefits. Refined grains, such as breads, cereals, and pastas, are enriched with vitamins and minerals that supply B vitamins, folic acid, and iron to our dietary patterns. Other refined and highly processed foods, such as cookies, cakes, and pastries, are energy-dense choices that are high in calories but contain few nutrients. For this reason, the *Dietary Guidelines* recommends making at least half of the grains you eat whole grains.

What did you eat to obtain the energy you are using right now? We will examine this question by focusing on the main nutrients the human body uses for fuel. These nutrients are carbohydrates (4 kcal per gram) and fats (9 kcal per gram). Although protein (4 kcal per gram) can be used for energy needs, the body typically prefers and reserves this nutrient for other key processes.

It is likely that you have recently consumed fruits, vegetables, dairy, and various grains. These foods all supply carbohydrates. Although some carbohydrate sources are more beneficial than others, carbohydrates should provide a significant portion of our dietary intake. Hence, we should strive to incorporate more nutritious and fiber-rich carbohydrates into our daily dietary patterns. As you will learn, naturally occurring sugars in whole fruit are metabolized differently than added sugars in highly processed foods (like candy). Indeed, almost all carbohydrate-rich foods, except pure sugar, provide essential nutrients and should generally constitute 45% to 65% of our daily calorie intake.

The beautiful chapter cover photo includes a diverse group of colorful carbohydrate sources that are rich in nutrients and phytochemicals. Now, let's take a closer look at carbohydrates.

# 4.1 Carbohydrates—Our Most Important Energy Source

Carbohydrates are a main fuel source for some cells, especially those in the brain, nervous system, and red blood cells. Muscles also rely on a dependable supply of carbohydrates to fuel intense physical activity. As mentioned, carbohydrates provide approximately 4 kcal per gram and are a readily available fuel for cells, in the form of both blood glucose and **glycogen** stored in the liver and muscles. The glycogen stored in the liver can be used to maintain blood glucose concentrations when you have not eaten for several hours or the food you eat does not supply enough carbohydrates. Regular intake of carbohydrates is important because liver glycogen stores can be depleted in about 18 hours if no carbohydrates are consumed. After that point, the body is forced to produce alternative fuel sources, either from the breakdown of protein or fat stores, depending on dietary patterns. To obtain adequate energy, the *Dietary Guidelines* recommends that 45% to 65% of the calories we consume each day come from carbohydrates.

As you will see in this chapter, whole grain products have greater overall health benefits than refined and highly processed forms of carbohydrate. Choosing whole grain carbohydrate sources and limiting the intake of refined carbohydrate sources contribute to a healthier dietary pattern. Let's explore this concept further as we learn about carbohydrates in detail.

Green plants synthesize most of the carbohydrates in our foods. Leaves capture the sun's solar energy in their cells and transform it into chemical energy. This energy is then stored in the chemical bonds of the carbohydrate glucose as it is produced from carbon dioxide in the air and water in the soil. This complex process is called **photosynthesis** (Fig. 4-1). Translated into simpler terms, 6 molecules of carbon dioxide combine with 6 molecules of water using energy from the sun to form 1 molecule of glucose and release 6 molecules of oxygen into the air. Converting solar energy into chemical bonds in the sugar is a key part of the process.

**glycogen** A carbohydrate made of multiple units of glucose with a highly branched structure. It is the storage form of glucose in humans and is synthesized (and stored) in the liver and muscles.

Shojiro Ishihara/sunabesyou/123RF

**photosynthesis** Process by which plants use energy from the sun to synthesize energy-yielding compounds, such as glucose.

**FIGURE 4-1** ▶ A summary of photosynthesis. Plants use carbon dioxide, water, and the sun's energy to produce glucose (sugar). Glucose is then stored in the plant and can undergo further metabolism to form starch and fiber in the plant. With the addition of nitrogen from soil or air, glucose can also be transformed into protein.

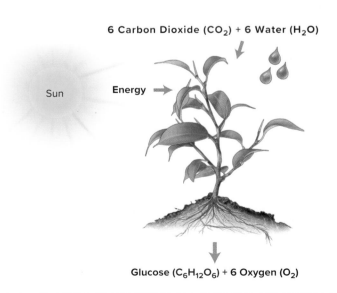

**6 Carbon Dioxide ($CO_2$) + 6 Water ($H_2O$)**

Sun

Energy →

Glucose ($C_6H_{12}O_6$) + 6 Oxygen ($O_2$)

**Photosynthesis Equation**

6 carbon dioxide + 6 water + solar energy → glucose + 6 oxygen

1. Why are carbohydrates considered our most valuable energy source?

2. What are the main components needed for photosynthesis?

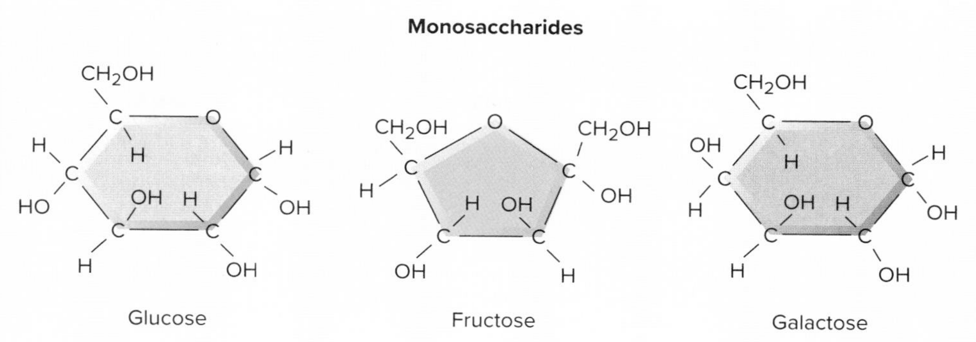

# 4.2 Forms of Carbohydrates

**Monosaccharides**

Glucose          Fructose          Galactose

**FIGURE 4-2** ▲ Chemical structure of monosaccharides.

As the name suggests, most carbohydrate molecules are composed of carbon, hydrogen, and oxygen atoms. Simple forms of carbohydrates are called **sugars**. **Simple carbohydrates** include **monosaccharides** and **disaccharides** (Fig. 4-2). Larger, more complex forms of carbohydrates, the **polysaccharides,** are primarily called starches or fibers, depending on their digestibility. Starches are digestible, whereas fibers are not. Polysaccharides will be discussed later in this chapter.

Simple carbohydrates contain only one (single) or two (double) sugar units and are called monosaccharides and disaccharides, respectively. These carbohydrates are quickly digested and absorbed. This rapid increase in blood sugar can result in a burst of energy that may be followed by fatigue as energy is depleted. Simple sugars are found in refined sugars, such as table sugar. These types of added sugars provide calories but lack vitamins, minerals, and fiber. This is why excessive added sugars contribute to weight gain. Recall that not all simple sugars are alike. On the other hand, naturally occurring simple sugars, found in fruit and milk, contribute key nutrients to our dietary pattern and should not be avoided. On the Nutrition Facts label, sugars are labeled as *Total Sugars* with *Added Sugars* listed as a subcategory.

## MONOSACCHARIDES: GLUCOSE, FRUCTOSE, AND GALACTOSE

Monosaccharides are the simple sugar units (*mono* means "one") that serve as the basic unit of all carbohydrate structures. The most common monosaccharides in foods are glucose, fructose, and galactose (Fig. 4-2). Although the structures vary between the monosaccharides, the chemical composition ($C_6H_{12}O_6$) remains the same in each.

**Glucose.** Glucose is the major monosaccharide found in the body. Glucose is also known as *dextrose,* and glucose in the bloodstream may be called *blood sugar*. Glucose is an important source of energy for human cells, although few foods contain glucose as their primary carbohydrate source. Most glucose comes from the digestion of starches and **sucrose** (common table sugar). Sucrose is made up of the monosaccharides glucose and fructose. For the most part, sugars and other carbohydrates in foods are eventually converted into glucose in the liver. This glucose then becomes available to serve as a source of fuel for cells.

**Fructose.** Also called *fruit sugar*, **fructose** is found naturally in fruits and forms half of each sucrose molecule. After it is consumed, fructose is absorbed by the small intestine and then transported to the liver, where it is quickly metabolized. Much is converted to glucose, but excess fructose may form other compounds, such as fat. Most of the free fructose in the food we eat comes from the use of **high-fructose corn syrup (HFCS)** in highly processed foods, including soft drinks, canned foods, cereals and baked goods, desserts, sweetened and flavored products (yogurt, condiments, jams, and jellies), candies, and many fast food items.

**Galactose.** The sugar **galactose** has nearly the same structure as glucose. Large quantities of pure galactose do not exist in nature. Instead, galactose is usually found bonded to glucose in **lactose,** a sugar found in milk and other milk products. During digestion, lactose is broken

**sugar** A simple carbohydrate with the chemical composition $(CH_2O)_n$. The basic unit of all sugars is *glucose*.

**monosaccharide** Simple sugar, such as glucose, that is not broken down further during digestion.

**disaccharide** Class of sugars formed by the chemical bonding of two monosaccharides.

**polysaccharides** Carbohydrates containing many glucose units, from 10 to 1000 or more. Also called *complex carbohydrates*.

**sucrose** Disaccharide composed of fructose bonded to glucose; also known as *table sugar*.

**fructose** A six-carbon monosaccharide that usually exists in a ring form; found in fruits and honey; also known as *fruit sugar*.

**high-fructose corn syrup (HFCS)** Corn syrup that has been manufactured to contain from 42% to 55% fructose.

**galactose** A six-carbon monosaccharide that usually exists in a ring form; closely related to glucose.

**lactose** A disaccharide consisting of glucose bonded to galactose; also known as *milk sugar*.

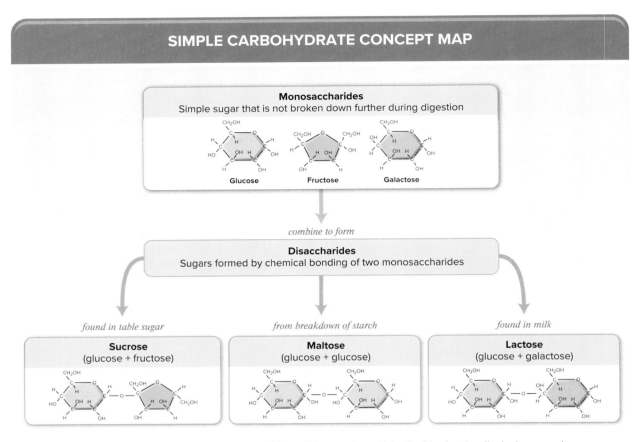

**SIMPLE CARBOHYDRATE CONCEPT MAP**

**Monosaccharides**
Simple sugar that is not broken down further during digestion

Glucose    Fructose    Galactose

*combine to form*

**Disaccharides**
Sugars formed by chemical bonding of two monosaccharides

*found in table sugar*    *from breakdown of starch*    *found in milk*

**Sucrose**
(glucose + fructose)

**Maltose**
(glucose + glucose)

**Lactose**
(glucose + galactose)

**FIGURE 4-3** ▲ This Simple Carbohydrate Concept Map will be continued later in this chapter displaying complex carbohydrates (Fig. 4-5). Here, we summarize the various forms, structures, and characteristics of the simple carbohydrates. Figure 4-7, located at the end of this section, brings the carbohydrate map together.

down to galactose and glucose and then absorbed. When galactose arrives at the liver, it is either transformed into glucose or further metabolized into glycogen, depending on the body's energy needs at the time. When required for milk production in the mammary gland of a lactating woman, galactose is resynthesized from glucose to help form the milk sugar lactose. Although milk consumption is recommended for lactating mothers, this amazing process produces milk for the baby even if mom doesn't consume milk!

**maltose** A disaccharide consisting of glucose bonded to glucose; also know as *malt sugar*.

### DISACCHARIDES: SUCROSE, LACTOSE, AND MALTOSE

Disaccharides are formed when two monosaccharides combine (*di* means "two"). The disaccharides in food are sucrose, lactose, and **maltose**. All contain glucose (Fig. 4-3).

**Sucrose.** Sucrose forms when the two sugars, glucose and fructose, bond together (Fig. 4-4). Sucrose is found naturally in sugarcane, sugar beets, honey, and maple sugar. These products are processed to varying degrees to make brown, white, and powdered sugars.

**Lactose.** Lactose forms when glucose bonds with galactose during the synthesis of milk. Therefore, our major food source for lactose is milk products. Section 4.4 on lactose maldigestion and lactose intolerance discusses the issues that result when a person can't readily digest lactose.

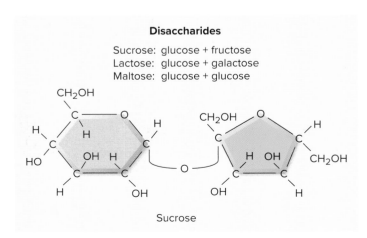

**Disaccharides**

Sucrose: glucose + fructose
Lactose: glucose + galactose
Maltose: glucose + glucose

Sucrose

**FIGURE 4-4** ▲ Chemical structure of the disaccharide sucrose.

**Maltose.** Maltose results when starch is broken down to just two glucose molecules bonded together. Maltose plays an important role in the beer and liquor industry. In the production of these alcoholic beverages, starches are converted to simpler carbohydrates by enzymes present in the grains. The breakdown products are then mixed with yeast cells in the absence of oxygen (anaerobic environment). The yeast cells convert most of the sugars to alcohol (ethanol) and carbon dioxide through a process called **fermentation.** Very little maltose remains in the final product. Small amounts of maltose can be found in some foods, including fruits, vegetables, and breads. Most maltose that we digest (in the small intestine) is produced during our own digestion of starch.

**fermentation** The conversion of carbohydrates to alcohols, acids, and carbon dioxide without the use of oxygen.

## POLYSACCHARIDES: STARCH, GLYCOGEN, AND FIBER

Complex carbohydrates are digested more slowly than simple carbohydrates and supply a consistent and steady release of sugar (glucose) into the bloodstream. As with simple sugars, some complex carbohydrates are better choices than others. Let's explore more about complex carbohydrates in Figure 4-5 and below.

**Starch.** In many foods, numerous single-sugar units are bonded together to form a chain known as a polysaccharide (*poly* means "many"). Polysaccharides, also called *complex carbohydrates* or *starch,* may contain 1000 or more glucose units and are found chiefly in grains, vegetables, and fruits. When the Nutrition Facts label on food products lists *Other Carbohydrates*, this primarily refers to the starch content.

Plants store carbohydrates in two forms of starch digestible by humans: **amylose** and **amylopectin.** Amylose, a long, straight chain of glucose units, comprises about 20% of the digestible starch found in vegetables, beans, breads, pasta, and rice. Amylopectin has a highly branched-chain structure and makes up the remaining 80% of digestible starches in the food we eat. **Cellulose** (a fiber) is another complex carbohydrate in plants. Although similar to amylose, it cannot be digested by humans.

**amylose** A digestible, straight-chain type of starch composed of glucose units.

**amylopectin** A digestible, branched-chain type of starch composed of glucose units.

**cellulose** An indigestible polysaccharide made up of glucose units and found in plant cell walls.

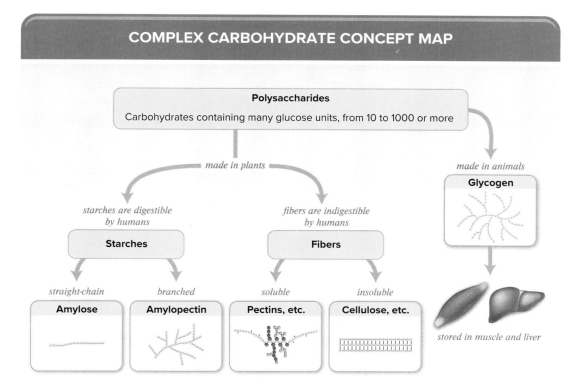

**FIGURE 4-5** ▲ This Complex Carbohydrate Concept Map summarizes the various forms, structures, and characteristics of the complex carbohydrates.

Root vegetables such as potatoes, yams, and tapioca are high in amylopectin starch. **Is amylopectin a digestible form of carbohydrate?**
C Squared Studios/Getty Images

The enzymes that break down starches to glucose and other related sugars act only at the end of a glucose chain. Amylopectin, because it is branched, provides many more sites (exposed ends) for action. Therefore, amylopectin is digested more rapidly and raises blood glucose much more readily than amylose.

**Glycogen.** As noted, animals—including humans—store glucose in the form of glycogen. Glycogen consists of a chain of glucose units with many branches, providing even more sites for enzyme action than amylopectin (Fig. 4-5). Because of its highly branched structure that can be broken down quickly, glycogen is an ideal storage form of carbohydrate in the body.

The liver and muscles are the major storage sites for glycogen. Because the amount of glucose immediately available in body fluids can provide only about 80 to 120 kcal, the carbohydrate energy stored as glycogen—amounting to about 1800 kcal—is extremely important. Of this 1800 kcal, liver glycogen (about 400 kcal) can readily contribute to blood glucose. Muscle glycogen stores (about 1400 kcal) cannot raise blood glucose but instead supply glucose for muscle use, especially during high-intensity and endurance exercise. Although animals store glycogen in their muscles, animal products such as meats, fish, and poultry are not good sources of carbohydrates because glycogen stores quickly degrade after the animal dies.

**Fiber.** Fiber is mainly made up of polysaccharides, but fibers differ from starches because the chemical bonds that join the individual sugar units together cannot be digested by enzymes in the gastrointestinal (GI) tract. This prevents the small intestine from absorbing the sugars because they cannot be released from the various fibers. Fiber is not a single substance but a group of substances with similar characteristics. The group is composed of the carbohydrates cellulose, **hemicelluloses, pectins, gums, β-glucans, inulin,** and **resistant starch,** as well as the noncarbohydrate **lignin.** In total, these constitute all the nonstarch polysaccharides in foods. Nutrition Facts labels combine the individual forms of fiber together under the term **dietary fiber.** Naturally occurring dietary fiber is found in beans, peas, lentils, fruits, nuts, seeds, vegetables, wheat bran, and whole grains (including whole oats, brown rice, popcorn, and quinoa) and foods made with whole grain ingredients (including breads, cereals, crackers, and pasta).

Cellulose, hemicelluloses, and lignin form the structural parts of plants (Fig. 4-6). Bran layers form the outer covering of all grains and are rich in hemicelluloses and lignin. **Whole** (i.e., unrefined) **grains** are good sources of bran fiber (Fig. 4-6).

**hemicellulose** A group of polysaccharides that are found in plant cell walls.

**pectin** Polysaccharides made of 300 to 1000 monosaccharides that are soluble viscous fibers and abundant in berries and other fruit.

**gums** Viscous polysaccharides often found in seeds and often used in the food industry for its thickening and stabilizing properties.

**β-glucan** Oats and barley are rich sources of these glucose polymers.

**inulin** A mixture of fructose chains that vary in length and occur naturally in plants.

**resistant starch** Indigestible dietary fiber sequestered in plant walls. Bananas and legumes are rich sources.

**lignin** A noncarbohydrate dietary fiber that is found in cell walls of woody plants and seeds.

**dietary fiber** Indigestible fiber found in food.

**whole grains** Grains containing the entire seed of the plant, including the bran, germ, and endosperm (starchy interior). Examples are whole wheat bread and brown rice.

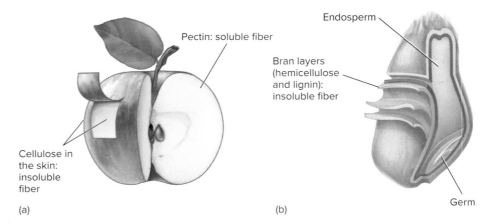

FIGURE 4-6 ▲ Soluble and insoluble fiber. (a) The skin of an apple consists of the insoluble fiber cellulose, which provides structure for the fruit. The soluble fiber pectin glues together the fruit cells. (b) The outside layer of a wheat kernel is made of layers of bran—primarily hemicellulose, a nonfermentable fiber—making this whole grain a good source of fiber. Overall, fruits, vegetables, whole grains, and beans are rich in fiber.

**TABLE 4-1** ■ **Classification of Dietary Fibers**

| Type | Description | Sources* | Physiological Effects |
|---|---|---|---|
| Insoluble, poorly fermented | Does not dissolve in water, does not trap water, and is poorly fermented | Cellulose (whole grains, bran, peas, root vegetables, beans, apple)<br>Lignin (vegetables, flour)<br><br>Hemicellulose (oats, wheat, barley, soy, rye, wheat, and barley) | Increases fecal bulk, may have constipating effect |
| Soluble, viscous, nonfermented | Dissolves in water and forms viscous gels | Psyllium (fortified foods) | Improves glycemic control, lowers blood cholesterol, and prevents constipation |
| Soluble, viscous, readily fermented | Dissolves in water and forms very viscous or visco-elastic gels | β-glucans (oats, barley, mushrooms)<br>Raw guar gum<br><br>Pectin (berries, citrus peel, apple pulp, and most fruit) | Improves glycemic control and lowers blood cholesterol |
| Soluble, nonviscous, readily fermented | Can dissolve in water, does not provide any health benefits associated with fiber viscosity | Inulin (artichokes, chicory root, rye, barley, bananas)<br>Oligosaccharides (cereals, vegetables, legumes, fruits)<br>Wheat dextrin<br>Resistant starch (peas, beans, lentils, potatoes, plantains) | Prebiotic effects with no known health benefits |

*All plant foods are a combination of both soluble and insoluble fiber in different proportions.
Source: Adapted from the Linus Pauling Institute

Because the majority of these fibers neither readily dissolve in water nor are easily metabolized by intestinal bacteria, they are called **insoluble.** Insoluble fiber, or *roughage*, is found in wheat bran, nuts, fruit skins, and some vegetables. This type fiber acts as a natural laxative because it speeds up the transit time of food through the GI tract.

Pectins, gums, and **mucilages** are contained around and inside plant cells. These fibers either dissolve or absorb water and are therefore called **soluble.** They also are readily fermented by bacteria in the large intestine. These fibers are primarily found in beans, oats, oat bran, and some fruits and vegetables. Soluble fiber slows the rate of absorption by attracting water into the GI tract, reduces blood cholesterol, promotes satiety, and controls blood glucose (Table 4-1).

Most foods contain mixtures of both soluble and insoluble fibers. Food labels do not generally distinguish between the two types. Types of fiber include: (1) *dietary fiber,* which describes the indigestible carbohydrates and lignin that are naturally occurring and intact in plants; and (2) **functional fiber,** which consists of the isolated indigestible carbohydrates that are added to food or extracted from a natural plant or animal sources or may be manufactured or synthesized because they have beneficial physiological effects in human beings. The commercially produced functional, or isolated fibers, include resistant starch, polydextrose, indigestible dextrins, and inulin in supplements or added to foods such as yogurt, cereal, bars, and bread. Figure 4-7 pulls together the entire carbohydrate concept map to help you visually display the connectedness of all carbohydrates. To determine your fiber intake, visit *Rate Your Plate: Estimate Your Fiber Intake* in Connect. The health benefits of many of these fibers are still unclear and are therefore a hot discussion topic.[1,2]

**insoluble fiber** A fiber that does not dissolve in water and may pass through the GI tract intact so is not a source of calories; also known as *roughage*.

**mucilage** A gelatinous substance of plants that contains protein and polysaccharides and is similar to plant gums.

**soluble fiber** A fiber that dissolves in water to form a thick gel-like substance and is broken down by bacteria in the large intestine, providing some calories.

**functional fiber** Indigestible carbohydrates that have beneficial physiological effects in humans.

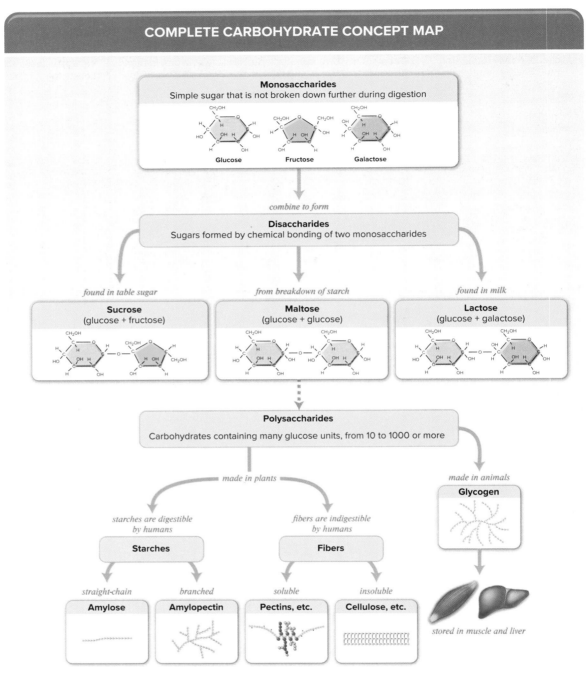

**FIGURE 4-7** ▲ This complete Complete Carbohydrate Concept Map summarizes the various forms, characteristics, and basic structures of the simple and complex carbohydrates.

## ✓ CONCEPT CHECK 4.2

1. What are the specific names and definitions of the monosaccharides and disaccharides, and what happens to them when they are digested and absorbed?

2. What is a polysaccharide, and what are the differences between the plant polysaccharides?

3. What is the name of the storage form of glucose? Where is this compound located in the body?

4. What makes fiber a very different kind of carbohydrate, and what is the difference between soluble, insoluble, and functional fiber?

# 4.3 Carbohydrates in Foods

Four of the groups on MyPlate—grains, vegetables, fruits, and dairy—contain the most nutrient-dense sources of carbohydrates (Fig. 4-8). A healthy, carbohydrate-rich dietary pattern emphasizes a variety of foods from these groups. Most dietary carbohydrate comes from starches. Because plants store glucose in the form of starches, plant-based foods (beans, potatoes, and grains used to make breads, cereals, and pasta) are the richest sources of starch. A dietary pattern rich in these starches also provides many micronutrients, phytochemicals, and fiber along with the carbohydrates. Soluble fibers (pectin, gums, and mucilages) are found in the skins and flesh of many fruits and berries; as thickeners and stabilizers in jams, yogurts, sauces, and fillings; and in products that contain psyllium and seaweed. Fiber is also available as a supplement or as an additive to certain foods (functional fiber) such that individuals with relatively low intakes of natural dietary fiber can obtain the health benefits of fiber. Be sure to discuss the pros and cons of fiber supplementation with your primary care provider or registered dietitian nutritionist (RDN).

Soft drinks are referred to as sugar-sweetened beverages due to their simple sugar content. **What other beverages contain added sugars?**
Andrew Bret Wallis/Stockbyte/Getty Images

Unfortunately, the top carbohydrate sources for U.S. adults include sugar-sweetened beverages, followed by desserts and sweet snacks. Sugar-sweetened beverages are typically sweetened with added sugars such as brown sugar, corn sweetener, corn syrup, dextrose, fructose, glucose, high-fructose corn syrup, honey, lactose, malt syrup, maltose, molasses, raw sugar, and sucrose. Examples of sugar-sweetened beverages include regular soda (not sugar free), fruit drinks, sports drinks, energy drinks, sweetened waters, and coffee and tea with added sugars. Remember that these sources provide many calories and few nutrients. This is why the *Dietary Guidelines* recommends limiting added sugars to less than 10% of total calories per day.

Many North Americans are paying more attention to carbohydrate intake and including more whole grain versions of breads, pasta, rice, and cereals, as well as fruits and vegetables in their dietary patterns. The decline in soft drink consumption in the United States is particularly encouraging, with consumers turning to water, and unsweetened tea and coffee instead. More tips for making smart beverage choices can be found in Figure 4-9.

It is important to understand the percentage of calories from carbohydrates when planning a healthy dietary pattern. The food sources that yield the highest percentage of calories from carbohydrates are table sugar, honey, jam, jelly, fruit, and plain baked potatoes. Cornflakes, rice, bread, and noodles are next, all containing at least 74% of calories as carbohydrates. Foods with moderate amounts of carbohydrate calories are peas,

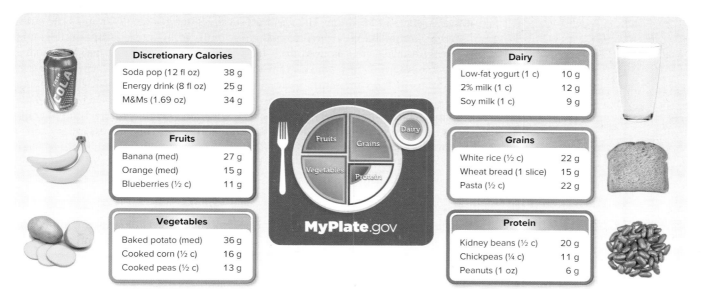

| Discretionary Calories | |
|---|---|
| Soda pop (12 fl oz) | 38 g |
| Energy drink (8 fl oz) | 25 g |
| M&Ms (1.69 oz) | 34 g |

| Fruits | |
|---|---|
| Banana (med) | 27 g |
| Orange (med) | 15 g |
| Blueberries (½ c) | 11 g |

| Vegetables | |
|---|---|
| Baked potato (med) | 36 g |
| Cooked corn (½ c) | 16 g |
| Cooked peas (½ c) | 13 g |

| Dairy | |
|---|---|
| Low-fat yogurt (1 c) | 10 g |
| 2% milk (1 c) | 12 g |
| Soy milk (1 c) | 9 g |

| Grains | |
|---|---|
| White rice (½ c) | 22 g |
| Wheat bread (1 slice) | 15 g |
| Pasta (½ c) | 22 g |

| Protein | |
|---|---|
| Kidney beans (½ c) | 20 g |
| Chickpeas (¼ c) | 11 g |
| Peanuts (1 oz) | 6 g |

**FIGURE 4-8** ▲ Food sources of carbohydrates. Overall, the richest sources of carbohydrates are discretionary calories (sugar-sweetened beverages), starchy vegetables (potatoes), and fruit (bananas). cola drink: scanrail/Getty Images; milk: McGraw-Hill Education; bananas: McGraw-Hill Education; wheat bread: Alex Cao/Photodisc/Getty Images; potato: Kaan Ates/Getty Images; kidney beans: zerbor/123RF; (MyPlate) U.S. Department of Agriculture

Source: U.S. Department of Agriculture, Agricultural Research Service. FoodData Central, 2019. fdc.nal.usda.gov.

**FIGURE 4-9 ▶** Smart beverage choices. Each day we make beverage choices that impact overall dietary patterns. Smoothies, energy drinks, sports beverages, and coffee-based gourmet drinks often contribute excess calories, added sugars, and sometimes fat to our dietary patterns. You can improve beverage choices by following these tips. McGraw-Hill Education (top); nesharm/Getty Images (middle); Cathy Yeulet/amenic181/123RF (bottom)

## Make Smart Beverage Choices

### During the Day

Keep a water bottle with you at all times.

Limit sugar-sweetened beverages. Try sugar-free sparkling or mineral water.

Drink water throughout the day and at meals.

Add lemon, cucumbers, oranges, strawberries, or herbs to vary the flavor.

Download an app to keep track of your water intake.

### Smoothie Smarts

Know your smoothie! Be sure you know all ingredients and amounts used to craft your smoothie.

Divide homemade smoothies into 2 servings and store one for later.

Check the calorie count and added sugars of smoothies before you buy.

Request a no-sugar added smoothie.

### Coffee Shop Options

Opt for low-fat milk or soy milk.

Always order the smallest size to save both calories and money.

Skip the cream and extra flavorings or consider sugar-free syrups.

Limit blended coffee drinks or select a light version.

A typical American lunch. **How does this lunch of a baked chicken and veggie burrito on whole grain tortilla with an apple, grapes, and a glass of low-fat milk compare to MyPlate?** chicken wrap: Lew Robertson/Stockbyte/Getty Images; glass of milk: NIPAPORN PANYACHAROEN/Shutterstock

broccoli, oatmeal, dry beans and other legumes, cream pies, French fries, and fat-free milk. In these foods, the carbohydrate content is diluted either by protein, as in the case of fat-free milk, or by fat, as in the case of cream pies. Foods with essentially no carbohydrates include beef, eggs, chicken, fish, vegetable oils, butter, and margarine (Fig. 4-8).

## WHOLE GRAINS

The *Dietary Guidelines* recommends that we consume at least half of all grains as whole grains. We can increase whole grain intake by replacing highly processed refined grains with whole grains and limiting the consumption of foods that contain refined grains, solid fats, added sugars, and sodium. The *Dietary Guidelines* defines whole grain as the entire grain seed or kernel made of three components: bran, germ, and endosperm (Fig. 4-6). When the term *whole grain* is used on a food package, it also means that the product contains a minimum of 51% whole grain ingredients by weight per serving. Examples of whole grains include amaranth, barley (not pearled), brown rice, buckwheat, bulgur, millet, oats, popcorn, quinoa, dark rye, whole-grain cornmeal, whole-wheat bread, whole-wheat chapati, whole-grain cereals and crackers, and wild rice. In contrast, refined grains have been milled, a process that removes the bran and germ. This process gives grains a finer texture and improves their shelf life, but it also removes dietary fiber, iron, and many B vitamins (Table 4-2). Some examples of refined grain products are white breads, refined-grain cereals and crackers, corn grits, cream of rice, cream of wheat, barley (pearled), masa, pasta, and white rice. Refined-grain choices should be enriched.

Although more fiber is one of the primary advantages of whole grains, many benefits of whole grains are thought to be due to the combined effects of several compounds. These compounds include fiber, minerals, trace minerals, vitamins, and an abundance of phytochemicals. These are mainly contained in the bran and germ parts of the grains. The *Dietary Guidelines* recommends consuming two to four servings of whole grains per day so that half of all the grains eaten are whole grains. Studies have shown that this is enough to impart numerous health benefits, including reducing risks of cardiovascular disease, diabetes, metabolic syndrome, some cancers, and obesity. Scientific evidence also supports the role of fiber in maintaining a healthy gut microbiota.

**TABLE 4-2 ■ Characteristics of Refined and Whole Grains**

| Refined Grains | Whole Grains |
|---|---|
| Contains endosperm only | Contains all components of grain seed or kernel (bran, germ, and endosperm) |
| Increased blood glucose response | Complex carbohydrate with slower glucose response |
| Lighter texture | Denser texture |
| Lower in fiber | Higher in fiber |
| Lower in nutrient density but enriched | Higher in vitamins, minerals, and antioxidants |

White whole wheat has the nutritional benefits of whole wheat but a milder taste, softer texture, and the lighter color of white bread. Traditional whole wheat is made from red wheat, which has a darker color and strongly flavored phenolic compounds. Switching to white whole wheat may be an acceptable option for those who prefer the taste and texture of white bread. **Based upon this information, is white whole wheat really a whole grain?** verastuchelova/123RF

Although average total grain intakes meets the *Dietary Guideline* recommendations, the intakes of refined grains exceeds the upper end of the recommended intake range for adults in both age groups. Whole grain intakes fall well below the recommendations at about 16% of total grain intake among adults on a given day[3] (Fig. 4-10). The reasons Americans are reluctant to consume whole grains include preferences in the taste, texture, cost, and availability of whole grains, compared to products made with refined flour.

In addition to our preference for refined grains, many consumers who are trying to choose a whole grain product are very confused by the deceptive marketing messages on the labels of grain products. For example, a label that says a cereal is *made with whole grains* does not guarantee that the cereal contains 100% whole grain. Terms such as *cracked wheat, stoneground wheat, enriched wheat flour, 12-grain,* and *multigrain* can be confusing because these products may contain little to no whole grains. Multigrain cereals may contain several grains, but many of them may be refined with just a small amount of whole grains. Some *whole grain breads* are actually white bread in disguise after brown coloring is added to enriched white flour.[4]

With all of the confusion, it is crucial to look beyond the front-of-the-package marketing claims and examine the list of ingredients. To confirm that products contain 100% whole grain, look for *whole* as the first word on the ingredient list. Sugary breakfast cereals that claim to be whole grain may list the first ingredient as a whole grain such as corn, rice, oat, or wheat, but then the next several ingredients may be various forms of sugar that together weigh more than the sum of the whole grains.[5]

Although whole grains make up only a small fraction of grains on grocery shelves, it is getting easier to find a variety of whole grain products. To help simplify the process even further, the Whole Grains Council developed the Whole Grain Stamp (Fig. 4-11). This stamp is on more than 13,000 different products in more than 61 countries.[6] Choosing products with the 100% whole grain stamp will help us reach the minimum goal of getting 3 ounce-equivalents of whole grains per day.[3] Schools are also doing their part to increase whole grain consumption. Foods must contain at least 50% whole grains to meet the whole grain–rich criteria for the federal school breakfast and lunch programs.[7] See Table 4-3 for information on some whole grains that are available, including their potential health benefits.

## VEGETABLES

Vegetables are a valuable source of carbohydrates in the form of starch and fiber. They are naturally low in fat and calories and come packed with many nutrients that are vital for health, including potassium, folate, vitamin A, and vitamin C. Potatoes, as featured in this chapter's *Farm to Fork,* contain many nutrients as well as a healthy dose of fiber. Eating the recommended amount of vegetables has been shown to improve weight and reduce the

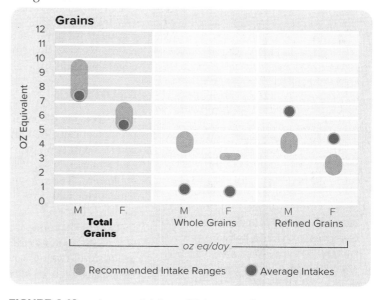

**FIGURE 4-10 ▲** Average Intakes of Subgroups Compared to Recommended Intake Ranges: Ages 31 Through 59.

Source: Average Intakes: Analysis of What We Eat in America, NHANES 2015-2016, day 1 dietary intake data, weighted. Recommended Intake Ranges: Healthy U.S.-Style Dietary Patterns

**FIGURE 4-11** ▶ The Whole Grains Council developed the Whole Grain stamps for use on grain products to identify whole grain foods. There are three versions of the stamp. (a) 100% stamp; (b) 50% stamp; (c) basic stamp. Wholegraincouncil.org

100% OF THE GRAIN IS WHOLE GRAIN
(a)

50% OR MORE OF THE GRAIN IS WHOLE GRAIN
(b)

EAT 48g OR MORE OF WHOLE GRAIN DAILY
(c)

| 100% Stamp | 50%+ Stamp | Basic Stamp |
|---|---|---|
| For products where ALL of the grain is whole grain. | For products where at least 50% of the grain is whole grain. | For products containing significant amounts of whole grain, but less than 50% of all grain is whole grain. |
| Minimum requirement: 16g whole grain per serving. (a full serving of whole grain) | Minimum requirement: 8g whole grain per serving. (one-half serving of whole grain) | Minimum requirement: 8g whole grain per serving. (one-half serving of whole grain) |

risk of several chronic diseases. As we discussed with whole grains, the fiber in vegetables may also reduce the risk of heart disease, obesity, cancers, and type 2 diabetes.

Within the *Dietary Guidelines*, any vegetable or 100% vegetable juice counts as a member of the vegetables group. Vegetables are organized into five subgroups: (1) dark-green vegetables; (2) starchy vegetables; (3) red and orange vegetables; (4) beans, peas, lentils: and (5) other vegetables based on their nutrient content. Vegetable choices should be selected from among the vegetable subgroups but may be raw or cooked; fresh, frozen, canned, or dried/dehydrated; and whole, cut up, or mashed.

Close to 90% of the U.S. population does not meet the *Dietary Guidelines* recommendations for vegetables. Vegetables, when consumed, are combined with high-sodium sources. The amount of vegetables you need depends on your age, sex, and level of physical activity. For example, recommended total daily amounts for women ages 19 to 50 years are 2½ cups. In general, 1 cup of raw or cooked vegetables or vegetable juice or 2 cups of raw leafy greens can be considered as a "1 cup equivalent" from the vegetables group. Recommended weekly amounts from each vegetable subgroup are given as amounts to eat weekly. For example, the recommendations for women ages 19 to 50 years are 1½ cups of dark-green vegetables, 5½ cups of red and orange vegetables, 1½ cups of beans, peas, and lentils, 5 cups of starchy vegetables, and 4 cups of other vegetables such as cauliflower or mushrooms weekly. See this Chapter's *Ask the RDN* for ideas on how to incorporate more vegetables and fruits into your dietary pattern using juicing.

## FRUITS

Fruits provide carbohydrates primarily in the form of natural sugar and fiber. Eating fruits provides health benefits similar to those discussed for vegetables. People who eat more fruits as part of an overall healthy dietary pattern are more likely to have lower body weight and a reduced risk of several chronic diseases. Dietary fiber from fruits helps reduce blood cholesterol levels, lower risk of disease, and promote proper bowel function. The fiber in fruits helps with weight maintenance by providing a feeling of fullness (satiety) with fewer calories. The MyPlate fruit group includes whole fruits and 100% fruit juice. Whole fruits include fresh, canned, frozen, and dried forms. It is recommended that at least half of the recommended amount of fruit should come from whole fruit, rather than 100% juice. Fruit juices should be 100% juice and always pasteurized. Diluting 100% juice with water (without added sugars) is an even healthier choice. When selecting canned fruit, choose options that are canned with 100% juice and without added sugars.

About 80% of the U.S. population does not meet the fruit recommendations. Although fruit is generally consumed in nutrient-dense forms such as bananas, apples, oranges, or grapes, some fruit is consumed in energy-dense forms such as fruit pies or similar desserts.

Achieving a nutrient-rich and sustainable food system will require collaborative efforts from individuals, communities, industry, organizations, and government. Supply and demand are bidirectional—a shift in local and sustainable food production requires a shift in our dietary patterns. Consumers must realize that our food choices impact more than just ourselves. A primarily plant-based dietary pattern is beneficial for ourselves and our planet.
Source: UN 2019 Climate Change and Land Report.

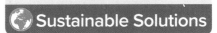

Sustainable Solutions

## TABLE 4-3 ■ Know Your Whole Grains

| Grain | Characteristics | Health Benefits |
|-------|-----------------|-----------------|
| Amaranth* | High-protein kernels that are tiny and have a peppery taste; gluten-free. | Contains a high level of complete protein, including lysine, an amino acid often missing in grains. |
| Barley | Highest in fiber; very slow cooking. Note: pearled barley is not technically a whole grain. | Barley may lower cholesterol even more effectively than oat fiber. Look for *whole barley*, *hulled barley*, or *hull-less barley*. |
| Buckwheat* | High levels of the antioxidant *rutin* and a high level of protein; gluten-free. | Rutin improves circulation and prevents LDL cholesterol from blocking blood vessels. |
| Bulgur* | Quick cooking time and mild flavor; often used in tabbouleh. | More fiber than quinoa, oats, millet, buckwheat, or corn. |
| Corn | Known for its sweet flavor; gluten-free. | Highest level of antioxidants of any grain or vegetable. Avoid labels that say *degerminated* and look for the words *whole corn*. |
| Einkhorn* | An ancient strain of wheat that has a strong hull. | May contain more protein, phosphorus, potassium, and beta-carotene than traditional wheat. |
| Farro | Also known as *emmer*, an ancient strain of wheat. Has a strong hull, chewy texture, and can be used to make pasta. | May be much higher in antioxidants than traditional wheat. Avoid pearled farro, and look for the words *whole farro* on labels. |
| Freekeh* | A hard wheat with a smoky flavor that is similar to bulgur. | Harvested while it is young, yielding a higher nutritional value. |
| Kamut | An heirloom grain with a rich, buttery taste. | Higher in protein and vitamin E than traditional wheat. Look for *whole kamut* in the ingredient list. |
| Kañiwa* | A cousin of quinoa that is high in protein; gluten-free. | May be higher in certain antioxidants than other whole grains. |
| Millet* | A versatile family of grains in white, gray, yellow, and red varieties; gluten-free. | Naturally high in protein and antioxidants and can help control blood sugar and cholesterol. |
| Oats | High in protein, oats are popular for breakfast. | Oat fiber is especially effective in lowering cholesterol. *Steel-cut oats* contain the entire oat kernel. |
| Quinoa* | Rich in high-quality protein, this is a small, light-colored round grain; gluten-free. | Quinoa contains complete protein, with all of the essential amino acids. |
| Rice | Many whole grain varieties, including brown, black, purple, or red. Brown rice is lower in fiber than other whole grains but rich in nutrients. White rice is refined; gluten-free. | One of the most easily digested grains, ideal for those on a restricted diet or who are gluten intolerant. Brown rice, and most other colored rice, is always whole. |
| Rye | Unusual among grains for the high level of fiber in its endosperm—not just in its bran. | Rye promotes a feeling of fullness, and it generally has a lower glycemic index than most other grains. Look for *whole rye* or *rye berries* in the ingredient list. |
| Sorghum* | Easily grown grain that can be eaten like popcorn, made into porridge, ground into flour, or brewed into beer; gluten-free. | Another gluten-free grain, popular among those with celiac disease. |
| Spelt | Variety of wheat that can be used as an alternative to traditional strains. | Spelt is higher in protein than common wheat. Look for *whole spelt* in the ingredient list. |
| Teff* | Versatile and easy to grow with a sweet, molasses-like flavor; gluten-free. | Twice the iron and three times the calcium of many other grains. |
| Triticale* | A hybrid of durum wheat and rye that is easily grown. | Shares many of the same health benefits as rye. |
| Wheat | Contains large amounts of gluten, a stretchy protein that enables bakers to create risen breads. | Look for *whole wheat* (in Canada, for the term *whole grain whole wheat*). |
| Wild Rice* | Not technically rice but a type of grass with a strong flavor; gluten-free. | Wild rice has twice the protein and fiber of brown rice. |

Source: Adapted from the Whole Grains Council, Whole grains A to Z, at http://wholegrainscouncil.org/whole-grains-101/whole-grains-a-to-z.
*When found on an ingredient list, it almost always indicates a whole grain

## FARM to FORK  Potatoes

DLeonis/Getty Images

Potatoes are one of the most productive and consumed crops around the world. This is good news because the average American eats more than 125 pounds of potatoes each year!

### Grow
- Potatoes can be easily grown in most climates with little maintenance.
- For the best results, plant one seed potato in the bottom third of a gallon container. As the plant grows, mound more soil/compost mix around the stem until you reach the top of the vessel. Stop watering for 2 weeks after the foliage dies.
- Harvest by hand, allow to dry for one day, then brush off soil, wash, and enjoy!

### Shop
- "New" potatoes (harvested early in the season) have thin skins and a waxy flesh. They cause a slower rise in blood glucose than "old" potatoes (harvested late in the season), which have thick skins and are typically used for baking.
- Of the most common varieties, russet potatoes contain the most phytochemicals. They are good sources of potassium, vitamin C, and several B vitamins. Sweet potatoes also contain these nutrients in addition to being rich in vitamin A and fiber.
- Try colorful novelty potatoes (red-, blue-, and black-skinned) with deep-colored flesh to get a more varied boost of phytochemicals.
- Consider buying organic potatoes to reduce pesticide residues.

### Store
- Store "new" potatoes in the refrigerator and eat within 1 week of purchase.
- Potatoes can be stored for months in a cool, dark, well-ventilated location.
- Green potatoes contain solanine, a poisonous compound that should be avoided.

### Prep
- Scrub potatoes before cooking, and eat with the skins to increase nutrient consumption by up to 50%.
- Most varieties of potatoes contain rapidly digested starch, which causes a sharp rise in blood glucose.
- For those concerned with blood glucose control, leftovers are better! Cook potatoes, refrigerate overnight, then reheat and eat them the next day. The cooler temperatures, postcooking, change the structure of the starch, which will delay a sharp rise in blood glucose.
- To add flavor and nutrients without extra fat, opt for toppings such as yogurt, vegetarian chili, salsa, herbs, and spices.
- Sadly, much of the U.S. potato consumption is in the form of French fries and potato chips. Try to limit consumption of these potato products and aim for less highly processed potatoes.

Source: Robinson J: Potatoes: From wild to fries. In *Eating on the Wild Side.* New York: Little, Brown and Company, 2013.

©Peter Madril

## DAIRY

Approximately 90% of the U.S. population does not meet the MyPlate dairy recommendations. Only 20% of adults consume milk on a daily basis.

The dairy food group contains fat-free and low-fat (1%) milk, yogurt, and cheese. For individuals who choose dairy alternatives, fortified soy beverages (commonly known as "soy milk") and soy yogurt—fortified with calcium, vitamin A, and vitamin D—are included as part of the dairy group as they are similar to milk and yogurt based upon their nutrient composition. Plant-based "milks" (e.g., almond, rice, coconut, oat, and hemp "milks") may contain calcium, but are not included as part of the dairy group because their overall nutritional content is not similar to dairy milk and fortified soy beverages. Therefore, they do not contribute to meeting the dairy group recommendations.

Dairy in the U.S. is often consumed in combination dishes with high amounts of sodium (e.g., cheeses on sandwiches, pizza, and pasta dishes) and saturated fat (e.g., high fat milks and yogurts) and added sugars such as flavored milk, ice cream, and sweetened yogurts. Foods made from milk that are primarily fat, such as cream cheese, cream, and butter, are not part of the dairy group. Some individuals experience the condition of lactose intolerance and maldigestion when they consume high-lactose foods (Table 4-4). Those suffering lactose intolerance can choose low-lactose and lactose-free dairy products. Typically, 1 cup of milk, yogurt, or soy milk (soy beverage); 1½ ounces of natural cheese; or 2 ounces of processed cheese are considered as "1 cup equivalent" from the dairy group.

### TABLE 4-4 ■ Lactose in Common Dairy Foods

| Food Product | Lactose (grams) |
|---|---|
| Cheese (1 oz, American) | 1.5 |
| Cheese (1 oz, Swiss) | 0 |
| Cottage cheese (½ cup) | 4 |
| Ice cream (½ cup) | 3 |
| Lactaid® milk (1 cup) | 0 |
| Milk (1 cup) | 11 |
| Sour cream (½ cup) | 4 |
| Yogurt (8 oz, Greek) | 7 |
| Yogurt (8 oz, 12 g protein) | 15 |

Source: USDA, ARS National Agriculture Library: Nutrient Data Laboratory. Lactose may vary by brand. Read labels carefully.

## NUTRITIVE SWEETENERS

The various substances that impart sweetness to foods fall into two broad classes: nutritive sweeteners, which can provide calories for the body, and alternative sweeteners, which, for the most part, provide no calories.[8] Alternative sweeteners are much sweeter on a per-gram basis than the nutritive sweeteners that provide calories. The taste and sweetness of sucrose (table sugar) make it the benchmark against which all other sweeteners are measured. Sucrose is obtained from sugarcane and sugar beet plants. Both sugars and sugar alcohols provide calories along with

## ASK THE RDN  Juicing

***Dear RDN:*** *Is juicing healthier than eating whole fruits or vegetables?*

Juicing has become a popular way to consume fruits and vegetables and can be a fun way to drink the fruits and vegetables you do not enjoy eating. It is also a smart way to get your vitamins and minerals if you have gas or cramping when you eat fiber-rich produce. The good news is that most of the vitamins, minerals, and phytochemicals remain in the juice that is extracted from the fruits and vegetables during the juicing process. Also, you control the amount of added sugars and preservatives in the end product. The downside of the juicing process is that the natural fiber found in whole fruits and vegetables is typically lost. Thus, the juice is not healthier than whole fruit. In fact, there is no scientific evidence to support claims that juice extracts provide any health benefits or that nutrients are better absorbed from juice. Remember that juices from both fruits and vegetables contain a significant amount of natural sugar and juicing concentrates the sugar and calories. If consumed in excess, those calories can quickly add up! For example, juicing three apples will result in 48 grams of sugar and 192 calories. To make your juice healthier, add some of the pulp back to obtain valuable fiber that can help you feel full. Even better, instead of a juicer, you can use an extractor (e.g., NutriBullet®), which pulverizes fruits, vegetables, and nuts into the consistency of juice. Finally, keep sugar, calories, and fiber content in mind when buying ready-made juices at trendy juice bars.

To your health,

***Anne M. Smith, PhD, RDN, LD***
Associate Professor Emeritus, The Ohio State University, Author of *Contemporary Nutrition*

Monty Soungpradith/
Open Image Studio LLC

There are many forms of sugar on the market. Together they contribute to our average daily intake of approximately 71 grams (17 teaspoons) of sugar. **How many calories are in 1 gram of sugar?**
C Squared Studios/Getty Images

sweetness. Sugars are found in many different food products, whereas sugar alcohols have rather limited uses.

**Sugars.** All of the monosaccharides (glucose, fructose, and galactose) and disaccharides (sucrose, lactose, and maltose) discussed earlier are designated *nutritive sweeteners* because they provide calories. The *Dietary Guidelines* recommend that we reduce the intake of calories from added sugars. Added sugars are defined as caloric sweeteners added to foods during processing or preparation or before consumption. It is estimated that Americans consume almost 270 calories (13%), of added sugars per day, far exceeding the recommendation of no more than 10% of total of calories per day.[9] As you might guess, almost 40% of added sugars come from sugar-sweetened beverages, including soft drinks, fruit drinks, sports and energy drinks, and sweetened coffee and tea. Desserts and sweet snacks are the next biggest source of added sugars.[9] Fortunately, the amount of added sugars must now be present on the Nutrition Facts label.

High-fructose corn syrup (HFCS) is a sweetener used in a wide variety of foods, from soft drinks to barbecue sauce. It is called *high-fructose* corn syrup because it contains up to 55% fructose, compared to sucrose, which contains only 50% fructose. HFCS is made by an enzymatic process that converts some of the glucose in cornstarch into fructose, which tastes sweeter than glucose. In the United States, corn is abundant and inexpensive compared to sugarcane or sugar beets, much of which is imported. Food manufacturers prefer HFCS because of its low cost and broad range of food-processing applications, and because it is easy to transport, has better shelf stability, and improves food properties. There has been much confusion and controversy surrounding the use and possible health effects of HFCS. After extensive review, the scientific community has concluded that there are no metabolic or endocrine differences between HFCS and sucrose related to obesity or any other adverse health outcome.[10]

Honey is a product of plant nectar that has been altered by bee enzymes. The enzymes break down much of the nectar's sucrose into fructose and glucose. Honey offers essentially the same nutritional value as other simple sugars—a source of energy and little else. However, honey is not safe for infants because it can contain spores of the bacterium *Clostridium*

*magnificent*

*microbiome*

One function of the gut microbiota is to consume carbohydrates. Microbiota-accessible carbohydrates (MACs) are indigestible complex carbohydrates found in fruits, vegetables, whole grains, and legumes that our microbiota feeds on. By consuming MACs, we can cultivate healthy microorganisms to improve overall health, reduce inflammation, support immunity, and even promote positive mental health.

▲ A variety of alternative sweeteners are available. Jill Braaten/McGraw-Hill

**sorbitol** Alcohol derivative of glucose that yields about 3 kcal/g but is slowly absorbed from the small intestine; used in some sugarless gums and dietetic foods.

**xylitol** Alcohol derivative of the five-carbon monosaccharide xylose. Absorbed more slowly than sucrose, xylitol supplies 40% fewer calories than table sugar.

**Acceptable Daily Intake (ADI)** Estimate of the amount of a sweetener that an individual can safely consume daily over a lifetime. ADIs are given as milligrams per kilogram of body weight per day.

**acesulfame-K** Alternative sweetener that yields no energy to the body; 200 times sweeter than sucrose.

*botulinum* that causes fatal foodborne illness. Unlike the acidic environment of an adult's stomach, which inhibits the growth of the bacteria, an infant's stomach does not produce much acid, making infants more susceptible to the threat that this bacterium poses.

Agave nectar comes from the same plant used to make tequila. Like high-fructose corn syrup, it's highly processed before it can be added to products. Agave contains about 60 calories per tablespoon compared to 40 calories per tablespoon for table sugar. Agave is sweeter than granular sugar, so you can use less to reduce calories.

In addition to sucrose and HFCS, brown sugar, turbinado sugar, honey, maple syrup, agave nectar, and other sugars are also added to foods. Brown sugar is essentially sucrose containing some molasses that is not completely removed from the sucrose during processing or is added to the sucrose crystals. Turbinado sugar is a partially refined version of raw sucrose that is often marketed as *raw sugar*. Maple syrup is made by boiling down and concentrating the sap from sugar maple trees. Because pure maple syrup is expensive, most pancake syrup is primarily corn syrup and HFCS with maple flavor added.

**Sugar Alcohols.** Food manufacturers and consumers have numerous options for obtaining sweetness while using less sugar and calories. Sugar alcohols are carbohydrates with a chemical structure that partially resembles both sugar and alcohol, but they don't contain ethanol. They are incompletely absorbed and metabolized by the body, and consequently contribute fewer calories than most sugars. This allows people with diabetes to enjoy the flavor of sweetness while controlling sugars; they also provide noncaloric or very-low-calorie sugar substitutes for persons trying to lose (or control) body weight.

Sugar alcohols, or polyols, such as **sorbitol** and **xylitol,** are used as nutritive sweeteners but contribute fewer calories (about 2.6 kcal per gram) than sugars. They are also absorbed and metabolized to glucose more slowly than are simple sugars. Because of this, they remain in the intestinal tract for a longer time and, in large quantities, can cause diarrhea. In fact, any product that may be consumed in amounts that result in a daily ingestion of 50 grams or more of sugar alcohols must bear the statement "Excess consumption may have a laxative effect" on the label.

Sugar alcohols must be listed on labels. If only one sugar alcohol is used in a product, its name must be listed; however, if two or more are used in one product, they are grouped together under the heading *sugar alcohols*. The caloric value of each sugar alcohol used in a food product is calculated so that when one reads the total amount of calories a product provides, it includes the sugar alcohols in the overall amount.

Sugar alcohols are used in sugarless gum, breath mints, and candy. Unlike sucrose, sugar alcohols are not readily metabolized by bacteria to acids in the mouth and thus do not promote tooth decay.

## ALTERNATIVE SWEETENERS

Alternative, artificial, high-intensity, or *nonnutritive* sweeteners yield few or no calories when consumed in amounts typically used in food products. They also are not metabolized by bacteria in the mouth, so they do not promote dental caries. A growing number of alternative sweeteners are currently available in the United States with an industry value of approximately $2.2 billion dollars.[8] Replacing added sugars with low- and no-calorie sweeteners may reduce overall calorie intake in the short-term and support weight management, yet questions still remain about their long-term weight effectiveness as a weight management strategy.[9]

For each sweetener, the FDA determines an **Acceptable Daily Intake (ADI)** guideline. ADIs are set at a level 100 times less than the level at which no harmful effects were noted in animal studies. Current evidence suggests that alternative sweeteners can be used safely by adults and children, and they are considered safe during pregnancy. Note that large, long-term studies have yet to be conducted in humans. A summary of alternative sweeteners can be found in Table 4-5 and additional information can be found at fda.gov/food/food-additives-petitions/high-intensity-sweeteners.

**Acesulfame-K.** **Acesulfame-K** is an organic acid linked to potassium (K). It is sold as Sunette® and can be used in baking. In the United States, it is currently approved for use as a general-purpose sweetener.

## TABLE 4-5 ■ Alternative Sweeteners

| Sweetener | Brand Names | Sweetness Intensity (Compared to Sucrose) | Acceptable Daily Intake (ADI) (mg/kg/day) | Approximate # Packets to Reach ADI* |
|---|---|---|---|---|
| Acesulfame Potassium (Ace-K) | Sweet One®, Sunett® | 200 × | 15 | 23 |
| Advantame | N/A | 20,000 × | 32.8 | 4,920 |
| Allulose | Dolcia Prima®, All-u-Lose® | 0.7 × | NS*** | Not determined |
| Aspartame | Nutrasweet®, Equal®, Sugar Twin® | 200 × | 50 | 75 |
| Neotame | Newtame® | 7,000–13,000 × | 0.3 | 23 (10,000 × sucrose) |
| Saccharin | Sweet and Low®, Sweet Twin®, Sweet'N Low®, Necta Sweet® | 200–700 × | 15 | 45 (400 × sucrose) |
| Luo Han Guo | Nectresse®, Monk Fruit in the Raw®, PureLo® | 100–250 × | NS*** | Not determined |
| Stevia | Truvia®, PureVia®, Enliten® | 200–400 × | 4** | 9 (300 × sucrose) |
| Sucralose | Splenda® | 600 × | 5 | 23 |

Source: Adapted from the Food and Drug Administration (FDA), Additional Information about High-Intensity Sweeteners Permitted for Use in Food in the United States, at https://www.fda.gov/food/ingredientspackaginglabeling/foodadditivesingredients/ucm397725.htm.
*Number of Tabletop Sweetener Packets a 60 kg (132 pound) person would need to consume to reach the ADI. Calculations assume 1 packet of high-intensity sweetener is as sweet as 2 teaspoons of sugar.
**ADI established by the Joint FAO/WHO Expert Committee on Food Additives (JECFA).
***NS means "not specified." A numerical ADI may not be deemed necessary for several reasons, including evidence of the ingredient's safety at levels well above the amounts needed to achieve the desired effect (e.g., as a sweetener) in food.

**Advantame.** A general-purpose sweetener and flavor enhancer, **advantame** is stable at higher temperatures and can be used as a tabletop sweetener as well as in cooking. Chemically, advantame is similar to aspartame but is much sweeter. Because only a small amount is needed to achieve the same level of sweetness (it is 20,000 times sweeter than sucrose), foods that contain advantame do not need to include alerts for people with PKU.

**Allulose.** Recently approved by the FDA, **allulose** (D-allulose or D-psicose) is a monosaccharide that is nearly identical to fructose in chemical structure and present in small quantities in natural products including wheat, figs, jackfruit, and raisins. Delivering about 0.2 kcal/g, or about 5% of the calories of sugar, this compound is absorbed in the small intestine but not metabolized, and excreted primarily in the urine. Its commercial uses include beverages, yogurt, ice cream, baked goods, and other typically high-calorie items. In terms of Nutrition Facts labeling, allulose is unique in that the FDA has exempted it from inclusion in the total and added sugars declaration, but it must be listed in the ingredient list, total carbohydrates, and total calories sections on a food label.

**Aspartame.** **Aspartame** is in widespread use throughout the world (typically packaged in blue packets, including Equal®). It has been approved for use in more than 90 countries, and its use has been endorsed by the World Health Organization, the American Medical Association, and the American Diabetes Association, among other reputable scientific groups.

The components of aspartame are the amino acids phenylalanine and aspartic acid, along with methanol. Recall that amino acids are the building blocks of proteins, so aspartame is more like a protein than a carbohydrate. Like protein, aspartame yields about 4 kcal per gram, but because it is about 200 times sweeter than sucrose, only a small amount is needed to obtain the desired sweetness. Like other proteins, however, aspartame is not heat stable, unless combined with other heat-stable sweeteners, and loses some sweetness if exposed to excessive heat. Aspartame is used in beverages, gelatin desserts, chewing gum, and toppings.

**advantame** Similar in structure to aspartame, this sweetener is 20,000 times sweeter than sucrose.

**allulose** Naturally occurring sugar in some food sources that is about 70% as sweet as sugar.

**aspartame** Alternative sweetener made of two amino acids and methanol; about 200 times sweeter than sucrose.

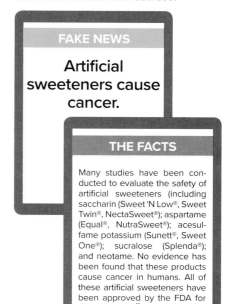

**FAKE NEWS**

Artificial sweeteners cause cancer.

**THE FACTS**

Many studies have been conducted to evaluate the safety of artificial sweeteners (including saccharin (Sweet'N Low®, Sweet Twin®, NectaSweet®); aspartame (Equal®, NutraSweet®); acesulfame potassium (Sunett®, Sweet One®); sucralose (Splenda®); and neotame. No evidence has been found that these products cause cancer in humans. All of these artificial sweeteners have been approved by the FDA for sale in the U.S.[11]

INGREDIENTS: SORBITOL, GUM BASE, MANNITOL, GLYCEROL, HYDROGENATED GLUCOSE SYRUP, XYLITOL, ARTIFICIAL AND NATURAL FLAVORS, ASPARTAME, RED 40, YELLOW 6 AND BHT (TO MAINTAIN FRESHNESS). PHENYLKETONURICS: CONTAINS PHENYLALANINE.

*Sugarless Gum*

Sugar alcohols and the alternative sweetener aspartame are used to sweeten this product. Note the warning for people with phenylketonuria (PKU) that this product is made with aspartame. **What is the significance behind this health warning?** Nancy R. Cohen/Getty Images

**phenylketonuria (PKU)** Disease caused by a genetic defect in the liver's ability to metabolize the amino acid phenylalanine; untreated, toxic by-products of phenylalanine build up in the body and lead to brain damage and severe health issues.

**neotame** General-purpose, nonnutritive sweetener that is approximately 7000 to 13,000 times sweeter than table sugar. It has a chemical structure similar to aspartame.

**saccharin** Alternative sweetener that yields no energy to the body; 200 to 700 times sweeter than sucrose.

**luo han guo** Extract of the monk fruit, this alternative sweetener is 100 to 250 times sweeter than sucrose.

**stevia** Alternative sweetener derived from South American shrub; 200 to 400 times sweeter than sucrose.

**sucralose** Alternative sweetener that has chlorines in place of 3 hydroxyl (—OH) groups on sucrose; 600 times sweeter than sucrose.

Persons with the rare disease called **phenylketonuria (PKU),** which interferes with the metabolism of phenylalanine, should avoid aspartame because of its high phenyl-alanine content. People with PKU will find a mandatory warning label on products containing aspartame.

**Neotame.** **Neotame** was approved by the FDA for use as a general-purpose sweetener but is used in few foods. Neotame is heat stable and can be used as a tabletop sweetener as well as in cooking. Neotame is safe for use by the general population, including children, pregnant and lactating women, and people with diabetes. Although similar to aspartame, neotame does not require labeling for people with PKU because it is not broken down in the body to individual amino acid components.

**Saccharin.** Discovered and used since 1879, **saccharin** represents about half of the alternative sweetener market in North America (typically packaged in pink packets, including Sweet 'N Low®).

**Luo Han Guo.** **Luo han guo** is an extract of the monk fruit. It was approved by the FDA and is sold as the sweeteners Nectresse™ and Monk Fruit in the Raw™.

**Stevia.** **Stevia,** sold as Truvia® and Sweet Leaf®, is an alternative sweetener derived from a South American plant and provides no energy. It has been used in teas and as a sweetener in Japan since the 1970s, and it is considered generally recognized as safe (GRAS) for use in foods.

**Sucralose.** **Sucralose** (Splenda®) is made by adding three chlorine molecules to sucrose. It cannot be broken down or absorbed, so it yields no calories. It can be used in cooking and baking because it does not break down under high heat conditions. Sucralose is approved as an additive to foods such as soft drinks, gum, baked goods, syrups, gelatins, frozen dairy desserts such as ice cream, jams, processed fruits, and fruit juices, and for tabletop use.

**To Sugar or Not to Sugar . . . That Is the Question!** There is much controversy surrounding which is a healthier option—consuming diet soft drinks or *natural* sugar-sweetened beverages. Like most of the field of nutrition, the answer is quite complex. In some studies, high consumption of artificial sweeteners has been linked with appetite alterations, obesity promotion, and even changes in gut microbes. On the other hand, we have definitive scientific evidence that large quantities of refined sugars (such as those found in soda and in sports and energy drinks) are related to dental caries, obesity and obesity-related chronic disease, and even alterations in brain function. Until we know more, reputable scientific organizations recommend that people replace sugar-sweetened and diet drinks with carbonated, plain, or unsweetened flavored water.[12]

### ✓ CONCEPT CHECK 4.3

1. Which food groups are the primary sources of carbohydrate in our dietary patterns?
2. What specific foods contain the highest percentage of calories from carbohydrates?
3. What are the most common nutritive sweeteners?
4. Which alternative sweeteners are approved for use in food?

## 4.4 Making Carbohydrates Available for Body Use

As discussed previously, simply consuming a food or beverage does not supply nutrients to body cells. Digestion and absorption must occur first.

## STARCH AND SUGAR DIGESTION

Food preparation can be viewed as the start of carbohydrate digestion because cooking softens tough connective structures in the fibrous parts of plants, such as broccoli stalks. When starches are heated, the starch granules swell as they soak up water, making them much easier to digest. All of these effects of cooking generally make carbohydrate-containing foods easier to chew, swallow, and break down during digestion.

The enzymatic digestion of starch begins in the mouth, when the saliva, which contains an enzyme called *salivary amylase*, mixes with carbohydrate-containing food products during the chewing of food. This amylase immediately breaks down starch into many smaller units, primarily disaccharides, such as maltose (Fig. 4-12). You can taste this conversion while chewing a saltine cracker. Prolonged chewing of the cracker causes it to taste sweeter as some starch breaks down into the sweeter disaccharides, such as maltose. Usually, food is in the mouth for such a short amount of time that this phase of digestion is negligible. In addition, once the food moves down the esophagus and reaches the stomach, the acidic environment inactivates salivary amylase.

When the carbohydrates reach the small intestine, the more alkaline environment of the intestine is better suited for further carbohydrate digestion. The pancreas releases

# Carbohydrate Digestion and Absorption

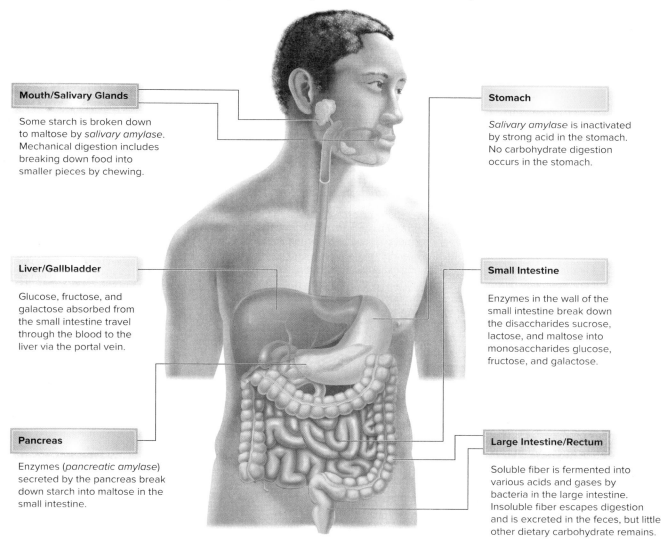

**Mouth/Salivary Glands**

Some starch is broken down to maltose by *salivary amylase*. Mechanical digestion includes breaking down food into smaller pieces by chewing.

**Stomach**

*Salivary amylase* is inactivated by strong acid in the stomach. No carbohydrate digestion occurs in the stomach.

**Liver/Gallbladder**

Glucose, fructose, and galactose absorbed from the small intestine travel through the blood to the liver via the portal vein.

**Small Intestine**

Enzymes in the wall of the small intestine break down the disaccharides sucrose, lactose, and maltose into monosaccharides glucose, fructose, and galactose.

**Pancreas**

Enzymes (*pancreatic amylase*) secreted by the pancreas break down starch into maltose in the small intestine.

**Large Intestine/Rectum**

Soluble fiber is fermented into various acids and gases by bacteria in the large intestine. Insoluble fiber escapes digestion and is excreted in the feces, but little other dietary carbohydrate remains.

**FIGURE 4-12** ▲ Carbohydrate digestion and absorption. Enzymes made by the mouth, pancreas, and small intestine participate in the process of digestion. Most carbohydrate digestion and absorption take place in the small intestine.

**maltase** An enzyme made by absorptive cells of the small intestine; this enzyme digests maltose to two glucose molecules.

**sucrase** An enzyme made by absorptive cells of the small intestine; this enzyme digests sucrose to glucose and fructose.

**lactase** An enzyme made by absorptive cells of the small intestine; this enzyme digests lactose to glucose and galactose.

**primary lactose maldigestion** Develops at about age 3 to 5 years when the production of the enzyme lactase decreases. When significant symptoms develop after lactose intake, it is then called *lactose intolerance.*

**secondary lactose maldigestion** Temporary condition in which lactase production is decreased in response to illness or surgery.

**congenital lactase deficiency** A rare birth defect resulting in the inability to produce lactase, such that a lactose-free diet is required from birth.

**lactose intolerance** A condition in which symptoms such as abdominal gas and bloating appear as a result of severe lactose maldigestion.

enzymes, such as *pancreatic amylase*, to aid the last stage of starch digestion. After amylase action, the original carbohydrates in a food are now present in the small intestine as the monosaccharides glucose and fructose, originally present as such in food, and disaccharides (maltose from starch breakdown, lactose mainly from dairy products, and sucrose from food and added at the table).

The disaccharides are digested to their single monosaccharide units once they reach the wall of the small intestine, where the specialized enzymes on the absorptive cells digest each disaccharide into monosaccharides. The enzyme **maltase** acts on maltose to produce two glucose molecules. **Sucrase** acts on sucrose to produce glucose and fructose. **Lactase** acts on lactose to produce glucose and galactose.

## LACTOSE MALDIGESTION AND LACTOSE INTOLERANCE

Deficient production of the enzyme lactase will impair the digestion of lactose. The most common form of this condition is **primary lactose maldigestion,** a normal pattern of physiology that often begins to develop around ages 3 to 5 years. This primary form of lactose maldigestion is estimated to be present in about 65% of the world's population, although not all of these individuals experience symptoms and most continue to produce some lactase.[13] **Secondary lactose maldigestion** is a temporary condition in which lactase production is decreased in response to another condition, such as intestinal diarrhea. Rarely, lactase production is absent from birth, a condition known as **congenital lactase deficiency.** Any of these types of lactose maldigestion can lead to symptoms of gas, abdominal bloating, cramps, and diarrhea when lactose is consumed. The bloating and gas are caused by bacterial fermentation of lactose in the large intestine. The diarrhea is caused by undigested lactose in the large intestine as it draws water from the circulatory system into the large intestine. When significant symptoms develop after lactose intake, it is then called **lactose intolerance.** It is important to note that lactose maldigestion and resultant lactose intolerance are not equivalent to a milk allergy.

Lactose intolerance is most prevalent in people of East Asian descent. It is also common in people of West African, Arab, Jewish, Greek, and Italian descent, and the occurrence

## CASE STUDY Problems with Milk Intake

Myeshia is a 19-year-old African-American female who recently read about the health benefits of calcium and decided to increase her intake of dairy products. To start, she drank a cup of 1% milk at lunch. Not long afterward, she experienced bloating, cramping, and increased gas production. She suspected that the culprit of this pain was the milk she consumed, especially because her parents and her sister complain of the same problem. She wanted to determine if other milk products were, in fact, the cause of her discomfort, so the next day she substituted a cup of yogurt for the glass of milk at lunch. Consuming the yogurt did not cause any pain.

Answer the following questions about Myeshia:

1. Why did Myeshia believe that she was sensitive to milk?
2. What component of milk is likely causing the problems that Myeshia experiences after drinking milk?
3. Why does this component cause intestinal discomfort in some individuals?
4. What is the name of this condition?
5. What groups of people are most likely to experience this condition?
6. Why did consuming yogurt not cause the same effects for Myeshia?
7. Are there any other products on the market that can replace regular milk or otherwise alleviate symptoms for individuals with this problem?
8. Can people with this condition ever drink regular milk?
9. What other foods should Myeshia include to supply the calcium, potassium, vitamin A, and vitamin D that are typically found in cow's milk?
10. Why do some individuals have trouble tolerating milk products during or immediately after an intestinal viral infection?

*Complete the Case Study. Responses to these questions can be provided by your instructor.*

▲ College students, like Myeshia, are wise to listen to their bodies. Thankfully, most college campuses have health clinics with dietitians who can help students create a dietary plan that works well for them. Chuckstock/Shutterstock

increases as people age. Many of these individuals can still consume moderate amounts of lactose with minimal or no gastrointestinal discomfort because of eventual lactose breakdown by bacteria in the large intestine. Studies have shown that nearly all individuals with decreased lactase production can tolerate ½ to 1 cup of milk with meals and that most individuals adapt to intestinal gas production resulting from the fermentation of lactose by bacteria in the large intestine. Thus, it is unnecessary for these people to totally restrict or avoid their intake of lactose-containing foods, such as milk and milk products, which are important for maintaining bone health. Obtaining enough calcium and vitamin D from food and beverages is much easier if milk and milk products are included in a dietary pattern.

Lactose maldigesters usually still produce some lactase. Combining lactose-containing foods with other foods is helpful because certain foods can have positive effects on rates of digestion. For example, fat in a meal slows digestion, leaving more time for lactase action. Hard cheese and yogurt are also more easily tolerated than milk. Much of the lactose is lost during the production of cheese, and the active bacteria cultures in yogurt digest the lactose with their lactase. In addition, products such as lactose-free or lactose-reduced milk (Lactaid® and Dairy Ease®) are made by treating regular milk with the lactase enzyme. Lactase supplements are also available to assist lactose maldigesters when they decide to consume products containing lactose. Many plant-based *milks*, including soy milk, almond milk, and rice milk, are naturally lactose free and can be used as an alternative to regular milk. Just be sure to read labels to see if these products are fortified with calcium and vitamin D.

Use of yogurt helps lactose maldigesters meet calcium needs. **What other foods might you recommend for lactose maldigesters?**
Anastasios71/Shutterstock

## CARBOHYDRATE ABSORPTION

Monosaccharides found naturally in foods and those formed as by-products of starch and disaccharide digestion in the mouth and small intestine generally follow an *active absorption* process. Recall that this is a process that requires a specific carrier and energy input for the substance to be taken up by the absorptive cells in the small intestine. Glucose and its close relative, galactose, undergo active absorption. They are pumped into the absorptive cells along with sodium.

Fructose is taken up by the absorptive cells via *facilitated diffusion*. In this case, a carrier is used, but no energy input is needed. This absorptive process is thus slower than that seen with glucose or galactose. So, large doses of fructose are not readily absorbed and can contribute to diarrhea as the monosaccharide remains in the small intestine and attracts water.

Once glucose, galactose, and fructose enter the absorptive cells, some fructose is metabolized into glucose. The single sugars in the absorptive cells are then transferred to the portal vein that goes directly to the liver. The liver then metabolizes those sugars by transforming the monosaccharides galactose and fructose into glucose and:

- releases it directly into the bloodstream for transport to organs such as the brain and kidneys and to muscles and adipose tissues;
- produces glycogen for storage of carbohydrate; and
- produces fat when carbohydrates are consumed in high amounts and overall calorie needs are exceeded.

Unless an individual has a condition that causes malabsorption or an intolerance to a carbohydrate such as lactose (or fructose), only a minor amount of some sugars (about 10%) escapes digestion. Any undigested carbohydrate travels to the large intestine. Some of that undigested carbohydrate (i.e., fermentable fiber) can be fermented by bacteria. The acids and gases produced by bacterial metabolism of the undigested carbohydrate are absorbed into the bloodstream. Scientists suspect that some of these products of bacterial metabolism promote the health of the large intestine by providing it with a source of calories.

## FIBER AND INTESTINAL HEALTH

Bacteria in the large intestine ferment soluble fibers into such products as acids and gases. The acids, once absorbed, also provide calories for the body. In this way, soluble fibers provide about 1.5 kcal per gram. Although the intestinal gas (flatulence) produced by this bacterial fermentation is not harmful, it can be painful and sometimes embarrassing. Over time, however, the body tends to adapt to a high-fiber dietary pattern, eventually producing less gas. Many gas-forming foods are good sources of soluble fiber.

Because insoluble fiber is an indigestible carbohydrate, it remains in the intestinal tract and supplies bulk to the feces, making elimination much easier. When enough fiber is consumed, the stool is large and soft because many types of plant fibers attract water. The larger size stimulates the intestinal muscles to contract, which aids elimination. Consequently, less pressure is necessary to expel the stool. When too little fiber is eaten, the opposite can occur: very little water is present in the feces, making it small and hard. Constipation may result, which forces one to exert excessive pressure in the large intestine during defecation. Hemorrhoids may also result from excessive straining during defecation.

Very high intake of fiber—for example, 60 grams per day—can also pose health risks and therefore should be followed only under the guidance of a primary care provider. Increased fluid intake is extremely important with a high-fiber dietary pattern. Inadequate fluid intake can leave the stool very hard and painful to eliminate. In more severe cases, the combination of excess fiber and insufficient fluid may contribute to blockages in the intestine, which may require surgery. Aside from problems with the passage of materials through the gastrointestinal tract, a high-fiber dietary pattern may also decrease the availability of nutrients. Certain components of fiber may bind to essential minerals, blocking them from being absorbed. For example, when fiber is consumed in large amounts, zinc, calcium, magnesium, and iron absorption may be hindered. Although there is no Upper Limit set for fiber, dangerous levels could occur with excessive fiber supplement use.

Many population studies have shown a link between increased fiber intake and a decrease in colon cancer development. Most research on dietary patterns and colon cancer focuses on the potential preventive effects of fruits, vegetables, whole grain breads and cereals, and beans. It is more advisable to increase fiber intake by using fiber-rich foods than by relying on fiber supplements. Overall, the health benefits to the colon that stem from a high-fiber dietary pattern are partially due to the nutrients that are commonly present in most high-fiber foods, such as vitamins, minerals, phytochemicals, and, in some cases, essential fatty acids.

A group of carbohydrates known as FODMAPs (fermentable oligo-, di-, and monosaccharides and polyols) may cause gastrointestinal symptoms such as gas, bloating, and diarrhea in some people. The FODMAPs include fructose, lactose, fructans (in wheat, onions, and garlic), galactans (in legumes), and polyols (sugar alcohols). Some individuals appear to be poor digesters of FODMAPs, thus these carbohydrates reach the large intestine without being digested, leading to bloating and gas. A low-FODMAP diet may provide relief for some people who experience gastrointestinal distress after meals, but this is not a cure-all. It is important to work with a gastroenterologist and RDN when limiting FODMAPs. Ideally, you should eliminate only those foods that trigger digestive problems; overly restrictive dietary patterns can lead to nutrient inadequacies.[14]

### ✓ CONCEPT CHECK 4.4

1. In what form are carbohydrates absorbed, and what happens to these compounds after absorption?
2. What are the names and locations of the enzymes that digest carbohydrates?
3. Why do some individuals feel discomfort after they consume large amounts of lactose? How can they avoid these symptoms?
4. What are the beneficial effects of fiber in the intestinal tract?

## 4.5 Putting Carbohydrates to Work in the Body

As discussed, all of the digestible carbohydrate that we eat is eventually converted into glucose. Glucose then goes on to function in body metabolism. The other sugars can generally be converted into glucose, and the starches are broken down to yield glucose, so the functions described here apply to most carbohydrates. The functions of glucose in the body start with supplying calories to fuel the body.

## PROVIDING ENERGY

The main function of glucose is to supply calories for use by the body. Certain tissues in the body, such as red blood cells, can use only glucose and other simple carbohydrate forms for fuel. Most parts of the brain and central nervous system also derive energy only from glucose, unless the dietary pattern contains almost no available glucose. In that case, the brain can use partial breakdown products of fat—called **ketone bodies**—for energy needs. Other body cells, including muscle cells, can use simple carbohydrates as fuel, but many of these cells can also use fat or protein for energy needs.

**ketone bodies** Partial breakdown products of fat that contain three or four carbons.

A dietary pattern that supplies enough digestible carbohydrates to prevent breakdown of proteins for energy needs is considered *protein sparing*. Under normal circumstances, digestible carbohydrates end up as blood glucose, and protein is reserved for functions such as building and maintaining muscles and vital organs. However, if you don't eat enough carbohydrates, your body is forced to make glucose from body proteins, draining the pool of amino acids available in cells for other critical functions. During long-term starvation, the continuous withdrawal of proteins from the muscles, heart, liver, kidneys, and other vital organs can result in weakness, poor function, and even failure of body systems.

The wasting of protein that occurs during long-term fasting can be life threatening. This has prompted companies that make formulas for rapid weight loss to include sufficient carbohydrates in the products to decrease protein breakdown and thereby protect vital tissues and organs, including the heart. Most of these very-low-calorie products are powders that can be mixed with different types of fluids and are consumed five or six times per day. When considering any weight-loss products, be sure that your total dietary pattern provides at least the RDA for carbohydrate.

In addition to the loss of protein, when you don't eat enough carbohydrates, the metabolism of fats is inefficient. In the absence of adequate carbohydrates, fats are not broken down completely in metabolism and instead form ketone bodies. This condition, known as **ketosis,** should be avoided because it disturbs the body's normal acid–base balance and leads to other health problems. This is a good reason to question the long-term safety of the low-carbohydrate diets that have been popular. Therapeutic ketogenic diets have proven effective for treating epilepsy for some individuals. Research is underway to evaluate keto diets in cancer therapies, sports nutrition, and weight loss.

**ketosis** The condition of having a high concentration of ketone bodies and related breakdown products in the bloodstream and tissues.

## REGULATING BLOOD GLUCOSE

Under normal circumstances, your blood glucose concentration is regulated within a narrow range. When carbohydrates are digested and taken up by the absorptive cells of the small intestine, the resulting monosaccharides are transported directly to the liver. One of the liver's roles, then, is to guard against excess glucose entering the bloodstream after a meal. The liver works together with the pancreas to regulate blood glucose.

When the concentration of glucose in the blood is high, such as during and immediately after a meal, the pancreas releases the hormone insulin into the bloodstream. Insulin delivers two different messages to various body cells to cause the level of glucose in the blood to fall. First, insulin directs muscle, adipose, and other cells to remove glucose from the bloodstream by taking it into those cells. Second, insulin directs the liver to store glucose as glycogen. By triggering both glycogen synthesis in the liver and glucose movement out of the bloodstream into certain cells, insulin keeps the concentration of glucose from rising too high in the blood (Fig. 4-13).

On the other hand, when you have not eaten for a few hours and blood glucose begins to fall, the pancreas releases the hormone **glucagon.** This hormone has the

FIGURE 4-13 ▶ Regulation of blood glucose. Insulin and glucagon are key factors in controlling blood glucose. When blood glucose rises above the normal range of 70 to 100 milligrams per deciliter (mg/dl) and blood glucose becomes elevated: (1) insulin is released from the pancreas (2) to lower it (3) and (4) blood glucose then falls back into the normal range (5). Inversely, when blood glucose falls below the normal range (6), glucagon is released (7), which has the opposite effect of insulin (8) and (9) and this then restores blood glucose to the normal range (10). Other hormones, such as epinephrine, norepinephrine, cortisol, and growth hormone, also affect blood glucose levels.

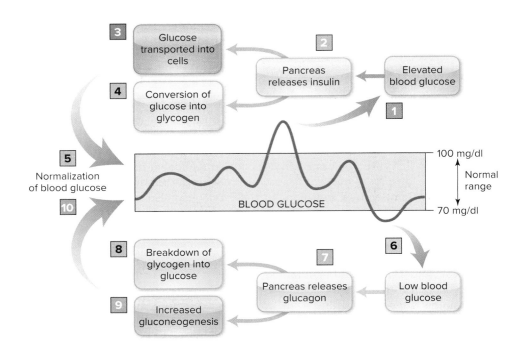

**hyperglycemia** High blood glucose, typically defined as above 125 mg/dl while in a fasted state.

**hypoglycemia** A condition caused by low levels of blood sugar that is often related to the treatment of diabetes and often defined by 70 mg/dl or less.

opposite effect of insulin. It prompts the breakdown of liver glycogen into glucose and the generation of glucose from noncarbohydrate substances, which is then released into the bloodstream to keep blood glucose from falling too low.

A different mechanism increases blood glucose during times of stress. Epinephrine (adrenaline) is the hormone responsible for the *flight or fight* response. Epinephrine is released in large amounts from the adrenal glands (located on top of each kidney) in response to a perceived threat, such as a car approaching head-on. These hormones cause glycogen in the liver to be quickly broken down into glucose. The resulting rapid flood of glucose from the liver into the bloodstream helps fuel quick mental and physical reactions.

This complex regulatory system is responsible for maintaining blood glucose within an acceptable range. It provides a safeguard against extremely high blood glucose **(hyperglycemia)** or low blood glucose **(hypoglycemia).** In essence, the actions of insulin on blood glucose are balanced by the actions of glucagon, epinephrine, and other hormones. If hormonal balance is not maintained, major changes in blood glucose concentrations occur. The disease type 1 diabetes is an example of the underproduction of insulin.

**Lifestyle Management and Blood Glucose Control.** Continued evidence supports lifestyle management techniques to improve blood glucose control. Directly aligning with the *Dietary Guidelines*, individuals are encouraged to consume a healthy dietary pattern rich in vegetables, fruits, whole grains, seafood, eggs, beans, peas, and lentils, unsalted nuts and seeds, fat-free and low-fat dairy products, and lean meats and poultry. Fat quality (selecting monounsaturated and polyunsaturated fats over saturated fats) has been found to be more important than fat quantity. Recommendations also encourage limiting or avoiding intake of sugar-sweetened beverages, reducing sodium to less than 2300 mg per day, and eating fatty fish at least two times (2 servings) per week. Note that dietary patterns that align to the majority of these guidelines include the Mediterranean, vegan, vegetarian, low-fat, low-carb, and DASH diets. To be safe, individuals with inadequate blood glucose control or those diagnosed with prediabetes or diabetes should work closely with an endocrinologist and registered dietitian to obtain individualized lifestyle recommendations. If insulin is prescribed, it may be necessary to count carbohydrates, protein, and fat until insulin dosing and blood glucose levels are controlled.

**Glycemic Response and Blood Glucose.** Our bodies react uniquely to different sources of carbohydrates. For example, a serving of a high-fiber food, such as black beans, results in lower blood glucose levels compared to the same size serving of mashed white potatoes. The effects of various foods on blood glucose are important to know because foods that result in a higher blood glucose level cause a larger release of insulin from the pancreas. When this higher insulin output occurs frequently, it leads to deleterious effects on the body. Some of these undesirable effects are high blood triglycerides, increased fat deposition in the adipose tissue, increased tendency for blood to clot, and increased fat synthesis in the liver. In addition, as insulin rapidly pulls glucose from the blood and into cells, hunger-regulating hormones are released, thus promoting increased refined carbohydrate intake. Over time, this increase in insulin output may also cause the muscles to become resistant to the action of insulin and eventually lead to type 2 diabetes in some people.

The **glycemic index (GI)** is a measurement of how the carbohydrate in a food raises blood glucose and has been used as a measure for planning diets for diabetics. Glycemic index is a ratio of the blood glucose response to a given food compared to the response to a reference such as glucose or white bread. Foods are ranked based on this comparison with a high GI food raising blood glucose more than a medium or low GI food.

The GI of a food is influenced by starch structure, fiber content, and food processing. In an effort to explain why glycemic responses to the same foods differ so greatly between people, scientists are now studying the interactions between individual physiological and genetic characteristics as well as the gut microbiome in regulation of individual glycemic responses.[15] Also keep in mind that the GI value only describes the type, not the amount, of carbohydrate in a food. The **glycemic load (GL)** measures both the quality and quantity of carbohydrates in meals. Knowing how different foods affect blood glucose can be helpful in meal planning. Portion sizes are important to manage in order to control blood glucose and maintain weight. Maintaining a healthy body weight and performing regular physical activity further reduces the effects of a high GI dietary pattern. The International Glycemic Index (GI) Database was developed to provide a public-use database of the glycemic index and glycemic load of foods.[16]

**glycemic index (GI)** The blood glucose response of a given food, compared to a standard (typically, glucose or white bread). Glycemic index is influenced by starch structure, fiber content, food processing, physical structure, and macronutrients.

**glycemic load (GL)** A measure of both the quality (GI value) and quantity (grams per serving) of a carbohydrate in a meal.

## FIBER: REDUCING CHOLESTEROL ABSORPTION AND OBESITY RISK

Aside from its role in maintaining bowel regularity, the consumption of fiber has many additional health benefits. A high intake of soluble fiber also inhibits absorption of cholesterol and cholesterol-rich bile acids from the small intestine, thereby reducing blood cholesterol and possibly reducing the risk of cardiovascular disease and gallstones. Recall good sources of soluble fiber include apples, bananas, oranges, carrots, barley, oats, and kidney beans. The beneficial bacteria in the large intestine degrade soluble fiber and produce certain fatty acids that probably also reduce cholesterol synthesis in the liver. In addition, the slower glucose absorption that occurs with dietary patterns high in soluble fiber is linked to a decrease in insulin release. One of the effects of insulin is to stimulate cholesterol synthesis in the liver, so this reduction in insulin may contribute to the ability of soluble fiber to lower blood cholesterol. Overall, a fiber-rich dietary pattern containing fruits, vegetables, beans, and whole grains is advocated as part of a strategy to reduce risk of cardiovascular disease (i.e., coronary heart disease and stroke). Again, this is something that a low-carbohydrate dietary pattern cannot promise.

A dietary pattern high in fiber helps control weight and reduces the risk of developing obesity.[17] Due to their bulky nature, high-fiber foods require more time to chew and move out of the stomach, and thus they fill us up without yielding many calories. Increasing intake of foods rich in fiber is one strategy for feeling satisfied or full after a meal.

Oatmeal is a rich source of soluble fiber. The FDA allows a health claim for the benefits of oatmeal to lower blood cholesterol because of the effects of this soluble fiber. **How does soluble fiber help to reduce blood cholesterol?** Alexis Joseph/ McGraw-Hill Education

1. What is the primary role of carbohydrates in the body?
2. How does the body respond when too little carbohydrate is consumed?
3. What are the mechanisms by which blood glucose levels are maintained within a narrow range?
4. What are some of the important functions of fiber?

# 4.6 Carbohydrate Needs

Whole grains contain all parts of the kernel. **What are the three components of the kernel?** Nancy R. Cohen/ Getty Images

The RDA for carbohydrates is 130 grams per day for adults. This is based on the amount needed to supply adequate glucose for the brain and nervous system, without having to rely on ketone bodies from incomplete fat breakdown as a calorie source. Somewhat exceeding this amount is fine; the *Dietary Guidelines* recommends that carbohydrate intake should range from 45% to 65% of total calorie intake. The Nutrition Facts label on foods uses 60% of calorie intake as the standard for recommended carbohydrate intake. This would be 300 grams of carbohydrate when consuming a dietary pattern of 2000 kcal.

Recommendations for carbohydrate consumption, however, emphasize the type of carbohydrates we should consume rather than just the total amount. Experts agree that one's carbohydrate intake should be based primarily on vegetables (dark green; red and orange; beans, peas, and lentils); fruits (especially whole fruit), and grains (at least half of which are whole grain), rather than on refined grains, starchy foods, and added sugars.

The *Dietary Guidelines* recommend that we choose fiber-rich fruits, vegetables, and whole grains often (Fig. 4-14). More specifically, 3 or more ounce equivalents of grains—roughly one-half of one's grains—should be whole. Whole grain is defined as the entire grain seed or kernel made of three components: the bran, germ, and endosperm, which must be in nearly the same relative proportions as the original grain if cracked, crushed, or flaked.[5]

## HOW MUCH FIBER DO WE NEED?

An Adequate Intake for fiber has been set based on the ability of fiber to reduce risk of disease. The Adequate Intake for fiber for adults is 25 grams per day for women and 38 grams per day for men. The goal is to provide at least 14 grams per 1000 kcal in a dietary pattern. After age 50, the Adequate Intake falls to 21 grams per day and 30 grams per day, respectively. The Daily Value used for fiber on food and supplement labels is 28 grams for a 2000-kcal dietary pattern. In North America, fiber intake remains well below the recommended values. More than 90% of women and 97% of men do not meet recommended intakes for dietary fiber.[9] This low intake is attributed to the lack of knowledge on the benefits of whole grains and the inability to recognize whole grain products at the time of purchase. Thus, most of us could benefit from increasing our fiber intake. At least 3 ounce equivalents of whole grains per day are recommended. Eating a high-fiber cereal (at least 3 grams of fiber per serving) for breakfast is one easy way to increase fiber intake (Fig. 4-15).[1,2] Other strategies include increasing dietary intakes of fruits, vegetables, and replacing refined grains with whole grains.

The *Rate Your Plate* activity in Connect shows a dietary pattern containing 25 or 38 grams of fiber within moderate calorie intakes. Dietary patterns to meet the fiber recommendations are possible and enjoyable if you incorporate plenty of whole wheat bread, fruits, vegetables, and beans. Use this activity to estimate the fiber content of your dietary pattern and determine *your* fiber score.

Remember that excessively high intakes of fiber can be unhealthy and that fluid intake must be increased with a high-fiber dietary pattern. A high-fiber dietary pattern may also decrease the absorption of essential minerals, especially zinc and iron.

**FIGURE 4-14** ◄ Carbohydrate-specific recommendations from the *Dietary Guidelines for Americans*.
Source: DietaryGuidelines.gov

Limit intake of calories from added sugars to less than 10% per day.

**Recommendations in the *Dietary Guidelines for Americans* regarding carbohydrate intake as part of a healthy eating pattern while staying within calorie needs**

Include grains, at least half of which are whole grain.

Include a variety of vegetables from all of the subgroups—dark green; red and orange; beans, peas, and lentils; starchy; and other.

Include fruits, especially whole fruit.

In the final analysis, keep in mind that any nutrient can lead to health problems when consumed in excess. High carbohydrate, high fiber, and low fat do not mean zero calories. Carbohydrates help moderate calorie intake in comparison with fats, but high-carbohydrate foods also contribute to total calorie intake.

## HOW MUCH SUGAR IS TOO MUCH?

The main problems with consuming an excess amount of sugar are that it provides discretionary calories and increases the risk for dental decay, weight gain, and other health issues.

**Dietary Quality Declines When Sugar Intake Is Excessive.** Overconsumption of sweet treats can leave little room for important, nutrient-dense foods, such as fruits and vegetables. Children and teenagers are at the highest risk for consuming too many discretionary calories in place of nutrients essential for growth. Many children and teenagers are drinking an excess of sugar-sweetened soft drinks and other sugar-containing beverages, including energy and sports drinks, and much less milk than ever before. Replacing sugar-laden drinks for milk can compromise bone health because milk contains calcium and vitamin D, which are both essential for bone health.

Remember that *added sugars* are sugars added to foods during processing and preparation. The *Dietary Guidelines* continue to strongly recommend that added sugars provide no more than 10% of total daily calorie intake. Dietary patterns

▲ When buying a bread labeled as *wheat bread*, most people think they are buying a whole wheat product. Because the flour is from the wheat plant, manufacturers can correctly list enriched white (refined) flour as wheat flour on food labels; however, if *whole wheat flour* is not listed first on the ingredient list, then the product is not primarily a whole wheat bread. Look for *100% whole grain* or *whole wheat* flour on the label for breads that are an excellent source of fiber. ninikas/iStock-photo/Getty Images

**FIGURE 4-15** ▶ Reading the Nutrition Facts on food labels helps us choose more nutritious foods. **Based on the information from these nutrition labels, which cereal is the better choice for breakfast?** Consider the amount of fiber in each cereal. Did the ingredients lists give you any clues? (Note: Ingredients are always listed in descending order by weight on a label.) When choosing a breakfast cereal, it is generally wise to focus on those that are rich sources of fiber. Sugar content can also be used for evaluation; strive for no more than 8 grams of added sugars per serving.

## Nutrition Facts

10 servings per container

| Serving size | | 1 cup (55g) |
|---|---|---|

| Amount per serving | Cereal | Cereal with ½ Cup Skim Milk |
|---|---|---|
| **Calories** | **170** | **210** |

| | % Daily Value** | |
|---|---|---|
| **Total Fat** 1.0g* | **2**% | **2**% |
| Saturated Fat 0g | **0**% | **0**% |
| *Trans* Fat 0g | | * |
| **Cholesterol** 0mg | **0**% | **0**% |
| **Sodium** 300mg | **13**% | **15**% |
| **Potassium** 340mg | **10**% | **16**% |
| **Total Carbohydrate** 43g | **14**% | **16**% |
| Dietary Fiber 7g | **28**% | **28**% |
| Total Sugars 16g | | |
| Includes 10g Added Sugars | **20**% | **20**% |
| **Protein** 4g | | |
| Vitamin D 1mcg, 2.5mcg | 10% | 25% |
| Calcium 20mg, 120mg | 2% | 12% |
| Iron 12mg, 12mg | 65% | 65% |
| Potassium 225mg, 402mg | 6% | 11% |
| Vitamin A 225mcg, 300mcg | 15% | 20% |
| Vitamin C 12mg, 13mg | 20% | 22% |
| Thiamin 0.3mg, 0.5mg | 25% | 30% |
| Riboflavin 0.4mg, 0.6mg | 25% | 35% |
| Niacin 5mg, 5mg | 25% | 25% |
| Vitamin B$_6$ 0.5mg, 0.5mg | 25% | 25% |
| Folic acid 120mcg, 120mcg | 30% | 30% |
| Vitamin B$_{12}$ 1.5mcg, 2.1mcg | 25% | 35% |
| Phosphorus 200mg, 300mg | 20% | 30% |
| Magnesium 80mg, 100mg | 20% | 25% |
| Zinc 3.7mg, 3.7mg | 25% | 25% |
| Copper 0.2mg, 0.2mg | 10% | 10% |

\* Amount in cereal. One half cup skim milk contributes an additional 40 calories, 65mg sodium, 6g total carbohydrate (6g sugars), and 4g protein.
\*\* The % Daily Value (DV) tells you how much a nutrient in a serving of food contributes to a daily diet. 2,000 calories a day is used for general nutrition advice.

## Nutrition Facts

17 servings per container

| Serving size | | ¾ cup (30g) |
|---|---|---|

| Amount per serving | Cereal | Cereal with ½ Cup Skim Milk |
|---|---|---|
| **Calories** | **170** | **210** |

| | % Daily Value** | |
|---|---|---|
| **Total Fat** 0g* | **0**% | **1**% |
| Saturated Fat 0g | **0**% | **1**% |
| *Trans* Fat 0g | | * |
| **Cholesterol** 0mg | **0**% | **1**% |
| **Sodium** 60mg | **2**% | **4**% |
| **Potassium** 80mg | **2**% | **8**% |
| **Total Carbohydrate** 35g | **9**% | **11**% |
| Dietary Fiber 1g | **4**% | **4**% |
| Total Sugars 20g | | |
| Includes 15g Added Sugars | **30**% | **30**% |
| **Protein** 3g | | |
| Vitamin D 1mcg, 2mcg | 10% | 20% |
| Calcium 0mg, 150mg | 0% | 15% |
| Iron 1.8mg, 1.8mg | 10% | 10% |
| Potassium 95mg, 272mg | 3% | 8% |
| Vitamin A 375mcg, 450mcg | 25% | 30% |
| Vitamin C 0mg, 1.2mg | 0% | 2% |
| Thiamin 0.3mg, 0.3mg | 25% | 25% |
| Riboflavin 0.4mg, 0.6mg | 25% | 35% |
| Niacin 5mg, 5mg | 25% | 25% |
| Vitamin B$_6$ 0.5mg, 0.5mg | 25% | 25% |
| Folic acid 100mcg, 100mcg | 25% | 25% |
| Vitamin B$_{12}$ 1.5mcg, 1.8mcg | 25% | 30% |
| Phosphorus 40mg, 150mg | 4% | 15% |
| Magnesium 16mg, 32mg | 4% | 8% |
| Zinc 1.5mg, 1.5mg | 10% | 10% |
| Copper 0.04mg, 0.04mg | 2% | 2% |

\* Amount in cereal. One-half cup skim milk contributes an additional 40 calories, 65mg sodium, 6g total carbohydrate (6g sugars), and 4g protein.
\*\* The % Daily Value (DV) tells you how much a nutrient in a serving of food contributes to a daily diet. 2,000 calories a day is used for general nutrition advice.

that go beyond this upper limit are likely to be deficient in vitamins and minerals. A moderate intake of about 10% of calorie intake corresponds to a maximum of approximately 50 grams (or 12.5 teaspoons) of sugars per day, based on a 2000 kcal dietary pattern. Because of the association between excessive consumption of sugars and several metabolic abnormalities and adverse health conditions, the American Heart Association recommends reductions in the intake of added sugars such that the upper limit of intake for most American women is no more than 100 kcal (25 grams) per day from added sugars and no more than 150 kcal (36 grams) per day for most American men.[18]

Most of the sugars we eat come from foods and beverages to which sugar has been added during processing and/or manufacturing. National consumption surveys indicate that the average American (age 1 and older) consumes approximately 266 kcal from added sugars per day, amounting to about 13% of calorie intake. Major sources of added sugars include sugar-sweetened beverages, desserts and sweet snacks, and sweetened coffee and tea (Fig. 4-16).

Supersizing sugar-rich beverages has also become common and has led to more sugar consumption; for example, in the 1950s, a typical serving size of a cola soft

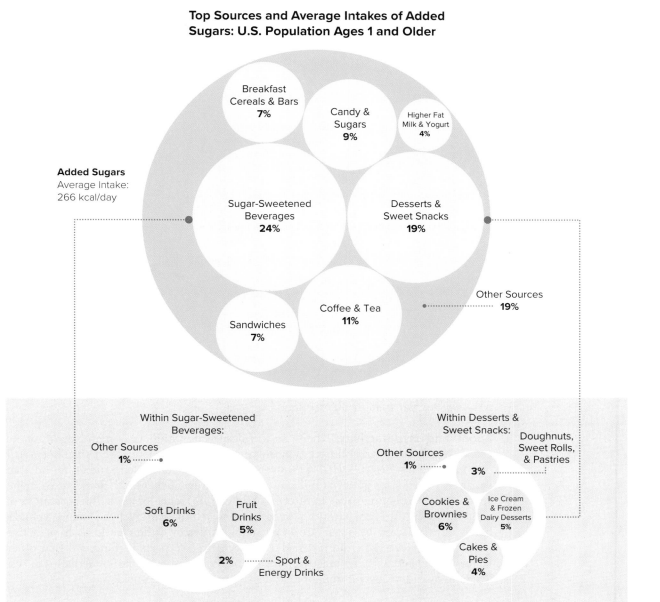

**FIGURE 4-16** ▲ Sources of added sugars in the dietary patterns of the U.S. population.

Data Source: Analysis of What We Eat in America, NHANES, 2013-2016, ages 1 and older, 2 days dietary intake data, weighted. As it appears in U. S. Department of Agriculture, Dietary Guidelines for Americans, 2020-2025. https://www.dietaryguidelines.gov/sites/default/files/2020-12/Dietary_Guidelines_for_Americans_2020-2025.pdf

**TABLE 4-6** ■ **Suggestions for Reducing Added Sugars**

**At the Grocery Store**

- Read the ingredients list on food labels. Look for all forms of added sugars. Added sugars are often hidden in tomato sauce, crackers, condiments, and salad dressings. Added sugars also hides under names such as high fructose corn syrup, invert sugar, sucrose, dried cane syrup, brown rice syrup, honey, molasses, and maple syrup. These can be listed separately on ingredients lists and add up!
- Buy unsweetened versions of foods typically high in sugar such as cereals, applesauce, yogurt, and canned fruit (in heavy syrup). Look for foods labeled *no added sugar* or *unsweetened*.
- Buy nuts and unsweetened dried fruits to replace candy for snacks.

**In the Kitchen**

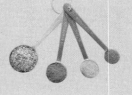

- Reduce the sugar in foods prepared at home. Try low-sugar recipes or adjust the sugar on your own. The amount of sugar can be decreased in recipes for foods such as pancakes, waffles, cookies, and cakes that will still taste great. Start by reducing the sugar gradually until you've decreased it by one-third or more.
- Add sweetness to foods with flavors and spices such as vanilla, citrus zest, cinnamon, cardamom, coriander, nutmeg, ginger, and mace.

**At the Table**

- Choose fewer foods high in sugar, such as prepared baked goods and sweet desserts. Reach for fresh fruit instead of cookies or candy for dessert and between-meal snacks.
- Include more protein, such as eggs and turkey, and healthy fats, such as nuts, seeds, and olive oil, to decrease the desire for sugar.
- Add less sugar to foods such as coffee, tea, cereal, and fruit. Reduce the use of white and brown sugars, honey, molasses, syrups, jams, and jellies. Cut back gradually to a quarter or half the amount. Use artificial sweeteners sparingly, as these may actually increase your desire for sweets.
- Substitute water for sugared soft drinks, sweet tea, coffee drinks, energy drinks, punches, and fruit juices.

drink was a 6.5-ounce bottle, and now a 20-ounce plastic bottle is a typical serving. This one change in serving size contributes an extra 164 kcal to the dietary pattern—all from added sugars. Most convenience stores now offer cups that will hold 64 ounces of soft drink. Health messages about the sugary beverages appear to be having some positive effects, resulting in a decrease in soft drink sales over the past decade. Market research indicates that youth are choosing more water, energy drinks, and coffee in place of soft drinks. Refer to Table 4-6 for suggestions on how to reduce added sugars in your diet, and read more about trends in added sugar consumption in *Newsworthy Nutrition* in this section.[19]

**Sugar and Hyperactivity.** There is a widespread notion that high sugar intake causes hyperactivity in children, typically part of the syndrome called *attention deficit hyperactivity disorder (ADHD)*. However, several well-controlled studies found that sugar in the dietary patterns did not affect children's behavior.[20] Some researchers suggest that expecting sugar to affect a child can influence parents' interpretation of what they see. A study of parents' perceptions showed that parents who believe a child's behavior is affected by sugar are more likely to perceive their child as hyperactive when they believe the child just had a sugary drink. Experts recommend that other factors associated with hyperactivity, including temperament, emotional disturbances, learning disorders, overstimulation, and sleep problems, be considered.[20]

**Sugar and Oral Health.** Sugars in the dietary pattern (and starches readily fermented in the mouth, such as crackers and white bread) also increase the risk of developing **dental caries.** Recall that caries, also known as cavities, are formed when sugars and other carbohydrates are metabolized, leading to the production of acids by bacteria that live in the mouth. These acids dissolve the tooth enamel and underlying structure.

**dental caries** Erosions in the surface of a tooth caused by acids made by bacteria as they metabolize sugars.

# Newsworthy Nutrition

### Americans are decreasing consumption of sugar-sweetened beverages

**INTRODUCTION:** Sugar-sweetened beverages (SSBs) are a significant source of added sugars and calories in U.S. dietary patterns for adults and children. **OBJECTIVE:** To provide national estimates for beverage consumption among children and adults in the United States. **METHODS:** This *cross-sectional* study utilized dietary data from 18,600 children (2 to 19 years) and from 27,652 adults (≥ 20 years) in the National Health and Nutrition Examination Survey (NHANES). Total beverage and SSB consumption were measured by 24-hour recall. **RESULTS:** From 2003 to 2014, per capita consumption of all beverages signficantly declined among children and adults. In the 2013–2014 survey wave, more than 60% of children and 50% of adults drank SSBs on any given day, which is significantly lower than the 2003–2004 cohort. The percentage of SSB drinkers and per capita consumption of SSBs were highest among African American, Mexican American, and non-Mexican Hispanic children, adolescents, and young adults for all study years. **CONCLUSION:** Beverage and SSB consumption declined for both children and adults from 2003 to 2014. The highest consumption levels were measured among African American, Mexican American, and non-Mexican Hispanic participants.

Source: Bleich SN and others: Trends in beverage consumption among children and adults, 2003–2014. *Obesity* 26:2, 2018.

Bacteria also use the sugars to make plaque, a sticky substance that both adheres acid-producing bacteria to teeth and diminishes the acid-neutralizing effect of saliva.

The worst offenders in terms of promoting dental caries are sticky and gummy foods high in sugars, such as caramel and gummies, because they stick to the teeth and supply the bacteria with a long-lived carbohydrate source. For this reason, snacking regularly on sugary foods is likely to cause caries because it gives the bacteria on the teeth a steady source of carbohydrates from which to continually make acid. Frequent consumption of liquid sugar sources (e.g., fruit juices, soda, sports beverages, energy drinks, and even many smoothies) can also cause dental caries. Chewing sugar-laden gum between meals is a prime example of a poor dental habit. Still, sugar-containing foods are not the only foods that promote acid production by bacteria in the mouth. As mentioned, if starch-containing foods (e.g., crackers and bread) are stuck to the teeth for a long time, the starch will be broken down to sugars by enzymes in the mouth; bacteria can then produce acid from these sugars. Overall, the sugar and starch contents of a food and its ability to remain in the mouth largely determine its potential to cause caries.

Fluoridated water and toothpaste are major factors in the prevention of dental caries in North American children due to fluoride's tooth-strengthening effect. Research has also indicated that certain foods—such as cheese, peanuts, and sugar-free chewing gum—can help reduce the amount of acid on teeth. In addition, rinsing the mouth after meals, drinking plenty of water, and eating healthy snacks reduce the acidity in the mouth. Certainly, good nutrition habits that do not present an overwhelming challenge to oral health (e.g., drinking plenty of water, chewing sugar-free gum), and routine visits to the dentist all contribute to improved dental health.

John and Mike are identical twins who like the same games, sports, and foods. However, John likes to chew sugar-free gum and Mike doesn't. At their last dental visit, John had no cavities but Mike had two. Mike wants to know why John, who chews sugar-free gum after eating, doesn't have cavities and he does. **How would you explain this to him?** Image Source Trading Ltd/ Shutterstock

## ✓ CONCEPT CHECK 4.6

1. What is the recommended intake of total carbohydrate per day?
2. How much fiber is recommended for each day?
3. How can we reduce our consumption of added sugars?
4. What is the link between sugar and oral health?

## Diabetes—When Blood Glucose Regulation Fails

BananaStock/Getty Images

Improper regulation of blood glucose results in either *hyperglycemia* (high blood glucose) or *hypoglycemia* (low blood glucose). High blood glucose is most commonly associated with diabetes (technically, *diabetes mellitus*), a disease that affects over 30 million adults, or 9% of the North American population 20 years or older.[21] It is estimated that over 28%, or 8 million, of these individuals do not know that they have the disease. Diabetes remains the seventh leading cause of death in the U.S. Diabetes is currently increasing in epidemic proportions in the U.S., with approximately 1.5 million new cases diagnosed in people age 18 or older annually. The American Diabetes Association (ADA) recommends testing fasting blood glucose in adults over age 45 every 3 years to screen for diabetes. Diabetes is often diagnosed using a **hemoglobin** A1c (HbA1c) test to diagnose diabetes with a threshold greater than 6.5%. The HbA1c is a more sensitive, long-term indicator of poor blood glucose control than the fasting blood glucose level. When blood glucose is too high, the glucose builds up in the blood and combines with hemoglobin (protein in red blood cells), making it glycated. The amount of glycated hemoglobin, or HbA1c, reflects the last several weeks or months of blood glucose levels.[22]

### Diabetes

There are two major forms of diabetes: **type 1** (formerly called insulin-dependent or juvenile-onset diabetes) and **type 2** (formerly called noninsulin-dependent or adult-onset diabetes) (Table 4-7). The change in names to type 1 and type 2 diabetes stems from the fact that many type 2 diabetics eventually must also rely on insulin injections as part of their treatment. In addition, many children today have type 2 diabetes. A third form, called gestational diabetes, occurs in some pregnant women. It is usually treated with an insulin regimen and dietary changes, and then resolves after delivery of the baby. However, women who have gestational diabetes during pregnancy are at high risk for developing type 2 diabetes later in life.

> **Symptoms of Diabetes**
> These symptoms may occur suddenly and include one or more of the following:
>
> | | |
> |---|---|
> | Drowsiness, lethargy | Increased appetite |
> | Extreme thirst | Stupor, unconsciousness |
> | Frequent urination | Sudden vision changes |
> | Fruity, sweet odor on breath | Sudden weight loss |
> | Heavy, labored breathing | Sugar in urine |

### Type 1 Diabetes

Type 1 diabetes often begins in late childhood, around ages 10 to 14 years, but can occur at any age. Approximately 1.3 million Americans have type 1 diabetes, and an estimated 40,000 people will be diagnosed with the disease each year in the U.S.[21] Children usually are admitted to the hospital with abnormally high blood glucose after eating, as well as evidence of ketosis.

The onset of type 1 diabetes is generally associated with decreased release of insulin from the pancreas. As insulin in the blood declines, blood glucose increases, especially after eating. Figure 4-17 shows a typical glucose response observed in a patient with this form of diabetes after consuming about 75 grams of glucose. When blood glucose levels are high, the kidneys let excess glucose spill into the urine, resulting in frequent urination that is high in sugar.

Most cases of type 1 diabetes begin with an immune system disorder, which causes destruction of the insulin-producing cells in the pancreas. The disease may stem from genetic, autoimmune, or environmental factors. Eventually, the pancreas loses its ability to synthesize insulin, and the clinical stage of the disease begins.

**hemoglobin** The iron-containing part of the red blood cell that carries oxygen to the cells and carbon dioxide away from the cells. The heme iron portion is also responsible for the red color of blood.

**type 1 diabetes** A form of diabetes characterized by total insulin deficiency due to destruction of insulin-producing cells of the pancreas. Insulin therapy is required.

**type 2 diabetes** A form of diabetes characterized by insulin resistance and often associated with obesity. Insulin therapy may be required in advanced stages of the disease.

**TABLE 4-7** ■ **Comparison of Type 1 and Type 2 Diabetes**

|  | Type 1 Diabetes | Type 2 Diabetes |
|---|---|---|
| **Occurrence** | 5% to 10% of cases of diabetes | 90% to 95% of cases of diabetes |
| **Cause** | Autoimmune destruction of the pancreas | Insulin resistance |
| **Risk factors** | Moderate genetic predisposition | Strong genetic predisposition<br>Obesity and physical inactivity<br>Ethnicity<br>Metabolic syndrome<br>Prediabetes |
| **Characteristics** | Distinct symptoms (frequent thirst, hunger, and urination)<br>Weight loss | Mild symptoms, especially in early phases of the disease (fatigue and nighttime urination) |
| **Treatment** | Insulin<br>Dietary pattern<br>Physical activity | Dietary pattern<br>Physical activity<br>Oral medications to lower blood glucose<br>Insulin (in advanced cases) |
| **Complications** | Cardiovascular disease<br>Kidney disease<br>Nerve disease<br>Blindness<br>Infections<br>Ketosis | Cardiovascular disease<br>Kidney disease<br>Nerve damage<br>Blindness<br>Infections<br>Ketosis in rare circumstances |
| **Monitoring** | Blood glucose<br>Urine ketones<br>HbA1c* | Blood glucose<br>HbA1c* |

*Hemoglobin A1c.

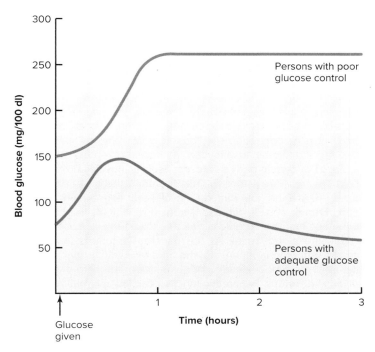

**FIGURE 4-17** ▲ Glucose tolerance test. A comparison of blood glucose concentrations after a 75-gram glucose load displaying a healthy blood glucose response (in blue) and a dysfunctional response (in red).

Hyperglycemia and other symptoms develop slowly and only after 90% or more of the insulin-secreting cells have been destroyed. HbA1c, fasting blood glucose, or an oral glucose tolerance test can be used to diagnose diabetes (Fig. 4-16). Remember that HbA1c is the recommended measure, and an HbA1c value of over 7% indicates poor blood glucose control.

Type 1 diabetes is treated primarily by insulin therapy, either with injections or with an *insulin infusion pump*. Advantages of using an insulin pump include eliminating individual insulin injections; delivering insulin more accurately than injections; improved stabilization of blood glucose levels; enabling more dietary and physical activity flexibility; reducing severe low blood glucose episodes; and eliminating unpredictable effects of intermediate- or long-acting insulin. Now, an amazing innovation is showing promise for management of type 1 diabetes. The *bionic pancreas* is a novel device that would fully automate blood glucose tracking and adjust the hormonal response accordingly. Several biotech companies are testing models.[23]

Medical nutrition therapy includes tailoring dietary pattern recommendations to the individual. It includes balancing carbohydrate intake with the insulin regimen and physical activity schedule to manage blood glucose levels. The amount and timing of carbohydrates eaten should be consistent from day to day to maintain blood glucose control. Insulin should be adjusted to match carbohydrate intake in persons who adjust their mealtime insulin doses or who are using an insulin pump. There are several methods available to estimate carbohydrate content of foods, including carbohydrate counting, exchange lists, and the glycemic index and glycemic load of foods.[24,25] If one does not eat often enough, the injected insulin can cause a severe drop in blood glucose or hypoglycemia because it acts on whatever glucose is available. Dietary patterns should be moderate or low in simple carbohydrates, include ample fiber and unsaturated fat, be low in saturated and *trans* fats, and supply an amount of calories to balance with needs. Providing adequate calories and nutrients to promote growth and development in children is crucial for young individuals with diabetes.[22,25,26]

The hormone imbalances that occur in people with untreated type 1 diabetes—primarily, not enough insulin—lead to mobilization of body fat, taken up by liver cells. Ketosis is the result because the fat is partially broken down to ketone bodies. Ketone bodies can rise excessively in the blood and eventually spill into the urine. These pull sodium and potassium ions as well as water into the urine. This series of events also causes frequent urination and can contribute to dehydration, ion imbalance, coma, and even death. Treatment includes provision of insulin, fluids, and minerals such as sodium and potassium.

Several degenerative complications, including cardiovascular disease, blindness, kidney disease, and nerve damage, result from poor blood glucose regulation, specifically long-term hyperglycemia. The high blood sugar concentration physically deteriorates small blood vessels (capillaries) and nerves. When improper nerve stimulation occurs in the intestinal tract, intermittent diarrhea and constipation result. Because of nerve deterioration in the extremities, many people with diabetes lose the sensation of pain associated with injuries or infections. They do not have as much pain, so they often delay treatment of hand or foot problems. This delay, combined with a rich environment for bacterial growth (bacteria thrive on glucose), sets the stage for damage and death of tissues in the extremities, sometimes leading to the need for amputation of feet and legs.

Current research has shown that aggressive treatment directed at keeping blood glucose within the normal range can slow the development of blood vessel and nerve complications of diabetes. A person with diabetes must work closely with a primary care provider and RDN to make the correct alterations in the dietary pattern and medications and to perform physical activity safely. Physical activity enhances glucose uptake by muscles independent of insulin action, which in turn can lower blood glucose. This outcome is beneficial, but people with type 1 diabetes need to be aware of their blood glucose response to physical activity and compensate appropriately to avoid hypoglycemia.

Insulin pumps alleviate the discomfort of injecting insulin under the skin multiple times per day. **How does this device work?** Oscar Gimeno Baldo/Alamy Images

### Prediabetes

Often people with type 2 diabetes did not develop the disease suddenly. It may develop for years before symptoms are noticed. Prediabetes is a condition in which the concentration of blood glucose drifts up higher than normal. By the time symptoms are noticeable, organs and tissues may already be damaged. Simple tests of your fasting blood glucose level or HbA1c can determine if you are prediabetic. Early detection of diabetes risk can help prevent diabetes if you make lifestyle changes. If you have a family history of diabetes or if your behaviors (being physically inactive and overweight, and having a poor dietary pattern) put you at risk, it is important to discover if your blood glucose is still in the prediabetic stage. Prediabetes, also called *impaired fasting glucose,* is diagnosed if the fasting blood glucose is 100 to 125 milligrams per deciliter or the HbA1c is 5.7% to 6.4%.

The Diabetes Prevention Plan (DPP) was a large, multisite study designed to improve glucose control and prevent the onset of diabetes. The DPP found that participants who lost even a modest amount of weight via dietary and physical activity changes sharply reduced their chances of developing diabetes. More impressive is that the DPP lifestyle intervention had health outcomes that were better than those in the medication-treated cohort.[27]

## Type 2 Diabetes

Type 2 diabetes usually begins after age 30. This is the most common type of diabetes, accounting for about 90% to 95% of the cases diagnosed in North America.[21] The disease is progressive and is present, in many cases, long before it is diagnosed. Hyperglycemia develops slowly such that the classic symptoms are not noticed in the early stages of the disease. Risk factors for type 2 diabetes are both genetic and environmental and include a family history of diabetes; older age; obesity, especially **upper-body obesity**; physical inactivity; prior history of gestational diabetes; prediabetes; and race or ethnicity. Latino/Hispanic Americans, African Americans, Asian Americans, Native Americans, and Pacific Islanders are at particular risk. The overall number of people affected is also on the rise, primarily because of widespread inactivity and obesity. There has also been a substantial increase in type 2 diabetes in children, due mostly to an increase in body fat in this population (coupled with limited physical activity).

Type 2 diabetes arises when the insulin receptors on the cell surfaces of certain body tissues, especially muscle and fat tissue, become *insulin resistant* (Fig. 4-18). During the onset of the disease, there is an abundance of insulin that is not used properly, and blood glucose is not readily transferred into cells. The person develops high blood glucose as a result of the glucose remaining in the bloodstream. As the pancreas attempts to increase insulin output to compensate, the beta cells in the pancreas lose the ability to produce sufficient quantities of the hormone. As the disease develops, pancreatic function can fail, eventually leading

**upper-body obesity** The type of obesity in which fat is stored primarily in the abdominal area; defined as a waist circumference more than 40 inches (102 centimeters) in men and more than 35 inches (88 centimeters) in women; closely associated with a high risk for cardiovascular disease, hypertension, and type 2 diabetes. Also known as *android obesity, visceral obesity* or *central obesity.*

## INSULIN RESISTANCE EXPLAINED

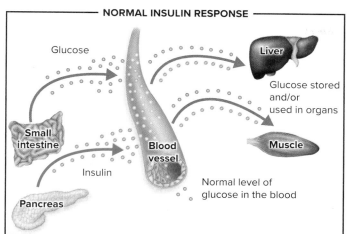
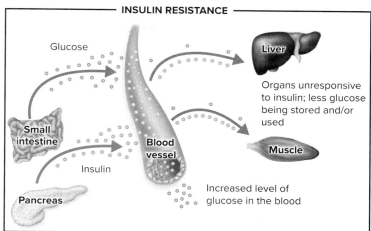

**FIGURE 4-18** ▲ During a normal insulin response (figure on left), glucose is absorbed from the small intestine after digestion into the bloodstream. As blood glucose levels rise, the pancreas secretes the hormone insulin to signal cells to take up glucose from the bloodstream. Insulin resistance (figure on right) occurs when the cells do not respond to normal insulin signals to take up glucose from the bloodstream. intestine: McGraw-Hill; liver: Macrovector/Shutterstock; pancreas: McGraw-Hill; muscle: McGraw-Hill; blood vessel: McGraw-Hill

to reduced insulin output. Because of the genetic link for type 2 diabetes, those who have a family history should schedule regular diabetes screening and be careful to avoid risk factors such as obesity and inactivity.

Most cases of type 2 diabetes (about 90%) are associated with overweight and obesity (especially with fat located in the abdominal region), but high blood glucose is not directly caused by the obesity. In fact, some lean people also develop this type of diabetes. Obesity associated with oversized adipose cells increases the risk for insulin resistance by the body as more fat is added to these cells during weight gain (Fig. 4-18).

Because type 2 diabetes is linked to obesity, achieving a healthy weight should be a primary goal of treatment, with even limited weight loss leading to better blood glucose regulation. Although many cases of type 2 diabetes can be relieved by reducing excess fat, many people struggle to lose weight. They remain affected with diabetes and may experience the degenerative complications seen in the type 1 form of the disease. Ketosis, however, is not usually seen in type 2 diabetes. Glucose-lowering medications and insulin are used as needed in patients with type 2 diabetes. New classes of drugs that mimic gut hormones are helping diabetic patients overcome the chronic problems that conventional treatments alone have been unable to control.

Regular patterns of meals and physical activity are important elements of therapy for type 2 diabetes. Physical activity helps the muscles take up more glucose. Medical nutrition therapy should emphasize overall calorie control, increased intakes of fiber-rich foods and fish, and reduced intake of added sugars and unhealthy fats. As in type 1 diabetes, it is critical that discussions include an RDN and primary care provider to create a tailored lifestyle management plan to meet the needs of each individual.[28] It is also important to consume the recommended 25 to 38 grams of fiber, with emphasis on soluble fiber sources, which will help regulate glucose.

Although sugar does not have to be completely eliminated, persons with diabetes will benefit from adhering to the recommendation to reduce consumption of added sugars. Because persons with diabetes are at increased risk of cardiovascular

disease, heart-healthy choices should also be included in the diabetic meal plan.

> For more information on diabetes, consult the following websites: www.diabetes.org and www.ndep.nih.gov.

## Hypoglycemia

People with diabetes who are taking insulin sometimes have hypoglycemia if they do not eat frequently enough. The first signs of diabetic hypoglycemia include shakiness, sweating, palpitations, anxiety, and hunger. Later symptoms are the result of insufficient glucose reaching the brain and include mental confusion, extreme fatigue, seizures, and unconsciousness. Symptoms should be treated immediately with consumption of glucose or food containing carbohydrate.

## Metabolic Syndrome

**Metabolic syndrome** is characterized by the presence of several risk factors for diabetes and cardiovascular disease. A person with metabolic syndrome must have at least three of the following metabolic risk factors (or be on medication to treat these risk factors) to be diagnosed with metabolic syndrome: a large waistline from abdominal obesity, high blood triglycerides, low HDL or *good* cholesterol, hypertension, and high fasting blood glucose (Fig. 4-19). Each aspect of metabolic syndrome is a unique health problem with its own treatment. In metabolic syndrome, however,

**metabolic syndrome** A condition in which a person has poor blood glucose regulation, hypertension, increased blood triglycerides, and other health problems. This condition is usually accompanied by obesity, lack of physical activity, and a dietary pattern high in refined carbohydrates. Also called *Syndrome X*.

### Diagnosis of Metabolic Syndrome

For an individual to be diagnosed with metabolic syndrome, he or she must have **three of the five risk factors** listed below.

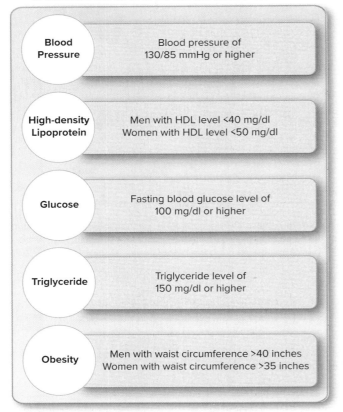

| Blood Pressure | Blood pressure of 130/85 mmHg or higher |
| High-density Lipoprotein | Men with HDL level <40 mg/dl<br>Women with HDL level <50 mg/dl |
| Glucose | Fasting blood glucose level of 100 mg/dl or higher |
| Triglyceride | Triglyceride level of 150 mg/dl or higher |
| Obesity | Men with waist circumference >40 inches<br>Women with waist circumference >35 inches |

**FIGURE 4-19** ▲ Metabolic syndrome is characterized by the presence of several risk factors for diabetes and cardiovascular disease.

▲ Regular physical activity is a key part of a plan to prevent (and control) type 2 diabetes. Ariel Skelley/Blend Images LLC

these risk factors are clustered together, making a person twice as likely to develop cardiovascular disease and five times more likely to develop diabetes.

It is generally accepted that one key element unifies all the aspects of metabolic syndrome is insulin resistance. As you learned, insulin is a hormone that directs tissues to pull glucose out of the blood and into cells for storage or fuel. With insulin resistance, the pancreas produces plenty of insulin, but the cells of the body do not respond to it effectively. Instead, excess glucose stays in the bloodstream. For a while, the pancreas may be able to compensate for the resistance of cells to insulin by overproducing insulin. Over time, however, the pancreas is unable to keep up the accelerated insulin production, and blood glucose levels remain elevated. With metabolic syndrome, blood glucose is not high enough to be classified as diabetes, but without intervention, it is likely to get worse and eventually lead to diabetes.

Genetics and aging contribute to the development of insulin resistance and the other elements of metabolic syndrome, but environmental factors, such as the dietary pattern and activity, play an important role. Body fatness, particularly abdominal obesity, is highly related to insulin resistance. Over 70% of adults in the U.S.

are overweight, almost 40% are obese, and these numbers continue to climb year after year. Increases in body weight among children and adolescents are of great concern because childhood obesity places them at high risk for these health problems. This increase in body weight has precipitated a dramatic surge in cardiovascular disease and diabetes risk: an estimated 50 million Americans now have metabolic syndrome.

The insulin resistance that precedes and contributes to type 2 diabetes also leads to several other components of the metabolic syndrome. The chief culprits contributing to the high blood triglycerides of metabolic syndrome are large meals rich in simple sugars and refined starches and low in fiber and whole grains, coupled with little to no physical activity. Nutrition and lifestyle changes are key strategies in addressing all of the unhealthy conditions of metabolic syndrome as a whole. Suggested interventions include:

- Decrease body weight. Even small improvements (e.g., 3% to 5% weight loss) for individuals who are overweight or obese can lessen disease risk. The most successful weight-loss and weight-maintenance programs include moderate dietary restriction combined with physical activity.
- Increase physical activity. To alleviate risks for chronic diseases, the *Physical Activity Guidelines* recommend at least 150 minutes of moderate-intensity physical activity each week.
- Limit solid fat consumption, especially *saturated* and *trans* fat sources. We will explore the different types and sources of fats and their effects later in the text.

### ✓ CONCEPT CHECK 4.7

1. List key differences between type 1 and type 2 diabetes.
2. Describe insulin resistance and how it contributes to diabetes.
3. List the risk factors for metabolic syndrome.

# Summary (Numbers refer to numbered sections in the chapter.)

**4.1** Carbohydrates are created in plants through photosynthesis. They are our main fuel source for body cells. Refined and highly processed products lack many of the health benefits provided by the carbohydrates found in whole grains, beans, fruits, and vegetables. The *Dietary Guidelines* suggest making at least half the grains you eat whole grains.

**4.2** The common monosaccharides in food are glucose, fructose, and galactose. The major disaccharides are sucrose (glucose + fructose), maltose (glucose + glucose), and lactose (glucose + galactose). When digested, these yield their component monosaccharides. Once these are absorbed from the small intestine and delivered to the liver, much of the fructose and galactose is converted into glucose.

One major group of polysaccharides consists of storage forms of glucose: starches in plants and glycogen in humans. These can be broken down by human digestive enzymes, releasing the glucose units. The main plant starches—straight-chain amylose and branched-chain amylopectin—are digested by enzymes in the mouth and small intestine. In humans, glycogen is synthesized in the liver and muscle tissue from glucose. Under the influence of hormones, liver glycogen is readily broken down to glucose, which can enter the bloodstream to fuel necessary cells.

Fiber is a group of indigestible polysaccharides including cellulose, hemicellulose, pectin, gum, and β-glucans, inulin, resistant starch, as well as the noncarbohydrate lignins. These substances are not broken down by human digestive enzymes. Dietary fiber can be categorized by their solubility, viscosity, and fermentation potential. Soluble fiber forms a gel in the intestines, reduces glucose and cholesterol absorption, promotes a sense of satiety, and serves as fuel for the microbiota. Insoluble fiber passes through the GI tract intact, adding bulk to the feces.

**4.3** Table sugar, honey, jelly, and fruit are some of the most concentrated sources of carbohydrates. Other high-carbohydrate foods, such as pie and fat-free milk, are diluted by either fat or protein. Nutritive (calorie-containing) sweeteners in food include sucrose, high-fructose corn syrup, brown sugar, and maple syrup. Several alternative sweeteners approved for use by the FDA include saccharin, aspartame, sucralose, neotame, stevia, allulose, and acesulfame-K.

**4.4** Some starch digestion occurs in the mouth with assistance from the enzyme amylase. Carbohydrate digestion is completed in the small intestine. Some plant fibers are digested by the bacteria present in the large intestine; undigested plant fibers become part of the feces. Monosaccharides in the intestinal contents mostly follow an active absorption process. They are then transported via the portal vein that leads directly to the liver.

The ability to digest lactose often diminishes with age. Undigested lactose travels to the large intestine, resulting in such symptoms as abdominal gas, pain, and diarrhea. The occurrence of severe symptoms after consuming lactose is called lactose intolerance. Most people with lactose maldigestion can tolerate cheese, yogurt, and moderate amounts of milk.

**4.5** Carbohydrates provide calories (4 kcal per gram), spare protein from being broken down for fuel, use of food and body protein for energy, and prevent ketosis. The RDA for carbohydrate is 130 grams per day. If carbohydrate intake is inadequate for the body's needs, protein is metabolized to provide glucose for energy needs. However, the price is loss of body protein, ketosis, and eventually a general body weakening. For this reason, low-carbohydrate diets are not recommended for extended periods.

Blood glucose concentration is regulated within a narrow range of 70 to 99 milligrams per deciliter. Insulin and glucagon are hormones that control blood glucose concentration. When we eat a meal, insulin promotes glucose uptake by cells. When fasting, glucagon promotes glucose release from glycogen stores in the liver.

**4.6** A goal of about half of calories as complex carbohydrates is a good one, with about 45% to 65% of total calories coming from carbohydrates in general. Foods to consume align with the *Dietary Guidelines*, which encourage fiber-rich fruits, vegetables, and whole grains. Specifically, 3 or more ounce equivalents of grains, roughly one-half of one's grains, should be whole. Added sugars should be limited to less than 10% of total calories.

Moderating sugar intake, especially between meals, reduces the risk of dental caries. Alternative sweeteners, such as aspartame, aid in reducing intake of sugars.

**4.7** Diabetes is characterized by a persistent high blood glucose concentration. Type 1 diabetes is caused by a lack of insulin, whereas type 2 diabetes is due to insulin resistance. Regular physical activity and a balanced meal plan that emphasizes fiber and limits added sugars and saturated fats are helpful in treating both type 1 and type 2 diabetes. Insulin is the main medication employed: it is required in type 1 diabetes and may be used in type 2 diabetes.

# Check Your Knowledge (Answers are available at the end of this question set.)

1. Dietary fiber
   a. raises blood cholesterol levels.
   b. speeds up transit time for food through the digestive tract.
   c. causes diverticulosis.
   d. causes constipation.

2. When the pancreas detects excess glucose, it releases the
   a. enzyme amylase.
   b. monosaccharide glucose.
   c. hormone insulin.
   d. hormone glucagon.

3. Cellulose is a(n)
   a. indigestible fiber.
   b. simple carbohydrate.
   c. energy-yielding nutrient.
   d. animal polysaccharide.

4. Digested white sugar is broken into _____ and _____.
   a. glucose, lactose     c. sucrose, maltose
   b. glucose, fructose     d. fructose, sucrose

5. Starch is a
   a. complex carbohydrate.     c. simple carbohydrate.
   b. fiber.                    d. gluten.

6. Fiber content of the dietary pattern can be increased by adding
   a. fresh fruits.             c. eggs.
   b. fish and poultry.         d. whole grains and cereals.
   e. both a and d.

7. Which form of diabetes is most common?
   a. type 1     c. type 3
   b. type 2     d. gestational

8. The recommended daily intake for fiber is approximately _____ grams.
   a. 5 to 10     c. 45 to 65
   b. 25 to 38    d. 75 to 109

9. Lactose intolerance is the result of
   a. drinking high-fat milk.
   b. eating a large amount of yogurt.
   c. low lactase activity.
   d. a high-fiber dietary pattern.

10. One of the components of metabolic syndrome is
    a. high HDL.
    b. high waist circumference.
    c. low blood sugar.
    d. low blood pressure.

11. During photosynthesis, glucose is stored in the plant and can undergo further metabolism to form starch and fiber in the plant. With the addition of _____ from soil or air, glucose can also be transformed into protein.
    a. oxygen     c. carbon
    b. nitrogen   d. hydrogen

Answer Key: 1. b (LO 4.5), 2. c (LO 4.5), 3. a (LO 4.2), 4. b (LO 4.2), 5. a (LO 4.2), 6. e (LO 4.3), 7. b (LO 4.7), 8. b (LO 4.6), 9. c (LO 4.4), 10. b (LO 4.7), 11. b (LO 4.1)

# Study Questions (Numbers refer to Learning Outcomes.)

1. Why do we need carbohydrates in our dietary pattern? **(LO 4.1)**

2. What are the three major monosaccharides and the three major disaccharides? Describe how each plays a part in the human dietary pattern. **(LO 4.2)**

3. Why are some foods that are high in carbohydrates, such as cookies and fat-free milk, not considered to be concentrated sources of carbohydrates? **(LO 4.3)**

4. List three alternatives to simple sugars for adding sweetness to the dietary pattern without adding calories. **(LO 4.3)**

5. Describe the digestion of the various types of carbohydrates in the body. **(LO 4.4)**

6. Describe the reason why some people are unable to tolerate high intakes of milk. **(LO 4.4)**

7. Outline the basic steps in blood glucose regulation, including the roles of insulin and glucagon. **(LO 4.5)**

8. What are the important roles that fiber plays in our dietary pattern? **(LO 4.5)**

9. Summarize current carbohydrate intake recommendations. **(LO 4.6)**

10. What, if any, are the proven ill effects of excessive sugar in the dietary pattern? **(LO 4.6)**

11. Type 1 diabetes is caused by a lack of insulin. What leads to type 2 diabetes? **(LO 4.7)**

# Further Readings

1. Collins K: Dietary fiber: Fiber—Increase amount and variety. *Today's Dietitian* 20(7):11, 2018.

2. Position of the Academy of Nutrition and Dietetics: Health implications of dietary fiber. *Journal of the Academy of Nutrition and Dietetics* 115:1861, 2015. DOI:10.1016/j.jand.2015.09.003)

3. Ahluwalia N and others: Contribution of whole grains to total grains intake among adults aged 20 and over: United States, 2013–2016. NCHS Data Brief, no 341. Hyattsville, MD: National Center for Health Statistics. 2019. Available at https://www.cdc.gov/nchs/data/databriefs/db341-h.pdf. Accessed January 20, 2020.

4. Oldways Whole Grain Council: Whole Grain Statistics. March 2018. Available at https://wholegrainscouncil.org/newsroom/whole-grain-statistics. Accessed January 20, 2020.

5. Schaeffer J: Boosting whole grain consumption. *Today's Dietitian* 15(2):33, 2013.

6. Oldways Whole Grain Council. Whole Grain Stamp. Available at https://wholegrainscouncil.org/whole-grain-stamp. Accessed January 1, 2020.

7. U.S. Department of Agriculture. Food and Nutrition Service. Nutrition Standards for School Meals. Available at https://www.fns.usda.gov/school-meals/nutrition-standards-school-meals. Accessed January 21, 2020.

8. Position of the Academy of Nutrition and Dietetics: Use of nutritive and nonnutritive sweeteners. *Journal of the Academy of Nutrition and Dietetics* 112:739, 2012. DOI:10.1016/j.jand.2012.03.009.

9. U.S. Department of Agriculture and U.S. Department of Health and Human Services. Dietary Guidelines for Americans, 2020-2025. 9th Edition. December 2020. Available at: DietaryGuidelines.gov.

10. Rippe JM and others: Relationship between added sugars consumption and chronic disease risk factors: Current understanding. *Nutrients* 4(8):11, 2016. DOI:10.3390/nu8110697.

11. National Institutes of Health. National Cancer Institute. Artificial Sweeteners and Cancer. Available at https://www.cancer.gov/about-cancer/causes-prevention/risk/diet/artificial-sweeteners-fact-sheet. Accessed January 18, 2020.

12. U.S. Department of Agriculture. Food and Drug Administration. How Sweet It Is: All About Sugar Substitutes. Available at https://www.fda.gov/consumers/consumer-updates/how-sweet-it-all-about-sugar-substitutes. Accessed January 21, 2020.

13. National Institutes of Health. U.S. National Library of Medicine. Lactose Intolerance. Available at https://ghr.nlm.nih.gov/condition/lactose-intolerance#statistics. Accessed January 19, 2020.

14. Dionne J and others. A systematic review and meta-analysis evaluating the efficacy of a gluten-free diet and a low FODMAPs diet in treating symptoms of irritable bowel syndrome. *Am J Gastroenterol* 113(9):1290–1300, 2018 Sep. DOI:10.1038/s41395-018-0195-4.

15. Mendes-Soares H and others: Assessment of a personalized approach to predicting postprandial glycemic responses to food among individuals without diabetes. *JAMA Newt Open* 2(2):e188102, 2019. DOI:10.1001/jamanetworkopen.2018.8102.

16. University of Sydney. GI Newsletter. Available at http://www.glycemicindex.com. Accessed January 21, 2020.

17. Seal CJ and others: Whole-grain foods and chronic disease: Evidence from epidemiological and intervention studies. *Proceedings of the Nutrition Society* 74(3):313, 2015. DOI:10.1017/S0029665115002104.

18. American Heart Association: Added Sugars. Available at Oldways Whole Grain Council: Whole Grain Statistics. March 2018. Available at https://wholegrainscouncil.org/newsroom/whole-grain-statistics. Accessed January 5, 2020.

19. Bleich SN and others: Trends in beverage consumption among children and adults, 2003–2014. *Obesity* 26:2, 2017. DOI:10.1002/oby.22622.

20. Academy of Nutrition and Dietetics. Sugar: Does It Really Cause Hyperactivity? Available at https://www.eatright.org/food/nutrition/dietary-guidelines-and-myplate/sugar-does-it-really-cause-hyperactivity. Accessed January 21, 2020.

21. American Diabetes Association. Statistics About Diabetes. Available at https://www.diabetes.org/resources/statistics/statistics-about-diabetes. Accessed January 25, 2020.

22. Standards of medical care in diabetes 2017: Summary of revisions. *Diabetes Care* 40(Suppl. 1):S4–S5, 2017. DOI: 10.2337/dc17-S003.

23. McAdams BH and others: An overview of insulin pumps and glucose sensors for the generalist. *Journal of Clinical Medicine* 4(5)1, 2016. DOI: 10.3390/jcm5010005.

24. Centers for Disease Control and Prevention, US Department of Health and Human Services: National Diabetes Statistics Report 2017. January 2018. Available at https://www.cdc.gov/features/diabetes-statistic-report/index.html. Accessed January 22, 2020.

25. Young-Hyman D and others: Psychosocial care for people with diabetes: A position statement of the American Diabetes Association. *Diabetes Care* 39.12: 2126–2140, 2016. DOI:10.2337/dc16-2053.

26. American Diabetes Association: Standards of medical care in diabetes—2018. *Diabetes Care* 36(1)14, 2018. DOI:10.2337/cd17-0119.

27. Centers for Disease Control and Prevention, US Department of Health and Human Services: National Diabetes Prevention Program. January 2018. Available at https://www.cdc.gov/diabetes/prevention/index.html. Accessed January 5, 2020.

28. Position of the Academy of Nutrition and Dietetics: The role of medical nutrition therapy and registered dietitian nutritionists in the prevention and treatment of prediabetes and type 2 diabetes. *Journal of the Academy of Nutrition and Dietetics* 118(2)343, 2018.

# Chapter 5

# Lipids

Julija Dmitrijeva/123RF

# Student Learning Outcomes

**Chapter 5 is designed to allow you to:**

**5.1** Understand the common properties of lipids.

**5.2** Describe the structures of the three forms of lipids: triglycerides, phospholipids, and sterols.

**5.3** Discuss the importance of the essential fatty acids.

**5.4** Identify food sources of saturated, monounsaturated, and polyunsaturated fatty acids; phospholipids; and sterols.

**5.5** Explain how lipids are digested and absorbed.

**5.6** Name the lipoproteins and classify them according to their functions.

**5.7** Describe the functions of the various forms of lipids in the body.

**5.8** Summarize current recommendations for fat intake.

**5.9** Characterize the relationship between lipids and cardiovascular disease.

## FAKE NEWS

### Eggs are bad for your heart.

## THE FACTS

Consuming too much saturated fat and cholesterol can increase your risk for cardiovascular disease. While eggs do provide some saturated fat and cholesterol, don't overlook their many nutritional benefits! Eggs are inexpensive sources of high-quality protein. Eggs also supply several micronutrients, such as vitamin D and choline, that tend to be low in the dietary patterns of most Americans. Research evidence indicates that eating an average of one egg per day (about 7 eggs per week) is reasonable for heart health.[1,2]

Should you be concerned about the amount and types of fat in your dietary pattern? Lipids perform vital functions, both in the body and in foods. Their presence in your dietary pattern is essential to good health. They are a source of fuel, they are part of every cell membrane, and they help to regulate body processes. However, some concern about fat intake is warranted. First, lipids are a concentrated source of energy; per gram, fats supply more than twice as much energy as carbohydrates or protein. In addition, certain types of fats may contribute to the risk of chronic diseases, such as **cardiovascular disease**. So, yes, the quantity and quality of fats in your dietary pattern deserve some attention.

Humans can survive with very little fat in their eating pattern. In fact, the body's need for the essential fatty acids can be met by daily consumption of only about 2 to 4 tablespoons of plant oil and two servings of fatty fish such as salmon or tuna per week. However, humans can thrive with a much more liberal fat intake. The Food and Nutrition Board suggests that fat intake should fall within the range of 20% to 35% of total calories for adults. Some experts suggest that a fat intake as high as 40% of calories is appropriate as long as the predominant type of fat is a healthy one, such as olive oil.

Let us look at lipids in detail—their forms, functions, and food sources. In this chapter, you will see how lipids are digested and absorbed and how specific types of lipids can influence your health—for better or for worse.

# 5.1 Lipids: A Chemistry Lesson

*Lipid* is a general term that includes **triglycerides, phospholipids,** and **sterols.** Food experts, such as chefs, refer to lipids that are solid at room temperature as *fats,* whereas lipids that are liquid at room temperature are called *oils.* To simplify our discussion, this chapter primarily uses the term *fat.* When necessary for clarity, the name of a specific lipid, such as cholesterol, will be used. This word use is consistent with the way many people use these terms.

As a class of nutrients, lipids share one main characteristic: they do not readily dissolve in water. Think of an oil-and-vinegar salad dressing. The oil is not soluble in the water-based vinegar; on standing, the two separate into distinct layers, with oil on the top and vinegar on the bottom.

In terms of chemistry, lipids are composed primarily of carbon, hydrogen, and oxygen. Compared to carbohydrates, lipids contain more carbon-hydrogen bonds and fewer carbon-oxygen bonds. Lots of chemical energy is released when those carbon-hydrogen bonds are broken down, so lipids yield more than twice as much energy (9 kcal per gram) as carbohydrates or proteins (4 kcal per gram).

The chemical structures of lipids are diverse. In this section, you will learn about the structures of three categories of lipids: triglycerides, phospholipids, and sterols.

## FATTY ACIDS AND TRIGLYCERIDES

Triglycerides are the primary form of lipids in the body and in foods. A triglyceride is composed of three fatty acids attached to a **glycerol** backbone. Each fatty acid is basically a long chain of carbons bonded together and flanked by hydrogens. At one end of a fatty acid (the alpha end) is an **acid group.** At the other end (the omega end) is a **methyl group** (Fig. 5-1).

Here is an important point to remember: the triglycerides in foods are not composed of just one type of fatty acid. Rather, dietary fats are complex mixtures of many different fatty acids, the combination of which provides each food its unique taste and smell.

**Fatty Acids Can Be Saturated or Unsaturated.** The terms *saturated* and *unsaturated* refer to the amount of hydrogen atoms that attach to the carbons in a fatty acid. Chemically speaking, a carbon atom can form four bonds. Within the carbon chain of a fatty acid, each carbon atom binds to two adjacent carbon atoms and one or two hydrogen atoms. The carbons that make up the chain of a **saturated fatty acid** are all connected to each other by single bonds. If only two of the carbon's bonds are used to join with adjacent carbon atoms, this leaves room to bind to two hydrogen atoms--the maximum amount. Just as a sponge can be saturated (full) with water, a saturated fatty acid, such as stearic acid, is saturated with hydrogen [see Fig. 5-1(a)]. The shape of saturated fatty acids is quite linear, so they can pack very close together. This close packing or stacking of saturated fatty acids gives them a solid texture at room temperature.

If the carbon chain of a fatty acid contains one or more double bonds, the carbon atoms involved in the double bond have fewer bonds to share with hydrogen, and the chain is said to be *unsaturated.* A fatty acid with only one double bond is **monounsaturated** [see Fig. 5-1(b)].

If two or more of the bonds between the carbons are double bonds, the fatty acid is even less saturated with hydrogens, and so it is **polyunsaturated** [see Fig. 5-1(c), (d)]. The double bonds in unsaturated fatty acids create kinks in the fatty acids' structure that keep them from packing closely together, so they are liquid at room temperature. Corn, soybean, sunflower, and safflower oils are rich in polyunsaturated fatty acids.

**Unsaturated Fatty Acids Differ in the Location of Their Double Bonds.** The location of the first double bond relative to the methyl end (also called the *omega* end) of an unsaturated fatty acid is another important feature. If the first double bond starts three carbons from the methyl end of the fatty acid, it is an **omega-3 (ω-3) fatty acid** [review Fig. 5-1(c)]. If it is located six carbons from the methyl end, it is an **omega-6 (ω-6) fatty acid** [review Fig. 5-1(d)]. An omega-9 fatty acid has its first double bond starting at the ninth carbon from the methyl end [review Fig. 5-1(b)]. Food sources of each of these types of fatty acids will be discussed in Section 5.2.

**triglyceride** The major form of lipid in the body and in food. It is composed of three fatty acids attached to glycerol.

**sterol** A compound containing a multi-ring (steroid) structure and a hydroxyl group (−OH). Cholesterol is a typical example.

**glycerol** A three-carbon alcohol used to form triglycerides.

The physical properties of fats and oils depend on their chemical structures. **Why is butter solid at room temperature, while olive oil and corn oil are liquid at room temperature?** Tetra Images/Getty Images

**acid group** In chemistry, a functional group that consists of a carbon atom that shares bonds with two oxygen atoms. This is the site where fatty acids are linked to glycerol to form triglycerides.

**methyl group** In chemistry, a carbon atom that shares bonds with three hydrogen atoms. The methyl group is the omega end of a fatty acid.

**saturated fatty acid** A fatty acid containing no carbon-carbon double bonds.

**monounsaturated fatty acid** A fatty acid containing one carbon-carbon double bond.

**polyunsaturated fatty acid** A fatty acid containing two or more carbon-carbon double bonds.

**omega-3 (ω-3) fatty acid** An unsaturated fatty acid with the first double bond on the third carbon from the methyl end (−CH₃).

**omega-6 (ω-6) fatty acid** An unsaturated fatty acid with the first double bond on the sixth carbon from the methyl end (−CH₃).

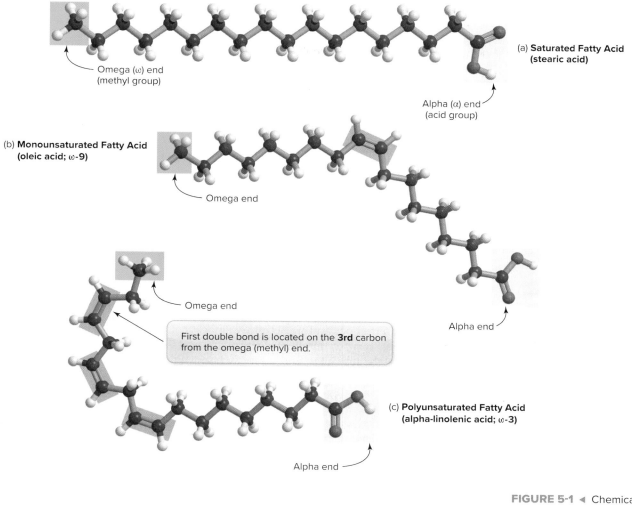

(a) **Saturated Fatty Acid (stearic acid)**

Omega (ω) end (methyl group)

Alpha (α) end (acid group)

(b) **Monounsaturated Fatty Acid (oleic acid; ω-9)**

Omega end

Alpha end

Omega end

First double bond is located on the **3rd** carbon from the omega (methyl) end.

Alpha end

(c) **Polyunsaturated Fatty Acid (alpha-linolenic acid; ω-3)**

Alpha end

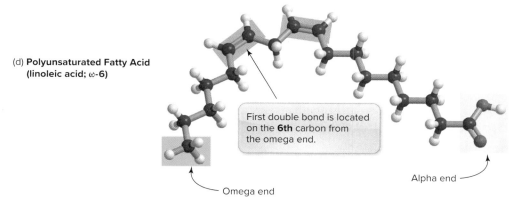

(d) **Polyunsaturated Fatty Acid (linoleic acid; ω-6)**

First double bond is located on the **6th** carbon from the omega end.

Alpha end

Omega end

**FIGURE 5-1** ◄ Chemical forms of saturated (a), monounsaturated (b), and polyunsaturated fatty acids (c and d). Each of the depicted fatty acids contains 18 carbons, but they differ from each other in the number and location of double bonds. The double bonds are shaded green. The linear shape of saturated fatty acids, as shown in (a), allows them to pack tightly together and form a solid at room temperature. In contrast, unsaturated fatty acids (b–d) have "kinks" where double bonds interrupt the carbon chain (Fig. 5-2). Thus, unsaturated fatty acids pack together only loosely and are usually liquid at room temperature.

**Unsaturated Fatty Acids Can Have *Cis* or *Trans* Configuration.** Unsaturated fatty acids, with their double bonds, can exist in two different structural forms: *cis* and *trans*. In nature, monounsaturated and polyunsaturated fatty acids usually are in the *cis* form (Fig. 5-2). In a ***cis* fatty acid,** the hydrogens are on the same side of the carbon-carbon double bond. In a ***trans* fatty acid,** the hydrogens are oriented on opposite sides of the carbon-carbon double bond. As shown in Figure 5-2, the *cis* bond causes the fatty acid's carbon chain to bend, whereas the *trans* bond allows the chain to remain more linear. This change in shape makes a *trans* unsaturated fatty acid look more like a saturated fatty acid. As you might guess, a

***cis* fatty acid** A form of an unsaturated fatty acid that has the hydrogens lying on the same side of the carbon-carbon double bond.

***trans* fatty acid** A form of an unsaturated fatty acid, usually a monounsaturated one when found in food, in which the hydrogens lie on opposite sides of the carbon-carbon double bond.

**FIGURE 5-2 ▶** *Cis* and *trans* fatty acids. In foods, *cis* fatty acids are much more common than *trans* fatty acids.

*Cis* means "*on the same side*." In a *cis* fatty acid, the hydrogens lie on the same side of the double bond. This causes a "kink" at that point in the fatty acid, typical of unsaturated fatty acids in foods.

*Trans* means "*on the other side*." In a *trans* fatty acid, the hydrogens lie on opposite sides of the double bond. This causes the fatty acid to exist in a linear form, like a saturated fatty acid.

**Oleic acid**

**Elaidic acid**

**linoleic acid** An essential omega-6 fatty acid with 18 carbons and two double bonds.

**alpha-linolenic acid** An essential omega-3 fatty acid with 18 carbons and 3 double bonds.

**essential fatty acids** Fatty acids that must be supplied by the diet to maintain health. Currently, only linoleic acid and alpha-linolenic acid are classified as essential.

**eicosanoids** A class of hormone compounds, including the prostaglandins, derived from the essential fatty acids. These signaling compounds are involved in cellular activity that affects practically all important functions in the body.

**eicosapentaenoic acid (EPA)** An omega-3 fatty acid with 20 carbons and 5 double bonds. It is present in large amounts in fatty fish and is slowly synthesized in the body from alpha-linolenic acid.

**docosahexaenoic acid (DHA)** An omega-3 fatty acid with 22 carbons and 6 double bonds. It is present in large amounts in fatty fish and is slowly synthesized in the body from alpha-linolenic acid. In the human body, high levels of DHA are found in the retina and brain.

**arachidonic acid (AA)** An omega-6 fatty acid made from linoleic acid with 20 carbon atoms and 4 double bonds.

*trans* fatty acid also acts more like a saturated fatty acid in the body. Remember this point when you learn about the health effects of various fatty acids in Sections 5.6 and 5.7.

**Two Fatty Acids Are Essential for Human Health.** The various classes of lipids have diverse functions in the body and are necessary for health. Of all the types of lipids found in foods, however, only two polyunsaturated fatty acids are essential. Remember, in nutrition, *essential* means that the substance is necessary for health, but cannot be made in the body, so it must be consumed as part of the dietary pattern. **Linoleic acid** (an omega-6 fatty acid) and **alpha-linolenic acid** (an omega-3 fatty acid) are the two **essential fatty acids** for humans. The essential fatty acids form body structures, perform important functions for the immune and nervous systems, and produce regulatory compounds, such as **eicosanoids** and hormones.

Many important *nonessential* fatty acids can be derived from the essential fatty acids (Fig. 5-3). Human enzymes can convert the two essential fatty acids to other long-chain polyunsaturated fatty acids, such as **eicosapentaenoic acid (EPA)** and **docosahexaenoic acid (DHA),** which are particularly important for proper function of the brain and nervous system. Because of its role in brain structure, DHA is especially important during pregnancy for fetal brain and nervous system development. Failing to consume enough of the essential fatty acids will limit the production of many nonessential fatty acids in the body. As you will learn in Section 5.6, a deficiency of essential fatty acids can severely impact the health of all body systems.

**Triglycerides Are the Main Form of Lipids in Foods and in the Body.** The lipids in foods and in body structures are mostly in the form of triglycerides. Although some fatty acids are transported in the bloodstream while attached to proteins, most fatty acids in the body are part of triglycerides.

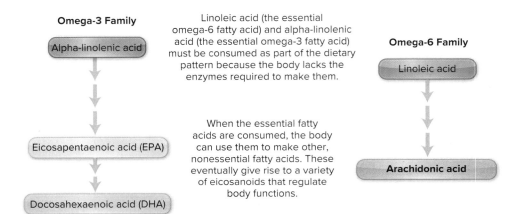

**FIGURE 5-3 ▶** The essential fatty acid (EFA) family.

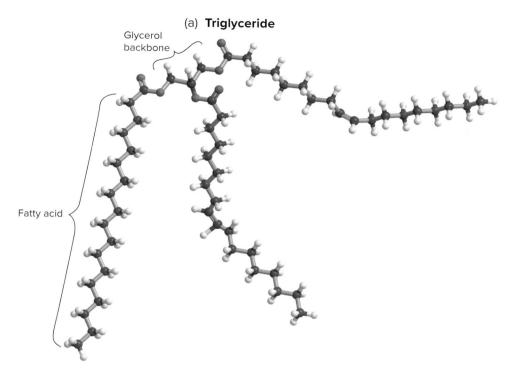

### (a) **Triglyceride**

Glycerol backbone

Fatty acid

**FIGURE 5-4** ◄ Chemical forms of common lipids: (a) triglyceride, (b) phospholipid (in this case, lecithin), and (c) sterol (in this case, cholesterol).

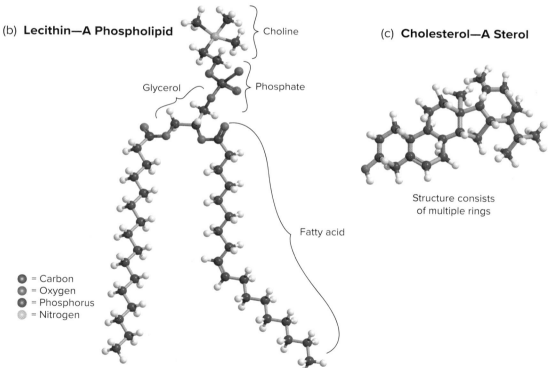

### (b) **Lecithin—A Phospholipid**

Choline

Glycerol

Phosphate

Fatty acid

● = Carbon
● = Oxygen
● = Phosphorus
● = Nitrogen

### (c) **Cholesterol—A Sterol**

Structure consists of multiple rings

Triglycerides contain a simple three-carbon alcohol, called glycerol, which serves as a backbone for three fatty acids [Fig. 5-4(a)]. Removing one fatty acid from a triglyceride forms a **diglyceride.** Removing two fatty acids from a triglyceride forms a **monoglyceride.** Later in this chapter you will see that before most dietary fats are absorbed, the two outer fatty acids are typically removed from the triglyceride during digestion in the small intestine. This produces a mixture of fatty acids and monoglycerides that can be absorbed into the intestinal cells. After absorption, the fatty acids and monoglycerides are mostly reformed into triglycerides inside body cells.

**diglyceride** A breakdown product of a triglyceride consisting of two fatty acids attached to a glycerol backbone.

**monoglyceride** A breakdown product of a triglyceride consisting of one fatty acid attached to a glycerol backbone.

## PHOSPHOLIPIDS

Like triglycerides, phospholipids are made of glycerol and fatty acids. However, at least one fatty acid is replaced with a compound containing phosphorus (and often other elements, such as nitrogen) [see Fig. 5-4(b)]. Many types of phospholipids exist in the body; they are part of the structure of every cell membrane. Phospholipids participate in fat digestion, absorption, and transport. The body is able to produce all the phospholipids it needs. **Lecithin** is one example of a phospholipid. Even though lecithin is sold as a dietary supplement and is used as an additive in many foods, phospholipids are not essential components of the dietary pattern.

**lecithin** A group of phospholipid compounds that are major components of cell membranes.

## STEROLS

Sterols are a class of lipids characterized by a multi-ringed structure that makes them structurally and functionally different from the other lipids already discussed [see Fig. 5-4(c)]. Although this waxy substance does not look much like the other lipids discussed so far, it is classified as a lipid because it does not readily dissolve in water. The most common example of a sterol is cholesterol, which is synthesized by the liver in animals and humans. Among other functions, cholesterol is a vital part of cell membranes and is used to form hormones and **bile acids**. It is important to note that the human body can make all the cholesterol it needs, so sterols are *not essential* components of the dietary pattern. Plants produce a different form of sterols called *phytosterols,* which may have benefits for human health (see Section 5.7).

The Lipids Concept Map (Fig. 5-5) summarizes the various forms of lipids.

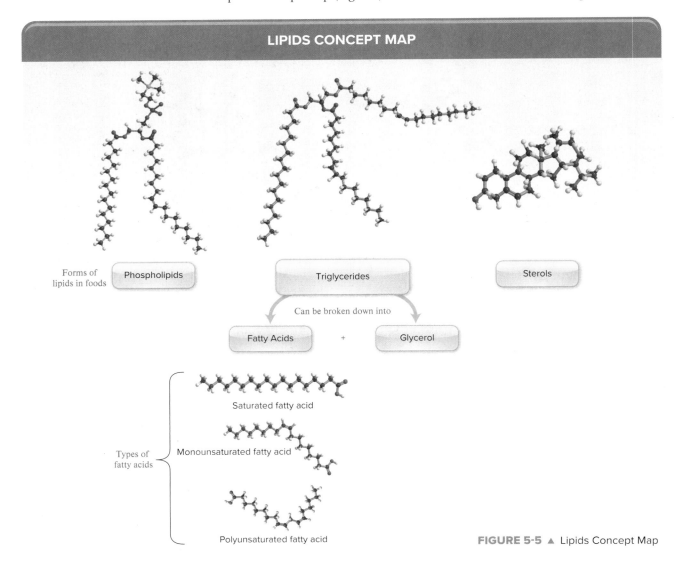

**FIGURE 5-5** ▲ Lipids Concept Map

✓ **CONCEPT CHECK 5.1**

1. What is the common property that all lipid compounds share?
2. Name three structural forms of lipids.
3. What is the structural difference between saturated and unsaturated fatty acids?
4. What is the structural difference between omega-3 and omega-6 fatty acids?
5. Which two fatty acids are essential?
6. How do triglycerides differ from phospholipids?
7. What are the main functions of cholesterol in the body?

# 5.2 Fats and Oils in Foods

## FATTY ACIDS AND TRIGLYCERIDES IN FOODS

Where is the fat on MyPlate? You'll notice there is no food group for fats and oils; it is assumed that fats and oils are incorporated into each food group. The fat content of foods may vary significantly because of fats and oils added during processing or food preparation. Figure 5-6 shows examples of food sources of fat.

In the protein foods group, there is a wide range of fat content. With less than 0.5 gram of fat per ½-cup serving, kidney beans are quite low in fat (4% of total kcal from fat). Lean cuts of meat, such as *loin* and *round* cuts of beef, have about 5 grams of fat per 3-ounce serving (25% of total kcal from fat). Nuts, marbled meats, and some processed meats are quite high in fat. Just 1 ounce of peanuts provides nearly 15 grams of fat (76% of total kcal from fat). A 3-ounce portion of prime rib has 29 grams of fat (76% of total kcal from fat). About 80% of the calories in processed meats, such as pepperoni and bacon, comes from fat.

In the dairy foods group, some items are very low in fat. A serving of 1 cup of skim milk or 1 cup of low-fat yogurt contains less than 1 gram of fat. Whole milk or yogurt made from whole milk contains 8 grams of fat per cup (50% of total kcal). With 9 grams of fat per slice, most types of cheese are fairly high in fat (74% of total kcal). Because it contributes so much to flavor and texture, the butterfat content of ice cream is the factor that determines quality of ice cream: standard ice cream has about 10 to 12 grams of fat per ½ cup (50% of total kcal), whereas premium ice cream has up to 18 grams of fat per ½ cup (60% of total kcal).

Most grains are quite low in fat. One slice of whole wheat bread, 1 cup of corn flakes cereal, and 1 cup of plain, cooked pasta, for example, provide just 1 gram of fat (about 5% to 10% of total kcal). However, we tend to add fat to grains to make them more palatable. Spreading 1 teaspoon of butter on a slice of bread adds about 4 grams of fat. When we make cookie dough or pie crust, we incorporate fat into flour to give it a delicious flavor and flaky texture. A slice of pie has about 5 grams of fat in the crust alone. Adding ½ cup of creamy alfredo sauce to your pasta will add 20 grams of fat (80% of total kcal from fat).

## FARM to FORK    Avocados

imagebroker/Alamy Stock Photo

Who knew? Avocados are not vegetables! Botanically, they are classified as berries, and like other berries, they are loaded with antioxidants and fiber.

### Grow
- Most avocados sold in North America are grown in California. The trees grow best in tropical or subtropical climates, but it is possible to grow hardy varieties of the plant in slightly cooler regions or indoors.

### Shop
- Most common are the Hass avocados. When ripe, they have dark green, almost black skin. A ripe avocado is soft at the top and yields only slightly when pressed in the middle.
- Another way to check for ripeness is to pluck the stem off the fruit. A vibrant green flesh is best. Light yellow (or if the stem is difficult to pluck) means the fruit is not ripe enough. Dark green or brown is too ripe.
- Don't buy avocados that are dented or mushy, or if you can feel the pit moving inside.

### Store
- Unripe avocados will ripen at room temperature. To promote ripening, place the avocado in a paper bag with a banana. Gases produced by the banana will help the avocado to ripen quickly.
- Ripe avocados will remain at peak quality for 2 or 3 days in the refrigerator.
- To prevent cut avocados from browning, sprinkle the cut surface with an acidic juice, such as lemon or lime, and store them in a plastic bag (remove as much air as possible) in the refrigerator.
- Storing cut avocados with sliced onions will prevent browning as well because volatile oils from the onion are very effective antioxidants.

### Prep
- Most often, avocados are served raw, chopped in flavorful salsas, or mashed in guacamole. Their mild flavor pairs well with the more defined flavors of onion, citrus, and fresh herbs.
- Avocados are rich in monounsaturated fatty acids, which gives them a creamy, delicate texture. Pureed avocados can substitute 1:1 for butter in recipes for baked goods. The mild flavor of the avocado will be imperceptible, and the final product will be lower in calories but higher in fiber and monounsaturated fats than the original recipe.
- Use half an avocado as a deliciously edible bowl for tuna salad.
- Search online for trendy ways to bake an egg in an avocado or grill it up with lean meats and veggies.

Source: Robinson J: Artichoke, asparagus, and avocados: Indulge! In *Eating on the Wild Side*. New York: Little, Brown and Company, 2013.

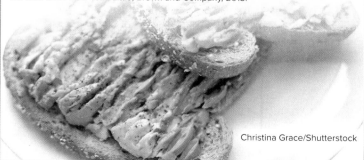

Christina Grace/Shutterstock

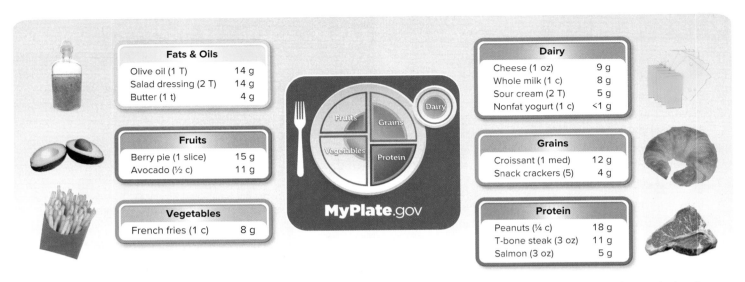

**FIGURE 5-6** ▲ Sources of fat from MyPlate. The fill of the background color (none, 1/3, 2/3, or completely covered) within each group in the plate indicates the average nutrient density for fat in that group. The fruit group and vegetable group are generally low in fat. In the other groups, both high-fat and low-fat choices are available. In general, any type of frying adds significant amounts of fat to a product, as with French fries and fried chicken. Careful reading of food labels can help you identify the fat content of foods. Italian dressing in a flask: McGraw-Hill Education; cheeses: Holly Curry/McGraw-Hill Education; avocado: lynx/iconotec.com/Glow Images; croissant: Ingram Publishing/Alamy Stock Photo; French fries in container: Comstock/Getty Images; frozen steak: Frank Bean/Getty Images; MyPlate: U.S. Department of Agriculture

Source: U.S. Department of Agriculture, Agricultural Research Service. FoodData Central, 2019. fdc.nal.usda.gov.

**long-chain fatty acid** A fatty acid that contains 12 or more carbons.

With less than 1 gram of fat per cup, most fruits and vegetables are low in fat. Avocados, discussed in *Farm to Fork*, are a notable exception. One whole avocado has about 30 grams of fat (82% of total kcal from fat). For most fruits and vegetables, however, what we add to these foods during preparation or processing can increase the fat content. Take the potato, for example. A plain, 3-ounce, baked potato is nearly devoid of fat. Adding a teaspoon of butter adds about 4 grams of fat. If we take that potato, slice it into fries, and deep-fry it in a vat of sizzling peanut oil, we add about 15 grams of fat (42% of total kcal). Likewise, adding just 2 tablespoons of ranch salad dressing incorporates 13 additional grams of fat into an otherwise fat-free pile of vegetables.

Foods that are high in fat are very energy dense, which means they pack a lot of calories into a small space. While you certainly should monitor the *total amount* of fat, the *type* of fat in foods is another important consideration when it comes to selecting a dietary pattern that promotes optimal health. Remember that triglycerides are composed of a mixture of fatty acids of various lengths and degrees of saturation.

Overall, a fat or an oil is classified as saturated, monounsaturated, or polyunsaturated based on the type of fatty acids present in the greatest concentration (Fig. 5-7). Fats in foods that contain primarily saturated fatty acids are solid at room temperature, especially if the fatty acids have long carbon chains (i.e., a **long-chain fatty acid**), as opposed to shorter versions. In contrast, fats containing primarily polyunsaturated or monounsaturated fatty acids (regardless of the length of the carbon chain) are usually liquid at room temperature. Most of the fatty acids in foods are long-chain fatty acids.

Animal fats are the chief contributors of saturated fatty acids to the North American dietary pattern. About 40% to 60% of total fat in dairy and meat products is in the form of saturated fatty acids. Palm oil and coconut oil are two plant oils that are rich sources of saturated fatty acids. Aside from palm oil and coconut oil, most plant oils are rich in unsaturated fatty acids, ranging from 73% to 94% of total fat. The major sources of monounsaturated fatty acids in the dietary pattern are canola oil, olive oil, and peanut oil. Corn, cottonseed, sunflower, soybean, and safflower oils contain mostly polyunsaturated fatty acids (54% to 77% of total fat).

**Food Sources of Essential Fatty Acids.** Rich food sources of linoleic acid—the essential omega-6 fatty acid—include safflower, sunflower, corn, and cottonseed oils, as well as nuts and seeds. In the North American dietary pattern, chicken is the leading source of linoleic acid, simply because it is so frequently consumed.

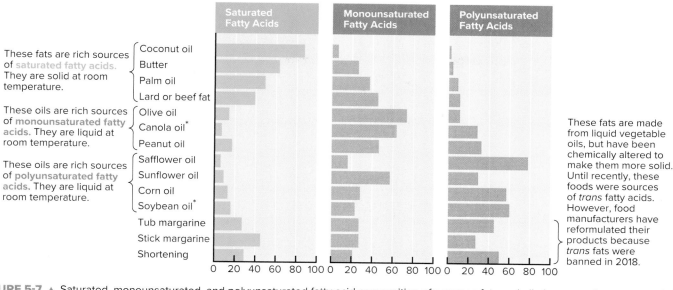

**FIGURE 5-7** ▲ Saturated, monounsaturated, and polyunsaturated fatty acid composition of common fats and oils (expressed as percent of all fatty acids in the product). Foods contain mixtures of fatty acids. *Rich source of the essential omega-3 fatty acid, alpha-linolenic acid.

Two nutrient-dense food sources of alpha-linolenic acid (the essential omega-3 fatty acid) are flaxseeds and walnuts.[3] Compared to other nuts and seeds, walnuts are one of the richest sources of alpha-linolenic acid (2.6 grams per 1-ounce serving, which is 14 walnut halves). In addition, walnuts are a rich source of plant sterols known to inhibit intestinal absorption of cholesterol. Other rich food sources of alpha-linolenic acid include oils from perilla seeds, chia seeds, canola seeds, and soybeans.

**Omega-3 Fatty Acids in Fish.** Looking at Figure 5-3, you can see that eicosapentaenoic acid (EPA) and docosahexaenoic acid (DHA) can be made in the body, so they are not essential fatty acids. However, by consuming food sources of EPA and DHA, you effectively skip a few rather slow and inefficient steps in the conversion of alpha-linolenic acid to EPA and DHA. These two omega-3 fatty acids are naturally found in fatty fish such as salmon, tuna, sardines, anchovies, striped bass, catfish, herring, mackerel, trout, and halibut (Table 5-1). The *Dietary Guidelines for Americans* and recommendations from the American Heart Association (AHA) encourage us to consume two servings of fatty fish per week

**TABLE 5-1** ■ Omega-3 Fatty Acids in Fish

| Food Item | Serving Size | Omega-3 Fatty Acids (g)* |
|---|---|---|
| Atlantic salmon | 3 oz | 1.8 |
| Anchovy | 3 oz | 1.7 |
| Sardines | 3 oz | 1.4 |
| Rainbow trout | 3 oz | 1.0 |
| Coho salmon | 3 oz | 0.9 |
| Bluefish | 3 oz | 0.8 |
| Striped bass | 3 oz | 0.8 |
| Tuna, white, canned | 3 oz | 0.7 |
| Halibut | 3 oz | 0.4 |
| Catfish | 3 oz | 0.2 |

*EPA + DHA

Source: U.S. Department of Agriculture, Agricultural Research Service. FoodData Central, 2019.

Public health authorities tell us to consume seafood twice per week. However, some populations of fish and shellfish are near extinction due to overfishing and damage to their habitats. How can we comply with these recommendations without jeopardizing the global supply of seafood for generations to come? Wild-capture fisheries in the U.S. must adhere to laws that limit overfishing and prevent damage to aquatic habitats. In addition, when managed responsibly, aquaculture (i.e., fish farming) can help to meet the global demand for seafood without harming the environment. Download a consumer guide about sustainable seafood at www.seafoodwatch.org. Also, look for the Aquaculture Stewardship Council (ASC)-certified and Marine Stewardship Council (MSC)-certified labels on packages of seafood sold in stores.

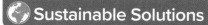

**Sustainable Solutions**

Flaxseeds are rich sources of alpha-linolenic acid, the essential omega-3 fatty acid. Whole flaxseeds must be chewed thoroughly or ground to improve digestibility and nutrient absorption. Flaxseed oil should be refrigerated to prevent it from turning rancid. **Why is flaxseed oil susceptible to rancidity?** Sergii Moskaliuk/123RF

**emulsifier** A compound that can suspend fat in water by isolating individual fat droplets, using a shell of water molecules or other substances to prevent the fat from coalescing.

for optimal cardiovascular health as well as improvements in brain health. An EPA/DHA supplement is useful for people who do not regularly consume fatty fish.

A word of caution: some types of fish can be a source of mercury, which is toxic in high amounts, especially during pregnancy, infancy, and early childhood (see Section 14.6). Fish species that are highest in mercury include shark, swordfish, king mackerel, tilefish, marlin, orange roughy, and bigeye (ahi) tuna. Salmon, sardines, herring, and albacore or yellowfin tuna are better choices because they are lower in mercury, yet they still provide heart-healthy omega-3 fatty acids. To limit exposure to mercury, vary your choices rather than always eating the same species of fish and limit overall intake to 12 ounces per week (two to three meals of fish or shellfish per week). Overall, research indicates that the benefits of fish intake, especially in reducing the risk of cardiovascular disease, outweigh the possible risks of mercury contamination.

## FOOD SOURCES OF PHOSPHOLIPIDS

Wheat germ, peanuts, egg yolks, soybeans, and organ meats are rich sources of phospholipids. Phospholipids such as lecithin, a component of egg yolks, are often added to salad dressing. Lecithin is used as an **emulsifier** in these and other products because of its ability to keep mixtures of lipids and water from separating (Fig. 5-8). Emulsifiers are added to salad dressings to keep the vegetable oil suspended in water. Likewise, eggs added to cake batters emulsify the fat with the milk.

## FOOD SOURCES OF STEROLS

There are two types of sterols in the food supply: (1) cholesterol and (2) phytosterols. Cholesterol is found only in foods of animal origin. Eggs, meats, and whole milk are our main dietary sources of cholesterol. One large egg yolk contains about 185 milligrams of

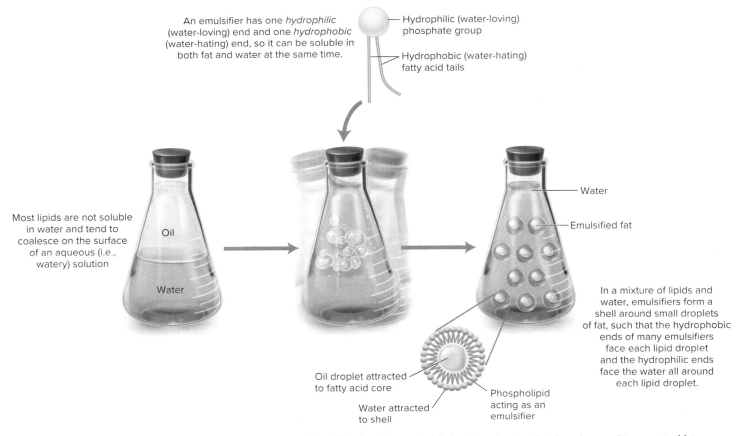

**FIGURE 5-8** ▲ Emulsifiers help to suspend lipids in water. Emulsification is important in food production, fat digestion, and transport of fats through the blood.

**TABLE 5-2 ■ Cholesterol Content of Foods**

| Food Item | Serving Size | Cholesterol (mg) |
|---|---|---|
| Beef brains, cooked | 3 oz | 2635 mg |
| Beef liver, braised | 3 oz | 334 mg |
| Egg yolk* | 1 large | 184 mg |
| Shrimp, cooked | 3 oz | 123 mg |
| Beef, pot roast, braised* | 3 oz | 89 mg |
| Pork loin, roasted | 3 oz | 69 mg |
| Chicken breast, skinless, roasted* | 3 oz | 85 mg |
| Trout, broiled | 3 oz | 63 mg |
| Ice cream chocolate, regular | 1 cup | 50 mg |
| Tuna, broiled | 3 oz | 42 mg |
| Hot dog | 1 each | 33 mg |
| Cheddar cheese* | 1 oz | 28 mg |
| Whole milk* | 1 cup | 24 mg |
| 1% milk | 1 cup | 12 mg |
| Fat-free milk | 1 cup | 5 mg |
| Egg white | 1 large | 0 mg |

*Leading dietary sources of cholesterol in American dietary patterns.

Source: U.S. Department of Agriculture, Agricultural Research Service, FoodData Central, 2019.

cholesterol. Because your body can make all the cholesterol it needs, cholesterol is not an essential nutrient. However, if you do consume dietary sources of cholesterol, it can be absorbed from your small intestine and perform all the same functions as the cholesterol produced by your liver.

Foods of plant origin are naturally cholesterol free, but they may contain phytosterols (e.g., plant sterols and plant stanols). Some phytosterols can help to lower blood cholesterol because they compete with cholesterol for absorptive sites in the small intestine and thereby limit cholesterol absorption. The highest sources of phytosterols include plant oils, nuts, seeds, beans, peas, lentils, and whole grains (refer to the discussion in Section 5.7 on medical interventions to lower blood lipids).

## USING FOOD LABELS TO IDENTIFY FAT

Some fat discussed so far is obvious: butter on bread, mayonnaise in potato salad, and marbling in raw meat. In some foods, however, fat is not immediately obvious. Foods that contain hidden fat include whole milk, pastries, cookies, cake, cheese, hot dogs, crackers, French fries, and ice cream. The fat (and calories) in these foods can add up rather quickly!

A good way to find out about the fat content of the foods you eat is on the food label (Fig. 5-9). The Nutrition Facts label displays total fat, saturated fat, and *trans* fat. You can use that information to select foods that comply with the recommendations of major health authorities, which urge Americans to limit saturated fat intake.

Also, check out the list of ingredients on the food label. Some examples of food ingredients that you may not immediately recognize as fats include tallow (beef fat), lard (pork fat), egg yolks or egg yolk solids, cream, cocoa butter, modified vegetable oils, or shortening. Conveniently, the label lists ingredients in order by weight in the product. If fat is one of the first ingredients listed, you are probably looking at a high-fat product.

**Nutrient Claims You May See on Food Labels**

**Fat**

- **Fat free:** less than 0.5 gram of fat per serving
- **Saturated fat free:** less than 0.5 gram of saturated fat per serving, and the level of *trans* fatty acids does not exceed 0.5 gram per serving
- **Low fat:** 3 grams of fat or less per serving and, if the serving is 30 grams or less or 2 tablespoons or less, per 50 grams of the food; 2% milk cannot be labeled low fat because it exceeds 3 grams per serving. *Reduced fat* is used instead.
- **Low saturated fat:** 1 gram of saturated fat or less per serving and not more than 15% of calories from saturated fatty acids
- **Reduced or less fat:** at least 25% less fat per serving than reference food
- **Reduced or less saturated fat:** at least 25% less saturated fat per serving than reference food

**Cholesterol**

- **Cholesterol free:** less than 2 milligrams of cholesterol and 2 grams or less of saturated fat per serving
- **Low cholesterol:** 20 milligrams or less of cholesterol and 2 grams or less of saturated fat per serving and, if the serving is 30 grams or less or 2 tablespoons or less, per 50 grams of the food
- **Reduced or less cholesterol:** at least 25% less of cholesterol and 2 grams or less of saturated fat per serving than reference food

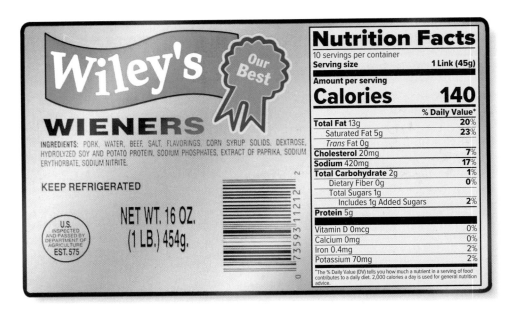

**FIGURE 5-9** ▲ Reading labels helps locate hidden fat. Who would think that wieners (hot dogs) can contain about 85% of food calories as fat? Looking at the hot dog does not suggest that almost all of its food calories come from fat, but the label shows otherwise. Let us do the math: 13 grams total fat × 9 kcal per gram of fat = 120 kcal from fat; 120 kcal/140 kcal per link = 0.86 or 86% kcal from fat.

## FAT IN FOOD PROVIDES SOME SATIETY, FLAVOR, AND TEXTURE

Does fat promote satiety? Fats tend to slow the process of digestion, which could help you to feel full after a meal. However, fats also improve the flavor and texture of foods, so they may stimulate increased food intake. Fat's effect on satiety may depend on the size and saturation of the fatty acids in foods. What everyone knows for sure is that fats contain more than twice the calories of carbohydrates and proteins. Therefore, a high-fat meal is likely to be a high-calorie meal.

Fat provides texture in foods. If you have ever taken a bite of high-quality milk chocolate, you probably agree that the feel of fat melting on the tongue is good. The fat in milk also provides a richness that fat-free milk lacks. The most tender cuts of meat are high in fat, visible as the marbling of meat. In addition, fat carries flavors in foods. Heating spices in oil intensifies the flavors of an Indian curry or a Mexican dish. You can see why reduced-fat foods seem less appealing than their full-fat counterparts!

## FAT-REPLACEMENT STRATEGIES FOR REDUCED-FAT FOODS

Manufacturers have introduced reduced-fat versions of numerous food products. The fat content of these alternatives ranges from 0% in fat-free Fig Newtons to about 75% of the original fat content in other products. However, the total calorie content of most fat-reduced products is not substantially lower than that of the regular products. When fat is removed from a product, what is used in its place? Sugar! It is difficult to reduce both the fat and sugar contents of a product at the same time and maintain flavor and texture. For this reason, many reduced-fat products (e.g., cakes and cookies) are still rich sources of energy. Use the Nutrition Facts label on the food package to choose the portion size that fits into your daily calorie needs.

To lower the fat in foods, manufacturers may replace some of the fat with water, protein (Simplesse®, Dairy-Lo®), or forms of carbohydrates such as starch derivatives (Z-Trim®), fiber (Maltrin®, Stellar™, Oatrim), and gums. In reduced-fat margarines, water replaces some of the fat. While regular margarines are 80% fat by weight

(11 grams per tablespoon), some reduced-fat margarines are as low as 30% fat by weight (4 grams per tablespoon). When used in recipes, the extra water added to these margarines can cause texture and volume changes in the finished product. Cookbooks can suggest alterations in recipes to compensate for the increased water content of these products.

Manufacturers also may use engineered fats, such as olestra (Olean®) and salatrim (Benefat®). These are made with fat and sucrose (table sugar) but provide few or no calories because they cannot be digested and/or absorbed well. The main problem with the fat replacer olestra is that it can bind to fat-soluble vitamins and reduce their absorption.

To date, fat replacements have had little impact on our eating patterns. This is partly because currently approved fat replacements lack versatility and are not used extensively by manufacturers. In addition, fat replacements are not practical for use in the foods that provide the most fat in our eating patterns: beef, cheese, whole milk, and pastries.

Canada has not approved the use of olestra in food products; the United States is the sole country that permits the use of this fat substitute in foods.

## RANCIDITY LIMITS SHELF LIFE OF FOODS

Decomposing oils emit a disagreeable odor, and they taste sour and stale. Foods with lots of unsaturated fatty acids (e.g., liquid plant oils) are most susceptible to **rancidity** because exposure to ultraviolet light, oxygen, and heat can break down the double bonds in these fatty acids when they are stored or heated to high temperatures. Foods most likely to become rancid are deep-fried foods and those with a large exposed surface area. The fat in fish is also susceptible to rancidity because it is highly polyunsaturated. Saturated fats can more readily resist these effects because they contain fewer carbon-carbon double bonds.

Rancidity is not a major problem for consumers because the odor and taste generally discourage us from eating enough decomposed fats to become sick. However, rancidity is a problem for the food manufacturing and restaurant industries because it can reduce a product's shelf life. How can food manufacturers prevent rancidity in food products?

For nearly a century, food manufacturers used a process called **hydrogenation** to make partially hydrogenated oils, which have extended shelf life. In hydrogenation, the addition of hydrogen to some of the double bonds in polyunsaturated fatty acids (liquid at room temperature) could make them more saturated (solid at room temperature), and therefore less likely to decompose. Unfortunately, this method of processing liquid plant oils to make them more solid had the unintended effect of producing some *trans* fatty acids. As mentioned, *trans* fatty acids are similar to saturated fatty acids in both shape and function. In fact, over several decades, research emerged to show that these artificial *trans* fats were even worse for cardiovascular health than saturated fats—they negatively impact blood lipids and increase inflammation in the body.

As these harmful effects on health became apparent, major health authorities, such as the World Health Organization, the AHA, and the *Dietary Guidelines* Advisory Committee, recommended minimizing or avoiding *trans* fats. Since 2006, federal regulations have required the disclosure of *trans* fat content on food labels to improve consumer awareness of *trans* fats in foods (review Fig. 5-9). A few city and state governments made laws to limit the use of *trans* fats by restaurants; some countries banned them altogether. In 2015, the FDA determined that partially hydrogenated oils—the main sources of *trans* fats in typical eating patterns—were no longer "generally recognized as safe" for use in human food products.[4] As of June 2018, partially hydrogenated oils have been banned from use in foods sold in grocery stores and restaurants in the United States.

Now that artificial *trans* fats have been banned, what are food manufacturers using to prevent rancidity in their food products? A different type of engineered fat called *interesterified fat* has been used in some food products to achieve the same shelf stability and food properties as partially hydrogenated oils. Interesterified fats are made by

**rancidity** Production of decomposed fatty acids that have an unpleasant flavor and odor.

**hydrogenation** The addition of hydrogen to a carbon-carbon double bond, producing a single carbon-carbon bond with two hydrogens attached to each carbon.

Some *trans* fatty acids occur naturally. Conjugated linoleic acid (CLA) is a family of fatty acids derived from linoleic acid. The bacteria that live in the rumens of some animals (cows, sheep, and goats, for example) produce *trans* fatty acids from the polyunsaturated fats in the grass the animals are fed. These natural *trans* fats eventually appear in foods such as beef, milk, and butter. CLA contains both *cis* and *trans* bonds, and the *trans* bond is in a different location compared to industrial *trans* fats. CLA may improve insulin levels in diabetics and decrease the risk of heart disease, cancer, and obesity—the very same diseases that industrial *trans* fats have been shown to increase. Isn't it amazing how a slight difference in the chemical structure of a fatty acid leads to vastly different health effects?

The *Dietary Guidelines* tell us to use plant oils instead of solid fats to prepare foods, but how can we keep them from spoiling too quickly? Store them in a cool, dark place, such as the pantry or refrigerator. This will limit exposure to sunlight or heat, which can cause unsaturated fats to break down. Liquid oils may solidify slightly in the refrigerator, but they will quickly return to liquid once they warm to room temperature. If you don't tend to use your oils very quickly, it would be prudent to buy smaller packages or even store them in the freezer. **Why is this bottle of olive oil packaged in a tinted glass bottle?** alexandr kornienko/123RF

**BHA, BHT** Butylated hydroxyanisole and butylated hydroxytoluene: two common synthetic antioxidants added to foods.

If the gallbladder is surgically removed (e.g., in cases of gallstone formation), bile will enter the small intestine directly from the liver. Moderate amounts of fat can still be digested adequately, but individuals may experience loose stools after eating high-fat meals if some fat reaches the large intestine unabsorbed.

enzymatically removing an unsaturated fatty acid from a triglyceride and replacing it with a saturated fatty acid to make a liquid plant oil into a more stable, solid fat without generating *trans* fatty acids. Other food manufacturers are using tropical plant oils, such as coconut oil or palm kernel oil, which are naturally high in saturated fatty acids. Some opt to use the oils from plants that have been genetically modified to produce fatty acids that are more shelf stable. On food labels, these appear as high-oleic soybean oil or high-oleic canola oil.[5]

Another way to prevent rancidity is to carefully package foods to protect them from exposure to light, heat, and oxygen. When you purchase frozen fish, you will notice that it has been vacuum packaged to limit exposure to oxygen. When high-fat snack foods, such as potato chips, are packaged, the bag is filled with an inert gas to prevent oxidation. Opaque packaging can limit exposure to light. Dark-colored glass or plastic packages are used for some plant oils and dietary supplements that contain oils. Addition of **antioxidants,** such as vitamin C, vitamin E, **BHA,** or **BHT,** can help to protect foods against rancidity by blocking the oxidation that causes fats to break down. Antioxidants are added to many processed foods that contain fat, such as salad dressings and cake mixes.

### ✓ CONCEPT CHECK 5.2

1. Name three foods for which at least 50% of total calories come from fat.
2. Where is cholesterol found in the food supply?
3. Which types of fat are used as emulsifiers, and what is their function in food?
4. What are some strategies used to produce reduced-fat foods?
5. How do fats become rancid, and how can this be prevented?

## 5.3 Making Lipids Available for Body Use

It is no secret that fats and oils make foods more appealing. Their presence in foods adds flavor, moistness, and texture. What happens to lipids once they are eaten? Let us take a closer look at the digestion, absorption, and uses of lipids in the body.

### DIGESTION

In the first phase of fat digestion, the stomach (and salivary glands to some extent) secretes the enzyme lipase. Salivary lipase and gastric lipase act primarily on triglycerides that have fatty acids with short chain lengths, such as those found in butterfat. The action of these lipase enzymes, however, is dwarfed by that of the lipase enzyme that is released from the pancreas to digest fats in the small intestine. Most of the triglycerides in a typical dietary pattern contain fatty acids with longer chain lengths; these are generally not digested until they reach the small intestine (Fig. 5-10).

In the small intestine, triglycerides are broken down by lipase into smaller products, namely monoglycerides (glycerol backbones with a single fatty acid attached) and fatty acids. Under the right circumstances, digestion is rapid and thorough. The "right" circumstances depend on the presence of **bile** from the gallbladder. Bile acids (components of bile) act to emulsify fats in the watery digestive juices. This is necessary because the lipid components of food are not soluble in chyme, which is mostly water; lipids tend to coalesce into large globules. Emulsification improves digestion and absorption because it separates large fat globules into smaller ones, thereby increasing the total surface area for lipase action (Fig. 5-11).

# Fat Digestion

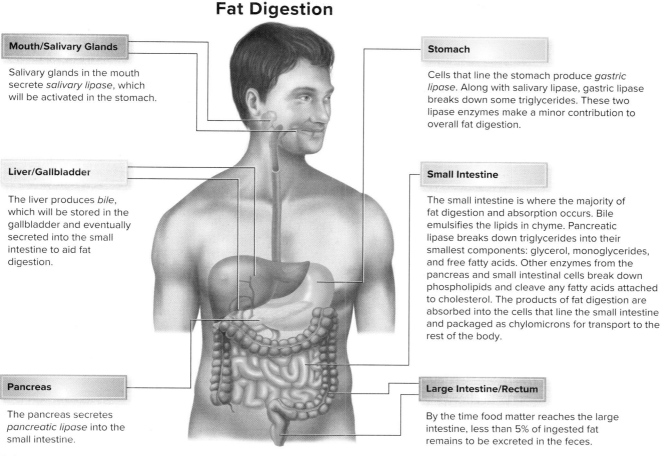

**Mouth/Salivary Glands**

Salivary glands in the mouth secrete *salivary lipase*, which will be activated in the stomach.

**Stomach**

Cells that line the stomach produce *gastric lipase*. Along with salivary lipase, gastric lipase breaks down some triglycerides. These two lipase enzymes make a minor contribution to overall fat digestion.

**Liver/Gallbladder**

The liver produces *bile*, which will be stored in the gallbladder and eventually secreted into the small intestine to aid fat digestion.

**Small Intestine**

The small intestine is where the majority of fat digestion and absorption occurs. Bile emulsifies the lipids in chyme. Pancreatic lipase breaks down triglycerides into their smallest components: glycerol, monoglycerides, and free fatty acids. Other enzymes from the pancreas and small intestinal cells break down phospholipids and cleave any fatty acids attached to cholesterol. The products of fat digestion are absorbed into the cells that line the small intestine and packaged as chylomicrons for transport to the rest of the body.

**Pancreas**

The pancreas secretes *pancreatic lipase* into the small intestine.

**Large Intestine/Rectum**

By the time food matter reaches the large intestine, less than 5% of ingested fat remains to be excreted in the feces.

**FIGURE 5-10** ▲ A summary of fat digestion. Section 3.8 covered general aspects of this process.

What happens to the bile acids after their work in fat digestion is complete? They get recycled! After participating in fat digestion, most bile acids are absorbed in the last segment of the small intestine and transported back to the liver. Through this process, approximately 98% of bile acids are recycled. This recycling of bile is called **enterohepatic circulation**. Only 1% to 2% of bile acids end up in the large intestine to be eliminated in the feces.

Phospholipid digestion is similar to triglyceride digestion. Enzymes from the pancreas and cells in the wall of the small intestine digest phospholipids. The eventual products are glycerol, fatty acids, and the remaining phosphorus-containing parts. With regard to cholesterol digestion, any cholesterol with a fatty acid attached is broken down to free cholesterol and fatty acids by enzymes released from the pancreas. Any fatty acids that are part of these structures could be broken down to yield energy, but contribution to overall calorie intake is miniscule compared to the energy derived from triglycerides.

## ABSORPTION

The products of fat digestion in the small intestine are monoglycerides, free fatty acids, and glycerol. These products diffuse into the absorptive cells of the small intestine. About 95% of dietary fat is absorbed in this way. The chain length of fatty acids dictates how they will be distributed to the rest of the body. If the chain length of a fatty acid is less than 12 carbon atoms, it is relatively soluble in water and small enough to be absorbed into the capillaries and transported via the portal vein directly to the liver. Longer-chain fatty acids are too large to be absorbed directly into portal circulation. These larger products of fat digestion are reformed into triglycerides within the absorptive cells of the small intestine, packaged as chylomicrons, and eventually enter circulation via the lymphatic system.

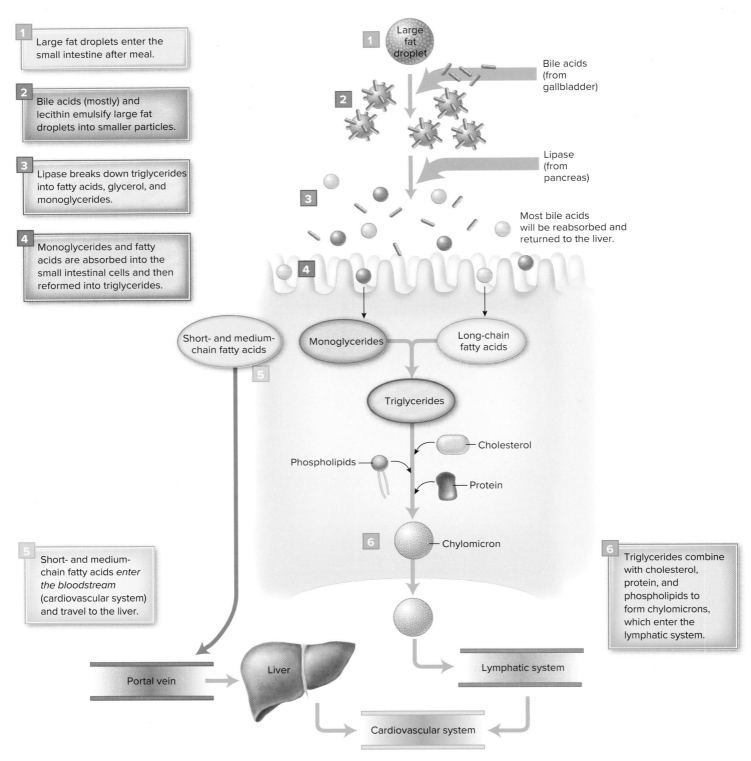

**1** Large fat droplets enter the small intestine after meal.

**2** Bile acids (mostly) and lecithin emulsify large fat droplets into smaller particles.

**3** Lipase breaks down triglycerides into fatty acids, glycerol, and monoglycerides.

**4** Monoglycerides and fatty acids are absorbed into the small intestinal cells and then reformed into triglycerides.

**5** Short- and medium-chain fatty acids *enter the bloodstream* (cardiovascular system) and travel to the liver.

**6** Triglycerides combine with cholesterol, protein, and phospholipids to form chylomicrons, which enter the lymphatic system.

**1** Large fat droplet

Bile acids (from gallbladder)

Lipase (from pancreas)

Most bile acids will be reabsorbed and returned to the liver.

Short- and medium-chain fatty acids

Monoglycerides

Long-chain fatty acids

Triglycerides

Phospholipids

Cholesterol

Protein

**6** Chylomicron

Portal vein

Liver

Lymphatic system

Cardiovascular system

**FIGURE 5-11** ▲ Summary of fat absorption. Bile acids facilitate the digestion of triglycerides by emulsifying fat droplets in the chyme in the small intestine. This allows for efficient action of pancreatic lipase and subsequent absorption of monoglycerides and fatty acids into the mucosal cells of the small intestine.

✓ **CONCEPT CHECK 5.3**

1. What enzymes are responsible for digestion of triglycerides?
2. What are the end products of fat digestion?
3. What are the differences between the absorption of long- versus short-chain fatty acids?

## 5.4 Carrying Lipids in the Bloodstream

As noted earlier, fat and water do not mix easily. This incompatibility presents a challenge for the transport of fats through the blood and lymph, which are mostly water. **Lipoproteins** serve as vehicles for transport of lipids from the small intestine and liver to the body tissues. They consist of a core of triglycerides and cholesterol surrounded by a shell of protein and phospholipids (Fig. 5-12).

Lipoproteins are classified into four groups, as detailed below (Table 5-3). Chylomicrons originate from the cells of the small intestine and are made of dietary fats. VLDLs, LDLs, and HDLs originate from the liver.

**lipoprotein** A compound found in the bloodstream containing a core of triglycerides and cholesterol surrounded by a shell of protein and phospholipids.

Proteins located in the shell help to identify the lipoprotein.

Triglycerides are housed in the core of the lipoprotein.

Phospholipids form the shell of the lipoprotein. Their phosphate heads are water-soluble, so they face the watery environment of the blood or lymph. Their fatty acid tails are oriented toward the lipid core of the lipoprotein.

Cholesterol is transported within the lipid core of the lipoprotein. Some cholesterol also lends structural support to the lipoprotein's outer shell.

**FIGURE 5-12** ◄ The structure of a lipoprotein. All the lipoproteins are made of triglycerides, cholesterol, phospholipids, and protein in various proportions. This unique structure allows fats to circulate in the water-based bloodstream.

**TABLE 5-3** ■ **Composition and Roles of the Major Lipoproteins in the Blood**

| Lipoprotein | Primary Component | Key Role |
|---|---|---|
| Chylomicron | Triglyceride | Carries dietary fat from the small intestine to cells |
| VLDL | Triglyceride | Carries lipids made and taken up by the liver to cells |
| LDL | Cholesterol | Carries cholesterol made by the liver and from other sources to cells |
| HDL | Protein | Contributes to cholesterol removal from cells and, in turn, excretion of it from the body |

Torbjorn Lagerwall/Alamy Stock Photo

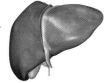
MedicalRF.com

☐ Triglyceride    ▨ Cholesterol    ▨ Phospholipid    ■ Protein

## CHYLOMICRONS TRANSPORT DIETARY FATS

**chylomicron** Lipoprotein that carries dietary fats (mainly triglycerides) from the small intestine to cells throughout the body. Chylomicrons are first absorbed into the lymphatic system, and then transported via the bloodstream.

Digestion of dietary fats results in a mixture of glycerol, monoglycerides, and fatty acids. Once these products are absorbed by the cells of the small intestine, they are reassembled into triglycerides. Then, the intestinal cells package the triglycerides into **chylomicrons.** Like the other lipoproteins, chylomicrons are composed of large droplets of triglycerides and cholesterol surrounded by a thin, water-soluble shell of phospholipids and protein (see Fig. 5-12). The water-soluble shell around a chylomicron allows the lipid to float freely in the lymph and blood, which are both water based. Some of the proteins in the shell may also help other cells identify the lipoprotein as a chylomicron.

Chylomicrons are loaded with dietary fat. They are the largest of the lipoproteins—too large to enter the capillaries that run through each villus. Therefore, they pass from the small intestinal cells into the **lacteals,** which are part of the lymphatic system. Chylomicrons travel through the lymph vessels, which eventually empty into the circulatory system through a duct near the heart.

**lipoprotein lipase** An enzyme attached to the cells that form the inner lining of blood vessels; it breaks down triglycerides into free fatty acids and glycerol.

Once a chylomicron enters the bloodstream, the triglycerides in its core are broken down into fatty acids and glycerol by yet another lipase enzyme. This one is called **lipoprotein lipase,** and it is attached to the inside walls of the blood vessels. As soon as the fatty acids are released to the bloodstream, they are absorbed by nearby cells. The remaining glycerol backbone circulates back to the liver. Muscle cells can immediately use the absorbed fatty acids for fuel. Adipose cells, on the other hand, tend to reassemble the fatty acids into triglycerides for storage.

**chylomicron remnant** Lipoprotein that remains after triglycerides have been removed from a chylomicron; composed of protein, phospholipids, and cholesterol.

What happens to the rest of the chylomicron after triglycerides have been removed? The leftover materials are called **chylomicron remnants.** Chylomicron remnants are removed from circulation by the liver, and their components are recycled to make other lipoproteins and bile acids.

## VLDLs AND LDLs TRANSPORT LIPIDS FROM THE LIVER TO THE BODY CELLS

The liver takes up various lipids from the blood. The liver also is the manufacturing site for lipids and cholesterol. The raw materials for lipid and cholesterol synthesis include free fatty acids taken up from the bloodstream, as well as carbon and hydrogen derived from carbohydrates, protein, and alcohol. The liver then must package the lipids it makes into lipoproteins for transport through the blood to body tissues.

**very-low-density lipoprotein (VLDL)** Lipoprotein that carries triglycerides (and cholesterol) from the liver to cells throughout the body.

First in our discussion of lipoproteins made by the liver are **very-low-density lipoproteins (VLDL).** These particles are composed of cholesterol and triglycerides surrounded by a water-soluble shell of phospholipids and protein. VLDLs are rich in triglycerides and thus are very low in density (Table 5-3). Once in the bloodstream, lipoprotein lipase on the inner surface of the blood vessels breaks down the triglycerides into fatty acids and glycerol. Fatty acids and glycerol are released from the VLDL into the bloodstream and are taken up by the body cells.

**low-density lipoprotein (LDL)** Lipoprotein that remains after most of the triglycerides have been removed from a VLDL; transports cholesterol from the liver to cells throughout the body. Elevated LDL is strongly linked to cardiovascular disease risk, so it is sometimes called *bad cholesterol.*

As its triglycerides are released, the VLDL becomes proportionately more dense and is known as a **low-density lipoprotein (LDL).** Now, most of what remains in the core of the lipoprotein is cholesterol. Thus, the role of LDL is to transport cholesterol to tissues. LDL particles are taken up from the bloodstream by specific receptors on cells, especially liver cells, and are then broken down. The cholesterol and protein components of LDL provide some of the building blocks necessary for cell growth and development, such as synthesis of cell membranes and hormones.

**high-density lipoprotein (HDL)** Lipoprotein that picks up cholesterol from dying cells and other sources and transfers it to the other lipoproteins in the bloodstream or directly to the liver. Higher HDL levels are associated with decreased risk for cardiovascular disease, so it is sometimes called *good cholesterol.*

## HDLs REMOVE CHOLESTEROL FROM THE BLOOD

The final group of lipoproteins, **high-density lipoproteins (HDL),** is a critical and beneficial participant in this process of lipid transport. Its high proportion of protein makes it the most dense lipoprotein. The liver synthesizes most of the HDL in the body (70% to 80%); the intestinal cells make the remainder. It roams the bloodstream, picking up cholesterol from dying cells and other sources. HDL donates the cholesterol primarily to other lipoproteins for transport back to the liver to be excreted. Some HDL travels directly back to the liver. This process is called **reverse cholesterol transport**.

**reverse cholesterol transport** Process by which HDL picks up cholesterol from the tissues and blood vessels and takes it to the liver for metabolism or excretion.

## "GOOD" AND "BAD" CHOLESTEROL IN THE BLOODSTREAM

HDL and LDL are often described as "good" and "bad" cholesterol, respectively. Many studies demonstrate that the amount of HDL in the bloodstream can closely predict the risk for cardiovascular disease. Risk increases with low HDL because little cholesterol is transported back to the liver and excreted. Compared to men, women tend to have high amounts of HDL, especially before **menopause.** High amounts of HDL slow the development of cardiovascular disease, so any cholesterol carried by HDL can be considered "good" cholesterol.

**menopause** The cessation of the menstrual cycle in women, usually beginning at about 50 years of age.

On the other hand, LDL is sometimes considered "bad" cholesterol. In our discussion of LDL, you learned that it is taken up by receptors on various cells. If LDL is not readily cleared from the bloodstream, **scavenger cells** in the arteries take up the lipoprotein, leading to a buildup of cholesterol in the blood vessels. This buildup, known as **atherosclerosis,** greatly increases the risk for cardiovascular disease. Low amounts of LDL are needed as part of routine body functions, but a high level of LDL cholesterol is a risk factor for cardiovascular disease. Read more about the link between LDL and cardiovascular disease in Section 5.7.

**scavenger cells** Specific form of white blood cells that can bury themselves in the artery wall and accumulate LDL. As these cells take up LDL, they contribute to the development of atherosclerosis.

**atherosclerosis** A buildup of fatty material (plaque) in the arteries, including the arteries that supply blood to the heart.

Here is an important point: the cholesterol in *foods* is not designated as "good" or "bad." It is only after cholesterol has been made or processed by the *liver* that it shows up in the bloodstream as part of LDL or HDL. Dietary patterns can certainly affect the lipoproteins in your blood. The cholesterol in your dietary pattern actually does not have a very big impact on the levels of LDL and HDL in your blood. If your body's regulatory mechanisms are working properly, when your dietary intake of cholesterol increases, your liver's synthesis of cholesterol should decrease. Research shows that *saturated fat* is the dietary lipid that has the biggest impact on blood lipid levels. Higher intakes of saturated fat tend to increase blood levels of LDL. This is why you hear so much dietary advice to limit saturated fat intake!

## ✓ CONCEPT CHECK 5.4

1. Describe how the structure of lipoproteins allows fats to be transported through the watery environment of the lymph and blood.
2. How are dietary fats packaged in the small intestine and transported?
3. Where are VLDLs made, and what do they contain?
4. Where do LDLs originate, and what is their role in the body?
5. Why are HDLs considered "good" cholesterol?

# 5.5 Roles of Lipids in the Body

Many vital body functions rely on lipids. They provide energy, serve as structural components of every cell, and regulate cellular processes.

## PROVIDING ENERGY

Fatty acids (part of triglycerides) are an important source of fuel for the body. Within cells, the breakdown of the chemical bonds between the carbon, hydrogen, and oxygen atoms in fatty acids releases energy that can be used to synthesize adenosine triphosphate (ATP), the source of energy for all cellular processes. Overall, about half of the energy used by the entire body at rest and during light activity comes from fatty acids; the remainder is derived mostly from carbohydrates.

▲ At rest or during light activity, the muscles use mostly fatty acids for fuel. mr.jerry/Flickr RF/Getty Images

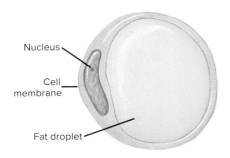

Nucleus

Cell membrane

Fat droplet

**FIGURE 5-13** ▲ An adipose cell.

## STORING ENERGY FOR LATER USE

We store energy mainly in the form of triglycerides. The body's ability to store fat is essentially limitless. Our fat storage sites, adipose cells (Fig. 5-13), can increase about 50 times in weight. If the amount of fat to be stored exceeds the ability of the existing cells to expand, the body can form new adipose cells.

An important advantage of using triglycerides to store energy in the body is that they can pack a lot of energy into a small space. Recall that fats yield, on average, 9 kcal per gram, whereas proteins and carbohydrates yield only about 4 kcal per gram. **Adipose tissue** contains about 80% lipid and only 20% water and protein. In contrast, muscle tissue is about 73% water. What if we had to store energy as muscle tissue? In addition, triglycerides are chemically stable, so they are not likely to react with other cell constituents. Fat is certainly a safe and efficient way to store energy.

## INSULATING AND PROTECTING THE BODY

The layer of fat that lies just below the skin serves as insulation to keep the body warm. Due to its low water content, adipose tissue does not conduct heat very well, so it slows the loss of heat from the body in a cold environment. In addition to insulating the body, fat tissue also surrounds and protects some organs (e.g., kidneys) from injury.

## DIGESTION, ABSORPTION, AND NUTRIENT TRANSPORT

Phospholipids and cholesterol participate in digestion and absorption. Cholesterol is the building block for bile acids, which you learned about in Section 3.8. Bile acids and phospholipids function as emulsifiers, which enable lipids to be suspended in water. During digestion, these emulsifiers help to break apart large droplets of fat into smaller droplets so that digestive enzymes can perform the work of breaking down the fats in a meal.

Lipids in food also carry fat-soluble vitamins to the small intestine and aid their absorption. People who absorb fat poorly (e.g., due to medical conditions or gastrointestinal surgery) are at risk for deficiencies of fat-soluble vitamins, especially vitamin K. A similar problem could result from frequent use of mineral oil as a laxative. The body cannot digest or absorb mineral oil; if taken with a meal, the undigested mineral oil carries fat-soluble vitamins from the meal through the GI tract and out of the body as part of feces. Fat malabsorption can impair mineral status as well. Within the GI tract, unabsorbed fatty acids can bind to some minerals, such as calcium and magnesium, and prevent them from being absorbed. As the fat is eliminated from the body, minerals will be lost, too.

Once absorbed, phospholipids are necessary for the transport of lipid-soluble materials through the lymph and blood. Recall that phospholipids form the outer shell of the lipoproteins: chylomicrons, VLDLs, LDLs, and HDLs.

## FORMING BODY STRUCTURES

Lipids are structural components of all cells. Phospholipids—with their water-soluble phosphate heads and fat-soluble fatty acid tails—have the unique ability to be simultaneously soluble in water and fat. In cell membranes, this property allows phospholipids to form a bilayer (Fig. 5-14). The phospholipids line up so that all the fatty acid tails point toward the interior of the membrane and all the phosphate heads face the watery environments on the inside (cytosol) or outside of the cell. The fatty acids, many of which are unsaturated, in the interior of the cell membrane lend fluidity to the cell membrane. Cholesterol, which is also present within the phospholipid bilayer, adds stability to the structure of cell membranes. Because it is made mostly of lipids, the cell membrane is impermeable to most water-soluble

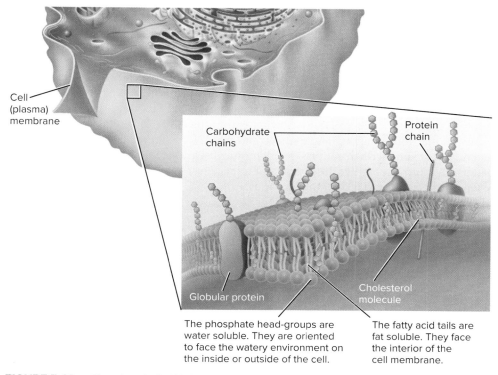

Cell
(plasma)
membrane

Carbohydrate
chains

Protein
chain

Globular protein

Cholesterol
molecule

The phosphate head-groups are
water soluble. They are oriented
to face the watery environment on
the inside or outside of the cell.

The fatty acid tails are
fat soluble. They face
the interior of the
cell membrane.

**FIGURE 5-14** ▲ The phospholipid bilayer. This cross-section of the cell membrane shows that phospholipids are the main components of cell membranes, forming a double layer (bilayer) of lipids.

substances (e.g., ions, glucose, amino acids, waste products). Thus, the movement of substances across the cell membrane often requires the help of proteins, which serve as gates, pumps, or channels.

Lipids are especially important for the structure of the cells of the nervous system. Although the exact composition changes throughout the life cycle, lipids make up about 60% of the weight of the human brain. Phospholipids and cholesterol are major components of the myelin sheath that surrounds and protects nerve cells. Myelin participates in rapid cell-to-cell communication within the nervous system.

## REGULATION AND COMMUNICATION

Lipids are potent regulators of chemical processes within the body. Sterols are needed for the synthesis of hormones, including estrogen, testosterone, and the active form of vitamin D. Individual fatty acids also orchestrate many aspects of human metabolism. As you learned in Section 5.1, cells can convert the essential omega-3 and omega-6 fatty acids into other fatty acids, which give rise to eicosanoids. Eicosanoids act as chemical messengers that direct growth and development, immune function, and the work of the central nervous system.

**COVID CORNER**

Although there is heightened interest in using omega-3 supplements to prevent or treat the extreme inflammatory response that accompanies COVID-19, there is insufficient evidence to recommend omega-3 fatty acid supplements to boost viral immunity among otherwise healthy adults. Incorporating food sources of omega-3 fatty acids, such as seafood, walnuts, and flaxseeds, into your dietary pattern is a safe way to support your immune system.

## ✓ CONCEPT CHECK 5.5

1. What are the functions of triglycerides in the body?
2. Where are phospholipids found in the body?
3. What are some compounds that are made from cholesterol in the body?

# 5.6 Recommendations for Fat Intake

## TOTAL FAT

There is no Recommended Dietary Allowance (RDA) for total fat for adults. Rather, the Food and Nutrition Board has set an Acceptable Macronutrient Distribution Range (AMDR) of 20% to 35% of total calories. This equates to 44 to 78 grams per day for a person who consumes 2000 kcal daily. Staying within the AMDR for total fat intake helps to ensure that you get enough fat to meet your needs for essential fatty acids, yet not so much fat that you increase your risks for chronic diseases. Table 5-4 illustrates a dietary pattern that provides either 20% or 30% of total calories as fat. Typically, the fat intake of Americans is around 33% of total calories.

Nut butters are plant products, so they are naturally cholesterol free and provide 2 or 3 grams of fiber per serving, which can lower blood cholesterol. Nuts are also a good source of unsaturated fats, which tend to promote heart health. Two tablespoons of peanut butter provide about 12 grams of unsaturated fats and only 3 grams of saturated fat. Be on the lookout for products with high sugar content. Nut butters naturally contain 1 or 2 grams of sugar, but processed varieties (e.g., cinnamon swirl) may contain up to 9 grams of sugar. **How can you determine which nut butters provide heart-healthy fats without added sugars?** Igor Dutina/iStock/Getty Images

**TABLE 5-4** ■ Daily Menu Examples Containing 2000 kcal and 30% or 20% of Calories as Fat

| 30% of Calories as Fat | | 20% of Calories as Fat | |
|---|---|---|---|
| Food | Fat (grams) | Food | Fat (grams) |
| **Breakfast** | | | |
| Orange juice, 1 cup | 0.5 | Orange juice, 1 cup | 0.5 |
| Shredded wheat, ¾ cup | 0.5 | Shredded wheat, 1 cup | 0.7 |
| Toasted whole grain bagel | 1.1 | Toasted whole grain bagel | 1.1 |
| Peanut butter, 3 teaspoons | 8.0 | Peanut butter, 3 teaspoons | 8.0 |
| 1% low-fat milk, 1 cup | 2.5 | Fat-free milk, 1 cup | 0.6 |
| **Lunch** | | | |
| Whole wheat bread, 2 slices | 2.4 | Whole wheat bread, 2 slices | 2.4 |
| Roast beef, 2 ounces | 4.9 | Light turkey roll, 2 ounces | 0.9 |
| Mustard, 3 teaspoons | 0.6 | Mustard, 3 teaspoons | 0.6 |
| Swiss cheese, 2 slices | 15.6 | Swiss cheese, 1 slice | 7.8 |
| Lettuce | — | Lettuce | — |
| Tomato | — | Tomato | — |
| Oatmeal cookie, 1 | 3.3 | Oatmeal cookie, 2 | 6.6 |
| **Snack** | | | |
| Apple | — | Apple | — |
| **Dinner** | | | |
| Chicken tenders frozen meal | 18.0 | Fat-free chicken tenders | — |
| Carrots, ½ cup | — | Carrots, ½ cup | — |
| Dinner roll, 1 | 2.0 | Dinner roll, 1 | 2.0 |
| Olive oil, 1.5 teaspoons | 6.8 | Olive oil, 1.5 teaspoons | 6.8 |
| Banana | 0.6 | Banana | 0.6 |
| 1% low-fat milk, 1 cup | 2.5 | Fat-free milk, 1 cup | 0.6 |
| **Snack** | | | |
| Raisins, 2 teaspoons | — | Raisins, ½ cup | — |
| Air-popped popcorn, 3 cups | 1.0 | Air-popped popcorn, 6 cups | 2.0 |
| Parmesan cheese, 2 tablespoons | 2.8 | Parmesan cheese, 2 tablespoons | 2.8 |
| **Totals** | **73.1** | | **44.0** |

Please note that fat recommendations for infants and children are different from those of adults (see Section 15.2). Youngsters are forming new tissue that requires fat, especially in the brain, so their intake of fat and cholesterol should not be greatly restricted.

## ESSENTIAL FATTY ACIDS

If humans fail to consume enough linoleic acid and alpha-linolenic acid, signs of essential fatty acid deficiency may appear. The first signs of essential fatty acid deficiency usually involve epithelial tissues because these cells are rapidly turned over; the skin becomes flaky and itchy, wound healing may be slow, and the individual may experience diarrhea. Over time, hair may fall out or lose its pigment. Due to the role of fatty acids in regulating the immune response, patients with essential fatty acid deficiency are more vulnerable to infections. Among children, growth may be restricted.[6]

The Food and Nutrition Board has set Adequate Intakes (AIs) for linoleic acid and alpha-linolenic acid. Table 5-5 lists the AIs for adults. To meet our dietary requirements for essential fatty acids, we only need to consume 2 to 4 tablespoons of plant oil each day, which equates to roughly 5% of our total calories. We can easily get that much from cooking oils, salad dressings, nuts and seeds, vegetables, and whole grain breads. Even a low-fat eating pattern will provide enough essential fatty acids if it follows a balanced plan such as MyPlate. Except in cases of GI tract diseases, eating disorders, or inadequate intravenous feeding, essential fatty acid deficiencies are quite rare in developed regions of the world.

**TABLE 5-5** ■ **Food and Nutrition Board Recommendations for Essential Fatty Acids**

| | Men (g/d) | Women (g/d) |
|---|---|---|
| Linoleic acid (omega-6) | 17 | 12 |
| Alpha-linolenic acid (omega-3) | 1.6 | 1.1 |

## SATURATED FAT

The *Dietary Guidelines* urge all Americans, starting at age 2, to limit saturated fat intake to less than 10% of total calories. Both the *Dietary Guidelines* and the AHA recommend replacing saturated fats with unsaturated fats by choosing a dietary pattern that emphasizes fruits, vegetables, whole grains, plant sources of protein (e.g., beans, peas, lentils, nuts, and seeds), fish, and lean animal proteins. These recommendations are based on evidence linking saturated fat intake with increased LDL levels in the blood, which tend to promote atherosclerosis.[7] Typical saturated fat intake among Americans is about 12% of total calories, so there is room for improvement in this area. Remember: fats that are solid at room temperature tend to be high in saturated fatty acids. Choosing nontropical plant oils, nuts, seeds, and fish (sources of unsaturated fatty acids) in place of animal sources of fat and tropical oils are effective strategies to reduce saturated fat intake.[8] Try *Rate Your Plate*: "Is Your Dietary Pattern High in Saturated Fat?" in Connect.

## CHOLESTEROL

About two-thirds of the cholesterol circulating through your body is made by body cells; only one-third comes from a typical dietary pattern. Each day, your body's cells produce approximately 875 milligrams of cholesterol. In addition to the cholesterol that cells make, Americans typically consume about 180 to 325 milligrams of cholesterol per day from animal-derived food products. Most of the time, cholesterol synthesis by the body is well regulated: if you eat more dietary cholesterol, the liver synthesizes less cholesterol.

Notice that the latest guidelines from the AHA set no specific limits on dietary cholesterol. This is because dietary cholesterol intake has little impact on blood cholesterol levels. Instead, evidence shows that saturated fat and *trans* fat have the greatest impact on blood lipids. That doesn't mean you should eat limitless quantities of cholesterol-rich foods. In fact, the *Dietary Guidelines* advise Americans to keep dietary cholesterol intake as low as possible while still meeting your needs for essential nutrients. As it turns out, rich food sources of cholesterol are usually sources of saturated fat as well. Thus, as you reduce the saturated fat in your eating pattern, you will likely reduce your cholesterol intake, too. The relationship between dietary lipids and cardiovascular disease risk will be discussed further in Section 5.7.

**Do fish oil capsules leave a nasty taste in your mouth?** First, make sure that your fish oil has not gone rancid. Pay attention to the expiration date on the package. Break open one capsule and smell it. If it smells foul, you should discard it. If your fish oil is fresh and you still experience "fish burps," you could try enteric-coated capsules, which will be digested after the capsules have passed the stomach. Freezing the capsules or taking them with a meal rather than on an empty stomach may also help to reduce the fishy aftertaste. Roman legoshyn/123RF

**ischemic stroke** Damage to part of the brain resulting from lack of blood flow to the brain.

**hemorrhagic stroke** Damage to part of the brain resulting from rupture of a blood vessel and subsequent bleeding within or over the internal surface of the brain.

## OMEGA-3 FATTY ACIDS

Have you heard advice to include seafood in your dietary pattern two times per week? The omega-3 fatty acids in seafood—namely EPA and DHA—help to regulate body processes that may promote heart health, protect brain health, and reduce the inflammation associated with rheumatoid arthritis.

How do these fatty acids influence health? After you eat foods containing fat, the fatty acids are taken up by your cells, where they can be incorporated into cell membranes or used to synthesize signaling molecules that influence blood clotting, inflammation, and heart rhythm. Indeed, many human studies show that people who eat fish once or twice a week (total weekly intake of 8 ounces) have lower risks for sudden cardiac death, coronary heart disease, and **ischemic stroke** compared to people who rarely eat fish. One to two servings per week of seafood that is not breaded or fried seems to be just the right amount for health benefits; higher intakes of seafood do not provide additional benefit.[9]

Although many people would benefit from consuming more omega-3 fatty acids, excessive intakes can be unhealthy, too. Certain groups of people, such as the Greenlandic Inuit, eat so much seafood that their normal blood-clotting ability can be impaired. This increases the risk for uncontrolled bleeding and may cause **hemorrhagic stroke.** An excessive intake of long-chain fatty acids from seafood can also depress immune function.

The *Dietary Guidelines* and the AHA Diet and Lifestyle Recommendations encourage Americans to eat 2 or more servings of fatty fish each week. Fish is not only a rich source of omega-3 fatty acids but also a valuable source of protein and trace elements that may protect the cardiovascular system. Broiled or baked fish is recommended rather than fried fish because frying may increase the ratio of omega-6 to omega-3 fatty acids and may produce *trans* fatty acids and oxidized lipid products that tend to increase cardiovascular disease risk.

Although consuming fish is thought to have greater benefits, fish oil capsules can be safely substituted for fish if a person does not like fish. Generally, about 1 gram of omega-3 fatty acids (about three capsules) from fish oil per day is recommended, especially for people with evidence of cardiovascular disease. The AHA also recently suggested that fish oil supplements (providing 2 to 4 grams of omega-3 fatty acids per day) could be employed to treat elevated blood triglycerides. However, a recent science advisory from the AHA noted there is insufficient evidence to recommend fish oil supplementation for prevention of cardiovascular diseases among the general population (i.e., individuals without existing cardiovascular disease).[10] Fish oil capsules should be limited for individuals who have bleeding disorders, take anticoagulant medications, or anticipate surgery because the extra omega-3 fatty acids may increase risk of uncontrollable bleeding and hemorrhagic stroke. Thus, for fish oil capsules, as well as other dietary supplements, it is important to follow the recommendations of your primary health care provider. Remember that fish oil supplements are not closely regulated by the FDA; the quality of these supplements is not standardized, and contaminants naturally present in the fish oil may not have been removed.

## EATING WELL FOR HEART HEALTH

Despite the potential to control many risk factors through dietary and lifestyle changes, cardiovascular disease remains the leading cause of death and disability worldwide. The AHA Diet and Lifestyle Recommendations (summarized in Table 5-6) aim to help Americans achieve and maintain a healthy body weight, blood lipid profile, blood pressure, as well as blood glucose levels. Most of the recommendations are meant for the general public; a few are targeted for people who are at high risk or who currently have cardiovascular disease.

In recent years, the Mediterranean diet has attracted a lot of attention because of the lower rates of chronic diseases seen among people following such a dietary plan.[11] Reduction of cardiovascular disease has been one of the most consistent results of the

**TABLE 5-6 ■ American Heart Association's *Diet and Lifestyle Recommendations***

Use up at least as many calories as you take in.

Eat a variety of nutritious foods from all the food groups, including:
• A variety of fruits and vegetables
• Whole grains
• Low-fat dairy products
• Skinless poultry and fish
• Nuts and legumes
• Nontropical vegetable oils

Eat less of the nutrient-poor foods:
• Limit foods and beverages that are high in calories but low in nutrients.
• Limit the amount of saturated fat, *trans* fat, and sodium you eat.

As you make daily food choices, base your eating pattern on these recommendations:
• Eat a variety of fresh, frozen, and canned fruits and vegetables without high-calorie sauces or added salt and sugars.
• Choose fiber-rich whole grains for most grain servings.
• Choose poultry and fish without the skin and prepare them in healthy ways.
• Eat a variety of fish (especially those rich in omega-3 fatty acids) at least twice a week.
• Select fat-free and low-fat dairy products.
• Omit foods containing partially hydrogenated vegetable oils to avoid *trans* fat.
• Limit saturated fat and *trans* fat and replace them with monounsaturated and polyunsaturated fats. If you need to lower blood cholesterol, reduce saturated fats to no more than 5% to 6% of total calories.
• Cut back on beverages and foods with added sugars.
• Choose foods with less sodium and prepare foods with little or no salt.
• If you drink alcohol, drink in moderation (no more than 1 drink per day for women and no more than 2 drinks per day for men).
• Follow the AHA recommendations when you eat out and keep an eye on portion sizes.

Source: Eckel RH et al.: 2013 AHA/ACC guideline on lifestyle management to reduce cardiovascular risk: A report of the American College of Cardiology/American Heart Association Task Force on Practice Guidelines. *Circulation* 129:S76, 2013.

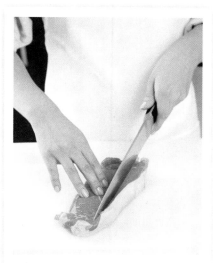

Trimming the visible fat from meats can help reduce saturated fat intake, but you cannot remove the marbling (streaks of fat running through the meat). **Is it possible to incorporate red meats into a heart-healthy dietary pattern?**
Chris Stein/Getty Images

Mediterranean diet. The major sources of fat in the Mediterranean diet include liberal amounts of olive oil compared to a small amount of animal fat (from animal flesh, eggs, and dairy products). In contrast, major sources of fat in the typical North American eating pattern include animal flesh, whole milk, pastries, cheese, margarine, and mayonnaise.

While dietary fat sources definitely play a role in prevention of chronic disease, it is important to remember that other aspects of one's lifestyle also contribute to disease risk. People who follow a Mediterranean diet also tend to consume moderate alcohol (usually in the form of red wine, which contains many antioxidants), eat plenty of whole grains and few highly processed carbohydrates, and are also more physically active than typical North Americans.[12]

An alternative plan for reduction of cardiovascular disease is Dr. Dean Ornish's purely vegetarian **(vegan)** dietary plan.[13] This dietary pattern is very low in fat, including only a scant quantity of vegetable oil used in cooking and the small amount of oils present in plant foods. Individuals restricting fat intake to 20% of calories should be monitored by a physician, as the resulting increase in carbohydrate intake can increase blood triglycerides in some people, which is not a healthful change. Over time, however, the initial problem of high blood triglycerides on a low-fat dietary pattern may self-correct. Among people following the Ornish plan, blood triglycerides initially increased but, within a year, fell to normal values as long as the individuals emphasized high-fiber carbohydrate sources, controlled (or improved) body weight, and followed a regular exercise program.

# Newsworthy Nutrition

## Effects of almond versus cracker snacks among college students

**INTRODUCTION:** During the transition to college, changes in dietary patterns, such as breakfast skipping, nutritionally inadequate food choices, and overall energy intake, may negatively impact the cardiometabolic health of young adults. Previous research has demonstrated a positive impact of nut consumption on risks for metabolic syndrome, cardiovascular disease, and type 2 diabetes. However, no intervention has examined the effects of nut consumption on health parameters among young adults, who are forming dietary habits that may persist throughout adulthood. **OBJECTIVES:** To evaluate the effects of snacks of almonds versus snacks of graham crackers on markers of cardiovascular health and blood glucose control among first-year college students. **METHODS:** Using a *randomized, controlled trial design,* 73 first-year college students were randomized to two groups: one group consumed a mid-morning almond snack (2 ounces) and the other group consumed a mid-morning graham cracker snack (5 sheets of crackers) daily for 8 weeks. Anthropometric measurements included body mass, height, waist circumference, and bioelectric impedance analysis of fat mass and fat-free mass. Biochemical analyses included glucose, insulin, and blood lipids, as well as several markers of insulin sensitivity and inflammation. **RESULTS:** Despite an increase in body mass over the 8-week period in both groups, almond snacking resulted in improved blood lipids and insulin sensitivity compared to cracker snacking. The almond snacking group experienced smaller reductions in HDL, 13% lower glucose area under the curve, and 34% lower insulin resistance index compared to the cracker snacking group. Both groups experienced improvements in fasting glucose and LDL. **CONCLUSION:** Snacking on almonds may improve cardiometabolic health among young adults. This is important because dietary behaviors established during young adulthood could influence health outcomes later in life.

Source: Dhillon G and others: Glucoregulatory and cardiometabolic profiles of almond vs. cracker snacking for 8 weeks in young adults: A randomized controlled trial. *Nutrients* 2018; 10:960. DOI:10.3390/nu10080960.

## FAT *QUALITY* VERSUS FAT *QUANTITY*

Weight control is a vitally important way to minimize risk for a variety of chronic diseases. Because it is a dense source of calories, many people seek to limit fat intake as a way to manage body weight. However, it is important to recognize that excess calories from any source—fat, carbohydrate, protein, or alcohol—contribute to weight gain. If weight loss is needed, understand that your overall calorie intake is more important than the specific macronutrient composition of your eating pattern.

As you have learned, the general consensus among nutrition experts is that fat should comprise 20% to 35% of total calories. When it comes to improving your dietary pattern, it is usually not necessary to drastically lower the *quantity* of fat in your dietary pattern. However, we should aim to choose more foods that provide unsaturated fats, including more omega-3 fatty acids, while choosing fewer sources of saturated and *trans* fats.[14] In short, most Americans need to improve their fat *quality.*

Simply selecting the reduced-fat versions of pastries, cookies, and cakes will not improve the quality of your dietary pattern. Often, extra sugar or salt has been added to these foods to compensate for losses of taste and texture. Instead, focus your efforts on choosing plant sources of fat more often than animal sources (Table 5-7). Fruits, vegetables, and whole grains are usually low in total fat and saturated fat; the (mostly unsaturated) fats they do provide are accompanied by vitamin E, vitamin K, and a variety of health-promoting phytochemicals.[8]

**TABLE 5-7** ■ **Tips for Cutting Back on Saturated Fats and *Trans* Fats**

|  | Eat Less of These Foods | Eat More of These Foods |
|---|---|---|
| Grains | Pasta dishes with cheese or cream sauces<br>Croissants<br>Pastries<br>Doughnuts<br>Pie crust | Whole grain breads<br>Whole grain pasta<br>Brown rice<br>Air-popped popcorn |
| Vegetables | French fries<br>Potato chips<br>Vegetables cooked in butter, cheese, or cream sauces | Fresh, frozen, baked, or steamed vegetables |
| Fruit | Fruit pies | Fresh, frozen, or canned fruits |
| Dairy | Whole milk<br>Ice cream<br>High-fat cheese<br>Cheesecake | Fat-free and reduced-fat milk<br>Low-fat frozen desserts (e.g., yogurt, sherbet, and ice milk)<br>Reduced-fat/part-skim cheese |
| Protein | Bacon<br>Sausage<br>Organ meats (e.g., liver)<br>Egg yolks | Fish<br>Skinless poultry<br>Lean cuts of meat (with fat trimmed away)<br>Beans, peas, and lentils<br>Egg whites/egg substitutes<br>Nuts and nut butters |
| Fat and Oils | Butter<br>Lard | Plant oils |

## CASE STUDY   Planning a Heart-Healthy Dietary Pattern

Jackie is a 21-year-old, health-conscious individual majoring in business. She recently learned that an eating pattern high in saturated fat can contribute to high blood cholesterol and that exercise is beneficial for the heart. Jackie now takes a brisk 30-minute walk each morning before going to class, and she has started to cut as much fat out of her dietary pattern as she can, replacing it mostly with carbohydrates. A typical day for Jackie now begins with a 2-cup bowl of Fruity Pebbles™ with 1 cup of skim milk and ½ cup of apple juice. For lunch, she might pack a turkey sandwich on white bread with lettuce, tomato, and mustard; a 1-ounce package of fat-free pretzels; and five reduced-fat vanilla wafers. Dinner could be a 2-cup portion of pasta with some olive oil and garlic mixed in, and a small iceberg lettuce salad with lemon juice squeezed over it. Her snacks are usually baked chips, low-fat cookies, fat-free frozen yogurt, or fat-free pretzels. She drinks five diet soft drinks throughout the day as her main beverage.

1. For heart health, is it necessary for Jackie to severely restrict her fat intake?
2. What types of fat should Jackie try to consume? Why are these types of fat the most desirable?
3. What types of foods has Jackie used to replace the fat in her dietary pattern?
4. What food groups are missing from her new dietary plan? How many servings should she be including from these food groups?
5. Is Jackie's new exercise routine appropriate?

*Complete the Case Study. Responses to these questions can be provided by your instructor.*

▲ Are there important food groups missing from Jackie's new dietary plan?
Stuart Pearce/Pixtal/age fotostock

## ASK THE RDN | Coconut Oil

**Dear RDN:** *My friend at the health food store said I should start using coconut oil in my cooking because it has all kinds of health benefits. Is coconut oil a good choice?*

Recently, coconut oil has been promoted for weight loss, heart health, cancer protection, and brain health. Indeed, there is some interesting research on the health effects of the particular fatty acids in coconut oil. At this point, however, there is little evidence to back up the claims that coconut oil is a miracle cure for anything.

Coconut oil is highly saturated. In fact, nearly 90% of the fatty acids in coconut oil are saturated, which is even more than butter or beef fat. This makes for a desirable product: it has that "melt in your mouth" quality and it is very shelf stable, which is attractive for food manufacturers and consumers alike. How do saturated fats impact your health? In this chapter, you have seen the recommendations from all the major health authorities urging Americans to limit their intakes of saturated fat. This is because saturated fatty acids have been shown to increase blood levels of LDL cholesterol, which are linked to atherosclerosis. Proponents of coconut oil point to research showing that the fatty acids in coconut oil, while they do raise LDL, also raise HDL levels. Recently, the AHA advised against the use of coconut oil (and other sources of saturated fatty acids) because of these effects on LDL levels.[8]

When it comes to weight loss, the producers of coconut oil focus on the size of the fatty acids in their product. Most dietary fats are long-chain fatty acids. About 60% of the fatty acids in coconut oil, however, are medium-chain fatty acids, which are more likely to be burned as fuel and less likely to be stored in adipose tissue. Does coconut oil help you lose weight? There are two small human studies that show a decrease in waist circumference with coconut oil, but there is not yet enough evidence to promote coconut oil as a weight-loss aid.

If you like coconut oil, *use it in moderation.* Like any dietary fat, coconut oil provides a lot of calories in a small portion. Simply adding coconut oil to your dietary pattern is not a wise choice because those extra fat calories can quickly add up. Extra calories from any source inevitably lead to weight gain. Use it instead of butter or margarine, aiming to keep your total intake of saturated fat less than 10% of total calories. Most of the time, though, choose unsaturated plant oils, which have proven benefits for heart health. Coconut oil is not a cure-all, but it can be included as part of an overall healthy dietary pattern that meets your calorie needs.

Tim Klontz, Klontz Photography

Yours in health,

**Angela Collene, MS, RDN, LD**

Lecturer, The Ohio State University, Author of *Contemporary Nutrition*

## ✓ CONCEPT CHECK 5.6

1. How does the percent of calories as fat in the typical North American dietary pattern compare to recommendations?

2. Describe the current recommendations for saturated fat intake.

3. What are the key features of the Mediterranean diet? Why is the Mediterranean diet promoted as a healthy dietary pattern?

# 5.7 Nutrition and Your Health
## Lipids and Cardiovascular Disease

Elena Nazarova/123RF

Cardiovascular disease is the major killer of North Americans. It includes a variety of diseases of the heart and blood vessels. The terms *coronary arterty disease*, *coronary heart disease*, or simply *heart disease* refer to cardiovascular disease that affects the heart. *Cerebrovascular disease* affects the brain and could lead to a stroke. *Peripheral artery disease* impacts the blood vessels that supply other areas of the body, such as the legs. Each year, more than 840,000 people die from cardiovascular disease in the United States. It is estimated that 1 person dies of a heart attack every 40 seconds in the United States.[15] Women generally lag about 10 years of age behind men in developing cardiovascular disease. Still, it eventually kills more women than any other disease.

## Development of Cardiovascular Disease

The symptoms of cardiovascular disease develop over many years and often do not become obvious until older adulthood. Nonetheless, autopsies of young adults under 20 years of age have shown that many of them had atherosclerotic **plaque** in their arteries. This finding indicates that plaque buildup can begin in childhood and continue throughout life, although it usually goes undetected for some time.[16]

The typical forms of cardiovascular disease—coronary heart disease and strokes—are associated with inadequate blood circulation in the heart and brain related to buildup of this plaque. Blood supplies the heart muscle, brain, and other body organs with oxygen and nutrients. When blood flow via the coronary arteries surrounding the heart is interrupted, the heart muscle can be damaged. A **heart attack** (also called *myocardial infarction*) may result. This may cause the heart to beat irregularly or to stop. About 25% of people do not survive their first heart attack. Similarly, if blood flow to parts of the brain is interrupted long enough, part of the brain dies, causing a **stroke** (also called *cerebrovascular accident*).

Sometimes, a heart attack can strike with the sudden force of a sledgehammer, with pain radiating up the neck or down the arm. Other times, it can sneak up at night, masquerading as indigestion, with slight pain or pressure in the chest. Crushing chest pain is a more common symptom in men.[17] Many times, the symptoms are so subtle in a woman that death occurs before she or the health professional realizes that a heart attack is taking place (see Fig. 5-15). If there is any suspicion that a heart attack is taking place, the person should first call 911 and then chew an aspirin (325 milligrams)

**plaque** A cholesterol-rich substance deposited in the blood vessels; it contains various white blood cells, smooth muscle cells, various proteins, cholesterol and other lipids, and eventually calcium.

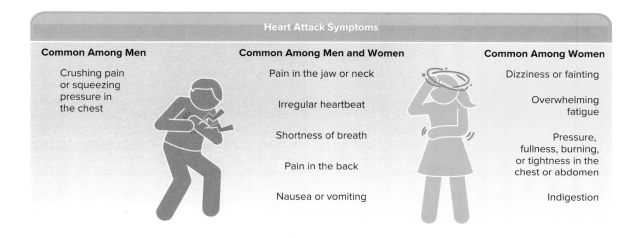

**FIGURE 5-15** ▲ Be aware of common warning signs of a heart attack. The signs may differ for men and women.

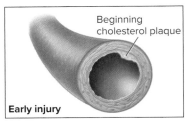

Cell lining (endothelium)

Muscle layer

Vessel opening

**Normal artery**

A healthy blood vessel is smooth and flexible. Blood can easily move through the vessel, and it is able to stretch to accommodate changes in blood pressure throughout the course of a day.

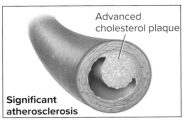

Beginning cholesterol plaque

**Early injury**

Injury to an artery wall begins the process of plaque formation. Possible causes for injury include smoking, diabetes, high blood pressure, high blood cholesterol, or infection.

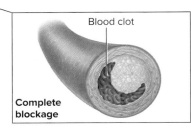

Advanced cholesterol plaque

**Significant atherosclerosis**

The initial injury is followed by a progressive buildup of plaque in the artery walls. The plaque consists of fatty materials (e.g., oxidized LDL), platelets, and minerals. Over time, the blood vessel narrows.

Blood clot

**Complete blockage**

A blood clot becomes lodged in the narrow blood vessel, blocking or greatly restricting blood flow to tissues. Nutrients and oxygen cannot reach the tissue, so it is damaged and may die. A heart attack results when this occurs in a blood vessel that supplies the heart muscle.

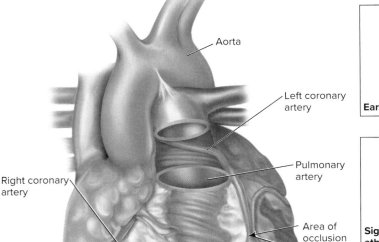

Aorta

Left coronary artery

Right coronary artery

Pulmonary artery

Area of occlusion

Area of muscle damage

**FIGURE 5-16** ▲ How atherosclerosis leads to heart attack.

thoroughly. Aspirin helps to reduce the blood clotting that leads to a heart attack.

Atherosclerosis probably first develops as part of an immune response to repair damage in a blood vessel (Fig. 5-16). A healthy blood vessel is smooth and flexible so that blood can easily move through it. What happens to cause damage to a blood vessel? Likely culprits include smoking, diabetes, high blood pressure, high blood cholesterol, infection—basically any process that causes inflammation in the body. (A laboratory test for C-reactive protein in the blood is used to detect inflammation.) Atherosclerosis can be seen in arteries throughout the body. The damage develops especially at points where an artery branches into two smaller vessels. A great deal of stress is placed on the vessel walls at these points due to changes in blood flow.

Once blood vessel damage has occurred, plaque continues to build up at the site of initial damage. The rate of plaque buildup is directly related to the amount of LDL in the blood. Specifically, oxidized LDL appears to be responsible for plaque

formation. Oxidized LDL has undergone changes that make it more likely to be taken up by scavenger cells in the arterial wall. The body also sends white blood cells called macrophages to the location of the cholesterol accumulation on the blood vessel wall. In an attempt to destroy it, the macrophage surrounds the fatty deposit and produces lipid-loaded **foam cells.** Over years, the blood vessels stiffen, so they cannot dilate or constrict to accommodate normal changes in blood pressure throughout the day. In addition, the buildup of plaque can restrict or block blood flow. Some plaques can become unstable and tear away from the artery. If they rupture, a blood clot will form inside the artery, and within minutes blood flow is cut off, resulting in a heart attack or stroke.

**foam cells** Lipid-loaded white blood cells that have surrounded large amounts of a fatty substance, usually cholesterol, on the blood vessel walls.

Blood clotting is a normal and necessary process that prevents blood loss in case of injury. Some people, however, develop disorders in which their blood forms clots too frequently. In areas of the blood vessels that are already partially blocked by plaque, a blood clot can cut off blood flow, leading to tissue damage or death. More than 95% of heart attacks are caused by total blockage of the coronary arteries due to a blood clot forming in an area of the artery already partially blocked by plaque.

Some common triggers for a heart attack include dehydration; acute emotional stress (e.g., getting fired from a job); strenuous physical activity when not otherwise physically fit (e.g., shoveling snow); waking during the night or getting up in the morning (linked to an abrupt increase in stress); and consuming large, high-fat meals (increases blood clotting).

## Risk Factors for Cardiovascular Disease

For a person at low risk of cardiovascular disease, the advice of health experts is to adhere to a balanced eating pattern (e.g., MyPlate), perform regular physical activity, have a complete fasting lipoprotein analysis performed at age 20 or beyond, and reevaluate risk factors every 4 to 6 years.

How do you know if you are at risk for cardiovascular disease? The AHA identifies several major risk factors, including age, sex, heredity, use of tobacco products, high blood cholesterol, high blood pressure, physical inactivity, excess body fat, and diabetes. Additional contributing factors include stress, alcohol, and dietary patterns that are high in calories, saturated fat, sodium, and added sugars.[18] See Figure 5-17 for additional details about risk factors for cardiovascular disease.

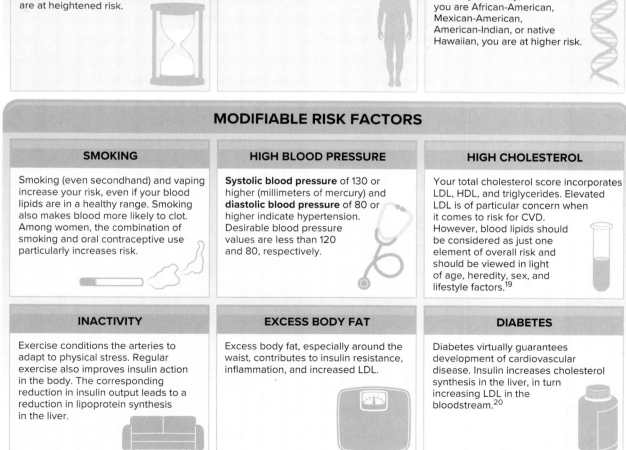

**FIGURE 5-17** ▲ Risk factors for cardiovascular disease.

Does this list of risk factors sound familiar? Many of these were introduced in Section 4.7 as part of the discussion of the **metabolic syndrome,** a cluster of risk factors that increases risk for heart disease, stroke, and type 2 diabetes. Recognize that a *risk factor* is not equivalent to a *cause* for disease. Nevertheless, the more of these risk factors you have, the greater your chances of ultimately developing cardiovascular disease. The good news: by following the diet and lifestyle recommendations of the AHA presented in Section 5.6 and staying physically active, you can control many of these risk factors.

## Lifestyle Modifications to Prevent Cardiovascular Disease

In Section 5.6, you read about recommendations to modify fat intake to promote heart health. For the general population, starting at age 2, the *Dietary Guidelines* and the AHA recommend limiting saturated fat intake to 10% of total kilocalories and avoiding *trans* fats. For people at risk of cardiovascular disease, the AHA recommends restricting saturated fat intake even further to 5% to 6% of total kilocalories. These recommendations are based on scientific evidence that links saturated and *trans* fat intake to higher levels of LDL, which promote plaque formation in blood vessels.[21]

Even while cutting back on saturated and *trans* fat, overall fat intake should generally fall within the range of 20% to 35% of total kilocalories. This means you shouldn't simply cut fat out of your eating pattern. Rather, you should choose monounsaturated and polyunsaturated sources of fat, such as fatty fish, plant oils, nuts, and seeds, instead of foods rich in saturated fats. The omega-3 polyunsaturated fatty acids in fish are particularly helpful for promoting heart health.

Remember that eating for heart health extends beyond a focus on dietary fat! Over the past few decades, evidence has accumulated that excessive intakes of added sugars have contributed to the high rate of cardiovascular disease. This is why it is important to replace saturated fat with unsaturated fat, rather than with simple carbohydrates. Excessive intakes of refined carbohydrates and added sugars promote high levels of insulin, inflammation, and increased blood cholesterol. As you just read, uncontrolled blood sugar among people with diabetes is a significant risk factor for cardiovascular disease. Review recommendations for healthy carbohydrate choices in Section 4.3.

You also read about the dangers of **oxidized** LDL—the form of LDL that contributes to plaque formation. Nutrients and phytochemicals that have antioxidant properties may reduce LDL oxidation. Fruits and vegetables are particularly rich in these compounds. Eating fruits and vegetables regularly is one positive step we can make to reduce plaque accumulation and slow the progression of cardiovascular disease. Some plant foods particularly helpful in this regard include beans, peas, lentils, nuts, dried plums (prunes), raisins, berries, plums, apples, cherries, oranges, grapes, spinach, broccoli, red bell peppers, and onions. Tea, coffee, and dark chocolate are also sources of antioxidants. Please note that *foods* are the best choices for antioxidants. The AHA does not support use of antioxidant supplements (such as vitamin E) to reduce cardiovascular disease risk. This is because large-scale studies have shown *no* decrease in cardiovascular disease risk with use of antioxidant supplements.

**oxidize** In the most basic sense, the loss of an electron or gain of an oxygen by a chemical substance. This change typically alters the shape and/or function of the substance.

Be sure to choose whole grains. Whole grains have more fiber, potassium, and magnesium but less added sugars than refined grains.

As you fill half your plate with fruits and vegetables, choose a variety of brightly colored produce. Fruits and vegetables provide fiber, potassium, magnesium, vitamin C, and vitamin E. Brightly colored selections, such as spinach and berries, are packed with disease-fighting phytochemicals. Cook vegetables with heart-healthy plant oils, such as olive oil, which are rich in unsaturated fatty acids. Use fresh or frozen fruits and vegetables to limit the amount of sodium in your dietary pattern.

Choose low-fat or fat-free dairy products as you aim to keep saturated fat intake under 10% of total calories (or 5% to 6% of total calories for individuals at risk of cardiovascular disease). Low-fat and fat-free dairy products provide just as much calcium and vitamin D as their full-fat counterparts.

For protein, choose mostly lean options. Substitute plant proteins for animal sources of protein several days per week. This meal illustrates a portion of grilled salmon. Fatty fish, like salmon, provide omega-3 fatty acids, which tend to decrease inflammation and promote healthy blood cholesterol levels.

**FIGURE 5-18** ▲ Use MyPlate with the AHA *Diet and Lifestyle Recommendations* to build a heart-healthy meal. Complement your plan for heart-healthy eating with an active lifestyle. Alexis Joseph/McGraw-Hill Education

Some plants contain natural cholesterol-lowering compounds called plant stanols or plant sterols. Rich food sources of plant sterols include wheat germ, sesame seeds, pistachios, and sunflower seeds. These compounds have been clinically shown to reduce LDL (bad) cholesterol, and products that contain these natural cholesterol reducers are backed by the following FDA-approved health claim: *Foods containing at least 0.4 gram per serving of plant sterols, eaten twice a day with meals for a daily total intake of at least 0.8 gram, as part of a diet low in saturated fat and cholesterol, may reduce the risk of heart disease.* CoroWise® is a leading brand of plant sterols. Products such as Smart Balance® margarines and Minute Maid Heart Wise® orange juice contain these plant sterols. The plant sterols work by reducing cholesterol absorption in the small intestine and lowering its return to the liver. The liver responds by taking up more cholesterol from the blood so it can continue to make bile acids. Studies show that 2 to 5 grams of plant sterols per day can reduce total blood cholesterol by 8% to 10% and LDL cholesterol by 9% to 14% (similar to what is seen with some cholesterol-lowering drugs).[22]

Besides dietary changes, exercise is an important way to modify risk for cardiovascular disease. The *Physical Activity Guidelines* state that benefits begin with about 90 minutes of moderate-intensity physical activity per week. In general, adults should strive to achieve 150 to 300 minutes of physical activity per week to reduce risks for chronic diseases. Both regular aerobic exercise and resistance exercise are recommended. Older adults and anyone with existing cardiovascular disease should seek physician approval before starting such a program.[23] The diet and lifestyle recommendations described above are summarized in Figure 5-18.

> Two approaches have been shown to reverse atherosclerosis in the body. One employs a vegan dietary pattern and other lifestyle changes that are part of the Dr. Dean Ornish program. The other employs aggressive LDL-lowering with medications.

## Medications to Lower Blood Lipids

For some people, eating and lifestyle changes are simply not enough to lower blood cholesterol. Fortunately, medications offer a more aggressive approach to treating high cholesterol.

Cholesterol-lowering medications may be appropriate for individuals who are at heightened risk for cardiovascular disease. Some factors to take into consideration include clinical evidence of atherosclerosis (see Fig. 5-16), very high LDL levels (e.g., ≥ 190 mg/dL), or preexisting diabetes or hypertension.

Medications work to lower blood cholesterol in several ways. Examples are listed in the *Medicine Cabinet* feature in this section. Keep in mind that medications may lead to adverse effects, especially on liver function, so physician monitoring is required. In addition, the cost of treatment with one of these drugs can vary widely, from as little as $12 per month to more than $500.[15]

## Medicine Cabinet

**Statins.** Most frequently prescribed cholesterol-lowering drugs that block a liver enzyme involved in cholesterol synthesis and thus reduce the amount of cholesterol in the blood. Examples: simvastatin (Zocor®), atorvastatin (Lipitor®), and rosuvastatin (Crestor®).

**Selective Cholesterol Absorption Inhibitors.** Just as the name would suggest, these drugs keep cholesterol from being absorbed from the small intestine. Cholesterol will still be produced by the liver, but the contribution from dietary cholesterol is minimized. Example: ezetimibe (Zetia®).

**PCSK9 inhibitors.** These biologic drugs bind to and inactivate an enzyme that blocks LDL cholesterol uptake by the liver. When the enzyme is inhibited, the liver takes up more LDL from the blood. Examples: alirocumab (Praluent®) and evolocumab (Repatha®).

**Resins.** These medications bind to bile acids in the intestine and are excreted in the feces, reducing their supply. This stimulates the liver to produce more bile acids, which uses more cholesterol and causes a decrease in blood cholesterol levels. Examples: cholestyramine (Questran®) and colesevelam (Welchol®).

**Fibrates.** These drugs lower blood triglycerides by decreasing the production of triglycerides by the liver. Example: gemfibrozil (Lopid®).

**Marine Oils and Derivatives.** Dietary supplementation with 2 to 4 grams per day of marine-derived polyunsaturated fatty acids, used under the supervision of a health care provider, can help to lower triglycerides. Chemically altered preparations of these fatty acids may also be prescribed. Example: Lovaza®.

**Combination Drugs.** Some pharmaceutical companies combine medications with different mechanisms of action. A statin drug (simvastatin) has been combined with another drug (ezetimibe) and is marketed as Vytorin®. While the statin reduces the cholesterol made by the liver, the ezetimibe helps to block the absorption of cholesterol from food.[22]

magnificent

**Could
your gut microbiota
influence your heart health?**
As gut microbes break down undigested food in the GI tract, they produce proteins and short-chain fatty acids
that can be absorbed into the bloodstream
and affect tissues throughout the body. For
example, some microbial enzymes can convert
dietary components (particularly those found in
animal sources of protein) into compounds that
promote inflammation, alter lipid metabolism in
the liver, and are associated with increased risk
of cardiovascular disease. Researchers are
investigating ways to alter the composition of the gut microbiota or the fuels
provided to these microbes to
modify cardiovascular outcomes.[24, 25]

microbiome

## Surgical Treatment for Cardiovascular Disease

The two most common surgical treatments for coronary artery block-age are percutaneous transluminal coronary angioplasty (PTCA) and coronary artery bypass graft (CABG). PTCA involves the insertion of a balloon catheter into an artery. Once it is advanced to the area of the blockage, the balloon is expanded to crush the buildup of plaque. Afterward, the blood vessel may be held open with metal mesh, called a stent. CABG involves the relocation of a large vein—from the leg, for example—to bypass the blocked blood vessel.

For more information on cardiovascular disease, see the website of the American Heart Association at www.heart.org or the heart disease section of Healthfinder at healthfinder.gov. This is a site created by the U.S. government for consumers. In addition, visit the website www.nhlbi.nih.gov.

### ✓ CONCEPT CHECK 5.7

1. List five risk factors for cardiovascular disease. Which of these are modifiable?

2. Define atherosclerosis. How does atherosclerosis cause a heart attack?

3. Identify three specific dietary strategies to lower your risk for cardiovascular disease.

## Summary (Numbers refer to numbered sections in the chapter.)

**5.1** Lipids are a group of compounds that do not dissolve in water. The three main forms of lipids are triglycerides, phospholipids, and sterols. Fatty acids are components of the chemical structure of lipids.

Saturated fatty acids contain no carbon-carbon double bonds, monounsaturated fatty acids contain one carbon-carbon double bond, and polyunsaturated fatty acids contain two or more carbon-carbon double bonds in the carbon chain. In omega-3 polyunsaturated fatty acids, the first of the carbon-carbon double bonds is located three carbons from the methyl end of the carbon chain. In omega-6 polyunsaturated fatty acids, the first carbon-carbon double bond counting from the methyl end occurs at the sixth carbon.

The essential fatty acids are linoleic acid (an omega-6 fatty acid) and alpha-linolenic acid (an omega-3 fatty acid). These must be included in the dietary pattern to maintain health.

Triglycerides are formed from a glycerol backbone with three fatty acids. Triglycerides rich in long-chain saturated fatty acids tend to be solid at room temperature, whereas those rich in

monounsaturated and polyunsaturated fatty acids are liquid at room temperature. Triglycerides are the major form of fat in both food and the body. They allow for efficient energy storage, protect certain organs, transport fat-soluble vitamins, and help insulate the body.

Phospholipids are derivatives of triglycerides in which one or two of the fatty acids are replaced by phosphorus-containing compounds. Sterols are lipids made of multiple carbon rings.

**5.2** Fats and oils have several functions as components of foods. Fats add flavor and texture to foods and provide some satiety after meals. Some phospholipids are used in foods as emulsifiers, which suspend fat in water.

Foods that are almost entirely fat include vegetable oils, butter, margarine, and mayonnaise. In the dairy group, whole milk and cheese are highest in fat. In the protein foods group, marbled meats, poultry with skin, cold water fish, and nuts are highest in fat. In the grains group, baked goods often have lots of fat added. In the fruits and vegetables groups, most foods are naturally low in fat, but fat may be added during food preparation (e.g., vegetable dip, butter).

Linoleic acid, the essential omega-6 fatty acid, is found in many plant oils, nuts, seeds, and poultry. Two rich sources of alpha-linolenic acid, the essential omega-3 fatty acid, are walnuts and flaxseed oil. Fatty fish are good sources of some nonessential omega-3 fatty acids, including eicosapentaenoic acid (EPA) and docosahexaenoic acid (DHA).

Food sources of phospholipids include organ meats, legumes (i.e., beans, peas, and lentils), eggs, and wheat germ. Cholesterol is only found in foods of animal origin, especially organ meats, egg yolks, and shellfish. Phytosterols are abundant in plant oils, nuts, seeds, and legumes.

**5.3** Fat digestion takes place primarily in the small intestine. First, bile emulsifies lipids in the chyme. Lipase released from the pancreas breaks down triglycerides into diglycerides, then into monoglycerides (glycerol backbone with single fatty acid attached) and fatty acids. Lipids are then taken up by the absorptive cells of the small intestine, packaged as chylomicrons, and eventually enter the lymphatic system before they pass into the bloodstream.

**5.4** Lipids are carried in the bloodstream as part of lipoproteins, which consist of a lipid core encased in a shell of protein, cholesterol, and phospholipids. There are four main types of lipoproteins. Chylomicrons are released from intestinal cells and carry lipids arising from dietary intake. Very-low-density lipoprotein (VLDL) and low-density lipoprotein (LDL) carry lipids both taken up by and synthesized in the liver. High-density lipoprotein (HDL) picks up cholesterol from cells and facilitates its transport back to the liver.

**5.5** Triglycerides are used for energy storage, insulation, and transportation of fat-soluble vitamins. Phospholipids are important parts of cell membranes, and some act as emulsifiers. Cholesterol forms vital biological compounds, such as hormones, cell membranes, and bile acids. Cells in the body make cholesterol whether we eat it or not. It is not an essential part of an adult's dietary pattern.

**5.6** The AMDR for fat is 20% to 35% of total calories. There is currently no RDA for total fat for adults, although Adequate Intakes (AIs) have been set for the essential fatty acids. Plant oils should contribute at least 5% of total calories to achieve the AIs proposed for essential fatty acids. Fatty fish are a rich source of omega-3 fatty acids and should be consumed at least twice per week.

The *Dietary Guidelines* advise Americans starting at 2 years of age, to limit saturated fat intake to less than 10% of total calories. For those who need to lower their blood cholesterol levels, the AHA recommends further restriction of saturated fat to 5% to 6% of total calories. *Trans* fat intake should be avoided; as of 2018, the use of partially hydrogenated oils (the leading dietary source of *trans* fatty acids) has been banned in the United States. Although there are no strict, numerical limits on dietary cholesterol intake for the general population, the *Dietary Guidelines* recommend keeping dietary cholesterol intake as low as possible while still meeting daily needs for essential nutrients.

**5.7** In the blood, elevated amounts of LDL and low amounts of HDL are strong predictors of risk for cardiovascular disease. Additional risk factors for the disease are smoking, hypertension, diabetes, obesity, and inactivity. Lifestyle modifications to improve heart health include choosing fish and plant oils instead of food sources of saturated fats; limiting added sugar intake; consuming plenty of fruits, vegetables, and whole grains to obtain antioxidants; incorporating phytosterols; and exercising regularly.

# Check Your Knowledge (Answers are available at the end of this question set.)

1. The main form of lipid found in the food we eat is
   a. cholesterol.
   b. phospholipids.
   c. triglycerides.
   d. diglycerides.

2. Which of the following is an essential fatty acid?
   a. Linoleic acid
   b. Oleic acid
   c. Docosahexaenoic acid
   d. Eicosapentaenoic acid

3. Which of the following foods are rich sources of saturated fatty acids?
   a. Olive oil, peanut oil, canola oil
   b. Palm oil, palm kernel oil, coconut oil
   c. Safflower oil, corn oil, soybean oil
   d. All of the above

4. Which of the following foods is the best source of omega-3 fatty acids?
   a. Fatty fish
   b. Peanut butter
   c. Lard and shortening
   d. Beef and other red meats

5. Lipoproteins are important for
   a. transport of fats in the blood and lymphatic system.
   b. synthesis of triglycerides.
   c. synthesis of adipose tissue.
   d. enzyme production.

6. Immediately after a meal, newly digested and absorbed dietary fats appear in the lymph and then the blood as part of which of the following?
   a. LDL
   b. HDL
   c. Chylomicrons
   d. Cholesterol

7. High blood concentrations of _____ decrease the risk for cardiovascular disease.
   a. low-density lipoproteins
   b. chylomicrons
   c. high-density lipoproteins
   d. cholesterol

8. _____ are unique among the lipids because they have one end that is soluble in water and one end that is soluble in fat.
   a. Saturated fatty acids
   b. Unsaturated fatty acids
   c. Sterols
   d. Phospholipids

9. Cholesterol is
   a. an essential nutrient.
   b. found in foods of plant origin.
   c. an important part of human cell membranes.
   d. all of the above.

10. The *Dietary Guidelines* recommend that Americans age 2 and older limit their intake of saturated fat to less than _____ of total calories.
    a. 5%
    b. 10%
    c. 35%
    d. 50%

10. b (LO 5.8)
5. a (LO 5.6), 6. c (LO 5.6), 7. c (LO 5.6), 8. d (LO 5.7), 9. c (LO 5.7),
**Answer Key: 1.** c (LO 5.2), 2. a (LO 5.3), 3. b (LO 5.4), 4. a (LO 5.4),

## Study Questions (Numbers refer to Learning Outcomes.)

1. Name a common property of all lipids. **(LO 5.1)**

2. Describe the chemical structures of saturated and unsaturated fatty acids and their different effects in both food and the human body. **(LO 5.2)**

3. Describe the significance of and possible uses for reduced-fat foods. **(LO 5.4)**

4. Describe the structures, origins, and roles of the four major blood lipoproteins. **(LO 5.6)**

5. What are two important functions of lipids in the human body? **(LO 5.7)**

6. Relate the need for omega-3 fatty acids in the dietary pattern to the recommendation to consume fatty fish at least twice a week. **(LO 5.8)**

7. What are the recommendations from various health care organizations regarding fat intake? What does this mean in terms of food choices? **(LO 5.8)**

8. Does the total cholesterol concentration in the bloodstream tell the whole story with respect to cardiovascular disease risk? **(LO 5.9)**

9. List the main risk factors for the development of cardiovascular disease. **(LO 5.9)**

10. Describe three lifestyle changes to decrease the risk of cardiovascular disease. **(LO 5.9)**

## Further Readings

1. Astrup A: Goodbye to the egg-white omelet—welcome back to the whole-egg omelet. *American Journal of Clinical Nutrition* 107:853, 2018.

2. American Heart Association News: Are eggs good for you or not? 2018. https://www.heart.org/en/news/2018/08/15/are-eggs-good-for-you-or-not. Accessed December 18, 2019.

3. Kris-Etherton PM and Fleming JA: Emerging nutrition science on fatty acids and cardiovascular disease: Nutritionists' perspectives. *Advances in Nutrition* 6(3):326S–337S, 2015.

4. Food and Drug Administration: *Trans* fat. 2018. https://www.fda.gov/food/food-additives-petitions/trans-fat. Accessed December 18, 2019.

5. Atchley C: Replacing PHOs with customizable alternatives. *Food Business News* 2018. https://www.foodbusinessnews.net/articles/12034-replacing-phos-with-customizable-alternatives. Accessed December 18, 2019.

6. Mogensen KM: Essential fatty acid deficiency. *Practical Gastroenterology* 2017; 41(6): 37. https://practicalgastro.com/2017/06/01/essential-fatty-acid-deficiency. Accessed December 18, 2019.

7. Eckel RH and others: 2013 AHA/ACC guideline on lifestyle management to reduce cardiovascular risk: A report of the American College of Cardiology/American Heart Association Task Force on Practice Guidelines. *Circulation* 2013; 129:S76.

8. Sacks FM and others: Dietary fats and cardiovascular disease: A presidential advisory from the American Heart Association. *Circulation* 2017; 135:e1.

9. Rimm EB and others: Seafood long-chain n-3 polyunsaturated fatty acids and cardiovascular disease: A science advisory from the American Heart Association. *Circulation* 2018; 138:e35. DOI:10.1161/CIR.0000000000000574.

10. Siscovick DS and others: Omega-3 polyunsaturated fatty acid (fish oil) supplementation and the prevention of clinical cardiovascular disease: A science advisory from the American Heart Association. *Circulation* 135:e867, 2017. DOI:10.1161/CIR.0000000000000482.

11. Shen J and others: Mediterranean dietary patterns and cardiovascular health. *Annual Review of Nutrition* 35:425, 2015.

12. Dennett C: Key ingredients of the Mediterranean diet—the nutritious sum of delicious parts. *Today's Dietitian* 18(5):28, 2016.

13. Palmer S: Low-fat vegan diets. *Today's Dietitian* 18(10):20, 2016.

14. Vannice G and others: Position of the Academy of Nutrition and Dietetics: Dietary fatty acids for healthy adults. *Journal of the Academy of Nutrition and Dietetics* 114:136, 2014.

15. Centers for Disease Control and Prevention: Heart disease facts. 2019. https://www.cdc.gov/heartdisease/facts.htm. Accessed December 18, 2019.

16. National Heart, Lung, and Blood Institute: Expert panel on integrated guidelines for cardiovascular health and risk reduction in children and adolescents: Summary report. *Pediatrics*. 128:S213–56, 2011. DOI:10.1542/peds.2009-2107C.

17. American Heart Association: What are the warning signs of heart attack? 2015. https://www.heart.org/-/media/data-import/downloadables/f/e/3/pe-abh-what-are-the-warning-signs-of-heart-attack-ucm_300319.pdf. Accessed December 18, 2019.

18. American Heart Association: Understand your risks to prevent a heart attack. 2016. https://www.heart.org/en/health-topics/heart-attack/understand-your-risks-to-prevent-a-heart-attack. Accessed December 18, 2019.

19. Grundy SM and others: AHA/ACC/AACVPR/AAPA/ABC/ACPM/ADA/AGS/APhA/ASPC/NLA/PCNA guideline on the management of blood cholesterol: A report of the American College of Cardiology/American Heart Association Task Force on Clinical Practice Guidelines. *Circulation* 2018. DOI:10.1161/CIR.0000000000000625.

20. American Diabetes Association: Cardiovascular disease and risk management: Standards of medical care in diabetes—2018. *Diabetes Care* 41:S86, 2018.

21. Arnett DK and others: 2019 ACC/AHA guideline on the primary prevention of cardiovascular disease: Part 1, lifestyle and behavioral factors. *JAMA Cardiology* 4:1043, 2019. DOI:10.1001/jamacardio.2019.2604.

22. American Heart Association: Cholesterol medications. 2018. https://www.heart.org/en/health-topics/cholesterol/prevention-and-treatment-of-high-cholesterol-hyperlipidemia/cholesterol-medications. Accessed December 18, 2019.

23. U.S. Department of Health and Human Services. *Physical Activity Guidelines for Americans*, 2nd edition. Washington, DC: U.S. Department of Health and Human Services. 2018.

24. Yoshida N and others: Gut microbiome and cardiovascular disease. *Diseases* 6:56, 2018. DOI:10.3390/diseases6030056.

25. Attye I and others: A crucial role for diet in the relationship between gut microbiota and cardiometabolic disease. *Annual Review of Medicine* 7i:10.1, 2020. DOI: 10.1146/annurev-med-062218-023720.

# Chapter 6

# Proteins

# Student Learning Outcomes

## FAKE NEWS

## It is impossible to eat too much protein.

### THE FACTS

It is possible to get too much of a good thing, including protein. Similar to consuming excess carbohydrate and fat, eating too much protein can also cause problems. Excess protein will more likely be converted to fat in the body rather than muscle. Too much protein on your plate also may be leaving little room for other important nutrients from high-quality carbohydrates, and fruits and vegetables. Visit Connect to learn more.

Consuming enough protein is vital for maintaining health. Proteins form important structures in the body, make up a key part of the blood, help regulate many body functions, and can fuel body cells.

Although North Americans generally eat more protein than is recommended, the timing of our protein consumption during the day may not be optimal for health. Our daily protein intake comes primarily at the evening meal and mostly from animal sources, such as meat, poultry, fish, eggs, milk, and cheese. In contrast, in the developing world, eating patterns can be deficient in protein.

Eating patterns that are mostly vegetarian still predominate in much of Asia and areas of Africa, and many North Americans practice some form of vegetarianism. In fact, North Americans are encouraged to consume a greater portion of their food, including protein, from plant sources. Over time, plant sources of protein such as nuts, seeds, and legumes have been sidelined by meats. Sources of plant proteins, however, offer a wealth of nutritional benefits—from lowering blood cholesterol to preventing certain forms of cancer.

We can benefit from eating more plant sources of proteins, but it takes some knowledge to get the right quantity and quality of protein at each meal to meet protein needs. This chapter takes a close look at protein, including the distribution of our protein intake throughout the day and the benefits of plant proteins in our eating patterns. It will also examine the benefits and potential risks of vegetarian eating patterns. Let's see why a detailed study of protein is worth your attention.

# 6.1 Amino Acids—Building Blocks of Proteins

Thousands of substances in the body are made of protein. Aside from water, proteins form the major part of lean body tissue, totaling about 17% of body weight. **Amino acids** are compounds that serve as the building blocks for proteins. They are chemically unique in that they contain nitrogen along with carbon, oxygen, and hydrogen. Plants combine nitrogen from the soil with carbon and other elements to form amino acids. They then link these amino acids to make proteins. Proteins are thus an essential part of a healthy eating pattern because they supply nitrogen in a form we can readily use—namely, amino acids.

Proteins are crucial to the *regulation* and *maintenance* of the body. Body functions such as blood clotting, fluid balance, hormone and enzyme production, vision, transport of many substances in the bloodstream, and cell repair require specific proteins. The body makes proteins in many configurations and sizes so that they can serve these greatly varied functions. Formation of these body proteins begins with amino acids from both the protein-containing foods we eat and those made from other compounds within the body. Proteins can also be broken down to *supply energy* for the body—on average, 4 kcal per gram.

Your body uses 20 different amino acids to function (Table 6-1). Although all of these commonly found amino acids are important, 11 (alanine, arginine, asparagine, aspartic acid, cysteine, glutamic acid, glutamine, glycine, proline, serine, and tyrosine) are considered **nonessential** with respect to our eating patterns. All tissues have some ability to make the nonessential amino acids as long as the right ingredients are present—the key factor being nitrogen that is already part of another amino acid. Therefore, it is not essential that these amino acids be consumed.

Amino acids are formed mostly of carbon, hydrogen, oxygen, and nitrogen. Figure 6-1 shows the structure of a generic amino acid and two examples of specific amino acids. The various amino acids used to make proteins are slight variations of the generic amino acid with different chemical makeups (see Appendix D). Each amino acid has an "acid" group, an "amino" group, and a "side" or R group specific to the amino acid.

The R group on some amino acids has a branched shape, like a tree. These so-called **branched-chain amino acids** are leucine, isoleucine, and valine. The branched-chain amino acids are the primary amino acids that promote and signal protein synthesis and turnover in muscles. Whey protein (from milk) is popular among strength-training athletes because it is particularly rich in branched-chain amino acids.

## ESSENTIAL AMINO ACIDS

The nine amino acids (histidine, isoleucine, leucine, lysine, methionine, phenylalanine, threonine, tryptophan, and valine) the body cannot make in sufficient amounts or at all are known as **essential.** The essential amino acids must be obtained from foods because body cells cannot make the needed carbon-based foundation of the amino acid, cannot put a nitrogen group on the needed carbon-based foundation, or just cannot do

**TABLE 6-1 ■ Amino Acids**

| Essential Amino Acids | Nonessential Amino Acids |
|---|---|
| Histidine | Alanine |
| Isoleucine* | Arginine |
| Leucine* | Asparagine |
| Lysine | Aspartic acid |
| Methionine | Cysteine |
| Phenylalanine | Glutamic acid |
| Threonine | Glutamine |
| Tryptophan | Glycine |
| Valine* | Proline |
| | Serine |
| | Tyrosine |

*A branched-chain amino acid.

**nonessential amino acids** Amino acids that can be synthesized by a healthy body in sufficient amounts; there are 11 nonessential amino acids. These are also called *dispensable amino acids.*

**branched-chain amino acids** Amino acids with a branching carbon backbone; these are leucine, isoleucine, and valine. All are essential amino acids.

**essential amino acids** The amino acids that cannot be synthesized by humans in sufficient amounts or at all and therefore must be included in the dietary pattern; there are nine essential amino acids. These are also called *indispensable amino acids.*

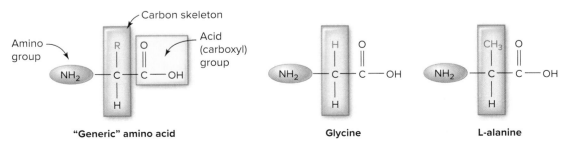

**FIGURE 6-1 ▲** Amino acid structure. The side chain (R) differentiates glycine (H) and alanine (CH₃).

the whole process fast enough to meet body needs. If you do not eat enough essential amino acids, your body struggles to conserve what essential amino acids it can. Eventually your body slows production of new proteins until, at some point, you will break protein down faster than you can make it. When that happens, health deteriorates.

Both nonessential and essential amino acids are present in foods that contain protein. Animal sources of protein, such as meat and dairy products, and plant sources of protein, such as beans, nuts, and seeds, can supply us with an adequate amount of both the essential and nonessential amino acid building blocks needed to maintain good health. The essential amino acid in smallest supply in a food or meal in relation to body needs becomes the limiting factor or **limiting amino acid** because it limits the amount of protein the body can synthesize. Typically, 50% of the amino acids in dietary proteins are essential. Fortunately, adults need only about 11% of their total protein requirement to be supplied by them.

The estimated requirements for essential amino acids for infants and preschool children are greater (40% of total protein intake) because of the needs of rapid growth and development; however, in later childhood, the need drops to 20%. Providing animal sources of protein, such as human milk or cow's milk–based formula for infants, or cow's milk for children, helps ensure that enough high-quality proteins are present. A major health risk for infants and children occurs in famine situations in which only one type of cereal grain, such as rice, is available, serving as a staple food and increasing the probability that one or more of the nine essential amino acids are lacking in the total dietary pattern.

**Conditionally Essential Amino Acids.** Some of the nonessential amino acids, which are usually synthesized in the body, can become **conditionally essential amino acids** during times of rapid growth, disease, or metabolic stress. For example, the need for amino acids to promote healing during recovery from surgery or burns is so high that synthesis of nonessential amino acids, especially arginine and glutamine, cannot keep up with demands. Studies have shown that supplementation with these nonessential amino acids is effective in improving the healing of surgical and burn wounds.[1]

In the genetic disorder phenylketonuria (PKU), a nonessential amino acid, tyrosine, becomes conditionally essential (Fig. 6-2). A person with PKU has a limited ability to convert the essential amino acid phenylalanine into tyrosine. In individuals with PKU, the activity of the enzyme used in processing much of our dietary phenylalanine to tyrosine is insufficient. The results are that: (1) tyrosine becomes essential (it must be consumed), whereas (2) phenylalanine that is consumed builds up to toxic levels in the blood. Elevated phenylalanine disrupts brain function, leading to intellectual disability. PKU is treated with a special diet that provides adequate protein, but that limits phenylalanine.

▲ Soy products such as soy milk, tofu, and edamame provide a plant source of all the essential amino acids. Image Source/ Glow Images

**limiting amino acid**  The essential amino acid in lowest concentration in a food or diet relative to body needs.

**conditionally essential amino acids**  Nonessential amino acids that cannot be made in adequate amounts to support the body's increased requirements during conditions of rapid growth, disease, or metabolic stress, and therefore become essential (i.e., must be obtained from food).

**PKU: Metabolism of Phenylalanine to Tyrosine Is Blocked**

*Phenylalanine hydroxylase*

**Phenylalanine (Phe)**
**(essential)**

**Tyrosine (Tyr)**
(becomes conditionally essential)

**FIGURE 6-2** ▲ In the genetic disorder phenylketonuria, tyrosine, a nonessential amino acid becomes a conditionally essential amino acid because its production from phenylalanine is blocked.

▲ All newborns are tested for phenylketonuria and some other amino acid disorders within the first few days of life using blood collected by a heel prick. If PKU is diagnosed, a special diet should begin as soon as possible after birth to ensure normal brain development.
Noor Haswan Noor Azman/Shutterstock

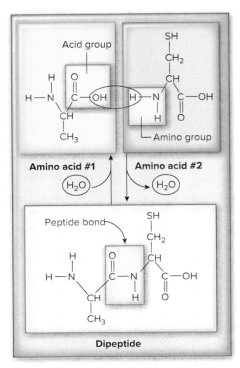

**FIGURE 6-3** ▲ Peptide bonds link amino acids. During synthesis of a peptide bond, a molecule of water is removed (dehydration). When peptide bonds are broken (as in digestion), a molecule of water is added (hydrolysis).

**peptide bond** A chemical bond formed between amino acids in a protein.

**polypeptide** A group of 10 to 2000 or more amino acids bonded together to form proteins.

**sickle cell disease** An illness that results from a malformation of the red blood cell because of an incorrect structure in part of its hemoglobin protein chains; also called *sickle cell anemia.*

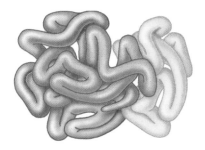

**FIGURE 6-4** ▲ Protein organization. Proteins often form a coiled shape, as shown by this drawing of the blood protein hemoglobin. This shape is dictated by the order of the amino acids in the protein chain. To get an idea of its size, consider that each teaspoon (5 milliliters) of blood contains about $10^{18}$ hemoglobin molecules. (One billion is $10^9$.)

✅ **CONCEPT CHECK 6.1**

1. What is the basic structure of an amino acid?
2. What is the difference between the essential and nonessential amino acids?
3. What are some examples of conditions in which a nonessential amino acid becomes essential?

# 6.2 Protein Organization and Synthesis

Within body cells, amino acids are linked together by chemical bonds—technically called **peptide bonds**—to form proteins (Fig. 6-3). Peptide bonds form between the amino group of one amino acid and the acid (carboxyl) group of another. Through peptide bonding of amino acids, cells can synthesize dipeptides (joining of two amino acids), tripeptides (joining of three amino acids), oligopeptides (joining of four to nine amino acids), and **polypeptides** (joining of 10 or more amino acids). Most proteins are polypeptides ranging from about 50 to 2000 amino acids. The body can synthesize many different proteins by linking together the 20 common types of amino acids with peptide bonds. These bonds are difficult to break, but heat, acids, enzymes, and other agents are able to do so during cooking and chemical digestion.

## PROTEIN ORGANIZATION

Through the bonding together of the 20 common types of amino acids into various combinations, cells can synthesize thousands of different proteins. The sequential order of the amino acids and the bonding between amino acid side chains then determines the protein's shape, which ultimately affects its function. Only correctly positioned amino acids can interact and fold properly to form the intended shape for the protein. The resulting unique, three-dimensional form, such as that shown for the protein hemoglobin in Figure 6-4, goes on to dictate the function of each particular protein. If it lacks the proper structure, a protein cannot function.

The **deoxyribonucleic acid (DNA)** present in the nucleus of the cell contains coded instructions for protein synthesis and organization, including which specific amino acids are to be placed in a protein and in which order. The relationship between DNA and the proteins eventually produced by a cell is very important. If the DNA code contains errors, this is a genetic defect and an incorrect amino acid will be added, resulting in an incorrect polypeptide chain being produced. Many diseases, including certain cancers, also stem from errors in the DNA code. Genetic engineering is now used to correct some **gene** defects in humans by placing the correct DNA code in the nucleus so that the correct protein can be made by the ribosomes.

**Sickle cell disease** (also called **sickle cell anemia**) is one example of an inherited genetic disease in which amino acids are out of order on a protein. In North America, people of African descent are especially prone to this genetic disease. Sickle cell anemia is not a nutritional disease but is caused by a mutation in the genetic code for hemoglobin, the protein depicted in Figure 6-4 that carries oxygen in red blood cells. The mutation causes the amino acid glutamic acid to be replaced with the amino acid valine. This error produces a profound change in hemoglobin structure. It can no longer form the shape needed to carry oxygen efficiently inside the red blood cell. Instead of forming normal circular disks, the red blood cells collapse into crescent (or sickle) shapes (see Fig. 6-5). Sickle red blood cells become hard and sticky, causing them to clog blood flow and break apart. This can cause severe bone and joint pain, abdominal pain, headache, convulsions, paralysis, and even death due to the lack of oxygen.

## PROTEIN SYNTHESIS

As we describe protein synthesis, let's use an analogy of preparing a recipe from an online database of recipes that you can access from your device. Whenever you want to prepare a meal, you go to the recipe database and download the desired recipe, which tells you all the ingredients you will need and the sequence of steps to prepare the foods. You must gather all the ingredients for the recipe on your kitchen counter, then set to work to prepare the recipe. In order for your recipe to turn out perfectly, it must download successfully, you must have all the ingredients, and you must follow the sequence of steps described in the recipe.

In protein synthesis, the database is the DNA, which is housed in the nucleus of the cell (online). Your cells must make hundreds of different proteins (foods), and DNA includes the genes (recipes) to make each and every protein your cells need. Protein synthesis, however, takes place in the cytoplasm of the cell (your kitchen), on the ribosomes (your kitchen counter), not in the nucleus (Fig. 6-6). Thus, the DNA code used for synthesis of a specific protein must be transferred from the nucleus to the cytoplasm. The download described in this analogy is akin to the process of transcription. Because the DNA cookbook cannot leave the nucleus to get to the "kitchen counter" (the ribosomes), the recipes for proteins must be transcribed into a form that can leave the nucleus. This transfer is the job of messenger **RNA** (mRNA), or cooks, that can leave the nucleus. During a step called **transcription**, the code (a gene) for a protein on a DNA sequence is transcribed into a single-stranded mRNA molecule (Fig. 6-6) that is ready to leave the nucleus.

Once in the cytoplasm, mRNA travels to the ribosomes. The ribosomes read the mRNA code, or recipe, and *translate* those instructions to produce a specific protein. During **translation**, amino acids are added one at a time to the growing polypeptide chain, according to the instructions on the mRNA. The amino acids that form a protein can be equated to the "ingredients" required for the recipe. Another key participant in protein synthesis, transfer RNA (tRNA), is responsible for bringing the specific amino acids to the

**FIGURE 6-5** ▲ An example of the consequences of errors in DNA coding of proteins. A normal circular red blood cell is shown on the right along with an abnormal sickle-shaped red blood cell on the left.
Sickle Cell Foundation of Georgia: Jackie George, Beverly Sinclair/photo by Janice Haney Carr/CDC

**transcription** The process by which the code or gene for a protein on a DNA sequence is copied into a single-stranded mRNA molecule that is ready to leave the nucleus.

**translation** The process of adding amino acids one at a time to a growing polypeptide chain, according to the instructions on the mRNA.

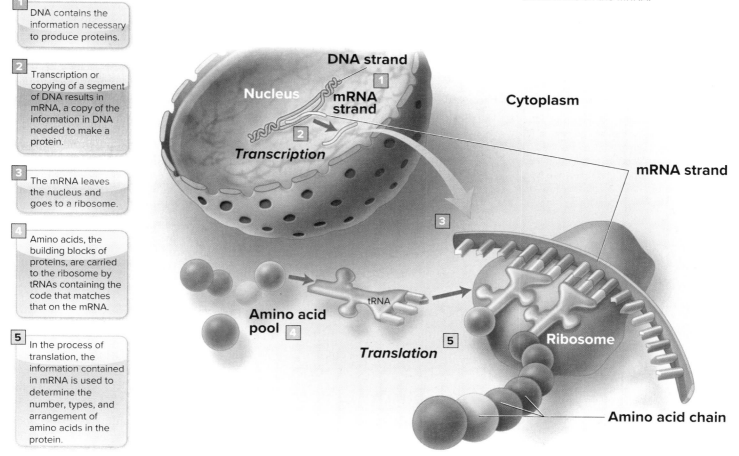

1 DNA contains the information necessary to produce proteins.

2 Transcription or copying of a segment of DNA results in mRNA, a copy of the information in DNA needed to make a protein.

3 The mRNA leaves the nucleus and goes to a ribosome.

4 Amino acids, the building blocks of proteins, are carried to the ribosome by tRNAs containing the code that matches that on the mRNA.

5 In the process of translation, the information contained in mRNA is used to determine the number, types, and arrangement of amino acids in the protein.

**FIGURE 6-6** ▲ Protein synthesis (simplified). Once the mRNA is fully read on the ribosome, the amino acids have been connected into the polypeptide, which is released from the ribosome into the cytoplasm. It generally is then processed further to become a functioning cell protein.

**epigenome** A network of chemical compounds surrounding DNA that modify the genome without altering the DNA sequences and have a role in determining which genes are active (expressed) or inactive (silenced) in a particular cell.

**epigenetics** Study of heritable changes in gene function that are independent of DNA sequence. For example, malnutrition during pregnancy may modify gene expression in the fetus and affect long-term body weight regulation in the offspring.

**nutritional genomics** Study of interactions between nutrition and genetics; includes nutrigenetics and nutrigenomics.

**denaturation** Alteration of a protein's three-dimensional structure, usually because of treatment by heat, enzymes, acid or alkaline solutions, or agitation.

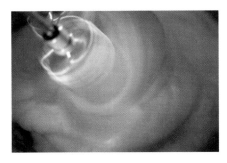

▲ Most of the protein in eggs is present in the egg white. Whipping egg whites is a form of agitation that changes the shape of the egg protein, resulting in the formation of a foam or meringue due to protein denaturation. I. Rozenbaum/PhotoAlto

ribosomes as needed during protein synthesis (review Fig. 6-6). Energy input is required to add each amino acid to the chain, making protein synthesis a significant user of calories.

Once synthesis of a polypeptide is complete, it twists and folds into the appropriate three-dimensional shape of the intended protein. These structural changes occur based on specific interactions among the amino acids on the polypeptide chain. The growth, development, and maintenance of cells, and ultimately of the entire organism, are directed by genetic mechanisms present in the cells. Many nutrients and other dietary components taken up by cells interact with our genes and affect gene expression and protein synthesis. While our human genome contains the code for the proteins that can be made by our bodies, our **epigenome** is an extra layer of instructions that can be altered by environmental and dietary factors and influence gene activity. **Epigenetics** refers to changes in gene expression caused by mechanisms other than changes in the underlying DNA sequence.

Through research, our understanding of the links between nutrition and genetics is becoming clearer. To some extent health professionals can now personalize nutrition recommendations based on their client's genetic information. Collectively, the interactions between genetics and nutrition are known as **nutritional genomics,** which is discussed in more detail in Section 6.8. In addition, genetic discoveries are leading to affordable genetic tests that predict our risk for disease, as well as new drugs that target disease processes at the genetic or molecular level.

## DENATURATION OF PROTEINS

Changing the shape of a protein often destroys its biological activity and thus its ability to function normally. Exposure to acid or alkaline substances, heat, or agitation (e.g., whipping egg whites) can alter a protein's structure, leaving it uncoiled or otherwise deformed. This process of altering the three-dimensional structure of a protein is called **denaturation**.

Denaturation of dietary proteins does not alter their nutritional value and is a necessary part of digestion and other body processes. The heat produced during cooking starts the denaturation of some proteins. After food is ingested, the secretion of stomach acid denatures many forms of proteins in foods, making it safer to eat. These include bacterial proteins, plant hormones, and many active enzymes. Denaturation also enhances digestion because the unraveling of the polypeptide chain increases its exposure to digestive enzymes. Denaturing proteins in some foods can also reduce their ability to cause allergic reactions.

Recognize that we need adequate intake of amino acids, especially essential ones, supplied by protein-rich foods to supply the building blocks for making proteins in the body. We dismantle ingested dietary proteins and use the amino acid building blocks to assemble the proteins we need.

✓ **CONCEPT CHECK 6.2**

1. Why is the amino acid order within a protein important?
2. What is the role of DNA in protein synthesis?
3. What are the steps of protein synthesis?
4. What are some of the ways a protein can become denatured?

# 6.3 Protein in Foods

## PROTEIN QUALITY OF FOODS

The quality of proteins can differ greatly according to their origin (animal or plant), individual amino acid composition, and level of amino acid bioactivity. While protein quality is an important consideration when planning our individual dietary patterns, at a population level the environmental effects of animal-protein versus plant-protein production must be considered. More sustainable protein sources should be considered as alternatives to popular meat-based sources.[2] Creating a sustainable dietary pattern to meet the needs of the growing population requires an evaluation of a food's ability to deliver nutrition. An important part of this evaluation is an accurate measure of protein quality. In 2013, the Food and Agriculture Organization of the United Nations (FAO) recommended a new, advanced method for assessing the quality of dietary proteins.[3] The Digestible Indispensable Amino Acid Score (DIAAS) determines amino acid digestibility at the end of the small intestine, which more accurately measures the amounts of amino acids absorbed by the body. The DIAAS method allows protein sources to be rated by their ability to supply amino acids for use by the body. For example, the DIAAS method has demonstrated the higher bioavailability of dairy proteins when compared to plant-based protein sources. Whole milk powder has a DIAAS score of 122, compared to scores of 64 for peas and 40 for wheat. When plant protein sources are combined, however, their DIAAS scores can increase. The individual scores for beans and rice are 60, but when combined the score increases to nearly 80.

For making protein claims on food items, the FAO has recommended the following protein quality scores: DIAAS < 75% is considered suboptimal and no protein quality claim is allowed. A protein or protein blend with a DIAAS of 75%–99% is considered good, but still does not provide an optimal (100%) supply of essential amino acids. A claim of excellent or high protein quality is allowed on a protein or protein blend with a DIAAS of 100% or more because it is considered to be an optimal provider of dietary essential amino acids. Recently, scientists have been calling for a more modernized definition of protein quality based not only on the protein in a food, but on the overall nutritional and environmental properties of the food.

Protein is found in all of the major food groups, but much of the protein we eat comes from animal sources in the dairy and protein sections of MyPlate (Fig. 6-7). Animal proteins contain ample amounts of all nine essential amino acids. Gelatin—made from the animal protein collagen—is an exception because it is low in essential amino acids. Generally, plant proteins do not match our need for essential amino acids as precisely as animal proteins. Many plant proteins, especially those found in grains, are low in one or more of the nine essential amino acids. Plant protein sources that do contain all the essential amino acids include soy, buckwheat, and the seeds of amaranth, chia, hemp, pumpkin, and quinoa. It is important, however, to consider the total

▲ High-quality protein is available from both animal and plant sources. Shown here are some common sources, including meat, fish, poultry, eggs, beans, and nuts.
D. Hurst/Alamy

**COVID CORNER**

In 2020, COVID-19 sent the livestock and meat industry way out of balance. COVID-19 sickened meat plant workers, forcing many meat processing plants to shut down and others to operate at reduced capacity. This slowdown resulted in a diminished supply of meat products that could not meet the high demand. At a time when customers were staying home and cooking more, they were faced with: (1) higher meat prices; (2) purchasing limits on fresh beef, chicken, and pork; and (3) a drop in the supply of frozen and processed meats. Turning to a plant-based dietary pattern is always a cost-efficient, healthful, and accessible alternative.

**FIGURE 6-7** ▲ Food sources of protein. The fill of the background color (none, 1/3, 2/3, or completely covered) within each food group on MyPlate indicates the average nutrient density for protein in that group. Overall, the dairy group provides much protein (7 to 14 grams per serving), as does the protein group (7 to 24 grams per serving). The fruits group provides little or no protein (about 1 gram per serving). Food choices from the vegetables group and grains group provide moderate amounts of protein (3 to 8 grams per serving). banana: Alamy Stock Photo; Brussel Sprouts: Pixtal/age footstock; dairy: Photodisc/Getty Images; oatmeal: McGraw-Hill Education; tuna: FotografiaBasica/E+/Getty Images; MyPlate: U.S. Department of Agriculture

Source: U.S. Department of Agriculture, Agricultural Research Service. FoodData Central, 2019. fdc.nal.usda.gov.

**high-quality proteins** Dietary proteins that contain ample amounts of all nine essential amino acids; also called *complete proteins*.

**lower-quality proteins** Dietary proteins that are low in or lack one or more essential amino acids; also called *incomplete proteins*.

**complementary proteins** Two food protein sources that make up for each other's inadequate supply of specific essential amino acids; together, they yield a sufficient amount of all nine and so provide high-quality (complete) protein for the diet.

amount of protein in these sources as well. For example, one tablespoon of chia seeds contains only 3 grams of protein.

"High-quality proteins" are those that are readily digestible and contain the essential amino acids in quantities that humans require. For the support of growth and maintenance, humans generally are able to use proteins from any single animal source more efficiently than from any single plant source. For this reason, animal proteins (except gelatin) are considered **high-quality** (also called **complete**) **proteins,** which contain sufficient amounts of the nine essential amino acids. The majority of individual plant sources of proteins are considered **lower-quality** (also called **incomplete**) **proteins** because their amino acid patterns can be quite different from human proteins. Thus, a single plant protein source, such as corn alone, cannot easily support human growth and maintenance. To obtain a sufficient amount of all nine essential amino acids, a variety of plant proteins needs to be consumed because most plant protein lacks adequate amounts of one or more essential amino acids.

An inadequate supply of just one of the essential amino acids prevents protein synthesis. This is known as the *all-or-none* principle: either all essential amino acids are available, or none can be used. When only lower-quality protein foods are consumed, the amount of the essential amino acids needed for protein synthesis may not be obtained. Therefore, a greater amount of lower-quality protein is needed to meet the demands of protein synthesis, compared to high-quality proteins. Once any of the essential amino acids is used up, further protein synthesis becomes impossible. The remaining amino acids are then used for energy needs or converted into carbohydrate or fat.

When two or more protein sources are combined in a meal or snack to compensate for deficiencies in their essential amino acid contents, the proteins are called **complementary proteins.** Meals with a variety of protein sources generally result in a complementary protein pattern. Figure 6-8 provides examples of plant group combinations in which the proteins complement each other based on their limiting amino acids. Many legumes, for example, are deficient in the essential amino acid methionine, whereas grains are limited in lysine. Eating a combination of legumes and grains, such as beans

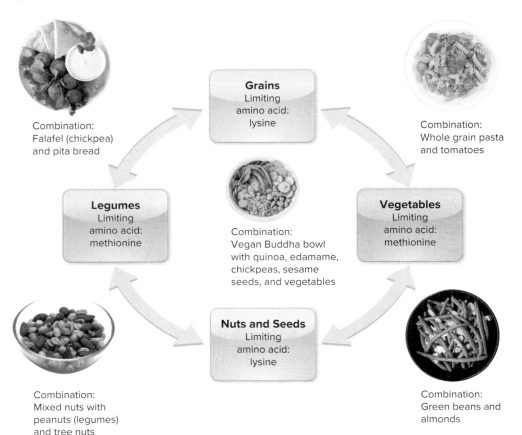

**FIGURE 6-8** ▶ Plant group combinations in which the proteins complement each other based on their limiting amino acids. falafel: Jennifer Barrow/123RF; Mixed nuts, L A Heusinkveld/Alamy; pasta: tomato, olegdudko/123RF; green beans with toasted almonds: Robyn Mackenzie/123RF; vegan Buddha bowl salad: Ekaterina Kondratova/123RF

and rice, will supply the body with adequate amounts of all essential amino acids. Likewise vegetables, which are limited in methionine, can be combined with nuts, which are limited in lysine. With these combinations, healthy adults should have little concern about obtaining enough of all nine essential amino acids. Even when following a plant-based dietary pattern, complementary proteins need not be consumed at the same meal. Meeting amino acid needs over the course of a day is a reasonable goal because there is a ready supply of amino acids present in body cells and in the blood.

In general, many North Americans are not reaping the full benefit of a plant-based eating pattern and would likely experience health benefits from replacing some animal sources of protein with plant sources of protein. Plant foods contribute fewer calories than most animal products but supply an ample amount of protein. Plant sources of proteins, especially legumes and nuts (Fig. 6-9), are a heart-healthy alternative to animal proteins because they contain very little saturated fat, aside from that added during processing or cooking.

## A CLOSER LOOK AT SOURCES OF PROTEINS

**Plant Proteins**   Plant-based protein sources include dairy alternatives, whole grains, nuts and seeds, and vegetables (Fig. 6-9). Per gram of protein, plant foods provide more magnesium, fiber, folate, vitamin E, iron (absorption is increased by the vitamin C also present), and zinc than animal sources of protein, and some calcium. Also, foods rich in phytochemicals help reduce risk of a wide variety of chronic diseases.

Legumes are a plant family with pods that contain a single row of seeds. Examples include garden and black-eyed peas, chickpeas, black beans, pinto beans, kidney beans, great northern beans, lentils, soybeans, and peanuts. Mature legume seeds—what we know as beans—make an impressive contribution to the protein, vitamin, mineral, and fiber content of a meal.[5] A ½-cup serving of legumes provides 100 to 150 kcal, 5 to 10 grams of protein, less than 1 gram of fat, and about 5 grams of fiber. Consumption of beans can lead to intestinal gas because our bodies lack the enzymes to break down certain carbohydrates in beans. Gas

**COVID CORNER**

A dietary pattern such as the Mediterranean diet is potentially beneficial against infections such as COVID-19 due to its effect on immune health.[6] The Mediterranean diet includes a high dietary intake of minimally processed fruit, vegetables, legumes, olive oil, whole grains, nuts, and monounsaturated fats; (2) low to moderate consumptions of fermented dairy products, fish, poultry, and wine; and (3) low consumptions of processed and red meats. Several foods associated with the Mediterranean diet contain vitamins C, D, and E, minerals such as zinc, copper, and calcium, and phytochemicals with potent anti-inflammatory and immunomodulatory properties.

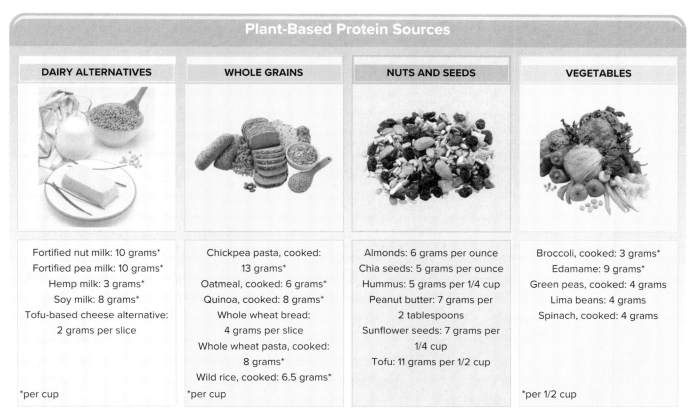

| DAIRY ALTERNATIVES | WHOLE GRAINS | NUTS AND SEEDS | VEGETABLES |
|---|---|---|---|
| Fortified nut milk: 10 grams* | Chickpea pasta, cooked: 13 grams* | Almonds: 6 grams per ounce | Broccoli, cooked: 3 grams* |
| Fortified pea milk: 10 grams* | Oatmeal, cooked: 6 grams* | Chia seeds: 5 grams per ounce | Edamame: 9 grams* |
| Hemp milk: 3 grams* | Quinoa, cooked: 8 grams* | Hummus: 5 grams per 1/4 cup | Green peas, cooked: 4 grams |
| Soy milk: 8 grams* | Whole wheat bread: 4 grams per slice | Peanut butter: 7 grams per 2 tablespoons | Lima beans: 4 grams |
| Tofu-based cheese alternative: 2 grams per slice | Whole wheat pasta, cooked: 8 grams* | Sunflower seeds: 7 grams per 1/4 cup | Spinach, cooked: 4 grams |
| | Wild rice, cooked: 6.5 grams* | Tofu: 11 grams per 1/2 cup | |
| *per cup | *per cup | | *per 1/2 cup |

**FIGURE 6-9** ▲ Plant-based proteins are found in a variety of food sources. dairy alternatives: Pixtal/AGE Fotostock; whole grains: Tetra Images/Getty Images; nuts and seeds: Thomas Northcut/Getty Images; vegetables: ©Pixtal

## FARM to FORK   Legumes

Legumes are seeds that grow in pods, such as beans, peas, and lentils. Beans have an oval or kidney shape, while peas are round, and lentils are flat disks.

### Grow

- Legumes have a beneficial relationship with bacteria in the soil. The bacteria take nitrogen from the soil and feed this nitrogen to the legumes. Legumes provide carbohydrates to the bacteria and nitrogen to support plants growing nearby.
- Pole beans can grow tall on trellises or in containers, making the most of limited garden spaces.
- Stagger the planting of beans to enjoy beans throughout the growing season.

### Shop

- Look for multicolored legumes—kidney beans, black beans, yellow peas, black-eyed peas, and lentils—to obtain abundant and varied phytochemicals.
- Buy canned beans. The process of heating during the canning process increases their nutritional value.
- Choose fresh or frozen pod peas to get the most fiber and antioxidants. Canned peas have lost up to 50% of their antioxidant content.

### Store

- Store fresh beans in a moisture-proof, airtight container to maintain freshness. Beans tend to get tough quickly after harvest.
- Canned beans remain highly nutritious over a long shelf life.
- To prevent dried beans from drying out further, keep them in a food-safe storage container with a tight lid and place in a cool, dry place away from sunlight.

### Prep

- Soak dry beans in water overnight prior to cooking. Discarding the water will help to reduce the flatulence that often occurs after eating legumes.
- During cooking, more than half of the antioxidants in dried beans will be leached into cooking water. Consume this water or allow the cooked beans to soak for 1 hour after cooking to help retain much of the nutrient content.
- Cook beans in a multifunction cooker or instant pot to save time, produce a tender product, and retain the most nutrients. If you prepare dried beans in a crockpot or slow cooker, make sure they are cooked adequately to deactivate the lectins.
- For the greatest convenience, use canned beans in cooking because they are higher in antioxidant content than fresh beans. Because canned beans are typically high in sodium, drain and rinse them under cold water to remove nearly half of the sodium.
- Most legumes are low in the essential amino acid methionine. Complement legumes with whole grains, which are a good source of methionine to achieve an eating pattern that provides all the essential amino acids.

Source: Robinson J: Legumes: Beans, peas and lentils. In *Eating on the Wild Side: The Missing Link to Optimum Health.* New York: Little, Brown and Company, 2013.

production can be decreased by soaking dry beans in water to leach the indigestible carbohydrates into the water, which then can be disposed.

Legumes are also naturally high in lectins, a class of proteins that bind carbohydrates and play a protective role in plants. Lectins have recently been wrongly blamed for causing a variety of health problems, including obesity, cancers, and inflammatory diseases. Although lectins are "natural pesticides" in plants, they are not toxic to humans because we cook legumes before consuming them. When legumes are cooked, fermented, sprouted, or processed for canning, the lectins bind to carbohydrates, which deactivates them and causes them to pass through the digestive tract and be eliminated. Eating raw or undercooked legumes would allow unbound lectins to attach to intestinal cells, resulting in vomiting, diarrhea, and abdominal pain. To prevent gastrointestinal distress, only consume beans that have been adequately cooked. Keep in mind that canned beans are already cooked. Read more about legumes in the *Farm to Fork* feature in this chapter.

Nuts and seeds are also excellent sources of plant protein. Commonly consumed nuts include almonds, cashews, pistachios, walnuts, and pecans. The defining characteristic of a nut is that it grows on a tree. Remember that peanuts, because they grow underground, are legumes. Seeds, including pumpkin, sesame, and sunflower seeds, are similar to nuts in nutrient composition. A 1-ounce serving of nuts or seeds generally supplies 160 to 190 kcal, 6 to 10 grams of protein, and 14 to 19 grams of fat. Although they are a dense source of calories, nuts and seeds make a powerful contribution to health when consumed in moderation. Chia and pumpkin seeds have the added advantage of being sources of high-quality protein with all of the essential amino acids.

In summary, plant proteins are a nutritious alternative to animal proteins. They are inexpensive, versatile, tasty, a colorful addition to your plate, and beneficial to health beyond their contribution of protein to the dietary pattern. Learning to substitute plant proteins in place of less healthy foods is one way to reduce your risk for many diseases. The impact of plant proteins on health is discussed in Section 6.9.

**Animal Proteins** Protein is found in all of the major food groups, but much of the protein we eat comes from animal sources in the dairy and protein sections of MyPlate (Fig. 6-7). For example, milk and eggs both contain very high-quality protein and are often used as standards against which other food proteins are compared. Milks with various fat contents, ranging from whole milk to fat-free milk, are rich sources of protein (8 grams per cup) and several other nutrients, including calcium and vitamin D. One egg has only 75 calories but 7 grams of high-quality protein. Due to the soaring costs of beef, the consumption of animal products declined from 2007 through 2014. Before the COVID-19 pandemic, that trend had recently reversed, however, with Americans eating record amounts of meat again, averaging about 222 pounds of red meat and poultry a year. There has also been a dramatic increase in meat consumption globally.[7] The greatest increases in consumption have occurred in East and Southeast Asia. Dairy production and consumption have also soared in China and India over the past few decades.

Dietary patterns that include high protein intakes are typically of concern only when these intakes rely heavily on animal sources of protein. An eating pattern that relies heavily on animal sources of protein is not recommended by the *Dietary Guidelines for Americans* or the American Heart Association because of its potential to increase the risk for cardiovascular and other diseases. Eating patterns rich in animal products are most likely low in beneficial substances found in plant sources, including fiber, some vitamins (e.g., folate), some minerals (e.g., magnesium), and phytochemicals, and they are high in substances such as saturated fat.

Although meat is one of the richest sources of protein, consumption of high levels of red and processed meat has been associated with an increased risk of colorectal cancer.[8] The review of more than 800 studies indicated that reducing consumption of processed meats can reduce the risk of colorectal cancer. Experts concluded that each 50-gram portion of processed meat eaten daily increases the risk of colorectal cancer by 18%.

An association also has been found between red meat consumption and deaths caused by cardiovascular disease and cancer.[9,10] This connection could be caused by the curing agents used to process meats such as ham and salami, as well as substances that form during cooking of meats at high temperatures. Any type of meat should be trimmed of all visible fat before cooking, especially grilling. The excessive fat or low-fiber contents of dietary patterns high in red meat may also be a contributing factor. These concerns could be avoided by focusing more on poultry, fish, nuts, legumes, and seeds to meet protein needs.

Red and processed meat has also been shown to be associated with increased risk of kidney disease.[11] Some researchers have expressed concern that a high-protein intake in general may overburden the kidneys by forcing them to excrete the extra nitrogen as **urea.** Also, animal proteins may contribute to kidney stone formation in certain individuals. A high-protein dietary pattern is not recommended for persons with limited kidney function such as those who have diabetes, early signs of kidney disease, or only one functioning kidney. There is some evidence that low-protein dietary patterns are helpful in slowing the decline in kidney function somewhat. High-protein dietary patterns increase urine output, which can lead to dehydration, especially in athletes.

Overall, the increases in demand and consumption of animal products have had a substantial impact on agriculture over the past three decades.[12] The "factory farms" have a significant environmental impact globally because of the large amount of land and water used, and the high level of waste produced, including agricultural greenhouse gas emissions.[13] These negative effects are in addition to the adverse health implications of consuming a high amount of animal products.

## FOOD PROTEIN ALLERGIES

Allergies occur when the immune system reacts to what it thinks is a foreign protein. In the case of food allergies, the immune system mistakes a food protein for a harmful invader. Overall, food allergies occur in up to 8% of children 4 years of age or younger and in up to 2% of adults. Eight foods account for 90% of food-related allergies (soy, peanuts, tree nuts, wheat, milk, eggs, fish, and shellfish; Fig. 6-10). The allergic reactions can range from a mild intolerance to fatal responses involving the cardiovascular and respiratory systems. Introducing allergenic foods such as peanut butter and eggs to infants as young as 4 to 6 months old is a new approach to combating food allergies.[15]

### ✓ CONCEPT CHECK 6.3

1. What types of foods contain high-quality proteins?
2. Why are complementary proteins important when pairing plant food sources?
3. What are the eight foods responsible for most food allergies?

▲ Animal-protein foods, such as roast beef (2 ounces) and Swiss cheese (1 ounce) on a bagel (3.5 inches), are typically our main sources of protein in North America. Although this sandwich provides 31 grams of protein, consumption of red meats such as roast beef has been associated with an increased risk of several chronic diseases, including colon cancer. Ingram Publishing/SuperStock

Because the Mediterranean diet is recognized as being low in meat, rich in fresh fruit and vegetables, and low in added sugar and saturated fatty acids, it is recommended as an alternative and sustainable dietary pattern. It has been advocated by the United Nations Food and Agricultural Organization (FAO) and could be a starting point for the creation of policies to support sustainable food production.[14]

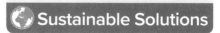

🌐 Sustainable Solutions

### Gluten Sensitivity

Gluten is a type of protein found in wheat, rye, and barley. Small peptides that arise from partial gluten digestion can be absorbed into the cells lining the small intestine and cause an inflammatory reaction in people with a genetic predisposition for celiac disease. Celiac disease is not a food allergy but an autoimmune response. Strict dietary avoidance of food products containing wheat, rye, and barley is the only proven way to manage the disease.[16]

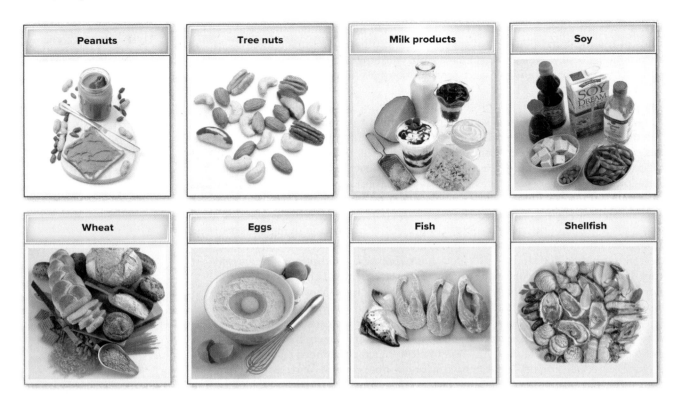

**FIGURE 6-10** ▲ Most common food allergens. peanut butter: Igor Dutina/iStock/Getty Images; tree nuts: Image Source/Glow Images; milk products, soy, wheat, eggs: Courtesy of Dennis Gottlieb; fish: Pixtal/age fotostock; shellfish: freeprod/123RF

# 6.4 Protein Digestion and Absorption

The digestion of most proteins begins with the cooking of food. Cooking unfolds (denatures) proteins (Fig. 6-11) and softens tough connective tissue in meat. The cooking process makes many protein-rich foods easier to chew and swallow, and facilitates their breakdown during later digestion and absorption. Cooking also makes many protein-rich foods, such as meats, eggs, fish, and poultry, much safer to eat.

## DIGESTION

The enzymatic digestion of protein begins in the stomach (Fig. 6-12). Thinking about food or chewing food stimulates the release of the hormone gastrin in the stomach. Gastrin then stimulates the stomach to produce acid and release pepsin. Proteins are first denatured by stomach acid. **Pepsin,** a major stomach enzyme for digesting proteins, then goes to work on the unraveled polypeptide chains. Pepsin can break only a few of the many peptide bonds found in these large polypeptide molecules, resulting in shorter chains of amino acids.

The partially digested proteins move from the stomach into the small intestine along with the rest of the nutrients and other substances in a meal (chyme). Once in the small intestine, the partially digested proteins (and any fats accompanying them) trigger the release of the hormone cholecystokinin (CCK) from the walls of the small intestine. CCK, in turn, travels through the bloodstream to the pancreas, where it causes the pancreas to release protein-splitting enzymes, such as **trypsin.** These digestive enzymes work in the small intestine to further divide the chains of amino acids into segments of two to three amino acids and some individual amino acids. Eventually, this mixture is digested into amino acids, using other enzymes from the lining of the small intestine and enzymes present in the absorptive cells themselves.

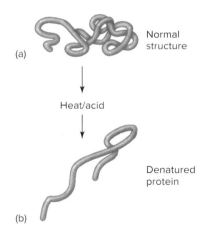

(a) Normal structure

Heat/acid

(b) Denatured protein

**FIGURE 6-11** ▲ Denaturation. (a) Protein showing typical coiled state. (b) Protein is partly uncoiled by heat or acid. This uncoiling can reduce biological activity and allow digestive enzymes to act on peptide bonds.

**pepsin** A protein-digesting enzyme produced by the stomach.

**trypsin** A protein-digesting enzyme secreted by the pancreas to act in the small intestine.

# Protein Digestion and Absorption

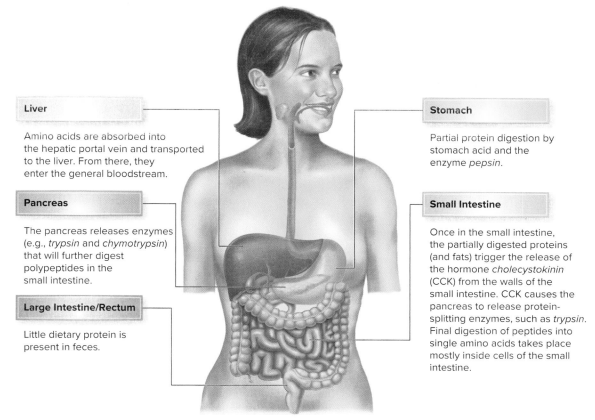

**Liver**

Amino acids are absorbed into the hepatic portal vein and transported to the liver. From there, they enter the general bloodstream.

**Pancreas**

The pancreas releases enzymes (e.g., *trypsin* and *chymotrypsin*) that will further digest polypeptides in the small intestine.

**Large Intestine/Rectum**

Little dietary protein is present in feces.

**Stomach**

Partial protein digestion by stomach acid and the enzyme *pepsin*.

**Small Intestine**

Once in the small intestine, the partially digested proteins (and fats) trigger the release of the hormone *cholecystokinin* (CCK) from the walls of the small intestine. CCK causes the pancreas to release protein-splitting enzymes, such as *trypsin*. Final digestion of peptides into single amino acids takes place mostly inside cells of the small intestine.

**FIGURE 6-12** ▲ A summary of protein digestion and absorption. Enzymatic protein digestion begins in the stomach and ends in the absorptive cells of the small intestine, where any remaining short groupings of amino acids are broken down into single amino acids. Stomach acid and enzymes contribute to protein digestion. Absorption from the intestinal lumen into the absorptive cells requires energy input.

## ABSORPTION

The short chains of amino acids and any individual amino acids in the small intestine are taken up by active transport into the absorptive cells lining the small intestine. Any remaining peptide bonds are broken inside intestinal cells to yield individual amino acids. They are water soluble, so the amino acids travel to the liver via the hepatic portal vein, which transports absorbed nutrients from the intestinal tract (see Fig. 6-12). In the liver, individual amino acids can undergo several modifications, depending on the needs of various body tissues. Individual amino acids may be: (1) combined into the proteins needed by specific cells; (2) broken down to meet energy needs; (3) released into the bloodstream; or (4) converted into nonessential amino acids, glucose, or fat. With excess protein intake, amino acids are converted into fat as a last resort.

Except during infancy, it is uncommon for intact proteins to be absorbed from the digestive tract. In infants up to 4 to 5 months of age, the gastrointestinal tract is somewhat permeable to small proteins, so some whole proteins can be absorbed. Because proteins from some foods (e.g., cow's milk and egg whites) may predispose an infant to food allergies, experts recommend waiting until an infant reaches 4 to 6 months of age to introduce solid foods.[15]

*magnificent*

Different types of protein sources can impact gut microbiota and obesity in both positive and negative ways. The high BCAA content of dairy proteins has a positive effect on gut microbiota and the prevention of obesity. Protein sources such as red meats and egg, however, contain compounds that the gut microbiota may convert into compounds such as trimethylamine and trimethylamine oxide (TMAO) that have been associated with increased risk for atherosclerosis and obesity.[17]

*microbiome*

✅ **CONCEPT CHECK 6.4**

1. Where and how does protein digestion begin?
2. What digestion steps take place in the stomach and small intestine?
3. What are the final products of protein digestion, and where do they go after absorption?

# 6.5 Putting Proteins to Work in the Body

Proteins function in many crucial ways in human metabolism and in the formation of body structures. Although we rely on foods to supply the amino acids needed to form these proteins, we also must eat enough carbohydrate and fat for food proteins to be used most efficiently. If we do not consume enough total calories to meet needs, proteins are broken down to supply energy to cells. This renders the amino acids unavailable for growth and repair of body tissues.

## PRODUCING VITAL BODY STRUCTURES

The amino acid pool in a cell can be used to form body proteins, as well as a variety of other compounds. Every cell contains protein. Muscles, connective tissue, mucus, blood-clotting factors, transport proteins in the bloodstream, lipoproteins, enzymes, immune antibodies, some hormones, visual pigments, and the support structure inside bones are all made of protein. Consuming excess protein does not enhance the synthesis of these body components, but eating too little protein can prevent it.

Most vital body proteins are in a constant state of breakdown, rebuilding, and repair. For example, the cells of the intestinal tract lining are constantly sloughed off. The digestive tract treats sloughed cells just like food particles, digesting them and absorbing their amino acids. In fact, most of the amino acids released throughout the body can be recycled to become part of the pool of amino acids available for the synthesis of future proteins.

Overall, **protein turnover** is a process by which a cell can respond to its changing environment by making proteins that are needed and disassembling proteins that are not needed. During a 24-hour period, an adult turns over (makes and degrades) about 250 grams of protein, recycling many of the amino acids. When you compare this to the 65 to 100 grams of protein typically consumed by adults in North America, recycled amino acids make an important contribution to total protein metabolism. If a person's dietary pattern is low in protein for a long period, the processes of rebuilding and repairing body proteins will slow down. Over time, skeletal muscles, blood proteins, and vital organs such as the heart and liver will lose protein and decrease in size or volume. Only the brain resists protein breakdown.

## REGULATORY FUNCTIONS

**Maintaining Fluid Balance.** Blood proteins help maintain body fluid balance. Normal blood pressure in the arteries forces blood into capillary beds. The blood fluid then moves from the **capillary beds** into the spaces between nearby cells (**extracellular spaces**) to provide nutrients to those cells (Fig. 6-13). Proteins in the bloodstream are too large, however, to move out of the capillary beds into the tissues. In fact, the presence of these proteins in the capillary beds attracts the proper amount of fluid back to the bloodstream, partially counteracting the force of blood pressure.

When protein intake is too low, the concentration of proteins in the bloodstream drops below normal. Excessive fluid then builds up in the surrounding tissues because the smaller amount of blood proteins is too weak to pull enough of the fluid back from the tissues into the bloodstream. As fluids accumulate in the tissues, the tissues swell, causing **edema.** Because edema can also be a symptom of other medical problems, an important step in diagnosing its cause is to measure the concentration of blood proteins.

Protein contributes to the structure and function of muscle. **Will eating protein in excess of requirements help this athlete build muscle mass?** Fancy Collection/SuperStock

**protein turnover** The process by which cells break down old proteins and resynthesize new proteins. In this way, the cell will have the proteins it needs to function at that time.

**capillary bed** Network of one-cell-thick vessels that create a junction between arterial and venous circulation. It is here that gas and nutrient exchange occurs between body cells and the blood.

**extracellular space** The space outside cells; contains one-third of body fluid.

**edema** The buildup of excess fluid in extracellular spaces of tissues.

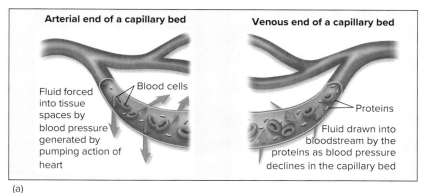

**FIGURE 6-13** ◄ Proteins help to maintain fluid balance. (a) As blood is pumped through the circulatory system, some fluid leaks out of the vessels. Normally, proteins in the blood draw water back into the blood vessels. (b) Without sufficient protein in the bloodstream, fluid remains in the tissues and edema develops. The lymph vessels of the lymphatic system also take up excess fluid that collects between cells and return it to the bloodstream.

### Arterial end of a capillary bed
Fluid forced into tissue spaces by blood pressure generated by pumping action of heart

Blood cells

(a)

### Venous end of a capillary bed
Proteins

Fluid drawn into bloodstream by the proteins as blood pressure declines in the capillary bed

### Normal tissue
Blood pressure is balanced by counteracting force of blood proteins.

### Swollen tissue (edema)
Blood pressure exceeds counteracting force of blood proteins, so fluid remains in the tissues.

(b)

The lymphatic system also helps maintain fluid balance by collecting excess fluid and substances, including proteins, from tissues and depositing them in the bloodstream.

**Contributing to Acid–Base Balance.** Proteins located in cell membranes help regulate acid–base balance in the blood by pumping chemical ions in and out of cells. This pumping of ions, among other factors, occurs in an effort to keep the blood slightly alkaline. In addition, some blood amino acids are especially good **buffers** in the bloodstream. Buffers are compounds that maintain acid–base conditions within a narrow range.

**Forming Hormones and Enzymes.** Many hormones, our internal body messengers, are proteins and therefore require amino acids for synthesis. The thyroid hormones are made from two molecules of only one type of amino acid: tyrosine. Insulin, on the other hand, is a hormone composed of 51 amino acid molecules. Almost all enzymes are proteins or have a protein component.

**Contributing to Immune Function.** Proteins are a key component of cells within the immune system. Antibodies, for example, are proteins produced by one type of white blood cell. These antibodies can bind to foreign proteins in the bloodstream—an important step in removing invaders from the body. Without sufficient dietary protein, the immune system lacks the materials needed to function properly. For example, a low-protein status can turn an infection such as measles into a fatal disease for a malnourished child.

### SOURCE OF ENERGY

**Forming Glucose.** If you do not consume enough carbohydrate to supply glucose, your liver (and kidneys, to a lesser extent) will be forced to make glucose from amino acids present in body tissues (Fig. 6-14). A fairly constant concentration of glucose must be maintained in the blood to supply energy for the brain, red blood cells, and nervous tissue. At rest, the brain uses about 19% of the body's energy requirements, and it gets most of that energy from glucose.

**buffer** Compounds that cause a solution to resist changes in acid–base conditions.

Neurotransmitters, released by nerve endings, are often derivatives of amino acids. This is true for dopamine and norepinephrine (both synthesized from the amino acid tyrosine) and serotonin (synthesized from the amino acid tryptophan). Most neurotransmitters are the size of a single amino acid.

**COVID CORNER**
Individuals with compromised immune systems are especially vulnerable to viral infections including COVID-19. During infections, adequate protein intake is important to maintain immune function, promote recovery, and prevent loss of lean body mass. To ensure production of immune factors, the dietary pattern, especially for the elderly and vegan populations, should include high-quality proteins. Patients who are ill with infections such as COVID-19 should consume about twice the amount of protein recommended for healthy individuals.

**FIGURE 6-14** ▶ Amino acid metabolism. The amino acid **pool** in a cell can be used to form body proteins, as well as a variety of other possible products. When the **carbon skeletons** of amino acids are metabolized to produce glucose or fat, ammonia ($NH_3$) is a resulting waste product. The ammonia is converted into urea and excreted in the urine.

**pool** The amount of a nutrient stored within the body that can be mobilized when needed.

**carbon skeleton** Amino acid structure that remains after the amino group ($-NH_2$) has been removed.

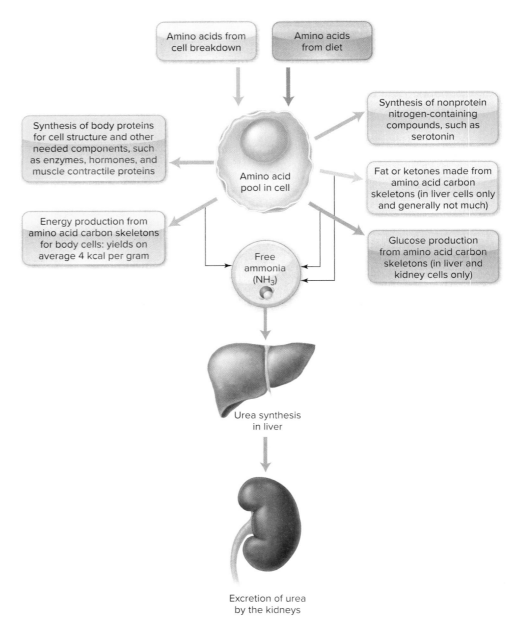

The vitamin niacin can be made from the amino acid tryptophan, illustrating another role of proteins.

Making some glucose from amino acids is normal. For example, when you skip breakfast and have not eaten since early the previous evening, glucose must be manufactured. In an extreme situation, however, such as in starvation, amino acids from muscle tissue are routinely converted into glucose, which decreases muscle tissue and can produce edema.

**Providing Energy.** Proteins supply little energy for a person at a healthy weight. Two situations in which a person does use protein to meet energy needs are during prolonged exercise and during calorie restriction, as with a weight-loss diet. In these cases, the amino group ($-NH_2$) from the amino acid is removed, and the remaining carbon skeleton is metabolized for energy needs (see Fig. 6-14). When the carbon skeletons of amino acids are metabolized to produce glucose or fat, ammonia ($NH_3$) is a resulting waste product. The ammonia is converted into urea and excreted in the urine. Under most conditions, cells primarily use fats and carbohydrates for energy needs. Although proteins contain the same amount of calories (on average, 4 kcal per gram) as carbohydrates, proteins are a costly source of calories, considering the amount of processing the liver and kidneys must perform to use this calorie source. The functions of proteins are summarized in Figure 6-15, the Protein Concept Map.

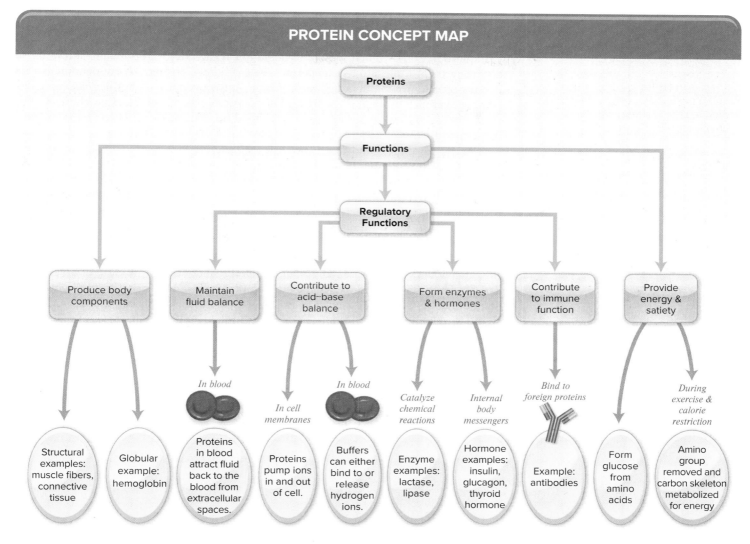

**FIGURE 6-15** ▲ Protein Concept Map illustrating the functions of protein throughout the body.

**Contributing to Satiety.** Many popular weight management diets recommend an increased amount of protein to enhance satiety and feelings of fullness. Studies of this phenomenon, however, have yielded different results depending on the timing and type of protein eaten. Increasing protein has shown the greatest effect on feelings of fullness when consumed prior to a meal.[18] This potential for protein to increase satiety and improve appetite control is greater than that of carbohydrates and/or fats. Several effective weight-loss diets include a percentage of calories from protein at the upper end of the Acceptable Macronutrient Distribution Range of 10% to 35% for protein. In general, these diets are appropriate if otherwise nutritionally sound, especially with regard to including a variety of foods from each food group.

## ✓ CONCEPT CHECK 6.5

1. Which body constituents are mainly proteins?
2. How many grams of protein does the body typically turn over each day?
3. If amino acids can be broken down to yield energy for cells, why do we need to convert some amino acids into glucose?

# 6.6 Protein Needs

The question of how much protein we need to eat each day for optimal health remains a controversial topic. Protein experts debate whether current recommendations for dietary protein are high enough, and if it is a misconception that Americans are consuming too much protein. We will begin our discussion of protein needs with the basics, then discuss the rationale for certain population groups to go beyond the current recommendations.

It is easy to understand that if you fail to consume an adequate amount of protein for weeks at a time, many metabolic processes slow down. This is because the body does not have enough amino acids available to build the proteins it needs. For example, the immune system no longer functions efficiently when it lacks key proteins, thereby increasing the risk of infections, disease, and death. Let's focus on how much protein (actually amino acids) we need to eat each day to prevent any decline in protein's metabolic functions.

People who are not growing or building muscle tissue need to eat only enough protein to match whatever they lose daily from protein breakdown. The amount of breakdown can be determined by measuring the amount of urea and other nitrogen-containing compounds in the urine, as well as losses of protein from feces, skin, hair, nails, and so on. In short, people need to balance protein intake with protein losses to maintain a state of **protein equilibrium,** also called *protein balance* (Fig. 6-16).

**protein equilibrium** A state in which protein intake is equal to related protein losses; the person is said to be in *protein balance.*

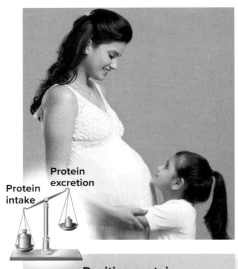

Protein excretion
Protein intake

### Positive protein balance

- Growth
- Pregnancy
- Recovery stage after illness, injury
- Athletic training**

(a)

Protein intake  Protein excretion

### Protein equilibrium

- Healthy adult meeting nutrient needs, notably protein, and calorie needs

(b)

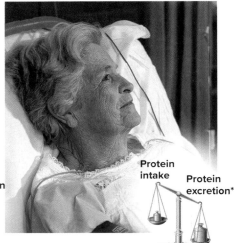

Protein intake
Protein excretion*

### Negative protein balance

- Inadequate protein intake (e.g., fasting, intestinal tract diseases)
- Inadequate calorie intake
- Fevers, burns, and infections
- Increased protein loss (e.g., kidney disease)

(c)

*Based on losses of urea and other nitrogen-containing compounds in the urine, as well as protein lost from feces, skin, hair, nails, and other minor routes.
**Only when additional lean body mass is being gained. Nevertheless, the athlete is probably already eating enough protein to support this extra protein synthesis; protein supplements are not needed.

**FIGURE 6-16** ▲ Protein balance in practical terms: (a) positive protein balance, (b) protein equilibrium, and (c) negative protein balance.
(left): Sudipta Halder/Getty Images; (middle) Nina Shannon/Getty Images; (right): Dynamic Graphics Group/Getty Images

When a body is growing or recovering from an illness or injury, it needs a **positive protein balance** to supply the raw materials required to build new tissues. To achieve this, a person must eat more protein daily than he or she loses. In addition, the hormones insulin, growth hormone, and testosterone all stimulate positive protein balance. Resistance exercise (weight training) that improves muscle protein synthesis also requires positive protein balance. Consuming less protein than needed leads to **negative protein balance,** such as when acute illness reduces the desire to eat, causing a person to lose more protein than he or she consumes.

For healthy people, the amount of dietary protein needed to maintain protein equilibrium (where intake equals losses) can be determined by increasing protein intake until it equals losses of protein and its related breakdown products (e.g., urea). Calorie needs must also be met so that amino acids are not diverted for use as energy.

The RDA for the amount of protein required for nearly all adults to maintain protein equilibrium is 0.8 gram of protein per kilogram of healthy body weight. Requirements are higher during periods of growth, such as pregnancy and infancy. An increased protein intake of more than 1.0 gram per kilogram of body weight per day also has been recommended for older adults as a result of the PROT-AGE Study.[19] Healthy weight is used in the calculation of protein needs because excess fat storage does not contribute much to protein needs. Calculations using the adult RDA are shown and estimate a requirement of about 56 grams of protein daily for a typical 70-kilogram (154-pound) man and about 46 grams of protein daily for a typical 57-kilogram (125-pound) woman.

**positive protein balance** A state in which protein intake exceeds related protein losses, as is needed during times of growth.

**negative protein balance** A state in which protein intake is less than related protein losses, such as often seen during acute illness.

### Calculation of Protein Recommendations for Adults

Convert weight from pounds to kilograms:

$$\frac{154 \text{ pounds}}{2.2 \text{ pounds/kilogram}} = 70 \text{ kilograms}$$

$$\frac{125 \text{ pounds}}{2.2 \text{ pounds/kilogram}} = 57 \text{ kilograms}$$

Calculate protein RDA:

$$70 \text{ kilograms} \times \frac{0.8 \text{ gram protein}}{\text{kilogram body weight}} = 56 \text{ grams}$$

$$57 \text{ kilograms} \times \frac{0.8 \text{ gram protein}}{\text{kilogram body weight}} = 46 \text{ grams}$$

The RDA for protein translates into about 10% of total calories. Many experts recommend up to 15% of total calories as protein to provide flexibility in planning a healthy eating pattern and to allow for the variety of protein-rich foods typically consumed. The *Dietary Guidelines* encourage the consumption of a variety of nutrient-dense protein foods from several subgroups including meats, poultry, and eggs; seafood; and nuts, seeds, and soy products (Fig 6-17). The upper range for protein intake is set at 35% of calories consumed. It is easy to meet these suggested daily protein needs, as given in Table 6-2. On a daily basis in North America, the typical man and woman consume about 100 and 65 grams of protein, respectively. Many consume more protein than the RDA recommends because they like many high-protein foods and can afford to buy them. Our bodies cannot store excess protein once it is consumed, so any excess amino acids not used for protein synthesis are stripped of the nitrogen-containing amino group and may be turned into glucose or ketone bodies (Fig. 6-14). Research also demonstrates that the timing of our protein intake throughout the day is important.

**Protein Requirements Per Meal.** Many adults have an unbalanced meal distribution of protein with more than 60% of daily protein consumed during a single evening meal and less than 15 grams at breakfast. Recent research, however, shows that distributing

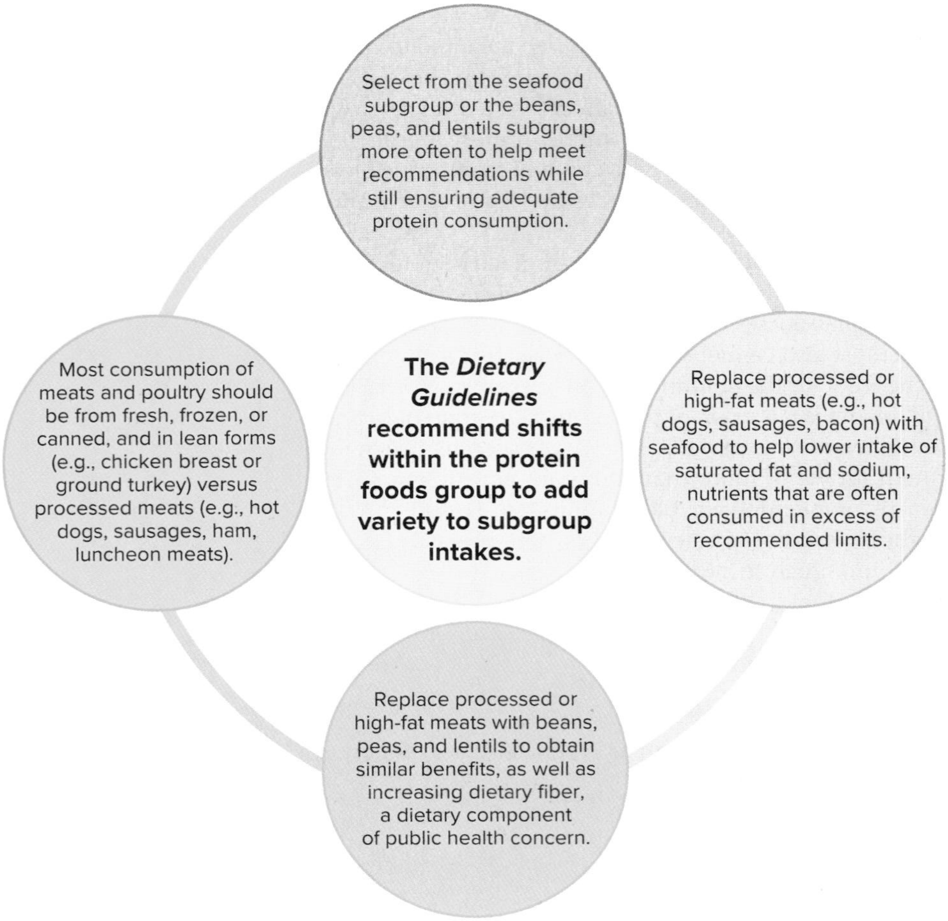

**FIGURE 6-17** ▶ Protein-specific recommendations from the *Dietary Guidelines*.

Source: U.S. Department of Agriculture and U.S. Department of Health and Human Services. *Dietary Guidelines for Americans, 2020–2025*. 9th Edition. December 2020. Available at https://www.dietaryguidelines.gov.

protein more evenly throughout the day is optimal for body functions.[20] Many protein functions, such as maintaining body composition and regulating glucose, are sensitive to the concentration of amino acids in blood and cells after meals. Meal-based responses to dietary protein have been linked to amino acids acting as signals. The best characterized of these postmeal amino acid signals is that of leucine stimulating the synthesis of skeletal muscle.

Studies have found that consuming at least 20 to 30 grams of protein at a given meal has positive effects on muscle protein synthesis, compared with spreading the same total amount of protein across multiple small meals. Researchers now recommend that adults consume at least 30 grams of protein at more than one meal in order to maintain healthy muscles and bones. Protein at breakfast is especially critical to regulate appetite and daily food intake and to replenish body proteins after an overnight fast.

**Protein Needs for Older Adults.** The concept of protein recommendations for each meal has also been studied in older adults.[19] Research shows that the responses to dietary protein decline with advancing age or reduced physical activity. The recommendation of the PROT-AGE Study Group is an increased daily protein intake of more than 1.0 gram per kilogram for older adults, while focusing on meal quantity and timing of protein.[19] Specific recommendations are for older adults to consume meals with greater than 20 grams of protein, including more than 2.2 grams of the branched-chain amino acid leucine, to optimize protein synthesis in skeletal muscle. Regarding physical activity, resistance exercise enhances the protein synthesis response, especially muscle growth, in older adults. Research indicates that with increases in resistance

**TABLE 6-2 ■ Protein Content of Sample Menus Containing 1600 and 2000 kcal**

| Menu | | 1600 kcal | | 2000 kcal | |
|---|---|---|---|---|---|
| | | Serving Size | Protein (g) | Serving Size | Protein (g) |
| **Breakfast** | | | | | |
| | Low-fat granola | ⅔ cup | 5 | ⅔ cup | 5 |
| | Blueberries | 1 cup | 1 | 1 cup | 1 |
| | Fat-free (skim) milk | 1 cup | 8.5 | 1 cup | 8.5 |
| | Coffee | 1 cup | 0 | 1 cup | 0 |
| **Lunch** | | | | | |
| | Broiled chicken breast | 3 ounces | 25 | 4 ounces | 33 |
| | Salad greens | 3 cups | 5 | 3 cups | 5 |
| | Croutons | ½ cup | 2 | ½ cup | 2 |
| | Low-fat salad dressing | 2 tbsp | 0 | 2 tbsp | 0 |
| **Dinner** | | | | | |
| | Fat-free (skim) milk | 1 cup | 8.5 | 1 cup | 8.5 |
| | Rice | 1 cup | 5 | 1.5 cups | 7.5 |
| | Shrimp | 4 large | 5 | 6 large | 7 |
| | Red beans | ½ cup | 21.5 | 1 cup | 43 |
| | Sweet red pepper | ½ cup | 0 | ½ cup | 0 |
| **Snack** | | | | | |
| | Woven wheat crackers | 6 crackers | 3 | 6 crackers | 3 |
| | Cheddar cheese | 1 ounce | 7 | 1 ounce | 7 |
| | Banana | ½ small | 0.5 | ½ small | 0.5 |
| | Total | | 97 | | 124 |

breakfast: Floortje/Getty Images; lunch: Olga Nayashkova/Shutterstock; dinner: Pixtal/age fotostock; snack: CWLawrence/Getty Images

exercise and protein intakes, older adults can achieve rates of muscle protein synthesis similar to young adults. In contrast, declining daily activity, including short-term bed rest due to hospitalization, illness, or injury, blunts the ability of amino acids to be incorporated into proteins and results in a significant loss of lean tissue in both young and older adults.

Mental stress, physical labor, and recreational weekend sports activities do not require an increase in the protein RDA. For some highly trained athletes, such as those participating in endurance or strength training, protein consumption may need to exceed the RDA. Many North Americans, especially men, already consume that much protein. Review the *Ask the RDN* entitled "Active Eating Advice" for ideas on including more protein in your meals throughout the day.

## AMINO ACID SUPPLEMENTS

Protein and amino acid supplements are used primarily by those trying to lose weight and by athletes hoping to build muscle. The branched-chain amino acids are especially popular with athletes looking to enhance their performance. Although the right amount of protein in the dietary pattern will aid athletic performance and help in weight control, consuming protein in the form of amino acid supplements is not considered safe. In Canada, the sale of individual amino acids to consumers is banned.

## ASK THE RDN | Active Eating Advice

*Dear RDN:* As a college student and athlete, my schedule is super busy with limited time for eating. What active eating advice could you give to college students on the go?

Busy lives with erratic schedules can present challenges when trying to fuel well to optimize academic and athletic performance. You may need to prioritize energy breaks in your day, pack food with you, and peruse offerings in the dining hall to make meal choices that work for you.

**Here are some recommendations to help you be in the know while on the go!**
1. *Prime Time:* Start your day with food, and be sure to surround workouts with fuel and fluid pre- and post-exercise.
2. *Think Your Drink:* Buying beverages can be quite costly and adds to the landfill. So instead, brew your own coffee or tea in your dorm room and bring a reusable thermos. This will save time and money. Carry your own water bottle, pack low-fat chocolate milk, or add a **protein isolate** to 100% juice or water to hydrate and also conserve money.
3. *Strive to compose a performance plate when you eat:* Start by filling half your plate with produce, such as salad, cooked vegetables, and/or fruit. Protein foods should make up 25% of your plate. Eggs, meat, poultry, fish, beans, soy foods, low-fat dairy foods, nuts, and seeds are all great options. If you are vegan, this means a good amount of beans, soy foods, lentils, peas, or seeds on the plate. Carbohydrates in the form of potatoes and whole grain forms of bread, cereal, rice, and pasta should cover at least 25% of the plate. Include some fat at meals through nuts or seeds, nut butters, hummus, salad dressing, or even avocado or guacamole.

**Here are some examples of on-the-go meals from the dining hall:**
- Breakfast burrito with eggs, black beans, salsa, and avocado in a whole grain tortilla with a glass of orange juice
- Rice bowl with greens, chicken, pineapple, nuts, and salad dressing
- Mongolian grill stir-fry of tofu, veggies, cashews, and sauce over noodles with an apple and peanut butter

**Try these, if you are making food to eat in your room or take with you:**
- Overnight oatmeal: ½ cup oats, ½ cup Greek yogurt, ½ cup low-fat milk, and then add in peanut butter, honey or syrup, a sliced banana, or dried fruit.
- DIY energy bites: In a bowl, mix together ½ cup peanut butter, ¼ cup honey, ½ cup dried fruit, 1 cup crispy rice cereal, ½ cup oats, ¼ cup peanuts. Roll into quarter-size balls and enjoy. Refrigerate or freeze the leftover energy bites for another time.
- Add a packet of tuna (for extra protein) and even some extra canned or frozen veggies to a can of cream-based or tomato-based clam chowder. Heat and have with crackers.
- Add microwaveable brown rice along with canned tomatoes and mushrooms to a can of lentil soup and heat, for an easy and filling lentil stew.

**Here are some foods to keep in your dorm room. They are shelf stable, don't take up a lot of space, minimize waste, and deliver on the taste.**

**PROTEIN**
Jerky: turkey or beef, or meat-based bars
Fruit and nut bars
Turkey pepperoni
Cans/pouches of tuna/salmon/turkey breast
Individual packets of whey, soy, or pea protein isolate
Protein bars: bars that provide at least 10 grams of protein per serving

**VEGETABLES**
Roasted beans (garbanzo, fava, broad)
Shelf-stable hummus
Vegetable chips (only ingredient is vegetables)
Freeze-dried bean soups

Lentil chips/crackers
Bean chips/snacks
Small cans of vegetable/tomato juice

**protein isolate** Protein powder that has been processed more than a protein concentrate to remove lower protein portions and collect pure protein fractions.

## ASK THE RDN *(continued)*

**GRAINS**

Unsweetened whole grain breakfast cereals
Oatmeal packets
Quinoa chips
Rice chips

Granola
Microwaveable rice/quinoa cups
Popcorn

**FRUITS**

Oranges
Apples/pears
Bananas

Unsweetened applesauce
Freeze-dried fruit

**NUTS/SEEDS**

Nuts and seeds are great, but be aware of the portion, which is *not* the entire bag or jar.
Seeds such as sunflower or pumpkin
Individual packets of nut butters (peanut, almond)
Nut and fruit bars

Eating on the go does not have to be a nutrition no-no. Explore options in campus food choices and also the time it will take to eat. Try to match your eating venues to your class sites, so you don't have to waste time trying to find a place to eat. Outfit your room with foods that you can eat fast but that will also last so you can eat well, learn well, and stay well.

Eat well to stay well and play well,

*Leslie J. Bonci, MPH, RD, CSSD, LDN*

Sports dietitian for collegiate, professional, Olympic and recreational athletes and performing artists

©Karen Meyers

Because the body's gastrointestinal system is adapted to handle whole proteins as a dietary source of amino acids, individual amino acid supplements can overwhelm the absorptive mechanisms in the small intestine. When amino acids are ingested as supplements, amino acid imbalances occur in the intestinal tract because an overload of chemically similar amino acids compete for absorption sites in the absorptive cells. For example, an excess of lysine can impair absorption of arginine because they are absorbed by the same transporter. The amino acids methionine, cysteine, and histidine are most likely to cause toxicity when consumed in large amounts. Due to this potential for imbalances and toxicities, the best advice is to stick to whole foods as sources of amino acids rather than supplements. Amino acid supplements also have a disagreeable odor and flavor and are much more expensive than food protein.

## ✓ CONCEPT CHECK 6.6

1. During what situations is the body in positive protein balance?
2. What is the RDA for protein for a person weighing 70 kilograms?
3. How much protein does the average American consume?

# 6.7 Protein-Calorie Malnutrition

Protein deficiency is rarely an isolated condition and usually accompanies a deficiency of calories and other nutrients resulting from insufficient food intake. In the developed world, alcohol use disorders can lead to cases of protein deficiency because of the low protein content of alcoholic beverages that make up a high percentage of calories. Protein and calorie malnutrition is a significant problem in hospitals worldwide, affecting patients from infancy through older adulthood. Malnutrition can be caused by the illnesses or injuries for which patients are admitted to the hospital and by the hospitalization itself.[21]

In developing areas of the world, people often have dietary patterns low in calories and protein. This state of undernutrition stunts the growth of children and makes them more susceptible to disease throughout life. People who consume too few of their calories as protein can eventually develop **protein-calorie malnutrition (PCM),** also referred to as *protein-energy malnutrition (PEM)* (Fig. 6-18). In its milder form, it is difficult to tell if a person with PCM is consuming too little calories or protein, or both. When an inadequate intake of nutrients, including protein, is combined with an existing disease, especially an infection, a form of malnutrition called **kwashiorkor** can develop. But if the nutrient deficiency—especially a calorie deficiency—becomes severe, a deficiency disease called **marasmus** can result. Both conditions are seen primarily in children but also may develop in adults, even in those hospitalized in North America. These two conditions form the tip of the iceberg with respect to states of undernutrition, and symptoms of these two conditions can even be present in the same person.

## KWASHIORKOR

*Kwashiorkor* is a word from Ghana that means "the disease that the first child gets when the new child comes." Infants in developing areas of the world are usually breastfed from birth. Often by the time the child reaches 1 to 1.5 years of age, the mother is pregnant or has already given birth again, and the newborn infant gets preference for breastfeeding. The older child's diet then abruptly changes from nutritious human milk to starchy roots and **gruel,** which are low in protein compared to their calorie content. Additionally, the foods are usually high in bulky plant fibers, which are very filling and

**protein-calorie malnutrition (PCM)** A condition resulting from regularly consuming insufficient amounts of calories and protein. The deficiency eventually results in body wasting, primarily of lean tissue, and an increased susceptibility to infections. Also known as *protein-energy malnutrition (PEM).*

**kwashiorkor** A form of protein-calorie malnutrition occurring primarily in young children who have an existing disease and consume a marginal amount of calories and insufficient protein in relation to needs. The child generally suffers from infections and exhibits edema, poor growth, weakness, and an increased susceptibility to further illness.

**marasmus** A form of protein-calorie malnutrition resulting from consuming a grossly insufficient amount of protein and calories. Victims have little or no fat stores, little muscle mass, and poor strength. Death from infections is common.

**gruel** A thin mixture of grains or legumes in milk or water.

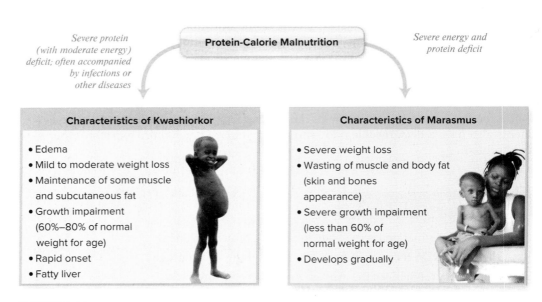

FIGURE 6-18 ▲ Classification of protein-calorie malnutrition occurring primarily in children (kwashiorkor and marasmus). (left): Christine Osborne Pictures/Alamy Stock Photo; (right): Phanie/Alamy Stock Photo

thus prevent the child from consuming enough food to meet calorie needs. Infections also raise calorie and protein needs, making it challenging to meet calorie needs, while protein consumption is grossly inadequate, and an increased amount is needed to combat infections. Many vitamin and mineral needs are also far from being fulfilled. Famine victims face similar problems.

The major symptoms of kwashiorkor are apathy, diarrhea, listlessness, failure to grow and gain weight, and withdrawal from the environment. These symptoms complicate other diseases present. For example, a condition such as measles, a disease that normally makes a well-nourished child ill for only a week or so, can become severely debilitating and even fatal when combined with kwashiorkor. Further symptoms of kwashiorkor are changes in hair color, potassium deficiency, flaky skin, fatty liver, reduced muscle mass, and massive edema in the abdomen and legs. The presence of edema in a child who has some subcutaneous fat (i.e., fat directly under the skin) is the hallmark of kwashiorkor (review Fig. 6-18). In addition, these children seldom move or cry. When you hold them, you feel the plumpness of edema, not muscle and fat tissue.

Many symptoms of kwashiorkor can be explained based on what we know about proteins. Proteins play important roles in fluid balance, lipoprotein transport, immune function, and production of tissues such as skin, cells lining the GI tract, and hair. Needless to say, children with an insufficient protein intake do not grow or mature normally. If children with kwashiorkor are helped in time—that is, if infections are treated and the dietary pattern becomes plentiful in protein, calories, and other essential nutrients—then the disease process reverses. They begin to grow again and may even show no signs of their previous condition, except perhaps shortness of stature. Unfortunately, by the time many of these children reach a hospital or care center, they already have severe infections. Despite the best care, they still die. Or, if they survive, they return home only to become ill again.

## MARASMUS

Marasmus (referred to as *protein-calorie malnutrition,* especially when experienced by older children and adults) typically occurs as an infant slowly starves to death. It is caused by dietary patterns containing minimal amounts of calories, as well as too little protein and other nutrients. The word *marasmus* means "to waste away" in Greek. Victims have a "skin-and-bones" appearance, with little or no subcutaneous fat (Fig. 6-18). Marasmus commonly develops in infants who either are not breastfed or have stopped breastfeeding in the early months. When people are poor and sanitation is lacking, bottle feeding often leads to marasmus. Often the weaning formula used is improperly prepared because of unsafe water or it is diluted with too much water because the parents cannot afford sufficient infant formula for the child's needs.

Marasmus in infants commonly occurs in the large cities of poverty-stricken countries. In cities, bottle feeding is often necessary because the infant is cared for by others when the mother is working or away from home. An infant with marasmus requires large amounts of calories and protein, similar to the needs of a **preterm** infant, and, unless the child receives them, full recovery from the disease may never occur. The majority of brain growth occurs between conception and the child's first birthday, with the brain growing at its highest rate after birth. If human milk or formula does not support brain growth during the first months of life, the brain may not grow to its full adult size, leading to diminished intellectual function. Both kwashiorkor and marasmus plague infants and children, often causing mortality rates in developing countries to be 10 to 20 times higher than in North America.

▲ Aims of the United Nations Sustainable Develophment Goals include drastically decreasing global poverty and hunger, including protein-calorie malnutrition, by 2030. Lissa Harrison

**preterm** Referring to an infant born before 37 weeks of gestation; also referred to as *premature.*

---

### ✔ CONCEPT CHECK 6.7

1. What are the characteristics of kwashiorkor and marasmus?
2. Why are bottle-fed infants at high risk for marasmus in poverty-stricken countries?

# 6.8 Nutrition and Genetics

## THE EMERGING FIELD OF NUTRITIONAL GENOMICS

We discuss nutrition and genetics in the protein chapter because the primary function of our genes is to produce proteins. The availability of genetic information is now enabling health professionals to personalize nutrition recommendations that can optimize nutritional status and improve the outcomes of nutrition-related diseases. It is evident that nutritional status can both *affect* and *be affected by* an individual's genetic makeup. The study of interactions between nutrition and genetics is known as nutritional genomics (Fig. 6-19). **Nutrigenetics** is the branch of nutritional genomics that examines how variations in genes can affect nutritional health. For example, the efficiency of absorption, metabolism, and excretion of a particular nutrient is controlled by genes. On the other hand, **nutrigenomics** refers to the many ways dietary components affect gene expression—particularly as it relates to development and treatment of nutrition-related diseases, such as cardiovascular disease. In this section we will examine each of these branches of nutritional genomics more closely.

Recall that nutrient recommendations, such as RDAs, are not absolute but are actually estimates of a level of intake that is likely to meet the needs of most (97% to 98%) of the population. For example, the RDA for folic acid (a B vitamin) is 400 micrograms per day. For most of the population, consuming this much folic acid from foods or supplemental sources will supply enough of the vitamin to optimize its functions in the body. There are certain subgroups of the population, however, that have dietary requirements for folic acid that are as much as 10 times higher than the RDA because of a genetic variation that alters the production of an enzyme necessary for amino acid metabolism. *Nutrigenetics* researchers are actively examining how genetic variations like this can affect individual nutrient requirements, how we can identify these people, and how we can personalize nutrition advice based on this knowledge.[22]

With *nutrigenomics*, researchers are interested in finding out how nutrients or other dietary components can influence gene expression, particularly as it relates to development of chronic diseases. Nutrigenomics research is now making it clear that

**nutrigenetics** A branch of nutritional genomics that studies of the effects of genes on nutritional health, such as variations in nutrient requirements and responsiveness to dietary modifications.

**nutrigenomics** A branch of nutritional genomics that studies how food impacts health through its interaction with our genes and its subsequent effect on gene expression.

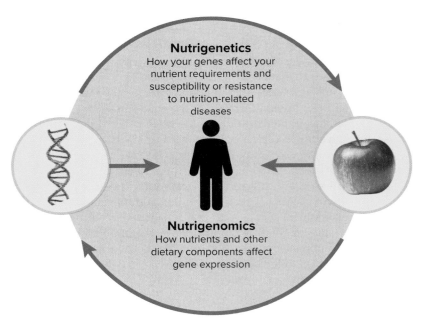

**Nutrigenetics**
How your genes affect your nutrient requirements and susceptibility or resistance to nutrition-related diseases

**Nutrigenomics**
How nutrients and other dietary components affect gene expression

**FIGURE 6-19** ▲ The study of nutritional genomics includes nutrigenetics and nutrigenomics. Nutrigenetics is the study of how genes dictate individual nutrient requirements and susceptibility or resistance to certain nutrition-related diseases. Nutrigenomics is the study of how nutrients and other dietary components can affect gene expression. DNA helix: Comstock/Stockbyte/Getty Images; red apple: Turnervisual/Getty Images

generalized nutrition recommendations may not apply to all individuals within a population group. Nutrients or other compounds consumed can turn certain genes on or off (like a light switch), thus manipulating the production of proteins that can affect—positively or negatively—the development or progression of diseases. Current areas of nutrigenomics research include obesity, diabetes, cardiovascular disease, celiac disease, cancer, osteoporosis, and Alzheimer's disease. With a better understanding of the interactions between genes and our eating patterns, it will not be long before dietary recommendations can be tailored to help those with genetically linked diseases.

## NUTRITIONAL DISEASES WITH A GENETIC LINK

Studies of families, including those with identical twins and adopted children, provide strong support for the effects of genetics in various disorders. In fact, family history of disease is considered to be an important risk factor in the development of many nutrition-related diseases.

▲ Studies of twins have provided strong evidence for the interaction between genes and dietary patterns and their combined effects on disease risk. Leland Bobbe/Getty Images

**Cardiovascular Disease.** There is strong evidence that cardiovascular disease is the result of gene–environment interactions. About one of every 500 people in North America has a defective gene that greatly delays cholesterol removal from the bloodstream. Elevated blood cholesterol is one of the risk factors for development of cardiovascular disease. Discoveries of gene and eating pattern interactions will allow for personalized treatment plans to manage blood lipids using medications and nutrition therapy that will help reduce disease risk and improve health outcomes. Although eating pattern modifications are important and can make a difference, medications and even surgery are often needed to fully address these problems.

**Obesity.** Most obese North Americans have at least one parent affected by obesity. This strongly suggests a genetic link. Findings from many human studies suggest that a variety of genes (likely 60 or more) are involved in the regulation of body weight. For example, specific gene variations have been linked to the propensity to overeat and alterations in metabolism.

Although some individuals may be genetically predisposed to store body fat, whether they do so depends on how many calories they consume relative to their needs. A common concept in nutrition is that how people live and the environmental factors that influence them allow each person's genetic potential to be expressed. Although not every person with a genetic tendency toward obesity becomes obese, those genetically predisposed to weight gain have a higher lifetime risk than individuals without a genetic predisposition to obesity.

**Diabetes.** Both type 1 and type 2 diabetes are influenced by genetics. Evidence for these genetic links comes from studies of families, including twins, and from the high incidence of diabetes among certain population groups (e.g., South Asians or Pima Indians). Diabetes, in fact, is a complex disease with more than 200 genes identified as possible causes. Only sensitive and expensive testing can identify who is at greatest risk. Type 2 diabetes, the most common form of diabetes (90% of all cases), is typically diagnosed after a person becomes obese, not before. In this case, a lifestyle leading to obesity affects genetic expression.

**Cancer.** It is estimated that about 10% of cancers have a genetic link. A much higher percentage (about 90%) of cancers are related to environmental and lifestyle factors. Body weight and eating patterns have been estimated to account for approximately 60% of all cancers. Other environmental influences on cancer risk include tobacco, sedentary lifestyles, family history, viruses, radiation, and toxic exposures.

▲ Genetic testing for disease susceptibility will be more common in the future as the genes that increase the risk of developing various diseases are isolated and decoded. Cultura RF/Getty Images

The following websites will help you gather more information about genetic conditions and testing:

**www.geneticalliance.org**

The Genetic Alliance is a nonprofit health advocacy group that provides a variety of publications for families about genetic conditions.

**http://learn.genetics.utah.edu**

The Genetic Science Learning Center maintained by the University of Utah provides an online tour of basic genetics.

**http://www.cancer.gov/publications /pdq/information-summaries/genetics**

The National Cancer Institute of the National Institutes of Health provides fact sheets pertaining to genetics and cancer for health professionals.

**https://www.genome.gov/26524162 /bringing-the-genomic-revolution-to-the -public**

The National Human Genome Research Institute website, from the National Institutes of Health, describes the latest research findings, discusses some ethical issues, and provides a talking glossary.

## YOUR GENETIC PROFILE

From this discussion, you can see that your genes can greatly influence your risk of developing certain diseases. By recognizing your potential for developing a particular disease, you can avoid behaviors and exposures that further raise your risk. Genetic testing can be valuable if it confirms that you carry a genetic mutation associated with a disease, and the results of testing might alter the course of treatment. Testing is also of interest when you do not know your family medical history or there are gaps in your family tree. Depending on the gene of interest, it typically costs under $1000 to have your DNA sequenced to reveal susceptibility for diseases or disorders. Many genetic tests are covered by health insurance plans, and the Genetic Information Nondiscrimination Act prohibits health insurers from raising premiums or denying coverage based on genetic information. DNA testing requires providing a DNA sample (blood, saliva, hair). Areas of interest in the genome are sequenced and read in a process known as genotyping. If you are considering genetic testing, it is best to consult a Certified Genetic Counselor. You can find one at nsgc.org.

Many of the genes involved in common diseases, such as diabetes, are still unknown. Although there are genetic tests available for some diseases, your family history of certain diseases is still a much better indicator of your genetic profile and risk of disease. Put together a family tree of illnesses and deaths by compiling a few key facts on your primary relatives: siblings, parents, aunts and uncles, and grandparents (ask your instructor about the *Rate Your Plate:* "Family Tree" activity in Connect). In general, the greater the number of your relatives who had a genetically transmitted disease and the closer they are related to you, the greater your risk. If there is a significant family history of a certain disease, lifestyle changes may help decrease your risk of developing this disease.

Figure 6-20 shows an example of a medical family tree (also called a *genogram*). Risk is high when two or more first-degree relatives in a family have a specific disease (first-degree relatives include one's biological parents, siblings, and offspring) or when a first-degree relative develops a disease before age 50 to 60 years. In the family depicted in Figure 6-20, prostate cancer killed the man's father. Knowing this, the man should be tested regularly for prostate cancer. His sisters should have frequent mammograms and breast exams because their mother died of breast cancer. Because heart attack and stroke are also common in the family, all the children should adopt a lifestyle that minimizes the risk of developing these conditions, such as avoiding excessive animal fat and salt intake. Colon cancer is also evident, so careful screening throughout life is important.

Information about our genetic makeup should increasingly influence our eating patterns and lifestyle choices. Throughout this text, we discuss "controllable" risk factors that could contribute to development of genetically linked diseases present in your family. This information will help you personalize nutrition advice based on your genetic background and identify and avoid the risk factors that could lead to the diseases present in your family.

## PERSONALIZING NUTRITION ADVICE

Nutrition professionals already recognize that dietary advice must be tailored to personal and cultural preferences. Research is now paving the way for even more personalized nutrition that incorporates the results of genetic testing when designing the most effective eating pattern for each person. There are genetic tests available for at least 1500 diseases and conditions. Many companies—most of them online— are already offering dietary advice and supplements based on direct-to-consumer genetic tests.

Remember that caution is always needed when evaluating nutrition information and claims. Some DNA-testing companies are responsible organizations, but many are

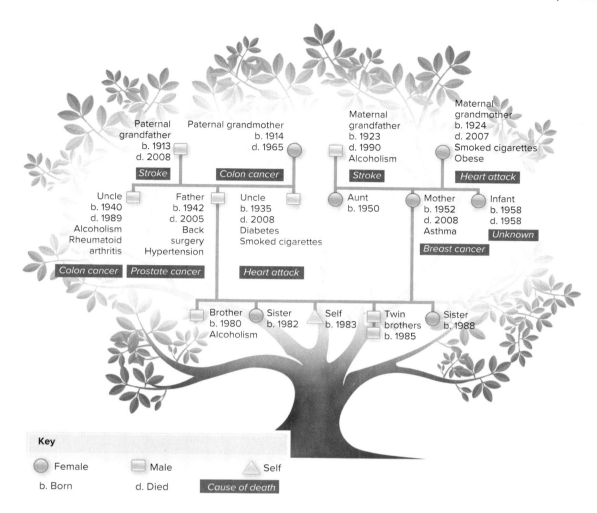

FIGURE 6-20 ▲ Example of a family tree for Justin, designated as "Self" at the trunk of the tree. The sex of each family member is identified by color (blue squares for males and red circles for females). Dates of birth (b) and death (d) are listed below each family member. If deceased, the cause of death is highlighted using white text against a red background. Other medical conditions the family members experienced are noted beneath each name. Create your own family tree of frequent diseases using the interactive tool available at https://phgkb.cdc.gov/FHH /html/index.html. Then show your family tree to your health care provider to discuss a more complete picture of what the information means for your health.

not. Marketing schemes may belittle the science of genetics to consumers. Even though there have been some great advancements in nutritional genomics in recent years, there is still much to learn. Not only is the science of nutritional genomics in its infancy, but application of this technology will require advanced training for health practitioners. Most health professionals agree that genetic testing complements carefully planned nutrient recommendations and dietary guidelines.

## ✓ CONCEPT CHECK 6.8

1. What is the difference between nutrigenetics and nutrigenomics?
2. List two nutrition-related diseases that are strongly affected by genetics.
3. Predict how nutritional genomics will affect nutrition recommendations in the future.

# 6.9 Nutrition and Your Health

## Vegetarian and Plant-Based Dietary Patterns

Mizina/Getty Images

Vegetarianism has evolved over the centuries from a necessity into an option. A 2019 study from the Vegetarian Resource Group found 4% of U.S. adults (10 million people) to be vegetarian, with only 2% or 5 million being strictly vegan.[23] Vegetarianism is popular among college students, and campus dining services offer vegetarian options at every meal. According to the 2019 poll, 6% of young adults (18 to 34 years of age) are vegetarian or vegan. In contrast, only 2% of adults 65 years or older are vegetarian. The growing popularity of vegetarian dietary patterns has prompted changes in the marketplace. Many restaurants offer vegetarian meals in response to the growing number of customers who want a vegetarian option when they eat out.

It is the position of the Academy of Nutrition and Dietetics that appropriately planned vegetarian, including vegan, diets are healthful, nutritionally adequate, and may provide health benefits for the prevention and treatment of certain diseases. These diets are appropriate for all stages of the life cycle.[24] There are many documented health benefits of following a vegetarian or plant-based eating pattern. Studies show that death rates from some chronic diseases, such as certain forms of cardiovascular disease, hypertension, many cancers, type 2 diabetes, and obesity, are lower for vegetarians than for nonvegetarians. Vegetarians often live longer, as shown in religious groups that practice vegetarianism. Other factors of healthful lifestyles, such as not smoking, abstaining from alcohol and drugs, and regular physical activity, are typical of vegetarians and probably partially account for the lower risks of chronic disease and longer lives seen in this population.

Advances in nutrition science help to ensure the nutritional adequacy of vegetarian dietary patterns. This information is important for vegetarians because an eating pattern of only plant-based foods has the potential to leave gaps between nutrient intake and nutrient needs. Nutrient deficiencies can diminish health at any life stage, but infants and children are at particular risk for stunting of growth and developmental delays. People who choose a vegetarian eating pattern can meet their nutritional needs by following a few basic rules and knowledgeably planning their meals.

The *Dietary Guidelines for Americans* and MyPlate emphasize that a healthy vegetarian dietary pattern can be achieved by incorporating protein foods from plants. In addition, the American Institute for Cancer Research promotes "The New American Plate," which includes plant-based foods covering two-thirds (or more) of the plate, leaving meat, fish, poultry, or low-fat dairy covering only one-third (or less) of the plate (aicr.org/cancer-prevention/healthy-eating/new-american-plate/).

### WHY DO PEOPLE BECOME VEGETARIANS?

People choose vegetarianism for a variety of reasons, including ethics, religion, economics, and health. Some believe that killing animals for food is unethical. Hindus and Trappist monks eat vegetarian meals as a practice of their religion. In North America, many Seventh-day Adventists base their practice of vegetarianism on biblical texts and believe it is a more healthful way to live.

▲ Meatless Monday (www.meatlessmonday.com) is a nonprofit initiative that began in the United States in 2003 to reduce dietary saturated fat and is now worldwide, active in over 40 countries. The Meatless Monday campaign recommends that we cut meat from our meals on Monday and thus encourages us to increase our consumption of fruits, vegetables, whole grains, and legumes. Ideally, this eating pattern will flow into other days of the week and translate into healthier eating habits. **Visit www.meatlessmonday.com and pick out one new meatless recipe to try this week.** Jonelle Weaver/Getty Images

Some advocates of vegetarianism base their food preference upon the inefficient use of animals as a source of protein. In the United States, nearly 70% of the grain crop is used for animal feed, and globally 35% of the grain harvest is used to produce animal protein. Although animals that humans consume do eat grasses that humans cannot digest, many also eat grains that humans can eat. When animals are raised on grains, it takes approximately 3 kilograms of grain to produce 1 kilogram of animal protein. In contrast, grass-fed livestock that eat from a pasture convert grass into protein more efficiently than those raised on grains.

People might also practice vegetarianism because it limits saturated fat and cholesterol intake, while encouraging a high intake of complex carbohydrates; vitamins A, C, and E; carotenoids; magnesium; and fiber.

## GOOD FOR DISEASE PREVENTION

**Heart Health.** Plant sources of proteins can positively impact heart health in several ways. First, the plant foods we eat contain no cholesterol or *trans* fat and little saturated fat. A vegan dietary pattern coupled with regular exercise and other lifestyle changes can lead to a reversal of atherosclerotic plaque in various arteries in the body. The major types of fat in plant foods are monounsaturated and polyunsaturated fats.

Beans and nuts contain soluble fiber, which binds to cholesterol in the small intestine and prevents it from being absorbed by the intestinal cells. Also, due to the activity of some phytochemicals, foods made from soybeans can lower production of cholesterol by the liver. The Food and Drug Administration (FDA) allows health claims for the cholesterol-lowering properties of soy foods, and the American Heart Association has recommended inclusion of some soy protein in the dietary patterns of people with high blood cholesterol. To list a health claim for soy on the label, a food product must have at least 6.25 grams of soy protein and less than 3 grams of fat, less than 1 gram of saturated fat, and less than 20 milligrams of cholesterol per serving. Based on inconsistent findings from recent studies of the ability of soy protein to lower LDL cholesterol, the FDA is proposing to downgrade the soy health claim from "authorized" to "qualified." The FDA is not questioning whether soy protein lowers cholesterol but, rather, by how much. Today soy products are readily available to consumers and can certainly play a role in reducing heart disease risk when combined with other cholesterol-reducing changes in the eating pattern.

There are several other heart-protective compounds in plant foods. Some of the phytochemicals may help to prevent blood clots and relax the blood vessels. Nuts are an especially good source of nutrients implicated in heart health, including vitamin E, folate, magnesium, and copper. Frequent consumption of nuts (about 1 ounce of nuts five times per week) is associated with a decreased risk of cardiovascular disease. The FDA allows a provisional health

▲ Plant proteins, such as those in soybeans, can be incorporated into one's dietary pattern in numerous ways, such as the edamame and tofu shown here. D. Hurst/Alamy Stock Photo

claim on food labels to link nuts with a reduced risk of developing cardiovascular disease.

**Cancer Prevention.** The numerous phytochemicals in plant foods are also thought to aid in preventing cancers of the breast, prostate, and colon. Many of the proposed anticancer effects of foods containing plant protein are through **antioxidant** mechanisms (see the *Newsworthy Nutrition* feature in this section).[25,26] In 2012, the American Institute for Cancer Research added soy to its list and online tool *AICR's Foods That Fight Cancer*™.[26] This website summarizes the current and emerging evidence on soy and cancer risk, which indicates that for all cancers, human studies show soy foods do not increase risk and in some cases may even lower it. This is especially good news for breast cancer patients and survivors who no longer need to worry about eating moderate amounts of soy foods. Walnuts are another plant source of protein that have been recognized as one of *AICR's Foods That Fight Cancer*™.[26]

**Diabetes Control.** Plants also may be particularly good sources of protein for people with diabetes or impaired glucose tolerance because the high fiber content of plant foods leads to a slower increase in blood glucose. Frequent nut consumption may reduce the risk of obesity and type 2 diabetes, as well as gallstones.

## TYPES OF PLANT-BASED EATING PATTERNS

There are several different types or degrees of plant-based eating (Fig. 6-21). Of the estimated 4% of American adults who call themselves vegetarians, only about 2% are total vegetarians, or **vegans,** who eat only plant foods. **Fruitarians** primarily eat fruits, nuts, honey, and vegetable oils. This plan is not recommended because it can lead to nutrient deficiencies in people of all ages. **Lactovegetarians**

**vegan** Referring to a dietary pattern that only includes foods of plant origin.

**fruitarian** Referring to a dietary pattern that primarily includes fruits, nuts, honey, and vegetable oils.

**lactovegetarian** Referring to a dietary pattern that primarily includes dairy products.

---

Plant-based diets are more environmentally sustainable than diets rich in animal products because they use fewer natural resources and are associated with much less environmental damage.

Sustainable Solutions

# Newsworthy Nutrition

## Vegetarian dietary patterns decrease risk of colorectal cancers

**INTRODUCTION:** Because colorectal cancers are a leading cause of cancer deaths, it is important to determine if eating patterns can decrease their risk. **OBJECTIVE:** The hypothesis of this study was that an association exists between vegetarian dietary patterns and the incidence of cancers of the colon and rectum. **METHODS:** This *cohort study* used, for 7 years, a validated quantitative food frequency questionnaire, of 77,659 Seventh-day Adventist men and women who were categorized into four vegetarian dietary patterns (vegan, lactoovovegetarian, pescovegetarian, or semivegetarian) or a nonvegetarian dietary pattern. State cancer registries were used to identify cases of colorectal cancer. **RESULTS:** There were 380 cases of colon cancer and 110 cases of rectal cancer during the 7-year follow-up of participants. As a group, vegetarians were 22% less likely to get colorectal cancers (19% for colon cancer; 29% for rectal cancer) than meat-eaters. When the risk was broken down by diet type, pescovegetarians had a 43% lower risk; vegans, 16%; lactoovovegetarians, 18%; and semivegetarians, 8% lower risk than meat-eaters. All of these effects were similar for men and women and for black and nonblack participants. **CONCLUSION:** These results support the hypothesis that vegetarian dietary patterns are associated with an overall lower incidence of colorectal cancers. Pescovegetarians had a much lower risk compared with nonvegetarians, suggesting that compounds in fish such as omega-3 fatty acids may play a protective role in the prevention of colorectal cancers.

Source: Orlich, MJ and others: Vegetarian dietary patterns and the risk of colorectal cancer, *Journal of American Medical Association Internal Medicine* 175:767, 2015.

and **ovovegetarians** allow dairy and egg products, respectively, in their plant-based eating pattern. **Lactoovovegetarians** eat dairy products and eggs. These inclusions make food planning easier because the dairy and eggs are rich in some nutrients, such as vitamin B-12 and calcium, that are missing or minimal in plants. The more variety in the dietary pattern, the easier it is to meet nutritional needs. A **pescovegetarian** diet, one that includes fish and other aquatic animal protein, has been associated with a lower risk of colon cancer (see the *Newsworthy Nutrition*).[25] A **pollovegetarian** diet includes chicken, turkey and other poultry. Anyone who goes meatless most of the time can call themselves a semivegetarian or a flexitarian.

**ovovegetarian** Referring to a dietary pattern that is primarily plant-based but also includes egg products.

**lactoovovegetarian** Referring to a dietary pattern that is primarily plant-based but also includes dairy products, and eggs.

**pescovegetarian** Referring to a dietary pattern that is primarily plant-based but also includes fish and other aquatic animal protein. Also called *pescatarian*.

**pollovegetarian** Referring to a dietary pattern that is primarily plant-based but also includes chicken, turkey, and other poultry.

## OPTIMIZING A VEGAN PLAN

Planning a vegan dietary pattern requires knowledge and creativity to yield high-quality protein and other key nutrients without animal products. When complementary proteins are eaten at the same meal or the next, the essential amino acids deficient in one protein source are supplied by those of another. Many legumes are deficient in the essential amino acid methionine, whereas grains are limited in lysine. Eating a combination of legumes and grains, such as beans and rice, will supply the body with adequate amounts of all essential amino acids (see Fig. 6-8). As with any meal plan, variety is an especially important characteristic of a nutritious vegan dietary pattern (Fig. 6-9). Table 6-3 lists vegetarian food plans, which emphasize grains, legumes, nuts, and seeds to help meet protein needs. To practice designing a vegetarian meal plan, ask your instructor about the *Rate Your Plate* activity: "Protein and the Vegetarian" in Connect.

Aside from amino acids, low intakes of certain micronutrients can be a problem for the vegan. At the forefront of nutritional concerns are riboflavin, vitamin B-12, iron, zinc, iodine, calcium, and vitamin D. Although use of a balanced multivitamin and mineral supplement can help, the following dietary advice should be implemented.

Riboflavin can be obtained from green leafy vegetables, whole grains, yeast, and legumes—components of most vegan plans. Vitamin B-12 only occurs naturally in animal foods. Vegans can prevent a vitamin B-12 deficiency by finding a reliable source of this vitamin, such as fortified soybean milk, ready-to-eat breakfast cereals fortified with vitamin B-12, and special nutritional yeast grown on media rich in vitamin B-12. Plants can contain soil or microbial contaminants that provide trace amounts of vitamin B-12, but these are negligible sources of the vitamin. Because the liver can store vitamin B-12 for about 4 years, it may take a long time for a vitamin B-12 deficiency to surface after removal of animal foods from the dietary pattern. If dietary vitamin B-12 inadequacy persists, deficiency can lead to anemia, nerve damage, and mental dysfunction. These deficiency consequences have been noted in the infants of vegetarian mothers whose breast milk was low in vitamin B-12.

For iron, the vegan can consume whole grains and ready-to-eat breakfast cereals, dried fruits, nuts, and legumes. The iron in these foods is not absorbed as well as iron in animal foods, but consuming these foods with a good source of vitamin C can enhance iron absorption. Cooking in iron pots and skillets can also add iron to food.

The vegan can find zinc in whole grains (especially ready-to-eat breakfast cereals), nuts, and legumes, but phytic acid and other substances in these foods limit zinc absorption. Breads are a good source of zinc because the leavening process (rising of the bread dough) reduces the influence of phytic acid. Iodized salt is a reliable source of iodine. It should be used instead of plain salt. Check the label to find out if your salt is iodized.

Of all nutrients, calcium and vitamin D are the most difficult to consume in sufficient quantities for vegans. Fortified foods including fortified soy milk, fortified orange juice, calcium-rich tofu, and certain ready-to-eat breakfast cereals and snacks are the vegan's best options for obtaining these nutrients. In addition to fortified foods, alternate sources of vitamin D include some mushroom varieties and regular sun exposure. Green leafy vegetables

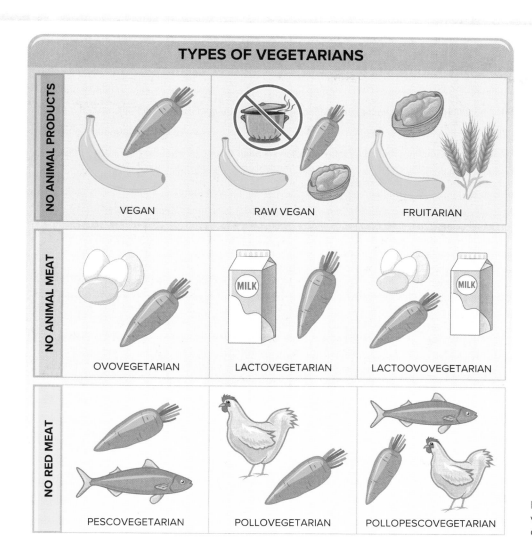

**FIGURE 6-21** ◄ The different types of vegetarians and the protein sources they consume.

**TABLE 6-3** ■ **Food Plan for Vegetarians Based on MyPlate**

| | MyPlate Servings | | |
|---|---|---|---|
| **Food Group** | **Lactovegetarian*** | **Vegan†** | **Key Nutrients Supplied‡** |
| Grains | 6–11 | 8–11 | Protein, thiamin, niacin, folate, vitamin E, zinc, magnesium, iron, and fiber |
| Beans and other legumes | 2–3 | 3 | Protein, vitamin B-6, zinc, magnesium, and fiber |
| Nuts, seeds | 2–3 | 3 | Protein, vitamin E, and magnesium |
| Vegetables | 3–5 (include 1 dark-green or leafy variety daily) | 4–6 (include 1 dark-green or leafy variety daily) | Vitamin A, vitamin C, folate, vitamin K, potassium, and magnesium |
| Fruits | 2–4 | 4 | Vitamin A, vitamin C, and folate |
| Dairy | 3 | ____ | Protein, riboflavin, vitamin D, vitamin B-12, and calcium |
| Fortified soy milk | ____ | 3 | |

*This plan contains about 75 grams of protein in 1650 kcal.

†This plan contains about 79 grams of protein in 1800 kcal.

‡One serving of vitamin- and mineral-enriched ready-to-eat breakfast cereal is recommended to meet possible nutrient gaps. Alternatively, a balanced multivitamin and mineral supplement can be used. Vegans also may benefit from the use of fortified soy milk to provide calcium, vitamin D, and vitamin B-12.

## ASK THE RDN   Plant-Based Eating

*Dear RDN: I am hearing more and more about the health benefits of a plant-based eating pattern. Can you give me some tips on replacing meat and dairy with high-quality plant proteins?*

Regularly consuming foods high in plant proteins, such as legumes (including tofu and other soybean products), whole grains, nuts, and seeds can help prevent and reverse a slew of chronic conditions, including cancers, diabetes, and heart disease. Plant foods are packed with fiber and phytochemicals that support immunity, combat inflammation, and promote healthy bacteria in our gut. As an added bonus, plant proteins are far more affordable, sustainable, and lower in terms of environmental impact than animal proteins.

The good news is that you don't have to swear off meat forever to reap these benefits. Research suggests that following a flexitarian diet (increasing plant-based foods and reducing, but not eliminating, animal foods) yields similar health benefits, like reduced risk of heart disease and diabetes. Eating less meat doesn't mean you're going to suffer from protein deficiency any time soon, either. It is important to note that protein is found in *almost all* foods; and It is nearly impossible not to get enough protein if you're eating enough calories.

In order to transition to a more plant-centric dietary pattern, start small. Overturning your entire eating pattern in a day can be a bit overwhelming initially. Instead of jumping to extremes, pick two small changes to implement each week. First, it may be swapping cow's milk with unsweetened almond or coconut milk. The great thing about nondairy beverages is that they're lower in calories, and some pack more calcium and vitamin D than dairy milk. Make your morning oatmeal with almond milk and stir in a tablespoon of peanut butter and chia seeds for a protein boost. Chia seeds are a hydrating powerhouse, made up of 20% protein and 25% fiber while absorbing up to 30 times their weight in water.

Did you know ¼ cup of pumpkin seeds has 7 grams of protein? Or that hemp seeds are the highest-protein seed, with 3 grams of protein per tablespoon? Peanuts boast the most protein in the nut category, with 7 grams per serving. For a tasty chocolate-banana shake, blend together 1 large frozen overripe banana, 1 tablespoon peanut butter, 1 tablespoon hemp seeds, 1 tablespoon cocoa powder, a handful of spinach, and 1 cup unsweetened vanilla almond milk. Breakfast is served!

For lunches, try power bowls made with a base of wild rice or quinoa, which yield 6 and 8 grams of protein per 1 cup serving, respectively. Top with ½ cup beans, chickpeas, or baked tempeh, a handful of arugula, avocado slices, a drizzle of tahini and lemon juice, and a sprinkle of hemp seeds for a calcium boost. If you're craving a sandwich, stuff a sprouted wheat wrap with ½ cup of black beans, a sprinkle of corn, salsa, avocado, crunchy romaine, and hot sauce.

Consider pasta night. Instead of refined white pasta, try one of the many bean or lentil-based noodles on the market. You can find spaghetti, fusilli, and penne made from black beans, lentils, or chickpeas that all boast 13 grams of protein or more per one cup serving. Stick with 100% whole grain pasta, and you've still got 8 grams of protein and 25% of the Daily Value of fiber per one cup serving. On top of pasta, instead of parmesan, sprinkle nutritional yeast, a cheesy-tasting inactive yeast that's packed with protein and vitamin B-12. Drizzle a tablespoon of tahini and a tablespoon of hemp seeds on your green salad for another 6 grams of protein.

For stir-fry night, swap the chicken for high-protein edamame, which you can usually find in the freezer section of your grocery store. Soy is not only a complete plant protein, but it also has a high concentration of branched-chain amino acids, which are beneficial to athletic performance. Many stores sell marinated tofu (or try tempeh) that's delicious in stir-fry as well. For a tasty peanut sauce, whisk together ¼ cup natural peanut butter, ¼ cup almond milk, 4 teaspoons honey, and 4 teaspoons reduced sodium soy sauce.

When you're craving chili, swap the meat for a couple cans of kidney beans. Adding sautéed mushrooms to the mix will up the umami factor and add meatiness. Boost spices like oregano and chili powder for extra flavor. High-plant-protein dinner is served!

When it comes to baking, experiment with nut- and seed-based flours. Peruse your favorite food blogs for chocolate chip cookies or banana bread made with almond flour or coconut flour for a protein boost. These versions are lower in carbohydrates and super moist thanks to the healthy fat content.

There's no doubt about it—plant proteins are trending *for good*. Do your health and wallet a favor and hop on the bandwagon!

Enjoy your plant proteins,

**Alexis Joseph, MS, RD, LD**

Dietitian, Nutrition Consultant, Founder of Hummusapien; Co-Owner of Alchemy Brands

©Raul Velasco

and nuts also contain calcium, but the mineral is either not well absorbed or not very plentiful from these sources. Dietary supplements are another option. Special planning of the eating pattern is always required because even a multivitamin and mineral supplement will not supply enough calcium to meet the needs for bone health.

Consuming adequate quantities of omega-3 fatty acids is yet another nutritional concern for vegetarians, especially vegans. Fish and fish oils, abundant sources of these heart-healthy fats, are omitted from many types of vegetarian plans. Alternative plant sources of omega-3 fatty acids include canola oil, soybean oil, seaweed, microalgae, flax seeds, chia seeds, and walnuts.

## CASE STUDY    Planning a Vegetarian Dietary Pattern

Jordan is a freshman in college. He lives in a campus residence hall and teaches martial arts in the afternoon. He eats two or three meals a day at the residence hall cafeteria and snacks between meals. Jordan and his roommate both decided to become vegetarians because they recently read an article on a fitness website describing the health benefits of a vegetarian dietary pattern. Yesterday, Jordan's vegetarian plan consisted of a Danish pastry for breakfast and a tomato-rice dish (no meat) with pretzels and a diet soft drink for lunch. In the afternoon, after his martial arts class, he had a milkshake and two cookies. At dinnertime, he had a vegetarian sub sandwich consisting of lettuce, sprouts, tomatoes, cucumbers, and cheese, with two glasses of fruit punch. In the evening, he had a bowl of popcorn.

▲ Has Jordan planned a healthy and nutritious vegetarian dietary pattern? Purestock/SuperStock

1. What type of health benefits can Jordan expect from following a well-planned vegetarian dietary pattern?
2. Evaluate Jordan's current dietary plan using the eating pattern principle of "balance." Are there any food groups he seems to be over- or under-consuming?
3. Which nutrients are missing in this current dietary plan?
4. Are there any food components in the current eating pattern that should be minimized or avoided?
5. How could he improve his new dietary pattern at each meal and snack to meet his nutritional needs and avoid undesirable food components?

*Complete the Case Study. Responses to these questions can be provided by your instructor.*

## SPECIAL CONCERNS FOR INFANTS AND CHILDREN

Infants and children, notoriously picky eaters in the first place, are at highest risk for nutrient deficiencies as a result of improperly planned vegetarian and vegan dietary patterns. With the use of complementary proteins and good sources of the problem nutrients just discussed, the calorie and nutrient needs of vegetarian and vegan infants and children can be met.[24] The most common nutritional concerns for vegetarian and vegan infants and children are deficiencies of iron, vitamin B-12, vitamin D, and calcium.

Vegetarian and vegan plans tend to be high in bulky, high-fiber, low-calorie foods that cause a feeling of fullness. While this is a welcome advantage for most adults, children have a small stomach capacity and relatively high nutrient needs, and thus may feel full before their calorie needs are met. The fiber content of a child's dietary pattern may need to be decreased by replacing high-fiber sources with some refined grain products, fruit juices, and peeled fruit. Other concentrated sources of calories for children who are vegetarian and vegan include fortified soy milk, nuts, dried fruits, and avocados.

▲ Children can safely enjoy vegetarian and vegan dietary patterns as long as certain adjustments are made to meet their age-specific nutritional needs. Brand X Pictures/Getty Images

## ✓ CONCEPT CHECK 6.9

1. What are some of the major health benefits of following a vegetarian eating pattern?
2. Describe the major types of plant-based eating patterns.
3. What are some nutritional concerns of following a vegan eating pattern?

# Summary (Numbers refer to numbered sections in the chapter.)

**6.1** Amino acids, the building blocks of proteins, contain a very usable form of nitrogen for humans. Of the 20 common types of amino acids found in food, nine must be consumed in food (essential) and the rest can be synthesized by the body (nonessential).

**6.2** Individual amino acids are bonded together to form proteins. The sequential order of amino acids determines the protein's ultimate shape and function. This order is directed by DNA in the cell nucleus. Diseases such as sickle cell anemia can occur when the amino acids are incorrect on a polypeptide chain. When the three-dimensional shape of a protein is unfolded—denatured—by treatment with heat, acid or alkaline solutions, agitation or other processes, the protein also loses its biological activity but not its nutritional value.

**6.3** High-quality (complete) protein is available from both animal and plant sources. The high quality of animal proteins means that they can be easily converted into body proteins. Excellent plant sources of protein include legumes, nuts, and seeds. High-quality protein foods contain ample amounts of all nine essential amino acids. Lower-quality (incomplete) protein foods lack sufficient amounts of one or more essential amino acids. This is typical of plant foods, especially cereal grains. Different types of plant foods eaten together often complement each other's amino acid deficits, thereby providing high-quality protein in the dietary pattern. Excessive intake of red meat, especially processed forms, has been linked to colon cancer and deaths caused by cardiovascular disease and cancers. Food allergies occur when the immune system mistakes a food protein for a harmful invader. Eight foods account for 90% of food-related allergies (soy, peanuts, tree nuts, wheat, milk, eggs, fish, and shellfish).

**6.4** Protein digestion begins in the stomach, where stomach acid and pepsin break down proteins into shorter polypeptide chains of amino acids. In the small intestine, these polypeptide chains eventually separate into amino acids in the absorptive cells. The free amino acids then travel via the portal vein to the liver. Some then enter the bloodstream from the liver.

**6.5** Important body components—such as muscles, connective tissue, transport proteins in the bloodstream, visual pigments, enzymes, some hormones, and immune cells—are made of proteins. These proteins are in a state of constant turnover. The carbon chains of proteins may be used to produce glucose (or fat) when necessary.

**6.6** The protein RDA for adults is 0.8 gram per kilogram of healthy body weight. This corresponds to 56 grams of protein daily for a 70-kilogram (154-pound) person and 46 grams per day for a 57-kilogram (125-pound) person. The North American dietary pattern generally supplies plenty of protein, with men consuming about 100 grams of protein daily and women consuming closer to 65 grams. These protein intakes are typically sufficient quality to support body functions. To maintain a state of protein balance, people need to balance protein intake with losses. A positive protein balance is needed to supply the raw materials required to build new tissues during growth or recovery from an illness or injury. To achieve this, a person must eat more protein daily than he or she loses. Consuming less protein than needed leads to negative protein balance, such as when acute illness reduces the desire to eat. Researchers now recommend that adults consume at least 30 grams of protein at each of three meals daily in order to maintain healthy muscles and bones.

For older adults, recent research indicates an increased daily protein requirement of more than 1.0 gram per kilogram. Due to the potential for imbalances and toxicities, the best practice to ensure adequacy is to stick to whole foods as sources of amino acids rather than supplements.

**6.7** Undernutrition can lead to protein-calorie malnutrition in the form of kwashiorkor or marasmus. Kwashiorkor results primarily from an inadequate protein intake in comparison with body needs, which often increase with concurrent disease and infection. Kwashiorkor often occurs when a child is weaned from human milk and fed mostly starchy gruels. Marasmus results from extreme starvation—a negligible intake of both protein and calories. Marasmus commonly occurs during famine, especially in infants.

**6.8** Nutritional genomics includes the study of how genes influence nutritional status (nutrigenetics) and how nutrients and other dietary components influence gene expression (nutrigenomics). Genograms and gene testing can be useful tools for identifying prevalence for disease. Personalized nutritional prescriptions can then be applied to promote optimal health.

**6.9** Vegetarian and other plant-based dietary patterns provide many health benefits, including lower risks of such chronic diseases as cardiovascular disease, diabetes, and certain cancers. The benefits associated with the plant-based dietary patterns appear to stem from the lower content of saturated fat and cholesterol and the higher amount of fiber, vitamins, minerals, and phytochemicals.

# Check Your Knowledge (Answers are available at the end of this question set.)

1. The "instructions" for making proteins are located in the
   a. cell membrane.
   b. cell nucleus.
   c. cytoplasm.
   d. lysosome.

2. A nutrient that could easily be deficient in the dietary pattern of a vegan would be
   a. vitamin C.
   b. folic acid.
   c. calcium.
   d. carbohydrate.

3. An example of protein complementation used in vegan dietary planning would be the combination of
   a. cereal and milk.
   b. bacon and eggs.
   c. rice and beans.
   d. macaroni and cheese.

4. If an essential amino acid is unavailable for protein synthesis,
   a. the cell will make the amino acid.
   b. protein synthesis will stop.
   c. the cell will continue to attach amino acids to the protein.
   d. the partially completed protein will be stored for later completion.

5. The study of how food impacts health through interaction with genes is
   a. epidemiology.
   b. nutrigenomics.
   c. nutrigenetics.
   d. genealogy.

6. Which of the following groups accounts for the differences among amino acids?
   a. Amine group
   b. Side chain
   c. Acid group
   d. Keto group

7. Absorption of amino acids primarily takes place in the
   a. stomach.
   b. liver.
   c. small intestine.
   d. large intestine.

8. Jack is not an athlete and weighs 176 pounds (80 kilograms). His daily protein requirement is _____ grams.
   a. 32        c. 64
   b. 40        d. 80

9. The basic building block of a protein is called a(n)
   a. fatty acid.
   b. monosaccharide.
   c. amino acid.
   d. gene.

10. Which of the following is true about protein intake of people in the United States?
   a. Most do not consume enough protein.
   b. Most consume the amount needed to balance losses.
   c. Athletes do not get enough protein without supplementation.
   d. Most consume more than is needed.

Answer Key: 1. b (LO 6.2), 2. c (LO 6.9), 3. c (LO 6.3), 4. b (LO 6.2), 5. b (LO 6.8), 6. b (LO 6.1), 7. c (LO 6.4), 8. c (LO 6.6), 9. c (LO 6.1), 10. d (LO 6.6)

# Study Questions (Numbers refer to Learning Outcomes)

1. Discuss the relative importance of consuming essential and nonessential amino acids. Why is it important for essential amino acids lost from the body to be consumed? (LO 6.1)

2. What is a limiting amino acid? Explain why this concept is a concern in a vegetarian dietary pattern. How can an individual following a vegetarian dietary pattern compensate for limiting amino acids in specific foods? (LO 6.1)

3. How are DNA and protein synthesis related? (LO 6.2)

4. Briefly describe the organization of proteins. How can this organization be altered or damaged? What might be a result of damaged protein organization? (LO 6.2)

5. Which eight foods are the major sources of proteins that cause food allergies? (LO 6.3)

6. What is the role of cholecystokinin (CCK) in protein digestion? (LO 6.4)

7. Describe four functions of proteins. Provide an example of how the structure of a protein relates to its function. (LO 6.5)

8. Describe how protein intake should be distributed throughout the day for more optimal function of protein. (LO 6.6)

9. Outline the major differences between kwashiorkor and marasmus. (LO 6.7)

10. What are the possible long-term effects of an inadequate intake of dietary protein among children between the ages of 6 months and 4 years? (LO 6.7)

11. Describe the nutrition-related diseases for which genetics or family history is considered to be an important risk factor. (LO 6.8)

12. Describe the different types of plant-based eating patterns and include the benefits and limitations of each. (LO 6.9)

# Further Readings

1. Ellinger S: Micronutrients, arginine, and glutamine: Does supplementation provide an efficient tool for prevention and treatment of different kinds of wounds? *Advances in Wound Care* 3(11):691, 2014. DOI:10.1089/wound.2013.0482.

2. Lonnie M and others: Protein for life: Review of optimal protein intake, sustainable dietary sources and the effect on appetite in aging adults. *Nutrients* 10(3):360, 2018. DOI:10.3390/nu10030360.

3. Food and Agriculture Organization of the United Nations: *Dietary Protein Quality Evaluation in Human Nutrition: Report of an FAO Expert Consultation.* FAO Food and Nutrition Paper 92, 2013.

4. Nadathur SR and others: *Sustainable Protein Sources.* London, UK: Elsevier, 2017.

5. Messina V: Nutritional and health benefits of dried beans. *American Journal of Clinical Nutrition* 100(suppl 1):437S, 2014. DOI:10.3945/ajcn.113.071472..

6. Zabetakis I and others: COVID-19: The inflammation link and the role of nutrition in potential mitigation. *Nutrients* 12, 1466, 2020. DOI:10.3390/nu12051466.

7. Oganization for Economic Cooperation and Development: Meat consumption (indicator), 2018, *OECD-FAO Agricultural Outlook (Edition 2017) Database, Agricultural Statistics.* DOI:10.1787/fa290fd0-en. Accessed on October 31, 2018.

8. Bouvard V and others: Carcinogenicity of consumption of red and processed meat. *Lancet Oncology* 16:1599, 2015. DOI:10.1016/S1470-2045(15)00444-1.

9. Pan A and others: Red meat consumption and mortality: Results from two prospective cohort studies. *Archives of Internal Medicine* 172:555, 2012. DOI:10.1001/archinternmed.2011.2287..

10. Micha R and others: Red and processed meat consumption and risk of incident coronary heart disease, stroke, and diabetes mellitus: A systemic review and meta-analysis. *Circulation* 121:2271, 2010. DOI:10.1161/CIRCULATIONAHA.109.924977..

11. Haring B and others: Dietary protein sources and risk for incident chronic kidney disease: Results from the Atherosclerosis Risk in Communities (ARIC) Study. *Journal of Renal Nutrition* 27:233, 2017. DOI:10.1053/j.jrn.2016.11.004.

12. Tilman D and Clark M: Global diets link environmental sustainability and human health. *Nature* 515:518, 2014. DOI:10.1038/nature13959.

13. Hilborn R and others: The environmental cost of animal source foods. *Frontiers in Ecology and the Environment* 16:329, 2018. DOI:10.1002/fee.1822.

14. Dernini S and others: Med Diet 4.0: The Mediterranean diet with four sustainable benefits. *Public Health Nutrition* 22:1, 2016. DOI:10.1017/S1368980016003177.

15. Fleischer DM and others: Primary prevention of allergic disease through nutritional interventions. *Journal of Allergy and Clinical Immunology: In Practice* 1:29, 2013. DOI:10.1016/j.jaip.2012.09.003.

16. Orenstein BW: Living gluten-free: Creating a healthful gluten-free kitchen. *Today's Dietitian* 17(2): 20, 2015.

17. Madsen L and others: Links between dietary protein sources, the gut microbiota, and obesity. *Frontiers in Physiology* 8:1047, 2017. DOI:10.3389/fphys.2017.01047.

18. Dhillon J and others: Effects of increased protein intake on fullness: A meta-analysis and its limitations. *Journal of the Academy of Nutrition and Dietetics* 116:968, 2016. DOI:10.1016/j.jand.2016.01.003.

19. Bauer J and others: Evidence-based recommendations for optimal dietary protein intake in older people: A position paper from the PROT-AGE Study Group. *Journal of the Medical Directors Association* 14:542, 2013. DOI:10.1016/j.jamda.2013.05.021.

20. Layman DK and others: Defining meal requirements for protein to optimize metabolic roles of amino acids. *American Journal of Clinical Nutrition* 101(suppl):1330S, 2015. DOI:10.3945/ajcn.114.084053.

21. Corkins MR and others: Malnutrition diagnoses in hospitalized patients: United States, 2010. *Journal of Parenteral and Enteral Nutrition* 38:186, 2014. DOI:10.1177/0148607113512154.

22. Dennett C: The future of nutrigenomics. *Today's Dietitian* 19(10):30, 2017.

23. Stahler C: How many people are vegan? How many eat vegan when eating out? The Vegetarian Resource Group website. https://www.vrg.org/nutshell/Polls/2019_adults_veg.htm. Accessed December 8, 2019.

24. Academy of Nutrition and Dietetics: Position of the Academy of Nutrition and Dietetics: Vegetarian diets. *Journal of the Academy of Nutrition and Dietetics* 116:1970, 2016. DOI:10.1016/j.jand.2016.09.025.

25. Orlich MJ and others: Vegetarian dietary patterns and the risk of colorectal cancers. *Journal of the American Medical Association Internal Medicine* 175:767, 2015. DOI:10.1001/jamainternmed.2015.59.

26. American Institute for Cancer Research: *AICR'S Foods That Fight Cancer*™, http://www.aicr.org/foods-that-fight-cancer. Accessed January 25, 2017.

# Chapter 7

# Energy Balance

# Student Learning Outcomes

**Chapter 7 is designed to allow you to:**

**7.1** Describe energy balance and the various uses of energy by the body.

**7.2** Compare methods to determine energy use by the body.

**7.3** Discuss methods for assessing and classifying body composition.

**7.4** Explain risk factors associated with overweight and obesity and related health consequences.

**7.5** List and discuss characteristics of a sound weight-loss program.

**7.6** Describe why reduced calorie intake is fundamental to weight loss and maintenance.

**7.7** Discuss why physical activity is a key component in weight loss and maintenance.

**7.8** Describe how modifying lifestyle behaviors fits into a sound and sustainable weight-loss program.

**7.9** Outline the pros and cons of various weight-loss methods for severe obesity.

**7.10** Discuss the causes and treatment of underweight.

**7.11** Evaluate popular weight-reduction methods, and determine which are safest and most successful.

---

## FAKE NEWS

## To lose weight, you should significantly cut calories.

## THE FACTS

When it comes to losing body fat, the calories you consume have a greater impact on weight loss than the amounts of macronutrients (carbohydrates, fat, or protein) in your food. Multiple studies have confirmed that a caloric deficit results in weight loss. With regard to consuming fewer calories, some experts suggest consuming less saturated fat; others suggest consuming less added sugars; others recommend consuming excessive protein. At this time, evidence shows that a nutrient-dense, primarily plant-based, high-fiber dietary pattern combined with regular physical activity is the most successful plan for long-term weight loss and maintenance.[1]

Almost 75% of adults are **overweight** or have **obesity**.[12] The highest rates of obesity is in adults ages 40 to 59. The **obesity** epidemic remains a crisis not only in the United States but also globally among affluent people and in developing countries where westernized dietary patterns (high in alcohol, saturated fat, added sugars, and sodium) are increasing in popularity while more sedentary lifestyles are adopted. Excess body fat increases the likelihood of many diet-related chronic conditions and diseases, such as cardiovascular disease, type 2 diabetes, and obesity, especially for inactive individuals.

It is clear that most diets fail before dieters reach a healthy weight range. Popular ("fad") diets are generally monotonous, highly restrictive, and short-lived. In addition, many of these diets may endanger some vulnerable populations, such as children, teenagers, women who are pregnant, and individuals with various health issues. A more logical approach is to change and adopt lifestyle behaviors that promote lifelong health and wellness.[2]

It is expected that without a national effort to promote effective new approaches to maintaining healthy weight, the current trends will continue (Fig. 7-1).[3] This chapter will guide you into a deeper understanding of the causes, consequences, and potential treatments of overweight and obesity.

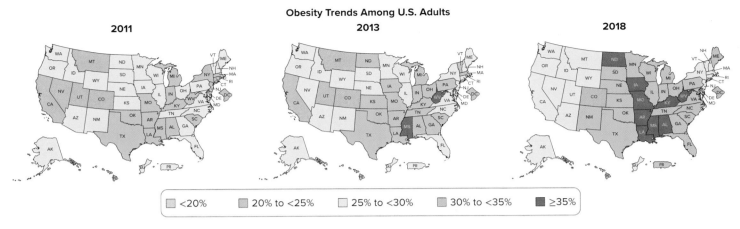

**FIGURE 7-1** ▲ Self-reported obesity prevalence by U.S. State and Territory; 2011, 2013, 2018.

Source: Centers for Disease Control and Prevention Behavioral Risk Factor Surveillance System

# 7.1 Energy Balance

We begin this chapter with good and bad news. The good news is that if you obtain and maintain a healthy body weight, you increase your chances of living a long and healthy life. The bad news is that almost 3/4 of all North American adults are overweight or obese.[4] Beginning in early to middle childhood, the majority of U.S. adults gain approximately 1.1 to 2.2 pounds per year. This modest weight gain over time contributes to the fact that 60% of adults are affected by obesity-related health consequences.

It is estimated that by 2030, there will be a 33% increase in obesity and a 130% increase in severe obesity. Yet if obesity rates stop climbing and remain at current levels, the savings in medical expenditures over the next two decades could be over $549 billion.[5]

There is no quick cure for overweight or obesity, despite what many weight-loss marketing advertisements claim. The most reliable and successful weight loss comes from a commitment to change and adopt lifestyle behaviors that improve dietary and physical activity patterns. A combination of improved energy balance, increased physical activity, and behavior modification is considered to be reliable treatment for overweight. And without a doubt, *prevention* of overweight and obesity is the most successful approach.

## POSITIVE AND NEGATIVE ENERGY BALANCE

**energy balance** The state in which energy (calorie) intake, in the form of food and beverages, matches the energy expended, primarily through basal metabolism and physical activity.

A healthy weight can result from understanding the important concept of **energy balance** (Fig. 7-2). Think of energy balance as an equation consisting of energy input and energy output:

> **Energy Balance**
> **Energy input = Energy output**
> (calories in from dietary intake)  (calories out from metabolism; digestion, absorption, and transport of nutrients; physical activity)

The balance of energy (or calories, measured in **kilocalories**) on the two sides of this equation can influence energy stores, especially the amount of fat stored in **adipose tissue.** When energy input is greater than energy output, the result is **positive energy balance.** The excess calories consumed are stored in the body, which results in weight gain. There are some situations in which positive energy balance is normal and healthy. During pregnancy and lactation (breastfeeding), a surplus of calories supports the developing fetus. Infants and children also require a positive energy

**positive energy balance** The state in which energy intake is greater than energy expended, generally resulting in weight gain.

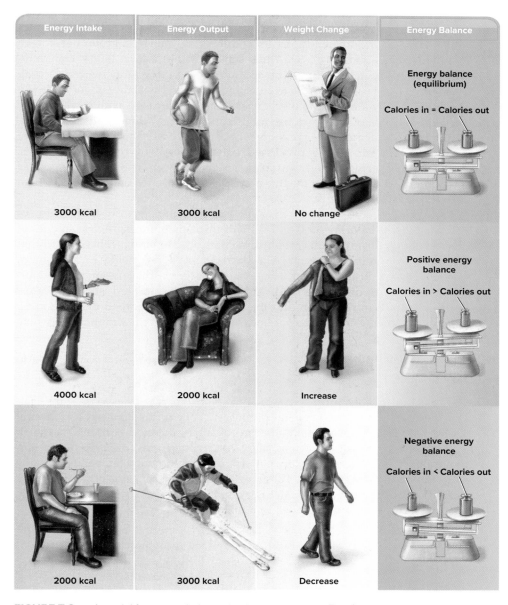

| Energy Intake | Energy Output | Weight Change | Energy Balance |
|---|---|---|---|
| 3000 kcal | 3000 kcal | No change | Energy balance (equilibrium) — Calories in = Calories out |
| 4000 kcal | 2000 kcal | Increase | Positive energy balance — Calories in > Calories out |
| 2000 kcal | 3000 kcal | Decrease | Negative energy balance — Calories in < Calories out |

**FIGURE 7-2** ▲ A model for energy balance: intake versus output. This figure depicts energy balance in practical terms.

balance for normal growth and development during youth and puberty. In adults, however, even a small positive energy balance can result in fat storage rather than muscle and bone and, over time, this can contribute to increased body fatness.

On the other hand, if energy input is less than energy output, there is a calorie deficit and **negative energy balance** results. A negative energy balance is necessary for weight loss. It is important to realize that when we lose weight, we typically lose some lean tissue in addition to adipose tissue.

The maintenance of energy balance while at an optimal weight substantially contributes to health and well-being by minimizing the risk of developing many common health problems associated with increased body fat. Adulthood is often a time of subtle weight gain that can lead to obesity and a greater risk of disease if left unchecked. Aging does not directly cause weight gain; rather, it often stems from a pattern of excess energy intake coupled with reduced physical activity and slower **metabolism.** Now, let us explore the factors that affect the energy balance equation.

**negative energy balance** The state in which energy intake is less than energy expended, resulting in weight loss.

*magnificent*

Think of the possibilities... what if you could regulate your appetite by simply flipping a switch! Now, researchers are doing just that with engineered mice! For some time, we have known that our brains influence hunger and appetite. Now we have growing evidence that our gut microbes are also involved. The gut-brain axis refers to the bidirectional communication between the GI tract and the brain that affects appetite and metabolism. Understanding the key mechanisms by which the gut microbiota can influence appetite and metabolism can provide clues to implement effective treatments for obesity, eating disorders, and other metabolic conditions.

*microbiome*

## ENERGY INTAKE

The *Dietary Guidelines* recommends focusing on meeting your food group needs with nutrient-dense foods and beverages, while staying within calorie limits. Determining the appropriate amount and type of food to meet our energy needs is a challenge for many of us. Our desire to consume food and the ability of our bodies to use it efficiently are survival mechanisms that have evolved with humans. The overabundance of energy-dense foods and beverages often leads to overconsumption of calories and excess stores of body fat. Given the wide availability of inexpensive and highly processed palatable food in vending machines, social gatherings, and fast-food restaurants—combined with *supersized* portions—it is no wonder that the average adult is over 10 pounds heavier than just a decade ago.

The number of calories in a food is determined with an instrument called a **bomb calorimeter** (Fig. 7-3). This instrument measures the amount of calories (kilocalories) from carbohydrate, fat, protein, and alcohol. Recall that carbohydrates and proteins each yield 4 kcal per gram, fats yield 9 kcal per gram, and alcohol yields 7 kcal per gram. These calorie estimates have been adjusted for: (1) the body's ability to digest the food; and (2) substances in food, such as fibrous plant parts, that burn in the bomb calorimeter but are not absorbed by our bodies, so they do not contribute calories to our body. The figures are then rounded to whole numbers.

## ENERGY OUTPUT

*Thermogenesis* is a metabolic process in which the body burns calories to produce heat. The body uses energy for three general purposes: (1) basal metabolism; (2) physical activity; and (3) digestion, absorption, and processing of ingested nutrients. A fourth minor form of energy output, known as **adaptive thermogenesis,** refers to production of heat in response to changes in dietary patterns or environmental temperature. Many factors affect our daily energy needs. These variables include age, sex, physical activity patterns, body composition, hormone status, sympathetic nervous system activity, body and ambient temperature, comorbidities, and medications (Fig. 7-4).

**bomb calorimeter** An instrument used to determine the calorie content of a food.

**adaptive thermogenesis** The ability of humans to regulate body temperature within narrow limits (thermoregulation) in response to changes in dietary patterns or environmental temperatures.

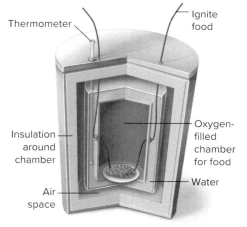

**FIGURE 7-3** ▲ Bomb calorimeters measure calorie content by igniting and burning a dried portion of food. The burning food raises the temperature of the water surrounding the chamber holding the food. The increase in water temperature indicates the number of calories in the food because 1 kcal equals the amount of heat needed to raise the temperature of 1 kilogram of water by 1 degree Celsius.

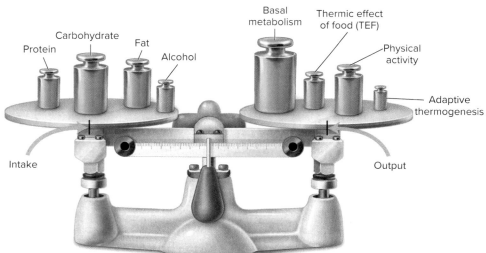

**FIGURE 7-4** ▲ The components of energy intake and expenditure. This figure incorporates the major variables that influence energy balance. Remember that alcohol is an additional source of energy. The size of each weight on the scale represents the relative contribution of that component to energy balance.

**Basal Metabolism.** **Basal metabolism** is expressed as basal metabolic rate (BMR) and represents the minimal amount of calories expended in a fasting state to keep a resting, awake body alive in a warm, quiet environment. For a sedentary person, basal metabolism accounts for about 60% to 80% of total energy use by the body. Some of the processes that utilize energy include the beating of the heart, respiration by the lungs, and the activity required by other organs, such as the liver, brain, and kidneys. It does not include energy used for physical activity or digestion, absorption, and processing of recently consumed nutrients. If the person is not fasting or completely rested, the term **resting metabolism** is used and expressed as resting metabolic rate (RMR). An individual's RMR is slightly higher than his or her BMR.

**basal metabolism** The minimal amount of calories the body uses to support itself in a fasting state when resting and awake in a warm, quiet environment. It amounts to roughly 1 kcal per kilogram per hour for men and 0.9 kcal per kilogram per hour for women; these values are often referred to as *basal metabolic rate (BMR)*.

**resting metabolism** The amount of calories the body uses when the person has not eaten in 4 hours and is resting (e.g., 15 to 30 minutes) and awake in a warm, quiet environment. It is usually slightly higher (~ 10%) than basal metabolism due to the more flexible testing criteria; often referred to as *resting metabolic rate (RMR)*.

### Estimation of Basal Metabolic Rate (BMR)

To see how basal metabolism contributes to energy needs, consider a 130-pound woman. First, knowing that there are 2.2 pounds (lb) for every kilogram (kg), convert her weight in pounds into kilograms:

$$130\ lb \div 2.2\ lb/kg = 59\ kg$$

Then, using a rough estimate of BMR of 0.9 kcal per kilogram per hour for an average female (note 1 kcal per kilogram per hour is used for an average male), calculate her BMR:

$$59\ kg \times 0.9\ kcal/kg = 53\ kcal/hr$$

Finally, use this hourly BMR to find her BMR for an entire day (24 hours):

$$53\ kcal/hr \times 24\ hr = 1272\ kcal/day$$

These calculations provide an estimate of basal metabolism, as it can vary as much as 25% to 30% among individuals.

**lean body mass** Body weight minus fat storage weight equals lean body mass. This includes organs such as the brain, muscles, and liver, as well as bone and blood and other body fluids.

Many factors influence BMR (Fig. 7-5). Of these, **lean body mass** (LBM) is most important. Persons with higher amounts of LBM have a higher BMR because lean tissue is more metabolically active than adipose tissue (fat mass). The lean tissue, therefore, requires more energy to support its activity. Although persons who are overweight or obese have an increased amount of body fat, they also typically have a high amount of LBM to support their weight and therefore a high BMR to go along with it (Fig. 7-6).

Many students are surprised to learn that a very low calorie intake decreases basal metabolism by about 10% to 20% (about 150 to 300 kcal per day) as the body shifts into a conservation or starvation mode. In addition, the effects of aging also make weight maintenance a challenge. As LBM slowly and steadily decreases, basal metabolism declines 1% to 2% for each decade past the age of 30. However, because physical activity helps maintain LBM, remaining active as one ages helps to preserve a higher basal metabolism and, in turn, aids in lifelong weight control.

**Energy for Physical Activity.** During intentional physical activity, your muscles burn calories to provide ample energy to fuel muscle contractions. During this thermogenic process, a considerable amount of the energy is lost as heat. The calorie expenditure from physical activity varies widely among people and is categorized as *exercise-induced thermogenesis (EAT)* and *nonexercise activity thermogenesis (NEAT)*.[6] Hence, EAT refers only to the energy burned during intentional physical activity. For persons who regularly engage in physical activity, EAT may account for 15% to 30% of TEE. NEAT is often categorized into

### Energy Output

| Basal Metabolic Rate (BMR) 60% to 80% of Total Energy Expenditure (TEE) | |
| --- | --- |
| **Factors That Increase BMR** Acute illness or injury Certain medical conditions Excess thyroid hormones Greater body surface area (e.g., tall height) Increased body temperature Lactation Lean body mass Periods of growth (pregnancy, infancy, adolescence) Post-exercise recovery Stimulant drugs (e.g., caffeine) Stress | |
| **Factors That Decrease BMR** Aging Insufficient thyroid hormone production Less body surface area (e.g., short stature) Starvation or very-low calorie diets | |
| **Physical Activity (PA)** 15% to 30% of TEE | **Thermic Effect of Food (TEF)** 8% to 15% of TEE |

**FIGURE 7-5** ▲ Contributions of BMR, activity, and TEF to energy output.

While a person (130-pound woman) is resting, the percentages of total energy use and corresponding energy use by various organs are approximately:[7]

| | | |
|---|---|---|
| Brain | 19% | 242 kcal/day |
| Skeletal muscle | 18% | 229 kcal/day |
| Liver | 27% | 343 kcal/day |
| Kidney | 10% | 127 kcal/day |
| Heart | 7% | 89 kcal/day |
| Other | 19% | 242 kcal/day |
| Total | 100% | 1272 kcal/day |

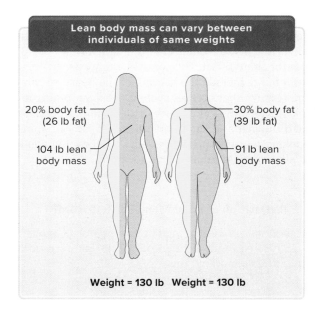

**FIGURE 7-6** ▲ Lean body mass (LBM), the most significant contributor to basal metabolic rate, varies greatly between individuals. Persons of the same body weight can have different amounts of LBM and body fat and, therefore, have varying energy needs.

**thermic effect of food (TEF)** The increase in metabolism that occurs during the digestion, absorption, and metabolism of energy-yielding nutrients. This typically represents 8% to 15% of calories consumed. Also called *diet-induced thermogenesis*.

A few foods, such as celery, have been hypothesized to use more calories for TEF than they contain, making them negative calorie foods. Despite its recurring popularity in fad diet plans, there is no scientific evidence supporting the idea that any food, including celery, is calorically negative. Although these foods still yield some calories, they remain excellent choices to include in a healthy dietary pattern. **Are celery, and other vegetables, considered a low or high energy-dense food option?** Ingram Publishing

three main components: (1) body posture; (2) ambulation; and (3) all other spontaneous movements, including fidgeting. Climbing stairs rather than riding the elevator, walking rather than riding an electric scooter to class, and standing at a workstation rather than sitting for long periods all increase physical activity and, hence, energy use. The increasing incidence of obesity in North America is partially the result of our increasingly sedentary lifestyles.

**Thermic Effect of Food.** In addition to basal metabolism and physical activity, the body uses energy to digest food and to absorb and metabolize the nutrients recently consumed. Energy used for these tasks is referred to as the **thermic effect of food (TEF)** or *diet-induced thermogenesis*. TEF is similar to a sales tax; it is like being charged about 8% to 15% for the total amount of calories you eat to cover the cost of processing that food. This *tax* equates to between 5 and 10 extra kcal for every 100 kcal needed for basal metabolism and physical activity. If your daily calorie intake was 3000 kcal, for example, TEF would account for 150 to 300 kcal. As with other components of energy output, the total amount can vary somewhat among individuals.

Food composition also influences TEF. For example, the TEF value for a protein-rich meal is 20% to 30% of the calories consumed, which is higher than that of a carbohydrate-rich (5% to 10%) or fat-rich (0% to 3%) meal. Lean protein foods, such as chicken breast, egg whites, and whitefish, have the highest TEF at almost 30%. This means that if you eat 100 kcal of chicken breast, almost 30 of those calories are burned off during digestion. This is because it takes more energy to metabolize amino acids (from protein sources) into fat than to convert glucose (from carbohydrate sources) into glycogen or transfer absorbed fat into adipose stores. In addition, large meals result in higher TEF values than the same amount of food eaten over many hours. The TEF value for alcohol is approximately 20%. In sum, consuming a nutritious dietary pattern and engaging in regular physical activity are still the best way to increase your metabolism and burn extra calories.

**Adaptive Thermogenesis.** Adaptive thermogenesis represents the increase in nonvoluntary physical activity triggered by reflex responses (versus intentional physical activity). For instance, body temperature is tightly regulated by the hypothalamus.

When your temperature begins to drop when exposed to cold temperatures, the hypothalamus signals your muscles to contract. These involuntary muscle contractions, known as *shivering thermogenesis*, serve as a defense mechanism to warm your body. Thus, exposure to colder climates can temporarily boost your metabolism. It is interesting to think that some activities, such as eating, include both voluntary and involuntary movement. Chewing is a voluntary movement, while the muscular contractions of the esophagus and intestine are not under conscious control.

**Brown adipose tissue** is a specialized form of adipose tissue that is brown in color and participates in thermogenesis. The brown appearance results from its greater number of mitochondria. Brown fat contributes to thermogenesis by releasing some of the energy from energy-yielding nutrients into the environment as heat instead of producing ATP. Hibernating animals use brown adipose tissue to generate heat to withstand a long winter. In infants, brown adipose tissue is metabolically active and contributes as much as 5% of body weight and regulates heat to protect their internal organs. Unlike the muscle-contracting shivering thermogenesis, *non-shivering thermogenesis* occurs in the brown adipose tissue. Compared to infants, adults have very little brown adipose tissue, and its role in adulthood remains poorly understood.

### ✓ CONCEPT CHECK 7.1

1. What are the main components of energy balance?
2. How is the energy content of food determined and expressed?
3. What are the main purposes for which the body uses energy?
4. List three factors that increase and three factors that decrease basal metabolic rate.

## 7.2 Determination of Energy Use by the Body

The amount of energy our body uses can be measured by both direct and indirect calorimetry or can be estimated based on height, weight, degree of physical activity, and age.

### DIRECT AND INDIRECT CALORIMETRY

**Direct calorimetry** measures the amount of body heat released by a person. The individual enters an insulated metabolic chamber, often the size of a very small bedroom, and over the course of 24 hours, any body heat released raises the temperature of a layer of water surrounding the chamber. A calorie (kilocalorie), as you recall, is related to the amount of heat required to raise the temperature of water. By measuring the water temperature in the direct calorimeter before and after the body releases heat, the energy expended can be calculated. Direct calorimetry works because almost all the energy used by the body eventually leaves as heat. However, because of its expense and complexity, direct calorimetry is rarely used.

A more commonly used method to estimate calorie needs is **indirect calorimetry.** This technique measures the respiratory gas exchange, which is the amount of oxygen a person consumes and the amount of carbon dioxide he or she expels (Fig. 7-7). A relationship exists between the body's use of energy and oxygen. For example, when metabolizing a mixed dietary pattern of the energy-yielding nutrients (carbohydrate, fat, and protein), the human body uses 1 liter of oxygen to yield about 4.85 kcal of energy.

Instruments to measure oxygen consumption for indirect calorimetry are widely used, relatively inexpensive, portable, and can vary considerably in terms of their accuracy. They can be mounted on carts (metabolic carts) or carried in a backpack while a person engages in physical activity to measure how many calories are burned during these activities. Tables presenting energy costs of various forms of physical activity rely

**brown adipose tissue** A specialized form of adipose (fat) tissue that produces large amounts of heat by metabolizing energy-yielding nutrients without synthesizing much useful energy for the body. The unused energy is released as heat.

As defined by the Food and Agriculture Organization (FAO), sustainable dietary patterns are those diets with low environmental impacts that contribute to food and nutrition security and to a healthy life for present and future generations. Sustainable diets are protective and respectful of biodiversity and ecosystems; culturally acceptable; accessible; economically fair and affordable; nutritionally adequate, safe, and healthy; while optimizing natural and human resources. More details to come in later chapters!

 Sustainable Solutions

**direct calorimetry** A method of determining a body's energy use by measuring heat released from the body. An insulated metabolic chamber is typically used.

**indirect calorimetry** A method to measure energy use by the body by measuring oxygen uptake and carbon dioxide output. Formulas are then used to convert this gas exchange value into energy use, estimating the proportion of energy nutrients that are being oxidized for energy in the fuel mix.

**FIGURE 7-7** ▲ Indirect calorimetry measures oxygen intake and carbon dioxide output from respirations to predict energy expended during activities. ©Sarah Rusnak

on information gained from indirect calorimetry studies. You will also see an estimation of calories burned during a workout on most exercise equipment and wearable fitness tracking devices.

## ESTIMATES OF ENERGY NEEDS

As previously mentioned, Chapter 2 contains a number of formulas to estimate energy needs, referred to as **Estimated Energy Requirements** (EER). Those for adults are shown below. As you calculate your own EER, remember to do multiplication and division before addition and subtraction. Note that the calories used for basal metabolism are already factored into these formulas.

### Estimated Energy Requirement (EER) Calculation

The variables in the formulas correspond to the following:

EER = Estimated Energy Requirement
AGE = Age in years
PA = Physical activity estimate (see table below)
WT = Weight in kilograms (pounds divided by 2.2)
HT = Height in meters (inches divided by 39.4)

*Estimated Energy Requirement Calculation for Men 19 Years and Older:*

$$EER = 662 - (9.53 \times AGE) + PA \times (15.91 \times WT + 539.6 \times HT)$$

*Estimated Energy Requirement Calculation for Women 19 Years and Older:*

$$EER = 354 - (6.91 \times AGE) + PA \times (9.36 \times WT + 726 \times HT)$$

**Track Your Energy Needs**

Have you ever wondered how many calories you burn each day? Use *NutritionCalc Plus* (available in Connect) to estimate your usual energy expenditure. On the Activities tab in *NutritionCalc Plus*, enter all your activities (everything from sleeping to studying to eating to working out) for a 24-hour period. Be sure your activities for each day total 1440 minutes (i.e., 24 hours). Doing this for one or more days will help you get a personalized estimate of your usual energy expenditure.

**Physical Activity (PA) Estimates**

| Activity Level | PA (Men*) | PA (Women*) |
|---|---|---|
| Sedentary (e.g., no physical activity) | 1.00 | 1.00 |
| Low activity (e.g., walks the equivalent of 2 miles per day at 3 to 4 mph) | 1.11 | 1.12 |
| Active (e.g., walks the equivalent of 7 miles per day at 3 to 4 mph) | 1.25 | 1.27 |
| Very active (e.g., walks the equivalent of 17 miles per day at 3 to 4 mph) | 1.48 | 1.45 |

*\*Men refers to biological males and women refers to biological females when estimating energy needs.*

### Estimating Energy Requirements

The following is a sample calculation for a man who is 25 years old, 5 feet 9 inches (1.75 meters), 154 pounds (70 kilograms), and has an active lifestyle. His EER is calculated as follows:

$$EER = 662 - (9.53 \times 25) + 1.25 \times (15.91 \times 70 + 539.6 \times 1.75) = 2997 \text{ kcal}$$

The next equation is a sample calculation for a woman who is 25 years old, 5 feet 4 inches (1.62 meters), 120 pounds (54.5 kilograms), and has an active lifestyle. Her EER is as follows:

$$EER = 354 - (6.91 \times 25) + 1.27 \times (9.36 \times 54.5 + 726 \times 1.62) = 2323 \text{ kcal}$$

▲ Physical activity, such as walking, is an important component of energy expenditure. Ronnie Kaufman/Blend Images LLC

You have determined the man's EER to be about 3000 kcal per day and the woman's EER to be about 2300 kcal per day. Remember that this is only an estimate; many other factors, such as genetics and hormones, can also affect actual energy needs.

✔ **CONCEPT CHECK 7.2**

1. What methods can be used to measure energy use by the body?
2. Estimated Energy Requirement (EER) can be calculated based on what five factors?

## 7.3 Assessing Body Weight

Numerous methods are used to establish a person's *healthy weight.* Several tables exist, generally based on weight-for-height criteria. When applied to a population, they provide adequate estimates of weight associated with health and longevity; however, they do not necessarily indicate the healthiest body weight for each individual. For example, athletes with more lean muscle mass but low fat mass have higher body weights (and BMI) than sedentary individuals.

Healthy body weight is based upon both personal and individual factors. Body weight must be considered in terms of overall health, not merely a mathematical calculation. Under the guidance of a registered dietitian nutritionist (RDN) or primary care provider, an individual can establish a *personal* healthy weight based on weight and medical history, body fat distribution patterns, physical activity, dietary patterns, family history, and current health status.

Red flags that your weight may be contributing to poor health or increasing your risk of disease include:

- Hypertension (high blood pressure)
- Hypercholesterolemia (high cholesterol)
- Family history of obesity or obesity-related diseases
- Upper-body (apple-shaped) android fat distribution
- Hyperglycemia (high blood sugar)

These criteria and current height/weight standards serve only as a rough estimate (Fig. 7-9). A healthy lifestyle may make a more important contribution to a person's overall health status than the number on the scale. Being fit and overweight are not necessarily mutually exclusive, and neither is *thin* synonymous with *healthy* if the person is not physically active.

| MyPlate Calorie Guidelines | | |
|---|---|---|
| **Children** | **Sedentary** ⟶ | **Active** |
| 2–3 years | 1000 ⟶ | 1400 |
| **Females** | **Sedentary** ⟶ | **Active** |
| 4–8 years | 1200 ⟶ | 1800 |
| 9–13 | 1400 ⟶ | 2200 |
| 14–18 | 1800 ⟶ | 2400 |
| 19–30 | 1800 ⟶ | 2400 |
| 31–50 | 1800 ⟶ | 2200 |
| 51+ | 1600 ⟶ | 2200 |
| **Males** | **Sedentary** ⟶ | **Active** |
| 4–8 years | 1200 ⟶ | 2000 |
| 9–13 | 1600 ⟶ | 2600 |
| 14–18 | 2000 ⟶ | 3200 |
| 19–30 | 2400 ⟶ | 3000 |
| 31–50 | 2200 ⟶ | 3000 |
| 51+ | 2000 ⟶ | 2800 |

**FIGURE 7-8** ▲ *Dietary Guidelines for Americans* recommendations for estimated calorie needs per day by age, sex, and physical activity level.

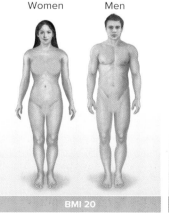

BMI 20

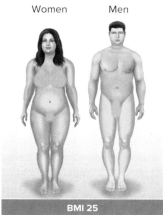

BMI 25

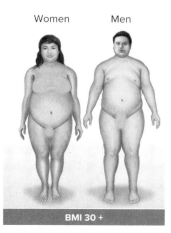

BMI 30 +

**FIGURE 7-9** ▲ Estimates of body shapes at different BMI classifications.

**TABLE 7-1** ■ **Body Mass Index Categories**

| Category | BMI |
|---|---|
| **Underweight** | **< 18.50** |
| Severe thinness | < 16.00 |
| Moderate thinness | 16.00–16.99 |
| Mild thinness | 17.00–18.49 |
| **Normal** | **18.50–24.99** |
| **Overweight** | **≥ 25.00** |
| Pre-obese | 25.00–29.99 |
| **Obese** | **≥ 30.00** |
| Obese class I | 30.00–34.99 |
| Obese class II | 35.00–39.99 |
| Obese class III | ≥ 40.00 |

BMI is a useful measure of weight-for-height standards and estimated body fat.
Source: Adapted from World Health Organization 1995, 2000, 2004, and 2018.

**body mass index (BMI)** Weight (in kilograms) divided by height (in meters) squared; a value of 25 and above indicates overweight, and a value of 30 and above indicates obesity.

# BODY MASS INDEX

Currently, **body mass index (BMI)** is the most widely used weight-for-height standard because it is a noninvasive clinical measurement related to risk of disease for most individuals.

**Calculating BMI**

$$\text{Body mass index} = \frac{\text{body weight (in kilograms)}}{\text{height}^2 \text{ (in meters)}}$$

$$\text{An alternate method for calculating BMI} = \frac{\text{weight (in pounds)} \times 703}{\text{height}^2 \text{ (in inches)}}$$

BMI weight classifications are shown in Table 7-1. A healthy weight for height is defined by a BMI between 18.5 and 24.9. Obesity-related health risks increase when BMI is 25 or more (Fig. 7-9). These are general cutoff values for the presence of overweight and obesity, respectively. Figure 7-10 lists the BMI for various heights and weights.

The concept of BMI is convenient to use because it is inexpensive, easy to obtain, and the values apply to both males and females. However, any weight-for-height standard remains a crude measure and does not consider body composition or fatness. Keep in mind that a BMI of 25 to 29.9 is a marker of *overweight* (compared to a standard population) and not necessarily a marker of excessive body fatness. As previously mentioned, many athletic individuals have a BMI greater than 25 because lean body mass is more dense than fat mass. Also, adults of very short stature (under 5 feet tall) may have a high BMI that may not necessarily reflect overweight or fatness. For these reasons, BMI should be used only as a screening tool. Adult BMIs should not be applied to children, growing adolescents, older individuals who are frail, women who are pregnant or lactating, and individuals with higher lean body mass. BMI is interpreted differently in these cohorts and will be discussed throughout the relevant chapters.

## ESTIMATING BODY COMPOSITION

If calorie intake exceeds calorie expenditure over time, overweight (and often obesity) will likely result. Often, obesity-related health issues eventually follow (Fig. 7-11). As mentioned, BMI values can be used as a convenient clinical tool to screen for overweight and obesity in adults. Remember, BMI is merely a ratio of weight and height; it is not a true assessment of body composition (i.e., the amount of lean mass versus fat mass). Medical experts, however, recommend that an individual's diagnosis of obesity should not be based primarily on body weight or BMI but, rather, on the total amount of fat in the body, the location of body fat, and the presence or absence of weight-related medical conditions. To determine your weight status, visit *Rate Your Plate*: "A Closer Look at Your Weight Status" in Connect.

Body composition varies widely among individuals. According to the American Council on Exercise, average amounts of body fat are about 18% to 24% for men and 25% to 31% for women. Men with over 25% body fat and women with over 32% body fat are

**Weight in pounds**

| Height | 120 | 130 | 140 | 150 | 160 | 170 | 180 | 190 | 200 | 210 | 220 | 230 | 240 | 250 |
|---|---|---|---|---|---|---|---|---|---|---|---|---|---|---|
| 4'6" | 29 | 31 | 34 | 36 | 39 | 41 | 43 | 46 | 48 | 51 | 53 | 56 | 58 | 60 |
| 4'8" | 27 | 29 | 31 | 34 | 36 | 38 | 40 | 43 | 45 | 47 | 49 | 51 | 52 | 56 |
| 4'10" | 25 | 27 | 29 | 31 | 34 | 36 | 38 | 40 | 42 | 44 | 46 | 48 | 50 | 52 |
| 5'0" | 23 | 25 | 27 | 29 | 31 | 33 | 35 | 37 | 39 | 41 | 43 | 45 | 47 | 49 |
| 5'2" | 22 | 24 | 26 | 27 | 29 | 31 | 33 | 35 | 37 | 38 | 40 | 42 | 44 | 46 |
| 5'4" | 21 | 22 | 24 | 26 | 28 | 29 | 31 | 33 | 34 | 36 | 38 | 40 | 41 | 43 |
| 5'6" | 19 | 21 | 23 | 24 | 26 | 27 | 29 | 31 | 32 | 34 | 36 | 37 | 39 | 40 |
| 5'8" | 18 | 20 | 21 | 23 | 24 | 26 | 27 | 29 | 30 | 32 | 34 | 35 | 37 | 38 |
| 5'10" | 17 | 19 | 20 | 22 | 23 | 24 | 26 | 27 | 29 | 30 | 32 | 33 | 35 | 36 |
| 6'0" | 16 | 18 | 19 | 20 | 22 | 23 | 24 | 26 | 27 | 28 | 30 | 31 | 33 | 34 |
| 6'2" | 15 | 17 | 18 | 19 | 21 | 22 | 23 | 24 | 26 | 27 | 28 | 30 | 31 | 32 |
| 6'4" | 15 | 16 | 17 | 18 | 20 | 21 | 22 | 23 | 24 | 26 | 27 | 28 | 29 | 30 |
| 6'6" | 14 | 15 | 16 | 17 | 19 | 20 | 21 | 22 | 23 | 24 | 25 | 27 | 28 | 29 |
| 6'8" | 13 | 14 | 15 | 17 | 18 | 19 | 20 | 21 | 22 | 23 | 24 | 26 | 26 | 28 |

Height in feet and inches

☐ Healthy weight  ☐ Overweight  ☐ Obese

Developed by the National Center for Health Statistics in collaboration with the National Center for Chronic Disease Prevention and Health Promotion

**FIGURE 7-10** ▲ Convenient height/weight table based on BMI. A healthy weight for height generally falls within a BMI range of 18.5 to 24.9 kg/m² (shown in green).

**FIGURE 7-11** ▲ The greater the degree of body fatness, the more likely and the more serious these health problems generally become. They are much more likely to appear in people who show an upper-body fat distribution pattern and/or are greater than twice their healthy body weight.

**FIGURE 7-12** ▲ Underwater (hydrostatic) weighing. In this technique, the subject exhales as much air as possible and then holds his or her breath and bends at the waist to become totally submerged. Once submerged, the underwater weight is recorded. Using this value, body volume can be calculated. David Madison/Getty Images

considered obese. A higher range of body fat percentage for women is physiologically acceptable to maintain reproductive functions, including estrogen production. Now, let's examine some tools used to measure body composition in clinical settings.

**Densitometry.** To measure body fat content accurately, both body weight and body volume of the person are used to calculate body density. Body weight is easy to measure on a conventional scale. Of the typical methods used to estimate body volume, **underwater (hydrostatic) weighing** remains quite accurate. This technique determines body volume using the difference between conventional body weight and body weight measured while submerged under water, along with the relative densities of fat tissue and lean tissue, using a mathematical formula. This procedure requires that an individual be totally submerged in a tank of water, with a trained technician directing the procedure (Fig. 7-12). **Air displacement** (Bod Pod®) is another method of determining body volume. Body volume is quantified by measuring the space a person takes up inside a measurement chamber (Fig. 7-13).

**Skinfold Measures. Skinfold measurements** are also a common anthropometric method to estimate total body fat content, although there are some limits to its accuracy. Clinicians use calipers to measure the fat layer directly under the skin at multiple sites and then plug these values into a mathematical formula (Fig. 7-14). Measurements are then compared to standards for males and females across different stages of the life cycle.

**underwater weighing** A method of estimating total body fat by weighing the individual on a standard scale, then weighing him or her again once submerged in water. The difference between the two weights is used to estimate total body volume. Also known as *hydrostatic weighing* or *hydrodensitometry*.

**air displacement** A method for estimating body composition that makes use of the volume of space taken up by a body inside a small chamber (Bod Pod®). This tool is also known as *air displacement plethysmography*.

**skinfold measurements** Skinfold or caliper testing is a common method to determine body fat percentage. This utilizes prediction equations that are population specific to estimate fat.

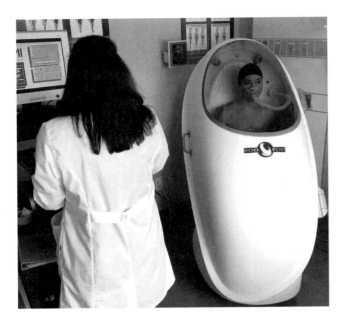

FIGURE 7-13 ▲ Bod Pod®. This device determines body volume based on the volume of displaced air, measured as a person sits in a sealed chamber. ©Sarah Rusnak

**bioelectrical impedance analysis (BIA)** The method to estimate total body fat that uses a low-energy electrical current. The more fat storage a person has, the more impedance (resistance) to electrical flow will be exhibited.

**dual energy X-ray absorptiometry (DXA)** A scientific tool used to measure bone mineral density and body composition.

**Bioelectrical Impedance Analysis.** The analysis of **bioelectrical impedance analysis (BIA)** is also used to estimate body fat. This procedure sends a painless, low-energy, and safe electrical current through the body to estimate body fat (Fig. 7-15). This estimation is based on the assumption that adipose (fat) tissue has proportionately greater electrical resistance than lean tissue. Within a few seconds, bioelectrical impedance analyzers convert body electrical resistance into an approximate estimate of total body fat, as long as hydration is adequate in the person being measured. Body composition monitors, better known as body fat calculators, which use bioelectric impedance, are now available for home use. These machines are similar in shape and use to bathroom scales, but their main purpose is to measure body fat. An electrical current passes easily through conductive foot pads and/or handheld electrodes. It is important to know that varying hydration levels can alter the results.

**Dual Energy X-ray Absorptiometry.** A more advanced determination of body fat content can be made using **dual energy X-ray absorptiometry (DXA).** DXA is considered one of the most accurate ways to determine body fat, but the equipment is expensive and not widely available. This X-ray system allows the clinician to separate body weight into distinct components: fat, fat-free soft tissue, and bone mineral. Additional software can determine body fat distribution. The typical whole-body scan requires about 10 to 25 minutes, and the dose of radiation is less than a chest X-ray. An assessment of bone mineral density and the risk of osteoporosis can also be made using DXA (Fig. 7-16).

There are other methods of assessing body composition, but those detailed here are the most commonly used in health clinics, fitness centers, and research. In the hands

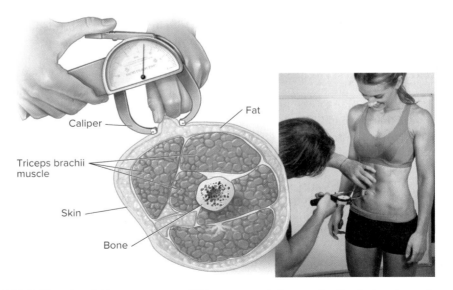

FIGURE 7-14 ▲ Skinfold measurements. With proper technique and calibrated equipment, skinfold measurements taken at various sites around the body can be used to predict body fat content in about 10 minutes. Measurements are made at several locations, including the triceps biceps, abdomen, and thigh. Ian Thraves/Alamy Stock Photo

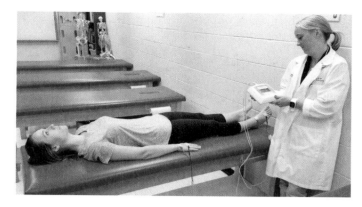

FIGURE 7-15 ▲ Bioelectrical impedance estimates total body fat in less than 5 minutes and is based on the principle that body fat resists the flow of electricity. This device (along with some home scales) sends an electrical current through the body and gives a percentage of body fat when it has completed its process. ©Sarah Rusnak

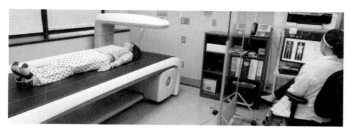

FIGURE 7-16 ▲ Dual energy X-ray absorptiometry (DXA). The scanner arm moves from head to toe and in doing so can determine body fat and bone density. DXA is currently considered the most accurate method for determining segmental body composition (as long as the person can fit under the arm of the instrument). The radiation dose is minimal. ©Sarah Rusnak

**upper-body obesity** The type of obesity in which fat is stored primarily in the abdominal area; defined as a waist circumference more than 40 inches (102 centimeters) in men and more than 35 inches (88 centimeters) in women; closely associated with a high risk for cardiovascular disease, hypertension, and type 2 diabetes. Also known as *android obesity, visceral obesity,* or *central obesity.*

**lower-body obesity** The type of obesity in which fat storage is primarily located in the buttocks and thigh area. Also known as *gynoid* or *gynecoid obesity.*

of a trained clinician, these assessments provide valuable information about body fat beyond simple measures of height and weight.

## BODY FAT DISTRIBUTION

In addition to the amount of fat we store, the location of that body fat is an important predictor of health risks. The apples and pears pictured on this chapter's opener are subtle reminders that shape matters when it comes to health risks. Some people store fat in upper-body areas, whereas others store fat lower on the body. Recall that **upper-body obesity,** characterized by a large abdomen or waist, is more often called *abdominal, visceral,* or *central* obesity and is related to insulin resistance and a fatty liver leading to chronic diseases. Because men typically develop upper-body obesity, it is also known as *android* obesity. While other adipose cells empty fat into general blood circulation, the fat released from abdominal adipose cells goes directly to the liver, by way of the portal vein. This influx of fat interferes with the liver's ability to use insulin and negatively affects lipoprotein metabolism by the liver. These upper-body adipose cells are not just storage depots; they are metabolically active cells that release many hormones and other cellular signals involved in long-term energy regulation. When they fill with excess fat, the cells become dysfunctional, resulting in inflammation, insulin resistance, and other adverse health conditions leading to chronic disease.

High testosterone levels encourage upper-body obesity, as do excessive alcohol intake and smoking. This pattern of fat storage is commonly known as the *apple shape* (large abdomen and small buttocks and thighs [Fig. 7-17]). Upper-body obesity is assessed by measuring the circumference of the abdomen at the waist. A waist circumference more than 40 inches (102 centimeters) in men and more than 35 inches (88 centimeters) in women indicates risk for upper-body obesity (Fig. 7-18). The combination of a large waist circumference and a BMI of 25 or more significantly increases health risks.

Estrogen and progesterone encourage **lower-body obesity**—the typical female pattern. The small abdomen and much larger buttocks and thighs give a *pear shape* appearance. Fat deposited in the lower body is not mobilized as easily as android fat cells and often resists being released. After menopause,

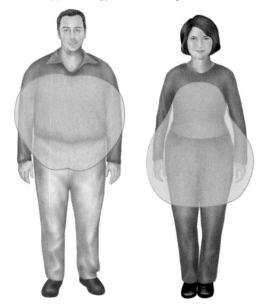

Upper-body fat distribution (android: apple shape)      Lower-body fat distribution (gynoid: pear shape)

FIGURE 7-17 ▲ Body fat stored primarily in the upper body (android) brings higher risks of obesity-related diseases than lower-body (gynoid) obesity. A waist circumference > 40 for men and > 35 for women indicates increased risk of disease.

**FIGURE 7-18** ▲ Waist circumference is an important measure of weight-related health risk. Note there are multiple protocols to measure waist circumference. ©Sarah Rusnak

blood estrogen levels fall, encouraging greater upper-body fat distribution and raising the risk of chronic disease for postmenopausal females.

## ✓ CONCEPT CHECK 7.3

1. How is body mass index (BMI) determined?
2. What are the BMI, body fat percentage, and waist circumference values for men and women that are associated with increased risk of health problems related to being overweight?
3. List three methods by which body fat content can be estimated.
4. Name four diseases associated with obesity.

# 7.4 Nature Versus Nurture

The energy imbalance that promotes weight gain is called an *obesogenic* environment and stems from many factors, including individual characteristics and behaviors in addition to one's environment, culture, and social and economic status.[5] Many studies of obesity attribute increasing trends to the growth of the global food system, including advancements in energy-dense food processing, persuasive marketing, and the widespread availability of more affordable and accessible energy-dense foods.

Both genetic (nature) and environmental (nurture) factors increase the risk for obesity (Table 7-2). The location of fat storage is strongly influenced by genetics. For example, research studies have found that offspring born to mothers affected by obesity are at heightened risk of obesity later in life. Consider the possibility that obesity is nurture allowing nature to express itself. Some people with obesity begin life with a slower basal metabolism, maintain a sedentary lifestyle, and consume highly refined, calorie-dense dietary patterns. These people are nurtured into gaining weight, promoting their natural genetic predisposition toward obesity. Even with a genetic tendency toward obesity, individuals can attain a healthier body weight by engaging in positive lifestyle behaviors with increased physical activity and decreased calorie consumption.

**identical twins** Two offspring that develop from a single ovum and sperm and, consequently, are born with the same genetic makeup.

## HOW DOES NATURE CONTRIBUTE TO WEIGHT MANAGEMENT OR MAINTENANCE?

Studies in pairs of **identical twins** provide insight into the contribution of nature (genetics) relative to obesity. Even when identical twins are raised apart with different environmental and behavioral influences, they tend to show similar weight gain patterns, both in overall weight and body fat distribution. A child with no biologic parent who is obese has approximately a 10% chance of becoming obese. When a child has one parent with obesity, that risk increases to 40%, and with two parents having obesity, risk soars to 80%.[8] Our genes play a role in our metabolic rate, fuel use, and differences in brain chemistry—all of which ultimately affect body weight.

We also inherit specific body types. Tall, thin people have an inherently easier time maintaining healthy body weight. This is likely due to the fact that BMR increases as body surface increases; therefore, people who are taller utilize more calories than people who are shorter, even at rest.

Humans have inherited a so-called *thrifty* metabolism or genotype that enables us to store fat readily.[9] In early human history, our genes adapted to an environment where

Studies in identical twins give us insight into the genetic contribution to obesity. **Does nature or nurture play a larger role in body weight?** Jose Manuel Gelpi Diaz/Alamy Stock Photo

**TABLE 7-2 ■ Factors That Encourage Excess Body Fat Stores and Obesity**

| Factor | How Fat Storage Is Affected |
|---|---|
| Age | Excess body fat is more common in adults and middle-age individuals due to loss of lean body mass and often a reduction in physical activity. |
| Basal metabolism | A low BMR due to factors such as genetics, thyroid problems, or energy restriction is linked to weight gain. |
| Childbearing years | A pattern of excessive weight gain during the childbearing years can occur to support a fetus. Fat stored during pregnancy to support lactation may not be fully lost in women who do not breastfeed. |
| Energy balance | Over time, dietary patterns consistent with positive energy balance promote storage of excess body fat. |
| Ethnicity | In some groups, higher body weight is socially acceptable, which may promote overeating. |
| Fat uptake by adipose tissue | Fat storage efficiency is high in some individuals with obesity and may even increase with weight loss. |
| Genetic predisposition | Genetic factors may affect metabolism, energy expenditure, deposition of adipose tissue or lean tissue, satiety, and the relative proportion of fuels used by the body. |
| Increased hunger sensations | Blunted satiety mechanisms may alter brain signals involved in food reward pathways. |
| Medications | Changes in hunger and appetite can be a side effect of many medications. |
| Menopause | Hormonal changes result in increased abdominal fat deposition. |
| Physical activity patterns | Sedentary behavior promotes positive energy balance and body fat storage. |
| Ratio of fat to lean tissue | A high ratio of fat mass to lean body mass is correlated with weight gain. |
| Residence | Regional environmental and lifestyle differences, such as calorie-laden diets and sedentary lifestyles, especially in the South and Midwest, are associated with higher rates of obesity. |
| Sex | Females have more fat mass than males due to less lean body mass and reduced surface area (height). |
| Social and behavioral factors | Obesity is associated with socioeconomic status; environment; social networks; binge eating; intakes of inexpensive, "supersized," high-saturated-fat food; sedentary lifestyles; increased screen time; smoking cessation; excessive alcohol intake; and frequency of meals eaten away from home. |
| Thermic effect of food | Some individuals are efficient metabolizers and thus expend fewer calories for digestion and absorption. |

food was sometimes scarce; thus, a metabolism that efficiently stored fat would have been a safeguard against starvation in times of famine. Now, with a constant overabundance of food, we require wise food choices and regular physical activity to maintain energy balance. Depending on genetic traits we inherit from our parents, some of us are more prone to weight gain in the modern food environment than others. Although we cannot overlook the impact of nurture, it appears that nature is a strong influencer of body weight and body composition.

**Influences on Weight.** The **set-point theory** of weight maintenance proposes that humans have a genetically predetermined body weight or body fat content, which the body closely regulates.[10] Several physiological changes that occur during calorie reduction and weight loss support this theory. For example, research suggests that the hypothalamus monitors the amount of body fat in humans and tries to keep that amount constant over time. The release and circulation of the hormone **leptin,** from adipose cells, promotes **satiety** and a sense of fullness, thus reducing **appetite.** As adipose cells increase in size and number, overall production of leptin increases, which should suppress appetite. If fat mass is reduced, leptin levels are reduced, so appetite should be increased. This system, however, is not foolproof. Research has shown that persons

**set-point theory**  The theory that changes in energy metabolism and appetite work to maintain a steady body weight throughout adulthood.

Body weight is influenced by many factors related to both nature and nurture. **Thinking back to your childhood, what influences do you think impacted your current weight?** szefei © 123RF.com

Shutterstock

The U.S. military acknowledges overweight and obesity as serious threats to our nation. Over 70% of potential recruits between the ages of 17 and 24 currently do not qualify for military service, and obesity accounts for over 30% of those disqualified. The majority of these young adults live in southern states that have the highest levels of obesity and lowest levels of physical fitness. Promoting healthy dietary and physical activity patterns is paramount to ensuring that all children grow up healthy, and that military enlistees are prepared to meet the strict eligibility requirements.[11]

who are overweight can have large amounts of leptin coming from the excess body fat, but their brains seem to be *leptin resistant* and are not receiving a functional signal to stop eating.

Thyroid hormone levels change in relation to body composition, too. When calorie intake is reduced, the blood concentration of thyroid hormones falls, which slows BMR. Also, the calorie cost of weight-bearing activity decreases, so that an activity that burned 100 kcal before weight loss may only burn 80 kcal after weight loss. Furthermore, as weight loss occurs, the body becomes more efficient at storing fat by increasing the activity of the enzyme lipoprotein lipase, which permits fat entry into cells. All of these changes protect the body from losing weight.

If a person overeats, in the short term BMR tends to increase. This causes some resistance to weight gain. In the long run, however, resistance to weight gain is much less than resistance to weight loss. When a person gains weight and stays at a stable weight for some time, the body tends to establish energy balance at a new set point.

Opponents of the set-point theory argue that weight does not remain constant throughout adulthood: the average person gains weight slowly through old age. Also, if an individual is placed in a different social, emotional, or physical environment, weight can be altered and maintained at a markedly higher or lower point. These arguments suggest that humans, rather than having a set point determined by genetics or the number of adipose cells, settle into a particular stable weight based on their circumstances, often regarded as a *settling point*.

## DOES NURTURE PLAY A ROLE?

Environmental factors, such as consuming an energy-dense dietary pattern and failing to meet physical activity guidelines, literally shape us. This seems reasonable when we consider that our gene pool has not significantly changed in the past century, yet the ranks of people having obesity have grown to epidemic proportions.

Some would argue that body weight similarities between family members stem more from learned behaviors than genetic similarities. Even couples, who have no genetic link, may behave similarly toward food and eventually assume similar degrees of leanness or fatness. Adult obesity is often correlated with childhood obesity.

Is poverty associated with obesity? Ironically, in developed nations the answer is *yes*. North Americans of lower socioeconomic status, especially mothers who are single, are more likely to be obese than those of higher socioeconomic status. Several social and behavioral factors promote fat storage and support the link between socioeconomic status and obesity. These factors may include social networks afflicted by overweight, a cultural/ethnic group that prefers higher body weight, a lifestyle that discourages healthy meals and adequate physical activity, easy availability of inexpensive energy-dense foods, limited access to fresh fruits and vegetables, excessive screen time, lack of adequate sleep, emotional stress, meals frequently eaten away from home, and limited access to health care.

✓ **CONCEPT CHECK 7.4**

1. Explain how body weight is influenced by nature (genetics).
2. What role does nurture (environment) play in determining body weight?
3. List three characteristics of an obesogenic environment.

# 7.5 Management of Overweight and Obesity

The decision to lose weight is an individual one. No person should be shamed or judged for the decisions they make about their own health. If someone elects to attempt weight-loss so they can improve their health, that decision is valid and should also be supported. Treatment of overweight and obesity should be long term and similar to that for any chronic disease. Effective weight loss maintenance requires sustainable lifestyle changes rather than quick fixes promoted by many popular *fad* diets. We often view a *diet* as something one attempts temporarily, only to resume prior behaviors once satisfactory results have been achieved. This is a primary reason so many people regain lost weight. Instead, an emphasis on healthy, active living with acceptable dietary patterns and behaviors will promote safe weight loss and sustained weight maintenance. Indeed, maintenance of a healthy weight requires lifelong changes in behaviors, not short-term and rapid weight loss as seen in so-called *crash diets*.

## LOSING BODY FAT

One pound of weight loss includes adipose tissue plus supporting lean tissues and fluids, and it represents approximately 3300 kcal per pound (about 7.2 kcal per gram). Because there are approximately 3500 kcal in 1 pound of fat, the past 50 years of weight-loss advice has centered on the notion that a deficit of approximately 500 kcal per day is required to lose 1 pound of fat tissue per week. This fairly simple *3500-kcal rule*, however, may be an inaccurate predictor of weight change over time, resulting in unrealistic expectations. When individuals lose weight, compensatory mechanisms kick in to prevent further weight loss and promote weight regain. To account for these variations, experts have developed new weight-loss prediction formulas that estimate a slower and more realistic pattern of weight loss that is not linear. Rather, weight loss often occurs most rapidly during the first six months after a period of negative energy balance and tapers off over time.

A web-based *body weight simulator* can help predict expected weight loss over time. The Body Weight Planner, which can be found at https://www.niddk.nih.gov/bwp, projects weight loss over time based on an individual's height, weight, sex, age, current calorie intake, calorie reduction, and activity level. As always, guidelines emphasize that the daily calorie deficit can come from decreased calorie intake, increased physical activity, or, ideally, a combination of both.

## WHAT TO LOOK FOR IN A SOUND WEIGHT-LOSS PLAN

Individuals can develop a weight-management plan by seeking advice from a health professional such as an RDN. Overall, a sound weight-loss program should include the

---

## FARM to FORK    Stone Fruits

offstocker/iStock/Getty Images

Stone fruits, such as peaches, nectarines, apricots, cherries, and plums, are soft-fleshed fruits with a hard (*stone*) seed. These fruits can contribute to a sound weight management plan while providing key nutrients.

### Grow

- The best time to plant stone fruit trees is in winter, when they are dormant. They prefer cold winters and warm, dry summers to produce flowers and fruit.
- Dwarf or semidwarf fruit trees allow those with small yards to grow stone fruits. Most trees produce their first viable crops after 3 to 4 years.
- Harvest stone fruits, such as peaches, at a U-pick orchard for fruit picked at the height of ripeness.

### Shop

- While the most flavorful stone fruits are locally grown and harvested when ripe, shipped fruits are often harvested prior to ripening and exposed to cold temperatures, resulting in mealy, leathery, and dry fruit.
- Choose fruit that is dent free and without bruises. Ripe stone fruit has a slight give when gently pressed between your palms.
- Select peaches and nectarines by their background color, not their blush. White-fleshed peaches and nectarines are higher in antioxidants than the yellow varieties, and the less common red flesh, or *blood peaches*, are the most nutritious.
- Apricots are one of the most nutritious stone fruits; deep orange and red dried apricots pack the most nutrients.
- Dark skins and flesh of plums have more anthocyanins. Dried plums (also known as *prunes*) are highly nutritious and rich in both soluble and insoluble fibers and have a reputation for relieving constipation.
- When selecting cherries, look for bright green and flexible stems. Darker cherries contain higher levels of anthocyanins and can reduce inflammation.

### Store

- For long-term storage of stone fruits, freezing preserves more antioxidants than canning. Before freezing, slice the fruit and sprinkle with lemon juice and sugar to retain the highest levels of nutrients and prevent browning.
- Store cherries in a microperforated plastic bag (with pin-size holes to allow moisture to escape) in the crisper drawer to allow gas exchange and decrease oxidation.

### Prep

- Dried plums (prunes) are an excellent source of antioxidants and have been linked to everything from reduced inflammation to bone health. To stew prunes, cover with water in a saucepan, bring to a boil, then reduce heat to simmer for 20 minutes. Add a sprinkle of sugar or slice of lemon for flavor.
- Eating the skin will provide the most fiber and nutrients. To de-fuzz stone fruits, wipe gently with a clean cloth.
- If frozen, thaw stone fruits in the microwave to retain the most antioxidants.

Source: Robinson J: Stone fruits: Time for a flavor revival, in *Eating on the Wild Side.* New York: Little, Brown and Company, 2013.

FotografiaBasica/E+/Getty Images

Fisher Photostudio/Shutterstock

The *Dietary Guidelines* provides the following recommendations related to "Focus on meeting food group needs with nutrient-dense foods and beverages, and stay within calorie limits."

- Meet nutritional needs primarily from foods and beverages—specifically, nutrient-dense foods and beverages.
- Consume nutrient-dense foods providing vitamins, minerals, and other health-promoting components with little to no added sugars, saturated fat, and sodium.
- A healthy dietary pattern consists of nutrient-dense forms of foods and beverages across all food groups, in recommended amounts, and within calorie limits.
- The core elements that make up a healthy dietary pattern include:
  - Vegetables of all types—dark green; red and orange; beans, peas, and lentils; starchy; and other vegetables
  - Fruits, especially whole fruit
  - Grains, at least half of which are whole grain
  - Dairy, including fat-free or low-fat milk, yogurt, and cheese, and/or lactose-free versions and fortified soy beverages and yogurt as alternatives
  - Protein foods, including lean meats, poultry, and eggs; seafood; beans, peas, and lentils; and nuts, seeds, and soy products
  - Oils, including vegetable oils and oils in food, such as seafood and nuts

**RATE OF WEIGHT LOSS**

- [ ] Encourages slow and steady weight loss, rather than rapid weight loss
- [ ] Sets goal of no more than 1 to 2 pounds of weight loss per week
- [ ] Includes a period of weight maintenance for a few months after about 10% of body weight is lost
- [ ] Evaluates need for further weight loss before more weight loss begins

**FLEXIBILITY**

- [ ] Supports participation in normal social activities (e.g., dining out, attending parties)
- [ ] Adapts the plan to individual habits and tastes

**DIETARY PATTERN**

- [ ] Meets nutrient needs and focuses on nutrient-dense options
- [ ] Includes common foods, with no foods being promoted as magical or special
- [ ] Uses MyPlate or a comparable food guide as a pattern for food choices

**BEHAVIOR CHANGE**

- [ ] Promotes reasonable changes that can be adopted
- [ ] Encourages social support
- [ ] Includes plans for relapse for those suffering setbacks
- [ ] Empowers individuals to manage behaviors
- [ ] Promotes self-monitoring practices such as keeping food diaries and setting goals

**OVERALL HEALTH**

- [ ] Requires screening by a primary care provider for people with existing health problems, those over 40 (men) to 50 (women) years of age who plan to increase physical activity substantially, and those who plan to lose weight rapidly
- [ ] Encourages regular physical activity, sufficient sleep, stress reduction, and other healthy changes in lifestyle
- [ ] Addresses underlying psychological weight issues, such as depression or stress

**FIGURE 7-19** ▲ Characteristics of a sound weight-loss plan. Use this checklist to evaluate any weight-loss plan before putting it into practice.

components listed in Figure 7-19. A one-sided approach that focuses only on restricting calories is a difficult plan of action. Instead, adding physical activity and an appropriate behavioral component will contribute to success in weight loss and eventual weight maintenance (Fig. 7-20).

## WEIGHT CONTROL IN PERSPECTIVE

Three principles point to the importance of preventing obesity. Public health strategies to address the current obesity problem must speak to all age groups. There is a particular need to focus on children and adolescents because patterns of excess weight and sedentary lifestyles developed during youth may form the basis for a lifetime of weight-related conditions and increased mortality. In the adult population, attention should

Control energy intake

Engage in positive lifestyle behaviors

Perform regular physical activity

**FIGURE 7-20** ◄ A successful weight-loss and maintenance strategy incorporates three interrelated components: (1) controlling energy intake; (2) performing regular physical activity; and (3) engaging in positive lifestyle behaviors. Without one component of the triad, weight loss and later maintenance become highly unlikely. (top) Sam Edwards/Age Fotostock; (left) Chrisgramly/Getty Images; (right) Oleksiy Rezin/Shutterstock

be directed toward weight management and maintenance by encouraging healthy dietary patterns and increasing physical activity.

1. **Improve diet quality** by emphasizing a dietary pattern rich in low energy-dense plant-focused foods and low-calorie beverages.
2. **Increase physical activity** to the equivalent of 150 to 300 minutes or more of moderate-intensity aerobic activity each week.
3. **Make positive behavior changes** to sustain lifestyle modifications promoting health.

### ✓ CONCEPT CHECK 7.5

1. What are the characteristics of an appropriate weight-control program?
2. What are the *Dietary Guidelines* related to "Choose a healthy eating pattern at an appropriate calorie level to help achieve and maintain a healthy body weight, support nutrient adequacy, and reduce the risk of chronic disease?"

## 7.6 Calorie Intake Is Essential for Weight Management

Following a healthy dietary pattern, engaging in regular physical activity, and managing body weight are critical during all life stages.

Losing and maintaining weight loss takes time and energy. Per day, the average female (ages 19 through 30) require about 1800 to 2400 kcal/day. Males in this age group need about 2400 to 3000 kcal/day. Calorie needs for adults ages 31 through 59 are generally lower with females requiring about 1600 to 2200 kcal/day and males requiring about 2200 to 3000 kcal/day.[12] A goal of losing 1 pound or so of stored fat per week may require initially limiting calorie intake by 500 kcal per day. Keep in mind that any calorie reduction will promote the loss of some lean mass along with fat tissue.

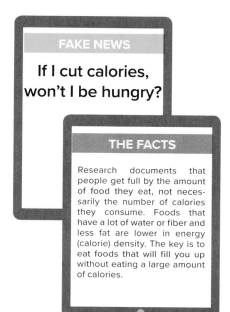

**FAKE NEWS**

### If I cut calories, won't I be hungry?

**THE FACTS**

Research documents that people get full by the amount of food they eat, not necessarily the number of calories they consume. Foods that have a lot of water or fiber and less fat are lower in energy (calorie) density. The key is to eat foods that will fill you up without eating a large amount of calories.

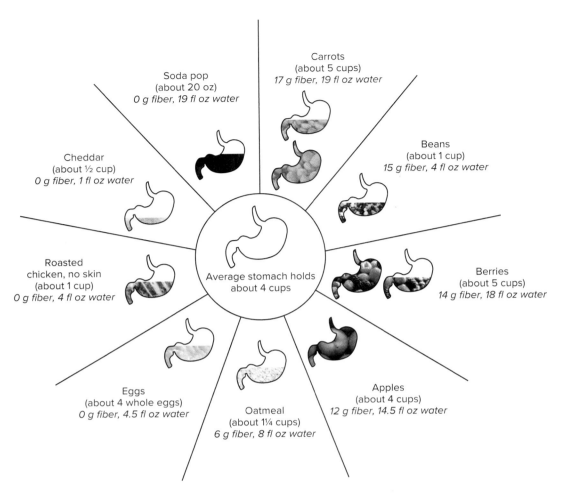

Soda pop (about 20 oz) 0 g fiber, 19 fl oz water

Carrots (about 5 cups) 17 g fiber, 19 fl oz water

Beans (about 1 cup) 15 g fiber, 4 fl oz water

Cheddar (about ½ cup) 0 g fiber, 1 fl oz water

Berries (about 5 cups) 14 g fiber, 18 fl oz water

Average stomach holds about 4 cups

Roasted chicken, no skin (about 1 cup) 0 g fiber, 4 fl oz water

Eggs (about 4 whole eggs) 0 g fiber, 4.5 fl oz water

Oatmeal (about 1¼ cups) 6 g fiber, 8 fl oz water

Apples (about 4 cups) 12 g fiber, 14.5 fl oz water

Portion control is another challenge that influences our calorie intake and requires a change in our approach to eating. As you learned in Chapter 2, the concept of **energy density** can help you choose more nutrient-rich foods with fewer calories per gram. With this technique, you can fill your plate with larger portions of foods with high **nutrient density** and low energy density (Fig. 7-21). Fruits and vegetables are great examples of low-energy-dense foods (low in calories but high in water, fiber, and key nutrients) that promote satiety. See this chapter's *Farm to Fork* for information on stone fruits.

One way to monitor calorie intake at the start of a weight-loss program is by reading Nutrition Facts labels. Label reading is critical because many foods are more energy dense than people realize (Fig. 7-22). With knowledge of current calorie intake, future food choices can be adjusted as needed. A recent report (see *Newsworthy Nutrition*) confirms that early dietary patterns may have a large effect on weight gain over the years. Finding what works is a process of trial and error.

Table 7-3 shows some simple strategies to reduce energy intake. As you should realize, it is best to consider healthy eating as a lifestyle pattern, rather than a fad diet or rapid weight-loss plan. Also, liquids deserve attention because liquid calories do not appear to stimulate satiety mechanisms to the same extent as solid foods.

## CONTROLLING HUNGER

A challenge to most weight-loss programs is to regulate appetite while eating less and engaging in more physical activity. Separating true **hunger** from habit or emotional eating is the first step toward controlling the drive to eat that can sabotage eating patterns. Hormones and your nervous system tell you when you are hungry. The blood hormones, along with an empty stomach, signal the brain that you are hungry. Likewise,

**Nutrition Facts**

3 servings per container

| Serving size | 2/3 cup (63g) |
|---|---|

**Amount per serving**

# Calories 110

| | % Daily Value* |
|---|---|
| **Total Fat** 4g | **5**% |
| Saturated Fat 3.5g | **18**% |
| *Trans* Fat 0g | |
| **Cholesterol** 0mg | **0**% |
| **Sodium** 150mg | **7**% |
| **Total Carbohydrate** 23g | **8**% |
| Dietary Fiber 3g | **11**% |
| Total Sugars 8g | |
| Added Sugars 6g | **12**% |
| Sugar Alcohol 4g | |
| **Protein** <1g | |
| Vitamin D 0mcg | 0% |
| Calcium 0mg | 0% |
| Iron 0mg | 0% |
| Potassium 0mg | 0% |

\* The % Daily Value (DV) tells you how much a nutrient in a serving of food contributes to a daily diet. 2,000 calories a day is used for general nutrition advice.

**Nutrition Facts**

3 servings per container

| Serving size | 2/3 cup (109g) |
|---|---|

**Amount per serving**

# Calories 350

| | % Daily Value* |
|---|---|
| **Total Fat** 22g | **29**% |
| Saturated Fat 14g | **68**% |
| *Trans* Fat 0g | |
| **Cholesterol** 30mg | **10**% |
| **Sodium** 55mg | **2**% |
| **Total Carbohydrate** 35g | **13**% |
| Dietary Fiber 2g | **5**% |
| Total Sugars 30g | |
| Added Sugars 25g | **50**% |
| Sugar Alcohol 4g | |
| **Protein** 4g | |
| Vitamin D 0.8mcg | 4% |
| Calcium 130mg | 10% |
| Iron 2.1mg | 10% |
| Potassium 250mg | 4% |

\* The % Daily Value (DV) tells you how much a nutrient in a serving of food contributes to a daily diet. 2,000 calories a day is used for general nutrition advice.

**FIGURE 7-22** ▲ Reading labels helps you choose more nutrient-dense foods and beverages. Which of these frozen desserts is the best choice, per ⅔-cup serving? The percent Daily Values are based on a 2000-kcal diet. ice cream: Mary-Jon Ludy/McGraw-Hill Education

nerves in the stomach signal the brain when you are full, but it can take up to 20 minutes for these satiety signals to reach the brain. The goal is to be hungry at mealtime but not so ravenous that you are tempted to binge. Eat slowly and stop eating when you are comfortably full. If you are craving food between meals, determine if you are feeling true hunger; if you are, then choose a small, high-fiber snack to satisfy you until the next meal. Drinking a glass of water can also help decrease hunger sensations between meals. Including lean protein (low-fat dairy, soy protein, or lean meat, fish, or chicken)

**TABLE 7-3** ■ **Saving Calories: Ideas for Getting Started**

| Save Kcals | By Choosing This | Instead of This |
|---|---|---|
| 45 kcal | 1 cup 1% milk | 1 cup whole milk |
| 50 kcal | 12 oz light beer | 12 oz regular beer |
| 80 kcal | 2 tbsp low-fat ranch dressing | 2 tbsp regular ranch dressing |
| 80 kcal | 2 slices whole wheat toast | 1 whole wheat bagel |
| 120 kcal | 1 cup broth-based vegetable soup | 1 cup cream-based vegetable soup |
| 120 kcal | 3 oz lean beef | 3 oz marbled beef |
| 120 kcal | 1 cup plain popcorn | 1 oz potato chips |
| 155 kcal | 12 oz water | 12 oz regular soft drink |
| 260 kcal | 1 apple | 1 slice apple pie |
| 270 kcal | 1 slice angel food cake | 1 slice cake with icing |
| 290 kcal | 1 English muffin | 1 Danish pastry |
| 380 kcal | 1 roasted chicken breast | 1 batter-fried chicken breast |

in meals and snacks will also keep hunger at bay. Eating high-volume foods that are rich in water and fiber will provide bulk with fewer calories, fill your stomach, and send satiety signals to the brain.

### CONQUERING THE WEIGHT-LOSS PLATEAU

It is important for anyone on a weight-loss program to know that healthy weight loss is slow and sometimes erratic, and it is normal to reach a weight-loss plateau. After losing weight for weeks, suddenly weight loss may stop. Fortunately, there are some strategies to overcome these plateaus and start obtaining results again.

There are several reasons why weight loss may stall. During the first part of a weight-loss program, individuals are typically losing fluid in addition to fat, causing a weight loss larger than the expected 1 to 2 pounds per week. Because a healthy weight-loss program is designed to cause fat loss rather than loss of muscle or fluid, your weight loss will begin to slow after the first week or so. Also the level of calorie deprivation needed to lose weight is hard to maintain, and you may begin to eat a few more calories. This *calorie creep* can contribute to the weight-loss plateau and eventually lead to weight gain. When this happens, it is important to go back to tracking your calories by recording what you eat and drink and weighing yourself frequently. Another possible reason for the weight-loss plateau is that your metabolism is adjusting to your lower calorie intake. In this case, it may be time to reduce your calories somewhat and drink plenty of water. Your metabolism may also be adapting to your physical activity routine. Varying the intensity of your workout routine, therefore, will help your muscles burn more calories and can help get you past the weight plateau. Strength training, along with the calorie-burning physical activities, is important to build muscle mass, which ultimately uses more calories for metabolism.

### ✓ CONCEPT CHECK 7.6

1. What MyPlate food groups are the best examples of low-energy-dense foods and promote satiety.
2. What mechanisms are involved in hunger control?
3. What are *calorie creep* and the *weight-loss plateau*, and how can they lead to weight gain?

## 7.7 Regular Physical Activity Promotes Weight Loss and Maintenance

Adults who engage in regular physically activity are healthier, feel better, and are less likely to develop chronic disease than adults who are inactive.[12] Regular physical activity can provides both immediate and long-term benefits ranging from improving sleep to reducing stress to improving bone health. All adults should move more and sit less, and any physical activity is better than none. The greatest health benefits are realized with at least 150 to 300 minutes of moderate-intensity aerobic activity each week. Muscle-strengthening activities, like lifting weights or doing push-ups, should be included at least 2 days each week.

Adding any of the activities in Table 7-4 to one's lifestyle can increase calorie expenditure, promote health, and result in weight loss and maintenance over time. Physical activity should be enjoyable, so it becomes part of a healthy pattern of

## Newsworthy Nutrition

### Plant-based dietary pattern promotes healthy body weight

**INTRODUCTION:** Healthy dietary patterns may be achieved with a plant-based diet, including a dietary regimen that encourages whole, plant-based foods in lieu of animal flesh, dairy products and eggs, as well as refined and highly processed foods. **OBJECTIVE:** The purpose of this *cohort study* was to assess correlations between diet quality and body weight in U.S. adults. The hypothesis of this study was that a plant-based diet would promote healthier body weights. **METHODS:** Over 146,000 participants from the Nurses' Health Study (NHS), Health Professionals Follow-Up Study (HPFS), and Nurses' Health Study II (NHS II) were assessed for the effect of diet quality over time on body weight. **RESULTS:** Healthy diet quality scores were correlated with greater intakes of fruits and vegetables, whole grains, and nuts. Subjects who had greater adherence to a plant-based dietary pattern, especially younger women or individuals who were overweight, gained significantly less weight over a 4-year period. **CONCLUSION:** Improving the quality of dietary patterns promotes less weight gain and a healthy body weight, especially in younger women or individuals who are overweight.

Source: Fung TT et al. Long-term change in diet quality is associated with body weight change in men and women. *Journal of Nutrition* 145:1850, 2015.

**TABLE 7-4 ■ Approximate Calorie Costs of Various Activities and Specific Calorie Costs Projected for a 150-Pound (68-kg) Person**

| Activity | Kcal per kg per Hour | Total kcal per Hour |
|---|---|---|
| Aerobic dance (casual) | 5.0 | 340 |
| Basketball | 6.0 | 408 |
| Cycling (12–13 mph) | 8.0 | 544 |
| Dancing | 3.0 | 204 |
| Frisbee | 3.0 | 204 |
| Housework | 3.0 | 204 |
| Jump rope | 8.0 | 544 |
| Rowing machine | 7.0 | 476 |
| Running (10-minute mile) | 10.0 | 680 |
| Sitting | 1.0 | 68 |
| Swimming | 7.0 | 476 |
| Tennis | 8.0 | 544 |
| Walking (3.5 mph, brisk pace) | 3.8 | 258 |
| Weight training | 3.0 | 204 |

The values above refer to total energy expenditure, including that needed to perform the physical activity, plus that needed for basal metabolism, the thermic effect of food, and thermogenesis. You can find the calorie costs of additional activities using the Physical Activity Calorie Counter at https://www.acefitness.org/education-and-resources/lifestyle/tools-calculators.

### Calorie Estimation on Fitness Machines

The control panel of fitness machines will typically display time, speed, distance covered, and calories burned. Time, speed, and distance are generally accurate values, but calories burned is a rough estimate based on the weight you enter before you start your workout. The calories burned are estimates that are not completely accurate because they do not consider factors other than weight, such as body fat percentage, fitness level, form, and fitness efficiency.

Treadmills and other fitness machines have been shown to overestimate calories burned by up to 20%. A heart rate monitor or fitness tracker is usually more accurate at estimating the number of calories you burn during a workout than are calorie estimators on stationary fitness machines.

Physical activity complements any weight-loss or maintenance plan. **How do you incorporate physical activity into your routine?** Zefa/Alamy Stock Photo

living. Although the goal is to achieve 150 to 300 minutes per week of moderate physical activity, everything counts! If you can't squeeze in 30 minutes a day, try shorter bouts of activity in 5- to 10-minute intervals. Muscle-strengthening activities will increase and retain lean body mass. As lean muscle mass increases, so will your BMR.

Unfortunately, opportunities to expend calories in our daily lives are diminishing as technology systematically eliminates the need to move our muscles. The easiest way to increase physical activity is to make it an enjoyable part of a daily routine. To start, one might pack a pair of athletic shoes and walk around the block after school or work or between classes every day. Other ideas are avoiding elevators in favor of stairs and parking the car farther away from your destination. Moving more and sitting less should be the goal!

Pedometers, cell phone apps, or wearable fitness trackers are relatively inexpensive devices that can help monitor activity and steps. A pedometer tracks steps and often distance. Fitness trackers are wearable devices that often track steps, distance, active minutes, heart rate, sleep patterns, and calorie expenditures throughout the day. Fitness trackers calculate calories by measuring heart rate, sweat rate, or heat loss and production. Like pedometers, these devices can motivate users to engage in more activity and reinforce positive behaviors. The *Physical Activity Guidelines for Americans* and related Move Your Way® resources have helpful information about physical activity and tips on how to get started at health.gov/paguidelines.

*magnificent*

Modifiable lifestyle factors, such as diet and physical activity, are key contributors to weight gain during the college years for many. Studies document that regular physical activity can enhance the number of beneficial microbial species, enrich microflora diversity, and improve the development of commensal bacteria. In addition, exercise is positively correlated with protein intake and creatine kinase levels. Together, these alterations are beneficial for the host.[13.]

*microbiome*

1. What should individuals remember about physical activity as part of a healthy lifestyle?

2. What are some simple ways one can incorporate more physical activity into a busy schedule?

# 7.8 Behavioral Strategies for Weight Management

Intensive behavioral interventions that use group sessions and incorporate changes in both dietary and physical activity patterns are most effective for individuals trying to lose and maintain weight.[14] Realistic weight-loss goals will keep you more focused and motivated and will help to guarantee success. The most helpful goals will focus on small and specific changes, such as engaging in physical activity for at least 150 minutes per week, drinking plenty of water, and eating five servings of fruits and vegetables a day, rather than solely focusing on weight or calories. The CDC website provides additional resources to support preventing weight gain (cdc.gov/healthyweight/ prevention/index .html) and losing weight (cdc.gov/healthyweight/ losing_weight/index.html).

The *Dietary Guidelines* identifies ways to empower people to make healthy *shifts* in their dietary patterns to encourage healthy weight management. This includes focusing on nutrient-dense foods and beverages within calorie limits. A healthy dietary pattern has little room for extra added sugars, saturated fat, sodium, or alcoholic beverages. Certainly, limited amounts of these discretionary items are permissible once your nutrient needs are met. The following limits are recommended by the *Dietary Guidelines*:

- Added sugars to less than 10% of calories per day starting at age 2.
- Saturated fat intake to less than 10% of calories per day starting at age 2.
- Sodium intake to less than 2300 milligrams per day and even less for children younger than 14.
- Alcoholic beverages, if consumed at all, should be limited to 2 drinks or less in a day for men and 1 drink or less in a day for women.

For most individuals, adopting a healthy dietary pattern begins with making simple substitutions that are manageable and fit within your personal budget, traditions, and cultural framework. Over time, these patterns will become habitual and allow for additional changes to promote a healthy lifestyle that can be maintained over each stage of life!

## MINDFUL AND INTUITIVE EATING

As emphasized throughout this chapter, becoming aware of your eating patterns and behavior triggers will ultimately help you achieve and maintain a healthy weight (Fig. 7-23). *Mindful eating* refers to being consciously aware of the entire eating experience, from food preparation to consumption, including recognizing and respecting everything in between (hunger cues, satiety, flavor, taste, texture, etc.). *Intuitive eating* encompasses the principles of mindful eating but also addresses the importance of rejecting the dieting mentality, respecting your body, coping with emotional eating, and nutrition without judgment.

The average person makes hundreds of food-related decisions every day. Many of us overeat in response to a constant barrage of triggers in our environment—social networks, sights, sounds, smells, schedules, and other prompts throughout our day. Try a few small nudges to get started in altering your own environment:

- Cover those desserts. There's some truth to the *out of sight, out of mind* philosophy.
- Keeping produce visible on the counter is a reminder to snack healthy. Families that keep a fruit bowl in the open tend to eat more produce.
- Some individuals may eat less when using smaller plates (Fig 7-24). Although not proven to work for everyone, it may be worth a try.[15]

Fruit is a great snack—high in nutrients and low in calories. **Do you recall what type of natural sugar is in fruit?** Dennis Gray/Cole Group/Getty Images

**The Blue Zone Power 9**

"Blue Zones" refer to areas where people live measurably longer lives than others. The five blue zones that have been identified are the Italian island of Sardinia; Okinawa, Japan; Loma Linda, California; Costa Rica's Nicoya Peninsula; and the Greek island of Ikaria. These centenarians share nine common behaviors:

1. Live in environments that nudge them into physical activity and movement without intentional thought.
2. Feel a sense of purpose in their lives.
3. Engage in stress-reduction behaviors (e.g., napping, socializing).
4. Eat their smallest meal in the late afternoon or early evening and don't eat more for the remainder of the day.
5. Consume many beans; animal meat is eaten only about five times per month in small amounts.
6. Drink alcohol moderately if at all and do not binge drink.
7. Belong to a faith-based community.
8. Prioritize families.
9. Interact in social circles that support physical and mental health. Smoking, obesity, depression, and other behaviors appear contagious.

Source: https://www.bluezones.com/2016/11/power-9.

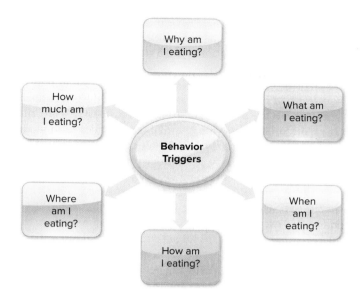

FIGURE 7-23 ▲ This figure captures the questions you can ask to help you identify and define your unique behavior triggers.

When changing behaviors, there is no one-size-fits-all approach. Experiment with different approaches and find what works for you!

## OTHER BEHAVIORAL STRATEGIES

Chain-breaking, stimulus control, cognitive restructuring, contingency management, and self-monitoring are examples of behavior modification strategies used by clinicians that can help put problem behaviors in perspective and organize the intervention into manageable steps.

**Chain-breaking** separates behaviors that tend to occur together (e.g., overindulging on potato chips while watching television). Although these activities do not have to occur together, they often do. Individuals need to identify and break these chain reactions.

**Stimulus control** puts us in charge of temptations. Options include covering tempting food with foil, removing energy-dense snacks from the kitchen counter, and avoiding bringing home especially tempting items. Provide a positive stimulus by keeping healthier snacks available to satisfy hunger and appetite.

**chain-breaking** Breaking the link between two or more behaviors that encourage overeating, such as snacking while watching television.

**stimulus control** Altering the environment to minimize the stimuli for eating, such as removing foods from sight by storing them in kitchen cabinets.

**Large dinner plate**          **Small dinner plate**

FIGURE 7-24 ▲ The dinner plate on the left is larger and makes the serving size of the food appear smaller. This optical illusion was first documented in 1865 and termed the *Delboeuf Illusion*. nesavinov/Shutterstock

**cognitive restructuring** Changing one's frame of mind regarding eating; for example, instead of using a difficult day as an excuse to overeat, a person would substitute other pleasures for rewards, such as a relaxing walk with a friend.

**contingency management** Forming a plan of action to respond to a situation in which overeating is likely, such as when snacks are within arm's reach at a party.

**self-monitoring** Tracking foods eaten and conditions affecting eating; actions are usually recorded in a diary, along with location, time, and state of mind. This is a tool to help people understand more about their eating behaviors.

**relapse prevention** A series of strategies used to help prevent and cope with weight-control lapses, such as recognizing high-risk situations and deciding beforehand on appropriate responses.

**weight bias** Negative attitudes toward, and beliefs about, others because of their weight that are often manifested by stereotypes or prejudice toward people with overweight and obesity.

**Cognitive restructuring** changes our frame of mind. For example, after a hard day, avoid using alcohol or comfort foods as quick relief for stress. Instead, plan for healthful, relaxing activities for stress reduction. Find a positive outlet such as taking a walk or catching up with a good friend.

Labeling some foods as *off limits* sets up an internal struggle to resist the urge to eat that food. This hopeless battle can make us feel deprived, and we eventually lose the fight. Managing food choices with the principle of moderation is best. If a favorite food becomes troublesome, place it off limits temporarily, until it can be enjoyed in moderation.

**Contingency management** prepares one for situations that may trigger overeating (e.g., when snacks are served at a party) or hinder physical activity (e.g., rain).

**Self-monitoring** can reveal problem eating behaviors—such as unconscious overeating—that may lead to weight gain. Records of dietary and physical activity behaviors can encourage new habits that will counteract unwanted behaviors. Obesity experts note this is a key behavioral tool to use in any behavior-change program. Several of these monitoring tools are available online and as apps for smartphones and other mobile devices.[17]

Overall, it's important to address specific barriers to success such as snacking, compulsive eating, and mealtime overeating. Modifying behaviors and changing thinking patterns are critical components of weight reduction and maintenance (Table 7-5). Without them, it is difficult to make lifelong lifestyle changes needed to achieve weight-control goals.

## RELAPSE PREVENTION

Preventing *relapse* is thought to be the hardest part of weight control—often harder than losing weight. Successful weight managers plan for lapses, do not panic, and take charge immediately. Modifying your language can include changing your internal language from "I ate that cookie; I'm a failure" to "I ate that cookie, but I did well to stop after only one!" When individuals lapse from their plan, newly learned behaviors should steer them back on track. Without a strong behavioral program for **relapse prevention** in place, a lapse frequently turns into a relapse and a potential collapse. Once a pattern of poor food choices begins, individuals may feel failure and stray farther from the plan. As the relapse lengthens, the entire plan collapses and falls short of the weight-loss goal. Losing weight is difficult. Overall, maintenance of weight loss is fostered by the *3 Ms*: motivation, movement, and monitoring.

## SOCIAL SUPPORT CAN PROMOTE BEHAVIORAL CHANGE

Healthy social support and networking are helpful in weight control. Family and friends can provide praise and encouragement. Unfortunately, your social network can also sabotage your efforts, so be aware of whom you can trust for support. An RDN can keep you accountable and help you navigate difficult situations. Long-term contact with a professional can be helpful for later weight maintenance. Groups of individuals attempting to lose weight or maintain losses can also provide empathetic support, either in person or online.

## SOCIETAL EFFORTS TO ADDRESS OBESITY

The incidence of obesity in the United States is now considered an *epidemic*. An epidemic is a public health problem, and such problems call for collective action. In fact, improvement in the health of our nation requires an approach that includes many sectors. Partnerships, programs, and policies that support healthy eating and active living must be coordinated. In an effort to eradicate **weight bias** and stigma associated with obesity, the Obesity Action Coalition (OAC) has joined forces with other obesity-focused organizations

Successful weight losers and maintainers from the National Weight Control Registry:

78% eat breakfast every day.
75% weigh themselves at least once a week.
62% watch less than 10 hours of TV per week.
90% exercise, on average, about 1 hr/day.

See http://www.nwcr.ws/ for details.

DNY59/E+/Getty Images

## TABLE 7-5 ■ Behavioral Tactics for Weight Loss

**Shopping**

1. Shop for food after eating so you are not hungry.
2. Shop from a list or use a shopping app.
3. Limit purchases of irresistible "problem" foods.
4. Shop for fresh foods around the perimeter of the store.
5. Avoid highly processed, ready-to-eat foods.

**Planning**

1. Plan meals in advance; keep healthy foods washed and prepared.
2. Substitute periods of physical activity for snacking.
3. Eat meals and snacks at scheduled times; don't skip meals.
4. Drink plenty of water throughout the day.

**Activities**

1. Store food out of sight to discourage impulsive eating.
2. Eat all food in a designated "dining" area; avoid buffet-type meals.
3. Keep serving dishes off the table, especially dishes of sauces and gravies.

Fisher Photostudio/Shutterstock

**Holidays and Parties**

1. Drink fewer alcoholic beverages; alternate water with alcohol.
2. Eat a low-calorie, high-fiber snack before parties.
3. Plan eating behavior before parties.
4. Practice polite ways to decline food.
5. Don't get discouraged by an occasional setback.

**Eating Behavior**

1. Put the fork down between bites; chew thoroughly and slowly before taking the next bite.
2. Leave some food on your plate.
3. Pause in the middle of the meal and evaluate satiety (feeling of fullness) signals.
4. Try not to be distracted while eating (e.g., reading, texting, watching TV).

**Rewards**

1. Plan specific rewards for positive behavior; allow for a small treat on physical activity days.
2. Solicit help from family and friends; engage in physical activity and healthy cooking together.
3. Use self-monitoring records (diet, physical activity, body weight) as basis for rewards. Many tracking apps are available.

Fisher Photostudio/Shutterstock

**Self-Monitoring**

1. Note the time and place of eating.
2. List the type and amount of food and beverages consumed.
3. Record who is present and how you feel.
4. Use the dietary intake diary to identify problem areas.
5. Use online or mobile apps to track your progress, including your new nutrition and health goals and habits.

**Cognitive Restructuring**

1. Avoid setting unreasonable goals; focus on small and manageable steps.
2. Focus on long-term progress, not occasional setbacks.
3. Avoid imperatives such as *always* and *never*.
4. Counter negative thoughts with positive restatements.

Fisher Photostudio/Shutterstock

**Portion Control**

1. Make healthy substitutions, such as small fries instead of large fries or add nuts or seeds instead of croutons to salads.
2. Think small. Order the entrée and share it with another person. Order a cup of soup instead of a bowl or an appetizer in place of an entrée.
3. Use a take-home box. Ask your server to pack half the entrée in a take-home box before bringing it to the table.

**Weight Bias**

**Understand Weight Bias**

Weight bias refers to the negative stereotypes, judgments, and discrimination often associated with individuals affected by overweight and obesity.

**Understand the Impact**

Weight bias often directly or indirectly results in inappropriate or inadequate treatment from members of society.

**Understand How You Can Help**

Using and promoting People-First Language helps to reframe the conversation to avoid weight bias.

**Understand People-First Language**

NO: "The obese man..."
YES: "The man affected by obesity..."

**FIGURE 7-25** ▲ Weight bias. The term *weight bias* refers to negative attitudes, judgments, stereotypes, beliefs, and discrimination targeting individuals because of their weight.

to raise awareness of the People-First Language initiative. *People-First Language* is not new. For many years, others with a chronic disease, including the mental health and disabilities communities, have embraced People-First Language (Fig. 7-25).

We all play an important role in promoting health within our communities. To make healthy eating a societal norm, changes must occur on multiple levels. For instance, providing access to safe, affordable, and healthy foods and beverages for all is a goal that will require collaborative efforts across all facets of society. Such changes are paramount in ensuring that Americans can achieve the recommendations set forth by the *Dietary Guidelines*.

The good news is that public, private, and nonprofit organizations have begun to work together to address and reverse this public health crisis. Food manufacturers are reformulating many products resulting in healthier options. Restaurants are now offering more plant-focused options with smaller portion sizes to align with the *Dietary Guidelines*. All of these small steps have an additive effect in nudging Americans in the right direction, so they can benefit from positive lifestyle behaviors.

✓ **CONCEPT CHECK 7.8**

1. What behavioral techniques are helpful in changing problem eating behaviors to improve weight-loss success?
2. List 5 behavioral tactics for weight loss.
3. What is weight bias? How can you help to reduce weight bias?

# 7.9 Professional Help for Weight Loss

A primary care provider can assist with setting up a weight-loss program. This professional is equipped to assess overall health and current weight status by examining health parameters such as blood pressure, blood lipids, and blood glucose that are affected by excess weight. Your primary care provider may recommend an RDN for a specific behavioral weight-loss plan and answers to nutrition-related questions. RDNs are uniquely qualified to help design a personalized weight-loss plan because they understand both food composition and the psychological importance of food. Sports dietitians and exercise physiologists can provide advice about programs to improve physical activity. The expense for such professional interventions is tax deductible in the United States in some cases and often covered by health insurance plans if prescribed by a primary care provider or covered by a student wellness plan.

Outside of the clinical setting, you can engage with a variety of weight-loss organizations. Some self-help groups, such as WW®, offer social support and education. Other programs are less desirable for the average person because they require expensive food purchases and don't encourage healthy food preparation. Often, the leaders of these programs are not credentialed dietitians or other qualified professionals. These programs also tend to be expensive because of their requirements for intense counseling or mandatory diet foods and supplements. The diet programs that require product purchases promote weight cycling, sometimes called *yo-yo dieting*, and ultimately result in rebound weight gain at a higher body fat level. In addition, the Federal Trade Commission has charged several commercial diet programs with misleading consumers through unsubstantiated weight-loss claims and deceptive testimonials.

## MEDICATIONS FOR WEIGHT LOSS

Candidates for medications to treat obesity, also called anti-obesity drugs or diet pills, include those with a BMI ≥ 30 or BMI ≥ 27 with at least one obesity-associated comorbid condition (e.g., type 2 diabetes, cardiovascular disease, hypertension) and who are motivated to lose weight. For specific populations, medication, also known as pharmacotherapy, may be considered to supplement lifestyle interventions to help achieve targeted weight loss and health goals. Drug therapy alone has not been found to be successful for long-term weight maintenance. Success with medications has been shown only in those who also modify their behavior, decrease calorie intake, and increase physical activity. For a drug to be considered effective in treating obesity, it must meet FDA guidelines and prove to be relatively safe (Table 7-6).[17]

One class of medication approved by the FDA for weight loss is orlistat (Xenical®). This medication reduces fat digestion by about 30% by inhibiting lipase enzyme action in the small intestine (Fig. 7-26). This cuts absorption of dietary fat by one-third for about 2 hours when taken along with a meal containing fat. This malabsorbed fat is excreted in the feces. Fat intake has to be controlled, however, because large amounts of fat in the feces cause numerous side effects, such as diarrhea, gas, bloating, and oily discharge. The malabsorbed fat also carries fat-soluble vitamins (vitamins A, D, E, K) into the feces, so the person taking orlistat must take a multivitamin and mineral supplement at bedtime. In this way, any micronutrients not absorbed during the day can be replaced. A low-dose form of orlistat (alli®) is available over the counter without a prescription.

Other approved weight-loss medications work in various ways to curb appetite. The FDA has approved a novel weight loss device called Plenity® for individuals with a BMI of 25 to 40. This is a hydrogel capsule made with cellulose (fiber) and citric acid. Plenity is designed to be taken with water before meals to absorb the water and increase satiety. The FDA cautions that Plenity should be used alongside diet and exercise. It may also be used with other weight-loss medications. Sometimes, primary care providers may prescribe medications that are not approved for weight loss but have weight loss as a side effect. Such an application is termed *off-label*. Over-the-counter medications and supplements are widely marketed as miracle cures for obesity, but in some cases they do more harm than good. Today more than ever, let the buyer beware concerning any purported weight-loss aid not prescribed by a primary care provider.

Although prescription medications can aid weight loss in some instances, they do not replace the need for reducing calorie intake, modifying behaviors, and increasing physical activity, both during and after therapy. Often, any weight loss during drug treatment can be attributed mostly to the individual's hard work at balancing calorie intake with calorie output.

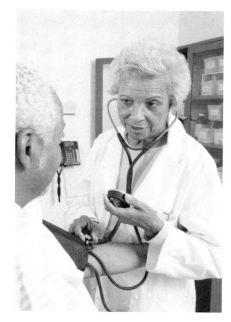

▲ All weight-loss programs should begin with a visit to your primary care provider.
sirtravelalot/Shutterstock

**TABLE 7-6 ■ Medications Approved for Obesity Treatment**

| Medication | Approval | Action | Possible Side Effects |
|---|---|---|---|
| Orlistat (Xenical® or alli® over-the-counter) | Over age 12<br><br>alli® – adults | Blocks fat absorption | Stomach pain, gas, diarrhea, leaky oily stools<br><br>Rare cases of severe liver injury; should not take with cyclosporine |
| Phentermine topiramate (Qsymia®) | Adults | Combination of appetite suppressor and migraine/seizure medication | Constipation, dizziness, dry mouth, taste changes, tingling of hands and feet, trouble sleeping<br><br>Do not use if pregnant or planning to become pregnant |
| Liraglutide (Saxenda®) | Adults | Slows gastric emptying and enhances satiety | Long-term effects remain unknown |
| Others:<br><br>Diethylpropion<br>Benzphetamine<br>Phendimetrazine | Adults | Alter brain chemicals to promote appetite suppression.<br><br>Approved by FDA for use up to 12 weeks only | Dry mouth, difficulty sleeping, dizziness, headache, feeling nervous, upset stomach, diarrhea, constipation, restlessness |

Sources: Food and Drug Administration (FDA) and https://www.niddk.nih.gov/health-information/weight-management/prescription-medications-treat-overweight-obesity.

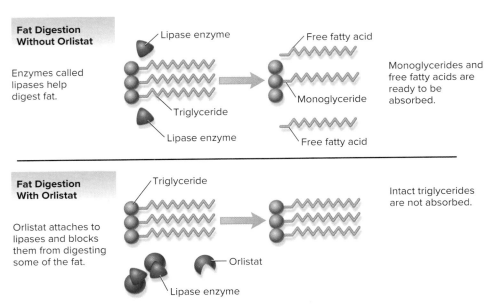

**FIGURE 7-26** ▲ Orlistat is a weight-loss drug that works in the digestive system to block digestion and subsequent absorption of about one-third of the fat in the food we eat. A low-dose form of this drug (alli®) is now available without a prescription.

## TREATMENT OF SEVERE OBESITY

Severe obesity, having a BMI ≥ 40 or weighing at least 100 pounds over healthy body weight (or twice one's healthy body weight), often requires professional treatment. Because of the serious health implications of severe obesity, drastic measures may be necessary. Such treatments are recommended only when traditional diets and medications fail. Drastic weight-loss procedures are not without side effects, both physical and psychological, making careful monitoring by a primary care provider a necessity.

**Very-Low-Calorie Diets.** If more traditional diets have failed, treating severe obesity with a **very-low-calorie diet (VLCD)** is possible, especially if the person has obesity-related diseases that are not well controlled (e.g., hypertension, type 2 diabetes). A VLCD can be dangerous because of its rapid weight loss and potential for severe health complications, including heart problems and gallstones. Often providing fewer than 800 kcal per day, all VLCD programs should be administered under strict medical supervision, as careful monitoring by a trained medical professional is crucial throughout this very restrictive form of weight loss.

Optifast® and other VLCD meal replacements are only available through medically supervised clinics. In general, these diets allow a person to consume only 400 to 800 kcal per day, often in liquid form. These diets were previously known as *protein-sparing modified fasts*. Of this amount, approximately 30 to 120 grams (120 to 480 kcal) are carbohydrates. The rest are typically from high-quality protein in the amount of 70 to 100 grams per day (280 to 400 kcal). Recall that **ketosis** is an accumulation of **ketone bodies** in the blood that occurs when fatty acids are broken down for fuel in the absence of carbohydrates. The main reasons for weight loss, however, are the minimal energy consumption and restricted food choices. About 2 to 4 pounds can be lost per week, primarily water and lean body mass losses. When physical activity and resistance training augment this diet, a greater loss of adipose tissue occurs.

Weight regain remains an all-too-common reality, especially without a behavioral and physical activity component. If behavioral therapy and physical activity supplement a long-term support program, maintenance of the weight loss is more likely but still difficult. Any program under consideration should include a detailed maintenance plan. Today, anti-obesity medications may also be included in this phase of the program.

**Intermittent Fasting.** The dietary practice known as *intermittent fasting*, where one cycles days of so-called normal eating with a day or days of eating little to nothing, is

**very-low-calorie diet (VLCD)** This diet allows a person fewer than 800 kcal per day, often in liquid form. Of this, 120 to 480 kcal are typically from carbohydrate, and the rest are mostly from high-quality protein.

alli® (orlistat) is an over-the-counter weight-loss drug that blocks fat digestion and absorption. **Why should people who take orlistat also take a dietary supplement of vitamins A, D, E, and K?** Jill Braaten/McGraw-Hill Education

gaining popularity. Research shows that intermittent fasting regimens yield weight-loss results that are similar to traditional weight-loss plans (i.e., moderate calorie restriction). Some individuals may experience improved blood sugar control and blood lipids. However, these plans may be hard to stick to over the long term and there are not many long-term studies to measure the impact on obesity-related diseases. More specific information on intermittent fasting, binge eating, and other weight-loss methods can be found in the *Ask the RDN* feature in this chapter.

**AspireAssist®.** The FDA has approved an obesity treatment device that uses a surgically placed tube to drain a portion of the stomach contents after every meal.[16] AspireAssist® is only intended for adults with obesity (BMI ≥ 35) who have failed to achieve and maintain weight loss through nonsurgical weight-loss therapy.

The device is inserted in the stomach surgically and connected to a disk-shaped port valve outside the body, flush against the abdomen. Approximately 20 to 30 minutes after a meal, the patient attaches tubing to the port valve and drains the contents. Once the valve is opened, it takes approximately 5 to 10 minutes to drain food matter through the tube and into the toilet. The device removes approximately 30% of the calories consumed.

Ongoing medical visits are necessary to monitor device use and weight loss, and to provide education and training. The device also contains a safety feature that tracks the number of drain cycles and stops after 115 cycles (approx. 5 to 6 weeks of therapy). This ensures the patient will return for frequent medical visits. Side effects include occasional indigestion, nausea, vomiting, constipation, and diarrhea.

**Bariatric Surgery. Bariatrics** is the medical specialty focusing on the treatment of obesity. Bariatric (or metabolic) surgery is considered for people with a BMI ≥ 30 and includes surgery aimed at promoting weight loss.[19] In addition to BMI, for screening the health care team also considers a person's nutrition and weight history, medical condition, motivation, age, and psychological status. Figure 7-27 provides details about the most common bariatric procedures. In 2007, gastric bands were the most common bariatric procedure. Now, the gastric sleeve is the most popular procedure.

The risks of bariatric surgery are serious and include both early and late postoperative complications, such as bleeding, blood clots, hernias, electrolyte imbalances, and severe infections. These risks depend on many factors related to the surgeon and facility, the patient, and the procedure. The procedures that are simply restrictive (e.g., adjustable gastric banding and sleeve gastrectomy) do not cause malabsorption and rarely affect bowel function. However, for those procedures that induce malabsorption (e.g., Roux-en-Y gastric bypass), nutrient deficiencies are of greater concern if the person is not adequately treated in the years following the surgery. Anemia and bone loss might then be the result.

Bariatric surgery is costly and may not be covered by medical insurance. The average cost for bariatric procedures is typically between $15,000 and $35,000. In addition, follow-up surgery is often needed after weight loss to correct stretched skin, previously filled with fat. Furthermore, the surgery necessitates major lifestyle changes, such as the need to plan frequent, small meals. Therefore, the dieter who has chosen this drastic approach to weight loss faces months of difficult adjustments.

Despite potential adverse effects, the benefits of bariatric surgery for those who are eligible usually outweigh the risks. Interestingly, research has uncovered that bariatric surgery may lead to long-term changes in gut bacteria (increased diversity and quantity) that contribute to weight loss. An increasing number of youth are turning to bariatric surgery to treat obesity. Table 7-7 details the pros and cons of the most common bariatric procedures for children. Weight-loss statistics vary by surgical method, but on average about 65% of people eventually lose and keep off 85% or more of excess body weight.[20] By no means is bariatric surgery a quick and easy fix for obesity, but with a serious commitment to permanent lifestyle changes and long-term follow-up with a health professional, these procedures can positively impact both quality and quantity of life.

**Youth and Adolescent Bariatric Surgery.** The American Academy of Pediatrics (AAP) recommends greater access to surgical treatments for severe obesity defined as a BMI of ≥ 120% of the 95th percentile for age and sex in youth. In the policy statement, the

**bariatrics** The medical specialty focusing on the treatment of obesity.

## ASK THE RDN Intermittent Fasting

*Dear RDN: I have had bad luck with traditional diets, but I've heard rave reviews of the "Fast Diet." Is an intermittent fasting program a good way to lose weight and keep it off?*

There are three basic methods of intermittent fasting: whole-day fasting, alternate-day fasting, and time-restricted feeding. The definition of *fasting* can vary from one plan to the next; some plans allow only calorie-free beverages and sugar-free gum during fasts, whereas other plans recommend reducing energy intake to 20% to 25% of your usual intake (i.e., a modified fast) on fast days. Plans also differ in recommendations for food choices during feeding phases; some prescribe specific calorie and nutrient goals, while others allow unrestricted food intake during feeding phases. Proponents of intermittent fasting recommend that a person experiment to find the method that best fits his or her lifestyle.

With whole-day fasting, like the "Fast Diet," you would eat normally on most days of the week but undertake a complete or modified fast 1 or 2 days each week. With alternate-day fasting, days of normal eating alternate with complete or modified fasts. For time-restricted feeding protocols, you would delay the first meal of the day to achieve a fast of 14 to 20 hours (including overnight), but an unrestricted feeding period would be allowed for the remaining hours of the day.

Besides weight loss, purported benefits of intermittent fasting include improved insulin sensitivity, enhanced ability to use fat for energy, decreased triglyceride levels, and reduced inflammation—changes that may potentially decrease risks for chronic diseases and lead to a longer life. Much of the research to date, however, has been conducted in animals, with few controlled studies in humans.

Reviews of current (although sparse) studies of the effects of intermittent fasting suggest that it may be a reasonable alternative to the traditional weight-loss method of moderate daily calorie restriction for some people.[17,18] Weight loss and improvements in metabolic effects are similar with these two weight-loss methods. However, intermittent fasting is not a good fit for everyone. If you struggle with blood sugar control (e.g., hypoglycemia or diabetes) or are pregnant, you should not attempt intermittent fasting. Quite predictably, some side effects of any fasting regimen include hunger, headaches, fatigue, and irritability. Some critics of intermittent fasting point out that underfeeding in this way could lead to nutrient deficiencies. We, as RDNs, can help you choose nutrient-dense foods during the feeding phases of any of these dietary patterns. Also, there is some concern that prolonged fasting may decrease your metabolic rate, which would make long-term weight maintenance very difficult. Because we have limited data on the long-term safety of intermittent fasting, it is not clear how much fasting is too much. Fasting too frequently may result in malnutrition, reduced immune function, organ damage, or eating disorders.

Wait—*eating disorders?*

Take a step back and look at the basic dietary patterns advocated here. Much like the dietary patterns of individuals with anorexia nervosa, intermittent-fasting programs involve calorie counting and extended periods of hunger. For some plans, foods are classified as "safe" or "off limits," which promotes an unhealthy, dichotomous view of foods and ties eating into emotions such as guilt, shame, and fear. Intermittent fasting programs that advise fasting (deprivation) followed by several hours or days of unrestricted eating (indulgence) may foster binge eating. Participants may feel as if they've failed themselves and others when they have trouble adhering to a strict regimen. For a person who already exhibits anxiety, depression, or obsessive traits, intermittent fasting may be a gateway to pathological dieting and eating disorders.

Although intermittent fasting may be a reasonable alternative to daily moderate calorie restriction for some people, more research is needed to determine which protocol is best and if mental and physical health are affected over the long term. If you are determined to try intermittent fasting, continue to seek the advice of an RDN to ensure that your dietary pattern still meets your nutritional needs. If you tend to obsess about food and body weight or you already have problems with depression or anxiety, it is important to steer clear of this dietary pattern. As you will learn later, eating disorders often begin with a simple diet. If efforts to control weight begin to interfere with daily activities and are linked to physiological and emotional changes, professional help will be needed to treat an eating disorder.

Fueling, not fasting,

**Angela Collene, MS, RDN, LD**
Lecturer, The Ohio State University, Author of *Contemporary Nutrition*

Tim Klontz, Klontz Photography

AAP details the health consequences of severe obesity leading to a dramatically shortened life expectancy for today's youth as compared to their parents. Research of adolescents and young adults who have undergone bariatric surgery have better long-term outcomes, including reductions in weight and chronic diseases. The AAP recommends pediatricians refer eligible youth to reputable multidisciplinary centers that have extensive pediatric bariatric surgical experience. The AAP also notes that access to bariatric surgery is often limited by a lack of insurance coverage, especially for youth from lower socioeconomic groups and racial and ethnic minorities. The AAP calls for primary care providers, governments, medical centers, and insurers to collaborate on strategies to improve access to bariatric surgery for those children and adolescents in need.[20]

# Select Types of Bariatric Surgery

### GASTRIC BYPASS

This procedure, also called Roux-en-Y gastric bypass, creates a small pouch that significantly limits the stomach volume. The stomach continues to make digestive juices, so this permits the digestive juices to flow to the small intestine. Because food now bypasses a portion of the small intestine, fewer calories and nutrients are absorbed.

Benefits: Greater weight loss than gastric band with no foreign objects used in the procedure.

Limitations: Increased risk of surgery-related issues with longer recovery. This procedure is difficult to reverse and nutrient deficiencies may occur.

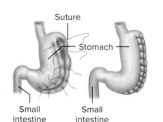

**Before surgery**       **After surgery**

1. Staples divide the stomach.

2. Part of the small intestine, called the jejunum, is connected to the pouch.

### LAPAROSCOPIC ADJUSTABLE GASTRIC BANDING

This procedure involves placing an inflatable band around the top portion of the stomach. This limits the space available for food and increases satiety. LAGB is often recommended for people who have tried other weight-loss plans without long-term success.

Benefits: The surgical procedure is rapid (30–60 minutes) and can be reversed. This surgery has the lowest risk of vitamin and mineral deficits.

Limitations: Results in less weight loss than other surgeries. The surgical procedure is challenging and requires multiple steps. In addition, the band may slip, so frequent follow-up is required.

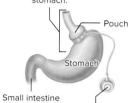

This procedure divides stomach into two sections. This creates a small pouch with a narrow opening that goes into the larger section of the stomach.

Port is used to adjust the gastric band after surgery.

### GASTRIC PLICATION

This is a restrictive procedure that shrinks the size of the stomach by suturing large folds in the stomach's lining. This reduces the stomach volume by approximately 80%, and increases satiety. The procedure typically takes between 40 minutes and 2 hours to complete.

Benefits: Increased weight loss over gastric band with no change in the intestines. This procedure has a relatively rapid recovery period.

Limitations: This procedure cannot be reversed, and there is an increased risk of acid reflux. As with most bariatric procedures, vitamin and mineral deficiencies are a serious concern.

### VERTICAL SLEEVE GASTRECTOMY

In this bariatric procedure, 80% to 85% of the stomach is removed to create a smaller stomach pouch. This limits the amount of food consumed and increases satiety. Vertical sleeve gastrectomy has most often been done on people who are too heavy to safely have other types of weight-loss surgery.

Benefits: This is a rapid surgical procedure, taking 30–60 minutes, that is safer than other procedures.

Limitations: This procedure cannot be reversed. Weight loss is typically slower than with other bariatric procedures.

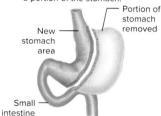

1. Using a video monitor to guide the instruments, surgeon removes a portion of the stomach.

2. The remaining portion of the stomach is closed using staples.

### ILEAL TRANSPOSITION

This metabolic procedure is often used for individuals who are overweight with type 2 diabetes. The technique relocates the distal part of the small intestine to the proximal part of the small intestine. This is a longer operation than other procedures (3–3.5 hours) and requires more advanced equipment, longer hospital stays, and higher costs than other commonly used, simpler procedures.

Benefits: Results in greater glycemic control than other procedures; considered key to managing the twin epidemics of obesity and diabetes.

Limitations: As with all invasive procedures, surgical complications are possible.

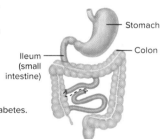

**FIGURE 7-27** ▲ Bariatric surgery promotes weight loss by altering the digestive tract anatomy, limiting the volume of food that can be consumed and digested. These surgical procedures are not appropriate for everyone and candidates must be screened carefully. In addition, many of the procedures are relatively new (e.g., intragastric balloon) and long-term effects remain unknown.

## TABLE 7-7 ■ Select Bariatric Surgeries Performed on Youth

|  | Adjustable Gastric Banding | Gastric Bypass | Sleeve Gastrectomy |
|---|---|---|---|
| Strengths | Low rate of complications and quicker recovery <br> Vitamin deficiencies are rare | Most frequent bariatric procedure <br> High success rate | Rapid surgical procedure <br> Considered safer than alternatives |
| Limitations | Weight loss not as rapid <br> May require replacement surgery | Longer recovery <br> Irreversible procedure | Weight loss not as rapid <br> Irreversible procedure |
| Possible Side Effects | Infection, bleeding, band slippage or erosion, stomach pouch enlargement, stoma blockage | Infection, bleeding, blood clots, bowel obstruction, "leaky" abdomen | Gastritis, heartburn, stomach ulcers, leaking from the surgical stitches on the stomach, intestinal blockage |

Source: American Society for Metabolic and Bariatric Surgery available at www.asmbs.org/patients/adolescent-obesity.

**Liposuction.** *Spot reducing* by using diet and physical activity is not possible. Problem local fat deposits can be reduced in size, however, using liposuction, or lipectomy. In this procedure, a pencil-thin tube is inserted into an incision in the skin, and the fat tissue, often in the buttocks and thigh area, is suctioned out through a tube and discarded. This invasive procedure involves risks, such as infection, permanent depressions in the skin, nerve damage, and blood clots, which can lead to kidney failure and sometimes death. The procedure is designed to help a person lose about 4 to 8 pounds per treatment. The costs vary greatly and depend upon site and clinician, but the average cost is approximately $3500 per procedure.

**Cryolipolysis.** Commonly referred to as fat freezing and patented under the name CoolSculpting™, cryolipolysis is a nonsurgical fat reduction procedure designed to reduce localized fat. The FDA has approved cryolipolysis to treat fat deposits beneath the chin, upper arms, inner and outer thighs, abdomen, hips, upper and lower back, and underneath the buttocks. This procedure is not targeted for people who are obese.

### ✓ CONCEPT CHECK 7.9

1. How restrictive is a very-low-calorie diet plan? Why is monitoring by a qualified health professional important?
2. What are the surgical options for people with severe obesity who have failed to lose weight with other weight-loss strategies?

## 7.10 Treatment of Underweight

**underweight** Ratio of weight to height that is lower than what is associated with optimal health. For adults, this is defined as BMI less than 18.5. For children, this is defined as BMI-for-age below the 5th percentile.

**Underweight** is defined by a BMI < 18.5 and can be caused by a variety of factors, such as cancer, infectious disease (e.g., tuberculosis), malabsorption or digestive tract disorders (e.g., Crohn's, inflammatory bowel disease), and excessive dieting or physical activity. Genetic factors may also lead to a higher RMR, a slight body frame, or both. Health problems associated with underweight include the loss of menstrual function (*amenorrhea*), low bone mass, complications with pregnancy and surgery, and slow recovery after illness. Significant underweight is also associated with increased death rates, especially when combined with smoking. We frequently hear about the risks of obesity but seldom of underweight. In our culture, being underweight is much more socially acceptable than being overweight or obese.

Sometimes being underweight requires medical intervention. A thorough physical exam should be obtained first to rule out hormonal imbalances, depression, cancer, infectious disease, digestive tract disorders, excessive physical activity, and other hidden diseases such as a serious eating disorder.

Yet the causes of underweight are not altogether different from the causes of obesity. Internal and external satiety-signal irregularities, the rate of metabolism, genetic factors, and psychological traits can all contribute to underweight.

In children who are growing, the high demand for calories to support physical activity and growth can also cause underweight. During growth spurts, active children and adolescents may not take the time to consume enough calories to support their needs. Moreover, gaining weight can be a formidable task for a person who is underweight. An extra 500 kcal per day or more may be required to gain weight, and this can prove challenging, especially for children who are active. Individuals attempting to gain weight may need to increase portion sizes and include frequent snacks.

When weight gain is necessary, one approach for treating adults is to gradually increase their consumption of calorie-dense foods, especially those high in vegetable fats. Nuts, seeds, and granola can be good calorie sources with low saturated fat content. Dried fruits and bananas are good fruit choices. Individuals who are underweight should replace sugar-free drinks with good calorie sources, such as 100% fruit juices and nutrient-rich smoothies.

Encouraging a regular meal and snack schedule also can aid in weight gain and maintenance. Sometimes individuals affected by underweight have experienced stress at work or school or have been too busy to eat. Making regular meals a priority may not only help them attain an appropriate weight but also may help with digestive disorders, such as constipation, sometimes associated with irregular eating patterns.

## HEALTHY WEIGHT GAIN

A combination of high-quality nutrition and strength training is needed to gain weight as muscle. Strength training slows muscle loss that comes with dieting and aging, increases the strength of your muscles and connective tissues, and increases bone density. When weight is lost, up to a quarter of the loss may come from muscle mass, which can slow basal metabolism. Strength training helps protect against lean body mass losses and rebuilds any muscle lost by dieting—or prevents it from being lost in the first place. The best bet when starting a strength-training program is to seek individualized counseling from a qualified sports dietitian or certified athletic trainer who can address personal goals and limitations and can help with alignment and execution of each exercise.

There are several things to consider when designing dietary patterns to accompany training. During a workout, it is normal for the body to break down some muscles due to the stress placed on them. Once you're finished strength training, you want to repair and build muscle again. It is very important to get proper nutrients into the body after a workout to promote recovery and muscle building. Immediately before and again after strength training, a serving of high-quality protein should be consumed to optimize performance and build lean muscle mass. It is also important to have some carbohydrate along with protein to increase the protein absorption, replenish glycogen stores, and provide future fuel for workouts. Low-fat chocolate milk has been shown to be a great source of protein, carbohydrate, and fluid immediately post-activity. Although a quick protein bar or shake is great when you're at the gym, it should not be the only source of protein. During meals, high-quality lean protein sources such as tuna, chicken, soy, and beans are recommended. Those who work out but eat nothing but food high in saturated fat and calories will gain fat in addition to any muscle gains. To gain lean muscle mass, one needs a balanced dietary pattern rich in protein and carbohydrates, including plenty of fruits, vegetables, and whole grains. The number of calories you require each day varies greatly and will depend on your weight, activity level, age, and muscle mass. If you are working out 3 days a week, you can eat about 15 kcal per pound of body weight. If you work out 5 days a week, you can increase that calorie count to 20 kcal per pound.

Individuals who are underweight should increase their consumption of calorie-dense foods, such as smoothies, that are also loaded with nutrients. **What are other good sources of nutritious foods that are calorie-dense?** Dynamic Graphics Group/Creatas/Alamy Stock Photo

The following are some specific tips for gaining healthy weight:
- Eat more frequently. Try five to six smaller meals during the day rather than two or three large ones.
- Choose nutrient-rich foods. Whole-grain breads, pastas and cereals, fruits and vegetables, dairy products, high-quality protein sources, and nuts and seeds are all wise choices. Try smoothies and shakes made with milk, fruit, and ground flaxseed. In some cases, a liquid meal replacement may be recommended.
- Drink fluids 30 minutes after a meal, not with it.
- Snack on nuts, nut butters, eggs, cheeses, dried fruits, and avocados to increase calories throughout the day.
- Add a bedtime snack, such as a peanut butter and jelly sandwich, smoothie, or a wrap with avocado, sliced vegetables, and lean proteins.
- Add extras to meals for more calories (cheese in casseroles, eggs in mixed dishes, and dried milk in soups and stews).
- Allow for an occasional treat but keep diet quality in mind. Try bran muffins, yogurt, and granola bars to satisfy the dessert craving.
- Engage in regular physical activity, especially strength training to enhance muscle gains and stimulate your appetite.

## ✓ CONCEPT CHECK 7.10

1. How is underweight defined, and what are some of its primary causes?
2. What are the components necessary to gain weight as muscle and not as fat?

## Popular Diets—Cause for Concern

BananaStock/PunchStock

Many individuals who are overweight try to help themselves by using the latest popular (also called *fad*) diets. But, as you will see, most of these fad diets do not help, and some can actually harm those who follow them. Research has shown that early dieting and unhealthful weight-control practices in adolescents can lead to an increased risk of weight gain, overweight, and eating disorders.

Most of these popular diets are nutritionally inadequate and include certain foods that people would not normally choose to consume in large amounts. To achieve weight loss and maintain it over time, experts agree that individuals should strive for reducing overall caloric intake in addition to increasing physical activity. People need a plan they can live with in the long run so that a healthy weight becomes permanent. The goal should be weight management over a lifetime, not immediate weight loss. Every popular diet leads to some immediate weight loss simply because daily intake is monitored and monotonous food choices are typically part of the plan.

People following diets often fall within a healthy BMI of 18.5 to 25. Rather than worrying about weight loss, these individuals should focus on a healthy lifestyle that allows for weight maintenance. Incorporating necessary lifestyle changes, especially regular physical activity, and learning to accept one's individual body characteristics should be the overriding goals.

The dieting mania can be viewed as mostly a social problem, stemming from unrealistic weight expectations and lack of appreciation for the natural variety in body shape and size. Not every person can or should look like a fashion model, nor can every man look like a Greek god, but all of us can strive for good health and, if physically possible, an active lifestyle.

### How to Recognize a Fad Diet

The criteria for evaluating weight-loss programs with regard to their safety and effectiveness were discussed earlier. In contrast, unreliable fad diets typically share some common characteristics:

1. They promote rapid weight loss. This is the primary temptation that attracts the dieter. As mentioned, this initial weight loss primarily results from water loss and lean muscle mass depletion, not adipose tissue depletion.
2. They often limit food selections and dictate specific rituals, such as eating only fruit for breakfast or cabbage soup every day.
3. They use testimonials from famous people and tie the diet to well-known cities, such as Beverly Hills or South Beach.
4. They present themselves as cure-alls. These diets claim to work for everyone, whatever the type of obesity or the person's genetic or environmental makeup.
5. They often recommend expensive supplements or meal replacements.
6. No attempts are made to change eating habits permanently. Dieters follow the diet until the desired weight is reached and then revert to old behaviors.
7. They are generally critical of and skeptical about the scientific community. The lack of a quick fix from medical and dietetic professionals has led some of the public to seek advice from those who appear to have the answer.
8. They claim that there is no need to exercise or engage in regular physical activity.

Probably the cruelest characteristic of these diets is that they essentially guarantee failure for the dieter. Fad diets are not designed for permanent weight loss. Habits are not changed, and the food selection is so limited that the person cannot follow the diet in the long run. Although dieters assume that they have lost fat, they often have lost mainly muscle mass and body fluids. As soon as they begin eating normally again, much of the lost tissue is replaced as fat mass. In a matter of weeks or months, most of the lost weight is back. The dieter appears to have failed, when actually the *diet* has failed. The gain and loss cycle is called weight cycling or *yo-yo* dieting. This whole scenario can add more blame and guilt, challenging the self-worth of the dieter. It can also come with some health costs, such as increased upper-body fat deposition. If someone needs help losing weight, consult a nutrition expert, such as an RDN. It is unfortunate that current trends suggest that people are spending more time and money on *quick fixes* than on such professional help.

## POPULAR DIET APPROACHES

A list of the best overall diet plans is presented in Table 7-8. Note that these plans follow the key recommendations of the *Dietary Guidelines* and include an abundance of plant-based foods rich in fiber, key nutrients, and phytochemicals. Dietary patterns that exclude entire food groups are not associated with long-term health benefits.

High-protein, low-carbohydrate diets continue to be a popular approach to losing weight. These diets typically recommend at least 30% to 50% of their total calories from protein and drastically restrict carbohydrate intake. Low carbohydrate intake leads to less glycogen synthesis and therefore less water in the body (about 3 grams of water are stored per gram of glycogen). As discussed, a very-low-carbohydrate intake also forces the liver to produce some glucose. The source of carbons for this glucose is mostly proteins from tissues such as muscle, resulting in loss of protein tissue, which is about 72% water. Essential ions, such as potassium, are also lost in the urine. In the initial stages of a low-carbohydrate diet, losses of glycogen stores, lean tissue, and water cause rapid weight loss. When a normal dietary pattern is resumed, the protein tissue is rebuilt and the weight is regained.

In addition, restricting carbohydrate causes your body to burn fat instead of carbohydrate for fuel. In theory, this burning of excess fat stores makes sense for weight loss. Remember, however, that when we burn fat without carbohydrate, it causes the body to go into the metabolic state called *ketosis*. For a person who is trying to lose weight, one benefit of ketone accumulation in body fluids is mild suppression of appetite. Increased excretion of electrolytes can cause nausea and headaches with some of the most restrictive plans. Taken to the extreme over time, ketosis can disrupt the body's acid–base balance and become quite dangerous.

For most dieters, a low-carb plan is such a major change from normal habits that it is very difficult to maintain. However, research indicates that low-carbohydrate diets may be an effective alternative to low-fat diets for some people.[21] Popular low-carb diets include the Ketogenic ("Keto") Diet, Atkins, Paleo, and Whole30.

### CARBOHYDRATE-FOCUSED DIETS

Rather than restricting carbohydrates, some diet plans emphasize eating plenty of fruits, vegetables, and whole grains, as well as limiting simple sugars and highly processed refined grains. In theory, these foods will cause a slow, steady rise and fall in blood sugar after a meal, which may help control hunger. They also provide lots of nutritious fiber and key nutrients to promote satiety.

### LOW-FAT DIETS

Very-low-fat diets contain approximately 5% to 10% of calories as fat and are often very high in carbohydrates. If followed consistently, these approaches lead to weight loss and may be helpful for reducing heart disease risks; however, they are difficult to follow. People are quickly bored with this type of diet because they cannot eat many of their favorite foods. These dieters eat primarily plant-based foods (grains, fruits, and vegetables), which many people cannot maintain for very long. Eventually, the person wants some foods higher in fat or protein. Furthermore, such diets may have too much carbohydrate for some people who have a family history of diabetes.[21]

**TABLE 7-8** ■ **Best Overall Diet Plans**

| Diet Plan | Rank |
|---|---|
| Mediterranean Diet | 1 |
| DASH Diet | 2 |
| Flexitarian Diet | 3 |

Source: U.S. News & World Report Best Diet Rankings 2021.

▲ In time, very-low-carbohydrate, high-protein diets typically leave a person fatigued and wanting more variety in meals, and so the diets are abandoned. Dropout rates are high on these diets and there is no evidence these types of diets are superior to any other diet plans that involve calorie restriction. Ernie Friedlander/Cole Group/Getty Images

### NOVELTY DIETS

Some novelty diets emphasize one food or food group and exclude almost all others. Other popular diet plans include a variety of detox regimens, fasting, or intermittent eating. Note that the Academy of Nutrition and Dietetics provides reviews of a variety of fad diet plans.[22]

### MEAL REPLACEMENTS

Meal replacements come in many forms, including beverages or formulas, frozen or shelf-stable entrees, and meal or snack bars. Most meal replacements are fortified with vitamins and minerals and are appropriate to replace one or two regular meals or snacks per day. Although they are not a magic bullet for weight loss, they have been shown to help some people lose weight. Advantages of these convenient products are that they provide portion- and calorie-controlled foods that can serve as a visual education on appropriate portion sizes. A disadvantage is that when dieters rely on foods selected and prepared by someone else, they do not learn behaviors necessary to select and prepare healthy foods on their own.

## QUACKERY IS COMMON WITH MANY POPULAR DIETS

Many popular diets fall under the category of *quackery*—people taking advantage of others. They usually involve a product or service that costs a considerable amount of money. Often, those offering the product or service don't realize that they are promoting quackery because they themselves were victims. For example, they tried the product and, by pure coincidence, felt it worked for them, so they promote it to all their friends and relatives.

Numerous other gimmicks for weight loss have come and gone and are likely to resurface. If in the future an important aid for weight loss is discovered, you can feel confident that reputable organizations, such as the National Academy of Medicine, Academy of Nutrition and Dietetics, or the Centers for Disease Control and Prevention (CDC) will report it. Quackwatch.com is an online resource for consumers to help identify quackery.

### ✓ CONCEPT CHECK 7.11

1. What are the common characteristics of a fad diet?
2. Why do popular diets often fail?
3. What is the primary mechanism by which all diets lead to weight loss?

---

## CASE STUDY Choosing a Weight-Management Program

Joe has a hectic schedule. During the day, he works full time at a warehouse distribution center filling orders. At night, three times a week, he attends class at the local community college in pursuit of computer certification. On weekends, he likes to watch sports on television, spend time with family and friends, and study. Joe has little time to exercise or think about what he eats—that is, convenience rules. He stops for coffee and a pastry on his way to work, has a burger or pizza for lunch at a fast-food restaurant, and for dinner picks up fried chicken or fish at the drive-through on his way to class. Unfortunately, over the past few years, Joe's weight has been climbing. He is 5 feet, 10 inches tall and weighs 200 pounds. Lately, he has frequently been short of breath during his shift at work. Watching a game on television a few nights ago, he saw an infomercial for a weight-loss supplement that promises to increase his energy level and allow him to continue to eat large portions of tasty foods but not gain weight. A famous actor supports the claim that this product allows one to eat at will and not gain weight. This claim is tempting to Joe.

▲ What changes can Joe make in his daily routine and diet to prevent weight gain?
Ryan McVay/Getty Images

1. Has Joe been experiencing positive or negative energy balance over the past few years? What is his current BMI?
2. What aspects of Joe's lifestyle (other than diet) are causing this effect on his energy balance?
3. What changes could Joe make to his dietary and physical activity patterns that would promote weight loss or maintenance?
4. Why should Joe be skeptical of the claims he heard about the weight-loss product in the infomercial?
5. Referring back to the characteristics of unreliable diets, what advice can you offer Joe for evaluating weight-loss programs?

*Complete the Case Study. Responses to these questions can be provided by your instructor.*

# Summary (Numbers refer to numbered sections in the chapter.)

**7.1** Energy balance considers energy intake and energy output. Negative energy balance occurs when energy output surpasses energy intake, resulting in weight loss. Positive energy balance occurs when calorie intake is greater than output, resulting in weight gain.

Basal metabolism, the thermic effect of food, physical activity, and adaptive thermogenesis account for total energy use by the body. Basal metabolism, which represents the minimum amount of calories required to keep the resting, awake body alive, is primarily affected by lean body mass, surface area, and thyroid hormone concentrations. Physical activity is energy use above that expended at rest. The thermic effect of food describes the increase in metabolism that facilitates digestion, absorption, and processing of the nutrients recently consumed. Adaptive thermogenesis includes nonvoluntary activities, such as shivering and fidgeting, that increase energy use and may counter extra calories from overeating. In a sedentary person, about 70% to 85% of energy use is accounted for by basal metabolism and the thermic effect of food.

**7.2** Energy use by the body can be measured as heat given off by direct calorimetry or as oxygen used by indirect calorimetry. A person's Estimated Energy Requirement (EER) can be calculated based on the following factors: sex, height, weight, age, and amount of physical activity.

**7.3** A body mass index (weight in kilograms ÷ height$^2$ in meters) of 18.5 to 24.9 is one measure of normal (healthy) weight. A healthy weight is best determined in conjunction with a thorough health evaluation by a clinician or dietitian. A body mass index of 25 to 29.9 represents overweight. Obesity is defined as a body mass index of 30 or more or a total body fat percentage over 25% (men) or 32% (women).

Fat distribution greatly determines health risks associated with obesity. Upper-body fat storage (android), as measured by a waist circumference greater than 40 inches (102 centimeters) for men or 35 inches (88 centimeters) for women, typically results in higher risks of hypertension, cardiovascular disease, and type 2 diabetes than lower-body fat storage (gynoid).

**7.4** Both genetic (nature) and environmental (nurture) factors can increase the risk of obesity. The set-point theory proposes that we have a genetically predetermined body weight or body fat content, which the body strives to regulate.

**7.5** A sound weight-loss and maintenance program should meet the individual's nutritional needs by emphasizing a wide variety of nutrient-dense, high-fiber foods; adapting to one's with readily obtainable foods; emphasizing regular physical activity; and stipulating the supervision by a clinician if weight is to be lost rapidly, if the person is obese (body mass index over 40), or if the individual is over the age of 40 (men) or 50 (women) and plans to perform substantially greater physical activity than recommended.

**7.6** For those choosing to lose weight, the evidence-based recommendations encourage an initial calorie deficit of approximately 500 kcal per day increased physical activity, or a combination of both. This should result in approximately 1 pound of weight loss per week initially. Note that weight change is not linear; it occurs most rapidly during the first year after a change in energy balance but tapers off over the next 2 years if changes in energy intake and physical activity are strictly maintained over time.

**7.7** Physical activity as part of a weight-management program should be focused on duration rather than just intensity. The recommendations promote 150 to 300 minutes of physical activity per week. Ideally, about 60 minutes of moderate-intensity physical activity should be part of each day to prevent adult weight gain.

**7.8** Behavior changes are a vital part of a weight-loss and management programs because individuals may have many behaviors that encourage an obesogenic environment. Specific behavior modification techniques, such as stimulus control and self-monitoring, can be used to help change lifestye behaviors.

**7.9** Medications to blunt appetite can aid weight loss. Orlistat (Xenical) reduces fat absorption from a meal when taken with the meal. Weight-loss drugs are reserved for those who are obese or have weight-related problems, and they must be administered under close physician supervision.

The treatment of severe obesity may include very-low-calorie diets containing 400 to 800 kcal per day or bariatric surgery, often to reduce stomach volume to approximately 30 milliliters (1 ounce). Both of these measures should be reserved for people who have failed at more conservative approaches to weight loss. They also require close medical supervision.

**7.10** Underweight can be caused by a variety of factors, such as excessive physical activity and genetics. Sometimes being underweight requires medical attention. A primary care provider should be consulted first to rule out underlying health issues. The underweight person may need to increase portion sizes, learn to incorporate calorie-dense foods frequently, and drink fluids between meals. In addition, encouraging a regular meal and snack schedule aids in both weight gain and weight maintenance.

**7.11** Many people who are overweight try popular fad diets that most often are not helpful and may actually be harmful. Unreliable diets typically share some common characteristics, including promoting quick weight loss, limiting food selections, using personal testimonials as proof, and requiring no physical activity.

# Check Your Knowledge (Answers are available at the end of this question set.)

1. An energy deficit of 500 kcal per day would result in a total weight loss of about 1 pound over a _____ period.
   a. 1-week
   b. 1-month
   c. 1-year
   d. 3-year

2. Thermic effect of food represents the energy cost of
   a. chewing food.
   b. peristalsis.
   c. basal metabolism.
   d. digesting, absorbing, and packaging nutrients.

3. A well-designed weight-loss program should
   a. increase physical activity.
   b. alter problem behaviors.
   c. reduce energy intake.
   d. include all of the above.

4. Which factor is associated with a lower basal metabolic rate
   a. stress.
   b. low calorie intake.
   c. fever.
   d. pregnancy.

5. The mechanism of bariatric surgery is to
   a. reduce stomach volume.
   b. slow transit time.
   c. surgically remove adipose tissue.
   d. prevent snacking.

6. Basal metabolism
   a. represents about 30% of total energy expenditure.
   b. is energy used to maintain heartbeat, respiration, other basic functions, and physical activities.
   c. represents about 60% to 80% of total calories used by the body during a day.
   d. includes energy to digest food.

7. It is recommended that adults should do the equivalent of _____ minutes of moderate-intensity aerobic activity each week for weight loss and to achieve and maintain a healthy body weight.
   a. 30 to 60
   b. 90 to 120
   c. 150 to 300
   d. 400 to 600

8. Probably the most important contributing factor for obesity rates today in the United States is
   a. food advertising.
   b. snacking practices.
   c. inactivity.
   d. eating processed food.

9. The major goal for weight reduction in the treatment of obesity is the loss of
   a. weight.
   b. body fat.
   c. body water.
   d. body protein.

10. For most adults, the greatest portion of their energy expenditure is for
    a. physical activity.
    b. sleeping.
    c. basal metabolism.
    d. the thermic effect of food.

**Answer Key:** 1. a (LO 7.6), 2. d (LO 7.2), 3. d (LO 7.6), 4. b (LO 7.1), 5. a (LO 7.10), 6. c (LO 7.1), 7. c (LO 7.8), 8. c (LO 7.5), 9. b (LO 7.6), 10. c (LO 7.1)

# Study Questions (Numbers refer to Learning Outcomes.)

1. How does energy imbalance, including the role of physical activity, lead to weight gain and obesity? **(LO 7.1)**

2. What are the five components that are included in calculating estimated energy requirements? **(LO 7.2)**

3. Define how a healthy weight may be determined. **(LO 7.3)**

4. Describe a practical tool to define obesity in a clinical setting. **(LO 7.3)**

5. List three health issues that are related to obesity and the mechanism or reason that each condition occurs. **(LO 7.3)**

6. Explain how nurture and nature can contribute to the development of obesity. What are the two most convincing pieces of evidence that both genetic and environmental factors play significant roles in the development of obesity? **(LO 7.4)**

7. List three key characteristics of a sound weight-loss program. **(LO 7.5)**

8. Why is the claim for quick, effortless weight loss by any method always misleading? **(LO 7.5)**

9. Why should obesity treatment be viewed as a lifelong commitment rather than a short diet or episode of weight loss? **(LO 7.5)**

10. List examples of positive behaviors that can lead to weight management. **(LO 7.8)**

11. What steps are important to remember when an underweight person wants to gain muscle but not fat? **(LO 7.10)**

# Further Readings

1. Johnston BC and others: Comparison of weight loss among named diet programs in overweight and obese adults: A meta-analysis. *Journal of the American Medical Association* 312(9):923, 2014. DOI:10.1001/jama.2014.10397.

2. Academy of Nutrition and Dietetics: Position of the Academy of Nutrition and Dietetics: Total diet approach to healthy eating. *Journal of the Academy of Nutrition and Dietetics* 113:307, 2014. DOI:10.1016/j.jand.2012.12.013.

3. Centers for Disease Control and Prevention: *Overweight and Obesity.* https://www.cdc.gov/obesity/data/prevalence-maps.html. Accessed January 15, 2020.

4. Ogden CL and others: Prevalence of obesity among adults and youth: United States, 2015–2016. NCHS data brief, no. 288. Hyattsville, MD: National Center for Health Statistics. 2017.

5. Bluher M: Obesity: Global epidemiology and pathogenesis. *Nature Reviews Endocrinology* 15:288, 2019. DOI:10.1038/s41574-019-0176-8.

6. Villablanca P and others: Nonexercise activity thermogenesis in obesity management. *Mayo Clinic Proceedings.* 90(4):509, 2015. DOI:https://doi.org/10.1016/j.mayocp.2015.02.001.

7. Passmore R and others: The chemical anatomy of the human body. In *Biochemical Disorders in Human Disease,* 3rd ed. London: Churchill, 1970, pp. 1–14.

8. Freedman DS and others: The relation of childhood BMI to adult adiposity: The Bogalusa Heart Study. *Pediatrics.* 115(1):22, 2005. DOI:https://doi.org/10.1542/peds.2004-0220.

9. Centers for Disease Control and Prevention, Geneomics and Precision Health: Behavior, environment, and genetic factors all have a role in causing people to be overweight and obese. Available at https://www.cdc.gov/genomics/resources/diseases/obesity/index.htm. Accessed January 8, 2020.

10. Obesity Action Coalition: Body Weight "Set Point"—what we know and what we don't know. Available at https://www.obesityaction.org/community/article-library/body-weight-set-point-what-we-know-and-what-we-dont-know. Accessed January 28, 2020.

11. Council for a Strong America: Mission Readiness Report. Unhealthy and unprepared: National security depends on promoting healthy lifestyles from an early age. https://www.strongnation.org/articles/737-unhealthy-and-unprepared. Accessed January 15, 2020.

12. Centers for Disease Control and Prevention. Healthy weight. Eat more, weigh less? Available at https://www.cdc.gov/healthyweight/healthy_eating/energy_density.html. Accessed January 22, 2020.

13. Mondo V and others: Exercise modifies the gut microbiota with positive health effects. *Oxidative Medicine and Cellular Longevity* 2017:3831972, 2017. DOI:10.1155/2017/3831972.

14. LeBlanc E and others: Behavioral and pharmacotherapy weight loss interventions to prevent obesity-related morbidity and mortality in adults: An updated systematic review for the U.S. Preventive Services Task Force. Rockville, MD: Agency for Healthcare Research and Quality (US); 2018 Sep. Report No.: 18-05239-EF-1. Available at https://www.ncbi.nlm.nih.gov/pubmed/30354042.

15. Peng M and others: How does plate size affect estimated satiation and intake for individuals in normal-weight and overweight groups? *Obesity Science & Practice* 3(3):282, 2017. DOI:10.1002/osp4.119.

16. U.S. Food and Drug Administration: FDA approves AspireAssist obesity device. https://www.fda.gov/newsevents/newsroom/pressannouncements/ucm506625.htm. Accessed January 25, 2020.

17. Webb D: Fasting regimens for weight loss. *Today's Dietitian* 2018; 20(2):34.

18. Patterson RE and others: Intermittent fasting and human metabolic health. *Journal of the Academy of Nutrition and Dietetics* 115:1203, 2015. DOI:10.1016/j.jand.2015.02.018.

19. American Society for Metabolic and Bariatric Surgery: ASMBS updated position statement on bariatric surgery in class i obesity. Available at https://asmbs.org/resources/asmbs-updated-position-statement-on-bariatric-surgery-in-class-i-obesity?sfns=mo. Accessed January 26, 2020.

20. Madura JA and others: Quick fix or long-term cure? Pros and cons of bariatric surgery. *F1000 Medicine Reports* 4:19, 2012. DOI:10.3410/M4-19.

21. Gardner CD and others: Effect of low-fat vs low-carbohydrate diet on 12-month weight loss in overweight adults and the association with genotype pattern or insulin secretion: The DIETFITS randomized clinical trial. *Journal of the American Medical Association* 319(7):667, 2018. DOI:10.1001/jama.2018.0245.

22. Academy of Nutrition and Dietetics: Fad diets through the years. Available at https://www.eatright.org/health/weight-loss/fad-diets/fad-diet-timeline?rdType=section_change&rdProj=nnm_redirects&rdInfo=food_resources_nnm. Accessed January 26, 2020.

# Chapter 8

# Vitamins and Phytochemicals

# Student Learning Outcomes

**Chapter 8 is designed to allow you to:**

**8.1** Describe the general characteristics of the fat- and water-soluble vitamins, absorption, and storage.

**8.2** List ways to preserve vitamins in foods and explain the sources and benefits of phytochemicals.

**8.3** Describe the roles of the fat-soluble vitamins (A, D, E, and K) in body defenses and bone health.

**8.4** Describe the roles of the B vitamins (thiamin, riboflavin, niacin, pantothenic acid, biotin, vitamin B-6, folate, and vitamin B-12) in energy metabolism and in blood and brain health; vitamin C in immune function; and choline in cell structures and metabolism.

**8.5** List the dietary sources and requirements for each fat- and water-soluble vitamin and choline as well as the dangers of exceeding the recommendations.

**8.6** Describe the signs and symptoms of vitamin deficiencies and risk factors leading to vitamin deficiencies.

**8.7** Evaluate dietary supplements, current recommendations, potential benefits and hazards of use.

**8.8** Describe modifiable risk factors that contribute to cancer risk and understand the role of food constituents in cancer prevention, treatment, and survivorship.

## FAKE NEWS

Vitamin-fortified foods are healthier because they provide essential vitamins without the need for a supplement.

## THE FACTS

Many manufacturers add vitamins to foods and beverages that are energy-dense and highly processed, such as cookies, candies, chips, and many snack foods. This marketing strategy has altered consumer purchasing behavior in favor of vitamin-fortified products. The fact is that added vitamins do not compensate for empty calories, added sugars, and saturated fats in any product. Consumers should strive to obtain a dietary pattern rich in vitamins from whole nutritious foods, such as fruits and vegetables.

Although the vitamins are essential nutrients, the amount of vitamins we need to prevent deficiency is quite small. Some people believe that consuming vitamins far in excess of their needs provides them with extra energy, protection from disease, and prolonged youth. They seem to think that if a little is good, then more must be better. More than half of the U.S. adult population have taken vitamin and/or mineral supplements on a regular basis, some at unsafe levels.

Vitamins are naturally found in plants and animals. Plants synthesize all the vitamins they need and are a healthy source of vitamins for animals. Animals vary in their ability to synthesize vitamins. For example, guinea pigs and humans are two of the few organisms unable to make their own supply of vitamin C.

Every major public health authority recommends that we increase our intake of fruits and vegetables. Which vitamins are especially found in fruits and vegetables? What are some other health-related attributes of plant-based foods in general? Which chronic diseases are associated with a poor intake of fruits and vegetables? Should we take a daily vitamin supplement if we do not include fruits and vegetables in our diet? This chapter provides some answers.

**FIGURE 8-1** ▶ Micronutrients contribute to many functions in the body. John Lund/ Getty Images

**Fluid and Electrolyte Balance**

Sodium
Potassium
Chloride
Phosphorus

**Energy Metabolism**

Thiamin
Riboflavin
Niacin
Pantothenic acid
Biotin
Vitamin B-12
Iodine
Chromium
Magnesium
Manganese
Molybdenum
Choline

**Bone Health**

Vitamin C
Vitamin D
Vitamin K
Calcium
Phosphorus
Magnesium
Fluoride
Boron
Silicon

**Body Defenses**

Vitamin A
Vitamin C
Vitamin D
Vitamin E
Carotenoids
Selenium
Copper
Iron
Magnesium
Manganese
Zinc

**Brain Health**

Thiamin
Riboflavin
Niacin
Vitamin B-6
Vitamin B-12
Folate
Choline
Vitamin C
Vitamin D
Vitamin E
Calcium
Iodine
Magnesium
Selenium
Iron
Zinc

**Blood Health**

Vitamin B-6
Vitamin B-12
Folate
Vitamin K
Iron
Zinc
Copper
Calcium

# 8.1 Vitamins: Vital Dietary Components

By definition, vitamins are essential organic (carbon-containing) substances needed in small amounts in the dietary pattern for normal function, growth, and maintenance of the body. All humans require the same essential vitamins, but the amount required can vary depending on age, sex, and presence of illness. Although vitamin requirements are very small, each vitamin is essential for one or more functions in the body (Fig. 8-1). Vitamins can be divided into two broad classes based on solubility: vitamins A, D, E, and K are **fat-soluble vitamins,** whereas the B vitamins and vitamin C are **water-soluble vitamins.** The B vitamins include thiamin, riboflavin, niacin, pantothenic acid, biotin, vitamin B-6, folate, and vitamin B-12. Choline is a vitamin-like nutrient but is not technically classified as a vitamin.

Vitamins are essential in human dietary patterns because they cannot be synthesized in the human body or produced in sufficient amounts. Notable exceptions to having a strict dietary need for a vitamin are vitamin A, which we can synthesize from certain pigments in plants; vitamin D, synthesized in the body if the skin is exposed to

**fat-soluble vitamins** Vitamins that dissolve in fat and some chemical compounds but not readily in water. These vitamins are A, D, E, and K.

**water-soluble vitamins** Vitamins that dissolve in water. These vitamins are the B vitamins and vitamin C.

adequate sunlight; niacin, synthesized from the amino acid tryptophan; and vitamin K, biotin, and others that are synthesized by the bacteria in the intestinal tract.

To be classified as a vitamin, a compound must meet the following criteria: (1) the body is unable to synthesize enough of the compound to maintain health; and (2) absence of the compound from the dietary pattern for a defined period produces deficiency symptoms. If caught in time, these symptoms can be quickly reversed when the compound is reintroduced. A compound does not qualify as a vitamin merely because the body cannot make it. Evidence must suggest that health eventually declines when the substance is not consumed.

As scientists began to identify various vitamins, related deficiency diseases such as **scurvy** (vitamin C) and rickets (vitamin D) were dramatically cured. For the most part, as the vitamins were discovered, they were named alphabetically: A, B, C, D, E, and so on. Later, many substances originally classified as vitamins were found not to be essential for humans and were removed from the list. Other vitamins, thought at first to be only one chemical, turned out to be several chemicals, so the alphabetical names had to be broken down by numbers (B-6, B-12, and so on).

In addition to their use in correcting deficiency diseases, a few vitamins have also proved useful in treating several nondeficiency diseases. These medical applications require administration of megadoses, amounts well above typical human needs for the vitamins. Understand that claimed benefits from use of vitamin supplements, especially intakes in excess of the Upper Level (if established), should be viewed critically because unproved claims are common. Remember, whenever you take a supplement at high doses, you are taking it at a pharmacological dose—that of a drug. Expect side effects as you would from any drug.

Vitamins isolated from foods (*natural*) or manufactured in the laboratory (*synthetic*) are the same chemical compounds and work equally well in the body. Contrary to claims on social media and in health food stores, natural vitamins isolated from foods are, with few exceptions, no different than those labeled synthetic. Of note, however, the natural form of vitamin E is much more potent than the synthetic form. In contrast, synthetic folic acid, the form of the vitamin added to ready-to-eat breakfast cereals and flour, is almost twice as potent as the natural vitamin form.[1]

## ABSORPTION AND STORAGE OF VITAMINS IN THE BODY

The fat-soluble vitamins (A, D, E, and K) are absorbed in the presence of dietary fat. These vitamins then travel with dietary fats as part of chylomicrons through the bloodstream to reach body cells. Special carriers in the bloodstream help distribute some of these vitamins. Fat-soluble vitamins are stored mostly in the liver and fatty tissues.

When fat absorption is efficient, about 40% to 90% of the fat-soluble vitamins are absorbed. Anything that interferes with normal digestion and absorption of fats, however, also interferes with fat-soluble vitamin absorption. For example, people with cystic fibrosis, a disease that often hampers fat absorption, may develop deficiencies of fat-soluble vitamins. Some medications, such as certain weight-loss drugs, also interfere with fat absorption. Unabsorbed fat carries these vitamins to the large intestine, and they are excreted in the feces. People with fat-malabsorption conditions are especially susceptible to vitamin K deficiency because body stores of vitamin K are lower than those of the other fat-soluble vitamins. Vitamin supplements, taken under a primary care provider's guidance, are part of the treatment for preventing a vitamin deficiency associated with fat malabsorption. Finally, people who use mineral oil as a laxative risk fat-soluble vitamin deficiencies. Fat-soluble vitamins dissolve in the mineral oil, but the intestine does not absorb mineral oil. Hence, the fat-soluble vitamins are eliminated with the mineral oil in the feces.

Water-soluble vitamins are handled much differently than fat-soluble vitamins. After being ingested, the B vitamins from food are first broken down from their active **coenzyme** forms into free vitamins in the stomach and small intestine. The vitamins are then absorbed, primarily in the small intestine. Typically, about 50% to 90% of the water-soluble vitamins in the diet are absorbed, which means they have relatively high

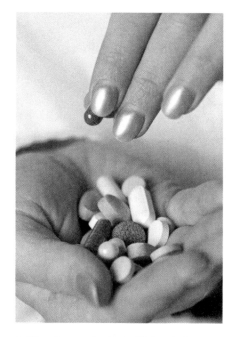

▲ Vitamins can be toxic if taken in large amounts as supplements. Liquid library/Getty Images

**coenzyme** An organic compound that combines with an inactive enzyme to form a catalytically active form. In this manner, coenzymes aid in enzyme function.

**bioavailability.** Water-soluble vitamins are transported to the liver via the hepatic portal vein and are distributed to body tissues. Once inside cells, the active coenzyme forms are resynthesized. Although some supplement manufacturers sell vitamins in their coenzyme forms, there is no benefit in consuming the coenzyme forms, as these are broken down during digestion and activated inside cells as needed.

Excretion of vitamins varies primarily on their solubility. Except for vitamin K, fat-soluble vitamins are not readily excreted from the body. Hence, toxicity can be an issue. Water-soluble vitamins are excreted based on **tissue saturation,** the degree to which the tissue vitamin stores are full. Tissue storage capacity is limited. As the tissues become saturated, the rate of excretion via the kidney increases sharply, preventing potential toxicity. Unlike other water-soluble vitamins, B-6 and B-12 are stored in the liver and not easily excreted in the urine.

In light of the limits of tissue saturation for many water-soluble vitamins, these vitamins should be consumed in the diet daily. However, an occasional lapse in the intake of water-soluble vitamins causes no harm. Symptoms of a vitamin deficiency occur only when that vitamin is lacking in one's dietary pattern, and the body stores are essentially exhausted. For example, for an average person, the dietary pattern must be devoid of thiamin for 10 to 14 days or lacking in vitamin C for 20 to 40 days before the first symptoms of deficiencies of these vitamins appear.

## VITAMIN TOXICITY

For most water-soluble vitamins, when you consume more than the RDA or AI, the kidneys efficiently filter the excess from the blood and excrete these compounds in urine. Notable exceptions are vitamin B-6 and vitamin B-12, which are stored in the liver. Although they are water-soluble, these two B vitamins may accumulate to toxic levels.

In contrast to the water-soluble vitamins, fat-soluble vitamins are not readily excreted, so some can easily accumulate in the body and cause toxic effects. Although a toxic effect from an excessive intake of any vitamin is theoretically possible, toxicity of the fat-soluble vitamin A is the most frequently observed. Vitamin A causes toxicity at intakes as little as 2 times the RDA. Vitamin E and the water-soluble vitamins niacin, vitamin B-6, and vitamin C can also cause toxic effects but only when consumed in very large amounts (15 to 100 times human needs). Overall, vitamins are unlikely to cause toxic effects unless taken in supplement (pill) form.

Some people believe that consuming vitamins far in excess of their needs provides them with extra energy, protection from disease, and prolonged youth. They seem to think that if a little is good, then more must be better. A *one-a-day* type of multivitamin and mineral supplement usually contains less than two times the Daily Values of its components, so daily use of these products is unlikely to cause toxic effects. However, consuming multiple pills, especially single-dose supplements such as vitamin A, can lead to toxicity.

## PRESERVATION OF VITAMINS IN FOODS

Good sources of vitamins can be found in all food groups, especially fruits and vegetables (Fig. 8-2). However, storage time and environmental factors can impact the vitamin content of foods. Fully ripe food contains more vitamins, but substantial amounts of vitamins can be lost from the time produce is harvested until it is eaten. Therefore, it is best to eat fresh produce as soon as possible after harvest. Food cooperatives, **community-supported agriculture (CSA),** and farmers' markets are great sources of freshly harvested fruits and vegetables. The water-soluble vitamins, particularly thiamin, vitamin C, and folate, can be destroyed with improper storage and excessive cooking. Heat, light, exposure to the air, cooking in water, and alkalinity are factors that can destroy vitamins. Some foods, like tomatoes, have improved bioavailability once cooked. See the *Ask the RDN* in this section for more details on raw versus cooked foods.

There are several steps you can take to preserve nutrients when you are purchasing, storing, and preparing fruits and vegetables (Table 8-1). Frozen vegetables and fruits

**Looking for a way to purchase fruits and vegetables at the grocery or farmers' market without wasting so many plastic bags?** Consider reusable produce bags. These environmentally friendly bags are durable, breathable, washable, and come in a variety of sizes. ©Anne Smith

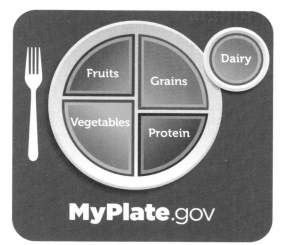

 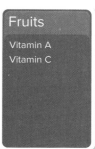  

| Grains | Vegetables | Fruits | Dairy | Protein |
|---|---|---|---|---|
| Folic acid | Folate | Vitamin A | Choline | Biotin |
| Niacin | Vitamin A | Vitamin C | Riboflavin | Choline |
| Riboflavin | Vitamin C | | Vitamin B-12 | Niacin |
| Thiamin | Vitamin K | | Vitamin D | Riboflavin |
| | | | | Thiamin |
| | | | | Vitamin B-6 |
| | | | | Vitamin B-12 |

**FIGURE 8-2** ◄ Certain food groups on MyPlate are especially rich sources of various vitamins and choline. Each may also be found in other MyPlate groups but in lower amounts. In addition to those vitamins listed here, pantothenic acid is present in moderate amounts in many groups, and vitamin E is abundant in plant oils.

MyPlate: U.S. Department of Agriculture

**TABLE 8-1** ■ **Tips for Preserving Vitamins in Fruits and Vegetables**

| Preservation Methods | Reason |
|---|---|
| Keep fruits and vegetables cool until eaten. | Enzymes in produce begin to degrade vitamins once harvested. Chilling slows down this process. |
| Refrigerate fruits and vegetables (except bananas, onions, potatoes, and tomatoes) in the vegetables drawer. | Nutrients keep best at temperatures near freezing, at high humidity, and away from air. |
| Trim, peel, and cut fruits and vegetables minimally and just prior to eating. | Oxygen breaks down vitamins faster when more of the food surface is exposed. Whenever possible, cook fruits and vegetables in their skins. |
| Microwave, steam, or stir-fry vegetables. | More nutrients are retained when there is minimal contact with water. |
| Minimize cooking time. | Prolonged cooking (slow simmering) and reheating reduce vitamin content. |
| Avoid adding fats to vegetables during cooking if you plan to discard the liquid. | Fat-soluble vitamins will be lost in discarded fat. If you want to add fats, do so after vegetables are fully cooked and drained. |
| Avoid adding baking soda to vegetables to enhance the green color. | Alkalinity destroys vitamin D, thiamin, and other vitamins. |
| Store canned and frozen fruits and vegetables carefully. | To protect canned foods, store them in a cool, dry location. To protect frozen foods, store them at 0°F (18°C) or colder. Eat within 12 months. |

Nutritional supplements are a growing industry in the U.S. **Do you think it is better to get nutrients from whole foods or dietary supplements?** Comstock Images

| ASK THE RDN | Raw Food Diet Plan |

**Dear RDN:** *My roommate follows a raw food diet and claims that cooking foods destroys key nutrients and makes foods toxic. Is this true?*

This highly restrictive diet includes consuming uncooked, unprocessed, mostly organic fruits, vegetables, and sprouted grains. Some may consume unpasteurized dairy products and raw eggs, meat, and fish. Advocates of this diet believe that heat destroys key nutrients and enzymes that are necessary to fight disease. Although some phytochemicals and nutrients are more bioavailable in raw form, there are many plant-based foods (e.g., tomatoes) that actually become more digestible once exposed to heat. In addition, many uncooked and unpasteurized products are sources of foodborne pathogens. Food poisoning is of particular concern for pregnant women, young children, older adults, immunocompromised individuals, and those with chronic disease.

Limitations: This diet is extremely difficult to maintain, especially when dining out, because of its many restrictions. Complete exclusion of food groups (e.g., dairy and meat) may lead to nutrient deficiencies, so supplements (e.g., iron, calcium, and vitamin B-12) may be necessary. Also, the cost and limited availability of organic foods may be prohibitive.

Strengths: This plan is compatible with vegetarian, vegan, and gluten-free dietary patterns. Raw food adopters often lose or maintain weight, given most of the foods allowed are low in calories and fat.

In sum, a plant-based dietary pattern is recommended for Americans; however, there is no evidence to support the nutritionally inadequate, extreme, and potentially harmful raw food practices.

Warm regards,

**Colleen Spees, PhD, RDN, LD, FAND**

Associate Professor, Researcher, The Ohio State University, and Author of *Contemporary Nutrition*

Ralphoto Studio

are often as nutrient-rich as freshly harvested ones because fruits and vegetables are typically frozen immediately after harvesting. As part of the freezing process, vegetables are quickly blanched in boiling water. Blanching destroys the enzymes that would otherwise degrade the vitamins over time. If produce will not be eaten within a few days of harvest, freezing is the best preservation method to retain nutrients.

## PHYTOCHEMICALS

In addition to the over 40 essential nutrients, there are thousands of other compounds in food. For many years, a great deal of nutrition research concentrated on gaining knowledge about carbohydrates, lipids, proteins, vitamins, and minerals. More recently, there is growing interest about the potential health benefits of other substances found in food. Foods that are sources of the chemicals that provide health benefits beyond being essential dietary nutrients are termed functional foods. Oatmeal is an example of a functional food as it contains soluble fiber that can lower cholesterol levels. Other foods are fortified or modified to improve health benefits. For example, some orange juice is fortified with calcium for bone health. Functional foods can be placed into two categories: (1) **zoochemicals,** health-promoting compounds found in animal food; and (2) **phytochemicals,** health-promoting compounds found in plant foods (*phyto* means "plant" in Greek). Examples of zoochemicals include omega-3 fatty acids in fish and prebiotics and probiotics in the GI tract of animals (including humans).

Phytochemicals are responsible for the unique colors, flavors, and odors observed in plants. For plants, phytochemicals serve as an environmental protective mechanism to help plants survive the elements (UV exposure, insects, and other predators).

**zoochemicals** Chemicals found in animal products that have health-protective actions.

Interestingly, these chemicals improve human health when dietary patterns high in plant foods are consumed. In addition to the carotenoids mentioned previously, some of the phytochemicals are listed in Table 8-2. Examples of phytochemicals include isoflavones in soy, sulforaphane in cruciferous vegetables, and resveratrol in grapes and wine. Other examples of foods that are rich sources of phytochemicals include fruits, vegetables, whole grains, peas, lentils, beans, herbs, spices, nuts, and seeds. The MyPlate food groups that contain phytochemicals include the fruits, vegetables, grains (whole grains), and protein (beans). See the *Farm to Fork* highlighting the many benefits of phytochemical-rich crucifer vegetables.

Thousands of phytochemicals have been identified yet few have been adequately studied. The full extent of these health benefits are just beginning to be elucidated. What we do know is that phytochemicals cannot be synthesized in the body, so we must obtain them from food; however, they are not considered essential nutrients because a deficiency disease is not observed when they are removed from the dietary pattern. Thus, even though you may see the word *phytonutrient* in some sources, *phytochemical* remains the most accurate term.

**Phytochemical Functions.** Although the study of the metabolic actions of phytochemicals is relatively new, numerous mechanistic protective health benefits of these plant-based foods have been noted, beyond those conferred by their vitamin and mineral contents:[2]

- Support the immune system
- Reduce inflammation and blood pressure
- Prevent DNA damage and aid in DNA repair
- Reduce oxidative damage to cells
- Promote cardiovascular, neurocognitive, eye, and bone health
- Regulate intracellular signaling of hormones and gene expression
- Activate insulin receptors
- Inhibit the initiation and proliferation of cancer, and stimulate spontaneous cell death
- Alter the absorption, production, and metabolism of cholesterol
- Mimic or inhibit hormones and enzymes
- Decrease the formation of blood clots

**Phytochemical Recommendations.** There are no specific dietary recommendations for the amount of phytochemicals that should be consumed, except that they should be consumed as food. At present, we know that phytochemicals have protective functions with minimal side effects when consumed naturally in a variety of foods. Therefore, it is wise to eat a wide variety of whole plant foods to obtain the optimum amount of macronutrients, vitamins, minerals, fiber, and phytochemicals. A combination of different plant foods will also increase the **antioxidant** capacity of the total dietary pattern, and there is no doubt that we need to consume a greater variety and quantity of antioxidant-rich foods.

## FARM to FORK  Crucifers

©Mary-Jon Ludy, Bowling Green State University, Garden of Hope Images

Cruciferous vegetables are cool-weather vegetables with four-petaled flowers that resemble a cross. The most common crucifers include broccoli, cauliflower, cabbage, bok choy, Brussels sprouts, and green leafy vegetables such as kale and arugula. Ounce for ounce, crucifers have it all—vitamins, minerals, fiber, and an abundance of disease-fighting phytochemicals.

### Grow

- The freshest and most nutrient-dense crucifers always come straight from the garden. Sadly, up to 80% of nutrients are lost in the transport from farm to fork.
- For crucifer container gardening, estimate 1 broccoli plant per 5 gallons of soil (see https://ofbf.org/2008/06/05/vegetable-container-gardens).
- Most crucifers require full sun and moist, fertile soil that's slightly acidic. Broccoli can germinate in soil with temperatures as low as 40°F, but check online for growing recommendations in your region.

### Shop

- Buy broccoli with dark green crowns, tight bud heads, and moist and firm stems. Avoid yellowing and dry produce that is not kept chilled.
- Intact heads of broccoli are not only more nutritious but less expensive than precut florets.

### Store

- To preserve the nutrients, antioxidants, and phytochemicals in crucifers, keep them cool and eat within days of harvest.
- If storing for any period, place in a plastic bag with about 20 pinprick holes (*micro-perforated* bag) in the crisper drawer.
- Prior to freezing broccoli, it is essential to *blanch* the produce to deactivate its enzymes (see http://ohioline.osu.edu/factsheet/HYG-5333).

### Prep

- In most cases, the leaves or flower buds of crucifers are eaten, but there are a few where either the roots or seeds are also eaten.
- Compounds known as glucosinolates are responsible for much of the bitterness and the health benefits offered by these amazing plants.
- Cooking and preservation expose crucifers to heat, oxygen, and light—all significantly decreasing nutrient and phytochemical content. Lower temperatures as well as drier and shorter cooking times reduce losses.
- Steaming for less than 5 minutes or sautéing with a small amount of extra virgin olive oil is recommended.

Source: Robinson J: Incredible crucifers: Tame their bitterness and reap the rewards, in *Eating on the Wild Side: The Missing Link to Optimum Health.* New York: Little, Brown and Company, 2013.

©Mary-Jon Ludy, Bowling Green State University, Garden of Hope Images

Blueberries are rich in health-promoting phytochemicals. They have been shown to have anticancer effects and therefore could be an important part of dietary cancer prevention strategies. **What is your favorite way to incorporate phytochemicals into your daily menu?** Lifesize/Getty Images

**TABLE 8-2 ■ Food Sources of Select Phytochemical Compounds**

| Food Sources | Phytochemical |
|---|---|
| Blueberries, strawberries, raspberries, grapes, apples, bananas, nuts | Polyphenols |
| Chili peppers | Capsaicin |
| Citrus fruit, onions, apples, grapes, red wine, tea, chocolate, tomatoes | Flavonoids |
| Cruciferous vegetables (broccoli, cabbage, kale) | Indoles |
| Cruciferous vegetables, especially broccoli | Isothiocyanates |
| Flaxseed, berries, whole grains | Lignans |
| Garlic, onions, leeks | Allyl sulfides/organosulfurs |
| Garlic, onions, licorice, legumes | Saponins |
| Grapes, peanuts, red wine | Resveratrol |
| Onions, bananas, oranges (small amounts) | Fructooligosaccharides |
| Orange, red, and yellow fruits and vegetables (egg yolks are a source as well) | Carotenoids (e.g., lycopene) |
| Oranges, lemons, grapefruit | Monoterpenes |
| Red, blue, and purple plants (blueberries, eggplant) | Anthocyanosides |
| Soybeans, other legumes | Isoflavones |
| Soybeans, other legumes, cucumbers, other fruits and vegetables | Phytosterols |
| Tea | Catechins |

## ✓ CONCEPT CHECK 8.1

1. Coenzyme Q (CoQ) is an organic compound that is required for the electron transport chain and has some antioxidant functions. This compound is synthesized within cells, and under most circumstances, the body synthesizes enough CoQ to meet its needs. Is CoQ a vitamin? Why or why not?

2. What is a megadose? Are there any negative consequences of consuming megadoses of vitamins? Are there any situations in which megadoses of vitamins are useful?

3. List at least three differences between fat-soluble and water-soluble vitamins.

4. List three ways to preserve vitamin content when storing, preparing, or cooking foods.

5. Give one example of a functional food.

6. List 5 foods that are rich sources of phytochemicals.

## 8.2 Vitamin A (Retinoids) and Carotenoids

Vitamin A was the first fat-soluble vitamin to be recognized as an important component of food essential for human health. Almost all (90%) of vitamin A is stored in the liver; the remaining 10% is in adipose tissue, kidneys, and the lungs. Either a deficiency or toxicity can cause severe problems, and there is a narrow range of optimal intakes between these two states.

Vitamin A is in a group of compounds known as **retinoids.** There are three active forms of vitamin A: (1) **retinol**; (2) **retinal**; and (3) **retinoic acid.** Collectively, these three are often called *preformed* vitamin A. They are only naturally present in animal products. When retinol is stored, it is *esterified* (joined to a fatty acid) and becomes **retinyl.** In supplements, you will often find vitamin A listed as *retinyl acetate* or *retinyl palmitate*.

Besides the retinoids, which come from animal sources, many foods of plant origin contain carotenoids, some of which can be converted into vitamin A in the body. Note that carotenoids are not essential nutrients; they are categorized as phytochemicals. Although hundreds of carotenoids have been identified, just three can be converted to retinol in the body: alpha carotene, **beta-carotene,** and beta-cryptoxanthin. Because these carotenoids can be turned into vitamin A, they are termed **provitamin A.** Of these three, only beta-carotene serves as a significant source of vitamin A. Neither the absorption of carotenoids nor the conversion of the provitamin A carotenoids into vitamin A is an efficient process. Other carotenoids that may play a role in human health but are not vitamin A precursors include lycopene, zeaxanthin, and lutein.

**retinoids** Chemical forms of preformed vitamin A found in animal foods.

**retinol** Alcohol form of vitamin A.

**retinal** Aldehyde form of vitamin A.

**retinoic acid** Acid form of vitamin A.

**retinyl** Storage form of vitamin A.

**beta-carotene** The orange-yellow pigment in carrots; beta-carotene is the only carotenoid that can be sufficiently absorbed and converted into retinol in the body.

**provitamin A** A substance that can be converted into vitamin A.

### FUNCTIONS OF VITAMIN A AND CAROTENOIDS

**Epithelial Cell Health and Immune Function.** For many years, vitamin A has been dubbed the *anti-infection* vitamin. Vitamin A maintains the health of epithelial cells, which line the surfaces of the lungs, intestines, stomach, vagina, urinary tract, and bladder, as well as those of the eyes and skin.[3] Retinoic acid is required for immature epithelial cells to develop into mature, functional epithelial cells. Epithelial tissues, such as those that form the skin and GI tract, serve as important barriers to infection. Vitamin A also supports the activity of certain immune system cells, specifically the T-lymphocytes, or T-cells.

**Eye Health and Vision.** The link between vitamin A and night vision has been known since ancient Egyptians used juice extracted from liver to cure **night blindness.** Vitamin A performs important functions in light–dark vision and, to a lesser extent, color vision. Light entering the eye reaches a lining called the **retina.** The retina consists of various cells, such as rods, cones, and nerve cells. Rods detect black and white, and they are responsible for night vision. Cones are responsible for color vision. Rods and cones require vitamin A for normal function. One form of vitamin A (retinal) allows certain cells in the eye to adjust to dim light (such as after seeing the headlights of an oncoming car; Fig. 8-3).

Some carotenoids are also important for vision. The macula (also known as the *macula lutea,* meaning "yellow spot") is in the central area of the retina and is responsible for the most detailed central vision. It contains the carotenoids lutein and zeaxanthin in high enough concentrations to impart a yellow color. *Age-related macular degeneration* (Fig. 8-4), the leading cause of blindness among older adults in North America, occurs because of changes in this macular area of the retina. Research studies indicate that higher intakes of these carotenoids may help to prevent or slow the progression of age-related macular degeneration.[4] Carotenoids may also decrease the risk of cataracts in the eyes. Although supplements containing carotenoids are marketed to older adults as a way to protect eye health, current research points to the safety and benefits of *dietary* sources of carotenoids. Overall, the richest sources of lutein and zeaxanthin are green leafy vegetables (Table 8-3).

**night blindness** Vitamin A deficiency disorder that results in loss of the ability to see under low-light conditions.

**retina** A light-sensitive lining in the back of the eye. It contains retinal.

**FIGURE 8-3** ▶ Vitamin A functions to maintain vision. Light enters the eye through the cornea and lens, and then hits the retina. The light reacts with vitamin A–containing rhodopsin, which is stored in the rod cells of the retina. Rod cells allow us to see black-and-white images. When light reacts with rhodopsin, retinal is cleaved from rhodopsin (bleaching), a process that stimulates an electrical impulse to the brain. A new molecule of vitamin A then combines with opsin to regenerate rhodopsin. The yellow background indicates the bleaching events that occur in the light; the gray background indicates the regenerative events that can occur in either light or dark conditions.

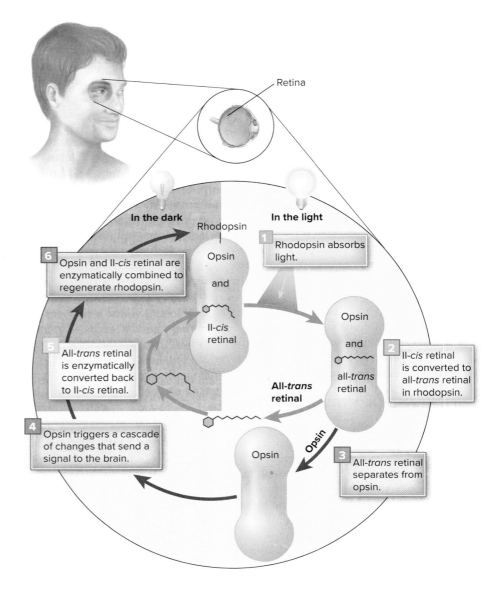

Retina

**In the dark**

Rhodopsin

Opsin

and

Il-*cis* retinal

**6** Opsin and Il-*cis* retinal are enzymatically combined to regenerate rhodopsin.

**5** All-*trans* retinal is enzymatically converted back to Il-*cis* retinal.

**4** Opsin triggers a cascade of changes that send a signal to the brain.

**All-*trans* retinal**

Opsin

Opsin

**In the light**

**1** Rhodopsin absorbs light.

Opsin

and

all-*trans* retinal

**2** Il-*cis* retinal is converted to all-*trans* retinal in rhodopsin.

**3** All-*trans* retinal separates from opsin.

**FIGURE 8-4** ▲ The blurry center of the image simulates the vision of a person with macular degeneration. ©National Eye Institute, National Institutes of Health

**TABLE 8-3** ■ **Vegetables Rich in Lutein and Zeaxanthin**

| Vegetables (serving size) | Lutein and Zeaxanthin (mg) |
|---|---|
| Broccoli (½ cup cooked) | 0.8 |
| Brussels sprouts (½ cup cooked) | 1.0 |
| Collard greens (½ cup cooked) | 3.9 |
| Kale (½ cup cooked) | 11.9 |
| Peas (½ cup cooked) | 1.9 |
| Romaine lettuce (1 cup raw) | 1.5 |
| Spinach (½ cup cooked) | 10.2 |
| Spinach (1 cup raw) | 3.1 |
| Swiss chard (½ cup cooked) | 8.0 |
| Zucchini (½ cup cooked) | 2.0 |

**Growth, Development, and Reproduction.** Vitamin A participates in the processes of growth, development, and reproduction in several ways. At the genetic level, vitamin A binds to receptors on DNA to increase synthesis of a variety of proteins. Some of these proteins are required for growth. During early fetal growth, vitamin A functions in the differentiation and maturation of cells that will ultimately form tissues and organs. For bones to grow and elongate, old bone must be remodeled (broken down) so that new bone can be formed. Vitamin A assists with the breakdown and formation of healthy bone tissue. Adequate intake of vitamin A is needed for reproduction; it aids in sperm production (associated with its epithelial role) and in a normal reproductive cycle for women.

**Cardiovascular Disease Prevention.** Carotenoids may play a role in preventing cardiovascular disease in individuals at high risk. This role may be linked to carotenoids' ability to inhibit the oxidation of low-density lipoproteins (LDLs). Until definitive studies are complete, many scientists recommend that we consume a total of at least five servings of a combination of fruits and vegetables per day as part of an overall effort to reduce the risk of cardiovascular disease. Other phytochemicals, including flavonoids, phenolic acids, and phytoestrogens, may have a more significant effect in cardiovascular disease prevention than carotenoids.[5] Until we have more answers, focus on an eating pattern rich in phytochemicals from whole foods.

**Cancer Prevention.** Vitamin A and the carotenoids have potential benefits but also potential dangers where cancer prevention is concerned. Vitamin A plays a role in cellular differentiation and embryonic development. Numerous studies have found that dietary patterns rich in provitamin A carotenoids are associated with a lower risk of skin, lung, bladder, and breast cancers. Still, because of the potential for toxicity, unsupervised use of megadose vitamin A or carotenoid supplements to reduce cancer risk is not advised and can be potentially dangerous.

## VITAMIN A DEFICIENCY

Typical American dietary patterns contain plentiful sources of preformed vitamin A (e.g., meats, eggs, and fortified milk) so most Americans are at low risk for vitamin A deficiency. However, in other parts of the world, vitamin A deficiency is a major public health concern. Across the globe, about one-third of children suffer from vitamin A deficiency.[6] Vitamin A deficiency results in three main problems: impaired vision, weakened immune function, and stunted growth. In severe cases, vitamin A deficiency contributes to death.

You learned about the important role of vitamin A in vision. In the early stages of vitamin A deficiency, the cells in the retina of the eye cannot quickly adjust from bright to dim light and the individual suffers from night blindness. Vitamin A deficiency can also lead to an eye disease called **xerophthalmia.** As vitamin A deficiency progresses, the cells that line the cornea of the eye (the clear window of the eye) lose the ability to produce mucus. The eye then becomes very dry. Eventually, dirt particles scratch and scar the dry surface of the eye. This can lead to accumulations of dead cells and secretions on the surface of the eye, which are called **Bitot's spots** (Fig. 8-5). Over time, impairments in the visual cycle and damage to the cornea lead to blindness. In fact, vitamin A deficiency is the leading cause of preventable blindness among children worldwide.[6]

So, for the eye, unhealthy epithelial tissue contributes to blindness. Lack of vitamin A also affects the integrity of other epithelial tissues, such as the skin, respiratory tract, and GI tract. **Hyperkeratosis** is a condition in which skin cells produce too much keratin, blocking the hair follicles and causing *gooseflesh* or *toadskin* appearance. The excessive keratin in these skin cells causes the skin to be hard and dry. The breakdown of epithelial tissues of the skin, respiratory tract, and GI tract also makes it easier for pathogens to enter the blood and cause infections. To further complicate immune function,

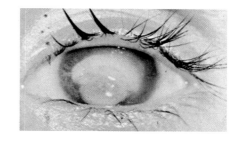

**FIGURE 8-5** ▲ Vitamin A deficiency eventually leads to blindness. The buildup of scar tissue on the surface of the eye leads to Bitot's spots. This problem is commonly seen today in Southeast Asia. Courtesy of Dr. Alfred Sommer

**xerophthalmia** Hardening of the cornea and drying of the surface of the eye, which can result in blindness as a result of vitamin A deficiency.

**Bitot's spots** Dry, foamy spots made from an accumulation of keratin (a protein) on the surface of the eye; caused by vitamin A deficiency.

**hyperkeratosis** A condition in which patches of skin become thicker, rougher, or drier than usual; a possible consequence of vitamin A deficiency.

**biofortification** Use of selective breeding or other biotechnology to enhance the nutrient content of crops.

Lycopene is the carotenoid that gives foods such as tomatoes, watermelon, guava, and pink grapefruit their reddish color. **If you want to absorb the most lycopene from tomatoes, is it better to eat them raw or cooked?** ©Sarah Rusnak

Inuits long knew and explorers soon learned to avoid eating the liver of polar bears. Just 4 ounces of polar bear liver will deliver a toxic dose of 1.36 million RAE of vitamin A. **How does this dose compare to the RDA?** M G Therin Weise/Getty Images

the ability of white blood cells to fight infection is impaired by lack of vitamin A. Vitamin A–deficient humans have an increased infection rate, but when they are supplemented with vitamin A, the immune response improves.

Vitamin A deficiency can also affect normal growth and development. This can lead to problems with fetal development during pregnancy. During infancy and childhood, vitamin A deficiency may lead to stunted growth.

Globally, attempts to address vitamin A deficiency include promoting breastfeeding, providing dietary supplements, and fortifying commonly consumed foods (e.g., sugar, margarine, and monosodium glutamate) with vitamin A. Most recently, **biofortification** of crops has become a viable technique to provide vitamin A to populations who need it most.[7]

## GETTING ENOUGH VITAMIN A AND CAROTENOIDS

Preformed vitamin A (e.g., retinol, retinal, and retinoic acid) is found in liver, fish, fish oils, fortified milk, butter, yogurt, and eggs (Fig. 8-6). Margarine and other plant oil spreads are also fortified with vitamin A. For individuals who choose dairy alternatives, fortified soy beverages (commonly known as "soy milk") and soy yogurt also contain vitamin A.

About 70% of the vitamin A in the typical North American dietary pattern comes from preformed vitamin A sources, whereas provitamin A carotenoids dominate the eating patterns among people in less developed areas of the world. The provitamin A carotenoids are mainly found in dark-green and yellow-orange vegetables and some fruits. Carrots, spinach and other leafy greens, winter squash, sweet potatoes, broccoli, mangoes, cantaloupe, peaches, and apricots are examples of such sources. Beta-carotene accounts for some of the orange color of carrots. Green vegetables also contain provitamin A. The yellow-orange beta-carotene is masked by dark-green chlorophyll pigments. Green, leafy vegetables, such as spinach and kale, have high concentrations of lutein and zeaxanthin. Tomato products contain significant amounts of lycopene. Cooking food and consuming food with small amounts of healthy fat improve the bioavailability of carotenoids. In raw fruits and vegetables, carotenoids are bound to proteins. Cooking disrupts this protein bond and frees the carotenoid for better absorption, and fat increases bioavailability.

The RDA for vitamin A is expressed in retinol activity equivalents (RAE). These RAE units consider the activity of both preformed vitamin A and the carotenoids that are converted into vitamin A in humans. The total RAE value for a food is calculated by adding the concentration of preformed vitamin A to the amount of provitamin A carotenoids in the food that will be converted into vitamin A. There is no separate DRI for beta-carotene or any of the other phytochemicals.

The dietary patterns of North American adults typically contain adequate vitamin A. Most adults in North America have liver reserves of vitamin A three to five times higher than needed to provide good health. Thus, the use of vitamin A supplements by most people is unnecessary. Populations in North America that may be at risk for vitamin A deficiency include those with low fruit and vegetable intakes (e.g., some children, older adults, and food-insecure individuals); people with alcoholism or liver disease; or people with severe fat malabsorption.

## AVOIDING TOO MUCH VITAMIN A AND CAROTENOIDS

Intakes in excess of the UL for vitamin A are linked to birth defects and liver toxicity. Other possible side effects include an increased risk of hip fracture and poor pregnancy outcomes.

During the early months of pregnancy, a high intake of preformed vitamin A is especially dangerous because it may cause fetal malformations and spontaneous abortions. This is because vitamin A binds to DNA and influences cell development. The Food and Drug Administration (FDA) recommends that women of childbearing age limit their overall intake of preformed vitamin A from food and dietary supplements to no

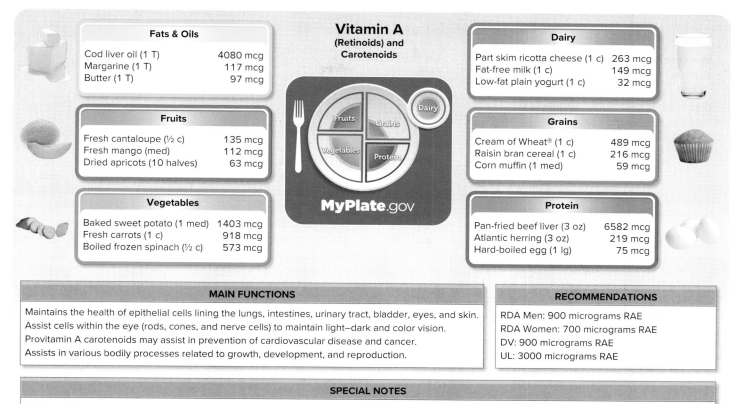

**FIGURE 8-6** ▲ Food sources of vitamin A and carotenoids. The fill of the background color (none, 1/3, 2/3, or completely covered) within each food group on MyPlate indicates the average nutrient density for vitamin A and provitamin A carotenoids in that group. The figure shows the vitamin A content of several foods from each food group. Overall, the fruits and vegetables groups provide many rich sources of carotenoids, whereas fortified dairy products and certain choices in the protein group are good sources of preformed vitamin A. The grains group also contains some foods that are nutrient dense because they are fortified with vitamin A. butter: 2/James Worrell/Ocean/CORBIS; milk: NIPAPORN PANYACHAROEN/Shutterstock; cantaloupe Renne Comet/National Cancer Institute; muffin: Anton Prado PHOTO/Shutterstock; sweet potato: lynx/iconotec.com/Glow Images; eggs: dancestrokes/123RF; MyPlate: U.S. Department of Agriculture

Sources: Office of Dietary Supplements, Dietary Supplements Fact Sheets, available from https://ods.od.nih.gov/factsheets/list-all; USDA FoodData Central, available from https://fdc.nal.usda.gov.

more than about 100% of the Daily Value. It is also important to limit consumption of rich food sources of preformed vitamin A, such as liver. In addition, medications that contain derivatives of vitamin A should be avoided during pregnancy. These precautions apply to women who are pregnant and also those who may possibly become pregnant; vitamin A is stored in the body for long periods, so women who ingest large amounts during the months before pregnancy place their **fetus** at risk.

In contrast, ingesting large amounts of vitamin A–yielding carotenoids does not appear to cause toxic effects. A high carotenoid concentration in the blood, *hypercarotenemia*, can occur if someone routinely consumes large amounts of carrots or takes pills containing beta-carotene (> 30 milligrams daily) or if infants eat an excessive amount of squash. The skin turns yellow-orange, particularly the palms of the hands and soles of the feet. It differs from jaundice, an accumulation of bilirubin caused by underlying disease, usually involving the liver. In jaundice, the yellow discoloration extends to the sclera (whites) of the eye, whereas in hypercarotenemia it does not. Hypercarotenemia does not appear to cause harm and disappears when carotenoid intake decreases. Dietary carotenoids do not produce toxic effects because: (1) their rate of conversion into vitamin A is relatively slow and regulated; and (2) the efficiency of carotenoid absorption from the small intestine decreases markedly as oral intake increases.

**fetus** The developing organism from about the beginning of the ninth week after conception until birth.

Biofortification increases the nutrient density of food crops through conventional plant breeding, enhanced agronomic practices, or biotechnology without sacrificing consumer and farmer preferences. So far, its application has been applied to iron-biofortification of beans, cowpea, and pearl millet; zinc-biofortification of maize, rice, and wheat; and pro-vitamin A carotenoid-biofortification of cassava, maize, rice, and sweet potato.

Source: WHO Biofortification of Crops with Minerals and Vitamins.

### 🌍 Sustainable Solutions

**7-dehydrocholesterol** Precursor of vitamin D found in the skin.

**vitamin D₃** Previtamin form of vitamin D synthesized in the skin and found naturally in some animal sources, including fish and egg yolks; also called *cholecalciferol*.

**25-hydroxyvitamin D₃** Intermediate form of vitamin D found in blood; also called *calcidiol* or *calcifediol* or abbreviated *25(OH)D₃*.

**1,25-dihydroxyvitamin D₃** Biologically active form of vitamin D that regulates blood calcium levels; also called *calcitriol* or abbreviated *1,25(OH)D₃*.

**vitamin D₂** Form found in nonanimal sources, such as in some mushrooms. Also synthetically produced and included in many supplements. Also called *ergocalciferol*.

### ✔ CONCEPT CHECK 8.2

1. What are the consequences of vitamin A deficiency?
2. How are the carotenoids related to vitamin A?
3. Name two carotenoids and identify a good food source of each.
4. List two rich food sources of retinoids.

# 8.3 Vitamin D (Calciferol or Calcitriol)

Vitamin D is a fat-soluble vitamin with two unique qualities. First, vitamin D is the only nutrient that is also a hormone. A hormone is a compound manufactured by one organ of the body that then enters the bloodstream and has a physiological effect on another organ or tissue. The cells that participate in the synthesis of vitamin D (skin, liver, and kidneys) are different from the cells that respond to vitamin D, namely bone and intestinal cells; therefore, vitamin D is considered a hormone.[8]

Second, vitamin D is the only nutrient that can be produced in the skin upon exposure to ultraviolet light. The human production of vitamin D begins when the ultraviolet B (UVB) rays of the sun convert a cholesterol precursor of vitamin D (**7-dehydrocholesterol**) in the skin into an inactive form of **vitamin D₃ (cholecalciferol).** As illustrated in Figure 8-7, this compound must be activated to **25-hydroxyvitamin D₃ (calcidiol or calcifediol)** in the liver and to **1,25-dihydroxyvitamin D₃ (calcitriol)** in the kidney before it can function as the vitamin D hormone. **Vitamin D₂ (ergocalciferol)** is a synthetic form of vitamin D that will be discussed later.

Our ability to absorb UVB rays and make vitamin D in the skin is affected by many factors. Dark skin pigmentation, geographic latitude, time of day, season of the year, weather conditions, genetics, and amount of body surface covered with clothing or sunscreen affect the skin's exposure to UVB rays and therefore influence vitamin D synthesis. Something as simple as complete cloud cover or severe pollution can reduce UVB rays by 50%. In addition, UVB rays will not penetrate glass. Aging reduces our ability to synthesize vitamin D—as much as 70% by age 70. Exposure of hands, face, and arms for about 10 minutes daily will support adequate vitamin D synthesis for most healthy children and adults. However, older adults and individuals with dark skin pigmentation require about three to five times this amount of sun exposure to synthesize an equivalent amount of vitamin D.

### FUNCTIONS OF VITAMIN D

**Blood Calcium Regulation.** The main function of vitamin D (calcitriol) is to maintain the normal range of calcium and phosphorus in the blood. Together with the hormones parathyroid hormone (PTH) and calcitonin, vitamin D closely maintains blood calcium in a narrow range. This tight regulation ensures that an appropriate amount of calcium is available to all cells. Vitamin D regulates calcium in three ways: (1) it influences the absorption of calcium and phosphorus from the small intestine; (2) in combination with PTH and calcitonin, it regulates calcium excretion via the kidney; and (3) it affects the deposition or withdrawal of minerals from the bones (Fig. 8-8).

**Gene Expression and Cell Growth.** The biological effects of vitamin D appear to extend beyond its role in bone health.[9] Vitamin D is involved in gene expression and cell growth; it binds to and subsequently affects cells of the immune system, brain and nervous system, parathyroid gland, pancreas, skin, muscles, and reproductive organs. In fact, vitamin D is considered one of the most potent regulators of cell growth, capable of influencing normal development of some cells (e.g., skin, colon, prostate,

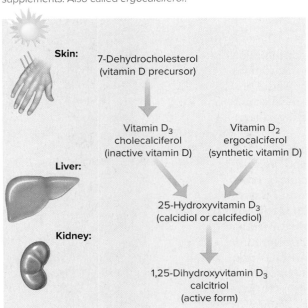

**FIGURE 8-7** ▲ A precursor to vitamin D is synthesized when skin is exposed to sunlight. Previtamin D must be further modified by the liver and kidney for maximal activity.

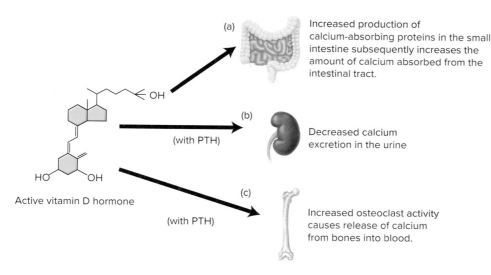

(a) Increased production of calcium-absorbing proteins in the small intestine subsequently increases the amount of calcium absorbed from the intestinal tract.

(b) (with PTH) Decreased calcium excretion in the urine

Active vitamin D hormone

(c) (with PTH) Increased osteoclast activity causes release of calcium from bones into blood.

**FIGURE 8-8** ◄ Vitamin D regulates blood calcium. When blood levels of calcium begin to drop from the normal range, PTH stimulates the synthesis of the most active form of vitamin D (calcitriol) by the kidney. Calcitriol acts at three different sites to increase blood calcium: (a) small intestine; (b) kidney; and (c) bone. When blood calcium levels increase above the normal range, PTH release is inhibited and calcitonin is released, which has the opposite effects of PTH.

and breast), which could in turn reduce cancer risk. Although the results of some studies suggest that vitamin D has a role in the prevention of several chronic diseases, an analysis of the existing research showed that a clear role of vitamin D does not exist for any of the 137 health outcomes or diseases studied.[9]

## VITAMIN D DEFICIENCY

When vitamin D levels are adequate, about 30% to 40% of dietary calcium is absorbed by the small intestine. If blood levels of vitamin D are low, the small intestine is able to absorb only about 10% to 15% of calcium consumed, which is not enough to maintain the calcium requirements for bone health and other functions. Subsequently, calcium and phosphorus deposition during bone synthesis is reduced, resulting in weaker bones that do not develop properly. Vitamin D deficiency can be traced to inadequate dietary intake, poor absorption (e.g., fat malabsorption in children with cystic fibrosis), specific genetic variants, altered metabolism (e.g., liver or kidney disease), or inadequate sun exposure. Vitamin D deficiency has been known to be a problem in individuals with dark skin and older adults. More recent research found that U.S. adults who were black, less educated, poor, obese, tobacco users, physically inactive, and infrequent milk consumers had a higher prevalence of vitamin D deficiency.[10]

Vitamin D deficiency can occur at any time, but when it occurs during infancy and early childhood, the resulting disease is known as **rickets**. The skeletal abnormalities of rickets include bowed legs, thick wrists and ankles, curvature of the spine, a pigeon chest (chest protrudes above the sternum), skull malformations, and pelvic deformities. Vitamin D deficiency is a concern among young infants who are not receiving: (1) vitamin D supplementation; (2) sufficient exposure to sunlight; or (3) food fortified with vitamin D (e.g., cereal and dairy products). In addition, consumption of non–cow's milk beverages with low vitamin D content has been associated with a decrease in blood vitamin D levels in early childhood compared to levels in children consuming cow's milk fortified with higher levels of vitamin D.[11] Studies show a high prevalence of vitamin D deficiency among children and adolescents worldwide.

**Osteomalacia,** which means "soft bone," is an adult disease comparable to rickets. It can result from inadequate calcium intake, inefficient calcium absorption in the intestine, or poor conservation of calcium by the kidneys. It occurs most commonly in people with kidney, stomach, gallbladder, or intestinal disease (especially when most of the intestine has been removed) and in people with cirrhosis of the liver. These diseases affect both vitamin D activation and calcium absorption, leading to a decrease in bone mineral density. Bones become porous and weak and break easily. Research shows that treatment with 10 to 20 micrograms per day of vitamin D, in conjunction with adequate dietary calcium, can reduce fracture risk in older adults. Thus, vitamin D is just as important as calcium when it comes to bone health.

▶ **Factors That Impair Vitamin D Status**
Aging
Dark skin pigmentation
Exclusive breastfeeding or low consumption of infant formula
Fat malabsorption
- Liver disease
- Cystic fibrosis
- Weight-loss medications
Inadequate dietary intake
Inadequate sun exposure
- Northern latitudes
- Concealing clothing (e.g., robes/veils)
- Air pollution (i.e., smog)
- Sunscreen with SPF > 8
- Excessive time spent indoors (e.g., due to health, work, or environmental conditions)
Kidney diseases
Liver diseases
Obesity

**rickets** A disease characterized by poor mineralization of newly synthesized bones because of low calcium content. Arising in infants and children, this deficiency is caused by insufficient amounts of vitamin D in the body.

**osteomalacia** Adult form of rickets. The bones have low mineral density and consequently are at risk for fracture.

## GETTING ENOUGH VITAMIN D

There are two forms of vitamin D in foods: vitamin $D_2$ and vitamin $D_3$. Vitamin D2 (ergocalciferol) is a synthetic product derived from the irradiation of plant sterols (ergosterol) and is used in some supplements. Vitamin $D_3$ (cholecalciferol)—the form synthesized in the human body and found naturally in a few foods—is more commonly used in supplements and fortified foods. Both forms of vitamin D must be modified by chemical reactions that occur in the kidney and liver before they can be active in the body.

Sunlight is the best source of vitamin D. Unlike with supplements, you can never receive a toxic dose. With just 10 minutes of exposure of the arms and legs to sunlight, it is estimated that we synthesize roughly 75 micrograms of vitamin D. To some extent, the vitamin D synthesized on sunny days can be stored "for a rainy day" in the liver and adipose cells. However, most people now limit sun exposure to decrease their risk of skin cancer. Unless you live in a year-round sunny climate and find yourself outside most days between 10:00 A.M. and 3:00 P.M., you are probably not meeting your vitamin D needs from sun exposure alone. Therefore, people residing in northern climates and with limited sun exposure in general should find dietary sources, especially in the winter months.

The RDA for vitamin D is based on a daily intake that is sufficient to maintain bone health and normal calcium metabolism, assuming minimal sun exposure. The *Dietary Guidelines* recommends that all children over age 9 and adults should choose more foods that provide vitamin D, a nutrient of concern in American dietary patterns.

Dietary sources of vitamin D are limited, with few foods being naturally high in vitamin D (Fig. 8-9). However, fortified foods are also an effective way to add vitamin D to your dietary pattern. Use of vitamin D–fortified milk began in the 1930s and effectively wiped out rickets in the United States. Because vitamin D is a fat-soluble vitamin, there is slightly more in a cup of whole milk (3.2 micrograms) compared to 1% (3 micrograms) or fat-free

▲ Fatty fish, such as salmon and trout, are considered the richest natural sources of vitamin D. Eggs are another natural source, with 1 large egg yolk delivering 1 microgram. Although butter, liver, and margarine contain some vitamin D, these foods are not considered significant sources because large servings must be eaten to obtain an appreciable amount of the vitamin. Catherine Markelov/happylark/123RF

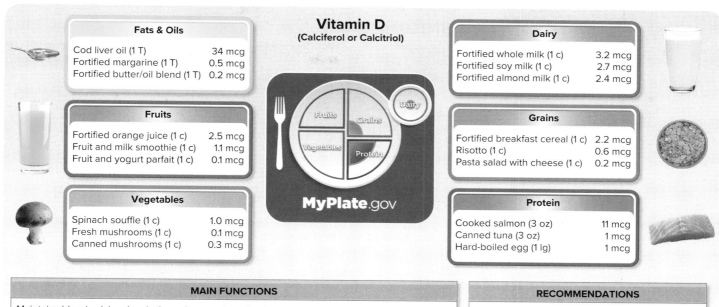

**Vitamin D**
(Calciferol or Calcitriol)

**Fats & Oils**

| | |
|---|---|
| Cod liver oil (1 T) | 34 mcg |
| Fortified margarine (1 T) | 0.5 mcg |
| Fortified butter/oil blend (1 T) | 0.2 mcg |

**Fruits**

| | |
|---|---|
| Fortified orange juice (1 c) | 2.5 mcg |
| Fruit and milk smoothie (1 c) | 1.1 mcg |
| Fruit and yogurt parfait (1 c) | 0.1 mcg |

**Vegetables**

| | |
|---|---|
| Spinach souffle (1 c) | 1.0 mcg |
| Fresh mushrooms (1 c) | 0.1 mcg |
| Canned mushrooms (1 c) | 0.3 mcg |

**Dairy**

| | |
|---|---|
| Fortified whole milk (1 c) | 3.2 mcg |
| Fortified soy milk (1 c) | 2.7 mcg |
| Fortified almond milk (1 c) | 2.4 mcg |

**Grains**

| | |
|---|---|
| Fortified breakfast cereal (1 c) | 2.2 mcg |
| Risotto (1 c) | 0.6 mcg |
| Pasta salad with cheese (1 c) | 0.2 mcg |

**Protein**

| | |
|---|---|
| Cooked salmon (3 oz) | 11 mcg |
| Canned tuna (3 oz) | 1 mcg |
| Hard-boiled egg (1 lg) | 1 mcg |

**MyPlate**.gov

**MAIN FUNCTIONS**

Maintains blood calcium levels through several mechanisms.
Contributes to bone health by affecting the deposition and withdrawal of minerals from bone.
Regulates cell growth and development, possibly reducing cancer risk.
May play a role in prevention of cardiovascular disease, diabetes, and hypertension.

**RECOMMENDATIONS**

RDA: 15 micrograms
DV: 20 micrograms
UL: 100 micrograms

**SPECIAL NOTES**

Vitamin D deficiency results in rickets, leading to abnormal bone development and characterized by bowing of the legs.
10–15 minutes of sun exposure to the arms and legs every day is sufficient to meet the body's vitamin D needs; however, many fall short of this.

**FIGURE 8-9** ▲ Food sources of vitamin D. The fill of the background color (none, 1/3, 2/3, or completely covered) within each food group on MyPlate indicates the average nutrient density for vitamin D in that group. The figure shows the vitamin D content of several foods in each food group. Overall, the richest sources of vitamin D are fish, fortified dairy products, and fortified breakfast cereals. Foods from the vegetables group are not a source of vitamin D, except for a select variety of mushrooms. oil: lumusphotography/iStock/Getty Images; milk: NIPAPORN PANYACHAROEN/ Shutterstock; juice: Sergei Vinogradov/seralexvi/123RF; cereal: Joe Belanger/iStock/Getty Images; mushrooms: lynx/iconotec.com/Glow Images; salmon: Foodcollection; MyPlate: U.S. Department of Agriculture

Sources: Office of Dietary Supplements, Dietary Supplements Fact Sheets, available from https://ods.od.nih.gov/factsheets/list-all; USDA FoodData Central, available from https:// fdc.nal.usda.gov.

(2.9 micrograms) milk. Consumption of at least 2 cups of vitamin D–fortified milk and soy beverages a day is advisable for people at risk for vitamin D deficiency. Ready-to-eat breakfast cereals are also fortified with vitamin D and other vitamins and minerals. A few provide about 2.2 micrograms per 1-cup serving, but most offer closer to 1 microgram per serving. You will almost double your intake of vitamin D if you add milk on top of your fortified cereal. Some brands of orange juice are now available fortified with about 2.5 micrograms of vitamin D per 1-cup serving.

It takes some planning to meet daily vitamin D requirements from food sources. If you work indoors, live in a northern location, have a family history of skin cancer, and are not eating fish every day, you may need a supplement. Supplements and fortified foods with vitamin $D_3$ are most effective at raising blood levels of vitamin D and reducing fracture risk.

◄ Solar radiation (UVB rays) on the skin is the most reliable way to maintain vitamin D status. In the absence of adequate sun exposure, especially in individuals with dark skin, the body must rely on dietary sources of vitamin D to meet needs. The melanin in dark skin reduces the skin's ability to produce vitamin D such that it may take anywhere from 30 minutes to 3 hours longer to get sufficient vitamin D, compared to lighter-skinned people. Dynamic Graphics Group/IT Stock Free/ Alamy Stock Photo

## FARM to FORK Mushrooms

barmalini/123RF

More than 2000 varieties of edible mushrooms are available in all shapes and sizes.[11] They are low in calories (about 20 calories in 1 cup of raw sliced mushrooms) and a good source of fiber. Most varieties are rich in the vitamins riboflavin, niacin, and pantothenic acid and the minerals potassium, selenium, and copper. Mushrooms also contain a vitamin D precursor, ergosterol, which is converted into vitamin D by exposing mushrooms to the sun's ultraviolet radiation. Mushrooms are a popular meat substitute because of their savory taste called umami.

### Grow
- Mushrooms can be cultivated at home or found in the wild (some wild mushrooms are poisonous, so never eat them unless you know they are safe). Mushrooms grow from very tiny spores in a growth mixture of sawdust, composted manure, or straw. This mixture, called spawn, supports growth of the threadlike roots, called mycelium. Mushroom kits are available, packed with mushroom mycelium growing on mushroom spawn.
- Mushrooms grow best in a dark, cool, moist, and humid environment, ideally in a basement or under the sink at home.
- White button mushrooms are the easiest types to grow at home. They should appear within 3 to 4 weeks and can be harvested every day for about 6 months. They are harvested when the caps open and the stalk can be cut from the stem with a sharp knife.

### Shop
- White button mushrooms are also the most common and least expensive mushrooms found in grocery stores. When purchasing mushrooms, choose those with a firm texture, even color, and tightly closed caps. Chanterelles are trumpet-shaped mushrooms and a good substitute for the more expensive morels.
- Crimini mushrooms are young portabella that look similar to the white button but are darker in color. The large portabella and shiitakes have a meaty taste and texture.
- Morels have a honeycomb-like shape and are available fresh in spring and summer, whereas dried morels are available year-round and are much less expensive.
- Porcinis are popular wild mushrooms with a distinct earthy, nutty flavor. Dried porcini mushrooms are less expensive and can be reconstituted and added to recipes.
- Some brands of mushrooms are now exposed to ultraviolet light to stimulate vitamin D production. Vitamin D–enhanced mushrooms providing 10 micrograms of vitamin D per 3-ounce serving (about 1 cup of diced mushrooms) are available in stores. They are usually labeled "UV treated" or "high in vitamin D" and cost more than regular mushrooms. In contrast, wild mushrooms such as chanterelles, maitake, and morels are naturally rich in vitamin D because they get natural sun exposure.

### Store
- Mushrooms are best when used within a few days of purchase. They should be stored in the refrigerator in their original package, a paper bag, or a damp cloth bag and used within 1 week. Storing mushrooms in a plastic bag will cause them to deteriorate quickly. Before preparing mushrooms, wipe them off with a clean, damp cloth or paper towel. Soaking fresh mushrooms in water will make them soggy. If they must be rinsed, do it lightly and dry gently.
- You can increase the vitamin D in your mushrooms by slicing them and placing the "gills" to face the sun for just 15 minutes. This will produce 5 to 20 micrograms of vitamin D (RDA is 15 micrograms) in 3 ounces of mushrooms and at least 90% of the vitamin is retained after storage and cooking.

### Prep
- Although mushrooms can be eaten raw, cooking mushrooms releases more of their nutrients and intensifies their color and flavor.
- Most mushrooms can be sautéed, grilled, or stir-fried. Searing mushrooms first in a dry pan browns the mushrooms and quickly drives off excess moisture. Adding a bit of butter and white wine or vinegar gives the mushrooms a nice sauce.
- Mushrooms can also be baked and filled with your favorite stuffing.

5second/123RF

During the time of growth from infancy through adolescence, consuming adequate vitamin D supports optimal bone mineralization. The American Academy of Pediatrics recommends that all infants, children, and adolescents should consume a minimum of 10 micrograms of vitamin D daily. Vitamin D supplementation is recommended for all infants (exclusively breastfed, partially breastfed, or formula fed) until vitamin D can be obtained from foods. Breast milk is a poor source of vitamin D, and the total intake of formula among young infants may not provide adequate vitamin D to meet needs. Keep in mind, however, that supplements must be used carefully, and under a primary care provider's guidance, to avoid vitamin D toxicity in the infant.

For adults over the age of 70, the RDA increases to 20 micrograms per day because of the reduced ability to absorb vitamin D from the intestine and decreased ability to synthesize it in the skin. A number of experts suggest older adults, especially those over age 70 who have limited sun exposure or dark skin, receive about 25 micrograms from a combination of vitamin D–fortified foods and a multivitamin and mineral supplement, with an individual supplement of vitamin D added if needed.[8]

Other population groups that have difficulty meeting vitamin D needs include individuals following a vegan dietary pattern and people with milk allergies or lactose intolerance. Vitamin D supplements or vitamin D–fortified soy milk or fruit juices are options for people who do not consume dairy products.

## AVOIDING TOO MUCH VITAMIN D

Too much vitamin D taken regularly can create serious health consequences in infants and children. Due to the role of vitamin D in calcium absorption, excretion, and release of calcium from bone, supplementation with high doses of vitamin D can cause calcium levels in the blood to increase above the normal range. The UL of 100 micrograms is based on the risk of overabsorption of calcium and eventual calcium deposits in the kidneys and other organs. Calcium deposits in organs can cause metabolic disturbances and cell death. Toxicity symptoms also include weakness, loss of appetite, diarrhea, vomiting, mental confusion, and increased urine output. Please note that vitamin D toxicity does not result from excessive exposure to the sun because the body regulates the amount made in the skin (i.e., as exposure to sunlight increases, the efficiency of vitamin D synthesis decreases).

### ✓ CONCEPT CHECK 8.3

1. Why is vitamin D sometimes not considered an *essential* nutrient?
2. What is the association between vitamin D and sun exposure?
3. Which two organs are involved in the activation of vitamin D in the body?
4. How does vitamin D work to maintain blood calcium levels?
5. What are some rich food sources of vitamin D?
6. Can vitamin D be toxic?

# 8.4 Vitamin E (Tocopherols)

In the 1920s, a fat-soluble compound was found to be essential for fertility in rats. This compound was named tocopherol from the Greek words *tokos,* meaning "birth," and *phero,* meaning "to bring forth." Later, this essential nutrient was named vitamin E. Vitamin E is a family of four tocopherols and four tocotrienols called alpha, beta, gamma, and delta. They differ in that tocopherols have a saturated side chain, whereas the tocotrienols have an unsaturated side chain. Tocotrienols have not been as extensively studied as tocopherols, but recent research explores their potential roles in prevention of cancers, diabetes,

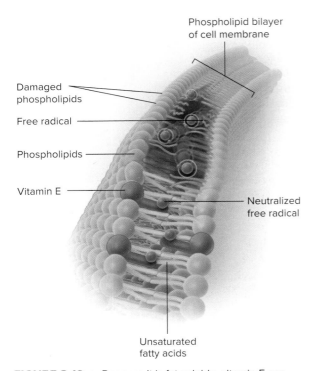

Phospholipid bilayer of cell membrane

Damaged phospholipids

Free radical

Phospholipids

Vitamin E

Neutralized free radical

Unsaturated fatty acids

**FIGURE 8-10** ▲ Because it is fat soluble, vitamin E can insert itself into cell membranes, where it helps stop free-radical chain reactions. If not interrupted, these reactions cause extensive oxidative damage to cells and, ultimately, cell death.

▲ The avocado in guacamole and the sunflower oil in the tortilla chip are good sources of vitamin E. Andrew Bret Wallis/ BananaStock/Getty Images

and cardiovascular diseases. Of these eight forms of vitamin E, alpha (α)-tocopherol is the most biologically active and the most potent.

## FUNCTIONS OF VITAMIN E

**Antioxidant.** The principal function of vitamin E in humans is as an anti-oxidant.[12] Vitamin E is a fat-soluble vitamin found primarily in adipose tissue and in the lipid bilayers of cell membranes (Fig. 8-10). Many of the lipids within these membranes are polyunsaturated fatty acids (PUFA), which are particularly susceptible to oxidative attack by free radicals. The formation of free radicals may destabilize the cell membrane, which may ultimately alter the ability of the cell to function properly. Vitamin E can donate electrons or hydrogen to free radicals found in membranes, thereby making them more stable. The antioxidant function of vitamin E appears to be critical in cells continually exposed to high levels of oxygen, particularly red blood cells and the cells lining the lungs.

Increasing vitamin E intake has been suggested as a way to prevent several chronic diseases that are linked to oxidative damage. For example, oxidized LDL cholesterol is a major component of the plaque that develops in arteries, which leads to atherosclerosis. Vitamin E is thought to attenuate the development of atherogenic plaque due to its ability to prevent or reduce the formation of oxidized LDL cholesterol.[13] In addition, oxidative damage to proteins in the eye leads to the development of cataracts. Oxidized proteins combine and precipitate in the lens, causing cloudiness and decreasing visual acuity. Insufficient consumption of antioxidants from foods increases one's risk of these diseases.

Experts do not know whether supplementation with megadoses of vitamin E can confer any significant protection against diseases linked to oxidative damage. The consensus among the scientific community is that the established benefits of lifestyle choices have a far greater effect than any proposed benefits of antioxidant supplementation. The consensus of scientific research organizations is that it is premature to recommend vitamin E supplements to the general population, based on current knowledge and the failure of large clinical trials to show any consistent benefit. In addition, the FDA has denied the request of the dietary supplement industry to make a health claim that vitamin E supplements reduce the risk of cardiovascular disease or cancer.

**Other Roles of Vitamin E.** Although vitamin E is essential for fertility in many animal species, it does not appear to serve this role in humans. It is, however, important for the formation of muscles and the central nervous system in early human development. Vitamin E has been shown to improve vitamin A absorption if the dietary intake of vitamin A is low. It also functions in the metabolism of iron within cells, and it helps maintain nervous tissue and immune function. Current research on vitamin E is focused on its ability to influence the expression of some genes involved in development of chronic diseases.[14]

## VITAMIN E DEFICIENCY

Specific population groups are especially susceptible to developing marginal vitamin E status. Infants who are born preterm tend to have low vitamin E stores because this vitamin is transferred from mother to baby during the late stages of pregnancy. Hence, the potential for oxidative damage, which could cause the cell membranes of red blood cells to break (hemolysis), is of particular concern for infants who are born preterm. The rapid growth of infants who are born preterm, coupled with the high oxygen needs of their immature lungs, greatly increases the stress on red blood cells. Special vitamin E–fortified formulas and supplements designed for infants who are born preterm compensate for this lack of vitamin E. Individuals who use tobacco products are another group at high

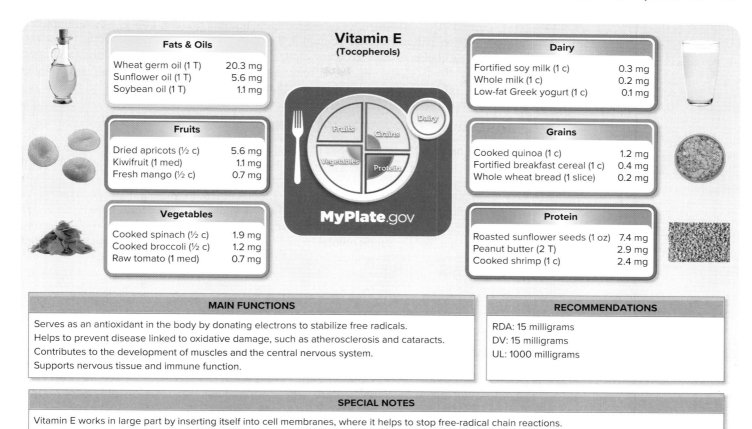

**Vitamin E**
(Tocopherols)

**Fats & Oils**

| | |
|---|---|
| Wheat germ oil (1 T) | 20.3 mg |
| Sunflower oil (1 T) | 5.6 mg |
| Soybean oil (1 T) | 1.1 mg |

**Fruits**

| | |
|---|---|
| Dried apricots (½ c) | 5.6 mg |
| Kiwifruit (1 med) | 1.1 mg |
| Fresh mango (½ c) | 0.7 mg |

**Vegetables**

| | |
|---|---|
| Cooked spinach (½ c) | 1.9 mg |
| Cooked broccoli (½ c) | 1.2 mg |
| Raw tomato (1 med) | 0.7 mg |

**Dairy**

| | |
|---|---|
| Fortified soy milk (1 c) | 0.3 mg |
| Whole milk (1 c) | 0.2 mg |
| Low-fat Greek yogurt (1 c) | 0.1 mg |

**Grains**

| | |
|---|---|
| Cooked quinoa (1 c) | 1.2 mg |
| Fortified breakfast cereal (1 c) | 0.4 mg |
| Whole wheat bread (1 slice) | 0.2 mg |

**Protein**

| | |
|---|---|
| Roasted sunflower seeds (1 oz) | 7.4 mg |
| Peanut butter (2 T) | 2.9 mg |
| Cooked shrimp (1 c) | 2.4 mg |

**MyPlate**.gov

**MAIN FUNCTIONS**

Serves as an antioxidant in the body by donating electrons to stabilize free radicals.
Helps to prevent disease linked to oxidative damage, such as atherosclerosis and cataracts.
Contributes to the development of muscles and the central nervous system.
Supports nervous tissue and immune function.

**RECOMMENDATIONS**

RDA: 15 milligrams
DV: 15 milligrams
UL: 1000 milligrams

**SPECIAL NOTES**

Vitamin E works in large part by inserting itself into cell membranes, where it helps to stop free-radical chain reactions.
Unlike the other fat-soluble vitamins, vitamin E is found naturally in many plant oils, including vegetable oil and soybean oil.

**FIGURE 8-11** ▲ Food sources of vitamin E. The fill of the background color (none, 1/3, 2/3, or completely covered) within each food group on MyPlate indicates the average nutrient density for vitamin E in that group. The figure shows the vitamin E content of several foods in each food group. Overall, the richest sources of vitamin E are nuts, seeds, plant oils, and quinoa. oil: Iconotec/Glow Images; milk: NIPAPORN PANYACHAROEN/Shutterstock; apricots: lynx/iconotec/Glowimages; cereal: Joe Belanger/iStock/Getty Images; spinach: Ingram Publishing; seeds: Glow Images; MyPlate: U.S. Department of Agriculture
Sources: Office of Dietary Supplements, Dietary Supplements Fact Sheets, available from https://ods.od.nih.gov/factsheets/list-all; USDA FoodData Central, available from https://fdc.nal.usda.gov.

risk for vitamin E deficiency, as smoking readily destroys vitamin E in the lungs. Others at risk of vitamin E deficiency include adults on very low-fat diets (< 15% total fat) or those with fat-malabsorption disorders.

## GETTING ENOUGH VITAMIN E

Because vitamin E is only synthesized by plants, plant products (especially the oils) are the best sources. In the North American dietary pattern nearly two-thirds of vitamin E is supplied by salad oils, margarines, spreads (low-fat margarine), and shortening (Fig. 8-11). Breakfast cereals fortified with vitamin E are good sources, but other than wheat germ, few other grain products provide much vitamin E. Milling of grains removes the germ, which contains the oils (mostly polyunsaturated fatty acids or PUFAs) and vitamin E. By removing the germ, the resulting grain product has less chance of spoiling (i.e., rancidity of the PUFAs) and thus a longer shelf life. Other good sources of vitamin E are nuts and seeds.

Because plant oils contain mostly unsaturated fatty acids, the relatively high amount of vitamin E in plant oils naturally protects these unsaturated lipids from oxidation. Animal products (meat, dairy, and eggs) and fish oils, on the other hand, contain almost no vitamin E. Vitamin E is susceptible to destruction by oxygen, metals, light, and heat, especially when oil is repeatedly reused in deep-fat frying; thus, the vitamin E content of a food depends on how it is harvested, processed, stored, and cooked.

The RDA of vitamin E for adults is 15 milligrams per day of alpha-tocopherol, the most active, natural form of vitamin E. This amount equals 22.4 milligrams of the less active, synthetic source. Typically, North American adults consume only about two-thirds of the RDA for vitamin E from food sources. On revised food and supplement labels, the Daily Value for vitamin E is 15 milligrams of alpha-tocopherol.

## AVOIDING TOO MUCH VITAMIN E

Unlike other fat-soluble vitamins, vitamin E is not as readily stored in the liver. It is stored in adipose tissue throughout the body. The UL for vitamin E is 1000 milligrams per day of supplemental alpha-tocopherol. Excessive intake of vitamin E can interfere with vitamin K's role in the clotting mechanism, leading to hemorrhage. The risk of insufficient blood clotting is especially high if vitamin E is taken in conjunction with anticoagulant medications (e.g., Coumadin® or high-dose aspirin). Always be cautious about using dietary supplements. In addition to the significant risk of drug interference and prolonged bleeding, vitamin E supplements can produce nausea, gastrointestinal distress, and diarrhea.

### ✓ CONCEPT CHECK 8.4

1. How does vitamin E work to prevent oxidative damage?
2. What are some rich food sources of vitamin E?
3. Why are infants who are born preterm, individuals who smoke, and people with fat-malabsorption disorders particularly susceptible to oxidative damage to cell membranes?
4. What are the possible results of vitamin E toxicity?

# 8.5 Vitamin K (Quinone)

A family of compounds known collectively as vitamin K is found in plants, plant oils, fish oils, and animal products. Vitamin K is also synthesized by bacteria in the human colon, which normally fulfills approximately 10% of human requirements. Vitamin K has three forms: (1) phylloquinone, the most abundant form of vitamin K, is synthesized by green plants; (2) menaquinone is synthesized by gut bacteria; and (3) menadione is the synthetic form found in supplements. Interestingly, the synthetic menadione form of vitamin K is twice as biologically available as the other two.[15]

## FUNCTIONS OF VITAMIN K

Vitamin K serves as a cofactor in chemical reactions that add carbon dioxide ($CO_2$) molecules to various proteins, thus enabling these proteins to bind to calcium. This is the biochemical basis for vitamin K's role in the life-and-death process of blood clotting. In the clotting cascade (Fig. 8-12), vitamin K imparts calcium-binding ability to seven different proteins, eventually leading to the conversion of soluble fibrinogen into insoluble fibrin (i.e., the clot). The "K" stands for *koagulation* in the language spoken by the Danish researchers who first noted the relationship between vitamin K and blood clotting.

Besides its role in blood clotting, vitamin K is also important for bone health. The calcium-binding protein in the bone, osteocalcin, depends upon vitamin K for its function in bone mineralization.

# HEMOSTASIS

Damage to tissues leads to bleeding and blood loss. If blood loss is significant, the delivery of oxygen and nutrients to the cells will be impaired. Therefore, the body has several efficient mechanisms to quickly stop bleeding. Collectively, the following processes that stop blood loss are called **hemostasis:**

1. **Vasoconstriction:** narrowing of the blood vessels, which limits blood flow to the damaged tissue.
2. Formation of a platelet plug: platelets stick to the damaged tissue and to each other, creating a temporary seal that stops bleeding.
3. **Coagulation** (blood clotting): process by which blood changes from a liquid to a solid. The end result is reinforcement of the platelet plug with the protein fibrin.

Coagulation of blood is a multistep process that involves the activation of a series of clotting factors that are always present in the plasma. Usually, platelets do not adhere to the inner lining of blood vessels and clotting factors remain in their inactive forms. This is important because inappropriate formation of blood clots could choke off the blood supply to living tissues. However, when activated by chemical signals released from damaged tissue, a cascade of reactions takes place in which one activated protein activates the next protein to its functional form, leading finally to the conversion of fibrinogen to fibrin, which forms a clot (Fig. 8-12). Individual

**hemostasis** The process of stopping blood loss.

**vasoconstriction** The narrowing of blood vessels. As part of hemostasis, this temporarily reduces the flow of blood to the site of injury while a clot is being formed.

**coagulation** Formation of a blood clot.

What happens to a clot after the damaged tissue has healed? Enzymes break down the clot, releasing some proteins into the bloodstream. Eventually, these protein waste products are picked up by the liver and kidneys, and their amino acids can be used to make other proteins.

**FIGURE 8-12** ◄ Vitamin K works to activate clotting factors, which are then able to bind to calcium. The calcium binding to clotting factors is necessary for clot formation. Steve Gschmeissner/Science Photo Library RF/Getty Images

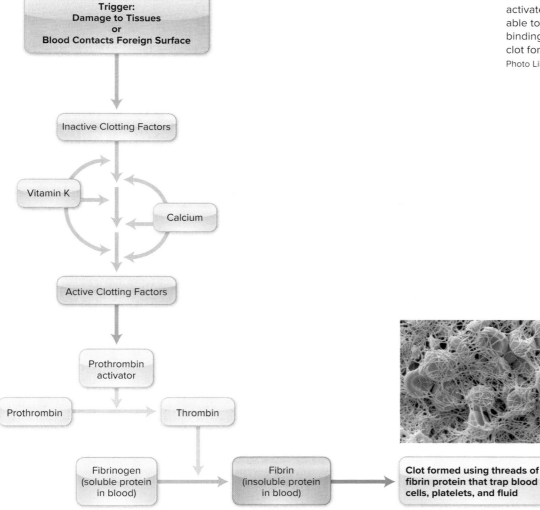

strands of fibrin join together to form a mesh that reinforces the platelet plug. At various points in the coagulation cascade, vitamin K and calcium are required as cofactors. Hemorrhage is essentially the opposite of hemostasis. In hemorrhage, blood loss cannot be controlled and the blood supply to body tissues becomes depleted. Hemorrhage could result from insufficient vitamin K. Excessive vitamin E could also lead to hemorrhage because, at high doses, vitamin E interferes with the functions of vitamin K. In addition, excessive intakes of omega-3 fatty acids could cause hemorrhage because these fatty acids lead to the production of chemical messengers that inhibit blood clotting.

Newborn infants are at highest risk for **vitamin K deficiency bleeding** because they are born with low stores of vitamin K. Furthermore, breast milk is quite low in vitamin K, and at birth a newborn's intestinal tract has not yet been populated with the bacteria that can synthesize vitamin K. Low vitamin K status in an infant sets the stage for serious bleeding problems if the infant is injured or needs surgery. Therefore, vitamin K is routinely administered by injection shortly after birth.[16] It is becoming more common, however, for parents to refuse vitamin K supplementation for their infants. In some cases, this has led to hemorrhagic complications, such as severe gastrointestinal or intracranial bleeding.[15]

Among adults, deficiencies of vitamin K can occur when a person takes antibiotics for an extended time. This destroys the bacteria that normally synthesize vitamin K. If vitamin K intake and/or absorption is low (e.g., among individuals with celiac disease) and the vitamin K–producing bacteria are knocked out by antibiotics, vitamin K deficiency may occur.

**vitamin K deficiency bleeding** Hemorrhage caused by a lack of vitamin K, especially among infants from birth to 6 months of age; also called *hemorrhagic disease of the newborn*.

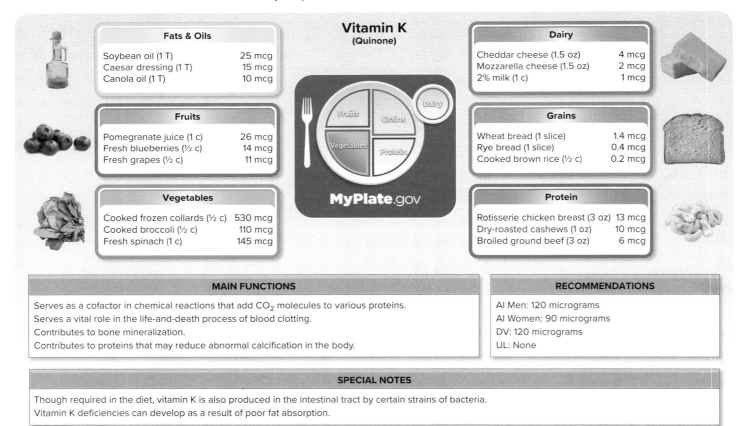

FIGURE 8-13 ▲ Food sources of vitamin K. The fill of the background color (none, 1/3, 2/3, or completely covered) within each food group on MyPlate indicates the average nutrient density for vitamin K in that group. The figure shows the vitamin K content of several foods in each food group. Overall, the richest sources of vitamin K are green, leafy vegetables. oil: Stockbyte/PunchStock; cheese: Brent Hofacker/Shutterstock; blueberries: photastic/Shutterstock; bread: Alex Cao/Photodisc/Getty Images; spinach: vvoennyy/123RF; cashews: Alistair Forrester Shankie/E+/Getty Images; MyPlate: U.S. Department of Agriculture

Sources: Office of Dietary Supplements, Dietary Supplements Fact Sheets, available from https://ods.od.nih.gov/factsheets/list-all; USDA FoodData Central, available from https://fdc.nal.usda.gov.

## VITAMIN K DEFICIENCY

Vitamin K deficiency is rare, but its consequences can be life-threatening. Even though it is a fat-soluble vitamin, the body does not store vitamin K to a great extent. Because of its important role in blood clotting, a deficiency of vitamin K can lead to excessive bleeding. In milder forms, this may cause easy bruising. In severe forms, vitamin K deficiency leads to hemorrhage, which can be fatal.

## GETTING ENOUGH VITAMIN K

Major food sources of the phylloquinone form of vitamin K are green leafy vegetables, broccoli, asparagus, and peas (Fig. 8-13). The menaquinone form of vitamin K is found in some meats, eggs, and dairy products, and it is the form synthesized by bacteria. Compared to that of plant sources, the nutrient density of vitamin K in foods of animal origin is rather low. Vitamin K is resistant to cooking losses.

As with other fat-soluble vitamins, absorption of vitamin K requires dietary fat and adequate liver and pancreatic secretions. Unlike other fat-soluble vitamins, though, not much vitamin K is stored in the body and excesses can be excreted via urine. Thus, a deficiency could develop rather quickly if dietary intake of this nutrient is insufficient, which can be a problem among older adults whose dietary patterns lack vegetables. However, because vitamin K is fairly widespread in foods and some can be synthesized by bacteria in the colon, deficiencies of this vitamin rarely occur. No reports of toxicity have been published.

**Medicine Cabinet**

People who are prone to develop blood clots may take anticoagulants or *blood thinners*. One commonly prescribed anticoagulant is Coumadin® (warfarin). This medication inhibits vitamin K–dependent coagulation factors. When taking Coumadin or similar drugs, it is important to keep dietary intake of vitamin K consistent from day to day.[15]

### ✓ CONCEPT CHECK 8.5

1. List and describe the three steps of hemostasis.
2. What is the role of vitamin K in hemostasis?
3. Why is it important for people who take Coumadin to monitor their dietary intake of vitamin K?

## 8.6 The Water-Soluble Vitamins and Choline

Regular consumption of good sources of the water-soluble vitamins is important. Most water-soluble vitamins are readily excreted from the body, with any excess generally ending up in the urine or stool and very little being stored. They dissolve in water, so large amounts of these vitamins can be lost during food processing and preparation. As emphasized earlier, vitamin content is best preserved by light cooking methods, such as stir-frying, steaming, and microwaving.

The B vitamins are thiamin, riboflavin, niacin, pantothenic acid, biotin, vitamin B-6, folate, and vitamin B-12. Choline is a related nutrient that has vitamin-like characteristics but currently is not classified as a vitamin. Vitamin C is also a water-soluble vitamin.

The B vitamins often occur together in the same foods, so a lack of one B vitamin may mean other B vitamins are also low in a diet. The B vitamins function as *coenzymes*, small molecules that interact with enzymes to enable the enzymes to function. In essence, the coenzymes contribute to enzyme activity (Fig. 8-14).

As coenzymes, the B vitamins play many key roles in metabolism. The metabolic pathways used by carbohydrates, fats, and amino acids all require input from B vitamins. Because of their role in energy metabolism, needs for many B vitamins increase somewhat as energy expenditure increases. Still, this is not a major concern because this increase in energy expenditure is usually accompanied by a corresponding increase in

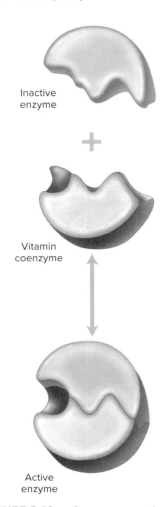

**FIGURE 8-14** ▲ Coenzymes, such as those formed from B vitamins, aid in the function of various enzymes. Without the coenzyme, the enzyme cannot function properly, and deficiency symptoms associated with the missing vitamin eventually appear. Health-food stores sell the coenzyme forms of some vitamins. These more expensive forms of vitamins are unnecessary. The body makes all the coenzymes it needs from vitamin precursors.

**anemia** A decreased oxygen-carrying capacity of the blood. This can be caused by many factors, such as iron deficiency or blood loss.

food intake, which contributes more B vitamins to a diet. Many B vitamins are interdependent because they participate in the same processes (Fig. 8-15). B vitamin–deficiency symptoms typically occur in the brain and nervous system, skin, and gastrointestinal (GI) tract. Cells in these tissues are metabolically active, and those in the skin and GI tract are also constantly being replaced.

After being ingested, the B vitamins are first broken down from their active coenzyme forms into free vitamins in the stomach and small intestine. The vitamins are then absorbed, primarily in the small intestine. Typically, about 50% to 90% of the B vitamins in the diet are absorbed, which means that they have relatively high bioavailability. Once inside cells, the active coenzyme forms are resynthesized. There is no need to consume the coenzyme forms themselves. Some vitamins are sold in their coenzyme forms, but these are broken down during digestion, and we activate them when needed.

## B VITAMIN INTAKES OF NORTH AMERICANS

The nutritional health of most North Americans with regard to the B vitamins is adequate. Typical dietary patterns contain plentiful and varied natural sources of these vitamins. In addition, many common foods, especially ready-to-eat breakfast cereals, are fortified with one or more of the B vitamins. In some developing countries, however, deficiencies of the B vitamins are more common and the resulting deficiency diseases pose significant threats to public health.

Because B vitamins are water soluble, very little is stored in the body and the excess that we eat ends up in the urine or stool. About 10% to 25% of these vitamins are lost from food during food processing and preparation because they dissolve in water. Light cooking methods, such as stir-frying, steaming, and microwaving, best preserve vitamin content.

Despite good B-vitamin status among North Americans, marginal deficiencies of these vitamins may occur in some cases, especially among older adults who eat small amounts of food and in other people with inadequate dietary patterns. In the short run, such a marginal deficiency in most people likely leads only to fatigue or other unspecified physical effects. Although the long-term effects of such marginal deficiencies are yet unknown, increased risks of cardiovascular disease, cancers, and cataracts of the eye are suspected. With rare exceptions, healthy adults do not develop the more serious B-vitamin–deficiency diseases from dietary inadequacy alone. The main exceptions are people with alcohol use disorders. The combination of extremely unbalanced eating patterns and alcohol-induced alterations of vitamin absorption and metabolism creates significant risks for serious nutrient deficiencies among people with alcohol use disorders.

## B VITAMINS IN GRAINS

The process of refining grains, such as processing white flour from whole wheat, leads to the loss of B vitamins as well as other vitamins and minerals. When grains are milled to make refined products, seeds are crushed and the germ, bran, and husk layers are discarded, leaving just the starch-containing endosperm in the refined grains. This starch is used to make white flour, bread, and cereal products. Unfortunately, many nutrients are lost in the discarded germ, bran, and husk materials. To counteract these losses in the United States, bread and cereal products made from milled grains are enriched with four B vitamins (thiamin, riboflavin, niacin, and folic acid) and with the mineral iron.

Food enrichment was initiated by federal legislation in the 1930s to help combat nutrient deficiencies such as pellagra (niacin deficiency) and iron-deficiency **anemia.** Folate was included on the list of nutrients required to be added to refined grain products in 1998. However, not all nutrients lost in milling are added back through enrichment. Products made with refined flour remain lower in several nutrients, including

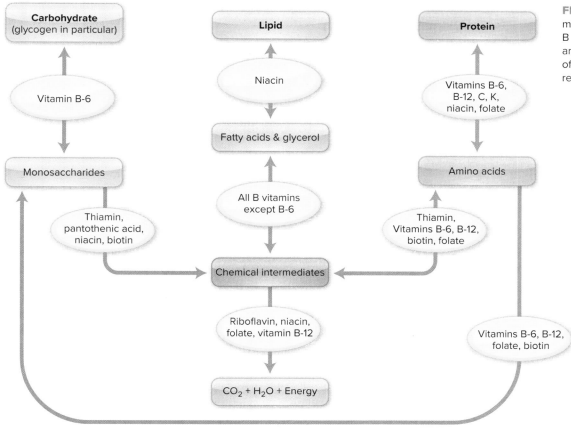

**FIGURE 8-15** ◄ Examples of metabolic pathways for which B vitamins and other vitamins are essential. The metabolism of energy-yielding nutrients requires vitamin input.

vitamins E and B-6, potassium, magnesium, and fiber, than products made with whole grains (Fig. 8-16). This lower nutrient density is why nutrition experts and the *Dietary Guidelines* advocate daily consumption of whole grain products, such as brown rice, oats, whole-wheat bread, whole-grain cereals and crackers, and wild rice, rather than refined grain products.

The following sections focus on the B vitamins and trace minerals involved in energy metabolism.

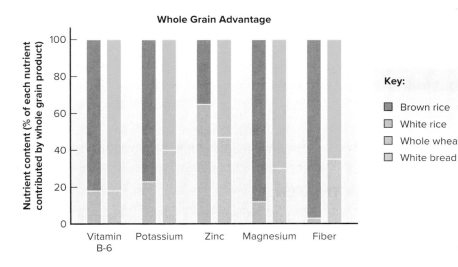

▲ Rapid cooking of vegetables in minimal fluids reduces the vitamins lost during cooking. Steaming is one effective method. C Squared Studios/Getty Images

**FIGURE 8-16** ◄ Comparison of the relative nutrient contents of refined versus whole grains. Nutrients are expressed as a percentage of the nutrient contribution of the whole grain product.

☑ **CONCEPT CHECK 8.6**

1. What body organs or tissues are most likely to show symptoms if there is a deficiency of B vitamins and why?
2. Why are B vitamins lost when foods are cooked in water?
3. What group of people is at high risk of B-vitamin deficiency?
4. What happens during the refining of grains that causes a decrease in nutrient density?

## 8.7 Thiamin (Vitamin B-1)

### FUNCTIONS OF THIAMIN

Thiamin was the first water-soluble vitamin to be discovered. One of its primary functions is to help release energy from carbohydrate. Its coenzyme form, thiamin pyrophosphate (TPP), participates in reactions in which carbon dioxide ($CO_2$) is released during the breakdown of carbohydrates and certain amino acids. These reactions are particularly important in the body's ATP-producing energy pathways. Thiamin also functions in chemical reactions that make RNA, DNA, and neurotransmitters.[17]

### THIAMIN DEFICIENCY

**beriberi** The thiamin-deficiency disorder characterized by muscle weakness, loss of appetite, nerve degeneration, and sometimes edema.

Thiamin-deficiency disease is called **beriberi,** a word that means "I can't, I can't" in the Sri Lanka language of Sinhalese. The symptoms include weakness, loss of appetite, irritability, nervous tingling throughout the body, poor arm and leg coordination, and deep muscle pain in the calves. This disease was described long before thiamin was discovered to be a vitamin in 1910. A person with beriberi often develops an enlarged heart and sometimes severe edema.

Beriberi is seen in areas where rice is a staple food and polished (white) rice is consumed rather than brown (whole grain) rice. In most parts of the world, even developing countries, white rice is preferred and is made by removing the bran and germ layer from brown rice. White rice is a poor source of thiamin, except for the enriched varieties sold in the United States and Canada.

Beriberi results when glucose, the primary fuel for brain and nerve cells, cannot be metabolized to release energy because of the lack of thiamin. Because the thiamin coenzyme participates in glucose metabolism, problems with functions that depend on glucose, such as brain and nerve action, are the first signs of a thiamin deficiency. Symptoms can develop in just 10 days without consuming adequate thiamin.

Alcohol abuse increases risk for thiamin deficiency. Absorption and use of thiamin are profoundly diminished and excretion is increased by consumption of alcohol. Furthermore, the low-quality eating pattern that often accompanies severe alcohol use disorders makes matters worse. There is limited storage in the body; therefore, an alcoholic binge lasting 1 to 2 weeks may quickly deplete already diminished amounts of the vitamin and result in deficiency symptoms. The beriberi associated with alcohol use disorders is also called *Wernicke-Korsakoff syndrome*. Other groups at risk of thiamin deficiency include adults who are older, are infected with human immunodeficiency virus, have diabetes, or have undergone bariatric surgery.

### GETTING ENOUGH THIAMIN

Average daily intakes for men exceed the DV by 50% or more, and women generally meet the RDA. Adults with low incomes and older people may barely meet their needs

*magnificent*

You have learned that the beneficial bacteria in the gut microbiome help us to digest our food, regulate our immune system, offer protection against diseases, and confer other health benefits. Did you also know that your microbiome is responsible for producing vitamins, including vitamin B-12, thiamin, riboflavin, and vitamin K?

*microbiome*

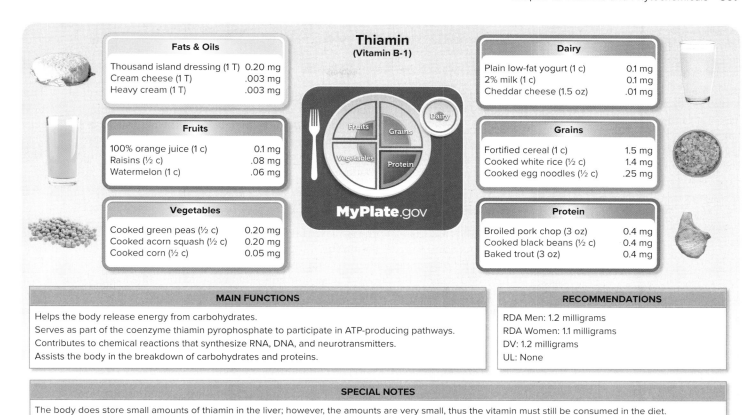

## Thiamin
(Vitamin B-1)

### Fats & Oils
| | |
|---|---|
| Thousand island dressing (1 T) | 0.20 mg |
| Cream cheese (1 T) | .003 mg |
| Heavy cream (1 T) | .003 mg |

### Fruits
| | |
|---|---|
| 100% orange juice (1 c) | 0.1 mg |
| Raisins (½ c) | .08 mg |
| Watermelon (1 c) | .06 mg |

### Vegetables
| | |
|---|---|
| Cooked green peas (½ c) | 0.20 mg |
| Cooked acorn squash (½ c) | 0.20 mg |
| Cooked corn (½ c) | 0.05 mg |

### Dairy
| | |
|---|---|
| Plain low-fat yogurt (1 c) | 0.1 mg |
| 2% milk (1 c) | 0.1 mg |
| Cheddar cheese (1.5 oz) | .01 mg |

### Grains
| | |
|---|---|
| Fortified cereal (1 c) | 1.5 mg |
| Cooked white rice (½ c) | 1.4 mg |
| Cooked egg noodles (½ c) | .25 mg |

### Protein
| | |
|---|---|
| Broiled pork chop (3 oz) | 0.4 mg |
| Cooked black beans (½ c) | 0.4 mg |
| Baked trout (3 oz) | 0.4 mg |

MyPlate.gov

| MAIN FUNCTIONS | RECOMMENDATIONS |
|---|---|
| Helps the body release energy from carbohydrates. | RDA Men: 1.2 milligrams |
| Serves as part of the coenzyme thiamin pyrophosphate to participate in ATP-producing pathways. | RDA Women: 1.1 milligrams |
| Contributes to chemical reactions that synthesize RNA, DNA, and neurotransmitters. | DV: 1.2 milligrams |
| Assists the body in the breakdown of carbohydrates and proteins. | UL: None |

**SPECIAL NOTES**

The body does store small amounts of thiamin in the liver; however, the amounts are very small, thus the vitamin must still be consumed in the diet.

Thiamin deficiency is known as beriberi and is characterized by weakness, loss of appetite, irritability, tingling in the body, poor coordination, and muscle pain.

**FIGURE 8-17** ▲ Food sources of thiamin. The fill of the background color (none, 1/3, 2/3, or completely covered) within each group on MyPlate indicates the average nutrient density for thiamin in that group. The figure shows thiamin content of several foods in each food group. Overall, the richest sources of thiamin are meats (especially pork), whole grains, and fortified breakfast cereals. cream cheese: Africa Studio/Shutterstock; milk: NIPAPORN PANYACHAROEN/Shutterstock; juice: Sergei Vinogradov/seralexvi/123RF; cereal: Joe Belanger/iStock/Getty Images; peas: Ingram Publishing/SuperStock; pork chop: FoodCollection; MyPlate: U.S. Department of Agriculture

Sources: Office of Dietary Supplements, Dietary Supplements Fact Sheets, available from https://ods.od.nih.gov/factsheets/list-all; USDA FoodData Central, available from https://fdc.nal.usda.gov.

for thiamin. Potential contributors to thiamin deficiency are eating patterns dominated by highly processed and unenriched foods, added sugars, and fat, as well as heavy alcohol intake combined with an inadequate eating pattern. No toxicity has been observed from the use of oral thiamin supplements, and thiamin is rapidly lost in the urine. Thus, no Upper Level (UL) has been set for thiamin.

Major sources of thiamin include pork products, whole grains, ready-to-eat breakfast cereals, enriched grains, green peas, orange juice, organ meats, and dried beans (Fig. 8-17). When considering the sections of MyPlate, the protein and grain groups contain the most foods that are nutrient-dense sources of thiamin.

### ✓ CONCEPT CHECK 8.7

1. How is thiamin involved in energy metabolism?
2. What body organs or tissues are most likely to show symptoms if there is a deficiency of thiamin?
3. What group of people is at high risk of thiamin deficiency?
4. What are some excellent sources of thiamin?

Pork is an excellent source of thiamin. **What food sources would be good sources of thiamin for an individual who excludes pork from their dietary pattern for religious reasons?** Ingram Publishing/SuperStock

**ariboflavinosis** A deficiency disease resulting from a riboflavin deficiency and often characterized by mouth sores, dermatitis, glossitis, and/or angular cheilitis.

**dermatitis** Condition that involves itchy, dry skin or a rash on swollen, reddened skin.

**glossitis** Inflammation and swelling of the tongue.

**angular cheilitis** Inflammation of the corners of the mouth with painful cracking; also called *cheilosis* or *angular stomatitis*.

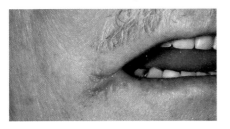

**FIGURE 8-18** ▲ *Angular cheilitis*, also called *cheilosis* or *angular stomatitis*, causes painful cracks at the corners of the mouth. Dr. P. Marazzi/Science Photo Library/Science Source

# 8.8 Riboflavin (Vitamin B-2)

## FUNCTIONS OF RIBOFLAVIN

Riboflavin derives its name from its yellow color (*flavus* means "yellow" in Latin). The coenzyme forms of riboflavin, flavin dinucleotide (FAD) and flavin mononucleotide (FMN), participate in many energy-yielding metabolic pathways, such as the breakdown of fatty acids. Some metabolism of vitamins and minerals also requires riboflavin. Indirectly, riboflavin also has an antioxidant role in the body through its support of the enzyme glutathione peroxidase.[18]

## RIBOFLAVIN DEFICIENCY

Symptoms associated with riboflavin deficiency (**ariboflavinosis**) include inflammation of the mouth and tongue, **dermatitis,** various eye disorders, sensitivity to the sun, and confusion. Riboflavin deficiency typically would occur jointly with deficiencies of niacin, thiamin, and vitamin B-6 because these nutrients often occur in the same foods. **Glossitis** is a painful soreness or inflammation of the tongue that can signal a deficiency of riboflavin, niacin, vitamin B-6, folate, or vitamin B-12. **Angular cheilitis,** also called cheilosis or angular stomatitis, is inflammation of the corners of the mouth that may cause painful cracking (Fig. 8-18). Both glossitis and angular cheilitis can be caused by other medical conditions; thus, further evaluation is required before diagnosing a nutrient deficiency. Such symptoms develop after approximately 2 months on a riboflavin-poor dietary pattern.

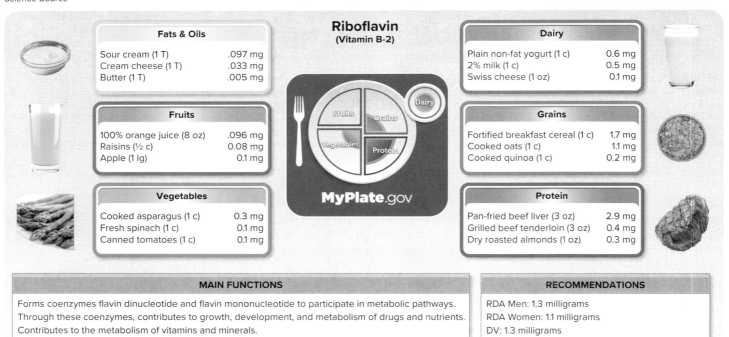

**Riboflavin (Vitamin B-2)**

| Fats & Oils | |
|---|---|
| Sour cream (1 T) | .097 mg |
| Cream cheese (1 T) | .033 mg |
| Butter (1 T) | .005 mg |

| Fruits | |
|---|---|
| 100% orange juice (8 oz) | .096 mg |
| Raisins (½ c) | 0.08 mg |
| Apple (1 lg) | 0.1 mg |

| Vegetables | |
|---|---|
| Cooked asparagus (1 c) | 0.3 mg |
| Fresh spinach (1 c) | 0.1 mg |
| Canned tomatoes (1 c) | 0.1 mg |

| Dairy | |
|---|---|
| Plain non-fat yogurt (1 c) | 0.6 mg |
| 2% milk (1 c) | 0.5 mg |
| Swiss cheese (1 oz) | 0.1 mg |

| Grains | |
|---|---|
| Fortified breakfast cereal (1 c) | 1.7 mg |
| Cooked oats (1 c) | 1.1 mg |
| Cooked quinoa (1 c) | 0.2 mg |

| Protein | |
|---|---|
| Pan-fried beef liver (3 oz) | 2.9 mg |
| Grilled beef tenderloin (3 oz) | 0.4 mg |
| Dry roasted almonds (1 oz) | 0.3 mg |

MyPlate.gov

| MAIN FUNCTIONS | RECOMMENDATIONS |
|---|---|
| Forms coenzymes flavin dinucleotide and flavin mononucleotide to participate in metabolic pathways. Through these coenzymes, contributes to growth, development, and metabolism of drugs and nutrients. Contributes to the metabolism of vitamins and minerals. Has an antioxidant role through its support of the enzyme glutathione peroxidase. | RDA Men: 1.3 milligrams RDA Women: 1.1 milligrams DV: 1.3 milligrams UL: None |

| SPECIAL NOTES |
|---|
| The body does store small amounts of riboflavin in the liver, heart, and kidneys; however, the amounts are very small, thus the vitamin must still be consumed in the diet. Riboflavin deficiency is known as ariboflavinosis and characterized by inflammation of the mouth and tongue, dermatitis, and cracking of skin around the mouth. |

**FIGURE 8-19** ▲ Food sources of riboflavin. The fill of the background color (none, 1/3, 2/3, or completely covered) within each group on MyPlate indicates the average nutrient density for riboflavin in that group. The figure shows riboflavin content of several foods in each food group. Overall, the richest sources of riboflavin are meats (especially liver), dairy products, and fortified breakfast cereals. Fruits are not particularly good sources of riboflavin. sour cream: Oleksandra Naumenko/Shutterstock; milk: NIPAPORN PANYACHAROEN/Shutterstock; juice: Sergei Vinogradov/seralexvi/123RF; cereal: Joe Belanger/iStock/Getty Images; asparagus Ingram Publishing; beef: MaraZe/Shutterstock; MyPlate: U.S. Department of Agriculture

Sources: Office of Dietary Supplements, Dietary Supplements Fact Sheets, available from https://ods.od.nih.gov/factsheets/list-all; USDA FoodData Central, available from https://fdc.nal.usda.gov.

## GETTING ENOUGH RIBOFLAVIN

On average, daily intakes of riboflavin are slightly above the RDA. As with thiamin, people with alcohol use disorders risk riboflavin deficiency because their eating patterns have low nutrient density. Although not common, others at risk for riboflavin deficiency are athletes who are vegetarian, individuals following a vegan eating plan, and women who are pregnant or lactating and their infants. No observable symptoms indicate that riboflavin taken in megadose supplements is toxic, so no UL has been set. Because we excrete excess riboflavin, riboflavin supplementation, including the riboflavin found in multivitamin supplements or in a heavily fortified breakfast cereal, can cause the urine to become bright yellow.

The grains, dairy, and protein groups of MyPlate contain the most nutrient-dense sources of riboflavin (Fig. 8-19). Foods that are major sources of riboflavin are ready-to-eat breakfast cereals, milk and dairy products, enriched grains, meat, and eggs. Vegetables such as asparagus, broccoli, and various greens are also good sources. Riboflavin is a relatively stable water-soluble vitamin; however, it is destroyed by light. Milk, therefore, is sold in paper or opaque plastic containers rather than clear glass to protect the riboflavin. In the United States, many meet the riboflavin recommendation by consuming fortified sources daily. Riboflavin can also be produced by bacteria in the large intestine. More of this riboflavin is absorbed by the large intestine after consumption of vegetable-based foods compared to meat-based foods.

### ✔ CONCEPT CHECK 8.8

1. How is riboflavin involved in energy metabolism?
2. What body organs or tissues are most likely to show symptoms if there is a deficiency of riboflavin?
3. What types of foods are the best sources of riboflavin?

## 8.9 Niacin (Vitamin B-3)

### FUNCTIONS OF NIACIN

Niacin functions in the body as one of two related compounds: nicotinic acid and nicotinamide. The coenzyme forms of niacin function in many cellular metabolic pathways. When you are generating energy (ATP) by burning carbohydrate and fat, a niacin coenzyme—nicotinamide dinucleotide (NAD) or nicotinamide dinucleotide phosphate (NADP)—is used. Anabolic pathways in the cell—those that make new compounds—also often use a niacin coenzyme.[19] This is especially true for fatty-acid synthesis.

One form of niacin, nicotinic acid, has been promoted as a natural method to lower blood lipids including LDL cholesterol; however, due to potential adverse side effects, its use is discouraged without supervision by a primary care provider.

### NIACIN DEFICIENCY

Because niacin coenzymes function in over 200 enzymatic reactions, niacin deficiency causes widespread problems in the body. Early symptoms include poor appetite, weight loss, and weakness. The distinct group of niacin-deficiency symptoms is known as **pellagra,** which means "rough or painful skin." The symptoms of the disease are **dementia,** diarrhea, and dermatitis. Left untreated, death often results.

Pellagra is the only dietary deficiency disease ever to reach epidemic proportions in the United States. It became a major problem in the southeastern United States in the late 1800s and persisted until the 1930s, when standards of living and eating patterns improved. Pellagra was particularly prevalent in populations that consumed corn as a major component of their eating pattern. Niacin in corn is bound by a protein that inhibits its absorption, making it less bioavailable. Soaking corn in an alkaline solution,

**pellagra** Niacin-deficiency disease characterized by dementia, diarrhea, and dermatitis, and possibly leading to death.

**dementia** A general loss or decrease in mental function.

▲ Corn is treated in an alkaline solution to release protein-bound niacin so it can be available in corn products, such as tortillas, taco shells, tortilla chips, and corn flour. mattjeacock/E+/Getty Images

such as lime water (water with calcium hydroxide), releases bound niacin and renders it more bioavailable. Hispanic people in North America never experienced much pellagra because they traditionally soak corn in lime water before making tortillas. Today, pellagra is rare in Western societies and is typically only seen associated with chronic alcohol use disorders in conjunction with poverty and malnutrition and in those with rare disorders of amino acid metabolism.

### GETTING ENOUGH NIACIN

We can obtain some preformed niacin from foods, and we can synthesize about 50% of the niacin required each day from the amino acid tryptophan. This synthesis requires two other vitamins (riboflavin and vitamin B-6) to function as coenzymes in this chemical conversion. In a dietary pattern, 60 milligrams of tryptophan yield about 1 milligram of niacin.

Intakes of niacin by adults are about double the RDA, not including the contribution from tryptophan. Tables of food composition values also ignore tryptophan contribution. The best food sources of niacin are found in the MyPlate protein group (Fig. 8-20). Major sources of niacin are tuna and other fish, poultry, peanuts, ready-to-eat cereals, beef, and asparagus. Coffee and tea also contribute some niacin to the dietary pattern. Niacin is heat stable; little is lost in cooking.

### AVOIDING TOO MUCH NIACIN

The UL for niacin pertains only to the nicotinic acid form. Side effects include headache, itching, and increased blood flow to the skin because of blood vessel dilation.

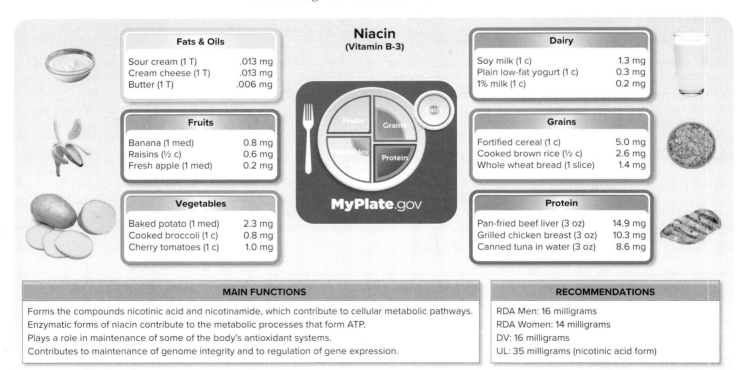

**Niacin**
(Vitamin B-3)

| Fats & Oils | |
|---|---|
| Sour cream (1 T) | .013 mg |
| Cream cheese (1 T) | .013 mg |
| Butter (1 T) | .006 mg |

| Fruits | |
|---|---|
| Banana (1 med) | 0.8 mg |
| Raisins (½ c) | 0.6 mg |
| Fresh apple (1 med) | 0.2 mg |

| Vegetables | |
|---|---|
| Baked potato (1 med) | 2.3 mg |
| Cooked broccoli (1 c) | 0.8 mg |
| Cherry tomatoes (1 c) | 1.0 mg |

| Dairy | |
|---|---|
| Soy milk (1 c) | 1.3 mg |
| Plain low-fat yogurt (1 c) | 0.3 mg |
| 1% milk (1 c) | 0.2 mg |

| Grains | |
|---|---|
| Fortified cereal (1 c) | 5.0 mg |
| Cooked brown rice (½ c) | 2.6 mg |
| Whole wheat bread (1 slice) | 1.4 mg |

| Protein | |
|---|---|
| Pan-fried beef liver (3 oz) | 14.9 mg |
| Grilled chicken breast (3 oz) | 10.3 mg |
| Canned tuna in water (3 oz) | 8.6 mg |

**MAIN FUNCTIONS**

Forms the compounds nicotinic acid and nicotinamide, which contribute to cellular metabolic pathways.
Enzymatic forms of niacin contribute to the metabolic processes that form ATP.
Plays a role in maintenance of some of the body's antioxidant systems.
Contributes to maintenance of genome integrity and to regulation of gene expression.

**RECOMMENDATIONS**

RDA Men: 16 milligrams
RDA Women: 14 milligrams
DV: 16 milligrams
UL: 35 milligrams (nicotinic acid form)

**SPECIAL NOTES**

Societies that rely on corn as a staple food were at high risk for niacin deficiency as the niacin in corn is bound by a protein and less bioavailable.
Niacin deficiency is known as pellagra and characterized by poor appetite, weight loss, weakness, dementia, diarrhea, and dermatitis.

**FIGURE 8-20** ▲ Food sources of niacin. The fill of the background color (none, 1/3, 2/3, or completely covered) within each group on MyPlate indicates the average nutrient density for niacin in that group. The figure shows niacin content of several foods in each food group. Overall, the richest sources of niacin are foods in the protein group and fortified breakfast cereals. Foods in the dairy group contain very little niacin, but the tryptophan in dairy foods can be converted into niacin. sour cream: Oleksandra Naumenko/Shutterstock; milk: NIPAPORN PANYACHAROEN/Shutterstock; banana: lynx/iconotec.com/Glow Images; cereal: Joe Belanger/iStock/Getty Images; potatoes Kaan Ates/Getty Images; chicken: Nycshooter/Michael Krinke/E+/Getty Images; MyPlate: U.S. Department of Agriculture

Sources: Office of Dietary Supplements, Dietary Supplements Fact Sheets, available from https://ods.od.nih.gov/factsheets/list-all; USDA FoodData Central, available from https://fdc.nal.usda.gov.

These symptoms are especially seen when intakes are above 100 milligrams per day. In the long run, GI tract and liver damage are possible, so any use of megadoses, including large doses recommended for treatment for cardiovascular disease, requires close medical monitoring.

✅ **CONCEPT CHECK 8.9**

1. How is niacin involved in energy metabolism?
2. What are the three distinct signs of a niacin deficiency?
3. What are some excellent sources of niacin?
4. What is the relationship between tryptophan and niacin?

## 8.10 Pantothenic Acid (Vitamin B-5)

### FUNCTIONS OF PANTOTHENIC ACID

Pantothenic acid is required for the synthesis of coenzyme A (CoA), a coenzyme in chemical reactions that allow the release of energy from carbohydrates, lipids, and protein. It also activates fatty acids, so they can yield energy and is used in the initial steps of fatty-acid synthesis.[20]

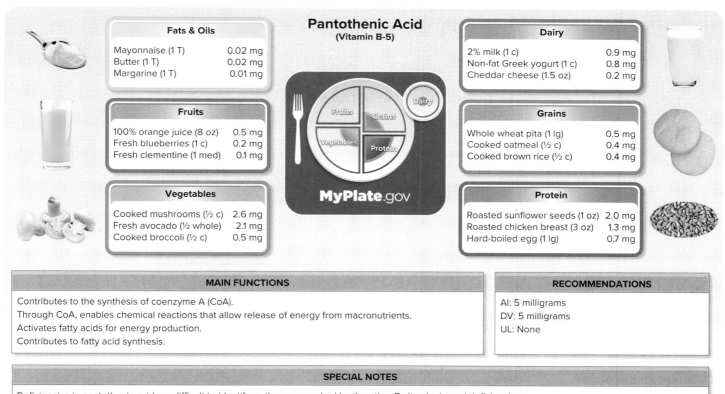

**Pantothenic Acid**
(Vitamin B-5)

**Fats & Oils**
| | |
|---|---|
| Mayonnaise (1 T) | 0.02 mg |
| Butter (1 T) | 0.02 mg |
| Margarine (1 T) | 0.01 mg |

**Fruits**
| | |
|---|---|
| 100% orange juice (8 oz) | 0.5 mg |
| Fresh blueberries (1 c) | 0.2 mg |
| Fresh clementine (1 med) | 0.1 mg |

**Vegetables**
| | |
|---|---|
| Cooked mushrooms (½ c) | 2.6 mg |
| Fresh avocado (½ whole) | 2.1 mg |
| Cooked broccoli (½ c) | 0.5 mg |

**Dairy**
| | |
|---|---|
| 2% milk (1 c) | 0.9 mg |
| Non-fat Greek yogurt (1 c) | 0.8 mg |
| Cheddar cheese (1.5 oz) | 0.2 mg |

**Grains**
| | |
|---|---|
| Whole wheat pita (1 lg) | 0.5 mg |
| Cooked oatmeal (½ c) | 0.4 mg |
| Cooked brown rice (½ c) | 0.4 mg |

**Protein**
| | |
|---|---|
| Roasted sunflower seeds (1 oz) | 2.0 mg |
| Roasted chicken breast (3 oz) | 1.3 mg |
| Hard-boiled egg (1 lg) | 0.7 mg |

**MyPlate**.gov

**MAIN FUNCTIONS**

Contributes to the synthesis of coenzyme A (CoA).
Through CoA, enables chemical reactions that allow release of energy from macronutrients.
Activates fatty acids for energy production.
Contributes to fatty acid synthesis.

**RECOMMENDATIONS**

AI: 5 milligrams
DV: 5 milligrams
UL: None

**SPECIAL NOTES**

Deficiencies in pantothenic acid are difficult to identify as they are masked by the other B vitamin–based deficiencies.
The intestinal microbiota produce small amounts of pantothenic acid, but it's not clear if this is absorbed into the body for utilization.

**FIGURE 8-21** ▲ Food sources of pantothenic acid. The fill of the background color (none, 1/3, 2/3, or completely covered) within each group on MyPlate indicates the average nutrient density for pantothenic acid in that group. The figure shows the pantothenic acid content of several foods in each food group. Overall, fortified foods and foods rich in protein are the best sources of pantothenic acid. mayo: Iconotec/Alamy Stock Photo; milk: NIPAPORN PANYACHAROEN/Shutterstock; juice: Sergei Vinogradov/seralexvi/123RF; pita bread: lynx/iconotec.com/Glowimages; mushrooms: Pixtal/age fotostock; seeds: Glow Images; MyPlate: U.S. Department of Agriculture

Sources: Office of Dietary Supplements, Dietary Supplements Fact Sheets, available from https://ods.od.nih.gov/factsheets/list-all; USDA FoodData Central, available from https://fdc.nal.usda.gov.

## PANTOTHENIC ACID DEFICIENCY

Pantothenic acid is so widespread in foods that a nutritional deficiency among healthy people with a varied eating pattern is unlikely. *Pantothen* means "from every side" in Greek. A deficiency of pantothenic acid might occur in alcohol use disorders along with a nutrient-deficient eating pattern. However, the symptoms would probably be hidden among deficiencies of thiamin, riboflavin, vitamin B-6, and folate, so the pantothenic acid deficiency might be unrecognizable.

## GETTING ENOUGH PANTOTHENIC ACID

The Adequate Intake (AI) set for pantothenic acid is 5 milligrams per day for adults. Average consumption is well in excess of this amount. Pantothenic acid is found in every food group, but the richest sources are sunflower seeds, mushrooms, peanuts, and eggs (Fig. 8-21). Other rich sources are meat, milk, and many vegetables. No observable toxicity is known for pantothenic acid, so no UL has been set.

### ✔ CONCEPT CHECK 8.10

1. What is the role of pantothenic acid in energy metabolism?
2. What are some rich sources of pantothenic acid?

# 8.11 Vitamin B-6 (Pyridoxine)

## FUNCTIONS OF VITAMIN B-6

Vitamin B-6 is known by its number, rather than its general name, and is a family of three structurally similar compounds. All can be converted into the active vitamin B-6 coenzyme, pyridoxal phosphate (PLP). The coenzyme forms of vitamin B-6 are needed for the activity of numerous enzymes involved in carbohydrate, protein, and lipid metabolism. One of the primary functions is as a coenzyme in over a hundred chemical reactions that involve the metabolism of amino acids and protein. The B-6 coenzyme, PLP, participates in reactions that allow the synthesis of nonessential (dispensable) amino acids by helping to split the nitrogen group ($-NH_2$) from an amino acid and making it available to another amino acid.

Vitamin B-6 also plays a role in **homocysteine** metabolism. Among the other important functions of vitamin B-6 are synthesis of neurotransmitters, such as serotonin and gamma aminobutyric acid (GABA); conversion of tryptophan to niacin; breakdown of stored glycogen to glucose; and synthesis of white blood cells and the heme portion of **hemoglobin.**[21]

**homocysteine** An amino acid that arises from the metabolism of methionine. Vitamin B-6, folate, vitamin B-12, and choline are required for its metabolism. Elevated levels are associated with an increased risk of cardiovascular disease.

## VITAMIN B-6 DEFICIENCY

Because of the role of vitamin B-6 in hemoglobin synthesis and amino acid metabolism, a deficiency in vitamin B-6 would affect multiple body systems, including the cardiovascular, immune, and nervous systems, as well as overall energy metabolism. Vitamin B-6 deficiency also results in widespread symptoms, including depression, vomiting, skin disorders, irritation of the nerves, anemia, and impaired immune response.

People with alcohol use disorders are susceptible to a vitamin B-6 deficiency. A metabolite formed in alcohol metabolism can displace the coenzyme form of vitamin B-6, increasing its tendency to be destroyed. In addition, alcohol decreases the absorption of vitamin B-6 and decreases the synthesis of its coenzyme form. Cirrhosis and hepatitis (both can accompany alcohol use disorders) also destroy healthy liver tissue. Thus, a cirrhotic liver cannot adequately metabolize vitamin B-6 or synthesize its coenzyme form.

## GETTING ENOUGH VITAMIN B-6

With their ample consumption of animal products, North Americans have an average daily consumption of vitamin B-6 that is greater than the RDA. There is some research to indicate that athletes may need slightly more vitamin B-6 than sedentary adults. The athlete's body processes large quantities of glycogen and protein, and the metabolism of these compounds requires vitamin B-6. However, unless athletes restrict their food intake, they are likely to consume plenty of this B vitamin.

Major sources of vitamin B-6 are animal products and fortified ready-to-eat breakfast cereals (Fig. 8-22). Other sources are vegetables and fruits such as potatoes, spinach, bananas, and cantaloupes. Overall, the protein group of MyPlate offers most of the food sources of vitamin B-6, and animal sources and fortified grain products are the most reliable because the vitamin B-6 they contain is more absorbable than that in plant foods. Vitamin B-6 is rather unstable; heating and freezing can easily destroy it.

▲ A baked potato is a good plant source of vitamin B-6. DNY59/E+/Getty Images

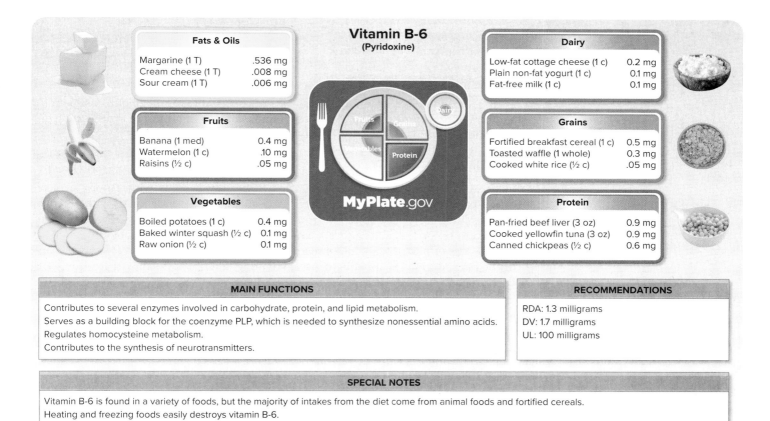

### Vitamin B-6
(Pyridoxine)

**Fats & Oils**

| | |
|---|---|
| Margarine (1 T) | .536 mg |
| Cream cheese (1 T) | .008 mg |
| Sour cream (1 T) | .006 mg |

**Fruits**

| | |
|---|---|
| Banana (1 med) | 0.4 mg |
| Watermelon (1 c) | .10 mg |
| Raisins (½ c) | .05 mg |

**Vegetables**

| | |
|---|---|
| Boiled potatoes (1 c) | 0.4 mg |
| Baked winter squash (½ c) | 0.1 mg |
| Raw onion (½ c) | 0.1 mg |

**Dairy**

| | |
|---|---|
| Low-fat cottage cheese (1 c) | 0.2 mg |
| Plain non-fat yogurt (1 c) | 0.1 mg |
| Fat-free milk (1 c) | 0.1 mg |

**Grains**

| | |
|---|---|
| Fortified breakfast cereal (1 c) | 0.5 mg |
| Toasted waffle (1 whole) | 0.3 mg |
| Cooked white rice (½ c) | .05 mg |

**Protein**

| | |
|---|---|
| Pan-fried beef liver (3 oz) | 0.9 mg |
| Cooked yellowfin tuna (3 oz) | 0.9 mg |
| Canned chickpeas (½ c) | 0.6 mg |

MyPlate.gov

**MAIN FUNCTIONS**

Contributes to several enzymes involved in carbohydrate, protein, and lipid metabolism.
Serves as a building block for the coenzyme PLP, which is needed to synthesize nonessential amino acids.
Regulates homocysteine metabolism.
Contributes to the synthesis of neurotransmitters.

**RECOMMENDATIONS**

RDA: 1.3 milligrams
DV: 1.7 milligrams
UL: 100 milligrams

**SPECIAL NOTES**

Vitamin B-6 is found in a variety of foods, but the majority of intakes from the diet come from animal foods and fortified cereals.
Heating and freezing foods easily destroys vitamin B-6.

**FIGURE 8-22** ▲ Food sources of vitamin B-6. The fill of the background color (none, 1/3, 2/3, or completely covered) within each group on MyPlate indicates the average nutrient density for vitamin B-6 in that group. The figure shows the vitamin B-6 content of several foods in each food group. Overall, the richest and most bioavailable sources of vitamin B-6 are animal sources of protein and fortified breakfast cereals. However, dairy foods are not a particularly good source of vitamin B-6. margarine: 2/James Worrell/Ocean/CORBIS; rice: Oleksandra Naumenko/Shutterstock; banana lynx/iconotec.com/Glow Images; cereal: Joe Belanger/iStock/Getty Images; potatoes: Kaan Ates/Getty Images; chickpeas: BRETT STEVENS/Cultura/Getty Images; MyPlate: U.S. Department of Agriculture

Sources: Office of Dietary Supplements, Dietary Supplements Fact Sheets, available from https://ods.od.nih.gov/factsheets/list-all; USDA FoodData Central, available from https://fdc.nal.usda.gov.

## AVOIDING TOO MUCH VITAMIN B-6

The UL for vitamin B-6 is 100 milligrams per day, based on the risk of developing nerve damage. Studies have shown that intakes of 2 to 6 grams of vitamin B-6 per day for 2 or more months can lead to irreversible nerve damage. Symptoms of vitamin B-6 toxicity include walking difficulties and hand and foot tingling and numbness. Some nerve damage in individual sensory neurons is probably reversible, but damage to the ganglia (where many nerve fibers converge) appears to be permanent. Other effects of excessive vitamin B-6 intake include painful, disfiguring skin lesions, photosensitivity, and gastrointestinal symptoms, such as nausea and heartburn. With 500-milligram tablets of vitamin B-6 available in health-food stores, taking a toxic dose is possible.[21]

### ✓ CONCEPT CHECK 8.11

1. What is the role of vitamin B-6 in energy metabolism and other body functions?
2. What are the primary sources of vitamin B-6?
3. Are vitamin B-6 supplements safe?

## 8.12 Biotin (Vitamin B-7)

### FUNCTIONS OF BIOTIN

In its coenzyme form, biotin aids in dozens of chemical reactions. Biotin assists in the addition of carbon dioxide to other compounds, a reaction critical in synthesizing glucose and fatty acids, as well as in breaking down certain amino acids.[22]

### BIOTIN DEFICIENCY

Biotin deficiency has never been reported in healthy persons with a normal mixed eating pattern. Groups at risk of biotin inadequacy include individuals with a genetic disorder that prevents free biotin from being released, individuals with chronic alcohol exposure, or pregnant and breastfeeding women. Symptoms of biotin deficiency include loss of hair, a scaly inflammation of the skin, changes in the tongue and lips, brittle nails, decreased appetite, nausea, vomiting, a form of anemia, depression, muscle pain and weakness, and poor growth.

### GETTING ENOUGH BIOTIN

Our food supply is thought to provide 40 to 60 micrograms per person per day, more than the AI. Biotin appears relatively nontoxic, thus no UL for biotin has been set.

Protein sources, such as egg yolks, peanuts, and cheese, are good sources of biotin (Fig. 8-23). The biotin content of food is typically not measured and therefore is often not available in food composition tables or nutrient databases. Because intestinal bacteria synthesize some biotin that you can absorb, a biotin deficiency is unlikely. Scientists are not sure how much of the bacteria-synthesized biotin in our intestines is absorbed, so we still need to consume biotin. If bacterial synthesis in the intestines is not sufficient, as in people who are missing a large part of the colon or who take antibiotics for many months, special attention must be paid to meeting biotin needs.

Biotin's bioavailability varies significantly among foods based on the food's biotin–protein complex. In raw egg whites, biotin is bound to avidin, which inhibits absorption of the vitamin. Consuming many raw egg whites can eventually lead to biotin-deficiency disease. Cooking, however, denatures the protein avidin in eggs so it cannot bind biotin. In addition to food safety concerns, this is one of the important reasons to avoid consuming raw eggs.[22]

▲ Eating a lot of raw eggs can lead to a biotin deficiency. Biotin is bound to avidin in raw egg whites, inhibiting its absorption. Cooking denatures avidin and releases biotin. Ingram Publishing

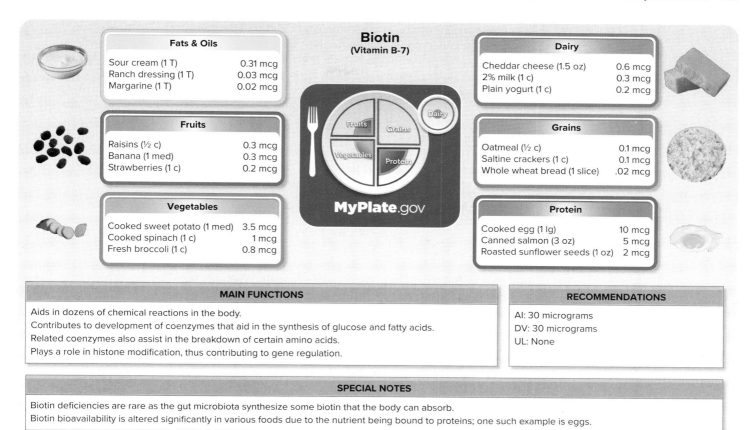

**FIGURE 8-23** ▲ Food sources of biotin. The fill of the background color (none, 1/3, 2/3, or completely covered) within each group on MyPlate indicates the average nutrient density for biotin in that group. The figure shows the biotin content of several foods in each food group. Overall, foods rich in protein are the best sources of biotin. Grains (even fortified varieties) contain very little biotin. sour cream: Oleksandra Naumenko/ Shutterstock; cheese: Brent Hofacker/Shutterstock; raisins: lynx/iconotec.com/Glowimages; oatmeal: McGraw-Hill Education; sweet potatoes lynx/iconotec.com/Glow Images; egg: McGraw-Hill Education; MyPlate: U.S. Department of Agriculture

Sources: Office of Dietary Supplements, Dietary Supplements Fact Sheets, available from https://ods.od.nih.gov/factsheets/list-all; USDA FoodData Central, available from https:// fdc.nal.usda.gov.

✓ **CONCEPT CHECK 8.12**

1. What is the role of biotin in energy metabolism?
2. What are the signs and symptoms of biotin deficiency?
3. What are the best sources of biotin?
4. Why does consumption of raw eggs lead to biotin deficiency?

# 8.13 Folate (Vitamin B-9)

The term *folate* is used to describe a variety of forms of this B vitamin found in foods and in the body. *Folic acid* is the synthetic form added to fortified foods and present in supplements.[23]

Folate recommendations for all but women of childbearing age are based on dietary folate equivalents (DFE). Synthetic folic acid, found in supplements and fortified foods, is more bioavailable than the folate that naturally occurs in food. The DFE unit takes into account these differences in bioavailability.

## FUNCTIONS OF FOLATE

A key role of the folate coenzyme is to supply or accept single carbon compounds. In this role, folate coenzymes aid in the synthesis of DNA and metabolism of amino acids. For example, folate works along with vitamin B-6 and vitamin B-12 to metabolize homocysteine.

▶ **Dietary Folate Equivalents (DFE)**

1 DFE = 1 microgram folate from food

= 0.6 microgram folic acid from fortified food

= 0.5 microgram folic acid from a supplement on an empty stomach

Folate also functions in the formation of neurotransmitters in the brain. Meeting the RDA for folate can improve symptoms of depression in some cases of mental illness.[24]

Research is also underway on the role of folate in cancer protection. Because folate aids in DNA synthesis, adequate folate status is important to maintain DNA integrity, including the control of certain cancer-promoting genes. Meeting the RDA for folate may be one way to reduce risk for some types of cancers. However, the influence of folate on cancer risk is complex; genetic variations in folate metabolism play a role. Folic acid supplements are not recommended for cancer prevention or treatment, but food sources of folate are a safe way to meet the RDA. However, studies of folate's effect on cancer risk vary based on the type of cancer. Furthermore, there are genetic variations in the way individuals metabolize folate.[25]

## FOLATE DEFICIENCY

One of the major results of a folate deficiency occurs in the early phases of red blood cell synthesis. With inadequate folate, immature red blood cells cannot divide because they cannot form new DNA. The cells grow progressively larger because they are still synthesizing protein and other cell parts to make new cells. When the time comes for the cells to divide, however, the amount of DNA is insufficient to form two nuclei. The cells thus remain in a large immature form, have a large nucleus, and are known as **macrocytes** or megaloblasts.

With the bone marrow of a folate-deficient person producing mostly immature macrocytes, few mature red blood cells arrive in the bloodstream. When fewer mature red blood cells are present, the blood's capacity to carry oxygen decreases, causing a condition known as **macrocytic anemia** (also called *megaloblastic* or large-cell anemia) (Fig. 8-24).

Clinicians focus on red blood cells as an indicator of folate status because they are easy to collect and examine. Folate deficiency, however, disrupts cell division throughout the entire body. Other symptoms of folate deficiency are inflammation of the tongue, diarrhea, poor growth, mental confusion, depression, and problems with nerve function. Another result of folate deficiency is elevated blood homocysteine levels, which have been associated with cardiovascular disease and are considered

**macrocyte** A large, immature red blood cell that results from the inability of the cell to divide normally; also called *megaloblast*.

**macrocytic anemia** Anemia characterized by the presence of abnormally large red blood cells; also called *megaloblasic anemia*.

**FIGURE 8-24** ▶ Macrocytic (megaloblastic) anemia occurs when red blood cells are unable to divide, leaving large, immature red blood cells. Either a folate or vitamin B-12 deficiency may cause this condition. Measurements of blood concentrations of both vitamins are taken to help determine the cause of the anemia.

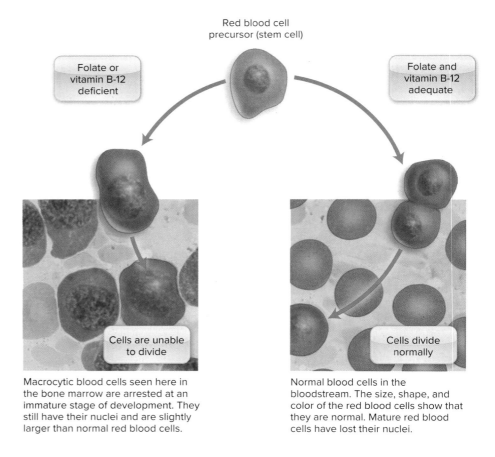

Macrocytic blood cells seen here in the bone marrow are arrested at an immature stage of development. They still have their nuclei and are slightly larger than normal red blood cells.

Normal blood cells in the bloodstream. The size, shape, and color of the red blood cells show that they are normal. Mature red blood cells have lost their nuclei.

an independent risk factor for atherosclerosis. Research is ongoing to examine the impact of folate (and other B vitamins) on cardiovascular disease risk.[26]

Maternal folate deficiency (along with a genetic abnormality related to folate metabolism) has been linked to the development of **neural tube defects** in the fetus. Examples of neural tube defects include **spina bifida** (spinal cord or spinal fluid bulge through the back) and **anencephaly** (absence of a major portion of the brain). Low maternal plasma folate and vitamin B-12 levels result in high plasma homocysteine levels and are risk factors for neural tube defects. Adequate folate status is crucial for all women of childbearing age, pregnant or not, because the neural tube closes within the first 28 days of pregnancy, a time when many women are not even aware that they are pregnant. Research suggests that the risk for neural tube defects can be decreased by folate supplementation for mothers-to-be.[27]

As they age, some adults may be at risk for folate deficiency due to a combination of inadequate folate intake and decreased absorption. Perhaps these people fail to consume sufficient amounts of fruits and vegetables because of limited financial resources or physical problems, such as poor dental health. In addition, folate deficiencies often occur with alcohol use disorders, due mostly to inadequate intake and impaired absorption. Symptoms of a folate-related anemia can alert a primary care provider to the possibility of alcohol use disorders.

**neural tube defect** A defect in the formation of the neural tube occurring during early fetal development. This type of defect results in various nervous system disorders, such as spina bifida. Folate deficiency in the pregnant woman increases the risk that the fetus will develop this disorder.

**spina bifida** Type of neural tube defect resulting from improper closure of the neural tube during embryonic development. The spinal cord or fluid may bulge outside the spinal column.

**anencephaly** Type of neural tube defect characterized by the absence of some or all of the brain and skull.

## GETTING ENOUGH FOLATE

Folate's name is derived from the Latin word *folium,* which means "foliage" or "leaves" Quite predictably, the richest sources of folate are green leafy vegetables. In addition, other vegetables, orange juice, dried beans, and organ meats are excellent sources of folate (Fig. 8-25). Fortified ready-to-eat breakfast cereals, bread, and milk are important sources of folic acid for many adults.

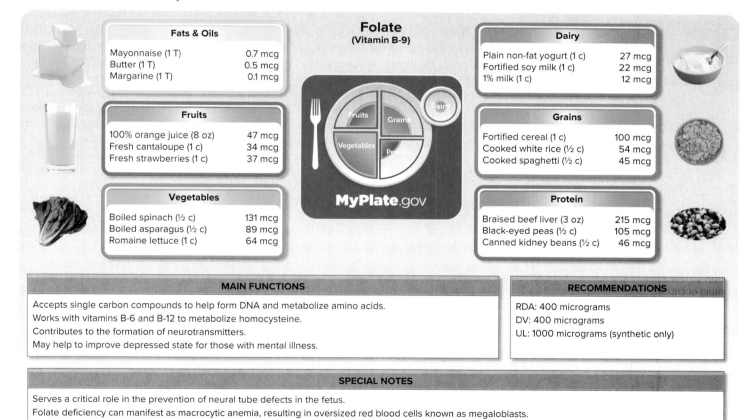

**Folate**
(Vitamin B-9)

| Fats & Oils | |
|---|---|
| Mayonnaise (1 T) | 0.7 mcg |
| Butter (1 T) | 0.5 mcg |
| Margarine (1 T) | 0.1 mcg |

| Fruits | |
|---|---|
| 100% orange juice (8 oz) | 47 mcg |
| Fresh cantaloupe (1 c) | 34 mcg |
| Fresh strawberries (1 c) | 37 mcg |

| Vegetables | |
|---|---|
| Boiled spinach (½ c) | 131 mcg |
| Boiled asparagus (½ c) | 89 mcg |
| Romaine lettuce (1 c) | 64 mcg |

| Dairy | |
|---|---|
| Plain non-fat yogurt (1 c) | 27 mcg |
| Fortified soy milk (1 c) | 22 mcg |
| 1% milk (1 c) | 12 mcg |

| Grains | |
|---|---|
| Fortified cereal (1 c) | 100 mcg |
| Cooked white rice (½ c) | 54 mcg |
| Cooked spaghetti (½ c) | 45 mcg |

| Protein | |
|---|---|
| Braised beef liver (3 oz) | 215 mcg |
| Black-eyed peas (½ c) | 105 mcg |
| Canned kidney beans (½ c) | 46 mcg |

**MyPlate**.gov

**MAIN FUNCTIONS**

Accepts single carbon compounds to help form DNA and metabolize amino acids.
Works with vitamins B-6 and B-12 to metabolize homocysteine.
Contributes to the formation of neurotransmitters.
May help to improve depressed state for those with mental illness.

**RECOMMENDATIONS**

RDA: 400 micrograms
DV: 400 micrograms
UL: 1000 micrograms (synthetic only)

**SPECIAL NOTES**

Serves a critical role in the prevention of neural tube defects in the fetus.
Folate deficiency can manifest as macrocytic anemia, resulting in oversized red blood cells known as megaloblasts.

**FIGURE 8-25** ▲ Food sources of folate. The fill of the background color (none, 1/3, 2/3, or completely covered) within each food group on MyPlate indicates the average nutrient density for folate in that group. The figure shows the folate content of several foods in each food group. Overall, the richest sources of folate are green, leafy vegetables and fortified grains. margarine: 2/James Worrell/Ocean/CORBIS; Greek yogurt: Anastasios71/Shutterstock; juice: Sergei Vinogradov/seralexvi/123RF; cereal: Joe Belanger/iStock/Getty Images; lettuce: Burke Triolo Productions/Getty Images; black-eyed peas: Lluis Real/AGE Fotostock; MyPlate: U.S. Department of Agriculture

Sources: Office of Dietary Supplements, Dietary Supplements Fact Sheets, available from https://ods.od.nih.gov/factsheets/list-all; USDA FoodData Central, available from https://fdc.nal.usda.gov.

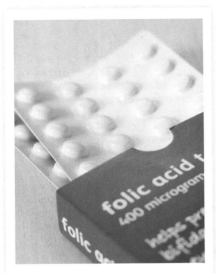

To lower the risk for neural tube defects, women of childbearing age should take a daily multivitamin and mineral supplement that contains 400 micrograms of folic acid. **How does the bioavailability of folic acid in supplements and fortified foods compare to the bioavailability of folate found naturally in foods?** Banana Stock/Getty Images

Folate is susceptible to destruction by heat and oxygen. The vitamin C present in some food sources of folate, such as orange juice, helps to reduce folate destruction, but food processing and preparation destroy 50% to 90% of the folate in food. This underscores the importance of regularly eating fresh fruits and raw or lightly cooked vegetables.

Pregnant women need extra folate (a total of 600 micrograms dietary folate equivalent [DFE]) to accommodate the increased rates of cell division and DNA synthesis in their bodies and in the developing fetus. A healthy eating pattern can supply this much. Still, prenatal care often includes a specially formulated multivitamin and mineral supplement enriched with folic acid to meet the higher RDA during pregnancy.

In 1998, the FDA mandated the fortification of refined grain products with folate with the aim of reducing birth defects of the spine. With this program, average intakes have increased by about 200 micrograms per day. Factors other than folate deficiency (e.g., genetics and environment) play a role in the development of neural tube defects, but studies have shown fortification of grain products with folic acid has decreased rates of neural tube defects in infants by an estimated 15% to 30% in the United States and up to 50% in other countries with higher background rates of neural tube defects.

The folic acid enrichment of grains has also been accompanied by a noticeable decline in cardiovascular risk, especially risk for stroke, due to a drop in blood homocysteine levels among U.S. adults. Supplements of folic acid, B-12, and B-6 have been promoted to help lower homocysteine and decrease cardiac and stroke risk. This is only likely to be effective for individuals who start out with elevated homocysteine levels. Taking pharmacological doses of folate is not recommended for people who have blood homocysteine levels in the normal range.

## AVOIDING TOO MUCH FOLATE

The UL for folate only refers to synthetic folic acid, because the absorption of the natural form of folate in food is limited. Some research indicates that too much folic acid could promote tumor development. Thus, even though folic acid fortification has been a public health success story for prevention of neural tube defects, there are concerns about the appropriate dose for the entire population.[28] In addition, large doses of folic acid can hide the signs of vitamin B-12 deficiency and therefore complicate its diagnosis. Specifically, regular consumption of large amounts of folate can prevent the appearance of an early warning sign of vitamin B-12 deficiency—enlarged red blood cells. For this reason, the FDA limits the amount of folic acid in supplements (for nonpregnant adults) to 400 micrograms.

### ✓ CONCEPT CHECK 8.13

1. Explain what happens to red blood cells in macrocytic anemia.
2. Why do folate needs increase from 400 micrograms to 600 micrograms per day for pregnant women?

## 8.14 Vitamin B-12 (Cobalamin or Cyanocobalamin)

Vitamin B-12 is a unique vitamin. Its structure is the largest of all the vitamins, and it is the only vitamin that contains a mineral as part of its structure. Unlike most water-soluble vitamins, B-12 can be stored to a significant extent in the liver, so it takes many months of an eating pattern devoid of vitamin B-12 for a deficiency to occur. Vitamin B-12 is only naturally found in foods of animal origin.[29] Finally, the means by which the body absorbs vitamin B-12 is complex; a problem at any of several steps of digestion and absorption could lead to deficiency. To illustrate the multistep process by which vitamin B-12 is absorbed, we will trace the path of a meal containing vitamin B-12 through the digestive tract (Fig. 8-26).

**Vitamin B-12 Digestion and Absorption**

**Mouth/Salivary Glands**

Salivary glands in the mouth produce *R-protein*.

**Liver/Gallbladder**

Some vitamin B-12 is stored in the liver.

**Pancreas**

The pancreas secretes *trypsin* into the small intestine.

**Stomach**

In the stomach, *HCl* and *pepsin* release vitamin B-12 from food protein. The free vitamin B-12 binds to R-protein (from the mouth). Parietal cells of the stomach secrete *intrinsic factor*.

**Small Intestine**

Trypsin (from the pancreas) releases vitamin B-12 from R-protein. The free vitamin B-12 links with intrinsic factor (from the stomach). When chyme reaches the ileum, the vitamin B-12/intrinsic factor complex is absorbed into the blood and binds with a transport protein.

**FIGURE 8-26** ▲ The absorption of vitamin B-12 requires several compounds produced in the mouth, stomach, and small intestine. Defects arising in the stomach or small intestine can interfere with absorption and result in vitamin B-12 deficiency.

In food, much of the vitamin B-12 is bound to protein and therefore cannot be absorbed. When food enters the mouth, **R-proteins** are secreted by the salivary glands. The bolus of food, including the R-proteins, travels down the esophagus to the stomach. Acid and enzymes present in the stomach release vitamin B-12 from food proteins, and the free vitamin B-12 then binds to R-protein. While food is in the stomach, the stomach cells release **intrinsic factor,** a protein-like compound. When the chyme reaches the duodenum, pancreatic enzymes release vitamin B-12 from R-proteins. The free vitamin B-12 then combines with intrinsic factor. The vitamin B-12–intrinsic factor complex travels the length of the small intestine to the ileum, where vitamin B-12 is finally absorbed.

If any of these steps fails or is altered, absorption of vitamin B-12 can drop to 1% to 2%. In these cases of malabsorption, the person usually takes monthly injections of vitamin B-12, uses nasal gels of the vitamin to bypass the need for intestinal absorption, or takes megadoses of a supplemental form (300 times the RDA). With this large dose, an adequate quantity of vitamin B-12 is able to cross the intestinal barrier via simple diffusion.

Most cases of vitamin B-12 deficiencies in healthy people result from defective absorption rather than from inadequate intakes. This is especially true for adults who are older. As we age, stomach acid production declines and our stomachs have a decreased ability to synthesize the intrinsic factor needed for adequate vitamin B-12 absorption.

**R-proteins** Proteins produced by the salivary glands that bind to free vitamin B-12 in the stomach and protect it from stomach acid.

▲ Vitamin B-12 is only naturally present in animal foods. With age, absorption of vitamin B-12 from food becomes less efficient, usually due to decreases in stomach-acid production. bitt24/Shutterstock

## FUNCTIONS OF VITAMIN B-12

The most important function of vitamin B-12 is participating in folate metabolism. Vitamin B-12 is required to convert folate coenzymes into the active forms of folate needed for metabolic reactions, such as DNA synthesis. Without vitamin B-12, reactions that require certain active forms of folate do not take place in the cell. Thus, a deficiency of vitamin B-12 can result in symptoms of a folate deficiency, including elevated homocysteine levels and macrocytic anemia.

Another vital function of vitamin B-12 is maintaining the **myelin** sheath that insulates neurons. Initial neurological symptoms of a vitamin B-12 deficiency include dysfunctional muscle control and impaired reflexes. Eventual destruction of the myelin sheath causes paralysis. If left untreated, vitamin B-12 deficiency eventually leads to death, mainly due to the destruction of nerves.

## VITAMIN B-12 DEFICIENCY

**pernicious anemia** The anemia that results from a lack of vitamin B-12 absorption; it is *pernicious* because of associated nerve degeneration that can result in eventual paralysis and death.

The fatal consequences of vitamin B-12 deficiency can be observed in a disease known as **pernicious anemia.** The word *pernicious* means "leading to death." Indeed, before the discovery that consuming large doses of raw liver—a rich source of vitamin B-12—could be used to treat this particular disease, many people did die from it. Pernicious anemia is characterized by macrocytic anemia, sore mouth, depression, back pain, apathy, and severe nerve degeneration. Nerve damage can lead to tingling in the extremities, weakness, paralysis, and eventually death from heart failure.

People with pernicious anemia usually do not lack vitamin B-12 in their dietary pattern. Instead, they suffer from an autoimmune disease that destroys the stomach cells which produce gastric acid and intrinsic factor. The resulting malabsorption of vitamin B-12 is responsible for all the symptoms of the disease. Because we are able to store some vitamin B-12, symptoms of nerve destruction do not develop until after about 3 years from the onset of the disease. Unfortunately, substantial nerve destruction often occurs before clinical signs of deficiency, such as anemia, are detected. The nerve destruction is irreversible.

Pernicious anemia, which affects about 2% of older adults, is the most common cause of vitamin B-12 malabsorption. Other causes are age-related atrophy of the acid-producing cells of the stomach and bacterial overgrowth in the small intestine. When acid production is low, bacteria normally present in the large intestine may colonize the small intestine and compete with our intestinal cells for vitamin B-12 absorption. Certain medications, such as antacids and metformin, also impair vitamin B-12 absorption.

Inadequate intake of vitamin B-12 is rarely responsible for deficiency, but it can occur. Vegan dietary patterns supply little vitamin B-12 unless they include vitamin B-12–fortified foods (e.g., soy milk) or supplements. Infants breastfed by vegetarian mothers are at risk for vitamin B-12 deficiency accompanied by anemia and long-term nervous system problems, such as diminished brain growth, degeneration of the spinal cord, and poor intellectual development. The problems may have their origins during pregnancy if the mother is deficient in vitamin B-12. Certainly, achieving an adequate vitamin B-12 intake is a key goal for individuals following a vegan eating pattern.

## GETTING ENOUGH VITAMIN B-12

Nutritional yeast is a rich source of vitamin B-12 for people who follow a vegetarian eating pattern. Usually, it is made of a dried strain of yeast called *Saccharomyces cerevisiae.* Its flavor is often described as cheesy or nutty. Mix it into sauces or soups. Sprinkle it over pasta, salad, or popcorn. **Why do people who follow a vegan dietary pattern need to be concerned about finding a good source of vitamin B-12?** Stephanie Frey/123RF

Vitamin B-12 compounds are originally synthesized by bacteria, fungi, and other lower organisms, then become incorporated into animal tissues when animals consume them. Organ meats (especially liver, kidneys, and heart) are especially rich sources of vitamin B-12. Other major sources of vitamin B-12 include meat, seafood, ready-to-eat breakfast cereals, milk, and eggs (Fig. 8-27). Adults over age 50 are often encouraged to seek a synthetic vitamin B-12 source to increase absolute absorption, which can be limited due to both reduced intrinsic factor and stomach-acid output. Synthetic vitamin B-12 is not food bound, so stomach acid is not required to release it from food. It will be more readily absorbed than the form found in food. Ready-to-eat breakfast cereals and dietary supplements are two possible synthetic sources.

On average, adults consume two times the RDA or more. This high intake provides the average omnivore with 2 to 3 years' storage of vitamin B-12 in the liver. A person would have to consume a dietary pattern essentially free of vitamin B-12 for approximately 20 years before exhibiting nerve destruction caused by a dietary deficiency. Still, vegans (who eat no animal products) should find a reliable source of vitamin B-12, such

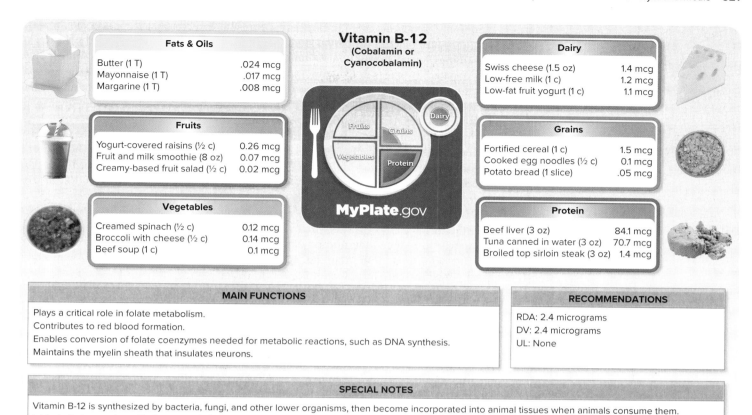

**FIGURE 8-27** ▲ Food sources of vitamin B-12. The fill of the background color (none, 1/3, 2/3, or completely covered) within each food group on MyPlate indicates the average nutrient density for vitamin B-12 in that group. The figure shows the vitamin B-12 content of several foods in each food group. Overall, foods of animal origin and fortified grains are the richest sources of vitamin B-12. Except for fortified grains, foods of plant origin do not contain vitamin B-12. butter: 2/James Worrell/Ocean/CORBIS; cheese: Comstock/Jupiter Images/Getty Images; smoothie: 5 second Studio/Shutterstock; cereal: Joe Belanger/iStock/Getty Images; soup: James Gathany/CDC; tuna: lynx/iconotec.com/Glow Images; MyPlate: U.S. Department of Agriculture

Sources: Office of Dietary Supplements, Dietary Supplements Fact Sheets, available from https://ods.od.nih.gov/factsheets/list-all; USDA FoodData Central, available from https://fdc.nal.usda.gov.

as fortified soy or rice milk, ready-to-eat breakfast cereals, and a form of yeast grown on media rich in vitamin B-12. Use of a multivitamin and mineral supplement containing vitamin B-12 is another option.[30] Vitamin B-12 supplements are essentially nontoxic, so no UL has been set.

### ✓ CONCEPT CHECK 8.14

1. How does a vitamin B-12 deficiency lead to macrocytic anemia?
2. What role does vitamin B-12 play in the health of the brain and nervous system?
3. Identify two population groups that are at risk for vitamin B-12 deficiency. Explain why these people are at risk.
4. List three rich food sources of vitamin B-12.

# 8.15 Vitamin C (Ascorbic Acid)

## FUNCTIONS OF VITAMIN C

**Supporting Body Defenses.** Vitamin C (also known as *ascorbic acid* or *ascorbate*) supports body defenses in two ways: it functions as an antioxidant, and it is necessary for the health of white blood cells.[31]

**nitrosamine** A carcinogen formed from nitrates and breakdown products of amino acids; associated with cancer risk.

Vitamin C functions as an antioxidant because it can readily accept and donate electrons. These antioxidant properties are thought to reduce the formation of cancer-causing **nitrosamines** in the stomach. Vitamin C also aids in the reactivation of vitamin E after it has donated an electron to a free radical. Population studies suggest that the antioxidant properties of vitamin C may be effective in the prevention of certain cancers and cataracts. The extent to which vitamin C functions in the reduction of specific diseases is debatable based on the scientific studies to date.

As part of its antioxidant role, vitamin C assists the immune system. Vitamin C protects the body's cells from the harmful oxidants that are generated as part of the immune response. Beyond its antioxidant activity, vitamin C is vital for the proper function of the immune system because it promotes the proliferation of white blood cells. Recall that certain white blood cells use oxidation reactions to destroy pathogens.

Can taking vitamin C fend off the common cold? Numerous well-designed, double-blind studies have failed to show that vitamin C *prevents* colds. Nevertheless, vitamin C does appear to reduce the duration of symptoms by a day or so and to lessen the severity of the symptoms.[32] For vitamin C to be effective, start taking it as soon as symptoms appear. Once the cold has taken hold, it is too late!

**Formation of Body Proteins.** The best understood function of vitamin C is its role in the synthesis of collagen. This protein is highly concentrated in connective tissue, bone, teeth, tendons, and blood vessels. The important function of vitamin C in the formation of connective tissue is exemplified in the early symptoms of a deficiency: pinpoint hemorrhages under the skin (Fig. 8-28), bleeding gums, and joint pain. Vitamin C is very important for wound healing; it strengthens structural tissues by increasing the cross-connections between amino acids found in collagen.

**FIGURE 8-28 ▲** Pinpoint hemorrhages of the skin—an early symptom of scurvy. The spots on the skin are caused by slight bleeding. The person may experience poor wound healing. These are signs of defective collagen synthesis. Dr. P. Marazzi/ Science Source

**Formation of Other Compounds.** Vitamin C has a specific function in the synthesis of numerous other compounds in the body. It is required for the synthesis of carnitine, a compound that transports fatty acids into the mitochondria so they can be used as fuel. In addition, it takes part in the formation of two neurotransmitters, serotonin and norepinephrine.

**Absorption of Iron.** Vitamin C enhances iron absorption by keeping iron in its most absorbable form, especially as the mineral travels through the alkaline environment of the small intestine. Consuming 75 milligrams (the amount of vitamin C in one orange) or more of vitamin C at a meal significantly increases absorption of the iron consumed at that meal. Increasing intake of vitamin C–rich foods is beneficial for those with poor iron status or for those who choose to limit iron-rich food sources. Iron deficiency is the most common nutritional deficiency worldwide.

## VITAMIN C DEFICIENCY

On long sea voyages before the mid-eighteenth century, half or more of sailing crews died due to scurvy, the vitamin C–deficiency disease. The symptoms of scurvy, which include bleeding gums, tooth loss, bruising, and scaly skin, illustrate the important function of vitamin C in the formation of connective tissue. Without vitamin C, the skin and blood vessels weaken and wounds will not heal. In 1740, the Englishman Dr. James Lind first showed that citrus fruits—two oranges and one lemon a day—could prevent the development of scurvy. Fifty years after Lind's discovery, daily rations for British sailors included limes (thus their nickname, *limeys*). Even after this discovery, scurvy continued to affect many people, including thousands who died during the American Civil War owing to inadequate intake of vitamin C.

## GETTING ENOUGH VITAMIN C

Fresh, ripe fruits and vegetables are loaded with vitamin C. Besides the foods listed in Figure 8-29, other citrus fruits, papayas, cauliflower, and other vegetables pack a lot of vitamin C into a low-calorie package. Ready-to-eat breakfast cereals,

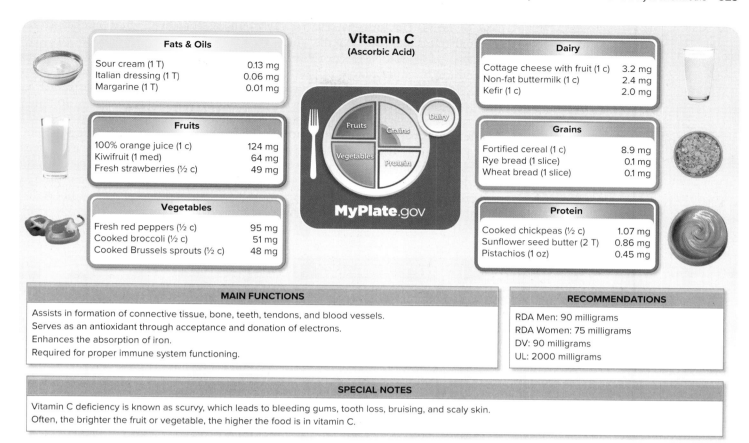

### Vitamin C (Ascorbic Acid)

**Fats & Oils**

| | |
|---|---|
| Sour cream (1 T) | 0.13 mg |
| Italian dressing (1 T) | 0.06 mg |
| Margarine (1 T) | 0.01 mg |

**Fruits**

| | |
|---|---|
| 100% orange juice (1 c) | 124 mg |
| Kiwifruit (1 med) | 64 mg |
| Fresh strawberries (½ c) | 49 mg |

**Vegetables**

| | |
|---|---|
| Fresh red peppers (½ c) | 95 mg |
| Cooked broccoli (½ c) | 51 mg |
| Cooked Brussels sprouts (½ c) | 48 mg |

**Dairy**

| | |
|---|---|
| Cottage cheese with fruit (1 c) | 3.2 mg |
| Non-fat buttermilk (1 c) | 2.4 mg |
| Kefir (1 c) | 2.0 mg |

**Grains**

| | |
|---|---|
| Fortified cereal (1 c) | 8.9 mg |
| Rye bread (1 slice) | 0.1 mg |
| Wheat bread (1 slice) | 0.1 mg |

**Protein**

| | |
|---|---|
| Cooked chickpeas (½ c) | 1.07 mg |
| Sunflower seed butter (2 T) | 0.86 mg |
| Pistachios (1 oz) | 0.45 mg |

**MAIN FUNCTIONS**

Assists in formation of connective tissue, bone, teeth, tendons, and blood vessels.
Serves as an antioxidant through acceptance and donation of electrons.
Enhances the absorption of iron.
Required for proper immune system functioning.

**RECOMMENDATIONS**

RDA Men: 90 milligrams
RDA Women: 75 milligrams
DV: 90 milligrams
UL: 2000 milligrams

**SPECIAL NOTES**

Vitamin C deficiency is known as scurvy, which leads to bleeding gums, tooth loss, bruising, and scaly skin.
Often, the brighter the fruit or vegetable, the higher the food is in vitamin C.

**FIGURE 8-29** ▲ Food sources of vitamin C. The fill of the background color (none, 1/3, 2/3, or completely covered) within each food group on MyPlate indicates the average nutrient density for vitamin C in that group. The figure shows the vitamin C content of several foods in each food group. Overall, fruits and vegetables are the richest sources of vitamin C. Foods in the dairy and protein groups are poor sources of vitamin C. sour cream: Oleksandra Naumenko/Shutterstock; buttermilk: FoodCollection; juice: Sergei Vinogradov/seralexvi/123RF; cereal: Joe Belanger/iStock/Getty Images; peppers: lynx/iconotec.com/Glow Images; sunflower butter: D. Hurst/Alamy; MyPlate: U.S. Department of Agriculture

Sources: Office of Dietary Supplements, Dietary Supplements Fact Sheets, available from https://ods.od.nih.gov/factsheets/list-all; USDA FoodData Central, available from https://fdc.nal.usda.gov.

potatoes, and fortified fruit drinks are also good sources of vitamin C. Five to nine servings of fruits and vegetables can easily provide enough vitamin C to meet the RDA. The brighter the fruit or vegetable, the higher it tends to be in vitamin C. For example, while a green bell pepper has 60 milligrams of vitamin C, a red bell pepper provides 95 milligrams of vitamin C. Keep in mind, however, that vitamin C is rapidly lost in processing and cooking as it is unstable in the presence of heat, iron, copper, or oxygen and is water soluble. Boiling fruits and vegetables can destroy much of the vitamin C or cause it to leach out of the food. Extended time on grocery store shelves or on your countertop at home will also decrease vitamin C content. Thus, you should consume fresh fruits and vegetables as soon after harvest as possible.

The RDA for vitamin C is 75 milligrams per day for adult women and 90 milligrams per day for adult men. According to the FDA's new food labeling rules, the Daily Value used on food and supplement labels is 90 milligrams. Tobacco users need to add an extra 35 milligrams per day to the RDA. The toxic by-products of cigarette smoke and the oxidizing agents found in tobacco products increase the need for the antioxidant action of vitamin C. Average daily consumption of vitamin C in the United States is in the range of 70 to 100 milligrams, and at this level of intake, absorption efficiency is about 80% to 90%.

## AVOIDING TOO MUCH VITAMIN C

The Upper Level (UL) of vitamin C is 2000 milligrams. Note that when vitamin C is consumed in large doses, the amount in excess of daily needs mostly ends up in the

▲ Citrus fruits are good sources of vitamin C. Maria Uspenskaya/Shutterstock

feces or urine. The kidneys start rapidly excreting vitamin C when intakes exceed 100 milligrams per day. As the amount ingested increases, absorption efficiency decreases precipitously—to approximately 50% with intake of 1000 milligrams per day and to 20% with intakes of 6000 milligrams daily. Regular consumption of more than 2000 milligrams per day may cause stomach inflammation and diarrhea. Even 1000-milligram supplement pills can cause some nausea and GI distress. Ingesting large amounts of vitamin C supplements is discouraged in people predisposed to kidney stones. Because vitamin C enhances the absorption of iron, vitamin C supplements are also not recommended for those who overabsorb iron or have excessive iron stores. High doses of vitamin C may interfere with medical tests for diabetes or blood in the feces. If you take vitamin C supplements at any dose, be sure to inform your primary care provider. Primary care providers may misdiagnose conditions if they do not realize the influence of large doses of vitamins on your medical test results.

## ✓ CONCEPT CHECK 8.15

1. How does the antioxidant role of vitamin C assist the immune system?
2. How do the signs of vitamin C deficiency relate to the many roles of the vitamin discussed in this chapter?
3. Why are fresh foods the best sources of vitamin C?

# 8.16 Choline and Other Vitamin-Like Compounds

▲ Choline is important for proper development of the fetal brain. Milk and other dairy products supply some choline. Vadim Guzhva/123RF

Our knowledge of micronutrients continues to evolve. The first vitamin was isolated in 1913—a mere century ago—and within 25 years, all the vitamins we know to be essential in the human dietary pattern were identified. For choline, a water-soluble compound that is important for many aspects of human health, the story is still unfolding.

In 1998, the National Academy of Medicine recognized choline as an essential nutrient. When the Dietary Reference Intakes were released in 2000, only limited research on the dietary requirements for choline existed. One study of male volunteers showed decreased choline stores and liver damage when they were fed choline-deficient intravenous nutrition solutions. Based on this human study, plus laboratory animal studies, choline has been deemed essential and *vitamin-like,* but it is not yet classified as a vitamin.[33]

### FUNCTIONS OF CHOLINE

Despite its lack of vitamin status, choline is needed by all cells and plays several important roles in the body.

**Cell Membrane Structure.** Choline is a precursor for several phospholipids. Phosphatidylcholine (also known as *lecithin*) accounts for about half of the phospholipids in cell membranes. Recall that phospholipids contribute to the flexibility of cell membranes and allow for the presence of both water- and fat-soluble compounds in cell membranes. With its role in cell membrane structure, choline is important for the health of every cell and particularly for the health of brain tissue, where it is present in high levels.

**Single-Carbon Metabolism.** Choline is a precursor for betaine, a compound that participates in many chemical reactions that involve the transfer of single-carbon groups in metabolism. Important examples of metabolic pathways that involve the transfer of single-carbon groups include the synthesis of neurotransmitters, modifications of DNA during embryonic development, and the metabolism of homocysteine. High levels of homocysteine in the blood are related to increased risk of heart disease. Betaine and the B vitamin folate both donate single-carbon groups to convert homocysteine to another compound, thus reducing levels of homocysteine in the blood.

Some research suggests a potential role of adequate choline for the prevention of birth defects. Choline's purported role in prevention of birth defects is similar to that of folate. Both folate and choline are involved in formation of DNA during embryonic development. Problems with DNA formation lead to birth defects. Indeed, animal studies show that maternal choline supplementation during critical stages of embryonic development can improve learning and memory in the offspring. In humans, as well, studies show that babies born to women with low choline intakes have four times higher rates of birth defects compared to babies born to women with high choline intakes. More recent studies, however, found no association between maternal blood concentrations of choline during pregnancy and risk of neural tube defects. More research is needed to determine whether choline supplementation would help prevent birth defects as has been demonstrated for folic acid.

**Nerve Function and Brain Development.** Choline is part of acetylcholine, a neurotransmitter associated with attention, learning, memory, muscle control, and many other functions. Sphingomyelin, a choline-containing phospholipid, is part of the myelin sheath that insulates nerve cells. As already mentioned, brain tissue is particularly high in choline. During pregnancy, the concentration of choline in amniotic fluid is high, supplying choline to the developing brain of the fetus. Animal studies demonstrate that choline deficiencies during pregnancy impair brain development, learning ability, and memory. The AI for choline is increased during pregnancy and breastfeeding to be available for proper brain development. Choline also may be useful for preventing or treating neurological disorders such as Alzheimer's disease.

**Lipid Transport.** As part of phospholipids, choline is a component of lipoproteins, which carry lipids through the blood. Choline deficiency leads to decreased production of lipid transport proteins, such as very-low-density lipoproteins (VLDLs). For certain, achieving adequate choline intake helps to avoid development of fatty liver. The inability of the liver to export fat to the rest of the body leads to the buildup of fat in the liver. A small amount of fat in the liver is normal, but excess fat leads to scarring of the liver tissue and eventual dysfunction. Fatty liver is a common cause of cirrhosis.

The roles of choline in lipid transport and homocysteine metabolism have implicated the nutrient in prevention of cardiovascular disease. However, research has also raised some concerns about possible negative associations between dietary choline intake and increased cardiovascular disease risk. Recent findings have shown that oral supplements of phosphatidylcholine, along with choline from foods such as eggs, are metabolized by the intestinal microbiota to produce an atherosclerosis-promoting compound that has been associated with an increased risk of major adverse cardiovascular events.[34]

▲ Eggs and other protein foods such as legumes, meat, and dairy products are natural sources of choline. John Thoeming/McGraw-Hill Education

## Choline

### Fats & Oils

| | |
|---|---|
| Mayonnaise (1 T) | 4.72 mg |
| Ranch dressing (1 T) | 4.82 mg |
| Vegetable oil (1 T) | 0.03 mg |

### Fruits

| | |
|---|---|
| Fresh tangerine (½ c) | 10 mg |
| Fresh sliced kiwifruit (½ c) | 7 mg |
| Fresh sliced apple (½ c) | 2 mg |

### Vegetables

| | |
|---|---|
| Cooked mushrooms (½ c) | 58 mg |
| Cooked Brussels sprouts (½ c) | 32 mg |
| Cooked broccoli (½ c) | 31 mg |

### Dairy

| | |
|---|---|
| 1% milk (1 c) | 43 mg |
| Plain non-fat yogurt (1 c) | 38 mg |
| Non-fat cottage cheese (1 c) | 26 mg |

### Grains

| | |
|---|---|
| Cooked quinoa (½ c) | 21 mg |
| Cooked egg noodles (½ c) | 20 mg |
| Long-grain brown rice (½ c) | 9.5 mg |

### Protein

| | |
|---|---|
| Beef liver (3 oz) | 356 mg |
| Hard-boiled egg (1 lg) | 147 mg |
| Roasted soy beans (½ c) | 107 mg |

**MyPlate**.gov

### MAIN FUNCTIONS

Serves as a precursor for several phospholipids that make up cell membranes.

Serves as a precursor for betaine, a compound that participates in many chemical reactions.

Plays critical roles in nerve function and brain development.

As part of phospholipids, contributes to lipid transport.

### RECOMMENDATIONS

AI Men: 550 milligrams
AI Women: 425 milligrams
DV: 550 milligrams
UL: 3500 milligrams

### SPECIAL NOTES

The AI increases for pregnant women to support brain development of the fetus.

Free choline is water-soluble, but is often a component of phospholipids, in which case it is fat-soluble and absorbed through the lymphatic system.

**FIGURE 8-30** ▲ Food sources of choline. The fill of the background color (none, 1/3, 2/3, or completely covered) within each group on MyPlate indicates the average nutrient density for choline in that group. The figure shows the choline content of several foods compared in each food group. Overall, foods that are rich sources of protein are good sources of choline. Grains and fruits are, in general, poor sources of choline. mayonnaise: Iconotec/Alamy Stock Photo; milk: NIPAPORN PANYACHAROEN/Shutterstock; mushroom: JIANG HONGYAN/Shutterstock; noodles: Joe Potato Photo/Getty Images; mushrooms: Pixtal/age fotostock; dancestrokes/123RF; MyPlate: U.S. Department of Agriculture

Sources: Office of Dietary Supplements, Dietary Supplements Fact Sheets, available from https://ods.od.nih.gov/factsheets/list-all; USDA FoodData Central, available from https://fdc.nal.usda.gov.

### GETTING ENOUGH CHOLINE

Choline is widely distributed in foods (Fig. 8-30). Soybeans, egg yolks, beef, shiitake mushrooms, almonds, and peanuts are good sources. In the U.S., most of our choline comes from eggs and other protein foods.[35] In addition to natural food sources, lecithin is often added to food products as an emulsifier during processing, so many other foods are sources of choline.

Eggs (with yolks) are by far the most nutrient-dense source of choline. One whole egg supplies about one-fourth of the daily choline needs in a 70-kcal package. Choline researchers suggest that an average of one egg per day would assist in achieving the AI for choline without raising risk for cardiovascular disease.

To some extent, choline also can be synthesized in the body by a process that involves other nutrients, such as folate and the amino acid methionine. If the body must synthesize choline to meet its needs, functional deficiencies of folate could result.

An AI for choline has been set for adults, but it is unknown whether a dietary supply is essential for infants or children. As noted, some choline can be synthesized in the body, but recent research indicates that synthesis by the body is not sufficient to meet the body's needs for choline. On average, Americans consume about 70% of the AI for choline. In addition, the AIs do not reflect wide genetic variation in individual choline requirements. Research suggests that at least half the population has genetic variations that increase dietary requirements for nutrients that serve in single-carbon metabolism, including choline and folate. Thus, even meeting the AI may not provide enough choline to support the body's needs for some people.[36]

The AI for choline increases during pregnancy (to 450 milligrams per day) and breast-feeding (to 550 milligrams per day) to support brain development of the fetus or infant. All prenatal vitamins do not contain choline; therefore, consumption of rich dietary sources of choline, such as eggs, is important for pregnant and breastfeeding women.

## AVOIDING TOO MUCH CHOLINE

The UL for adults is set at 3.5 grams per day. Routinely exceeding the UL will result in a fishy body odor and low blood pressure.

## OTHER VITAMIN-LIKE COMPOUNDS

A variety of vitamin-like compounds are found in the body. These include the following:

- Carnitine, needed to transport fatty acids into cell mitochondria
- Inositol, part of cell membranes
- Taurine, part of bile acids
- Lipoic acid, which participates in carbohydrate metabolism and acts as an antioxidant

These vitamin-like compounds can be synthesized by cells using common building blocks, such as amino acids and glucose. Our diets are also a source. In disease states or periods of active growth, the synthesis of vitamin-like compounds may not meet needs, so dietary intake can be crucial. The needs for vitamin-like compounds in certain groups of individuals, such as preterm infants, are being investigated. Although promoted and sold by health-food stores, these vitamin-like compounds need not be included in the diet of the average healthy adult.

---

### ✓ CONCEPT CHECK 8.16

1. Describe three functions of choline in the human body.
2. List three ways to incorporate more choline into the diet.
3. Is it necessary to take dietary supplements of vitamin-like compounds, such as carnitine and taurine? Why or why not?

---

# 8.17 Dietary Supplements—Who Needs Them?

The terms *multivitamin* and *mineral supplement* (MVM) have been mentioned many times so far in this textbook. Often, these and other supplements are marketed as cures for anything and everything. This cure-all approach is promoted by the supplement industry and countless health-food stores, pharmacies, and supermarkets.

According to the Dietary Supplement Health and Education Act of 1994 (DSHEA), a supplement in the United States is a product intended to supplement the diet that bears or contains one or more of the following ingredients:

- A vitamin
- A mineral
- An herb or another botanical
- An amino acid
- A dietary substance to supplement the diet, which could be an extract or a combination of the first four ingredients in this list

The definition is broad and covers a wide variety of nutritional substances. The use of dietary supplements is a common practice among North Americans and generates about $35 billion annually for the industry in the United States (Fig. 8-31).

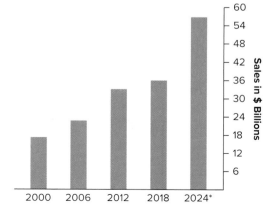

**FIGURE 8-31** ▲ The dietary supplement industry is a growing multibillion-dollar business in the United States.
* Projected sales.

# Newsworthy Nutrition

### Increased emergency department visits for dietary supplement users

**INTRODUCTION:** Intake of dietary supplements, including herbs, vitamins, and mineral supplements, continues to increase in the United States, yet few studies have investigated the safety of these products. **OBJECTIVE:** The aim of this *cohort study* was to evaluate adverse events directly related to dietary supplement intake that resulted in emergency department visits. **METHODS:** Surveillance data were collected from 63 nationally representative emergency departments from 2004 to 2013. **RESULTS:** Over 23,000 emergency department visits per year were attributed to adverse events related to dietary supplement intake, resulting in 2154 hospitalizations annually. Of emergency department visits, 28% related to supplement use involving young adults (ages 20 to 34). After excluding dietary supplement intake by unsupervised children, 66% of supplement-related emergency department visits involved herbal or complementary nutritional products and 32% involved micronutrients. Herbals and complementary nutritional products for weight loss and enhanced energy were prevalent. These products resulted in complaints of heart palpitations, chest pain, or tachycardia. **CONCLUSION:** In light of the fact that the dietary supplement industry is not well regulated, supplement users must be extremely cautious and stay well informed of the potential health risks.

Source: Geller A and others: Emergency department visits for adverse events related to dietary supplements, *New England Journal of Medicine* 15:373, 2015.

---

## ASK THE RDN    Supplements in Perspective

*Dear RDN:* *There's so much talk about nutrition to reduce risk of chronic diseases like cancer, diabetes, and heart disease. Are dietary supplements smart insurance to make sure I get enough protective nutrients?*

A dietary supplement can be valuable to fill a specific nutrient gap, but routine use of micronutrient supplements as a blanket approach to prevent chronic disease is not recommended.[37]

The first step in a decision about supplementation is to assess whether there is a gap between dietary intake and nutrient needs that is unlikely to be filled by a realistic change in eating patterns. The people who are most likely to have a gap are those with elevated nutrient needs (due to malabsorption, aging, chronic diseases, medications, growth, pregnancy, or lactation) or those with limited ability to consume enough food (due to poor appetite, illness, avoidance of one or more major food groups, or severe calorie restriction). Although some genetic mutations may affect digestion, absorption, or metabolism in a subset of individuals, the use of genetic testing to support supplementation is premature.[37]

The idea of a supplement as "insurance" suggests that "If some is good, more is better" for protecting health. But this is not what research shows. Laboratory studies in cells or in animals show mechanisms through which a nutrient could potentially be protective against a disease. And observational studies show greater risk of a disease in people with vitamin or mineral deficiencies compared to people with optimal nutritional status. Current recommendations and position statements of major health organizations conclude that human research does not support dietary supplements as an effective strategy to reduce risk of cancer, heart disease, diabetes, or other chronic diseases.[37,38] For example, laboratory studies support the potential for vitamin D to inhibit cancer development and slow cancer progression. Limited evidence from population-level studies link low blood levels of vitamin D with greater risk of colorectal cancer.[38] However, in recent randomized controlled studies, supplementation with vitamin D raised blood levels, but it did not demonstrate a reduction in colorectal or total cancer risk.[39] Based upon the evidence, using supplements to push considerably beyond your RDA for vitamin D may not protect your health.

When it comes to vitamins and minerals, many people assume that *more is better*. Actually, studies that examine relationships between nutrient intakes and disease rates indicate that benefits are seen when people with the lowest intakes start meeting the recommendations for nutrients. If a person has a deficiency, then increasing intake of a nutrient is likely to help. However, exceeding the RDA or AI for a nutrient does not offer additional benefits. For example, supplements of vitamin E and selenium at doses originally thought to be protective against chronic disease were found to increase risk of high-grade prostate cancer in a controlled intervention trial called SELECT (ask your instructor about the "Research Collection" activity in Connect). And you might expect that people exposed to asbestos or tobacco smoke would benefit from an antioxidant supplement like beta-carotene. However, in controlled trials, doing so increased risk of lung cancer in these men.[39]

In recent years, we have come to understand that overall dietary patterns are much more important than individual nutrients. In other words, a healthy dietary pattern is worth more than the sum of its parts. A supplement of an isolated vitamin or mineral can be helpful to fill a specific nutrient gap that's been identified, but it does not appear to have the same benefit as a food which provides that nutrient along with other nutrients, phytochemicals, and fiber.

To your health,

**Karen Collins, MS, RDN, CDN, FAND**

Consultant dietitian specializing in cancer prevention and cardiometabolic health

Karen Collins

Supplements can be sold without proof that they are safe and effective. Unless the FDA has evidence that a supplement is inherently dangerous or marketed with an illegal claim, it does not regulate such products closely (see *Newsworthy Nutrition*). The vitamin folate is an exception. The FDA has limited resources to police supplement manufacturers and has to act against these manufacturers one at a time. Thus, we cannot rely on the FDA to protect us from any dangers associated with vitamin and mineral supplement overuse and misuse. We bear that responsibility ourselves, with the help of professional advice from a primary care provider or registered dietitian nutritionist (RDN).

The supplement makers can make broad claims about their products under the *structure or function* provision of the law. The products, however, cannot claim to prevent, treat, or cure a disease. Menopause in women and aging are not diseases per se, so products alleging to treat symptoms of these conditions can be marketed without FDA approval. For example, a product that claims to treat hot flashes arising during menopause can be sold without any evidence to prove that the product works, but a product that claims to decrease the risk of cardiovascular disease by reducing blood cholesterol must have results from scientific studies that justify the claim.

Why do people take supplements? Frequently given reasons include:

- Fill nutrient gaps in the dietary pattern
- Increase "energy"
- Maintain overall health
- Prevent disease (cancer, osteoporosis, etc.)
- Reduce stress
- Reduce susceptibility to health conditions (e.g., colds)

## SHOULD YOU TAKE A SUPPLEMENT?

Multivitamin and mineral supplements (MVMs) are popularly regarded as a simple backup plan or an insurance policy, even for people who consciously try to maintain a balanced dietary pattern. Users aim to prevent nutrient deficiencies or chronic diseases by filling any gaps between dietary intake and nutrient needs. However, evidence to support the widespread use of MVMs is lacking. While there is little risk of harm from consuming a balanced MVM that supplies no more than 100% of the Daily Value for the nutrients it contains, most studies indicate no discernible advantage.[37] The National Institutes of Health, in its *State-of-the-Science Report*, concluded that the present evidence is insufficient to recommend either for or against the use of MVMs by Americans to prevent chronic disease.[37]

Do specific vitamin or mineral supplements provide any benefit? Only a few studies of vitamin and mineral supplements demonstrate beneficial effects for the prevention of deficiencies or chronic diseases. For example, postmenopausal women may benefit from taking calcium and vitamin D supplements to increase bone mineral density and decrease fracture risk. The *Dietary Guidelines* recommends that beverage supplements not replace regular food intake unless instructed by a health professional. Table 8-4 outlines the population groups that are most likely to benefit from taking dietary supplements (see *Ask the RDN*).

While there may be moderate benefits of consuming dietary supplements for specific subgroups, uninformed use of supplements can be risky. Indeed, most cases of nutrient toxicity are a result of supplement use. High doses of one nutrient can affect absorption or metabolism of other nutrients. For example, excessive zinc intake can inhibit copper absorption, and large amounts of folate can mask signs and symptoms of a vitamin B-12 deficiency. In addition, some supplements can interfere with medications. For instance, high intakes of vitamin K or vitamin E alter the action of anticlotting medications, vitamin B-6 can offset the action of L-dopa (used in treating Parkinson's disease), and large doses of vitamin C might interfere with certain cancer therapy regimens.

For most Americans, finding ways to incorporate the recommended servings of fruits, vegetables, and whole grains into the dietary pattern is the safest and healthiest way to ensure nutrient adequacy.[5,37] Many of the health-promoting effects of foods cannot be found in a bottle. Few or no phytochemicals or fiber are present in most supplements. Multivitamin and mineral supplements also contain little calcium to keep the pill size small. Furthermore, the oxide forms of magnesium, zinc, and copper used in many supplements

Top Five Dietary Supplements
1. Multivitamins
2. Calcium
3. Vitamin D
4. Vitamin C
5. Protein

Source: Council for Responsible Nutrition.

Believing that supplements provide the nutrition her body needs, Janice regularly takes numerous supplements while paying relatively little attention to daily food choices. **How would you explain to her that this practice may lead to health problems?** Ryan McVay/Getty Images

Websites to help you evaluate supplements:
- ods.od.nih.gov
- www.acsh.org
- www.eatright.org
- www.ncahf.org
- www.quackwatch.com
- www.usp.org/dietary-supplements/overview

**TABLE 8-4 ▪ Who Is Most Likely to Benefit from Dietary Supplements?**

| Type of Supplement | Who May Benefit |
|---|---|
| Multivitamin/mineral supplement | People on restrictive diets (< 1200 kcal per day), vegans, vegetarians |
| | People with suboptimal diets (e.g., in cases of food insecurity or *picky* eaters) |
| | People with malabsorptive diseases |
| | People who take medications that interfere with nutrient absorption or metabolism |
| | Older adults (over 50 years of age) |
| | Pregnant women or those of childbearing age |
| Various B vitamins | People who abuse alcohol |
| Folic acid | Women of childbearing age (especially during pregnancy and breastfeeding) |
| Vitamin B-12 | Older adults |
| | Strict vegans |
| Vitamin C | People who use tobacco |
| Vitamin D | People with limited dairy intake (due to allergies or lactose intolerance) |
| | People with limited exposure to sunlight (e.g., all infants, many African Americans, and some older adults) |
| | Strict vegans |
| Vitamin E | People who follow diets low in fat (especially plant oils) |
| Vitamin K | Newborns |
| Calcium | Strict vegans |
| | Older adults with bone loss |
| Fluoride | Some older infants and children (as directed by a dentist) |
| Iron | Women with excessive bleeding during menstruation |
| | Women who are pregnant |
| | Strict vegans |
| Zinc | Strict vegans |

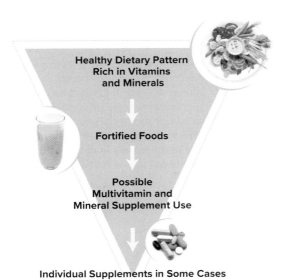

**FIGURE 8-32** ▲ Supplement savvy—an approach to the use of nutrient supplements. Emphasizing a healthy dietary pattern rich in vitamins and minerals is always the first option. glass of orange juice: Stockbyte/Getty Images; plate of fresh vegetables: C Squared Studios/Getty Images; multi drugs: Don Wilkie/Getty Images

are not as well absorbed as forms found in foods. Overall, supplement use cannot fix an inadequate dietary pattern in all respects.

As illustrated in Figure 8-32, when it comes to improving nutrient intake, choose foods rich in vitamins and minerals before considering dietary supplements. First, you should assess your current dietary patterns. MyPlate is a tool consumers can use to plan a healthy dietary pattern. If nutrient gaps still remain, identify food sources that can help. For example, fortified, ready-to-eat breakfast cereals supply a variety of micronutrients, including vitamin E, folic acid, vitamin B-6, and highly absorbable forms of vitamin B-12. Other fortified foods, such as calcium-fortified orange juice, can also be helpful. Be aware of portion sizes of highly fortified foods, however, as multiple servings could lead to excessive intakes of some nutrients, such as vitamin A, iron, and synthetic folic acid. Last, if supplement use is desired, educate yourself and discuss it with your primary care provider or RDN.

## WHICH SUPPLEMENT SHOULD YOU CHOOSE?

If, after consulting with your primary care provider or dietitian, you decide to take a multivitamin and mineral supplement, start by choosing a nationally recognized brand (from a supermarket or pharmacy) that contains about 100% of the Daily Values for the nutrients present. A multivitamin and mineral supplement should generally be taken with or just after meals to maximize absorption. Make sure also that intake from the total of this supplement, any other supplements used, and highly fortified foods (such as ready-to-eat breakfast cereals) provides no more than the Upper Level for each vitamin

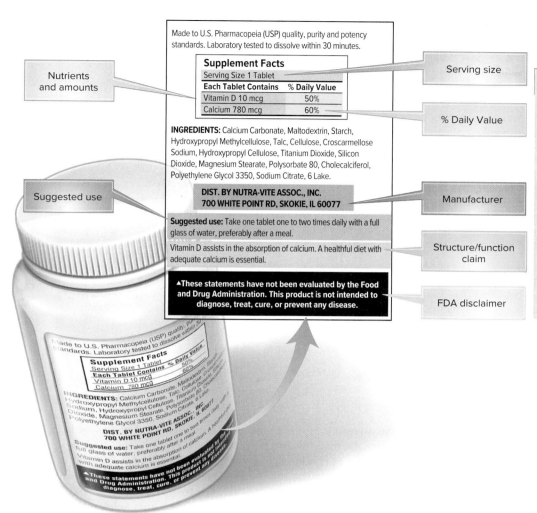

Made to U.S. Pharmacopeia (USP) quality, purity and potency standards. Laboratory tested to dissolve within 30 minutes.

**Nutrients and amounts**

**Serving size**

| Supplement Facts | |
|---|---|
| Serving Size 1 Tablet | |
| **Each Tablet Contains** | **% Daily Value** |
| Vitamin D 10 mcg | 50% |
| Calcium 780 mcg | 60% |

**% Daily Value**

**INGREDIENTS:** Calcium Carbonate, Maltodextrin, Starch, Hydroxypropyl Methylcellulose, Talc, Cellulose, Croscarmellose Sodium, Hydroxypropyl Cellulose, Titanium Dioxide, Silicon Dioxide, Magnesium Stearate, Polysorbate 80, Cholecalciferol, Polyethylene Glycol 3350, Sodium Citrate, 6 Lake.

**DIST. BY NUTRA-VITE ASSOC., INC.
700 WHITE POINT RD, SKOKIE, IL 60077**

**Manufacturer**

**Suggested use**

**Suggested use:** Take one tablet one to two times daily with a full glass of water, preferably after a meal.

Vitamin D assists in the absorption of calcium. A healthful diet with adequate calcium is essential.

**Structure/function claim**

▲These statements have not been evaluated by the Food and Drug Administration. This product is not intended to diagnose, treat, cure, or prevent any disease.

**FDA disclaimer**

**FIGURE 8-33** ▲ Nutrient supplements display a nutrition label different from that of foods. This Supplement Facts label must list the ingredient(s), amount(s) per serving, serving size, suggested use, and % Daily Value if one has been established. In addition, this label includes structure/ function claims, which are not mandatory elements of the supplement label. When structure/ function claims are made, however, the label also must include the FDA warning that these claims have not been evaluated by the agency.

**COVID CORNER**

Although there has been much buzz in the media about dietary supplements "boosting" the immune system to fight COVID-19, currently there is no evidence to support this statement. Unfortunately, social media wellness influencers (without any professional training) are pushing near-lethal doses of some vitamins, minerals, botanicals, herbs, and extracts. The bottom line is that the key to maintaining a healthy immune system is adopting and maintaining healthy patterns over time. This includes consuming nutritious dietary patterns, obtaining adequate sleep, and engaging in regular physical activity.

▲ If you must take vitamins, look for the USP symbol on your vitamin or mineral supplement to ensure quality and safety.
David Tietz/McGraw-Hill Education

and mineral. This is especially important with regard to preformed vitamin A intake. Two exceptions are: (1) older adults should make sure any product used is low in iron or iron free to avoid possible iron overload; and (2) somewhat exceeding the Upper Level for vitamin D is likely a safe practice for adults. Read the labels carefully to be sure of what is being taken (Fig. 8-33). Because research on a variety of nutrient supplements has revealed a lack of product quality, the FDA now requires supplement makers to test the identity, purity, strength, and composition of all their products. As an extra protection, select supplements bear the logo of the United States Pharmacopeial Convention (USP). The USP is an independent, nonprofit group of scientists who review products for strength, quality, purity, packaging, labeling, speed of dissolution, and shelf-stability. The USP designation on a supplement label indicates that the product has been evaluated and meets professionally accepted standards of supplement quality. Other supplement testing organizations include NSF International, National Product Association (NPA), and ConsumerLab. Although these organizations do not guarantee that a product has therapeutic value, nor do they test each batch of supplements, their seals indicate that the product contains the amount

In the dietary supplements aisle of the grocery store, the choices are endless—and expensive. Julie, a college sophomore, just read the Academy of Nutrition and Dietetics' position paper on nutrient supplementation for her class. She learned that dietary supplements, such as a balanced multivitamin and mineral supplement, can be a good backup plan to ensure adequate nutrition, but the jury is still out when it comes to demonstrating a benefit of dietary supplements for long-term health. The majority of Americans regularly take nutrient supplements, but it is usually the people who already have a healthy dietary pattern who take them. Getting more than the recommended amount of a nutrient does not confer additional health benefits. In fact, too much of some vitamins and minerals can lead to toxicity.

Julie decides she would rather focus on getting her nutrients from whole foods. How can she get the most vitamins and minerals out of the foods she eats?

▲ Julie is shopping for nutrient-dense foods. After educating herself about the benefits of eating whole foods, she has realized most of these items are located around the periphery of her grocery store. Adam Melnyk/Shutterstock

1. What factors can damage or reduce vitamins in food?
2. To maximize vitamin content, what should Julie keep in mind as she selects fresh produce for purchase?
3. How does food processing affect vitamin and mineral content? Does it make a difference if Julie chooses products with whole grains or refined grains?
4. When storing fruits and vegetables in her apartment, what steps can Julie take to minimize nutrient losses?
5. Which cooking methods are best for preserving vitamin content?

*Complete the Case Study. Responses to these questions can be provided by your instructor.*

of the active ingredient advertised on the label and is free from dangerous or toxic substances, such as bacteria, arsenic, or lead.

Another consideration in choosing a supplement is avoiding superfluous ingredients, such as added sugars, para-aminobenzoic acid (PABA), hesperidin complex, inositol, bee pollen, and lecithins. These are not needed in our dietary patterns. They are especially common in expensive supplements sold in health-food stores and online. In addition, use of l-tryptophan and high doses of beta-carotene or fish oils is discouraged. The National Institutes of Health, Office of Dietary Supplements provides the My Dietary Supplement and Medicine Record, to help individuals track supplement and medicine use. See https://ods.od.nih.gov/pubs /DietarySupplementandMedicineRecord.pdf for more information.

**✓ CONCEPT CHECK 8.17**

1. Name four types of ingredients that are classified as dietary supplements by the Dietary Supplement Health and Education Act.
2. Identify three potential risks from use of dietary supplements.
3. Describe three situations in which use of dietary supplements is warranted.

# 8.18 Nutrition and Your Health

## Nutrition and Cancer

Mitch Hrdlicka/Getty Images

Cancers are the second leading cause of death for North American adults. It is estimated that more than almost 1700 people die each day of cancer in the United States, accounting for one in four deaths.[40] Cancer-related medical expenses exceed $80 billion each year and are projected to increase as newer, more sophisticated treatments become available. The top four cancers, causing more than 50% of cancer deaths, are breast, lung, prostate, and colorectal.[41]

Cancer includes over 100 distinct and complex diseases; each differs in the types of cells affected and, in some cases, in the factors contributing to cancer development and progression. For example, environmental exposures and genetic factors leading to skin cancer often differ from those leading to breast cancer. Similarly, the treatments for the different types of cancer must also vary.

## Cancer Progression

Cancer begins (initiation) when a DNA mutation occurs in a cell. Promotion and progression of cancer refer to the stages in which mutated cells multiply and proliferate uncontrollably. Without prompt and effective treatment, cancer continues to grow and spread. Most cancers take the form of tumors, although not all tumors are cancers. A **tumor** is spontaneous new tissue growth that appears to serve no physiological purpose. It can be either **benign,** like a wart, or **malignant,** like most lung cancers. The terms *malignant tumor* and *malignant neoplasm* are synonymous with cancer.

Whereas benign (noncancerous) tumors are dangerous if their presence interferes with normal bodily functions, malignant (cancerous) tumors invade surrounding structures, including blood vessels, the lymph system, and nervous tissue. Cancer can spread, or **metastasize,** to distant sites via the blood and lymphatic circulation, thereby producing invasive tumors in almost any part of the body. Cancer metastasis is much more difficult to treat as each new cancer takes on distinct characteristics in response to its new environment. The fact that uncontrolled cancers spread explains why early detection and targeted treatment are critical. Cancers that are often diagnosed in the early stages, mainly due to widespread screening programs, are those in the colon, breast, and cervix.

## Early Detection of Cancer

The CAUTION acronym listed here is a useful aid for remembering many early warning signs for cancer. Unexplained weight loss can be an additional symptom that should not be ignored.

- **C**hange in bowel or bladder habits
- **A** sore that does not heal
- **U**nusual bleeding or discharge
- **T**hickening or lump in the breast or elsewhere
- **I**ndigestion or difficulty in swallowing
- **O**bvious change in a wart or mole
- **N**agging cough or hoarseness

Routine screenings are important for early detection of cancer. The American Cancer Society publishes current cancer screening guidelines based upon expert recommendations and research. Families with genetic predispositions to cancer should consult their primary care provider to discuss enhanced surveillance.

## Factors That Influence the Development of Cancer

Genetics, environment, and lifestyle are potent forces that influence the risk for developing cancer. Of cancers, 5% to 10% are thought to be inherited and 90% to 95% are related to environmental factors. Genetic predispositions to cancers are most prevalent in the colon, breast, and prostate cancers. Modifiable lifestyle and environmental exposures explain the huge variation in cancer rates from country to country. Excessive body fatness and dietary patterns account for over half of all environmentally related cancers.

**tumor** Mass of cells; may be cancerous (malignant) or noncancerous (benign).

**benign** Noncancerous; tumors that do not spread.

**malignant** Malicious; in reference to a tumor, the property of invading surrounding tissues and spreading to distant sites.

**metastasize** The spreading of disease from one part of the body to another, even to parts of the body that are remote from the site of the original tumor. Cancer cells can spread via blood vessels, the lymphatic system, or direct growth of the tumor.

## FAKE NEWS

# Sugar feeds cancer.

### THE FACTS

No. Although cancer cells consume more glucose than non-cancer cells, no research has proven that eating sugar will make your cancer more aggressive or grow faster. However, a high-sugar diet, over time, can certainly contribute to weight gain. Recall that obesity is associated with an increased risk of developing several types of cancer.

Source: National Cancer Institute, *Common Cancer Myths and Misconceptions*.

in tumor development, regardless of the macronutrient composition of the dietary pattern. Currently, calorie restriction appears to be the most effective technique for preventing cancer in laboratory animals. Unfortunately, it is difficult for humans to reduce dietary calories to 70% of usual intake. While the data obtained from animal studies are interesting, understand that severe calorie restrictions are not feasible and sustainable for most individuals. In addition, once cancer is present, calorie restriction may no longer be helpful.

### CANCER-FIGHTING FOODS

Research shows that foods of plant origin protect against a range of cancers. This could be due to a variety of nutrients, phytochemicals, and plant other compounds (Table 8-5). Besides containing key vitamins and minerals, which strengthen our immune system, fruits, vegetables, nuts, and seeds are good sources of biologically active phytochemicals. Foods containing fiber are also linked to a reduced risk of cancer. Fiber is thought to speed up *gut transit time*, or the length of time it takes food to move through the digestive system.

Although no single food or food component can fully protect you against cancer, evidence clearly shows that dietary patterns that are rich in a variety of plant foods help to reduce the risk for many cancers. MyPlate is aimed at disease prevention. It is likely that consumption of a wide variety of plant-based foods and adequate physical activity result in a dose effect that, together, are more potent and protective than either in isolation.[43]

## Evidence-Based Recommendations for Cancer Prevention

Given the devastating toll of cancer treatment and lack of a definitive cure, efforts at prevention are of prime importance. Several health organizations have issued their own sets of dietary and lifestyle guidelines for cancer prevention and survivorship. Here, we present the most recent recommendations of the American Cancer Society, which are consistent with the recommendations of other cancer organizations.[44]

Although we have little control over our genetic risk factors for cancer, we have tremendous influence over our lifestyle behaviors, especially with regard to smoking, alcohol intake, physical activity, UV exposure, and dietary patterns. Indeed, over one-third of all cancers in North America are due to use of tobacco products. About half of the cancers of the mouth, pharynx, and larynx are associated with heavy use of alcohol. The combined use of both alcohol and tobacco products increases cancer risks higher than either alone.

### BODY FATNESS LINKED TO CANCER RISK

An estimated one of every three cancer deaths in the United States is linked to excess body fat, suboptimal nutrition, and inadequate physical activity. Of these, excess body fat appears to have the greatest impact on cancer risk. This includes a greater risk for at least 13 different types of cancer.[42]

There are several ways excess body fat can influence cancer risk. Excess body fat stimulates secretion of hormones (insulin from the pancreas and estrogen from adipose cells) and other proteins that promote systemic inflammation and oxidative stress, which contribute to carcinogenesis.

A strong link exists between cancer risk and excess calories in the dietary pattern. In animal experiments, restricting total calorie intake to about 70% of usual intake results in about a 40% reduction

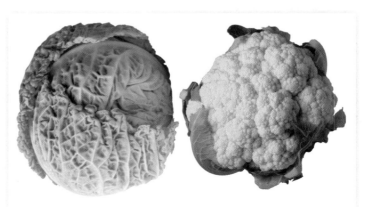

A dietary pattern that includes cruciferous vegetables, such as cabbage and cauliflower, is related to lower risk of developing cancer. **What components of these foods may support your body's defense against cancer?** C Squared Studios/Getty Images

**TABLE 8-5 ■ Food Constituents Associated with Cancer**

| Constituent | Dietary Sources | Action |
| --- | --- | --- |
| **Protective when consumed within recommended ranges\*** | | |
| Calcium | Milk products, green vegetables | Slows cell division in the colon and binds bile acids and free fatty acids, thus reducing colon cancer risk. |
| Carotenoids | Fruits, vegetables | Antioxidant-like properties; possibly influences cell metabolism. |
| Conjugated linoleic acid | Milk products, meats | May inhibit tumor development and act as an antioxidant. |
| Fiber-rich foods | Fruits, vegetables, whole grain breads and cereals, beans, nuts | Colon and rectal cancer risk may be decreased by accelerating intestinal transit or binding carcinogens such that they are excreted. |
| Flavonoids, indoles, phenols, and other phytochemicals | Vegetables, especially cabbage, cauliflower, broccoli, Brussels sprouts, garlic, onions, tea | May reduce cancer in the stomach and other organs. |
| Folate | Fruits, vegetables, whole grains | Encourages normal cell development; reduces the risk of colon cancer. |
| Omega-3 fatty acids | Cold-water fish, such as salmon and tuna | May inhibit tumor growth. |
| Selenium | Meats, whole grains | Part of antioxidant system that inhibits tumor growth and kills cancer cells. |
| Soy products | Tofu, soy milk, tempeh, soy nuts | Phytic acid possibly binds carcinogens in the intestinal tract; genistein component possibly reduces growth and metastasis of malignant cells. |
| Vitamin A | Liver, fortified milk, fruits, vegetables | Encourages normal cell development. |
| Vitamin C | Fruits, vegetables | Can block conversion of nitrites and nitrates to potent carcinogens; antioxidant. |
| Vitamin D | Fortified milk, fatty fish | Increases production of a protein that suppresses cell growth. |
| Vitamin E | Whole grains, vegetable oils, green leafy vegetables | Prevents formation of nitrosamines; antioxidant. |
| **Possibly carcinogenic** | | |
| Alcohol | Beer, wine, liquor | Contributes to cancers of the throat, liver, bladder, breast, and colon (especially if the person does not consume enough folate). |
| Benzo(a)pyrene and other heterocyclic amines | Charcoal-broiled foods, especially meats | Linked to stomach and colon cancer; to limit this risk, trim fat from meat before cooking, cut barbecuing time by partially cooking meat (such as in a microwave oven) prior to grilling, and don't consume blackened parts. |
| Excessive calorie intake | All macronutrients can contribute. | Excess fat mass leading to obesity; increased synthesis of estrogen and other sex hormones, which may increase the risk for some cancer; high insulin levels (as a result of insulin resistance) may promote tumor growth. |
| High glycemic load carbohydrates | Cookies, cakes, sugared soft drinks, candy | Insulin surges associated with these foods may increase tumor growth, such as in the colon. |
| Multiring compounds: aflatoxin | Formed when mold is present on peanuts or grains. | May alter DNA structure and inhibit its ability to properly respond to physiologic controls; aflatoxin in particular is linked to liver cancer. |
| Nitrites, nitrates | Cured meats, especially ham, bacon, and sausages | Under very high temperatures will bind to amino acid derivatives to form nitrosamines, potent carcinogens. |
| Saturated fats | Meats, high-fat milk and milk products, animal fats, and vegetable oils | The strongest evidence is for excessive saturated and polyunsaturated fat intake; saturated fat is linked to an increased risk of prostate cancer. |

\*Many of the actions listed for these possibly protective agents are speculative and have been verified only by experimental animal studies. The best evidence supports obtaining these nutrients and other food constituents from foods. The U.S. Preventive Services Task Force supports this statement, noting there is no clear evidence that nutrient supplements provide the same benefits.

## AMERICAN CANCER SOCIETY GUIDELINE FOR DIET AND PHYSICAL ACTIVITY FOR CANCER PREVENTION

1. Achieve and maintain a healthy body weight throughout life.
   - Keep body weight within the healthy range and avoid weight gain in adult life.
2. Be physically active.
   - Adults should engage in 150–300 min of moderate-intensity physical activity per week, or 75–150 min of vigorous-intensity physical activity, or an equivalent combination; achieving or exceeding the upper limit of 300 min is optimal.
   - Limit sedentary behavior, such as sitting, lying down, and watching television, and other forms of screen-based entertainment.
3. Follow a healthy eating pattern at all ages.
   - A healthy eating pattern includes:
     ◦ Foods that are high in nutrients in amounts that help achieve and maintain a healthy body weight;
     ◦ A variety of vegetables—dark green, red, and orange, fiber-rich legumes (beans and peas), and others;
     ◦ Fruits, especially whole fruits with a variety of colors; and
     ◦ Whole grains.
   - A healthy eating pattern limits or does not include:
     ◦ Red and processed meats;
     ◦ Sugar-sweetened beverages; or
     ◦ Highly processed foods and refined grain products.
4. It is best not to drink alcohol.
   - People who do choose to drink alcohol should limit their consumption to no more than 1 drink per day for women and 2 drinks per day for men.

Recommendation for Community Action
- Public, private, and community organizations should work collaboratively at national, state, and local levels to develop, advocate for, and implement policy and environmental changes that increase access to affordable, nutritious foods; provide safe, enjoyable, and accessible opportunities for physical activity; and limit alcohol for all individuals.

## Nutrition Concerns During Active Cancer Treatment

Eating and lifestyle changes can exert a powerful influence on the risk for developing cancer, but it is important to note that they are no substitute for preventive screening and appropriate medical care. Once cancer has developed, dietary and lifestyle changes will not be adequate to prevent cancer growth or metastasis. Nutrition concerns during active cancer treatment (chemotherapy, immunotherapy, radiation, surgery) vary depending on the site and stage of the cancer, but the overall goals of nutrition therapy are to minimize weight fluctuations and prevent nutrient deficiencies.

Weight loss, particularly loss of muscle mass, is a major concern during cancer treatment because malnutrition can interrupt treatment regimens and impede recovery. Common effects of cancer and/or cancer treatments include fatigue, mouth sores, dry mouth, taste abnormalities, nausea, and diarrhea—all of which can lead to inadequate food intake.

During active treatment, the most appropriate food choices are those that the cancer patient craves and can tolerate. Although food choices vary widely based on each patient's individual symptoms, cool, nonacidic liquids and soft, mildly flavored foods are generally well accepted. Small, frequent meals and foods with high nutrient and calorie density should be emphasized to meet calorie and protein needs. Often, liquid nutritional supplements are warranted during this time. Because many cancer patients are immunocompromised as a result of their treatment, safe food handling practices are extremely important.[40] Individuals in active cancer treatment should seek the advice of an experienced RDN.

To learn more about cancer, review these sources of credible cancer information on the Internet:
American Cancer Society: www.cancer.org
American Institute for Cancer Research: www.aicr.org
National Cancer Institute: www.cancer.gov

### ✔ CONCEPT CHECK 8.18

1. List the components of the cancer CAUTION acronym.
2. Describe the top two modifiable risk factors associated with cancer risk.
3. What are the physical activity recommendations for cancer prevention?

## Summary (Numbers refer to numbered sections in the chapter.)

**8.1** Vitamins are organic substances required in small amounts in the dietary pattern for growth, function, and body maintenance. These can be categorized as fat soluble (vitamins A, D, E, and K) or water soluble (B vitamins and vitamin C). Vitamins cannot be synthesized by the body in adequate amounts to support health, and absence of a vitamin from the diet leads to the development of a deficiency disease. Fat-soluble vitamins require dietary fat for absorption and are carried by lipoproteins in the blood. Vitamin toxicity is most likely to occur from megadoses of fat-soluble vitamins because they are readily stored in the body. Intakes of water-soluble vitamins that exceed the storage ability of tissues are typically excreted in urine. Some vitamins are susceptible to destruction by light, heat, air, or alkalinity, or may be lost from foods in cooking water or fats. Functional foods, such as oatmeal, provide health benefits beyond basic nutrition. Functional foods often contain large amounts of plant-derived compounds known as phytochemicals. These nonessential compounds are what provides plants with their unique color, odor, and flavor. Human consumption is linked to numerous protective health benefits.

**8.2** Vitamin A maintains the health of epithelial tissues and is responsible for the function of mucus-secreting cells. It is also vitally important for normal vision. Vitamin A is found in its active forms in meats, fortified dairy products, fish, and eggs, and it can be derived from carotenoids in a variety of fruits and vegetables. Carotenoids are phytochemicals that can be converted into vitamin A in the body. Although carotenoids are not essential nutrients, some have health-promoting qualities for humans. In addition to their contribution to vitamin A intake, carotenoids are powerful antioxidants. The antioxidant abilities of several carotenoids are linked to prevention of macular degeneration, cataracts, cardiovascular disease, and cancer. Carotenoids are plentiful in dark-green and orange vegetables.

**8.3** Vitamin D is both a hormone and a vitamin. Human skin synthesizes it using sunshine and a cholesterol-like substance. If we do not spend enough time in the sun, foods such as fish and fortified milk can supply the vitamin. The active hormone form of vitamin D helps regulate blood calcium in part by increasing calcium absorption from the intestine. Infants and children who do not get enough vitamin D may develop rickets, and adults with inadequate amounts in the body develop osteomalacia. Older people and infants often need a supplemental source. Toxicity may lead to calcification of soft tissue, weakness, and gastrointestinal disturbances.

**8.4** Vitamin E functions primarily as an antioxidant and is found in plant oils. By donating electrons to electron-seeking, free-radical (oxidizing) compounds, it neutralizes them. This effect shields cell membranes and red blood cells from breakdown. Claims are made about the curative powers of vitamin E, but more information is needed before megadose vitamin E recommendations for healthy adults can be made with certainty.

**8.5** Vitamin K is essential for blood clotting and imparts calcium-binding ability to various proteins, including those in bone. Hemostasis describes the mechanisms that prevent blood loss, such as vasoconstriction, formation of a platelet plug, and blood coagulation. Some vitamin K absorbed each day comes from bacterial synthesis in the intestine, but most comes from foods, primarily green, leafy vegetables.

**8.6** The B vitamins yield no energy directly, but they contribute to energy-yielding chemical reactions in the body by virtue of their coenzyme functions. B vitamins are highly bioavailable. North American diets are typically adequate in B vitamins except in cases of food insecurity, metabolic disorders, or alcoholism. Whole grains are more nutrient-dense sources of B vitamins (as well as other nutrients) than refined grains. Several B vitamins function as coenzymes in energy metabolism.

**8.7** Thiamin's coenzyme form is involved in the metabolism of carbohydrates and proteins as well as the synthesis of RNA, DNA, and neurotransmitters. Rich food sources of thiamin include pork, enriched or fortified grain products, and milk. Beriberi, the thiamin-deficiency disease, leads to muscle weakness and nerve damage. Thiamin toxicity is unknown, and no UL has been set.

**8.8** The coenzymes of riboflavin participate in the catabolism of fatty acids, metabolism of other vitamins and minerals, and antioxidant activity of glutathione peroxidase. Dairy products, enriched and fortified grain products, meat, and eggs are rich food sources of riboflavin. Symptoms of ariboflavinosis include glossitis and angular cheilitis. There is no evidence of toxicity with high doses of riboflavin; no UL has been set.

**8.9** Niacin's coenzymes function in many synthetic reactions, especially fatty-acid synthesis. Rich food sources include seafood, poultry, meats, peanuts, and enriched or fortified grains. Pellagra, the disease of niacin deficiency, results in dermatitis, diarrhea, dementia, and, eventually, death. Megadoses of niacin have been used to lower blood lipids, but they cause side effects, such as flushing of the skin.

**8.10** Vitamin B-6 coenzymes activate many enzymes of carbohydrate, lipid, and, especially, protein metabolism. They also help synthesize neurotransmitters and participate in homocysteine metabolism. Rich food sources include animal products and enriched or fortified grain products, as well as some fruits and vegetables. A deficiency of vitamin B-6 leads to headaches, depression, gastrointestinal symptoms, skin disorders, nerve problems, anemia, and impaired immunity. Vitamin B-6 toxicity can result in nerve damage.

**8.11** Pantothenic acid functions as a coenzyme in reactions that yield energy from carbohydrates, lipids, and protein, as well as fatty-acid synthesis. It is widely distributed among foods, with sunflower seeds, mushrooms, peanuts, and eggs among the richest sources. A deficiency of pantothenic acid is unlikely, but symptoms would be similar to those seen with deficiencies of other B vitamins. There is no known toxicity and no UL for pantothenic acid.

**8.12** Biotin's coenzyme form aids in reactions that synthesize glucose and fatty acids and in the metabolism of amino acids. Egg yolks, peanuts, and cheese provide dietary biotin, but this vitamin is also synthesized by bacteria in the intestines. Consuming raw egg whites may lead to a biotin deficiency because avidin in egg whites binds biotin and reduces its bioavailability. Biotin deficiency can lead to inflammation of the skin and mouth, gastrointestinal symptoms, muscle pain and weakness, poor growth, and anemia. No UL has been set for biotin as no toxicity has ever been observed.

**8.13** Folate plays an important role in DNA synthesis and homocysteine metabolism. Symptoms of a deficiency include generally poor cell division in various areas of the body, macrocytic anemia, tongue inflammation, diarrhea, and poor growth. Pregnancy puts high demands for folate on the body; deficiency during the first month of pregnancy can result in neural tube defects in offspring. A deficiency can also occur in people with alcoholism. Food sources are leafy vegetables, organ meats, and orange juice.

**8.14** Vitamin B-12 is needed to metabolize folate and homocysteine, and to maintain the insulation surrounding nerves. Absorption of vitamin B-12 is a complex process that requires a salivary protein, adequate stomach acid production, and an intrinsic factor produced by the stomach. A deficiency, which results in anemia and nerve degeneration, most likely results from poor absorption of vitamin B-12 rather than poor dietary intake. Pernicious anemia is one condition that can impair vitamin B-12 absorption. Vitamin B-12 is found in foods of animal origin, fortified foods, and supplements.

**8.15** Vitamin C is a potent antioxidant and also promotes the production of white blood cells to support immune function. In addition, vitamin C takes part in the synthesis of collagen, carnitine, and neurotransmitters, and it can modestly enhance iron absorption.

A vitamin C deficiency results in scurvy, evidenced by pinpoint hemorrhages in the skin, bleeding gums, and joint pain. Smoking increases the possibility of vitamin C deficiency. Fresh fruits and vegetables, especially citrus fruits, are good sources. Vitamin C also modestly enhances iron absorption. A great amount of vitamin C is lost in storage and cooking. Deficiencies can occur in people with alcoholism and those whose diets lack sufficient fruits and vegetables.

**8.16** Choline is an essential nutrient but is not classified as a vitamin. As a component of phospholipids, it is important for cell membrane structure, myelination of nerves, and lipid transport. Like folate, choline plays a role in single-carbon metabolism, which has implications for prevention of birth defects, cancer, and heart disease. Egg yolks, meats, dairy products, soybeans, and nuts are good food sources of choline.

**8.17** To meet nutrient needs and prevent chronic disease, whole foods should be emphasized, but occasionally, dietary supplements may be necessary. For example, women of childbearing age, older adults, vegans, and people with malabsorptive diseases are most likely to benefit from dietary supplements. Consumers should educate themselves about possible benefits and risks and discuss concerns with their primary care provider.

The Dietary Supplement Health and Education Act of 1994 (DSHEA) is a federal statute that defines and regulates dietary supplements in the U.S.

**8.18** Given the toll of cancer treatment and lack of a definitive cure, efforts at prevention are key. A variety of dietary changes will reduce your risk for cancer. Start by making sure that your dietary pattern is moderate in calories, added sugars, and saturated fat content and that you consume many fruits and vegetables, whole grains, beans, peas, lentils, some fish, and low-fat dairy products. In addition, remain physically active; avoid obesity; consume alcohol in moderation (if at all); and limit intake of animal fat and salt-cured, smoked, and nitrate-cured foods.

# Check Your Knowledge (Answers to the following questions are below)

1. Vitamins are classified as
   a. organic and inorganic.
   b. fat soluble and water soluble.
   c. essential and nonessential.
   d. elements and compounds.

2. A vitamin synthesized by bacteria in the intestine is
   a. A.
   b. D.
   c. E.
   d. K.

3. A deficiency of vitamin A can lead to the disease called
   a. xerophthalmia.
   b. osteomalacia.
   c. scurvy.
   d. pellagra.

4. Vitamin D is called the sunshine vitamin because
   a. it is available in orange juice.
   b. exposure to sunlight converts a precursor into vitamin D.
   c. it can be destroyed by exposure to sunlight.
   d. it is an ingredient in sunscreen.

5. Vitamin E functions as
   a. a coenzyme.
   b. a hormone.
   c. an antioxidant.
   d. a peroxide.

6. Bowed legs, an enlarged and misshapen head, and enlarged knee joints in children are all symptoms of
   a. rickets.
   b. xerophthalmia.
   c. osteoporosis.
   d. vitamin D toxicity.

7. A deficient intake of _____ has been shown to increase the risk of having a baby with a neural tube defect such as spina bifida.
   a. vitamin A
   b. vitamin C
   c. vitamin E
   d. folate

8. Vitamin C is necessary for the production of
   a. stomach acid.
   b. collagen.
   c. insulin.
   d. clotting factors.

9. B vitamins, including thiamin, riboflavin, and niacin, are called the "energy" vitamins because they
   a. can be broken down to provide energy.
   b. are ingredients in energy drinks such as Powerade.
   c. are part of coenzymes needed for release of energy from carbohydrates, fats, and proteins.
   d. are needed in large amounts by competitive athletes.

10. Noodles, spaghetti, and bread are made from wheat flour that is enriched with all of the following nutrients except
    a. vitamin B-6.
    b. thiamin.
    c. niacin.
    d. riboflavin.

11. Which of the B vitamins is sensitive to and can be degraded by light?
    a. Riboflavin
    b. Niacin
    c. Thiamin
    d. Pantothenic acid

12. Niacin can be synthesized in the body from the amino acid
    a. tyrosine.
    b. tryptophan.
    c. phenylalanine.
    d. glutamine.

13. Avidin, a component of raw egg whites, may decrease the absorption of
    a. biotin.
    b. thiamin.
    c. iron.
    d. riboflavin.

14. Choline is an important component of
    a. cholesterol.
    b. an antioxidant.
    c. a phospholipid.
    d. proteins.

15. Which of the following meals is most compatible with American Cancer Society's guideline for cancer prevention?
    a. Flame-broiled chicken breast, baked potato, and mixed vegetables
    b. Poached salmon, steamed broccoli, and corn on the cob
    c. Baked ham, sweet potato casserole, and spinach salad
    d. Cheese pizza and bread sticks with marinara sauce

16. Vitamins are *not* destroyed by exposure to which of the following?
    a. Heat
    b. Light
    c. Water
    d. Oxygen

17. As a safeguard for supplements, look for supplements that have the _____ logo to indicate the product has been independently evaluated and meets quality standards.
    a. U.S. Pharmacopeial Convention (USP)
    b. FDA Nutrition Facts Label for Supplements
    c. USDA Organic Label
    d. NIH Office of Dietary Supplement Standards (ODSS)

Answer Key: 1. b (LO 8.1), 2. d (LO 8.3), 3. a (LO 8.6), 4. b (LO 8.3), 5. c (LO 8.3), 6. a (LO 8.6), 7. d (LO 8.6), 8. b (LO 8.4), 9. c (LO 8.4), 10. a (LO 8.5), 11. a (LO 8.6), 12. b (LO 8.4), 13. a (LO 8.5), 14. c (LO 8.5), 15. b (LO 8.8), 16. d (LO 8.2), 17. a (LO 8.7)

## Study Questions (Numbers refer to Learning Outcomes.)

1. Why is the risk of toxicity greater with the fat-soluble vitamins A and D than with water-soluble vitamins in general? **(LO 8.1)**

2. Diets rich in phytochemicals have been associated with which protective health benefits? **(LO 8.2)**

3. How would you determine which fruits and vegetables displayed in the produce section of your supermarket are likely to provide plenty of carotenoids? **(LO 8.2)**

4. What is the primary function of the vitamin D hormone? Which groups of people likely need to supplement their diets with vitamin D, and on what do you base your answer? **(LO 8.3)**

5. Describe how vitamin E functions as an antioxidant. **(LO 8.3)**

6. Describe how the RDA, DV, and UL for vitamin B-6 should be used in everyday life. **(LO 8.4)**

7. Is it necessary for North Americans to consume a great excess of vitamin C to avoid the possibility of a deficiency? Do vitamin C intakes well above the RDA have any negative consequences? **(LO 8.4)**

8. Why is choline not considered a vitamin? **(LO 8.4)**

9. What are the best food sources for thiamin? **(LO 8.5)**

10. What are the signs of a riboflavin deficiency? **(LO 8.6)**

11. Milling (refining) grains removes which vitamins and minerals? Which of these are replaced during processing? **(LO 8.6)**

12. Describe the three signs of the niacin deficiency pellagra. **(LO 8.6)**

13. Why does the consumption of raw eggs lead to a biotin deficiency? **(LO 8.6)**

14. Why does FDA limit the amount of folate that may be included in supplements and fortified foods? **(LO 8.7)**

15. List five modifiable behaviors recommended for cancer prevention. **(LO 8.8)**

## Further Readings

1. ADA Reports: Position of the Academy of Nutrition and Dietetics: Micronutrient supplementation. *Journal of the Academy of Nutrition and Dietetics* 118(11):2162, 2018. DOI:1.10.1016/j.jand.2018.07.022.

2. Harvard Medical School, Harvard Health Publishing: Fill up on phytochemicals. Available at https://www.health.harvard.edu/staying-healthy/fill-up-on-phytochemicals. Accessed January 30, 2020.

3. U.S. Department of Health & Human Services, National Institutes of Health, Office of Dietary Supplements: Vitamin A fact sheet for health professionals. Available at https://ods.od.nih.gov/factsheets/VitaminA-HealthProfessional. Accessed January 20, 2020.

4. Chapman NA and others: Role of diet and food intake in age-related macular degeneration: A systematic review. *Clinical and Experimental Ophthalmology* 47:106, 2019. DOI:10.1111/ceo.13343.

5. Aune D and others: Dietary intake and blood concentrations of antioxidants and the risk of cardiovascular disease, total cancer, and all-cause mortality: a systematic review and dose-response meta-analysis of prospective studies. *American Journal of Clinical Nutrition.* 108(5):1069, 2018. DOI:10.1093/ajcn/nqy097.

6. UNICEF for Every Child: Vitamin A. Available at https://data.unicef.org/topic/nutrition/vitamin-a-deficiency. Accessed January 13, 2020.

7. Bouis HE and Saltzman A: Improving nutrition through biofortification: A review of evidence from HarvestPlus, 2003 through 2016. *Global Food Security* 12:49, 2017. DOI:10.1016/j.gfs.2017.01.009.

8. U.S. Department of Health & Human Services, National Institutes of Health, Office of Dietary Supplements: Vitamin D fact sheet for health professionals. Available at https://ods.od.nih.gov/factsheets/VitaminD-HealthProfessional. Accessed January 20, 2020.

9. Dunne S and Bell JA: Vitamin D's role in health—deterministic or indeterminate? *Today's Dietitian* 16(7):48, 2014.

10. Theodoratou E and others: Vitamin D and multiple health outcomes: Umbrella review of systematic reviews and meta-analyses of observational studies and randomised trials. *British Medical Journal* 348:g2035, 2014. DOI:10.1136/bmj.g2035.

11. Liu X and others: Vitamin D deficiency and insufficiency among US adults: Prevalence, predictors and clinical implications. *British Journal of Nutrition* 119(8):928, 2018. DOI:10.1017/S0007114518000491.

12. U.S. Department of Health & Human Services, National Institutes of Health, Office of Dietary Supplements, Vitamin E fact sheet for health professionals. Available at https://ods.od.nih.gov/factsheets/VitaminE-HealthProfessional. Accessed January 22, 2020.

13. Ashor AW and others: Effect of vitamin C and vitamin E supplementation on endothelial function: A systematic review and meta-analysis of randomised controlled trials. *British Journal of Nutrition* 113:1182, 2015. DOI:10.1017/S0007114515000227.

14. Khadangi F and Azzi A: Vitamin E—The next 100 years. *IUBMB Life* 9999:1, 2018. DOI:10.1002/iub.1990.

15. U.S. Department of Health & Human Services, National Institutes of Health, Office of Dietary Supplements: Vitamin K fact sheet for health professionals. Available at https://ods.od.nih.gov/factsheets/VitaminK-HealthProfessional. Accessed January 22, 2020.

16. Centers for Disease Control and Prevention: Facts about vitamin K deficiency bleeding. Available at https://www.cdc.gov/ncbddd/vitamink/facts.html. Accessed January 30, 2020.

17. U.S. Department of Health & Human Services, National Institutes of Health, Office of Dietary Supplements: Thiamin fact sheet for health professionals. Available at https://ods.od.nih.gov/factsheets/Thiamin-HealthProfessional. Accessed January 22, 2020.

18. U.S. Department of Health & Human Services, National Institutes of Health, Office of Dietary Supplements: Riboflavin fact sheet for health professionals. Available at https://ods.od.nih.gov/factsheets/Riboflavin-HealthProfessional. Accessed January 21, 2020.

19. U.S. Department of Health & Human Services, National Institutes of Health, Office of Dietary Supplements: Niacin fact sheet for health professionals. Available at https://ods.od.nih.gov/factsheets/Niacin-HealthProfessional. Accessed January 24, 2020.

20. U.S. Department of Health & Human Services, National Institutes of Health, Office of Dietary Supplements: Pantothenic acid fact sheet for health professionals. Available at https://ods.od.nih.gov/factsheets/PantothenicAcid-HealthProfessional. Accessed January 26, 2020.

21. U.S. Department of Health & Human Services, National Institutes of Health, Office of Dietary Supplements: Vitamin B-6 fact sheet for health professionals. Available at https://ods.od.nih.gov/factsheets/VitaminB6-HealthProfessional. Accessed January 26, 2020.

22. U.S. Department of Health & Human Services, National Institutes of Health, Office of Dietary Supplements: Biotin fact sheet for health professionals. Available at https://ods.od.nih.gov/factsheets/Biotin-HealthProfessional. Accessed January 26, 2020.

23. U.S. Department of Health & Human Services, National Institutes of Health, Office of Dietary Supplements: Folate fact sheet for health professionals. Available at https://ods.od.nih.gov/factsheets/Folate-HealthProfessional. Accessed January 26, 2020.

24. Zheng Y and Cantley LC: Toward a better understanding of folate metabolism in health and disease. *Journal of Experimental Medicine* 216:253, 2019. DOI:10.1084/jem.20181965.

25. Pieroth R and others: Folate and its impact on cancer risk. *Current Nutrition Reports* 7:70, 2018. DOI:10.1007/s13668-018-0237-y.

26. Jayedi A and others: Intake of vitamin B6, folate, and vitamin B12 and risk of coronary heart disease: A systematic review and dose-response meta-analysis of prospective cohort studies. *Critical Reviews in Food Science and Nutrition* 2019;59(16):2697–2707. DOI:10.1080/10408398.2018.1511967.

27. Peker E and others: The levels of vitamin B12, folate and homocysteine in mothers and their babies with neural tube defects. *Journal of Maternal and Fetal Neonatal Medicine* 23:1, 2015. DOI:10.3109/14767058.2015.1109620.

28. Kim Y: Folate and cancer: A tale of Dr. Jekyll and Mr. Hyde? *American Journal of Clinical Nutrition* 107:139, 2018. DOI:10.1093/ajcn/nqx076.

29. U.S. Department of Health & Human Services, National Institutes of Health, Office of Dietary Supplements: Vitamin B-12 fact sheet for health professionals. Available at https://ods.od.nih.gov/factsheets/VitaminB12-HealthProfessional. Accessed January 26, 2020.

30. Palmer S: Vitamin B-12 and the vegan diet. *Today's Dietitian* 20:38, 2018.

31. U.S. Department of Health & Human Services, National Institutes of Health, Office of Dietary Supplements: Vitamin C fact sheet for health professionals. Available at https://ods.od.nih.gov/factsheets/VitaminC-HealthProfessional. Accessed January 26, 2020.

32. Hemila H and Chalker E: Vitamin C for preventing and treating the common cold. *Cochrane Database of Systematic Reviews,* 2013. DOI:10.1002/14651858.CD000980.pub4.

33. U.S. Department of Health & Human Services, National Institutes of Health, Office of Dietary Supplements: Choline fact sheet for health professionals. Available at https://ods.od.nih.gov/factsheets/Choline-HealthProfessional. Accessed January 26, 2020.

34. Meyer KA and Shea JW: Dietary choline and betaine and risk of CVD: A systematic review and meta-analysis of prospective studies. *Nutrients* 9:711, 2017. DOI:10.3390/nu9070711.

35. Wallace TC and Fulgoni VL: Usual choline intakes are associated with egg and protein food consumption in the United States. *Nutrients* 9(8):839, 2017. DOI:10.3390/nu9080839.

36. Weisenberger J: Choline under the microscope. *Today's Dietitian* 19(11):36, 2017.

37. World Cancer Research Fund/American Institute for Cancer Research: Diet, nutrition, physical activity and cancer: A global perspective. Continuous Update Project Expert Report 2018. Available at www.dietandcancerreport.org. Accessed January 15, 2020.

38. Manson JE and others: Vitamin D supplements and prevention of cancer and cardiovascular cisease. *New England Journal of Medicine* 380:33–44, 2019. DOI:10.1056/NEJMoa1809944.

39. U.S. Department of Agriculture and U.S. Department of Health and Human Services. Dietary Guidelines for Americans, 2020-2025. 9th Edition. December 2020. Available at: DietaryGuidelines.gov.

40. Siegel RL, Miller KD, and Jemal A: Cancer statistics, 2019. *CA: A Cancer Journal for Clinicians* 69:7, 2019. DOI:10.3322/caac.21551.

41. American Cancer Society: Economic impact of cancer. Updated January 3, 2018. https://www.cancer.org/cancer/cancer-basics/economic-impact-of-cancer.html. Accessed January 16, 2020.

42. U.S. Cancer Statistics Working Group: U.S. Cancer Statistics Data Visualizations Tool, based on November 2017 submission data (1999–2015): U.S. Department of Health and Human Services, Centers for Disease Control and Prevention and National Cancer Institute. Available at http://www.cdc.gov/cancer/dataviz. Accessed January 30, 2020.

43. Rock CL and others: American Cancer Society guideline for diet and physical activity for cancer prevention: *CA: A Cancer Journal for Clinicians* 0:1–27, 2020. DOI:10.3322/caac.21591.

44. American Institute for Cancer Research: Ten recommendations for cancer prevention. Available at http://www.aicr.org/reduce-your-cancer-risk/recommendations-for-cancer-prevention. Accessed January 30, 2020.

# Chapter 9

# Water and Minerals

Sarah Rusnak

# Student Learning Outcomes

## Chapter 9 is designed to allow you to:

**9.1** Understand the functions of water in the body, the regulation of fluid balance, and the health consequences of fluid imbalance, as well as list recommended intakes and sources of water.

**9.2** Describe the general characteristics of the major and trace minerals, their absorption and storage, the dangers of mineral toxicities, and ways to preserve minerals in foods.

**9.3** Describe the roles of sodium, potassium, and chloride in controlling fluid balance, acid–base balance, and nerve impulse transmission.

**9.4** Describe the role of calcium, phosphorus, magnesium, and fluoride in body functions including bone health, as well as the process of osteoporosis development and prevention.

**9.5** Describe the functions of the trace minerals including the role of iron in blood health, zinc in immune function, iodine in thyroid metabolism, and chromium in glucose metabolism.

**9.6** List some of the best dietary sources for each major and trace mineral.

**9.7** List the dietary requirements for each major and trace mineral, as well as the dangers of exceeding these recommendations.

**9.8** Describe the signs and symptoms of mineral deficiency and conditions that lead to a deficiency.

**9.9** Describe factors that can contribute to the development of hypertension and strategies to lower blood pressure.

## FAKE NEWS

Everyone needs to drink eight glasses of water a day.

## THE FACTS

The claim to drink eight 8-ounce (1 cup) glasses of water per day has probably been accepted for a long time because it is easy to remember. Eight glasses is a half gallon of water or about 2 liters. The claim, otherwise known as the 8 x 8 rule, is not supported by scientific evidence. The National Academy of Medicine recommendation is somewhat greater at 9 cups (2.7 liters) per day for women and 13 cups (3.7 liters) per day for men, but it is a guideline for *total beverage* consumption, not just water. Water is not the only beverage that contributes to fluid balance, and most foods are full of water. The truth is that water intake need varies by individual and can be partially guided by thirst. Body size, body composition, activity level, and climate are all factors that impact fluid needs. Drink when you are thirsty and drink more during hot weather and exercise.

Water ($H_2O$)—the most abundant molecule in the human body and the most versatile medium for a variety of chemical reactions—constitutes the major portion of the human body. Without water, biological processes necessary to life would cease in a matter of days. We must replenish water regularly because the body does not store it *per se.* Fluids must be consumed daily to replenish what is lost through respiration (lungs), perspiration (skin), and excretion (urine and feces). We recognize this constant demand for water as thirst, but the body has several intricate mechanisms to ensure fluid conservation. Maintenance of fluid balance relies on strict control of levels of dissolved minerals inside and outside the cells. These dissolved minerals—sodium, chloride, potassium, and phosphorus—are called electrolytes. Not only do they regulate the distribution of water throughout the body, they are also involved in maintenance of acid–base balance and in the conduction of nerve impulses. In this chapter, you will learn about the importance of water and how the electrolytes work together to regulate fluid balance, acid–base balance, and nerve function.

Like water, many minerals are vital to health. They are considered inorganic because they are typically not bonded to carbon atoms. In addition to water balance, minerals are key participants in body metabolism, muscle movement, body growth, and other wide-ranging processes (Fig. 9-1). We also know that some mineral deficiencies can cause severe health problems.

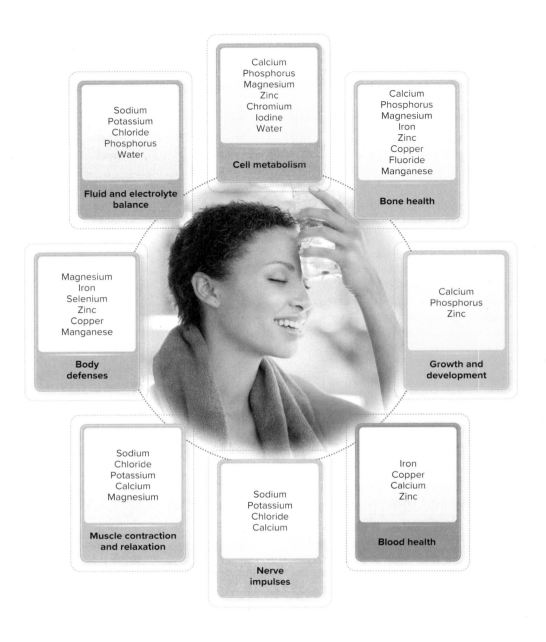

## 9.1 Water

Life as we know it could not exist without water. Every cell, tissue, and organ contains some water. Overall, water comprises 50% to 70% of the human body (Fig. 9-2). Indeed, water is essential for life. Whereas humans can live for several weeks without food, we cannot survive for more than a few days without water. The simple $H_2O$ molecule has some amazing properties: it is a versatile solvent, a coolant, and a lubricant (Fig. 9-3). These properties make water uniquely suited to carry out a number of essential roles in the human body.

### WATER IS THE UNIVERSAL SOLVENT

Water is often called the "universal solvent" because so many different solutes can be dissolved in it. This property of water makes it: (1) an ideal transport vehicle for nutrients and wastes; and (2) a medium for many chemical reactions of human metabolism.

**Water Transports Nutrients and Wastes.** The majority of the nutrients we consume—carbohydrates, proteins, minerals, and many vitamins—are water soluble. Lipids, although they are not soluble in water, can be surrounded by a layer of water-soluble protein so that they can be dispersed throughout the water-based environment within and around cells and tissues. As the primary component of blood and lymph, water acts to transport nutrients to all the cells of the body.

The metabolism of nutrients generates some waste products, most of which can dissolve in water and exit the body as part of urine. For example, when proteins are broken down as fuel, the nitrogen portion of amino acids cannot be used for energy production. The liver converts these nitrogenous waste products into urea. In addition, when we consume more than we need of some nutrients, such as sodium, the excess can be dissolved in water and excreted in urine. Typical urine output ranges from 1 to 2 liters per day, depending mostly on our intakes of fluid, protein, and sodium.

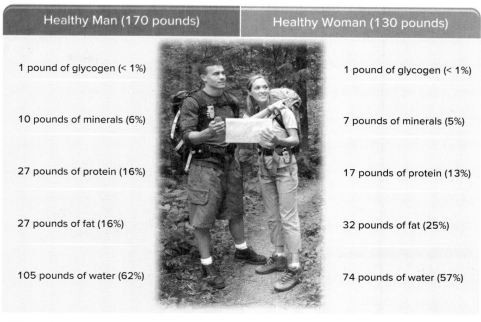

| Healthy Man (170 pounds) | Healthy Woman (130 pounds) |
| --- | --- |
| 1 pound of glycogen (< 1%) | 1 pound of glycogen (< 1%) |
| 10 pounds of minerals (6%) | 7 pounds of minerals (5%) |
| 27 pounds of protein (16%) | 17 pounds of protein (13%) |
| 27 pounds of fat (16%) | 32 pounds of fat (25%) |
| 105 pounds of water (62%) | 74 pounds of water (57%) |

**FIGURE 9-2** ▲ This figure shows that, compared to other body constituents, water typically is the biggest contributor to our body weight. Where is this water? Although the percentages vary for men and women, the main constituent of the body is water. The percentage of water varies tremendously among tissues. For example, muscle is 73% water, adipose tissue is 10% to 20% water, and bone contains approximately 20% water. As the fat content of the body increases, the percentage of lean muscle decreases, and subsequently the percentage of body water decreases. When body composition measurements are performed on extremely lean athletes, the percentage of body water can be around 70%. Digital Vision/Getty Images

**Water Is a Medium for Chemical Reactions.** Because so many compounds dissolve in water, it provides a medium in which chemical reactions take place in the body. Furthermore, water itself ($H_2O$) is an important participant in many chemical reactions. When carbohydrates, lipids, and proteins are metabolized as sources of energy, water is one of the by-products. In fact, this **metabolic water** (1 cup or more per day) contributes to the maintenance of fluid balance in the body.

**metabolic water** Water formed as a by-product of carbohydrate, lipid, and protein metabolism.

## WATER CONTRIBUTES TO BODY TEMPERATURE REGULATION

Water has a great ability to hold heat, so changes in water temperature occur very slowly. Because the human body is 50% to 70% water, it holds heat efficiently and a lot of energy is required to change body temperature. Water molecules are polar (charged), so they are attracted to each other. This attractive force is strong, and requires energy from heating to separate the water molecules.

When overheated, the body secretes fluids in the form of perspiration, which evaporates through skin pores, releasing heat energy. So, as the water in perspiration evaporates, heat energy is removed from the skin, cooling the body in the process (Fig. 9-4). In response to an increased body temperature, blood vessels in the skin become larger, allowing greater water loss through perspiration. Each quart (approximately 1 liter or 2 pounds) of perspiration that evaporates represents approximately 600 kcal.

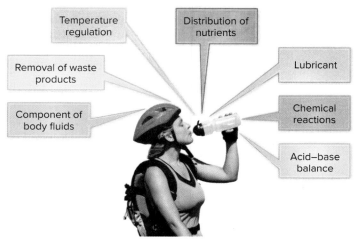

Temperature regulation

Distribution of nutrients

Removal of waste products

Lubricant

Component of body fluids

Chemical reactions

Acid–base balance

**FIGURE 9-3** ▲ Water has many essential functions in the body.
Ron Chapple/Thinkstock Images/Getty Images

**FIGURE 9-4** ▶ Body temperature is reduced when heat is transported from the body through the bloodstream to the surface of the skin. As perspiration evaporates from the surface of the skin, heat is dissipated. This cools the blood, which circulates back to the body, reducing body temperature.

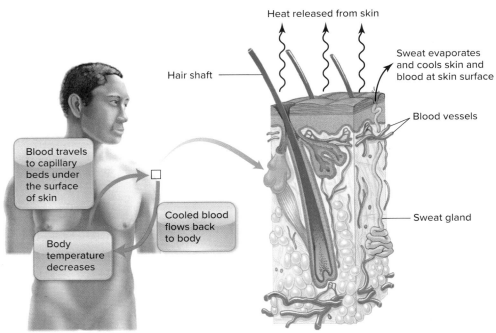

Heat released from skin

Hair shaft

Sweat evaporates and cools skin and blood at skin surface

Blood vessels

Sweat gland

Blood travels to capillary beds under the surface of skin

Cooled blood flows back to body

Body temperature decreases

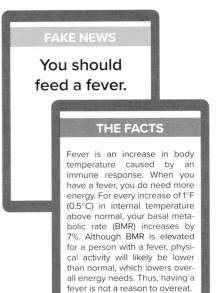

**FAKE NEWS**

## You should feed a fever.

### THE FACTS

Fever is an increase in body temperature caused by an immune response. When you have a fever, you do need more energy. For every increase of 1°F (0.5°C) in internal temperature above normal, your basal metabolic rate (BMR) increases by 7%. Although BMR is elevated for a person with a fever, physical activity will likely be lower than normal, which lowers overall energy needs. Thus, having a fever is not a reason to overeat.

▲ Tears are one example of water's role as a lubricant. This fluid allows the eyeball to move smoothly in its socket and helps to flush away foreign particles. Lack of tear production is a symptom of dehydration. Danil Chepko/123RF

When carbohydrates, lipids, and proteins are used by cells in the body, energy is released in the form of heat. About 60% of the chemical energy in food is turned into body heat; the other 40% is converted into forms of energy that cells can use (ATP). Almost all of that energy eventually leaves the body in the form of heat. Perspiration is the primary way to prevent this rise in body temperature. If this heat could not be dissipated, the body temperature would rise enough to prevent enzyme systems from functioning efficiently, ultimately leading to death.

## WATER MOISTENS, LUBRICATES, AND CUSHIONS

The body secretes many fluids that are primarily water. Water-based secretions are produced by the digestive tract, respiratory tract, urogenital tract, eyes, and skin. Saliva acts as a lubricant, allowing food to pass through the esophagus to the stomach. Mucus provides a protective fluid coating throughout the digestive tract. The lungs are coated with a layer of mucus that provides an important immunologic function. Water helps form the lubricant found in knees and other joints of the body. The spinal cord and brain are cushioned by cerebrospinal fluid. Water is also the basis of amniotic fluid, which functions as a shock absorber surrounding the growing fetus in the mother's womb. Without adequate availability of water, the ability of the body to produce these critical secretions will be limited.

## THE WATER BALANCING ACT

Despite its critical importance for human survival, water is not stored in the body. It is continually lost through respiration (lungs), perspiration (skin), and excretion (urine and feces). Through mechanisms that monitor blood pressure and the concentration of solutes in body fluids, the nervous, endocrine, digestive, and urinary systems work together elegantly to maintain fluid balance and support life.

**Water Intake.** The Adequate Intake (AI) for *total* water is 2.7 liters (11 cups) for adult women and 3.7 liters (15 cups) for adult men. This amount is based primarily on average water intake from fluids and foods. For *fluid* alone, this corresponds to about 2.2 liters (9 cups) for women and about 3 liters (13 cups) for men.

Fluid intake—including water, fruit juice, coffee, tea, soft drinks, milk, and even alcoholic beverages—makes the biggest contribution to our total water needs. For the woman in Figure 9-5, fluid intake adds up to about 9 cups. In addition, nearly all foods contain water. Many fruits and vegetables are more than 80% water, and many meats contain at least 50% water (Fig. 9-6). In Figure 9-5, water from foods supplies another 2 cups. As mentioned previously, the body produces 250 to 350 milliliters (1 to 1½ cups) of water each day as a by-product of the chemical reactions used to metabolize energy. For the woman in Figure 9-5, metabolic water amounts to 1.25 cups. However, the amount of metabolic water produced can double in physically active people.

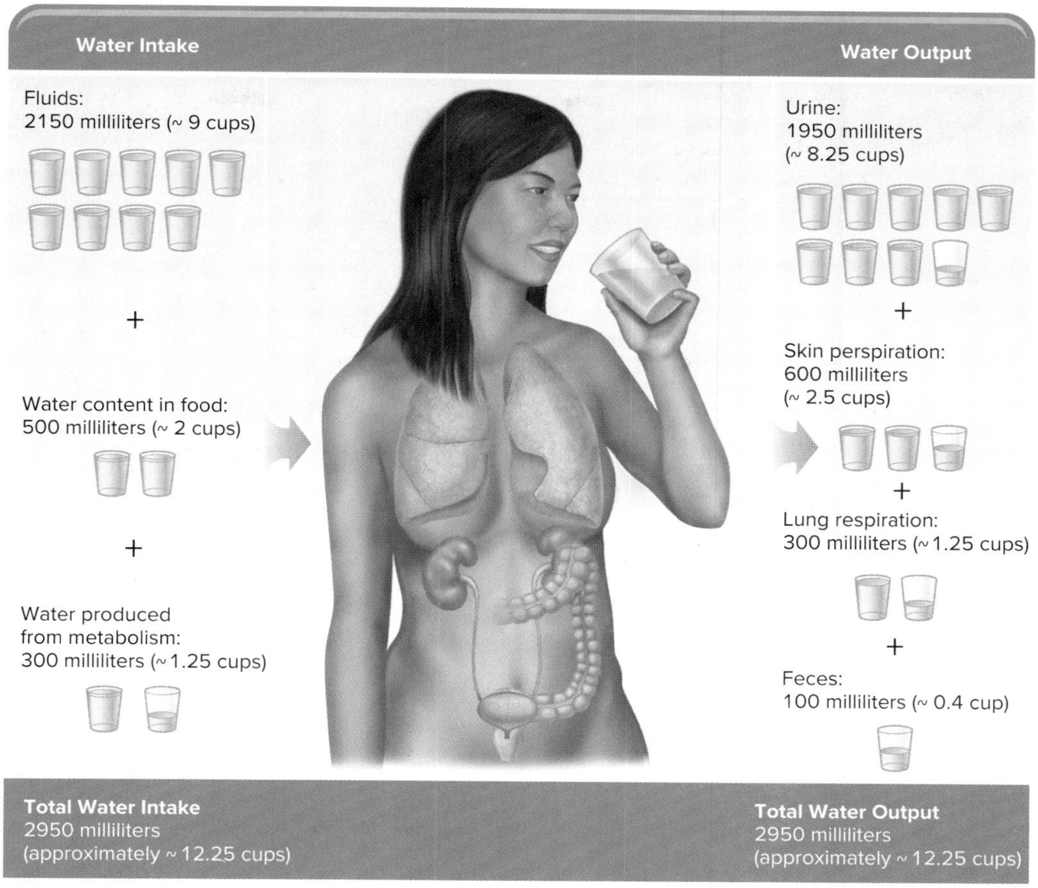

**Water Output.** Usually, urinary excretion of water accounts for the greatest source of output. Average urinary water loss per day is approximately 1950 milliliters (8.5 cups), but it may vary based on intake of fluids, protein, and sodium. Removal of waste products requires at least 500 milliliters (2 cups) of urine production per day. Urine output consistently below this level is often a sign of chronic **dehydration** due to low fluid intake.

**Water Intake**

Fluids: 2150 milliliters (~ 9 cups)

+

Water content in food: 500 milliliters (~ 2 cups)

+

Water produced from metabolism: 300 milliliters (~1.25 cups)

**Total Water Intake**
2950 milliliters
(approximately ~ 12.25 cups)

**Water Output**

Urine: 1950 milliliters (~ 8.25 cups)

+

Skin perspiration: 600 milliliters (~ 2.5 cups)

+

Lung respiration: 300 milliliters (~1.25 cups)

+

Feces: 100 milliliters (~ 0.4 cup)

**Total Water Output**
2950 milliliters
(approximately ~12.25 cups)

**FIGURE 9-5** ▲ Estimate of water balance—intake versus output—in a woman. We primarily maintain body fluids at an optimum amount by adjusting water output to intake. As you can see with this woman, most water comes from the liquids we consume. Some comes from the moisture in foods, and the remainder is manufactured during metabolism. Water output includes that lost via urine, skin, lungs, and feces.

**dehydration** A harmful condition in which water intake is inadequate to replace losses.

Water is lost through the skin in the form of perspiration. On days of low physical activity, these losses amount to about 1 liter. Under hot, humid conditions or with strenuous physical activity, losses can be much greater than 1 liter per day. Some water is also lost from the lungs in the form of water vapor in exhaled air. Together with perspiration, the fluid lost through lung respiration is sometimes called "insensible" water loss because it is difficult to measure.

A relatively small amount of water is lost daily in the feces. When we consider the large amount of water used to lubricate the digestive tract, the loss of only 100 milliliters (0.5 cup) of water each day through the feces is remarkable. In addition to the variable amount of water ingested in the form of foods and fluids, about 8000 milliliters (34 cups) of water enter the digestive tract daily through secretions from the mouth, stomach, intestine, pancreas, and other organs. The small intestine absorbs most of this water, while the colon takes up a lesser but still important amount. The kidneys also greatly conserve water. They can reabsorb as much as 97% of the water filtered each day. The volumes of water intake and output shown in Figure 9-5 are estimates. Caffeine and alcohol intake, ambient temperature, humidity, altitude, and physical activity will influence water loss.

**Fluid Conservation.** The blood pressure and the concentration of solutes in the blood are closely monitored by receptors in the kidneys, blood vessels, and brain. Once the

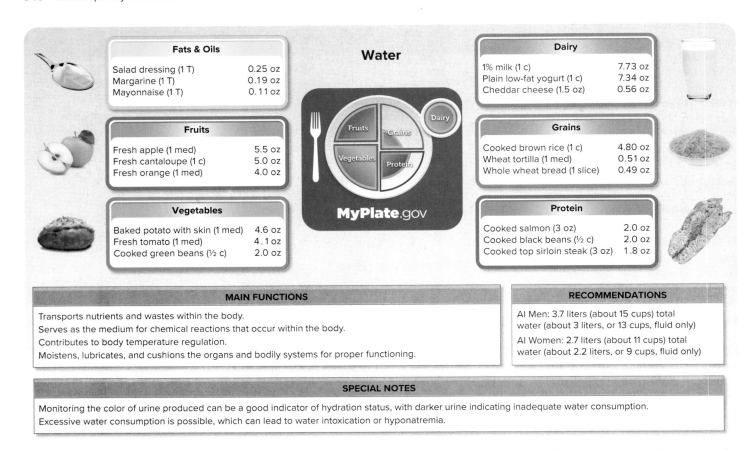

**FIGURE 9-6** ▲ Sources, functions, and recommendations for water. On MyPlate, the fill of the background color (none, 1/3, 2/3, or completely covered) within each group on the plate indicates the average nutrient density for water in that group. Overall, the vegetables, fruits, dairy, and protein groups contain many foods that are nutrient-dense sources of water. Although not depicted on MyPlate, all beverages are nearly 100% water. Fats and oils, on the other hand, have almost no water. mayonnaise: Iconotec/Alamy Stock Photo; ripe green apple: Roman Samokhin/Shutterstock; baked potato: DNY59/E+/Getty Images; glass of milk: NIPAPORN PANYACHAROEN/Shutterstock; brown rice: Yellow Cat/Shutterstock; salmon: Zoran Kolundzija/E+/Getty images; MyPlate: U.S. Department of Agriculture

Sources: Office of Dietary Supplements, Dietary Supplements Fact Sheets, available from https://ods.od.nih.gov/factsheets/list-all; USDA FoodData Central, available from https://fdc.nal.usda.gov.

**antidiuretic hormone** A hormone secreted by the pituitary gland when blood concentration of solutes is high. It causes the kidneys to decrease water excretion, which increases blood volume.

**angiotensin** A hormone produced by the liver and activated by enzymes from the kidneys. It signals the adrenal glands to produce aldosterone and also directs the kidneys to conserve sodium (and therefore water). Both of these actions have the effect of increasing blood volume.

**aldosterone** A hormone produced by the adrenal glands when blood volume is low. It acts on the kidneys to conserve sodium (and therefore water) to increase blood volume.

**hemoconcentration** Decrease in plasma volume, causing an increase in the concentration of red blood cells and other constituents of the blood.

body registers a shortage of available water, it increases fluid conservation (Fig. 9-7). Hormones that participate in this process are **antidiuretic hormone (ADH), angiotensin,** and **aldosterone.** The pituitary gland, located in the brain, senses the concentration of solutes in the blood. When blood concentration of solutes is high, the pituitary gland releases ADH. The kidneys respond to ADH by reducing urine production and output. ADH also causes blood vessel constriction, which acts to raise blood pressure. Meanwhile, the kidneys possess receptors that monitor blood pressure. Low blood pressure triggers the release of an enzyme that activates angiotensin and, eventually, aldosterone, two hormones that signal the kidneys to retain more sodium and, in turn, more water via osmosis. As sodium and water are retained, blood pressure increases back to normal.

**Dehydration.** Despite mechanisms that work to conserve water, fluid continues to be lost via the feces, skin, and lungs. Those losses must be replaced. In addition, there is a limit to how concentrated urine can become. Eventually, if fluid is not consumed, the body becomes dehydrated and suffers ill effects.

By the time a person loses 1% to 2% of body weight in fluids, he or she will be thirsty and experience loss of appetite and **hemoconcentration** (Fig. 9-8). Even this small water deficit can cause one to feel tired and dizzy and to experience headaches. At a 4% loss of body weight from fluids, muscles lose significant strength and endurance, and central nervous system function is negatively affected (e.g., memory and reaction time are compromised and one becomes impatient). By the time body weight is reduced by 10% from fluid loss, heat tolerance is decreased and weakness results. Ultimately, dehydration

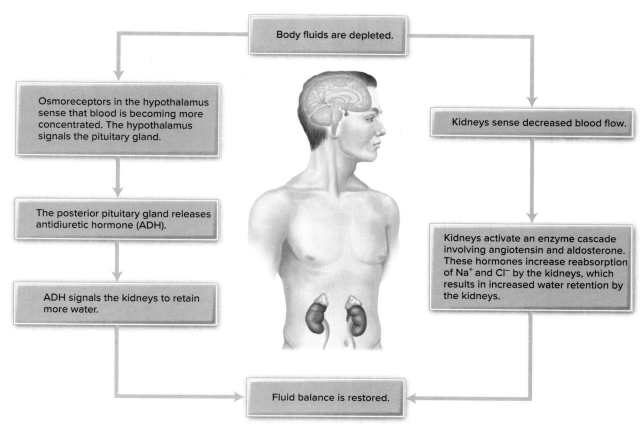

**FIGURE 9-7** ▲ When body fluids are depleted, hormonal signals from the pituitary gland and the kidneys work together to increase fluid retention by the kidneys and thereby restore fluid balance.

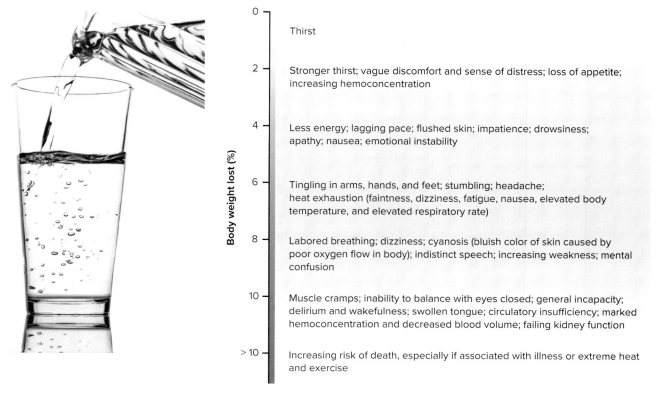

**FIGURE 9-8** ▲ This list of dehydration effects ranges from thirst to risk of death, depending on the extent of water weight lost, shown on the left as a percentage. glass of water: McGraw-Hill Education

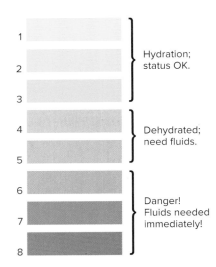

| 1 | ]  |
| 2 | } Hydration; status OK. |
| 3 | ] |
| 4 | } Dehydrated; need fluids. |
| 5 | ] |
| 6 | ] |
| 7 | } Danger! Fluids needed immediately! |
| 8 | ] |

**FIGURE 9-9** ▲ Monitoring the color of urine is a good gauge of hydration.

**COVID CORNER**

You may have heard the suggestion to drink water frequently (e.g., every 15 minutes) to help prevent COVID-19 infection. The basis of this advice is the assumption that washing the virus down the esophagus will prevent it from getting to the lungs. The practice is ineffective, however, because the respiratory system is directly exposed to thousands of viruses through respiratory droplets in the air. Through the airborne droplets and touching your nose, many of the viruses make their way into the nose long before you start drinking water.

**water intoxication** Potentially fatal condition that occurs with a high intake of water, which results in a severe dilution of the blood and other fluid compartments.

**hyponatremia** Dangerously low blood sodium level.

**hard water** Water that contains high levels of calcium and magnesium.

**soft water** Water that contains a high level of sodium.

will lead to kidney failure, coma, and death. Dehydration is a contributing factor to the development of heatstroke, a serious condition. Performing strenuous physical activity, including athletic training, in hot, humid conditions can lead to dehydration and the inability to regulate body temperature. Heart rate is increased and the skin becomes dry. Unassisted, the individual will become unconscious and die. Adequate fluid intake and, if possible, avoidance of physical activity in hot, humid conditions are the best steps to prevent heat illness.

Another potential consequence of inadequate fluid intake is kidney stones. When urine production is lower than about 500 milliliters (approximately 2 cups) per day, the urine is concentrated, increasing the risk of kidney stone formation in susceptible people (generally men). Kidney stones form from minerals and other substances that have precipitated out of the urine and accumulate in the kidney.

The simplest way to determine if water intake is adequate is to observe urine color (Fig. 9-9). If hydration is adequate, urine should be clear or pale yellow (like lemonade); concentrated urine is dark yellow (like apple juice). Urine color, however, can be influenced by consuming supplements (especially some B vitamins), medications, and certain food. Lots of carotenoid-containing items (e.g., carrot juice, pumpkin, winter squash) can tint the urine orange. Eating a large amount of fava beans or rhubarb will turn urine dark brown. A reddish or pinkish urine results from eating too many beets or blackberries. Asparagus not only makes the urine smell like asparagus but can also turn it a bit green.

## IS THIRST A GOOD INDICATOR OF HYDRATION STATUS?

If you do not drink enough water, your brain communicates the need to drink by signaling thirst (Fig. 9-8). In most cases, drinking fluids in response to the thirst sensation will result in adequate hydration. However, the thirst mechanism can lag behind actual water loss during prolonged physical activity and illness, as well as in older adults. Children who are ill, especially those with fever, vomiting, diarrhea, and increased perspiration, as well as older persons, often need to be reminded to drink plenty of fluids because their thirst sensation may be less pronounced.

As you will learn in Chapter 10, athletes need to monitor fluid status. They should weigh themselves before and after training sessions to determine their rate of water loss and, thus, their water needs. However, the body can absorb only about 60% of the water consumed—thus, athletes should drink about 50% more than what they lose through sweat in a workout. The goal is to consume 2 to 3 cups of fluid for every pound lost.

## CAN A PERSON CONSUME TOO MUCH WATER?

Even though the kidneys of a healthy person can produce up to 15 liters of urine per day, it is possible to drink too much water. As water intake increases above what is needed, kidneys process the excess fluid and excrete a more dilute urine. If water intake far exceeds the kidneys' processing ability, overhydration and dilution of blood sodium result. This condition is commonly known as **water intoxication** or, more accurately, **hyponatremia.** Endurance athletes exercising for prolonged times and drinking large volumes of water to replace sweat losses are especially at risk. Using sport drinks will help replace the sodium lost in sweat. Water intoxication can happen to healthy people when they drink a great deal of water in a very short period of time. Severe hyponatremia results along with extreme and rapid blood dilution, causing tissue swelling. The heartbeat becomes irregular, allowing fluid to enter the lungs; the brain and nerves swell, causing severe headaches, confusion, seizure, and coma. Unless water is restricted and a concentrated salt solution administered under close medical monitoring, the person will die.

## SOURCES OF WATER

**What Is the Difference Between Hard and Soft Water?** In North America, 89% of homes have hard water. **Hard water** contains relatively high levels of the minerals calcium and magnesium, whereas **soft water** is high in sodium. Naturally occurring soft water is found in the Pacific North and Northwest, New England, South Atlantic–Gulf States, and Hawaii. Hard water can be converted into soft water through the use of a commercial water softener. As water travels through the water softener, calcium and magnesium exchange with sodium found in the water softener device. The water that exits the softener has a low calcium and magnesium content; however, the sodium content is high. The additional intake of sodium from soft water is undesirable for people restricting sodium intake for reasons such as **hypertension**. The additional intake of calcium and magnesium afforded by consuming hard water would be more beneficial than increasing sodium intake through the use of softened water.

**Is Bottled Water Healthier Than Tap Water?** Bottled water is a popular alternative to tap water; in 2019, Americans consumed 45 gallons of bottled water per person. Bottled water is currently a $67 billion industry, with the average American spending $205 a year on this form of hydration. Many people choose bottled water because they believe it is less likely to be contaminated with pathogens or impurities than tap water. Much of the bottled water produced in the United States, however, is actually processed municipal tap water. There are some differences in water treatment methods: rather than using chlorine to disinfect water, most bottled water is treated with ozone, which does not impart a flavor to the water. Although the Environmental Protection Agency regulates and monitors public water supplies, and the Food and Drug Administration regulates bottled water, the standards for quality and contaminant levels are identical for both.

Consumers should be aware of FDA definitions for types of bottled water. *Artesian water* must come from a confined aquifer. *Springwater* must flow naturally to the surface. *Purified water* is produced through an approved process such as distillation or reverse osmosis.

Beyond the level of contaminants, there are some definite differences between bottled and tap water. A small amount of fluoride is added to municipal water supplies to prevent dental caries. Very few bottled water manufacturers add the mineral fluoride to the water. People who drink primarily bottled water should include regular tap water throughout the day in order to receive the benefits of fluoride. Trips to the drinking fountain or making coffee or tea with tap water should suffice.

Beyond its impact on the environment, the use of plastics may pose additional threats to human health. Drinking water from a freshly washed or newly opened bottle is fine. But plastic, like the food we eat, has a shelf life. Over time, the chemicals that make up plastic break down and can leach into the liquid inside the container. Temperature, age of the bottle, acidity of the contents, and type of plastic (recycling code) all make a difference. Age of the consumer makes a difference, too, as babies and young children are more susceptible to problems than adults. Figure 9-10 lists several guidelines to ensure the safety of the bottled water you drink.

It is well known that America has some of the cleanest, safest tap water in the world. While following the recommendations to drink more water, hydrate with tap or home-filtered water in a reusable container.

**Is Sparkling or Seltzer Water Harmful to Teeth?** Sparkling or seltzer waters have become popular and healthier alternatives to sugar-sweetened sodas and diet sodas. Sparkling water is simply water with carbon dioxide pumped into it. This carbonated water becomes more acidic because the carbon dioxide turns into carbonic acid. Pure water has a neutral pH of 7, but seltzer or sparkling water has a pH between 3 and 4 and is acidic enough to erode tooth enamel. Beverages with a pH below 4 are considered erosive, whereas those with a pH of 2 to 3, including colas, are extremely erosive. Drinking sparkling water with food increases the pH of the combination in your mouth, making it less erosive. Club soda and carbonated mineral waters, such as plain San Pellegrino® or Perrier®, are also not erosive because the natural or added minerals raise the pH to about 5. In contrast, if citric acid, a component of many "flavors," is added to sparkling water, the pH drops. Unfortunately, the citric acid does not have to be listed separately on labels, making it difficult for consumers to know it is present.

▶ Water that comes out of your refrigerator dispenser or through a filter on your faucet typically runs through a charcoal (carbon) filter. The carbon attracts compounds present in tap water to remove off-flavors. Importantly, it does not remove fluoride, a mineral essential to help fight tooth decay.

One of the big differences between bottled water and tap water is the way it is delivered. Tap water travels through pipes into your home, but bottled water requires extra packaging and more costly methods of transport and storage. The large amount of plastic used to package over a billion gallons of bottled water consumed in the United States per year creates huge concerns related to energy use, recycling, and solid-waste disposal. To eliminate the use of disposable plastic water bottles, use a reusable bottle, preferably stainless steel, with a wide mouth for easy cleaning.

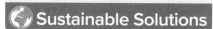

Sustainable Solutions

Production Perig/Shutterstock

▶ Since the 1930s, BPA has been used to make clear plastic bottles. Unfortunately, this organic compound can leach into a bottle's contents when the plastic is exposed to acidic or hot conditions. BPA is considered an endocrine disrupter. At low doses, it can mimic the body's own hormones. Thus, there is concern about chronic exposure for infants and young children. BPA has been banned for use in Canada and many European countries. In the United States, the FDA has banned the use of BPA in the manufacture of baby bottles and sippy cups. Many manufacturers of reusable water bottles are switching to "BPA free" plastics.

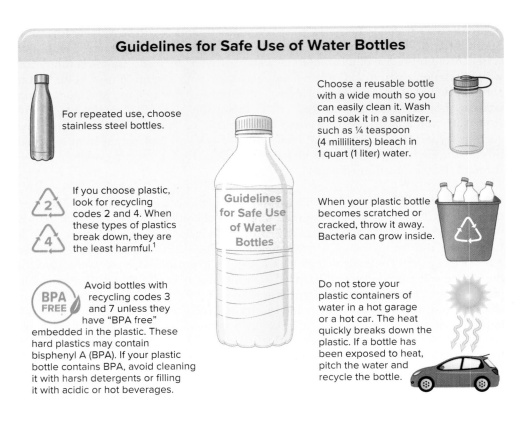

## Guidelines for Safe Use of Water Bottles

For repeated use, choose stainless steel bottles.

If you choose plastic, look for recycling codes 2 and 4. When these types of plastics break down, they are the least harmful.[1]

BPA FREE

Avoid bottles with recycling codes 3 and 7 unless they have "BPA free" embedded in the plastic. These hard plastics may contain bisphenyl A (BPA). If your plastic bottle contains BPA, avoid cleaning it with harsh detergents or filling it with acidic or hot beverages.

Guidelines for Safe Use of Water Bottles

Choose a reusable bottle with a wide mouth so you can easily clean it. Wash and soak it in a sanitizer, such as ¼ teaspoon (4 milliliters) bleach in 1 quart (1 liter) water.

When your plastic bottle becomes scratched or cracked, throw it away. Bacteria can grow inside.

Do not store your plastic containers of water in a hot garage or a hot car. The heat quickly breaks down the plastic. If a bottle has been exposed to heat, pitch the water and recycle the bottle.

**intracellular fluid** Fluid contained within a cell; it represents about two-thirds of body fluid.

**FIGURE 9-10** ▲ Following the guidelines listed here for the safe use of water bottles will protect the consumer from the ingestion of harmful plastics and microorganisms, while saving the environment from the pollution caused by disposable water bottles.

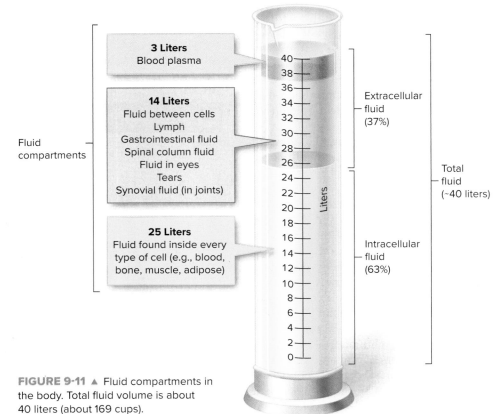

Fluid compartments

**3 Liters**
Blood plasma

**14 Liters**
Fluid between cells
Lymph
Gastrointestinal fluid
Spinal column fluid
Fluid in eyes
Tears
Synovial fluid (in joints)

**25 Liters**
Fluid found inside every type of cell (e.g., blood, bone, muscle, adipose)

Extracellular fluid (37%)

Total fluid (~40 liters)

Intracellular fluid (63%)

**FIGURE 9-11** ▲ Fluid compartments in the body. Total fluid volume is about 40 liters (about 169 cups).

Because its acid content can soften tooth enamel, wait at least 30 minutes to brush your teeth after drinking sparkling water. To decrease the time acid is in contact with your teeth, drink sparkling water with food rather than sipping it all day. Because saliva is needed to neutralize acid, avoid sparkling water if you have dry mouth.[2]

## ELECTROLYTES REGULATE FLUID BALANCE

Each cell of the body is surrounded by a membrane that allows water to pass freely through it. Water found inside the cell membrane is part of the **intracellular fluid.** Intracellular fluid accounts for 63% of fluid in the body. The remaining body fluid is found outside of cells in one of two extracellular spaces: (1) the fluid portion of blood (plasma) and lymph, accounting for 7% of body fluid; or (2) the fluid between cells (interstitial), making up 30% of body fluid (Fig. 9-11).

Although water can freely travel into and out of cells across the cell membrane, the body controls the amount of water in the intracellular and extracellular compartments mainly by controlling ion concentrations. **Ions** are minerals that dissolve in water and are either positively (+) or negatively (−) charged. These charged ions allow the transfer of electrical current, so they are called **electrolytes**. Four electrolytes predominate: sodium ($Na^+$) and chloride ($Cl^-$) are primarily found in the **extracellular fluid,** and potassium ($K^+$) and phosphate ($PO_4^-$) are the principal electrolytes in the intracellular fluid.

The term **osmosis** is used to describe the passage of water through a membrane from an area of lower electrolyte concentration to an area of higher electrolyte concentration. Where ions move, water follows passively. Under normal conditions, the concentration of electrolytes on either side of the cell membrane is controlled in such a way that the intracellular fluid and extracellular fluid are **isotonic.** Thus, water movement into and out of the cell is in equilibrium (Fig. 9-12). However, if the intracellular concentration of electrolytes is greater than the extracellular concentration—that is, **hypertonic**—water will flow freely into the cell. If too much water flows in, the cell can burst, similar to filling a balloon with too much air. The opposite can also occur. When the intracellular concentration of electrolytes is relatively low or **hypotonic** compared to the extracellular environment, water will exit the cell, leading to cell shrinkage.

The principle of osmosis is used by the digestive tract to absorb water from the colon. Water from beverages, foods, and intestinal tract secretions makes the contents of the intestinal tract high in water as it enters the colon from the small intestine. Cells that line the colon actively absorb sodium. Water follows sodium, causing a large amount of water to be reabsorbed in the colon. As a result, daily water loss from feces is low—approximately 100 milliliters (0.5 cup).

**ion**  A positively or negatively charged atom.

**extracellular fluid**  Fluid found outside the cells; it represents about one-third of body fluid.

**osmosis**  The passage of water through a membrane from a less concentrated compartment to a more concentrated compartment.

**isotonic**  Having equal concentration of solutes.

**hypertonic**  Having high concentration of solutes.

**hypotonic**  Having low concentration of solutes.

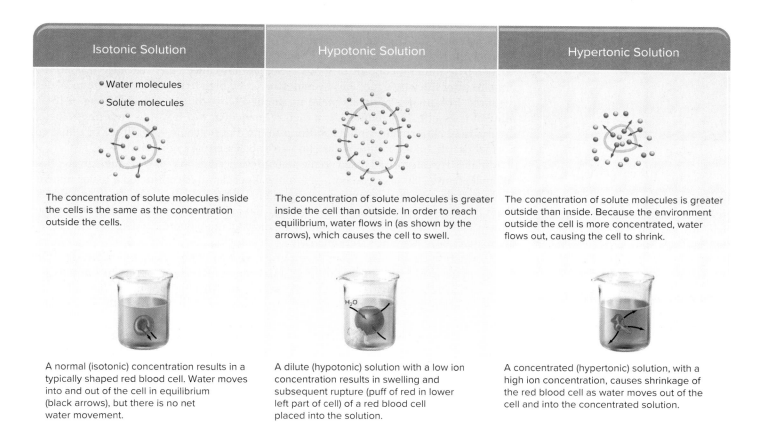

| Isotonic Solution | Hypotonic Solution | Hypertonic Solution |
|---|---|---|
| • Water molecules<br>• Solute molecules | | |
| The concentration of solute molecules inside the cells is the same as the concentration outside the cells. | The concentration of solute molecules is greater inside the cell than outside. In order to reach equilibrium, water flows in (as shown by the arrows), which causes the cell to swell. | The concentration of solute molecules is greater outside than inside. Because the environment outside the cell is more concentrated, water flows out, causing the cell to shrink. |
| A normal (isotonic) concentration results in a typically shaped red blood cell. Water moves into and out of the cell in equilibrium (black arrows), but there is no net water movement. | A dilute (hypotonic) solution with a low ion concentration results in swelling and subsequent rupture (puff of red in lower left part of cell) of a red blood cell placed into the solution. | A concentrated (hypertonic) solution, with a high ion concentration, causes shrinkage of the red blood cell as water moves out of the cell and into the concentrated solution. |

**FIGURE 9-12** ▲ Effects of various ion concentrations in a fluid on human cells. This shows the process of osmosis. Fluid is shifting in and out of the red blood cell in response to changing ion concentrations in the fluid surrounding the cells.

During illness, when large volumes of fluid are lost through vomiting and diarrhea, osmosis can lead to a life-threatening condition. Fluid losses from the gastrointestinal tract result in an increased concentration of electrolytes in the extracellular space. As intracellular water exits the cells in an attempt to dilute the extracellular fluid, cells shrink and lose their ability to function normally. In the heart, this imbalance can lead to a decreased ability of the heart to pump blood and, ultimately, to cardiac failure. Infants and older adults are particularly susceptible to the effects of severe dehydration.

### ✓ CONCEPT CHECK 9.1

1. Why is it significant that water is the "universal solvent"?
2. Describe how water regulates body temperature.
3. Provide two examples of water's role as a lubricant.
4. List the components of water intake and water output.
5. What is the AI for fluid intake for adult men? For adult women?
6. Examine Figure 9-7 and describe the hormonal regulation of water balance in your own words.
7. List two situations in which thirst is *not* a reliable indicator of fluid needs.
8. What is water intoxication?
9. List the guidelines for the safe use of water bottles.
10. Define *osmosis*.

## 9.2 Minerals: Essential Elements for Health

While vitamins are compounds consisting of many elements (e.g., carbon, oxygen, and hydrogen), minerals are individual chemical elements. They cannot be broken down further. The mineral content of foods is sometimes called "ash" because it is all that remains after the whole food has been destroyed by high temperatures or chemical degradation. In humans, minerals make up about 4% of adult body weight (Fig. 9-13). A mineral is essential for humans if a dietary inadequacy of it results in a physiological or structural abnormality and its addition to the diet prevents such illness or reinstates normal health. Sixteen minerals are known to be essential in the diet.

Minerals are categorized based on the amount we need per day. If we require greater than 100 milligrams (1/50 of a teaspoon) of a mineral per day, it is considered a major mineral. These include calcium, phosphorus, magnesium, sulfur, sodium, potassium, and chloride. Trace minerals are required at levels less than 100 milligrams per day. Nine essential trace minerals (iron, zinc, copper, iodine, selenium, molybdenum, fluoride, manganese, and chromium) have been identified for humans.

Information about trace minerals is a growing area of knowledge in nutrition. With the exceptions of iron and iodine, the importance of trace minerals to humans has been recognized only within the last 50 years or so. Although we need 100 milligrams or less of each trace mineral daily, they are as essential to good health as are major minerals.

In some cases, discovering the importance of a trace mineral reads like a detective story, and the evidence is still unfolding. In 1961, researchers linked dwarfism in Middle Eastern villagers to a zinc deficiency. Other scientists recognized that a rare form of heart disease in an isolated area of China was linked to a selenium deficiency. In North America, some trace mineral deficiencies were first observed in the late 1960s and early 1970s, when the minerals were not added to synthetic formulas used for intravenous feeding.

▶ Sulfur is unique among the major minerals because it has no known dietary requirement. Proteins supply all the sulfur we need. It is found in many important compounds in the body, such as some amino acids (e.g., methionine) and vitamins (e.g., biotin and thiamin). Sulfur helps to maintain acid–base balance in the body and is an important part of the liver's drug-detoxifying pathways.

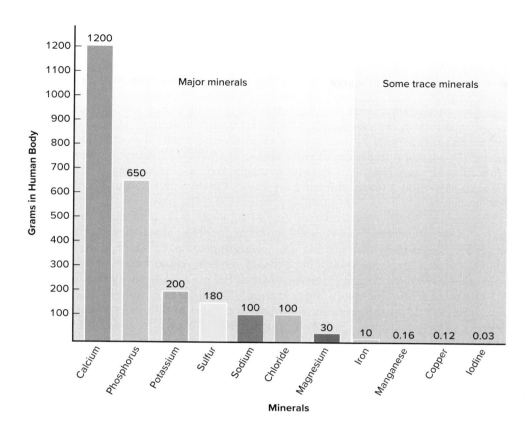

**FIGURE 9-13** ◄ Approximate amounts of various minerals present in the average human body. Other trace minerals of nutritional importance not listed include chromium, fluoride, molybdenum, selenium, and zinc.

It is difficult to define precisely our trace mineral needs because we need only minute amounts. Highly sophisticated technology is required to measure such small amounts in both food and body tissues.

There are several additional trace minerals (sometimes called **ultratrace minerals**) found in the human body, but many of them have no known requirements. These include arsenic, boron, nickel, silicon, and vanadium.

**ultratrace mineral** A mineral present in the human diet in trace amounts but that has not been shown to be essential to human health.

## ABSORPTION AND STORAGE OF MINERALS IN THE BODY

Foods offer us a plentiful supply of many minerals, but the ability of our bodies to absorb and use them varies. The **bioavailability** of minerals depends on many factors, including many nonmineral components of foods. Age, gender, genetic variables, nutritional status, and diet will affect mineral absorption and bioavailability. Numerous prescription drugs also adversely affect mineral absorption. The mineral content listed in a food composition table is a starting point for estimating the contribution the food will make to our mineral needs.

Components of fiber, such as **phytic acid (phytate)** and **oxalic acid (oxalate),** can limit absorption of some minerals by binding to them. Spinach, for example, contains plenty of calcium, but only about 5% (compared to the average 25% bioavailability of calcium from other foods) of it can be absorbed because of the high concentration of oxalic acid in spinach, which binds calcium. High-fiber diets—particularly those in excess of current recommendations of 25 (adult women) to 38 grams (adult men) of fiber per day—can decrease the absorption of iron, zinc, and possibly other minerals.

Many minerals, such as magnesium, calcium, iron, and copper, are of similar sizes and electrical charges (+2 charge). Having similar sizes and the same electrical charge causes these minerals to compete with each other for absorption; therefore, an excess of one mineral decreases the absorption and metabolism of other minerals. For example,

**phytic acid** A constituent of plant fibers that binds positive ions to its multiple phosphate groups; also called *phytate*.

**oxalic acid** An organic acid found in spinach, rhubarb, and sweet potatoes that can depress the absorption of certain minerals present in the food, such as calcium; also called *oxalate*.

▲ The zinc deficiencies found among some Middle Eastern populations are attributed partly to their consumption of unleavened breads, resulting in low bioavailability of dietary zinc. If grains are leavened with yeast, as they are in leavened bread making, enzymes produced by the yeast can break some of the bonds between phytic acid and minerals. This increases mineral absorption. Robyn Mackenzie/Shutterstock

**COVID CORNER**

While there is no good data on the effects of mineral supplements on the risk or severity of COVID-19, there is evidence indicating that supplements of several nutrients can reduce risk or severity of some viral infections, especially in persons deficient in those nutrients. Therefore, taking a standard multi-supplement reflecting the RDA for nutrients is reasonable in individuals with inadequate dietary sources. No supplements contain all the benefits provided by healthy foods, however, so supplements should not be used as substitutes for a good dietary pattern. As always, megadose supplements (many times the RDA) can be harmful and are discouraged. Consumers were warned to avoid any supplements promoting wild health claims and the FDA monitored and warned companies selling fraudulent products claiming to prevent, diagnose, treat, or cure COVID-19.

a large intake of zinc decreases copper absorption. This becomes an issue with the use of individual mineral supplements, which should be avoided unless a dietary deficiency or medical condition specifically warrants it. Food sources pose little risk for these mineral interactions, giving us another reason to emphasize foods in meeting nutrient needs.

Several beneficial vitamin-mineral interactions occur during nutrient absorption and metabolism. When consumed in conjunction with vitamin C, absorption of certain forms of iron—such as that in plant products—improves. The active form of the vitamin D hormone improves calcium absorption. Many vitamins require specific minerals to act as components in their structure and as cofactors for their function. For example, without magnesium or manganese, the thiamin coenzyme cannot function efficiently.

The average North American diet derives minerals from both plant and animal sources. Overall, minerals from animal products are better absorbed than those from plants because binders such as fiber are not present to hinder absorption. Also, the mineral content of plants greatly depends on mineral concentrations of the soil in which they are grown. Vegans must be aware of the potentially poor mineral content of some plant foods and choose some concentrated sources of minerals. Soil conditions have less of an influence on the mineral content of animal products because livestock usually consume a variety of plant products grown from soils of differing mineral contents.

Like vitamins, the majority of the minerals are absorbed in the small intestine. Minor amounts may be absorbed in the stomach, and some sodium and potassium are absorbed in the large intestine. After minerals are absorbed, some travel freely in the bloodstream, but many are carried by specific transport proteins to their sites of action or storage. Calcium is one example of a mineral that can travel as an ion in the blood or bound to a blood protein called albumin. Iron, on the other hand, has damaging effects in its unbound form, so it is transported bound to proteins, such as transferrin.

Minerals are stored in various tissues throughout the body. Some minerals must remain in the bloodstream to maintain fluid balance and supply body functions. Others, such as calcium, phosphorus, magnesium, and fluoride, are stored mainly in bones. Iron, copper, zinc, and many trace minerals are stored in the liver. Still others are stored in muscle tissue, organs, or glands.

## MINERAL TOXICITIES

Excessive mineral intake, especially of trace minerals such as iron and copper, can have toxic results. For many trace minerals, the gap between just enough and too much is small. Taking minerals in supplement form poses the biggest threat for mineral toxicity, whereas food sources are unlikely culprits. Mineral supplements exceeding current standards for mineral needs—especially those that supply more than 100% of the Daily Values on supplement labels—should be taken only under a primary care provider's supervision. The Daily Values for minerals are typically greater than our current standards (e.g., Recommended Dietary Allowances [RDA]). Without close monitoring, doses of minerals should not exceed any Upper Level set on a long-term basis.

The potential for toxicity is not the only reason to be cautious about the use of mineral supplements. Harmful interactions with other nutrients are possible as well as contamination of mineral supplements—with lead, for example. Use of brands approved by the United States Pharmacopeial Convention (USP) lessens this risk.

## PRESERVATION OF MINERALS IN FOODS

Minerals are found in plant and animal foods, but as you previously read, the bioavailability of minerals varies widely. Minerals are not typically lost from animal sources during processing, storage, or cooking; but for plant sources, significant amounts may be lost during food processing. When grains are refined, the final products have lost the majority of their vitamin E, many B vitamins, and trace minerals. The more refined a plant food, as in the case of white flour, the lower its mineral content. During the enrichment of refined grain products, iron is the only mineral added, whereas the selenium, zinc, copper, and other minerals lost during refinement are not replaced. Following the recommendation of the MyPlate Plan to "make half your grains whole" will effectively improve the mineral content of an eating pattern

## ✓ CONCEPT CHECK 9.2

1. Are ultratrace minerals essential for humans? List three examples of ultratrace minerals.
2. Should people take individual mineral supplements? Why or why not?
3. Where are minerals stored in the body?

# 9.3 Sodium

Sodium (Na) is best known as one of the elements in common table salt. Salt is nutritionally important and was once a highly valued chemical compound! Historically, salt was treasured because of its ability to preserve foods. It was so valuable, it was used as a form of payment, and the word *salary* comes from the Latin word for salt. Salt is 40% sodium and 60% chloride by weight (the chemical symbol Na represents the Latin term for sodium, *natrium*). One teaspoon of salt contains 2400 milligrams of sodium. Although salt was once difficult to find, it is now overly abundant in our food supply. Indeed, with Americans on the high side of dietary requirements for sodium, reducing sodium in our dietary patterns is the focus of major public health campaigns.

## FUNCTIONS OF SODIUM

When sodium chloride (NaCl) is dissolved in water, the chemical bond holding the two atoms together breaks and the charged ions, $Na^+$ and $Cl^-$, are released. These electrolytes, as well as others, attract water. The concentration of intracellular and extracellular water is controlled by the concentration of the electrolytes. Fluid balance is maintained by moving or actively pumping sodium ions where more water is needed. Sodium ions also function in nerve impulse conduction and absorption of some nutrients (e.g., glucose).

Although sodium consumption varies tremendously from day to day, even meal to meal (unless strictly controlled for health reasons), our blood levels vary only slightly. The kidneys function as a sodium filter. If blood sodium is low, sodium is secreted back into the blood as it flows through the kidney, resulting in a decreased urine output. Conversely, if our blood sodium levels are too high, the sodium is filtered out by the kidneys and excreted into the urine. When this excess sodium is removed, water follows, resulting in greater urine output. Without drinking extra water, dehydration can result. Fortunately, high-sodium (salty) foods make us thirsty and drive us to drink more fluids.

## SODIUM DEFICIENCY

A low sodium intake, coupled with excessive perspiration and persistent vomiting or diarrhea, has the ability to deplete the body of sodium. This state can lead to muscle cramps, nausea, vomiting, dizziness, and later shock and coma. The likelihood of this occurring is low, because the kidneys are efficient at conserving sodium under conditions of low sodium status.

**Table salt, kosher salt, sea salt—which one is best?** Choosing the best salt really comes down to personal preference or what works best in a recipe.

- *Table salt* is typically mined from the earth and processed into fine grains. These fine, evenly shaped grains can be measured precisely when a recipe needs an exact amount of salt. Most of the table salt sold in North America is fortified with iodine, an essential nutrient for thyroid function.

- *Kosher salt*, also called coarse salt, is chunky and its large grains are easy to see when applying it by hand. It gets its name from its role in *koshering*, also known as curing or pickling.

- *Sea salt* is made through the evaporation of seawater. It is minimally processed, which gives it a coarser texture. It may contain traces of magnesium, calcium, and potassium, giving it some color and flavor variations. Flaky sea salt can be used as the last touch in dishes where the flakes of salt create tiny explosions of salty flavor. Although many consumers prefer the taste and texture of sea salt to table salt, when it comes to heart health, there is no significant difference in sodium or chloride content.

I. Rozenbaum & F. Cirou/PhotoAlto

Perspiration contains about two-thirds the sodium concentration found in blood or about 1 gram of sodium per liter. When excessive perspiration leads to weight loss that exceeds 2% to 3% of total body weight (or about 5 to 6 pounds), sodium losses should raise concern. Even then, adding salt to food or selecting some salty foods, such as soup or crackers, is sufficient to restore body sodium for most people. Athletes who perspire for hours during endurance activities need to consume sodium from electrolyte-replacement or sports drinks during exercise to avoid depletion of sodium (hyponatremia).

## GETTING ENOUGH SODIUM

Nearly all the sodium we consume is absorbed by the digestive tract. About 77% of the sodium we consume is in the form of salt added during food manufacturing and food preparation at restaurants. Sodium added while cooking or at the table at home provides about 11% of our intake, and naturally occurring sodium in foods provides the remaining 12% (Fig. 9-14). Most unprocessed foods are relatively low in sodium; milk is one exception (about 120 milligrams per cup).

The AI for sodium is 1500 milligrams for adults. Adults typically consume 2300 to 4700 milligrams of sodium, however, from dietary patterns that include highly processed foods, dining out, and salt added while cooking. The more we consume highly processed and restaurant food, the higher our sodium intake. Conversely, the more unprocessed and home-cooked meals prepared, the more control a person has over his or her sodium intake. If we ate only unprocessed foods and added no salt, we would consume about 500 milligrams of sodium per day. Major contributors of sodium in the foods we eat are bread and rolls, pizza, sandwiches, cold cuts and cured meats, cheese, soups, burritos and tacos, savory snacks (chips, popcorn, pretzels, snack mixes, and crackers), chicken, and eggs and omelets. When dietary sodium must be restricted, it is important to check the sodium content listed on food labels to monitor sodium intake.

Our kidneys help us adapt to wide variations in dietary sodium intakes, with today's sodium intake found in tomorrow's urine. However, approximately 10% to 15% of adults are *salt sensitive* or, more specifically, *sodium sensitive*—that is, sodium intake has a direct effect on their blood pressure. As their sodium intake increases, so does their blood pressure. Among individuals who are sodium sensitive, lowering sodium intake to about 2000 milligrams per day can often decrease blood pressure. Groups that appear to be especially affected are African Americans, Asian Americans, and people who have diabetes and/or are overweight. Lifestyle factors, such as being overweight and inactive, are the major contributors to the development of high blood pressure or hypertension.

The National Academies of Sciences recently revised the Dietary Reference Intakes for both sodium and potassium. The sodium DRI, which is an AI, is now

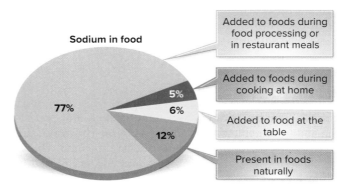

**FIGURE 9-14** ▶ Americans consume salt largely as a result of the addition of sodium during food processing.

1500 milligrams per day for ages 14 and older. The sodium AI for children ages 1 to 13 decreased to 1200 milligrams (see Appendix G for a full list of the sodium DRIs). A new DRI category, Chronic Disease Risk Reduction Intake (CDRR), was also introduced for sodium. This new category is historic in that it is the first time that overuse of a nutrient is being tied to chronic disease. There is now sufficient evidence that a reduction in sodium intake has a beneficial effect on cardiovascular disease risk, hypertension risk, **systolic blood pressure**, and **diastolic blood pressure**. Reducing sodium intake may also help maintain a healthy calcium status, as sodium intake greater than about 2000 milligrams per day may cause calcium to be lost along with the sodium in the urine.

Adopting a reduced-salt dietary pattern is a significant lifestyle change for most people because many typical food choices will have to be limited (Fig. 9-15). At first, foods may taste bland, but eventually you will perceive flavor as the taste receptors in the tongue become more sensitive to the salt content of foods. It takes 6 to 8 weeks to retrain your taste buds to sense sodium at a lower level. Slowly reducing sodium intake by substituting lemon juice, herbs, and spices will allow you to become accustomed to a dietary pattern that contains minimal amounts of salt. Many cookbooks and online sources offer excellent recipes for flavorful dishes.

**systolic blood pressure** The pressure on blood vessels when the heart beats, squeezing and pushing blood through the arteries to the rest of the body.

**diastolic blood pressure** The pressure in the arteries when the heart rests between beats and fills with blood and receives oxygen.

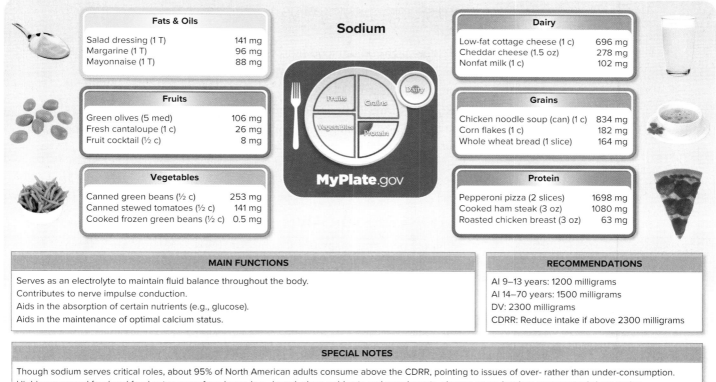

| **Fats & Oils** | |
| --- | --- |
| Salad dressing (1 T) | 141 mg |
| Margarine (1 T) | 96 mg |
| Mayonnaise (1 T) | 88 mg |

| **Fruits** | |
| --- | --- |
| Green olives (5 med) | 106 mg |
| Fresh cantaloupe (1 c) | 26 mg |
| Fruit cocktail (½ c) | 8 mg |

| **Vegetables** | |
| --- | --- |
| Canned green beans (½ c) | 253 mg |
| Canned stewed tomatoes (½ c) | 141 mg |
| Cooked frozen green beans (½ c) | 0.5 mg |

**Sodium**

MyPlate.gov

| **Dairy** | |
| --- | --- |
| Low-fat cottage cheese (1 c) | 696 mg |
| Cheddar cheese (1.5 oz) | 278 mg |
| Nonfat milk (1 c) | 102 mg |

| **Grains** | |
| --- | --- |
| Chicken noodle soup (can) (1 c) | 834 mg |
| Corn flakes (1 c) | 182 mg |
| Whole wheat bread (1 slice) | 164 mg |

| **Protein** | |
| --- | --- |
| Pepperoni pizza (2 slices) | 1698 mg |
| Cooked ham steak (3 oz) | 1080 mg |
| Roasted chicken breast (3 oz) | 63 mg |

| **MAIN FUNCTIONS** |
| --- |
| Serves as an electrolyte to maintain fluid balance throughout the body. |
| Contributes to nerve impulse conduction. |
| Aids in the absorption of certain nutrients (e.g., glucose). |
| Aids in the maintenance of optimal calcium status. |

| **RECOMMENDATIONS** |
| --- |
| AI 9–13 years: 1200 milligrams |
| AI 14–70 years: 1500 milligrams |
| DV: 2300 milligrams |
| CDRR: Reduce intake if above 2300 milligrams |

| **SPECIAL NOTES** |
| --- |
| Though sodium serves critical roles, about 95% of North American adults consume above the CDRR, pointing to issues of over- rather than under-consumption. Highly processed food and food eaten away from home (e.g., bread, pizza, cold cuts and cured meats, cheese, soup, burritos, savory snacks) are major contributors of sodium. |

**FIGURE 9-15** ▲ Food sources, functions, and recommendations for sodium. The fill of the background color (none, 1/3, 2/3, or completely covered) within each food group on MyPlate indicates the average nutrient density for sodium for *natural, unprocessed foods* in that group. The figure shows the sodium content of several natural and processed foods from each food group. Overall, the dairy group is the only food group that provides much sodium in its natural form. Food processing adds significant sodium to foods such as canned vegetables and cured meats.

mayonnaise: Iconotec/Alamy Stock Photo; green olives: Iconotec/Glow Images; green beans: ncognet0/Getty Images; glass of milk: NIPAPORN PANYACHAROEN/Shutterstock; chicken soup: ma-k/E+/Getty Images; pepperoni pizza slice: Burke/Triolo/Brand X Pictures/Getty Images; MyPlate: U.S. Department of Agriculture

Sources: Office of Dietary Supplements, Dietary Supplements Fact Sheets, available from https://ods.od.nih.gov/factsheets/list-all; USDA FoodData Central, available from https://fdc.nal.usda.gov/

**Sandwiches are a top source of sodium with 21% of our average intake of sodium coming from them. This includes chicken and turkey sandwiches, burritos and tacos, hot dogs, breakfast sandwiches and peanut butter and jelly sandwiches. How could you reduce the sodium content of this meal of a grilled cheese sandwich, chips, and tomato soup?** Matthew Antonino/123RF

## AVOIDING TOO MUCH SODIUM

The CDRR recommendation for individuals, ages 14 and older, is to reduce sodium intakes if they are above 2300 mg per day. Reducing sodium intakes that exceed the CDRR are expected to reduce risk for cardiovascular disease and high blood pressure within the healthy population. Current research also links excessive sodium consumption to overweight and obesity. As salt intake increases, fluid intake also increases; if high-calorie beverages are chosen, weight gain may follow.[3] About 97% of North American males and 84% of females have sodium intakes that exceed the CDRR. It must be noted that the Daily Value (DV) of 2300 milligrams equals the CDRR for sodium. Most Americans, including college-age students, should reduce sodium intakes to < 2300 milligrams per day. A healthier goal is to aim for the AI of 1500 milligrams.[4] To lower sodium intake, the *Dietary Guidelines* recommend cooking at home more often; using the Nutrition Facts label to choose products with less sodium, reduced sodium, or no-salt-added; and flavoring foods with herbs and spices instead of salt based on personal and cultural foodways. Deli meats and cheeses are very high in sodium. The sodium in breads, pizza, crackers, and snack foods add up quickly because we typically eat so much of them. Also, reducing sodium while at the same time eating foods rich in potassium, such as fruits and vegetables, might achieve greater health benefits than reducing sodium alone. In sum, choosing fewer highly processed foods and preparing most of your meals at home will result in a healthy, yet achievable, sodium intake. Ask your instructor about the *Rate Your Plate* activity, "How High Is Your Sodium Intake?" available in Connect.

✓ **CONCEPT CHECK 9.3**

1. List three high-sodium food sources in your dietary pattern.
2. Which organ regulates the amount of sodium in your blood?
3. Define sodium sensitivity.
4. Define the term CDRR and explain why it is important.
5. List three specific changes you could make to your dietary pattern to decrease your sodium intake.

## 9.4 Potassium

### FUNCTIONS OF POTASSIUM

Potassium (K) performs many of the same functions as sodium, such as water balance and nerve impulse transmission. (The chemical symbol K represents the Latin term *kalium.*) All membranes contain an energy-dependent pump that can transfer sodium from inside to outside the cell. When sodium ($Na^+$) is actively pumped out of the cell, potassium ($K^+$) enters the cell in an attempt to balance the loss of the positively charged sodium ions. That makes potassium the principal positively charged ion inside cells. Intracellular fluids contain 95% of the potassium in the body. Higher potassium intake is associated with *lower* rather than higher blood pressure values.

### POTASSIUM DEFICIENCY

▲ Fruits and vegetables are rich sources of potassium. Mary Jon Ludy/McGraw-Hill Education

**diuretic** A substance that increases urinary fluid excretion.

Low blood potassium, also known as *hypokalemia,* is a life-threatening problem. Symptoms often include a loss of appetite, muscle cramps, confusion, and constipation. Eventually, the heart beats irregularly, decreasing its capacity to pump blood.

Hypokalemia can result from malnutrition and inadequate food intake, but it is most commonly seen with chronic diarrhea or vomiting, or as a side effect of medications, including laxatives and some **diuretics**. Vulnerable populations include people with certain eating disorders (see Chapter 11) or alcohol use disorders (see Chapter 1). Other populations at increased risk for potassium deficiency include people on very low-calorie diets and athletes who exercise for prolonged periods. In these situations, more potassium-rich foods should be consumed to compensate for potentially low body potassium. People who take potassium-wasting diuretics (water pills) for high blood pressure need to carefully monitor their dietary intake of potassium. Increased intake of fruits and vegetables or potassium chloride supplements are typically prescribed by their primary care providers.

## GETTING ENOUGH POTASSIUM

Unprocessed foods are rich sources of potassium, including fruits, vegetables, milk, whole grains, legumes, and meats (Fig. 9-16). An easy guide is to remember that the more processed your food, the higher it is in sodium and the lower it is in potassium. Major contributors of potassium to the adult dietary pattern include milk, potatoes, beef, coffee, tomatoes, and orange juice. Bananas are a popular fruit and excellent source of potassium. Read more about them in the *Farm to Fork* feature.

Approximately 90% of the potassium consumed is absorbed, but dietary patterns are more likely to be lower in potassium than sodium because we add salt to our food, not potassium. New DRIs for potassium recommend that most people should consume between 2000 and 3400 milligrams of potassium per day (Fig. 9-16). Many of us need to increase potassium intake, preferably by increasing fruit and vegetable intake.

> ▶ **Potassium: A Dietary Component of Public Health Concern for Underconsumption**
>
> According to the *Dietary Guidelines* potassium is considered a dietary component of public health concern for the general U.S. population because low intakes are associated with health concerns. If a healthy dietary pattern is consumed, amounts of potassium can meet recommendations. The *Dietary Guidelines* encourage individuals to make shifts to increase the intake of vegetables, fruits, beans, whole grains, and dairy to move intakes of potassium closer to recommendations.

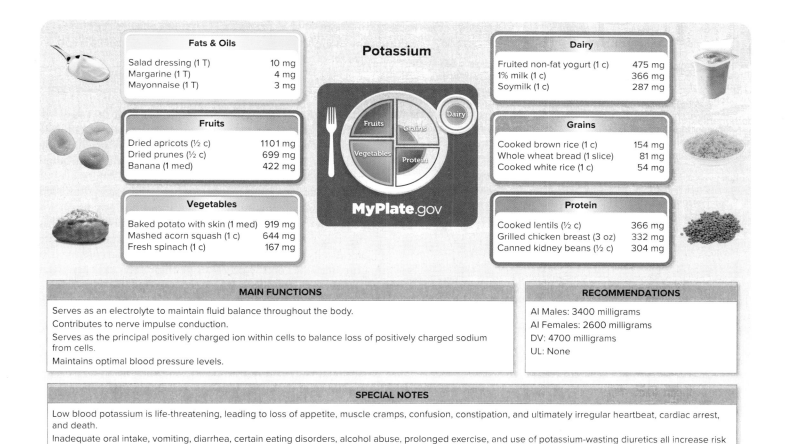

**FIGURE 9-16** ▲ Food sources, functions, and recommendations for potassium. The fill of the background color (none, 1/3, 2/3, or completely covered) within each food group on MyPlate indicates the average nutrient density for potassium in that group. The figure shows the potassium content of several foods in each food group. Overall, the richest sources of potassium are unprocessed foods of plant origin, such as fruits, vegetables, and beans. mayonnaise: Iconotec/Alamy Stock Photo; dried apricots: lynx/iconotec/Glowimages; baked potato: DNY59/E+/Getty Images; yogurt in a container: Ingram Publishing/SuperStock; brown rice: Yellow Cat/Shutterstock; heap of lentils: asterix0597/Getty Images; MyPlate: U.S Department of Agriculture

Sources: Office of Dietary Supplements, Dietary Supplements Fact Sheets, available from https://ods.od.nih.gov/factsheets/list-all; USDA FoodData Central, available from https://fdc.nal.usda.gov/

## FARM to FORK   Bananas

Corey Hochachka/Design Pics

When you enjoy tropical fruits, you are eating globally. The most common tropical fruits are bananas, pineapples, mangos, and papayas—imported from Ecuador, Costa Rica, Mexico, and Hawaii. Bananas are the most consumed fruit in the United States—more than apples and oranges combined!

### Grow

• Bananas grow on "banana palms," which are actually a perennial herb rather than a tree.
• Bananas do not grow from a seed but instead from a bulb or rhizome. After planting a banana palm in a subtropical environment, it can take up to 12 months to enjoy the fruits of your labor.
• Because bananas grow in tropical climates, they are not seasonal fruits but rather continue to grow all year long.

### Shop

• There are two main categories of bananas. Plantains (or cooking bananas) are starchy, and Cavendish (or dessert bananas) are the sweet, yellow bananas most commonly sold at the supermarket.
• To increase your antioxidant content when eating bananas, look beyond Cavendish bananas. Red bananas and niños, or Lady Fingers, provide greater vitamin C, potassium, calcium, manganese, carotenoids, and zinc than the common Cavendish.

### Store

• Bananas continue to ripen after harvest and are often harvested when immature and green. These bananas ripen in 5 to 7 days at room temperature.
• Store ripe bananas in the refrigerator. Although their skins will turn brown, the flesh will stay fresher for days.

### Prep

• Plantains are the main carbohydrate source for 20 million people globally. Harvested when green, plantains are skinned, then steamed, baked, or fried.
• In the Caribbean, fried plantains are a staple. For a healthier recipe, try baked plantains. Use ripe plantains, spotted with brown or black spots, and slice after peeling. Coat a nonstick baking sheet with cooking spray and lay the plantains in rows. Cook at 450°F for 15 minutes, flipping frequently.
• Celebrate bananas as the perfect on-the-go snack, lunch box treat, or sliced treat on top of cereal, yogurt, or whole wheat pancakes.

Source: Robinson J: Tropical fruits: Make the most of eating globally, in *Eating on the Wild Side*. New York: Little, Brown and Company, 2013.

lynx/iconotec.com/Glow Images

## AVOIDING TOO MUCH POTASSIUM

If the kidneys function normally, typical potassium intakes from food will not lead to toxicity. Thus, no Upper Level or CDRR for potassium has been set. When the kidneys function poorly, potassium builds in the blood, inhibiting heart function and leading to a slowed heartbeat. If left untreated, the heart eventually stops beating, resulting in a cardiac arrest and death. Therefore, in cases of kidney failure or kidney disease, close monitoring of the levels of potassium in blood and in the dietary pattern becomes critical. Experts suggest being cautious with potassium supplements.

### ✓ CONCEPT CHECK 9.4

1. List two functions of potassium in the body.
2. How is potassium intake related to blood pressure?
3. List three specific changes you could make to your eating pattern to increase your potassium intake.

## 9.5 Chloride

### FUNCTIONS OF CHLORIDE

Chloride ($Cl^-$) is a negative ion found primarily in the extracellular fluid. Along with sodium and potassium, chloride helps to regulate fluid balance in the body. In fact, chloride itself may be partially responsible for increases in blood pressure that accompany high-salt diets.

Chloride ions are also a component of the acid produced in the stomach (hydrochloric acid) and are important for overall maintenance of acid–base balance in the body. This electrolyte also is used during immune responses as white blood cells attack foreign cells. In addition, nervous system function relies on the presence of chloride.

### CHLORIDE DEFICIENCY

Low levels of chloride in the blood can lead to a disturbance of the body's acid–base balance. A chloride deficiency is unlikely, however, because our dietary salt intake is so high. Frequent and lengthy bouts of vomiting, if coupled with a poor intake of chloride, can contribute to a deficiency because stomach secretions contain a lot of chloride. Individuals with bulimia or severe cases of gastroenteritis are at risk for chloride deficiency. In addition, low chloride levels could occur as a side effect of some medications, such as diuretics or laxatives.

### GETTING ENOUGH CHLORIDE

When it comes to sources of chloride, it is important to make the distinction between the chloride ion, which is vital for body functions, and chlorine ($Cl_2$), which is a poisonous gas. *Chlorine* is used to disinfect municipal water supplies. A small amount of chlorine may remain in tap water, but the substance evaporates quickly. Municipal and well water usually contains some *chloride* (leached from the earth) as well, but water does not represent a significant source of chloride.

A few fruits and vegetables, such as seaweed, celery, tomatoes, and olives, are naturally good sources of chloride. Most of our dietary chloride, however, comes from salt added to foods. Knowing a food's salt content allows for a close prediction of its chloride content. Salt is 60% chloride by weight.

Like sodium, nearly all the chloride consumed is efficiently absorbed. The AI for chloride is based on the 40:60 ratio of sodium to chloride in salt (1500 milligrams of sodium: 2300 milligrams of chloride). If the average adult consumes about 9 grams of salt daily, that yields 5.4 grams (5400 milligrams) of chloride. The principal route of excretion is the kidneys, although some chloride is lost in perspiration.

### AVOIDING TOO MUCH CHLORIDE

The average adult typically consumes an excess of chloride. Because chloride has a role in raising blood pressure, it is important that aging adults consciously control salt intake to decrease risk of developing hypertension. Learning at a young age to select foods lower in sodium chloride is the best way to start.

▶ **Chloride**
AI: 2300 milligrams
DV: 3400 milligrams
UL: 3600 milligrams

▲ Olives are a naturally good source of chloride. I. Rozenbaum & F. Cirou/PhotoAlto

### ✓ CONCEPT CHECK 9.5

1. List two functions of chloride in the body.
2. How is chloride intake related to blood pressure?

## 9.6 Calcium

### FUNCTIONS OF CALCIUM

**Bone Growth, Maintenance, and Repair.** Calcium (Ca) represents 40% of all the minerals present in the body and equals about 2.5 pounds (1200 grams) in the average person (Fig. 9-13). All cells require calcium to function; however, more than 99% of the calcium in the body is used for growth, development, and maintenance of bones. Calcium is the main component of **hydroxyapatite,** the crystalline compound responsible for the structure and hardness of bone.

**Muscle and Nerve Function.** Less than 1% of calcium is found in blood, but this circulating calcium is critical to supply the needs of cells other than bone cells. Forming and maintaining bone are calcium's major roles in the body, but it is critical for other processes as well. Muscle contraction is activated by calcium release and the flow of calcium along the surface of the muscle cell. In nerve transmission, calcium assists in the release of neurotransmitters and permits the flow of ions in and out of nerve cells.

**Additional Functions and Health Benefits.** Calcium helps regulate cellular metabolism by influencing the activities of various enzymes and hormonal responses. Calcium also functions in the maintenance of cell membrane integrity, normal blood pressure, regulation of glucose concentration, and **cellular differentiation.** Calcium is essential for formation of a blood clot. It is the tight regulation of the concentration of calcium in the blood that keeps these processes going, even if a person fails to consume enough dietary calcium from day to day. In its 2018 report, the World Cancer Research Fund reported convincing evidence that dietary calcium is related to a decrease in colorectal cancer and a probable decrease in breast cancer (in premenopausal women only).[5]

**hydroxyapatite** Crystalline compound containing calcium, phosphorus, and sometimes fluoride, also known as *bone mineral.*

**cellular differentiation** The process of a less-specialized cell becoming a more specialized type. An example is when stem cells in the bone marrow become red and white blood cells.

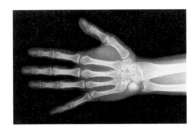

▲ Ninety-nine percent of calcium in the body is in bones. itsmejust/Shutterstock

**parathyroid hormone (PTH)** A hormone made by the parathyroid glands that increases synthesis of the vitamin D hormone and aids calcium release from bone and calcium conservation by the kidneys, among other functions.

**tetany** A body condition marked by sharp contraction of muscles and failure to relax afterward; usually caused by abnormal calcium metabolism.

Supplements, however, were not recommended for decreasing cancer risk. Overall, the benefits of a dietary pattern providing adequate calcium extend far beyond bone health.

## CALCIUM DEFICIENCY

The body tightly regulates blood calcium concentration within a narrow range, regardless of dietary intake (Fig. 9-17). If dietary calcium intake is inadequate and blood calcium concentration begins to decrease, three hormonally controlled actions are stimulated to reestablish normal levels of calcium in the blood: (1) bones release calcium; (2) intestines absorb more calcium; and (3) the kidneys retain more calcium in the blood. **Parathyroid hormone** (**PTH**, from the parathyroid gland) and vitamin D are released in response to low blood calcium. Because of this tight hormonal regulation, insufficient dietary calcium is not likely to result in low blood calcium. Rather, kidney diseases, hormonal abnormality, or medications are the likely culprits. If blood calcium falls below a critical point, muscles cannot relax after contraction and nerve function is disrupted. The result is a condition called **tetany** in which muscles become stiff or twitch involuntarily.

You can see that the skeleton does more than simply provide the framework of the body; it also functions as a bank to which calcium can be added and from which calcium can be withdrawn. Only about 1% of the calcium in bone is available at any time for this purpose. Over time, however, bone loss due to inadequate calcium intake and/or absorption may occur, though slowly. It takes many years of calcium loss from bones to show up as clinical symptoms. By not meeting calcium needs, some people, especially women, are setting the stage for **osteoporosis** and future bone fractures.

## OSTEOPOROSIS

In the U.S., 54 million adults age 50 and older—more than one-half of the total U.S. adult population—are affected by osteoporosis and low bone mass.[6] Of those affected, 10.2 million have osteoporosis and another 43.4 million have low bone mass. Of those with osteoporosis, it is estimated that 7.7 million are non-Hispanic white, 0.5 million are

**FIGURE 9-17** ▶ Blood calcium levels are tightly regulated by hormones that influence the bones, intestine, and kidneys.

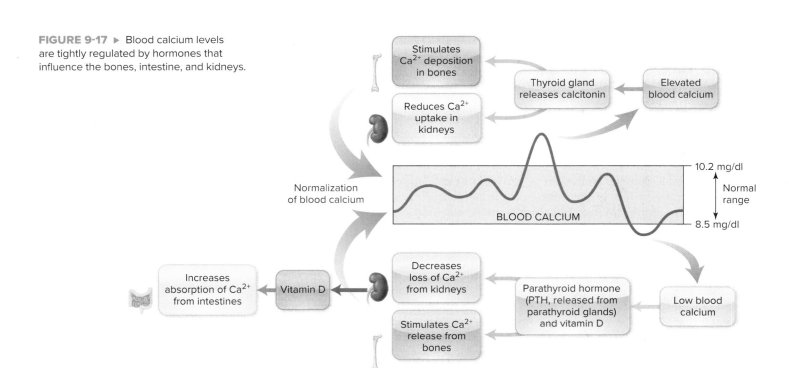

non-Hispanic black, and 0.6 million are Mexican American. The number of adults over 50 with osteoporosis or low bone mass is expected to reach 71.2 million by 2030.

Osteoporosis is responsible for approximately 2 million bone fractures per year in the United States, including nearly 300,000 broken hips. The health care costs associated with these fractures are projected to reach more than $25 billion by the year 2025, with the most rapid increases among minority populations. Hip fractures are regarded as devastating. They often result in loss of mobility and the need for long-term care. Women experience 80% of hip fractures, and the average age for hip fracture is 80 years.[7] Only 40% of people with hip fractures regain their earlier level of independence. It is estimated that a year after fracturing a hip, 90% of those who needed no assistance climbing stairs before the fracture will not be able to climb five stairs; 66% will need help to get on or off a toilet; 50% will not be able to raise themselves from a chair; 31% will need assistance to get out of bed; and 20% will not be able to put on a pair of pants by themselves.

Hip fracture is associated with significant mortality. It is estimated that 20% to 30% of the 300,000 Americans age 65 or older who fracture a hip each year will die within 12 months. Other types of fractures or even fear of fracture due to osteoporosis can affect quality of life. Vertebral fractures, especially if there are several of them, cause significant pain, reduced lung function, loss of height, and a curved spine. Movement can be restricted and gait altered, increasing the risk and fear of falls and/or more fractures.

Women tend to lose 1% to 3% of their bone mass each year after menopause. Men also lose bone mass as they age, but the loss is more gradual. When the bone-demineralizing activities of **osteoclasts** exceed the bone-building activities of **osteoblasts,** bone mass declines. This is a normal part of aging and does not always lead to unhealthy bones. However, if bone mass is low when this process begins, even moderate bone demineralization can lead to a condition called **osteopenia.** As more bone is lost, the entire matrix of the bone tissue also begins to break down. When this occurs, osteoporosis is likely to result. About 25% of women older than age 50 develop or have osteoporosis. Among people older than age 80, osteoporosis becomes the rule, not the exception. Although young adults may not feel threatened by osteoporosis, to older adults, the diagnosis presents the reality of life permanently changed.

**Type 1 and Type 2 Osteoporosis.** There are two types of osteoporosis. **Type 1 osteoporosis,** also called postmenopausal osteoporosis, typically appears in women between 50 and 60 years of age. This type of osteoporosis is directly linked to decreased estrogen concentrations that occur at menopause. Type 1 osteoporosis most dramatically affects trabecular bone, as this type of bone undergoes faster remodeling than cortical bone (Fig. 9-18).

Trabecular bone has a much higher density of osteoblast and osteoclast cells than cortical bone. The osteoblast cells require estrogen for maximal activity. After menopause, osteoblast activity decreases; however, osteoclast activity remains high. The result is greater bone resorption than resynthesis. Minerals are released but not reincorporated into bone. The locations where the osteoclasts left pockets for osteoblasts to rebuild the collagen mineral matrix will now become holes. A woman can lose 20% to 30% of trabecular bone and 5% to 10% of cortical bone between ages 50 and 60, unless intervention occurs (Fig. 9-19). The bones at greatest risk for osteoporotic fractures are the trabecular-rich bones of the pelvis (7% of fractures), vertebrae of the spine (27%), and portions of long bones such as the wrist (19%). People with either form of osteoporosis can lose significant height

▶ Individuals with cancer, especially those with breast and prostate cancer, have an increased risk of osteoporosis. The cancer as well as cancer therapies can cause bone loss and fractures. These patients need to be given advice on the changes they can make in their eating, physical activity, and lifestyle patterns that will strengthen their bones.

**osteoclast** Bone cells that break down bone and subsequently release bone minerals into the blood.

**osteoblast** Bone cells that initiate the synthesis of new bone.

**osteopenia** A bone disease defined by low mineral density.

**type 1 osteoporosis** Porous trabecular bone characterized by rapid bone demineralization following menopause.

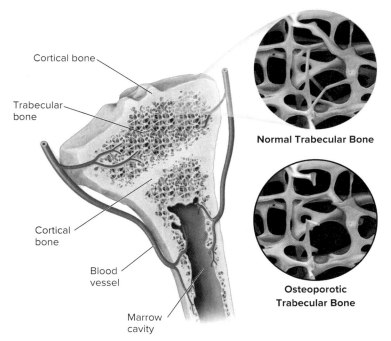

**FIGURE 9-18** ▲ Normal and osteoporotic bone. Cortical bone forms the shafts of bones and their outer mineral covering. Trabecular bone supports the outer shell of cortical bone in various bones of the body. Note how the osteoporotic bone has much less trabecular bone. This leads to a more fragile bone and is not reversible to any major extent with current therapies.

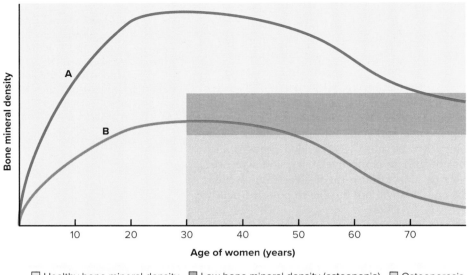

**Woman A** had developed a high peak bone mass by age 30. Her bone loss was slow and steady between ages 30 and 50 and sped up somewhat after age 50 because of the effects of menopause. At age 75, the woman had a healthy bone mineral density value and did not show evidence of osteoporosis.

**Woman B** achieved a low peak bone mass by age 30 but experienced the same rate of bone loss as Woman A. By age 50, she already had low bone mineral density, and by age 70 kyphosis and spinal fractures had occurred (see Fig. 9-20).

☐ Healthy bone mineral density  ☐ Low bone mineral density (osteopenia)  ☐ Osteoporosis

**FIGURE 9-19** ▲ The relationship between peak bone mass and the ultimate risk of developing osteoporosis and related bone fractures.
Source: Davies JH and others: Bone mass acquisition in healthy children. *Archives of Disease in Childhood* 90:373, 2005.

and experience severe pain, especially in the vertebrae. A woman may lose an inch or more in height as the bone is demineralized (see Fig. 9-20).

**type 2 osteoporosis** Porous trabecular and cortical bone observed in men and women after the age of 70.

**Type 2 osteoporosis** tends to be diagnosed later in life (70 to 75 years of age). Type 2 osteoporosis is a result of breakdown of both cortical and trabecular bone. It is due to a combination of dietary and age-related factors; low dietary intake of bone-building nutrients compounds the problems resulting from the decreased ability to absorb or metabolize nutrients.

**FIGURE 9-20** ▶ A young woman with healthy bones and an older woman with osteoporosis. Osteoporotic bones have less supportive structure, so osteoporosis generally leads to loss of height, distorted body shape, fractures, and possibly loss of teeth. Monitoring changes in adult height is one way to detect early evidence of osteoporosis. Kyphosis, or curvature of the upper spine, seen here in the woman on the right who is older, results from demineralization of the vertebrae. This can lead to both physical and emotional pain. Kyphosis occurs in both men and women.

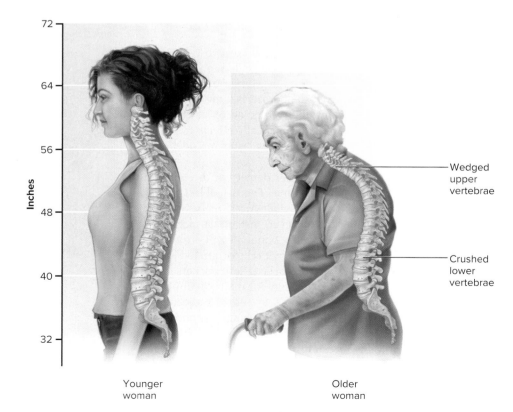

Both men and women with osteoporosis can develop a curvature in the upper spine called **kyphosis** (see Fig. 9-20). Kyphosis is a major concern because the bending of the spine may decrease the volume of the chest cavity, resulting in difficulty breathing, abdominal pain, decreased appetite, and premature satiety. As discussed, osteoporotic bone is also more susceptible to fracture following a fall.

According to the National Osteoporosis Foundation guidelines, adults diagnosed with osteoporosis should first be counseled on risk factor reduction.[8] Daily patterns of adequate calcium and vitamin D consumption as well as physical activity should be stressed. Primary care providers have guidelines regarding pharmacological (drug) intervention for patients. These drugs are indicated for postmenopausal women and men over 50 who meet specific standards for level of risk of future fracture and medical history.[9] Current medication options are **bisphosphonates** and para-thyroid hormone. Women may also use calcium and hormone replacement therapy (see *Medicine Cabinet* for a list of osteoporosis medications).

The incidence of osteoporosis increases as a person ages. Its significance as a personal and public health problem is intensifying as the U.S. population ages. It appears, however, that a large percentage of the cases of osteoporosis can be prevented. The key to prevention is to build dense bones during the first 30 years of life and then limit the amount of bone loss in adulthood. People who achieve a higher peak bone mass early in life have more calcium to lose before bones become weak and fracture easily (Fig. 9-19). Males experience osteoporosis less often than females do because they typically have a higher peak bone mass and therefore have more bone mass to lose.

**Bone Health Assessment.** Before the 1990s, a routine X-ray could detect osteoporosis, but not until 30% or more of bone mass was already lost. A hairline fracture of the vertebrae also was used to diagnose osteoporosis, but neither of these tools predicted the risk of osteoporosis. Today, we have tools that can quantitate bone mass and bone density, and, subsequently, the likelihood of a person developing bone disease.

The most accurate test for assessing bone density is the central dual energy X-ray absorptiometry (DXA) measurement of the hip and spine. The central DXA procedure is simple, painless, safe, noninvasive, and generally takes less than 15 minutes. The ability of the bone to block the path of a low-level X-ray is used as a measure of bone mineral density. A very low dose of radiation is used for the DXA—about one-tenth of the exposure from a chest X-ray. The hip and spine are measured because these sites are commonly affected by osteoporosis and are likely to result in more serious injuries.

The National Osteoporosis Foundation recommendations call for DXA testing for the following groups of people:

- All women age 65 and older and men age 70 and older
- Younger postmenopausal women and men (ages 50 to 69) who have risk factors
- Women going through perimenopause (transitioning into menopause) who are at risk due to low body weight, have prior low-trauma fracture, or take high-risk medications, such as steroids
- Adults with fracture after age 50
- Adults with a health condition (e.g., rheumatoid arthritis, Crohn's disease, asthma) for which they take steroids for a prolonged period
- Anyone being considered for medication for osteoporosis or receiving therapy for osteoporosis

**kyphosis** Abnormally increased bending of the spine; commonly known as *dowager's hump.*

**bisphosphonates** Drugs that bind minerals and prevent osteoclast breakdown of bone. Examples are alendronate (Fosamax) and risedronate (Actonel).

## Medicine Cabinet

### Types of Osteoporosis Medications

**Antiresorptive agents** are used to prevent bone loss and decrease the risk of bone fractures. They slow bone loss that occurs during the breakdown phase of the remodeling cycle. These medicines slow bone loss without affecting bone synthesis so that bone density can increase.

- Bisphosphonates
  - Alendronate (Fosamax® and Fosamax Plus D®)
  - Ibandronate (Boniva®)
  - Risedronate (Actonel®, Actonel® with Calcium, and Atelvia™)
  - Zoledronic acid (Reclast®)
- Calcitonin (Fortical® and Miacalcin®)
- Denosumab (Prolia®)
- Estrogen therapy
- Estrogen agonists/antagonists (Evista®)

**Anabolic agents** are used to increase the rate of bone formation and decrease the risk of fractures.

- Teriparatide (Forteo®)
- Abaloparatide (Tymlos)
- Romosozumab-aqqg (Evenity)

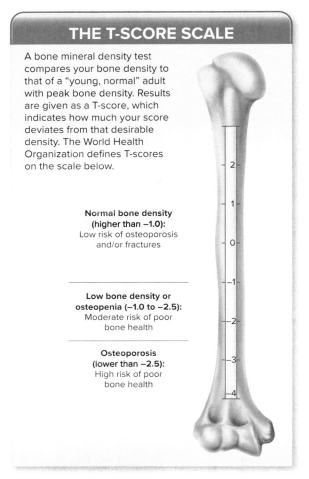

## THE T-SCORE SCALE

A bone mineral density test compares your bone density to that of a "young, normal" adult with peak bone density. Results are given as a T-score, which indicates how much your score deviates from that desirable density. The World Health Organization defines T-scores on the scale below.

**Normal bone density (higher than –1.0):**
Low risk of osteoporosis and/or fractures

**Low bone density or osteopenia (–1.0 to –2.5):**
Moderate risk of poor bone health

**Osteoporosis (lower than –2.5):**
High risk of poor bone health

**FIGURE 9-21** ▲ This classification for diagnosing osteoporosis is used by the World Health Organization.

▲ Weight-bearing physical activity, such as walking or running, is associated with increased bone density. Female athletes, however, must maintain an adequate energy intake to maintain estrogen levels, which stimulates bone formation. Derek E. Rothchild/Getty Images

From the DXA measurement of bone density, a T-score is generated, which compares the observed bone density to that of a person at peak bone mass (e.g., age 30). The T-score is interpreted as shown in Figure 9-21.

When DXA measurements of the hip and spine are not feasible due to portability, expense, or severe obesity, several peripheral measurements of bone density can be used as screening methods at various sites on the body. Peripheral DXA uses the same technology as central DXA but scans only the ankle or wrist. Peripheral quantitative computed tomography (pQCT) generates a three-dimensional scan of the radius in the arm or the tibia in the lower leg. It measures bone mineral density and can provide information on cortical and trabecular bone density, area, and thickness, as well as some measurements of bone strength. The quantitative ultrasound (QUS) technique uses sound waves to measure the density of bone in the heel, shin, and kneecap. These peripheral tests are sometimes used at health fairs, medical offices, and research settings. Although they cannot be used to diagnose osteoporosis, they can inform the health care provider if additional testing (i.e., central DXA) is warranted.

Whether or not you have had your bone mineral density tested, you can use the online Fracture Risk Assessment Tool (FRAX) to estimate your risk of fracture. You can calculate your FRAX score at http://www.shef.ac.uk/FRAX. The National Osteoporosis Foundation Guide recommends that doctors consider prescribing medication if your risk of a hip fracture is at least 3% over the next 10 years. The incidence of osteoporosis in many other countries is far less than in the United States. In reviewing Tables 9-1 and 9-2, you can see that the factors related to the greater incidence in the United States are mainly related to dietary patterns and other lifestyle behaviors. That means that osteoporosis can largely be prevented.

The synthesis, maintenance, and repair of bone requires several vitamins, minerals, and hormones. The key nutrients include vitamin D and the minerals calcium, phosphorus, magnesium, and fluoride. Other nutrients that support bone health include vitamins C and K, boron, and silicon. In addition, adequate protein must be consumed, especially in the elderly, to reduce bone loss.

## GETTING ENOUGH CALCIUM

Bone growth, development, and maintenance require an adequate calcium intake. Calcium requires an acidic environment in the gastrointestinal tract to be absorbed efficiently (Fig. 9-22). Absorption occurs primarily in the upper part of the small intestine. This area tends to remain somewhat acidic because it receives the stomach contents.

**TABLE 9-1** ■ **Biological Factors Associated with Bone Status**

| Biological Factors | Effect on Bone Status |
|---|---|
| Sex | Women have lower bone mass and density than men. |
| Age | Bone loss often begins after age 30. |
| Race | Individuals of Caucasian or Asian heritage are at greater risk for poor bone health than individuals of African descent. |
| Frame size | People with "small bones" have a lower bone mass. |
| Estrogen | Women at menopause and beyond should consider use of current medical therapies to reduce bone loss linked to the fall in estrogen output. |

**TABLE 9-2 ■ Modifiable Lifestyle Factors Associated with Bone Status**

| Lifestyle Factors | Call to Action |
| --- | --- |
| Adequate dietary pattern containing an appropriate amount of nutrients | Follow the MyPlate Plan or the DASH diet with special emphasis on adequate amounts of fruits, vegetables, and low-fat and fat-free dairy products.<br>Consider use of fortified foods (or supplements) to make up for specific nutrient shortfalls, such as vitamin D and calcium. If deficient, consult your primary care provider and a registered dietitian nutritionist (RDN). |
| Healthy body weight | Maintain a healthy body weight (BMI of 18.5–24.9) to support bone health. |
| Normal menses | During childbearing years, seek medical advice if menses cease for more than 3 months (such as in cases of anorexia nervosa or extreme athletic training). |
| Movements that create impact and muscle-loading forces on bone. | Perform weight-bearing activity as this contributes to bone maintenance, whereas bed rest and a sedentary lifestyle promote bone loss.<br>Strength training, especially upper body, is helpful to bone maintenance. |
| Smoking | Smoking lowers estrogen synthesis in women. Cessation is advised. Passive exposure is a risk. |
| Medications | Some medications (e.g., thyroid hormone, cortisol, diuretics) stimulate urinary calcium excretion.<br>Some medications (e.g., alcohol, diuretics, cancer medications) stimulate urinary excretion of magnesium. |
| Excessive intake of protein, phosphorus, sodium, caffeine, wheat bran, or alcohol | Moderate intake of these dietary constituents is recommended. Problems primarily arise when excessive intakes of these nutrients are combined with inadequate calcium consumption.<br>Excessive soft drink consumption is especially discouraged. |
| Inadequate UV-B exposure | If sunlight exposure is limited (< 10 to 15 minutes per day without sunscreen), focus on food to meet current RDA for vitamin D. Consult your primary care provider before using vitamin D supplements. |

After the first section of the small intestine, secretions from the pancreas enter the small intestine and the pH becomes neutral to slightly basic, causing calcium absorption to decrease. Efficient calcium absorption in the upper small intestine also depends on the presence of the active form of vitamin D (Fig. 9-22). Adults absorb about 30% of the calcium in the foods eaten, but during times when the body needs extra calcium, such as in infancy and pregnancy, absorption increases to as high as 60%. The normal aging process negatively influences the absorption efficiency of calcium. Due to decreased synthesis, absorption, and activation of vitamin D, as well as decreased acid secretion in the stomach, people over age 40 have a harder time meeting their needs for calcium (Fig. 9-22).

The Recommended Dietary Allowance for calcium is 1000 milligrams per day for adults through 50 years of age. For women older than age 50, and for men over age 70, the RDA increases to 1200 milligrams per day. The RDA is based on the amount of calcium needed each day to offset calcium losses in urine, feces, and other fluids, such as sweat. The RDA for young people ages 9 to 18 (1300 milligrams per day) includes an additional amount to allow for increases in bone mass during growth and development.

Calcium is found in both plant and animal foods. Overall, dairy products provide about 75% of the calcium in North American dietary patterns. In fact, fat-free milk is the most nutrient-dense (milligrams per kcal) source of this bone-building nutrient. Calcium tends to have high bioavailability from dairy products because they contain vitamin D and lactose, which enhance calcium absorption (Fig. 9-22). An exception is cottage cheese, in which most calcium is bound and unavailable for absorption. Bread, rolls, crackers, and foods made with milk products are secondary contributors. Other calcium sources are leafy greens (such as kale), almonds, some legumes, sardines, and canned salmon (Fig. 9-23).

The calcium intakes of many Americans fall short of meeting the RDA. During late childhood and adolescence—a critical time for accretion of bone mass—sugar- and

▶ **Calcium: A Dietary Component of Public Health Concern for Underconsumption**

According to the *Dietary Guidelines* calcium is considered a dietary component of public health concern for the general U.S. population because low intakes are associated with health concerns. Close to 30 percent of men and 60 percent of women older than age 19 years do not consume enough calcium. The *Dietary Guidelines* recommend individuals consume adequate amounts of foods with calcium during adult years to promote optimal bone health and prevent the onset of osteoporosis. A healthy dietary pattern with nutrient-dense, calcium-rich foods, such as low-fat milk and yogurt and fortified soy alternatives and canned sardines and salmon, can help adults better meet intake recommendations.

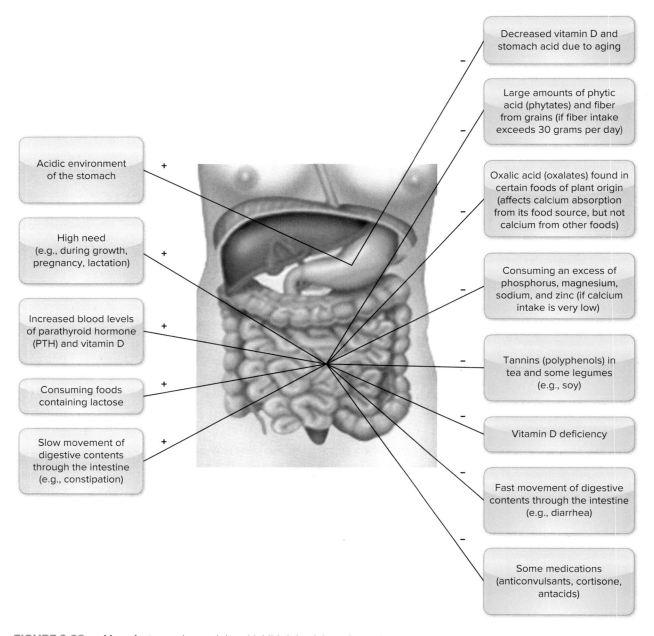

Decreased vitamin D and stomach acid due to aging −

Large amounts of phytic acid (phytates) and fiber from grains (if fiber intake exceeds 30 grams per day) −

Oxalic acid (oxalates) found in certain foods of plant origin (affects calcium absorption from its food source, but not calcium from other foods) −

Consuming an excess of phosphorus, magnesium, sodium, and zinc (if calcium intake is very low) −

Tannins (polyphenols) in tea and some legumes (e.g., soy) −

Vitamin D deficiency −

Fast movement of digestive contents through the intestine (e.g., diarrhea) −

Some medications (anticonvulsants, cortisone, antacids) −

+ Acidic environment of the stomach

+ High need (e.g., during growth, pregnancy, lactation)

+ Increased blood levels of parathyroid hormone (PTH) and vitamin D

+ Consuming foods containing lactose

+ Slow movement of digestive contents through the intestine (e.g., constipation)

**FIGURE 9-22** ▲ Many factors enhance (+) and inhibit (−) calcium absorption.

caffeine- laden beverages are frequently chosen instead of milk or dairy alternatives (see *Ask the RDN* in this section). This calcium deficit increases the risk for osteoporosis later in life. Into adulthood, average daily calcium intakes are approximately 800 milligrams for women and 1000 milligrams for men. About half of adult women in the United States consume less than 60% of the recommended intake of calcium. Some of this is due to their perception that dairy products are high in calories (although many reduced-fat dairy products are available). A growing number of Americans are choosing plant-based dietary patterns, including a vegan dietary pattern, which eliminates dairy products. Other adults simply lose their taste for milk as they age. In addition, lactose intolerance becomes more prevalent as people age.

Through the production of more products fortified with calcium, food and beverage companies are responding to consumers' desire to increase consumption of calcium.

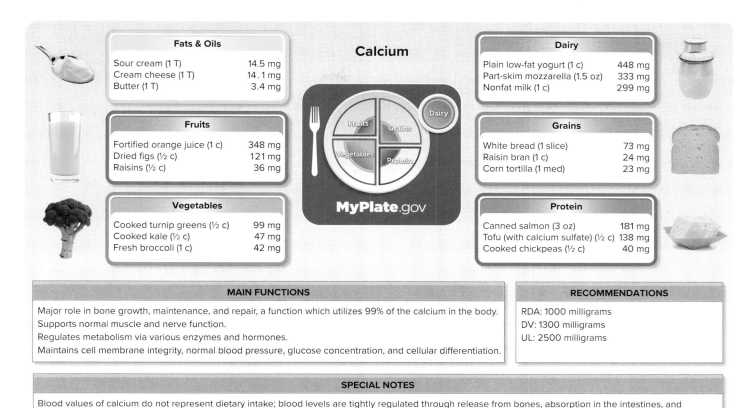

## Calcium

**Fats & Oils**

| | |
|---|---|
| Sour cream (1 T) | 14.5 mg |
| Cream cheese (1 T) | 14.1 mg |
| Butter (1 T) | 3.4 mg |

**Fruits**

| | |
|---|---|
| Fortified orange juice (1 c) | 348 mg |
| Dried figs (½ c) | 121 mg |
| Raisins (½ c) | 36 mg |

**Vegetables**

| | |
|---|---|
| Cooked turnip greens (½ c) | 99 mg |
| Cooked kale (½ c) | 47 mg |
| Fresh broccoli (1 c) | 42 mg |

**Dairy**

| | |
|---|---|
| Plain low-fat yogurt (1 c) | 448 mg |
| Part-skim mozzarella (1.5 oz) | 333 mg |
| Nonfat milk (1 c) | 299 mg |

**Grains**

| | |
|---|---|
| White bread (1 slice) | 73 mg |
| Raisin bran (1 c) | 24 mg |
| Corn tortilla (1 med) | 23 mg |

**Protein**

| | |
|---|---|
| Canned salmon (3 oz) | 181 mg |
| Tofu (with calcium sulfate) (½ c) | 138 mg |
| Cooked chickpeas (½ c) | 40 mg |

**MyPlate.gov**

**MAIN FUNCTIONS**

Major role in bone growth, maintenance, and repair, a function which utilizes 99% of the calcium in the body.
Supports normal muscle and nerve function.
Regulates metabolism via various enzymes and hormones.
Maintains cell membrane integrity, normal blood pressure, glucose concentration, and cellular differentiation.

**RECOMMENDATIONS**

RDA: 1000 milligrams
DV: 1300 milligrams
UL: 2500 milligrams

**SPECIAL NOTES**

Blood values of calcium do not represent dietary intake; blood levels are tightly regulated through release from bones, absorption in the intestines, and retention by the kidneys.
Inadequate calcium over time can release excessive amounts of the mineral from bone and contribute to the development of osteoporosis.

**FIGURE 9-23** ▲ Food sources, functions, and recommendations for calcium. The fill of the background color (none, 1/3, 2/3, or completely covered) within each food group on MyPlate indicates the average nutrient density for calcium in that group. The figure shows the calcium content of several foods in each food group. Overall, the richest sources of calcium are dairy foods (and dairy alternatives), legumes, green leafy vegetables, and fortified foods. mayonnaise: Iconotec/Alamy Stock Photo; orange juice: Sergei Vinogradov/seralexvi/123RF; broccoli: lynx/iconotec.com/Glow Images; two jars of yogurt, Foodcollection; slice of bread: Ingram Publishing/Age Fotostock; tofu: chengyuzheng/iStock/Getty Images; MyPlate: U.S. Department of Agriculture Sources: Office of Dietary Supplements, Dietary Supplements Fact Sheets, available from https://ods.od.nih.gov/factsheets/list-all; USDA FoodData Central, available from https://fdc.nal.usda.gov.

Calcium-fortified foods such as orange juice, breakfast cereals, breakfast bars, waffles, and soy products provide considerable amounts of calcium. In fact, an 8-ounce glass of calcium-fortified orange juice can provide up to 350 milligrams of calcium, whereas 8 ounces of milk has 300. Many dairy alternatives that look, feel, and taste like typical dairy foods are available in the grocery store today and are reviewed in the next section. Another source of calcium is soybean curd (tofu) if it is made with calcium carbonate (check the label). It is easy to assess the calcium content of foods because it is among the nutrients required to be listed on the Nutrition Facts label. The Daily Value (DV) for calcium used for food and supplement labels is 1000 milligrams.

To estimate your calcium intake, use the rule of 300s. Count 300 milligrams for the calcium provided by foods scattered throughout your eating pattern. Add another 300 milligrams to that for every cup of milk or yogurt or 1.5 ounces of cheese. If you eat a lot of tofu, almonds, or sardines or drink calcium-fortified beverages, use Figure 9-23 and diet-analysis software to get a more accurate calculation of calcium intake.

Keep in mind, however, that many factors can influence the bioavailability of calcium. Calcium absorption can be reduced by the presence of oxalates, tannins, and phytic acid (Fig. 9-22). These compounds chelate (chemically bind) calcium in the digestive tract. Oxalates are found in sweet potatoes, collard greens, spinach, and rhubarb. It is estimated that a person would have to consume over eight servings (8 cups) of spinach to absorb the same amount of calcium present in one serving (1 cup) of milk.

▲ Chickpeas (garbanzo beans) are a good plant source of calcium at 80 milligrams per cup. A cup of hummus made from chickpeas provides 93.5 milligrams of calcium. L. Mouton/PhotoAlto

**CASE STUDY** Worried About Grandma

▲ Grace is thinking about her family history of osteoporosis and exploring ways she can decrease her risk factors and promote her bone health. Dennis Wise/ Getty Images

Grace, a 23-year-old woman of Korean descent, is in her final year of nursing school in Boston. She also works 20 hours per week at a local pharmacy. Grace is worried after a phone call she just received from her mom. While out on her walk yesterday, Grandmother Kyon caught her foot on an uneven section of the sidewalk, fell, and broke her hip. The doctor diagnosed Grandmother Kyon with osteoporosis, and the family is worried about the long recovery ahead. As a nursing student, Grace knows how devastating a hip fracture can be. She also knows that osteoporosis can run in families. Grace resolves to do whatever she can to learn about osteoporosis and strengthen her bones now.

She starts by searching the Internet for a website that can help her determine if she is at high risk for osteoporosis. The New York State Department of Health Osteoporosis Education Prevention Program (http://www.health.ny.gov/publications /1988) offers a list of risk factors for osteoporosis. For herself, Grace sees a few risk factors that she cannot change, but there are a few that are glaringly obvious. She uses this website to "bone up" on ways to promote her bone health while she is still young.

Grace discovers she is probably pretty low in vitamin D intake. She also does not have much free time for physical activity. She cooks traditional Korean dishes for half of her meals but eats fast food or sandwiches for the rest. Daily, she takes a multivitamin/ mineral pill plus one calcium carbonate pill. She generally takes them when she brushes her teeth at night before bed. On weekends, she gets out with friends or her boyfriend and has a glass or two of wine. Grace does not smoke.

Answer the following questions about Grace's situation.

1. Go to the URL that Grace visited and click on the link for "Risk Factors for Osteoporosis." What risk factors do you see for Grace? What additional questions would you ask Grace to assess her risk for osteoporosis?
2. Grace went to the section "Keeping Your Bones and Teeth Strong for Life" to determine key actions she can take to strengthen her bones now. Which ones does she already do? Which lifestyle actions could Grace integrate into her daily life right now?
3. What are the environmental conditions that inhibit Grace from adequately synthesizing vitamin D? How can she overcome some of them?
4. What are some calcium and vitamin D sources in traditional Korean dishes? (See http://www.pbs.org/hiddenkorea/food.htm for information about traditional Korean food.)
5. Give Grace some tips to increase her calcium and vitamin D intake at fast-food restaurants without significantly increasing costs or calories.
6. Would you recommend any changes to her supplement regimen?

*Complete the Case Study. Responses to these questions can be provided by instructor.*

Oxalates bind only the calcium in the food they are in; oxalate-containing foods do not affect the calcium availability from other foods. This does not hold true for phytates or tannins, which decrease calcium absorption from the entire meal. Tea and some legumes are rich sources of tannins. Phytates are found in whole grains, raw beans, and nuts. Dietary patterns high in dietary fiber reduce mineral absorption. With this in mind, individuals consuming a vegan diet should take extra care to include good sources of calcium in their dietary patterns. Estimate your calcium intake using the *Rate Your Plate* activity, "Working for Denser Bones," in Connect.

## DAIRY ALTERNATIVES: WHY GO DAIRY FREE?

There are many reasons why you may choose to avoid dairy products. These include following a vegan dietary pattern or just trying to increase your consumption of plant-based foods and beverages. You may have heard that a dairy-free diet will lessen your digestive problems or increase weight loss. Or you may need to avoid dairy products because you have an allergy to milk protein or intolerance to lactose. Whatever the

## ASK THE RDN Caffeine and Bone Health

*Dear RDN: I look forward to the caffeine boost I get from a cup of coffee in the morning and late afternoon, and I enjoy a diet cola with lunch. Could the caffeine in these drinks be damaging my bones and leading to osteoporosis?*

Coffee and colas are major sources of caffeine. An 8-ounce cup of coffee usually contains at least 95 milligrams of caffeine, and a 12-ounce cola typically has about 30 milligrams. High intakes of caffeine are known to decrease intestinal calcium absorption and simultaneously increase the amount of calcium excreted in the urine, suggesting that excessive use of caffeine can eventually contribute to bone loss. Research has confirmed that consuming 330 milligrams or more of caffeine daily (the amount found in about four cups of coffee) increases the risk of brittle or broken bones. The evidence shows both coffee and colas having effects on bone density that could contribute to osteoporosis. It is interesting that these associations appear only in women and are not linked to the consumption of black tea, which also contains caffeine. In fact, older women who drink tea have been shown to have greater bone density than women who do not drink tea at all. While the tannins in tea are known to reduce the absorption of calcium, studies have suggested that other phytochemicals in tea, such as flavonoids, may provide a protective effect against bone loss due to their estrogen-like properties.

In addition to caffeine, regular and diet colas also contain phosphoric acid that can affect bones. Phosphorus is widespread in our food supply, both naturally and as an additive, raising concern that we may actually be getting too much of a good thing. Getting more than the RDA for phosphorus is generally considered safe if you consume adequate calcium and have healthy kidney function. However, high phosphorus intakes can contribute to bone loss when they are combined with a low calcium intake, creating an imbalance in the calcium-to-phosphorus ratio. This situation is more common when you regularly substitute soft drinks for milk.

The bottom line is that you can enjoy coffee and cola as long as you get enough calcium to meet your body's needs and make up for any caffeine-induced calcium losses. Be sure and include calcium-rich foods in your dietary pattern and consult your primary care provider or dietitian to discuss calcium supplementation if your intake falls short of the recommendations. Most important, make sure that your coffee and colas are not replacing milk and calcium-fortified juice consumption that promotes bone health. Practice moderation with your caffeine consumption to prevent caffeine from interfering with calcium absorption and excretion. Less than two cans of cola per day is advised, and keep your coffee habit to no more than three (8-ounce) cups a day. Adding fat-free milk to your coffee can help offset any calcium losses.

Start your day with café au lait,

*Anne M. Smith, PhD, RDN, LD*

Associate Professor Emeritus, The Ohio State University, Author of *Contemporary Nutrition*

Monty Soungpradith/
Open Image Studio LLC

reasons for avoiding milk and other dairy products, you must understand the nutritional implications, especially related to bone health. The typical American dietary pattern relies on dairy products to provide several essential nutrients, including protein, calcium, potassium, and vitamin D. If you do not eat dairy products, it is critical that you know how to make healthy food substitutions to get the nutrients you need.[10]

Quantity and quality must be considered when finding a substitute for the protein found in dairy foods. Dairy protein is *complete protein*, meaning it provides all of the essential amino acids necessary for protein synthesis in the body. Quinoa and soy foods (e.g., soy milk, tofu, or edamame) are excellent plant protein choices because they also contain all of the essential amino acids. Rice products are used, too, but rice protein is not a complete protein source.

It is always preferable to consume most of your nutrients in whole foods because they will be more easily absorbed by your body. The concentration of calcium in nondairy foods is significantly less than found in dairy products, making it difficult to get the calcium and other dairy nutrients just from whole dairy-alternative foods. Leafy greens and broccoli are nondairy sources of calcium, but you would have to eat 8 cups of cooked spinach or 3 cups of cooked Swiss chard to get the amount of calcium (300 milligrams) in 1 cup of milk. Chia seeds, flaxseeds, and sesame seeds, as well as quinoa, also are rich plant sources of calcium. Foods fortified with calcium, such as soy milk and some types of orange juice, are usually necessary to fill in the gaps and bring calcium intake up to the RDA.

Exposure to sunlight for 10 minutes per day will ensure most of us are getting a full day's supply of vitamin D. Because this is not always possible, fortified breakfast cereals and orange juice are good nondairy sources of vitamin D.

Because vitamin B-12 is found almost exclusively in meat and dairy foods, fortified foods will be a likely option when avoiding dairy. Foods fortified with vitamin B-12 include breakfast cereals, soy milk, and meat substitutes.

## DAIRY-FREE FOOD OPTIONS

There are many dairy alternatives available in the grocery store today. These are products that look, feel, and taste like typical dairy foods but are not made from the milk of animals. They include dairy-free milks, sour creams, cheeses, and ice creams. Not all of these substitutes, however, have the same nutritional profiles as the original milk-based product. It is very important, therefore, to know what nutrients you are looking for and then to read labels carefully to compare products. The following are examples of the dairy alternatives you can find in stores or online.

**Milks.** Dairy milk alternatives are now widely available and growing in popularity. In 2019, the sales of nondairy milk beverages were nearly $1.8 billion. Growth in the market has been led by almond ($1.3 billion sales in 2019) and coconut milk products. At the same time, interest in soy milk has waned. Most of the milk alternatives are healthy and delicious but not exact nutritional replicas of cow's milk (Table 9-3). Although fortified with calcium and vitamin D, most contain little if any protein because the protein remains in the pulp fraction that is strained out and discarded during the production of the plant-based milk. Several of the milk substitutes have unique health benefits. For example, almond milk is rich in vitamin E, magnesium, and potassium, and soy milk is a good source of protein.

*Soy milk* was one of the first and most widely available dairy-free milk substitutes in grocery stores. In the European Union, this product is commonly called "soy drink" because the term *milk* has been restricted to mammary secretions. Soy milk is made by soaking and grinding dry soybeans with water to make an emulsion of oil, water, and protein. Soy milk contains the highest quantity of protein of all the dairy alternatives, with 6 to 10 grams per cup, and it is also the best quality in that it contains all of the essential amino acids. It has about the same amount of protein, fat, and carbohydrate as cow's milk. It contains omega-3 fatty acids, fiber, magnesium, and manganese; and most brands are fortified with calcium (450 milligrams per cup), riboflavin, and vitamins A, D, and B-12. There are many brands of soy milk, including unsweetened and sweetened varieties, available at grocery and health food stores. Soy milk is a popular product for individuals who consume a vegan diet because it is plant based, as well as for people with lactose intolerance because it is lactose free. Soy milk can be coagulated to produce tofu just as dairy milk can be made into cheese.

*Rice milk* is another popular dairy alternative. It is made from boiled rice, brown rice syrup, and brown rice starch. It is naturally sweeter than cow's milk because the rice carbohydrates are broken down enzymatically into sugars during production. Additional

▲ It is not just dairy! If you are lactose intolerant or just do not like dairy, you can still find calcium in foods. Soy and rice milks as well as some types of orange juice are fortified with calcium. Food Passionates/Corbis

**TABLE 9-3** ■ **How Nondairy Milk Beverages Compare to the Nutrient Content of Cow's Milk**

| Nutrients per 1-Cup Serving | 2% Milk | Soy Milk | Rice Milk | Almond Milk | Coconut Milk | Coconut Milk Beverage | Oat Milk | Flax Milk | Cashew Milk |
|---|---|---|---|---|---|---|---|---|---|
| Calories | 103 | 90 | 120 | 40–60 | 552 | 70 | 130 | 50 | 60 |
| Total Fat (grams) | 2.4 | 3.5 | 2.5 | 3 | 57 | 4.5 | 2.5 | 2.5 | 2.5 |
| Protein (grams) | 8 | 6 | 1 | 1 | 5 | 0 | 4 | 0 | < 1 |
| Calcium (milligrams) | 246 | 450 | 300 | 200 | 40 | 100 | 350 | 300 | 450 |

sweeteners and flavors, such as vanilla, chocolate, or almond, may also be added to some varieties. Compared to cow's milk, rice milk is much higher in carbohydrates (24 grams per cup) and has significantly less protein (1 gram per cup) and calcium, as well as no lactose. Most brands are fortified to contain about 300 milligrams of calcium per cup (23% of Daily Value), as well as iron, riboflavin, and vitamins A, D, and B-12. Rice milk is probably the most hypoallergenic of all the dairy alternatives and therefore is a good choice for people who are allergic to cow's milk or soy, in addition to those who are lactose intolerant. It is also a popular dairy substitute for individuals who consume a vegan diet.

*Almond milk* is made from ground almonds and water, and it has become the most popular dairy alternative. On the positive side, a cup of almond milk contains about 200 milligrams of calcium (15% of Daily Value), 2.5 micrograms of vitamin D (12.5% of Daily Value), and 10 milligrams of vitamin E (67% of Daily Value). Almond milk is much lower in calories, however, than cow's milk. There are sweetened (60 calories per cup) and unsweetened varieties (40 calories per cup). Almond milk also contains very little protein (1 gram per cup), carbohydrate (2 grams per cup), and fiber (1 gram per cup), and it has a fat content similar to that of rice milk (3 grams per cup). Because almond milk is low in fat and calories and contains some essential vitamins and minerals, it can be used as a milk substitute by those who would like to lose weight. Many find the taste of almond milk more acceptable compared to other dairy substitutes.

▲ Almonds are a natural source of calcium: 1 ounce contains 80 milligrams.
Ingram Publishing/SuperStock

*Coconut milk* is the liquid that is made from grated coconut meat. It has a high oil content and therefore is much higher in calories, fat, and saturated fat than other milks and milk substitutes. A 1-cup serving of canned coconut milk contains 552 calories and 57 grams of fat, of which 51 grams are saturated fat. Although it contains very little calcium, vitamin D, and other nutrients involved in bone health, coconut milk does contain about 22% of the Daily Value for iron. There are also coconut milk beverages available that compare closely to soy, rice, and almond milks. These beverages are a mixture of coconut cream and water, and per cup they have only 4.5 grams of fat, 0 grams of protein, 70 calories, and 100 milligrams of calcium (8% of Daily Value).

*Potato, oat, hemp, flax, and cashew milks* are some more recent dairy-free alternatives. Potato milk, made simply from potatoes and water, is high in carbohydrates and low in protein. It is usually fortified with calcium and vitamins.

▲ Coconut milk is loaded with calories because of its high fat content, and contains very little calcium and vitamin D.
Bernatskaya Oxana/Shutterstock

Oat milk, like rice milk, is considered a "grain milk," or nondairy beverage. It is made from oat groats mixed with water and potentially other grains and beans. It is low in calories (130 calories per cup) and fat but high in fiber and iron. It typically does not contain calcium and is low in vitamin D (5% of Daily Value), so it would not be a good choice for maintaining bone health.

Hemp milk or hemp seed milk is made from hulled hemp seeds that are soaked and ground in water. The beverage is creamy and has a nutty flavor. A 1-cup serving contains about 160 calories, 24 grams of carbohydrate, 5 grams of fat, and 3 grams of protein, and it is fortified with calcium (38% of Daily Value) and vitamin D (15% of Daily Value).

Flax milk is cold-pressed flax oil mixed with filtered water. One cup unsweetened has 50 calories and provides 1200 milligrams of omega-3s but contains no protein. Flax milks are typically fortified with vitamin A, vitamin D, vitamin B-12, and calcium.

Cashew milk is one of the most recent additions to the plant-based beverages industry. Although it has only 60 calories per cup and no saturated fat or cholesterol, it is creamier than skim milk and has nearly twice the calcium of cow's milk.

**Cheeses.** Finding a dairy-free cheese that has the same taste, texture, and cooking qualities of regular cheeses is a challenge. Soy cheeses are the easiest to find, but nut-based cheeses made from cashews and macadamia nuts are also available in grocery stores and online. Soy cream cheeses are available in grocery stores and are popular dairy substitutes because their taste and texture are similar to those of regular cream cheese. Likewise, soy-based sour creams are the most prevalent dairy-free sour creams. Recipes are available for cashew-based vegan cream cheese at https://www.hummusapien.com /vegan-cashew-cream-cheese-recipe.

**Ice Cream, Yogurt, Smoothies, and Whipped Cream.** Soy-based ice creams are widely available in stores, and rice-based ice creams are also popular because of their sweetness. Almond and coconut ice creams also are available, and most sorbets and fruit-based sherbets are dairy free. As with most of the dairy-free products, soy-based yogurts are the most readily available in stores and are the only yogurt alternatives that will provide protein. Coconut- and nut-based yogurts can also be found in grocery stores, and more consumers are looking for nondairy probiotic sources. Dairy-free smoothies are available. Rice milk is recommended for homemade smoothies because of its sweetness, whereas coconut and almond milk can provide their distinctive flavors. Vegan smoothie recipes can be found at https://www.hummusapien.com/healthy-smoothies-recipes. Soy-, rice-, coconut-, and nut-based whipped toppings are also available but may contain added sugars and saturated fats. Less processed whipping cream can also be made at home from coconut milk, almonds, or cashews.

**Dairy-Free Shopping Guide: Things to Consider**

- Replacing dairy milk with a plant-milk alternative can be part of a plant-based eating plan.
- Look for the unsweetened varieties.
- Choose plant-based milks that are fortified with vitamin D and calcium.
- Remember that most plant-based milks have negligible protein. Only soy milk is similar to dairy milk in terms of protein. Be aware that almond milk, which is extremely low in protein, is becoming popular in coffee shops as a dairy-free creamer.
- Look for a dairy alternative with a short ingredient list (e.g., nuts, water, salt, and spices). Some dairy-free milks are full of additives, including stabilizers and thickening agents such as sugar, gums, inulin, and carrageenan.

## AVOIDING TOO MUCH CALCIUM

The Upper Level (UL) for calcium intake is 2500 milligrams per day for young adults, based on the observation that greater intakes increase the risk for some forms of kidney stones. Excessive calcium intakes by some people can also cause high blood and urinary calcium concentrations, irritability, headache, kidney failure, soft tissue calcification, and decreased absorption of other minerals.

**Calcium Supplementation.** There has been much debate over the effectiveness of calcium and vitamin D supplementation for maintaining bone health. A recent review of studies found that calcium supplementation can be a double-edged sword.[11] It has been shown to promote bone formation and therefore prevent osteoporosis, but it may also have a negative impact. While achieving the RDAs for calcium and vitamin D through foods and beverages has been found to improve bone mineral density and reduce rate of fractures, use of supplements may actually impair skeletal health and worsen cardiovascular health. One study found that a calcium intake beyond the RDA for elderly women and men, usually achieved by calcium supplements, did not provide any benefit for hip or lumbar spine bone mineral density in older adults.[12] The results of other studies, including the large Women's Health Initiative trial, have shown an increase in the rate of heart attacks and possibly stroke among older adults taking calcium supplements with or without vitamin D.[13] A recent meta-analysis of cohort studies and randomized controlled trials also determined that calcium supplements are linked to adverse cardiovascular events, especially heart attacks (see *Newsworthy Nutrition*).

On the positive side, another report from the Women's Health Initiative study indicates that long-term use of a daily calcium and vitamin D supplement that is close to the RDA results in a substantial reduction in the risk of hip fracture among postmenopausal women.[14] These authors also reported that this level of calcium and vitamin D supplementation did not result in an increase in other chronic diseases, including heart disease. The more positive effects of supplementation appear to happen when the level of total calcium and vitamin D intake is kept very close to the RDA. Therefore, taking 1000 milligrams of calcium carbonate or calcium citrate daily as a supplement in divided doses (about 500 milligrams per tablet) is probably safe in many instances.

So which is better: calcium from food or supplements? The National Osteoporosis Foundation and other experts agree that we should consume the recommended amounts of calcium and vitamin D from foods first and that more research is needed to better comprehend the benefits and risks associated with calcium and vitamin D supplementation. Modification of eating habits to include foods that are good sources of calcium is a better plan of action and appears to be the safest means to prevent osteoporosis without jeopardizing heart health. Foods that contain calcium also supply other vitamins, minerals, phytochemicals, and fats needed to support health. Problems associated with excessive consumption of calcium, such as constipation, are not likely when foods are the primary sources of calcium.

In an effort to provide guidance for the public, the U.S. Preventive Services Task Force reviewed current research studies on the use of vitamin D and calcium supplements to prevent fractures and issued recommendations.[15] The recommendations apply to adult men and women who live at home. They do not apply to those living in assisted living or skilled nursing facilities or who have been diagnosed with osteoporosis or vitamin D deficiency. The Task Force concluded that there is not enough current evidence to determine if vitamin D and calcium supplementation, alone or combined, are beneficial or harmful for the prevention of fractures (see adjacent box).

▶ *Vitamin D, Calcium, or Combined Supplementation for the Primary Prevention of Fractures in Community-Dwelling Adults* **Task Force Conclusions:**[15]

(1) Current evidence is insufficient to assess the balance of the benefits and harms of vitamin D and calcium supplementation, alone or combined, for the primary prevention of fractures in community-dwelling, asymptomatic men and premenopausal women.

(2) Current evidence is insufficient to assess the balance of the benefits and harms of daily supplementation with doses greater than 400 IU (10 micrograms) of vitamin D and greater than 1000 mg of calcium for the primary prevention of fractures in community-dwelling, postmenopausal women.

(3) The Task Force recommends against daily supplementation with 400 IU (10 micrograms) or less of vitamin D and 1000 mg or less of calcium for the primary prevention of fractures in community-dwelling, postmenopausal women. These recommendations do not apply to persons with a history of osteoporotic fractures, increased risk for falls, or a diagnosis of osteoporosis or vitamin D deficiency.

# Newsworthy Nutrition

## Calcium supplements and cardiovascular disease risk

**INTRODUCTION:** Although several studies have shown a beneficial effect of calcium intake on cardiovascular effects, others have shown that calcium intake, especially from calcium supplements, is associated with increased mortality or the risk of heart attack and stroke. **OBJECTIVE:** The goal of this study was to explore the associations between calcium from dietary and supplemental intakes and cardiovascular disease (CVD) risks. **METHODS:** The study design was a systematic review and meta-analysis of 16 randomized controlled trials and 26 prospective cohort studies of dietary or supplemental intake of calcium, with or without vitamin D, and cardiovascular outcomes. Data was from PubMed, Cochrane Central, Scopus, and Web of Science, published up to March 2019. **RESULTS:** Results of cohort studies indicated that there were no associations between dietary calcium intakes and the risk of CVD, coronary heart disease (CHD), and stroke, for intakes ranging from 200 to 1500 mg/day. Results showed that calcium supplements, ranging from 1000 to 1400 mg/day, did not increase the risk of CVD and stroke; however, the risk of CHD increased by 20% and the risk of heart attack increased by 21% with the use of oral calcium supplements. **CONCLUSIONS:** Keeping in mind that very high calcium intakes are difficult if not impossible to achieve by dietary sources alone, the authors conclude that calcium intake from dietary sources does not increase the risk of CVD, and they suggest that adequate dietary calcium intakes are beneficial to cardiovascular protection. They conclude that calcium supplements might raise CHD risk, especially heart attack, and therefore the concerns regarding potential adverse cardiovascular risks are related to the use of calcium supplements.

Source: Yang C and others: The evidence and controversy between dietary calcium intake and calcium supplementation and the risk of cardiovascular disease: A systematic review and meta-analysis of cohort studies and randomized controlled trials. *Journal of the American College of Nutrition* 18:1, 2019. DOI:10.1080/07315724.2019.1649219.

**TABLE 9-4  ■  Calcium Supplement Comparisons**

|  | Calcium Carbonate | Calcium Citrate |
|---|---|---|
| **Calcium content** | 40% | 21% |
| **Forms** | Tablets, chewable tablets, soft chews | Pills (can be quite large) Liquid (sometimes easier to tolerate) |
| **Cost** | Least expensive, most common form | Most expensive |
| **Bioavailability and meal timing** | Needs acid environment in stomach, so take with acid food or take with meals | Best absorbed; does not need acid environment to be absorbed |

Tablet or liquid calcium supplements with the United States Pharmacopeia (USP) symbol are considered the safest. FDA has cautioned the public on the use of calcium supplements from dolomite, bone meal, coral, or oyster shell because of the potential for unhealthy levels of environmental contaminants, especially lead. **Why are supplements containing calcium citrate more desirable than those containing calcium carbonate?** Isadora Getty Buyou/Image Source

Keeping in mind the recommendations just discussed, increasing calcium intake through the use of a calcium supplement may be beneficial if you have a milk allergy, do not like milk, are ovovegetarian, vegan, lactose intolerant, or cannot incorporate enough calcium-containing foods into your dietary pattern.[16] Always look for a supplement with added vitamin D, as it enhances calcium uptake. This additional vitamin D typically does not add to the cost of the supplement. Table 9-4 compares the two most common forms of calcium supplements. Calcium carbonate should be taken with meals because it requires an acid environment in the stomach to dissolve and maximize calcium absorption. Calcium citrate is indicated for people who cannot remember to take calcium carbonate with meals and for those who have low-acid stomach conditions, such as people who take acid-reducing medications for ulcers or reflux, or those who have had surgery for obesity reduction.

Calcium supplements have side effects, including gas, bloating, or constipation. Distributing small-dose supplements throughout the day, taking it with meals, or even changing the brand of supplement may alleviate some problems. Intake of calcium from supplements and/or food above 500 milligrams at any one time significantly reduces the percentage absorbed.

With calcium supplements, interactions with other minerals are a concern. There is evidence that calcium supplements may decrease the absorption of zinc, iron, and other minerals. An effect of calcium supplementation on iron absorption is possible; however, this appears to be small over the long term. To be safe, people using a calcium supplement on a regular basis should notify their primary care provider of the practice. Calcium supplements can also interfere with the body's ability to absorb certain antibiotics. If your doctor prescribes antibiotics, especially tetracycline, be sure to talk with your pharmacist about the timing of your supplement, medication, and meals.

## ✓ CONCEPT CHECK 9.6

1. What percentage of calcium in the body is found in bone and teeth?
2. What are the two types of osteoporosis, and how do they differ?
3. What type of bone is most affected by osteoporosis?
4. Why is the achievement of peak bone mass as a young adult so important in preventing osteoporosis?
5. What are some current treatments for osteoporosis?
6. What is the most accurate test for bone density, and how is it done?
7. According to the National Osteoporosis Foundation, who should have their bone density checked?
8. Beyond its role in bone health, what are some other critical functions of calcium?
9. What role does vitamin D play in calcium metabolism?
10. What factors reduce calcium absorption?

# 9.7 Phosphorus

## FUNCTIONS OF PHOSPHORUS

Phosphorus (P) is the second most abundant mineral in the body. Approximately 85% of phosphorus is found as a component of hydroxyapatite crystals that provide the functional component of bone and teeth. The remaining 15% of phosphorus is in the soft tissues, blood, and extracellular fluid. Phosphorus is part of DNA and RNA, the genetic material present in every cell. Because DNA and RNA are responsible for protein synthesis, phosphorus is critical for cellular replication and growth. Phosphorus is also a primary component of adenosine triphosphate (ATP), the energy molecule that fuels body functions. Phosphorus is essential for the activation and deactivation of certain enzymes, and many of the B vitamins are functional only when a phosphate group is attached.

A major class of lipids, the phospholipids, also contains phosphorus. Recall that phospholipids are the principal structural component of cell membranes, making up approximately 60% of membranes. These phospholipid membranes regulate the transport of nutrients and waste products into and out of cells. Phosphorus also serves as a buffer to maintain blood pH. Lastly, phosphorus (in the form of the phosphate ion) is the principal negatively charged ion in intracellular fluid and thus is essential for maintenance of fluid balance.

## PHOSPHORUS DEFICIENCY

Deficiencies of phosphorus are unlikely in healthy adults because it is widespread in food and beverages and efficiently absorbed. If intakes of phosphorus are inadequate, the kidney compensates by increasing the reabsorption of phosphorus to prevent blood phosphorus levels from decreasing. Dietary phosphorus deficiency usually occurs only in cases of near-total starvation, including anorexia nervosa. Some health conditions, such as diabetes and alcohol use disorders, as well as medications, such as antacids and diuretics, can cause body phosphorus levels to fall. Marginal phosphorus status can be found in infants born preterm, individuals consuming a vegan dietary pattern, older people with nutrient-poor eating patterns, and people with long-term bouts of diarrhea, as often occurs in Crohn's disease and celiac disease. Symptoms of phosphorus deficiency include loss of appetite, anemia, muscle weakness, bone pain, fragile bones, increased susceptibility to infection, numbness and tingling of the extremities, difficulty walking, and irregular breathing.[17]

## GETTING ENOUGH PHOSPHORUS

In contrast to calcium, phosphorus is naturally abundant in many foods and beverages. Milk, cheese, meat, and bread provide most of the phosphorus in the adult dietary pattern. Nuts, fish, breakfast cereals, bran, and eggs are also good sources (Fig. 9-24). About 20% to 30% of dietary phosphorus comes from food additives, especially in baked goods, cheeses, processed meats, and many soft drinks (about 75 milligrams per 12 ounces). As a food additive, phosphorus is considered a GRAS (generally recognized as safe) substance, and its function is to increase water binding and taste. Phosphoric acid, which gives a tangy, sour taste, will also significantly lower the pH of a food or beverage, such as a soft drink that has a pH less than 3. Absorption of phosphorus is generally high, ranging from 55% to 80%. Phosphorus absorption from grains, however, is reduced because of the high phytic acid content. Vitamin D enhances phosphorus absorption.

▲ Trail mix is rich in phosphorus.
imagebroker/Alamy Stock Photo

The RDA for phosphorus is 700 milligrams for adults (Fig. 9-24). The recommendation is higher (1250 milligrams per day) for young people ages 9 to 18 years to support growth and development. Average daily adult consumption is about 1000 to 1600 milligrams.

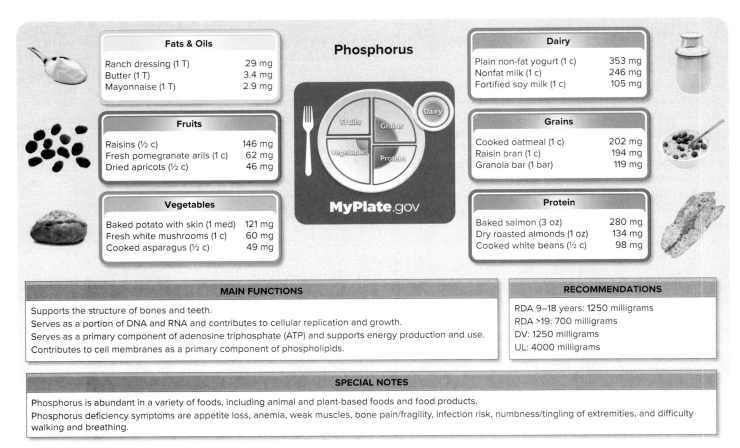

| Fats & Oils | |
|---|---|
| Ranch dressing (1 T) | 29 mg |
| Butter (1 T) | 3.4 mg |
| Mayonnaise (1 T) | 2.9 mg |

**Phosphorus**

| Dairy | |
|---|---|
| Plain non-fat yogurt (1 c) | 353 mg |
| Nonfat milk (1 c) | 246 mg |
| Fortified soy milk (1 c) | 105 mg |

| Fruits | |
|---|---|
| Raisins (½ c) | 146 mg |
| Fresh pomegranate arils (1 c) | 62 mg |
| Dried apricots (½ c) | 46 mg |

| Grains | |
|---|---|
| Cooked oatmeal (1 c) | 202 mg |
| Raisin bran (1 c) | 194 mg |
| Granola bar (1 bar) | 119 mg |

| Vegetables | |
|---|---|
| Baked potato with skin (1 med) | 121 mg |
| Fresh white mushrooms (1 c) | 60 mg |
| Cooked asparagus (½ c) | 49 mg |

| Protein | |
|---|---|
| Baked salmon (3 oz) | 280 mg |
| Dry roasted almonds (1 oz) | 134 mg |
| Cooked white beans (½ c) | 98 mg |

**MyPlate.gov**

| MAIN FUNCTIONS |
|---|
| Supports the structure of bones and teeth.<br>Serves as a portion of DNA and RNA and contributes to cellular replication and growth.<br>Serves as a primary component of adenosine triphosphate (ATP) and supports energy production and use.<br>Contributes to cell membranes as a primary component of phospholipids. |

| RECOMMENDATIONS |
|---|
| RDA 9–18 years: 1250 milligrams<br>RDA >19: 700 milligrams<br>DV: 1250 milligrams<br>UL: 4000 milligrams |

| SPECIAL NOTES |
|---|
| Phosphorus is abundant in a variety of foods, including animal and plant-based foods and food products.<br>Phosphorus deficiency symptoms are appetite loss, anemia, weak muscles, bone pain/fragility, infection risk, numbness/tingling of extremities, and difficulty walking and breathing. |

**FIGURE 9-24** ▲ Food sources, functions, and recommendations for phosphorus. The fill of the background color (none, 1/3, 2/3, or completely covered) within each food group on MyPlate indicates the average nutrient density for phosphorus in that group. The figure shows the phosphorus content of several foods in each food group. Overall, the richest sources of phosphorus are dairy products and protein foods.

mayonnaise: Iconotec/Alamy Stock Photo; raisins: lynx/iconotec/Glowimages; baked potato: DNY59/E+/Getty Images; two jars of yogurt: Foodcollection; oatmeal and blueberries: Lucy Stein/Glow Images; salmon: Zoran Kolundzija/E+/Getty images; MyPlate: U.S. Department of Agriculture

Sources: Office of Dietary Supplements, Dietary Supplements Fact Sheets, available from https://ods.od.nih.gov/factsheets/list-all; USDA FoodData Central, available from https://fdc.nal.usda.gov.

## AVOIDING TOO MUCH PHOSPHORUS

The UL for phosphorus intake is 4000 milligrams per day for adults up to 70 years and 3000 milligrams per day for those older than 70 years. Intakes greater than this can result in mineralization of soft tissues. Phosphorus levels in the blood are regulated primarily by the kidneys, and these organs are particularly sensitive to phosphorus toxicity. High intakes can lead to serious problems in people with certain kidney diseases. In addition, a high phosphorus intake coupled with a low calcium intake can cause a chronic imbalance in the calcium-to-phosphorus ratio in the dietary pattern and contribute to bone loss. This situation most likely arises when the RDA for calcium is not met, as can occur in adolescents and adults who regularly substitute soft drinks for milk or otherwise undercomsume calcium.

### ✓ CONCEPT CHECK 9.7

1. What effect does vitamin D have on phosphorus absorption?
2. What are the key functions of phosphorus beyond bone health?
3. Is a deficiency of phosphorus likely? Why or why not?
4. What are the primary food sources of phosphorus?
5. What are the risks of excess intake of phosphorus?

# 9.8 Magnesium

## FUNCTIONS OF MAGNESIUM

Magnesium (Mg) has many functions, some of which are related to bone health. Magnesium is similar to calcium and phosphorus in that most of the magnesium in the body (60%) is found in bones. Magnesium in bones provides rigidity, and it functions as a storage site that can be drawn upon by other tissues when dietary intake is inadequate. Magnesium is also required for the synthesis of vitamin D in the liver. It also promotes resistance to tooth decay by stabilizing calcium in tooth enamel.

Beyond its role in bone health, magnesium is important for nerve and heart function and aids in many enzyme reactions. Magnesium functions to relax muscles after contraction. Over 300 enzymes use magnesium, and many energy-yielding compounds in cells require magnesium to function properly (e.g., ATP). Magnesium plays a critical role in the synthesis of DNA and protein.

Other possible benefits of magnesium in relation to cardiovascular disease include decreasing blood pressure by dilating arteries and preventing heart abnormalities. People with cardiovascular disease should closely monitor magnesium intake, especially because they are often on medications, such as diuretics, that reduce magnesium levels. A dietary pattern that includes food sources that are rich in magnesium and calcium is associated with lower risk of type 2 diabetes in some populations.

## MAGNESIUM DEFICIENCY

In humans, a magnesium deficiency causes an irregular heartbeat, sometimes accompanied by weakness, muscle pain, disorientation, and seizures. In terms of bone health, low magnesium levels disrupt the hormonal regulation of blood calcium by parathyroid hormone and affect the activity of vitamin D (Fig. 9-17).[18] You might expect that magnesium deficiency would result in diminished bone mass, but to date this has only been observed in animals. There is some evidence, however, that magnesium supplementation may improve bone density in postmenopausal women.

Magnesium deficiency develops very slowly. Not only is the mineral present in foods of both plant and animal origin, but the kidneys also are very efficient at retaining magnesium. Poor magnesium status is most commonly found among people with abnormal kidney function, whether as a result of kidney disease or as a side effect of certain diuretics. Alcohol use disorders also can increase the risk of deficiency because dietary intake may be poor and because alcohol increases magnesium excretion in the urine. The disorientation and weakness associated with alcohol use disorders closely resembles the behavior of people with low blood magnesium. In addition, people with malabsorptive diseases (e.g., Crohn's disease), heavy perspiration, or prolonged bouts of diarrhea or vomiting are susceptible to low blood levels of magnesium.

## GETTING ENOUGH MAGNESIUM

Magnesium is found in the plant pigment chlorophyll. Rich sources of magnesium are plant products, such as squash, whole grains (especially wheat bran), beans, nuts, seeds, and broccoli (Fig. 9-25). Animal products (e.g., milk and meats) and chocolate supply some magnesium, although not as much as foods of plant origin. Another source of magnesium is hard tap water, which contains a high mineral content.

The adult RDA for magnesium of 310 to 420 milligrams is based on the amount needed to offset daily losses. Adult men consume on average 320 milligrams daily, whereas women consume closer to 220 milligrams daily, suggesting that many of us should improve our intakes of magnesium-rich foods, such as whole grain breads and

▲ Magnesium is in good supply in dark chocolate (64 milligrams in 1 ounce) and nuts (74 milligrams in 1 ounce of cashews) shown here. CDL Creative Studio/Shutterstock

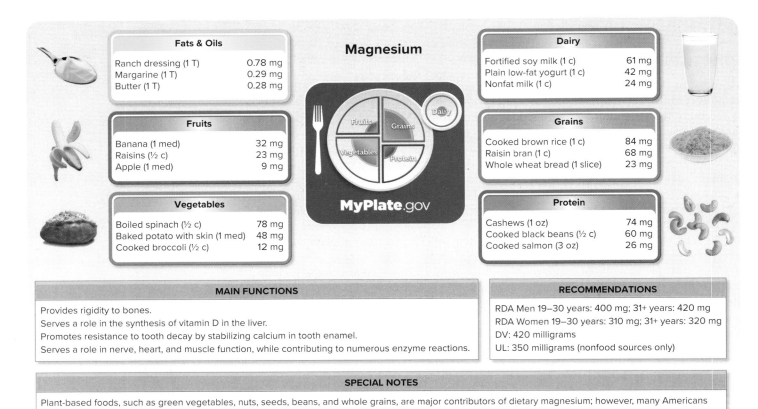

**FIGURE 9-25** ▲ Food sources, functions, and recommendations for magnesium. The fill of the background color (none, 1/3, 2/3, or completely covered) within each food group on MyPlate indicates the average nutrient density for magnesium in that group. The figure shows the magnesium content of several foods in each food group. Overall, the richest sources of magnesium are vegetables and whole grains.

mayonnaise: Iconotec/Alamy Stock Photo; banana: lynx/iconotec.com/Glow Images; baked potato: DNY59/E+/Getty Images; glass of milk: NIPAPORN PANYACHAROEN/Shutterstock; brown rice in bowl: Yellow Cat/Shutterstock; cashew nuts: C Squared Studios/Getty Images; MyPlate: U.S. Department of Agriculture

Sources: Office of Dietary Supplements, Dietary Supplements Fact Sheets, available from https://ods.od.nih.gov/factsheets/list-all; USDA FoodData Central, available from https://fdc.nal.usda.gov.

cereals. The refined grain products that dominate the dietary patterns of many North Americans are poor sources of this mineral because refining reduces the magnesium content by as much as 80%. This low value also reflects poor intake of green and other brightly colored vegetables. Speak to your primary care provider or an RDN if you are concerned about your magnesium intake. If dietary intake of magnesium is inadequate, a balanced multivitamin and mineral supplement containing approximately 100 milligrams of magnesium can help close the gap between intake and needs. Dietary patterns very high in phosphorus or fiber (phytate) limit intestinal absorption of magnesium, as do dietary patterns too low in protein.

## AVOIDING TOO MUCH MAGNESIUM

The UL for magnesium intake is 350 milligrams per day, based on the risk of higher intakes causing diarrhea. This guideline refers only to *nonfood* sources such as antacids, laxatives, or supplements.[18] Food sources are not known to cause toxicity. Magnesium toxicity especially occurs in people who have kidney failure or who overuse over-the-counter medications that contain magnesium, such as certain antacids and laxatives (e.g., milk of magnesia). Older people are at particular risk, as kidney function may be compromised.

## ✓ CONCEPT CHECK 9.8

1. What are the key functions of magnesium other than bone health?
2. What are the primary food sources of magnesium?
3. Who is at greatest risk of developing a magnesium deficiency?
4. When is magnesium toxicity most likely to occur?

# 9.9 Iron

Iron (Fe) is a trace mineral. The total amount of iron in the body is only 10 grams, which is about as heavy as two nickels. Even though we only need to eat small amounts of this trace mineral each day, iron has some mighty important functions.

## FUNCTIONS OF IRON

Iron is part of the protein **hemoglobin** in red blood cells and **myoglobin** in muscle cells. Hemoglobin molecules in red blood cells transport oxygen ($O_2$) from the lungs to cells and then transport carbon dioxide ($CO_2$) from cells to the lungs for excretion. In addition, iron is used as part of many enzymes and other proteins and compounds that cells use in energy production. Iron also is needed for brain and immune function, drug detoxification in the liver, and synthesis of collagen for bone health.

## IRON DEFICIENCY

Iron deficiency is the most common nutrient deficiency worldwide. About 30% of the world's population is anemic, and about half of these cases are caused by iron deficiency.

When neither the dietary pattern nor body stores can supply the iron needed for hemoglobin synthesis, the concentration of hemoglobin in red blood cells decreases. The percentage of blood made up of red blood cells **(hematocrit)** as well as the hemoglobin concentration are used to assess iron status. Other measures of iron status include the concentration of iron (serum iron) and iron-containing proteins (**ferritin** or **transferrin**) in blood.

When hematocrit and hemoglobin fall, an iron deficiency is suspected. In severe deficiency, hemoglobin and hematocrit fall so low that the amount of oxygen carried in the bloodstream is decreased. This condition is called *iron-deficiency anemia*.

Iron deficiency can be categorized into three stages:

- **Stage 1:** Iron stores become depleted, but no physiological impairment is observed.
- **Stage 2:** The amount of iron in transferrin is depleted; some physiological impairment occurs. Heme production is decreased, and activities of enzymes that require iron as a cofactor are limited.
- **Stage 3 (iron-deficiency anemia):** Red blood cells are small (microcytic), pale (hypochromic), and reduced in number; oxygen-carrying capacity of red blood cells declines.

Clinical symptoms of iron-deficiency anemia are associated with the lack of oxygen getting to the tissues. An individual with iron-deficiency anemia may experience pale skin, fatigue upon exertion, poor temperature regulation (always cold, especially toes and fingers), loss of appetite, and apathy. Poor iron stores may decrease learning ability, attention span, work performance, and immune status even before a person is anemic. Children with chronic anemia have abnormal cognitive development.

It is important to note that many more people have an iron deficiency *without* anemia (stages 1 or 2) than have iron-deficiency anemia (stage 3). Their blood hemoglobin values are still normal, but they have no stores to draw from in times of pregnancy or illness, and basic functioning may be marginally impaired. That could mean anything from chronic fatigue to difficulties staying mentally alert.

**myoglobin** Iron-containing protein that binds oxygen in muscle tissue.

**hematocrit** The percentage of blood made up of red blood cells.

**ferritin** A protein that stores iron and releases it in a controlled manner; acts as a buffer against iron deficiency and iron overload.

**transferrin** Iron-binding protein; controls the level of free iron in blood.

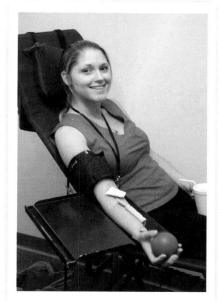

Red blood cells contain about two-thirds of the body's total iron supply. Each time blood is donated, about 10% of total blood volume is sacrificed, which removes about 7% of the body's iron supply. Over the next few weeks, the red blood cells will be replaced, so healthy people can usually donate blood two to four times a year without harmful consequences. **How do blood banks screen potential donors for anemia?**

David H. Lewis/E-plus/Getty Images

It is important to understand that many conditions lead to an anemic state; iron-deficiency anemia is the most prevalent nutrient deficiency worldwide. Probably about 10% of North Americans in high-risk categories have iron-deficiency anemia. This appears most often during infancy, the preschool years, and puberty. Growth—with accompanying expansion of blood volume and muscle mass—increases iron needs, and some individuals are unable to consume enough iron to meet the body's requirements. Women are vulnerable to anemia during childbearing years due to menstrual blood loss. Anemia is also found in pregnant women because blood volume expands during pregnancy and extra iron is needed to synthesize red blood cells for the mother and the fetus.[19] Iron-deficiency anemia can also be caused by blood loss from ulcers, colon cancer, or hemorrhoids. Endurance athletes may have increased iron requirements due to increased blood loss in feces and urine and chronic lysis of red blood cells in the feet due to the trauma of running. Additional risk factors for iron-deficiency anemia include inappropriately planned vegan dietary patterns, extreme dietary restrictions (e.g., eating disorders), and frequent blood donation.

To cure iron-deficiency anemia, a person needs to take iron supplements.[20] A primary care provider should also find the cause so that the anemia does not reoccur. Changes in the dietary pattern may *prevent* iron-deficiency anemia, but supplemental iron is the only reliable *cure* once it has developed. Supplements must be taken for 3 to 6 months or perhaps longer. Hemoglobin levels respond quickly to dietary changes and supplementation, but stopping supplements too soon means that iron stores (blood, bone marrow, etc.) will not be fully replenished. Remember, it takes longer than 1 month to become anemic, so it will take longer than 1 month to cure it.

## ABSORPTION AND DISTRIBUTION OF IRON

Overall, iron absorption depends on the following factors: (1) the person's iron status; (2) its form in food; (3) the acidity of the GI tract; and (4) other dietary components consumed with iron-containing foods. Controlling iron levels in the body is important because there is a narrow gap between just enough and too much iron. As you've learned, too little iron can impair oxygen transport. To prevent deficiency, the human body highly conserves iron. Except for bleeding associated with menstruation, injury, or childbirth, body loss of iron is minimal. Approximately 90% is recovered and reused every day. On the other hand, too much iron in the body is also extremely damaging. It can accumulate in organs and promote oxidative damage. To avoid toxicity, iron absorption from the small intestine is tightly regulated.

The most important factor influencing iron absorption is the body's need for iron. Iron needs are increased during pregnancy and growth. At high altitudes, the lower oxygen concentration of the air causes an increase in the hemoglobin concentration of blood and thus an increase in iron needs.

The principal mechanism to regulate iron content in the body is tight control of absorption. High doses of iron can still be toxic, but absorption is carefully regulated under most conditions. In general, healthy people with adequate iron stores absorb between 5% and 15% of dietary iron, which is quite low compared to other nutrients. When iron stores are inadequate or needs are high due to growth or pregnancy, the main protein that carries iron (transferrin) more readily binds iron, shifting it from the intestinal cells into the bloodstream. Absorption efficiency in times of need can be as high as 50%. On the other hand, if iron stores are adequate and the iron-binding protein in the blood is fully saturated with iron, absorption from the intestinal cells is minimal—as low as 2%. The iron remains in the intestinal cells, and it will be excreted in the feces when those intestinal cells slough off, which occurs every 5 to 6 days.

Another major influence on iron absorption is the form of iron in the food. **Heme iron,** derived from hemoglobin and myoglobin, comprises 40% of the iron in meat, fish, and poultry (MFP). Absorption of heme iron ranges from about 15% to 35%. Almost nothing affects its absorption. **Nonheme iron,** on the other hand, is subject to many

*magnificent*

Several *in vitro* or "test-tube" studies have shown that iron significantly affects the intestinal microbiota. Gut microbiota favorably modify dietary nonheme iron, leading to an increase in iron availability. In addition, *Bifidobacteriaceae* can bind iron present in the large intestine, which limits the formation of free radicals synthesized in the presence of iron and reduces the risk of colorectal cancer. On the negative side, high levels of iron in the intestine can promote the development of pathogenic microorganisms.[21]

*microbiome*

**heme iron** Iron provided from animal tissues in the form of hemoglobin and myoglobin. Approximately 40% of the iron in meat, fish, and poultry is heme iron; it is readily absorbed.

**nonheme iron** Iron provided from plant sources, supplements, and animal tissues other than in the forms of hemoglobin and myoglobin. Nonheme iron is less efficiently absorbed than heme iron; absorption is closely dependent on body needs.

**TABLE 9-5 ■ Dietary Factors That Affect Nonheme Iron Absorption**

| Nonheme Enhancers | Nonheme Inhibitors |
|---|---|
| Vitamin C<br>• Add marinara sauce to your spaghetti noodles.<br>MFP (meat, fish, poultry) meat protein<br>• Add some tuna to your snack of crackers. | Tannins (found in tea)<br>• Drink tea between meals rather than with meals. (Does not apply to herbal tea)<br>Oxalates (spinach, rhubarb, and chard)<br>Phytates (whole grains, bran, and soybean)<br>Megadoses of calcium |

conditions that can either enhance or inhibit its absorption, which ranges from 2% to 8%. Table 9-5 summarizes dietary factors that affect bioavailability of nonheme iron. Nonheme iron makes up 60% of iron in MFP and 100% of the iron found in dairy, eggs, fruit, vegetables, grains, fortified foods, and supplements. Because most of our dietary iron is nonheme iron, our overall dietary iron absorption is 5% to 15%.

Acidity also affects iron absorption: an acidic environment solubilizes iron and keeps it in a form that can be readily absorbed. Therefore, any medication or health condition that lowers acid production of the stomach can decrease iron absorption. For example, acid-reducing medications that people take to control heartburn or ulcers can impair iron absorption. Also, as people age, gastric acid secretion may decline. This puts older adults at risk for iron-deficiency anemia.

Lastly, other micronutrients affect iron absorption and availability. Large doses of zinc or calcium can compete with iron for absorption in the small intestine. If you need to take an iron supplement, it is best to take it at least 2 hours before or after a calcium-rich meal or supplement. In contrast, vitamin C is a powerful enhancer of iron absorption. Doses of 75 milligrams of vitamin C can increase nonheme absorption by 4%—a lot for nonheme. If you want to get the most iron out of your dietary supplement, take it with a glass of orange juice. Research also indicates that adequate vitamin A and copper intakes promote good iron status due to their roles in iron absorption, transport, and metabolism.[20]

## GETTING ENOUGH IRON

The major iron sources in the adult eating pattern are ready-to-eat breakfast cereals, beans, and animal products (Fig. 9-26). Animal sources contain approximately 40% heme iron, the most bioavailable form. Iron is added to flour during the enrichment process. Other iron sources are peas and legumes, but the absorption of nonheme iron found in these products is relatively low. Milk and eggs are not good sources of iron. A common cause of iron-deficiency anemia in children is high consumption of milk coupled with insufficient meat intake. Individuals who follow a vegan dietary pattern are particularly susceptible to iron-deficiency anemia because of their lack of dietary heme iron. Ask your instructor about the *Rate Your Plate* activity in Connect to learn more about food sources of iron and other nutrients involved in blood health.

The adult RDA is based on a 10% absorption rate to cover average losses of about 0.8 milligram per day. For women of reproductive age, menstrual losses are an average of about 1 gram of additional iron per day. Thus, iron is the only nutrient for which women have higher requirements than men. Most women do not consume the recommended 18 milligrams of iron daily. The average daily amount consumed by women is closer to 13 milligrams, while in men it is about 18 milligrams per day. Women of reproductive age can close this gap between average daily intakes and needs by seeking out iron-fortified foods, such as ready-to-eat breakfast cereals that contain at least 50% of the DV. Use of a balanced multivitamin and mineral supplement containing up to 100% of the DV for iron is another option. Consuming more than the RDA for iron is not advised unless recommended by a primary care provider.

Iron deficiency is the most common nutrient deficiency. **Why are pregnant women and preschool-age children at risk for iron-deficiency anemia?** Tanya Constantine/ Blend Images/Getty Images

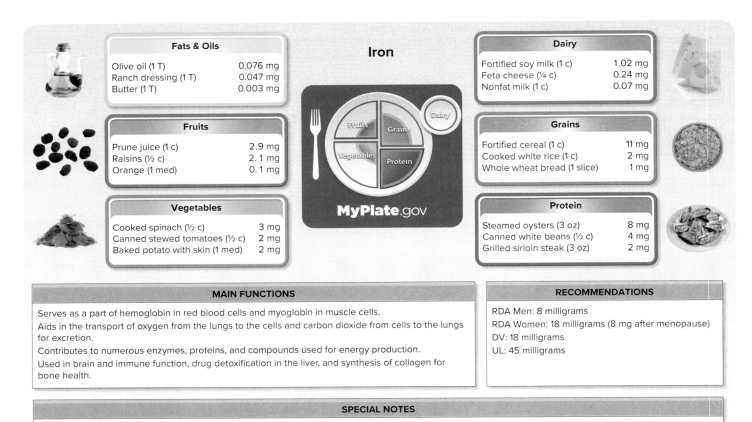

**Iron**

| Fats & Oils | |
|---|---|
| Olive oil (1 T) | 0.076 mg |
| Ranch dressing (1 T) | 0.047 mg |
| Butter (1 T) | 0.003 mg |

| Fruits | |
|---|---|
| Prune juice (1 c) | 2.9 mg |
| Raisins (½ c) | 2.1 mg |
| Orange (1 med) | 0.1 mg |

| Vegetables | |
|---|---|
| Cooked spinach (½ c) | 3 mg |
| Canned stewed tomatoes (½ c) | 2 mg |
| Baked potato with skin (1 med) | 2 mg |

| Dairy | |
|---|---|
| Fortified soy milk (1 c) | 1.02 mg |
| Feta cheese (¼ c) | 0.24 mg |
| Nonfat milk (1 c) | 0.07 mg |

| Grains | |
|---|---|
| Fortified cereal (1 c) | 11 mg |
| Cooked white rice (1 c) | 2 mg |
| Whole wheat bread (1 slice) | 1 mg |

| Protein | |
|---|---|
| Steamed oysters (3 oz) | 8 mg |
| Canned white beans (½ c) | 4 mg |
| Grilled sirloin steak (3 oz) | 2 mg |

**MAIN FUNCTIONS**

Serves as a part of hemoglobin in red blood cells and myoglobin in muscle cells.

Aids in the transport of oxygen from the lungs to the cells and carbon dioxide from cells to the lungs for excretion.

Contributes to numerous enzymes, proteins, and compounds used for energy production.

Used in brain and immune function, drug detoxification in the liver, and synthesis of collagen for bone health.

**RECOMMENDATIONS**

RDA Men: 8 milligrams

RDA Women: 18 milligrams (8 mg after menopause)

DV: 18 milligrams

UL: 45 milligrams

**SPECIAL NOTES**

Iron deficiency is the most common nutrient deficiency worldwide.

Although an individual's iron status impacts iron absorption, the heme iron found in animal foods is more bioavailable than the nonheme form found in plant-based foods.

**FIGURE 9-26** ▲ Food sources, functions, and recommendations for iron. The fill of the background color (none, 1/3, 2/3, or completely covered) within each food group on MyPlate indicates the average nutrient density for iron in that group. The figure shows the iron content of several foods in each food group. Overall, the richest sources of iron are meats, legumes, and fortified grain products. olive oil: Iconotec/Glow Images; raisins: lynx/iconotec.com/Glow Images; pile of spinach: Ingram Publishing; sheep's cheese (feta): Foodcollection; bowl of cereal: Joe Belanger/iStock/Getty Images; oyster plate: lynx/iconotec.com/Glow Images; MyPlate: U.S. Department of Agriculture

Sources: Office of Dietary Supplements, Dietary Supplements Fact Sheets, available from https://ods.od.nih.gov/factsheets/list-all; USDA FoodData Central, available from https://fdc.nal.usda.gov.

## AVOIDING TOO MUCH IRON

The UL for iron is 45 milligrams per day. Higher amounts can lead to stomach irritation. Although iron overload is not as common as iron deficiency, the consequences can be dire. Children are the most likely victims of iron poisoning (acute toxicity) because supplements, which often look a lot like candy, may be easily accessible on kitchen tables and from cabinets. Even a large single dose of 60 milligrams of iron can be life threatening to a 1-year-old. The FDA requires that all iron supplements carry a warning about toxicity. Since the introduction of that warning label in the 1990s, cases of accidental iron poisoning among children have been greatly reduced.

Iron toxicity accompanies hereditary **hemochromatosis,** a genetic disease. It is associated with a substantial increase in iron absorption from both food and supplements. The harshest effects are seen in iron-storing organs such as the liver and heart. Some iron is also deposited in the pancreas and muscles. Blood levels of iron remain high too, which increases the likelihood of infections and may promote cardiovascular disease.

Hereditary hemochromatosis occurs when a person carries two dysfunctional copies of a particular gene. People with one dysfunctional gene and one functional gene (i.e., carriers) may also absorb too much dietary iron, but not to the same extent as

**hemochromatosis** A disorder of iron metabolism characterized by increased iron absorption and deposition in the liver and heart. This eventually poisons the cells in those organs.

those with two dysfunctional genes. About 5% to 10% of North Americans of Northern European descent are carriers of hemochromatosis. Approximately 1 in 250 North Americans has both hemochromatosis genes. These numbers are high, considering that many primary care providers regard hemochromatosis as a rare disease and therefore do not routinely test for it.

Anyone who has a blood relative (including uncles, aunts, and cousins) who has hemochromatosis or is a carrier should be screened for iron overload. At your next visit to a primary care provider, ask for a transferrin saturation test to assess iron stores. A ferritin test may also be added to assess your stores. Hemochromatosis can go undetected until a person reaches age 50 to 60, so some experts recommend screening for anyone over the age of 20.

If the disease goes untreated, iron accumulates and serious health problems may result: arthritis, heart disease, diabetes, liver disease, gallbladder disease, some cancers, hypothyroidism, reproductive dysfunction, and depression. Even with iron overload, the person may have anemia due to damage to the bone marrow or liver. Treatment of hemochromatosis is relatively easy, but it must be monitored consistently. **Therapeutic phlebotomy,** as a blood donation, to remove excess iron is essential. One must be very careful about dietary choices. Few sources of heme iron should be eaten, and supplements with iron or vitamin C should be avoided. Highly fortified breakfast cereals must also be avoided.

FDA requires all dietary supplements with 30 milligrams or more of elemental iron to carry a warning label about accidental poisoning. **What are the risks of consuming too much iron?** ©Angela Collene

**therapeutic phlebotomy** Periodic blood removal, as a blood donation, for the purpose of ridding the body of excess iron.

## ☑ CONCEPT CHECK 9.9

1. List three symptoms of iron deficiency. How do these symptoms relate to the roles of iron in the body?
2. What are heme and nonheme iron? What can you do to enhance your absorption of nonheme iron?
3. What is hemochromatosis?

---

## CASE STUDY  Anemia

Anita is a 60-year-old woman who prides herself on taking charge of her health. Her daily physical activities include walking her dog, playing tennis, or participating in a tai chi class at the senior center. She follows the *Dietary Guidelines*, choosing a variety of whole grains; eating at least five servings per day of fruits and vegetables; and keeping her saturated fat and sodium intakes to a minimum. She eats lean sources of protein, choosing poultry, fish, or vegetable sources of protein instead of red meat. She maintains a healthy body weight, has never had high blood pressure or high blood sugar, and does not take medications or supplements. Last week, she went to a blood drive at her church with the intent of donating blood, but she was turned away because her hematocrit (a measure of the percentage of red blood cells in the blood) was slightly below the requirements for donation. Anita was surprised because she has never had a problem donating blood before. The nurse told Anita that her low hematocrit level was indicative of anemia, which has many possible causes. As Anita thought about it, she realized that she had been feeling more tired than usual.

▲ Anita has been feeling tired lately after playing tennis. How could her fatigue be related to her dietary pattern? Big Cheese Photo/SuperStock

1. Iron deficiency is the most common form of anemia. What role does iron play in the health of red blood cells? From the description, is it possible that Anita has low iron stores? What dietary changes could Anita make to improve her iron status?
2. Folate deficiency may lead to anemia. What role does folate play in the health of red blood cells? From the description, is it likely that Anita is deficient in folate?
3. Low vitamin B-12 may result in anemia. What role does vitamin B-12 play in the health of red blood cells? Suggest an explanation for why Anita may have low vitamin B-12 status even with adequate dietary intake of vitamin B-12. What dietary changes could Anita make to improve her vitamin B-12 status?
4. What should Anita do now that she knows she is anemic? Will changes to her food choices suffice to resolve the problem?

*Complete the Case Study. Responses to these questions can be provided by your instructor.*

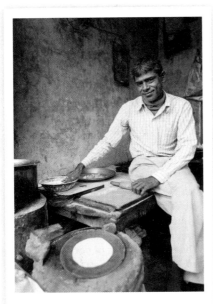

Zinc deficiency has been associated with the consumption of unleavened bread in Middle Eastern countries. **What is the name of the plant compound in grains that binds to zinc and decreases its bioavailability?**

hadynyah/Getty Images

# 9.10 Zinc

Zinc (Zn) deficiency was first recognized in the early 1960s in Egypt and Iran, where it was linked to growth retardation and poor sexual development. Even though the zinc content of the diets of people in these areas was fairly high, absorption of the mineral was limited by the phytic acid in unleavened bread. Parasite infestation and the practice of eating clay and other parts of soil also contributed to the severe zinc deficiency.

## FUNCTIONS OF ZINC

Approximately 200 enzymes require zinc as a cofactor for activity. Adequate zinc intake is necessary to support many physiological functions:

- DNA synthesis and function
- Protein metabolism, wound healing, and growth
- Development of bones and reproductive organs
- Storage, release, and function of insulin
- Cell membrane structure and function
- Component of superoxide dismutase, an enzyme that aids in the prevention of oxidative damage to cells (zinc, therefore, has an indirect **antioxidant** function)
- Immune function through white blood cell formation

It is important to note that although zinc is important for immune function, intakes in excess of the RDA do not provide any extra benefit towards immunity. In fact, chronic excessive intakes of zinc can actually depress immune function. Zinc supplementation may be useful to slow the progression of macular degeneration of the eye and reduce the risk for developing certain forms of cancer.[22]

## ZINC DEFICIENCY

Symptoms of adult zinc deficiency include an acnelike rash, diarrhea, lack of appetite, delayed wound healing, impaired immunity, reduced sense of taste (metallic-like) and smell, and hair loss. In children and adolescents with zinc deficiency, growth, sexual development, and learning ability may also be hampered.

## GETTING ENOUGH ZINC

Protein-rich diets, especially those that include many animal sources of protein, are high in zinc. The average North American consumes 10 to 14 milligrams of zinc per day, about 80% of which is provided by meat, fish, poultry, fortified cereal, and dairy products (Fig. 9-27). There are no indications of moderate or severe zinc deficiencies in an otherwise healthy adult population. It is likely, however, that some North Americans—especially children with food insecurity, vegans, and older people with alcohol use disorders—have marginal zinc status. These and other people who show deterioration in taste sensation, recurring infections, poor growth, or depressed wound healing should have their zinc status checked.

Overall, about 40% of dietary zinc is absorbed. Absorption efficiency depends on the body's need for zinc and the form of the mineral in foods. When zinc status is poor, absorption of the mineral increases. The zinc found in animal foods is better absorbed than that found in plants. Worldwide, however, most people rely on unfortified cereal grains (low in zinc) as their source of protein, calories, and zinc. Phytic acid in plant foods binds to zinc and limits its availability. Adding yeast to grains (leavening) breaks down phytic acid, increasing zinc bioavailability from leavened grain products. In populations that consume mainly unleavened bread, zinc deficiency can be a problem.

The form generally used in multivitamin and mineral supplements (zinc oxide) is not as well-absorbed as zinc found naturally in foods but still contributes to meeting zinc needs. High-dose calcium supplementation decreases zinc availability if taken too close to mealtime. Finally, zinc competes with copper and iron for absorption,

## Zinc

**Fats & Oils**

| | |
|---|---|
| Cream cheese (1 T) | 0.07 mg |
| Sour cream (1 T) | 0.05 mg |
| Ranch dressing (1 T) | 0.03 mg |

**Fruits**

| | |
|---|---|
| Fresh blackberries (1 c) | 0.76 mg |
| Dried apricots (½ c) | 0.25 mg |
| Banana (1 med) | 0.18 mg |

**Vegetables**

| | |
|---|---|
| Cooked asparagus (½ c) | 0.6 mg |
| Cooked frozen peas (½ c) | 0.5 mg |
| Cooked zucchini (½ c) | 0.4 mg |

**MyPlate.gov**

**Dairy**

| | |
|---|---|
| Swiss cheese (1.5 oz) | 1.8 mg |
| Fruited low-fat yogurt (1 c) | 1.8 mg |
| Nonfat milk (1 c ) | 1 mg |

**Grains**

| | |
|---|---|
| Fortified cereal (1 c) | 5 mg |
| Cooked quinoa (1 c) | 2 mg |
| Cooked oatmeal (1 c) | 1 mg |

**Protein**

| | |
|---|---|
| Steamed oysters (3 oz) | 65.7 mg |
| Beef chuck roast (3 oz) | 7 mg |
| Dry roasted cashews (1 oz) | 1.6 mg |

**MAIN FUNCTIONS**

Serves as a cofactor in approximately 200 enzymes.

Supports numerous physiological functions, including DNA synthesis and function, protein metabolism.

Supports normal growth and development, including wound healing and development of bones and certain organs.

Serves as a component of one of the body's natural antioxidant enzyme, superoxide dismutase.

**RECOMMENDATIONS**

RDA Men: 11 milligrams

RDA Women: 8 milligrams

DV: 11 milligrams

UL: 40 mg

**SPECIAL NOTES**

Absorption efficiency of zinc depends on the body's needs and the form of mineral provided.

Major sources of zinc in most dietary patterns include animal-based products, such as meat and fish, which is better absorbed in the body.

**FIGURE 9-27** ▲ Food sources, functions, and recommendations for zinc. The fill of the background color (none, 1/3, 2/3, or completely covered) within each food group on MyPlate indicates the average nutrient density for zinc in that group. The figure shows the zinc content of several foods. Overall, the richest sources of zinc are in the protein group. bagel with cream cheese: Renee Comet/National Cancer Institute (NCI); fresh blackberries: Vitalina Rybakova/Shutterstock; asparagus: Ingram Publishing/SuperStock; wedge of swiss cheese: Comstock/Jupiter Images/Getty Images; bowl of cereal: Joe Belanger/iStock/Getty Images; oyster plate: lynx/iconotec.com/Glow Images; MyPlate: U.S. Department of Agriculture

Sources: Office of Dietary Supplements, Dietary Supplements Fact Sheets, available from https://ods.od.nih.gov/factsheets/list-all; USDA FoodData Central, available from https://fdc.nal.usda.gov.

and vice versa, when supplemental sources are consumed. Supplements with more than 100% of the Daily Value for individual minerals are not recommended without medical supervision.

## AVOIDING TOO MUCH ZINC

Excessive zinc intake over time can lead to problems by interfering with copper metabolism. The interference with copper metabolism is the basis for setting the UL. Zinc toxicity can occur from zinc supplements and overconsumption of zinc-fortified foods. A person using megadose supplementation should be under close medical supervision and take a supplement containing copper (2 milligrams per day). Zinc intakes over 100 milligrams result in diarrhea, cramps, nausea, vomiting, and loss of appetite. Intakes consistently over 2000 milligrams per day can lead to depressed immune function and decreased high-density lipoproteins.

## ✓ CONCEPT CHECK 9.10

1. List three good sources of zinc.
2. What are the consequences of zinc deficiency?

Pixtal/age fotostock

## FARM to FORK — Onions and Garlic

Onions and garlic, along with shallots, scallions, chives, and leeks, are part of the allium family. The alliums have long been associated with health and medicinal properties. The hot, pungent flavor of alliums comes from thiosulfinates, compounds that contain the mineral sulfur. Some of these compounds have been shown to have antiviral and antibacterial properties.

### Grow
- The most common garlic grown in America, the California silverskin, is very productive. Plant one clove and a new head grows with 16 cloves.
- Several varieties of onions are grown, including white, yellow, red, pearl, and the sweet onions, such as Vidalia.
- Farmers have cultivated larger and sweeter varieties of onions that are popular but have much lower health benefits than wild varieties.
- Green onions, or scallions, are one of the most nutritious alliums and are easy to grow even in small gardens.

### Shop
- Buy garlic bulbs that are plump and tightly encased in their papery outer wrapping. If the outer skin is loose or frayed, the bulbs may be dried or moldy.
- Purchase onions with their papery skin intact; this outer skin preserves the juiciness of the onion and protects it from mold and fungal infections.

### Store
- Garlic can be stored for 1 or 2 months but becomes more pungent the longer it is stored.
- Store sweet onions and garlic on a shelf in the refrigerator to keep them freshest. Keep them out of the crisper drawer where the high humidity will cause them to sprout. Other onions can be stored in a net bag in a cool, dark location such as an unheated room or basement.

### Prep
- Before cooking, garlic should be sliced, chopped, or minced, and then allowed to rest for 10 minutes. This will maximize the production of the phytochemical, allicin, before heat destroys the enzyme that creates it.
- Onions can be cooked as soon as you slice or chop them without losing any health benefits. All cooking methods, except boiling, increase the quercetin content of onions. Cooking also makes the hottest onions taste mild and sweet.

Source: Robinson J. Alliums: All things to all people. In *Eating on the Wild Side*. New York: Little, Brown and Company, 2013.

Milovan Radmanovac/123RF

## 9.11 Selenium

### FUNCTIONS OF SELENIUM

Selenium (Se) is a trace mineral that exists in many absorbable chemical forms. Selenium's best-understood role is aiding the activity of one of the body's antioxidant enzymes, glutathione peroxidase. Glutathione peroxidase converts potentially damaging peroxides (e.g., hydrogen peroxide) into water. As part of this antioxidant enzyme, selenium spares vitamin E and helps maintain cell-membrane integrity. Selenium is also part of an enzyme that is essential for the activation of **thyroid hormone**.

### SELENIUM DEFICIENCY

Selenium content of foods is strongly dependent on the selenium content of the soil where plants are raised or animals graze. Worldwide, only one region—Keshan County in China—has such low soil selenium levels that selenium deficiencies result. Selenium deficiency symptoms in humans include muscle pain and wasting, and a certain form of heart damage. Also, due to its role in thyroid hormone metabolism, selenium deficiency may impair thyroid function, thereby limiting growth. In China's Keshan County, unless children and adults receive selenium supplements, they develop characteristic muscle and heart disorders associated with inadequate selenium intake.

Low blood levels of selenium have been linked to an increased incidence of some forms of cancer, specifically prostate cancer. Although selenium could prove to have a role in prevention of cancers in those with low or marginal selenium stores, it is premature to recommend selenium supplementation for this purpose. Animal studies in this area are conflicting. Current studies examine the interaction of selenium and vitamin E on gene expression in some cancers.

### GETTING ENOUGH SELENIUM

Overall, the major selenium contributors to the adult dietary pattern are animal and grain products. Fish, shellfish, meat (especially organ meats), and eggs are good animal sources of selenium (Fig. 9-28). Brazil nuts, as well as grains and seeds grown in soils containing selenium, are good plant sources. Some geographic regions with low-selenium soil in North America include the Northeast, Pacific, Southwest, and coastal plain of the Southeast in the United States, along with the north-central and eastern regions in Canada. If you consume a variety of foods from many geographic areas, it is unlikely that your eating pattern is deficient in selenium. Garlic and onions can be rich sources of selenium because they have the ability to accumulate it from soil, which may increase their cancer-fighting potential. Read more about getting the most from garlic and onions in *Farm to Fork*.

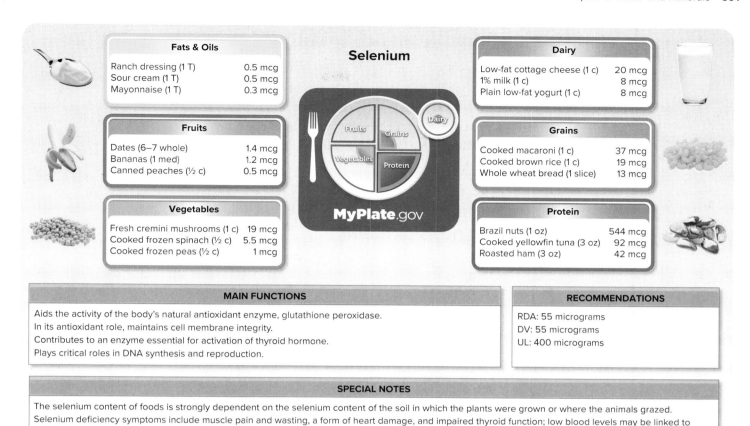

**Selenium**

### Fats & Oils
| | |
|---|---|
| Ranch dressing (1 T) | 0.5 mcg |
| Sour cream (1 T) | 0.5 mcg |
| Mayonnaise (1 T) | 0.3 mcg |

### Fruits
| | |
|---|---|
| Dates (6–7 whole) | 1.4 mcg |
| Bananas (1 med) | 1.2 mcg |
| Canned peaches (½ c) | 0.5 mcg |

### Vegetables
| | |
|---|---|
| Fresh cremini mushrooms (1 c) | 19 mcg |
| Cooked frozen spinach (½ c) | 5.5 mcg |
| Cooked frozen peas (½ c) | 1 mcg |

### Dairy
| | |
|---|---|
| Low-fat cottage cheese (1 c) | 20 mcg |
| 1% milk (1 c) | 8 mcg |
| Plain low-fat yogurt (1 c) | 8 mcg |

### Grains
| | |
|---|---|
| Cooked macaroni (1 c) | 37 mcg |
| Cooked brown rice (1 c) | 19 mcg |
| Whole wheat bread (1 slice) | 13 mcg |

### Protein
| | |
|---|---|
| Brazil nuts (1 oz) | 544 mcg |
| Cooked yellowfin tuna (3 oz) | 92 mcg |
| Roasted ham (3 oz) | 42 mcg |

**MyPlate.gov**

**MAIN FUNCTIONS**

Aids the activity of the body's natural antioxidant enzyme, glutathione peroxidase.
In its antioxidant role, maintains cell membrane integrity.
Contributes to an enzyme essential for activation of thyroid hormone.
Plays critical roles in DNA synthesis and reproduction.

**RECOMMENDATIONS**

RDA: 55 micrograms
DV: 55 micrograms
UL: 400 micrograms

**SPECIAL NOTES**

The selenium content of foods is strongly dependent on the selenium content of the soil in which the plants were grown or where the animals grazed. Selenium deficiency symptoms include muscle pain and wasting, a form of heart damage, and impaired thyroid function; low blood levels may be linked to some cancers.

**FIGURE 9-28** ▲ Food sources, functions, and recommendations for selenium. The fill of the background color (none, 1/3, 2/3, or completely covered) within each food group on MyPlate indicates the average nutrient density for selenium in that group. The figure shows the selenium content of several foods. Overall, the richest sources of selenium are found in the protein foods and grains groups. mayonnaise: Iconotec/Alamy Stock Photo; banana: lynx/iconotec.com/Glow Images; pile of peas: Ingram Publishing/SuperStock; glass of milk: NIPAPORN PANYACHAROEN/Shutterstock; pile of pasta gomiti: olgaman/Getty Images; raw brazil nuts: 4kodiak/Getty Images; MyPlate: U.S. Department of Agriculture

Sources: Office of Dietary Supplements, Dietary Supplements Fact Sheets, available from https://ods.od.nih.gov/factsheets/list-all; USDA FoodData Central, available from https://fdc.nal.usda.gov.

The RDA for selenium is 55 micrograms per day for adults (Fig. 9-28). This intake maximizes the activity of selenium-dependent enzymes. On updated food and supplement labels, the Daily Value is also 55 micrograms. Most adults meet the RDA, consuming on average 105 micrograms each day.

## AVOIDING TOO MUCH SELENIUM

High concentrations of selenium are rarely found in food, with the exception of Brazil nuts. Therefore, selenium toxicity has not been reported from eating food. Excessive selenium supplementation for an extended period has been shown to be toxic. The UL for selenium is 400 micrograms per day for adults. This is based on overt signs of selenium toxicity, such as hair loss, weakness, nausea, vomiting, and cirrhosis. Because Brazil nuts are such a concentrated source of selenium, you should not consume them every day, and when you do eat them, limit your portion size.

This portion of 10 nuts contains 960 micrograms of selenium. **How does this compare to the UL for selenium?** 4kodiak/Getty Images

**goiter** An enlargement of the thyroid gland; this is often caused by insufficient iodine in the dietary pattern.

**congenital hypothyroidism** The stunting of body growth and poor development in the offspring that result from inadequate maternal intake of iodine during pregnancy that impairs thyroid hormone synthesis (formerly called *cretinism*).

## ✅ CONCEPT CHECK 9.11

1. How does selenium participate in the body's antioxidant defenses?
2. What other functions does selenium play in the body?
3. What are the signs of a selenium deficiency?
4. What food groups are the best sources of selenium?
5. What are the signs of selenium toxicity?

# 9.12 Iodine

## FUNCTIONS OF IODINE

The thyroid gland actively accumulates and traps iodine (I) from the bloodstream to support thyroid hormone synthesis. Thyroid hormones are synthesized using iodine and the amino acid tyrosine. Because these hormones help regulate metabolic rate and promote growth and development throughout the body, iodine adequacy is important for overall energy metabolism.

## IODINE DEFICIENCY

If a person's iodine intake is insufficient, the thyroid gland enlarges as it attempts to take up more iodine from the bloodstream. This eventually leads to **goiter**. Simple goiter is a painless condition but, if uncorrected, can lead to pressure on the trachea (windpipe), which may cause difficulty in breathing. Although iodine can prevent goiter formation, it does not significantly shrink a goiter once it has formed. Surgical removal may be required in severe cases.

If a woman has an iodine-deficient dietary pattern during the early months of her pregnancy, the fetus suffers iodine deficiency because the available iodine is used by the mother's body. The infant may be born with stunted growth and developmental delays that collectively are known as **congenital hypothyroidism** (formerly called cretinism). This condition appeared in North America before iodine fortification of table salt began. Today, congenital hypothyroidism still appears in Europe, Africa, Latin America, and Asia. Marginal iodine status has been detected recently in women of childbearing years in the United States. Recent studies have shown a link between mild iodine deficiency during pregnancy and decreased IQ in offspring in the United Kingdom and Australia.[27,28]

During World War I, men drafted into the military from areas such as the Great Lakes Region of the United States had a much higher rate of goiter than did men from other areas of the country. The soils in these areas have low iodine contents. In the 1920s, a researcher in Ohio found that low doses of iodine given to children over a 4-year period could prevent goiter. That finding led to the addition of iodine to salt beginning in the 1920s, the first time a nutrient was purposely added to food to prevent a disease.

In the United States, salt can be purchased either iodized or plain. Check for this on the label when you purchase salt. Whereas many nations, such as Canada, require iodine fortification of salt, some areas of Europe, such as northern Italy, have yet to adopt an iodine-fortification program, despite having very low soil levels of iodine. People in these areas, especially women, still suffer from goiter, as do people in areas of Latin America, the Indian subcontinent, Southeast Asia, and Africa. With about 2 billion people worldwide at risk of iodine deficiency, and approximately 800 million of these people suffering the effects of iodine deficiency, eradication of iodine deficiency is a goal of many health-related organizations worldwide.

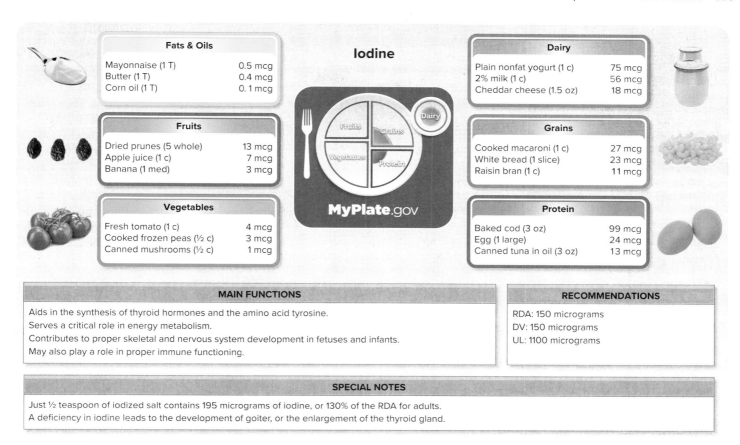

### Iodine

**MyPlate.gov**

| Fats & Oils | |
|---|---|
| Mayonnaise (1 T) | 0.5 mcg |
| Butter (1 T) | 0.4 mcg |
| Corn oil (1 T) | 0.1 mcg |

| Fruits | |
|---|---|
| Dried prunes (5 whole) | 13 mcg |
| Apple juice (1 c) | 7 mcg |
| Banana (1 med) | 3 mcg |

| Vegetables | |
|---|---|
| Fresh tomato (1 c) | 4 mcg |
| Cooked frozen peas (½ c) | 3 mcg |
| Canned mushrooms (½ c) | 1 mcg |

| Dairy | |
|---|---|
| Plain nonfat yogurt (1 c) | 75 mcg |
| 2% milk (1 c) | 56 mcg |
| Cheddar cheese (1.5 oz) | 18 mcg |

| Grains | |
|---|---|
| Cooked macaroni (1 c) | 27 mcg |
| White bread (1 slice) | 23 mcg |
| Raisin bran (1 c) | 11 mcg |

| Protein | |
|---|---|
| Baked cod (3 oz) | 99 mcg |
| Egg (1 large) | 24 mcg |
| Canned tuna in oil (3 oz) | 13 mcg |

**MAIN FUNCTIONS**

Aids in the synthesis of thyroid hormones and the amino acid tyrosine.
Serves a critical role in energy metabolism.
Contributes to proper skeletal and nervous system development in fetuses and infants.
May also play a role in proper immune functioning.

**RECOMMENDATIONS**

RDA: 150 micrograms
DV: 150 micrograms
UL: 1100 micrograms

**SPECIAL NOTES**

Just ½ teaspoon of iodized salt contains 195 micrograms of iodine, or 130% of the RDA for adults.
A deficiency in iodine leads to the development of goiter, or the enlargement of the thyroid gland.

**FIGURE 9-29** ▲ Food sources, functions, and recommendations for iodine. The fill of the background color (none, 1/3, 2/3, or completely covered) within each food group on MyPlate indicates the average nutrient density for iodine in that group. The figure shows the iodine content of several foods. Overall, the richest sources of iodine are iodized salt (added to foods in any group), seafood and seaweed, and dairy products. Fruits and vegetables other than seaweed are poor sources of iodine. mayonnaise: Iconotec/Alamy Stock Photo; three prunes: I. Rozenbaum & F. Cirou/PhotoAlto; tomato branch: Tim UR/Shutterstock; two jars of yogurt: Foodcollection; pile of pasta: olgaman/Getty Images; eggs: clubfoto/E+/Getty Images; MyPlate: U.S. Department of Agriculture

Sources: Office of Dietary Supplements, Dietary Supplements Fact Sheets, available from https://ods.od.nih.gov/factsheets/list-all; USDA FoodData Central, available from https://fdc.nal.usda.gov/

## GETTING ENOUGH IODINE

Iodized salt, dairy products, and grain products contain various forms of iodine (Fig. 9-29). Sea salt and kosher salt, however, are not typically iodized. The RDA for iodine (Fig. 9-29) was set to support thyroid gland function. This is the same as the DV used on food and supplement labels. A half teaspoon of iodine-fortified salt (about 2 grams) supplies that amount. Most North American adults consume more iodine than the RDA—an estimated 190 to 300 micrograms daily, not including that from use of iodized salt at the table. This extra amount of iodine adds up, because dairies use it as a sterilizing agent, bakeries use it as a dough conditioner, food producers use it as part of food colorants, and it is added to salt. There is concern, however, that vegans may not consume enough unless iodized salt is used. Iodized salt, dairy products, and grain products contain various forms of iodine (Fig. 9-29). Sea salt and kosher salt, however, are not typically iodized.

## AVOIDING TOO MUCH IODINE

The UL for iodine is 1.1 milligrams per day. When high amounts of iodine are consumed, thyroid hormone synthesis is inhibited, as in a deficiency. This can appear in people who eat a lot of seaweed because some seaweeds contain as much as 1% iodine by weight. Total iodine intake then can add up to 60 to 130 times the RDA.

▲ This woman has an enlargement of the thyroid gland (also known as a goiter) caused by insufficient iodine in the diet. Scott Camazine/Science Source

✓ **CONCEPT CHECK 9.12**

1. What is the role of iodine in thyroid metabolism?
2. What are the effects of iodine deficiency?
3. What is a goiter?
4. Is salt always a good source of iodine?

## **9.13** Copper

Copper (Cu) and iron are similar in terms of food sources, absorption, and functions. Copper is a component of blood. In the body, it is found in highest concentration in the liver, brain, heart, kidneys, and muscles. **Ceruloplasmin** is the name of the protein that carries most of the body's copper in the blood.

**ceruloplasmin** Copper-containing protein in the blood; functions in the transport of iron.

### FUNCTIONS OF COPPER

Copper is a cofactor for many enzymes, including some involved in the body's antioxidant defenses. Copper serves as a cofactor for superoxide dismutase, an enzyme that defends the body against free-radical damage. Copper also has a role in the function of enzymes that create cross-links in connective tissue proteins, such as the collagen in bone. Another very important role of copper is as a cofactor in the last stage of energy metabolism, which converts the energy stored in carbohydrates, fats, and proteins into ATP.

Copper is important for blood health because one of its roles is making iron available for the formation of red blood cells. Copper is part of three different enzymes that assist in the transport of iron out of intestinal cells, through the blood, and to the bone marrow, where iron is incorporated into hemoglobin. Copper is also a cofactor for enzymes involved in blood clotting, immune system function, and blood lipoprotein metabolism. In addition, copper is needed for brain health through its role in enzymes involved in nerve myelination and neurotransmitter synthesis.

▶ A genetic disease called **Menkes syndrome** decreases the amount of copper available to the brain and nervous system. Babies born with Menkes syndrome suffer from nervous system disorders, weak muscle tone, and delays in physical and cognitive development related to the lack of copper-containing enzymes that help to form nervous tissue and synthesize neurotransmitters. They usually do not live past the age of 3.

**Menkes syndrome** An inherited X-linked recessive pattern disorder that affects copper levels in the body.

### COPPER DEFICIENCY

Considering the many roles of copper, it is not surprising that copper deficiency affects so many different body systems. Symptoms of copper deficiency include a form of anemia, weakened immunity, bone loss, poor growth, and some forms of cardiovascular disease.

The groups most likely to develop copper deficiencies are infants born preterm and people recovering from intestinal surgery. A copper deficiency can also result from the overuse of zinc supplementation, because zinc and copper compete with each other for absorption.

### GETTING ENOUGH COPPER

Rich sources of copper include liver, legumes, seeds, whole grain breads and cereals, and cocoa (Fig. 9-30). Milk and dairy products, fruits, and vegetables are generally poor sources of copper. Also, the form of copper typically found in multivitamin and mineral supplements (copper oxide) is not readily absorbed. It is best to rely on food sources to meet copper needs.

Dark chocolate is a rich source of copper. One ounce of dark chocolate candy provides about 500 micrograms of copper. **Is it likely that anyone would suffer from copper toxicity because of eating too much chocolate?**
Baiba Opule/Baibaz/123RF

Copper absorption is highly variable; as copper intake increases, the mineral is absorbed less efficiently. Absorption takes place in the stomach and upper small intestine. Excess copper is not stored to a great extent, so when intake exceeds needs, the liver incorporates it into bile, which is excreted as part of the feces. Phytates, fiber, and excessive zinc and iron supplements may all interfere with copper absorption.

The copper status of adults appears to be good: the average adult intake is about 1 milligram for women and 1.6 milligrams for men per day. However, sensitive laboratory tests to determine copper status are lacking.

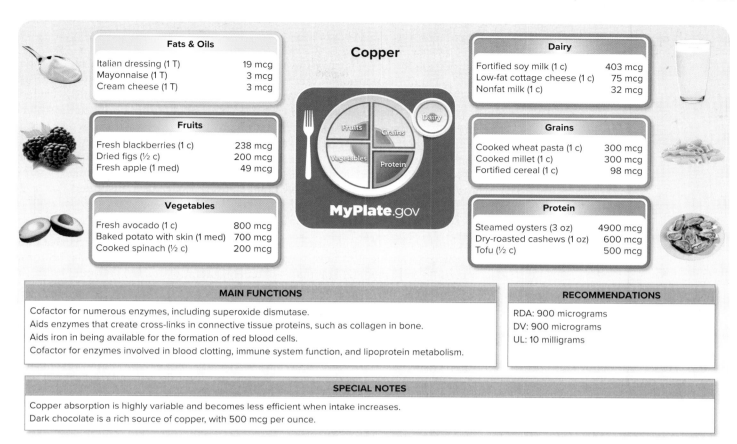

**FIGURE 9-30** ▲ Food sources, functions, and recommendations for copper. The fill of the background color (none, 1/3, 2/3, or completely covered) within each food group on MyPlate indicates the average nutrient density for copper in that group. The figure shows the copper content of several foods in each food group. Overall, the richest sources of copper are found in the protein foods and grains groups. mayonnaise: Iconotec/ Alamy Stock Photo; fresh blackberries: Vitalina Rybakova/Shutterstock; avocado: lynx/iconotec.com/Glow Images; glass of milk: NIPAPORN PANYACHAROEN/Shutterstock; wholemeal pasta: HandmadePictures/iStock/Getty Images; oyster plate: lynx/iconotec.com/Glow Images; MyPlate: U.S. Department of Agriculture

Sources: Office of Dietary Supplements, Dietary Supplements Fact Sheets, available from https://ods.od.nih.gov/factsheets/list-all; USDA FoodData Central, available from https://fdc.nal.usda.gov.

## AVOIDING TOO MUCH COPPER

A single dose of copper greater than 10 milligrams can cause toxicity. Consequences of copper toxicity include GI distress, vomiting blood, tarry feces, and damage to the liver and kidneys. Toxicity cannot occur with food, only supplements or excessive exposure to copper salts used in agriculture.

**Wilson's disease** is a genetic disease in which the liver cannot synthesize ceruloplasmin. In turn, copper accumulates in tissues, such as lungs and liver. People with Wilson's disease suffer damage to the liver and nervous system. A primary treatment for Wilson's disease is a vegan dietary pattern, as fruits and vegetables are low in copper. Researchers are currently studying how excess copper in the blood may influence the development of Alzheimer's and Parkinson's diseases.

**Wilson's disease** A genetic disorder that results in accumulation of copper in the tissues; characterized by damage to the liver, nervous system, and other organs.

### ✓ CONCEPT CHECK 9.13

1. List three functions of copper.
2. Describe some interactions among iron, zinc, and copper in the body.
3. What changes to the dietary pattern will be required for a person with Wilson's disease?

## 9.14 Fluoride

The fluoride ion (F⁻) is the form of this trace mineral essential for human health. Nearly all (about 95%) of the fluoride in the body is found in the teeth and skeleton. Dentists in the early 1900s noticed a lower rate of dental caries (cavities) in areas of the United States, particularly the Southwest, that contained high amounts of fluoride in the water. The amounts of fluoride were sometimes so high that small brown spots developed on the teeth (mottling). Even though mottled teeth were discolored, they contained few dental caries. Experiments in the early 1940s showed that fluoride in the water decreased the incidence of dental caries by 20% to 80% in children. Fluoridation of public water supplies in many parts of the United States has since been instituted.[29]

### FUNCTIONS OF FLUORIDE

Fluoride functions in the following ways to prevent dental caries: (1) incorporates into tooth structure, causing it to be stronger and more resistant to acid degradation from bacteria found in plaque; (2) stimulates remineralization of enamel and inhibits tooth demineralization; and (3) has antibacterial effect on acid-producing microorganisms found in plaque.

### GETTING ENOUGH FLUORIDE

The list of foods that are good sources of fluoride is rather short: marine fish, clams, lobster, crab, shrimp, tea, and seaweed. Most of our fluoride actually comes from oral hygiene products and the fluoridated water supply. Numerous products are available to apply fluoride to teeth topically. These include gels applied at a dentist's office, toothpaste, and mouth rinses for everyday use. Fluoride is also available in supplement form, although use should be directed by a dentist or primary care provider. The most economical method of distributing fluoride is to add the mineral to the community's drinking water. When water fluoridation and fluoridated topical products are used in combination, the reductions in dental caries are additive.

In a few areas of the world, the fluoride content of groundwater is naturally high, but most groundwater supplies contain low levels of fluoride. In the 1950s, after researchers established a connection between fluoride and rates of dental caries, communities in the United States began adding fluoride to the municipal water supply to achieve a fluoride level of 0.7 to 1.2 milligrams per liter. (The lower levels are for communities in hotter climates, where total water consumption is higher.) About two-thirds of North Americans currently consume fluoridated water, with these policies made by individual municipalities. Because most people now have ample access to oral hygiene products with fluoride, the level of water fluoridation has been lowered to just 0.7 milligram per liter.

The AI for fluoride for adults is 3.1 to 3.8 milligrams per day. This range of intake provides the benefits of resistance to dental caries without causing ill effects. As described, typical fluoridated water contains about 1 milligram per liter, which works out to about 0.25 milligram per cup. In communities without fluoridated water (e.g., those that rely on private well water), use of fluoride-containing oral hygiene products or dietary supplements is of heightened importance for combating dental decay.

Note that fluoride is generally not added to bottled water. Frequent use of bottled water or a household reverse osmosis water purification system significantly restricts fluoride intake. A refrigerator or Brita® filter does not remove fluoride.

### AVOIDING TOO MUCH FLUORIDE

The UL for fluoride is set at 1.3 to 2.2 milligrams per day for young children and 10 milligrams per day for children over 9 years of age and adults, based on skeletal and tooth damage seen with higher doses. Children may develop **fluorosis** if they swallow large amounts of fluoride toothpaste as part of daily tooth care. Fluorosis leads to stained

▶ **Fluoride**
AI: 3.1 to 3.8 milligrams
UL:
 Young children: 1.3 to 2.2 milligrams
 > 9 years: 10 milligrams

▶ To find a list of brands of bottled water that contain fluoride, check out the website of the International Bottled Water Association at www.bottledwater.org /fluoride.

**fluorosis** Discoloration of tooth enamel sometimes accompanied with pitting due to consuming a large amount of fluoride for an extended period.

and pitted teeth and can permanently damage teeth if it occurs during tooth development (first decade of life) (Fig. 9-31). Not swallowing toothpaste and limiting the amount used to "pea" size are the best ways to prevent this problem. In addition, children under 6 years should have tooth brushing supervised by an adult and should never use fluoride mouthwash. In adults, fluorosis is associated with hip fractures, weak or stiff joints, and chronic stomach inflammation.

There have been opponents to the fluoridation of public water supplies. Some people argue that water fluoridation standards were set at a time when much of the population did not have adequate access to fluoride-containing oral hygiene products, and that addition of fluoride to the water supply is no longer necessary. Other critics claim that chronic exposure to fluoridated water is linked to a variety of health ailments affecting the skeletal, nervous, or endocrine systems. At this time, there is little scientific evidence to support claims that water fluoridation at current levels has adverse health effects other than dental fluorosis. The recommendations for the level of water fluoridation aim to take advantage of the oral health benefits of fluoride while limiting unwanted health effects, including fluorosis. The CDC provides more information on fluoridation at https://www.cdc.gov/fluoridation/index.html.

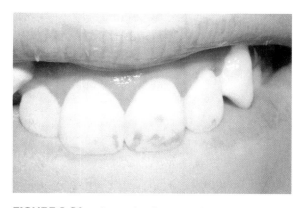

**FIGURE 9-31** ▲ Example of mottling (brown spots) on teeth called fluorosis and caused by overexposure to fluoride. Children can develop fluorosis as a result of swallowing large amounts of fluoride toothpaste. Paul Casamassimo, DDS, MS

### ✓ CONCEPT CHECK 9.14

1. How does fluoride help reduce the development of dental caries?
2. What are our primary sources of fluoride?
3. What are the risks of excessive fluoride intake?

## 9.15 Chromium

### FUNCTIONS OF CHROMIUM

Chromium (Cr) enhances the function of insulin, so it is required for glucose uptake into cells. The mineral is involved in the metabolism of lipids and proteins as well, although the exact mechanisms are not known. Chromium supplements have been promoted for building muscle mass and for weight loss, but there is not much evidence to support these claims.

### CHROMIUM DEFICIENCY

A chromium deficiency is characterized by impaired blood glucose control and elevated blood cholesterol and triglycerides. Low or marginal chromium intakes may contribute to an increased risk for developing type 2 diabetes, but opinions are mixed on the true degree of this effect. Chromium deficiency appears in people maintained on intravenous nutrition solutions not supplemented with chromium and in children with malnutrition. Marginal deficiencies may go undetected because sensitive measures of chromium status are not available.

### GETTING ENOUGH CHROMIUM

Specific data regarding the chromium content of various foods are scant, and most food-composition tables do not include values for this trace mineral. Meat, whole grain products, eggs, mushrooms, nuts, beer, and spices are relatively good sources of chromium. Brewer's yeast is also a very good source.

▲ Mushrooms are a good source of chromium. Pixtal/age fotostock

▶ **Chromium**
AI
   Men: 35 micrograms
   Women: 25 micrograms
DV: 35 micrograms
UL: none

Chromium absorption is quite low: only 0.4% to 2.5% of the amount consumed. Absorption is enhanced by vitamin C and niacin. Any unabsorbed chromium is excreted in the feces. Once absorbed, it is stored in the liver, spleen, soft tissue, and bone, and it is excreted via urine. Certain conditions can enhance urinary excretion of chromium: eating patterns high in simple sugars (more than 35% of total calories), significant infection, acute prolonged exercise, pregnancy and lactation, and major physical trauma. If chromium intakes are already low, these states potentially can lead to deficiency.

The Adequate Intake (AI) for chromium is 25 to 35 micrograms per day, based on the amount present in a balanced dietary pattern (see box). Average adult intakes in North America are estimated at about 30 micrograms per day but could be somewhat higher.

No UL for chromium has been set because toxicity from food sources has not been observed. Chromium toxicity, however, has been reported in people exposed to industrial waste and in painters who use art supplies with high chromium content. Liver damage and lung cancer can result. In general, aim for no more than the DV when taking any dietary supplements unless your primary care provider recommends otherwise.

✓ **CONCEPT CHECK 9.15**

1. How is chromium involved in carbohydrate metabolism?
2. What conditions can increase the loss of chromium in the urine?
3. Which foods are considered the best sources of chromium?

## 9.16 Other Trace Minerals

### MANGANESE

The mineral manganese (Mn) is easily confused with magnesium (Mg). Not only are their names similar, but they also often substitute for each other in metabolic processes. As a participant in energy metabolism, manganese is required as a cofactor for synthesis of glucose and metabolism of some amino acids. Manganese is also needed by some enzymes, especially superoxide dismutase used in free-radical metabolism. Manganese is also important in bone formation.

Manganese deficiency does not develop in humans unless the mineral is purposely removed from the diet. Animals on manganese-deficient diets suffer alterations in brain function, bone formation, and reproduction. If human dietary patterns were low in manganese, these symptoms would probably appear as well. As it happens, our need for manganese is very low, and our eating patterns tend to be adequate in this trace mineral.

The AI for manganese is 1.8 to 2.3 milligrams to offset daily losses. Average intakes fall within this range. The DV used on food and supplement labels is 2 milligrams. Good food sources of manganese are nuts, rice, oats and other whole grains, beans, and leafy vegetables. Manganese is toxic at high doses. Supplements are not recommended because large doses can decrease absorption of other minerals. People with low iron stores must avoid manganese supplements or risk worsening anemia. The UL is 11 milligrams per day. This value is based on the development of nerve damage. Miners who have inhaled dust fumes high in manganese experience symptoms that mimic Parkinson's disease, including cognitive and muscular dysfunction.

▲ Beans are sources of manganese and molybdenum. Pixtal/AGE Fotostock

▶ **Manganese**
AI
  Men: 2.3 milligrams
  Women: 1.8 milligrams
DV: 2 milligrams
UL: 11 milligrams

## MOLYBDENUM

Several human enzymes use molybdenum (Mo), including some involved in metabolism of amino acids that contain sulfur. No molybdenum deficiency has been reported in people who consume food and beverages orally. Deficiency symptoms have appeared in people maintained on intravenous nutrition devoid of this trace mineral. Symptoms include increased heart and respiratory rates, night blindness, mental confusion, edema, and weakness.

Good food sources of molybdenum include milk and dairy products, beans, whole grains, and nuts. The RDA for molybdenum is 45 micrograms to offset daily losses. The DV used on food and supplement labels is 75 micrograms. Our daily intakes average 76 micrograms (for women) and 109 micrograms (for men). The Upper Level for molybdenum is 2 milligrams per day. When consumed in high doses, molybdenum causes toxicity in laboratory animals, resulting in weight loss and decreased growth. Toxicity risk in humans is quite low.

▶ **Molybdenum**
RDA: 45 micrograms
DV: 75 micrograms
UL: 2 milligrams

☑ **CONCEPT CHECK 9.16**

1. What are the primary functions of manganese and molybdenum in metabolism of nutrients?

2. What are the good food sources of manganese and molybdenum?

3. Why is a deficiency of manganese or molybdenum possible when patients are fed intravenously?

## Minerals and Hypertension

Sam Edwards/Glow Images

The American College of Cardiology and American Heart Association revised blood pressure (BP) guidelines in 2017 (Figure 9-32). These guidelines are intended to serve as a resource for health care providers and the general public while emphasizing the importance of earlier detection and treatment of this potentially deadly condition. Hypertension is defined as sustained systolic pressure exceeding 130 mmHg or diastolic blood pressure exceeding 80 mmHg. Most cases of hypertension (about 95%) have no clear-cut cause. Such cases are classified as **primary or essential hypertension.** Kidney disease, sleep-disordered breathing (sleep apnea), and other causes often lead to the other 5% of cases, classified as **secondary hypertension.** African Americans and Asian Americans are more likely than Caucasians to develop hypertension and to do so earlier in life.

Unless blood pressure is periodically measured, the development of hypertension is easily overlooked. By lowering the cutoff for hypertension, the guidelines promote earlier detection and treatment to reduce the complications of hypertension. It is anticipated that the guidelines will result in more diagnoses of hypertension, especially in those under age 45.

## How High Is High?

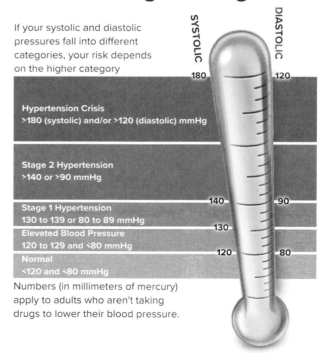

If your systolic and diastolic pressures fall into different categories, your risk depends on the higher category

**Hypertension Crisis**
>180 (systolic) and/or >120 (diastolic) mmHg

**Stage 2 Hypertension**
>140 or >90 mmHg

**Stage 1 Hypertension**
130 to 139 or 80 to 89 mmHg

**Elevated Blood Pressure**
120 to 129 and <80 mmHg

**Normal**
<120 and <80 mmHg

Numbers (in millimeters of mercury) apply to adults who aren't taking drugs to lower their blood pressure.

In the United States, 75 million, or an estimated one in three adults, has hypertension. Over the age of 65, the number rises to one in every two adults. Only about half of the cases are being treated. Blood pressure is expressed by two numbers. The higher number represents systolic blood pressure, the pressure in the arteries when the heart muscle is contracting and pumping blood into the arteries. Optimal systolic blood pressure is less than 120 mmHg. The second value is diastolic blood pressure, the artery pressure when the heart is relaxed. Optimal diastolic blood pressure is less than 80 mmHg. Elevations in both systolic and diastolic blood pressure are strong predictors of disease (Fig. 9-32).

**primary hypertension** Blood pressure of 130/80 mmHg or higher with no identified cause; also called *essential hypertension.*

**secondary hypertension** Blood pressure of 130/80 mmHg or higher as a result of disease (e.g., kidney dysfunction or sleep apnea) or drug use.

**FIGURE 9-32** ▲ Guidelines for the prevention, detection, evaluation, and management of high blood pressure in adults. Values listed for each category are for systolic and diastolic, respectively. Hypertension is now defined as a BP ≥ 130 (systolic) or 80 (diastolic) mmHg.

## Why Control Blood Pressure?

Because it usually does not cause symptoms, hypertension is described as a silent disorder. Blood pressure must be controlled mainly to prevent cardiovascular disease, kidney disease, strokes and related declines in brain function, poor blood circulation in the legs, problems with vision, and sudden death. These conditions are much more likely to be found in individuals with hypertension than in people with normal blood pressure. Smoking and elevated blood lipoproteins make these diseases even more likely. Individuals with hypertension need to be diagnosed and treated as soon as possible, as the condition generally progresses to a more serious stage over time and even resists therapy if it persists for years. The benefits of controlling blood pressure were confirmed by the results of a randomized study of patients at high risk for cardiovascular events.[30] Decreasing the systolic blood pressure of these patients to less than 120 mmHg, as compared with less than 140 mmHg, resulted in lower rates of fatal and nonfatal major cardiovascular events and death from any cause.

## Contributors to Hypertension

Because we do not know the cause of 95% of the cases of hypertension, we can identify only risk factors that contribute to its development. A *family history of hypertension* is a risk factor, especially if both parents have (or had) the condition. In addition, blood pressure can increase as a person ages. Some increase is caused by atherosclerosis. As plaque builds up in the arteries, the arteries become less flexible and cannot expand. When vessels remain rigid, blood pressure remains high. Eventually, the plaque begins to make the problem worse by decreasing the blood supply to the kidneys, which decreases their ability to control blood volume and, in turn, blood pressure.

*Excess body fat* greatly increases the risk of having hypertension. Overall, obesity is considered the number-one lifestyle factor related to hypertension. This is especially the case in minority populations. Additional blood vessels develop to support excess tissue in overweight and obese individuals, and these extra miles of associated blood vessels increase work by the heart and also blood pressure. Hypertension is also linked to obesity if elevated blood insulin levels result from insulin-resistant adipose cells. This increased insulin level augments sodium retention in the body and accelerates atherosclerosis. In such cases, a weight loss of as little as 10 to 15 pounds often can help treat hypertension.

*Inactivity* is considered the number-two lifestyle factor related to hypertension. If an obese person can engage in regular physical activity (at least 5 days per week for 30 to 60 minutes) and lose weight, blood pressure often returns to normal.

*Excess alcohol intake* is responsible for about 10% of all cases of hypertension, especially in middle-age males and among African Americans. When hypertension is caused by excessive alcohol intake, it is usually reversible. A sensible alcohol intake for people with hypertension is no more than two drinks per day for men and no more than one drink per day for women and all older adults.

In some people, particularly African Americans and seniors affected by overweight, blood pressure is especially sensitive to sodium. In these people, *excess salt* leads to fluid retention by the kidneys and a corresponding increase in blood volume, resulting in increased blood pressure. It is not clear whether sodium or chloride is more responsible for the effect. Because most of our sodium comes in the form of salt, if one reduces sodium intake, chloride intake naturally falls. For the most part, a recommendation to consume less sodium is equivalent to a call for less salt in the dietary pattern. Only some North Americans are susceptible to increases in blood pressure from salt intake, so it is only the number-four lifestyle factor related to hypertension. It is unfortunate that salt intake receives the major portion of public attention with regard to hypertension. Efforts to prevent hypertension should also focus on obesity, inactivity, and excessive consumption of alcohol.

## Other Minerals and Blood Pressure

Minerals such as calcium, potassium, and magnesium also deserve attention when it comes to prevention and treatment of hypertension. Studies show that altering a dietary pattern to include these minerals and be low in salt can decrease blood pressure within days, especially among African Americans. The response is similar to that seen with commonly used medications. The dietary pattern is called the Dietary Approaches to Stop Hypertension (DASH) diet (Table 9-6). The diet is rich in calcium, potassium, and magnesium and low in salt. Both magnesium and calcium are important for healthy blood pressure because they help blood vessels relax. It takes a standard MyPlate Daily Food Plan; adds one to two extra vegetables and fruits servings; and emphasizes consumption of nuts, seeds, or legumes (beans) 4 to 5 days of the week. In DASH studies, participants also consumed no more than 3 grams of sodium and no more than one to two alcoholic drinks per day. A DASH 2 dietary trial tested three daily sodium intakes (3300 milligrams, 2400 milligrams, and 1500 milligrams). Blood pressure steadily declined in people on the DASH diet as their sodium intake declined. Overall, the DASH diet is seen as a total dietary approach to treating hypertension, with the many

**TABLE 9-6** ■ **What Is the DASH Diet?**
**The DASH diet is low in fat and sodium and rich in fruits, vegetables, and low-fat dairy products. Here is the breakdown:**

| Per Day | Per Week |
|---|---|
| 6 to 8 servings of grains and grain products | — |
| 4 to 5 servings of fruit | — |
| 4 to 5 servings of vegetables | — |
| 2 to 3 servings of low-fat or fat-free dairy products | — |
| 2 or fewer servings of meats, poultry, and fish | 4 to 5 servings of nuts, seeds, or legumes |
| 2 to 3 servings of fats/oils | 5 or fewer servings of sweets and added sugars |

# Newsworthy Nutrition

## Low-potassium diets lead to arterial stiffness

**INTRODUCTION:** Studies have linked low dietary potassium intake with cardiovascular diseases, including hypertension and stroke; however, the underlying mechanisms are unknown. **OBJECTIVE:** The researchers hypothesized that potassium promotes blood flow by preventing the buildup of calcium in the smooth muscle cells of arteries. The objective of the study was to discover how dietary potassium affects arterial calcification. **METHODS:** In this animal study, three groups of atherosclerosis-prone mice were fed a high-fat, high-cholesterol diet that contained normal, low, or high levels of potassium for 30 weeks. Calcification was determined using tissue staining and ultrasound imaging methods. **RESULTS:** Mice fed a low-potassium diet had increased vascular calcification and artery stiffness, whereas a high-potassium diet reduced calcification and stiffness. Reduction in potassium also increased intracellular calcium, which activated proteins often associated with bone cells. These bone-associated proteins promoted vascular smooth muscle cell calcification, suggesting a direct effect of low potassium on the artery calcification. **CONCLUSIONS:** The mouse study found mechanisms that could help explain how low potassium intakes in the diet cause the calcification and stiffness of arteries that are linked to high blood pressure, heart disease, and stroke and lead to better therapies to control vascular disease.

Source: Sun Y and others: Dietary potassium regulates vascular calcification and arterial stiffness. *The Journal of Clinical Investigation Insight* 2(19):e94920, 2017. DOI:10.1172/jci.insight.94920.

healthful practices of the dietary pattern being responsible for the reduction in blood pressure. The DASH diet was created over 20 years ago, and it has stood the test of time. Numerous studies have shown that it consistently lowers blood pressure in diverse populations with elevated blood pressure or hypertension.[31] A panel of health experts has named the DASH diet one of the best overall eating plans, ranking it second after the Mediterranean Diet on the *U.S. News & World Report's* 2020 Best Diets list.

Studies have linked low dietary potassium intake with cardiovascular diseases, including hypertension and stroke. A recent mouse study found that low potassium intakes in the diet cause the calcification and stiffness of arteries that are linked to high blood pressure, heart disease, and stroke (see *Newsworthy Nutrition*). Other studies also show a reduction in stroke risk among people with a dietary pattern rich in fruits, vegetables, and vitamin C (recall that fruits and vegetables are rich in vitamin C). Overall, a dietary pattern low in salt and rich in low-fat and fat-free dairy products, fruits, vegetables, whole grains, and some nuts can substantially reduce hypertension and stroke risk in many people, especially those with hypertension.[32]

## Medications to Treat Hypertension

There are a variety of high blood pressure medications, otherwise known as antihypertensives. The different classes of these drugs are summarized in the *Medicine Cabinet* feature in this section. Diuretics, or "water pills," are one class that works to reduce blood volume (and therefore blood pressure) by increasing fluid output in the urine. Other medications act by slowing heart rate or by causing relaxation of the small muscles lining the blood vessels. A combination of two or more medications is commonly required to treat hypertension that does not respond to nutrition and lifestyle therapy.

# Medicine Cabinet

*Diuretics* are commonly prescribed to lower blood pressure. Diuretics cause the kidneys to excrete more urine but at the same time may increase urinary excretion of minerals and decrease blood levels of potassium, magnesium, and zinc. Those taking diuretics need to carefully monitor their dietary intake of these minerals, especially potassium, and increase intake of fruits and vegetables or potassium chloride supplements as prescribed by primary care providers.

Examples:

- Hydrochlorothiazide (Microzide®)
- Furosemide (Lasix®)

*Beta-blockers* decrease the heart's rate, as well as its workload and blood output.

Examples:

- Atenolol (Tenormin®)
- Metoprolol tartrate (Lopressor®)

*ACE (angiotensin-converting enzyme) inhibitors* cause the body to produce less angiotensin, which causes the blood vessels to relax and open up and thus lower blood pressure.

Examples:

- Benazepril hydrochloride (Lotensin®)
- Lisinopril (Prinivel®, Zestril®)

Source: American Heart Association. Prevention and treatment of high blood pressure, http://www.heart.org.

| Advice | Details | Drop in Systolic Blood Pressure |
|---|---|---|
| **Lose excess weight** | For every 20 pounds you lose (if BMI > 25) | 5 points |
| **Adopt a DASH eating plan** | Eat a lower-fat diet rich in vegetables, fruits, and low-fat dairy foods | 11 points |
| **Exercise daily** | Get 90–150 min/wk of aerobic activity (such as brisk walking) | 5 to 8 points |
| **Limit sodium** | < 1500 mg/d is optimal goal but at least 1000 mg/d reduction in most adults | 5 to 6 points |
| **Enhance intake of dietary potassium** | Consume 3500–5000 mg per day, preferably from a dietary pattern rich in potassium | 4 to 5 points |
| **Limit alcohol** | Have no more than 2 drinks per day for men, 1 drink per day for women<br>(1 drink = 12 oz beer, 5 oz wine, or 1.5 oz 80-proof whiskey) | 4 points |

**FIGURE 9-33** ▲ What works? If your blood pressure is high, here's how much lifestyle changes can lower it.

Source: *2017 Guideline for the Prevention, Detection, Evaluation, and Management of High Blood Pressure in Adults.* https://www.acc.org/~/media/Non-Clinical/Files-PDFs-Excel-MS-Word-etc/Guidelines/2017/Guidelines_Made_Simple_2017_HBP.pdf.

## Prevention of Hypertension

Many of the risk factors for hypertension and stroke are controllable, and appropriate lifestyle changes can reduce a person's risk (Fig. 9-33), depending on the severity of the hypertension. Experts recommend that those with hypertension lower blood pressure through diet and lifestyle changes before resorting to blood pressure medications.

With nearly half of U.S. adults diagnosed with hypertension, and only about one in four of them having their hypertension under control, *The Surgeon General's Call to Action to Control Hypertension (Call to Action)* was released in 2020 by the CDC.[33,34] The aim of the *Call to Action* is to avert the negative health effects of hypertension by identifying evidence-based interventions that can be implemented, adapted, and expanded in diverse settings across the United States. The *Call to Action* identifies three goals to improve hypertension control across the United States, and each goal is supported by strategies to achieve success:

**Goal 1.** Make hypertension control a national priority.

- Increase awareness of health risks
- Recognize economic burden
- Eliminate disparities

**Goal 2.** Ensure that the places where people live, learn, work, and play support hypertension control.

- Promote physical activity opportunities
- Promote healthy food opportunities
- Connect to lifestyle change resources

**Goal 3.** Optimize patient care for hypertension.

- Use standardized treatment approaches
- Promote team-based care
- Empower and equip patients
- Recognize and reward clinicians

**COVID CORNER**

Hypertension, along with diabetes and obesity, has been identified as one of the strongest risk factors for suffering a severe case of COVID-19 infection. In New York City, about 57 percent of those hospitalized with COVID-19 had hypertension. The fact that hypertension is more prevalent in African Americans, occurring in 40 percent of black adults compared with 28 percent of white and Hispanic adults, is one reason why African Americans have been hit harder by COVID-19.

**✓ CONCEPT CHECK 9.17**

1. What are the systolic and diastolic blood pressure values for Stage 1 and Stage 2 hypertension?
2. How does the DASH diet differ from the MyPlate recommendations?
3. What lifestyle changes can help lower high blood pressure?

## Summary (Numbers refer to numbered sections in the chapter.)

**9.1** Water constitutes 50% to 70% of the human body. Its unique chemical properties enable it to dissolve substances as well as serve as a medium for chemical reactions, temperature regulation, and lubrication. Water also helps regulate the acid–base balance in the body.

For adults, daily water needs are estimated at 9 cups (women) to 13 cups (men) per day; fluid intake contributes to meeting this need. Hormones participate in the process of fluid conservation. Receptors in the kidneys, blood vessels, and brain monitor blood pressure and solute concentration in the blood. Dehydration leads to kidney failure, coma, and death. The amount of water in the intracellular and extracellular compartments is controlled mainly by ion concentrations.

**9.2** Minerals are categorized based on the amount we need per day. If we require greater than 100 milligrams of a mineral per day, it is considered a major mineral; otherwise it is considered a trace mineral. Many minerals are vital for sustaining life. For humans, animal products are the most bioavailable sources of most minerals. Supplements of minerals exceeding 100% of the Daily Values should be taken only under a primary care provider's supervision. Toxicity and nutrient interactions are especially likely if the Upper Level (when set) is exceeded on a long-term basis.

**9.3** Sodium, the major positive ion found outside cells, is vital in fluid balance and nerve impulse transmission. The North American diet provides abundant sodium through processed foods and table salt. About 10% to 15% of the adult population, such as overweight people, is especially sodium-sensitive and at risk for developing hypertension from consuming excessive sodium.

**9.4** Potassium, the major positive ion found inside cells, has a similar function to sodium. Milk, fruits, and vegetables are good sources. Low blood potassium, hypokalemia, is a life-threatening problem that is most commonly seen with chronic diarrhea or vomiting, or as a side effect of medications, including laxatives and some diuretics.

**9.5** Chloride is the major negative ion found outside cells. It is important in digestion as part of stomach acid and in immune and nerve functions. Table salt supplies most of the chloride in our diets.

**9.6** Calcium forms a part of bone structure and plays a role in blood clotting, muscle contraction, nerve transmission, and cell metabolism. Calcium absorption is enhanced by stomach acid and the active vitamin D hormone. Dairy products are important calcium sources, but many dairy alternatives such as soy or almond milk are available. Deficient calcium intake decreases bone mineralization, ultimately leading to osteopenia and osteoporosis. Women are particularly at risk for developing osteoporosis as they age. Numerous lifestyle and medical options can help reduce this risk, including an adequate intake of calcium and many other minerals.

**9.7** Phosphorus aids enzyme function and forms part of key metabolic compounds, cell membranes, and bone. It is efficiently absorbed, and deficiencies are rare, although there is concern about possible poor intake by some older women. Good food sources are dairy products, bakery products, and meats.

**9.8** Magnesium is a mineral found mostly in plant food sources. It is important for nerve and heart function and as an activator for many enzymes. Whole grain breads and cereals (bran portion), vegetables, nuts, seeds, milk, and meats are good food sources.

**9.9** Iron absorption depends mainly on the form of iron present and the body's need for it. Heme iron from animal sources is better absorbed than the nonheme iron obtained primarily from plant sources. Consuming vitamin C or meat simultaneously with nonheme iron increases absorption. Iron operates mainly in synthesizing hemoglobin and myoglobin and in the action of the immune system. Women are at highest risk for developing iron deficiency, which decreases blood hemoglobin and hematocrit. When this condition is severe, iron-deficiency anemia develops. This decreases the amount of oxygen carried in the blood. Iron toxicity usually results from a genetic disorder called hemochromatosis. This disease causes overabsorption and accumulation of iron, which can result in severe liver and heart damage.

**9.10** Zinc aids in the action of up to 200 enzymes important for growth, development, cell membrane structure and function, immune function, antioxidant protection, wound healing, and taste. A zinc deficiency results in poor growth, loss of appetite, reduced sense of taste and smell, hair loss, and a persistent rash. Zinc is best absorbed from animal sources. The richest sources of zinc are oysters, shrimp, crab, and beef. Good plant sources are whole grains, peanuts, and beans.

**9.11** An important role of selenium is decreasing the action of free-radical (oxidizing) compounds. In this way, selenium acts along with vitamin E in providing antioxidant protection. Muscle pain, muscle wasting, and a form of heart damage may result from a selenium deficiency. Meats, eggs, fish, and shellfish are good animal sources of selenium. Good plant sources include grains and seeds.

**9.12** Iodine forms part of the thyroid hormones. A lack of dietary iodine results in the development of an enlarged thyroid gland or goiter. Iodized salt is a major food source.

**9.13** Copper is important for iron metabolism, cross-linking of connective tissue, and other functions, such as enzymes that provide antioxidant protection. A copper deficiency can result in a form of anemia. Copper is found mainly in liver, seafood, cocoa, legumes, and whole grains.

**9.14** Fluoride as part of regular dietary intake or toothpaste use makes teeth resistant to dental caries. Most North Americans receive the bulk of their fluoride from fluoridated water and oral hygiene products.

**9.15** Chromium aids in the action of the hormone insulin. Chromium deficiency results in impaired blood glucose control. Egg yolks, meats, and whole grains are good sources of chromium.

**9.16** Manganese and molybdenum are used by various enzymes. One enzyme that uses manganese provides antioxidant protection. Clear deficiencies in otherwise healthy people are rarely seen for these nutrients. Human needs for other trace minerals are so low that deficiencies are uncommon.

**9.17** Hypertension is defined as sustained systolic pressure exceeding 130 mmHg or diastolic blood pressure exceeding 80 mmHg. Controlling weight and alcohol intake; exercising regularly; decreasing salt intake; and ensuring adequate potassium, magnesium, and calcium in the diet all can play a part in controlling high blood pressure.

# Check Your Knowledge (Answers to the following questions are below.)

1. Dietary heme iron is derived from
   a. elemental iron in food.
   b. animal flesh.
   c. breakfast cereal.
   d. vegetables.

2. Chloride is
   a. a component of hydrochloric acid.
   b. an intracellular fluid ion.
   c. a positively charged ion.
   d. converted to chlorine in the intestinal tract.

3. Minerals involved in fluid balance are
   a. calcium and magnesium.
   b. copper and iron.
   c. calcium and phosphorus.
   d. sodium and potassium.

4. In a situation where there is an insufficient intake of dietary iodine, the thyroid-stimulating hormone promotes the enlargement of the thyroid gland. This condition is called
   a. Graves' disease.
   b. goiter.
   c. hyperparathyroidism.
   d. congenital hypothyroidism.

5. Ninety-nine percent of the calcium in the body is found in
   a. intracellular fluid.
   b. bones and teeth.
   c. nerve cells.
   d. the liver.

6. At the end of long bones, inside the spinal vertebrae, and inside the flat bones of the pelvis is a spongy type of bone known as _____ bone.
   a. cortical
   b. osteoporotic
   c. trabecular
   d. compact

7. Which compartment contains the greatest amount of body fluid?
   a. Intracellular
   b. Extracellular
   c. They contain the same amount.

8. The primary function of sodium is to maintain
   a. bone mineral content.
   b. hemoglobin concentration.
   c. immune function.
   d. fluid distribution.

9. Hypertension is defined as a blood pressure greater than
   a. 110/60.
   b. 120/65.
   c. 130/80.
   d. 190/80.

10. Which of the following individuals are most likely to develop osteoporosis?
    a. Premenopausal women athletes
    b. Women taking estrogen replacement therapy
    c. Slender, inactive women who smoke
    d. Women who eat a lot of high-fat dairy products

Answer Key: 1. b (LO 9.6), 2. a (LO 9.3), 3. d (LO 9.3), 4. b (LO 9.5), 5. b (LO 9.4), 6. c (LO 9.4), 7. a (LO 9.1), 8. d (LO 9.3), 9. c (LO 9.9), 10. c (LO 9.4)

# Study Questions (Numbers refer to Learning Outcomes.)

1. Approximately how much water do you need each day to stay healthy? Identify at least two situations that increase the need for water. Then list three sources of water in the average person's dietary pattern. **(LO 9.1)**

2. Identify four factors that influence the bioavailability of minerals from food. **(LO 9.2)**

3. What is the relationship between sodium and water balance, and how is that relationship monitored as well as maintained in the body? **(LO 9.3)**

4. List three sources of dietary calcium. Identify two factors that negatively influence the absorption of calcium. Identify two factors that positively influence the absorption of calcium. **(LO 9.6)**

5. Describe two methods that can be used to assess bone density. What demographic groups should have bone density measured? **(LO 9.4)**

6. What is the UL for calcium and what is the advice regarding the use of calcium supplements? **(LO 9.7)**

7. List three roles of magnesium in the body. Identify two chronic diseases that may be affected by magnesium status. **(LO 9.4)**

8. Describe the symptoms of iron-deficiency anemia, and explain possible reasons they occur. **(LO 9.8)**

9. What is the relationship between iodine, the thyroid gland, and energy metabolism? **(LO 9.5)**

10. Describe the functions of fluoride in the body. List three sources of fluoride. **(LO 9.4, 9.6)**

11. Explain the function of chromium in carbohydrate metabolism. **(LO 9.5)**

12. List three dietary strategies to lower blood pressure. **(LO 9.9)**

# Further Readings

1. Seaman G: Plastics by the numbers. *Eartheasy*. May 2, 2012. https://learn.eartheasy.com/articles/plastics-by-the-numbers.

2. Dow C: Is your seltzer habit harming your teeth? *Nutrition Action Newsletter,* p.10, June 2018.

3. Grimes CA and others: Dietary salt intake, sugar-sweetened beverage consumption, and obesity risk. *Pediatrics* 131:14, 2013. DOI:10.1542/peds.2012-1628.

4. Collins K: Sodium reduction—How low should clients go for optimal health? *Today's Dietitian* 20(2):18, 2018.

5. World Cancer Research Fund International. *Diet, Nutrition, Physical Activity and Cancer: a Global Perspective—The Third Expert Report*. London, UK: World Cancer Research Fund International, 2018. Available at https://www.wcrf.org/dietandcancer.

6. Wright NC and others: The recent prevalence of osteoporosis and low bone mass in the United States based on bone mineral density at the femoral neck or lumbar spine. *Journal of Bone Mineral Research* 29:2520, 2014. DOI:10.1002/jbmr.2269.

7. LeBlanc KE and others; Hip fracture: Diagnosis, treatment, and secondary prevention. *American Family Physician* 89:945, 2014.

8. National Osteoporosis Foundation: *2014 clinician's guide to prevention and treatment of osteoporosis*. Washington, DC: National Osteoporosis Foundation. www.nof.org. Accessed January 3, 2020.

9. Ishtiag S and others: Treatment of post-menopausal osteoporosis: Beyond bisphosphonates. *Journal of Endocrinological Investigation* 38:13, 2015. DOI:10.1007/s40618-014-0152-z.

10. Schaeffer J: What's new in the dairy-free aisle. *Today's Dietitian* 17(2):30, 2015.

11. Kelvin L and others: The good, the bad, and the ugly of calcium supplementation: A review of calcium intake on human health. *Clinical Interventions in Aging* 13:2443, 2018. DOI:10.2147/CIA.S157523.

12. Anderson JJ and others: Calcium intakes and femoral and lumbar bone density of elderly U.S. men and women: National Health and Nutrition Examination Survey 2005–2006 analysis. *Journal of Clinical Endocrinology and Metabolism* 97:4531, 2012. DOI:10.1210/jc.2012–1407.

13. Reid IR and others: Cardiovascular complications of calcium supplements. *Journal of Cellular Biochemistry* 116:494, 2015. DOI:10.1002/jcb.25028.

14. Prentice RL and others: Health risks and benefits from calcium and vitamin D supplementation: Women's Health Initiative clinical trial and cohort study. *Osteoporosis International* 24(2):567, 2013. DOI:10.1007/s00198-012-2224-2.

15. U.S. Preventive Services Task Force: Vitamin D, calcium, or combined supplementation for the primary prevention of fractures in community-dwelling adults: US Preventive Services Task Force Recommendation Statement. *Journal of the American Medical Association* 319:1592, 2018. DOI:10.1001/jama.2018.3185.

16. Office of Dietary Supplements, National Institutes of Health: Calcium fact sheet for health professionals. September 26, 2018. http://ods.od.nih.gov/factsheets/calcium; and Magnesium fact sheet for health professionals. September 26, 2018. http://ods.od.nih.gov/factsheets/magnesium. Accessed January 3, 2020.

17. Linus Pauling Institute Micronutrient Information Center: Phosphorus. June 2014. http://lpi.oregonstate.edu/mic/minerals/phosphorus. Accessed January 3, 2020.

18. Uwitonze AM and Razzaque MS: Role of magnesium in vitamin D activation and function. *The Journal of the American Osteopathic Association* 118:181, 2018. DOI:10.7556/jaoa.2018.037.

19. Office of Dietary Supplements, National Institutes of Health: Iron—Health professional fact sheet. Available at https://ods.od.nih.gov/factsheets/Iron-HealthProfessional. Accessed February 11, 2019.

20. Hurrell R and Egli I: Iron bioavailability and dietary reference values. *American Journal of Clinical Nutrition* 91:1461S, 2010. DOI:10.3945/ajcn.2010.28674F.

21. Skrypnik K and Suliburska J: Association between the gut microbiota and mineral metabolism. *Journal of the Science of Food and Agriculture* 98:2449, 2018. DOI:10.1002/jsfa.8724.

22. Zelman K: Micronutrients: Zinc. *Food and Nutrition Magazine* 8:14, 2019.

23. Zabetakis I and others: COVID-19: The inflammation link and the role of nutrition in potential mitigation. *Nutrients* 12, 1466, 2020. DOI:10.3390/nu12051466.

24. Science M and others: Zinc for the treatment of the common cold: A systematic review and meta-analysis of randomized controlled trials. *Canadian Medical Association Journal* 2012, 184, E551–E561.

25. McCarty MF and DiNicolantonio JJ: Nutraceuticals have potential for boosting the type 1 interferon response to RNA viruses including influenza and coronavirus. *Progress in Cardiovascular Diseases* 2020. Available online: https://www.ncbi.nlm.nih.gov/pubmed/32061635 (accessed on 1 April 2020).

26. Rayman MP and others: Association between regional selenium status and reported outcome of COVID-19 cases in China. *The American Journal of Clinical Nutrition*, 111(6):1297, 2020; DOI: 10.1093/ajcn/nqaa095.

27. Bath SC and others: Effect of inadequate iodine status in UK pregnant women on cognitive outcomes in their children: Results from the Avon Longitudinal Study of Parents and Children (ALSPAC). *Lancet* 382:331, 2013. DOI:10.1016/S0140-6736(13)60436-5.

28. Hynes KL and others: Mild iodine deficiency during pregnancy is associated with reduced educational outcomes in the offspring: 9-year follow-up of the Gestational Iodine Cohort. *Journal of Clinical Endocrinology Metabolism* 98:1954, 2013. DOI:10.1210/jc.2012-4249.

29. Rugg-Gunn AJ and Do L: Effectiveness of water fluoridation in caries prevention. *Community Dentistry and Oral Epidemiology* 240(Suppl. 2):55, 2012. DOI:10.1111/j.1600-0528.2012.00721.

30. SPRINT Research Group: A randomized trial of intensive versus standard blood-pressure control. *New England Journal of Medicine* 373:2103, 2015. DOI:10.1056/NEJMoa1511939.

31. Steinberg D and others: The DASH diet, 20 years later. *Journal of the American Medical Association* 317:1529, 2017. DOI:10.1001/jama.2017.1628.

32. Aburto NJ and others: Effect of lower sodium intake on health: Systematic review and meta-analyses. *British Medical Journal* 346:f1326, 2013. DOI:doi.org/10.1136/bmj.f1326.

33. Ritchey MD and others: Potential need for expanded pharmacologic treatment and lifestyle modification services under the 2017 ACC/AHA Hypertension Guideline. *The Journal of Clinical Hypertension* 20(10):1377, 2018. DOI:10.1111/jch.13364.

34. U.S. Department of Health and Human Services. *The Surgeon General's Call to Action to Control Hypertension*. Washington, DC: U.S. Department of Health and Human Services, Office of the Surgeon General; 2020. Available at www.surgeongeneral.gov.

# Chapter 10

# Nutrition: Fitness and Sports

Erik Isakson/Blend Images LLC

**Chapter 10 is designed to allow you to:**

**10.1** List positive health-related outcomes of a physically active lifestyle.

**10.2** List key elements of a sound fitness regimen.

**10.3** Describe the use of carbohydrates, fat, and protein to meet energy needs during various activities.

**10.4** Differentiate between anaerobic and aerobic uses of glucose and identify advantages and disadvantages of each.

**10.5** Explain how muscles and related organs adapt to an increase in physical activity.

**10.6** Describe how to estimate an athlete's nutrient needs and discuss the general principles for meeting overall nutrient requirements in the training diet.

**10.7** Examine problems associated with dehydration and outline the importance of maintaining optimal fluid status during physical activity.

**10.8** Understand how athletes can optimize performance by consuming foods and fluids before, during, and after exercise.

**10.9** List several ergogenic aids and describe their effects on an athlete's performance.

## FAKE NEWS

### The more protein you eat, the more muscle you will gain.

## THE FACTS

Although protein does impact muscle mass, the fact is that the vast majority of athletes and college students meet their daily protein. Protein beyond what is required equates to extra calories that most do not need. It is best to time your protein intake in relation to your workouts. For instance, consume ample protein before or immediately after a workout for muscle repair and recovery. Specific protein and nutrient recommendations are discussed in detail in this chapter and vary by age, type of activity, intensity, and other factors.

Source: Nestlé Nutrition.

Are you an individual looking to start a personal fitness program? Do you participate in recreational sports? Or are you a competitive athlete who would like to take your performance to the next level? One thing is certain: at every level of fitness, sensible nutrition and physical activity complement each other in the pursuit of wellness.

Some individuals fall into the trap of thinking that participation in physical activity alleviates the need to pay close attention to eating patterns. As you will learn in this chapter, adequate dietary patterns and hydration have a significant impact on physical performance.

Most athletes are on the lookout for any advantage that might enhance performance and give them the winning edge. For this reason, athletes are likely targets for nutrition quackery and misinformation. As you sort fact from fiction, be sure of this: long-term health promotion and maintenance should include adherence to the evidence-based dietary and physical activity guidelines.

Overall, making informed choices about foods, beverages, and dietary supplements can optimize many aspects of physical performance, from preventing fatigue to gaining muscle to recovering from workouts. In this chapter, you will discover how all physical activity benefits both physical and mental health and how nutrition relates to optimal physical performance. This information is important for the everyday, recreational athlete who engages in physical activity to manage weight or just to have fun. A sports dietitian can assist in guiding individuals in setting and achieving their own nutrition and physical activity goals.

USDA

**physical activity** Any movement of skeletal muscles that requires energy.

**exercise** Physical activities that are planned, repetitive, and intended to improve physical fitness.

**physical fitness** The ability to perform moderate to vigorous activity without undue fatigue.

**FIGURE 10-1** ▶ The benefits of physical activity for adults. Note that the down arrows represent risk reduction or decreases in factors, the up arrows represent gains or increases in factors. Rubberball/Getty Images
Source: *2018 Physical Activity Guidelines*

# 10.1 Introduction to Physical Fitness

Randi, Chandra, and Marlie used to chat over a cup of coffee after their 8:00 A.M. nutrition class, but now they meet at the student recreation center for a workout. Randi listens to nutrition podcasts as she power walks around the track. Chandra always takes the stairs to her dorm room on the fourth floor. Marlie reviews her course notes on the recumbent bike. What do these students have in common? They each have found inventive ways to incorporate physical activity into their busy college schedules.

Without a doubt, the benefits of **physical activity** outweigh the risks for most Americans. The potential benefits of regular physical activity include improved heart health, sleep patterns, and body composition. Physical activity also can reduce stress and improve clinical parameters such as blood pressure, blood cholesterol, blood glucose regulation, and immune function. In addition, physical activity aids in weight control, both by raising resting energy expenditure and by increasing overall energy expenditure. In fact, as physical fitness improves, so does the ability to mobilize fat as a source of energy. See Figure 10-1 for a closer look at these and other benefits of a physically active lifestyle.[1]

Note that physical activity and exercise are not synonymous. Physical activity refers to any movement of skeletal muscles that requires energy. It includes exercise, sports, as well as all the simple, unplanned activities of daily living, such as raking the yard, walking up and down the stairs, and carrying bags of groceries into the apartment. **Exercise** specifically refers to those physical activities that are planned, repetitive, and intended to improve **physical fitness.** Examples of exercise include scheduled walking, biking, swimming, team sports, and running.

The U.S. Department of Health and Human Services' (DHHS) *Physical Activity Guidelines for Americans* recommends that all individuals avoid inactivity.[2] Any physical activity is better than none, and people who participate in even small amounts of physical activity gain health benefits. The key physical activity recommendations include the following:

- For substantial health benefits, adults should do at least 150 to 300 minutes (2½ to 5 hours) per week of **moderate-intensity,** or 75 to 150 minutes (1¼ to 2½ hours) per week of **vigorous-intensity aerobic physical activity,** or an equivalent combination of moderate- and vigorous-intensity aerobic activity. Preferably, aerobic activity should be spread throughout the week.
- For additional and more extensive health benefits, adults should increase their physical activity beyond 300 minutes (5 hours) per week of moderate-intensity or 150 minutes per week of vigorous-intensity aerobic physical activity, or an equivalent combination of moderate- and vigorous-intensity activity.
- Adults should also include **muscle-strengthening activities** that involve all major muscle groups on 2 or more days a week.
- *Move more and sit less.*

Complying with the *Physical Activity Guidelines* is the first step toward managing weight, reducing risks for chronic diseases, and improving physical fitness. Unfortunately, many adults in the U.S. lead sedentary lifestyles. Only about 20% of American adults consistently achieve the level of physical activity set forth by the *Physical Activity Guidelines.*[2]

**moderate-intensity aerobic physical activity** Aerobic activity that increases a person's heart rate and breathing to some extent. Examples include brisk walking, dancing, swimming, or bicycling.

**vigorous-intensity aerobic physical activity** Aerobic activity that greatly increases a person's heart rate and breathing. Examples include jogging, tennis, swimming or bicycling uphill.

**muscle-strengthening activity** Physical activity that increases skeletal muscle strength, power, endurance, and mass. Examples include lifting free weights, using weight machines, and calisthenics (e.g., push-ups).

### ✓ CONCEPT CHECK 10.1

1. Differentiate between physical activity and exercise.
2. How much physical activity is recommended by the *Physical Activity Guidelines for Americans* to reduce risks for chronic diseases?
3. List five benefits of regular physical activity.

## 10.2 Achieving and Maintaining Physical Fitness

Many of the recommendations that appear later in this chapter focus on enhancing the performance of highly competitive athletes; however, not many students are elite athletes. Furthermore, in a health profession, most clients are at a beginner or intermediate level of fitness. Certainly, proper nutrition supports physical performance at all levels. This section outlines how to get started with a plan to achieve physical fitness.

### ASSESS YOUR CURRENT LEVEL OF FITNESS

Start by assessing your current level of fitness. In some cases, it is beneficial to seek medical advice before getting started. Men age 40 years or older and women age 50 years or older, anyone who has been inactive for many years, or those who have an existing health problem should discuss their fitness goals with their primary care provider before altering physical activity patterns.[4] Health concerns that may require medical evaluation before beginning a fitness program include obesity, cardiovascular disease (or family history), hypertension, diabetes (or family history), chest pains, shortness of breath, history of dizzy spells, respiratory ailments, osteoporosis, hypertension, or

*magnificent*

Ongoing research confirms that exercise is associated with beneficial changes in gut microbial composition and metabolites independent of diet in both rodents and humans. Intense activities have also been linked to regulating oxidative stress and inflammatory responses in addition to improvements in metabolism and energy expenditure.[3] Another reason the microbiome is truly magnificent!

*microbiome*

arthritis. If you are unsure, you can take the Physical Activity Readiness Questionnaire (PAR-Q), found at www.health.harvard.edu/PAR-Q.[4] This tool can help you decide if you need to see your primary care provider first. Even if you do not have preexisting medical problems, enlisting the aid of a sports dietitian can help you to determine a safe starting point and establish realistic fitness goals.

## ASSESS YOUR LEVEL OF PHYSICAL ACTIVITY

There are four levels of aerobic physical activity:[2]

1. *Inactive* means that you are not getting moderate- to vigorous-intensity activities beyond daily life activities.
2. *Insufficiently active* is engaging in some moderate- or vigorous-intensity physical activity but less than 150 minutes of moderate-intensity physical activity per week or 75 minutes of vigorous-intensity physical activity (or the equivalent in combination). This level is less than the target range for meeting the key physical guidelines for adults.
3. *Active* physical activity is doing the equivalent of 150 minutes to 300 minutes of moderate-intensity physical activity per week. This level meets the key guideline target range for adults.
4. *Highly active* means you are meeting the equivalent of more than 300 minutes of moderate-intensity physical activity per week. This level exceeds the key guideline target range for adults.

## SET YOUR PHYSICAL ACTIVITY GOALS

For any behavior change, goal setting will enhance your success. Sports dietitians are trained professionals who can help you set reasonable physical activity goals based on your current level of fitness. A Board Certification as a Specialist in Sports Dietetics (CSSD) credential is the premier professional sports nutrition credential in the U.S. CSSDs are registered dietitian nutritionists (RDNs) who provide safe, effective, evidence-based nutrition services for health, fitness, and athletic performance.

Wearable fitness devices to measure physical movement are now more accurate and less expensive than ever. These can help motivate you on your quest to become more physically active. Devices include pedometers that count steps and accelerometers that measure trunk or limb movement. Smartphone app accelerometers, such as wrist watches, have become the norm. Many of these devices use multi-sensor systems that measure steps, often paired with global positioning systems (GPS) that provide estimates of speed and distance. Many now include heart rate monitors and track intensity of movements. We should all strive for a lifespan approach to physical activity. This means we should try to be active for life. These devices may help you reinforce positive behaviors for some people.[2]

## PLAN YOUR PROGRAM

Although your goal may focus on just one aspect of fitness, such as being able to bench-press your body weight, a balanced fitness program will include various types of activities, such as aerobics, muscle- and bone-strengthening, balance training, flexibility, and more.

Many physical activity experts use the *FITT principle* to design a fitness program. FITT stands for *frequency, intensity, time,* and *type* of activity. Frequency typically refers to the number of sessions or bouts of moderate to vigorous physical activity per day or per week. Relative intensity refers to the ease or difficulty with which an individual performs physical activity. Type (also called *mode*) is your choice of activity, such as walking or running. Table 10-1 summarizes the American College of Sports Medicine (ACSM) recommendations for planning a general fitness program.

**TABLE 10-1** ■ **Elements of a Well-Rounded Fitness Program**

| | Aerobic Fitness | Muscular Fitness | Flexibility |
|---|---|---|---|
| **Frequency** | 3 to 5 days per week | 2 to 3 days per week | 2 to 3 days per week |
| **Intensity** | 60% to 85% of MHR[a] or RPE[b] of 4 or higher (Fig. 10-3) | 70% to 80% of 1 RM[c] (lower for endurance and higher for strength) | To point of tension |
| **Time** | 150 minutes per week | 2 to 3 sets of 8 to 12 repetitions or 1 set of 8 to 12 repetitions to fatigue | 2 to 4 repetitions of 8 to 12 different exercises, held for 10 to 30 seconds each |
| **Type** | Brisk walking, running, cycling, swimming, basketball, tennis, and soccer | Bench press, squat, biceps curl, and abdominal crunch | Hamstring stretch, shoulder reach, and side bend |

[a]Maximal heart rate; [b]Rating of perceived exertion; [c]1-repetition maximum
Source: American College of Sports Medicine.

**Cardiorespiratory Endurance.** As you learned earlier, *aerobic* means "with oxygen." Aerobic activity is intense enough and long enough to maintain or improve your cardiorespiratory fitness.[2] There are three components of aerobic activity: (1) *Intensity* or how hard a person works to complete the activity (usually described as moderate or vigorous); (2) *frequency* or how often a person engages in aerobic activity; and (3) *duration* or how long a person engages in the activity for one session. The ability to perform aerobic activity depends on the health of your heart and lungs—the organ systems that provide oxygen to the cells of the body. Aerobic activities usually form the backbone of a fitness program. Indeed, many of the benefits of fitness are direct effects of aerobic training.

Remember that heart rate is just an estimate of workout intensity. Medications, such as those for hypertension and other health conditions, may impact heart rate. If you

**Measuring Your Heart Rate**

Most wearable fitness trackers now contain automatic heart rate monitors. If you do not own one, you can manually check your heart rate by following the following steps:

1. Stop and count your heartbeats for 10 seconds.
2. Multiply that number by 6 to determine your heartbeats per minute.

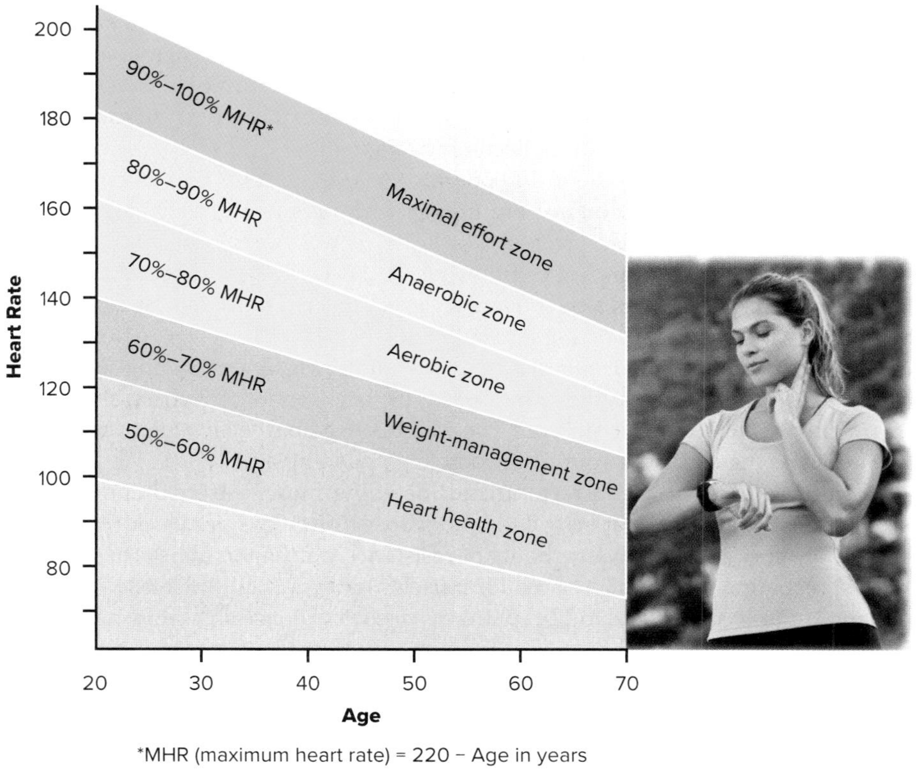

**Calculating Your Target Heart Rate**

1. Determine your age-predicted maximum heart rate (MHR, Fig. 10-2) but subtracting your age from 220:

For a 20-year-old person, MHR equals 200 beats per minute (220 − 20 (age) = 200 beats per minute (bpm).

2. At the initiation of an aerobic exercise program, aim for about 50% to 70% of MHR as a target heart rate (THR):

200 bpm × 0.5 (50%) = 100 bmp

200 bpm MHR × 0.7 (70%) = 140 bpm

As you progress and become more physically fit, you can work up to a higher THR. For intermediate exercisers, 60% to 75% of MHR is recommended. For trained exercisers, 70% to 85% MHR is suitable.

*MHR (maximum heart rate) = 220 − Age in years

**FIGURE 10-2** ▲ Heart rate training chart. This chart shows the number of heartbeats per minute that corresponds to various exercise intensities. PeopleImages/Getty Images

**FIGURE 10-3** ▶ Rating of Perceived Exertion Scale. When engaging in physical activity, a rating of a 5 or 6 produces noticeable increases in breathing rate and heart rate. A level of 7 or 8 produces large increases in a person's breathing and heart rate.[2]

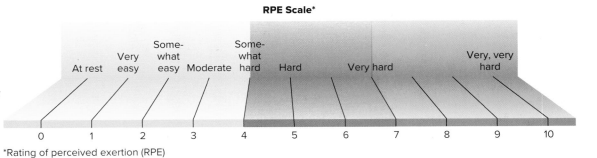

**RPE Scale***

At rest  Very easy  Some-what easy  Moderate  Some-what hard  Hard  Very hard  Very, very hard

0   1   2   3   4   5   6   7   8   9   10

*Rating of perceived exertion (RPE)

Try the *talk test* to determine your level of physical activity.

Generally, if you are engaging in moderate-intensity aerobic activity, you should be able to talk, but not sing, during the activity.

If you are engaging in vigorous-intensity activity, you probably cannot say more than a few words without pausing for a breath.[2]

have health concerns, a primary care provider can help to personalize your safe target heart rate zone.

Another way of determining the intensity of activities is the *Rating of Perceived Exertion (RPE) Scale*. The Borg system includes a range of 0 to 10, with each number corresponding to a subjective feeling of exertion. For example, the number 0 is "nothing at all" (e.g., sitting), and the number 10 is considered close to maximal effort or "very, very heavy" (e.g., all-out sprint; Fig. 10-3).

When using the 10-point RPE scale, the goal is to aim for a minimum rating of 4, which corresponds to the beginning of "somewhat hard." This is the point at which you begin to obtain significant benefits. If you are hovering around an RPE of 4, you should still be able to talk somewhat comfortably to a workout partner (*talk test*).

**Musculoskeletal Fitness.** Muscles are strengthened by the principles of overload, adaptation, progression, and specificity. *Overload* indicates that a resistance is applied on a regular basis and manageable to handle. The muscles *adapt* to this new load and become stronger. Over time, *progression* is possible as the body adapts. These improvements in strength are *specific* to the overloaded muscles.[2] Muscle-strengthening activities include three components: (1) *intensity* or how much force or weight is applied relative to how much you can lift; (2) *frequency* or how often you engage in muscle-strengthening activities; and (3) *sets and repetitions* or how many times you complete the muscle-strengthening activity. *Muscular endurance* refers to the ability of the muscle to perform repeated, submaximal contractions over time without becoming fatigued. An athlete training for muscular endurance may bench-press 80 to 100 pounds for several sets of 8 to 12 repetitions. Both muscular strength and endurance are important aspects of muscular fitness that relate to health for athletes of all levels. *Muscular power* combines strength with speed for explosive movements such as jumping or throwing. Power is a crucial aspect of muscular fitness for many athletes. Studies also show that developing muscular power can help to improve function and balance among older adults.

Overall, muscular fitness is developed by performing resistance training for all the major muscle groups of the body, including legs, hips, back, abdomen, chest, shoulders, and arms. This resistance may come from free weights (e.g., barbells), weight machines (e.g., leg press), or the weight of your own body (e.g., push-ups).

The *Physical Activity Guidelines* recommend including muscle-strengthening activities in your fitness program on at least 2 to 3 nonconsecutive days per week (Fig. 10-4). Taking a day or more to rest between bouts of resistance training or alternating groups of muscles is recommended to allow time for muscles to recover and increase in size.

To enhance muscle strength, 8 to 12 repetitions of each exercise should be performed to volitional fatigue. One set of 8 to 12 repetitions is effective at increasing muscular strength; limited evidence suggests that 2 or 3 sets are more effective.[2]

**Bone-Strengthening Activities.** Also referred to as *weight bearing* or *weight loading*, these activities produce force on the bones to promote bone growth and strength.[2] Jumping, running, brisk walking, and weight lifting are examples of bone-strengthening activities.

**Balance Activities.** These activities improve the ability to resist forces that cause falls while a person is stable or moving.[2]

**Multicomponent Physical Activities.** The concept of physical fitness has been operationalized as a multicomponent construct that includes cardiorespiratory endurance (aerobic power), musculoskeletal fitness, flexibility, balance, and speed of movement.[2] Dancing, tai chi, gardening, and sports are considered multicomponent physical activities.

**Flexibility Activities.** *Flexibility* is an often overlooked aspect of physical health and tends to decline with age. Flexibility, often called stretching, refers to the ability to move a joint through its full range of motion. Poor flexibility is often linked to chronic pain, especially in the lower back. Contrary to popular belief, research studies do not clearly support a role of flexibility training in preventing injury or muscle soreness from aerobic or strength-training activities. However, gains in flexibility can improve balance and stability, thereby reducing risks of falls and injuries.

**Warm-Up and Cool-Down.** Be sure to plan for adequate warm-up and cool-down periods as part of your physical activity routine. Begin by warming up with low-intensity exercises, such as walking, slow jogging, or any low-intensity performance of the anticipated activity. This warms up your muscles so that muscle filaments slide over one another more easily, which increases range of motion and flexibility and decreases the risk of injury. It is also thought to lower cardiovascular risks, particularly among people who are not accustomed to regular exercise. During cool-down, slow down to low-intensity activity followed by stretching. The same physical activities performed during warm-up are appropriate. Although the cool-down does not actually prevent muscle soreness, it does reduce the dizziness or light-headedness that can occur with an abrupt end to a vigorous workout.

**Get Started.** For sedentary people who are otherwise healthy, gradual **progression** toward meeting the physical activity guidelines is recommended. During the first phase of a fitness program to promote health, you should begin to incorporate short periods of physical activity into your daily routine. This includes brisk walking, taking the stairs instead of the elevator, dancing, hiking, and other activities. A sensible goal is at least 30 to 60 minutes of this moderate type of physical activity on most (and preferably all) days of the week. If there is not much time for activity, you can still obtain significant benefits from performing shorter sessions of increased intensity. Remember that any activity is better than no activity. Aim to move more and sit less!

**Stick with It!** Even after all the effort of assessing baseline fitness, setting goals, designing a program, and getting started, it turns out that maintaining a physical activity program may be the hardest part! To stick with a fitness program, experts recommend the following:

- Forget occasional setbacks; focus on the long-term benefits to your health.
- Frequently revisit both short- and long-term goals.
- Include friends and family members for additional motivation.
- Reward yourself when achieving goals.
- Set aside a specific time each day for movement; set physical activity calendar appointments.
- Start slowly and progress incrementally.
- Vary your activities to keep it fresh; rotate indoor and outdoor activities.

**Josie's Fitness Plan**

**Monday**
Spinning class (45 min)
Stretching routine (20 min)

**Tuesday**
Walking (30 min)
Weight machines (30 min)

**Wednesday**
Spinning class (45 min)
Stretching routine (20 min)

**Thursday**
Walking (30 min)
Weight machines (30 min)

**Friday**
Yoga (30 min)

**Saturday**
Trail hiking (60 min)

**Sunday**
Rest

**FIGURE 10-4** ▲ Sample fitness plan. Josie's fitness plan incorporates aerobic, strength, and flexibility exercises. John Lund/Sam Diephuis/Blend Images LLC

**progression** Incremental increase in frequency, intensity, and time spent in each type of physical activity over several weeks or months.

Good social support is a key to maintaining behavioral change. **Which friends or family members would help you achieve your fitness goals?** Ariel Skelley/Getty Images

**✓ CONCEPT CHECK 10.2**

1. List the four levels of aerobic physical activity.

2. Why is goal setting important for the success of a physical activity program?

3. Differentiate between muscular strength, muscular endurance, and muscular power.

4. Provide three tips for a person who needs help maintaining a fitness program.

# 10.3 Energy Sources for Active Muscles

Like other cells, muscle cells cannot directly use the energy released from breaking down **glucose** or triglycerides. Muscle cells need a specific form of energy for contraction. Body cells must first convert the chemical energy in carbohydrates, proteins, and fats into **adenosine triphosphate (ATP).**

The chemical bonds between phosphates in ATP and related molecules are high-energy bonds. Using the energy obtained from food, cells make ATP from its breakdown product **adenosine diphosphate (ADP)** and a phosphate group (abbreviated $P_i$). Conversely, to release energy from ATP, cells partially break down the compound into ADP and $P_i$ (Fig. 10-5). The released energy is used for many cell functions.

**adenosine diphosphate (ADP)** A breakdown product of ATP. ADP is synthesized into ATP using energy from foodstuffs and a phosphate group; abbreviated *Pi*.

## ANAEROBIC METABOLISM SUPPLIES ENERGY FOR SHORT BURSTS OF INTENSE ACTIVITY

**Stored ATP.** Essentially, ATP is the immediate source of energy for body functions. The primary goal in the use of any fuel, whether carbohydrate, fat, or protein, is to make ATP. A resting muscle cell contains only a small amount of ATP that can be used immediately. This amount of ATP could keep the muscle working maximally for only about 2 seconds if no resupply of ATP were possible. Fortunately, the cells have various mechanisms to resupply ATP. Overall, cells must constantly and repeatedly use and then reform ATP, using a variety of energy sources.

**Phosphocreatine.** As soon as ATP stored in muscle cells begins to be used, another high-energy compound, **phosphocreatine (PCr),** is used to resupply ATP. An enzyme in the muscle cell is activated to split PCr into phosphate and **creatine** (Fig. 10-6). This releases energy that can be used to reform ATP from its breakdown products. If no other source of energy for ATP resupply were available, PCr could probably maintain maximal muscle contractions for about 15 seconds.

**phosphocreatine (PCr)** A high-energy compound that can be used to reform ATP. It is used primarily during bursts of activity, such as lifting and jumping.

**creatine** An organic (i.e., carbon-containing) molecule in muscle cells that serves as part of a high-energy compound (termed *creatine phosphate* or *phosphocreatine*) capable of synthesizing ATP from ADP.

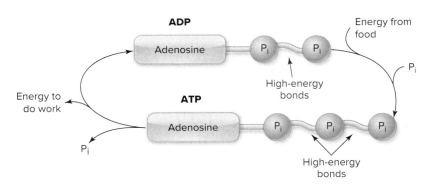

**FIGURE 10-5** ▲ Food energy is stored in the chemical bonds between the phosphate groups in ATP. When a phosphate group is cleaved from ATP, energy to perform work is released. The product of the breakdown of ATP is adenosine diphosphate (ADP).

The main advantage of PCr is that it can be activated instantly and can replenish ATP at rates fast enough to meet the energy demands of the fastest and most powerful actions, including jumping, lifting, throwing, and sprinting. The disadvantage of PCr is that not much of it is made and stored in the muscles. Strength-training athletes sometimes use creatine supplements in an effort to increase PCr in muscles.

**Anaerobic Glucose Breakdown.** Carbohydrates are an important fuel for muscles. The most useful form of carbohydrate fuel is the simple sugar *glucose*, available to all cells from the bloodstream. As you will recall, glucose is stored as **glycogen** in the liver and muscle cells. Blood glucose is maintained by the breakdown of liver glycogen. Breakdown of glycogen stored in a specific muscle also helps meet the carbohydrate demand of that muscle, but the actual amount of glycogen stored in muscle is limited (about 350 grams for all the muscles in the body yielding approximately 1400 kcal).

Whether oxygen is available or not, glucose can be broken down to a three-carbon compound called **pyruvate** (sometimes called *pyruvic acid*) in a pathway that releases some energy. However, when the oxygen supply in the muscle is limited (**anaerobic** conditions), glucose cannot be fully broken down. The pyruvate accumulates in the muscle and is then converted to **lactate** (also called *lactic acid*). Only about 5% of the total amount of ATP that could be formed from complete breakdown of glucose is released through this anaerobic process (Fig. 10-7).

The advantage of anaerobic glucose breakdown is that it is the fastest way to resupply ATP, other than PCr breakdown. It therefore provides most of the energy needed for events that require a quick burst of energy, ranging from about 30 seconds to 2 minutes. Examples of activities that primarily rely on anaerobic glucose breakdown include sprinting 400 meters or swimming 100 meters.

The two major disadvantages of the anaerobic process are: (1) the high rate of ATP production cannot be sustained for long periods; and (2) the rapid accumulation of lactate increases the acidity of the muscle. Normally, the pH of the muscle is about 7.1. Intense activity that relies on anaerobic glucose breakdown can decrease the pH of muscle tissue to about 6.5. Acidity inhibits the activities of key enzymes in the muscle cells, slowing anaerobic ATP production and causing short-term fatigue. In addition, the acidity leads to a net potassium loss from muscle cells, which also contributes to fatigue. During physical activity, you need to choose a sustainable pace according to the goals of the activity.

Before long, muscle cells release the accumulating lactate into the bloodstream. The liver (and the kidneys to some extent) takes up the lactate and resynthesizes it into glucose. Glucose can then reenter the bloodstream, where it is available for cell uptake and breakdown. Individuals vary in their ability to clear lactate from the muscles and recycle it. Physical training may improve the ability of the body to remove and recycle lactate.

## AEROBIC METABOLISM FUELS PROLONGED, LOWER-INTENSITY ACTIVITY

**Carbohydrates.** If plenty of oxygen is available in the muscle (aerobic conditions), such as when the physical activity is of low to moderate intensity, the bulk of the pyruvate is shuttled to the mitochondria of the cell, where it is fully metabolized into carbon dioxide ($CO_2$) and water ($H_2O$) (Fig. 10-8). This aerobic breakdown of glucose yields approximately 95% of the ATP made from complete glucose metabolism (glucose → $CO_2$ + $H_2O$).

Aerobic glucose breakdown supplies more ATP than the anaerobic process, but it releases the energy more slowly. This slower rate of aerobic energy supply can be sustained for hours. One reason is that the products are carbon dioxide and water, not lactate. Aerobic glucose breakdown makes a major energy contribution to activities that last anywhere from 2 minutes to several hours. Examples of such activities include jogging or distance swimming (Table 10-2).

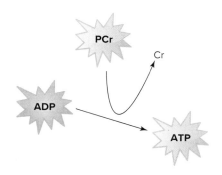

**FIGURE 10-6** ▲ Phosphocreatine (PCr) reacts with ADP to yield ATP plus creatine (Cr).

**pyruvate** A three-carbon compound formed during glucose metabolism; also called *pyruvic acid*.

**lactate** A three-carbon acid formed during anaerobic cell metabolism; a partial breakdown product of glucose; also called *lactic acid*.

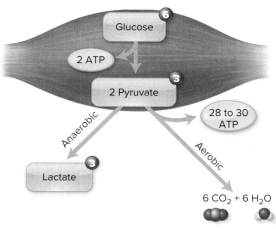

**FIGURE 10-7** ▲ ATP yield from aerobic versus anaerobic glucose use. Encircled numbers indicate the number of carbons in each molecule.

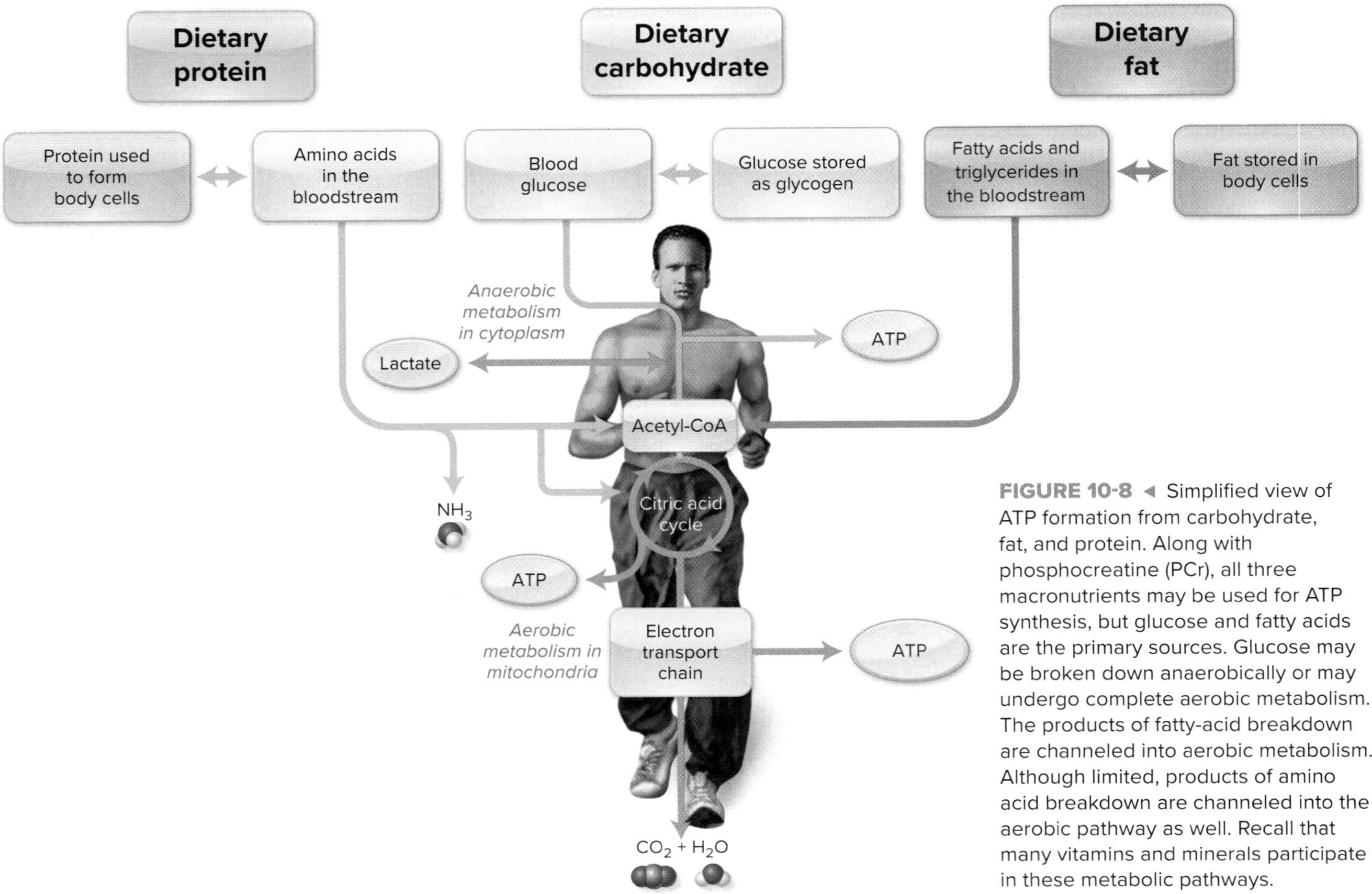

**FIGURE 10-8** ◄ Simplified view of ATP formation from carbohydrate, fat, and protein. Along with phosphocreatine (PCr), all three macronutrients may be used for ATP synthesis, but glucose and fatty acids are the primary sources. Glucose may be broken down anaerobically or may undergo complete aerobic metabolism. The products of fatty-acid breakdown are channeled into aerobic metabolism. Although limited, products of amino acid breakdown are channeled into the aerobic pathway as well. Recall that many vitamins and minerals participate in these metabolic pathways.

**TABLE 10-2** ■ **Energy Sources Used by Resting and Working Muscle Cells**

| Energy Source* | When in Use | Activity |
|---|---|---|
| ATP | At all times | All types |
| Phosphocreatine (PCr) | All physical activity initially; short bursts up to 10 seconds | Throws, jumps, or 100- to 400-m sprints |
| Carbohydrate (anaerobic) | High-intensity activity, especially lasting 30 seconds to 2 minutes | 200-yard (about 200-meter) sprint |
| Carbohydrate (aerobic) | Physical activity lasting 2 minutes to several hours; the higher the intensity, the greater the use | Basketball, swimming, jogging, power walking |
| Fat (aerobic) | Activities lasting more than a few minutes; greater amounts are used at lower intensities | Long-distance running or cycling; much of the fuel used in a 30-minute brisk walk |
| Protein (aerobic) | Low amount used during all activities; slightly more in endurance exercise, especially when carbohydrate fuel is depleted | Long-distance running |

* At any given time, more than one source is used. The relative amount of use differs during various activities.

Endurance athletes sometimes reach a point in an event at which extreme physical and mental fatigue sets in; it feels impossible to stand up, let alone continue competing. Long-distance runners call this phenomenon *hitting the wall*, and cyclists sometimes refer to it as *bonking*. This occurs because muscle glycogen has been depleted and blood glucose has begun to decline during exercise, leading to deterioration of both physical and mental function. As discussed in more detail later, maximizing glycogen storage before physical activity, supplying carbohydrates during activity, and replenishing glycogen stores between events can help athletes avoid the game-stopping effects of glycogen depletion.

**Fat.** When fat stores in body tissues begin to be broken down for energy, each triglyceride first yields three fatty acids (*tri-*) and a glycerol molecule (*glyceride*). The majority of the stored energy is found in the fatty acids. During physical activity, fatty acids are released from various adipose tissue depots into the bloodstream and travel to the muscles, where they are taken into each cell and broken down aerobically to carbon dioxide and water (Fig. 10-9). Some of the fat stored in muscles (intramuscular triglycerides) also is used, especially as activity increases from a low to a moderate pace.

Fat is an advantageous fuel for muscles: we generally have plenty of it stored, and it is a concentrated source of energy. For a given weight of fuel, fat supplies more than twice as much energy as carbohydrate does. However, the ability of muscles to use fat for fuel depends on the intensity of the activity. During intense, brief exercise, muscles may not be able to use much fat. The reason for this is that some of the steps involved in fat breakdown cannot occur fast enough to meet the ATP demands of short-duration, high-intensity exercise. However, fat becomes a progressively more important energy source as duration increases, especially when physical activity remains at a low or moderate (aerobic) rate for more than 20 minutes (Fig. 10-10).

For lengthy activities at a moderate pace (e.g., hiking) or even sitting at a desk for 8 hours a day, fat supplies about 70% to 90% of the energy required. Carbohydrate use is much less. As intensity increases, carbohydrate use goes up and fat use decreases.

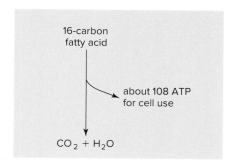

**FIGURE 10-9** ▲ ATP yield from aerobic fatty acid utilization.

▲ Fatty acids are recruited from all over the body, not necessarily from fat stored near the active muscles. *This is why spot reducing does not work.* Physical activity can tone the muscles near adipose tissue but does not preferentially use those stores. Comstock/Getty Images

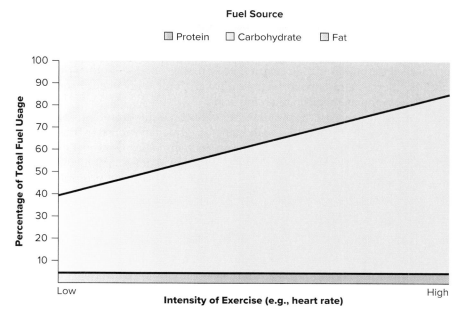

**FIGURE 10-10** ▲ Generalized relationship between fuel use and exercise intensity. At rest or during light activity, nearly equal amounts of carbohydrate and fat are used to generate ATP. As activity intensifies (e.g., sprinting), anaerobic processes supply quick fuel, so the relative proportion of carbohydrates used for fuel increases. Except during endurance activities, very little protein is used for fuel.

# Newsworthy Nutrition

### Branched-chain amino acids impact health and lifespan in mice

Amino acids have long been advocated by the sports communities for their muscle building benefits. Branched-chain amino acids (BCAAs) are a group of three essential amino acids: leucine, isoleucine, and valine, most commonly found in red meat and dairy. **INTRODUCTION:** Elevated BCAAs are associated with obesity and insulin resistance. While delivering muscle-building benefits, excessive consumption of BCAAs may reduce lifespan, negatively impact mood, and lead to weight gain. **OBJECTIVE:** The aim of this *animal study* was to evaluate the long-term effects of BCAA exposure on appetite, weight, mood, and lifespan. **METHODS:** Mice (312 male and female C57BL/6J) were fed dietary BCAAs in amounts of 200%, 50%, or 20% of the typical intakes throughout their life. **RESULTS:** Mice who were fed the highest amount of BCAAs increased their blood BCAA levels, which reduced serotonin levels in the brain. These levels of BCAA resulted in obesity and a shortened lifespan. **CONCLUSION:** The results of this study show that diets high in protein and low in carbohydrates had detrimental effects for health of mice in mid to late life, and also led to a shortened lifespan. Based on this information, athletes should vary protein sources in order to get a variety of essential amino acids through a healthy and balanced diet rich in fiber, vitamins, and minerals.

Source: Solon-Biet SM and others: Branched chain amino acids impact health and lifespan indirectly via amino acid balance and appetite control. *Nature Metabolism* 1(5):532, 2019. DOI: 10.1038/s42255-019-0059-2.

During a 5-mile run at a moderate pace, muscles use about a 50:50 ratio of fat to carbohydrate. In comparison, for a sprint, the contribution of fat to resupply ATP is minimal. To summarize, remember that the only fast-paced (anaerobic) fuel we eat is carbohydrate; slow and steady (aerobic) activity uses fat in addition to carbohydrate.

**Protein.** Although amino acids derived from protein can be used to fuel muscles, their contribution is relatively small, compared with that of carbohydrate and fat. Most protein is reserved for building and repairing body tissues and for synthesizing important enzymes, hormones, and transporters. As a rough guide, only about 5% of the body's ATP comes from the metabolism of amino acids.

During endurance activities, proteins can contribute importantly to energy needs, perhaps as much as 10% to 15%, especially as glycogen stores in the muscle are exhausted. Most of the energy supplied from protein comes from metabolism of the **branched-chain amino acids**: leucine, isoleucine, and valine. A healthy dietary pattern provides ample branched-chain amino acids to supply this amount of fuel; protein or amino acid supplements are rarely needed (see *Newsworthy Nutrition*).

## PHYSICAL TRAINING AFFECTS FUEL USE

As people start exercising regularly, they experience a *training effect*. Initially, these individuals might be able to engage in physical activity for 20 minutes before tiring. Months later, activities can be extended to an hour before they become fatigued. The training effect results from changes in the ability of exercising cells to use food fuel to generate ATP.

Almost immediately after a person begins a physical activity program, both aerobic and strength training improve the insulin sensitivity of cells. In other words, more glucose can be transported from the bloodstream into the cells, where it can be broken down either anaerobically or aerobically. Improved blood glucose management is an added benefit for preventing or treating metabolic syndrome or type 2 diabetes.

Marty started going to the gym about 8 weeks ago. At first, he noticed that he began huffing and puffing about 7 minutes into his basketball game. Now, however, he can play for about 25 minutes without tiring. **What is a possible explanation for this ability to work out longer?** Realistic Reflections

Endurance aerobic activities also increase the ability of muscles to store glycogen. This highly branched polymer of glucose can be broken down into individual glucose units when the energy needs of the cell are high or when blood glucose levels start to drop. Increasing glycogen storage will help to delay fatigue during prolonged activity.

Fat is a concentrated source of calories; complete oxidation of a long-chain fatty acid yields more than three times as much ATP as metabolism of glucose. Many endurance athletes attempt to train their muscles to readily use fat for fuel and thereby conserve muscle glycogen. Later in this chapter, you will learn more about this technique, known as **fat adaptation** or *ketoadaptation*.

Protein use becomes more efficient with training, too. Endurance training increases the ability of muscle cells to use branched-chain amino acids for fuel during prolonged activity. However, the ability to use carbohydrates and fats for fuel is also increased; as long as the diet is adequate in carbohydrates and fat, most protein is spared for muscle synthesis and repair.

In addition, training increases the number of mitochondria within muscle cells. Recall that mitochondria are the powerhouses of the cells; this is where glucose and fat are broken down aerobically to generate ATP. With more mitochondria, muscle cells can use carbohydrates and fat more efficiently.

Overall, the cardiovascular and respiratory systems become more efficient at providing oxygen to the cells of the body. Plasma volume increases shortly after a training program is started, and red blood cell volume eventually increases as well. The heart pumps more blood with each contraction. Training also increases the number of capillaries in muscles, which increases oxygen supply to the muscles. Meanwhile, lung capacity increases, so more oxygen is available. The increased supply of oxygen translates into more efficient aerobic metabolism of carbohydrates and fats. Thus, lactate production from anaerobic glucose metabolism decreases. Lactate contributes to short-term muscle fatigue, so the less lactate produced, the longer the activity can be sustained.

Through all these adaptations, physical training improves the ability of cells to convert food energy into fuel for physical activity.

**fat adaptation** Manipulating the diet and physical training regimen so that muscles become more efficient at metabolizing fat as fuel during aerobic activity. Also known as *ketoadaptation*.

### ✓ CONCEPT CHECK 10.3

1. Describe one process used to resupply ATP during a short, intense burst of activity, such as a 100-meter running sprint.

2. How does the ATP yield of anaerobic breakdown of glucose compare to that of aerobic breakdown of glucose?

3. Why is fat a useful source of energy during exercise? Name three types of activity during which fat supplies 50% or more of fuel.

4. Is protein a useful source of energy during physical activity? Why or why not?

## 10.4 Nutrient Recommendations for Active Adults and Athletes

Athletic training and genetic makeup are two important determinants of athletic performance. Although a healthy dietary pattern can't substitute for either factor, it can help to enhance and maximize an athlete's potential. On the other hand, suboptimal dietary patterns can seriously reduce performance.[5] As highlighted in earlier chapters, the *Dietary Guidelines* notes several nutrients of public health concern given their inadequate consumption in the U.S. Active adults should consume nutrient-dense sources of dietary fiber, vitamin D, calcium, iron, and potassium for optimal performance.

▲ Intense athletic training may require thousands of additional calories. This increased food intake should easily provide ample protein and other nutrients to support activity. Ty Milford/Aurora Open/ Getty Images

## CALORIES

The daily calorie needs of athletes are highly individualized: genetics, hormones, age, sex, height, weight, temperature, altitude, stress, physical health, medications, body composition, and training level influence energy expenditure. A small, female gymnast may need 1800 kcal daily to sustain her training regimen without losing body weight, whereas a large, muscular football player may need 4000 kcal per day. Because athletes are such a heterogeneous group, there is no perfect equation to estimate their daily calorie needs. Even for nonathletes, the *Estimated Energy Requirement* (EER) equation provides only a rough approximation. However, you can use the EER equation as a starting point and individualize recommendations based on trial and error.

An estimate of the calories required to sustain moderate activity is 5 to 8 kcal per minute. The calories required for sports training or competition, then, must be added to those used to carry on normal activities. For example, consider a 135-pound young woman who requires 2200 kcal per day to fuel her normal activities. If she starts teaching two 45-minute dance classes each day, she will need about 500 extra kcal (a total of 2700 kcal per day) to maintain her current body weight. If an athlete experiences daily fatigue, the first consideration should be whether he or she is consuming enough food. Up to six meals per day may be needed, including one before each workout.

How can we know if an athlete is getting enough calories? Consulting with a sports dietitian would help answer this question. Estimating daily intake from a dietary intake log kept by the athlete is one way. Another option is to estimate the athlete's body fat percentage via body composition measures such as those detailed earlier (e.g., bioelectrical impedance analysis, DXA etc.). Body fat should be the typical amount found for athletes in the specific sport practiced. The recommended body fat percentage for athletes varies based on the sex of the athlete and the sport itself. The best way to determine if an athlete is consuming enough calories is to monitor body weight changes over time. If body weight starts to fall unintentionally, calories should be increased; if weight rises and it is because of increases in body fat, the athlete should reduce calories or increase physical activity. Again, a sports dietitian is an excellent resource for estimating individual nutrient needs for athletes.

Body mass index (BMI) is not an appropriate surrogate for assessing body fat for athletes. Body composition assessments are preferred for determining body fatness. If assessments of body composition show that an athlete needs to reduce body fat or gain weight, a sports dietitian should be consulted to provide professional recommendations related to balancing dietary and physical activity patterns to achieve optimal performance status.

Historically, athletes who competed in sports with weight classes (e.g., wrestlers, boxers) would try to lose weight before a competition. Many of the methods used to cut weight are unhealthy and dangerous. It is important to remember that losing as little as 2% of body weight by **dehydration** can adversely affect physical and mental performance, especially in hot weather. A pattern of repeated weight loss or gain of more than 5% of body weight by dehydration carries risk of kidney dysfunction and heat-related illness. Death is also a possibility.

To discourage such unhealthy practices and prevent future deaths, the National Collegiate Athletic Association (NCAA) and many states have authorized physicians or athletic trainers to set safe weight and body fat content minimums in weight-class sports. Under current guidelines, athletes wishing to lose weight should slowly descend to the desired weight class by not losing more than 1.5% of body weight per week.[6]

**relative energy deficiency in sport (RED-S)** A syndrome of altered metabolism, immune function, and mental health caused by low energy availability in athletes, which may be due to unintentional failure to meet the high energy demands of sports or intentional restriction of energy intake to control weight.

**Relative Energy Deficiency in Sport (RED-S).** The *female athlete triad syndrome* has historically been linked to low energy availability (LEA), but new insights have expanded the concept to be more inclusive of male athletes and extend beyond bone and menstrual health. Hence, the concept of **relative energy deficiency in sport (RED-S)** addresses the full range of concerns to include amenorrhea (females), reduced testosterone levels and libido (males), suboptimal bone health, increased risk of illness and injuries, gastrointestinal disturbances, cardiovascular disease, impaired training capacity, and

## Effects of Relative Energy Deficiency in Sport (RED-S)

**Cardiovascular system**
- Irregular heart rhythm
- Fainting
- Anemia

**Nervous system**
- Decreased coordination
- Irritability
- Depression

**Endocrine system**
- Impaired growth
- Decreased BMR

**Low Energy Availability**

**Digestive system**
- Esophagitis
- Ulcers
- Constipation

**Musculoskeletal system**
- Low bone mass
- Muscle injuries
- Decreased strength
- Decreased endurance

**Immune system**
- Frequent illness

**Reproductive system**
- Amenorrhea
- Infertility

**FIGURE 10-11** ▲ Relative energy deficiency in sport occurs when the athlete has low energy availability (with or without an eating disorder), which negatively impacts multiple body systems. Long-term health is at risk; thus, prevention and early treatment are crucial. Photos: (a) Lane Oatey/Blue Jean Images/Getty Images; (b) Aaron Amat/Shutterstock

poor performance. The highest prevalence of LEA involves weight-sensitive endurance sports. Athletes at greatest risk for disordered eating often follow misguided weight-loss programs and fail to recognize increased energy expenditure associated with training and competition. Treatment for these at-risk athletes includes increasing energy availability and often requires a comprehensive care team approach that includes a sports physician, sports dietitian, physiologist, and psychologist (Fig. 10-11).[7]

What is *energy availability*? Recall that you learned about energy balance. Energy input comes from foods and beverages. Components of energy output include basal metabolism, physical activity, and the thermic effect of food. For athletes, the proportion of energy expenditure devoted to physical activity may be quite high. Energy availability is the amount of energy left after the demands of physical activity have been met. If the amount of energy supplied by the dietary pattern is insufficient to cover total energy needs (i.e., low energy availability), some basic metabolic functions will suffer.

## CARBOHYDRATES

Anyone who exercises vigorously, especially for more than 1 hour per day on a regular basis, should adhere to an eating pattern that includes moderate to high amounts of carbohydrate. Numerous servings of varied grains, starchy vegetables, and fruits provide enough carbohydrate to maintain adequate liver and muscle glycogen stores.

Although it has been known for some time that the role of carbohydrate intake within (during) activity serves as an additional substrate for the muscle and the brain, there is now scientific evidence that carbohydrate consumption during physical activity provides additional performance benefits. Specifically, carbohydrate intake appears to stimulate areas of the brain that control both pacing and the reward system by

Fruits, vegetables, and whole grains should form the foundation of an athlete's dietary pattern. **What macronutrient contributes the majority of fuel from these food sources?**
Alexis Joseph/McGraw-Hill Education

## FARM to FORK    Carrots and Beets

Alexis Joseph/McGraw-Hill Education

Carrots and beets are root crops with a wide range of nutrients, flavors, and health-promoting benefits. Red beet juice has also been touted to enhance athletic performance.

### Grow

- The wild ancestor of carrots was purple, but most carrots grown in the United States are orange, a good indicator of the nutrients and other phytochemicals they contain, especially beta-carotene. Farmers are again producing purple carrot varieties, which are sweeter and higher in beta-carotene and the purple pigments, anthocyanins.
- Red beets are high in betalins, phytochemicals that may reduce the risk of cancer and other diseases.

### Shop

- For the freshest and sweetest carrots and beets, buy them with the green tops still attached. They will be at most only a few weeks old. Carrots and beets without tops can be several weeks to months old.
- Refrain from buying baby or frozen carrots. Baby carrots originate from misshapen mature carrots that have been whittled down to a smaller uniform size, and the remaining inner core is not as nutritious as the outer part that is discarded. The peeling, processing, and freeze/thaw cycle of frozen carrots destroys about half of their antioxidant value.
- The processing of beets into canned beets renders them more nutritious with more antioxidant value.

### Store

- Cut the tops off of your fresh carrots so they retain their moisture. To protect carrots from the ethylene gas produced by other fruits and vegetables, store them in a sealed plastic bag in the refrigerator.
- Remove beet greens and store beet roots unwrapped in the crisper drawer of the refrigerator and use them within 2 weeks.

### Prep

- Carrots and beets are more nutritious cooked. The heat breaks down cell walls and makes nutrients more bioavailable. Scrub carrots and beets, and cook them whole.
- Carrots can be sautéed or steamed to preserve more nutrients and sweetness. Eat carrots with oil or fat to allow best absorption of the fat-soluble beta-carotene, a precursor of vitamin A.
- Beets can be steamed, microwaved, or roasted. The skin helps retain the water-soluble nutrients while cooking and can be slipped off when cool. Beets will stain hands and surfaces, including wooden cutting boards. Use rubber gloves to avoid beet stains on your hands.
- Eating beets with mustard, horseradish, or vinegar will disguise their earthly flavor.

Source: Robinson J: The other root crops: Carrots, beets and sweet potatoes. In *Eating on the Wild Side.* New York: Little, Brown and Company, 2013.

L. Mouton/PhotoAlto

enhancing communications with mouth and gut receptors. Termed *mouth sensing,* this may provide more rationale for frequent intakes of carbs during longer events. Depletion of carbohydrate ranks just behind depletion of fluid and electrolytes as a major cause of fatigue and poor performance. Optimal carbohydrate intake depends on the size of the athlete and the type of physical activity. Recall that your weight in pounds divided by 2.2 converts your weight into kilograms. For light or skill-based sports, 3 to 5 grams of carbohydrate per kilogram of body weight will suffice. For physical activity of moderate intensity, consume 5 to 7 grams of carbohydrate per kilogram. Athletes who train for several hours per day may need up to 12 grams of carbohydrates per kilogram of body weight, which may add up to 600 grams per day (or more). Attention to carbohydrate intake is especially important when performing multiple training bouts in one day (e.g., two-a-day swim practices) or heavy training on successive days (e.g., cross-country running).

Athletes should obtain at least 45% to 65% of their total energy needs from carbohydrate,[8] especially if activity duration is expected to exceed 2 hours and total caloric intake is about 3000 kcal per day or less. Eating patterns providing 4000 to 5000 kcal per day can be as low as 50% carbohydrate, as these will still provide sufficient carbohydrate (e.g., 500 to 600 grams or so per day).

Besides the total amount of carbohydrates, the *quality* of carbohydrates can certainly influence mental and physical performance. Sports dietitians emphasize the difference between a high-carbohydrate meal and a high-carbohydrate/high-fat meal. Before endurance events, such as marathons or triathlons, some athletes seek to increase their carbohydrate reserves by eating foods such as potato chips, French fries, banana cream pie, and pastries. Although such foods provide carbohydrate, these highly processed foods also contain excessive amounts of added sugars, saturated fats, and sodium. See the *Farm to Fork* feature on carrots and beets to read about beet juice and athletic performance. Also be sure to check the Nutrition Facts label for carbohydrate content (Table 10-3). Consuming a moderate amount of fiber during the final day of training is a good precaution to reduce the chances of bloating and intestinal gas during the next day's event. It is also essential to hydrate properly when consuming additional fiber.

### FAT

A dietary pattern containing up to 35% of calories from fat is generally recommended for athletes.[8] Rich sources of monounsaturated fat, such as olive and canola oil, should be emphasized, and saturated fat intake should be limited. Consuming ≤ 20% of energy intake from fat does not enhance performance, and extreme restriction of fat intake may limit the food range needed to meet overall health and performance goals. At the other end of the spectrum,

TABLE 10-3 ■ Carbohydrate-Rich Foods

**Starches = 15 Grams Carbohydrate (80 kcal)**

| | |
|---|---|
| Dry breakfast cereal,* ½–¾ cup | Baked potato, ¼ large |
| Cooked breakfast cereal, ½ cup | Bagel, ¼ (or 4 ounces) |
| Cooked rice, ⅓ cup | Bread, 1 slice |
| Cooked pasta, ⅓ cup | Pretzels, ¾ ounce |
| Cooked corn, ½ cup | Saltine crackers, 6 |
| Cooked dry beans, ½ cup | Pancake, 4-inch diameter, 1 |

**Vegetables = 5 Grams Carbohydrate (25 kcal)**

| | |
|---|---|
| Cooked vegetables, ½ cup | Examples: carrots, green beans, broccoli, |
| Raw vegetables, 1 cup | cauliflower, spinach, tomatoes, and vegetable |
| Vegetable juice, ½ cup | juice |

**Fruits = 15 Grams Carbohydrate (60 kcal)**

| | |
|---|---|
| Canned fruit or berries, ½ cup | Grapes, 17 |
| 100% fruit juice, ½ cup | Grapefruit, ½ |
| Apple or orange, 1 small | Peach, 1 |
| Banana, 1 small | Watermelon cubes, 1¼ cups |

**Dairy = 12 Grams Carbohydrate (100 to 150 kcal)**

| | |
|---|---|
| Milk, 1 cup | Soy milk, 1 cup |
| Plain low-fat yogurt, ⅔ cup | |

**Sweets = 15 Grams Carbohydrate (variable kcal)**

| | |
|---|---|
| Cake, 2-inch square | Ice cream, ½ cup |
| Cookies, 2 small | Sherbet, ½ cup |

*The carbohydrate content of dry cereal varies widely. Check the labels of the ones you choose and adjust the serving size accordingly.

Source: Modified from *Choose Your Foods: Food Lists for Diabetes* by the American Diabetes Association and Academy of Nutrition and Dietetics.

extremely high-fat, carbohydrate-restricted diets are not supported by the current evidence.

## PROTEIN

Whether the recommended daily protein allowances for the general population, set at 0.8 to 1.0 g/kg, is appropriate for athletes remains a point of controversy.[8] For athletes, many experts recommend protein intake within the range of 1.2 to 2.0 grams of protein per kilogram of body weight, spacing modest amounts of high-quality protein (0.3 g/kg) after physical activity and throughout the day. Sports dietitians contend that the RDAs have been set to prevent deficiency among the general population, not to optimize physical performance among athletes.[8]

The optimal protein intake for an athlete will vary based on the athlete's activity and level of fitness. Athletes who are just beginning a strength-training program are likely to need the most protein to supply the building blocks for synthesis of new muscle tissue. Once the desired muscle mass is achieved, daily protein intake need not exceed 1.2 grams per kilogram of body weight. Some strength-training athletes tend to aim for excessive protein intakes, sometimes as high as 3 or 4 grams of protein per kilogram of body weight. There remains conflicting evidence that protein intake above 2.0 grams per kilogram of body weight will benefit the athlete. Protein intakes above this amount result in an increased use of amino acids for energy needs; no further increase in muscle protein synthesis is seen. To practice evaluating an athlete's protein needs, visit *Rate Your Plate*: "A Student Athlete's Dietary Pattern" in Connect.

Unless an athlete follows a low-calorie diet, the recommended range of 1.2 to 2.0 grams of protein per kilogram of body weight can be met by eating a variety of foods.

## ASK THE RDN  The No-Meat Athlete

*Dear RDN:* For a variety of reasons, I am considering switching to a vegan eating pattern. However, I am a sprinter on the college women's swimming and diving team, and I want to make sure that a vegan diet would not affect my athletic performance. Can a "no-meat athlete" compete?

A well-planned plant-based diet can absolutely support optimal athletic performance. From better weight management to reduced risk for chronic diseases, vegan diets offer many health benefits. However, some nutrient needs may be difficult to meet with a plant-based diet. These include protein, iron, calcium, and vitamin B-12.

Protein intakes of athletes who follow a vegan dietary pattern, especially young women, are often low. You may need more than the RDA for protein because, first, you are an athlete and, second, the quality of plant proteins is generally lower than that of animal proteins. Protein recommendations for athletes who follow a vegetarian or vegan dietary pattern are 1.3 to 1.8 grams per kilogram of body weight. If you consume adequate calories, you can meet your protein needs without protein or amino acid supplements, but the *quality* of your protein matters. Be sure to consume a variety of plant proteins (legumes, grains, nuts) throughout the day to obtain all of the essential amino acids.

Iron delivers oxygen to exercising muscles and helps to convert food fuel into ATP—two processes that impact athletic performance. Iron needs can be difficult to meet even with omnivorous diets, so iron intake deserves special attention. Remember that plant (nonheme) sources of iron are not as well absorbed as animal (heme) sources, but foods with vitamin C can boost absorption of nonheme iron. Have your iron status checked periodically. If your iron stores are low, discuss options with your primary care provider or registered dietitian nutritionist (RDN).

Lean female athletes place themselves at risk for early osteoporosis if their dietary patterns are low in calcium (and vitamin D). A vegan eating pattern eliminates animal products, including dairy foods, which are excellent sources of these bone-building nutrients.

A deficiency of vitamin B-12 could also affect red blood cell health and possibly nerve function. Unlike iron and calcium, there are no plant-based alternatives for vitamin B-12. Use of fortified foods and/or dietary supplements will be crucial to meeting your vitamin B-12 needs.

One more concern is that this highly restrictive dietary change—even though it seems like a healthful choice—could promote disordered eating behaviors. As you move toward a plant-based dietary pattern, make sure your overall energy intake is adequate to support your needs for competitive training. If your weight drops too low, it could significantly affect your athletic performance and long-term bone health.

Start slowly and follow these general recommendations: Begin your day with a bowl of fortified whole grain breakfast cereal. Check the label for brands of cereal that provide at least 20% of the DV for iron and vitamin B-12. Top your cereal with almond milk (or a similar dairy alternative) that provides at least one-third of the DV for calcium. Also include calcium-fortified dairy alternatives as a substitute whenever you would have chosen dairy products. Enjoy a glass of calcium-fortified 100% orange juice with breakfast to enhance absorption of iron from your cereal. Throughout the day, be sure to choose plentiful and varied sources of plant proteins, including legumes, nuts, seeds, and whole grains. Try soy and quinoa, two versatile plant proteins that contain all essential amino acids. Last, monitor changes in your weight and performance. If your calorie intake is too low, consider incorporating more nuts, nut butters, avocados, and hummus into your meals and snacks, as these are energy-dense sources of healthy fats.

These are just a few basic ideas to help you remain healthy and compete at your highest level. To tailor your food intake to your specific needs, seek the expert advice of your team's sports dietitian at your university health center.

To your record-breaking success!

*Angela Collene, MS, RDN, LD*
Lecturer, The Ohio State University, Author of *Contemporary Nutrition*

Tim Klontz, Klontz Photography

To illustrate, a 117-pound (53-kilogram) woman performing moderate-intensity endurance activity can consume 58 grams of protein (53 × 1.1) during a single day by including 3 ounces of chicken (a small chicken breast), 3 ounces of tuna, and two glasses of low-fat milk. Similarly, a 170-pound (77-kilogram) man who aims to gain muscle mass through strength training needs to consume only 6 ounces of chicken (a large chicken breast), ½ cup of cooked beans, a 6-ounce can of tuna, and three glasses of milk to achieve an intake of 154 grams of protein (77 × 2.0) in a day. For both athletes, these calculations do not include the protein present in the grains or vegetables they will also eat. By simply meeting their calorie needs, many athletes easily meet their protein requirements.

Despite marketing claims, high-quality whole food proteins are superior to supplements for the maintenance, repair, and synthesis of muscle. The protein foods group comprises a broad group of foods from both animal and plant sources, and includes several subgroups: meats, poultry, and eggs; seafood; and nuts, seeds, and soy products. Beans, peas, and lentils may be considered a part of the protein foods group as well.

The consumption of milk-based protein, especially after resistance training, is effective in increasing muscle strength and promotes positive results in body composition. Although supplements are usually expensive and unnecessary, many athletes choose to use protein powders (e.g., whey, casein, or soy) to add additional protein to their diet. Whey protein is especially popular among strength-trained athletes. It is particularly rich in leucine, an essential branched-chain amino acid that has been shown in some studies to stimulate gains in muscle mass during strength training.[9] Yet experts still recommend that individuals attempt to meet their protein needs via whole foods including plant-based proteins (see *Ask the RDN*).

Consuming excessive amounts of protein has drawbacks. It increases calcium loss somewhat in the urine. It also leads to increased urine production, which could increase dehydration. Excess animal protein also may lead to kidney stones in people with a history of this or other kidney problems. Last but not least, a dietary pattern that is so focused on animal protein may leave the athlete short on carbohydrates, leading to fatigue and poor athletic performance.

## VITAMINS AND MINERALS

Compared to the needs of sedentary adults, vitamin and mineral needs for athletes are the same or slightly higher. At this time, there are not enough data to support separate Dietary Reference Intakes (DRIs) specific to athletes for any of the micronutrients. Athletes usually have high calorie intakes, so they tend to consume plenty of vitamins and minerals. An exception is athletes consuming low-calorie diets (1200 kcal or less), as seen with some female athletes participating in events in which maintaining a low body weight is crucial. These eating patterns may not meet B vitamin and other micronutrient needs. Athletes who follow a vegan dietary pattern may also fall short on some nutrients. In these cases, consuming fortified foods, such as ready-to-eat breakfast cereals, or a balanced multivitamin and mineral supplement may be recommended. You can discuss options with a sports dietitian or your primary care provider. Vegetarian athletes, also discussed later in this chapter, are encouraged to consume generous amounts of fruits, vegetables, whole grains, and plant proteins (nuts, seeds, soy products, etc.) that contribute adequate fiber, phytochemicals, antioxidants, and other nutrients.

**B Vitamins Support Energy Metabolism and Red Blood Cell Health.** Recall that coenzyme forms of B vitamins facilitate chemical reactions that generate ATP from carbohydrates, proteins, and fats. Some B vitamins are involved in processes that build and repair tissue. Although no separate DRIs have been set, athletes may need more than the current RDA for some B vitamins, such as riboflavin and vitamin B-6.

Furthermore, physical performance is highly dependent on the availability of oxygen to active muscles. Folate, vitamin B-6, and vitamin B-12 are involved in the formation of healthy red blood cells, which transport oxygen to all body tissues.

As you can imagine, an inadequate supply of B vitamins could impair an athlete's physical performance. Yet deficiencies of B vitamins are not very common. As athletes consume greater quantities of food to meet their increased calorie needs, they typically consume enough B vitamins from food sources to support energy metabolism and red blood cell health. Taking more than the RDA for B vitamins is not likely to enhance performance.

On the other hand, for a person with a diagnosed vitamin or mineral deficiency, supplementation could improve athletic performance. At-risk populations include athletes who are vegan or older (vitamin B-12), female athletes of childbearing age (folate), and any athlete who restricts dietary intake to control body weight (variety of micronutrients). In these cases, fortified foods or a balanced multivitamin and mineral supplement may be beneficial to overall health and athletic performance.

**Antioxidants May Prevent Oxidative Damage.** Physical activity leads to increased production of free radicals. The presence of some free radicals in muscle tissue is

Is there an optimal dose of protein? Researchers suggest a dose of at least 30 grams of protein per meal is ideal to promote muscle synthesis.[13] Rather than simply focusing on total protein intake, athletes should consume foods that provide 30 to 45 grams of high-quality protein per meal spaced throughout the day. **How many sources of high-quality protein can you spot on this plate?** asife/Shutterstock

**Do strength-trained athletes need to ingest tuna, chicken, and lean beef at every meal to build muscle?** Patrik Baboumian is a strongman competitor and former bodybuilder who follows a vegan diet. The plant-based protein sources that built his 250-pound physique included beans, peas, lentils, nuts, seeds, and soy. Baboumian was a world record–holder for the log lift in his weight class and was named Germany's strongest man. David Cooper/ Toronto Star via Getty Images

actually beneficial for muscle contraction and adaptation to exercise. However, excessive free radicals can lead to fatigue and cell damage.

Athletes' needs for antioxidants such as vitamin E and vitamin C may be somewhat greater because of the potential protection these nutrients provide. However, there is evidence that antioxidant system activity increases in the body as training progresses. The use of large doses of antioxidants is not currently an accepted part of the dietary guidance for athletes. Food groups rich in antioxidants, include vegetables, fruits, and grains.[5]

**Optimal Iron Status Improves Performance.** Iron is involved in red blood cell production, oxygen transport, and energy production, so a deficiency of this mineral can noticeably detract from optimal athletic performance. Some of the consequences of iron deficiency include weakness, fatigue, and decreased work capacity. The potential causes for iron deficiency in athletes vary.[10] As in the general population, female athletes are most susceptible to low iron status due to monthly menstrual losses. Restrictive eating plans, such as low-calorie and vegan eating plans, are likely to be lower in iron. Distance runners should pay special attention to iron intake because their intense workouts may lead to gastrointestinal bleeding.

A less threatening concern is **sports anemia.** At the start of an endurance training regimen, plasma volume expands, but the synthesis of additional red blood cells is slower to increase. This results in a temporary dilution of the blood; even if iron stores are adequate, blood iron tests may appear low. Sports anemia is not detrimental to performance, but it is hard to differentiate between sports anemia and true anemia. If iron status is low and not replenished, iron-deficiency anemia can markedly impair endurance performance.

Although true iron-deficiency anemia (a depressed blood **hemoglobin** level) is not that common among athletes, some studies suggest that iron deficiency *without* anemia may have a negative impact on physical activity and performance. Recall that iron deficiency occurs long before anemia is detected clinically. As body stores of iron are depleted, body processes that use iron, such as energy-yielding reactions, are impaired.

It is a good idea for athletes (especially women who are premenopausal) to have their iron status checked at the beginning of a training season and at least once midseason.[10] Current evidence suggests that as many as half of female athletes may be iron deficient. To identify iron deficiency without anemia among athletes, many experts advocate serum ferritin testing. Ferritin is an iron transport protein; low serum ferritin levels indicate low iron stores even before red blood cell health is affected.

Any blood test indicating low iron status—sports anemia or not—warrants follow-up. A primary care provider will need to determine the cause of iron depletion. Whatever the cause, once depleted, iron stores can take months to replenish. Dietary sources are typically not enough to correct iron-deficiency anemia; supplementation (under a primary care provider's supervision only) is required. Athletes must be especially careful to meet iron needs because preventing iron deficiency is a lot simpler than treating it.

Knowing that iron is required for red blood cell synthesis, athletes may be tempted to self-prescribe iron supplements in an attempt to boost the oxygen-carrying capacity of the blood. However, indiscriminate use of iron supplements for people with normal hemoglobin and serum ferritin levels is never advised.[10] Research does not support a benefit of iron supplementation on athletic performance for athletes with normal iron status. Furthermore, liver damage and increased rates of heart disease and some forms of cancer are possible consequences of iron toxicity. A safer alternative would be to have iron status checked periodically. In addition, monitor dietary patterns to become aware of usual iron intakes. If dietary iron intake is low, incorporate more food sources of heme iron and pair nonheme sources with vitamin C to enhance absorption. Avoid drinking tea or iced tea with meals since this may inhibit iron absorption. The decision to use an iron supplement is best left to a primary care provider.

**sports anemia** Exercise-induced iron-deficiency anemia caused by plasma volume expansion, low hemoglobin synthesis, or increased destruction of red blood cells.

▲ Weight-restricted athletes should make sure they are consuming enough calcium, protein, and other essential nutrients. Juice Images Ltd/Getty Images

**Calcium Intake Is Important, Especially in Women.** Athletes, especially women trying to lose weight by restricting their intake of dairy products, can have marginal or low dietary intakes of calcium. This practice compromises bone health. Of still greater concern are female athletes who have stopped menstruating because their arduous training and low body fat interfere with the normal secretion of reproductive hormones. Female athletes who do not menstruate regularly are more likely to suffer **stress fractures** during training and will be susceptible to bone injuries throughout life. The negative impacts of low dietary calcium intake and irregular menses in female athletes outweigh the benefits of weight-bearing activities on bone density. Increasing energy intake to restore body weight and body fat stores is important to correct hormonal imbalances and prevent further bone loss.

Female athletes whose menstrual cycles become irregular should consult a primary care provider or sports dietitian. Decreasing the amount of training or increasing energy intake and body weight often restores regular menstrual cycles. If irregular menstrual cycles persist, severe bone loss (much of which is not reversible) and osteoporosis can result. Extra calcium in the diet does not necessarily compensate for these damaging effects of menstrual irregularities, but inadequate dietary calcium can make matters worse.

**Vitamin D.** Recent studies have documented a relationship between vitamin D status and injury prevention, improved neuromuscular function, enhanced muscle size, decreased inflammation, and reduced risk of stress fractures. Athletes who live at latitudes above the 35th parallel or who predominantly train indoors are at higher risk for vitamin D deficiency. Although vitamin D status is important for athletes and nonathletes alike, current data do not support vitamin D as an **ergogenic** aid. More data are needed to elucidate the role of vitamin D in athletics. Those who have a history of stress fractures, joint or overtraining injuries, or muscle pain and weakness should consult their primary care provider or sports dietitian.

## FLUID

Fluid needs for an average adult are about 9 to 11 cups per day for women and 13 to 15 cups per day for men. Athletes generally need even more water to regulate body temperature. Heat production in contracting muscles can rise 15 to 20 times above that of resting muscles. Unless this heat is quickly dissipated, heat exhaustion, heat cramps, and potentially fatal heatstroke may ensue.

Fluid and electrolyte needs vary widely, based on differences in genetics, body mass, environmental conditions, level of training, and event duration. Because fluid needs are highly individualized and dynamic, it is difficult to make general recommendations for fluid replacement.[5] Rather, an athlete should consume enough fluid to prevent short-term (i.e., fluid-related) changes in body weight. The American College of Sports Medicine recommends losing *no more than 2%* of body weight during physical activity, especially in hot weather.[5] A football player wearing equipment in hot weather can lose this much within 30 minutes. However, it is important to remember this advice even when sweating can go unnoticed, such as when swimming or during the winter.

To determine fluid needs, the athlete should first calculate 2% of his or her body weight. Next, it is useful to know the body's hourly sweat rate, which can be calculated by comparing weight loss during the activity to the amount of fluid consumed. This will require some self-monitoring of pre- and post-workout weight and fluid intake during workouts. For reference, sweat rates during prolonged activities can range from 3 to 8 cups (750 to 2000 mL) per hour. If weight change cannot be monitored, urine color is another measure of hydration status. Urine color should be no more yellow than lemonade.

## BEFORE, DURING, AND AFTER PHYSICAL ACTIVITY

The fluid plan that suits most athletes needs to be tailored to meet the athlete's tolerance and experience, opportunities for drinking fluids throughout activities, and

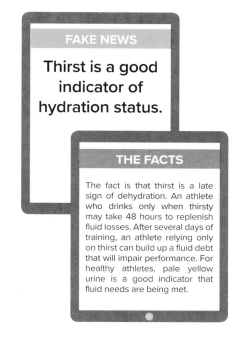

**FAKE NEWS**

Thirst is a good indicator of hydration status.

**THE FACTS**

The fact is that thirst is a late sign of dehydration. An athlete who drinks only when thirsty may take 48 hours to replenish fluid losses. After several days of training, an athlete relying only on thirst can build up a fluid debt that will impair performance. For healthy athletes, pale yellow urine is a good indicator that fluid needs are being met.

**stress fracture** A fracture that occurs from repeated jarring of a bone. Common sites include bones of the foot and shins.

**ergogenic** A mechanical, nutritional, psychological, pharmacological, or physiological substance or treatment intended to directly improve physical performance.

Annually in America, 25 billion styrofoam coffee cups and 50 billion water bottles are used with 75% ending up in landfills. Do your part to make a difference by recycling single-use items, carrying reusable water bottles, and purchasing only what you need and in bulk when possible.

Source: Earth Day 2020 Fact Sheet.

 Sustainable Solutions

consumption of other nutrients (e.g., carbohydrate) in liquid form. The following fluid-replacement approach for before, during, and after workouts can meet an athlete's fluid needs in most cases.[5]

- Drink 5 to 10 mL/kg of body weight (about 1.5 to 3 cups for a 150-pound male) of water in the 2 to 4 hours before exercise.
- During events lasting more than 30 minutes, athletes should consume fluid to prevent dehydration (losses of > 2% body weight). Recommendations for marathon runners suggest about 1.5 to 3.5 cups (400 to 800 milliliters) per hour to prevent dehydration. Football players wearing equipment for two-a-day practices during the heat of August may need even more than 800 milliliters per hour to prevent dehydration. The best plan is to determine individual rate of fluid losses during training and plan accordingly. In many cases, athletes, especially children and teenagers, need to be reminded to consume fluids during physical activity.
- Within 4 to 6 hours after exercise, about 2 to 3 cups of fluid should be consumed for every pound lost. It is important that weight be restored before the next exercise period. Skipping fluids before or during events will almost certainly impair performance.

**Sports Drinks.** For sports that require less than 60 minutes of continuous exertion or when total weight loss is less than 5 to 6 pounds, the primary concern is replacing the water lost in sweat. When continuous exercise extends beyond 60 minutes, electrolyte (especially sodium) and carbohydrate replacement becomes increasingly important.

Use of sports drinks (Fig. 10-12) during long bouts of continuous exercise—especially in hot weather—offers several advantages for athletes:

- *Water* increases blood volume to allow for efficient cooling and transport of fuels and waste products to and from cells.
- *Carbohydrates* supply glucose to muscles as they become depleted of glycogen and also add flavor, which encourages athletes to drink.
- *Electrolytes* in sports drinks help to maintain blood volume, enhance the absorption of water and carbohydrate from the intestine, and stimulate thirst.

Overall, the decision to use a sports drink hinges primarily on the duration, type, and intensity of the activity. As the projected duration of continuous activity approaches 60 minutes or longer, the advantages of sports drinks over plain water emerge. However, athletes should experiment with sports drinks during practice, instead of trying them for the first time during competition. Also, both athletes and nonathletes should recognize that sports drinks can be a source of excess added sugars and calories that may result in undesired weight gain.

**Energy Drinks.** The popularity of caffeine-containing energy drinks has surged in recent years. Some studies show that caffeine may improve athletic performance during endurance events (e.g., cycling) or sports that require a high level of mental alertness (e.g., archery). However, excessive caffeine consumption can lead to shakiness, nervousness, anxiety, nausea, irregular heart rate, elevated blood pressure, and insomnia. Chronic intakes of energy drinks can lead to fatigue and headaches. The high added sugar content in energy drinks also can lead to weight gain. In addition, the **diuretic** effect of caffeine may not support optimal hydration, particularly for athletes who are not accustomed to caffeine. Compare the caffeine and calorie contents of several top-selling energy drinks in Table 10-4.

**Alcohol.** Excessive alcohol consumption, consistent with binge drinking patterns, is observed among some athletes, particularly in team sports. Besides the nonnutritive calorie load of alcohol (7 kcal/g), alcohol has negative effects on performance and recovery from physical activity. A solid body of evidence cautions against consumption of alcohol before, during, or after physical activities.

**TABLE 10-4** ■ **Caffeine, Calorie, and Sugar Content of Popular Energy Drinks**

| Energy Drink | Serving | Caffeine (mg) | Energy (kcal) | Sugars (g) |
|---|---|---|---|---|
| Full Throttle® | 16 fl oz | 164 | 228 | 60 |
| Monster Energy® | 16 fl oz | 164 | 233 | 54 |
| Monster Lo-Carb® | 16 fl oz | 144 | 24 | 7 |
| Monster Zero Sugar® | 16 fl oz | 140 | 0 | 0 |
| NOS® | 16 fl oz | 169 | 218 | 56 |
| Red Bull® | 8.4 fl oz | 75 | 112 | 27 |
| Red Bull Sugar Free® | 8.4 fl oz | 76 | 13 | 0 |
| Red Bull The Blue Edition® | 8.4 fl oz | 80 | 110 | 27 |
| Rockstar® | 16 fl oz | 164 | 288 | 61 |

Source: USDA FoodData Central

**Heat-Related Illness.** As environmental temperature rises above 95°F (35°C), virtually all body heat is lost through the evaporation of sweat from the skin. As humidity rises, especially above 75%, evaporation slows and sweating is insufficient to cool the body. The result is rapid fatigue, increased work for the heart, and difficulty with prolonged exertion. Heat-related injuries—heat exhaustion, heat cramps, and heatstroke—can be deadly (Table 10-5). To decrease the risk of developing heat-related injuries, watch for rapid body-weight changes (2% or more of body weight), replace lost fluids, and avoid exercising under extremely hot, humid conditions.

**FIGURE 10-12** ▲ Most sports drinks for fluid and electrolyte replacement typically contain added sugars plus sodium and potassium. The various sugars in this product total 14 grams of carbohydrates per cup (240-mL) serving. Sports drinks typically contain about 6% to 8% sugar. Drinks with a sugar content above 8% to 10%, such as soft drinks or fruit juices, may cause gastrointestinal distress, fail to empty from the stomach rapidly, and are not recommended for fluid repletion during sport.

**TABLE 10-5** ■ **Heat-Related Illnesses**

| Heat-Related Illness | Symptoms | Recommended Treatment |
|---|---|---|
| **Heat exhaustion** is the first stage of heat-related illness that occurs after depletion of blood volume from fluid loss by the body. | Profuse sweating<br>Headache<br>Dizziness<br>Nausea<br>Muscle weakness<br>Visual disturbances<br>Flushed skin<br>Hyperthermia<br>Heat cramps | Move to cool environment.<br>Remove excess clothing.<br>Cool the skin with ice packs or cold water.<br>Replenish lost fluids and electrolytes. |
| **Heat cramps** are a frequent complication of heat exhaustion. They usually occur in people who have experienced significant sweat losses from exercising for several hours in a hot environment and have consumed a large volume of water without replacing electrolytes. | Painful skeletal muscle cramps<br>Involuntary muscle spasms | Replenish lost fluids and electrolytes immediately. |
| **Heatstroke** can occur when internal body temperature reaches about 104°F. Sweating often ceases and blood circulation is greatly reduced. Nervous system damage may ensue, and death results in 10% of cases. | Hyperthermia<br>Hot dry skin<br>Nausea<br>Confusion<br>Irritability<br>Poor coordination<br>Fainting<br>Seizures<br>Coma | Cool the skin with ice packs or cold water.<br><br><br>*SEEK PROFESSIONAL MEDICAL ATTENTION IMMEDIATELY.* |

**Water Intoxication.** It is also possible for some athletes to drink so much water that they develop **water intoxication (hyponatremia).** Recall that hyponatremia causes cardiovascular and neurological problems and can lead to death. Symptoms of hyponatremia include bloating, puffiness, weight gain, nausea, vomiting, headache, confusion, delirium, seizures, respiratory distress, loss of consciousness, and possibly death.

✓ **CONCEPT CHECK 10.4**

1. What does it mean to *cut weight* before a competition? How might this affect physical performance?

2. Greta, a point guard on the women's basketball team, complains of chronic fatigue. Describe three nutritional concerns you would investigate.

3. During one day of preseason training for football, David loses 7 pounds as a result of sweat losses. How much fluid should he drink to rehydrate after practice?

# 10.5 Recommendations for Endurance, Strength, and Power Athletes

You have learned about ways in which nutrient needs can be universally affected by participation in sports. The definition of sports, however, is broad, and each individual is unique. Endurance athletes, who need to fuel activity that lasts several hours, should take a different approach to nutrition than strength and power athletes, who focus on gains in muscle mass. In this section, we present specific nutrition strategies for endurance, strength, and power athletes.

## ENDURANCE ATHLETES: STRATEGIES TO DELAY OR PREVENT FATIGUE

**carbohydrate loading** A process in which a high-carbohydrate diet is consumed for several days before an athletic event while tapering exercise duration in an attempt to increase muscle glycogen stores.

The overarching goal for endurance athletes is to consume adequate carbohydrates and fluids. *Before* the event, endurance athletes should focus on maximizing muscle and liver glycogen stores, which will later be used to fuel muscles and maintain blood glucose. *During* an event, the goal is to prevent dehydration and glycogen depletion, as both of these conditions lead to fatigue and detract from physical performance. *After* an event, muscle glycogen stores need to be replenished, damaged muscle tissue must be repaired, and hydration should be restored.[11]

**Maximize Glycogen Stores Before the Event.** For athletes who compete in continuous, intense aerobic events lasting more than 60 to 90 minutes (or in shorter events taking place more than once within a 24-hour period), a regimen of **carbohydrate loading** can maximize the amount of energy stored in the form of muscle glycogen for the event. In one regimen, during the week prior to the event, the athlete gradually reduces the intensity and duration of exercise (*tapering*) while simultaneously increasing the percentage of total calories supplied by carbohydrates. Shorter carbohydrate-loading regimens (e.g., 1 or 2 days before an event) may also be effective.

For example, consider the carbohydrate-loading schedule of a 25-year-old man preparing for a marathon. His typical calorie needs are about 3500 kcal per day. Six days before competition, he completes a final, hard workout of 60 minutes. On that day, carbohydrates contribute 45% to 50% of his total calorie intake. As he goes through the rest of the week, the duration of his workouts decreases to 40 minutes and then to about 20 minutes by the end of the week. Meanwhile, he increases the amount of carbohydrate in his eating plan to reach 70% to 80% of total calorie intake as the week continues (Table 10-6). Total calorie intake should decrease as exercise time decreases

▲ Carbohydrate loading is appropriate only for endurance activities such as a long-distance race. Fuse/Corbis/Getty Images

**TABLE 10-6 ■ Sample Carbohydrate-Loading Regimen**

| Days Before Competition | 6 | 5 | 4 | 3 | 2 | 1 |
|---|---|---|---|---|---|---|
| Physical activity time (minutes) | 60 | 40 | 40 | 20 | 20 | Rest |
| Carbohydrate intake (grams) | 450 | 450 | 450 | 600 | 600 | 600 |

**Carbohydrate Loading May Be Beneficial for These Activities**

- Marathons
- Long-distance swimming
- Cross-country skiing
- 30-kilometer runs
- Triathlons
- Tournament-play basketball
- Cycling time trials

throughout the week. On the final day before competition, he rests while maintaining the high carbohydrate intake.

This carbohydrate-loading technique usually increases muscle glycogen stores by 50% to 90% over typical conditions (i.e., when dietary carbohydrate constitutes only about 50% of total calorie intake). A potential disadvantage of carbohydrate loading is that additional water (about 3 grams) is incorporated into the muscles along with each gram of glycogen. Although the additional water aids in maintaining hydration for some individuals, this additional water weight and related muscle stiffness detract from their sports performance.

Athletes considering a carbohydrate-loading regimen should test it out during training (and well before an important competition) to experience its effects on their performance. They can then determine whether it is worth the effort. Currently, expert advice is shifting away from such regimented carbohydrate loading in favor of supplying carbohydrates during the event (along with a consistent dietary pattern high in carbohydrate).

Even if an endurance athlete chooses not to practice a strict carbohydrate-loading regimen, a meal should be eaten 1 to 4 hours before an endurance event to top off muscle and liver glycogen stores, prevent hunger during the event, and provide extra fluid. A pre-event meal should consist primarily of carbohydrate, contain moderate fat or fiber, and include high-quality protein (Table 10-7). The longer the period before an event, the larger the meal can be, because there will be more time available for digestion. Anything consumed within 1 hour before an event should be blended or liquid to promote more rapid stomach emptying. Examples include diluted 100% juice, low-fat smoothies, and sports drinks.

**TABLE 10-7 ■ High-Carbohydrate Pre-Event Meals**

| Breakfast | |
|---|---|
| Cheerios,® ¾ cup<br>Reduced-fat milk, 1 cup<br>Blueberry muffin, 1<br>Orange juice, 4 ounces | 748 kcal<br>110 grams (59%) carbohydrate |
| or | |
| Low-fat fruit yogurt, 1 cup<br>Plain bagel, ½<br>Apple juice, 4 ounces<br>Peanut butter (for bagel), 1 tbsp | 541 kcal<br>92 grams (68%) carbohydrate |

| Lunch or Dinner | |
|---|---|
| Broiled chicken, 3 ounces<br>Rice, 1½ cups<br>Steamed zucchini, 1 cup<br>Low-fat chocolate milk, 1 cup<br>Jello®, ½ cup | 709 kcal<br>116 grams (64%) carbohydrate |
| or | |
| Spaghetti noodles, 2 cups<br>Spaghetti sauce, 1 cup<br>Reduced-fat milk, 1½ cups<br>Green beans, 1 cup | 709 kcal<br>132 grams (66%) carbohydrate |

With regard to the timing of pre-activity meals, a general guide is to allow 4 hours for a big meal (about 1200 kcal), 3 hours for a moderate meal (about 800 to 900 kcal), 2 hours for a light meal (about 400 to 600 kcal), and an hour or less for a snack (about 300 kcal).

▲ Pre-event meals may require a higher proportion of grains than suggested by MyPlate to boost carbohydrate content. Choose starchy vegetables and grain-based snacks to help top off glycogen stores. cobraphotography/Shutterstock

Many power athletes utilize a training technique called **periodization,** in which physical stresses on the body change throughout the year:

- Early in the training season, athletes work on building aerobic endurance.
- After gains in aerobic capacity have been achieved, the focus shifts to building strength, power, and sport-specific skills.
- During the competitive season, daily workouts are scaled back, but activity is intense and of long duration on game days.
- In the off-season, athletes continue to work out to stay in shape, but the volume is certainly lower than it was in season.

Athletes taking part in periodized training will use the full spectrum of energy systems we have discussed. Nutrition recommendations should also be periodized to match such dynamic training plans.[8]

**periodization** Cycling the volume, intensity, and activities of workouts throughout the training season.

Carbohydrate-rich food choices for a pre-event meal include spaghetti, muffins, bagels, pancakes with fresh fruit topping, oatmeal with fruit, a baked potato topped with a small amount of low-fat cheese, toasted bread with jam, bananas, or low-sugar breakfast cereals with reduced-fat milk. Liquid meal-replacement formulas, such as instant breakfast drinks, also can be used. Foods especially rich in fiber can be eaten the previous day to help empty the colon before an event, but they should not be eaten the night before or on the morning of the event. Avoid fatty or fried foods, such as sausage, bacon, sauces, and gravies. Some foods (e.g., dairy products) may cause gastrointestinal upset. Athletes should experiment with the size, timing, and composition of pre-event meals during training to determine what is best tolerated.

**Fat Adaptation.** Traditionally, high-carbohydrate dietary patterns have been the nutrition norm for endurance athletes. However, an alternative approach in training, known as *fat adaptation*, is becoming more popular among endurance athletes.

Of all the energy-yielding nutrients, carbohydrates are utilized most rapidly to fuel exercising muscles. When athletes are following a high-carbohydrate dietary pattern or practice carbohydrate loading before an endurance event, they ensure that muscle and liver glycogen will be available to muscles throughout the race. Even after carbohydrate loading, however, the total amount of energy available from muscle glycogen is limited to 2000 to 3000 kcal.

In comparison, the supply of energy from triglycerides stored in the muscle and adipose tissue is virtually limitless (50,000 to 100,000 kcal). Recall that, depending on intensity, about half of the energy for endurance events comes from fat. The metabolism of fat for energy occurs more slowly, but it provides more than twice as many calories per gram as carbohydrates or protein.

With fat adaptation, rather than following a traditional high-carbohydrate dietary pattern (about 65% of calories from carbohydrates and about 20% from fat) during the days leading up to an event, endurance athletes replace much of the carbohydrates with fat. For example, a high-fat training eating plan might consist of just 25% of calories from carbohydrate with 60% to 70% of calories from fat. The rationale is that high-carbohydrate dietary patterns, especially those with many simple sugars and refined grains, boost insulin secretion, which inhibits the breakdown of fat. By lowering carbohydrates and increasing the fat content of the diet, the cells will adapt to greater use of fat for fuel. If the athlete uses more fat for fuel during an endurance event, muscle glycogen might be spared, so that those stored carbohydrates would be available for a burst of speed at the end of the race.[12]

Research comparing the effects of high-carbohydrate or high-fat training diets on athletic performance has yielded mixed results. One possible explanation is that the muscles of *fat-adapted* athletes are able to break down more fat for fuel during physical activity at low or moderate intensities, but the low carbohydrate intake depletes glycogen stores, so higher-intensity activity is impaired. More information can be found at https://www.sportsdietitians.com.au.

At this time, there is not enough consistent evidence to support a recommendation for high-fat diets for athletes. Although high-fat diets can lead to more fat oxidation and lower rates of muscle glycogen utilization, no improvements in overall performance are reported. Current evidence does support a performance-enhancing effect of carbohydrate ingestion before and during physical activity. To avoid compromising training performance, athletes exercising at a moderate to high intensity or engaging in competition should include quality carbohydrates in their dietary pattern that matches their training and body composition goals.

**Replenish Fuel During the Event.** We have already established the importance of consuming adequate fluids during endurance activities. For continuous sporting events longer than 60 minutes, consumption of carbohydrates and electrolytes during activity can also improve athletic performance. Prolonged exercise depletes muscle glycogen stores and may transiently lower blood glucose, leading to physical and mental fatigue. One way to avoid *hitting the wall* is to maintain normal blood glucose concentrations by carbohydrate feedings during the activity. During short events (e.g., 30 minutes or so), carbohydrate intake is not as important because the muscles do not take up much blood glucose during short-term activities, relying instead primarily on their glycogen stores for fuel.

A general guideline for endurance events is to consume 30 to 60 grams of carbohydrate per hour. A trend in sports nutrition is to use multiple sources of carbohydrates (e.g., glucose, fructose, and maltodextrin) with different routes and rates of absorption to maximize the supply of glucose to cells and lessen the risk of gastrointestinal distress.[5]

Some experts suggest that consuming protein with carbohydrate during exercise provides added benefit. This is an area of ongoing research, but there is not enough evidence at this time to support a clear recommendation for ingesting protein during exercise. Some products formulated for consumption during exercise do contain amino acids, so you can try various formulations to see what works best for you.

Compared to carbohydrates, fat is more slowly digested, absorbed, and metabolized. Thus, although fat serves as fuel during prolonged aerobic activity, consumption of fat during activity is unlikely to improve athletic performance and is likely to cause gastrointestinal distress.

Sports drinks are a good source of carbohydrate calories during continuous endurance events lasting over 60 minutes. Sports drinks usually contain about 14 grams of carbohydrate per 8-ounce serving. They supply the necessary fluid, electrolytes, and carbohydrates to keep an athlete performing at his or her best in endurance or ultra-endurance events.

As an alternative to sports drinks, some athletes use carbohydrate gels or chews. Gels and chews are formulated with one or more sugars or starches to rapidly supply about 25 grams of carbohydrate per serving (Table 10-8). In addition, they provide electrolytes to replenish those lost in sweat. Some of these products also may contain certain amino acids, vitamins, caffeine, or herbal ingredients. An advantage of gels compared to energy bars or sports drinks is they are convenient to carry.

Popular energy bars typically provide about 180 to 250 kcal and anywhere from 2 to 45 grams of carbohydrate. The wide range of carbohydrate content in energy bars is due to a variety of marketing trends in the sports supplement industry. If a bar is chosen, look for one with about 40 grams of carbohydrate and no more than 10 grams of protein, 4 grams of fat, and 5 grams of fiber. The bars are fortified with vitamins and minerals in amounts ranging from about 25% to 100% of typical human needs. Outside of sporting events, some people use energy bars as a quick and convenient meal or snack. Keep in mind that energy bars contain a concentrated source of calories. Whole foods are always preferred for snacking.

Check the label on all of these products to gauge the amount of gel or bar that provides 30 to 60 grams of carbohydrate per hour. In addition, remember that any carbohydrate-containing food must be accompanied by fluid to ensure adequate hydration. At a minimum, one serving of any of these products will cost at least $1, and some

▲ Sports gels and chews are often used to provide fuel during athletic events.
McGraw-Hill Education. Mark Dierker/McGraw-Hill

**TABLE 10-8 ■ Energy and Macronutrient Content of Energy Bars, Gels, and Chews**

| Energy Bars | Serving (oz) | Energy (kcal) | Carbohydrates (g) | Fiber (g) | Protein (g) | Fat (g) |
|---|---|---|---|---|---|---|
| Atkins® (chocolate coconut) | 1.41 | 170 | 19 | 9 | 4 | 12 |
| Clif® Bar (crunchy peanut butter) | 2.4 | 260 | 40 | 4 | 11 | 7 |
| KIND® (dark chocolate cherry cashew) | 1.41 | 160 | 22 | 6 | 4 | 10 |
| Luna® (lemon zest) | 1.69 | 190 | 28 | 3 | 8 | 6 |
| Larabar® (banana bread) | 1.59 | 200 | 24 | 4 | 5 | 10 |
| **Gels & Chews** | | | | | | |
| Clif® Shot (vanilla) | 1.1 | 100 | 25 | 0 | 0 | 0 |
| GU™ Energy Gel (lemon sublime) | 1.1 | 100 | 23 | 0 | 0 | 0 |
| Huma Chia Energy Gel® (strawberry) | 1.52 | 100 | 22 | 2 | 1 | 0.5 |
| PowerBar® Gel (strawberry banana) | 1.44 | 110 | 27 | 0 | 0 | 0 |

Overall, choosing energy bars is preferable to choosing candy bars and packaged desserts. An additional concern is that micronutrient toxicity (e.g., vitamin A) might occur if numerous bars are eaten in a day, as many are highly fortified.

brands with all natural or organic ingredients cost as much as $5. Are sports drinks, energy bars, and gels worth the price? Critics suggest that these products are essentially the nutritional equivalent of a cup of low-fat yogurt and a piece of fruit (Table 10-8). For an athlete on a tight budget, a small bag of graham crackers or jelly beans could just as easily provide a quick shot of glucose during a race. With a little bit of time and an Internet connection, you can even find recipes to make your own sports drinks and energy bars at home for a fraction of the cost of name-brand products.

**After Physical Activity, Replenish Glycogen and Fluid.** After prolonged aerobic activities, muscle and liver glycogen stores will be depleted. An athlete's need to pay special attention to nutrition during recovery depends on the type of workout completed and the timing of the next workout. For example, multiple events in the same day will require rapid restoration of glycogen stores. However, if an athlete will be able to rest for 1 or 2 days before the next exercise session, immediate consumption of a post-workout meal is not as crucial.

To rapidly restore glycogen stores, carbohydrate-rich foods providing approximately 1.0 to 1.2 grams of carbohydrate per kilogram of body weight should be consumed shortly after continuous (endurance) physical activity. Immediately after exercise is when glycogen synthesis is greatest because the muscles are insulin-sensitive at this point. Foods such as fruit, fruit juice, bread, or a sports drink contribute to rapid restoration of glycogen stores.

Although carbohydrate intake is the most important factor for replenishing glycogen after endurance activities, adding an appropriate amount of high-quality protein during recovery can be helpful for stimulating glucose uptake and repairing damaged muscle tissue. The current recommendation for protein after endurance exercise is 0.25 to 0.3 gram of protein per kilogram of body weight.

Fluid and electrolyte (i.e., sodium and potassium) intake is another essential component of recovery for an endurance athlete, especially if two workouts a day are performed or if the environment is hot and humid. Specialized recovery drinks containing carbohydrates, amino acids, and electrolytes are available, but if food and fluid intake is sufficient to restore weight loss, it generally will also supply enough electrolytes to meet needs during recovery from endurance activities.

## STRENGTH AND POWER ATHLETES: STRATEGIES TO ENHANCE MUSCLE GAIN

Strength training—improving the maximal force that can be exerted by the muscle—should be part of any well-rounded physical activity program. Resistance training may utilize free weights, specialized weight machines, or one's own body weight. As previously mentioned, a workout typically includes 8 to 12 different muscle-strengthening activities that target all the major muscle groups of the body. Most of the people you see lifting weights in the gym are probably performing several sets of 8 to 12 repetitions each, lifting about 50% of the maximum weight they could lift (1 repetition maximum [RM]). This type of workout improves muscular *endurance,* which is an important part of muscular fitness for overall health. However, to truly build muscular *strength,* athletes need to work against greater resistance (around 80% of 1 RM) over fewer repetitions. For a few athletes, such as those who participate in weight-lifting or body-building competitions, muscular strength is the focus of training.

Muscular power combines strength with speed, improving the ability to apply force quickly. Examples of power sports include middle-distance running, gridiron football, rowing, and swimming. In reality, many sports and everyday activities involve muscular power: jumping for a rebound in basketball, delivering a roundhouse kick to an

## ASK THE RDN    Boosting your Metabolism

*Dear RDN: I am concerned that my "slow metabolism" may be making it difficult to maintain a healthy weight. Are there things I can do that will increase my metabolism and help prevent my weight from creeping up?*

Metabolism is an important component of weight control. Recall that your basal metabolic rate (BMR) determines how quickly your body burns energy to maintain basic functions such as heartbeat and breathing. You may have inherited a "slow" metabolism, and after age 40 BMR steadily declines for all of us. This is a major contributor to age-related weight gain. Gaining fat mass while losing lean mass will slow metabolism even more.

Good news! You can speed up your metabolism, by incorporating the following tips into your lifestyle.

1.  After a good night's *sleep* (7 to 9 hours), simply *eating breakfast* is one of the most effective ways to improve your metabolism. While we sleep, BMR slows down to conserve energy. Depriving yourself of sleep, however, can lead to an overall slower metabolic rate because of negative effects on insulin and stress hormone levels, appetite, and your body's ability to metabolize sugar. Eating a healthy breakfast in the morning jump-starts your metabolism, breaking it out of its fasting metabolic state.[12] Research shows that breakfast eaters are more likely to have a healthy BMI (under 25). A breakfast that combines complex carbohydrates, protein, and a little fat will get your metabolism going and will keep you satisfied until lunch.[13]

▲ Eating smaller meals more often, drinking plenty of water, and exercising to stimulate and build muscle will all help to increase your metabolism. PM78/iStock/Getty Images

2.  Believe it or not, *eating more often* and snacking can actually help boost your metabolism. As you go through your day, eating a small meal or snack every 3 to 4 hours will help you to burn more calories. In contrast, eating large meals, skipping meals, following very-low-calorie diets, and fasting are not good strategies to boost your metabolism. These actions often decrease overall calorie intake but at the expense of a slower metabolism. Although very-low-calorie diets or intermittent fasting[14] may help you lose weight initially, they will eventually fail because they are difficult to maintain and burn muscle, not fat.

3.  *Eating more protein-rich foods* will boost metabolism because more calories are burned digesting protein compared to fat or carbohydrates. Protein-rich foods, including lean meats, tofu, nuts, seeds, beans, eggs, and low-fat dairy products, will also provide building blocks for the growth and maintenance of muscle.[13]

4.  Muscle is more metabolically active than fat; every pound of muscle requires about 6 kcal a day just to sustain it, compared to only 2 kcal per pound of fat. Therefore, *building muscle* can cause a long-term increase in your metabolism. Resting metabolic rate (the calories your body burns even when you are at rest) is much higher in people with more muscle mass. Strength training not only leads to muscle growth but also raises your energy expenditure for the whole day. *Aerobic exercise* will increase your metabolism, not only during a workout but also for a few hours afterward. Including **high-intensity interval training** (HIIT) will result in a greater, more sustained rise in resting metabolic rate compared to low- or moderate-intensity workouts.

5.  As you are eating more protein and getting more physical activity, remember to *stay hydrated.* Even mild dehydration will slow your metabolism. Individuals who drink more water (eight or more glasses a day) burn more calories than those who drink less. Drinking a glass of water or other unsweetened beverage before every meal and snack will help you stay hydrated. Drinking caffeine-containing beverages such as coffee, tea, or energy drinks may cause a short-term rise in your metabolic rate. However, "energy drinks" should not be part of your metabolism-boosting program (see *What the Dietitian Chose* in Connect for more on energy drinks). The high levels of caffeine and various other ingredients may cause more harm than good.

In a nutshell, to speed up your metabolism, get plenty of sleep, start your day with breakfast, eat smaller frequent meals, drink plenty of water, and exercise to stimulate and build muscle.

*Kicking it up a notch,*

**Anne M. Smith, PhD, RDN, LD**
Associate Professor Emeritus, The Ohio State University, Author of *Contemporary Nutrition*

Monty Soungpradith/
Open Image Studio LLC

**high-intensity interval training (HIIT)**  Short, intense aerobic workout that alternates short periods of intense anaerobic exercise with less-intense recovery periods to provide improved athletic capacity and condition, improved glucose metabolism, and improved fat burning.

▲ Low-fat chocolate milk is the go-to recovery drink for many athletes. This 2-cup serving of low-fat chocolate milk is a tasty vehicle for 49 grams of carbohydrate and 17 grams of protein. Pixtal/SuperStock

▲ A special nutrition issue that concerns some strength-trained athletes is muscle dysmorphia, informally called *bigorexia*. In this disorder, individuals see themselves as being too thin, even though they are more muscular than average. People who suffer from muscle dysmorphia may practice disordered eating behaviors or use steroids to achieve high levels of muscularity. Ingram Publishing

opponent in martial arts, or driving the ball down the fairway in a round of golf are examples of muscular power in sports.

For strength and power athletes, some sports nutrition alterations are warranted. First, calorie needs will be higher due to the additional lean mass and high-volume training routines of these athletes. Recall that the primary types of fuel for strength and power moves are phosphocreatine (PCr) and carbohydrates for the brief bursts of activity, with fat providing energy during the resting stages. Very little protein is used as fuel during resistance activities. Second, there will be some extra emphasis on protein intake in the recovery phase.[15] Low-fat chocolate milk is a great option to replenish the body!

Strength and power athletes tend to be extremely focused on consuming adequate protein to support muscle protein synthesis. Strength-training athletes in the early phases of training have the highest estimated protein needs of any athletes. Once desired muscle mass has been achieved, protein requirements for maintenance of muscular strength decrease slightly. Meeting these recommendations for protein intake optimizes muscle protein synthesis, but consuming more than the recommended range of protein intake does not appear to offer advantage and could be detrimental. Recall that excess amino acids are used as fuel or stored as fat; they do not directly translate into increased muscular strength.

**Before and During Strength and Power Training, Focus on Calories, Carbohydrates, and Fluids.** Adequate hydration supports optimal athletic performance; strength and power athletes are no exception. Checking the urine color or urine specific gravity is a good indication of fluid status. If an athlete is poorly hydrated before an event, water or a sports drink should be sufficient to restore hydration.

Similar to nutrition strategies for endurance athletes, adequate carbohydrate ingestion in the days leading up to and hours immediately before physical activity has been shown to enhance performance for strength and power events, too. Athletes who perform many repetitions with moderate resistance will use more of their muscle glycogen stores than athletes who perform fewer repetitions with high resistance. Overall, research has shown that consuming approximately 4 to 7 grams of carbohydrates per kilogram of body weight per day is appropriate for strength and power training. The optimal rate of carbohydrate ingestion before and during resistance activities has not yet been established, but some research indicates that 1 to 4 grams of carbohydrate per kilogram of body weight in a pre-event meal or beverage will enhance work capacity during resistance workouts.

In strength and power sports, many athletes also use creatine supplements to increase levels of phosphocreatine in muscles. Recall that phosphocreatine is used to resupply ATP during short, intense bursts of activity. When phosphocreatine stores are increased, muscle glycogen may be preserved.[15]

Focus on achieving adequate hydration and maximizing muscle glycogen before activities because there may not be an opportunity to replenish fluids and carbohydrates during strength or power competitions. During extended training sessions, however, supplying fluids and carbohydrates will enhance both physical and mental performance.

While a few experts advocate ingesting protein before or during a resistance workout in an effort to promote muscle protein synthesis, the bulk of evidence points to emphasizing protein during the recovery period for optimal performance.

For athletes (and adults in general), fat intake should fall into the range of 20% to 35% of overall calorie intake. Dietary analysis shows that usual fat intake of resistance-trained athletes is slightly higher than recommended, probably because too many of their choices of protein-rich foods (e.g., meats and dairy products) are also rich sources of fat. If fat intake is above 35% of total calories, replacing the excess fat with carbohydrates would have a favorable effect on protein balance. This is because insulin, secreted in response to glucose in the blood, triggers uptake of amino acids by cells, which provides materials for protein synthesis within the cells.

**TABLE 10-9** ■ **Sample Recovery Meals**

**Option 1:** 491 kcal, 69 grams carbohydrate, 31 grams protein, 10 grams fat
   Bagel, 1 regular
   Deli turkey, 1 ounce
   Mozzarella cheese, 1 ounce
   Low-fat milk, 1 cup

**Option 2:** 571 kcal, 81 grams carbohydrate, 38 grams protein, 12 grams fat
   Low-fat flavored Greek yogurt, 16 ounces
   Banana, 1 medium

**Option 3:** (plant-based): 289 kcal, 52 grams carbohydrate, 13 grams protein, 4 grams fat
   Chickpeas, ½ cup
   Quinoa, ½ cup
   Tomatoes, ⅔ cup canned

**After Strength and Power Activities, Consuming Carbohydrates and Protein Promotes Recovery.** Those first few hours after resistance training, according to many researchers, are the best time to provide carbohydrates and protein to replenish muscle glycogen and promote muscle repair and synthesis. Right after exercise, the cells are insulin sensitive, so they rapidly take up glucose from the blood and store it as glycogen. General ACSM guidelines are to consume 1.0 to 1.2 grams of carbohydrate per kilogram of body weight to restore muscle glycogen. To promote muscle protein synthesis along with glycogen restoration, many experts recommend intakes at the upper end of that range (e.g., 1.2 to 1.5 grams of carbohydrates per kilogram of body weight) shortly after training.[5] The presence of certain amino acids further stimulates insulin secretion to enhance the uptake of glucose and synthesis of glycogen.

To promote gains in muscle mass, many experts recommend 0.25 to 0.30 gram of high-quality protein per kilogram of body weight within the first 1 or 2 hours after exercise to maximize protein synthesis.[11] Novice strength-training athletes who are seeking to gain muscle mass have the highest requirements for protein. With advanced training, the rate of protein turnover during exercise decreases. Therefore, well-trained strength athletes require less protein to repair and maintain muscles than their untrained counterparts. Some amino acids (e.g., leucine) may stimulate the metabolic pathways that lead to synthesis of muscle protein. The process of muscle protein synthesis not only requires amino acids as building blocks, of course, but also depends on carbohydrate as a source of energy.

Overall, recovery from resistance training requires a combination (3:1 ratio) of carbohydrates and high-quality protein. For a 154-pound (70-kilogram) athlete, this corresponds to about 70 grams of carbohydrate and 25 grams of protein in each 2-hour interval. Table 10-9 provides several options for recovery meals for athletes.

**Special Dietary Patterns for Athletes.** Some athletes adopt special dietary patterns to align with their culture, beliefs, or perceptions that these will improve health or physical performance. Vegetarian; low fermentable oligo-, di-, monosaccharides and polyols (FODMAPs); gluten free; and fasting are the most popular special diets of athletes. Note that dietary restrictions often do more harm than good for this population. For instance, no scientific evidence has documented any direct benefits of avoidance of gluten with healthy athletes. Gluten-free diets are associated with a reduction of FODMAPs, which may improve adverse gastrointestinal symptoms. Vegetarian dietary patterns can support the nutritional demands of athletes, but careful planning and execution are required to ensure proper intakes of energy and nutrients of concern (e.g., iron).[14] When adherence to a special dietary pattern is a necessity (e.g., during Ramadan), advice from an RDN is recommended to guide the athlete on nutrition strategies for optimal health and performance.[16]

▲ Athletes are subject to all the same nutritional challenges as the general public: overreliance on convenience foods, abundance of nutrition misinformation, temptations to eat out of boredom or for emotional comfort. Furthermore, they must adapt to the seasonal demands of their sport and maintain exhausting training and travel schedules. ©George Postalakis

## CONCLUDING REMARKS

Nutritional strategies have the potential to optimize athletic performance. Here, we have presented several generalized guidelines to plan nutritionally adequate dietary patterns that optimize energy stores, ensure hydration, and give athletes a competitive edge. We have stressed the importance of a food-first philosophy for carbohydrates and fluids pre-event, within-event, and between-events activity, as well as protein for muscle recovery. Above all, recognize that each athlete is unique. Remember that genetics can impact nutrition requirements. Each type of activity demands its own set of energy sources. Sports vary in training regimens, duration, and opportunities to acquire nourishment before, during, and between events. Even within a particular sport, each player's position has its own physical demands, which can alter nutritional needs.[14] Finally, personal taste preferences and gastrointestinal tolerance will dictate adherence to any nutrition plan. As you pursue your own physical activity pattern, start with your solid foundation of knowledge about nutrient needs, but be attentive to your concerns, be adaptable, and always continue to learn.

## CASE STUDY  Planning a Training Diet

Michael is a 6-foot, 185-pound male who is training for a 10K run coming up in 3 weeks. He has read a lot about sports nutrition and especially about the importance of eating more carbohydrates while in training. He also has been struggling to keep his weight in a range that he feels contributes to better speed and endurance. Consequently, he is also trying to eat as little fat as possible. Unfortunately, over the past week, his workouts in the afternoon have not met his expectations. His run times are slower, and he shows signs of fatigue after just 20 minutes into his training program.

His breakfast yesterday was a large bagel with cream cheese and orange juice. For lunch, he had a small salad with fat-free dressing, a large plate of pasta with marinara sauce and broccoli, and a diet soft drink. For dinner, he had a small broiled chicken breast, a cup of rice, some carrots, and iced tea. Later, he snacked on fat-free pretzels.

1. Is a high-carbohydrate dietary pattern a good idea during Michael's training?
2. Are there any important components missing in Michael's diet? Are missing components contributing to his fatigue?
3. Describe some changes that should be made in Michael's eating plan, including specific foods that should be emphasized.
4. How should fluid needs be met during Michael's workouts?
5. Should Michael focus on fueling his body before, during, or after workouts?

*Complete the Case Study. Responses to these questions can be provided by your instructor.*

▲ Michael, a senior studying engineering, is wise to seek nutrition advice in preparing for his first 10K. Comstock/Getty Images

## ✓ CONCEPT CHECK 10.5

1. Which nutrient(s) should be emphasized in a pre-event meal for an endurance athlete? Provide an example of a suitable pre-event meal for a long-distance cyclist.

2. What is carbohydrate loading? List three sports for which carbohydrate loading could enhance performance.

3. Why is a combination of carbohydrate and protein recommended for recovery after resistance training? Suggest a suitable recovery meal.

# 10.6 Nutrition and Your Health
## Ergogenic Aids and Athletic Performance

▲ The demands of collegiate athletics are great. Elite athletes should be encouraged to gain the competitive edge with proper rest, fluids, stress reduction, and healthy dietary patterns versus opting for ergogenic aids and unnecessary supplementation. Colleen Spees

Extreme diet manipulation to improve athletic performance is not a recent innovation. Today's athletes are as likely as their predecessors to experiment with any substance that promises a competitive advantage. The U.S. sports nutrition supplement market now exceeds $11 billion, accounting for more than 35% of the global market. Caffeine, creatine, nitrate/beetroot juice, beta-alanine, and bicarbonate are just some of the substances used by athletes in hopes of gaining an ergogenic edge.[17]

Based on what is known, today's athletes can benefit from scientific evidence documenting the ergogenic properties of a few dietary substances. These ergogenic aids include sufficient water and electrolytes, adequate carbohydrates, and a balanced and varied dietary pattern consistent with MyPlate. Protein and amino acid supplements are often not needed as the vast majority of athletes can easily meet their protein needs from a food-first approach. In general, nutrient supplements should only be used to meet a specific dietary shortcoming, such as an inadequate iron intake. Beyond the proven benefits of the nutrition strategies presented in this chapter, Table 10-10 provides a list of well-studied ergogenic aids.

Dietary supplements rumored to enhance athletic performance require careful evaluation and monitoring. Overall, there is little scientific evidence to support the efficacy of many substances that are touted as performance-enhancing aids. Of these, many are useless, and some are dangerous enough to promote organ damage (Table 10-11). The liver and kidneys are particularly susceptible to damage because these organs help detoxify harmful compounds.[15] Athletes should be skeptical of any substance until its ergogenic effect is scientifically validated. The Food and Drug Administration (FDA) has a limited ability to regulate these dietary supplements, and the manufacturing processes for dietary supplements are not as tightly regulated by the FDA as they are for prescription drugs.

Some supplements contain substances that will cause athletes to test positive for various banned substances. For instance, creatine is not a substance banned by the National Collegiate Athletic Association (NCAA) since it is classified as a nutritional supplement. However, positive drug tests have occurred with some creatine supplements because they may contain other nonpermissible substances banned by the NCAA. Studies consistently show that many supplements do not contain the substance and/or the amount listed on the label. Not only must athletes determine whether there is evidence that a dietary supplement is safe and effective, but they must also question if the dietary supplement contains what it is supposed to contain.

### National Collegiate Athletic Association (NCAA) and Nutrition Supplements

The NCAA has developed lists of *permissible* and *nonpermissible* nutritional supplements for athletic departments to provide to student athletes. The NCAA has issued a warning advising students to discuss their use of *any* dietary supplement with their team medical staff to avoid unknowingly ingesting banned substances. Following are a few key examples:

**Select Permissible Nutritional Supplements**
- Carbohydrate boosters
- Electrolyte/carbohydrate replacement drinks
- Energy bars
- Vitamins and minerals

**Select Nonpermissible Drug Classes**
- Alcohol and beta blockers (rifle only)
- Anabolic agents
- Beta-2 agonists
- Cannabinoids
- Diuretics and masking agents
- Growth factor, related substances, and mimetics
- Hormone and metabolic modulators (anti-estrogens)
- Narcotics
- Peptide hormones
- Stimulants

For a complete explanation of the NCAA's rules regarding dietary supplements, see www.ncaa.org.

### Sports Food

*Electrolyte supplements* may be used for rehydration or hydration by replacing electrolytes lost in sweat.

*Liquid meals* may provide a quick and convenient source of carbohydrate, protein, and nutrients when it is eating is not practical.

*Protein supplements* may provide a quick and convenient source of easily digested, high-quality protein.

*Sports drinks* may be used for hydration and fueling strategies for longer or high-quality training sessions or longer races.

*Sports gels/confectionery* may be for fueling strategies during longer training sessions and races.

Source: International Association of Athletics Federations Consensus Statement 2019: Nutrition for Athletics.

**TABLE 10-10 ■ Commonly Used Supplements for Athletes**

| Substance/Practice | Purported Benefits | Potential Risks |
|---|---|---|
| Antioxidants | Decreased levels of free radicals | Increased risk of cancers for certain supplements, nutrient-specific toxicities |
| Beta-Alanine | Increased muscle carnosine, a protein that neutralizes acidic compounds that contribute to muscle fatigue during high-intensity activity | Flushing and feelings of "pins and needles" |
| Beta-Hydroxy Beta-Methylbutyric Acid (HMB) | Decreased muscle damage, speeds up recovery from intense physical activity | Itching, abdominal pain, and constipation |
| Branched-Chain Amino Acids (BCAA) | Increased energy delivered to muscles during activity to increase muscle size and strength | Increased levels of ammonia in the blood, fatigue, loss of motor coordination, digestive discomfort, nausea, vomiting, and diarrhea |
| Caffeine | Increased vigilance and mental alertness; improved endurance; reduced perception of fatigue; enhanced lipolysis and fast oxidation | High doses may cause insomnia, nervousness, restlessness, digestive discomfort, nausea, vomiting, rapid heart rate, increased respirations, tremors, delirium, convulsions, and increased urination. |
| Calcium | Maintains strength of bones and teeth | High doses may cause digestive discomfort, kidney stones, and heart problems. |
| Collagen/Gelatin | Alleviate joint pain | Digestive discomfort, constipation, anorexia, and skin itching |
| Creatine | Increased lean mass; improved short-term performance; aid muscle recovery; delay muscle fatigue | Possible heat intolerance, fever, dehydration, reduced blood volume, electrolyte imbalances, digestive discomfort, and muscle cramping |
| Curcumin/Turmeric | Muscle repair, decreased soreness, and decreased inflammation | Digestive discomfort, constipation, indigestion, diarrhea, abdominal distension, acid reflux, nausea, and vomiting |
| Fish Oil | Decreased muscle soreness | Fish burps, heartburn, acid reflux, nausea, diarrhea, and skin rash |
| Iron | Overcome deficiency that can lead to fatigue and irritability | High doses can cause digestive discomfort, nausea, vomiting, constipation, and diarrhea. |
| Magnesium | Support testosterone production | Digestive discomfort, nausea, vomiting, and diarrhea |
| Nitric Oxide Boosters (arginine, beetroot juice, citrulline) | Improved blood flow to muscles and enhanced performance by improving oxygen consumption to deliver nutrients to muscles | Drop in blood pressure, dizziness, lightheadedness, and loss of balance |
| Protein Powder | Provide necessary protein in absence of whole foods, increased lean mass production | High doses can cause nausea, thirst, bloating, cramps, diarrhea, reduced appetite, and fatigue. Overreliance on supplements over food can increase potential for poor intakes of other key nutrients. |
| Probiotics | Improved gut health with mental and physical benefits | Digestive discomfort, rash, itching; infections in certain high-risk individuals |
| Ketones | Support fat burning | Shakiness and abnormal heartbeat |
| Sodium Bicarbonate (baking soda) | Neutralize acidic compounds that contribute to muscle fatigue | Digestive discomfort, diarrhea, and vomiting |
| Tart Cherry | Reduced stress on the body from heavy bouts of training | Digestive discomfort and diarrhea |
| Vitamins and Minerals | Provide necessary nutrition in absence of a balanced diet | Potential for nutrient-specific toxicity and related side effects (e.g., nausea, vomiting, organ damage); overreliance on supplements over food can increase potential for poor intakes of other key nutrients |
| Vitamin D | Increased skeletal muscle function, decreased recovery time, improved power, and support of testosterone production | High doses can cause toxicities, leading to weakness, fatigue, sleepiness, headaches, loss of appetite, dry mouth, metallic taste, digestive discomfort, weight loss, and seizures. |

Sources: Adapted from Natural Medicines Comprehensive Database (http://naturaldatabase.therapeuticresearch.com) and Dietary Supplements for Exercise and Athletic Performance, National Institutes of Health, Office of Dietary Supplements (https://ods.od.nih.gov/factsheets/ExerciseAndAthleticPerformance-Consumer).

**TABLE 10-11  ■  Dangerous, Banned, or Illegal Substances and Practices**

| Substance/Practice | Purported Use | Risks |
|---|---|---|
| Anabolic Agents (e.g., testosterone) | Increase muscle mass and strength | Liver cysts; increased risk of heart disease, hypertension, reproductive dysfunction; depression, sleep disturbances, mood swings |
| Alcohol and Beta Blockers | Decrease anxiety and allow for muscle relaxation | Changes in blood sugar, symptoms of heart failure |
| Anti-Estrogens (e.g., anastrozole) | Improved physique, reduced estrogen and maximized testosterone production, masked signs of anabolic steroid use | Weakness, headaches, sweating, stomach pain, nausea/vomiting, poor appetite, weight gain, joint/bone/muscle pain, mood changes, depression |
| Beta-2 Agonists (e.g., bambuterol) | Increased muscle strength, power; improved lung function | Chest pain, dizziness, dry mouth, headache, changes in blood pressure, muscle cramps, rapid heartbeat |
| Blood Doping | Enhanced aerobic capacity by increasing red blood cells | Blood thickening that strains heart |
| Diuretics and Masking Agents (e.g., furosemide) | Rapid weight loss and masked presence of other banned substances | Rapid depletion of electrolytes, fatigue, dizziness, or muscle cramps |
| Gene Doping | Increased muscle mass, fat burning, and endurance | Fatal immune responses, blood thickening, death |
| Illicit Drugs | Relaxation, stress relief, pain management | Increased heart rate, dizziness, slowed reaction time, overdose, death |
| Local Anesthetics | Pain management and injury recovery | Delayed recovery, worsening of injuries, damage to muscle or tendons |
| Peptide Hormones and Analogues (e.g., growth hormone) | Increased muscle mass and fat metabolism | Uncontrolled growth of the heart and other internal organs; death |
| Stimulants (e.g., caffeine, ephedrine) | Increased muscle strength and power, promote mental alertness, weight loss | Heart palpitations, anxiety, and death |

Source: NCAA Banned Drugs List (http://www.ncaa.org/2018-19-ncaa-banned-drugs-list).

Even substances whose ergogenic effects have been supported by scientific evidence should be used with extreme caution, as the testing conditions may not match those of the intended use. Careful judgment should be exercised when it comes to using the appropriate dose of supplements or using multiple types of supplements concurrently.[18]

Rather than waiting for a magic bullet to enhance performance, athletes are advised to concentrate their efforts on improving their training routines and sport techniques, while adhering to a well-balanced dietary pattern embracing whole foods as described in this chapter.

**✓ CONCEPT CHECK 10.6**

1. Name two permissible and two nonpermissible ergogenic aids.

2. What are some risks associated with commonly used ergogenic aids?

# Summary (Numbers refer to numbered sections in the chapter.)

**10.1** A gradual increase in regular physical activity is recommended for all persons. Benefits include improvements in cardiovascular health, gastrointestinal function, blood glucose regulation, mental health, quality of life, and sleep; reduced risk of certain cancers and enhanced muscle and bone strength.

**10.2** The *Physical Activity Guidelines for Americans* advise adults to do 150 to 300 minutes of moderate-intensity or 75 to 150 minutes of vigorous-intensity aerobic physical activity per week. In addition, adults should perform muscle-strengthening activities and flexibility activities at least twice per week. Workouts should allow time for warm-up exercises to increase blood flow and warm the muscles and then end with cool-down exercises, including stretching. Ultimately, any activity is better than no activity.

**10.3** Human metabolic pathways extract chemical energy from carbohydrate, fat, and protein to yield ATP. Phosphocreatine is a high-energy compound that can be used to resupply ATP during short, intense activities. The mix of macronutrients used for fuel depends on the intensity and duration of activity: short-term, intense activities primarily use carbohydrate for fuel, whereas low- or moderate-intensity endurance activities use more fat for fuel. Protein makes a minor contribution as a fuel source.

**10.4** To support physical activity, athletes require 5 to 8 kcal per minute of activity above energy needs for a sedentary person. Monitoring weight changes over time is a good way to assess the adequacy of energy intake. Athletes should obtain energy from a varied dietary pattern that includes sources of carbohydrates (3 to

12 grams of carbohydrate per kilogram of body weight; usually 60% of total energy), protein (1.2 to 2.0 grams of protein per kilogram, depending on the type of training), and fat (up to 35% of energy, focusing on vegetable oils instead of solid fats). The increased overall food intake of athletes typically furnishes adequate vitamins and minerals. Some micronutrients of concern are iron and calcium, especially for women. Athletes should drink fluid before, during, and after physical activity (approximately 2 to 3 cups per pound lost). Sports drinks may help replace fluid, electrolytes, and carbohydrates lost during continuous workouts that last beyond 60 minutes.

**10.5** Endurance athletes can delay or prevent fatigue by consuming enough fluids, electrolytes, and carbohydrates before, during, and after events. In addition, protein in the post-activity period will aid muscle recovery. In addition to these strategies to maintain hydration and muscle glycogen stores, athletes who train to develop strength or power should place special emphasis on protein during the recovery period.

**10.6** Athletes can benefit from ergogenic properties of sufficient water, electrolytes, and carbohydrates, and a balanced and varied dietary pattern consistent with the *Dietary Guidelines* and MyPlate. Although some ergogenic aids may be useful for enhancing athletic performance, extreme caution should be practiced as many ergogenic aids are dangerous. Protein and amino acid supplements are not necessary because athletes almost always meet protein needs from whole foods.

# Check Your Knowledge (Answers are available at the end of this question set.)

1. An energy-rich compound, phosphocreatine (PCr), is found in _____ tissue.
   a. adipose
   b. muscle
   c. liver
   d. kidney

2. A physical activity program should include
   a. aerobic activities 5 days per week.
   b. strength-training activities 2 to 3 days per week.
   c. stretching exercises 2 to 3 days per week.
   d. all of these.

3. During muscle-building regimens, athletes should consume _____ grams of protein per kilogram body weight.
   a. 0.5 to 0.7
   b. 0.8
   c. 1 to 2
   d. 2 to 2.5

4. Which of these foods is the best choice for carbohydrate loading before endurance events?
   a. Potato chips
   b. French fries
   c. High-fiber cereal
   d. Rice

5. As the body adapts to regular exercise, the *training effect* results in
   a. decreased blood flow to muscles.
   b. increased lactate production.
   c. decreased muscle triglyceride content.
   d. decreased resting heart rate.

6. A physically active lifestyle leads to
   a. increased bone strength.
   b. decreased risk of colon cancer.
   c. reduced anxiety and depression.
   d. all of these.

7. Approximately how many cups of fluid are required to replace each pound of weight lost during an athletic event or workout?
   a. 0.5 to 0.75 cup
   b. 1 to 1.5 cups
   c. 2 to 3 cups
   d. 4 to 5 cups

8. The benefit of a sports drink is to provide
   a. water to hydrate.
   b. electrolytes to enhance water absorption in the intestine and maintain blood volume.
   c. carbohydrate for energy.
   d. all of these.

9. Compared to anaerobic glucose metabolism, aerobic glucose metabolism produces more
   a. lactate.
   b. ATP.
   c. phosphocreatine.
   d. fatty acids.

10. Caffeine is used as an ergogenic aid by some athletes because it is thought to
    a. decrease fatigue.
    b. decrease the buildup of lactate.
    c. serve as an energy source.
    d. increase muscle mass and strength.

Answer Key: 1. b (LO 10.3), 2. d (LO 10.2), 3. c (LO 10.6), 4. d (LO 10.8), 5. d (LO 10.5), 6. d (LO 10.1), 7. c (LO 10.7), 8. d (LO 10.7), 9. b (LO 10.7), 10. a (LO 10.9)

# Study Questions (Numbers refer to Learning Outcomes.)

1. How does greater physical fitness contribute to better overall health? Explain the process. **(LO 10.1)**

2. You have set a goal to increase muscle mass and decrease body fat. Plan a weekly fitness regimen using the FITT principle. **(LO 10.2)**

3. How are carbohydrates, fat, and protein used to supply energy during a 100-meter sprint? During a weight-lifting session? During a 3-mile walk? **(LO 10.3)**

4. What is the difference between anaerobic and aerobic activities? Explain why aerobic metabolism is increased by a regular fitness routine. **(LO 10.4)**

5. Is fat from adipose tissue used as an energy source during physical activity? If so, when? **(LO 10.5)**

6. What are some typical measures used to assess whether an athlete's calorie intake is adequate? **(LO 10.6)**

7. List five nutrients of specific concern for athletes and the appropriate food sources from which these nutrients can be obtained. **(LO 10.6)**

8. Your neighbor is planning to run a 5-kilometer race. Summarize for her what you have learned about fluid intake before, during, and after the event. **(LO 10.7)**

9. You plan to participate in a half-marathon. Plan your menu for the day of the event, being sure to include appropriate levels of macronutrients and fluids before, during, and after the athletic event. **(LO 10.8)**

10. Should competitive athletes take amino acid supplements? Why or why not? **(LO 10.9)**

# Further Readings

1. Centers for Disease Control and Prevention: Active People, Healthy Nation: At a Glance. Washington, DC. Available at https://www.cdc.gov/physicalactivity/downloads/Active_People_Healthy_Nation_at-a-glance_082018_508.pdf. Accessed January 22, 2020.

2. U.S. Department of Health and Human Services: *2018 Physical Activity Guidelines for Americans,* 2nd edition, 2018. Washington, DC: Office of Disease Prevention and Health Promotion. Available at http://www.health.gov/paguidelines. Accessed January 12, 2020.

3. Mach N and others: Endurance exercise and gut microbiota: A review. *Journal of Sport and Health Science* 6(2):170, 2017. DOI:10.1016/j.jshs.2016.05.001.

4. Harvard Medical School: Healthbeat: Do you need to see a doctor before starting your exercise program? Available at https://www.health.harvard.edu/healthbeat/do-you-need-to-see-a-doctor-before-starting-your-exercise-program. Accessed January 14, 2020.

5. Thomas DT and others: Position of the Academy of Nutrition and Dietetics, Dietitians of Canada, and the American College of Sports Medicine: Nutrition and athletic performance. *Journal of the Academy of Nutrition and Dietetics* 116:501, 2016. DOI:10.1016/j.jand.2015.12.006.

6. National Collegiate Athletic Association (NCAA): 2019–20 and 2020–21 NCAA Wrestling Rules. Available at http://www.ncaapublications.com/productdownloads/WR20.pdf. Accessed January 29, 2020.

7. Daily JP and Stumbo JR: Female athlete triad. *Primary Care: Clinics in Office Practice* 45:615, 2018. DOI:10.1016/j.pop.2018.07.004.

8. U.S. Department of Agriculture and U.S. Department of Health and Human Services. Dietary Guidelines for Americans, 2020-2025. 9th Edition. December 2020. Available at: DietaryGuidelines.gov.

9. Loenneke JP and others: Per meal dose and frequency of protein consumption is associated with lean mass and muscle performance. *Clinical Nutrition* 35(6)1506, 2016. DOI:10.1016/j.clnu.2016.04.002.

10. U.S. Department of Health and Human Services: Iron: Fact sheet for health professionals, Washington, DC, 2018. Available at https://ods.od.nih.gov/factsheets/Iron-HealthProfessional. Accessed January 26, 2020.

11. Kerksick CM and others: International Society of Sports Nutrition position stand: Nutrient timing. *Journal of the International Society of Sports Nutrition* 29(14)33, 2017. DOI:10.1186/1550-2783-5-17.

12. Sievert K and others: Effect of breakfast on weight and energy intake: Systematic review and meta-analysis of randomised controlled trials. *British Medical Journal* 364:142, 2019. DOI:10.1136/bmj.142.

13. Giancoli AN: Breakfast: Protein power. *Today's Dietitian* 18(9):4, 2016.

14. Patterson RE and others: Intermittent fasting and human metabolic health. *Journal of the Academy of Nutrition and Dietetics* 115(8):1203, 2015. DOI:10.1016/j.jand.2015.02.018.

15. Burke LM and others: International Association of Athletics Federations Consensus Statement 2019: Nutrition for Athletics. *Int J Sport Nutr Exerc Metab* 1;29(2):73, 2019. DOI:10.1123/ijsnem.2019-0065.

16. Lis DM and others: Dietary practices adopted by track-and-field athletes: Gluten-free, low FODMAP, vegetarian, and fasting. *International Journal of Sport Nutrition and Exercise Metabolism* 29, 2019. DOI:10.1123/ijsnem .2018-0309.

17. U.S. Department of Agriculture, National Agricultural Library: Ergogenic Aids. Available at https://www.nal.usda.gov/fnic/ergogenic-aids. Accessed February 28, 2020.

18. U.S. Department of Agriculture, National Agricultural Library: Nutrition for Athletes. Available at https://www.nal.usda.gov/fnic /nutrition-athletes. Accessed January 23, 2020.

# Chapter 11 › Eating Disorders

# Student Learning Outcomes

**Chapter 11 is designed to allow you to:**

**11.1** Contrast healthy attitudes toward uses of food with behavior patterns that could lead to unhealthy uses of food.

**11.2** Describe current hypotheses about the origins of eating disorders.

**11.3** List physical and mental characteristics of anorexia nervosa, and outline current best practices for its treatment.

**11.4** List physical and mental characteristics of bulimia nervosa, and outline current best practices for its treatment.

**11.5** Enumerate physical and mental characteristics of binge eating disorder, and outline current best practices for its treatment.

**11.6** Describe pica and other specified feeding and eating disorders.

**11.7** Discuss other patterns of disordered eating that are seen in clinical practice, but are not formally diagnosed as eating disorders.

**11.8** Describe strategies to reduce the development of eating disorders.

## FAKE NEWS

# Dysfunctional parent-child relationships cause eating disorders.

## THE FACTS

Historically, controlling parents were thought to cause anorexia nervosa, while unstructured, neglectful family relationships were implicated in the development of bulimia nervosa. However, placing the blame on family members is counterproductive in the treatment of eating disorders. Research has revealed that at least 50% of the risk for eating disorders is genetic. Indeed, eating disorders are biologically based mental disorders.[1] Family discord may trigger maladaptive coping mechanisms, but family members can be a vital ally in the treatment of eating disorders.

Many of us occasionally eat until we're stuffed and uncomfortable, such as at a holiday dinner. Faced with savory and tempting foods, we find that we cannot easily stop eating. Usually we forgive ourselves, vowing not to overeat the next time. Nevertheless, many of us struggle with controlling our food intake and body weight. The combination of too many instances of simple overeating and too little physical activity eventually leads to progressive weight gain.

The wellness community's focus on obesity tends to upstage all other nutrition-related issues. However, the consequences of eating disorders are just as serious as obesity; indeed, if left untreated, eating disorders can be fatal. What is most alarming about these disorders—anorexia nervosa, bulimia nervosa, binge eating disorder, and others—is the increasing number of cases reported each year.

It is widely accepted that eating disorders arise as a result of interactions between brain biology and environmental influences. Successful treatment of eating disorders, therefore, is complex and must go beyond nutritional therapy. Keep in mind that eating disorders are not restricted to any socioeconomic class or ethnicity. They can also strike at any age in either females or males. Let us examine the causes, effects, and treatments of these conditions and consider strategies for prevention.

▲ Maintaining an ultraslim body type is an all-too-common goal in today's culture. The media and the fashion world bombard us with body images that are unrealistic for most people. ©Lars A. Niki

**endorphins** Natural body tranquilizers that function in pain reduction and may be involved in the feeding response.

**disordered eating** Mild and short-term changes in eating patterns that occur in relation to a stressful event, an illness, or a desire to modify one's dietary pattern for a variety of health and personal appearance reasons.

# 11.1 From Ordered to Disordered Eating Habits

Eating—a completely instinctive behavior for animals—serves an extraordinary number of physiological, psychological, social, and cultural purposes for humans. Eating practices may take on religious meanings; signify bonds within families and ethnic or racial groups; and provide a way to demonstrate affection, concern, prestige, or class values. Within the family, preparing, sharing, and even withholding food may be a means of expressing love or power.

We are bombarded daily with our society's portrayal of the "ideal" body. Dieting is promoted to achieve this ideal body—eternally young and physically fit. Television programs, billboard and Internet advertisements, video games, magazines, movies, and social media suggest that an ultraslim body will bring happiness, love, admiration, and success. This fantasy notion is contradictory to the high rate of obesity in our society. In response to this social pressure, some individuals take an extreme approach: the pathological pursuit of weight control or weight loss.

Early in life, we develop images of "acceptable" and "unacceptable" body types. Of the attributes that constitute attractiveness, many people view body weight as the most important, perhaps because we can exert some control over body weight. Fatness is the most dreaded deviation from our cultural ideals of body image, the one most derided and shunned. In a study about weight bias conducted by researchers at Yale University, nearly half of survey respondents said they would rather give up 1 year of life than live with obesity.[2]

It can be difficult to resist comparing your body to what the popular media portrays as "ideal." For some people, the disparity between their own body image and the perceived ideal may be enough to trigger an eating disorder.

## FOOD: MORE THAN JUST A SOURCE OF NUTRIENTS

From birth, we link food with personal and emotional experiences. As infants, we associate milk with security and warmth, so the breast or bottle becomes a source of comfort. Even as we age, most people continue to derive comfort and great pleasure from food. This is both a biological and a psychological phenomenon. Indeed, eating can stimulate the release of certain **neurotransmitters** (e.g., serotonin) and natural opioids (including **endorphins**), which produce a sense of calm and euphoria in the human body. Thus, in times of great stress, some people turn to food for its naturally calming, druglike effect.

Food may also be used as a reward or a bribe. Perhaps some of the following comments sound familiar:

- You can have your dessert if you eat five more bites of your vegetables.
- You cannot play until you clean your plate.
- I will eat the broccoli if you let me use my cell phone.
- If you love me, you will eat your dinner.

On the surface, using food as a reward or bribe seems harmless. Eventually, however, this practice may encourage both caregivers and children to use food to achieve goals other than satisfying hunger and nutrient needs. Regularly using food as a bargaining chip can contribute to abnormal eating patterns. Carried to the extreme, these patterns can lead to **disordered eating.**

Disordered eating refers to short-term, mild changes in eating patterns that occur in response to a stressful event, an illness, or even a desire to modify food intake for a variety of health and personal appearance reasons. The problem may be no more than a bad habit, a style of eating adapted from friends or family members, or an aspect of preparing for athletic competition. While disordered eating can lead to weight fluctuations or nutrient deficiencies, it rarely requires in-depth professional attention.

If, however, disordered eating becomes sustained or distressing, starts to interfere with everyday activities, or is linked to negative physiological changes, then it may require professional intervention.

## ORIGINS OF EATING DISORDERS

Given the common practice of dieting, it can sometimes be difficult to draw a definitive line between disordered eating and an **eating disorder.** Indeed, many eating disorders start with a simple diet. Eating disorders then go on to involve physiological changes associated with food restriction, binge eating, purging, and/or fluctuations in weight. They also involve feelings of distress or extreme concern about body shape or weight. Eating disorders are not due to a failure of will or behavior; rather, they are real, treatable medical illnesses in which certain maladaptive patterns of eating take on a life of their own.

People who suffer from eating disorders can experience a wide range of health complications, including heart conditions and kidney failure, which may lead to death. Thus, it is important to recognize the signs of eating disorders and treat them early. The three main types of eating disorders are **anorexia nervosa, bulimia nervosa,** and **binge eating disorder.** Although it is convenient to label patients with a clear-cut diagnosis, the various types of eating disorders have more similarities than distinctions, particularly in their underlying biological and psychological processes.

Specific criteria from the *Diagnostic and Statistical Manual of Mental Disorders,* 5th Edition (DSM-5), are used by clinicians to diagnose eating disorders. As you read about the different eating disorders in this chapter, look for the relevant diagnostic criteria in Tables 11-1, 11-2, and 11-3. Keep in mind that individuals may exhibit a few symptoms of eating disorders but not enough to warrant a formal diagnosis. These people may be classified as having **subthreshold eating disorders.** Also, some people show characteristics of more than one eating disorder or may migrate from one disorder to another over time. Indeed, about half of women diagnosed as having anorexia nervosa eventually develop symptoms of bulimia.[3] Still, appreciating the differences between the disorders helps us to understand the various approaches to prevention and treatment.

Over the years, researchers have theorized that dysfunctional family interactions, especially between parents and adolescents, precipitate eating disorders. While unhealthy family relationships may lead to some emotional distress, there is little scientific evidence that family functioning is a primary cause for eating disorders. In fact, insinuating that the family has caused a person's eating disorder leads to feelings of guilt and shame within the family that may actually hinder efforts at treatment.[4]

Scientists now recognize that genes bear much of the blame for eating disorders.[1] Twin studies have shown that when one of a set of identical twins (i.e., who develop from a single fertilized egg, so they have the same genome) has an eating disorder, it is more likely that the other twin will also have an eating disorder—more likely than in sets of fraternal twins. Genetic factors appear to account for an estimated 50% to 83% of the overall risk for developing an eating disorder.

A variety of genes could be involved in the development of eating disorders, including those responsible for the synthesis of hormones and neurotransmitters involved in weight regulation and eating behaviors (review Section 7.4). In about 80% of cases, eating disorders co-occur with other psychological disorders, such as anxiety disorders, clinical depression, and substance use disorders.[5] It appears that genes influence the brain biology that determines how we perceive ourselves and respond to food stimuli or stress. In response to environmental triggers, some individuals may resort to self-destructive coping mechanisms, including disordered eating.

Because stressful life events may precipitate an eating disorder in genetically predisposed individuals, there is a strong association between eating disorders and a history of abuse. In fact, a history of physical and sexual abuse is about twice as common among people with eating disorders compared to the population as a whole.[6] Other stressful

---

**eating disorder** Severe alterations in eating patterns linked to physiological changes. The alterations are associated with food restriction, binge eating, inappropriate compensatory behaviors, and fluctuations in weight. They also involve a number of emotional and cognitive changes that affect the way a person perceives and experiences his or her body.

**anorexia nervosa** An eating disorder characterized by extreme restriction of energy intake relative to requirements, leading to significantly low body weight.

**bulimia nervosa** An eating disorder characterized by recurrent episodes of binge eating followed by inappropriate compensatory behaviors to prevent weight gain.

**binge eating disorder** An eating disorder characterized by recurrent episodes of binge eating that are associated with marked distress and lack of control over behavior, but not followed by inappropriate compensatory behaviors to prevent weight gain.

**subthreshold eating disorder** A clinically recognized eating disorder that meets some, but not all of the criteria for diagnosis of anorexia nervosa, bulimia nervosa, or binge eating disorder.

▶ **Progression from Ordered to Disordered Eating**

Attention to hunger and satiety signals; limitation of calorie intake to restore weight to a healthful level

↓

Some disordered eating habits begin as weight loss is attempted, such as very restrictive eating.

↓

Clinically evident eating disorder recognized

life events, such as wartime military service, the death of a loved one, or constant social pressures to achieve thinness, are potential triggers for eating disorders.

In addition to genetics, researchers are studying epigenetics (heritable changes in gene expression that do not involve changes to the underlying DNA) as it relates to the development of eating disorders.[7] Some studies suggest that maternal stress or hormone levels during pregnancy can predict the future development of an eating disorder in their children by affecting how the child's genes function.[8]

Overall, genes may set the stage for development of an eating disorder, but environmental factors also play a role. Identifying specific genes or epigenetic markers linked to eating disorders eventually could help in tailoring prevention and treatment efforts for at-risk individuals. However, the counseling that is part of current therapy will still be of value.

## THE CHANGING FACE OF EATING DISORDERS

If you were asked to paint a picture of a person with an eating disorder, who would you depict? The predominant stereotype is that eating disorders typically affect young, white females of middle or upper socioeconomic status. However, the face of eating disorders is changing.

When it comes to anorexia and bulimia, women do outnumber men by about 9 to 1. Perhaps social pressures can account for part of this disparity: in the media, women are held to standards of unnatural thinness, whereas the image conveyed for men is big and muscular. However, men are affected by eating disorders as well. Among men, exercise status and sexual orientation are factors that particularly influence the development of anorexia and bulimia.[9] Among males, athletes are more prone than nonathletes to develop these eating disorders, especially those who participate in sports that require weight classes (e.g., boxers, wrestlers, and jockeys) or where judging is partly based on aesthetics (e.g., ice skating, diving, or dancing).[10] With reference to sexual orientation, the prevalence of eating disorders in gay men is two to three times that in heterosexual men.[11] Certain types of eating disorders are more common among men. While men make up less than 10% of cases of anorexia and bulimia, they account for about 40% of cases of binge eating disorder.

Eating disorders typically develop during adolescence or young adulthood. Adolescence is a period of turbulent sexual and social tensions. At this time of life, teenagers establish their own identities. While declaring independence, they often seek acceptance and support from peers and parents and react strongly to how they think others perceive them. At the same time, their bodies are changing, and much of the change is beyond their control. This is a time when extreme dieting practices may take root. It is alarming that eating disorders are being diagnosed at earlier ages.[12] Note that calorie restriction is not always evidenced as weight loss; stunting (failure to grow in height) and delayed sexual maturation also could be signs of eating disorders among children and adolescents.

While much of the focus has been on youth, middle-aged and older adults are not immune to the devastating effects of eating disorders. Although eating disorders rarely make their first appearance late in adulthood, it may not be until later in adulthood that people who have suffered from eating disorders for years finally seek treatment. Furthermore, some adults who had recovered may relapse into former disordered eating practices. Research reveals that disordered eating behaviors and dissatisfaction with body weight and shape are quite common among older women. In one study, binge eating was reported by 3.5% of a community-based sample of women over age 50; purging behaviors, such as excessive exercise, were employed by 7.8% of the women.[13] Negative body image can have a dramatic impact on self-esteem and overall quality of life at any age.

Until recently, most researchers have reported that eating disorders primarily affect middle- and upper-class white women. Now, studies show greater similarities in the rates of body dissatisfaction and disordered eating behaviors across ethnic and cultural groups. Perhaps minorities with eating disorders have been less likely to seek help in the past due to fear of shame or stigma, lack of resources, or language barriers. Also, health care workers may have been less likely to diagnose non-whites as having eating disorders.

Individuals who participate in sports with weight classes, such as wrestling, may practice disordered eating behaviors in order to gain a competitive advantage over other athletes in a lower weight class. **How do eating disorders impact sports performance?** Rubberball/Getty Images

Previously, it seemed that non-white cultures were more accepting of larger body shapes, but mainstream pressures for thinness now cut across cultural lines.

Do you know someone who is at risk for an eating disorder? If so, suggest that the person seek a professional evaluation because the sooner treatment begins, the better the chances for recovery. (See *Rate Your Plate* in Connect for a simple screening questionnaire for eating disorders.) However, do not try to diagnose eating disorders in your friends or family members. Only a trained professional can exclude other possible diseases and correctly diagnose an eating disorder. Once an eating disorder is diagnosed, immediate treatment is advisable. As a friend, the best you can do is to encourage an affected person to seek professional help. Such help is commonly available at student health centers and student guidance/counseling facilities on college campuses.

▶ Since 1922, the BMI values of Miss America winners have steadily decreased; during the last three decades, most winners had a BMI in the "underweight" range (less than 18.5).

## ✓ CONCEPT CHECK 11.1

1. Differentiate between disordered eating and an eating disorder.
2. Describe how genetics and environment interact in the development of eating disorders.
3. Why are eating disorders more common among adolescents than other age groups?

## 11.2 Anorexia Nervosa

Anorexia nervosa is an eating disorder characterized by extreme weight loss, an irrational fear of weight gain, and a distorted body image. These three criteria are outlined in Table 11-1 and described in detail in this section. Anorexia nervosa affects an estimated 0.8% of American adults (0.12% of men and 1.42% of women) at some point during their lives.[14]

First, people with anorexia nervosa severely restrict energy intake relative to requirements. The term *anorexia* implies a loss of appetite; however, a denial of one's appetite more accurately describes the behavior of people with anorexia nervosa. Low energy intake leads to a body weight that is significantly less than expected when compared to others of the same age, sex, stage of physical development, and activity level. While low body weight (i.e., less than 85% of expected body weight for a given age and sex or BMI of less than 17) may indicate anorexia nervosa, a variety of other medical conditions could also result in low body weight. The next two diagnostic criteria set anorexia nervosa apart from other problems related to inadequate food intake or low body weight.

The second key criterion for diagnosis of anorexia nervosa involves an intense fear of gaining weight or becoming overweight or obese. Some individuals with eating disorders may deny a fear of weight gain, so persistent behaviors that interfere with weight gain are also included in this criterion. To be diagnosed with anorexia nervosa, an individual must have experienced fear of weight gain or practiced behaviors to prevent weight gain at least 75% of the days in the last 3 months.

**TABLE 11-1 ■ Diagnostic Criteria for Anorexia Nervosa**

A. Extreme dietary restriction that leads to significantly low body weight

B. Overwhelming distress about weight gain (or avoidance of behaviors that may lead to weight gain) despite having a low body weight

C. Disturbed perception of one's own body weight or shape, overemphasis on body weight or shape in determining self-worth, or failure to recognize the dangers of extremely low body weight

Source: American Psychiatric Association, *Diagnostic and Statistical Manual for Mental Disorders*, 5e, 2013, American Psychiatric Association.

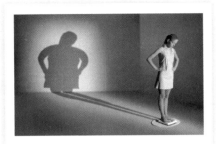

For people with eating disorders, the difference between the real and desired body images may be too difficult to accept. **See the website womenshealth.gov/body-image to learn more about body image.** Steve Niedo Photography/Stone/Getty Images

Third, as depicted in the photo on this page, individuals with anorexia nervosa have a distorted body image. The term *nervosa* refers to an unhealthy obsession. These individuals irrationally believe they are fat, even though others constantly comment on their thin physique. Some people with anorexia realize they are thin but continue to be haunted by certain areas of their bodies that they believe to be fat (such as thighs, buttocks, and stomach). Even though extremely low body weight results in severe health effects, as described later in this section, people with anorexia do not acknowledge the problem. They may persist in efforts at weight loss and try to thwart efforts of family members and medical professionals to increase their body weight to a healthy level.

Note that there are two subtypes of anorexia nervosa: a *restricting type* and a *binge eating/ purging type*. All individuals with anorexia nervosa severely restrict their calorie intake to achieve or maintain a low body weight; those with the binge eating/purging type of anorexia nervosa also engage in episodes of binge eating followed by compensatory behaviors to rid the body of calories. One difference between bulimia nervosa and anorexia nervosa of the binge eating/purging subtype is that individuals with anorexia nervosa have low body weight (e.g., BMI < 17), whereas individuals with bulimia tend to have normal or high BMI.

## COMMON BEHAVIORS OF ANOREXIA NERVOSA

Individuals who develop anorexia share some common traits. Take a young woman, for example, who is described by parents and teachers as responsible, meticulous, and obedient. She holds herself to high standards of performance and appearance. She is competitive and perfectionistic. At home, she may not allow clutter in her bedroom. Clinicians note that after a physical examination, she folds her examination gown very carefully and cleans up the examination room before leaving. As mentioned, both genetic and environmental factors contribute to both our self-perception and our responses to stress. These obsessive personality traits—rigidity and perfectionism—are related to the same brain biology that predicts development of eating disorders.

Anorexia nervosa may begin as a simple attempt to lose weight. A comment from a well-meaning friend, relative, or coach suggesting that the person seems to be gaining weight or is too fat may be all that is needed. The stress of having to maintain a certain weight to look attractive or perform better can also lead to disordered eating. Abusive experiences, a difficult breakup, or the stress of leaving home for college are examples of triggers for extreme dieting. Changing one's appearance might be viewed as a way to avoid future conflict or ensure success in a new situation.

Extreme dietary rules severely limit nutritional intake among people with anorexia nervosa. **What happens to your basal metabolic rate (BMR) when you severely restrict your calorie intake?** zinkevych/123RF

Still, losing weight does not help people deal with anger, grief, anxiety, or depression. If these psychological issues are not addressed, individuals may intensify efforts to lose weight "to look even better," rather than work through unresolved psychological concerns. At first, dieting becomes the life focus. Such persons may derive a sense of achievement from their "success" at controlling body weight or perceive improvements in other areas of their lives. What began as a diet leads to very abnormal self-perceptions and eating habits, such as cutting a pea in half before eating it.

Extreme dieting is the most important predictor of an eating disorder. (Adolescents expressing concern about their weight should be advised to focus on physical activity, which does not appear to impart a risk for subsequent problems.) Once dieting begins, a person developing anorexia nervosa does not stop. As the disorder progresses, the range of foods eaten may narrow; the list of "safe foods" shortens, whereas the list of "unsafe foods" gets longer. Abnormal habits include hiding and storing food or spreading food around a plate to make it look as if much has been eaten. A person with anorexia nervosa may cook a large meal and watch others eat it while refusing to eat anything or may insist on having different meals from the rest of the family. Frequent weighing and body checking—such as measuring the width of the thigh—are common.[15]

As mentioned, among some people with anorexia nervosa, disordered eating behaviors may eventually include bingeing on large amounts of food in a short time and/or inappropriate behaviors to compensate for the large number of calories consumed. Covered further in Section 11.3, compensatory behaviors (sometimes called *purging*) include vomiting, using **laxatives** or **diuretics**, and excessive exercise. Thus, people with anorexia nervosa may exist in a state of continuous semistarvation or may alternate between periods of starvation and periods of bingeing and purging.

A state of semistarvation can cause depression, irritability, and hostility. People with anorexia may be excessively critical of themselves and usually withdraw from family and friends. Despite a strong drive for perfection, performance in school, sports, and work begins to deteriorate. (See Section 11.8 for a glimpse of the inner turmoil of a woman with anorexia nervosa.)

Ultimately, a person with anorexia nervosa eats fewer calories than are required to maintain a healthy body weight. This can vary considerably from person to person, but 600 to 800 kcal daily is not unusual. In place of food, the person may consume up to 20 cans of diet soft drinks and chew many pieces of sugarless gum each day.

## PHYSICAL EFFECTS OF ANOREXIA NERVOSA

The state of semistarvation disturbs many body systems as it forces the body to conserve energy stores as much as possible. Many of the following complications can be reversed by returning to a healthy weight, provided the duration of the semistarvation has not been too long.[17] (Many of these are illustrated in Figure 11-1.)

▶ A disturbing Internet trend is the attempt to promote eating disorders as a way of life. Some individuals with eating disorders have personified their illness into a role model named "Ana," who tells them what to eat and mocks them when they don't lose weight. Similarly, pro-"Mia" sites provide tips and encouragement for people with bulimia (e.g., how to induce vomiting and cover up evidence of compensatory behaviors). Pro-Ana and pro-Mia websites reject the serious health risks of eating disorders and instead dispense unsafe "thinspiration" to vulnerable individuals.[16]

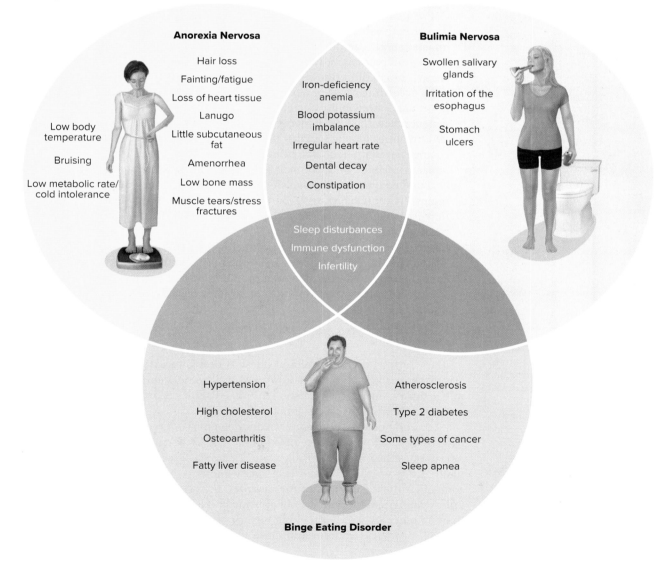

**Anorexia Nervosa**
- Hair loss
- Fainting/fatigue
- Loss of heart tissue
- Lanugo
- Little subcutaneous fat
- Amenorrhea
- Low bone mass
- Muscle tears/stress fractures
- Low body temperature
- Bruising
- Low metabolic rate/cold intolerance

**Bulimia Nervosa**
- Swollen salivary glands
- Irritation of the esophagus
- Stomach ulcers

Iron-deficiency anemia
Blood potassium imbalance
Irregular heart rate
Dental decay
Constipation

Sleep disturbances
Immune dysfunction
Infertility

**Binge Eating Disorder**
- Hypertension
- High cholesterol
- Osteoarthritis
- Fatty liver disease
- Atherosclerosis
- Type 2 diabetes
- Some types of cancer
- Sleep apnea

**FIGURE 11-1** ▲ Physical effects of eating disorders. Although this figure contains many potential consequences, it is not an exhaustive list. Some of these physical effects can also serve as warning signs that a problem exists.

**lanugo** Downlike hair that appears after a person has lost much body fat through semistarvation. The hair stands erect and traps air, acting as insulation for the body to compensate for the relative lack of body fat, which usually functions as insulation.

**amenorrhea** Absence of menstrual periods in a woman of reproductive age.

▲ In 2010, the death of French fashion model Isabelle Caro increased awareness of the serious nature of anorexia nervosa. In 2006, the 5-foot, 4-inch model was hospitalized when she slipped into a coma at her lowest weight of 55 pounds. After surviving the coma, she resolved to speak out against dieting in the fashion industry and posed for a controversial billboard ad under the words "No Anorexia." She also authored a book about her 15-year struggle with an eating disorder. Her weight had increased to at least 80 pounds by the filming of her interview for *National Geographic's Taboo: Beauty* documentary, but she died, at age 28, before it aired. Ernesto Ruscio/Getty Images

- Lowered body temperature and cold intolerance from loss of insulating fat layer
- Slowed metabolic rate from decreased synthesis of thyroid hormones
- Decreased heart rate as metabolism slows, leading to premature fatigue, fainting, and an overwhelming need for sleep (Other changes in heart function may occur as well, including loss of heart tissue, irregular heart rhythm, and low blood pressure.)
- Iron-deficiency anemia, which leads to further weakness and fatigue
- Rough, dry, scaly, cracked, and/or cold skin, which may show multiple bruises because of the loss of the protective fat layer normally present under the skin
- Low white blood cell count, which increases the risk of infection—one cause of death in people with anorexia nervosa
- Abnormal feeling of fullness or bloating, which can last for several hours after eating
- Loss of hair from the head
- Appearance of **lanugo**—downy hairs on the body that trap air to partially counteract heat loss that occurs with loss of fat tissue
- Impaired swallowing ability due to loss of muscle tissue in the pharynx, which could lead to respiratory infections from aspiration of food and liquids into the lungs
- Constipation due to deterioration of the gastrointestinal tract and abuse of laxatives (In extreme cases, impaired motility can cause rupture of the GI tract, leading to infection and even death.)
- Low blood potassium, worsened by potassium losses during vomiting and use of some types of diuretics, which increases the risk of heart rhythm disturbances—another leading cause of death in people with anorexia nervosa
- Low energy availability decreases production of sex hormones, such as estrogen. **Amenorrhea**, the loss of menstrual periods, may be the first sign of endocrine abnormalities among females (see *Ask the RDN* in this section).
- Changes in brain size, blood flow to the brain, and neurotransmitter function, all of which contribute to depression and complicate treatment
- Osteopenia or osteoporosis, which is evident in 85% of women and at least 25% of men with anorexia nervosa (Bone loss is due to decreased body weight and lean mass, several related hormonal changes, and prolonged use of some antidepressant medications.)
- Eventual loss of teeth caused by acid erosion of tooth enamel if frequent vomiting occurs (Tooth loss, along with low bone mass, can be lasting signs of the disease, even if other physical and mental problems are resolved.)
- Muscle tears and stress fractures in athletes because of decreased bone and muscle mass
- Sleep disturbances
- Depression

Many of the psychological and physical problems associated with anorexia nervosa arise from insufficient calorie intake, as well as deficiencies of nutrients such as thiamin, calcium, and iron. A person with this disorder is psychologically and physically ill and needs immediate professional help.

About one-quarter of those with anorexia nervosa recover within 6 years, but many will struggle with the disease throughout life. Among all psychiatric diseases, anorexia nervosa has the highest mortality rate: 2% to 5% of people with anorexia nervosa eventually die from the disease—from heart ailments, infections, or suicide. The longer someone suffers from this eating disorder, the poorer the chances for complete recovery. A young patient with a brief episode and a supportive family has a better outlook than an older patient with a long history of disordered eating and no family support. Overall, prompt treatment and close long-term follow-up improve the chances for a successful recovery.

## ASK THE RDN

# Not Having a Period is NOT Normal

***Dear RDN:*** *I have been diligently tracking my weight on a daily basis since I was a freshman in high school. I am happy that I can keep my weight just under 100 pounds. Should I be concerned that I have not had my period since starting college two years ago?*

Yes, you should be concerned about missing your period! Contrary to popular belief, not having a regular period, also known as amenorrhea, is not normal. It also means your body isn't producing enough estrogen, and low levels of estrogen may lead to infertility and premature osteoporosis or other health conditions.

So, what *is* a "normal" period? Well, that depends on the woman. A normal cycle, counted from the first day of your last period to the first day of the next period, is 25 to 35 days. A period can be as short as two days or as long as a week. If you're someone who has had three-day periods since you began menstruating, that's your normal. If your pattern suddenly changes, you should be concerned.

There are some medical conditions that may prevent or affect menstruation, including polycystic ovary syndrome (see Section 14.1), pelvic inflammatory disease, and genetic abnormalities. See your health care provider to rule out these possibilities. Most likely, the loss of your period is the result of a nutrition or lifestyle change.

Now let's talk birth control. If you're on the pill or any other kind of hormonal birth control, your period isn't a "real" period. Birth control has varying concentrations of synthetic female hormones. The absence of these hormones (when you take the sugar pills in your pack) induces a period. You cannot heal amenorrhea with birth control. For that reason, it is not recommended to be on birth control to regain a natural period, especially if you're recovering from an eating disorder. Keep in mind that your body will need time to restore its metabolic function, hormone levels, and natural set point where your body can freely menstruate. After coming off birth control, it can take anywhere from 3 to 6 months for your true period to normalize.

There are four factors that typically affect menstruation: nutrition, exercise, stress, and sleep. When we underfuel our bodies and/or overexercise, menstruation takes a back seat. Your body can't focus on your reproductive system when it doesn't have enough energy to maintain vital functions, like breathing and heartbeat. Normal menstrual function typically requires an energy intake of about 30 kcal per kilogram of body weight (or about 13 to 14 kcal per pound),[18] but every body has its own threshold to menstruate. Undernourishment—eating too few calories and/or consuming a suboptimal dietary pattern—can cause your period to stop. If low energy availability is the cause, slowly increasing caloric intake and eating a wider variety of foods should help induce a period. Seek the help of an RDN who specializes in eating disorders to help guide you through the process.

Respect your own body's threshold for exercise. If you love to run but suspect you may be putting too much stress on your body, perhaps alternate running days with other activities or decrease your mileage. If you're in a high-stress phase of life where you're dealing with anxiety, school demands, and working extra hours, intense exercise may place more demands on your body. Practicing meditation or yoga, taking walks, visiting with friends, or engaging in low-intensity physical activity would be wise choices until things settle down. Exercise in a way that's appropriate for *your* body and respect rest days.

If nutrition and exercise are not to blame for missed periods, stress could be the issue. When you're under significant stress, your body releases stress hormones, such as cortisol. When this happens, it's best to slow down and work on self-care and activities that bring you joy and relaxation.

Last, but certainly not least, a lack of sleep is one of the most overlooked factors when it comes to your health. Research suggests that sleep dysregulation may suppress reproductive hormones and affect frequency of periods. If you struggle to get at least 7 hours of sleep each night, create a nighttime routine that facilitates more quality sleep.

©Raul Velasco

Menstruation matters—period.

**Alexis Joseph, MS, RD, LD**

Dietitian, Nutrition Consultant, Founder of Hummusapien, Co-Owner of Alchemy Brands

Early treatment for an eating disorder, such as anorexia nervosa, improves chances of success. **What resources are available on your campus for students who are struggling with eating disorders?** Tetra Images/Getty Images

*magnificent*

Beneficial gut microbes produce compounds that regulate mood, appetite, and gastrointestinal symptoms. Among individuals with anorexia nervosa, beneficial probiotic microorganisms are reduced in number and activity, both before and after nutritional rehabilitation. To induce weight gain, the refeeding protocol for treatment of individuals with anorexia nervosa must include high-fat foods. However, food sources of probiotics (e.g., yogurt and other fermented foods) and prebiotics (e.g., fruits, vegetables, and whole grains) can also help the beneficial microorganisms in the GI tract to flourish and potentially improve physical and mental health outcomes for individuals in recovery for eating disorders.[20]

*microbiome*

## TREATMENT FOR THE PERSON WITH ANOREXIA NERVOSA

People with anorexia often sink into shells of isolation and fear. They deny that a problem exists. Frequently, their friends and family members meet with them to confront the problem in a loving way. This is called an *intervention.* They present evidence of the problem and encourage immediate treatment.

Treatment requires a multidisciplinary team of experienced professionals (e.g., physicians, psychiatrists, psychologists, registered dietitian nutritionists [RDNs], nurses, occupational therapists, and social workers). An ideal setting is an eating disorders clinic in a medical center. Outpatient therapy generally begins first. This may be extended to 3 to 5 days per week. Day hospitalization (6 to 12 hours per day) is another option. Total hospitalization is necessary once a person falls below 75% of expected weight (i.e., how much an individual would weigh at a healthy BMI), experiences medical problems, and/or exhibits severe psychological problems or suicidal risk. Still, even in the most skilled hands at the finest facilities, efforts may fail. This tells us that the prevention of anorexia nervosa is of utmost importance.[19]

The health care team must gain the cooperation and trust of the person with anorexia and work together to restore body weight, correct medical complications, and treat mental illness. However, the individual who has been barely existing in a state of semistarvation cannot focus on much besides food. Dreams and even morbid thoughts about food will interfere with therapy until the person regains sufficient weight. Currently, the average time for recovery from anorexia nervosa is 7 years. Many insurance companies cover only a fraction of the estimated $150,000 cost of treatment.

**Nutrition Therapy.** The ultimate goal of nutrition therapy is to restore body weight to a healthy range. The individual must increase oral food intake and gain enough weight to raise the metabolic rate to normal and reverse as many physical signs of the disease as possible. The refeeding plan is designed first to minimize or stop any further weight loss. Then, the focus shifts to restoring appropriate eating behaviors. After this, the expectation can be switched to gaining weight; 2 to 3 pounds per week is appropriate. Tube and/or intravenous (IV) feeding is used only if immediate renourishment is required, as this can frighten the person and cause him or her to distrust medical staff.

Persons with anorexia nervosa need considerable reassurance during the refeeding process because of uncomfortable and unfamiliar effects—such as bloating and increased body temperature. These symptoms occur because of starvation-related changes in GI tract function and will usually resolve over time, but they can be frightening for the person recovering from anorexia. Rapid changes in electrolytes and minerals in the blood associated with refeeding—especially potassium, phosphorus, and magnesium—can be dangerous. Therefore, monitoring blood levels of these minerals is of critical importance during the process of incorporating more food into the dietary pattern.

In addition to helping the person with anorexia nervosa reach and maintain adequate nutritional status, the RDN provides accurate nutrition information throughout the treatment, promotes a healthy attitude toward food, and helps the person learn to eat in response to natural hunger and satiety cues. The focus then turns to identifying healthy and adequate food choices that promote weight gain to achieve a BMI of 20 or more for adults or BMI-for-age between the 5th and 85th percentile for children (see Section 15.1). For children and adolescents, additional energy needs for growth must be considered.

As noted, nutrient deficiencies are commonly observed in patients with anorexia nervosa. As treatment progresses, the health care team will work with the patient to fully restore nutritional status. A multivitamin and mineral supplement will be added, as well as enough calcium to raise intake to about 1500 milligrams per day.

Bone loss will not likely be completely restored, even if nutrition is adequate. Adolescent patients, in particular, may have missed a critical time for accrual of bone mass or gains in height. However, supplementation with calcium and vitamin D is still necessary to prevent further bone loss.

Despite their extreme need for nutritional intervention, surrendering control over eating can be scary and frustrating for individuals with anorexia nervosa. These patients may be very resistant to therapy. They may try to hide weight loss by wearing many layers of clothes, putting coins in their pockets, or drinking many glasses of water before stepping on the scale. Excessive physical activity may also impede weight gain. At many treatment centers, moderate bed rest is used in the early stages of treatment to help promote weight gain. Overall, experienced professional help is the key for effectively treating anorexia nervosa.[21]

**Psychological Therapy.** Once the immediate physical problems of anorexia nervosa are addressed, the focus of treatment shifts to the underlying emotional problems that preceded the eating disorder. If therapists can discover the psychological conflicts that triggered the disorder, they can develop more effective treatment strategies. Education about the medical consequences of semistarvation also may be helpful. A key aspect of psychological treatment is showing affected individuals how to regain control of other facets of their lives and cope with difficult situations. As eating evolves into a normal routine, they can turn to previously neglected activities.

Family-based treatment (usually 6 to 12 months) is the preferred method of psychological treatment for anorexia nervosa among *adolescents*.[22] Importantly, family-based treatment for anorexia nervosa absolves parents of blame for the eating disorder. Early treatment focuses on ways the family can help the person with anorexia achieve a healthy body weight. Eventually, responsibility for eating and weight control will be transferred back to the patient. Beyond eating behaviors, family-based treatment for anorexia nervosa also helps the patient to establish healthy relationships with parents and other family members.

At this time, there is not enough evidence to promote one particular type of psychological therapy over another for *adults* with anorexia nervosa. Therapists may use **cognitive behavioral therapy,** which involves helping the person confront and change irrational beliefs about body image, eating, relationships, and weight. However, after subsisting in a starved state for months or years, the brain chemistry of a person with anorexia nervosa is so altered that attempts at cognitive restructuring are usually not effective in the early stages of treatment. Underlying issues that may have triggered the eating disorder, such as sexual abuse, also must be identified and addressed by the therapist. Guided self-help groups for people with eating disorders, as well as their families and friends, represent additional nonthreatening first steps into treatment.

**cognitive behavioral therapy** Psychological therapy in which the person's assumptions about dieting, body weight, and related issues are confronted. New ways of thinking are explored and then practiced by the person. In this way, an individual can learn new ways to control disordered eating behaviors and related life stress.

**Pharmacological Therapy.** There are no medications approved by the U.S. Food and Drug Administration (FDA) specifically for the treatment of anorexia nervosa. Rather, *food* is the treatment of choice for people with this disorder. Generally, medications are not effective in managing the primary symptoms of anorexia nervosa. Fluoxetine (Prozac®) and related medications may stabilize recovery once 85% of expected body weight has been attained. These medications work by prolonging serotonin activity in the brain, which in turn regulates mood and feelings of satiety. A variety of other pharmacological agents, such as olanzapine (Zyprexa®), may have some role in treating mood changes, anxiety, or psychotic symptoms associated with anorexia nervosa, but they have limited value unless weight gain is also achieved.[23]

With professional help, many people with anorexia nervosa can regain a balanced dietary pattern. Although they may not be totally cured, recovering individuals no longer depend on unusual eating habits to cope with daily problems. They recover a sense of normalcy in their lives. Longer follow-up—sometimes several years—is associated with better outcomes. Recovery rates are around 20% to 30% for short-term therapy but increase to 70% to 80% with 8 years of follow-up. No universal approach exists because each case is unique. Establishing a strong relationship with either a therapist or another supportive person is especially important to recovery. As they learn alternative coping mechanisms, individuals with anorexia nervosa can relinquish dysfunctional relationships with food and instead develop healthy personal relationships.

> ✓ **CONCEPT CHECK 11.2**
>
> 1. Identify the three diagnostic criteria for anorexia nervosa.
> 2. List five physical effects of anorexia nervosa.
> 3. Describe elements of nutritional, psychological, and pharmacological therapy for anorexia nervosa.

# 11.3 Bulimia Nervosa

Literally translated, *bulimia* means ravenous (oxlike) hunger. This eating disorder is characterized by recurrent episodes of binge eating followed by some type of compensatory behavior to prevent weight gain (see Table 11-2). As with anorexia nervosa, individuals with bulimia nervosa overvalue body weight and shape.

**Binge eating** is defined as consuming an abnormally large amount of food within a short time period (e.g., 2 hours). Notably, binges are characterized by a lack of control over the food consumed. **Compensatory behaviors** (also known as *purging*) used to rid the body of excess calories consumed during a binge may include vomiting; misuse of laxatives, diuretics, or enemas; or excessive exercise. For a diagnosis of bulimia nervosa, binge eating followed by inappropriate compensatory behaviors must take place at least once per week over a period of 3 months or more.[24]

It is likely that many people with bulimic behaviors are never diagnosed. People with bulimia nervosa lead secret lives, hiding their abnormal eating habits. Moreover, it can be difficult to recognize the disorder based on appearance because people with bulimia nervosa are usually at or slightly above normal weight. By rough estimate, bulimia nervosa affects about 0.28% of U.S. adults (0.08% of men and 0.46% of women) at some point during their lives.[14] About 10% of cases occur in men. However, most diagnoses of bulimia nervosa rely on self-reports, so the disorder may be much more widespread than commonly thought.

**binge eating** Consuming an abnormally large amount of food within a short time period (e.g., 2 hours).

**compensatory behaviors** Actions taken to rid the body of excess calories and/or to alleviate guilt or anxiety associated with a binge; examples include vomiting, misuse of laxatives, or excessive exercise.

**TABLE 11-2 ■ Diagnostic Criteria for Bulimia Nervosa**

| |
|---|
| A. Repeated binge eating, characterized by:<br>  1. Eating a large amount of food in a short period of time (e.g., within 2 hours)<br>  2. Experiencing a loss of control over eating during binges |
| B. Repeated use of unsafe means of preventing weight gain (e.g., self-induced vomiting; inappropriate use of laxatives, diuretics, enemas, or other medications; fasting; or excessive exercise) |
| C. Binge-compensate cycles occur at least one time per week for 3 months |
| D. Undue influence of body weight or shape on self-evaluation |
| E. Behaviors are distinct from the binge eating/purging subtype of anorexia nervosa |

Source: American Psychiatric Association, *Diagnostic and Statistical Manual for Mental Disorders*, 5e, 2013, American Psychiatric Association.

## COMMON BEHAVIORS OF BULIMIA NERVOSA

Bulimia nervosa involves episodes of binge eating followed by various means to rid the body of excess calories. Susceptible people often have genetic factors and lifestyle patterns that predispose them to becoming overweight, and many have tried multiple weight-reduction diets in the past. A person with bulimia nervosa may think of food constantly; however, unlike the person with anorexia nervosa, who turns away from food when faced with problems, the person with bulimia nervosa turns toward food in critical situations. Also, unlike those with anorexia nervosa, people with bulimia nervosa recognize their behavior as abnormal.

Individuals with bulimia nervosa tend to be impulsive, which is sometimes expressed in behaviors such as stealing, increased sexual activity, drug and alcohol abuse (see *Newsworthy Nutrition* in this section), self-mutilation, or attempted suicide.[25] Some experts have suggested that part of the problem may actually arise from an inability to control responses to impulse and desire. Approximately half of the people with bulimia nervosa struggle with depression. Many people with bulimia nervosa report a history of sexual abuse. They appear competent to outsiders, while they actually feel out of control, ashamed, and frustrated.

For food intake to qualify as a binge, an atypically large amount of food must be consumed in a short time and the person must exhibit a lack of control over the behavior. Among sufferers of bulimia nervosa, bingeing often alternates with attempts to rigidly restrict food intake. Elaborate food rules are common, such as avoiding all sweets. Thus, eating just one cookie or doughnut may cause an individual with this disorder to feel as though he has broken a rule. At that point, in the mind of a person with bulimia, the objectionable food must be eliminated. Usually this leads to further overeating, partly because it is easier to regurgitate a large amount of food than a small amount.

Binge-compensate cycles may be practiced daily, weekly, or across longer intervals. A specific time often is set aside. Most binge eating occurs at night, when other people are less likely to interrupt, and usually lasts from 30 minutes to 2 hours. A binge can be triggered by stress, boredom, loneliness, depression, or any combination thereof. It often follows a period of calorie restriction and thus may be linked to intense hunger. A binge is not at all like normal eating; once begun, it seems to propel itself. The person not only loses control but generally does not even taste or enjoy food during a binge. This separates the practice from overeating.

Most commonly, individuals with bulimia consume cakes, cookies, ice cream, and other high-carbohydrate convenience foods during binges because these foods can be purged relatively easily and comfortably by vomiting. In a single binge, foods supplying 3000 kcal or more may be eaten. Compensatory behaviors follow in hopes that no weight will be gained; however, even when vomiting follows the binge, up to 75% of the calories taken in are still absorbed, inevitably causing some weight gain. When laxatives or enemas are used, about 90% of the calories are absorbed, as laxatives act in the large intestine, beyond the point of most nutrient absorption. The belief that purging soon after bingeing will prevent excessive calorie absorption and weight gain is a misconception.

At the onset of bulimia nervosa, some individuals may induce vomiting by placing their fingers deep into the mouth to trigger the gag reflex. They may bite down on their fingers inadvertently, resulting in bite marks and scars around the knuckles. These marks on the knuckles, known as **Russell's sign,** are a characteristic sign of this disorder (Fig. 11-2). Once the disease is established, however, a person may be able to vomit simply by contracting the abdominal muscles. Vomiting may also occur spontaneously.

Another way a person with bulimia may attempt to compensate for a binge is by engaging in excessive exercise to expend a large amount of calories. In this practice, referred to as *debting*, individuals try to estimate the amount of calories eaten during a

▶ Bingeing and purging (via vomiting) were evident in pre-Christian Roman times but were practiced in a group setting. The eating disorder bulimia nervosa is generally practiced in private. It was first described in the medical literature in 1979.

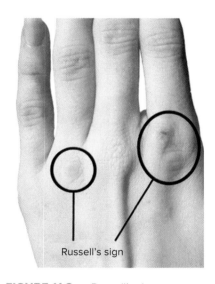

Russell's sign

**FIGURE 11-2** ▲ Russell's sign may indicate bulimia nervosa. It may appear as abrasions, calluses, or scars on the knuckles caused by trauma from the teeth while using the fingers to trigger the gag reflex to induce vomiting. ScienceSource

**Russell's sign**  Evidence of abrasion that appears on the knuckles of a person who repeatedly induces vomiting by using the fingers to trigger the gag reflex in the back of the throat; named after the psychiatrist who first identified bulimia nervosa.

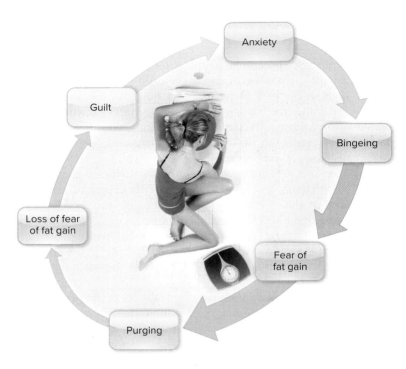

FIGURE 11-3 ▲ Bulimia nervosa's vicious cycle of obsession. Peter Dazeley/Getty Images

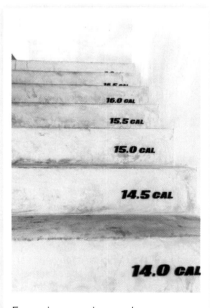

Excessive exercise can be one component of bulimia if it is used as a way to offset the calorie intake from a binge. **How can you tell if exercise is *excessive*?** ronnarong/123RF

binge and then work out to burn off the excess calories. Exercise is considered excessive when it is done at inappropriate times or in inappropriate settings, or when a person does it despite injury or other medical complications.

People with bulimia nervosa are not proud of their behavior. After a binge, they usually feel guilty and depressed. Over time, they experience low self-esteem, feel hopeless about their situation, and are caught in a vicious cycle of obsession (Fig. 11-3).[25] Compulsive lying, shoplifting to obtain food, and drug abuse can further intensify these feelings. A person discovered in the act of bingeing by a friend or family member may lash out and order the intruder to leave. Sufferers gradually distance themselves from others, spending more time preoccupied by and engaging in bingeing and compensating.

Because individuals with bulimia nervosa attempt to hide their behaviors, it may be difficult to identify them early in the disease process, when treatment is likely to be most effective. An early warning sign of bulimia is frequent trips to the bathroom during or after meals. To cover the sounds of vomiting, these individuals may run the bathroom fan or turn on the shower. Despite efforts to disguise the behavior with air fresheners, mouthwash, or breath mints, there may be a lingering odor of vomit. Be suspicious of packages or receipts for laxatives, diuretics, diet pills, or enemas. People who use exercise to compensate for binges are usually preoccupied with their workout schedule or might seem extremely distressed when they are unable to work out. If you suspect someone is falling victim to bulimia nervosa, encourage the individual to get professional help. Early intervention can prevent some of the serious physical health effects described next.

## PHYSICAL EFFECTS OF BULIMIA NERVOSA

Repeated vomiting is a physically destructive method of purging. Indeed, the majority of health problems associated with bulimia nervosa, as noted here, arise from vomiting:[24,26]

• Small blood vessels, including those in the eyes and nose, may burst due to the pressure of vomiting. This can lead to bloodshot eyes or nosebleeds.

- Repeated exposure of teeth to stomach acid causes demineralization (Fig. 11-4), making the teeth painful and sensitive to heat, cold, and acids. Eventually, the teeth may decay severely, erode away from fillings, and fall out. Dental professionals are sometimes the first health professionals to notice signs of bulimia nervosa.
- Blood potassium can drop significantly with regular vomiting or the use of certain diuretics or laxatives. This can disturb the heart's rhythm and even produce sudden death.
- Salivary glands may swell as a result of irritation or infection from persistent vomiting.
- Ulcers, bleeding, and **perforations** may develop in the stomach or esophagus.
- Constipation may result as a complication of frequent laxative use.

Bulimia nervosa is a serious and potentially deadly mental disorder. The mortality rate for patients with bulimia nervosa is estimated around 0.4%. Potential causes for death include suicide, cardiac arrest, or infections.

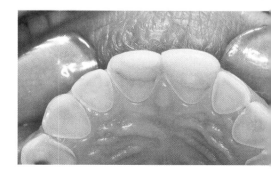

**FIGURE 11-4** ▲ Dental erosion resulting from frequent episodes of self-induced vomiting by a patient with bulimia nervosa. Courtesy of Carl M. Allen, DDS, MSD

## TREATMENT FOR THE PERSON WITH BULIMIA NERVOSA

Therapy for bulimia nervosa, as for anorexia nervosa, requires a team of experienced clinicians.[22] Individuals with bulimia nervosa are less likely than those with anorexia nervosa to enter treatment in a state of semistarvation. However, if a person with bulimia nervosa has lost significant weight, this must be treated before psychological treatment begins. Although clinicians have yet to reach consensus on the best therapy for bulimia nervosa, they generally agree that treatment should last at least 16 weeks. Hospitalization may be indicated in cases of extreme laxative abuse, regular vomiting, substance abuse, and depression, especially if there is evidence of self-harm.

**Nutritional Therapy.** Nutritional counseling by an RDN has two main goals: reestablishing regular eating habits and correcting misconceptions about food.

# Newsworthy Nutrition

## Substance use among adolescents with eating disorders

**INTRODUCTION:** Previous research suggests a relationship between eating disorders and substance use. **OBJECTIVE:** To examine the relationship between eating disorders and substance use among adolescents. **METHODS:** Between 2001 and 2012, researchers collected data from 290 adolescents (ages 12 to 18) with eating disorders who sought treatment at the University of Chicago Medical Center. **RESULTS:** Analyses showed that 24.6% of adolescents with anorexia nervosa and 48.7% of adolescents with bulimia nervosa had used substances. Alcohol was most commonly used, followed by cannabis, tobacco, and other illicit drugs. Binge drinking was reported by 27.5% of those who used alcohol. Substance use in this cohort was lower than use reported for the general population of adolescents, but this may be due to underreporting or the relatively young sample. It is notable that substance use was twice as common among adolescents with bulimia compared to anorexia. The frequency of binge-purge cycles was directly related to substance use among adolescents with bulimia nervosa. **CONCLUSION:** Substance use may co-occur with eating disorders and should be considered during assessment and treatment of adolescents with eating disorders. The impulsive and risky nature of alcohol, tobacco, and other drug use is consistent with some behavioral traits of bulimia nervosa, which may be related to disruptions in neuroendocrine pathways. It is imperative to identify these behaviors early because the coexistence of eating disorders with substance use greatly increases health risks for adolescents.

Source: Mann AP and others: Factors associated with substance use in adolescents with eating disorders, *Journal of Adolescent Health* 55:182, 2014.

In general, the focus is not on stopping bingeing and purging per se but on developing regular eating habits. Once this is achieved, the binge-purge cycle should start to break down. Initially, the RDN must help the person with bulimia nervosa to decrease the amount of food consumed in a binge. This will decrease the risk of esophageal tears from purging by vomiting. A decrease in the frequency of this type of purging will also decrease damage to the teeth. Next, to develop a normal eating pattern, some specialists encourage their patients to develop daily meal plans and keep a food diary in which they record food intake, internal sensations of hunger, environmental factors that precipitate binges, and thoughts and feelings that accompany binge-compensate cycles.

Avoidance of binge foods may be recommended early in treatment. Setting time limits for the completion of meals and snacks can also be important for people who struggle with binge eating. Many individuals with bulimia eat quickly and have trouble interpreting satiety cues. Suggesting that they put their fork down after each bite is a behavioral technique that a therapist might try during recovery for bulimia. (In comparison, many people with anorexia eat in an excessively slow manner—for example, taking 1 hour to eat a muffin cut into tiny, bite-size pieces.)

Providing accurate information about bulimia nervosa and its consequences can help affected individuals to recognize the need to change the behavior. Over the long term, individuals are discouraged from following strict rules about dieting because this mimics the typical obsessive attitudes associated with bulimia nervosa. Rather, they must learn to honor internal hunger and satiety cues and to choose a variety of foods from each food group.

▶ The binge-purge cycle can create an initial state of euphoria in the person. Giving up this euphoria has been equated to giving up an addiction.

**Psychological Therapy.** The objectives of psychotherapy are to improve people's self-acceptance and to help them be less concerned about body weight. Cognitive behavioral therapy is generally used.[25] Psychotherapy helps correct the all-or-none mentality that is typical of people with bulimia nervosa: "If I eat one cookie, I'm a failure and might as well binge." The premise of this psychotherapy is that if abnormal attitudes and beliefs can be altered, normal eating will follow. In addition, the therapist guides the person in establishing food habits that will minimize bingeing: avoid fasting, eat regular meals, and use alternative methods—other than eating—to cope with stressful situations. Another goal of therapy is to help the person with bulimia to accept some depression and self-doubt as normal. Group therapy can be useful to foster strong social support.

**Pharmacological Therapy.** Although pharmacological agents should not be used as the sole treatment for bulimia nervosa, studies indicate that some medications may be beneficial in conjunction with nutritional and psychological therapies. Fluoxetine (Prozac®) is the only antidepressant that has been approved by the FDA for use in the treatment of bulimia nervosa. It can help to increase feelings of satiety after eating and thereby reduce the frequency of binges. (Women of childbearing age should be cautious because fluoxetine has been linked to increased rates of miscarriages and other complications of pregnancy.)[24] Some clinicians may also prescribe lisdexamfetamine (Vyvanse®), which is approved for treatment of binge eating disorder. Although this drug is not currently FDA-approved for treatment of bulimia nervosa, research shows it may help to reduce binge frequency. Physicians also may prescribe other similar forms of antidepressants, other classes of psychiatric medications, or certain antiseizure medications (e.g., topiramate [Topamax®]).

People with bulimia nervosa must recognize that they have a serious disorder that can have grave medical complications if not treated. Relapse is likely, so therapy should be long term. As seen with anorexia, earlier intervention and longer duration of treatment are associated with better outcomes. After just 1 year of therapy, about one-fourth of patients with bulimia nervosa experience recovery, but most will require many years of therapy. Those with bulimia nervosa need professional help because of their depression and high risk for suicide. About 50% of people with bulimia nervosa recover completely from the disorder. Others continue to struggle with it to varying degrees for a lifetime. Such a difficult course of treatment underscores the need for prevention.

1. Describe the eating behaviors of bulimia nervosa. What triggers them? What distinguishes them from simple overeating at a holiday meal?

2. What is a compensatory behavior? List at least three examples of inappropriate compensatory behaviors.

3. Describe at least three physical effects of bulimia nervosa.

4. Outline the basic components of nutritional therapy for people with bulimia nervosa. How do psychological and pharmacological therapies contribute to recovery?

# 11.4 Binge Eating Disorder

First officially described in 1994, binge eating disorder is a growing, complex, and serious problem. Generally, binge eating disorder can be defined as binge eating episodes not accompanied by compensatory behaviors (as seen in bulimia nervosa) at least one time per week, on average, for at least 3 months. The diagnostic criteria for binge eating disorder are listed in Table 11-3.

While anorexia and bulimia disproportionately affect females, about 40% of people with binge eating disorder are males. For adults, the lifetime prevalence of binge eating disorder is estimated to be about 1.25% for women and about 0.42% for men.[14] Among the general U.S. population, about 4 million have this disorder. However, many more people in the general population are likely to have less severe forms of the disorder that do not meet all the criteria described in Table 11-3. The number of cases of binge eating disorder is far greater than that of either anorexia nervosa or bulimia nervosa. This disorder is most common among individuals with severe obesity and those with a long history of yo-yo dieting, although obesity is not a criterion for having binge eating disorder. Compared to anorexia and bulimia, binge eating disorder tends to be diagnosed later in life, usually in adults in their late 40s or early 50s.

Like anorexia and bulimia, binge eating disorder seems to arise when a person with a genetic predisposition for the disorder is faced with an environmental trigger, such as emotional stress. Individuals with binge eating disorder may have disruptions in neurotransmitter function or altered activity in the part of the brain that responds to rewarding stimuli. There is a strong association between binge eating disorder and other psychological disorders, especially anxiety, depression, and addiction.[6,7,27]

▶ Up to 25% of college-age women exhibit some degree of binge eating. Such behavior may have a negative impact on physical appearance, health, social life, and academics.

**TABLE 11-3 ■ Diagnostic Criteria for Binge Eating Disorder**

A. Recurrent binge eating, characterized by:
1. Eating a large amount of food in a short period of time (e.g., within 2 hours)
2. Experiencing a loss of control over eating during binges

B. Episodes of binge eating are associated with at least three of the following:
1. Rapid rate of eating
2. Continuing to eat beyond feelings of fullness
3. Overeating in the absence of hunger
4. Eating alone to avoid embarrassment
5. Feelings of self-disgust, depression, or guilt after overeating

C. Extreme distress about binge eating

D. Binges occur at least one time per week for 3 months.

E. Behaviors are distinct from bulimia nervosa and anorexia nervosa.

Source: American Psychiatric Association, *Diagnostic and Statistical Manual for Mental Disorders*, 5e, 2013, American Psychiatric Association.

Some experts describe binge eating disorder as an addiction to food. **What are some similarities between binge eating and other addictive behaviors, such as smoking or gambling?** Ryan McVay/Photodisc/Getty Images

## COMMON BEHAVIORS OF BINGE EATING DISORDER

As described in Section 11.3, a binge refers to the uncontrolled consumption of an unusually large amount of food within a discrete period of time. A binge can include any food, but most often consists of foods that carry the social stigma of "junk" or "bad" foods—ice cream, cookies, potato chips, and similar snack foods. During binges, food is eaten without regard to biological need and often in a recurrent, ritualized fashion. Unlike those with bulimia nervosa, people with binge eating disorder do not attempt to purge the excess calories.

Binge eating is usually triggered by negative emotions, such as stress, anxiety, loneliness, grief, or anger.[27] In fact, almost half of individuals with severe binge eating disorder exhibit clinical depression. Often, people with binge eating disorder have not learned to express or appropriately deal with their feelings, so they turn to food to cope with stress or meet emotional needs. Those who regularly practice binge eating may grow up nurturing others instead of themselves, avoiding their own feelings, and taking little time for themselves. Unfortunately, unresolved conflicts and unmet emotional needs will resurface. To make matters worse, binge eating itself brings added feelings of guilt, embarrassment, and shame.

Typically, binge eaters isolate themselves and eat large quantities of a favorite food—such as a whole pizza in one sitting—when an emotional setback occurs. Other people with this disorder eat food continually over an extended period, called *grazing*. For instance, someone with a stressful or frustrating job might come home every night and graze until bedtime.

Many people with binge eating disorder have struggled to lose weight throughout their lives. As noted in Section 7.6, overly restrictive diets can lead to hunger and a sense of deprivation that trigger binge eating. People with binge eating disorder tend to perceive themselves as hungry more often than normal. During periods when little food is eaten, they get very hungry and obsessive about food. Restricting favorite foods, such as chocolate, leads to feelings of deprivation. When these individuals finally give themselves permission to eat a forbidden food or loosen up a rigid meal plan, they feel driven to eat in a compulsive, uncontrolled way. People with binge eating disorder usually began this cycle of strict dieting alternating with binge eating during adolescence or in their early twenties and have had little success with traditional weight-control programs.

## PHYSICAL EFFECTS OF BINGE EATING DISORDER

Although obesity is not among the criteria for diagnosis of binge eating disorder, about 70% of those with the disorder have a BMI of 30 or higher. The physical effects of the disorder stem from the comorbid conditions of obesity, which were summarized in Figure 7-11.[27] The most deadly physical effects are listed here:

- Hypertension from excess body weight and high sodium intake
- Elevated cholesterol levels, which contribute to atherosclerosis (Binge eating may have a more severe effect on blood lipids than overeating by grazing throughout the day because large meals are linked to high insulin and high triglycerides.)
- Cardiovascular disease, which contributes to deaths from heart attacks and strokes
- Type 2 diabetes, which is strongly linked to obesity

## TREATMENT FOR THE PERSON WITH BINGE EATING DISORDER

Many participants in organized weight-control programs—perhaps as many as half—have binge eating disorder.[28] This means RDNs and others who lead weight-management programs may be the first to identify binge eating disorder among their clients. As with anorexia and bulimia, binge eating disorder truly is a psychological problem with nutritional consequences. The success of traditional weight-loss therapies for people with binge eating disorder has been poor because these approaches fail to address the underlying psychological causes of the disorder. While people with binge eating disorder will likely benefit from increased nutrition knowledge, first addressing the psychological needs of individuals with binge eating disorder is essential to successful treatment. Unless they find a way to manage or overcome negative emotions that underlie eating behaviors, the success of attempts to improve food choices and increase physical activity will be short lived.

The primary goal of treatment for people with binge eating disorder is to decrease and eventually eliminate episodes of binge eating. Losing excess body weight, alleviating comorbid conditions of obesity, and improving psychological disturbances are secondary goals of therapy.[29]

**Psychological Therapy.** Similar to treatments for bulimia nervosa, there is growing evidence that cognitive behavioral therapy techniques are useful for overcoming binge eating disorder. Many people with binge eating disorder may experience difficulty in identifying personal emotional needs and expressing emotions. This problem is a common predisposing factor in binge eating, so communication issues should be addressed during treatment. Binge eaters often must be helped to recognize their buried emotions in anxiety-producing situations and then encouraged to share them with their therapist or therapy group. Learning simple but appropriate phrases to say to oneself can help stop bingeing when the desire is strong. Even if negative situations cannot be changed, people must learn how to adapt to and bear with them effectively—not through self-destructive binge eating behaviors.

Cognitive behavioral therapy can be delivered in many forms. One-on-one sessions with a therapist and group therapy are the most common methods. Self-help groups such as Overeaters Anonymous® also have value. Their treatment philosophy, which parallels that of Alcoholics Anonymous®, is to create an environment of encouragement and accountability to overcome this eating disorder. Recently, even web-based adaptations of cognitive behavioral therapy have proved to be useful.

Although psychological therapy has been valuable for correcting binge eating behavior, it is not always successful at inducing weight loss among a population of overweight and obese individuals who may have one or more other health issues related to obesity. Thus, nutrition therapy is an important part of treatment for binge eating disorder.[29]

**Nutrition Therapy.** Once effective coping mechanisms are learned, the RDN can educate the patient on developing normal eating patterns and making healthful food choices. First, those with binge eating disorder must learn to eat in response to hunger—a biological signal—rather than in response to emotional needs or external factors (such as the time of day, boredom, or the simple presence of food). Counselors often direct binge eaters to record their perceptions of physical hunger throughout the day and at the beginning and end of every meal. Individuals with binge eating disorder must learn to respond to a prescribed amount of fullness at each meal.

Individuals recovering from binge eating disorder should initially avoid weight-loss diets because feelings of food deprivation can trigger binge eating. Even if exposure to favorite binge foods is limited in the early stages of treatment, many experts feel that learning to eat all foods—but in moderation—is an effective long-term goal for people with binge eating disorder. This practice can prevent the feelings of desperation and deprivation that come from limiting particular foods.

**Pharmacological Therapy.** Psychological and nutritional therapies are useful tools to treat binge eating disorder, but they are not 100% effective. Thus, there is growing interest in drug therapy.[29] The first drug to be approved for treatment of binge eating disorder is lisdexamfetamine dimesylate (Vyvanse®), a stimulant that has been used to treat attention deficit hyperactivity disorder. Some antidepressants (e.g., fluoxetine [Prozac®] and duloxetine [Cymbalta®]) and antiseizure medications (e.g., topamirate [Topamax®]) are successful in reducing binge eating. Despite their usefulness for reducing binge eating, these medications still may not induce significant weight loss. Orlistat (Xenical®) and phentermine (Adipex-P®), discussed in Section 7.9, can assist with weight-loss efforts after binge eating is under control. Other weight-loss medications have yet to be extensively tested for treatment of binge eating disorder.

**COVID CORNER**

▶ As communities try to cope with the COVID-19 pandemic with physical distancing, individuals struggling with eating disorders are at heightened risk. Increased anxiety and depression may trigger maladaptive coping behaviors, such as binge eating. In addition, lack of access to appropriate health care due to fear or limited financial resources may disrupt or compromise professional therapy for people with eating disorders. Telehealth appointments and virtual support groups may be useful for some, yet many individuals with eating disorders may be hesitant or unable to reach out for help with recovery.[30]

Given the similarities between binge eating and other addictive behaviors, medications used to treat substance abuse (e.g., naltrexone) are being studied for use among patients with binge eating disorder. Another novel therapy that is currently being studied is Namenda®, which is used in patients with Alzheimer's disease.

Overall, people who have binge eating disorder are usually unsuccessful in controlling it on their own. Furthermore, unrecognized binge eating disorder will undermine the success of weight-loss therapies among many overweight and obese individuals who do seek professional help. Health professionals can screen weight-loss clients for binge eating disorder by asking questions about eating patterns, feelings of loss of control over eating behaviors, and feelings of guilt after eating, and then refer affected individuals to appropriate treatments.

### ✓ CONCEPT CHECK 11.4

1. What distinguishes binge eating disorder from bulimia nervosa?
2. List at least three health effects that may result from binge eating disorder.
3. Why do people with binge eating disorder have little success in traditional weight-loss programs?

### CASE STUDY  Eating Disorders—Steps to Recovery

At age 16, Sarah suddenly became self-conscious about her body when her peers teased her about being overweight. She began exercising to an aerobics video for an hour each day and found that she had success in losing weight; this was the beginning of her obsession to be thin. Next, Sarah turned to eating less food to lose even more weight and began eliminating certain foods from her eating pattern, such as candy and meat. She increased her water and vegetable intake and chewed sugarless gum to curb her appetite. Once she began dieting, it felt impossible to stop. She enjoyed having a high degree of self-control over her body. Still, Sarah was literally obsessed with food, even staring at others while they were eating a meal. She occasionally cooked large meals and then refused to eat all but a few bites. By the time Sarah was 19 years old and 5 feet 6 inches tall, her weight had dropped from 150 to 105 pounds. Her family was concerned about her weight status, demanding that she go to her primary care provider for an evaluation. Sarah was not happy about this idea because she worried the doctors would force her to eat and gain back unwanted weight, but she believed that her family would stop pestering her if she went. Sarah did not think she had a problem; she thought she was still grotesquely overweight. She did notice, however, that she always felt cold and was concerned that she had not menstruated in a year.

Individuals with eating disorders suffer from a distorted body image. Even though she had lost significant weight, Sarah still believed she was overweight. Ted Foxx/Alamy Stock Photo

1. Sarah appears to have an eating disorder. Which eating disorder best describes her behavior?
2. List the behaviors that Sarah developed between ages 16 and 19 that are signs of the development of this eating disorder.
3. What physical symptoms of this disorder does Sarah have (review Fig. 11-1)?
4. Outline the therapies you think the primary care provider will recommend for Sarah. Where could she go for the therapy she needs? Which types of professionals would be involved?
5. Do you think Sarah has developed any vitamin or mineral deficiencies? Which ones would be most likely? How could these deficiencies be treated?
6. What is the likelihood that she will fully recover from her condition?

*Complete the Case Study. Responses to these questions can be provided by your instructor.*

# 11.5 Other Eating Disorders

Besides those already covered, there are several lesser-known types of eating disorders. Some of these, such as **pica** (discussed next), rumination disorder, and avoidant/restrictive food intake disorder (discussed in Section 15.4), are distinct eating disorders with their own sets of diagnostic criteria. Others fall under the *DSM-5* category of *Other Specified Feeding or Eating Disorders,* which encompasses several disorders that do not meet all of the criteria for diagnosis of anorexia, bulimia, or binge eating disorder. These include several categories of subthreshold eating disorders, **purging disorder,** and **night eating syndrome.**

## PICA

Pica is a disorder in which a person persistently eats nonnutritive, nonfood substances over a period of at least 1 month. A few examples of nonfood substances ingested by people with pica are clay, dirt, ice, chalk, or wood. Pica could lead to some serious health consequences, including microbial infections, poisoning from toxins present in the nonfood material, gastrointestinal blockages, or nutrient deficiencies (in cases when nonfood substances displace nutritive foods in the dietary pattern).[31] Pica tends to co-occur with other mental disorders, such as autism spectrum disorder or obsessive-compulsive disorder.

## OTHER SPECIFIED FEEDING OR EATING DISORDERS

**Subthreshold Eating Disorders.** Individuals who meet some but not all of the criteria for diagnosis with anorexia nervosa, bulimia nervosa, or binge eating disorder may fall into one of five subthreshold classifications.[15] For example, **atypical anorexia nervosa** describes a person who meets most of the criteria for diagnosis of anorexia nervosa but whose weight is still within a normal range. This could occur if an overweight person has just begun severely restricting calories. Despite significant weight loss, BMI may still fall within the healthy range of 18.5 to 24.9. Other diagnoses in this category include cases of bulimia nervosa or binge eating disorder in which episodes of binge eating occur less than once per week **(bulimia nervosa of low frequency** or **binge eating disorder of low frequency)** or have taken place for less than 3 months **(bulimia nervosa of limited duration** or **binge eating disorder of limited duration).**

**Purging Disorder.** Purging disorder is the name given to the behavior of people who repeatedly purge (i.e., vomit) to promote weight loss even in the absence of binge eating.[32] This disorder is related to body dissatisfaction, anxiety, and depression. The physical effects of purging disorder are the same as those of bulimia nervosa and include dental problems, mouth sores, damage to the esophagus, constipation, dehydration, electrolyte imbalances, and overall malnutrition.

**Night Eating Syndrome.** Night eating syndrome is characterized by recurrent episodes of night eating, manifested by eating after awakening from sleep or by excessive food consumption after the evening meal. With night eating syndrome, the person is fully aware of and able to recall the behavior, which results in marked distress.[33] This eating disorder should not be confused with *sleep-related eating disorder,* a sleep disorder (like sleepwalking) in which individuals have only partial or no memory of nocturnal food intake.

Although night eating syndrome was first observed among patients with obesity, it also occurs among people within a healthy BMI range. It has been estimated to occur in 1.5% of the general population and in 8.9% of patients treated in obesity clinics; it tends to co-occur with other disorders. Some typical signs and symptoms of night eating syndrome include:

- Not feeling hungry in the morning and delaying the first meal until several hours after waking

**pica** The practice of eating nonfood items, such as dirt, laundry starch, or clay.

**purging disorder** An eating disorder characterized by repeated purging (e.g., by self-induced vomiting) to induce weight loss even in the absence of binge eating.

**night eating syndrome** An eating disorder characterized by consumption of a large volume of food in the late evening and nocturnal awakenings with ingestion of food.

**atypical anorexia nervosa** A subthreshold eating disorder in which a person meets most of the criteria for diagnosis of anorexia nervosa, except BMI is within a normal range.

**bulimia nervosa of low frequency** A subthreshold eating disorder in which a person meets all of the criteria for diagnosis of bulimia nervosa, except the frequency of binge-compensate cycles is less than once per week.

**binge eating disorder of low frequency** A subthreshold eating disorder in which a person meets all of the criteria for diagnosis of binge eating disorder, except the frequency of binges is less than once per week.

**bulimia nervosa of limited duration** A subthreshold eating disorder in which a person meets all of the criteria for diagnosis of bulimia nervosa, except the duration of the disordered eating behavior is less than 3 months.

**binge eating disorder of limited duration** A subthreshold eating disorder in which a person meets all of the criteria for diagnosis of binge eating disorder, except the duration of the disordered eating behavior is less than 3 months.

- Overeating in the evening with more than 25% of daily food intake consumed after dinner
- Difficulty falling asleep and needing to eat something to help fall asleep faster
- Waking at least once during the night with a need to eat to be able to fall asleep again
- Feelings of guilt and shame regarding eating behaviors
- Feeling depressed, especially at night

Research shows that the circadian rhythm (your body's 24-hour clock) of food intake appears to be disturbed in night eating syndrome. Studies have also shown that night eating syndrome is prevalent among outpatients with sleep apnea, restless leg syndrome, or some psychiatric conditions. Behavioral changes, such as establishing and monitoring the sleep-wake schedule and doing regular physical activity, can help. Symptoms are significantly improved with use of the antidepressant sertraline (Zoloft®).

## ✓ CONCEPT CHECK 11.5

1. List three ways pica could harm health.
2. Describe three different cases in which a person would be diagnosed with subthreshold eating disorders.
3. How is purging disorder similar to bulimia nervosa? How do these two disorders differ?
4. List three characteristics of night eating syndrome.

# 11.6 Additional Disordered Eating Patterns

There are several other patterns of disordered eating that have not yet been classified by *DSM-5*, but health care providers report encountering them in clinical practice. Continued research will be required to set standardized diagnostic criteria and establish evidence-based treatments.

## ORTHOREXIA

**orthorexia** A proposed psychological disorder characterized by an obsession with proper or healthful eating.

**muscle dysmorphia** A psychological disorder in which an individual perceives himself or herself as too thin, desires a highly muscular physique, and engages in obsessive behaviors to gain muscle mass.

Orthorexia describes a condition in which a person is unduly concerned with eating only healthful foods. **When does a healthy concern about nutrition become an unhealthy obsession?** Cathy Yeulet/ Hemera/Getty Images

As you have learned about nutrition in this course, you may have changed some of your own food choices. Obviously, a focus on healthy eating can help with weight management and prevention of disease, but when strict food rules begin to interfere with everyday life, they can become pathological. Some people worry excessively about the availability of food they permit themselves to eat, such as low-fat, sugar-free, or organic food choices. Although not currently classified as an eating disorder, **orthorexia** describes an emerging condition in which healthful eating becomes an obsession.[34,35] The term comes from Greek words meaning "straight or proper appetite." Unlike cases of anorexia or bulimia, orthorexia does not usually originate in the drive for thinness. Rather, a need for perfection or purity lies at the heart of orthorexia. Such extreme dietary perfectionism may be related to obsessive-compulsive disorder. Criteria to diagnose orthorexia have been proposed. These include eating behaviors that interfere with normal activities of living and social activities, feelings of guilt or anxiety when eating "forbidden" foods, and physical signs of malnutrition.[36]

## MUSCLE DYSMORPHIA

**Muscle dysmorphia** has gained attention as a psychological condition with many similarities to other eating disorders.[10,37] First identified among male bodybuilders in the 1990s, muscle dysmorphia was initially termed "reverse anorexia." Men (and some women) with this disorder perceive themselves as *too thin,* rather than too fat, and are preoccupied with strict weightlifting and diet regimens to achieve a high level of muscularity. Among male high school students and college-age men, in particular, body dissatisfaction may lead to disordered eating and exercise practices.

In muscle dysmorphia, numerous hours are devoted to working out at the gym; planning and eating meals that fit into a certain macronutrient ratio; and keeping meticulous records of exercise, body measurements, and food intake. The dietary practices (e.g., high-volume eating, protein intakes of up to 5 grams per kilogram of body weight), and use of unproven ergogenic aids or anabolic steroids can result in physical impairment. Such exercise and eating routines also interfere with social, occupational, and recreational activities. People with muscle dysmorphia may avoid social contact, eating in restaurants, and being seen without clothes because of distress over appearing too thin. Some experts support recognition of muscle dysmorphia as a distinct eating disorder because of its many similarities with anorexia nervosa and a tendency for people with muscle dysmorphia to cross over to other forms of eating disorders over time.

## EATING DISORDERS AND DIABETES

Individuals with diabetes—especially adolescents and young adults—are at heightened risk for eating disorders.[38,39] As you learned in Chapter 4, when insulin is absent or when cells are insulin resistant, cells are unable to use glucose for energy. Weight loss occurs because carbohydrate calories essentially are wasted. After a person starts insulin or insulin-sensitizing therapy, cells are able to utilize glucose and store fat, so weight gain is a common side effect. Individuals with diabetes may also become overwhelmed by a constant focus on achieving perfect numbers: not just body weight but also blood glucose measurements, insulin dosages, calories, carbohydrates, and perhaps physical activity.

Research shows that disordered eating behaviors are twice as common among patients with type 1 diabetes as in those without disease.[40] Approximately one in three teens with type 1 diabetes admits to intentionally skipping doses of insulin to induce weight loss.[41] This practice, colloquially dubbed *diabulimia,* can lead to severe hyperglycemia and its myriad consequences, which include eye damage, kidney damage, diabetic coma, or death. On the other hand, about one in five teens with type 1 diabetes uses overdoses of insulin to compensate for episodes of binge eating, a practice that could lead to dangerously low levels of blood sugar. Among adolescents with type 2 diabetes, up to 25% show signs of binge eating disorder. For youth with diabetes, evidence of poor blood sugar control, frequently missed clinic visits, and the presence of depression can be warning signs of eating disorders.[42]

## DISORDERED EATING AND BINGE DRINKING

A pattern of inappropriate compensatory behaviors to avoid weight gain from consuming alcohol—informally called *drunkorexia*—has recently gained recognition in the media and medical literature. This pattern of disordered eating involves compensatory behaviors (e.g., calorie restriction, self-induced vomiting, or excessive exercise) used to offset the calories consumed during episodes of binge drinking. These compensatory behaviors may occur before or after binge drinking. Recall from Chapter 1, binge drinking is defined as consuming 4 or more drinks (women) or 5 or more drinks (men) within a 2-hour period.

As with the other patterns of disordered eating described in this section, drunkorexia has not yet been recognized as a distinct eating disorder, nor does it have standard diagnostic criteria. However, several researchers have proposed defining drunkorexia in much the same way *DSM-5* defines bulimia nervosa: a pattern of binge drinking accompanied by inappropriate compensatory mechanisms to rid the body of excess calories that occurs at least once per week for at least 3 months.[43] In light of its similarities with bulimia nervosa, some experts suggest this pattern should be named *alcoholimia.*

Researchers have just begun to examine the etiology of drunkorexia. Like binge eating, binge drinking may be a maladaptive coping mechanism that occurs in response to extreme stress, grief, or traumatic life events. Body dissatisfaction and fear of weight gain also play a role. The behaviors appear to be more common among women than men and are practiced mainly by adolescents and young adults. Reported motivations for these behaviors include both avoidance of weight gain and rapid induction of intoxication.[44]

The physical effects of drunkorexia could be severe. Malnutrition could result from repeated episodes of calorie restriction. Dehydration and electrolyte imbalances induced by vomiting or laxative abuse can threaten heart function. Overlapping these disordered behaviors with binge drinking can only intensify the dangerous effects of excessive alcohol. In the short term, this could include impaired judgment, increased risk of accidents, or alcohol poisoning. Repeated over time, these behaviors can harm relationships, diminish work or academic performance, and damage the liver and nervous system. Likewise, by altering the absorption, metabolism, and excretion of nutrients, heavy drinking may worsen the effects of malnutrition. Indeed, the combination of disordered eating and binge drinking is a double-edged sword.

### ✓ CONCEPT CHECK 11.6

1. How would you distinguish between healthy eating and orthorexia?
2. How is muscle dysmorphia similar to anorexia nervosa? How is it different?
3. For an individual with type 1 diabetes, what are the long-term harmful effects of misusing insulin to prevent weight gain?

## 11.7 Prevention of Eating Disorders

A key to developing and maintaining healthful eating behavior is to realize that some concern about eating, health, and weight is normal. It is also normal to experience variation in what we eat, how we feel, and even how much we weigh. For example, it is common to experience some minimal weight change (up to 2 to 3 pounds) throughout the day and even more over the course of a week. A large weight fluctuation or ongoing weight gain or weight loss is more likely to indicate a problem. If you notice a significant change in your eating habits, how you feel, or your body weight, it is a good idea to consult your primary care provider. Treating physical and emotional problems early helps lead you to peace of mind and good health.

Many people begin to form opinions about food, nutrition, health, weight, and body image prior to or during puberty. Parents, friends, and professionals working with children and young adults should consider the following advice for preventing eating disorders:

- Discourage restrictive dieting and meal skipping. Fasting is also discouraged (except for religious occasions).
- Provide information about normal changes that occur during puberty.
- Correct myths about nutrition, healthy body weight, and approaches to weight loss.
- Carefully phrase any weight-related recommendations and comments.
- Do not overemphasize numbers on a scale. Instead, teach the basics of proper nutrition and regular physical activity in school and at home.
- Encourage normal expression of disruptive emotions.
- Encourage children to eat when they are hungry and stop eating when they are full.
- Provide adolescents with an appropriate, but not unlimited, degree of independence, choice, responsibility, and self-accountability for their actions.
- Increase self-acceptance and appreciation of the power and pleasure emerging from one's body.
- Enhance tolerance for diversity in body weight and shape.

▶ **These websites provide further information on eating disorders:**
- Academy for Eating Disorders, www.aedweb.org
- The National Eating Disorders Association, www.nationaleatingdisorders.org
- National Institute of Mental Health's concise review of eating disorders, https://www.nimh.nih.gov/health/topics/eating-disorders/index.shtml

- Build respectful environments and supportive relationships.
- Encourage coaches to be sensitive to weight and body-image issues among athletes.
- Emphasize that thinness is not necessarily associated with better athletic performance.
- Support programs for eating disorder screening and prevention at high schools and colleges.

Our society as a whole can benefit from a fresh focus on nutritious food practices and a healthful outlook toward food and body weight (see *Farm to Fork* in this section). Not only is treatment of eating disorders far more difficult than prevention, but these disorders also have devastating effects on the entire family. For this reason, caregivers and health care professionals must emphasize the importance of an overall healthful dietary pattern that focuses on moderation, as opposed to restriction and perfection.

Overall, the challenge facing many North Americans is achieving a healthy body weight without excessive dieting. A growing number of health professionals support a nondiet approach to weight management. This means adopting and maintaining sensible eating habits, a physically active lifestyle, and realistic and positive attitudes and emotions while practicing creative ways to handle stress. Certain cultural ideals of beauty can trigger eating disorders; increased acceptance of diverse shapes and sizes can help to reduce the pressures predisposing some people to various types of disordered eating behavior. For example, the Health at Every Size® approach aims to fight discrimination against individuals based on body shape and size and asserts that each person should be free to find his or her natural weight. Women who combine careers and motherhood are saying that they have more important things to worry about. Trendsetters in the fashion industry are tolerating more curves. We cannot change the genes that predispose people to eating disorders (at least not yet!), but we can make a difference in the environmental triggers that set these devastating disorders in motion.

### ✓ CONCEPT CHECK 11.7

1. Why is prevention of eating disorders so important?
2. Mr. Thomas, a high school health teacher, wants to try to prevent young adults from falling into the discouraging traps of anorexia nervosa and bulimia nervosa. What are some of the topics and issues he should discuss with students in his health classes?

## FARM to FORK   Apples

twixx/123RF

An apple a day keeps the doctor away? Evidence does support this old adage! With more than 7000 varieties to choose from, apples offer a range of colors, flavors, and phytochemicals.

### Grow
- Small apples (< 2 inches in diameter) are called crabapples. These are quite tart, but they are edible and actually contain more phytochemicals than larger apples.
- Apple trees can be grown from seeds, but the end-product can be quite unpredictable. To ensure a desirable harvest, get a 2- or 3-year-old plant from a nursery.

### Shop
- Common varieties with high phytochemical content include Cortland, Fuji, Granny Smith, and Honeycrisp. For the best phytochemical content, shop for varieties such as Haralson, Liberty, Northern Spy, and Ozark Gold at farmers' markets or specialty stores.
- Among the red apples, look for those with deep red color on all sides. This is an indication that the apples were exposed to lots of sunlight while growing. Sun exposure increases the phytochemical content of the fruit.
- When it comes to controlling calories, whole fruit is a better choice than fruit juice, but if you do shop for apple juice, choose a cloudy bottle. Filtering juice removes many healthy phytochemicals. If you hold the bottle up to the light, the juice should not be see-through and there should be some sediment at the bottom.

### Store
- Store apples in the refrigerator, preferably in the crisper drawer, set to high humidity. These will last about 10 times longer than apples stored at room temperature.
- Apples harvested early in the season (July and August) can be stored for just 2 or 3 weeks before spoiling, but apples harvested late in the season (September, October, and November) store well for several months.

### Prep
- Don't throw the skin in the compost bin! Many nutrients and phytochemicals are concentrated in the apple's skin. Scrubbing the fruit under running water is a good way to remove pesticide residues. If you can afford them, organic apples would be a great choice for reducing pesticide exposure.
- Apple skins can be puréed in a food processor and incorporated into recipes for baked treats and applesauce.
- To prevent sliced apples from browning, sprinkle with lemon juice. The citric acid inhibits the oxidase enzymes that cause browning.
- What can you do with crabapples? These small, tart fruits are excellent for use in making jams and jellies or sauces to cook with meats.

Source: Robinson J. Apples: From potent medicine to mild-mannered clones, in *Eating on the Wild Side*. New York: Little, Brown and Company, 2013.

Alexis Joseph/McGraw-Hill Education

## Eating Disorder Reflections

eggeeggjiew/123RF

## Reflections from a Woman with Anorexia Nervosa

It was the spring of my freshman year of high school, and I had just turned 15 years old. I was determined to land a leading role in the upcoming high school musical, *West Side Story*. I thought I should lose some weight to appear more attractive to the student director, Shawn, so I decided to give up "junk" food. The next day, my friend Sandra looked at my lunch, spread out neatly on a napkin before me, and squawked, "Dill pickles?! Who brings dill pickles for lunch in a Ziploc bag?" The other girls at the table fell into a fit of hysterics. "Casting for *West Side Story* is coming up," I said, "and I gave up 'junk' food to try to lose a few extra pounds." One of my friends thought it would be funny to give me one M&M—just to smell. Aren't they funny? So I put it in a little plastic container and kept it in my backpack as a reminder. Every once in a while, I did smell it.

For the next few weeks, there were times when I would find myself cracking open the refrigerator door and just staring at what I knew to be a deliciously crunchy, crisp, and cold Kit Kat® bar in the dairy bin. I didn't eat it, though. At the mall, my friends Bridgette and Nora wanted to stop and get a cinnamon bun. They chided me, but I didn't budge. The cinnamon bun smelled so good. But as I sat opposite them in the food court and watched them overdramatize its ooey-gooey goodness, I felt a sense of pride that I could make a decision and stick with it. I could see that they were jealous of my willpower.

When Easter came around, I took a look at the contents of the Easter basket my mom insisted on preparing and turned up my nose at it. I had proven to myself that I could resist temptation . . . why stop now?

After the musical, I started running with my friend Laura. She was getting in shape for the next season of field hockey. After school, we met in the locker room, changed out of our school clothes, and out we went. The running helped. Every morning, just after going to the bathroom and before getting any breakfast, I would pop onto the scale in my mom's bathroom. One hundred and fifteen pounds and still going. At 5 feet 7 inches, that wasn't too bad.

Cheese and butter had made it to the "no" list by the time I was 16 and down to 105 pounds. Fat-free was my mantra. For my sixteenth birthday, my friends threw a little surprise party for me. Nora, knowing I would put up a fight, made me a cake. "It's your birthday! You can have a piece of cake!" I politely said no, that I would cut it for everyone else, but I really didn't want any. They pestered me, and Nora started to feel offended, so finally I took a few bites, so she wouldn't burst into tears. It had been so long since I'd had so much sugar. I felt bloated and sick. I ate nothing for the rest of the day and only six saltines, an apple, and two stalks of celery the next day. Those foods were on the "yes" list. Salads also were okay but only with salt and vinegar. I told my parents that the dissections in biology class had given me a distaste for meat, but really, I just didn't want all those calories. For a while, I craved food day and night, but slowly, I was getting better and better at holding my ground.

By my senior year, I was skipping lunches altogether, opting instead to hang out in the library and read over my AP bio text. "Where were you at lunch today?" Bridgette would ask later. "Oh, I had some reading to do. The AP exam is going to be tough." At 100 pounds, I was getting closer to finding out what "tough" really meant.

Even though Laura moved out of state, I didn't give up on exercising. Now, my mom's stair stepper in the basement was my favorite. I'd take my biology notes, prop them up in front of me, and step-step-step until I had burned 400 kcal. I felt so efficient knowing that I could multitask. Sometimes I'd exercise twice a day. As senior year wore on, though, it got harder and harder to get up in the morning and put on my tennis shoes. And then one morning, in the shower, I just collapsed under the stream of hot water.

I ended up in this hospital bed with an IV in my arm. At 92 pounds, my body was starving. As it turns out, if you don't give your body enough fuel, you start to cannibalize yourself, in a sense. My body had been so hungry, my muscles had been wasting away, and the episode in the shower was due to a problem with my heart. It's a problem I have created . . . not my parents or my distant group of friends . . . only me. My mom was there, next to me, caressing the arm with the IV tube, putting her whole life on hold because of me. Isn't this what I'd wanted—to be in control of my own destiny?

Where do I go from here?

## Thoughts of a Woman with Bulimia Nervosa

I am wide awake and immediately out of bed. I think back to the night before when I made a new list of what I wanted to get done. My husband is not far behind me on his way into the bathroom to get ready for work. Maybe I can sneak onto the scale before he notices me. I am already in my private world. I am overjoyed when the scale says that I am the same weight as I was the night before, and I can feel that slightly hungry feeling. Maybe it will stop today, maybe today everything will change. What were the projects I wanted to do?

We eat the same breakfast, except that I take no butter on my toast, no cream in my coffee, and never take seconds (until he gets out the door). Today, I am going to be really good, which means eating certain predetermined portions of food and not taking one more bite than I think I am allowed. I am careful to see that I don't take more than he does. I can feel the tension building. I wish he'd hurry up and leave, so I can get going!

As soon as he shuts the door, I try to involve myself with one of the myriad responsibilities on my list. But I hate them all! I just want to crawl into a hole. I don't want to do anything. I'd rather eat. I am alone, I am nervous, I am no good, I always do everything wrong anyway. I am not in control, I can't make it through the day—I know it. It has been the same for so long.

I remember the starchy cereal I ate for breakfast. I am back into the bathroom and onto the scale. It measures the same, but I don't want to stay the same! I want to be thinner! I look in the mirror. I think my thighs are ugly and deformed looking. I see a lumpy, clumsy, pear-shaped wimp. I feel frustrated, trapped in this body, and I don't know what to do about it.

I float to the refrigerator knowing exactly what is inside. I begin with last night's brownies. I always begin with the sweets. At first I try to make it look like nothing is missing, but my appetite is huge and I resolve to make another batch of brownies to replace the one I'm devouring. I know there is half of a bag of cookies in the trash, thrown out the night before, and I dig them out and polish them off. I drink some milk so my vomiting will be smoother. I like the full feeling I get after downing a big glass. I get out six pieces of bread and toast one side in the broiler, turn them over and cover them with butter and put them under the broiler again till they are bubbling. I take all six pieces on a plate to the television and go back for a bowl of cereal and a banana. Before the last toast is finished, I am already preparing the next batch of six more pieces. I might have another brownie or five from the new batch, and a couple large bowlfuls of ice cream, yogurt, or cottage cheese. My stomach is stretched into a huge ball below my rib cage. I know I'll have to go into the bathroom soon, but I want to postpone it. I am in never-never land. I am waiting, feeling the pressure, pacing the floor in and out of rooms. Time is passing. Time is passing. It is almost time.

Bulimic episodes add to the despair felt in this disorder. **What are some strategies recommended by clinicians to develop normal eating habits?** Royalty-Free/Corbis

I wander aimlessly through the living room and kitchen once more, tidying, making the whole house neat and put back together. Finally, I make the turn into the bathroom. I brace my feet, pull my hair back and stick my finger down my throat, stroking twice. I get up a huge gush of food. Three times, four, and another stream of partially digested food. I can see everything come back. I am glad to see those brownies, because they are SO fattening. The rhythm of the emptying is broken and my head is beginning to hurt. I stand up feeling dizzy, empty, and weak.

The whole episode has taken about an hour.

## ✓ CONCEPT CHECK 11.8

1. Which of the diagnostic criteria can you identify in the narrative about the young woman with anorexia nervosa?

2. What are the defining characteristics of a binge? Are these characteristics evident in the narrative about the woman with bulimia nervosa?

# Summary (Numbers refer to numbered sections in the chapter.)

**11.1** Disordered eating encompasses mild and short-term changes in eating patterns that occur as a result of life stress, illness, or a desire to change body weight. When carried to the extreme, disordered eating may progress to an eating disorder in which severe changes in eating patterns have lasting and detrimental effects. Current research on the origins of eating disorders indicates that genetic factors dictate brain biology, which affects how certain individuals perceive their bodies and respond to life stresses. Thus, a person who is genetically predisposed to eating disorders may use disordered eating behaviors to cope with feelings of depression, anger, or guilt. The three main types of eating disorders are anorexia nervosa, bulimia nervosa, and binge eating disorder.

**11.2** Anorexia nervosa is characterized by extreme weight loss, a distorted body image, and an irrational fear of weight gain and obesity. Weight loss is achieved primarily by severely restricting food intake. Physical consequences include a profound decrease in body weight and body fat, heart irregularities, iron-deficiency anemia, impaired immunity, digestive dysfunction, and loss of menstrual periods. Treatment of anorexia nervosa includes increasing food intake to support weight gain. Family-based treatment can help individuals with anorexia nervosa establish healthy eating behaviors and body image.

**11.3** Similar to anorexia nervosa, bulimia is characterized by placing too much value on body weight and shape. However, the disordered eating patterns of individuals with bulimia nervosa involve recurrent binge eating followed by compensatory behaviors. Binge eating is consuming an abnormally large amount of food within a short time period. A person with bulimia nervosa has lack of control over bingeing behaviors and feels extremely distressed after a binge. Inappropriate compensatory behaviors used to rid the body of excess calories include vomiting or misusing laxatives, diuretics, or enemas. Alternatively, fasting and excessive exercise may be used. Vomiting as a means of purging is especially destructive to the body; it can cause severe tooth decay, stomach ulcers, irritation of the esophagus, low blood potassium, and other problems. Treatment of bulimia nervosa includes psychological and nutritional counseling. Prozac® is the only FDA-approved medication for treatment of bulimia nervosa, but other antidepressants and antiseizure medications are sometimes prescribed.

**11.4** Binge eating disorder is the most widespread eating disorder. It tends to be diagnosed in middle age and affects men and women nearly equally. It is characterized by recurrent episodes of bingeing, which cause marked distress, but are not followed by compensatory behaviors. About 70% of people with binge eating disorder are obese. The health effects that stem from binge eating disorder are comorbid conditions of obesity, including hypertension, high blood cholesterol, cardiovascular disease, and type 2 diabetes. Treatment involves cognitive behavioral therapy and nutrition counseling. Vyvanse® is the only FDA-approved drug for treatment of binge eating disorder. Antidepressants and other medications may also be useful.

**11.5** Pica is an eating disorder in which a person persistently ingests nonnutritive, nonfood items such as clay, dirt, or ice. Subthreshold eating disorders (e.g., atypical anorexia nervosa, bulimia nervosa of low frequency or limited duration, or binge eating disorder of low frequency or limited duration) describe the behaviors of individuals who meet some but not all of the criteria for diagnosis of anorexia nervosa, bulimia nervosa, or binge eating disorder. People with purging disorder practice purging behaviors to achieve weight loss, but they do not exhibit the binge eating behaviors typical of bulimia nervosa. Individuals with night eating syndrome consume more than 25% of daily food intake after dinner, may have difficulty falling asleep without eating, and wake up at least once during the night to consume food.

**11.6** Several additional disordered eating patterns that are not currently classified as eating disorders include orthorexia, reverse anorexia, diabulimia, and drunkorexia. Orthorexia refers to an obsession with healthy eating such that overly restrictive food choices reduce the quality of life and cause physical harm. Muscle dysmorphia could be described as "reverse anorexia"; a person views him- or herself as less muscular than desired and resorts to patterns of disordered eating and obsessive exercise to achieve a muscular body shape. Individuals with type 1 diabetes are at increased risk for disordered behaviors, such as the misuse of diabetes medication to regulate body weight. Some people with type 1 diabetes skip doses of insulin to induce weight loss but then suffer the consequences of hyperglycemia. Alternatively, overdoses of insulin or glucose-lowering medications could be used to counter the effects of a binge. This practice could result in hypoglycemia. Drunkorexia refers to compensatory caloric restriction to counter the calories consumed during binge drinking.

**11.7** Prevention of eating disorders is crucial because treatments are expensive, lengthy, and not 100% effective. Encouraging healthy attitudes about eating and exercise from a young age will aid in preventing the development of eating disorders. Those who work closely with children and young adults should encourage acceptance of diversity in body sizes and carefully phrase comments about body weight. Helping children to develop healthy ways to cope with emotions is also important.

**11.8** Anorexia nervosa and bulimia nervosa are both biologically based disorders that are characterized by undue influence of body weight or shape on one's self-evaluation. The coping mechanisms (i.e., extreme dietary restriction versus binge-compensate cycles) used by people with these disorders can vary considerably.

# Check Your Knowledge (Answers are available at the end of this question set.)

1. For 3 weeks leading up to her friend's wedding, Teresa skipped meals and restricted her food intake to 800 kcal per day so that she could fit into her bridesmaid dress. After the wedding, she resumed eating 2200 kcal per day. This is an example of
   a. disordered eating.
   b. an eating disorder.
   c. size acceptance.
   d. muscle dysmorphia.

2. Factors that could contribute to the development of eating disorders include
   a. genetics.
   b. social pressures to be thin.
   c. sexual abuse.
   d. all of these.

3. Anorexia nervosa can be defined as
   a. compulsive eating.
   b. hyperactivity.
   c. denial of appetite.
   d. purging.

4. The most likely long-term health consequence of anorexia nervosa is
   a. fractures resulting from bone loss.
   b. atherosclerotic heart disease.
   c. esophageal ulcers.
   d. cancer.

5. A serious health consequence related to low estrogen levels among women of reproductive age is
   a. low energy availability.
   b. low basal metabolic rate.
   c. low bone mineral density.
   d. insomnia.

6. Cuts or calluses on the fingers or knuckles caused by using the fingers to trigger the gag reflex are known as
   a. lanugo.
   b. parotid signs.
   c. Russell's sign.
   d. Barrett's esophagus.

7. The *most life-threatening* health risk from frequent vomiting due to bulimia nervosa is
   a. a drop in blood potassium.
   b. constipation.
   c. weight gain.
   d. swollen salivary glands.

8. Binge eating disorder can be characterized as
   a. bingeing accompanied by purging.
   b. an obsession with healthy eating.
   c. eating to avoid feeling and dealing with emotional pain.
   d. the early phase of bulimia nervosa.

9. Night eating syndrome is characterized by
   a. eating dinner but no breakfast or lunch.
   b. the need to eat to fall asleep.
   c. waking at night to purge by vomiting.
   d. consuming all of the daily calories at night.

10. If you were assigned to speak to a group of middle school students about healthy eating, which message would be best?
    a. Ask students to sort various snack ideas into "good" or "bad" groups.
    b. Illustrate how many minutes of exercise are needed to burn the calories in various snacks.
    c. Advise kids to restrict favorite treats (e.g., ice cream), except as a reward for reaching a goal, such as getting a good grade on a test.
    d. Emphasize that students should eat when they are hungry and stop eating when they are full.

Answer Key: 1. a (LO 11.1), 2. d (LO 11.2), 3. c (LO 11.3), 4. a (LO 11.3), 5. c (LO 11.3), 6. c (LO 11.4), 7. a (LO 11.4), 8. c (LO 11.5), 9. b (LO 11.6), 10. d (LO 11.8)

# Study Questions (Numbers refer to Learning Outcomes)

1. Based on your knowledge of good nutrition and sound dietary habits, answer the following questions:
   a. How can repeated bingeing and purging lead to significant nutrient deficiencies?
   b. How can significant nutrient deficiencies contribute to major health problems in later life?
   c. A friend asks you if it is okay to "cleanse" the body by only drinking fruit juice for a week. What is your response? (LO 11.1)

2. Explain the role of excessive exercise in eating disorders. (LO 11.1)

3. Provide an example of the way society contributes to development of eating disorders. (LO 11.2)

4. Describe some differences between the types and rates of eating disorders among males and females. What contributes to these disparities? (LO 11.2)

5. What are the typical characteristics of a person with anorexia nervosa? What may influence a person to begin rigid, self-imposed dietary patterns? (LO 11.3)

6. List the detrimental physical effects of bulimia nervosa. Describe important goals of the psychological and nutrition therapy used to treat patients with bulimia nervosa. (LO 11.4)

7. How does binge eating disorder differ from bulimia nervosa? Describe factors that contribute to the development of binge eating disorder. (LO 11.5)

8. What medications are currently used in the treatment of anorexia nervosa, bulimia nervosa, and binge eating disorder? (LOs 11.3–11.5)

9. What is orthorexia? How may social media contribute to this pattern of disordered eating? (LO 11.7)

10. Provide two recommendations to reduce the problem of eating disorders in our society. (LO 11.8)

# Further Readings

1. Klump K and others: Academy for Eating Disorders position paper: Eating disorders are serious mental illnesses. *International Journal of Eating Disorders* 42:97, 2009. DOI:10.1002/eat.20589.

2. Schwartz MB and others: The influence of one's own body weight on implicit and explicit anti-fat bias. *Obesity* 14:440, 2006. DOI:10.1038/oby.2006.58.

3. Eddy KT and others: Diagnostic crossover in anorexia nervosa and bulimia nervosa: implications for DSM-V. *American Journal of Psychiatry* 165:245, 2008. DOI:10.1176/appi.ajp.2007.07060951.

4. Le Grange D and others: Academy of Eating Disorders position paper: The role of the family in eating disorders. *International Journal of Eating Disorders* 43:1, 2010. DOI:10.1002/eat.20751.

5. Steinglass JE, Berner LA, and Attia E: Cognitive neuroscience of eating disorders. *Psychiatric Clinics of North America* 42:75, 2019. DOI:10.1016/j.psc.2018.10.008.

6. Mitchison D and Hay P: The epidemiology of eating disorders: Genetic, environmental, and societal factors. *Journal of Clinical Epidemiology* 6:89, 2014. DOI:10.2147/CLEP.S40841.

7. Campbell IC and others: Eating disorders, gene-environment interactions, and epigenetics. *Neuroscience & Biobehavioral Reviews* 35:784, 2011. DOI:10.1016/j.neubiorev.2010.09.012.

8. Rivera HM and others: The role of maternal obesity in the risk of neuropsychiatric disorders. *Frontiers in Neuroscience* 9:194, 2015. DOI:10.3389/fnins.2015.00194.

9. Limbers CA, Cohen LA, and Gray BA: Eating disorders in adolescent and young adult males: Prevalence, diagnosis, and treatment strategies. *Adolescent Health, Medicine, and Therapeutics* 9:111, 2018. DOI:10.2147/AHMT.S147480.

10. Mitchison D and Mond J: Epidemiology of eating disorders, eating disordered behaviour, and body image disturbance in males: A narrative review. *Journal of Eating Disorders* 3:20, 2015. DOI:10.1186/s40337-015-0058-y.

11. Matthews-Ewald MR and others: Sexual orientation and disordered eating behaviors among self-identified male and female college students. *Eating Behaviors* 15:441, 2014. DOI:10.1016/j.eatbeh.2014.05.002.

12. Campbell K and Peebles R: Eating disorders in children and adolescents: State of the art review. *Pediatrics* 134:582, 2014. DOI:10.1542/peds.2014-0194.

13. Luca A and others: Eating disorders in late-life. *Aging and Disease* 6:48, 2015. DOI:10.14336/AD.2014.0124.

14. Udo T and Grilo CM: Prevalence and correlates of DSM-5-defined eating disorders in a nationally representative sample of U.S. adults. *Biological Psychiatry* 84:345, 2018. DOI:10.1016/j.biopsych.2018.03.014.

15. Rowe E: Early detection of eating disorders in general practice. *Australian Family Physician* 46:833, 2017.

16. Borzekowski DLG and others: e-Ana and e-Mia: A content analysis of pro-eating disorder websites. *American Journal of Public Health* 100:1526, 2010. DOI:10.2105/AJPH.2009.172700.

17. Mehler PS and Brown C: Anorexia nervosa—medical complications. *Journal of Eating Disorders* 3:11, 2015. DOI:10.1186/s40337-015-0040-8.

18. Daily JP and Stumbo JR: Female athlete triad. *Primary Care: Clinics in Office Practice* 45:615, 2018. DOI:10.1016/j.pop.2018.07.004.

19. Ozier AD and Henry BW: Position of the American Dietetic Association: Nutrition intervention in the treatment of eating disorders. *Journal of the American Dietetic Association* 111:1236, 2011. DOI:10.1016/j.jada.2011.06.016.

20. Ruusunen A and others: The gut microbiome in anorexia nervosa: Relevance for nutritional rehabilitation. *Psychopharmacology* 236:1545, 2019. DOI:10.1007/s00213-018-5159-2.

21. Zipfel S and others: Anorexia nervosa: Etiology, assessment, and treatment. *Lancet Psychiatry* 2:1099, 2015. DOI:10.1016/S2215-0366(15)00356-9.

22. Waterhous T and others: Practice paper of the American Dietetic Association: Nutrition intervention in the treatment of eating disorders. *Journal of the American Dietetic Association* 111:1261, 2011.

23. Marvanova M and Gramith K: Role of antidepressants in the treatment of adults with anorexia nervosa. *Mental Health Clinics* 8:127, 2018. DOI:10.9740/mhc.2018.05.127.

24. Castillo M and Weiselberg E: Bulimia nervosa/purging disorder. *Current Problems in Pediatric and Adolescent Health Care* 47:85, 2017. DOI:10.1016/j.cppeds.2017.02.004.

25. Wade TD: Recent research on bulimia nervosa. *Psychiatric Clinics of North America* 42:21, 2019. DOI:10.1016/j.psc.2018.10.002.

26. Mehler PS and Rylander M: Bulimia nervosa—medical complications. *Journal of Eating Disorders* 3:12, 2015. DOI:10.1186/s40337-015-0044-4.

27. Guerdjikova AI and others: Binge eating disorder. *Psychiatric Clinics of North America* 40:255, 2017. DOI:10.1016/j.psc.2017.01.003.

28. Da Luz FQ and others: Obesity with comorbid eating disorder: Associated health risks and treatment approaches. *Nutrients* 10:829, 2018. DOI:10.3390/nu10070829.

29. Brownley KA and others: Pharmacological approaches to the management of binge eating disorder. *Drugs* 75:9, 2015. DOI:10.1007/s40265-014-0327-0.

30. Touyz S, Lacey H, and Hay P. Eating disorders in the time of COVID-19. *Journal of Eating Disorders* 8:19, 2020. DOI:10.1186/s40337-020-00295-3.

31. Miao D, Young SL, and Golden CD: A meta-analysis of pica and micronutrient status. *American Journal of Human Biology* 27:84, 2015. DOI:10.1002/ajhb.22598.

32. Murray SB and Anderson LK: Deconstructing "atypical" eating disorders: An overview of emerging eating disorder phenotypes. *Current Psychiatry Reports* 17:86, 2015. DOI:10.1007/s11920-015-0624-7.

33. Allison KC, Spaeth A, and Hopkins CM: Sleep and eating disorders. *Current Psychiatry Reports* 18:92, 2016. DOI:10.1007/s11920-016-0728-8.

34. Dunn TM and Bratman S: On orthorexia nervosa: A review of the literature and proposed diagnostic criteria. *Eating Behaviors* 21:11–17, 2016. DOI:10.1016/j.eatbeh.2015.12.006.

35. Getz L: Understanding orthorexia—when healthful eating becomes an obsession. *Today's Dietitian* 16:18, 2014.

36. Cena H and others: Definition and diagnostic criteria for orthorexia nervosa: A narrative review of the literature. *Eating and Weight Disorders*, 2018. DOI:10.1007/s40519-018-0606-y.

37. Baghurst T: Muscle dysmorphia and male body image: Signs and symptoms. *SCAN's Pulse* 36(1):5, 2017.

38. Colton PA and others: Eating disorders in girls and women with type 1 diabetes: A longitudinal study of prevalence, onset, remission, and recurrence. *Diabetes Care* 38:1212, 2015. DOI:10.2337/dc14-2646.

39. Clery P and others: Systematic review and meta-analysis of the efficacy of interventions for people with type 1 diabetes mellitus and disordered eating. *Diabetic Medicine* 34:1667, 2017. DOI:10.1111/dme.13509.

40. Staite E and others: "Diabulimia" through the lens of social media: A qualitative review and analysis of online blogs by people with type 1 diabetes mellitus and eating disorders. *Diabetic Medicine* 35:1329, 2018. DOI:10.1111/dme.13700.

41. Wisting L and others: Prevalence of disturbed eating behavior and associated symptoms of anxiety and depression among adult males and females with type 1 diabetes. *Journal of Eating Disorders* 6:28, 2018. DOI:10.1186/s40337-018-0209-z.

42. Cecilia-Costa R, Volkening LK, and Laffel LM: Factors associated with disordered eating behaviours in adolescents with type 1 diabetes. *Diabetic Medicine*, 2018. DOI:10.1111/dme.13890.

43. Thompson-Memmer C, Glassman T, and Diehr A: Drunkorexia: A new term and diagnostic criteria. *Journal of American College Health* 67:620, 2019. DOI:10.1080/07448481.2018.1500470.

44. Laghi F and others: Psychological characteristics and eating attitudes in adolescents with drunkorexia behavior: An exploratory study. *Eating and Weight Disorders–Studies on Anorexia, Bulimia and Obesity*, 2019. DOI:10.1007/s40519-019-00675-y.

# Chapter 12

# Protecting Our Food Supply

# Student Learning Outcomes

**Chapter 12 is designed to allow you to:**

**12.1** Understand the effects of conventional and sustainable agriculture on our food choices, including the use of organic farming and food biotechnology; explain the benefits of locally grown foods and community supported agriculture.

**12.2** Understand the reasons behind pesticide use, the possible long-term health implications, and their safety limits.

**12.3** List types and common sources of viruses, bacteria, fungi, and parasites that can make their way into food; compare and contrast food preservation methods.

**12.4** Understand the foodborne illnesses caused by bacteria, viruses, and parasites.

**12.5** Describe the main reasons for using chemical additives in foods, the general classes of additives, and the functions of each class; identify natural substances in foods that can cause illness and the consequences of their ingestion.

**12.6** Describe the procedures that can be used to limit the risk of foodborne illness.

## FAKE NEWS

Organic products, including produce, have more nutritional value beyond conventionally grown produce, which makes them worth the extra cost.

## THE FACTS

Current research is not sufficient to recommend organic over conventional produce on the basis of nutrient content. Produce has its highest vitamin and phytochemical content when it is harvested at peak ripeness. To maximize the nutrient content of your produce either grow your own, buy it locally at a farmers' market, or use canned or frozen fruits and vegetables because they are harvested close to ripeness.

This chapter focuses on where food comes from, how it was grown, and the overall safety of our food supply. During the last century, agriculture, the primary way our food supply is produced, has seen many changes. There have been tremendous rises in productivity, with conventional agriculture emphasizing large yields and low costs. Unfortunately our food system results in an increasing amount of wasted food, one of the leading drivers of climate change. On a positive note, there is a growing trend toward sustainability and an understanding of the long-term environmental impact of agricultural practices. Consumers are demanding increased transparency in the food supply and are driving up demand for organic and locally grown products.

Although warnings about the safety of food and water appear everywhere, experts agree that North Americans enjoy one of the safest food supplies in the world, especially if foods are stored and prepared properly. Although tremendous progress has been made regarding food safety, microorganisms and certain chemicals in foods still pose a health risk. As always, the nutritional and health benefits of food must be balanced against any food-related hazards. This chapter focuses on these hazards: how real they are and how you can minimize their effect on your life. You bear some responsibility for this; government agencies and industry can only do so much. The *Dietary Guidelines* encourage us to prepare and store foods safely. Which foods pose the greatest risk for foodborne illness? Is any food safe after being stored in the refrigerator for 6 months? Are food additives and pesticides a growing concern? Are there ways we can produce and eat food more sustainably? This chapter provides some answers.

# 12.1 Food Production Choices

During the last century, agriculture—the production of food and livestock—has seen tremendous rises in productivity as a result of human labor being replaced by automated technologies, selective animal and plant breeding, and synthetic fertilizers and pesticides. At one time, nearly everyone was involved in food production. Today, less than one in three people around the globe (less than 2% in the United States) is now involved in farming. The number of farms in the United States has been dropping steadily over the last decade. While there were about 2.2 million U.S. farms in 2007, there were just over 2 million farms in 2018.[1] Approximately 570 million farms, primarily run by individuals, produce about 80% of the world's food. The majority of large farms are in the Americas, whereas smaller farms predominate in Asia. Numerous advances in agricultural sciences are affecting our food supply; of particular note are organic food production, food biotechnology, and sustainable agriculture. Many of these new developments in agriculture are aimed at reducing the overall **carbon footprint** (carbon dioxide and methane emissions) generated as crops move from the farm to the fork. At the same time, a growing area of concern relative to the economic and environmental impacts on our food supply is the issue of **food waste.**

## ORGANIC FOODS

The production of **organic foods** relies on farming practices such as **biological pest management,** composting, manure application, and crop rotation to maintain healthy soil, water, crops, and animals. In contrast, synthetic pesticides, fertilizers, and hormones; antibiotics; sewage sludge (used as fertilizer); genetic engineering; and irradiation are not permitted in the production of organic foods. However, many synthetic substances and natural pesticides may be used in organic crop production. See the U.S. Organic Regulations at https://www.ams.usda.gov/grades-standards/organic-standards for more information. Additionally, organic meat, poultry, eggs, and dairy products must come from animals allowed to graze outdoors and consume only organic feed.

Interest in personal and environmental health has contributed to the increasing availability and sales of organic foods. Organic foods are increasingly available in supermarkets, specialty stores, farmers' markets, and restaurants.[2] Consumers can select organic fruits, vegetables, grains, dairy products, meats, eggs, and many processed foods, including sauces and condiments, breakfast cereals, cookies, and snack chips. Direct marketing of farm products through farmers' markets is a growing sales outlet for organic products nationwide. According to the Organic Trade Association, U.S. sales of organic foods exceed $47 billion annually. Canada's organic food and beverage market has also tripled, reaching $5.4 billion annually. Despite this rapid growth, only 4.2% of foods sold are organic. Organic foods, because they cost more to grow and produce, are more expensive than comparable conventional foods.

The Organic Foods Production Act of 1990 established standards for the production of foods that bear the USDA Organic seal (Fig. 12-1). Foods labeled and marketed as organic must be grown on farms that are certified by the U.S. Department of Agriculture (USDA) as following all of the rules established in the 1990 act. Products labeled *100% organic* may only contain organically produced ingredients and processing aids, excluding water and salt. No other ingredients or additives are permitted. Foods made from multiple ingredients (e.g., breakfast cereal) can be labeled as *organic* if at least 95% of their ingredients (by weight) meet organic standards. The term *made with organic ingredients* can be used if at least 70% of the ingredients are organic. Small organic producers and farmers with sales less than $5000 per year are exempt from the certification regulation. Some farmers use organic production methods but choose not to be USDA certified. Their foods cannot be labeled as organic, but many of these farmers market and sell to those seeking organic foods.

There has been an increase in USDA support and funding for research, cost-share assistance, and other organic food programs since national organic standards were

**carbon footprint** The greenhouse gas emissions caused by an organization, event, product, or individual.

**food waste** Food that is edible or fit for consumption which is being discarded as plate waste by consumers and by retailers due to color or appearance.

**organic food** Food grown without use of pesticides, synthetic fertilizers, sewage sludge, genetically modified organisms, antibiotics, hormones, or ionizing radiation.

**biological pest management** Control of agricultural pests by using natural predators, parasites, or pathogens. For example, ladybugs can be used to control an aphid infestation.

**FIGURE 12-1** ▲ The USDA organic seal identifies organic foods grown on USDA-certified organic farms. USDA

implemented. The organic food market grew exponentially in 2009, when the USDA offered $50 million in new funding to encourage greater production of organic food in the United States. With the additional financial support for farmers and ranchers, the number of USDA-certified organic operations in the United States increased to over 24,500 out of a global total of over 37,000. Because most stores now offer organic products, consumers have the opportunity to compare products and prices. Increased availability and use of coupons, the proliferation of private label and store brands of organic products, and better-value products offered by major organic brands all have contributed to increased sales.

**Organic Foods and Health.**  Consumers may choose to eat organic foods for a variety of reasons. Some are trying to decrease their synthetic pesticide exposure and protect the environment. Organic produce typically carries fewer pesticide residues than conventional produce; however, the residues on both organic and nonorganic produce are tested annually by the USDA and remain well below government safety thresholds. Cautious consumers may consider organic as a wise choice for vulnerable populations (e.g., young children, seniors) and opt for organic foods to encourage environmentally friendly sustainable agriculture practices.

Some consumers believe that eating organic food will improve the overall nutritional quality of their dietary intake (see the *Fake or Fact* feature). In terms of nutritional quality, a large study examined half a century of scientific evidence about the nutrient content of organic and conventional foods. The researchers concluded that organic and conventional foods are not significantly different in their nutrient content or nutritional value.[3] At this time, it is not possible to recommend organic foods over conventional foods based on nutrient content alone: both can meet nutritional needs. A healthy dose of common sense also is important; an *organic* label does not change a less healthy food into a more healthy food. For example, organic potato chips have the same calorie and fat content as conventional potato chips.

One concern raised about organic foods is that food safety may be jeopardized because animal manures used for fertilizers may contaminate food with pathogens. Although reports of outbreaks of foodborne illness linked to organically grown foods have been increasing (see *Newsworthy Nutrition* in this section), research has not shown that certified organic food has higher contamination with bacterial pathogens. To avoid exposure to potential pathogens, consumers should carefully wash or scrub all produce—organic and conventional—under running water. This safe food handling practice is critical for individuals with depressed immune systems.

Unlike the term *organic,* the term *natural* is not regulated by any federal agency. Products labeled as "natural" are generally those derived from natural ingredients, such as a plant source, which retain their native properties in the finished product. Meat or poultry labeled "natural" is expected to be minimally processed and contain no artificial flavoring, coloring, chemical preservative, or other artificial or synthetic ingredients. Unfortunately, few regulations are in place to ensure adherence to the policies, and there is debate over what constitutes *minimally processed.* Although all organic products fit this definition of natural, not all natural products are necessarily organic. Also, many "natural" products, such as manure and arsenic, are detrimental to health and harmful if ingested.

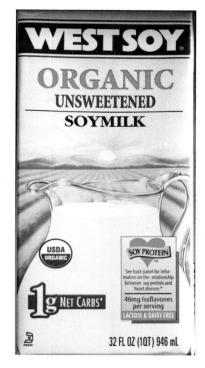

▲ USDA organic products have strict production and labeling requirements. Organic products must be: (1) produced without excluded methods (e.g., genetic engineering, ionizing radiation, or sewage sludge); (2) produced using allowed substances; and (3) overseen by a USDA National Organic Program—accredited certifying agent, following all USDA organic regulations. Andrew Resek/McGraw-Hill Education

## FOOD BIOTECHNOLOGY

The ability of humans to modify natural resources has enabled us to improve the production and yield of many important foods. Traditional biotechnology is almost as old as agriculture itself. The first farmer to improve stocks by selectively breeding the best bull with the best cows was implementing biotechnology. The first baker to use yeast to make bread rise took advantage of biotechnology.

By the 1930s, biotechnology made possible the selective breeding of improved plant hybrids. As a result, corn production in the United States quickly doubled. Through similar methods, agricultural wheat was crossed with wild grasses to confer

# Newsworthy Nutrition

### Increase in foodborne illnesses associated with organic foods

**INTRODUCTION:** Consumers often choose organic foods based on the perception that they are safer than conventionally produced foods. Organic standards, however, do not directly protect against microbial contamination. **OBJECTIVE:** The aim of this study was to identify the number of foodborne illness outbreaks linked to organic foods. **METHODS:** In this *epidemiological study*, researchers reviewed outbreaks of foodborne illness reported to the Foodborne Disease Outbreak Surveillance System of the Centers for Disease Control and Prevention (CDC) where the implicated food was reported to be organic. For each outbreak, information collected included the year, state, number of illnesses, pathogen, and implicated food. **RESULTS:** From 1992 to 2014, 18 outbreaks were identified that were caused by organic foods. These outbreaks resulted in 779 illnesses, 258 hospitalizations, and 3 deaths; 56% of outbreaks occurred in the last 5 years of the study. *Salmonella* and *E. coli* O157:H7 were the most commonly occurring pathogens, resulting in 44% and 33% of the outbreaks, respectively. Produce items were implicated in eight of the outbreaks; unpasteurized dairy products in four; and eggs, nut and seed products, and multi-ingredient foods in two outbreaks each. Foods that were definitely or likely USDA-certified organic were linked to 15 (83%) outbreaks. These results indicate that an increase in foodborne illness outbreaks associated with organic foods has paralleled increases in organic food production and consumption in recent years. This study was unable to compare the risk of outbreaks due to organic foods with that of conventional foods because data on food production method are not systematically collected as part of foodborne outbreak surveillance. **CONCLUSION:** The authors conclude that consumers should focus attention on food safety regardless of whether foods are produced organically or conventionally and that consumers should be especially aware of the risk of milk and produce consumed raw, including organic.

Source: Harvey RR and others: Foodborne disease outbreaks associated with organic foods in the United States. *Journal of Food Protection* 79:1953, 2016.

**genetic engineering** Manipulation of the genetic makeup of any organism with recombinant DNA technology. This includes DNA insertion, deletion, modification, or replacement. Also referred to as *gene editing* or *genetic editing*.

**genetically modified organism (GMO)** Any organism created by genetic engineering.

more desirable properties, such as greater yield, increased resistance to mildew and bacterial diseases, and tolerance to salt or adverse climatic conditions. Another type of biotechnology uses hormones rather than breeding. In the last decade, Canadian salmon have been treated with a hormone that allows them to mature three times faster than normal—without changing the fish in any other way. In general terms, biotechnology can be understood as the use of living things—plants, animals, bacteria—to manufacture novel products.

Biotechnology used in agriculture includes several methods that directly modify products. It differs from traditional methods because it directly changes some of the genetic material (DNA) of organisms to improve characteristics. Crossbreeding of plants or animals is no longer the only tool. Development of **genetic engineering** began in the 1970s. The field now features a wide range of cell and subcell techniques for the synthesis and placement of genetic material into organisms. In comparison to modern biotechnology, conventional breeding is inefficient and has inconsistent results; biotechnology uses genetic material more precisely. Scientists select the traits they desire and genetically engineer or introduce the gene that produces that trait into plants or animals. The new organisms are called **genetically modified organisms (GMOs).**

The primary category where biotechnology is applied is the addition of a unique characteristic, called an *input trait*, to a crop. These enhanced input traits include herbicide (weed killer) tolerance, insect and virus protection, and tolerance to environmental stressors such as drought. Other categories are value-added *output traits*, such as plant oils with increased levels of omega-3 fatty acids, and crops that produce pharmaceuticals. Scientists have engineered plants that thrive with fewer pesticides, potatoes that can be stored longer, and apples that do not turn brown when cut or sliced. In addition, biotechnology allows scientists to create fruits and grains with greater amounts of nutrients such as beta-carotene (e.g., *golden rice*) and vitamins E and C. Biotechnology is being used cautiously and conservatively, so the benefits are subtle. The ultimate benefits, however, could be significant in developing countries given the technology, training, and access.

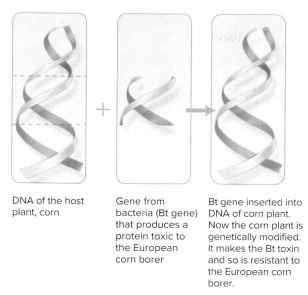

DNA of the host plant, corn

Gene from bacteria (Bt gene) that produces a protein toxic to the European corn borer

Bt gene inserted into DNA of corn plant. Now the corn plant is genetically modified. It makes the Bt toxin and so is resistant to the European corn borer.

**FIGURE 12-2** ▲ Biotechnology involves various techniques for transferring foreign DNA into an organism. In this diagram, a sample of DNA is cleaved out of a larger DNA fragment and inserted into the DNA of a host cell. Thus, the host cell contains new genetic information, with the potential of providing the cell with new capabilities. For corn, this could mean resistance to the European corn borer, a plant predator that feeds on corn and attacks hundreds of crops. In another application, bacteria can be engineered to produce the human form of the hormone insulin.

Few consumers realize that over 90% of corn and soybeans produced in the United States have been genetically engineered to either resist certain insects, thereby reducing pesticide use, and/or survive when sprayed with herbicides that kill surrounding weeds (https://www.ers.usda.gov/data-products/adoption-of-genetically-engineered-crops-in-the-us/recent-trends-in-ge-adoption.aspx). Papaya and sugar beet plants have been genetically engineered for viral resistance. Corn has been genetically altered by inserting a gene from the bacterium *Bacillus thuringiensis,* usually referred to as the *Bt gene,* into corn DNA (Fig. 12-2). The gene allows the corn plant to make a protein lethal to predator caterpillars that destroy the crop. The Bt protein in the corn is present in low concentrations with no effect on humans; it is digested along with the other proteins in corn. In fact, for many years, organic farmers have used the Bt bacteria directly on plants to destroy pests without changing the DNA of the plant.

The U.S. Food and Drug Administration (FDA) and the National Academy of Sciences are confident that approved varieties of genetically engineered foods are safe to consume. The USDA has established the National Bioengineered Food Disclosure Standard (NBFDS) to provide a mandatory disclosure standard, by which uniform information for bioengineered foods is provided to consumers. This standard defines "bioengineering" to mean any food that: (1) contains genetic material that has been modified through in vitro recombinant DNA techniques; and (2) for which the modification could not otherwise be obtained through conventional breeding or not found in nature. The rule makes no mention of whether crops produced use other gene-editing techniques. Figure 12-3 displays the USDA labels that must appear on foods meeting the definition by January 1, 2022. The *Ask the RDN* feature in this section contains more information on GMOs.

New breeding techniques (NBTs) now genetically modify both crops and animals more precisely. Using these technologies, experts can enhance, silence, insert, or remove target characteristics. CRISPR-Cas9 (Clustered Regulatory Interspaced Short Palindromic Repeats) is a powerful gene-editing technique. It is a natural bacterial defense system that scientists have reprogrammed to precisely target and edit DNA. CRISPR uses a molecular scissors that snips away or inserts specific traits found

The sustainable practice of **hydroponics**, as seen on our cover image, is an alternative agricultural practice that involves growing plants in soilless and nutrient-rich root mediums in a variety of controlled environments, including gutters, pipes (as pictured growing vertically on the cover), and other space-saving and inexpensive containers. The benefits of hydroponics include rapid plant growth with greater yields, reduced food and water waste, plants free of weeds and soilborne diseases, and the flexibility to farm in small spaces—perfect for students and urban dwellers! While we acknowledge that hydroponic farming is not meant to replace traditional farming techniques, it is a piece of the puzzle of finding sustainable solutions in a changing ecosystem.

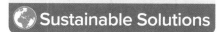 Sustainable Solutions

**hydroponics** A horticultural method of growing plants without soil, by using mineral nutrient solutions in a water solvent.

## ASK THE RDN  GMO and Organic Foods

*Dear RDN: I have started to see many more items in the grocery store labeled as "organic" and/or "nonGMO." These are frequently more expensive than the traditional versions. Is there enough evidence about the positive effects of organic and non-GMO foods to justify the added expense?*

This is a complicated question to unpack and the science is constantly evolving and changing, but here's what we know right now:

*On organic foods:* The USDA organic program, operated through the Agricultural Marketing Service (AMS), provides guidance and a framework for farmers, ranchers, and growers on how crops, orchards, and animals can be grown and raised in order to meet the USDA organic standards.[4] The green and white organic symbol (Fig 12-1), displayed on packaging or as a label on produce, is a marketing symbol, and those using it pay a fee and must be able to provide documentation that products and ingredients have been grown or raised according to USDA organic standards. The symbol is not meant to provide any information regarding nutrition, food safety, or health of the crop, animals, ingredients, or product.

Despite occasional headlines to the contrary, there is still no conclusive proof that eating foods that bear the USDA organic symbol or were grown or raised using organic methods will automatically result in better health or nutrition. As is the case with any food item, much depends on the crop or product, how it is prepared, and how much is consumed. Often individuals equate the organic symbol with "healthier" attributes, though the organic symbol appears on a variety of products representing varying levels of processing, including such snack items as chips and cookies, beverages, and sodas, as well as canned and frozen items. While many equate organic with "pesticide free," this is not necessarily the case. Organic farmers can, and often do, use pesticides (herbicides, fungicides, insecticides) if proactive integrative pest management are not effective, but these pesticides must meet the USDA organic standards. Whether the fruit or vegetable is raised using the USDA organic standards, some organic methods, or nonorganic methods, the amounts of pesticide residue in samples tested by the USDA are well below acceptable standards for virtually all fruits and vegetables. Overall, the health benefits of eating fruits and vegetables far outweigh the risk of pesticide residue. Be sure to see the *Pesticide Calculator* at https://www.safefruitsandveggies.com.

*On "GMO" foods:* There are no specific "GMO foods." The commonly used term *GMO,* taken to mean *genetically modified organism,* is in and of itself subject to dispute, and currently the USDA has used the term *bioengineered.* Rather than "GMO foods," there are *genetically engineered crops.* At this time there are only about 10 crops that are considered genetically engineered or bioengineered.[5] The traits of these GMO crops range from pest resistance so fewer pesticides can be used, to drought or flood tolerance so crops can better survive adverse conditions.

As technology advances and climate, economic, and health issues create demand, the number of crops (and even animals and seafood) that are developed using bioengineering is expected to increase. Some of the current crops are used to make clothing (cotton), provide feed for animals (corn, soy, and alfalfa), or become ingredients in foods (soy, sugar beets, corn). GMO products undergo years of research and testing before being released commercially. There have been no medically verified or reported illnesses or diseases in humans or animals linked to the consumption of GMO crops or ingredients in foods derived from currently available GMO crops. In addition, new technology, such as genome editing, will likely result in more advances in the use of technology for plant and animal breeding. It is expected that this will result in regulatory challenges in terms of how these products will be identified or labeled.

The National Academies of Sciences, Engineering, and Medicine (NAS) released a report addressing the safety of genetically engineered crops.[6] After interviewing subject matter experts, taking public comments, and reviewing hundreds of studies, the NAS concluded that it "*found no substantiated evidence of a difference in risks to human health between currently commercialized genetically engineered (GE) crops and conventionally bred crops, nor did it find conclusive cause-and-effect evidence of environmental problems from the GE crops.*" The Center for Science in the Public Interest (CSPI), a nonprofit consumer advocacy group, reached similar conclusions in its Biotechnology Report.[7]

*The bottom line:* Our health is affected by many factors in addition to our food choices. These factors include our genes, lifestyle choices, exercise and activity, stress, drug and alcohol use, economic status, and access to diagnostic tests, screening, and medical care. Food items that are labeled *non-GMO* or *organic* as a marketing strategy do not necessarily guarantee a better nutritional profile or lead to better health for those who purchase and consume them. Follow #FactsNotFear.

All the best,

**Leah McGrath, RDN, LDN**

Retail (supermarket) dietitian and former Army officer/dietitian

©Leah McGrath

naturally in the species. At any given time, there are hundreds of CRISPR crops in development with advantageous traits. NBTs and other gene-editing techniques offer advantages over traditional transgenic methods. NBTs have allowed product engineering that has been quickly approved by regulatory systems, including a non-browning Arctic® apple.

Public response to use of biotechnology remains mixed. The biggest debate in the United States surrounds the potential environmental hazards of introducing

**FIGURE 12-3** ▲ Retail food products that are bioengineered or contain bioengineered ingredients will carry one of these bioengineered labels. Regulated entities have several disclosure options: text, symbol, electronic or digital link, and/or text message. A phone number or web address is available to small food manufacturers or for small and very small packages. United States Department of Agriculture

genes from one species to another. Some challengers of these transgenic technologies even question the reduction in pesticide use that accompanies the cultivation of genetically modified crops. Although the use of genetically modified crops may reduce the need for environmentally harmful activities, such as applying pesticides to crops, critics point out that seeds produced with additional insecticide, such as the Bt protein, may lead to insect resistance to the toxic compound. Use of traditional pesticides has always required prudent application, in part to avoid the same type of insect resistance. In addition, accidental release of genetically modified animals, such as fish, may go on to harm wild varieties.

A global analysis of over 145 studies of biotechnology crops over the past 20 years confirmed the significant benefits of biotech crops. The study found that, on average, GMO technology has reduced chemical pesticide use by over 35%, increased crop yield by over 20%, and increased farmer profits by almost 70%. This study estimated the value added to be over $130 billion.[8]

## SUSTAINABLE AGRICULTURE

Conventional agriculture focuses on maximizing production through the use of large acreages, powerful machines, chemicals to control pests, and synthetic fertilizers to boost growth. A culture of sustainability has emerged, however, including a clear trend for sustainable food choices manufactured in an environmentally responsible way. Sustainable agriculture is an integrated system of plant and animal production that promises, over the long term, to have the following results:

- Satisfy human food needs
- Enhance environmental quality
- Efficiently use nonrenewable resources
- Sustain the economic viability of farm operations
- Enhance the quality of life for farmers and society as a whole

In areas such as South America, successful sustainable practices have increased productivity. Sustainable farming practices include the following:

- *Crop rotation,* which protects the soil by reducing nutrient depletion of the soil
- *Intercropping,* or the growing of two or more crops in proximity, which encourages plants to thrive in varying soil characteristics
- *Step farming,* also known as *terrace farming*, which increases productivity by enabling planting on hillsides by terracing slopes to hold water for a long duration and retain the topsoil more effectively

Sustainable farming using a terraced rice field in China. **What are the benefits of terrace farming?**
Bilderbuch/Design Pics

▲ The locavore movement is based on the assumption that local products are more nutritious and taste better, and it encourages consumers to buy from farmers' markets or produce their own food. Arina P. Habich/Shutterstock

**Sustainable Living and Eating.** Adhering to a *lifestyle of health and sustainability* (LOHAS) describes a growing demographic group focused on sustainable living. Many of today's college students are joining this market segment and developing behaviors associated with social responsibility. These consumers are driving changes in many areas, including the food industry. The food industry has responded with a move toward *green* initiatives that should be sustainable for the long term. Slow Food USA is an example of a nonprofit group dedicated to creating a framework for a deeper environmental connection to our food and aiming to inspire and empower Americans to build a food system that is sustainable, healthy, and delicious.

Eating sustainably includes reducing the amount of food we waste in an effort to save money and resources.[9] While the global population is growing and resources are becoming scarce, Americans are throwing away an enormous amount of food.[10] Along with all that uneaten food go wasted resources, including water, fertilizer, farmland, and energy. See *Ask the RDN* on wasted food in Chapter 13, which includes the food waste hierarchy that can be used to manage food surplus and food waste.[11]

**Sustainable Seafood.** Seafood choices become more complex when we consider the issue of overfishing and protecting endangered species of fish.[12] An *overfished* species is a population whose survival is jeopardized due to harvesting at a rate that exceeds the replenishing of stock. The good news is that fish production, whether farmed or wild caught, has a lower environmental cost compared to the production of meats. Fewer greenhouse gases are emitted, fewer chemicals and antibiotics are used, and fewer pounds of protein in feed are used than in beef, pork, or poultry production.

Fortunately, the United States has rigorous standards and closely monitors its fishing and aquaculture (fish farming) operations. The National Oceanic and Atmospheric Administration Fisheries establishes strict fishing catch levels in U.S. waters. As a result, when we buy U.S. wild-caught or farmed fish, we are making a sustainable choice. Look for the words "U.S. seafood product" on fish labels to ensure that the fish or shellfish has been sustainably harvested. In addition, groups such as the Natural Resources Defense Council (www.nrdc.org) and Seafood Watch (www.seafoodwatch.org) regularly update lists of acceptable fish.

The sustainable solution, however, is not that simple, with over 90% of the seafood we eat in the U.S. coming from international sources. About half of this seafood is from Southeast Asia, and a significant portion of the seafood caught by American fishermen is exported overseas for processing and then imported back to the U.S. Americans also have a limited seafood palate, with only 10 species making up 80% of the seafood we eat. Fortunately, four of the most popular fish—shrimp, salmon, tilapia, and pangasius (a genus of catfish)—are largely raised by certified and sustainable aquaculture operations. For example, salmon must be farmed to supply two-thirds of the 350 million tons of salmon that Americans consume annually. While many consumers perceive that wild-caught fish are the more environmentally friendly variety, many unregulated wild-caught fishing harvests have reached their peak capacity and are threatening future global seafood supplies. While sustainable seafood sourcing remains complex, both farmed and wild-caught fish can be healthy, sustainable, and economical choices.

## LOCALLY GROWN FOODS

Consumers are demanding increased transparency in the food supply, and local food helps answer questions about where food comes from and how it was grown. Retailers are using the *locally grown* label to respond to consumer desires for fresh, safe products that also support small, local farmers and help the environment. Local products typically provide fresher options, do not have the added costs of long transportation, and thus use less fossil fuel. Food-service establishments are also placing greater emphasis on supporting local producers, encouraging a farm-to-fork approach.

Typically, consumers have had access to locally grown farm-fresh produce at farmers' markets. These markets are also an integral part of the way that urban communities are linked to farms and continue to gain popularity. There are almost 9000 farmers' markets listed in the USDA's *National Farmers Market Directory*. Visit https://www.ams .usda.gov/local-food-directories/farmersmarkets to find a farmers' market near you.

The interest in "local" foods has become such a phenomenon that the term *locavore* was the 2007 Word of the Year in the *New Oxford American Dictionary*. A **locavore** is defined as someone who eats food grown or produced locally or within a certain radius from home, such as 50, 100, or 150 miles. The locavore movement has gained prominence due to consumers' food-safety concerns and the search for local, sustainable foods. It also encourages consumers to buy from farmers' markets or produce their own food, with the argument that fresh, local products are more nutritious and taste better.

There is no evidence, however, that locally grown products are safer. Although many small producers have proper food-safety practices, they often lack the expensive food-safety audits that are more common among big producers. Food-safety auditors evaluate evidence of insects on produce, sanitation practices, and similar food-safety criteria. Undetected foodborne illness outbreaks are more likely with "local" products delivered in small quantities and sold in a small area. Local products are not necessarily pesticide free and may not be cheaper, given that smaller growers lack the economic advantages of bigger growers.

Unlike organic products, there are no federal regulations specifying the meaning of "locally grown." Searchable databases and mapping resources such as MarketMaker (https://foodmarketmaker.com) are available to connect growers with buyers, restaurants with distributors, and consumers with local farmers' markets. These tools make it easier for people to find and sell locally grown foods. Positive attitudes toward organic, local, and sustainable food production practices are on the rise and appear to be increasing the conversations around the quality of dietary patterns. For example, a study of college students in Minnesota showed that students who put a high importance on alternative food production methods had a higher-quality dietary pattern. They consumed more fruits and vegetables and dietary fiber, fewer added sugars and sugar-sweetened drinks, and less fat.[13]

**locavore**  Someone who eats food grown or produced locally or within a certain radius such as 50, 100, or 150 miles.

## COMMUNITY-SUPPORTED AGRICULTURE

Consumers are not only taking comfort in knowing where their food comes from but also are becoming interested in community connections with local and regional farmers. Stemming from the interest in locally grown food, there is growing national support for local food collaboratives and community-supported agriculture. *Community-supported agriculture (CSA)* programs involve a partnership between local food producers and consumers. During each growing season, CSA farmers offer a share of foods to individuals, families, or companies that have pledged support to the CSA either financially and/or by working for the CSA.

Another example of a farm-community partnership is the National Farm to School Network, a nonprofit effort to connect farmers with nearby school (K–12) cafeterias. The objectives of this program are to serve healthy meals in school cafeterias; improve student nutrition; provide agriculture, health, and nutrition education opportunities; and support local and regional farmers. Since 1997, this program has grown from only 6 local programs to over 40,000 in all 50 states, incorporating the local bounty into their menus. Administrators of the program have found that if children can meet the farmer who actually grew the food, they are much more likely to eat it.

Farmers participating in community-supported agriculture (CSA) and the National Farm to School Network offer other avenues for exposing customers to locally grown foods. **Why is it important to expose children to fresh produce at an early age?** Sarah Rutter

### ✓ CONCEPT CHECK 12.1

1. What are the basic requirements for foods to be labeled as *organic?*
2. What two characteristics define a food as "bioengineered" according to the National Bioengineered Food Disclosure Standard (NBFDS)?
3. What is the definition of *sustainable agriculture?*
4. What are some of the programs that support local agriculture and youth nutrition?

▲ There are risks and benefits associated with pesticide use. The greatest short-term risk is in rural communities, where exposure is more direct.

Jeff Vanuga/USDA Natural Resources Conservation Service

▶ One of the problems with pesticides is they create new pests when they destroy the predators (spiders, wasps, and beetles) that naturally keep most plant-feeding insect populations in check. In the United States, spider mites and the cotton bollworm were merely nuisances until pesticides decimated their predators.

Fruits and vegetables grown without use of synthetic pesticides are available and may bear an "organic" label. These products generally are more expensive than conventional produce. **What are some reasons that the costs outweigh the benefits of pesticide use?**

Nancy R. Cohen/Getty Images

# 12.2 Environmental Contaminants in Food

A variety of environmental contaminants can be found in foods. Aside from pesticide residues, other potential contaminants are listed in Table 12-1. An approach to minimize exposure to environmental contaminants includes following safe food handing procedures and consuming a wide variety of foods in moderation.

## PESTICIDES

Pesticides used in food production yield both beneficial and unwanted effects. Most health authorities believe that the benefits far outweigh the risks. Pesticides help ensure a safe and adequate food supply and help make foods available at reasonable cost. Nevertheless, many consumers consider organic foods safer than conventional foods based on the lack of pesticide use. The *Ask the RDN* feature in this chapter helps decipher the current evidence.

Most concern about pesticide residues in food appropriately focuses on long-term toxicity because the amounts of residue present, if any, are extremely small. These low concentrations found in foods are not known to produce adverse effects in the short term, although harm has been caused by high amounts that result from accidents or misuse. For humans, pesticides pose a danger mainly in their cumulative effects, so their threats to health are difficult to determine. The contamination of underground water supplies and destruction of wildlife habitats indicate that pesticide use should be reduced. The U.S. federal government and many farmers are working toward that end. The use of biotechnology to reduce pesticide use is one alternative.

## WHAT IS A PESTICIDE?

Federal law defines a pesticide as any substance or mixture of substances intended to prevent, destroy, repel, or mitigate any pest. The built-in toxic properties of pesticides lead to the possibility that other, nontarget organisms, including humans, might also be harmed. The term *pesticide* tends to be used as a generic reference to many types of products, including insecticides (to kill insects), herbicides (to kill weeds), rodenticides (to kill rodents), and fungicides (to control fungi, mold, and mildew). A pesticide product may be chemical or bacterial, natural or synthetic.

For agriculture, the Environmental Protection Agency (EPA) allows about 10,000 pesticides to be used, containing some 300 active ingredients. The EPA reports that over 760 million pounds of agricultural pesticide are used annually, with herbicides being the most widely used type.

Once a pesticide is applied, it can turn up in a number of unintended and unwanted places. It may be carried in the air and dust by wind currents, remain in the soil attached to soil particles, be taken up by organisms in the soil, decompose to other compounds, be taken up by plant roots, enter the groundwater, or invade aquatic habitats. Each is a route to the food chain; some are more direct than others. Another concern is that pesticides are the probable cause of *massive colony collapse disorder* (CCD). CCD occurs when bees disappear from the colony and then die off en masse. This has been an ongoing concern, and is a critical issue for our food supply because one-third of all foods and beverages come from crops pollinated by honeybees.

## WHY USE PESTICIDES?

The primary reason for using pesticides is economic: the use of agricultural chemicals increases production and lowers the cost of food, at least in the short run. Many farmers believe that it would be impossible to stay in business without pesticides. It is estimated that for every $1 that is spent on pesticides to increase crop yields, $4 is saved in crops. Farmers also rely more on pesticides to produce cosmetically attractive fruits and vegetables.

**TABLE 12-1 ■ Potential Environmental Contaminants in Our Food Supply**

| Chemical Substance | Sources | Toxic Effects | Preventive Measures |
|---|---|---|---|
| Acrylamide | Fried foods rich in carbohydrate cooked at high temperatures for extended periods, such as French fries and potato chips | Known carcinogen for laboratory animals; acrylamide ingestion has not been proven to be carcinogenic in humans. | Limit intake of deep-fat fried foods rich in carbohydrate. |
| Bisphenol A (BPA) | Leaching of BPA from plastic food and beverage packages and from resins coating the inside of some metal food cans | Reproductive and developmental defects in animals | Use BPA-free bottles and cups. Do not microwave plastic containers, avoid plastic packaging with recycle codes 3 or 7, and reduce use of canned foods. |
| Cadmium | Plants grown in soil rich in cadmium. Clams, shellfish, tobacco smoke. Occupational exposure in some cases | Kidney disease. Liver disease. Bone deformities. Lung disease (when inhaled) | Consume a wide variety of foods, including seafood sources. |
| Dioxin | Trash-burning incinerators. Bottom-feeding fish from the Great Lakes. Fat from animals exposed to dioxin via water or soil | Abnormal reproduction and fetal/infant development. Immune suppression. Cancer in laboratory animals | Pay attention to warnings of dioxin risks from local fish and limit intake as suggested. Consume a variety of fish from diverse water sources. |
| Lead | Contaminated water supply. Lead-based paint chips and dust in older homes. Occupational exposure (e.g., radiator repair). Fruit juices and pickled vegetables stored in galvanized or tin containers or leaded glass. Some solder used in joining copper pipes in older homes. Mexican pottery dishes. Some imported herbal remedies. Leaded glass decanters | Anemia. Kidney disease. Nervous system damage (fatigue and changes in behavior). Learning impairments in childhood (even from mild lead exposure) | Avoid paint chips and related dust in older homes; see information about programs to reduce lead-based paint hazards in homes at www.hud.gov/offices/lead. Meet iron and calcium needs to reduce lead absorption. Store fruit juices and pickled vegetables in glass, plastic, or waxed paper containers. Let water for cooking or drinking run 1 minute or so if off for more than 2 hours, and use only cold water for cooking; do not soften drinking water. Do not store alcoholic beverages in leaded glass. |
| Mercury | Swordfish, shark, king mackerel, and tilefish. Fresh and canned albacore tuna is also a possible source (light chunk tuna is very low in mercury). | Reduced fetal/child development; birth defects; toxic to nervous system | Consume affected food no more than once per week (no more than two times per week for albacore tuna). Pregnant and breastfeeding women as well as young children should avoid these species of fish, but some albacore tuna consumption is fine. Two to three fish meals per week is appropriate for pregnant (and nursing) women if different types of fish are eaten. |
| Polychlorinated biphenyls (PCBs) | Fish from the Great Lakes and Hudson River Valley. Farmed salmon are a possible source. | Cancer in laboratory animals. Potential for liver, immune, and reproductive disorders | Pay attention to warnings of PCB contamination from local fish and limit intake as suggested on the fishing license or on state advisories. Choose a variety of fish from diverse water sources. |
| Urethane | Alcoholic beverages such as sherry, bourbon, sake, and fruit brandies | Cancer in laboratory animals | Avoid generous amounts of typical sources. |

Source: Adapted from NIH US Library of Medicine TOXNET

## REGULATION OF PESTICIDES

The responsibility for ensuring that residues of pesticides in foods are below amounts that pose a danger to health is shared by the FDA, the EPA, and the Food Safety and Inspection Service (FSIS) of the USDA in the United States. Table 12-2 lists the roles of various food protection agencies. The FDA is responsible for enforcing pesticide tolerances in all foods except meat, poultry, and certain egg products, which are monitored by the USDA. A newly proposed pesticide must be tested extensively, perhaps over 10 years or more, before it is approved for use. The EPA must decide that the pesticide causes no unreasonable adverse effects on people and the environment and that benefits of use outweigh the risks. The FDA tests thousands of raw products each year for pesticide residues. Note that a pesticide residue is considered illegal in this case if it has not approved for use on the crop in question or if the amount used exceeds the allowed tolerance.

## HOW SAFE ARE PESTICIDES?

Dangers from exposure to pesticides through food depend on how potent the chemical toxin is, how concentrated it is in the food, how much and how frequently it is eaten, and the consumer's resistance or susceptibility to the substance. As part of the FDA's Total Diet Study (TDS), typical foods are analyzed for elements, pesticides, and industrial chemicals four times per year. Specific foods are also analyzed for mercury. For rural counties in the U.S., the incidence of lymph, genital, brain, and digestive-tract

**TABLE 12-2 ▪ Agencies Responsible for Monitoring the U.S. Food Supply***

| Agency | Responsibilities |
|---|---|
| U.S. Department of Agriculture (USDA) Food Safety and Inspection Service (FSIS) www.fsis.usda.gov | The FSIS ensures that the nation's commercial supply of meat, poultry, and egg products is safe, wholesome, and correctly labeled and packaged. |
| Food and Drug Administration (FDA) www.fda.gov | Protects consumers against impure, unsafe, and fraudulently labeled products Sets standards for specific foods The FDA Center for Food Safety and Applied Nutrition (CFSAN) regulates foods other than the meat, poultry, and egg products regulated by the FSIS. |
| Centers for Disease Control and Prevention (CDC) National Outbreak Reporting System (NORS) https://wwwn.cdc.gov/norsdashboard | Leads federal efforts to gather data on foodborne illnesses, investigate foodborne illnesses and outbreaks, and monitor the effectiveness of prevention and control efforts in reducing foodborne illnesses Plays a key role in building state and local health department capacity to support foodborne disease surveillance and outbreak response |
| Environmental Protection Agency (EPA) www.epa.gov | Regulates pesticides Establishes water-quality standards |
| National Marine Fisheries Service or NOAA Fisheries www.nmfs.noaa.gov | Domestic and international conservation and management of living marine resources Voluntary seafood inspection program; can use official mark to show federal inspection |
| Bureau of Alcohol, Tobacco, Firearms and Explosives (ATF) www.atf.gov | Enforces laws on alcoholic beverages |
| State and local governments www.FoodSafety.gov | Regulate milk safety Monitor food industry within their borders Inspect food-related establishments |

*Government agencies responsible for monitoring food safety in Canada and the specific laws followed can be found at http://www.inspection.gc.ca.

cancers increases with higher-than-average pesticide exposure. Respiratory cancer cases increase with greater insecticide use. In general when comparing cancer incidence and death rates in rural and urban America, new cases of cancers of the lung, colon, and cervix as well as death rates from lung, colorectal, prostate, and cervical cancers have been higher in rural America.[14]

Interestingly, research shows that the cancer risk from pesticide residues is hundreds of times less than the risk from eating such common foods as peanut butter, brown mustard, and basil. Recall that plants naturally manufacture toxic substances to defend themselves against insects, birds, and grazing animals (including humans). When plants are stressed or damaged, they produce even more of these toxins. Because of this, many foods contain naturally occurring chemicals considered toxic, and some possibly carcinogenic.

The USDA Pesticide Data Program collects data on pesticide residues in food, particularly foods most likely consumed by infants and children. For the most recent samples tested, less than 1% of domestic and about 3% of imported samples had residues over tolerance levels. These findings are consistent with previous evidence showing that, in general, pesticide residues in food are well below EPA tolerances, confirming the safety of the food supply relative to pesticide residues. Visit the EPA website (https://www.epa.gov) for more information about pesticides and food.

▲ The FDA's yearly evaluation of a **market basket** of typical foods shows that pesticide residues are minimal in most foods. C Squared Studios/Getty Images

**market basket** Food the FDA buys, prepares, and analyzes as part of the ongoing Total Diet Study, which monitors levels of about 800 contaminants and nutrients in the average U.S. dietary pattern; the number varies slightly from year to year. About 280 kinds of foods and beverages from representative areas of the country are included four times a year.

## PERSONAL ACTION

The FDA and other scientific organizations report that the hazards of pesticides are extremely low and in the short run are much less dangerous than the hazards of foodborne illness that arise in our own kitchens. We can encourage farmers to use fewer pesticides to reduce exposure to our foods and water supplies, but we will have to settle for produce that is not perfect in appearance or that has been grown with the aid of biotechnology. Additional advice for limiting exposure to pesticides is found in Table 12-3. Choosing organic produce is one way to reduce your exposure to synthetic pesticides. One study found that people who report they "often or always" buy organic produce had significantly less insecticides in their urine samples, even though they reported eating 70% more servings of fruits and vegetables per day than adults reporting they "rarely or never" purchase organic produce.[15]

**TABLE 12-3 ■ What You Can Do to Reduce Exposure to Pesticides**

**WASH:** Wash and scrub all fresh fruits and vegetables thoroughly under running water to remove bacteria and soil. Running water has an abrasive effect that soaking does not. This will help remove bacteria and traces of chemicals from the surface of fruits and vegetables and dirt from crevices. Antibacterial washing products are not necessary.

**PEEL AND TRIM:** Peel fruits and vegetables when possible to reduce dirt, bacteria, and pesticides. Discard outer leaves of leafy vegetables. Trim fat from meat and skin from poultry and fish because some pesticides are fat-soluble and can accumulate in the fatty tissues of animals.

**SELECT A VARIETY:** Eat a wide variety of foods from different sources to provide a better mix of nutrients and reduce your likelihood of exposure to a single pesticide.

**CHOOSE ORGANIC:** Choose organically grown foods to reduce exposure to synthetic pesticides, but keep in mind that organic regulations allow for natural pesticides.

**USE INSECT REPELLENTS SAFELY:** Read the label for pesticide safety information and apply insect repellents safely. See www.epa.gov/pesticides/factsheets/pest_ti.htm for more pesticide safety tips.

Source: Adapted from U.S. Environmental Protection Agency.

## FARM to FORK    Melons

Pixtal/AGE Fotostock

Americans eat, on average, 26 pounds of melons a year, making them one of our favorite fruits. They are about 95% water, so they are a great source of fluid but a diluted source of other nutrients. Most contain a reasonable amount of vitamin C and are a juicy, low-calorie treat. Honeydew and casaba melons are the sweetest melons, but also the least nutritious.

### Grow

- Most melons available in the summer are grown in the U.S. The majority of melons sold in spring, fall, and winter have been imported from Mexico.
- Seedless watermelons are the most popular, making up 50% of the world market.
- Watermelons can grow in your backyard garden when daytime temperatures are between 70°F and 90°F and nighttime temperatures stay above 60°F.

### Shop

- Fully ripe melons with deep-colored flesh are the most nutritious and delicious. The darker the red flesh of watermelons, the greater the lycopene content, whereas deep orange cantaloupe flesh is higher in overall carotenoid content.
- Melons presectioned into halves, quarters, or wedges are typically fresh and allow you to see the inside color before you buy them.
- Small watermelons are more nutritious than the large varieties.
- To find a ripe watermelon, look for one that has lost its gloss, has a yellow "ground spot," and has a deep sound when you thump it.
- A ripe cantaloupe will have a slight depression or "innie" at its stem end.

### Store

- Storing a watermelon at room temperature for a few days will increase its antioxidant value.
- Eat ripe cantaloupes as soon as possible. They will keep for up to five days in the drawer of the refrigerator.

### Prep

- Scrub the outside of melons to remove any harmful bacteria on the outside. Although cantaloupes are virtually free of pesticides, they can harbor more bacteria because of their "netted" surface and therefore need vigorous rinsing.
- Once you have sliced open a melon, cover uneaten portions and refrigerate to inhibit the growth of bacteria. Eat within a day or two of slicing.

Source: Robinson J: Melons: Light in flavor and nutrition. In *Eating on the Wild Side*. Little Brown, New York, 2013.

MIXA/Getty Images

Each year, the Environmental Working Group (EWG) releases the "Dirty Dozen" and "Clean 15" lists of food they claim are most/least likely to contain pesticide residues. Recently researchers have discovered that the list makes some claims that are not based on scientific evidence. More importantly, the USDA and FDA have released reports showing that both organic food and conventional food are safe when handled appropriately and that such independent lists appear to be inducing fear and reducing consumer produce intake. According to these data, 99% of residues on the tested produce, if present at all, were well below safe threshold levels set by the EPA. In addition, 50% of foods sampled had no detectable residues at all. Fortunately, we have decades of high-quality nutrition studies documenting the numerous health benefits of consuming a dietary pattern rich in plant-based products. The Pesticide Residue Calculator (https://www.safefruitsandveggies.com/pesticide-calculator) mentioned in *Ask the RDN* shows that a child could eat more than 180 servings of strawberries per day and still remain below the known harmful pesticide residue levels. To reduce the risk of foodborne illness and remove most pesticide residues, concerned consumers should simply wash their fruits and vegetables.

## ENVIRONMENTAL CONTAMINANTS IN FISH

The presence of the environmental contaminants mercury and polychlorinated biphenyls (PCBs) in fish has caused some confusion regarding the risks and benefits of fish consumption. In our previous discussion of the benefits of omega-3 fatty acids, it was recommended that we include cold-water fatty fish, such as salmon or tuna, in our eating pattern about two to three times a week (8 to 12 oz per week for adults).[16] Conversely, you may have heard recommendations to eat less fish because they are a source of environmental contaminants. Balancing the benefits and risks of consuming fish is tricky, and not all experts agree. For example, advice from the FDA and EPA indicates that salmon is safe to eat, even during pregnancy, because it is low in mercury.

Mercury and PCBs are by-products of industrial processes and accumulate in fish tissue. PCBs were banned from use in 1979, but environmental levels have been decreasing very slowly and therefore still persist in our food supply, especially in seafood. The contaminants become more concentrated in bigger fish as they eat smaller, contaminated fish. Fish are of primary concern because they are the only predators we eat regularly. The National Academy of Medicine, Food and Agriculture Organization, FDA, and EPA all have issued similar guidelines for fish consumption. These groups advise pregnant women to eat up to 12 ounces of low-mercury fish per week and to avoid the four highest-mercury fish, which are swordfish, shark, tilefish, and king mackerel. For other adults, the basic recommendation is to "eat fish" but to vary the source to reduce the risk of chronic exposure to the same contaminants.

North Americans typically do not eat enough fish to cause concern about high intakes of environmental contaminants. On average, we consume only about 4 ounces of seafood per week. Around 80% of that is shrimp, canned tuna, salmon, and white fish, which are relatively low in environmental contaminants. Most Americans would benefit from eating more fish—a rich source of omega-3 fatty acids. Research shows that the risk of dying from heart disease is about 50% greater among people who do not eat fish compared to those who eat one or two servings of fatty fish each week. Overall, it appears that the benefits of consuming fish twice per week outweigh the potential risks. Pregnant women should follow the FDA/EPA guidelines, and the rest of us should eat a variety of types of fish, focusing on the smaller, fatty fish at the bottom of the food chain.

## AGROTERRORISM

Agroterrorism is the deliberate introduction of harmful agents, biological and otherwise, into the food supply chain with the intent of causing actual or perceived damage. The potential target areas for agroterrorism are typically farm animals and livestock, plant crops, and the food processing, distribution, and retailing system. In response to acts of terrorism in 2001, the U.S. Congress passed the Public Health Security and Bioterrorism Preparedness and Response Act (Bioterrorism Act). The Food Safety Modernization Act (FSMA) followed in 2011, giving the FDA increased power to monitor and control food in the United States. Although the United States has not been the victim of an agroterrorism attack, there remain potential vulnerabilities within our agricultural and food processing systems. The goals of the Bioterrorism Act and FSMA include the establishment of a process for regulators, scientists, and public health officials to improve the defensive position of the agriculture industry and to reduce the threat of agroterrorism.

This wild Alaska salmon is a top choice among types of fish based on its high nutritional value and low mercury levels. **What types of fish have the highest levels of mercury?** Digital Vision/Getty Images

### ✔ CONCEPT CHECK 12.2

1. What are the benefits of pesticide use?
2. What agencies regulate the use of pesticides?
3. What environmental contaminants can be found in fish, and which fish are most likely to contain these toxins?
4. What can you do to reduce your exposure to pesticides?

## **12.3** Food Preservation and Safety

During the early stages of urbanization in North America, contaminated water and food, especially milk, were responsible for large outbreaks of devastating human diseases. These experiences led to the development of procedures for purifying water, treating sewage, and **pasteurizing** milk. Since that time, safe water and milk have become more universally available, yet not always accessible. The greatest health risk from food today is contamination of a variety of foods by **viruses** and **bacteria** and, to a lesser extent, by various forms of **fungi** and **parasites.** These microorganisms can all cause **foodborne illness.** The Centers for Disease Control and Prevention (CDC) estimates that foodborne illness affects about one in six Americans each year. We generally have a safe food supply, but there are occasional instances of foodborne illnesses.

Microbial contamination of food is, by far, the more important issue for our short-term health. Americans are also concerned about health risks from chemicals such as food additives, although they cause less than 5% of all cases of foodborne illness in North America.

**pasteurizing**  The process of heating food products to kill pathogenic microorganisms and reduce the total number of bacteria.

**virus**  One of the smallest known types of infectious agents, many of which cause disease in humans. A virus is essentially a piece of genetic material surrounded by a coat of protein. Viruses do not metabolize, grow, or move by themselves. They reproduce only with the aid of a living cellular host.

**bacteria**  Single-cell microorganisms; some produce toxins, which cause illness in humans. Bacteria can be carried by water, animals, and people. They survive on skin, clothes, and hair and thrive in foods at room temperature. Some can live without oxygen and survive by means of spore formation.

**fungi**  Simple parasitic life forms, including molds, mildews, yeasts, and mushrooms. They live on dead or decaying organic matter. Fungi can grow as single cells, like yeast, or as a multicellular colony, as seen with molds.

**parasite**  An organism that lives in or on another organism and derives nourishment from it.

**foodborne illness**  Sickness caused by the ingestion of food containing harmful substances.

## EFFECTS OF FOODBORNE ILLNESS

According to the CDC, foodborne illnesses cause nearly 50 million illnesses, 130,000 hospitalizations, and 3000 deaths in the United States each year.[17,18] The numbers of foodborne disease outbreaks by state are shown in Figure 12-4. Those most susceptible to foodborne illness include the following:

- Infants and children
- Older adults
- Those with liver disease, diabetes, **HIV** infection (and AIDS), or cancer
- Patients recovering from surgery
- Women who are pregnant
- People taking immunosuppressant agents (e.g., transplant patients)

Some bouts of foodborne illness, especially when coupled with ongoing health problems, are lengthy and lead to food allergies, seizures, blood poisoning (from **toxins** or microorganisms in the bloodstream), or other illnesses. Foodborne illnesses often result from the unsafe handling of food at home, so we each bear some responsibility for preventing them.[19,20] You cannot usually tell that a particular food contains harmful microorganisms by taste, smell, or sight; therefore, you might not even suspect that food has caused your distress. In fact, your last case of diarrhea may have been caused by foodborne illness.

In response to the significant, largely preventable, public health risks of foodborne illness, the FDA Food Safety Modernization Act strengthened the food-safety system, enabling the FDA to better protect public health. It allows the FDA to focus on prevention of food-safety problems before they occur. The law provides new tools for inspection and compliance and for holding imported foods to the same standards as domestic foods.[21] The law directs the FDA to build a national food-safety system that is integrated and in partnership with state and local authorities. Several government agencies are at work on problems regarding food safety (Table 12-2). In addition, the CDC established the National Outbreak Reporting System (NORS) to alert the public of outbreaks (Fig. 12-5). Of course, the work of these agencies does not substitute for individual safety efforts.

## WHY IS FOODBORNE ILLNESS SO COMMON?

Foodborne illness is carried or transmitted to people by food. Most foodborne illnesses are transmitted through food in which microorganisms are able to grow rapidly.

**human immunodeficiency virus (HIV)** A virus that attacks the body's immune cells, which normally fight infection. As HIV progresses, immune function may be severely compromised, leading to acquired immune deficiency syndrome (AIDS).

**toxins** Poisonous compounds produced by an organism that can cause disease.

**FIGURE 12-4** ▶ Foodborne disease outbreaks, 2017. The color of each state represents the rate of foodborne disease outbreaks per 1 million people. The numbers on the map represent the number of confirmed outbreaks in each state in one year.

Source: https://wwwn.cdc.gov/norsdashboard/

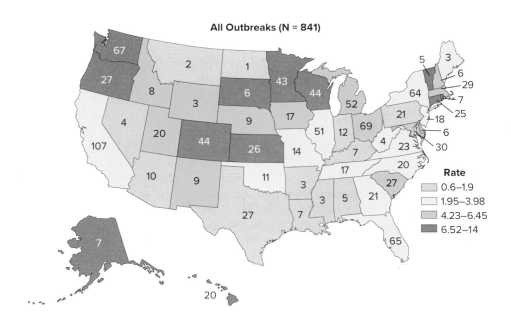

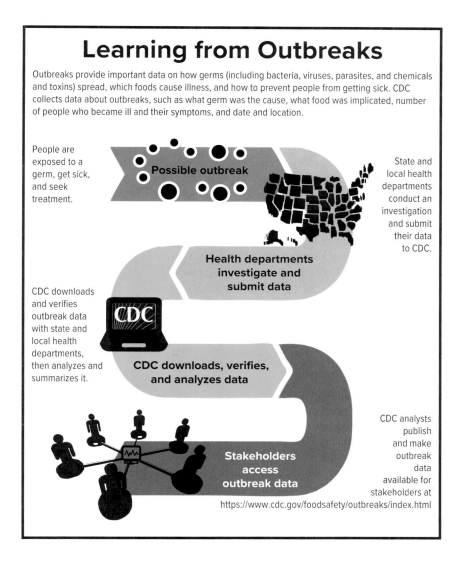

## Learning from Outbreaks

Outbreaks provide important data on how germs (including bacteria, viruses, parasites, and chemicals and toxins) spread, which foods cause illness, and how to prevent people from getting sick. CDC collects data about outbreaks, such as what germ was the cause, what food was implicated, number of people who became ill and their symptoms, and date and location.

People are exposed to a germ, get sick, and seek treatment.

**Possible outbreak**

State and local health departments conduct an investigation and submit their data to CDC.

**Health departments investigate and submit data**

CDC downloads and verifies outbreak data with state and local health departments, then analyzes and summarizes it.

**CDC downloads, verifies, and analyzes data**

CDC analysts publish and make outbreak data available for stakeholders at

**Stakeholders access outbreak data**

https://www.cdc.gov/foodsafety/outbreaks/index.html

**FIGURE 12-5** ◄ Learning from outbreaks. The CDC is responsible for tracking outbreaks of foodborne illnesses to gather key information and keep the public informed.

Source: Centers for Disease Control and Prevention, Steps in a Foodborne Outbreak Investigation, Retrieved from https://www.cdc.gov/foodsafety/outbreaks/investigating-outbreaks/investigations/index.html

▶ **Microorganisms are able to grow rapidly in foods that are:**

- generally moist.
- rich in protein.
- have a neutral or slightly acidic pH.

Unfortunately, this includes many of the foods we eat every day, such as meats, eggs, and dairy products.

The U.S. food industry tries whenever possible to prolong the shelf life of food products; however, a longer shelf life allows more time for bacteria in foods to multiply. Some bacteria even grow at refrigeration temperatures. Partially cooked—and some fully cooked—products pose a greater risk because refrigerated storage may only slow, not prevent, bacterial growth. Furthermore, we now know that in addition to providing a good growth medium for microorganisms, food (especially seafood) is also a carrier of many microorganisms.

The following consumer, environmental, and industry trends have increased the risk of contracting foodborne illness:

- Greater consumption of raw or undercooked animal products
- More foods prepared in kitchens outside the home
- Consumption of more imported ready-to-eat foods
- Centralized food production where food is prepared off-site for distribution
- Increased use of antibiotics in animal feeds
- More medications used that suppress the ability to combat foodborne infectious agents
- A continuing increase in the number of older adults in the population

Restaurants that choose to serve raw or undercooked foods of animal origin are required to have a consumer advisory on the menu warning customers of the health risks of consuming such foods. Think of a recent outbreak of foodborne illness you heard about in the news. **What foods were implicated?** alisali/123RF

- Increased number of severe storms and natural disasters resulting in power outages and contamination of water supplies
- Shipping foods between multiple locations in the supply chain

With the high number of two-income families in the U.S., many people look for convenient and easy-to-prepare foods. Many grocery stores provide an alternative to cooking at home by offering a variety of prepared foods from the meat departments, salad bars, and bakeries. Supermarkets offer take-home meal kits or entrées that can be served immediately or reheated. The foods are usually prepared in central kitchens or processing plants and shipped to individual stores. This centralization of food production by the food processing and restaurant industry increases the risk of foodborne illness. If a food product is contaminated in a central processing plant, consumers over a wide area can suffer foodborne illness. For example, in 1994 a contaminated ice cream mix used in a Schwan's® ice cream plant resulted in 224,000 suspected cases of *Salmonella* bacterial infections. The contamination was caused by an ice cream pre-mix that had been delivered to Schwan's in a truck that had not been properly washed after carrying raw, unpasteurized eggs.

Greater consumption of ready-to-eat foods imported from foreign countries is another cause of increased foodborne illness in North America. Almost 20% of all food consumed in the U.S. is imported, including approximately 95% of fish and shellfish, 50% of fresh fruits, and 20% of fresh vegetables. Due to this increase, U.S. authorities are reexamining inspection procedures for these imports. For example, a 2012 outbreak of *Listeria* was caused by contaminated ricotta cheese imported from Italy.

The use of antibiotics in animal feeds has also had an impact on outbreaks and the severity of cases of foodborne illness. It is good to know that animals pass the antibiotics through their systems before they are slaughtered and that animal products to be used for human consumption are tested for antibiotics. The real danger from such widespread use of antibiotics is that it encourages the development of antibiotic-resistant strains of bacteria. In other words, the animals may be a reservoir of pathogens that can grow even when exposed to typical antibiotic medicines. This issue of resistant bacteria in food-producing animals is receiving considerable attention by scientists. The CDC encourages the judicious use of antibiotics in humans and animals because both uses contribute to the emergence, persistence, and spread of antibiotic-resistant bacteria. The good news is the significant decrease in the amount of antibiotics sold recently, without an increase in animal health issues or food safety problems. The FDA reports that sales and distribution of medically important antibiotics for use in U.S. livestock decreased by 33% from 2016 through 2017, and by 43% since sales peaked in 2015. There has been a decrease in sales for most of the drug classes used in food-producing animals. The biggest decrease has been in sales of antibiotics used for growth promotion, which dropped from 5.7 million kilograms in 2016 to 0 kilograms in 2017. The greatest decline in antibiotic sales was seen in chickens.[22]

Every decade, the list of microorganisms suspected of causing foodborne illness gets longer. One reason more cases of foodborne disease are reported now is that health care providers are more likely to suspect foodborne contaminants as a cause of illness.

## FOOD PRESERVATION—PAST, PRESENT, AND FUTURE

For centuries, salt, sugar, smoke, fermentation, and drying have been used to preserve food. Ancient Romans used sulfites to disinfect wine containers and preserve wine. In the age of exploration,

Antibiotics may be given to animals, including turkeys, to prevent disease and increase feed efficiency. When antibiotics are used, a "withdrawal" period is required to ensure birds are free from any residues prior to slaughter. The USDA's Food Safety and Inspection Service randomly samples turkeys at slaughter to test for residues. **Are animals containing low levels of antibiotic residues permitted into the food supply?** Ingram Publishing/SuperStock

*magnificent*

*microbiome*

**Fermentation** can result in foods that contain live, probiotic organisms. The fermented foods that deliver the most beneficial bacteria to the gastrointestinal tract are kefir and yogurt. These dairy products have probiotic capabilities and have been associated with reduced inflammation and improved gut health. Most cheeses as well as *non-heated* kimchi and sauerkraut, kombucha, and miso also provide large numbers of live probiotic bacteria. Pickles that are fermented with salt, not vinegar, contain probiotics and are found in the refrigerated section of the grocery store. The live organisms in many other fermented foods, including sourdough bread and canned fermented vegetables such as sauerkraut, are destroyed by heat treatment. Beer and wine are also fermented, but the microbes used are filtered out of the finished product. Consume at least one serving of fermented foods daily to support your gut health. When adding fermented foods to your meals remember to mix them in at the end to avoid destroying the beneficial probiotics with heat.

European adventurers traveling to the New World salted their meat to preserve it. Most preserving methods work on the principle of decreasing water content (Table 12-4). Bacteria need abundant stores of water to grow; yeasts and molds can grow with less water, but some is still necessary. Decreasing the water content of some high-moisture foods, however, causes them to lose essential characteristics. Fermentation is an ancient preservation method using selected bacteria or yeast to ferment or pickle foods, thereby producing pickles, sauerkraut, yogurt, and wine from cucumbers, cabbage, milk, and grape juice, respectively. Check out the *Magnificent Microbiome* feature to see how these fermented foods provide a source of probiotics in the GI tract.

Today, we can add pasteurization, cooking, sterilization, refrigeration, freezing, canning, chemical preservation, and food **irradiation** to the list of food preservation techniques. Two specific methods of food sterilization, **aseptic processing** and **ultra-high temperature (UHT) processing,** are especially useful for liquid foods, such as fruit juices. With aseptic packaging and UHT processing, boxes of sterile milk, smoothies, and juices can remain unrefrigerated on supermarket shelves, free of microbial growth, for many years.

Food irradiation is an FDA-approved preservation technique that has dramatically improved food safety because it takes minimal doses of radiation to control pathogens such as *E. coli* O157:H7 and *Salmonella.* Irradiation is approved for use with eggs (still in the shell), seeds, meats, spices, dry vegetable seasonings, and fresh fruits and vegetables. The **radiation** energy used does not make the food radioactive. The energy essentially passes through the food, as in microwave cooking, and no radioactive residues are left behind.

**irradiation** A process in which radiation energy is applied to foods, creating compounds (free radicals) within the food that destroy cell membranes, break down DNA, link proteins, limit enzyme activity, and alter a variety of other proteins and cell functions of microorganisms that can lead to food spoilage. This process does not make the food radioactive.

**aseptic processing** A method by which food and container are separately and simultaneously sterilized; it allows manufacturers to produce boxes of milk that can be stored at room temperature.

**ultra-high temperature (UHT) processing** Method of sterilizing food by heating it above 275°F (135°C) for 2 to 5 seconds. Also called *ultra-heat treatment.*

**radiation** Literally, energy that is emitted from a center in all directions. Various forms of radiation energy include X-rays and ultraviolet rays from the sun.

**TABLE 12-4 ■ Food Preservation Techniques**

| **Historic Methods** | |
| --- | --- |
| Salt, sugar | These bind and reduce the water available to microorganisms. |
| Smoke | Heat kills microbes; chemicals in the smoke act as preservatives; water evaporates through drying. |
| Fermentation | Bacteria or yeast make acids and alcohol; minimizes growth of other bacteria and yeast. |
| Drying | Evaporates water. |
| **Modern Methods** | |
| Pasteurization | Moderately high (62°C to 100°C [144°F to 212°F]) temperatures are used for about 15 to 30 minutes to inactivate certain enzymes and kill microorganisms, especially in milk. |
| Refrigeration | Household refrigeration (typically at or below 40°F) slows down the deteriorative effects of microorganisms and enzymes. |
| Freezing | Freezing stops the growth of microorganisms, which do not grow when the temperature of the food is below 15°F (−10°C). |
| Canning | Food is heated in containers to a temperature that destroys microorganisms. Heating also causes air to be driven out of the container, forming a vacuum seal that prevents air and microorganisms from getting back into the product. |
| Chemical preservation | Preservation is usually based on the combined or synergistic activity of several additives. Certain preservatives have been used for centuries and include salt, sugar, acids, alcohols, and components of smoke. Some other chemicals used include sulfur dioxide, benzoic acid, sorbic acid, and formic acid. |
| Irradiation | Radiation energy passes through food and controls growth of insects, bacteria, fungi, and parasites by breaking chemical bonds, destroying cell walls and cell membranes, breaking down DNA. |
| Sterilization: Aseptic processing | Food and package are sterilized separately before the food enters the package. |
| Sterilization: Ultra-high temperature processing | Food is sterilized by heating it above 275°F (135°C) for 2 to 5 seconds. |

**FIGURE 12-6** ▲ The Radura® international label denotes prior irradiation of the food product. tatadonets/123RF

Irradiated food, except for dried seasonings, must be labeled with the international food irradiation symbol, the Radura (Fig. 12-6), and a statement that the product has been treated by irradiation. Some consumers have expressed concern that irradiation destroys nutrient content or makes food radioactive, but extensive studies confirm that irradiation does not compromise nutritional quality or noticeably change the taste, texture, or appearance of food. Keep in mind that even when foods, especially meats, have been irradiated, it is still important to follow basic food-safety procedures, as later contamination during food preparation is possible.

## ✓ CONCEPT CHECK 12.3

1. What are the annual rates of hospitalizations and deaths as a result of foodborne illnesses in the United States?
2. What lifestyle changes have made foodborne illness so common today?
3. Choose three agencies that bear some responsibility for monitoring the safety of our food supply and describe their specific roles.
4. What food preservation techniques have been used for centuries?
5. What is the UHT technique, and what types of foods are preserved with this process?
6. Why is irradiation considered a safe technique to preserve food?

# 12.4 Foodborne Illness Caused by Microorganisms

Most cases of foodborne illness are caused by specific viruses, bacteria, and parasites. Prions—proteins involved in maintaining nerve cell function—can also turn infectious and lead to diseases such as bovine spongiform encephalopathy, better known as mad cow disease.

## BACTERIA

Bacteria are single-cell organisms found in the food we eat, the water we drink, and the air we breathe. Many types of bacteria cause foodborne illness, including *Bacillus, Campylobacter, Clostridium, Escherichia, Listeria, Vibrio, Salmonella,* and *Staphylococcus* (Table 12-5). Bacteria are everywhere: each teaspoon of soil contains about 2 billion bacteria. Luckily, only a small number of all bacteria types pose a threat.

Bacteria can cause foodborne illness in three ways:

1. **Foodborne infection.** Foodborne bacteria directly invade the intestinal wall.
2. **Toxin-mediated infection.** Foodborne bacteria produce a harmful toxin as they colonize the GI tract.
3. **Foodborne intoxication.** Bacteria secrete a toxin into food before it is eaten, which causes harm to humans after the food is ingested.

The main way to distinguish an infectious route from an intoxication is time: if symptoms appear in 4 hours or less, it is an intoxication. *Salmonella,* for example, causes an infection because the bacteria cause the illness. *Clostridium botulinum, Staphylococcus aureus,* and *Bacillus cereus* produce toxins and therefore cause illness from intoxication. In addition, whereas most strains of *E. coli* are harmless, *E. coli* O157:H7 (Fig. 12-7) and O104:H4 produce a toxin that can cause severe illness, including severe bloody diarrhea and the potentially fatal kidney complication known as **hemolytic uremic syndrome (HUS).** Between 2009 and 2018, the

**foodborne infection** Occurs when a person eats food containing harmful microorganisms, which then grow in the intestinal tract and cause illness.

**toxin-mediated infection** Occurs when a person eats food containing harmful bacteria. While in the intestinal tract, the bacteria produce toxins that cause illness.

**foodborne intoxication** Results when a person eats food containing toxins that cause illness.

**hemolytic uremic syndrome (HUS)** Disease characterized by anemia caused by destruction of red blood cells (hemolytic), acute kidney failure (uremic), and a low platelet count.

FDA and CDC identified 40 foodborne outbreaks of toxin-producing E. coli infections in the U.S. that were confirmed or suspected to be linked to leafy greens. As a result of these recurrences, the FDA has developed the 2020 Leafy Greens Action STEC (Shiga toxin-producing *E. coli*) Plan to advance work in the prevention, response, and addressing of knowledge gaps in this area (https://www.fda.gov/food /foodborne-pathogens/2020-leafy-greens-stec-action-plan).

Bacterial foodborne illnesses typically cause gastrointestinal symptoms such as vomiting, diarrhea, and abdominal cramps. *Salmonella, Listeria, E. coli* O157:H7 and O104:H4, and *Campylobacter* are the bacterial foodborne illnesses of particular interest because they are the ones most often associated with death. *E. coli* O157:H7 and O104:H4 have caused deaths when HUS has developed. Listeriosis is of particular concern for pregnant women because they are about 20 times more likely to get this infection than other healthy adults. Listeriosis can cause preterm birth, spontaneous abortion, or stillbirth because the *Listeria* bacteria can cross the placenta and infect the fetus.

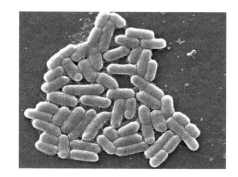

**FIGURE 12-7** ▲ An electron micrograph of *E. coli* bacteria, strain *E. coli* O157:H7, magnified 6836x. Although most strains of *E. coli* are harmless and live in the intestines of healthy humans and animals, this strain produces a toxin that causes severe illness. Janice Haney Carr/CDC

**TABLE 12-5** ■ **Bacterial Causes of Foodborne Illness**

| Bacteria | Illness and Food Sources | Outbreaks | Additional Information |
|---|---|---|---|
| *Campylobacter jejuni*<br><br>I. Rozenbaum/F. Cirou/Photo Alto | **Onset:** 2 to 5 days<br><br>**Symptoms:** Diarrhea, cramps, fever, and vomiting<br><br>**Duration:** 2 to 10 days<br><br>**Sources:** Raw and undercooked poultry, unpasteurized milk, contaminated water | In March 2015, raw milk produced in San Benito County, California, tested positive for *Campylobacter* and was implicated in illnesses in six northern California residents diagnosed with campylobacteriosis. | Estimated 1.3 million infections/year; produces a toxin that destroys intestinal mucosal surfaces; can cause Guillain-Barré syndrome, a rare neurological disorder that causes paralysis |
| *Clostridium botulinum*<br><br>Kari Marttila/Alamy Stock Photo | **Onset:** 12 to 72 hours<br><br>**Symptoms:** Vomiting, diarrhea, blurred vision, double vision, difficulty swallowing, muscle weakness; can result in respiratory failure and death<br><br>**Duration:** Variable, days to weeks<br><br>**Sources:** Improperly canned foods, especially home-canned vegetables, fermented fish, improperly stored baked potatoes | In April 2015, the largest botulism outbreak in the U.S. in nearly 40 years occurred among 77 persons who ate a church potluck meal in Ohio. Consumption of homemade potato salad prepared with improperly home-canned potatoes was implicated. Of the 29 patients, 25 received botulinum antitoxin, 11 required breathing tubes, and one patient died of respiratory failure shortly after arriving at the hospital.[23] | Estimated 205 cases/year; caused by a neurotoxin; grows only in the absence of air in nonacidic foods; incorrect home canning causes most botulism, but in 2007 commercially canned chili sauce caused an outbreak; honey can contain botulism spores and should not be given to infants younger than 1 year of age |
| *Clostridium perfringens*<br><br>Ingram Publishing/SuperStock | **Onset:** 8 to 16 hours<br><br>**Symptoms:** Intense abdominal cramps, watery diarrhea; more serious in elderly or ill persons<br><br>**Duration:** Usually 24 hours<br><br>**Sources:** Meats, poultry, gravy, dried or precooked foods, and/or improperly stored foods | In 2010, 42 residents and 12 staff members at a Louisiana state psychiatric hospital experienced vomiting, abdominal cramps, and diarrhea. Three patients died within 24 hours. Chicken that was cooked about 24 hours before serving and not cooled properly was associated with the illness. *C. perfringens* enterotoxin was detected in the chicken. | Estimated 1 million cases/year; anaerobic bacteria widespread in soil and water; multiplies rapidly in prepared foods, such as meats, casseroles, and gravies, held for extended time at room temperature |

*(continued)*

**TABLE 12-5 ■ Bacterial Causes of Foodborne Illness (*continued*)**

| Bacteria | Illness and Food Sources | Outbreaks | Additional Information |
|---|---|---|---|
| ***Escherichia coli* (O157:H7, O104:H4, and other strains)**<br><br>John A. Rizzo/Getty Images | **Onset:** 1 to 8 days<br><br>**Symptoms:** Severe (often bloody) diarrhea, abdominal pain, and vomiting. Usually little or no fever is present. More common in children 4 years or younger. Can lead to kidney failure.<br><br>**Duration:** 5 to 10 days<br><br>**Sources:** Undercooked beef (especially hamburger), unpasteurized milk and juice, raw fruits and vegetables (e.g. sprouts), and contaminated water | In 2019, a multi-state outbreak of *E. coli* O157:H7 infections occurred among 167 people from 27 states. Romaine lettuce from the Salinas Valley growing region was the likely source of this outbreak that resulted in 85 people hospitalized, including 15 people who developed kidney failure. No deaths were reported.<br><br>In 2015, a meal item or ingredient served at Chipotle® restaurants in nine states was the likely source of an outbreak of *E. coli* O157:H7. Of the 53 people infected, 20 were hospitalized. | Leading cause of bloody diarrhea in the United States; estimated 95,000 illnesses/year; lives in the intestine of healthy cattle; cattle and cattle manure are chief sources; illness caused by the powerful Shiga toxin made by the bacteria (so called because toxin is virtually identical to that produced by *Shigella dysenteria* toxin); petting zoos, lakes, and swimming pools can contain pathogenic *E. coli*. |
| ***Listeria monocytogenes***<br><br>Digital Vision/Getty Images | **Onset:** 9 to 48 hours for gastrointestinal symptoms, 2 to 6 weeks for invasive disease<br><br>**Symptoms:** Fever, muscle aches, and nausea or diarrhea. During pregnancy, women may have mild flu-like illness, and infection can lead to preterm delivery or stillbirth.<br><br>**Duration:** Variable, days to weeks<br><br>**Sources:** Unpasteurized milk, soft cheeses made with unpasteurized milk, ready-to-eat deli meats | From 2017 to 2019, hard-boiled eggs produced at an Almark Foods facility in Georgia were linked to one death in Texas and four people were hospitalized in five states. In 2015, Blue Bell Creameries® products were implicated in an outbreak of listeriosis in 10 people from four states. All 10 victims were hospitalized. Five of the people contracted listeriosis while hospitalized for unrelated problems. While in the hospital, all had consumed milkshakes made with Blue Bell ice cream. Three of these people died as a result of being infected by *Listeria*. | Estimated 1600 cases with 260 fatalities per year; widespread in soil and water and can be carried in healthy animals; grows at refrigeration temperatures; about one-third of cases occur during pregnancy; high-risk persons should avoid uncooked deli meats, soft cheeses (e.g., feta, Brie, and Camembert), blue-veined cheeses, cheeses made from unpasteurized milk (e.g., queso blanco), refrigerated meat spreads or pâtés, uncooked refrigerated smoked fish. |
| ***Salmonella* species**<br><br>Clement Mok/Photodisc/Getty Images | **Onset:** 6 to 48 hours<br><br>**Symptoms:** Diarrhea, fever, abdominal cramps, vomiting; can be fatal in infants, the elderly, and those with impaired immune systems<br><br>**Duration:** 4 to 7 days<br><br>**Sources:** Eggs, poultry, meat, unpasteurized milk or juice, cheese, contaminated raw fruits and vegetables | In 2020, 869 people across 47 states were infected with *Salmonella newport* linked to onions from Thomson International, Inc., and resulted in at least 116 hospitalizations. In 2018, 420 individuals across 16 states were infected with *Salmonella newport* linked to ground beef packaged in Arizona. More than 90 were hospitalized. | Estimated 1.2 million infections per year; bacteria live in the intestines of animals and humans; food is contaminated by infected water and feces; three strains of *Salmonella* account for almost 50% of cases; *S. enteritidis* infects the ovaries of healthy hens and contaminates eggs; almost 20% of cases are from eating undercooked eggs or egg-containing dishes; reptiles, such as turtles, also spread the disease. |
| ***Shigella* species**<br><br>Comstock Images/PictureQuest | **Onset:** 4 to 7 days<br><br>**Symptoms:** Abdominal cramps, fever, and diarrhea. Stools may contain blood and mucus.<br><br>**Duration:** 24 to 48 hours<br><br>**Sources:** Raw produce, contaminated drinking water, uncooked foods, other foods contaminated by infected food handlers with poor hygiene | In August 2018 an outbreak of shigellosis occurred at a wedding in Oregon. Contaminated asparagus was the likely source, with 112 people sickened and 10 hospitalized.<br><br>In 2015, 190 people developed shigellosis after eating at a restaurant in San Jose, California. At least 11 of the *Shigella* victims were treated in hospital intensive-care units. | Estimated 500,000 cases/year in U.S.; humans and primates are the only sources; common in day care centers and custodial institutions from poor hygiene; traveler's diarrhea often caused by *S. dysenteriae* |

**TABLE 12-5** ■ **Bacterial Causes of Foodborne Illness (*continued*)**

| Bacteria | Illness and Food Sources | Outbreaks | Additional Information |
|---|---|---|---|
| ***Staphylococcus aureus***<br><br>Michael Lamotte/Cole Group/Getty Images | **Onset:** 1 to 6 hours<br><br>**Symptoms:** Sudden onset of severe nausea and vomiting. Abdominal cramps. Diarrhea and fever may be present.<br><br>**Duration:** 24 to 48 hours<br><br>**Sources:** Unrefrigerated or improperly refrigerated meats, potato and egg salads, cream pastries | In June 2015, 86 children were sickened and 30 were hospitalized as a result of S. *aureus* toxin found in several food products served at Sunnyside Child Care Centers in Montgomery, Alabama. | Estimated 241,148 cases per year; bacteria live on skin and within nasal passages of up to 25% of people; can be passed to foods; multiplies rapidly when contaminated foods are held for extended time at room temperature; illness caused by a heat-resistant toxin that cannot be destroyed by cooking |
| ***Vibrio parahaemolyticus***<br><br>I. Rozenbaum/F. Cirou/Photo Alto | **Onset:** 1 to 7 days<br><br>**Symptoms:** Watery (occasionally bloody) diarrhea, abdominal cramps, nausea, vomiting, fever<br><br>**Duration:** 2 to 5 days<br><br>**Sources:** Undercooked or raw seafood, such as shellfish | In 2018, a multistate outbreak of *Vibrio parahaemolyticus* affected 26 individuals from 7 states and the District of Columbia. The source was linked to eating fresh crab meat imported from Venezuela. | Estimated 100 cases/year; found in coastal waters; more infections in summer; those with impaired immune systems and liver disease at higher risk of infection; fatality rate of 50% with bloodstream infection |
| ***Yersinia enterocolitica***<br><br>Yotrak Butda/123RF | **Onset:** 4 to 7 days<br><br>**Symptoms:** Fever, abdominal pain, and diarrhea (often bloody)<br><br>**Duration:** Lasts 1–3 weeks or longer<br><br>**Sources:** Raw or undercooked pork, particularly pork intestines (chitterlings); tofu; water; unpasteurized milk | In 2011, an outbreak of *Y. enterocolitica* was associated with drinking milk or eating ice cream made by a dairy in Pennsylvania. Sixteen individuals were affected. *Y. enterocolitica* was found in an unopened container and was isolated from homemade yogurt made from the dairy's milk. | Estimated 117,000 illnesses, 640 hospitalizations, and 35 deaths in the U.S. per year. Children under 5 years at greater risk; bacteria live mainly in pigs but can be found in other animals. |

Source: Adapted from USDA, available at https://www.fda.gov/food/resourcesforyou/consumers/ucm103263.htm

**Effects of Temperature: The Danger Zone.** To proliferate, bacteria require nutrients, water, and warmth. Most grow best in **danger zone** temperatures of 40°F to 140°F (4.4°C to 60°C, Fig. 12-8). Pathogenic bacteria typically do not multiply when food is held at temperatures above 140°F (60°C) or stored at safe refrigeration temperatures, 32°F to 40°F (0°C to 4.4°C). One important exception is *Listeria* bacteria, which can multiply at refrigeration temperatures. Also note that high temperatures can kill toxin-producing bacteria, but any toxin produced in the food will not be inactivated by high temperatures. Most pathogenic bacteria also require oxygen for growth, but *Clostridium botulinum* and *Clostridium perfringens* grow only in anaerobic (oxygen-free) environments, such as those found in tightly sealed cans and jars. Food acidity can affect bacterial growth, too. Although most bacteria do not grow well in acidic environments, some, such as disease-causing *E. coli,* can grow in acidic foods, such as fruit juice.

## VIRUSES

Viruses, like bacteria, are widely dispersed in nature. Unlike bacteria, however, viruses can reproduce only after invading body cells, such as those that line the intestines.

*magnificent*

Norovirus researchers are looking to the gut microbiome for some answers. Early animal studies have uncovered that some cells in the biome enhance norovirus replication and provide a safe haven that allows some people to remain contagious for weeks, even after symptoms dissipate. Scientists are investigating ways to manipulate the gut environment or microbiome itself in order to stimulate the immune system in ways that could shut down norovirus infection.[24]

*microbiome*

**danger zone** Temperature range (40°F to 140°F) where bacteria grow most rapidly.

▲ Raw shellfish, especially bivalves (e.g., oysters and clams), present a particular risk related to foodborne viral disease. These animals filter-feed, a process that concentrates viruses, bacteria, and toxins present in the water as it is filtered for food. Adequate cooking of shellfish will kill viruses and bacteria, but toxins may not be affected. It is important to buy shellfish from reliable sources that have harvested these foods from safe areas. duybox/123RF

### COVID CORNER

People do not need to worry about contracting COVID-19 from food or packaging. Coronaviruses cannot multiply in food. They only survive and multiply in a live animal or human host. The COVID-19 virus causes illness by infecting the respiratory tract, not the gut. The virus is transmitted through person-to-person contact and through direct contact with respiratory droplets from an infected person's cough or sneeze. Outbreaks of COVID-19 have been traced only to contact with other people and not from eating contaminated food. So, keep eating fruits and vegetables and wash them under running water. Also, there is no need to disinfect food packaging, but always wash hands after handling food packages and before eating. For more information visit https://www.who.int/emergencies/diseases/novel-coronavirus-2019.

Experts speculate that about 70% of foodborne illness cases go undiagnosed because they result from viral causes, and there is no easy way to test for these pathogens. Table 12-6 describes the two most common viral causes of foodborne illness, norovirus and hepatitis A, along with typical food sources and symptoms and outbreaks of the illnesses they cause.

Norovirus is the leading cause of foodborne illness, causing almost 60% of illnesses in the U.S. annually. The origin of its name came from Norwalk, Ohio, where the virus was first isolated after a 1968 outbreak. Each year, norovirus results in almost

**TABLE 12-6** ■ **Viral Causes of Foodborne Illness**

| Viruses | Illness and Sources | Outbreaks | Additional Information |
|---|---|---|---|
| **Hepatitis A virus**<br><br>Iconotec/Glow Images | **Onset:** 15 to 50 days<br><br>**Symptoms:** Diarrhea, dark urine, jaundice, and flu-like symptoms (i.e., fever, headache, nausea, and abdominal pain)<br><br>**Duration:** Variable, 2 weeks to 3 months<br><br>**Sources:** Raw produce, contaminated drinking water, uncooked foods and cooked foods that are not reheated after contact with an infected food handler; shellfish from contaminated waters | In 2013, 162 people were sickened, with 71 hospitalized in 10 states, by pomegranate arils found in frozen berries contaminated with hepatitis A.<br><br>In 2003, over 500 adults in the United States contracted hepatitis A after eating raw green onions in a restaurant. These were contaminated during growth in Mexico and not properly washed by food-service workers. | Estimated to cause over 4000 cases per year, but rates have declined by more than 95% since the hepatitis A vaccine became available. Infected food handlers contaminate food and transmit the disease; children and young adults are most susceptible; immunoglobulin given within 1 week to those exposed to hepatitis A virus can also decrease infection. |
| **Norovirus (Norwalk and Norwalk-like viruses), human rotavirus**<br><br>lynx/iconotec.com/Glow Images | **Onset:** 12 to 48 hours<br><br>**Symptoms:** Nausea, vomiting, abdominal cramping, diarrhea, fever, headache. Diarrhea is more prevalent in adults, vomiting more common in children.<br><br>**Duration:** 12 to 60 hours<br><br>**Sources:** Raw produce, contaminated drinking water, uncooked foods and cooked foods that are not reheated after contact with an infected food handler; shellfish from contaminated waters | In 2016, at least 120 Boston College students were sickened by Norovirus after eating at a Chipotle® restaurant. In 2019, 10 outbreaks were linked to Norovirus on seven different cruise lines, affecting 1110 passengers and 114 crew members. The largest outbreak in 2019 occurred on Royal Caribbean's *Oasis of the Seas*, where 561 passengers (9%) and 31 crew (1.5%) reported being ill. | Estimated to cause over 20 million cases of gastroenteritis, 71,000 hospitalizations, and 800 deaths per year. Most of these outbreaks occur in food-service settings when infected food-service workers touch ready-to-eat foods, such as raw fruits and vegetables; Noroviruses are very infectious—as few as 10 to 100 particles can lead to infection; workers with Norovirus symptoms should not work until symptoms are resolved. |

Source: Adapted from USDA, available at https://www.fda.gov/food/resourcesforyou/consumers/ucm103263.htm.

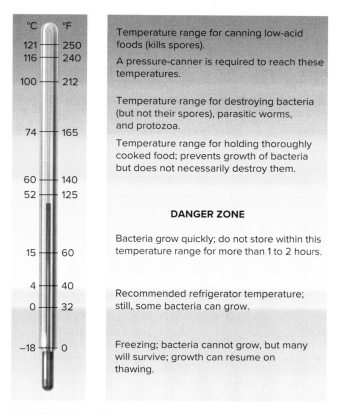

**FIGURE 12-8** ▲ Effects of temperature on microbes that cause foodborne illness. Source: USDA Food Safety and Inspection Service, Safe Minimum Internal Temperature Chart, www.fsis.usda.gov

2 million outpatient clinic visits, 400,000 emergency department visits, and 800 deaths, mainly in young children and seniors. This virus is also responsible for over 90% of the diarrheal disease outbreaks on cruise ships, and almost 1 million pediatric medical care visits annually. It causes an illness commonly misdiagnosed as the "stomach flu." Norovirus infection typically has a sudden onset and usually a short duration of only 12 to 60 hours. These microorganisms are hardy and survive freezing, relatively high temperatures, as well as chlorination up to 10 parts per million. The most commonly reported norovirus outbreaks from food contamination are at restaurants (Fig. 12-9).

## PROTOZOAN AND HELMINTH PARASITES

Table 12-7 describes common parasites and typical food sources and symptoms of the illnesses they cause. Parasitic infections spread via person-to-person contact and contaminated food, water, and soil. Parasites live in or on another organism, known as the host, from which they absorb nutrients. Humans may serve as hosts to parasites. These tiny ravagers rob millions of people around the globe of their health and, in some cases, their lives. Those hardest hit live in tropical countries where poor sanitation fosters the growth of parasites.

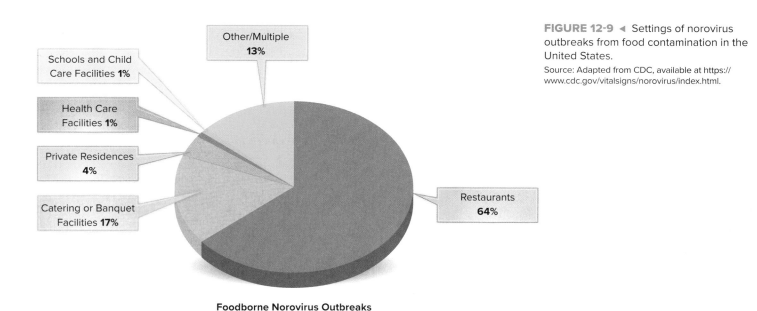

**Foodborne Norovirus Outbreaks**

**FIGURE 12-9** ◄ Settings of norovirus outbreaks from food contamination in the United States.

Source: Adapted from CDC, available at https://www.cdc.gov/vitalsigns/norovirus/index.html.

## TABLE 12-7 ■ Parasitic Causes of Foodborne Illness

| Parasite | Typical Food Sources | Illness | Additional Information |
|---|---|---|---|
| **Anisakis nematodes** <br><br> Pixtal/AGE Fotostock | Contaminated raw or under-cooked fish | **Onset:** 12 hours or less <br><br> **Symptoms:** Abdominal pain, nausea, vomiting, abdominal distention, diarrhea, blood and mucus in stool, and mild fever | Caused by eating the larvae of roundworms; the infection is more common where raw fish is routinely consumed. |
| **Cryptosporidium** <br><br> Ingram Publishing | Uncooked food or food contaminated by an ill food handler after cooking; contaminated drinking water | **Onset:** 2 to 10 days <br><br> **Symptoms:** Diarrhea, stomach cramps, upset stomach, slight fever <br><br> **Duration:** 1 to 2 weeks in otherwise healthy people | The largest U.S. outbreak was in 1993 in Milwaukee, with more than 443,000 persons affected; in August 2014, 11 persons in Idaho contracted cryptosporidiosis from drinking raw goat's milk. An increasing number of outbreaks are occurring in U.S. water parks, community swimming pools, and spas. |
| **Cyclospora cayetanensis** <br><br> hadynyah/Getty Images | Contaminated food or water | **Onset:** 1 to 14 days <br><br> **Symptoms:** Diarrhea (usually watery), loss of appetite, substantial loss of weight, stomach cramps, nausea, vomiting, fatigue <br><br> **Duration:** Weeks to months | Most common in tropical and subtropical areas, but since 1990, about a dozen outbreaks, affecting 3600 people, have occurred in the United States and Canada. |
| **Tapeworms** <br><br> Ingram Publishing | Contaminated raw beef, pork, and fish | **Symptoms:** Abdominal discomfort and diarrhea | Tapeworm larvae can get into the muscles of their hosts and cause infection when raw or under-cooked meat (from an infected animal) is eaten. |
| **Toxoplasma gondii** <br><br> Tetra Images/Alamy Stock Photo | Contaminated raw or under-cooked meat and unwashed fruits and vegetables | **Onset:** 5 to 20 days <br><br> **Symptoms:** Most people are asymptomatic; those with symptoms have fever, headache, sore muscles, and diarrhea; can be fatal to the fetus of pregnant women | Parasite is spread to humans from animals, including cats, the main reservoir of the disease; humans acquire the disease from ingesting contaminated meat or from fecal contamination from handling cat litter. |

**TABLE 12-7 ■ Parasitic Causes of Foodborne Illness (*continued*)**

| Parasite | Typical Food Sources | Illness | Additional Information |
|---|---|---|---|
| ***Trichinella spiralis***<br><br>Foodcollection | Contaminated raw or under-cooked pork or wild game | **Onset:** Weeks to months<br><br>**Symptoms:** GI symptoms followed by muscle weakness, fluid retention in the face, fever, and flu-like symptoms | The number of trichinosis infections has decreased greatly because pigs are now less likely to harbor this parasite; cooking pork to 160°F (72°C) will kill *Trichinella*, as will freezing it for 3 days at −4°F (−20°C). |

Source: Adapted from CDC, available at https://www.cdc.gov/parasites/az/index.html#t.

The more than 80 foodborne parasites known to affect humans include mainly **protozoa,** such as *Cryptosporidium* and *Cyclospora,* and **helminths,** such as tapeworms and the roundworm *Trichinella spiralis.*

**protozoa** (singular, protozoan) One-celled animals that are more complex than bacteria. Disease-causing protozoa can be spread through food and water.

**helminth** Parasitic worm that can contaminate food, water, feces, animals, and other substances.

### ✓ CONCEPT CHECK 12.4

1. What is the temperature "danger zone" in which bacteria can grow rapidly?
2. What type of microorganisms pose the greatest risk for foodborne illness?
3. What is the setting for the most norovirus outbreaks from contaminated food?

# 12.5 Food Additives

By the time you see a food item on the market shelf, it usually contains substances added to make it taste better, increase its nutrient content or shelf life, or make it easier to process. Yet other substances may have accidentally found their way into the foods you buy. All of these extraneous substances are known as **additives,** and, although some may be beneficial, others, such as sulfites, may be harmful for some people. All purposefully added substances must be evaluated by the FDA. To evaluate the safety of an additive, the FDA considers: (1) the substance composition and properties; (2) typical amount consumed; (3) immediate and long-term effects on health; and (4) a variety of safety factors.

**additives** Substances added to foods, either intentionally or incidentally.

Food additives are classified into two types: **direct food additives** (intentionally added to foods) and **indirect food additives** (incidentally added as contaminants). Both types of agents are regulated by the FDA in the United States. Currently, more than 3900 different substances are directly added to foods. As many as 3200 other substances enter foods as indirect contaminants. This includes substances that may reasonably be expected to enter food through contact with processing equipment or packaging materials.

**direct food additives** Additives knowingly (intentionally) incorporated into food products by manufacturers.

**indirect food additives** Additives that appear in food products incidentally, from environmental contamination of food ingredients or during the manufacturing process.

### WHY ARE FOOD ADDITIVES USED?

Most additives are used to reduce food spoilage. Common food additives serve the general function of **preservatives,** which can extend the shelf life of some foods. For example, **antioxidants** (e.g., vitamin E and sulfites) prevent discoloration caused by exposure to oxygen and enzymes. Antimicrobial additives (e.g., potassium sorbate) retard the growth of microbes in food products. Table 12-8 describes common food additives in detail.

**preservatives** Compounds that extend the shelf life of foods by inhibiting microbial growth or minimizing the destructive effect of oxygen and metals.

**TABLE 12-8** ■ Types of Food Additives—Sources and Related Health Concerns

| Food Additive Class | Attributes | Health Risks |
|---|---|---|
| **Acidic or alkaline agents,** such as citric acid, calcium lactate, and sodium hydroxide | Acids impart a tart taste to foods; inhibit mold growth; lessen discoloration and rancidity; and reduce the risk of botulism in naturally low-acid vegetables. Alkaline agents improve flavor by neutralizing acids produced during fermentation. | No known health risks when used properly. |
| **Nonnutritive sweeteners** (see Table 4-5) | Sweeten foods without adding more than a few calories | Moderate use considered safe (except for use of aspartame by people with the disease phenylketonuria (PKU). |
| **Anticaking and free-flow agents,** such as calcium silicate<br><br>Karen Appleyard/Alamy Stock Photo | Absorb moisture to keep table salt and powdered food products free-flowing | No known health risks when used properly. |
| **Antimicrobial agents,** such as salt and sodium benzoate | Inhibit mold and fungal growth | Salt increases the risk of developing hypertension, especially in sodium-sensitive individuals. No known health risks from other agents when used properly. |
| **Antioxidants,** such as BHA (butylated hydroxyanisole), BHT (butylated hydroxytoluene), vitamin E, vitamin C, and sulfites | Delay food discolorations from oxygen exposure; reduce rancidity from the breakdown of fats; maintain the color of deli meats; prevent the formation of cancer-causing nitrosamines | Approximately 10% of people have a sulfite intolerance. Symptoms include difficulty breathing, wheezing, hives, diarrhea, abdominal pain, cramps, and dizziness. Sulfites are used as preservatives in some foods and beverages. |
| **Color additives,** such as tartrazine<br><br>Jules Frazier/Getty Images | Make foods more visually appealing | Tartrazine (Yellow Dye No. 5) can cause allergic symptoms such as hives and nasal discharge, especially in people allergic to aspirin. FDA requires all forms of synthetic colors used in a food to be listed on its label. |
| **Curing and pickling agents,** such as salt, nitrates, and nitrites<br><br>Renee Comet/National Cancer Institute | Nitrates and nitrites act as preservatives, especially to prevent the growth of *Clostridium botulinum,* often used in conjunction with salt | Salt increases the risk of developing hypertension, especially in sodium-sensitive individuals. Nitrate and nitrite consumption has been associated with synthesis of nitrosamines. Some nitrosamines are known carcinogens.<br><br>The American Cancer Society and other organizations recommend avoiding processed meats to lower cancer risk based on evidence that regular consumption of processed meats increases colorectal cancer risk. |

**TABLE 12-8** ■ **Types of Food Additives—Sources and Related Health Concerns** (*continued*)

| Food Additive Class | Attributes | Health Risks |
|---|---|---|
| **Emulsifiers,** such as monoglycerides and lecithins Viktor1/Shutterstock | Suspend fat in water to improve uniformity, smoothness, and body of foods such as baked goods, ice cream, and mayonnaise | No known health risks when used properly. |
| **Fat replacements,** such as maltodextrins, emulsifiers, fiber, modified food starch, and engineered fats (see Section 5.3) | Limit calorie content of foods by replacing some of the fat content | Generally no known health risks when used properly. Possible loss of fat-soluble vitamins, and GI distress if used in excess. |
| **Flavor and flavoring agents,** such as natural and artificial flavors, sugar, and corn syrup | Impart more or improve flavor of foods | Sugar and corn syrup can increase risk for dental caries. Generally no known health risks for flavoring agents when used properly. Possible weight gain and its comorbid conditions from excess calories. |
| **Flavor enhancers,** such as monosodium glutamate (MSG) and salt Superstock/Corbis | Help bring out the natural flavor of foods, such as meats | Some people (especially infants) are sensitive to the glutamate in MSG and after exposure experience flushing, chest pain, facial pressure, dizziness, sweating, rapid heart rate, nausea, vomiting, increase in blood pressure, and headache. Affected individuals should look for the word *glutamate* on food labels, especially on labels for isolated protein, yeast extract, bouillon, and soup stock. |
| **Humectants,** such as glycerol, propylene glycol, and sorbitol | Retain more moisture, texture, and fresh flavor in foods such as candies, shredded coconut, and marshmallows | No known health risk when used properly. |
| **Leavening agents,** such as yeast, baking powder, and baking soda | Introduce carbon dioxide into food products | No known health risk when used properly. |
| **Maturing and bleaching agents,** such as bromates, peroxides, and ammonium chloride | Shorten the time needed for maturation of flour to become usable for baking products | No known health risk when used properly. |
| **Nutrient supplements,** such as vitamin A, vitamin D, and iodine | Enhance the nutrient content of foods such as margarine, milk, and ready-to-eat breakfast cereals | No known health risk if intake from the supplemental sources combined with natural food sources does not exceed the Upper Level for a particular nutrient. |
| **Sequestrants,** such as EDTA and citric acid | Bind free ions to prevent them from causing rancidity in products containing fat | No known health risk when used properly. |

(*continued*)

Contemporary Nutrition

**TABLE 12-8 ■ Types of Food Additives—Sources and Related Health Concerns (*continued*)**

| Food Additive Class | Attributes | Health Risks |
|---|---|---|
| **Stabilizers and thickeners,** such as pectins, gums, gelatins, and agars <br><br> Lauri Patterson/E+/Getty Images | Impart a smooth texture and uniform color and flavor to candies, frozen desserts, chocolate milk, and beverages containing alternative sweeteners; prevent evaporation and deterioration of flavorings used in cakes, puddings, and gelatin mixes | No known health risk when used properly. |

Source: Adapted from FDA, available at https://www.fda.gov/Food/IngredientsPackagingLabeling/FoodAdditivesIngredients/ucm094211.htm.

▶ Sugar, salt, corn syrup, and citric acid constitute 98% of all additives (by weight) used in food processing.

**sequestrants** Compounds that bind free metal ions. By so doing, they reduce the ability of ions to cause rancidity in foods containing fat.

**generally recognized as safe (GRAS)** A list of food additives that in 1958 were considered safe for human consumption. The FDA continues to bear responsibility for proving they are not safe and can remove unsafe products from the list.

Additives are also used to reduce food spoilage caused by the activity of some enzymes that leads to undesirable changes in color and flavor in foods. This type of food spoilage occurs when enzymes in a food react to oxygen—for example, when apple and peach slices darken or turn rust color as they are exposed to air. Antioxidants are a type of preservative that slows the action of oxygen-requiring enzymes on food surfaces. These preservatives include vitamins E and C and a variety of sulfites.

Without the use of some food additives, it would be impossible to produce massive quantities of foods and safely distribute them nationwide or worldwide, as is now done. Despite consumer concerns about the safety of food additives, many have been extensively studied and proven safe when the FDA guidelines for their use are followed.

## THE GRAS LIST

The Federal Food, Drug, and Cosmetic Act requires that any substance which is intentionally added to food is subject to review and approval by the FDA before it is used. Substances are exempted from review if they are generally recognized by qualified experts as having been adequately shown to be safe under the conditions of its intended use. In 1958, all food additives used in the United States and considered safe at that time were put on a **generally recognized as safe (GRAS)** list. The U.S. Congress established the GRAS list because it believed manufacturers did not need to prove the safety of substances that had been used for a long time and were already generally recognized as safe.

A substance can be removed from the GRAS list if the FDA can prove that it does *not* belong on the list. A few substances, such as cyclamates, failed the review process and were removed from the list. Red Dye No. 3 was removed because it was found to be linked to cancer. Many chemicals on the GRAS list have not yet been rigorously tested, primarily because of expense and because they have long histories of use without evidence of toxicity or their chemical characteristics do not suggest that they are potential health hazards. Substances may be added to the GRAS list if there are enough data to establish that the substance is safe under the conditions of its intended use.

In the past, the American Heart Association (AHA) and other experts have questioned the appropriateness of the GRAS listing for salt. They have suggested that sodium, one of the two components of salt, has negative health consequences and therefore does not meet the "safe" requirement of the GRAS. The AHA has advocated for the FDA to amend the GRAS listing for sodium chloride in an effort to reduce sodium content in processed foods. Hearing these concerns, the FDA issued a ruling that formalized the documentation of GRAS determinations and notifications. This ruling strengthened the oversight of food ingredients and included the consideration of both research results and expert opinion when making final determinations.[25]

## SYNTHETIC CHEMICALS

Although human endeavors contribute some toxins to foods, such as synthetic pesticides and industrial chemicals, nature's poisons are often even more potent and widespread. Nothing about a natural product makes it inherently safer than a synthetic product. Many synthetic products are laboratory copies of chemicals that also occur in nature. Some cancer researchers estimate that we ingest at least 10,000 times more (by weight) natural toxins produced by plants than we do synthetic pesticide residues. Plants produce these toxins to protect themselves from predators and disease-causing organisms. Some of these plant toxins are the beneficial phytochemicals we have already discussed. This comparison does not make synthetic chemicals any less toxic, but it does put them in perspective.

Last, toxicity is related to dosage. Consider vitamin E, often added to food to prevent rancidity of fats. This chemical is safe when used within certain limits. However, high doses have been associated with health problems, such as interfering with vitamin K activity in the body. Thus, even well-known, commonly used chemicals can be toxic in some circumstances and at high concentrations.

## TESTS OF FOOD ADDITIVES FOR SAFETY

Food additives are tested by the FDA for safety on at least two animal species, usually rats and mice. Scientists determine the highest dose of the additive that produces no *observable effects* in the animals. These doses are proportionately much higher than humans ever consume. The maximum dosage that produced no observable effects is then divided by at least 100 to establish a conservative margin of safety for human use. This 100-fold margin is used because it is assumed that we are at least 10 times more sensitive to food additives than laboratory animals and that any one person might be 10 times more sensitive than another. This conservative estimate essentially ensures that the food additive in question will cause no harmful health effects in humans.

One important exception applies to the procedure for testing direct food additives: if an additive is shown to cause cancer, even though only in high doses, no margin of safety is allowed. The food additive cannot be used because it would violate the **Delaney Clause.** This clause prohibits intentionally adding to foods a compound introduced after 1958 that causes cancer at any level of exposure. Evidence for cancer could come from either laboratory animal or human studies. A few exceptions to this clause, including the curing and pickling agents, nitrites and nitrates, are allowed.

Indirect food additives are another matter. The FDA cannot ban various industrial chemicals, pesticide residues, and mold toxins from foods, even though some of these contaminants increase cancer risk. These products are not purposely added to foods. The FDA sets an acceptable level for these substances. An incidental substance found in a food cannot contribute to more than one cancer case during the lifetimes of 1 million people. If a higher risk exists, the amount of the compound in a food must be reduced until the guideline is met.

In general, if you consume a variety of foods in moderation, the chances of food additives jeopardizing your health are minimal. Pay attention to your body. If you suspect an intolerance or a sensitivity, consult your health care provider for further evaluation. Remember that you are more likely to suffer in the short run from foodborne illness due to the microbial contamination of food than from consuming additives.

## APPROVAL FOR A NEW FOOD ADDITIVE

Before a new food additive can be added to foods, the FDA must approve its use. Besides rigorously testing an additive to establish its safety margins, manufacturers must give the FDA information that: (1) identifies the new additive; (2) gives its chemical

**Delaney Clause** A clause to the 1958 Food Additives Amendment of the Pure Food and Drug Act in the United States that prevents the intentional (direct) addition to foods of a compound shown to cause cancer in laboratory animals or humans.

▶ **Key terms often used by toxicologists:**

| toxicology | Scientific study of harmful substances |
|---|---|
| **safety** | Relative certainty that a substance will not cause injury |
| **hazard** | Chance that injury will result from use of a substance |
| **toxicity** | Capacity of a substance to produce injury or illness at some dosage |

composition; (3) states how it is manufactured; and (4) specifies laboratory methods used to measure its presence in the food supply at the amount of intended use.

Manufacturers must also offer proof that the additive will accomplish its intended purpose in a food, that it is safe, and that it is to be used in no higher amount than needed. Additives cannot be used to hide defective food ingredients, such as rancid oils; to deceive customers; or to replace good manufacturing practices. A manufacturer must establish that the ingredient is necessary for producing a specific food product.

**Worldwide Differences in Approval.** Despite these guidelines, many activists and public health watchdogs are not satisfied with the FDA procedures for regulating and monitoring the safety of food additives. These groups have urged the FDA and food manufacturers to stop the use of various chemicals until their safety can be more fully determined. A major incentive for these requests is the fact that many of the chemicals used in the United States are illegal to use as food additives in the countries of the European Union (EU) and others such as Brazil, Canada, India, and Japan.

Worldwide, there are differences in the approaches countries take to the approval of food additives. A key difference is that other countries do not rely on a GRAS list of compounds. An element that distinguishes the approach of the European Union (EU) from that of the United States is what the EU calls the *precautionary principle*. The EU believes that protective action or "precaution" should be taken when substantial, credible evidence of danger to human or environmental health is available, despite continuing scientific uncertainty. In contrast, the FDA's approach is that proof of harm must be demonstrated before regulatory action is taken. FDA approval is also unique in that when making determinations about additive safety, the FDA often relies on studies performed by the companies seeking approval.

As an example, the artificial colors Red Dye No. 40 and Yellow Dyes No. 5 and No. 6 are allowed in the United States but have been taken off the market in the United Kingdom. In the rest of Europe, products that contain these dyes must carry labels warning of their potential adverse effect on children's attention and behavior. These differences stem from varying conclusions made by authorities in the United States and Europe after considering the results of a study published in 2007.[26] The study found that artificial colors or a sodium benzoate preservative (or both) resulted in increased hyperactivity in children. The study persuaded British authorities to ban use of these dyes as food additives, whereas the EU chose to require warning labels on products that contain them. In the United States, the colors remain in use because the FDA found the study inconclusive based on the fact that it examined effects of a mixture of additives rather than individual colorings.

Several large companies and retailers have policies voluntarily barring some approved additives from their products. The sandwich chain Subway voluntarily discontinued the use of the approved dough conditioner azodicarbonamide, whose breakdown product, urethane, raised health concerns. Panera Bread went even further, announcing that foods served in its cafés are free of artificial additives. If you are concerned about the additives in your food sources, you can easily avoid most of them by consuming unprocessed whole foods such as produce, lean meats, whole grains, and low-fat dairy. However, no evidence shows that this will necessarily make you healthier, nor can you avoid all additives. Because some additives are used even in whole foods, it amounts to a personal decision. For more on food additives, visit *Rate Your Plate*, "Take a Closer Look at Food Additives" in Connect.

## NATURAL SUBSTANCES IN FOODS THAT CAN CAUSE ILLNESS

Foods contain a variety of naturally occurring substances that can cause illness. Table 12-9 shows some of the more important examples and the problems they can cause.

People have coexisted for centuries with these naturally occurring substances and have learned to avoid some of them and limit intake of others. They pose little health risk because we have developed cooking and food preparation methods to limit the potency of harmful substances, such as thiaminase. Spices are used in such small amounts that

Your choice to consume fresh rather than processed foods will lower your intake of food additives. **Where can you usually find fresh produce, lean meats, and low-fat dairy in a grocery store?** Rob Melnychuk/Getty Images

**TABLE 12-9** ■ **Examples of Natural Substances in Foods That Can Cause Illness**

| Substance | Source | Effect |
|---|---|---|
| Avidin | Raw egg whites (cooking destroys avidin) | Binds the vitamin biotin in a way that prevents its absorption, so a biotin deficiency may ultimately develop over the long term |
| Mushroom toxins | Some species of mushrooms such as the jack-o'-lantern (shown here) | Stomach upset, dizziness, hallucinations, and other neurological symptoms. More lethal varieties can cause liver and kidney failure, coma, and even death. The FDA regulates commercially grown and harvested mushrooms. There are no systematic controls on individual gatherers harvesting wild species, except in Illinois and Michigan. |
| Oxalic acid | Spinach, strawberries, sesame seeds | Binds calcium and iron in the foods and so limits absorption of these nutrients |
| Safrole | Sassafras, mace, and nutmeg | Causes cancer when consumed in high doses |
| Senna or comfrey | Herbal teas | Diarrhea and liver damage |
| Solanine | Potato shoots and flesh when it has been stressed by exposure to light or pests; indicated by green spots on potato skins | Inhibits the action of neurotransmitters |
| Tetrodotoxin | Puffer fish | Causes respiratory paralysis |
| Thiaminase | Raw fish, clams, and mussels | Destroys the vitamin thiamin |

AwakenedEye/Getty Images

Ingram Publishing/Alamy Images

Source: Adapted from CDC. Available at https://www.cdc.gov/biomonitoring/nutritional_indicators.html.

health risks do not result. Farmers know potatoes must be stored in the dark so that solanine will not be synthesized. Nevertheless, it is important to understand that some potentially harmful chemicals in foods occur naturally.

## IS CAFFEINE A CAUSE FOR CONCERN?

Why all the controversy over a cup of coffee? Researchers have spent a great deal of time on the study of caffeine, the substance of greatest concern in the favorite beverage of many. Caffeine is found naturally in the leaves, seeds, or fruits of more than 60 plants, including coffee beans, cacao beans, kola nuts, guarana berries, and tea. Caffeine is a stimulant found as a natural or added ingredient in many beverages and chocolate. On average, we consume 64% of our caffeine intake as coffee, 16% as tea, 18% as soft drinks, and less than 1% from energy drinks (Table 12-10). For teenagers and young adults, this ratio is often relatively higher for soft drinks and lower for coffee.

Caffeine is not often consumed by itself. With the popularity of trendy coffee shops that serve everything from mocha java to flavored lattes, it is difficult to separate the effects of caffeine intake from those of cream, sugar, alternative sweeteners, and flavorings. Although a 6-ounce cup of black coffee contains just 2 kcal, adding cream and sugar increases the calorie count significantly. Adding just one small creamer package of half-and-half will give you an extra 20 kcal; liquid nondairy creamer adds 20 kcal; and

Because gourmet coffee drinks are consumed several times a day, it is difficult to separate the effects of caffeine from those of cream, sugar, chocolate, and other flavorings. **Do you know how many calories are in the typical gourmet coffee beverage?** Ingram Publishing/SuperStock

### TABLE 12-10 ■ Caffeine Content of Common Sources

| Source and Amount | Milligrams of Caffeine | |
|---|---|---|
| | Typical | Range* |
| **Chocolate-flavored beverages** (8 fl oz) | | |
| Chocolate milk beverage | 2 | 2–7 |
| Hot cocoa beverage | 5 | 3–25 |
| **Chocolate foods** (1 oz) | | |
| Chocolate-flavored syrup | 2 | 0–5 |
| Dark chocolate, semisweet | 19 | 5–35 |
| Milk chocolate | 6 | 1–15 |
| **Coffee** (8 fl oz) | | |
| Decaffeinated, brewed | 2 | 2–25 |
| Espresso (1 fl oz serving) | 63 | 50–150 |
| Regular, brewed, drip method | 95 | 45–190 |
| **Energy drinks** (8.4 fl oz) | | |
| Monster® | 75 | 75 |
| Red Bull® | 80 | 80 |
| **Soft drinks** (8 fl oz) | | |
| Coca-Cola® | 22 | 22–23 |
| Mountain Dew® | 36 | 20–40 |
| **Teas** (8 fl oz) | | |
| Brewed, black tea | 47 | 40–50 |
| Brewed, green tea | 28 | 8–30 |
| Iced | 25 | 15–95 |
| Instant | 40 | 40–80 |

*For the coffee and tea products, the range varies due to brewing method, plant variety, brand of product, and so on. For a more extensive list of caffeine content of foods and drugs, visit www.cspinet.org/new/cafchart.htm.

Source: USDA.

▲ A cup of tea typically contains half the amount of caffeine in coffee. Opt out of using tea in trendy plastic tea bags, however. Micro- and nanoplastic particles have been found in tea steeped in these plastic tea bags compared to tea brewed in paper tea bags or in a reusable loose-leaf tea infuser. John A. Rizzo/Getty Images

The caffeine content of coffee varies with brewing method. Coffee made with a French press, shown here, is not filtered. **What risks are associated with unfiltered coffee consumption?** lynx/iconotec.com/Glow Images

1 teaspoon of sugar adds about 16 kcal. So what is the health-conscious coffee drinker to think? Let us explore the myths and facts of caffeine intake.

Caffeine does not accumulate in the body and is normally excreted within several hours following consumption. Caffeine stimulates the central nervous system and can cause anxiety, increased heart rate, insomnia, increased urination (possibly resulting in dehydration), diarrhea, and gastrointestinal upset in high doses. Those already suffering from ulcers or heartburn may experience irritation because caffeine relaxes the lower esophageal sphincter and increases acid production. Those who have anxiety or panic attacks may find that caffeine worsens their symptoms. Some people need only a little caffeine to feel such effects, and the threshold for children is likely even lower than that for adults.

Withdrawal symptoms are also real. Former coffee drinkers may experience headache, nausea, and depression for a short time after discontinuing use. These symptoms can be expected to peak at 20 to 48 hours following the last intake of caffeine. Withdrawal symptoms often occur for those trying to quit as little as a cup of coffee per day. Slow tapering of use over a few days is recommended to avoid these problems.

Are there more serious consequences of consuming caffeine regularly? Although it has been hypothesized that caffeine consumption can lead to certain types of cancer, the association of caffeine with cancer has not been established in the literature. In fact, regular coffee consumption has been linked to a decreased risk of head and neck, colorectal, breast, prostate, endometrial, and liver cancers.[27]

Heavy coffee consumption does increase blood pressure for a short period of time, and coffee consumption has been linked to increased LDL-cholesterol and triglycerides in the blood. This association was found to be caused specifically by cafestol and kahweol, two oils in ground coffee. However, filtered and instant coffees do not contain the harmful oils. It is prudent, though, to limit the amount of coffee in general, especially from French coffee presses and from espresso, as these beverages are not filtered.

Heavy caffeine use does mildly increase the amount of calcium excreted in urine. For this reason, it is important that heavy coffee drinkers ensure their overall dietary patterns contain adequate calcium sources. Women are thought to be at higher risk for a variety of deleterious effects with caffeine consumption, including miscarriages, osteoporosis, and birth defects in their offspring. Some studies show a higher likelihood for miscarriages in women consuming more than five 8-ounce cups of coffee per day (about 500 milligrams of caffeine). The position of the American College of Obstetricians and Gynecologists is that moderate caffeine intake—less than 200 milligrams a day—won't increase the risk of miscarriage, preterm birth, or birth defects.[28]

In contrast to these potential harmful effects of caffeine, coffee consumption is effective for treating migraines, likely effective for improving mental alertness, and possibly effective for improving memory, pain, Parkinson's disease, athletic performance, and glucose metabolism in diabetes.[29] Swedish scientists recently found that women ages 40 to 83 who consumed more than a cup of coffee per day for 10 years had a 22% to 25% lower risk of stroke.

Though the debate over caffeine will likely continue as long as North Americans drink coffee, research does not support many old misconceptions about caffeine. These studies are reinforcing the idea of moderation. A prudent dose of caffeine is 200 to 300 milligrams (about 2 to 3 cups of regular, brewed coffee) per day.

## ✔ CONCEPT CHECK 12.5

1. Choose three common food additives and state their purpose in food production.
2. What is the GRAS list?
3. How do scientists determine safe limits for the amount of additives that are allowed in foods?
4. What is the purpose of the Delaney Clause?
5. Pick three examples of naturally occurring substances and describe their harmful effects.
6. What is the typical caffeine content of 8-ounce cups of coffee and tea?
7. What are some of the negative effects of excess caffeine on the body?
8. Moderate coffee consumption has been linked to a decreased risk of which disorders?

# 12.6 Nutrition and Your Health
## Preventing Foodborne Illness

Ingram Publishing

You can greatly reduce the risk of foodborne illness by following some important rules.[20,30] It is a long list because many risky habits need to be addressed.

## Purchasing Food

- When shopping, select frozen foods and perishable foods such as meat, poultry, or fish last. Always place fresh meat, poultry, and fish in separate plastic bags, so that drippings do not contaminate other foods in the shopping cart. Do not let groceries sit in a warm car; this allows bacteria to grow. Also consider using recyclable insulated grocery bags to transport your cold items. Get the perishable foods such as meat, eggs, and dairy products home quickly and promptly refrigerate or freeze them.
- Do not buy or use food from damaged containers that leak, bulge, or are severely dented, or from jars that are cracked or have loose or bulging lids. Do not taste or use food that has a foul odor or spurts liquid when the can is opened; the deadly *Clostridium botulinum* toxin may be present.
- Purchase only pasteurized milk and cheese (check the label). This is especially important for pregnant women because highly toxic bacteria and viruses that can harm the fetus thrive in unpasteurized milk.
- Purchase only the amount of produce needed for a week's time. The longer you keep fruits and vegetables, the more time is available for bacteria to grow.
- When purchasing precut produce or bagged salad greens, avoid those that look slimy, discolored, or dry; these are signs of improper holding temperatures.
- Observe sell-by and expiration dates on food labels, and do not buy products that are near or past these dates.

- Follow food recalls online at https://www.fsis.usda.gov /wps/portal/fsis/topics/recalls-and-public-health-alerts /current-recalls-and-alerts. Note that a Class I recall means that there is a *reasonable probability* that consuming the food will cause serious health consequences or death.

## Preparing Food

- Thoroughly wash your hands for 20 to 30 seconds with warm, soapy water before and after handling food. This practice is especially important when handling raw meat, fish, poultry, and eggs; after using the bathroom; after playing with pets; or after changing diapers. The four F's of contamination include fingers, foods, feces, and flies.

Washing hands thoroughly (for at least 20 to 30 seconds) with warm water and soap should be the first step in food preparation. **What are the four F's of food contamination?** Dave & Les Jacobs/Getty Images

- Make sure counters, cutting boards, dishes, and other equipment are thoroughly washed, rinsed, sanitized, and air-dried before use. Be especially careful to use hot, soapy water to wash surfaces and equipment that come in contact with raw meat, fish, poultry, and eggs as soon as possible to remove *Salmonella* bacteria that may be present. Otherwise, bacteria on the surfaces will infect the next foods that come in contact with the surface, a process called **cross-contamination.** In addition, replace sponges and wash kitchen towels frequently. Microwaving sponges for 30 to 60 seconds helps rid them of live bacteria.
- If possible, cut foods to be eaten raw on a clean cutting board reserved for that purpose. Then clean this cutting board using hot, soapy water. If the same board must be used for both meat and other foods, cut any potentially contaminated items, such as meat, last. After cutting the meat, wash the cutting

**cross-contamination**  Process by which bacteria or other microorganisms are unintentionally transferred from one substance or object to another, with harmful effect.

board thoroughly. The FDA recommends cutting boards with unmarred surfaces made of easy-to-clean, nonporous materials, such as plastic, marble, or glass. Wooden boards should be made of a nonabsorbent hardwood, such as oak, maple, or bamboo, and have no obvious seams or cracks. Recently, bamboo has become popular for cutting boards because its dense wood resists knife scarring and water penetration, leaving bacteria without a place to multiply. Keep a separate wooden cutting board for chopping produce and slicing bread to prevent cross-contamination. Furthermore, the FDA recommends that all cutting boards be replaced when they become streaked with hard-to-clean grooves or cuts, which may harbor bacteria. In addition, cutting boards should be sanitized once a week in a dilute bleach solution. Flood the board with the solution, let it sit for a few minutes, then rinse thoroughly.

- Ignore the 5-second rule. Food that has fallen on the floor picks up bacteria immediately upon contact.
- When thawing foods, do so in the refrigerator, under cold potable running water, or in a microwave oven. Also, cook foods immediately after thawing under cold water or in the microwave. Never let frozen foods thaw unrefrigerated all day or night. Also, marinate food in the refrigerator.
- Avoid coughing or sneezing over foods, even when you are healthy. Cover cuts on hands with a sterile bandage. This helps stop *Staphylococcus* bacteria from contaminating food.
- Carefully wash fresh fruit and vegetables under running water to remove dirt and bacteria clinging to the surface, using a vegetable brush if the skin is to be eaten. People have become ill from *Salmonella* introduced while cutting melons that were contaminated with surface bacteria. The bacteria were on the outside of the melons and oranges.
- Completely remove moldy portions of food, or do not eat the food. If a food is covered in mold, discard the food. Mold growth is prevented by properly storing food at cold temperatures and using the food promptly. Also discard soft foods with high moisture content such as bread, yogurt, soft cheeses, and deli meats if there are spots of mold on them. It is safe to trim off any moldy spots of dense foods such as hard cheeses or firm fruits and vegetables.
- Use refrigerated ground meat and patties in 1 to 2 days and frozen meat and patties within 3 to 4 months.

▶ **The World Health Organization's Golden Rules for Safe Food Preparation:**
1. Choose foods processed for safety.
2. Cook food thoroughly.
3. Eat cooked foods immediately.
4. Store cooked foods carefully.
5. Reheat cooked foods thoroughly.
6. Avoid contact between raw and cooked foods.
7. Wash hands repeatedly.
8. Keep all kitchen surfaces meticulously clean.
9. Protect foods from insects, rodents, and other animals.
10. Use pure water.

## Cooking Food

- Cook food thoroughly and use a bimetallic thermometer to check for doneness, especially for fresh beef and fish (145°F [63°C]), pork (145°F [63°C]), and poultry (165°F [74°C]). Minimal internal temperatures for doneness are shown in Figure 12-10. Eggs should be cooked until the yolk and white are hard. The FDA does not recommend that eggs be prepared sunny-side up. Alfalfa sprouts and other types of sprouts should be cooked until they are steaming. Cooking is by far the most reliable way to destroy foodborne viruses and bacteria, such as norovirus and toxic strains of *E. coli*. Freezing only temporarily halts viral and bacterial growth.

▶ The USDA simplified the rules of foodborne illness prevention into four actions as part of its food-safety program (www.foodsafety.gov):
1. **Clean.** Wash hands and surfaces often.
2. **Separate.** Don't cross-contaminate.
3. **Cook.** Cook to proper temperatures.
4. **Chill.** Refrigerate promptly.

▲ Check Your Steps: 4 Steps to Food Safety.
Source: FoodSafety.gov

▶ **To reduce the risk of bacteria surviving during microwave cooking:**
- Cover food with glass or ceramic when possible to decrease evaporation and heat the surface.
- Stir and rotate food at least once or twice for even cooking. Then, allow microwaved food to stand, covered, after heating is completed to help cook the exterior and equalize the temperature throughout.
- Use the oven temperature probe or a meat thermometer to check that food is done. Insert it at several spots.
- If thawing meat in the microwave, use the oven's defrost setting. Ice crystals in frozen foods are not heated well by the microwave oven and can create cold spots, which later cook more slowly.

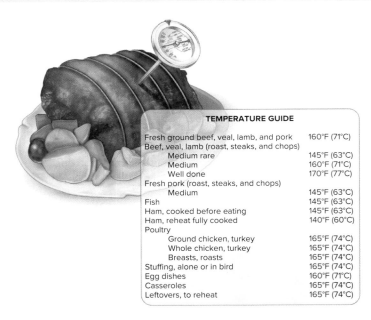

| TEMPERATURE GUIDE | |
|---|---|
| Fresh ground beef, veal, lamb, and pork | 160°F (71°C) |
| Beef, veal, lamb (roast, steaks, and chops) | |
|     Medium rare | 145°F (63°C) |
|     Medium | 160°F (71°C) |
|     Well done | 170°F (77°C) |
| Fresh pork (roast, steaks, and chops) | |
|     Medium | 145°F (63°C) |
| Fish | 145°F (63°C) |
| Ham, cooked before eating | 145°F (63°C) |
| Ham, reheat fully cooked | 140°F (60°C) |
| Poultry | |
|     Ground chicken, turkey | 165°F (74°C) |
|     Whole chicken, turkey | 165°F (74°C) |
|     Breasts, roasts | 165°F (74°C) |
| Stuffing, alone or in bird | 165°F (74°C) |
| Egg dishes | 160°F (71°C) |
| Casseroles | 165°F (74°C) |
| Leftovers, to reheat | 165°F (74°C) |

**FIGURE 12-10** ▲ Minimum internal temperatures when cooking or reheating foods. Source: USDA Food Safety and Inspection Service, Safe Minimum Internal Temperature Chart, www.fsis.usda.gov.

- A general precaution is to not eat raw animal products. As noted, many restaurants now include an advisory on menus stating that an increased risk of foodborne illness is associated with eating undercooked eggs. As long as restaurants provide this warning on their menus, however, they are allowed to cook eggs to any temperature requested by the consumer. The FDA warns us not to consume homemade ice cream, eggnog, and mayonnaise if made with unpasteurized, raw eggs because of the risk of *Salmonella* foodborne illness. The pasteurization of eggs or egg products kills *Salmonella* bacteria. Consuming raw seafood, especially oysters, also poses a risk of foodborne illness. Properly cooked seafood should flake easily and/or be opaque or dull and firm. If it is translucent or shiny, it is not done.
- Cook stuffing separately from poultry (or stuff immediately before cooking, and then transfer the stuffing to a clean bowl immediately after cooking). Make sure the stuffing reaches 165°F (74°C). *Salmonella* is the major concern with poultry.
- Once a food is cooked, consume it right away, or cool it to 70°F (21°C) within 2 hours, and then make sure any leftovers are cooled to 40°F (4.4°C) within 4 hours. If it is not to be eaten immediately, in hot weather (80°F and above) make sure that this cooling is done within 1 hour. Do this by separating the food into as many shallow pans as needed to provide a large surface area for cooling. Be careful not to recontaminate cooked food by contact with raw meat or juices from hands, cutting boards, or other dirty utensils.
- Serve meat, poultry, and fish on a clean plate—never the same plate used to hold the raw product. For example, when grilling hamburgers, do not put cooked items on the same plate used to carry the raw product out to the grill.

- For outdoor cooking, cook food completely at the picnic site, with no partial cooking in advance.

## Storing and Reheating Cooked Food

- Keep foods out of the *danger zone* (see Fig. 12-8) by keeping hot foods hot and cold foods cold. Hold food below 40°F (4.4°C)

▶ The FoodKeeper app (http://foodsafety.gov) will guide you on appropriate storage of foods. Developed by the Food Safety and Inspection Service, this resource is also available as a mobile app.

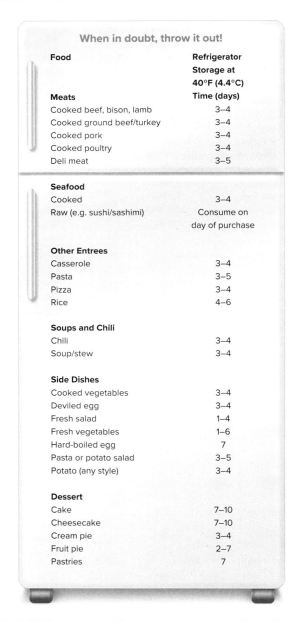

| When in doubt, throw it out! | |
|---|---|
| **Food** | **Refrigerator Storage at 40°F (4.4°C) Time (days)** |
| **Meats** | |
| Cooked beef, bison, lamb | 3–4 |
| Cooked ground beef/turkey | 3–4 |
| Cooked pork | 3–4 |
| Cooked poultry | 3–4 |
| Deli meat | 3–5 |
| **Seafood** | |
| Cooked | 3–4 |
| Raw (e.g. sushi/sashimi) | Consume on day of purchase |
| **Other Entrees** | |
| Casserole | 3–4 |
| Pasta | 3–5 |
| Pizza | 3–4 |
| Rice | 4–6 |
| **Soups and Chili** | |
| Chili | 3–4 |
| Soup/stew | 3–4 |
| **Side Dishes** | |
| Cooked vegetables | 3–4 |
| Deviled egg | 3–4 |
| Fresh salad | 1–4 |
| Fresh vegetables | 1–6 |
| Hard-boiled egg | 7 |
| Pasta or potato salad | 3–5 |
| Potato (any style) | 3–4 |
| **Dessert** | |
| Cake | 7–10 |
| Cheesecake | 7–10 |
| Cream pie | 3–4 |
| Fruit pie | 2–7 |
| Pastries | 7 |

**FIGURE 12-11** ▲ Length of time to keep leftovers safely in the refrigerator. Source: Visit www.foodsafety.gov for more tips on keeping your food safe.

or above 140°F (60°C). Foodborne microorganisms thrive in more moderate temperatures (60°F to 110°F [16°C to 43°C]). Some microorganisms can even grow in the refrigerator. Store dry food at 60°F to 70°F (16°C to 21°C).

- Reheat leftovers to 165°F (74°C); reheat gravy to a rolling boil to kill *Clostridium perfringens* bacteria, which may be present. Merely reheating to a good eating temperature is not sufficient to kill harmful bacteria.
- Store peeled or cut-up produce, such as melon balls, in the refrigerator.
- Keep leftovers in the refrigerator only for the recommended length of time (Fig. 12-11).
- Make sure the refrigerator stays below 40°F (4.4°C). Either use a refrigerator thermometer or keep it as cold as possible without freezing milk and lettuce.
- When the power goes out, keep the freezer and refrigerator doors closed as much as possible. Food can stay cold in an unopened refrigerator for about 4 hours; after 4 hours without power, discard perishable foods such as milk, meat, leftovers, and deli meats. Unopened freezers will keep food frozen for 2 days if full and 1 day if half full. Meat, poultry, and seafood can be refrozen if the freezer has not risen above 40°F (4.4°C).

Cross-contamination is not only a threat during food preparation; it can also become a problem during food storage. Make sure all foods, including leftovers, are contained and covered in the refrigerator to prevent drippings from uncooked and potentially hazardous foods from tainting other foods. It is a good idea to store foods likely to pose risk of foodborne illness on lower shelves of the refrigerator, beneath other foods to be eaten raw.

Raw fish dishes, such as sushi, can be safe for most people to eat if they are made with very fresh fish that has been commercially frozen and then thawed. The freezing is important to eliminate potential health risks from parasites. The FDA recommends that the fish be frozen to an internal temperature of −10°F (−23°C) for 7 days. If you choose to eat uncooked fish, purchase the fish from reputable establishments that have high standards for quality and sanitation. If you are at high risk for foodborne illness, it is wise to avoid raw fish products.

When it comes to keeping food safe, keep those four main topics of clean, separate, cook, and chill in mind. Keep your hands, surfaces, and utensils clean. Separate raw foods from cooked and ready-to-eat foods at all stages of food preparation, cooking, and storage to prevent cross-contamination. Use a food thermometer to ensure that you have cooked foods to the proper temperatures to kill harmful pathogens. Finally, chill the leftovers promptly, being sure that foods are not kept in the danger zone for more than two hours. These simple strategies will help to keep you safe from foodborne illness.

## ✓ CONCEPT CHECK 12.6

1. What is cross-contamination and how can you avoid it?
2. What are the four actionable items recommended by the USDA to prevent foodborne illnesses?
3. What are some safe food handling practices to implement during a power outage?

## CASE STUDY Preventing Foodborne Illness at Gatherings

Nicole attended a gathering of her coworkers on a warm Saturday in July. The theme of the party was international dining. Nicole and her husband brought an Argentinian dish: potato and beef empanadas. They followed the recipe and cooking time carefully, removing the dish from the oven at 1:00 P.M. and keeping it warm by wrapping the pan in a towel. They traveled in their car to the party and set the dish out on the buffet table at 3:00 P.M. Dinner was to be served at 4:00 P.M. However, the guests were enjoying themselves so much that no one began to eat until 6:00 P.M. Nicole made sure she sampled the empanadas that she and her husband made, but her husband did not. She also had some salad, garlic bread, and a sweet dessert made with coconut.

The couple returned home at 11:00 P.M. and went to bed. At about 2:00 A.M., Nicole knew something was wrong. She had severe abdominal pain and had to make a dash to the bathroom. She spent most of the next 3 hours in the bathroom with severe diarrhea. By dawn, the diarrhea subsided and she started to feel better. After a few cups of tea and a light breakfast, she was feeling like herself by noon. On Monday at work, she discovered that several of her coworkers also had diarrhea on Saturday night.

Nicole and her husband enjoying the international dining summer celebration with her coworkers. Digital Vision/ Getty Images

1. Based on her symptoms, what type of foodborne illness did Nicole likely contract?
2. Why is the beef a likely suspect for this type of foodborne illness?
3. Why is consuming food at large gatherings risky?
4. What precautions for avoiding foodborne illness were ignored by Nicole and the rest of the people at the party?
5. How could this scenario be rewritten to substantially reduce the risk of foodborne illness?

*Complete the Case Study. Responses to these questions can be provided by your instructor.*

## Summary (Numbers refer to numbered sections in the chapter.)

**12.1** Conventional agriculture emphasizes large yields and low costs, but the more recent trend toward sustainability considers the long-term environmental impact of agricultural practices. Consumers are driving up demand for organic and locally grown products.

**12.2** A variety of environmental contaminants and pesticide residues can be found in foods. It is helpful to know which foods pose the greatest risks and act accordingly to reduce exposure, such as washing fruits and vegetables before use.

**12.3** Infants, children, older adults, patients post-surgery, individuals who are immunosuppressed, and women who are pregnant are most susceptible to foodborne illness. The risk of foodborne illness has increased as more of our foods are prepared outside of the home. In the past, salt, sugar, smoke, fermentation, and drying were used to protect against foodborne illness. Today, careful cooking, pasteurization, irradiation, attention to food temperature, and thorough handwashing provide additional insurance.

**12.4** Viruses, bacteria, and other microorganisms in food pose the greatest risk for foodborne illness. Major causes of foodborne illness are norovirus and the bacteria *Campylobacter jejuni, Salmonella, Staphylococcus aureus,* and *Clostridium perfringens.* In addition, such bacteria as *Clostridium botulinum, Listeria monocytogenes,* and *Escherichia coli* have been found to cause illness.

**12.5** Food additives are used primarily to extend shelf life by preventing microbial growth and the destruction of food components by oxygen, metals, and other substances. Food additives are classified as those directly (intentionally) added to foods and those that indirectly (incidentally) appear in foods. Under its jurisdiction in the United States, the Delaney Clause allows the FDA to ban the use of any direct food additive that increases cancer risk. Toxic substances occur naturally in a variety of foods, such as green potatoes, raw fish, mushrooms, and raw egg whites. Cooking foods limits their toxic effects in some cases; others are best to avoid altogether, such as toxic mushroom species and the green parts of potatoes.

**12.6** Safe food handling can be summed up in four easy steps: (1) *clean* hands and surfaces often; (2) *separate* raw and ready-to-eat foods to prevent cross-contamination; (3) *cook* (and reheat) potentially hazardous foods thoroughly, measuring temperature with a food thermometer; and (4) *chill* foods by refrigerating promptly after eating. Two of these practices focus on food temperature; foods should not be held in the "danger zone" (40°F to 140°F; 4.4°C to 60°C) for more than 2 hours.

## Check Your Knowledge (Answers are available at the end of this question set.)

1. Nitrite prevents the growth of
   a. *Clostridium botulinum.*
   b. *Escherichia coli.*
   c. *Staphylococcus aureus.*
   d. yeasts.

2. Substances used to preserve foods by lowering the pH are
   a. smoke and irradiation.
   b. baking powder and baking soda.
   c. salt and sugar.
   d. vinegar and citric acid.

3. Food additives widely used for many years without apparent ill effects are on the _____ list.
   a. FDA
   b. GRAS
   c. USDA
   d. Delaney

4. The four actions that are part of the USDA food-safety program are clean, _____, cook, and chill.
   a. sterilize
   b. separate
   c. pasteurize
   d. freeze

5. *Salmonella* bacteria are usually spread via
   a. raw meats, poultry, and eggs.
   b. pickled vegetables.
   c. home-canned vegetables.
   d. raw vegetables.

6. It is unwise to thaw meats or poultry
   a. in a microwave oven.
   b. in the refrigerator.
   c. under cool running water.
   d. at room temperature.

7. Milk that can remain on supermarket shelves, free of microbial growth, for many years has been processed by which of the following methods?
   a. Use of humectants
   b. Using antibiotics in animal feed
   c. Use of sequestrants
   d. Aseptic processing

8. Those at greatest risk for foodborne illness include
   a. pregnant women.
   b. infants and children.
   c. immunosuppressed individuals.
   d. all of these individuals.

9. Pasteurization involves the
   a. exposure of food to high temperatures for short periods to destroy harmful microorganisms.
   b. exposure of food to heat to inactivate enzymes that cause undesirable effects in foods during storage.
   c. fortification of foods with vitamins A and D.
   d. use of irradiation to destroy certain pathogens in foods.

10. Food can be kept for long periods by adding salt or sugar because these substances
    a. make the food too acidic for spoilage to occur.
    b. bind to water, thereby making it unavailable to the microorganisms.
    c. effectively kill microorganisms.
    d. dissolve the cell walls in plant foods.

Answer Key: 1. a (LO 12.5), 2. d (LO 12.3), 3. b (LO 12.5), 4. b (LO 12.6), 5. a (LO 12.3), 6. d (LO 12.6), 7. d (LO 12.3), 8. d (LO 12.4), 9. a (LO 12.3), 10. b (LO 12.5)

## Study Questions (Numbers refer to Learning Outcomes.)

1. Describe some of the advances in agricultural science that are positively affecting our food supply. **(LO 12.1)**

2. Describe four recommendations for reducing the risk of ill effects from environmental contaminants. **(LO 12.2)**

3. What three trends in food purchasing and production have led to a greater number of cases of foodborne illness? **(LO 12.3)**

4. Which types of foods are most likely to be involved in foodborne illness? Why are they prone to contamination? **(LO 12.3)**

5. Identify three major classes of microorganisms responsible for foodborne illness. **(LO 12.4)**

6. Define the term *food additive,* and give examples of four direct or intentional food additives. What are their specific functions in foods? What is their relationship to the GRAS list? **(LO 12.5)**

7. Describe the federal process that governs the use of food additives, including the Delaney Clause. **(LO 12.5)**

8. Put into perspective the benefits and risks of using additives in food. Point out an easy way to reduce the consumption of food additives. Do you think this is worth the effort in terms of maintaining health? Why or why not? **(LO 12.5)**

9. Name some substances that occur naturally in foods but may cause illness. **(LO 12.5)**

10. List four techniques other than thorough cooking that are important in preventing foodborne illness. **(LO 12.6)**

# Further Readings

1. Statista Research Department: *U.S. Agriculture—Statistics & Facts.* Updated April 15, 2019. Available at https://www.statista.com /topics/1126/us-agriculture. Accessed January 10, 2020.

2. U.S. Department of Agriculture Economic Research Service: *Organic Market Overview.* Updated October 09, 2019. Available at https://www .ers.usda.gov/topics/natural-resources-environment/organic-agriculture /organic-market-overview.aspx. Accessed January 10, 2020.

3. Smith-Spangler C and others: Are organic foods safer or healthier than conventional alternatives? A systematic review. *Annals of Internal Medicine* 157:348, 2012. https://doi.org/10.7326/0003-4819-157-5-201209040-00007.

4. Agricultural Marketing Service, U.S. Department of Agriculture: *Organic Standards.* https://www.ams.usda.gov/grades-standards/organic -standards. Accessed January 13, 2020.

5. Genetic Literacy Project: *GMO FAQ.* https://gmo.geneticliteracyproject .org/FAQ/what-is-crisprcas9-and-other-new-breeding-technologies-nbts. Accessed January 18, 2020.

6. National Academies of Sciences, Engineering, and Medicine: *Genetically Engineered Crops: Experiences and Prospects 2016.* Report in Brief available at https://www.nap.edu/resource/23395/GE-crops-report-brief.pdf. Accessed January 10, 2020.

7. Center for Science in the Public Interest: *Biotechnology.* Available at https://cspinet .org/protecting-our-health/biotechnology. Accessed January 10, 2020.

8. Klümper W and Qaim M: A meta-analysis of the impacts of genetically modified crops. *PLoS One* 9(11):e111629, 2014. DOI:10.1371/journal. pone.0111629.

9. Vogliano C and Brown K: The state of America's wasted food and opportunities to make a difference. *Journal of the Academy of Nutrition and Dietetics* 116:1199, 2016. DOI:doi.org/10.1016/j.jand.2016.01.022.

10. Gunders D: National Resource Defense Fund: Wasted: Second Edition of NRDC's Landmark Food Waste Report. Updated August 17, 2017. Available at https://www.nrdc.org/sites/default/files/wasted-2017-report.pdf. Accessed January 13, 2020.

11. Papargyropoulou E and others: The food waste hierarchy as a framework for the management of food surplus and food waste. *Journal of Cleaner Production* 76:106, 2014. DOI:doi.org/10.1016/j.jclepro.2014.04.020.

12. U.S. Department of Commerce, National Oceanic and Atmospheric Administration (NOAA): *New record: Number of overfished stocks in the U.S. reaches all time low.* Updated May 17, 2018. Available at https://www .noaa.gov/media-release/new-record-number-of-overfished-stocks-in-us -reaches-all-time-low. Accessed January 13, 2020.

13. Pelletier JE and others: Positive attitudes toward organic, local, and sustainable foods are associated with higher dietary quality among young adults. *Journal of the Academy of Nutrition and Dietetics* 113(1): 127, 2013. DOI:10.1016/j.jand.2012.08.021.

14. Henley SJ and others: Invasive cancer incidence, 2004–2013, and deaths, 2006–2015, in nonmetropolitan and metropolitan counties—United States. *Morbidity and Mortality Weekly Report Surveillance Summaries* 66(No. SS-14):1, 2017. DOI:10.15585/mmwr.ss6614a1external icon.

15. Curl CL and others: Estimating pesticide exposure from dietary intake and organic food choices: The Multi-Ethnic Study of Atherosclerosis (MESA). *Environmental Health Perspectives* 123:475, 2015. DOI:10.1289/ehp.1408197.

16. U.S. Food and Drug Administration: *Advice About Eating Fish.* Updated July, 2019. Available at https://www.fda.gov/food/consumers/advice -about-eating-fish. Accessed January 18, 2020.

17. Painter JA and others: Attribution of foodborne illnesses, hospitalizations, and deaths to food commodities by using outbreak data, United States, 1998–2008. *Emerging Infectious Diseases Journal* 19(3):407, 2013. DOI:10.3201/eid1903.111866.

18. Centers for Disease Control and Prevention: *CDC estimates of foodborne illness in the United States.* Available at www.cdc.gov/foodborneburden. Accessed January 18, 2020.

19. U.S. Department of Agriculture, Food Safety and Inspection Service: *Cleanliness helps prevent foodborne illness.* Updated December 2, 2016. https://www.fsis.usda.gov/wps/portal/fsis/topics/food-safety-education /get-answers/food-safety-fact-sheets/safe-food-handling/cleanliness-helps -prevent-foodborne-illness/CT_Index. Accessed January 18, 2020.

20. Carothers M: Are you and your food prepared for a power outage? *USDA Blog.* February 21, 2017. Available at https://www.usda.gov /media/blog/2016/09/26/are-you-and-your-food-prepared-power-outage. Acccessed January 19, 2019.

21. Thatte D: The food safety modernization act in a nutshell. *National Institute of Standards and Technology Blog,* October 17, 2019. Available at https://www.nist.gov/blogs/manufacturing-innovation-blog/food-safety -modernization-act-nutshell. Accessed January 14, 2020.

22. Dall C: FDA reports major drop in antibiotics for food animals. *Center for Infectious Disease Research and Policy News,* December 19, 2018. Available at http://www.cidrap.umn.edu/news-perspective/2018/12/fda-reports -major-drop-antibiotics-food-animals. Accessed January 16, 2020.

23. McCarty CL: Notes from the field: Large outbreak of botulism associated with a church potluck meal—Ohio, 2015. *CDC Morbidity and Mortality Weekly Report* 64(29):802, 2015.

24. Grau KR and others: The intestinal regionalization of acute norovirus infection is regulated by the microbiota via bile acid-mediated priming of type III interferon. *Nature Microbiology.* 5:84, 2020 DOI:10.1038/ s41564-019-0602-7.

25. U.S. Department of Agriculture: Food: FDA issues final rule on food ingredients that may be "generally recognized as safe." Updated December 12, 2017. Available at https://www.fda.gov/food/newsevents /constituentupdates/ucm516332.htm. Accessed January 20, 2019.

26. McCann D and others: Food additives and hyperactive behaviour in 3-year-old and 8-/9-year-old children in the community: A randomised, double-blinded, placebo-controlled trial. *Lancet* 370(9598):1560, 2007. DOI:10.1016/S0140-6736(07)61306-3.

27. Kennedy OJ and others: Coffee, including caffeinated and decaffeinated coffee, and the risk of hepatocellular carcinoma: A systematic review and dose–response meta-analysis. *British Medical Journal Open* 7(5):e013739. DOI:10.1136/bmjopen-2016-013739.

28. Committee on Obstetric Practice: Moderate caffeine consumption during pregnancy. Committee Opinion No. 462. American College of Obstetricians and Gynecologists. *Obstetrics and Gynecology* 116:467, 2010. DOI:10.1097/AOG.0b013e3181eeb2a1.

29. Reis CEG and others: Effects of coffee consumption on glucose metabolism: A systematic review of clinical trials. *Journal of Traditional and Complementary Medicine* 9(3):184, 2019. DOI:10.1016/j.jtcme.2018.01.001

30. U.S. Department of Health & Human Services: *Storage times for the refrigerator and freezer.* Updated April 12, 2019. Available at https://www.foodsafety .gov/keep/charts/storagetimes.html. Accessed January 18, 2020.

# Chapter 13

# Global Nutrition

# Student Learning Outcomes

**Chapter 13 is designed to allow you to:**

**13.1** Define and characterize the terms *food insecure, nutrition security, hunger,* and *malnutrition.*

**13.2** Examine malnutrition in the United States, and highlight several programs established to combat suboptimal nutrition.

**13.3** Examine global nutrition, and evaluate the factors related to health outcomes.

**13.4** Outline possible solutions to malnutrition in the developing world.

**13.5** Evaluate the consequences of malnutrition during critical periods in a person's life.

## FAKE NEWS

### There's just not enough food on the planet to feed everyone.

## THE FACTS

Food production is an efficient process. The world is currently producing enough food to feed the global population (or 10 billion people). Sadly, approximately ⅓ of annual global human food production gets lost or wasted. These losses account for about $680 billion in industrialized countries and $310 billion in developing countries. Global food losses and waste per year are approximately 30% for cereals; 45% for root crops, fruits, and vegetables; 20% for oil seeds, meat, and dairy; plus 35% for fish. Annually, consumers in more affluent countries waste almost as much food as the entire net food production of sub-Saharan Africa. This chapter's *Ask the RDN* provides ideas for reducing food waste.

Source: FAO Global Policy Forum.

Global nutrition has never been more important than it is today. Currently, nearly one in nine individuals are chronically undernourished—too hungry to lead a productive, active life. The consequences of **undernutrition** are widespread—even though there is enough food available to sufficiently feed all of us. As depicted in the beautiful and bountiful image on this chapter's cover page, plentiful food is the source of not only nourishment, but also comfort and sustainability.

Within the developing world, over 820 million people suffer from hunger. Chronic **malnutrition** contributes to lowered resistance to disease, infection, and death, with children under 5 years being the most susceptible.

This chapter examines global nutrition and focuses on the problem of malnutrition, contributing factors, and potential solutions. If we are to eradicate this issue, we all have to understand the problem. We must begin to assume responsibility today, not tomorrow, to work on solutions to hunger both close to home and in faraway nations. The United Nations has led the charge to adopt and promote a set of goals to alleviate global poverty, protect the planet, and ensure prosperity for all. These 17 *Sustainable Development Goals* comprise the core of the global agenda. To coordinate and achieve this initiative, stakeholders, governments, the private sector, society, and individuals (just like you) will be called on to support these sustainable goals. In Section 2, we will discuss malnutrition in the U.S. and the *Dietary Guidelines* recommendations to customize and enjoy food and beverages that reflect personal preferences, cultural traditions, and budgetary considerations.

## 13.1 World Hunger

Uncertainty regarding the source of one's next meal remains a daily experience for 2 billion people around the world.[1] This situation is troubling, considering that agriculture worldwide produces more than enough food to meet the energy requirements of each of the planet's 7.6 billion people. Even with this abundance, the United Nations Food and Agriculture Organization (FAO) reports that over 26% of people are still unable to access enough food to lead active, healthy lives. Hence, these individuals are **food insecure.** Availability, access, utilization, and stability are the pillars of food security, which exists when people have physical, social, and economic access to sufficient, safe, and nutritious food to meet their dietary needs and food preferences for an active and healthy life. These dimensions are defined in detail in Figure 13-1.[1]

Food security is actually a component of a larger concept called **nutrition security.** The FAO defines nutrition security as secure access to an appropriately nutritious diet (i.e., protein, carbohydrate, fat, vitamins, minerals, and water), coupled with a sanitary environment and adequate health services and care, in order to ensure a healthy and active life for all household members.[2] The combined term *food and nutrition security* can be used to emphasize both food and health.[1]

Food insecurity and malnutrition exist in virtually every nation in the world (Fig. 13-2). The prevalence of undernutrition is most common in regions of the developing world, where the majority of people suffering from food insecurity, hunger, or malnutrition are poor. There has been global interest in poverty and hunger for decades, thus development of the Sustainable Development Goals (SDGs) have been set forth by the United Nations (UN). These goals and targets will be discussed later in this chapter.[3]

### HUNGER

Recall that the physiological state that results when not enough food is eaten to meet energy needs is called *hunger.* This sensation is often described as an uneasiness, discomfort, weakness, or pain caused by a lack of food. The medical and societal costs of the malnutrition that can result from chronic hunger include high rates of **preterm** births, intellectual disabilities, inadequate or stunted growth and development, poor academic performance, decreased work output, chronic disease, and many other preventable issues.

Although malnutrition (both undernutrition and overnutrition) does occur in North America, it is not always a result of extreme poverty over a large segment of the population. Instead, there are often specific causes, such as an eating disorder,

**food insecure** The state of being without reliable access to a sufficient quantity of affordable, nutritious food.

**nutrition security** Secure access to a nutritious diet coupled with a sanitary environment and adequate health services and care.

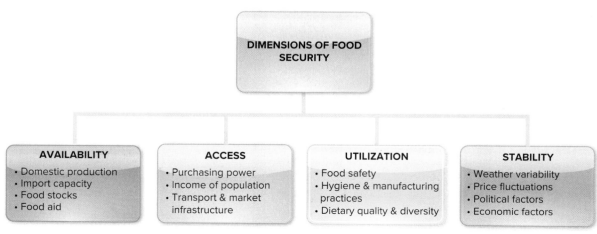

**FIGURE 13-1** ▲ The four dimensions of food security.
Source: FAO, http://www.fao.org/docrep/013/al936e/al936e00.pdf.

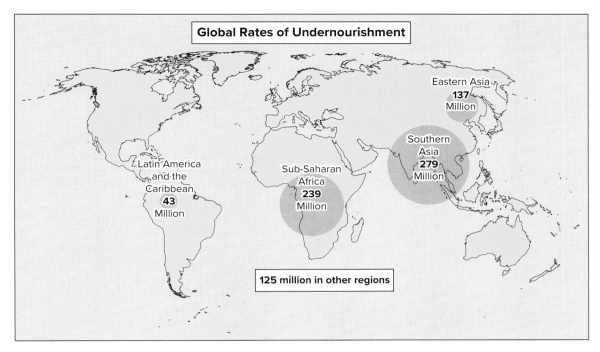

**FIGURE 13-2** ▲ The world's most undernourished people are found in Southern Asia, followed by sub-Saharan Africa, Eastern Asia, and Latin America and the Caribbean.
Source: Food and Agriculture Organization of the United Nations.

alcohol use disorders, unavailable or inadequate caregiving, or homelessness that contribute to this issue.[2]

Fortunately, there are *food safety nets*, such as food pantries and soup kitchens to help those in need, though sometimes there remain transportation and other barriers to food access for many in need. Food insecurity, as measured in the United States using the U.S. Department of Agriculture (USDA) Household Food Security Survey Module is also a problem for many vulnerable households.[4] There are four levels of food security: *high food security, marginal food security, low food security*, and *very low food security*. Approximately 11% of U.S. households (14 million households) are food insecure. Worldwide, the FAO identifies five levels of food security (Fig. 13-3). The FAO's Integrated Food Security Phase Classifications (IPC) are based on global mortality, malnutrition, food and water access and availability, dietary diversity, coping strategies, and livelihood assets.[1]

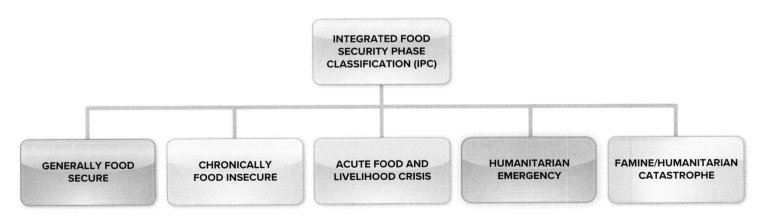

**FIGURE 13-3** ▲ Integrated food security phase classifications.
Source: FAO, http://www.fao.org/docrep/013/al936e/al936e00.pdf.

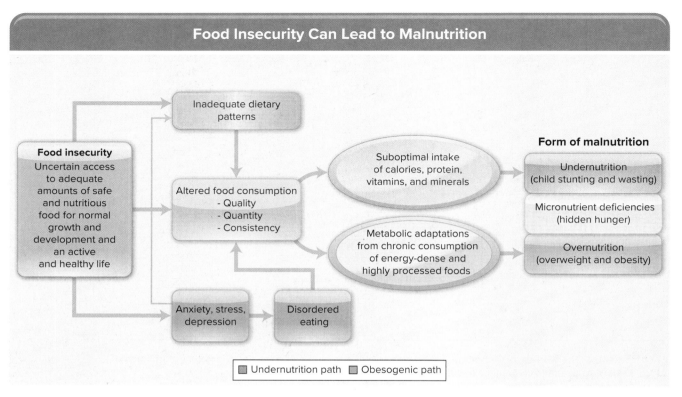

**FIGURE 13-4** ▲ Food insecurity can contribute to wasting, stunting, and micronutrient deficiencies. A dietary pattern characterized by insufficient intake of calories, protein, vitamins, and minerals will impede fetal, infant, and child growth and development. Such diets contribute to maternal undernutrition and consequently to higher risk of low birth weight, which in turn are both risk factors for child stunting. The stress of living with food insecurity can also have a negative effect on the nutrition of infants by compromising breastfeeding. Source: Food and Agriculture Organization of the United Nations.

**FAKE NEWS**

*Malnutrition* means the same thing as *hunger.*

**THE FACTS**

Malnutrition can occur with or without hunger. Malnutrition occurs when a dietary pattern is suboptimal and results over time in a diet that is lacking in nutrients. Also note that malnutrition includes both undernutrition and overnutrition.

**federal poverty level (FPL)** An economic measure used to determine if an individual or family qualifies for specific federal benefits and programs; also called *poverty line*.

Poverty is a primary driver of malnutrition. In the United States, child poverty affects over 15 million children in families living below the federal poverty level. Poverty guidelines are established to establish criteria to qualify for federal support. The **federal poverty level** (FPL) accounts for family size and is updated annually. For example, a family of four having an income less than $25,700 per year would qualify for support under these guidelines. Research, however, estimates that families need an income of about twice that level to cover basic expenses. Research consistently documents that poverty impairs a child's ability to learn and contributes to inadequate growth, behavioral problems, and poor health (Fig. 13-4). Fortunately, the United States does offer food assistance programs for low-income families and children. These programs help shield children from hunger and will be discussed later in this chapter.

## MALNUTRITION AND MICRONUTRIENT DEFICIENCIES

*Malnutrition* is a condition of impaired development or function caused by either a chronic deficiency or excess in calorie and/or nutrient intake. When food supplies are low and the population is large, undernutrition is common, leading to nutritional deficiency diseases, such as **goiter** (iodine deficiency), **anemia** (iron deficiency), and **xerophthalmia** (vitamin A deficiency). However, when the food supply is ample or overabundant, energy-dense food choices coupled with an excessive intake can lead to overnutrition, obesity, and its related chronic diseases, such as type 2 diabetes.

Undernutrition is the most common form of malnutrition among the poor in developing countries. Undernutrition is also the primary cause of specific nutrient deficiencies that can result in muscle **wasting**, blindness, scurvy, pellagra, beriberi, anemia, rickets, goiter, and a host of other problems (Table 13-1). Recall that the term **hidden**

## TABLE 13-1 ■ Nutrient-Deficiency Diseases That Commonly Accompany Undernutrition

| Disease and Key Nutrient Involved* | Typical Effects of Deficiency | Sources of Nutrients Involved |
|---|---|---|
| **Ariboflavinosis** Riboflavin | Inflammation of oral cavity; nervous system disorders | Milk, mushrooms, spinach, liver, and enriched grains |
| **Beriberi** Thiamin | Nerve degeneration, altered muscle coordination, and cardiovascular problems | Sunflower seeds, pork, whole and enriched grains, and dried beans |
| **Goiter** Iodine | Enlarged thyroid gland in teenagers and adults, possible mental retardation, and congenital hypothyroidism | Iodized salt and saltwater fish |
| **Iron-deficiency anemia** Iron | Reduced work output, delayed growth, and increased health risk in pregnancy | Meats, seafood, broccoli, peas, bran, whole grains, and enriched products |
| **Macrocytic anemia** Folate | Enlarged red blood cells, fatigue, and weakness | Green leafy vegetables, legumes, oranges, and liver |
| **Pellagra** Niacin | Diarrhea, dermatitis, dementia, and death | Mushrooms, bran, tuna, chicken, beef, peanuts, and whole and enriched grains |
| **Rickets** Vitamin D | Poorly calcified bones, bowed legs, and other bone deformities | Fortified milk, fish oils, and sun exposure |
| **Scurvy** Vitamin C | Delayed wound healing, internal bleeding, and abnormal formation of bones and teeth | Citrus fruits, strawberries, broccoli, and tomatoes |
| **Xerophthalmia** Vitamin A | Blindness from chronic eye infections, dryness, restricted growth, and keratinization of epithelial tissues | Fortified milk, sweet potatoes, spinach, greens, carrots, cantaloupe, and apricots |

*Although the nutrients are listed separately to illustrate the important role of each one, often two or more nutrition-deficiency diseases are simultaneously found in an undernourished person in the developing world.

**hunger** affects a large proportion of the world population. This refers to those with micronutrient (vitamin and mineral) deficiencies, often with no observable signs or symptoms. Recall that protein-calorie malnutrition (PCM) is a form of undernutrition caused by an extremely deficient intake of calories or protein and is generally accompanied by illness. Kwashiorkor and marasmus are both forms of PCM.

The most critical micronutrients missing from diets worldwide (Fig. 13-5) are iron, vitamin A, iodine, zinc, and various B vitamins (e.g., folate), as well as selenium and vitamin C. About 3 billion people, mostly in the developing world, are affected by iron and zinc deficiencies. With poor iron status, cognitive development will likely be impaired, particularly if prolonged deficiency occurs during early infancy. UNICEF estimates that over 6 billion people around the globe now consume iodized salt, representing a colossal large-scale food fortification achievement. Vitamin A deficiency predominantly affects pregnant women and young children. Although severe vitamin A deficiency, which causes blindness, is on the decline, up to 500,000 preschool-age children are still blinded by it each year, with half of them likely to die within 12 months of losing their sight. The United Nations International Children's Emergency Fund (UNICEF) reports that the lives of one in three children could be spared annually in the developing world if vitamin A supplements were provided a few times each year. The annual cost per child is approximately 2 cents.

## FAMINE

**Famine** can also lead to chronic hunger. Periods of famine are characterized by large-scale loss of life, climate change, social disruption, economic chaos, and political conflicts. As a result of these extreme events, the affected community may experience the following: a

**wasting** Low weight for height (thinness) that typically indicates a recent and severe process of weight loss, often associated with acute starvation or severe disease.

**famine** An extreme shortage of food, which leads to massive starvation in a population; often associated with crop failures, war, and political unrest.

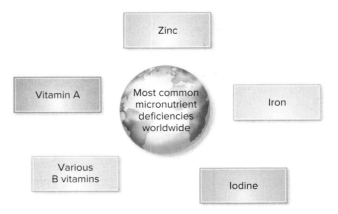

**FIGURE 13-5** ▲ Critical micronutrient deficiencies worldwide.

downward spiral characterized by human distress; sales of land, livestock, and other farm assets; mass migration; division and impoverishment of the poorest families; crime; and humanitarian crises. The most at-risk countries for humanitarian catastrophes are ranked by the International Rescue Committee (IRC). The top 10 countries account for about 50% of internally displaced individuals, almost 70% of refugees, and are experiencing the worst humanitarian crises in the world. More on war and famine is discussed later in this chapter.

Special focused and global efforts are needed to eradicate the fundamental causes of famine. These causes vary by region and decade, but the most common cause of famine is crop failure. The most obvious reasons are extreme weather conditions such as floods or drought, war, and civil strife.

## GENERAL EFFECTS OF SEMISTARVATION

In the initial stages, the results of undernutrition from semistarvation are often so mild that observable physical symptoms are absent and blood tests do not usually detect these slight metabolic changes. Even in the absence of detectable symptoms, however, undernourishment may affect the ability to work, learn, reproduce, and recover from illnesses or injuries. Recall that as tissues continue to be depleted of nutrients, blood tests eventually detect biochemical changes, such as a drop in blood hemoglobin concentration. Physical symptoms, such as body weakness or fatigue, tend to follow with further depletion. Finally, the full-blown symptoms of the deficiency are recognizable, such as **stunting** or blindness (vitamin A deficiency).

When a few people in a population develop a severe deficiency, this may represent only the tip of the iceberg. Typically, a much greater number of individuals in that area may be suffering milder degrees of undernutrition. These deficiencies should not, therefore, be dismissed as trivial, especially in the developing world. In many low- and even middle-income countries, it is typical for a combination of micronutrient deficiencies to occur together. These are caused by any number of factors, including dietary patterns of poor nutritional quality related to seasonal variation in food availability, low bioavailability of nutrients from plant sources, cultural food practices, and poverty. It is becoming clear that combined deficiencies of specific vitamins and minerals (especially iron and zinc) can significantly reduce work performance, even in the absence of obvious physical symptoms. This resulting state of ill health, in turn, diminishes the ability of individuals, communities, and even whole countries to perform at peak levels of physical and mental capacity. Because nutritional requirements are high during periods of rapid growth, pregnant women, infants, and children are especially vulnerable to the effects of undernutrition. Studies of pregnant women and children in developing countries have proven the negative impact of micronutrient deficiencies on birth size, length of gestation, growth, and intellectual development.

Added to their lack of nourishment, the inhabitants of poverty-stricken countries must also contend with recurrent infections, contaminated water sources, poor sanitation, extreme weather conditions, and regular exposure to infectious diseases. Deficiencies of micronutrients, especially iron and zinc, can lead to reduced immune function and thereby increase the risk of diseases, such as diarrhea and pneumonia.

Strategies to address the multiple nutrient deficiencies in developing countries have included supplementation and fortification of ready-to-use foods and beverages. Although supplementation has been the most widely practiced intervention, public health experts believe that fortification of a commonly consumed food could be a single, cost-effective intervention strategy to target a larger population. Studies of micronutrient supplementation during pregnancy have shown an increase in birth weight and a reduction in **low birth weights** but no impact on preterm births or perinatal mortality. In children, micronutrient supplementation with three or more nutrients has resulted in increases in height and weight.[5]

**stunting** Impaired growth and development characterized by low height-for-age.

▶ The effects of hunger are diverse, widespread, and devastating. Characteristics include decreased or impaired:
- Concentration
- Energy and strength
- Immunity
- Infant birth weights
- Learning and development
- Mental health
- Productivity

Nikhil Gangavane/thefinalmiracle/123RF

**low birth weight (LBW)** Referring to any infant weighing less than 2.5 kilograms (5.5 pounds) at birth; most commonly results from preterm birth.

☑ **CONCEPT CHECK 13.1**

1. What is the definition of *food insecure?*
2. Name two types of malnutrition?
3. List three health consequences of chronic hunger?
4. What is the primary cause of undernutrition?
5. At which stages of the life cycle is undernutrition especially damaging?

# 13.2 Malnutrition in the United States

Roughly 38 million (11.8%) people in the United States are living at or below the poverty level.[6] Of these, approximately 57% of households report having to choose between purchasing food and housing costs, 69% between food and utilities, 66% between food and medicine, and 67% between food and access to transportation.[7] While housing and utility costs, medical care, and transportation fares are nonnegotiable, a person may choose to eat less. Food is one of the few optional items in a household budget. The short-term consequences of eating less may be less dramatic than getting evicted, but the long-term cumulative effects are significant. A healthy eating pattern will only be achieved when adequate resources and supports exist in places where individuals live, work, and gather.

## HELPING THE HUNGRY IN THE UNITED STATES

Until the twentieth century, individuals and a wide variety of charitable, often faith-based organizations, provided most of the help to poor, undernourished people in the United States. Early programs rarely distributed direct cash payments to people in need because these were thought to reduce recipients' motivation to improve their circumstances or engage in negative behaviors, such as smoking, that contributed to their poverty. Beginning in the early 1900s, the involvement of local, county, and state governments in providing assistance to the poor has steadily increased (Table 13-2).

**Federal Food Assistance.** After observing extensive hunger and poverty during his presidential campaign in the 1960s, President John F. Kennedy revitalized the Food Stamp Program, which had begun two decades earlier, and expanded commodity distribution programs. Renamed the Supplemental Nutrition Assistance Program (SNAP), this $68 billion program helps low-income people and families buy food. SNAP allows recipients to use an electronic benefit transfer (EBT) card to purchase food and garden seeds—but excludes tobacco, cleaning items, alcoholic beverages, and nonedible products—at stores authorized to accept them. The average recipient of SNAP receives about $130 a month. Over 34 million participate in the program. About 80% of SNAP households include a child, older individuals, or those with a disability.[8]

A USDA report assessed the shopping cart of SNAP participants and found the top item purchased with SNAP dollars was soft drinks. The report compared SNAP households and non-SNAP households. While SNAP participants purchased slightly more junk food and fewer vegetables, both groups purchased an abundant amount of sweetened drinks, candy, ice cream, and potato chips. SNAP households spent about 40 cents of each dollar on meat, fruits, vegetables, milk, eggs, and bread. Another 40 cents were spent on cereal, prepared foods, dairy products, rice, and beans. The final 20 cents was spent on highly processed and low-nutrient-dense foods such as sweetened beverages, desserts, salty snacks, candy, and sugar.[8]

**Other Food Assistance.** The U.S. Congress established the National School Lunch Program in the 1940s and the School Breakfast Program in the 1960s. School breakfast and lunch programs enable low-income students—12 million for breakfast and 20 million

▶ **Historical Research on Undernutrition**

In the 1940s, a group of researchers led by Dr. Ansel Keys examined the general effects of undernutrition on adults. Previously healthy men were fed a diet averaging about 1800 kcal daily for 6 months. During this time, the men lost an average of 24% of their body weight. After about 3 months, the participants complained of fatigue, muscle soreness, irritability, intolerance to cold, and hunger pains. They exhibited a lack of ambition, self-discipline, and concentration and were often moody, apathetic, and depressed. Their heart rate and muscle tone decreased, and they developed edema. When the men were permitted to eat normally again, feelings of recurrent hunger and fatigue persisted, even after 12 weeks of rehabilitation. Full recovery required about 8 months. This study helps us understand the general state of undernourished adults worldwide.

Source: Keys A: Will you starve that they be better fed? *Brochure*, May 27, 1944.

**TABLE 13-2** ■ Federal and Non-Federal Programs That Supply Food for People in the United States

| Program | Eligibility | Description | Website |
|---|---|---|---|
| **Child Nutrition Programs** | | | |
| Child and Adult Care Food Program | Child-care program participants; income guidelines match National School Lunch Program | Reimbursement for meals supplied at sites that follow USDA guidelines | https://www.fns.usda.gov/cacfp /child-and-adult-care-food-program |
| Fresh Fruit and Vegetable Program | Low-income elementary schools | Free fresh fruits and vegetables | https://www.fns.usda.gov/ffvp /fresh-fruit-and-vegetable-program |
| National School Lunch Program | Low-income school-age children | Provides nutritionally balanced, low-cost, or free lunches to school-age children | https://www.fns.usda.gov/nslp /national-school-lunch-program-nslp |
| School Breakfast Program | Low-income school-age children | Free or reduced-price breakfast program to school-age children | https://www.fns.usda.gov/sbp /school-breakfast-program-sbp |
| Special Milk Program | Low-income school-age children | Provides milk in schools and child-care institutions | https://www.fns.usda.gov/smp /special-milk-program |
| Summer Food Service Program | Low-income neighborhood or participation in a program | Free meals and snacks for children when school is not in session | https://www.fns.usda.gov/sfsp /summer-food-service-program |
| Team Nutrition | Schools that participate in other USDA programs | Delivers nutrition to children and caregivers through extensive networks and channels | https://www.fns.usda.gov/tn |
| **Food Distribution Programs** | | | |
| Commodity Supplemental Food Program | Low-income individuals at least 60 years old | Supplements the diet with USDA surplus foods | https://www.fns.usda.gov/csfp /commodity-supplemental-food -program |
| Community Food Systems | Schools that participate in existing USDA programs | Services include helping nutrition program operators incorporate local foods into food service, as well as the Summer Food Service Program and Child and Adult Care Food Program | https://www.fns.usda.gov/cfs |
| Food Distribution Program on Indian Reservations (FDPIR) | Low-income American Indian and non-American Indians living on reservations or members of recognized tribes | Alternative to SNAP; distributes monthly food packages; includes nutrition education | https://www.fns.usda.gov/fdpir /food-distribution-program-indian reservations-fdpir |
| Emergency Food Assistance Program (TEFAP) | Low-income families | Nutrition assistance via USDA food commodity distributions | https://www.fns.usda.gov/tefap /emergency-food-assistance -program-tefap |
| Senior Farmers' Market Nutrition Program | Low-income households | Awards grants to provide coupons for eligible foods at farmers' markets, roadside stands, and community agriculture programs | https://www.fns.usda.gov/sfmnp /senior-farmers-market-nutrition -program-sfmnp |
| Special Supplemental Nutrition Program for Women, Infants, and Children (WIC) | Low-income pregnant/lactating women, infants, and children under 5 years old and at nutritional risk | Provides coupons to purchase nutritious foods; includes nutrition education | https://www.fns.usda.gov/wic /women-infants-and-children-wic |
| Supplemental Nutrition Assistance Program (SNAP) | Low-income families | Electronic benefit transfer cards given to purchase foods; amount based on household size and income | https://www.fns.usda.gov/snap /supplemental-nutrition-assistance -program-snap |
| SNAP Nutrition Education (SNAP-Ed) | Families using or eligible for SNAP | Nutrition education and physical activity via community partnerships | https://www.fns.usda.gov/snap /supplemental-nutrition-assistance -program-education-snap-ed |

**TABLE 13-2** ■ **Federal and Non-Federal Programs That Supply Food for People in the United States (*continued*)**

| Program | Eligibility | Description | Website |
|---|---|---|---|
| USDA Department of Defense Fresh Fruit and Vegetable Program | Low-income elementary schools | USDA Foods entitlement dollars for fresh produce | https://www.fns.usda.gov/fdd/usdadod-fresh-fruit-and-vegetable-program |
| USDA Foods in Schools | Participants in National School Lunch Program, Child and Adult Care Food Program, or Summer Food Service Program | Supports purchase of 100% American-grown and -produced foods | https://www.fns.usda.gov/fdd/schoolscn-usda-foods-programs |
| USDA Foods Processing | Eligible state and distributing recipient agencies | Permits raw donated foods processed into convenient, ready-to-use end-products | https://www.fns.usda.gov/fdd/usda-foods-processing-home |
| WIC Farmers Market Nutrition Program | WIC-eligible participants | Provide fresh, unprepared, locally grown fruits and vegetables | https://www.fns.usda.gov/fmnp/wic-farmers-market-nutrition-program-fmnp |
| **Other Federally Funded Programs and Resources** | | | |
| Expanded Food and Nutrition Education Program (EFNEP) | Low-income families | Offers nutrition education and programming to limited-resource families and children | https://nifa.usda.gov/program/about-efnep |
| Farm-to-School | Participants of the National School Lunch, Summer Food Service, or Child and Adult Care Food Program | Helps operators incorporate more local foods into meal programs | https://www.fns.usda.gov/farmtoschool/farm-school |
| **Community-Based Programs and Other Organizations and Resources** | | | |
| Congressional Hunger Center | | Advocates for public policy with emphasis on issues related to hunger and poverty | https://www.hungercenter.org |
| Feeding America | Individual food banks | Network and database of select national food banks | https://www.feedingamerica.org |
| Food Research and Action Center (FRAC) | | Conducts research, policy initiatives, and provider training to reduce poverty and undernutrition | http://www.frac.org |
| Gleaning and Food Recovery | | Focuses on gleaning to promote gathering leftover produce and food recovery at local, state, regional, and national levels | https://www.nal.usda.gov/aglaw/gleaning-and-food-recovery |
| Meals on Wheels Association of America | Adults age 60 or over | Home-delivered meals, as well as congregate meals, for those without mobility concerns | https://www.mealsonwheelsamerica.org |
| Mazon: A Jewish Response to Hunger | | Focus on hunger-related policy and improving food access | http://www.mazon.org |
| Share Our Strength | Varies based on program | Supports delivery of nutrition and cooking education | https://www.shareourstrength.org |
| WHY (World Hunger Year) | | Supports grassroots efforts, community programs, and advocacy efforts to raise awareness about hunger | https://whyhunger.org |

Source: Position of the Academy of Nutrition and Dietetics: Food Insecurity in the United States.

This man is hoping to find work to fill his basic need of food. **What may be feasible for you to help this person find employment, food, and/or shelter?** DebbiSmirnoff/Getty Images

for lunch—to receive meals free or at reduced cost if certain income guidelines are met.[9] The basis of eligibility for free and reduced-price meals can be determined through participation in other programs (categorical eligibility), by schools (direct certification), or by household income (income eligibility).

The Older Americans Act (OAA) authorizes nutrition services programs to reduce hunger and food insecurity, enhance socialization, promote health and well-being, and delay adverse health conditions for older individuals. Over 11 million (one in six) older adults are served annually by OAA programs such as congregate and home-delivered meals.[10]

In 1972, the Special Supplemental Nutrition Program for Women, Infants, and Children (WIC) was authorized. This program provides electronic vouchers to pay for food purchases and nutrition education to low-income, nutritionally at-risk women who are pregnant or lactating and their young children. It serves 6 million women and children each month and remains the most successful federally funded nutrition program to date based upon health outcomes.[9]

Government food assistance programs are like a *safety net:* they are strong, yet porous. These programs were originally known as *welfare* and were designed to provide short-term relief for Americans suffering financial hardship. Now, many low- and middle-income U.S. households rely on food pantries and federal assistance continually and for the majority of their nutritional needs. It is estimated that one in seven Americans—that is, 46 million people—rely on food pantries and emergency meal service programs to feed themselves and their households.[11] Statistics show that most people requesting emergency food assistance are members of families. This is not surprising considering that households with children report a higher rate of food insecurity than households without children, and the rate is even higher for households with children headed by single parents.

Undernutrition in North America is a much more subtle problem than in developing countries. To the untrained eye, undernourished children may just seem lean, when, in fact, their growth is being stunted by insufficient nutrients. More likely, though, children from food-insecure households are prone to be overweight. The coexistence of undernutrition with overweight or obesity is referred to as the *double burden of malnutrition*. This form of malnutrition may be the result of considerable reliance on highly processed convenience foods that provide excessive calories, saturated fat, and added sugars. The availability of cooking facilities also affects nutrient intake among the poor. Without adequate cooking facilities, people may buy expensive convenience foods that require little to no preparation. These are typically highly processed foods that provide calories but are often lacking in nutrients. Malnutrition in either form (underweight or overweight) leads to significant health issues and greater mortality (Fig. 13-6).

## SOCIOECOLOGICAL FACTORS RELATED TO MALNUTRITION

In the United States, poverty, psychosocial factors, and functional limitations all increase the risk of malnutrition. Extreme poverty used to be considered the root cause of hunger and undernutrition, but we now understand that hunger and undernutrition are also major causes of poverty. There appears to be a bidirectional relationship between these factors that is much more complex than once considered.

**Poverty.** Poverty is determined in the United States using poverty thresholds that are issued each year by the Census Bureau. Persons are categorized as poor if they have an income less than that deemed sufficient to purchase the basic needs of food, shelter, clothing, and other essentials. Poverty, however, is complex and does not mean the same thing to all people. Poverty can be characterized as **situational poverty** or **generational poverty.** Households can fall into situational poverty as the result of dire or unexpected circumstances, such as getting laid off from a job or a family member needing expensive medical treatment. Some have linked situational poverty to the *seven D's:* divorce, death, disease, downsizing, disablement, disasters, and debt. **Situational poverty** may be episodic or cyclical, may affect entire communities as a

▲ Food insecurity is part of the North American landscape. A *safety net* of programs exists but fails to serve many households in need.
Source: USDA.

**situational poverty** State of poverty caused by specific circumstances such as death, illness, divorce, or some catastrophe.

**generational poverty** Chronic state of poverty lasting for two generations or longer.

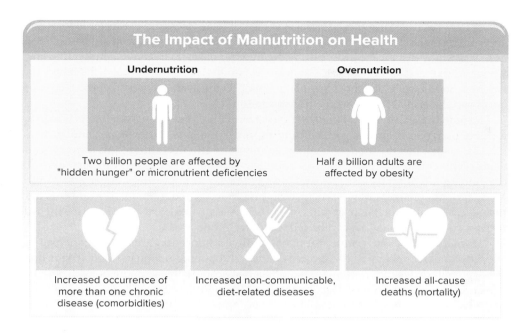

## The Impact of Malnutrition on Health

**Undernutrition**

Two billion people are affected by "hidden hunger" or micronutrient deficiencies

**Overnutrition**

Half a billion adults are affected by obesity

Increased occurrence of more than one chronic disease (comorbidities)

Increased non-communicable, diet-related diseases

Increased all-cause deaths (mortality)

**FIGURE 13-6** ◄ Malnutrition includes both undernutrition and overnutrition and occurs when a person's diet contains too few or too many nutrients. Source: FAO of the United Nations.

result of economic downturns, and may be temporarily alleviated by social service programs. In contrast, **generational poverty** refers to a culture of persistent poverty passed from parents to children, with two or more generations living in poverty. Generational poverty is more complex because the cultures and traditions of previous generations are passed along to future generations and often include an undervaluation of the benefits of education and a lack of confidence in one's abilities when it comes to changing one's economic circumstances. Clearly, a major driving force behind poverty is underemployment.

**Homelessness.** The growing shortage of affordable housing and ongoing poverty are the two trends largely responsible for homelessness. The homelessness created by the gap between the number of affordable housing units and the number of people needing them impacts the health and well-being of all individuals. Children affected by homelessness are most vulnerable and are twice as likely to experience hunger compared to children residing in a home.

Over 550,000 people experience homelessness—meaning they sleep outside or in an emergency shelter, transitional housing, or permanent supportive housing (PSH) on any given night. Over 37,500 of these are veterans.[13] Often, homelessness is secondary to an unexpected financial crisis that prevents individuals from retaining their housing. People in poverty must make very difficult choices when limited resources force competition between housing, food, child care, health care, and education. Because housing absorbs a high percentage of income, it is often sacrificed.[14]

*Chronic homelessness* is defined as involving either long-term and/or repeated occurrences of homelessness combined with physical and/or mental disability. Individuals who are chronically homeless are more likely to be older, male, non-white, and experiencing long-term unemployment. Without access to health care, they may cycle through emergency departments and inpatient beds, detox programs, jails, prisons, and psychiatric institutions, all at high public expense. Although progress has been made in addressing chronic homelessness in the last decade, much work still needs to be done to mitigate this societal issue.

**Access to Healthy Food.** Access to affordable and nutritious foods from supermarkets, grocery stores, or other retailers is a challenge for many Americans, making it harder for them to follow a balanced, nutrient-dense dietary pattern. Low-income, low-access areas with inadequate availability of nutritious foods were originally called **food deserts** and more recently are termed *healthy food priority areas*. USDA's online *Food*

**COVID CORNER**

Before COVID-19, food insecurity was already a serious issue in the U.S. Sadly, the pandemic has driven many households into a state of situational poverty, thus exponentially increasing food insecurity levels across the nation. Globally, about 80% of the world's poorest and most food insecure individuals reside in remote and rural areas, often relying on small-scale agriculture for survival. The COVID crisis highlighted many areas of inequality and gaps in our food system that must be addressed in order to transform it into a more sustainable and resilient network. These colossal efforts will require global collaborations and widespread support to be successful. For more information on the impact of COVID on food security, read the United Nation's Ensuring Food Security in the Era of COVID-19 online.

▲ Almost 70% of those experiencing unsheltered homelessness are men or boys.[12]
Ruben Sanchez @lostintv/Getty Images

**food desert** Areas where residents do not live near supermarkets or other food retailers that carry affordable and nutritious food.

Food pantries and soup kitchens are important sources of nutrients for a growing number of people in the United States. **Do you know how to locate your local food pantry?**

Ariel Skelley/Getty Images

*Access Research Atlas* presents a spatial overview of food access indicators for low-income and other neighborhoods.[15] Measures include: accessibility to sources of healthy food, as measured by distance to a store or by the number of stores in an area; individual-level resources that may affect accessibility, such as family income or vehicle availability; and neighborhood-level indicators of resources, such as the average income of the neighborhood and the availability of public transportation.[15]

The federal Healthy Food Financing Initiative (HFFI) provides government financing for developing and equipping grocery stores, small retailers, corner stores, and farmers' markets selling healthy food in low-income communities. In addition, this initiative provides employment and business opportunities in underserved areas. Farmers' markets are gaining attention as a simple and local approach in impacting hunger and health. Many farmers' markets now participate in programs for older adults, WIC, SNAP, and the SNAP Double Dollar program, where shoppers using their SNAP EBT card will receive matching funds that can be redeemed for produce at that market.

For many years, government-funded food assistance programs have helped to alleviate undernutrition in the United States.[11] Obtaining adequate food to not only survive, but thrive, is a tremendous challenge, especially when people experiencing homelessness are forced to live outside. Although many people assume that food pantries and soup kitchens are abundant and accessible for those in need, there are many barriers to accessing these resources. The majority of food pantries are staffed by volunteers and dependent upon food donations. They often have limited hours of operation, require appointments well in advance, often require proof of residence, and limit the amount of food that patrons can take home per month. Individuals who are homeless often lack proper identification and proof of residence, do not have cooking facilities and equipment needed for food preparation, and do not have adequate storage to prevent foodborne illnesses and practice safe food-handling procedures. Food availability through soup kitchens is limited in most cities and absent in most rural areas.

An additional challenge is that a growing number of cities have taken strides to restrict or ban the act of sharing food with the homeless. These laws are concerning, considering that most cities do not have adequate food resources to meet the needs of their poor and homeless. Advocates and food providers are eager to work with cities and other government agencies to help address the problems of hunger and homelessness by improving access to federal food benefits and other food resources. State governments have also begun moving to protect the right of individuals and groups to share excess food with others.

## CULTURAL TRADITIONS, PERSONAL PREFERENCES, AND BUDGET

Exposure to a variety of foods and beverages, early in life, promotes a child's willingness to consume and enjoy a variety of foods throughout each life stage. Establishing and maintaining a healthy dietary pattern, aligning with individual and cultural preferences, should be a priority. *Cultural foodways* refers to the intersection of food in culture, traditions, and history. Indeed, culture has a significant influence on both individual and family food and beverage choices. The *Dietary Guidelines* highlights that nutrient-dense and culturally relevant foods are part of all of the food groups. For instance, spices and herbs can improve food flavor while reducing added sugars, saturated fat, and sodium. A healthy dietary pattern can be affordable and fit within budgetary constraints. Strategies for success requires some advanced planning; consideration of regional and seasonal food availability; and incorporation of a variety of fresh, frozen, dried, and canned options as low-cost options to meet nutrient needs within a budget.

### ✓ CONCEPT CHECK 13.2

1. Name two food assistance programs in the United States.
2. What factors influence the presence of poverty, hunger, and malnutrition?
3. What is the difference between situational and generational poverty?

# 13.3 Malnutrition in the Developing World

Malnutrition in the developing world is also tied to poverty, and any sustainable solution must address this issue. However, such countries have a multitude of problems so complex and interrelated that they cannot be treated separately. Programs that have proved helpful in the United States are only a starting point. The major obstacles challenging those seeking a solution are illustrated in Figure 13-7 and described in the next few sections.

## FOOD-TO-POPULATION RATIO

The world has over 7.5 billion inhabitants. Population growth exceeds economic growth in much of the developing world, and as a result, poverty is on the rise globally. This disrupts the balance in the food-to-population ratio, tipping it toward food shortages. To ensure a decent life for a widening segment of humanity, many experts suggest that the growth in the earth's most vulnerable populations should slow.

According to the FAO, there are over 820 million individuals affected by hunger in the world, and 98% reside in developing countries.[1] More than 75% of these people live in rural areas of Africa and Asia and remain totally dependent on local agriculture for all of their nutritional needs. Unfortunately, major food and health disparities continue to exist between developed and developing countries, among the rich and the poor, and even within the family unit (e.g., males may be fed before females).

Economists estimate that world food production will continue to increase more rapidly than the world population. This will come at a high cost in terms of the water, fertilizer, and chemicals (pesticides, fungicides) needed to allow for adequate food production. In the short term, the primary problem is not related to production but rather to the supply chain—both distribution and use. This reality is even more dire in poverty-stricken areas of developing nations. Unfortunately, many vulnerable people do not have the income to purchase their own farmland or to have access to nonlocal sources.

Most viable farmland in the world is already in use, and because of inadequate farming practices or competing land-use demands, the number of farmable acres is decreasing annually. For these reasons, the world's *sustainable* food output—an amount that does not deplete the earth's resources—is now running well behind food consumption.

Historical records and artifacts dating back over 1.5 million years in South Africa provide evidence that humans have farmed and consumed insects. This practice is known as entomophagy and is seen as a delicacy in some cultures. As a high-protein food source, insect farming is a form of sustainable agriculture requiring minimal inputs compared to other high-protein food sources. In addition to protein, insects are high in iron, calcium, and B vitamins and low in both carbohydrates and fat.

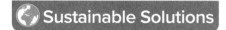 Sustainable Solutions

▲ This young Maasai woman, with her baby on her back, lives in Kenya, which is in a region with low contraceptive rates and the highest rates of maternal death.
Britta Kasholm-Tengve/Getty Images

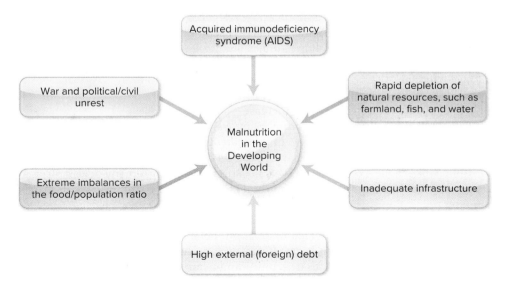

**FIGURE 13-7** ▲ Many factors contribute to malnutrition in the developing world. Any solutions to the problem must take these multifaceted factors into consideration.

▲ This mother, holding her 27-month-old malnourished child at a camp for internally displaced persons in Sudan, is dependent on assistance from relief organizations that are often not welcome in the war-torn country.

Source: USAID

**global maternal mortality ratio** The number of maternal deaths per 100,000 live births.

**Top 10 Countries at Risk for a Humanitarian Crisis**

1. Yemen
2. Democratic Republic of the Congo (DRC)
3. Syria
4. Nigeria
5. Venezuela
6. Afghanistan
7. South Sudan
8. Burkina Faso
9. Somalia
10. Central African Republic (CAR)

Source: International Rescue Committee

**Global Hunger Index (GHI)** A tool that tracks the state of hunger worldwide and calls attention to locations where immediate hunger relief is urgently needed.

This discrepancy suggests that food production in less-developed countries will not keep up with population growth and will soon lag behind.

Responsible family planning can contribute to the reduction of poverty and hunger and can *decrease* the number of maternal and childhood deaths. On a daily basis, approximately 810 women die from preventable causes related to pregnancy and childbirth; 94% of these deaths occur in developing countries.[16] In addition, more than 214 million women in developing countries lack access to modern contraception.[17] Global organizations are working to: (1) empower women and couples in developing countries to lead healthier lives by increasing access to family planning and maternal health products and services; and (2) address policy, financing, delivery, and sociocultural barriers to women who desire access to such information, services, and supplies. Although strides have been made, the **global maternal mortality ratio** remains unacceptable.

Promoting breastfeeding also contributes to infant health and birth control. Women who do not breastfeed generally begin to ovulate again within a month or so after childbirth. Although it is not a completely reliable method of contraception, exclusively breastfeeding an infant may delay ovulation after childbirth, thereby lowering the likelihood of fertilization. When births are widely spaced, not only do fewer total births occur, but the mother also has a longer period in which to physically recover from pregnancy, and the infant receives feeding priority for a longer period of time. One potential contraindication for breastfeeding, however, is the **human immunodeficiency virus (HIV)**. When mothers are infected with HIV, the risk of transferring the virus through human milk is about 10%. Depending on the circumstances, this risk may outweigh the benefits of breastfeeding.

Increasing per capita income, improving the standard of living, and improving education, especially for women in developing nations, can contribute to successful long-term solutions to excessive population growth. Global fertility is projected to decline to 2.4 children birthed per woman by 2030.[18] Yet the major concern remains whether there are enough resources worldwide to raise per capita income and provide enough education to slow total global population growth.

## WAR AND POLITICAL/CIVIL UNREST

Global military spending hovers around $1.8 trillion.[19] Five major contributors to military expenditures are the United States, China, Saudi Arabia, India, and France; they account for 60% of military spending worldwide. Aside from the economic impact of military spending, civil disruptions and wars continue to set back the progress of addressing poverty and contribute to massive undernutrition. Globally, 1 in every 108 humans is now either a refugee, internally displaced, or seeking political or religious asylum. The UN Refugee Agency reports that worldwide over 70 million people have been forcibly displaced; over 50% of these refugees are children.[20] As war rages on, health, education, and public services continue to decline for affected populations. All the while, poverty and malnutrition increase.

Sadly, food has become a weapon in many wars. At the local level, fields are often mined and water wells intentionally contaminated. Even when food is readily available and accessible, political divisions may impede its distribution to the point that undernutrition will plague countries for years. Especially during emergencies, programs designed to aid the hungry have been undermined by unstable administrations, corruption, and political influence. During this chaos, relief agencies may be caught between warring factions and those attempting to alleviate suffering.

To assess hunger, the International Food Policy Research Institute annually calculates the **Global Hunger Index (GHI)**, a tool designed to measure and track hunger worldwide. The GHI uses a multidimensional approach to assess hunger that includes population undernourishment, child wasting, child stunting, and child mortality.[21]

During the 1960s and 1970s, the problem of undernutrition in developing countries was perceived as a technical one: how to produce enough food for the growing world population. The problem is now seen as largely political: how to achieve cooperation among and within nations, so that gains in food production and infrastructure are not weapons of war. The best answer may lie in a combination of approaches: finding technical solutions to address chronic hunger and poverty, and resolving political crises that push developing nations into a state of nutrition despair and chaos.

## AGRICULTURE AND THE RAPID DEPLETION OF NATURAL RESOURCES

As we continue to deplete the earth's resources, population control grows increasingly critical. Agriculture production is approaching its limits in many areas worldwide. Environmentally unsustainable farming methods have been undermining food production, especially in developing countries. In addition, climate change is increasingly linked to hunger and poverty. Climate change patterns, characterized by drought, flooding, and severe storms, require significant alterations in farming practices.

The **green revolution** was a phenomenon that began in the 1960s when crop yields rose dramatically in some countries, such as the Philippines, India, and Mexico. The increased use of fertilizers and irrigation, and the development of superior crops through careful plant breeding, made this boost in agricultural production possible. The movement toward practices in **sustainable agriculture** and the development of new crops through food **biotechnology** have begun to improve crop yields on the shrinking amount of farmable land. Crops that are pest or chemical resistant or biofortified have been produced as a result of novel biotechnology. Controversy still surrounds the use of this technology, and more research is warranted to assess the effects of these crops on human health and the environment.

The world's fertile farmland is increasingly threatened by increased land and soil degradation. Other areas of the world remain uncultivated or ungrazed and cannot sustain farming because they are too rocky, steep, infertile, dry, wet, or inaccessible. Agriculture is the largest user of freshwater resources. China produces approximately 24% of the world's grain and feeds about 20% of the world's population with less than 10% of the world's arable land. Yet China is also plagued with a growing scarcity of fresh water. In the future, billions of people are expected to face ongoing water shortages.[22]

In recent years, the amount of fish caught worldwide has leveled off as fish consumption has increased to about 17% of animal consumption globally. Fish, once considered the poor person's protein, are quickly becoming depleted in our oceans. Fish farming simply cannot compensate for the degree of reduction in wild fish populations. Coastal fisheries are overexploited, and in the past 30 years, stocks have declined by almost 65%.[23] Yet 400 million people in Southeast Asia and Africa must depend on fish for their protein and nutritional needs. The impact of globalization can affect the dietary patterns and subsequent health of locals. In Micronesia, obesity rates increased as local fish exports rose. This is an example of dietary patterns becoming altered in response to global trade resulting in an influx of U.S. commodities that include many highly processed and energy-dense foods.

If the world population continues to expand at its current rate, we face the potential threats of serious famine, disease, and death. A profound change is needed if we are to nourish an estimated 9.7 billion inhabitants of planet Earth by 2050. The FAO *State of Food Insecurity in the World* report calls for global sustainability support and practices. Food systems should sustain the environment while providing adequate, healthy food. If food production is to keep up with the expanding population, immediate action is needed to protect the earth's already deteriorated environment from further destruction.[1]

**green revolution** Refers to increases in crop yields that accompanied the introduction of new agricultural technologies in less-developed countries. The key technologies were high-yielding, disease-resistant strains of rice, wheat, and corn; greater use of fertilizer and water; and improved cultivation practices.

**sustainable agriculture** Agricultural system that provides a secure living for farm families; maintains the natural environment and resources; supports the rural community; and offers respect and fair treatment to all involved, from farm workers to consumers to the animals raised for food.

**biotechnology** A collection of processes that involves the use of biological systems for altering and, ideally, improving the characteristics of plants, animals, and other forms of life.

(a)

(b)

▲ Rich agricultural resources in North America, as seen in these wheat fields (a), contrast with gardens being tended by a farmer on a floating island in Myanmar (b).

(a) Aleksandar Dickov/Shutterstock; (b) Getty Images

**Sustainable Development Goals for Clean Water and Sanitation**

**Worldwide, over 2 billion individuals lack access to safe water. Of those...**

**785 million do not have access to basic water services**

**144 million obtain water from streams or lakes**

**263 million travel over 30 minutes per trip to obtain water**

**FIGURE 13-8** ▲ Water quality and safety are paramount to optimal human development and well-being. Access to safe water is one of the most effective ways to promote health and reduce poverty. Source: World Health Organization (WHO).

## INADEQUATE SHELTER AND SANITATION

When people die from undernutrition, other factors, such as inadequate shelter and sanitation, almost always contribute. No one can survive without clean water, nor can we stay healthy for any length of time without proper sanitation and shelter. Many more people suffer from the effects of poor sanitation and an unreliable water supply in developing countries than are affected by war and political/civil unrest. Without *wa*ter, sanitation, and *h*ygiene (*WASH*), sustainable development remains impossible.

Fortunately, organizations such as UNICEF are working in more than a hundred countries across the globe to improve water supplies and sanitation facilities in schools and communities and to promote safe hygiene practices. The UNICEF WASH programs are designed to support the Sustainable Development Goal for clean water and sanitation, which aims to achieve universal and equitable access to safe and affordable drinking water for all by 2030. This goal can't come soon enough as about 2.2 billion people still lack access to safe water (Fig. 13-8).[24]

The tremendous movement of individuals to urban settings has caused a population redistribution that has challenged the capacity for shelter and sanitation. In 1950, 30% of the world's population resided in urban areas, but by 2050, over 65% of the world's population is projected to be urban. It is expected that in the next 30 years, most urban population growth will be in cities of developing countries, reaching almost 5.2 billion in 2050. People go to the cities to find employment and resources the countryside can no longer provide. In developing countries, the poor make up most of the urban population, and their needs for housing and community services often exceed available governmental resources. About 25% of current urban dwellers live in slum-like conditions worldwide.[3]

Most poverty-stricken urban residents live in overcrowded, self-made shelters, which lack a safe and adequate water supply and are only partially served by public utilities. The WHO/UNICEF Joint Monitoring Program for Water Supply and Sanitation has estimated that 4.4 billion people lack safely managed sanitation, and 2.1 billion lack safe and readily available water at home.[25] The shantytowns and ghettos of the developing world are often worse than the rural areas left behind. The urban poor also need resources to purchase food, so they often subsist on food even more meager than the

▲ Women in developing countries are forced to spend large parts of their day fetching water. 68/Ocean/Corbis

homegrown rural fare. Making matters worse, haphazard shelters often lack facilities to protect food from spoilage or damage by insects and rodents. This inability to protect food supplies in some developing countries leads to the loss of as much as 40% of all perishable foods.[1]

The shift from rural to urban life takes its greatest toll on infants and children. WHO/UNICEF reports that inadequate access to safe water and sanitation services, combined with poor hygiene practices, kills and sickens almost 300,000 under age 5 annually.[26] Contributing to this issue is the fact that infants are often weaned early from breast milk to infant formula, partly because the mother must find employment. Mothers may also be influenced by advertisements depicting images of sophisticated, formula-feeding women. Unfortunately, because infant formulas are expensive, food-insecure parents may try to conserve the formula by either overdiluting the mixture or reducing the feeding volume. In addition, the water supply may not be safe, so the prepared formula is also likely to be contaminated with pathogens or contaminants. Human milk, in contrast, is much more economical, hygienic, readily available, and nutritious. It also provides infants with immunity to some ailments. In most situations, breastfeeding should be promoted as the safest and most nutritious choice for infant feeding.

Poor sanitation also creates a critical public health problem and, along with under-nutrition, particularly raises the risk of infection. Inadequate sanitation is another example of the inferior infrastructure in the developing world. Human urine and feces are two of the most dangerous substances encountered in overcrowded urban areas. In addition, rotting garbage and associated insect and rodent infestations are potent sources of disease-causing organisms commonly seen in urban areas of the developing world. The inability to dispose of the massive numbers of deceased people (and animals) resulting from disease and wars causes additional sanitation problems. In some developing countries, diarrheal diseases account for almost 10% of all deaths in children younger than 5 years of age.[27] Additional repercussions of poor sanitation are that children are denied access to education because their schools lack private and decent sanitation facilities.

## THE IMPACT OF AIDS WORLDWIDE

Nutrition security is impacted greatly in developing countries by the high prevalence of HIV infection and **acquired immunodeficiency syndrome (AIDS)**. HIV/AIDS impairs absorption of nutrients, increases nutrient requirements, and decreases the capacity to work. About 38 million people around the world are infected with HIV/AIDS. Cases in sub-Saharan Africa account for almost 65% of the global total of new HIV/AIDS infections.[28]

An individual can be infected with HIV through contact with bodily fluids including blood, semen, vaginal secretions, and human milk. Thus, the virus can be transmitted through sexual contact, through blood-to-blood contact, and from mother to child during pregnancy, delivery, or breastfeeding. The virus has a very limited ability to exist outside the body. Once infected with HIV, the individual is said to be *HIV positive*. If untreated, the viral disease progresses over the next few years, and the individual develops opportunistic infections with symptoms such as diarrhea, lung disease, weight loss, and a form of cancer. Once the individual has developed these symptoms, that person is said to have AIDS. Without treatment, an individual will likely die from AIDS within 4 to 5 years.

Ensuring healthy lives and promoting the well-being for all is one of the Sustainable Development Goals. This specific goal proposes to end the epidemics of AIDS, tuberculosis, malaria, and other tropical diseases and to combat hepatitis, water-borne diseases, and other communicable diseases by 2030. Several countries have noted modest declines in the number of children newly infected with HIV, but the Middle East and North Africa have seen no reduction in the number

▼ Blood loss caused by intestinal and bloodborne parasite infections is a common cause of anemia among poor populations, especially when people do not wear shoes. Parasites, such as hookworms, can easily penetrate the soles of the feet and legs and enter the bloodstream. Although hookworm disease has been largely eradicated in the United States and other industrialized nations through improved sanitation, it continues to plague more than one-eighth of the world's population, mostly in tropical regions.

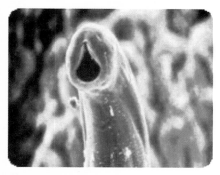

Source: CDC

**acquired immunodeficiency syndrome (AIDS)** Late-stage HIV when the body's immune system is badly damaged because of the virus.

▲ A particularly sad consequence of AIDS in Africa is the number of AIDS orphans—children whose parents have both died of AIDS. These orphans then become responsible for the care of their siblings and other family members. RachelKolokoffHopper/Shutterstock

of newly infected children. Reductions in the number of adults who are newly infected with HIV, as well as the direct use of preventive antiretroviral medications, appear to be the reason for the reductions in the number of children acquiring HIV/AIDS.

**Toward Universal Treatment.** Although no vaccine is available to prevent AIDS, the latest antiretroviral therapy (ART) can significantly slow the progression of the disease. Providing AIDS drugs to women who are pregnant is also an effective preventive measure. The goal of therapy for women who are pregnant is to maximize viral suppression to reduce the transmission of HIV to the fetus and newborn.

Although ART is effective, it is very costly. Drug companies and governments have worked successfully to lower the cost of AIDS drugs for developing nations. Most countries that aspire to expand access to treatment set a goal of providing ART to around 80% of those in need. Although life-saving ART is reaching more individuals, global treatment is still reaching only 62% of those who need it, meaning millions of people are still waiting to be treated. Progress to the goal of ensuring access to HIV treatment has been slower for children than for adults. Sadly, just 54% of HIV-infected children had access to antiretroviral therapy.[29]

**Nutrition and HIV/AIDS.** Although adequate nutrition cannot prevent or cure HIV infection or AIDS, nutritional status can affect the progression of the disease. A focus on eliminating or reducing malnutrition can significantly slow disease progression and severity and can improve longevity. A dietary pattern adequate in energy, protein, and micronutrients can lessen the impact of infections associated with AIDS. Low levels of vitamins A and E can contribute to a more rapid onset of symptoms, including body wasting and fever. Maintenance of an optimal nutritional status should be an integral focus of the treatment for AIDS. Daily use of a balanced vitamin and mineral supplement has also been shown to slow health declines in people with HIV/AIDS. Nutritional counseling by a registered dietitian nutritionist (RDN) is paramount in assessing nutritional status, providing personalized education and recommendations, and monitoring outcomes. The main goals of nutritional treatment are to achieve and maintain a healthy body weight, body composition, and optimal nutritional status; minimize nutrition-related side effects and complications; improve quality of life; and expand access to nutrition services.[30]

✓ **CONCEPT CHECK 13.3**

1. Why does malnutrition continue despite an adequate food supply in many areas?
2. How have war and declines in natural resources contributed to malnutrition in developing countries?
3. What effects have overpopulation and increased urbanization had on nutrition and disease risk?
4. What role does nutrition play in combating AIDS?

# 13.4 Reducing Malnutrition in the Developing World

Reducing malnutrition in the developing world is quite a complex goal and takes considerable time to accomplish. It has been a common practice for affluent nations to supply famine-stricken areas with direct food aid; however, this is not a long-term

# THE GLOBAL GOALS
## For Sustainable Development

**FIGURE 13-9** ▲ Achieving these targets will contribute to achievement of the United Nations Sustainable Development Goals (SDGs). Goals 2 and 3 aim to end hunger, achieve food security, improve nutrition, address all forms of malnutrition for all age groups, and ensure health and well-being for all at every stage of life. In total, 12 of the 17 SDGs have indicators that are highly relevant to nutrition.

solution. Although it reduces the number of deaths from famine, it can also reduce incentives for local production by driving down food prices. In addition, the affected countries may have little or no means of transporting the food to those who need it most, and the donated foods may not be culturally acceptable or nutrient dense. In the short run or during a crisis, there is no choice: aid must be provided because people are starving.[4] Still, improving the infrastructure for poor people, especially rural people, must be the long-term focus.

With the adoption of the Sustainable Development Goals (SDGs) by all United Nations member states, world leaders committed to address the many dimensions of extreme poverty and create a better life for those in need by 2030 (Fig. 13-9). Pertinent hunger-related goals include reducing by half the proportion of men, women, and children of all ages living in poverty; eradicating extreme poverty for all, currently measured as people living on less than $1.90 a day; and ensuring that the poor and the vulnerable have equal rights to economic resources, as well as access to basic services, ownership and control over land and other forms of property, inheritance, natural resources, appropriate new technology, and financial services.

Twelve of the seventeen SDGs contain nutrition-related metrics. Without substantial and sustained investments in adequate nutrition, the SDGs will certainly fail. As the SDGs emphasize, malnutrition results not only from a lack of nutritious and safe food but also from an intertwined and complete series of factors including health care, sociodemographics, education, water, sanitation, hygiene, food access, available resources, human empowerment, and more.

**SDG Report Card**

The 2020 SDG Report highlights progress had been made in some key areas including maternal and child health, increasing global access to electricity, and expanding representation of women in government.

Unfortunately, these gains were offset by increasing food insecurity, continued environmental deterioration, and persistent inequalities. The impact of COVID-19 caused further disruption to SDG progress, especially for our poorest and most vulnerable populations. This includes children, women, older individuals, persons suffering disabilities, migrants, and refugees.

For more details, visit https://www.un.org/sustainabledevelopment/progress-report/.

▲ Food security is fostered by communities raising and distributing locally grown food. ©Erin Koran

## DEVELOPMENT TAILORED TO LOCAL CONDITIONS

Although world food supplies have grown in recent years, an increase in malnutrition has resulted from inadequate food distribution and access. In addition, millions of farmers are losing access to resources they need to be self-reliant. There is a growing realization that unless economic opportunities can be created as part of a plan for sustainable development, rural residents who do not own land will flock to the over-crowded cities.

For the most part, the sustainable solutions lie in helping people meet their own needs and directing them to resources and employment opportunities. History has shown that the provision of credit—along with education and training, food storage facilities, and marketing support—allows rural people to actively participate in their success, which benefits their families and communities. Understanding and appreciating the local conditions, including crop rotation, utilization of varieties of plants, and crop diversification, are essential to understanding barriers to success. Agroforestry initiatives and promotion of local and culturally sensitive foods are needed. Communities that have significant land and water limitations may require the intervention and assistance of local authorities.

One U.S. program, the Peace Corps, has helped to improve conditions in developing nations for over 50 years by providing education, distributing food and medical supplies, and building structures for local use. The aim of the Peace Corps is to help create independent, self-sustaining economies around the world. Becoming a Peace Corps volunteer is a meaningful way to make a difference around the world. Learn more about this program at www.peacecorps.gov.

Suitable technologies for processing, preserving, marketing, and distributing nutritious local staples also should be fostered, so that small farmers can flourish. Education and training on how to use whole foods to create healthful eating plans adds further benefit. Supplementing indigenous foods with nutrients that are in short supply, such as iron, various B vitamins, zinc, and iodine, also deserves consideration. One such program involves enriching rice with vitamin A in various parts of the world. Later in this section, we examine the role of biotechnology in improving nutrient quality and other characteristics of plants and animals, another positive step in reducing malnutrition. In addition, advances in water purification must be employed.

Another SDG aims to ensure environmental sustainability, which may include promoting extensive land ownership and thus increasing the availability of food. If food resources are concentrated among a minority of people, as often happens with unequal land ownership, food is unlikely to be equally distributed unless efficient transportation systems are in place. Small-scale industrial development is another way to create meaningful employment and purchasing power for vast numbers of the rural poor.

Raising the economic status of impoverished people by employing them is as important as expanding the food supply. If an increase in food supply is achieved without an accompanying rise in employment, there may be no long-term change in the number of undernourished people. Although food prices may fall with increased mechanization, use of fertilizers, and other modern farming technologies, these advances can also displace people from jobs, a result that may harm, rather than help, the population.

## IMPROVING EQUALITY

▲ This school teacher and her students are in a classroom in a village in Senegal, Africa. Eliminating disparity is a goal of the Sustainable Development Goals. See https://www.nutritionintl.org for more on this issue. Andia/UIG/Getty Images

Women living in poverty are a special concern. In addition to working longer hours than men, they grow most of the food for family consumption and account for 39% of the labor force in the informal sector of the economy and an increasing proportion in the formal sector.[3] Of the 3 billion people in the world living on less than $2 a day, 70% are women. Clearly, economic opportunities for women and education regarding family planning must be augmented. Empowerment of women is critical to improve the level of food and nutrition security, increase the production and distribution of food and other agricultural products, and enhance living conditions in general (see *Newsworthy Nutrition*).

Goal 5 of the SDGs aims to achieve equality and empower all women and girls. Increasing women's access to education, information and communication technologies, economic resources, and governance reduces poverty, promotes development, achieves equality, protects women's human rights, and eliminates violence against women. Although the proportion of women who are paid to work in jobs outside of agriculture is increasing, women continue to experience significant gaps in terms of poverty, labor market and wages, and participation in private and public decision making. There is still plenty of room for improvement.

## SUSTAINABLE AGRICULTURE

Over the years, changes in agricultural practices have positively impacted the availability of food around the world. Along with the positive effects on farming, however, negative results have occurred. Most significant among these are the depletion of topsoil, contamination of groundwater, decline of family farms, neglect of living and working conditions for farm laborers, increasing costs of production, and lack of integration of economic and social conditions in rural communities.

The concept of **sustainable development** has been embraced as economic growth that will reduce poverty while, at the same time, it will protect the environment and preserve natural capital. The UN cites economic development, social development, and environmental protection as the *reinforcing pillars* of sustainable development. The role of the agriculture industry in promoting practices that contribute to environmental and social issues has been at the center of discussions of **sustainable intensification** in agriculture

**sustainable development** Economic growth that will simultaneously reduce poverty, protect the environment, and preserve natural capital.

**sustainable intensification** Agricultural practices that consider whole landscapes, territories, and ecosystems to optimize resource utilization and management.

## FARM to FORK   Pineapples

9comeback/Shutterstock

Luscious tropical fruits, like pineapples, not only provide key nutrients but also an economic advantage for developing countries. After bananas, pineapples are the most consumed tropical fruit in the U.S. Typically imported from Costa Rica and Hawaii, pineapples have a long history of being bred for certain traits, resulting in the sweet and low-acidic varieties available today. James Dole, the most famous pineapple industrialist, moved to Hawaii in 1899 and started the first pineapple plantation.

### Grow

- To plant a pineapple at home, slice off the pineapple crown (top) from a fresh pineapple. Remove all fresh fruit to avoid rot. Make thin slices in the stalk, displaying a ring of brown root dots. Remove the lower leaves on the stalk to expose approximately 1" of bare stalk. Allow the stalk to dry well for a few days. Plant the pineapple 1" deep in a pot with fast-draining potting mix. Water very lightly and place in a bright, sunny location. The pineapple should root in 1 to 3 months. Repot as needed to encourage growth.

### Shop

- When shopping for pineapple, select the golden sweet varieties. Although sweet pineapples have 25% more sugar, they have 135% more beta-carotene and 350% more vitamin C than traditional Cayenne pineapples. To select the freshest pineapple, look for dark green crown leaves that are difficult to pluck and free of browning or fading.

### Store

- Pineapples are harvested ripe and do not continue to ripen, so they should be kept refrigerated and eaten within 4 days. When purchasing tropical fruits, look for *fair trade* fruits, signifying that the fruit was grown under environmentally friendly conditions with attention to small-scale producers.

### Prep

- Both pineapple juice and flesh are used in cuisines around the globe. Pineapple slices or chunks are used in desserts, fruit salads, and savory dishes such as pizza toppings or grilled with a hamburger. Crushed pineapple is often added to yogurt, jams, and ice cream. Pineapple juice is often consumed straight, added to smoothies, and as a main ingredient in piña coladas. Pineapple juice can also serve as a meat marinade and tenderizer.

Source: Robinson J: Tropical fruits: Make the most of eating globally, in *Eating on the Wild Side*. New York: Little, Brown and Company, 2013.

Pixtal/SuperStock

# Newsworthy Nutrition

## Maternal depression and child severe acute malnutrition: A case-control study

**INTRODUCTION:** Maternal depression is the leading cause of disability and adversely affects the health of mothers and growth of their infants. **OBJECTIVE:** The aim of this *case-control study* was to assess maternal depression and its impact on malnutrition in Kenyan children ages 6 to 60 months. **METHODS:** A matched case-control study design was conducted with children admitted to a Kenya hospital with severe acute malnutrition. The controls were normal-weight children who were age, acute diagnosis, and sex matched. Mothers of both groups were assessed for depression using the PHQ-9 questionnaire. Logistic regression was used to compare the odds of maternal depression in cases and controls, controlling for other factors associated with youth malnutrition. **RESULTS:** The prevalence of moderate to severe maternal depression of malnourished children was high (64%) compared to mothers of normal weight children (5%). In multivariate analyses, the odds of maternal depression were higher in cases than in controls (adjusted OR = 53.5, 95% CI = 8.5–338.3), as were the odds of having very low income (adjusted OR = 77.6 95% CI = 5.8–1033.2). **CONCLUSION:** Kenyan mothers whose children were diagnosed with malnutrition carry a significant mental health burden. The authors recommend implementation of interventions that offer social support, mental health counseling, strategies to address food insecurity, and economic wellness training for mothers of malnourished children.

Source: Haithar S and others: Maternal depression and child severe acute malnutrition: A case-control study from Kenya. *BMC Pediatrics* 18:289, 2018. DOI: https://doi.org/10.1186/s12887-018-1261-1.

*magnificent*

Like humans, soil also has a vast and dynamic microbiome. This vibrant living community has a profound effect on the life and death of plants. The soil microbiome is a rich reservoir of biodiversity that has enormous functional potential in health care, agriculture, food production, and climate regulation.

Source: Plant Soil Microbial Community Consortium, NC State.

*microbiome*

**regenerative agriculture** A system of farming and grazing practices that aims to reverse climate change by rebuilding soil organic matter and restoring degraded soil biodiversity, resulting in both carbon drawdown and improving the water cycle.

**hydroponics** This type of agriculture involves growing plants in a nutrient solution root medium in a controlled, soilless environment.

(Fig. 13-10).[31] As a result, interest in alternative farming practices has grown and reinforced a movement toward sustainable agriculture. Sustainable agriculture flows directly from the SDGs to ensure sustainable consumption and production patterns globally. Meeting these targets depends on integration of several goals, including environmental health, economic profitability, and social and economic equity. Sustainable agriculture addresses many environmental and social concerns and offers innovative and economically viable opportunities for many in the food system, including growers, laborers, consumers, and policy makers. It is gaining support and acceptance from conventional farmers in many countries.

Sustainable agriculture involves maintaining or enhancing the land and natural resources for use long into the future. It also requires consideration for human resources, including working and living conditions of laborers, the needs of rural communities, and consumer health and safety. The potential of sustainability is best understood when the consequences of farming practices on both human communities and the environment are considered. Many farmers around the globe are transitioning to sustainable agriculture by taking small, realistic steps based on their personal goals and family economics. Reaching the goal of worldwide sustainable agriculture requires participation by all stakeholders, including farmers, laborers, retailers, consumers, researchers, and policy makers.

**Regenerative agriculture** is a system of farming that focuses on restoring degraded soils, increasing biodiversity, improving watersheds, and enhancing the ecosystem. The principles and practices strive to capture carbon in the soil and above ground to counter atmospheric carbon accumulation. Critics of regenerative agriculture claim that it is a watered-down version of organic farming promoted by industry to appease environmentally conscious consumers. Ultimately, more rigorous research is needed to fully assess the long-term impact of various farming practices on the health of plants and the planet.

As depicted on the cover, a novel area of agricultural technology that is gaining momentum is **hydroponics**. Implemented correctly, this soilless method of plant production can grow plants 25% faster and produce up to 30% greater yield as compared

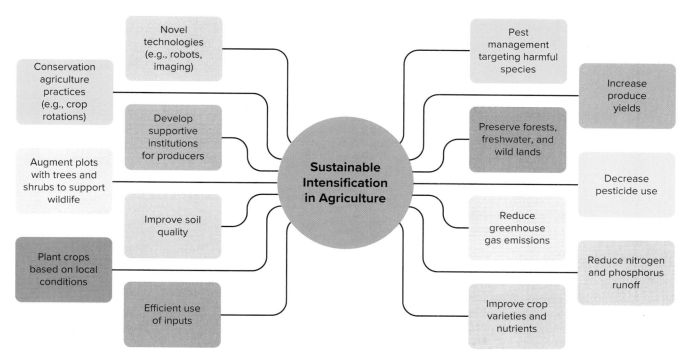

**FIGURE 13-10** ▲ Goals of sustainable intensification of agriculture needed to efficiently produce enough food for the world's population using environmentally friendly techniques.
Source: Adapted from FAO

to traditional soil-grown plants. These significant benefits are possible through intense control of the nutrient solution, water, and pH levels. In addition, hydroponics is environmentally friendly and reduces the waste and pollution often associated with soil runoff. Yet despite these positives, there remain limitations to hydroponics as well. Initially, system start-ups require extensive training, time, space, and money. Starting small is a wise choice for novice growers.

## ROLE OF THE NEW BIOTECHNOLOGY IN THE DEVELOPING WORLD

Each year, up to 17 million farmers in 26 countries have planted over 190 million hectares of biotech crops. The top four GMOs include corn, soybean, canola, and cotton. The 75 million U.S. biotech hectares include maize, soybeans, cotton, canola, sugar beets, alfalfa, papaya, squash, potatoes, and apples. These crops help to alleviate poverty by improving the economic situation of over 16 million small farmers.[32]

Whether applications of genetic engineering will help to significantly reduce malnutrition in the developing world remains to be seen. Unless price cuts accompany the increased production, only landowners and suppliers of biotechnology will enjoy the benefits. Small farmers may benefit if they can afford to purchase the genetically modified seeds. This point deserves emphasis: the person who cannot afford to buy enough food today will still face that same predicament in the future. As with most innovations, the more successful farmers—often those with larger farms—will adopt new biotechnology first. Because of this, the present trend toward fewer and larger farms will continue in the developing world, a movement that undermines the solution to one of the most pressing undernutrition issues. Furthermore, biotechnology does not promise dramatic increases in the production of most grains and cassava, the primary food resources in developing parts of the world.

With the introduction of more drought- and pest-resistant crops, as well as self-fertilizing crops, agricultural biotechnology may help to reduce world hunger. Perhaps the most promising potential of genetically modified foods lies within the realm

**Biofortification** is one way to provide nutrients to deficient populations. *Fortification* refers to adding nutrients to foods during processing. In contrast, biofortification uses plant breeding or biotechnology to grow plants with increased levels of certain nutrients. Biofortified crops, such as maize, sweet potatoes, and rice, can be cultivated to provide rich sources of vitamin A in areas where vitamin A deficiencies are common. **What advantages does biofortification offer compared to other public health efforts to correct undernutrition?** Shutterstock

## ASK THE RDN  Wasted Food

*Dear RDN:* I hear food waste is a problem, but how bad is it really?

Food waste—or wasted food—is a seemingly innocent act that is collectively creating one of the gravest issues of the twenty-first century. Producing food to feed our growing global population requires immense amounts of resources, including land, petroleum, agrochemical inputs, fresh water, and labor. These inputs make our food system one of the leading drivers of climate change, biodiversity loss, and deforestation.

Today, nearly 40% of the global food supply ends up in the landfill. When wasted food enters the landfill, it creates methane gas, which is 25 times more powerful than carbon dioxide at raising the earth's temperature. In fact, if food waste were a country, it would be the third-largest greenhouse gas–emitting country in the world, behind the U.S. and China.

Depending on the country, wasted food occurs in different paths along the food supply chain (*see Food Recovery Hierarchy*). In developing countries, wasted food often occurs closer to the beginning of food production. This is related to a lack of adequate storage facilities, transportation routes, or refrigeration. In developed regions like North America, wasted food typically occurs at the end of the food supply chain. This means most food in America is wasted in retail settings like grocery stores or restaurants, or with us—the consumers.

Reducing wasted food not only cuts down on environmental stress but also saves us money. The following are six simple ways to start reducing wasted food today:

1. **Plan ahead.** Shopping with meals in mind can help reduce impulse purchases, leading to a more efficient use of the food we purchase.
2. **Understand dates posted on product labels.** Contrary to popular belief, few foods contain expiration dates (infant foods are the exception). Most foods contain *best by*, *sell by*, or *best if used by* dates. These dates are indicative of quality, not safety. Most often, food is perfectly safe to eat after these dates. Check www.stilltasty.com for a free shelf-life guide for nearly every food product.
3. **Store produce properly.** Some fruits and vegetables give off natural gases, which expedite the natural ripening process of neighboring fruits. Refer to the *Farm to Fork* features in this book and consult a fruit and vegetable storing guide to determine the best possible place to store your produce.[33] See https://www.fruitsandveggiesmorematters.org/fruit-and-vegetable-storage-101 for more information.
4. **Learn to love your freezer.** Frozen foods are a great option for extending the shelf life of food. Frozen bananas or berries make great additions to smoothies. Additionally, purchasing frozen foods can also be cost saving, as they are often less expensive and just as nutritious as their fresh counterparts.
5. **Donate it.** It is hard to prevent all food from entering the landfill. If faced with large amounts of leftover food, such as uneaten and safe leftovers from a catered event, these foods can often be donated to local food pantries for hungry neighbors. Check with www.ampleharvest.org to locate a food pantry near you.
6. **Compost.** While increasing food donations and reducing the amount of wasted food are preferred, composting is a great option when compared to throwing food in the landfill. Twenty cities around the world, including San Francisco and Seattle, have integrated composting services into their weekly municipal pickups, similar to trash or recycling. If your city does not have this option, home composting is easier than ever, as the Internet is home to numerous how-to guides. Composting food helps prevent the methane gas emissions that occur when it enters landfills. Compost is also a nutrient-rich soil improvement, which helps improve the productivity of gardens.

Cheers to a healthier and more sustainable food system,

*Chris Vogliano, MS, RDN, PhD Candidate*

Globally recognized public health dietitian and expert in food security

### Food Recovery Hierarchy

Most Preferred

**Source Reduction**
Reduce the volume of surplus food generated

**Feed Hungry People**
Donate extra food to food banks, soup kitchens, and shelters

**Feed Animals**
Divert food scraps to animal feed

**Industrial Uses**
Provide waste oils for rendering and fuel conversion and food scraps for digestion to recover energy

**Composting**
Create a nutrient-rich soil amendment

**Landfill/Incineration**
Last resort to disposal

Least Preferred

EPA Food Recovery Hierarchy. https://www.epa.gov/sustainable-management-food/food-recovery-hierarchy

©Melissa Olson

of plant breeding for micronutrients. If they have access to farming resources to increase the micronutrient composition of crops, developing countries will have a tool to treat and prevent nutrient deficiencies. In addition, greater yields for indigenous plants, such as tomatoes that tolerate high soil salinity, are another hopeful outcome. Biotechnology will likely be a useful tool against the complex scourge of world undernutrition. Improved crops produced by this technology, together with political and other efforts, can contribute to success in the battle against worldwide malnutrition.

Another exciting bioengineering technique includes *biofortification*. This process can improve the nutritional quality of crops while the plants are growing rather than through conventional fortification during food processing. Examples of successful biofortification of crops with nutrients include iron (rice, beans, sweet potatoes, cassava, and legumes), zinc (wheat, rice, beans, sweet potatoes, and maize), provitamin A carotenoid (sweet potatoes, maize, and cassava), and amino acid and protein (sorghum and cassava). UNICEF estimates that biofortification could impact over 5 million children under age 5 suffering from iron deficiency.[34]

Any technology comes with risks. The public has long been opposed to "unnatural" products or processes perceived as harmful to the environment. Concern exists regarding potential cumulative effects of biotechnology over time. With these concerns in mind, the FDA carefully examines all products developed using this technology and will enforce labeling of potential allergens that may be newly present in food altered by biotechnology.

## CONCLUDING THOUGHTS

The economic loss from malnutrition is staggering, and the amount of human pain and suffering it causes is incalculable. With all the international relief efforts and assistance from governments and private organizations combined, the battle is far from over. Read more about the progress toward the Sustainable Development Goals and the ongoing challenges of hunger in several regions of the world, even as poverty has decreased, at https://sustainabledevelopment.un.org.[3]

Ultimately, the rapid depletion of world resources, massive debt incurred by poorer countries, threat of danger to more prosperous countries nearby, and toll of war and famine all affect the population's physical and economic well-being. Planet Earth has enough food and the technical expertise to alleviate hunger. With world leaders affirming the UN Sustainable Development Goals and working to achieve its targets by 2030, we have a promising coordinated political effort that is making progress toward lasting reductions in hunger and optimal health for all.

Stan and his friends have read about various relief efforts to help undernourished people in developing countries, especially the emergency food aid programs for famine-ravaged areas. Yet many of these efforts appear temporary, and they wonder what long-range approaches might help alleviate the problem of undernutrition. **What suggestions would you give them about possible long-term solutions for undernutrition in developing countries?** ZouZou/Shutterstock

### ✓ CONCEPT CHECK 13.4

1. Why is development tailored to local conditions important in combating global hunger?
2. What is the definition of *sustainable agriculture?*
3. Describe one way biotechnology could help to relieve global malnutrition.

## Malnutrition at Critical Life Stages

Andrew Holbrooke/Corbis via Getty Images

## Critical Life Stages When Malnutrition Is Particularly Devastating

Prolonged malnutrition is detrimental to many aspects of human health, resulting in increased maternal, infant, and child mortality; loss of parents (especially linked to AIDS); exploitation of women; reduced work capacity; reduced intellectual and social development; and overall human suffering, especially during a famine. It is particularly damaging during periods of growth and old age. A conceptual framework of malnutrition lists the immediate causes as inadequate dietary intake and unsatisfactory health. These underlying causes relate to families and include lack of access to food, inadequate care for women and children, and inadequate health services. Effective strategies to combat malnutrition should embrace a life cycle approach. Public health programs should be complementary and comprehensive across vulnerable periods of the whole reproductive cycle.

### PREGNANCY

Undernutrition poses the greatest health risk during pregnancy. Every day, approximately 810 women die from preventable causes, with 94% occurring in low- and middle-income countries.[16] A pregnant woman needs extra nutrients to meet both her own needs and those of her developing offspring. Nourishing the fetus may deplete maternal stores of nutrients. Maternal iron-deficiency anemia is one possible consequence of malnutrition during pregnancy.

Birth rates in Africa are the highest in the world. In Niger, for example, a woman gives birth to an average of nearly seven children. Coupled with chronic malnutrition, a woman's risk of dying increases with every pregnancy and birth. This results in a strong connection between a high fertility rate and high maternal mortality rates. Although most of us consider pregnancy and childbirth as natural parts of life, for women living in developing countries and without adequate access to health care, pregnancy and childbirth complications are among the leading causes of death. The cumulative effect of successive pregnancies does not allow the mother to recover key nutrients, such as iron and folate, lost during pregnancy and breastfeeding. Yet some progress is being made. Since 1990, the maternal mortality rate has been almost 60%. In South Asia, the maternal mortality rate has declined by 65%, and in sub-Saharan Africa it has fallen by almost 40%.[16] The new SDG is to reduce the global maternal mortality ratio to less than 70 per 100,000 live births by 2030.

### FETAL DEVELOPMENT AND INFANT STAGES

An unborn baby faces major health risks from malnutrition during gestation. To support growth and development of the brain and other body tissues, a growing fetus requires a rich supply of

▲ This Lobi tribeswoman, standing with her six children in a grassy field in Burkina Faso, Africa, is an illustration of the typical birth rate in this region. Maternal nutrient levels become depleted with successive pregnancies followed by breastfeeding. Lissa Harrison

protein, lipids, vitamins, and minerals. When these nutrient needs are not met, it is more likely that the infant will be born before 37 weeks of gestation, well before the ideal 40 weeks. Preterm birth and low birth weight (5.5 pounds or less) contribute to 60% to 80% of all neonatal deaths.[35] If the infant survives, abnormal growth and development can result. In extreme cases, low-birth-weight infants face exponentially greater risk of dying before their first birthday, primarily due to complications of reduced lung development. When low birth weight is accompanied by other physical abnormalities, medical costs may exceed hundreds of thousands of dollars.

Worldwide, the incidence of infants born with low birth weight is around 15%.[36] Much higher rates, however, occur in developing countries. Undernutrition is a major contributor to these high rates of low birth weight in developing countries. About 8% of infants born in the United States have low birth weights. These low birth weights account for 60% to 80% of all deaths of infants.[35] Pregnancy during the teenage years, when the mother's body is still growing, contributes to low birth weight worldwide.

## CHILDHOOD

Early childhood, when growth is rapid, is another period when malnutrition is extremely risky. The greatest impact of undernutrition occurs in the first 1000 days of life: from conception to a child's second birthday. Irreversible damage is the result if optimal nutrition is not available at critical times in development. The central nervous system—including the brain—is highly vulnerable because of rapid growth through early childhood. After the preschool years, brain growth and development slow dramatically. Nutritional deprivation, especially in early infancy, can lead to permanent brain impairment. Without an effective intervention, it is projected that ongoing undernutrition could leave more than 1 billion children with mental impairments by 2020.[37]

In general, children living in poverty are at the greatest risk for nutrient deficiencies and their subsequent consequences. The WHO reports that nearly 150 million suffer from *stunted growth.* Iron-deficiency anemia is also more common among low-income children. This deficiency can lead to fatigue, reduced stamina, stunted growth, impaired motor development, and learning problems. Undernutrition in childhood can also weaken immune function and increase the risk of infection when nutrients such as protein, vitamin A, and zinc are low in a dietary pattern. Clearly, malnutrition and illness have a cyclical relationship. Not only does undernutrition lead to illness, but diarrheal and other infectious diseases increase malnutrition. For this reason, millions of children in developing countries are dying from the combination of malnutrition and infection. Conversely, when missing nutrients, such as vitamin A and iron, are restored to a child's diet, improvements in health are evident.

Reducing wasting in children under 5 years of age and addressing the nutritional needs of adolescent girls are SDGs. Death rates for children under 5 years of age have already been reduced to 1 in 26 children in 2018 compared to 1 in 11 in 1990. Despite this progress, over 5 million children under 5 years of age still died in 2018.[38] Furthermore, the progress made in recent decades has been unequally distributed across regions and countries and within countries. India (18%) and Nigeria (13%) account for one-third of all under-age-5 deaths. These statistics indicate that undernutrition is one of the most important challenges to overcome because it is estimated to be an underlying cause in as many as 45% of the preventable deaths among children under age 5.[35]

## LATER YEARS

The WHO predicts a significant increase in the number of people age 65 years or older, from an estimate of 524 million in 2010 to almost 1.5 billion in 2050. Most of this increase in global aging will be in developing countries. Older adults, especially older women living alone in poverty, are at risk for undernutrition. All older adults require nutrient-dense foods in amounts dependent on their current health and degree of physical activity. Many older adults have fixed incomes and incur significant medical costs over time, so food often becomes a low-priority item. In addition, depression, social isolation, and declining physical and mental health can compound the problem of undernutrition in older adults.

▲ This senior tribeswoman in Burkina Faso, Africa, with an ivory piercing in her upper lip, reminds us that the population of people aged 65 years or older will increase exponentially in developing countries. Lissa Harrison

## ✓ CONCEPT CHECK 13.5

1. What are some underlying causes of malnutrition?
2. What two factors contribute to up to 80% of all neonatal deaths?
3. During childhood, when is the period when undernutrition is most impactful?

## CASE STUDY  Undernutrition During Childhood

Jamal traveled to the Philippines with his church group last summer. During their stay, the group helped build shelters in a village where, a few weeks before, a storm had destroyed several houses. Jamal noticed that many of the children were very short, much shorter than the children in his neighborhood in the United States. His group worked in a remote, low-elevation area where the storm and subsequent flooding had caused the most damage. On several occasions, he noticed young mothers crouched on curbs or in doorways, holding their children. These children rarely moved; they appeared pale and listless. In contrast to the children Jamal's group had met at a church in the capital city, most of the children in this village were not active and lively. One evening, a nurse from the local clinic came to speak to Jamal's group. She said that many children in this area do not get enough to eat and that malnutrition was rampant. She considered the recent storm a blessing in disguise, hoping it would spur the Philippine government to send supplies to the village, particularly food and medicines. Jamal was shocked by such a degree of suffering. He wonders why children in the Philippines can be starving to death while so many children in his hometown in the United States have more food than they need.

▲ Family members from a small village in the Philippines are coping with recent flood waters encroaching on their living space. What are some potential health issues associated with living in these conditions?
Digital Vision/Getty Images

1. Which nutrients are likely to be deficient in the diets of these children?
2. Which nutrient deficiencies contribute to *stunted* growth?
3. Which nutrient deficiencies may be reducing immune defenses?
4. Are these children likely to be consuming adequate calories? What effect does inadequate consumption of calories have on growth?
5. How might the recent flooding impact the health of the children in the village?

*Complete the Case Study. Responses to these questions can be provided by your instructor.*

## Summary (Numbers refer to numbered sections in the chapter.)

**13.1** Poverty is commonly linked to chronic or episodic undernutrition. Malnutrition includes both undernutrition and overnutrition and can occur when the food supply is either scarce or abundant. The resulting deficiency conditions and degenerative diseases contribute to poor health.

Undernutrition is the most common form of malnutrition in developing countries. It results from inadequate intake, absorption, or use of nutrients or food energy. Nutrient deficiencies impact all body systems; infectious diseases run rampant because the immune system is depressed.

Malnutrition diminishes both physical and mental capabilities. In poor countries, this is worsened by recurrent infections, unsanitary conditions, extreme weather, inadequate shelter, and exposure to pathogens and diseases.

**13.2** In North America, famine is not seen, but food insecurity and malnutrition remain concerns, due in large part to poverty. Single mothers and their children are more likely to live in poverty. Soup kitchens, SNAP benefits, the school lunch and breakfast programs, WIC, and other programs focus on improving the

nutritional health of individuals at risk. When adequately funded, these programs have proved effective in reducing malnutrition and improving food security.

**13.3** Multiple factors contribute to the problem of undernutrition in the developing world. In densely populated countries, food resources, as well as the means for distributing food, may be inadequate. Environmentally unsustainable farming methods hamper future efforts to grow food. Limited water availability hinders food production. Natural disasters, urbanization, war, and disease all contribute to the major problem of undernutrition.

**13.4** Proposed solutions to world undernutrition must consider multiple interacting factors, many thoroughly embedded in cultural traditions. Family planning efforts, for example, may not succeed until life expectancy increases. Through education and training, efforts should be made to upgrade farming methods, improve crops, limit pregnancies, encourage breastfeeding, and improve sanitation and hygiene.

Direct food aid is only a short-term solution. In what may appear to be a step backward, many experts recommend more subsistence-level sustainable farming. Small-scale industrial development is another way to create meaningful employment and purchasing power for vast numbers of the rural poor. Various biotechnology applications may also prove beneficial. The UN's 17 Sustainable Development Goals focus on global solutions.

**13.5** The greatest risk of malnutrition occurs during critical periods of growth and development: gestation, infancy, and childhood. Low birth weight is a leading cause of infant deaths worldwide. Many developmental problems are caused by nutritional deprivation during critical periods of brain growth. People in their senior years are also at great risk.

## Check Your Knowledge (Answers are available at the end of this question set.)

1. There are an estimated _____ chronically undernourished people in the world.
   a. 14 million
   b. 1 billion
   c. 3 billion
   d. 6 billion

2. The number-one killer of children in developing countries is
   a. xerophthalmia.
   b. iron-deficiency anemia.
   c. iodine deficiency.
   d. diarrhea.

3. The human organism is particularly susceptible to the effects of undernutrition during
   a. pregnancy.
   b. infancy.
   c. childhood.
   d. all of these stages.

4. Which is a barrier to solving undernutrition in the developing world?
   a. ill-defined solutions.
   b. poor infrastructure.
   c. a lack of manpower.
   d. a lack of interest.

5. Many of the child deaths each year in developing countries could be prevented if
   a. technology were improved.
   b. doctors were more specialized.
   c. mothers would learn more about nutrition.
   d. sanitation and hygiene were improved.

6. The Supplemental Nutrition Assistance Program (SNAP) allows
   a. families with low income to buy surplus food at government stores with government-issued electronic benefit transfer cards.
   b. individuals with low income to purchase food, cleaning supplies, alcoholic beverages, and anything else sold in supermarkets with electronic benefit transfer cards.
   c. individuals with low income to turn in government-issued electronic benefit transfer cards for cash to buy food.
   d. individuals with low income to purchase food and seeds with government-issued electronic benefit transfer cards.

7. A long-term solution to world hunger is
   a. the green revolution.
   b. cash crops.
   c. jobs and self-sufficiency.
   d. government and private aid.

8. How many U.S. crops are genetically engineered?
   a. 0
   b. Fewer than 12
   c. More than 50
   d. All U.S. crops

9. Genetically modified soybeans make up approximately _____% of the soybeans grown in the United States.
   a. 25
   b. 50
   c. 75
   d. 90

10. The National Bioengineered Food Disclosure Law established a national mandatory standard for disclosing foods that are bioengineered.
    a. True
    b. False

Answer Key: 1. b (LO 13.1), 2. d (LO 13.3), 3. d (LO 13.5), 4. b (LO 13.3), 5. d (LO 13.3), 6. d (LO 13.2), 7. c (LO 13.2), 8. b (LO 13.4), 9. d (LO 13.4), 10. a (LO 13.4)

## Study Questions (Numbers refer to Learning Outcomes.)

1. Describe the difference between malnutrition, overnutrition, and undernutrition. (**LO 13.1**)

2. Name three nutrients often lacking in the diets of people who lack access to an adequate, safe food supply. What effects can be expected with each deficiency? (**LO 13.1**)

3. What do you believe are the major factors contributing to malnutrition in affluent nations, such as the United States? What are some possible solutions to this problem? (**LO 13.2**)

4. What federal programs are available to address the problem of malnutrition in the United States? (**LO 13.2**)

5. Outline how war and civil unrest in developing countries have worsened issues of chronic hunger over the past few years. (**LO 13.3**)

6. How important is population control in addressing the problem of world hunger now and in the future? (**LO 13.3**)

7. Why is solving the problem of malnutrition a key factor in the ability of developing countries to reach their full potential? (**LO 13.3**)

8. Describe how sustainable agriculture and biotechnology can improve food availability worldwide. (**LO 13.4**)

9. List three long-term consequences of malnutrition during fetal development or infancy. (**LO 13.5**)

10. Related to nutrition, what are the critical stages of life when individuals are most vulnerable? (**LO 13.5**)

## Further Readings

1. FAO, IFAD, UNICEF, WFP, and WHO: *The State of Food Security and Nutrition in the World 2019.* Available at https://www.unicef.org /media/55921/file/SOFI-2019-full-report.pdf. Accessed January 24, 2020.

2. Academy of Nutrition and Dietetics: Position of the Academy of Nutrition and Dietetics: Nutrition security in developing nations: Sustainable food, water, and health. *Journal of the Academy of Nutrition and Dietetics* 113:581, 2013. DOI:10.1016/j.jand.2013.01.025.

3. United Nations: Sustainable development goals: 17 goals to transform our world. Available at http://www.un.org/sustainabledevelopment. Accessed January 15, 2020.

4. Economic Research Service, U.S. Department of Agriculture: Food security in the U.S. survey tools. Available at https://www.ers.usda .gov/topics/food-nutrition-assistance/food-security-in-the-us/survey -tools. Accessed January 25, 2020.

5. Centers for Disease Control and Prevention: Micronutrient Facts. Available at https://www.cdc.gov/nutrition/micronutrient-malnutrition /micronutrients/index.html. Accessed January 15, 2020.

6. U.S. Census Bureau: Poverty in the U.S. Available at http://www.census .gov/topics/income-poverty/poverty.html. Accessed January 15, 2020.

7. Feeding America: Hunger in America. Available at https://www .feedingamerica.org/hunger-in-america/impact-of-hunger. Accessed January 22, 2020.

8. U.S. Department of Agriculture: Foods typically purchased by Supplemental Nutrition Assistance Program (SNAP) households. Available at https://www.fns.usda.gov/snap/foods-typically-purchased -supplemental-nutrition-assistance-program-snap-households. Accessed January 15, 2020.

9. U.S. Department of Agriculture Food and Nutrition Service: Program Information Report, U.S. Summary FY 2018–FY 2019. Available at https://fns-prod.azureedge.net/sites/default/files/data-files/Keydata -August-2019.pdf. Accessed January 20, 2020.

10. National Council on Aging: Older Americans Act. Available at https://www.ncoa.org/public-policy-action/older-americans-act. Accessed January 20, 2020.

11. Academy of Nutrition and Dietetics Position of the Academy of Nutrition and Dietetics: Food insecurity in the United States. *Journal of the Academy of Nutrition and Dietetics* 117:1991, 2017. DOI: 10.1016/j.jand.2017.09.027.

12. National Alliance to End Homelessness: The Demographic Data Project. Available at https://endhomelessness.org/resource/the-demograhic-data -project. Accessed January 15, 2020.

13. The U.S. Department of Housing and Urban Development, Office of Community Planning and Development: The 2018 Annual Homeless Assessment Report (AHAR) to Congress. Available at https://files.hudexchange .info/resources/documents/2018-AHAR-Part-1.pdf?te=1&nl=california -today&emc=edit_ca_20191214. Accessed January 30, 2020.

14. United States Interagency Council on Homelessness: Homelessness in America: Focus on Chronic Homelessness Among People with Disabilities. Available at https://www.usich.gov/resources/uploads /asset_library/Homelessness-in-America-Focus-on-chronic.pdf. Accessed January 25, 2020.

15. Economic Research Service, U.S. Department of Agriculture: Food access research atlas. Available at http://www.ers.usda.gov/data-products /food-access-research-atlas.aspx. Accessed January 15, 2020.

16. World Health Organization (WHO): Maternal mortality. Available at https://www.who.int/news-room/fact-sheets/detail/maternal-mortality. Accessed January 11, 2020.

17. World Health Organization (WHO): Family planning/contraception. Available at https://www.who.int/news-room/fact-sheets/detail/family -planning-contraception. Accessed January 24, 2020.

18. United Nations: World Population Prospects: The 2019 Revision. Available at https://data.un.org/Data.aspx?d=PopDiv&f=variableID%3A54. Accessed January 16, 2020.

19. Stockholm International Peace Research Institute (SIPRI): World Military Expenses. Available at https://www.sipri.org/media/press -release/2019/world-military-expenditure-grows-18-trillion-2018. Accessed January 29, 2020.

20. United Nations Refugee Agency: Figures at a Glance. Available at https:// www.unhcr.org/en-us/figures-at-a-glance.html. Accessed January 26, 2020.

21. Global Hunger Index: Global Hunger Index by Severity. Available at https://www.globalhungerindex.org/results.html. Accessed January 10, 2020.

22. FAO of the United Nations: FAO in China. Available at http://www.fao.org/china/fao-in-china/china-at-a-glance/en. Accessed January 19, 2020.

23. FAO of the United Nations: The State of World Fisheries and Aquaculture 2018— Meeting the Sustainable Development Goal. Available at http://www.fao.org/3/i9540en/i9540en.pdf. Accessed at January 22, 2020.

24. World Health Organization (WHO): Drinking water. Available at https://www.who.int/news-room/fact-sheets/detail/drinking-water. Accessed January 22, 2020.

25. UNICEF: Water, Sanitation and Hygiene. Available at https://www.unicef.org/wash. Accessed January 29, 2020.

26. World Health Organization (WHO): Sanitation. Available at https://www.who.int/news-room/fact-sheets/detail/sanitation. Accessed January 28, 2020.

27. UNICEF: Diarrhoeal Disease. Available at https://data.unicef.org/topic/child-health/diarrhoeal-disease. Accessed January 12, 2020.

28. World Health Organization (WHO): Global Health Observatory (GHO) Data HIV/AIDS. Available at https://www.who.int/gho/hiv/en. Accessed January 4, 2020.

29. HIV.gov: U.S. Statistics Fast Facts. Available at https://www.hiv.gov/hiv-basics/overview/data-and-trends/statistics. Accessed January 12, 2020.

30. Academy of Nutrition and Dietetics: Practice Paper: Nutrition intervention and human immunodeficiency virus infection. *Journal of the Academy of Nutrition and Dietetics* 118:486, 2018.

31. FAO of the United Nations: Sustainable Intensification of Crop Production. Available at http://www.fao.org/policy-support/policy-themes/sustainable-intensification-crop-production/en. Accessed January 22, 2020.

32. International Service for the Acquisition of Agri-biotech Applications (ISAAA): Pocket K No. 16: Biotech Crop Highlights in 2018. Available at https://www.isaaa.org/resources/publications/pocketk/16. Accessed January 15, 2020.

33. U.S. Department of Agriculture Food and Nutrition Service: Storing Fresh Produce. Available at https://www.fns.usda.gov/storing-fresh-produce. Accessed January 25, 2020.

34. UNICEF: The State of the World's Children 2019. Children, Food and Nutrition: Growing Well in a Changing World. Available at https://www.unicef.org/media/60806/file/SOWC-2019.pdf. Accessed January 10, 2020.

35. World Health Organization (WHO): Maternal, newborn, child and adolescent health. Available at https://www.who.int/maternal_child_adolescent/newborns/en. Accessed January 15, 2020.

36. UNICEF: Low Birth Weight. Available at https://data.unicef.org/topic/nutrition/low-birthweight. Accessed January 24, 2020.

37. UNICEF: The State of the World's Children 2019. Children, food and nutrition. Available at https://data.unicef.org/resources/state-of-the-worlds-children-2019. Accessed January 17, 2020.

38. UNICEF: Under Five Mortality. Available at https://data.unicef.org/topic/child-survival/under-five-mortality. Accessed January 25, 2020.

# Chapter 14

# Nutrition During Pregnancy and Breastfeeding

# Student Learning Outcomes

**Chapter 14 is designed to allow you to:**

**14.1** Describe how nutrition affects fertility.

**14.2** Summarize the physiological changes of pregnancy and how these changes affect key nutrient requirements.

**14.3** Define "success" in terms of positive health outcomes during pregnancy, and identify lifestyle factors that promote a successful pregnancy for both the mother and the infant.

**14.4** Specify optimal ranges of weight gain during pregnancy for women with low, healthy, or high prepregnancy BMI.

**14.5** Outline guidelines for physical activity during pregnancy.

**14.6** Describe how dietary changes can prevent or alleviate some discomforts and complications of pregnancy.

**14.7** Summarize the physiological processes involved in breastfeeding and how breastfeeding affects the nutritional requirements of a woman.

**14.8** Design an adequate, balanced meal plan for a pregnant or breastfeeding woman based on MyPlate.

**14.9** List several advantages of breastfeeding for both the mother and the infant.

**14.10** Relate the nutritional status of the parents to the risk of birth defects in the child.

## FAKE NEWS

# Food cravings are the body's way of getting needed nutrients during pregnancy.

## THE FACTS

Food cravings during pregnancy may be related to hormonal changes or may simply stem from family or cultural traditions. It's okay to give in to pregnancy cravings in moderation. In fact, some researchers think this may be a subtle way women cope with stress and anxiety during pregnancy.[1] However, there is no evidence that consuming the foods craved during pregnancy will correct nutrient deficiencies.

Some women crave nonfood items (e.g., chalk or clay). This practice, known as *pica* (see Section 11.5), is an eating disorder. It could be either a cause or a consequence of nutritional problems, such as deficiencies of iron or zinc.[2]

The responsibility of nourishing and protecting a child is exhilarating and, at the same time, intimidating. Parents' desire to produce a healthy baby can arouse interest in nutrition and health information. They usually want to do everything possible to maximize their chances that their newborn will thrive.

Despite the best intentions, the infant mortality rate in North America is higher than that of many other industrialized nations. In Canada, about 5 of every 1000 infants per year die before their first birthday, and in the U.S. it is almost 6 infants. (For comparison, the rate of infant mortality in Sweden is less than 3 of every 1000 infants.) In addition, about 11% of pregnant women in the U.S. receive inadequate prenatal care. These are alarming statistics for two countries that have such a high per capita expenditure for health care compared to many other countries in the world.

Some aspects of fetal and newborn health are beyond our control. Still, conscious decisions about social, health, environmental, and nutritional factors during pregnancy significantly affect the baby's future. Choosing to breastfeed the infant adds further benefits. Let us examine how eating well during pregnancy and breastfeeding can help a baby to have a healthy start in life.

# **14.1** Nutrition and Fertility

"Surprise, we're pregnant!" Considering the fact that only about half of all pregnancies are planned, a positive pregnancy test can be shocking news. Even when planned, women often do not suspect they are pregnant during the first few weeks after conception. They may not seek medical attention until 2 to 3 months after conception. Without fanfare, though, the **embryo** grows and develops daily. As you will learn in this chapter, the mother's nutritional status will affect the health of her baby before and long after birth. For that reason, the health and nutrition habits of a woman who is trying to become pregnant—or has the potential to become pregnant—are vitally important.

For up to 15% of couples who are planning for pregnancy, however, that pregnancy test shows a negative result month after month. **Infertility** refers to the inability of a couple to conceive after 1 year of unprotected intercourse. There are numerous causes for male and female infertility, many of which are outside the couple's control. In some cases, however, modifying nutrition and other lifestyle behaviors can improve a couple's chances of conceiving a child.

The nutritional status of both the mother- and the father-to-be can affect the likelihood of conception.[3] Some of the nutritional factors discussed in this chapter affect the hormone levels involved in reproduction. Others directly affect the viability of the egg or the sperm. Thus far, research clearly supports a link between body fat and fertility, and some evidence points to roles of certain dietary fats, carbohydrates, antioxidant nutrients, B vitamins, zinc, iron, and especially overall dietary patterns.

**ENERGY BALANCE**

Recall from Section 7.1 that energy balance refers to matching calorie intake to energy expenditure. Positive energy balance describes a situation in which the amount of calories consumed exceeds the amount of calories required to support basic body processes and physical activity. Sustained over time, positive energy balance leads to gains in both lean mass and adipose tissue. Negative energy balance occurs when calorie intake falls short of calorie needs. Prolonged negative energy balance leads to loss of both lean and adipose tissue. At either extreme, prolonged energy imbalance can impair fertility.

It is important to note that adipose tissue serves as more than just a storage depot for energy; it also produces estrogen and other hormones and cellular signaling molecules with widespread effects. For example, leptin, a hormone produced primarily by adipose tissue, affects appetite, metabolic rate, immune function, growth, and reproduction.

In terms of energy, reproductive function is costly. Synthesis of reproductive hormones, maintenance of normal menstrual cycles, pregnancy, and breastfeeding require calories. With negative energy balance, little energy is available to maintain normal reproductive function. Consequently, many women with underweight BMI experience amenorrhea, a sign of impaired ovulation. Some causes of low energy availability include undernutrition that stems from **food insecurity,** eating disorders, or high levels of athletic training. During World War II, for example, a famine in Holland drastically cut the calorie intake of women to about 1000 kcal per day. Many women became amenorrheic, and birth rates declined by about 50% during that time period. Studies in female athletes indicate that energy intake of at least 30 kcal per kilogram of lean mass is needed for normal reproductive function among women. For men, low body fat can decrease sex drive and sperm count.

On the opposite end of the spectrum, prolonged positive energy balance also decreases fertility. Excess adipose tissue affects the availability of reproductive

**embryo** In humans, the developing offspring in utero from about the beginning of the third week to the end of the eighth week after conception.

**infertility** Inability of a couple to conceive after 1 year of unprotected intercourse.

hormones and induces insulin resistance. For women, these endocrine changes impair the success of ovulation and implantation. In fact, excess body fat is thought to cause about 25% of problems with ovulation that lead to infertility. Among men, excess body fat increases estrogen levels and decreases testosterone. Also, extra fat tissue increases the temperature of the testicular area. The changes in hormones and temperature result in lower sperm production. Obesity is also associated with increased oxidative stress, which damages DNA in both the egg and the sperm. For adults who are overweight or obese, losing just 5% to 10% of body weight can increase the chances of conception.

## HORMONAL BALANCE

As we discuss the ways energy imbalances influence fertility, it is important to mention **polycystic ovary syndrome (PCOS).** Women with PCOS have hormonal imbalances that may lead to problems with ovulation, unusually high levels of androgen hormones (e.g., testosterone), and the presence of many tiny cysts that surround the ovaries like a strand of pearls. All women secrete some testosterone, but women with PCOS secrete more than normal. The high levels of male hormones lead to some of the signs and symptoms of PCOS: excess hair growth on the face, acne, and a tendency to deposit fat around the waist. Insulin resistance is another common feature of the syndrome. Thus, women with PCOS are at higher risk for diabetes, high blood pressure, and cardiovascular disease. Many women with PCOS experience irregular or absent periods, difficulty becoming pregnant, and higher-than-average rates of **spontaneous abortions.** Indeed, PCOS is the leading cause of female infertility.[4]

Many dietary and lifestyle changes have been studied to see how they can alter the course of PCOS. To date, evidence most clearly indicates that weight loss is important to improve metabolic and fertility issues among overweight women with PCOS. If an overweight woman with PCOS loses just 5% to 10% of her body weight, her chances of conception improve.[5] Daily physical activity, known to improve insulin sensitivity, is a key component of any weight-management strategy.

In addition to managing body weight, the quality and quantity of carbohydrates may make a difference in controlling PCOS and improving fertility. Some experts recommend reducing carbohydrate intake to the lower end of the range recommended by the Food and Nutrition Board (about 45% of total kcal) and choosing low glycemic-index carbohydrates. For example, women with PCOS are urged to choose whole grains instead of refined grains and whole fruits and vegetables rather than juices, and to steer clear of sugar-sweetened beverages.[6] Beware of unproven advice to reduce carbohydrate intake below the RDA during pregnancy. Severely restricting carbohydrate choices could limit the intake of important nutrients, such as fiber and B vitamins.

## KEY NUTRIENTS

**Vitamins.** A variety of micronutrients may contribute to fertility, but folic acid tops the list for both men and women.[3] Folate is involved in DNA synthesis and the metabolism of **homocysteine.** For no other cells is proper DNA synthesis so important as for the egg and sperm, which transmit genetic information from one generation to the next. Foods such as leafy green vegetables, strawberries, and orange juice are sources of natural folate. The synthetic form, folic acid, can be found in fortified foods (e.g., ready-to-eat breakfast cereals) and dietary supplements.

The chemical reactions of metabolism produce free radicals (molecules with unpaired electrons) that can damage cell membranes and DNA. The body has some antioxidant mechanisms that limit the activity of free radicals, but when the production of free radicals exceeds the antioxidant capacity of the body, oxidative damage to cells is likely. Free radicals can damage egg and (especially) sperm cells and can

**polycystic ovary syndrome (PCOS)** A condition of hormonal imbalance (e.g., elevated testosterone and insulin) in a woman that can lead to infertility, weight gain in the abdominal region, excessive growth of body hair, and acne.

**spontaneous abortion** Cessation of pregnancy and expulsion of the embryo or nonviable fetus prior to 20 weeks of gestation. This is the result of natural causes, such as a genetic defect or developmental problem; also called *miscarriage.*

affect how well a fertilized egg implants and matures. Research studies show that dietary patterns rich in antioxidants—vitamin E, vitamin C, selenium, zinc, beta-carotene, and some other plant pigments—are linked to improved fertility for both men and women.

Routine intake of a daily multivitamin and mineral supplement has been linked to improved fertility in many studies, but food sources of nutrients appear to offer multiple benefits beyond vitamin and mineral content.[7] Foods of plant origin—including brightly colored fruits and vegetables, whole grains, and plant oils—are rich sources of vitamins linked to improved fertility.

**Minerals.** Iron and zinc are two minerals that have been linked to fertility. Zinc appears to be especially important for male fertility. Not only is it involved as a cofactor in antioxidant reactions that could protect sperm from oxidative damage, but zinc also is required for normal sexual maturation and production of sperm and reproductive hormones. Men with poor zinc status may have poor sperm quality (e.g., low sperm production, impaired sperm motility, and/or damaged DNA). Studies have shown that zinc supplements can improve sperm quality in men who are deficient in zinc.[8] However, a recent trial showed that zinc supplementation (in combination with folic acid supplementation) in men did not improve rates of live births or markers of sperm quality.[9] Rather than relying on supplements, men should strive to consume a *dietary pattern* that meets the RDA for zinc.

For women, iron and zinc are needed for normal ovulation. Data from a large observational study of nurses showed that iron intake before conception was linked to improved ovulatory function and therefore better fertility. Interestingly, in this study, higher intakes of nonheme iron (from plant sources) were specifically related to improved fertility[10] (see *Ask the RDN* in this section).

**Dietary Fat.** As recommended for the general population of healthy adults, men and women who are trying to conceive should try to limit sources of saturated fats (and *trans* fats, which were recently banned from use in food manufacturing in the United States). For women, a dietary pattern rich in saturated fat promotes insulin resistance and impairs ovulation. For men, high intakes of saturated fat are linked to poor sperm quality.[3] Instead of consuming the typical American fare of pizza and fast foods, men and women should emphasize plant oils and fish oils, which provide more unsaturated fats. Among men experiencing infertility, boosting intakes of **omega-3 fatty acids,** which are the type of polyunsaturated fatty acids found in fish oils and walnuts, can improve sperm quality.

## ALCOHOL

To date, research does not demonstrate a clear relationship between alcohol consumption and fertility.[3] However, in Section 14.8, you will learn about the devastating effects of alcohol on the developing fetus in utero. Given the fact that many women do not realize they are pregnant until several weeks after conception, it seems prudent to avoid alcohol while trying to conceive.

When trying to conceive, should you switch to decaf? There has been some concern that caffeine could decrease fertility, but the evidence is mixed. In some studies, intake of more than 4 cups of coffee (about 500 mg caffeine) has been linked to decreased fertility, but most studies show no clear link between moderate caffeine intake and fertility.[3]

**trimesters** Three 13- to 14-week periods into which the normal pregnancy (on average, 40 weeks) is divided somewhat arbitrarily for purposes of discussion and analysis. Development of the offspring, however, is continuous throughout pregnancy, with no specific physiological markers demarcating the transition from one trimester to the next.

### ✓ CONCEPT CHECK 14.1

1. How is energy balance related to fertility?
2. What hormonal changes are characteristic of polycystic ovary syndrome? List two dietary or lifestyle modifications that may help a woman with PCOS.
3. List three nutrients that have been linked to fertility.

## ASK THE RDN  The Fertility Diet

**Dear RDN:** *My husband and I have been trying to conceive for 2 years. We have an appointment to see a fertility specialist, but I recently found* The Fertility Diet. *Is this a good resource for couples who are trying to get pregnant?*

*The Fertility Diet,* written in 2007 by scientists from the Harvard School of Public Health, sums up findings from the Nurses' Health Study II regarding relationships between dietary patterns, lifestyle, and pregnancy success.[11] Based on extensive data collected from 18,555 women who were trying to conceive during the study, the authors consolidated their conclusions into 10 recommendations:

1.  Avoid *trans* fats.
2.  Use more unsaturated fats.
3.  Eat more vegetable protein.
4.  Choose "low and slow" carbohydrates (e.g., whole grains, foods with low glycemic index).
5.  Temporarily trade in low-fat and fat-free dairy products for full-fat versions.
6.  Use a daily multivitamin and mineral supplement that contains folic acid.
7.  Ensure adequate iron intake from sources other than red meats.
8.  Choose water for hydration; coffee, tea, and alcohol can be used in moderation, but sugar-sweetened beverages should be avoided.
9.  Lose 5% to 10% of body weight (if overweight).
10. Incorporate moderate exercise.

With the exception of tip #5, all of these recommendations are in line with current dietary advice for Americans from major health authorities, including the *Dietary Guidelines*. For women who are trying to get pregnant, in particular, these recommendations are in agreement with the latest research on dietary patterns and fertility covered in this section.

What about #5? Grab an ice cream scoop? Public health messages have been telling us to choose low-fat and fat-free dairy foods for decades. Why might women seeking to become pregnant benefit from drinking whole milk and eating ice cream? The authors hypothesize that dairy fat, which does contain many hormones from lactating and sometimes pregnant cows, may influence the woman's hormone levels in a way that improves fertility. This is an intriguing hypothesis and one that certainly warrants further investigation. Notably, in their chapter on dairy choices, the authors also point out that dairy products may not be the best way for Americans to obtain calcium, in general. Instead, they advocate plant sources of calcium, dietary supplements, and a daily dose of sunshine to support bone health.

In addition to the tips offered in *The Fertility Diet,* I would also encourage you to find ways to include at least five servings per day of fruits and vegetables. These foods have much to offer in terms of the micronutrients that are needed to support a healthy pregnancy and can help with blood glucose control and weight management.

Research demonstrates that insulin resistance underlies many cases of infertility. Dietary changes that improve insulin sensitivity—losing excess body fat, choosing healthy fats, avoiding dramatic rises and falls in blood sugar throughout the day—can optimize a woman's chances of ovulating normally and becoming pregnant. Furthermore, adequate intakes of several key micronutrients—iron and folic acid—are vital for growth and development of the fetus.

Keep in mind that the Nurses' Health Study (like most research on dietary patterns and fertility) is an observational study. Even though these dietary patterns are associated with improved fertility, they have not been tested as part of a controlled experiment. Also, the authors provide the rationale for these dietary and lifestyle strategies to counter ovulatory infertility only; dietary changes are less likely to help women with infertility due to structural abnormalities of the reproductive system or other health conditions. In other words, neither *The Fertility Diet* nor any other dietary or lifestyle change can guarantee a healthy pregnancy; however, this chapter does offer the best of what we know to enhance your chances of success.

In summary, *The Fertility Diet* provides excellent nutrition advice, backed by evidence from the Nurses' Health Study and many other research studies, that can be adopted by the whole family. The authors have a knack for explaining complex scientific concepts in terms nonscientists can understand. Furthermore, the book includes practical advice, meal plans, and recipes to help women translate science into food choices.

Wishing you pink or blue,

**Angela Collene, MS, RDN, LD**
Lecturer, The Ohio State University, Author of *Contemporary Nutrition*

Tim Klontz, Klontz Photography

## 14.2 Prenatal Growth and Development

The length of a normal pregnancy is 38 to 42 weeks, measured from the first day of the woman's last menstrual period. For purposes of discussion, the duration of pregnancy is commonly divided into three periods, called **trimesters.** For 8 weeks after conception, a human embryo develops from a fertilized **ovum** into a **fetus.**

**ovum** The egg cell from which a fetus eventually develops if the egg is fertilized by a sperm cell.

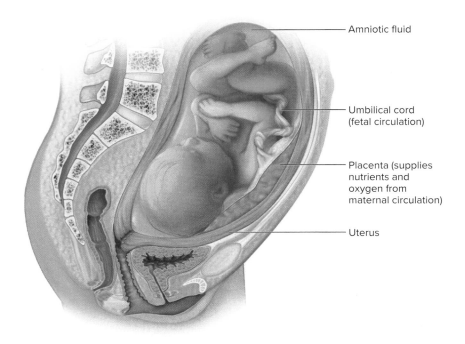

Amniotic fluid

Umbilical cord
(fetal circulation)

Placenta (supplies
nutrients and
oxygen from
maternal circulation)

Uterus

**placenta** An organ that forms in the uterus in pregnant women. Through this organ, oxygen and nutrients from the mother's blood are transferred to the fetus, and fetal wastes are removed. The placenta also releases hormones that maintain the state of pregnancy.

**zygote** The fertilized ovum; the cell resulting from the union of an egg cell (ovum) and sperm until it divides.

Until birth, the mother nourishes the fetus via a **placenta,** an organ that forms in her uterus to accommodate the growth and development of the fetus (Fig. 14-1). The role of the placenta is to exchange nutrients, oxygen, and other gases between the mother and fetus and to eliminate waste products. This occurs through a network of capillaries that bring the fetal blood close to the maternal blood supply, but the two blood supplies do not mix.

## EARLY GROWTH—THE FIRST TRIMESTER IS A CRITICAL TIME

In the formation of the human organism, egg and sperm unite to produce the **zygote** (Fig. 14-2). From this point, the reproductive process occurs rapidly:

- Within 30 hours: the zygote divides in half to form two cells.
- Within 4 days: cell number climbs to 128 cells.
- At 14 days: the group of cells is called an embryo.
- Within 35 days: heart is beating, embryo is 1/30 of an inch (8 millimeters) long, eyes and limb buds are clearly visible.
- At 8 weeks: the embryo is known as a fetus.
- At 13 weeks (end of first trimester): most organs are formed, and the fetus can move.

Growth begins within a day of conception with a rapid increase in cell number. This type of growth dominates embryonic and early fetal development. The newly formed cells then begin to grow larger. Further growth is a mix of increases in cell number and cell size. By the end of 13 weeks—the first trimester—most organs are formed and the fetus can move (see Fig. 14-2).

As the offspring develops, nutritional deficiencies, toxicities, and other harmful environmental exposures have the potential to damage organ systems. For example, adverse reactions to medications, high intakes of vitamin A, exposure to radiation, or trauma can alter or arrest the current phase of fetal development, and the effects may last a lifetime (review Fig. 14-2). The most critical time for these potential problems is during the first trimester. Most spontaneous abortions happen at this time. One-half or more of pregnancies end in this way, often so early that a woman does not even realize she was pregnant. (An additional 15% to 20% are lost before normal delivery.) Early spontaneous abortions usually result from a genetic defect or fatal error in fetal development.

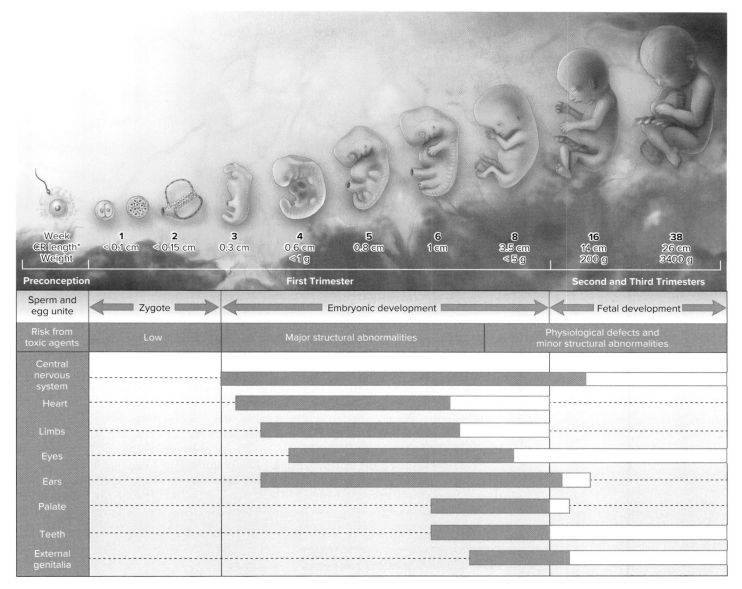

| Week | 1 | 2 | 3 | 4 | 5 | 6 | 8 | 16 | 38 |
|------|---|---|---|---|---|---|---|----|-----|
| CR length* | <0.1 cm | <0.15 cm | 0.3 cm | 0.6 cm | 0.8 cm | 1 cm | 3.5 cm | 14 cm | 26 cm |
| Weight | | | | <1 g | | | <5 g | 200 g | 3400 g |

| | Preconception | | First Trimester | | Second and Third Trimesters |
|---|---|---|---|---|---|
| Sperm and egg unite | ← Zygote → | | ← Embryonic development → | ← Fetal development → | |
| Risk from toxic agents | Low | | Major structural abnormalities | | Physiological defects and minor structural abnormalities |
| Central nervous system | | | | | |
| Heart | | | | | |
| Limbs | | | | | |
| Eyes | | | | | |
| Ears | | | | | |
| Palate | | | | | |
| Teeth | | | | | |
| External genitalia | | | | | |

**FIGURE 14-2** ▲ Harmful effects of toxic agents during pregnancy. Vulnerable periods of fetal development are indicated with purple bars. The purple shading indicates the time of greatest risk to the organ. The most serious damage to the fetus from exposure to toxins is likely to occur during the first 8 weeks after conception, two-thirds of the way through the first trimester. As the white bars in the chart show, however, damage to vital parts of the body—including the eyes, brain, and genitals—can also occur during the later months of pregnancy.

*CR length = measurement from the crown (top of the head) to the rump (lowest part of the buttocks)

Smoking (any form of nicotine or marijuana), alcohol abuse, use of aspirin and NSAIDs, and illicit drug use raise the risk for spontaneous abortion.

A woman should avoid substances that may harm the developing fetus, especially during the first trimester. This holds true, as well, for the time when a woman is trying to become pregnant. As previously mentioned, she is unlikely to be aware of her pregnancy for at least a few weeks. In addition, the fetus develops so rapidly during the first trimester that, if an essential nutrient is not available, the fetus may be affected even before evidence of the nutrient deficiency appears in the mother.

For this reason, the *quality*—rather than the *quantity*—of the woman's eating pattern is most important during the first trimester. In other words, the mother should consume the same amount of calories as she did before she became pregnant, but she should focus on choosing more nutrient-dense foods. Although some women lose their appetite and feel nauseated during the first trimester, they should be careful to meet nutrient needs as much as possible.

## SECOND TRIMESTER

By the beginning of the second trimester (i.e., 13 weeks' **gestation**), a fetus weighs about 1 ounce and is about the size of a lemon. Arms, hands, fingers, legs, feet, and toes are fully formed. The fetus has ears and begins to form tooth sockets in its jawbone. Organs continue to grow and mature, and, with a stethoscope or Doppler instrument, clinicians can detect the fetal heartbeat. Most bones are distinctly evident throughout the body. Eventually, the fetus begins to look more like an infant. It may suck its thumb and kick strongly enough to be felt by the mother. As shown in Figure 14-2, the fetus can still be affected by exposure to toxins, but not to the degree seen in the first trimester.

During the second trimester, the mother's breast weight typically increases by approximately 30% due to the development of milk-producing cells and the deposition of 2 to 4 pounds of fat for **lactation.** This stored fat serves as a reservoir for the extra energy that will be needed to produce breast milk.

## THIRD TRIMESTER

By the beginning of the third trimester, a fetus weighs about 2 to 3 pounds and is about as big as a head of lettuce. The third trimester is a crucial time for fetal growth. The fetus will double in length and increase its weight by three to four times. The third trimester is the time when many nutrients are transferred from the mother to the fetus. An infant born after only about 26 weeks of gestation has a good chance of surviving if cared for in a nursery for high-risk newborns. However, the infant will have low stores of minerals (e.g., iron and calcium), fat, and fat-soluble vitamins that normally accumulate during the last month of gestation. This and other medical problems, such as a poor ability to suck and swallow, complicate nutritional care for **preterm** infants.

By 40 weeks, the fetus usually weighs 7 to 9 pounds (3 to 4 kilograms) and is about 20 inches (50 centimeters) long. Soft spots (fontanels) on top of the head indicate where the skull bones are growing together. These bones close by the time the baby reaches 12 to 18 months of age.

**gestation** The period of intrauterine development of offspring, from conception to birth; in humans, normal gestation is 38 to 42 weeks.

**lactation** The period of milk secretion following pregnancy; typically called *breastfeeding.*

### ✓ CONCEPT CHECK 14.2

1. What is the role of the placenta?
2. Describe how risks for fetal malformations vary throughout pregnancy.
3. During which trimester are most organs being formed?
4. During which trimester does fetal size increase the most?

# **14.3** Success in Pregnancy

The goal of pregnancy is to achieve optimal health for both the baby and the mother. For the mother, a successful pregnancy is one in which her physical and emotional health is protected so that she can return to her prepregnancy health status. For the infant, two widely accepted criteria are: (1) a gestation period longer than 37 weeks and (2) a birth weight greater than 5.5 pounds (2.5 kilograms). Sufficient lung development, likely to have occurred by 37 weeks of gestation, is critical to the survival of a newborn. The longer the gestation (up to 42 weeks), the greater the ultimate birth weight and maturation state, leading to fewer medical problems and better quality of life for the infant.

As you read this chapter, you will notice frequent references to lifelong effects of maternal nutrition, physical activity, and other lifestyle practices on the child. This is called the *developmental origins of health and disease hypothesis* or, more simply, the **fetal origins hypothesis.** Emerging evidence links prenatal influences, such as famines, fasting, exposure to alcohol, environmental pollution, and even maternal stress to the child's risks for disease later in life (see *Newsworthy Nutrition* in Section 14.4). As we learn more about how environmental factors modify the ways our genes are expressed, it becomes increasingly evident that many aspects of our physical and mental health are initially programmed while we are yet in utero. What's more, these epigenetic changes may be passed down from generation to generation.[12]

Overall, a successful pregnancy is the outcome of many complex gene–environment interactions. The decisions both parents make today can affect the health of their child for years to come. Although a mother's decisions, practices, and precautions before conception and during pregnancy contribute to the health of her fetus during all three trimesters, she cannot guarantee her fetus good health because some genetic and environmental factors are beyond her control. She and others involved in the pregnancy should not hold an unrealistic illusion of total control.

On average, a healthy newborn weighs about 7.5 pounds and is 20 inches long. **Low-birth-weight (LBW)** infants are those weighing less than 5.5 pounds (2.5 kilograms) at birth. In the United States, 1 in 12 newborns is born LBW. Most commonly, LBW is associated with preterm birth. Medical costs during the first year of life for LBW infants are higher than those for normal-weight infants. In fact, hospital-related costs of caring for a preterm and/or LBW infant are more than $55,000 for the first year of life in the United States. For comparison, average medical costs during the first year of life for a healthy, **full-term** infant are around $5000. Overall, preterm births cost employers more than $12 billion in excess health care costs each year.[13] (See the March of Dimes website at www.marchofdimes.org for additional information on preterm birth.)

Full-term and preterm infants who weigh less than the expected weight for their duration of gestation as a result of insufficient growth are described as **small for gestational age (SGA).** Thus, a full-term infant weighing less than 5.5 pounds at birth is SGA but not preterm, whereas a preterm infant born at 30 weeks of gestation is probably LBW without being SGA. Infants who are SGA are more likely than normal-weight infants to have medical complications, including problems with feeding, blood glucose control, temperature regulation, growth, and development in the weeks after birth.

## PRENATAL CARE AND COUNSELING

Adequate prenatal care helps to promote success in pregnancy. Ideally, women should receive examinations and counseling *before* becoming pregnant and should continue regular prenatal care throughout pregnancy. If prenatal care is inadequate, delayed, or absent, untreated maternal nutritional deficiencies can deprive a developing fetus of needed nutrients. In addition, untreated health conditions, such as **anemia, AIDS, hypertension, diabetes,** or depression, must be carefully addressed to minimize complications during pregnancy. Treating ongoing infections will also decrease risks of fetal damage. Without prenatal care, a woman is three times more likely to deliver an LBW baby—one who will be 40 times more likely to die during the first 4 weeks of life than a normal-birth-weight infant. Although the ideal time to start prenatal care is before conception, about 20% of women in the United States receive *no* prenatal care throughout the first trimester—a critical time to positively influence the outcome of pregnancy.

Eating patterns cannot be predicted from income, education, or lifestyle. Although some women already have healthy dietary patterns, most can benefit from nutritional advice. All should be reminded of behaviors that may harm the growing fetus, such as extreme calorie restriction or fasting. By focusing on appropriate prenatal care, nutrient intake, and healthy behaviors, parents give their fetus—and later, their infant—the best chance of thriving. Overall, the chances of having a healthy baby are maximized with education, an adequate eating pattern, and early and consistent prenatal medical care.

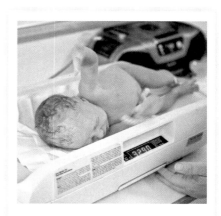

This newborn weighs 3290 grams (about 7 pounds, 3 ounces). **Is this a healthy birth weight?** Don Bayley/Getty Images

**fetal origins hypothesis** A theory that links nutritional and other environmental factors during gestation to the future health of the offspring.

**low birth weight (LBW)** Referring to any infant weighing less than 2.5 kilograms (5.5 pounds) at birth; most commonly results from preterm birth.

**full term** Referring to an infant born from 37 up to 42 weeks of gestation; also called *term*.

**small for gestational age (SGA)** Referring to infants who weigh less than the expected weight for their length of gestation. This corresponds to less than 2.5 kilograms (5.5 pounds) in a full-term newborn.

▲ To ensure optimal health and rapid treatment of medical conditions that develop during pregnancy, a pregnant woman should consult with her health care provider on a regular basis. Ideally, this consultation should begin before she becomes pregnant. JGI/Tom Grill/Blend Images/CORBIS

TABLE 14-1 ■ Effects of Teenage Pregnancy on Mother and Child

| For Mother | For Child |
|---|---|
| ⬆ Depression and other mental health problems | ⬆ Premature birth |
| ⬆ Use of illicit drugs and alcohol | ⬆ Infant mortality |
| ⬆ Poverty and reliance on public assistance | ⬆ Hospital admissions during childhood |
| ⬆ Single parenthood | ⬆ Rates of imprisonment during adulthood |
| ⬇ Graduation rates from high school and college | ⬇ Birth weight |
| | ⬇ Academic performance |
| | ⬇ Nutritional status |

## EFFECTS OF MATERNAL AGE

The age of the mother is another factor that determines pregnancy outcome. The ideal age for pregnancy is between 20 and 35 years of age. Outside that age range—at either extreme—complications are more likely to arise. The rates of teen pregnancy have declined since 1990; still, approximately 210,000 babies are born to teen mothers in the United States each year—the highest rate of any industrialized country. Teen pregnancy increases risk for negative outcomes for both mother and child (Table 14-1) and costs taxpayers an estimated $9.4 billion each year.[14]

Teens who are pregnant frequently exhibit a variety of risk factors that can complicate pregnancy and pose risk to the fetus.[15] For instance, teenagers are more likely than adult women to be underweight at the start of pregnancy and to gain too little weight during pregnancy. In addition, their bodies may lack the physical maturity needed to carry a fetus safely. Even with prenatal care, 10% of children born to teenage mothers are of low birth weight and 14.5% are preterm.

Advanced maternal age also poses special risks for pregnancy.[16] The likelihood of LBW and preterm delivery increases modestly, but progressively, with maternal age beyond 35 years. Given close monitoring, however, a woman older than 35 years has an excellent chance of giving birth to a healthy infant.

## CLOSELY SPACED AND MULTIPLE BIRTHS

The interval between the birth of one child and subsequent conception may affect the outcome of the latter pregnancy. Infants conceived less than 12 months after a previous birth have increased risks for preterm birth, LBW, SGA, birth defects, and developmental problems compared to those conceived at least 18 months following the birth of their older siblings. These poor outcomes are probably related to insufficient time to rebuild nutrient stores depleted by the previous pregnancy. Similarly, multiple births (i.e., twins) increase the risk for preterm birth, LBW, and other medical complications for the mother and the offspring. The American College of Obstetricians and Gynecologists recommends that women should wait until 18 months after giving birth before getting pregnant again.[17]

## EXPOSURE TO TOXIC CHEMICALS

Figure 14-2 depicts how and when toxic agents can harm the developing fetus. In preparation for and during pregnancy, the mother should undoubtedly eliminate alcohol, nicotine, and illicit drugs (e.g., marijuana). During organ development, exposure to toxic chemicals can cause malformations. The adverse effects of fetal exposure to alcohol during gestation are discussed in Section 14.8. Smoking is linked to preterm birth and appears to increase the risk of birth defects, sudden infant death, childhood cancer, and disabilities.

Illicit drug use is particularly harmful during pregnancy. Many chemicals in recreational drugs cross the placenta and affect the fetus, whose detoxification systems (e.g., liver and kidneys) are immature. Marijuana, the most common recreational drug used during the reproductive years, can result in reduced blood flow to the uterus and placenta, leading to poor fetal growth. Risks of low birth weight and preterm birth are increased for infants whose mothers used marijuana during pregnancy. Use of psychoactive drugs, such as opioids, cocaine, and methamphetamines, can restrict fetal growth and brain development, leading to lifelong physical and mental disabilities.

Even prescribed and common over-the-counter medications could have harmful effects on the developing fetus. Problem drugs include aspirin (especially when used chronically or heavily), hormone ointments, nose drops and related "cold" medications, rectal suppositories, weight-control pills, antidepressants, and medications prescribed for preexisting illnesses. Some herbal therapies also have the potential to damage the fetus. Lower doses and/or safer alternatives should be substituted when a woman is planning to or has become pregnant.[18]

In addition to these toxic chemicals, health hazards in the mother's environment, including job-related hazards and exposure to X-rays, should be minimized.

## FOOD SAFETY

Foodborne illness during any stage of life is a concern, but one type of foodborne illness that poses particular danger for pregnant women is caused by the bacterium *Listeria monocytogenes* (review Section 12.4).[19] Infection with this microorganism typically causes mild flu-like symptoms, such as fever, headache, and vomiting, about 7 to 30 days after exposure. However, pregnant women and their offspring may suffer more severe complications, including fetal death. Unpasteurized milk, soft cheeses made from raw milk (e.g., Brie, Camembert, feta, and blue cheeses), and some raw vegetables (e.g., cabbage and sprouts) can be sources of *Listeria* organisms, so it is especially important that pregnant women (and other people at high risk for infection) avoid these products. Experts advise consuming only pasteurized milk products and cooking meat, poultry, and seafood thoroughly to kill *Listeria* and other foodborne organisms. It is unsafe in pregnancy to eat any raw meats or other raw animal products, uncooked hot dogs, or undercooked poultry. These food safety recommendations are included in Chapter 5 of the *Dietary Guidelines,* which you can download from www.dietaryguidelines.gov.

The U.S. Department of Agriculture (USDA) warns pregnant women to thoroughly cook (e.g., microwave) all ready-to-eat meats, including hot dogs and cold cuts. **Why should pregnant women be extra careful about properly handling these foods?** Carolyn Taylor Photography/Stockbyte/Getty Images

## PREPREGNANCY BMI

Women should aim to achieve a healthy body weight prior to becoming pregnant. Infants born to women who begin pregnancy substantially above or below a healthy BMI are more likely to experience problems than those born to women who begin pregnancy at a healthy BMI.[20,21] For instance, babies born to women who are obese are at increased risk of birth defects, death in the first few weeks after birth, and obesity in childhood. Many pregnant women who are obese experience high blood pressure, diabetes, and more difficult deliveries.

At the other extreme, women who begin pregnancy underweight (BMI under 18.5) are more likely to have infants who are low birth weight and preterm compared to women at a normal weight. These differences may be because underweight women tend to have lighter placentas and lower nutrient stores, especially iron, than heavier women, which could negatively affect fetal growth and development. A woman who is underweight can improve her nutrient stores and pregnancy outcome by gaining weight before pregnancy or gaining extra weight during pregnancy.

## NUTRITIONAL STATUS

Is attention to good nutrition worth the effort? Yes! Extensive research suggests that an adequate vitamin and mineral intake at least 8 weeks before conception and throughout pregnancy can improve outcomes of pregnancy.[22] Extra nutrients and calories support fetal growth and development, of course, but also undergird the synthesis of maternal tissues. Her uterus and breasts grow, the placenta develops, her total blood volume increases, the heart and kidneys work harder, and stores of body fat increase. In particular, meeting folate needs (400 micrograms of synthetic folic acid per day) helps to prevent birth defects such as **neural tube defects** (see Section 14.8) and decrease the risk of preterm delivery. Low intakes of calcium and iron or excessive intakes of vitamin A also are causes for concern during pregnancy.

Although it is difficult to predict the degree to which poor nutrition will affect each pregnancy, a daily eating pattern containing only 1000 kcal has been shown to greatly restrict fetal growth and development. Increased maternal and infant death rates seen in famine-stricken areas of Africa provide further evidence.

Genetic background can explain little of the observed differences in birth weight between developed and developing countries. Environmental factors, including nutritional factors, are important. The worse the nutritional status of the mother at the beginning of pregnancy, the more valuable a healthy prenatal dietary pattern and/or use of prenatal supplements are in improving the course and outcome of her pregnancy.

## NUTRITION ASSISTANCE FOR FAMILIES AFFECTED BY FOOD INSECURITY

Poverty impacts pregnancy in many ways. Families of low socioeconomic status tend to receive inadequate health care. A lack of education, limited access to health care, or scant financial resources may contribute to poor health practices, such as dietary patterns that fail to meet the mother's increased nutrient requirements.

Several U.S. government programs provide high-quality health care and foods to reduce infant mortality and improve nutritional status of pregnant women and their children. These government-subsidized programs are designed to alleviate the negative impact of poverty, insufficient education, and inadequate nutrient intake on pregnancy outcomes. An example of such a program is the Special Supplemental Nutrition Program for Women, Infants, and Children (WIC), described in Section 13.2. This program offers health assessments, education, and electronic benefit transfer (EBT) cards to purchase foods that supply high-quality protein, calcium, iron, and vitamins A and C to pregnant women, infants, and children (up to 5 years of age) from eligible populations. The WIC program is available in all areas of the United States and has a staff trained to promote healthy behaviors during pregnancy. Participation in WIC has been shown to improve nutrient intakes of mothers, infants, and children.[23] More than 8 million women, infants, and young children are already benefiting from this program, yet many eligible individuals are not participating.

### ✓ CONCEPT CHECK 14.3

1. In one or two sentences, how would you define success in pregnancy?
2. Define the terms *preterm, low birth weight,* and *small for gestational age.*
3. How is maternal age related to the outcome of pregnancy?
4. Is attention to good nutrition during pregnancy worth the effort? Why or why not?

## 14.4 Increased Nutrient Needs to Support Pregnancy

Pregnancy is a time of increased nutrient requirements. Some general principles are true for most women with regard to increased nutrient needs. However, recognize that each expectant mom will benefit from individualized counseling tailored to her unique nutritional and health needs.

### CALORIE NEEDS

To support the growth and development of the fetus, pregnant women need to increase their calorie intake. Calorie needs during the first trimester are essentially the same as for women who are not pregnant. However, during the second and third trimesters, it is

In the United States, low-income pregnant women and their children benefit from federal nutrition assistance programs:

Special Supplemental Nutrition Program for Women, Infants, and Children (WIC)
**www.fns.usda.gov/wic**

Supplemental Nutrition Assistance Program (SNAP)
**www.fns.usda.gov/snap /supplemental-nutrition -assistance-program**

Food Distribution Program on Indian Reservations (FDPIR)
**www.fns.usda.gov/fdpir/food -distribution-program-indian -reservations**

UpperCut Images/Getty Images

necessary for a pregnant woman to consume approximately 350 to 450 kcal more per day than her prepregnancy needs (Table 14-2).[24]

Rather than seeing this as an opportunity to fill up on sugary desserts or fat-filled snacks, pregnant women should consume these extra calories in the form of nutrient-dense foods. For example, throughout the day, about six whole wheat crackers, 1 ounce of cheese, and ½ cup of fat-free or low-fat milk would supply the extra calories (and also some calcium). Although she "eats for two," a woman who is pregnant must not double her normal calorie intake. The "eating for two" concept refers more appropriately to increased needs for several vitamins and minerals. Micronutrient needs are increased by up to 50% during pregnancy, whereas calorie needs only increase by about 20% (and only during the second and third trimesters).

Keep in mind that the estimated energy needs shown in Table 14-2 are just a starting point. Changes in weight over time are the best way to evaluate the adequacy of energy intake. Individual variations in metabolic rate, prepregnancy weight status, and physical activity level may also affect a pregnant woman's daily energy needs. Women who are pregnant with multiple fetuses may need even more than 450 extra kcal per day.

## STAYING ACTIVE DURING PREGNANCY

Pregnancy is not the time to begin an intense fitness regimen, but women can generally take part in most low- or moderate-intensity activities during pregnancy. In fact, the *Physical Activity Guidelines* and current advice from the American College of Obstetrics and Gynecology recommend that pregnant women perform at least 30 minutes per day of moderate-intensity physical activity. Walking, cycling, swimming, or light aerobics for at least 150 minutes per week is generally advised. Such activity may prevent pregnancy complications and promote an easier delivery.[25,26,27] Some research indicates that regular physical activity during pregnancy lowers a woman's risk of developing gestational diabetes by 50% and preeclampsia by 40%. These disorders of pregnancy are discussed further in Section 14.6. Figure 14-3 illustrates the broad range of benefits of exercise during pregnancy for both the mother and the baby.

Women who were highly active prior to pregnancy can maintain their activities as long as they remain healthy and review plans with their health care providers. Adequate fluid intake before, during, and after physical activity is of heightened importance to regulate

**TABLE 14-2 ■ Estimated Energy Needs During Pregnancy**

| Trimester | Estimated Energy Needs |
|---|---|
| 1st Trimester | Prepregnancy EER + 0 kcal |
| 2nd Trimester | Prepregnancy EER + 340 kcal |
| 3rd Trimester | Prepregnancy EER + 452 kcal |

Source: Data from *Dietary Reference Intakes for Energy, Carbohydrate, Fiber, Fat, Fatty Acids, Cholesterol, Protein, and Amino Acids*, Copyright 2005 by the Institute of Medicine. National Academies Press, Washington, DC.

**Benefits of Physical Activity During Pregnancy**

| Benefits for Mother | Benefits for Baby |
|---|---|
| Prevents excessive weight gain during pregnancy | **During Gestation** |
| Improved cardiovascular function | Decreased resting heart rate |
| Lower risk for gestational diabetes | Healthier placenta |
| Lower risk for gestational hypertension | Increased amniotic fluid |
| Reduced bone loss associated with pregnancy | Possible improvements in brain development |
| Less edema in the legs and feet | Longer gestation |
| Better sleep | **After Gestation** |
| Decreased back pain | Lower birth weight |
| Improved satisfaction with body image | Leaner BMI during childhood |

FIGURE 14-3 ◄ Benefits of physical activity during pregnancy for the mother and baby. Tanya Constantine/Blend Images LLC

body temperature, because heat stress can be harmful to the developing fetus. Activities with inherent risk of falls or abdominal trauma—downhill skiing, horseback riding, and contact sports (e.g., soccer and basketball)—can potentially harm the fetus and should be avoided. In addition, pregnant women should avoid activities that require excess straining (e.g., heavy weightlifting) or exposure to extremes in air pressure (e.g., SCUBA diving).

Women with high-risk pregnancies, such as those experiencing premature labor contractions, may need to restrict physical activity. To ensure optimal health for both herself and her infant, a pregnant woman should first consult her primary care provider about physical activity and possible limitations.

## OPTIMAL WEIGHT GAIN

Healthy prepregnancy weight and appropriate weight gain during pregnancy are excellent predictors of pregnancy outcome.[28,29] The mother's dietary pattern should allow for approximately 2 to 4 pounds (0.9 to 1.8 kilograms) of weight gain during the first trimester and then a subsequent weight gain of 0.8 to 1 pound (0.4 to 0.5 kilogram) weekly during the second and third trimesters (Fig. 14-4). A healthy goal for total weight gain for a woman of normal weight (based on prepregnancy BMI; Table 14-3) averages about 25 to 35 pounds (11.5 to 16 kilograms).

For women who begin pregnancy with a low BMI, the goal for total weight gain increases to 28 to 40 pounds (12.5 to 18 kilograms). The goal decreases to a total of 15 to 25 pounds (7 to 11.5 kilograms) for women who are overweight at conception. Target weight gain for women who are obese at conception is a total of 11 to 20 pounds (5 to 9 kilograms). Figure 14-4 shows why the typical recommendation begins at 25 pounds.

A total weight gain of between 25 and 35 pounds for a woman starting pregnancy at normal BMI has repeatedly been shown to yield optimal health for both mother and fetus if gestation lasts at least 38 weeks. The weight gain should yield a birth weight of 7.5 pounds (3.5 kilograms). Although some extra weight gain during pregnancy is usually not harmful (about 5 to 10 pounds), it can set the stage for a pattern of weight gain

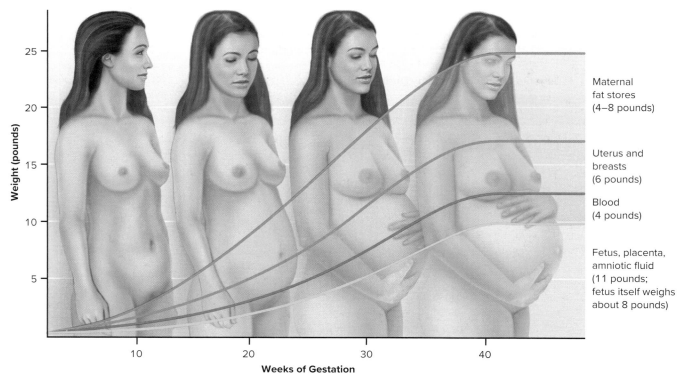

**FIGURE 14-4** ▲ The components of weight gain in pregnancy. A total weight gain of 25 to 35 pounds throughout the whole pregnancy is recommended for most women. The various components shown in this figure total about 25 pounds.

**TABLE 14-3** ■ Recommended Weight Gain in Pregnancy Based on Prepregnancy Body Mass Index (BMI)

| Prepregnancy BMI Category | Total Weight Gain* | |
|---|---|---|
| | pounds | kilograms |
| Low (BMI less than 18.5) | 28 to 40 | 12.5 to 18 |
| Normal (BMI 18.5 to 24.9) | 25 to 35 | 11.5 to 16 |
| High (BMI 25.0 to 29.9) | 15 to 25 | 7 to 11.5 |
| Obese (BMI greater than 30.0) | 11 to 20 | 5 to 9 |

*The listed values are for pregnancies with one fetus. For women of normal prepregnancy BMI carrying twins, the range is 37 to 54 pounds (17 to 24.5 kilograms), or less for heavier women.

Source: Data from *Weight Gain During Pregnancy: Reexamining the Guidelines,* Copyright 2009 by the Institute of Medicine and National Research Council of the National Academies. National Academies Press, Washington, DC.

during the childbearing years if the mother does not return to her approximate prepregnancy weight after delivery.

Women who are carrying more than one fetus need to gain additional weight to support fetal growth and development. There are provisional guidelines for women carrying multiple fetuses. Women of normal prepregnancy BMI carrying twins should aim to gain within the range of 37 to 54 pounds, whereas women who are overweight or obese and carrying twins should gain less (a total of 31 to 50 pounds or a total of 25 to 42 pounds, respectively).

Overweight and obesity contribute to complications during pregnancy. Excess maternal body fat increases risk for diabetes, hypertension, blood clots, and spontaneous abortions during pregnancy. After childbirth, lasting effects of excess gestational weight gain for the mother include increased BMI, central body fat distribution, and elevated blood pressure. For the baby, there is a greater chance of birth defects and **fetal macrosomia**, in which the fetus grows larger than average in utero. Compared to infants of normal birth weight, larger infants require surgical delivery (i.e., Cesarean sections) more often. Over the long term, excessive maternal weight gain during pregnancy has been linked to increased risk for obesity and metabolic syndrome in the child.

Gestational weight gain is a key issue in prenatal care and a concern of many mothers-to-be. Even after the 2009 release of the weight gain guidelines summarized in Table 14-3, many women report receiving *no* guidance regarding weight gain from health care providers before or during pregnancy. Considering the extensive consequences of either inadequate or excessive weight gain during gestation, health professionals should take a more active role in educating pregnant women about what weight changes to expect, the consequences of too little or too much weight gain, and how to make adjustments if the weight gain trajectory veers off course.[30,31,32] Weight gain during pregnancy should generally follow the pattern in Figure 14-4. Weekly monitoring of weight changes, especially on a chart that shows expected weight gains, can help a pregnant woman to assess how much to adjust her food intake and physical activity.

Realistic information about increased calorie requirements should be provided. Pregnant women need to understand that "eating for two" is more about increasing the *quality* of their dietary pattern (i.e., choosing nutrient-dense foods) rather than the *quantity* of foods consumed. Furthermore, successful weight management strategies involve a behavioral component; some women need to learn skills, such as self-monitoring of weight, dietary intake, and physical activity. Online tools and smartphone apps (e.g., Sprout Pregnancy) offer specialized tools to make self-monitoring tasks simple and social.

If a woman deviates from the recommended pattern of weight gain, she should make appropriate adjustments, but weight loss during pregnancy is never advised. For example, if a woman begins to gain too much weight during her pregnancy, she should *not* lose weight to get back on track. Even if a woman gains 35 pounds in the first 7 months

**fetal macrosomia** A condition in which an infant grows excessively large (e.g., birth weight > 4000 grams) in utero, usually as a consequence of maternal hyperglycemia.

Check out the pregnancy weight gain calculator at www.MyPlate.gov.

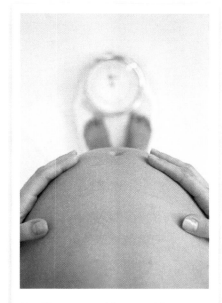

Health care providers must be on the lookout for women who exhibit disordered eating behaviors during pregnancy. Although it is not yet recognized as a distinct diagnosis, *pregorexia* is gaining attention among practitioners who work with expectant moms. Some women are exceptionally concerned about weight gain and may restrict food intake, exercise excessively, or engage in other methods of purging excess calories to prevent weight gain.[33] **What are the risks of inadequate weight gain during pregnancy?** John Slater/Getty Images

# Newsworthy Nutrition

## Maternal dietary pattern during pregnancy is linked to cognitive development of children

**INTRODUCTION:** Neurological development begins in utero and continues throughout childhood. The third trimester is a particularly important time during which the supply of nutrients, such as B vitamins, iron, zinc, and long-chain fatty acids, supports optimal development of the brain and nervous system. Maternal dietary patterns during gestation could influence the intellectual ability of offspring throughout life. **OBJECTIVE:** To examine the relationship between maternal dietary patterns during gestation and IQ of offspring at 8 years of age. **METHODS:** The Avon Longitudinal Study of Parents and Children is a prospective longitudinal study that includes more than 13,000 mother–offspring pairs in the United Kingdom who have been followed from pregnancy, through childbirth, and now into the offspring's young adulthood. The analyses presented here include 12,039 children with IQ data and whose mothers had complete dietary data. The researchers used cluster analysis to group the mothers into three distinct dietary patterns: fruits and vegetables, meat and potatoes, and white bread and coffee. (The dietary patterns differed in other ways, but these were the distinguishing features.) **RESULTS:** After adjusting for confounding variables (e.g., maternal education, various socioeconomic factors, smoking, alcohol use, prepregnancy BMI, and the child's eating pattern), children of mothers in the fruits and vegetables cluster had higher mean IQ scores at 8 years of age than children of mothers in the meat and potatoes or white bread and coffee groups. **CONCLUSION:** Consuming a nutrient-rich dietary pattern during pregnancy offers benefits for the neurological development of the offspring. This study is unique because it examines the influence of maternal dietary patterns, as opposed to individual nutrients, on offspring neurological development.

Source: Freitas-Vilela AA and others: Maternal dietary patterns during pregnancy and intelligence quotients in the offspring at 8 years of age: Findings from the ALSPAC cohort. *Maternal & Child Nutrition* 14:e12431, 2018. DOI:10.1111/mcn.12431.

of pregnancy, she must still gain more during the last 2 months. She should, however, slow the increase in weight to parallel the rise on the prenatal weight gain chart. In other words, the sources of the unnecessary calories should be found and minimized. On the other hand, if a woman has not gained the desired weight by a given point in pregnancy, she should not gain the needed weight rapidly. Instead, she should slowly gain a little more weight than the typical pattern to meet the goal by the end of the pregnancy. An RDN can help make any needed adjustments.

## PROTEIN NEEDS

The RDA for protein increases to 1.1 grams per kilogram per day, which equates to an additional 25 grams per day during pregnancy[24] (Fig. 14-5). All women should check to make sure they are eating enough protein as well as enough calories (so that protein can be spared for synthesis of new tissue). However, many women already consume protein in excess of their needs and therefore do not need to focus specifically on increasing protein intake. Small changes are usually all that is necessary. For example, simply adding a cup of fat-free milk to the daily menu adds 90 nutrient-dense kilocalories and 8 grams of protein.

## CARBOHYDRATE NEEDS

The RDA for carbohydrate increases to 175 grams daily, primarily to prevent ketosis.[24] Ketone bodies, a by-product of metabolism of fat for energy, are poorly used by the fetal brain and may impair fetal brain development. Carbohydrate intakes of most women, pregnant or not, already fulfill this requirement (see Fig. 14-5).

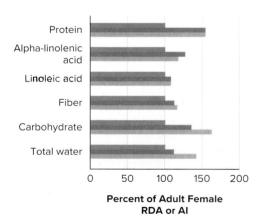

**FIGURE 14-5** ▲ Relative macronutrient and water requirements for pregnancy and breastfeeding. Note that there is no RDA or AI for total fat; needs are based on 20% to 35% of overall energy intake.

## LIPID NEEDS

Fat intake should increase proportionally with calorie intake during pregnancy to maintain around 20% to 35% of total calories from fat.[24] Pregnancy is not a time for a low-fat diet, as lipids are a source of extra calories and **essential fatty acids** needed during pregnancy. Recommendations for the types of lipids needed during pregnancy are generally the same as for females who are not pregnant: saturated fat should contribute no more than 10% of total calories and *trans* fat should be avoided.

During pregnancy, it is particularly important to make sure to consume adequate essential fatty acids: linoleic acid (omega-6) and alpha-linolenic acid (omega-3). Essential fatty acids cannot be synthesized in the body and must be consumed. For the developing fetus, essential fatty acids are required for growth, brain development, and eye development. Children whose mothers consumed sources of omega-3 fatty acids during pregnancy tend to have better sleep patterns, better speech development, higher intelligence, and fewer behavioral problems as they grow and develop.[34] For mothers, a healthy dietary pattern that provides adequate omega-3 fatty acids during pregnancy may help to reduce postpartum depression.[35]

Recommendations are slightly increased during pregnancy to 13 grams per day of linoleic acid and 1.4 grams per day of alpha-linolenic acid. These needs can be met by consuming 2 to 4 tablespoons per day of plant oils. However, usual intakes of omega-3 fatty acids among women—pregnant or not—fall short of recommendations.[36] Consumption of two to three servings (8 to 12 ounces) per week of fish is recommended for meeting needs for essential fatty acids.[37] Women who choose not to eat fish can also obtain the same omega-3 fatty acids found in fish from specially raised eggs (it's in the chicken feed!) or fish oil supplements. For supplements, consumers should choose brands that have been distilled to remove environmental contaminants.[38] (See Section 14.8 for a discussion of mercury in fish.)

Eating 8 to 12 ounces (2 to 3 servings) of fish per week will help pregnant or breastfeeding women ensure that their infants receive ample omega-3 fatty acids. It is important, however, to avoid those fish likely to be contaminated with environmental pollutants. **Which fish are likely to be highest in mercury?** David Papazian/Getty Images

## FLUID NEEDS

During pregnancy, women need extra water to support increased plasma volume, heart function, kidney function, and to ensure an adequate amount of amniotic fluid, which surrounds and protects the growing fetus. Inadequate fluid intake contributes to constipation, a common complaint during pregnancy. More severe dehydration during pregnancy can cause electrolyte imbalances and restrict the transfer of oxygen, nutrients, and wastes between the mother and fetus. The Adequate Intake for total water increases by 0.3 liter (1¼ cups) above prepregnancy needs to 3 liters (about 12½ cups) per day.[24] Remember that total water includes water from foods and beverages. Increasing fluid intake to about 10 cups per day is sufficient for most pregnant women. More fluid may be needed by women with high levels of physical activity and women who live in hot or dry climates. Thus, fluid intake goals should be individualized. An easy way to monitor hydration status is to check the color of the urine: pale yellow or straw-colored urine indicates adequate hydration.[39]

Water is the best choice for hydration, but some nutrient-dense beverages may also be included. Low-fat or fat-free milk would help to ensure adequate calcium and vitamin D intake. Although it can also be a source of added sugars, 100% fruit juice does provide some vitamins and minerals. (In general, it is better to choose whole fruit instead of fruit juice.) Unsweetened coffee and tea can also contribute to meeting fluid needs, but most authorities recommend that pregnant women limit caffeine intake to 200 milligrams per day (equivalent to about 2 cups of coffee).

Even though pregnant women do have room for some extra calories, they should not fill up on sugary beverages. High intakes of sugar-sweetened beverages promote excessive weight gain and complicate blood sugar control for the mother. There is evidence, as well, that intake of sugar-sweetened beverages may have effects on body weight and disease risk later in life for the offspring.[40,41]

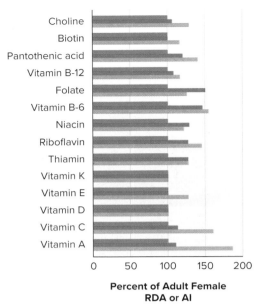

**FIGURE 14-6** ▲ Relative vitamin requirements for pregnancy and breastfeeding.

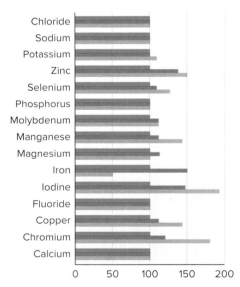

**FIGURE 14-7** ▲ Relative mineral requirements for pregnancy and breastfeeding.

## VITAMIN NEEDS

Vitamin needs increase from prepregnancy RDAs/AIs by up to 30% for most of the B vitamins, by 45% for vitamin B-6, and by 50% for folate (Fig. 14-6). Vitamin A needs only increase by 10%, so a specific focus on this vitamin is not needed.[24] Recall that excess amounts of vitamin A are harmful to the developing fetus.

**Folate.** The extra amounts of vitamin B-6 and other B vitamins (except folate) needed in the dietary pattern are easily met via nutrient-dense food choices, such as a serving of a typical ready-to-eat breakfast cereal and some lean animal protein sources. Folate needs, however, often merit strategic menu planning and possible vitamin supplementation.[12] The synthesis of DNA, and therefore cell division, requires folate, so this nutrient is especially crucial during pregnancy. Ultimately, both fetal and maternal health depend on an ample supply of folate. Red blood cell formation, which requires folate, increases during pregnancy. Serious folate-related anemia therefore can result if folate intake is inadequate. The RDA for folate increases during pregnancy to 600 micrograms DFE per day (review Section 8.13 for calculation of DFE). This is a critical goal in the nutritional care of a woman who is pregnant. Increasing folate intakes to meet 600 micrograms DFE per day during pregnancy can be achieved through dietary sources (see *Farm to Fork* in this section), a supplemental source of folic acid, or a combination of both. Choosing foods that are rich in synthetic folic acid, such as from ready-to-eat breakfast cereals or meal replacement bars (look for approximately 50% to 100% of the Daily Value), is especially helpful in meeting folate needs. Recall that synthetic folic acid is much more easily absorbed than the various forms of folate found naturally in foods.[22]

**Vitamin D.** Emerging evidence indicates that low maternal levels of vitamin D during pregnancy affect multiple health parameters in the mother and the offspring. About 54% of black women and 42% of white women have insufficient blood levels of the active form of vitamin D, even though many take a daily supplement containing 10 micrograms of vitamin D. Aside from its well-known roles in bone health and the incidence of rickets, poor vitamin D status is implicated in serious complications of pregnancy, including higher rates of blood sugar and blood pressure abnormalities during pregnancy and poor growth of the fetus.[42] Vitamin D's role in immune regulation is evidenced by higher rates of respiratory infections and mother-to-child transmission of HIV when vitamin D is deficient. Because of the ability of vitamin D to modulate gene expression, vitamin D status is critically important during early fetal development through infancy. Diseases that develop later in childhood or adulthood, such as **type 1 diabetes,** multiple sclerosis, allergies, asthma, schizophrenia, and certain types of cancer, are associated with low vitamin D status during gestation. The RDA for women during pregnancy is 15 micrograms of vitamin D daily (the same as for adult females who are not pregnant). Including 3 servings per day of dairy products and 8 to 12 ounces of seafood per week, as recommended by the Dietary Guidelines, would go a long way toward meeting the RDA for vitamin D. On average, pregnant women consume about half of the recommended servings of dairy foods and seafood.[43] Currently, there is no recommendation for universal vitamin D screening or supplementation, but the American College of Obstetricians and Gynecologists advises 25 to 50 micrograms per day for individuals with vitamin D deficiency is a safe level of supplemental vitamin D during pregnancy.[44]

## MINERAL NEEDS

Mineral needs generally increase during pregnancy, especially the requirements for iodine, iron, and zinc (Fig. 14-7).[12,45] Calcium needs do not increase but still may deserve special attention because many women find it difficult to meet their calcium needs.

**Iodine.** Pregnant women need extra iodine (RDA of 220 micrograms per day) to support **thyroid hormone** synthesis (for the mother and the developing fetus) and fetal brain development.[24] If a mother is deficient in iodine during pregnancy, she may develop a goiter (see Section 9.12), and her child may suffer from **congenital hypo-thyroidism** (formerly called *cretinism*), a devastating birth defect. Section 14.8 provides further information on this and other nutrition-related birth defects. Typical iodine intakes are plentiful if the woman uses iodized salt.

**Iron.** Extra iron (RDA of 27 milligrams per day) is needed to synthesize a greater amount of **hemoglobin** during pregnancy and to establish iron stores for the fetus.[24] Less than 5% of women begin pregnancy with clinically diagnosed iron-deficiency anemia, but an estimated one-third of pregnant women have poor iron stores, so heavy demands for iron during pregnancy cannot be met.[46] The consequences of iron-deficiency anemia—especially during the first trimester—can be severe. Negative outcomes include preterm delivery, LBW infants, and increased risk for fetal death in the first weeks after birth.

To meet the increased RDA during pregnancy, women often need a supplemental source of iron, especially if they do not consume iron-fortified foods, such as highly fortified breakfast cereals containing close to 100% of the Daily Value for iron (18 milligrams). The American College of Obstetricians and Gynecologists recommends screening for iron deficiency for all pregnant women and provision of iron supplements for women with iron deficiency. Most prenatal supplements contain iron. A potential pitfall is that iron supplements can decrease appetite and can cause nausea and constipation. To alleviate these problems, it may be helpful to take these supplements between meals or just before going to bed. Milk, coffee, or tea should not be consumed with an iron supplement because these beverages have compounds that interfere with iron absorption. On the other hand, eating foods rich in vitamin C helps increase iron absorption. Pregnant women who are not anemic may wait until the second trimester, when pregnancy-related nausea generally lessens, to start prenatal supplements if gastrointestinal side effects are a problem.

**Zinc.** The RDA for zinc increases from 8 to 11 milligrams per day during pregnancy.[24] Zinc is involved in protein synthesis and the function of many enzymes. Zinc deficiency during pregnancy has been linked to preterm and LBW births. The zinc intakes of pregnant women in the United States are generally adequate, but low-income women and those who follow vegan or vegetarian dietary patterns are more susceptible to low zinc status. Also, because iron and zinc compete for absorption, high levels of iron supplementation during pregnancy may impair zinc absorption. Incorporating lean animal proteins or a fortified ready-to-eat breakfast cereal in the dietary pattern is a good way to obtain enough extra zinc during pregnancy.

## USE OF PRENATAL VITAMIN AND MINERAL SUPPLEMENTS

With the exceptions of folate, iron, and vitamin D, the vitamin and mineral intakes of pregnant women in developed countries are generally adequate.[47] Research evidence supports

## FARM to FORK    Greens

Mary-Jon Ludy

Leafy green vegetables are good sources of folate, a vital nutrient for the prevention of neural tube defects.

### Grow
- Even if you don't have space for gardening, it's easy to grow lettuce and other greens in containers on your patio or balcony. Garden-fresh greens will be ready to eat in 30 to 60 days.
- Seed catalogs offer a much wider variety of greens than you can find in the grocery store. Find greens that will grow where you live, and try something new and nutritious!

### Shop
- To identify greens that are highest in phytochemicals and nutrients, look for the deepest colors. Dark green, red, purple, and reddish-brown leaves have the most disease-fighting phytochemicals and the highest levels of micronutrients, including folate.
- Choose lettuces with leaves that are loose and open. These lettuce leaves produce more phytochemicals to protect themselves from the sun. Lettuces such as iceberg, with tightly wrapped leaves, provide only a fraction of the nutrients found in more colorful, loosely packed leaves.
- Fresh heads of lettuce or bunches of spinach tend to last longer than precut, prewashed greens sold in bags and plastic containers.

### Store
- Greens need just the right amount of humidity and oxygen to stay fresh. After you bring home fresh greens, pull the lettuce apart, soak it in cold water for about 10 minutes, and spin it dry before storing. This increases the moisture content within the leaves but keeps the outside of the leaves dry, delaying spoilage.
- Use a pin to prick 10 to 20 holes in a gallon-size plastic bag. Store your greens in the perforated bag in the crisper drawer of your refrigerator to maintain optimal freshness. The bag helps the greens to retain their moisture. The tiny holes allow for some exchange of gases.

### Prep
- Even though plants with the darkest colors have the most nutritional value, some people do not tolerate their somewhat bitter or peppery flavors. Harvesting the leaves early (i.e., baby lettuce greens) will yield a milder salad. Try mixing some bitter greens, such as arugula, with milder varieties, such as romaine or Bibb lettuce. Add sweet flavors, such as chopped fruit or a touch of honey in your salad dressing, to balance the stronger flavors.
- Don't skimp on the dressing! Many of the healthy compounds in these plants are fat-soluble. You will absorb more of the vitamin E and phytochemicals if you consume some fat as part of your meal. Olive oil is an excellent choice for homemade salad dressing.
- Boiling causes more than half of the nutrients to leach out of the vegetables and into the cooking water. If you cook your greens, sauté, steam, or microwave them.

Source: Robinson J. From wild greens to iceberg lettuce: Breeding out the medicine, in *Eating on the Wild Side*. New York: Little, Brown and Company, 2013.

D. Hurst/Alamy Stock Photo

routine supplementation with folic acid for the prevention of neural tube defects during pregnancy. Supplementation with iron is beneficial specifically for women at risk of iron-deficiency anemia.[48] Beyond these nutrients, some studies indicate that use of a multivitamin and mineral supplement is advantageous for reducing the number of LBW and SGA births.[49] More research is needed to know if prenatal supplements are effective for reducing many other pregnancy complications.[50] To date, there is not enough evidence to recommend prenatal multivitamin and mineral supplements for all pregnant women, but they are prescribed routinely by most health care providers. Some supplements formulated for pregnancy are sold over the counter, while others are dispensed by prescription because of their high synthetic folic acid content (1000 micrograms), which could pose problems for others, such as people who are older. These supplements are high in iron (27 milligrams per pill). There is no evidence that use of such supplements causes significant health problems in pregnancy, with the possible exception of combined amounts of supplemental and dietary vitamin A (see Section 14.8).

Instances when prenatal supplements may especially contribute to a successful pregnancy are those involving women living in poverty, teenagers, women with an inadequate dietary pattern, women carrying multiple fetuses, women who smoke or use alcohol or illegal drugs, and women who follow a vegan dietary pattern. In other cases, healthy eating patterns can provide the needed nutrients. When choosing a multivitamin, discuss options with a primary care provider and rely on brands that display the USP symbol on their label, signifying that the supplement meets the content, quality, purity, and safety standards of the United States Pharmacopeial Convention (see Section 8.17). Of course, avoid megadoses of any nutrient. Skip supplements containing herbs, enzymes, and amino acids. Many of these ingredients have not been evaluated for safety during pregnancy or breastfeeding and may be toxic to the fetus. Furthermore, discard supplements that are past the expiration date, as some ingredients lose potency over time.

### ✓ CONCEPT CHECK 14.4

1. Danica is a moderately active, 25-year-old woman who required 2200 kcal per day before she became pregnant. How many kilocalories will she need each day during her first, second, and third trimesters?

2. What is the optimal range of weight gain during pregnancy for a woman who begins pregnancy at a healthy BMI? How does this differ for a woman who begins pregnancy underweight? Overweight? Obese?

3. List two nutrients that may need to be supplemented in the dietary pattern of a pregnant woman and the reason for each.

## 14.5 Eating Patterns for Pregnant Women

One dietary approach to support a successful pregnancy is based on MyPlate. For an active 24-year-old woman, about 2200 kcal is recommended during the first trimester (the same as recommended for such a woman when not pregnant). The plan should include 7 ounce equivalents from the grains group, 6 ounce equivalents from the protein foods group, 3 cups from the dairy group (or dairy alternatives), 3 cups of vegetables, and 2 cups of fruit. In addition, 6 teaspoons per day of plant oil, used in cooking or as salad dressing, will supply essential fatty acids. Figure 14-8 shows a balanced dietary pattern, based on MyPlate, for a woman in the first trimester. By selecting nutrient-dense foods from the food groups, the dietary pattern should be moderate in sodium (no more than 2300 milligrams per day), saturated fat (limit to 10% of total calories), and added sugars (limit to 10% of total calories).

In the second and third trimesters, about 2600 kcal is recommended for this woman. The plan should now include an additional 2 ounce equivalents from the grains group,

This lunch of spinach salad has several sources of nonheme iron. **Is iced tea a good beverage choice? Why or why not?** TheCrimsonMonkey/ Getty Images

**FIGURE 14-8 ▶** Daily meal plan for a pregnant woman based on MyPlate. Servings per day for the first trimester (2200 kcal per day) are shown in dark colors. Additional servings per day for the second and third trimesters (2600 kcal per day) are shown in lighter colors.

Source: United States Department of Agriculture, Find Your Healthy Eating Style, retrieved from https://www.myplate.gov/myplate-plan/results/2600-calories-ages-14-plus

### Grains

Choices from the grains group should focus on whole grains and enriched foods. One ounce of a whole grain, ready-to-eat breakfast cereal significantly contributes to meeting many vitamin and mineral needs.

### Protein Foods

Foods from the protein group help to provide the extra iron and zinc needed during pregnancy. Within the protein foods group, most people would benefit from selecting plant-based sources more often.

### Vegetables

One cup of vegetables should be a green vegetable or other rich source of folate.

### Fruit

One cup of fruit each day should be a good vitamin C source.

### Dairy

Choices from the dairy group should include low-fat or fat-free versions of milk, yogurt, and cheese. These foods are good sources of protein, calcium, and vitamin D. For those who cannot tolerate dairy products, soy milk, other dairy alternatives, and calcium-fortified foods can be useful to meet nutrient needs.

**cups**     **cups**     **ounce equivalents**     **ounce equivalents**     **cups**

an extra ½ ounce equivalent from the protein group, an additional ½ cup from the vegetables group, and 2 additional teaspoons of plant oil throughout the day. These additional servings are shown in lighter color in Figure 14-8.

Table 14-4 illustrates one daily menu based on the 2600-kcal plan for pregnancy for women in the second or third trimesters. This menu meets the extra nutrient needs associated with pregnancy. Women who need to consume more than this—and some do for various reasons—should incorporate additional fruits, vegetables, and whole grain breads and cereals, not nutrient-poor snacks and sugar-sweetened beverages. Try the "Rate Your Plate" activity in Connect to apply your knowledge of meal planning for pregnant women.

Moms-to-be can personalize MyPlate for each trimester at www.myplate.gov/myplate-plan.

## PREGNANT VEGETARIANS

Women who are either lactoovovegetarians or lactovegetarians generally do not face special difficulties in meeting their nutritional needs during pregnancy. Although many people assume that vegetarian dietary patterns are deficient in protein, as long as a woman meets her energy needs, she is likely to obtain adequate protein. Women who follow a vegetarian dietary pattern during pregnancy should plan to consume a variety of plant sources of protein to get all of the **essential amino acids.**

On the other hand, for a woman who follows a vegan dietary pattern during pregnancy, careful meal planning during preconception and pregnancy is crucial to ensure sufficient iron, zinc, calcium, vitamin D, vitamin B-12, and omega-3 fatty acids, in addition to protein.[51] The basic vegan dietary pattern described in Section 6.8 should be modified to include more whole grains, beans, nuts, and seeds to supply the necessary extra amounts of some of these nutrients. As mentioned, use of a prenatal multivitamin

# 578  Contemporary Nutrition

**TABLE 14-4 ■ Sample 2600-kcal Daily Menu That Meets Nutritional Needs During Pregnancy and Breastfeeding for Most Women\***

| | Vitamin B-6 | Folate | Iron | Zinc | Calcium |
|---|:---:|:---:|:---:|:---:|:---:|
| **Breakfast** | | | | | |
| 1 cup multigrain cereal | ✓ | ✓ | ✓ | ✓ | ✓ |
| 1 cup orange juice | | ✓ | | | |
| 1 cup fat-free milk | ✓ | | | | ✓ |
| **Snack** | | | | | |
| 2 tbsp peanut butter | ✓ | ✓ | ✓ | ✓ | |
| 2 stalks of celery | | ✓ | | | |
| 1 slice whole wheat toast | | ✓ | ✓ | ✓ | |
| 1 cup plain low-fat yogurt | ✓ | | | | ✓ |
| ½ cup strawberries | | ✓ | | | |
| **Lunch** | | | | | |
| 2 cups spinach and fruit salad with 2 tbsp oil and vinegar dressing | | ✓ | | | ✓ |
| 2 slices whole wheat toast | | ✓ | ✓ | ✓ | |
| 1½ ounces provolone cheese | ✓ | | | | ✓ |
| **Snack** | | | | | |
| 5 whole wheat crackers | | ✓ | ✓ | ✓ | |
| 1 cup grape juice | | | | | |
| **Dinner** | | | | | |
| 2 bean burritos | ✓ | ✓ | ✓ | ✓ | ✓ |
| 1 cup cooked broccoli | | ✓ | | | ✓ |
| 1 tsp soft margarine | | | | | |
| Unsweetened iced tea | | | | | |
| **Snack** | | | | | |
| Granola bar (2 ounces) | | | | | |
| ½ banana | ✓ | ✓ | ✓ | ✓ | |

*This dietary plan meets nutrient needs for pregnancy and breastfeeding. Check marks indicate that the menu item is a good source of the nutrient, whereas the lack of a check indicates a poor source of the nutrient. The vitamin- and mineral-fortified breakfast cereal used in this example makes an important contribution to meeting nutrient needs. Fluids can be added as desired. Total intake of fluids, such as water, should be 10 cups per day for pregnant women or about 13 cups per day for breastfeeding women.

and mineral supplement also is generally advocated to help fill micronutrient gaps. However, although these are high in iron, this is not true for calcium (200 milligrams per pill). If iron and calcium supplements are used, they should not be taken together to avoid possible competition for absorption.

## ✓ CONCEPT CHECK 14.5

1. In Chapter 5, review the Mediterranean diet and how it compares to MyPlate. Is the Mediterranean diet suitable to meet the nutrient needs of a pregnant woman?
2. Modify the sample 2600-kcal daily menu in Table 14-4 so that it would be suitable for a woman who follows a vegan dietary pattern during pregnancy.
3. Michaela tells you about her pregnancy craving for ice cream. She says she has been eating one or two ice cream bars after lunch every day and usually stops for a milkshake at a fast-food restaurant on her way home from work. What nutrition information would you share with her?

## CASE STUDY  Eating for Two

Lily and her husband have just found out that Lily is pregnant with their first child. She is 25 years old, weighs 135 pounds, and is 67 inches tall. Lily has been reading everything she can find on pregnancy because she knows that her health is important to the success of her pregnancy.

Lily knows she should avoid alcohol, especially because alcohol is potentially toxic to the growing fetus. Lily is not a smoker, does not take any medications, and limits her coffee intake to 4 cups a day and soft drink intake to 3 colas per day. Based on her reading, she has decided to breastfeed her infant and has already inquired about childbirth classes. She has modified her dietary pattern to include some extra protein, along with more fruits and vegetables. She has also started taking an over-the-counter vitamin and mineral supplement. Lily has always kept in good shape, and she is admittedly worried about gaining too much weight during pregnancy. Recently, she started a running program 5 days a week, and she plans to continue running throughout her pregnancy.

1. What recommendations do you have regarding Lily's use of dietary supplements?
2. What is Lily doing to prevent neural tube defects? What else could she do?
3. What recommendations would you make regarding Lily's caffeine consumption?
4. Should Lily include fish as a source of protein in her meals? Why or why not?
5. Constipation is a common complaint during pregnancy. What suggestions do you have to help Lily avoid this health concern?
6. What information would you share with Lily about appropriate weight gain during pregnancy?

*Complete the Case Study. Responses to these questions can be provided by your instructor.*

▲ Lily and her husband want to do everything they can to ensure a healthy pregnancy. Hill Street Studios/Blend Images LLC

## 14.6 Nutrition-Related Concerns During Pregnancy

During pregnancy, fetal needs for oxygen and nutrients as well as excretion of waste products increase the burden on the mother's lungs, heart, and kidneys. Although a mother's organ systems work efficiently, some discomfort may accompany the changes her body undergoes to accommodate the fetus.

### HEARTBURN

Hormones (such as progesterone) produced by the placenta relax smooth muscles in the uterus and the gastrointestinal tract. This often causes heartburn as the **lower esophageal sphincter** relaxes, allowing stomach acid to reflux into the esophagus (review Section 3.11). To prevent or reduce heartburn, the woman should avoid lying down for at least three hours after eating, choose meals that are lower in fat so that foods pass more quickly from the stomach into the small intestine, and avoid spicy foods, which can worsen symptoms. She should also consume most liquids between (rather than with) meals to limit the volume of stomach contents; this helps to decrease some of the pressure inside the stomach that encourages reflux. Women with more severe cases should consult a primary care provider about using antacids or related medications.[52]

### CONSTIPATION AND HEMORRHOIDS

The hormone-induced relaxation of muscles in the GI tract may also lead to constipation. This is especially likely to develop late in pregnancy, when the fetus competes with the GI tract for space in the abdominal cavity. To alleviate constipation, a woman should focus on consuming adequate fluid and fiber and should perform regular physical activity. The AI for fiber in pregnancy is 28 grams, slightly more than for the nonpregnant woman. Fluid needs are 10 cups per day. The extra iron in prenatal supplements

Sandy, 5 months pregnant, has been having heartburn and difficult bowel movements. As a student of nutrition, you understand the digestive system and the role of nutrition in health. **What dietary and lifestyle strategies might you suggest to Sandy to relieve her digestive complaints?** antoniodiaz/123RF

may also contribute to constipation. If this is the case, a clinician can reevaluate the woman's needs for supplemental iron. Alleviating constipation can also help to prevent hemorrhoids, which affect about one-third of women during pregnancy.

## EDEMA

Placental hormones cause various body tissues to retain fluid during pregnancy. Blood volume also greatly expands during pregnancy. The extra fluid may cause some swelling (edema), especially in the extremities. There is no reason to restrict salt severely or use **diuretics** to limit mild edema. Mild edema during pregnancy simply reflects normal and expected changes in plasma volume and kidney function. However, the edema may limit physical activity late in pregnancy and occasionally requires a woman to elevate her feet or wear compression stockings to control the symptoms. Overall, edema is only a mild nuisance unless it is accompanied by hypertension and the appearance of extra protein in the urine (see "Hypertensive Disorders of Pregnancy" later in this section).

## NAUSEA AND VOMITING OF PREGNANCY

About 70% to 85% of women experience nausea during the early stages of pregnancy. Nausea and vomiting of pregnancy are probably related to changes in GI motility induced by pregnancy-related hormones. Although commonly called "morning sickness," pregnancy-related nausea may occur at any time and may persist all day. It is often the first signal to a woman that she is pregnant.

To help control mild nausea during pregnancy, women can try the following: avoiding foods that are likely to trigger nausea (e.g., greasy, spicy, or acidic foods); cooking with good ventilation to dissipate nauseating smells; eating saltine crackers or dry cereal before getting out of bed; avoiding large fluid intake early in the morning; and eating smaller, more frequent meals instead of two or three large meals per day. The iron in prenatal supplements triggers nausea in some women, so changing the type of supplement used or postponing use until the second trimester may provide relief in some cases. If a woman thinks her morning sickness is exacerbated by her prenatal supplement, she should talk with her primary care provider about switching to another supplement.

Usually, nausea is mild and stops after the first trimester. However, in up to 3% of pregnancies, nausea and vomiting are prolonged and severe enough to cause weight loss, dehydration, and dangerous electrolyte imbalances. In these cases, the preceding practices offer little relief; medical attention is needed to protect the health of the mother and her growing baby.

## ANEMIA

To supply the fetus with nutrients and oxygen, the mother's blood volume expands by approximately 50%. The number of red blood cells, however, increases by only 20% to 30%, and this occurs more gradually. As a result, a woman has a lower ratio of red blood cells to total blood volume in her system during pregnancy. This hemodilution is known as **physiological anemia.** It is a normal response to pregnancy, rather than the result of inadequate nutrient intake.

However, if iron stores and/or dietary iron intake are not sufficient to meet her increased needs during pregnancy, the woman may develop iron-deficiency anemia. The *Dietary Guidelines* identify iron as a nutrient of public health concern because several population groups, including pregnant women, are likely to have low intakes of this nutrient. Iron-deficiency anemia requires medical attention at any stage of the life cycle, but consequences for the fetus are especially severe (see Section 14.4).[53] Maternal anemia is associated with LBW, preterm birth, and infant mortality. To identify problems with iron status during pregnancy, all women should be screened for iron deficiency at least once per trimester. Lean meats, beans, and fortified grains are nutrient-dense sources of iron that should be emphasized to prevent anemia during pregnancy. Importantly, once anemia has developed, *dietary supplements will be needed to replenish iron status;* dietary changes alone are not sufficient to correct iron deficiency.

Some studies show a potential benefit of foods or dietary supplements made with ginger to alleviate nausea and vomiting of pregnancy for some women. **However, why is it important to consult your primary care provider before taking dietary supplements?** Toltek/iStock/Getty Images

**physiological anemia** The normal decrease in red blood cell concentration in the blood due to increased blood volume during pregnancy; also called *hemodilution.*

## GESTATIONAL DIABETES

Hormones synthesized by the placenta decrease the efficiency of insulin. This leads to a mild increase in blood glucose, which is normal and helps supply calories to the fetus. If the rise in blood glucose becomes excessive, this leads to **gestational diabetes,** often beginning between weeks 20 and 28, particularly in women who have a family history of diabetes or who had a prepregnancy BMI in the obese range. Other risk factors include maternal age over 35 and gestational diabetes in a prior pregnancy.

In North America, gestational diabetes develops in as many as 10% of pregnancies (estimates vary based on the diagnostic criteria used). Today, women with risk factors for **type 2 diabetes** (e.g., obesity, family history) should be screened for undiagnosed type 2 diabetes at the first prenatal visit. Pregnant women without type 2 diabetes should be screened for gestational diabetes between 24 and 28 weeks. Various testing methods are used but typically involve being given an oral dose of glucose, then testing blood glucose levels one, two, or three hours later.[54] If gestational diabetes is detected, a special dietary pattern that distributes carbohydrates throughout the day must be implemented. Carbohydrate choices should be mostly whole, unprocessed grains, vegetables, fruits, and beans, which have a lower impact on blood glucose than refined grains or foods with lots of added sugars. Sometimes insulin injections or oral medications are also needed. Regular physical activity also helps to control blood glucose.

The primary risk of uncontrolled diabetes during pregnancy is fetal macrosomia, which means that the fetus grows to a large size in utero. This is a result of the oversupply of glucose from maternal circulation coupled with an increased production of insulin by the fetus, which allows fetal tissues to take up an increased amount of building materials for growth. The mother may require a Cesarean section if the fetus is too large for a vaginal delivery. Another threat is that the infant may have low blood glucose at birth because of the tendency to produce extra insulin that began during gestation. Other concerns are the potential for early delivery, low iron stores, and increased risk of birth trauma and malformations.

Although gestational diabetes often disappears after the infant's birth, it increases the mother's risk of developing diabetes later in life, especially if she fails to maintain a healthy body weight. **Hyperglycemia** during gestation may have some long-term repercussions for the child as well. Studies show that infants of mothers with gestational diabetes may also have higher risks of developing obesity, metabolic syndrome, and type 2 diabetes as they grow to adulthood. For all these reasons, proper control of blood glucose during pregnancy is extremely important.

Researchers are working to understand how breastfeeding may be helpful for women who had gestational diabetes. Some studies indicate that breastfeeding may lower the woman's risk of developing type 2 diabetes later in life.[55]

## HYPERTENSIVE DISORDERS OF PREGNANCY

Hypertension occurs in about 6% to 8% of pregnancies in the United States.[56] Sometimes, women enter pregnancy with chronic hypertension. However, when hypertension first appears after 20 weeks of gestation, it is termed **gestational hypertension** (formerly called *pregnancy-induced hypertension*).

About half of women with gestational hypertension eventually develop **preeclampsia** (mild form) or **eclampsia** (severe form). Early symptoms include a rise in blood pressure, excess protein in the urine, edema, changes in blood clotting, headache, and visual disturbances. Very severe effects, including convulsions, can occur in the second and third trimesters. If not controlled, eclampsia eventually damages the liver and kidneys, and the mother and fetus may die.

The causes for hypertensive disorders of pregnancy are not well understood but likely involve interactions among genetics, certain environmental or lifestyle influences, and abnormal function of the placenta. The populations most at risk for these disorders are women under age 17 or over age 35, women who are overweight or obese, and those who have had multiple-birth pregnancies. A family history of gestational hypertension in the mother's or father's side of the family, diabetes, African-American race, and a first pregnancy also raise risk.

**gestational diabetes** A high blood glucose concentration that develops during pregnancy and returns to normal after birth; one cause is the placental production of hormones that antagonize the regulation of blood glucose by insulin.

**gestational hypertension** Blood pressure of 140/90 mmHg or higher that is first diagnosed after 20 weeks of gestation. This may evolve into preeclampsia or eclampsia. Note that the current cutoff for gestational hypertension is slightly higher than the recently updated threshold for diagnosis of chronic hypertension among the general population.

**preeclampsia** A form of gestational hypertension characterized by protein in the urine.

**eclampsia** A severe form of gestational hypertension characterized by protein in the urine and seizures; formerly called *toxemia*.

▲ The Baby-Friendly Hospital Initiative promotes and supports breastfeeding. Guidelines include training hospital personnel to assist mothers with lactation, initiating breastfeeding within 30 minutes of birth, allowing infants to "room in" with mothers, not offering formula or pacifiers to newborn babies, and having a plan to support breastfeeding moms after they leave the hospital. Simplefoto,Tyler Olson/leaf/123RF

Gestational hypertension resolves once the pregnancy ends, making delivery the most reliable treatment for the mother. However, if eclampsia develops before the fetus is ready to be born, medical intervention is necessary. Bed rest and administration of magnesium sulfate are the most effective treatment methods. Magnesium likely acts to relax blood vessels and so leads to a reduction in blood pressure. There is good evidence that adequate intakes of calcium and vitamin D are involved in reducing incidence of gestational hypertension. There is interest in the use of antioxidants to prevent or treat gestational hypertension, but current evidence does not support the use of antioxidant supplements for preventing these disorders.[57,58] Severe sodium restrictions are not recommended for prevention or treatment of gestational hypertension (see *Fake News*). Several other treatments are under study. In the meantime, the best advice is to comply with the key recommendations of the *Dietary Guidelines,* aiming to meet nutrient needs with a variety of foods.

**✓ CONCEPT CHECK 14.6**

1. What dietary and lifestyle strategies would you suggest to a pregnant woman who complains of morning sickness?
2. Define *gestational diabetes.* What are the potential consequences of this disorder for the mother and baby?
3. Differentiate between chronic hypertension and gestational hypertension. Have any dietary supplements been shown to reliably prevent or treat preeclampsia?

# 14.7 Breastfeeding

Breastfeeding provides the new infant optimal nutrition from the very start of life. The *Dietary Guidelines,* the Academy of Nutrition and Dietetics (AND), and the American Academy of Pediatrics (AAP) recommend breastfeeding *exclusively* (i.e., with no other foods or beverages) for the first 6 months, followed by the combination of breastfeeding and infant foods until 1 year.[61,62] The World Health Organization goes beyond that to recommend breastfeeding (along with age-appropriate solid foods; see Section 15.3) for at least 2 years. Surveys show that about 80% of North American mothers breastfeed their infants in the hospital. By 6 months, about half of mothers are still breastfeeding their infants, but only about 25% are exclusively breastfeeding. At 1 year of age, about one-third of mothers are still breastfeeding (in addition to providing solid foods).[63]

Women who choose to breastfeed usually find it an enjoyable, special time in their lives that strengthens the bond with their new infant. Although bottle feeding with an infant formula is also safe for infants, it cannot replicate all of the benefits derived from breastfeeding. If a woman does not breastfeed her child, milk production ceases within a few days after birth.

## PLANNING TO BREASTFEED

Almost all women are physically capable of breastfeeding their children (see the later section "Medical Conditions Precluding Breastfeeding" for exceptions). In most cases, problems encountered in breastfeeding are due to a lack of knowledge or support. Anatomical problems in breasts, such as inverted nipples, can be corrected during pregnancy. Breast size generally increases during pregnancy and is no indication of success in breastfeeding. Most women notice a dramatic increase in the size and weight of their breasts by the third or fourth day of breastfeeding. If these changes do not occur, a woman needs to seek advice from her primary care provider or a **lactation consultant**.

Breastfed infants must be followed closely by caregivers over the first few days of life to ensure that feeding and weight gain are proceeding normally. Monitoring is especially important with a mother's first child because the mother will be inexperienced with the technique of breastfeeding. The infant is at risk of undernutrition and dehydration if feeding does not proceed smoothly.

Although it is the most natural way to feed a newborn child, the *technique* does not always come naturally. Gathering information on breastfeeding, what obstacles to anticipate, and how to respond to such obstacles will help new mothers and their babies to succeed with breastfeeding. It also helps to have an experienced friend, family member, or a professional lactation consultant to call for advice when questions arise.

## HUMAN MILK PRODUCTION

During pregnancy, breast weight increases by 1 to 2 pounds due to the action of placental hormones that stimulate the formation of milk-producing cells. Milk will be produced in the **lobules** (Fig. 14-9). A cluster of lobules forms a lobe. Ducts lead from each lobe to the nipple, where milk is ejected from the breast.

Most of the protein found in human milk is synthesized by breast tissue. Some proteins also enter the milk directly from the mother's bloodstream. These proteins include immune factors (e.g., antibodies) and enzymes. Some of the fats in human milk come from what the mother has eaten; other fats are synthesized by breast tissue. The sugar galactose is synthesized in the breast, whereas glucose enters from the mother's bloodstream. Together, these sugars form lactose, the main carbohydrate in human milk.

Lactation is regulated by two hormones: prolactin and oxytocin. **Prolactin** is the hormone that stimulates milk production in the lobules of the breast. As milk is produced between feedings, it is stored in the lobules. An important brain–breast connection—commonly called the **let-down reflex**—is necessary to release the milk stored in the lobules. This reflex occurs in response to the infant suckling at the breast. Stimulation of the nerves in the nipple area signals the pituitary gland to release **oxytocin,** which allows the lobules to contract and let down (release) stored milk (Fig. 14-10). The milk then travels via ducts to the nipple. The mother may feel a tingling sensation shortly before milk flow begins. Throughout the feeding, the suckling of the infant further stimulates prolactin release from the pituitary gland, so milk synthesis within the lobules continues. The more the infant suckles, the more milk is produced. As the quantity of milk production increases, the quality does not change. Because of this, even twins (and triplets) can be breastfed adequately.

If the let-down reflex does not operate properly, the infant will not receive enough milk. The infant gets frustrated, which then frustrates the mother. The let-down reflex is easily inhibited by nervous tension, a lack of confidence, and fatigue. Mothers should be especially aware of the link between tension and a weak let-down reflex. They need to find a relaxed environment where they can breastfeed.

After a few weeks, the mother's let-down reflex becomes automatic. Milk ejection can be triggered just by thinking about her infant or seeing or hearing another one cry. At first, however, the process can be a bit bewildering. A mother cannot directly measure the amount of milk the infant takes in, so she may fear that she is not adequately nourishing the infant.

As a general rule, a well-nourished breastfed infant should: (1) have 3 to 5 wet diapers per day by 3 to 5 days of age and 4 to 6 wet diapers per day thereafter; (2) show a normal pattern of weight gain; and (3) pass at least one or two stools per day that look like lumpy mustard. In addition, softening of the breast during the feeding indicates that milk is being removed from the breast. Parents who sense their infant is not consuming enough milk should consult their primary care provider immediately because dehydration can develop rapidly. It is normal for newborn infants to lose some weight (up to 10% of birth weight) in the first few days after birth. However, losing more than 10% of birth weight indicates a feeding problem that requires intervention.[64]

It generally takes 2 to 3 weeks to fully establish the feeding routine: infant and mother begin to feel comfortable, the milk supply

**lactation consultant** Health care professional (often a registered nurse or RDN) with special training to provide education and support for breastfeeding mothers and their infants.

**lobules** Saclike structures in the breast that store milk; also called *alveoli.*

**prolactin** A hormone secreted by the pituitary gland that stimulates the synthesis of milk in the breast.

**let-down reflex** A reflex stimulated by infant suckling that causes the release (ejection) of milk from milk ducts in the mother's breasts; also called *milk ejection reflex.*

**oxytocin** A hormone secreted by the pituitary gland. It causes contraction of the musclelike cells surrounding the ducts of the breasts and the smooth muscle of the uterus.

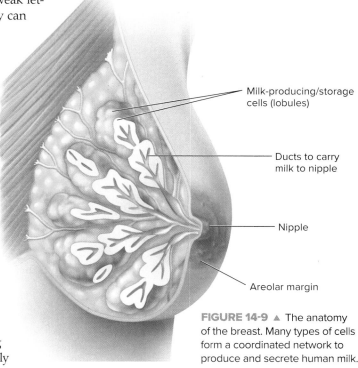

Milk-producing/storage cells (lobules)

Ducts to carry milk to nipple

Nipple

Areolar margin

**FIGURE 14-9** ▲ The anatomy of the breast. Many types of cells form a coordinated network to produce and secrete human milk.

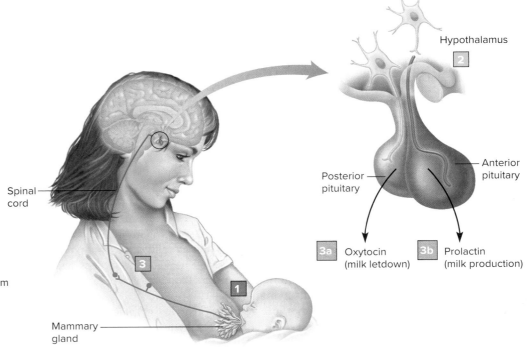

1. Suckling stimulates nerves in the nipple and areola that transmit impulses to the hypothalamus.

2. In response, the hypothalamus stimulates the posterior pituitary to release oxytocin and the anterior pituitary to release prolactin.

3. Oxytocin stimulates lobules in the breast to let down (release) milk from storage. Prolactin stimulates additional milk production.

**FIGURE 14-10** ▲ Let-down reflex. Suckling sets into motion the sequence of events that lead to milk let-down, the flow of milk into ducts of the breast. Both milk production and let-down are regulated by hormones produced by the pituitary gland.

*magnificent*

**Did you know that human breast milk is not sterile? Along with human milk oligosaccharides, the unique collection of microorganisms from the mother's breast helps to colonize the infant's GI tract.**

*microbiome*

**colostrum** The first fluid secreted by the breast during late pregnancy and the first few days after birth. This thick fluid is rich in immune factors and protein.

**human milk oligosaccharides** Small, indigestible carbohydrates made in the human breast from lactose and other simple sugars.

meets infant demand, and initial nipple soreness disappears. Establishing the breastfeeding routine requires patience, but the rewards are numerous. The adjustments are easier if supplemental formula feedings are not introduced until breastfeeding is well established, after at least 3 to 4 weeks. Then it is fine if a supplemental bottle or two of infant formula per day is needed, but supplemental feedings will decrease the infant's demand for the breast, and thereby decrease milk production.

## NUTRITIONAL QUALITIES OF HUMAN MILK

Human milk is different in composition from cow's milk. Unless altered (i.e., to make infant formula), cow's milk should never be used in infant feeding until the infant is at least 12 months old. Unaltered cow's milk is too high in minerals and protein and does not contain enough carbohydrate to meet an infant's nutritional needs. In addition, the major protein in cow's milk is harder for an infant to digest than the major proteins in human milk. The proteins in cow's milk may also spur allergies in some infants.

**Colostrum.** At the end of pregnancy, the first fluid made by the human breast is **colostrum.** This thick, yellowish fluid may leak from the breast during late pregnancy and is produced in earnest for a few days after birth. Colostrum differs from mature human milk in that it contains a higher concentration of antibodies, immune system cells, and growth factors, some of which pass intact through the lining of the infant's immature GI tract into the bloodstream. The first few months of life are the only time when we can readily absorb whole proteins, such as these immune factors, across the GI tract. These immune factors and cells protect the infant from some GI tract diseases and other infectious disorders, compensating for the infant's immature immune system during the first few months of life. These protective qualities of colostrum are an important reason for mothers to nurse their infants if even for just the first few weeks after birth.

Colostrum is particularly rich in **human milk oligosaccharides.** These small carbohydrates cannot be digested by human enzymes, but they help to establish a healthy gut microbiota by promoting the growth of beneficial bacteria, including *Lactobacillus bifidus.* These beneficial bacteria limit the growth of potentially toxic bacteria in the intestine. Human milk oligosaccharides are thought to promote the health of the infant's gut and stimulate normal immune function. This is one way breastfeeding supports the infant's digestive health.[65]

**Mature Milk.** Human milk composition gradually changes from colostrum to mature milk several days after delivery. Human milk looks very different from cow's milk. (Table 15-3 provides a direct comparison.) Human milk is thin and almost watery in appearance and often has a slightly bluish tinge. Its nutritional qualities, however, are impressive!

Human milk's proteins form a soft, light curd in the infant's stomach and are easy to digest. Some human milk proteins offer immune protection. Still others bind iron, reducing the growth of some bacteria that can cause diarrhea.

The lipids in human breast milk are high in linoleic acid and cholesterol, which are needed for brain development. Breast milk also contains long-chain omega-3 fatty acids, such as docosahexaenoic acid. This polyunsaturated fatty acid is used for the synthesis of tissues in the central nervous system, especially in the brain and the retina of the eye.

The fat composition of human milk changes during each feeding. The consistency of milk released at the start of a feeding (*fore milk*) resembles that of skim milk. The milk released after 10 to 20 minutes of feeding (*hind milk*) is much higher in fat—like cream. Babies need to nurse long enough (e.g., a total of 20 or more minutes) to get the calories in the rich hind milk to be satisfied between feedings and to grow well. The overall calorie content of infant formulas has been based on that of human milk (on average, 67 kcal per 100 milliliters).

Human milk composition also allows for adequate fluid status of the infant, provided the baby is exclusively breastfed. Caregivers often wonder if the infant needs additional water if stressed by hot weather, diarrhea, vomiting, or fever. The AAP advises against supplemental water or juice during the first 6 months of life. The practice may unnecessarily introduce pathogens or allergens. Excessive water can lead to brain disorders, low blood sodium, and other problems. Thus, before 6 months of age, supplemental fluids should be given only with a clinician's guidance.

**Human Milk for Preterm Infants.** Can a mother breastfeed her preterm infant? In some cases, human milk is the most desirable form of nourishment, depending on infant weight and length of gestation. Feeding of human milk to preterm infants has been linked to lower infant mortality, decreased risk of infections, reduced stays in the neonatal intensive care unit, fewer hospital readmissions, better growth, and improved brain development.[62]

Breastfeeding a preterm infant, however, demands a high level of dedication from caregivers. Preterm infants sometimes have a weak or uncoordinated suck-and-swallow reflex, so it may be necessary to express milk from the breast and feed that milk to the infant through a tube until the infant's feeding ability improves. Fortification of the milk with calcium, phosphorus, sodium, and protein is often necessary to meet the needs of a rapidly growing preterm infant. In some cases, special feeding problems may prevent the use of human milk or necessitate supplementing it with specialized formula. Sometimes intravenous feeding is the only option. Working as a team, the pediatrician, neonatal nurses, and RDN will guide the parents in this decision.

## EATING PATTERN FOR WOMEN WHO BREASTFEED

Nutrient needs for a woman who breastfeeds change to some extent from those of the pregnant woman in the second and third trimester (review Fig. 14-6 and Fig. 14-7 and see the DRI summary tables in Appendix G). The DRIs for folate and iron decrease while the DRIs for calories; vitamins A, E, and C; riboflavin; copper; chromium; iodine; manganese; selenium; and zinc increase. Still, these increased needs of the breastfeeding woman will be met by the general dietary plan proposed for a woman in the latter stages of pregnancy:

- 3 cups from the dairy group (which includes fortified soy beverages), or use of calcium-fortified foods to make up for any gap between calcium intake and need
- 6½ ounce equivalents from the protein group
- 3½ cups from the vegetables group
- 2 cups from the fruits group
- 9 ounce equivalents from the grains group
- 8 teaspoons of vegetable oil

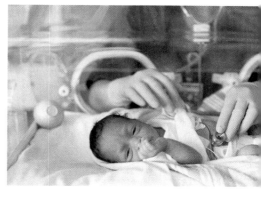

▲ If human milk is used to feed the preterm infant, fortification of the milk with certain nutrients is often needed. ERproductions Ltd./Blend Images LLC

**FAKE NEWS**

A woman must avoid consuming potential food allergens during pregnancy and breastfeeding to reduce the risk for food allergies in her infant.

**THE FACTS**

There is no evidence that maternal dietary restrictions (e.g., peanuts, eggs, and fish) during pregnancy or breastfeeding prevent food allergies in infants.[66]

**TABLE 14-5 ■ Estimated Energy Needs During Lactation**

| Months Postpartum | Estimated Energy Needs |
|---|---|
| 0 up to 6 months | Prepregnancy EER + 330 kcal |
| 6 up to 12 months | Prepregnancy EER + 400 kcal |

Source: Data from *Dietary Reference Intakes for Energy, Carbohydrate, Fiber, Fat, Fatty Acids, Cholesterol, Protein, and Amino Acids,* Copyright 2005 by the Institute of Medicine. National Academies Press, Washington, DC.

Table 14-4 provides a sample menu. As in pregnancy, a serving of fortified ready-to-eat breakfast cereal (or use of a balanced multivitamin and mineral supplement) is helpful to meet extra nutrient needs. And, as mentioned for pregnancy, during breastfeeding, women should consume 8 to 12 ounces of low-mercury fish per week (or 1 gram per day of omega-3 fatty acids from a fish oil supplement) because the omega-3 fatty acids present in fish are secreted into breast milk and are likely to be important for development of the infant's nervous system.[37]

Depending on the infant's demand for milk (which varies over time as the infant grows and begins to consume solid foods), milk production may require 500 kcal (or more) every day. The recommended energy intake during lactation includes an extra 330 to 400 kcal per day above prepregnancy recommendations (Table 14-5). The difference between the energy needed for milk production and the recommended intake—about 170 kcal during the first 6 months postpartum—should allow for a gradual loss of the extra body fat accumulated during pregnancy.

After giving birth, women are often eager to shed the excess "baby fat." Breastfeeding, however, is no time for crash diets. A gradual weight loss of 1 to 4 pounds per month by the nursing mother is appropriate. At significantly greater rates of weight loss—when calories are restricted to less than about 1500 kcal per day—milk output decreases. A reasonable approach for a breastfeeding mother is to follow a dietary plan that supplies at least 1800 kcal per day; has moderate fat content; and includes a variety of nutrient-dense foods from all food groups.

Hydration is especially important during breastfeeding; the woman should drink fluids every time her infant nurses. Drinking about 13 cups of fluids per day encourages ample milk production. Detrimental habits, such as smoking cigarettes or drinking more than two alcoholic beverages a day, can decrease milk output. (Even very small amounts of alcohol can have a deleterious effect on milk output in some women.)

## EXPRESSING AND STORING HUMAN MILK

There are many reasons why lactating women may wish to express their milk and store it for later use. Infants who are born preterm or who have certain medical conditions may have difficulty latching on to the breast for effective feedings. In these cases, expressed breast milk can be fed to the infant by bottle or tube to provide nutritional and immunological benefits beyond what infant formula can provide. For mothers with low milk supply, pumping can stimulate the breast to increase milk production. Even for healthy mother/infant dyads, the ability to let another caregiver feed the infant allows the mother some freedom to spend a few hours away from home, return to work outside the home, or simply get a full night's sleep! Some women also donate milk to milk banks for use by infants whose mothers cannot breastfeed.

Women can express their milk by hand (i.e., manual expression) into a sterile plastic bottle or nursing bag. Several styles of breast pumps are also available: manually operated or electric, single or double. For a woman who wishes to return to work while continuing to breastfeed her infant, an electric double pump is a worthwhile investment. The cost for these supplies may be covered by insurance or may be tax-deductible.

Expressing and storing human milk requires careful sanitation and rapid chilling to ensure safety for the infant. Hands, surfaces, and all pumping equipment should be thoroughly washed before expressing milk. Pumps should not be shared with others. Any parts (e.g., tubing) that appear contaminated with mold should be replaced immediately. Freshly expressed human milk is safe at room temperature for up to 4 hours or can be kept in the refrigerator for up to 4 days. For longer-term storage, human milk will maintain its highest quality in the freezer for up to six months but can be kept up to a year. Sterile nursing bags are a practical, space-saving option for mothers who plan to keep a sizable supply of milk in the freezer. Thawed milk should be used within 24 hours. Any milk leftover after a feeding should be used within two hours or discarded because bacteria and enzymes from the infant's saliva may make the milk unsafe.[67] See Chapter 15 for more information on infant feeding techniques.

Breast milk can be expressed by the mother using a manual, battery-operated, or electric (shown) breast pump. In 2010, an amendment to the Fair Labor Standards Act mandated that most U.S. employers allow break time and a private setting for breastfeeding mothers to express milk. **Why would it be advantageous for employers to support breastfeeding women at work?** Courtesy of Medela, LLC

Expressing milk may seem awkward at first; it takes some practice! The breastfed infant may need some time to adapt to drinking human milk from a bottle, as well. After 1 month or so, the breastfeeding routine should be well established, and the infant can easily transition between breast and bottle feedings. It is worth the effort to be able to provide optimal nutrition for the infant even when the mother cannot be present.

## ADVANTAGES OF BREASTFEEDING

With appropriate training and support, the vast majority of women are capable of breastfeeding, and their infants would benefit from it (see Table 14-6).[62] Nonetheless, some circumstances may make breastfeeding impractical or undesirable for a woman. Mothers who do not want to breastfeed their infants should not feel pressured to do so. Breastfeeding provides advantages, but none so great that a woman who decides to feed her infant an infant formula should feel she is compromising her infant's well-being.

Human milk is tailored to meet infant nutrient needs for the first 4 to 6 months of life. Beyond meeting nutritional needs, there are many other benefits of breastfeeding.

**Fewer Infections.** Due in part to the antibodies in human milk, breastfeeding reduces the infant's overall risk of developing infections. Infants who are breastfed also have fewer ear infections (otitis media) because they do not sleep with a bottle in their mouth. Experts strongly discourage allowing an infant to sleep with a bottle in the mouth; milk can pool in the mouth, throat, and inner ear, creating a growth medium for bacteria, which can lead to ear infections and dental caries. By reducing these common ailments, parents can decrease discomfort for the infant, avoid related trips to the doctor, and prevent possible hearing loss.

**Lower Risk for Noncommunicable Diseases.** Research now links breastfeeding with reduced risks for many obesity-related, chronic diseases. Infants who are breastfed may learn to self-regulate food intake and avoid overeating, which may explain the connection to lowered risk for obesity and type 2 diabetes among adults who were breastfed as infants. The immunologic benefits of breastfeeding seem to be involved in lowered rates of type 1 diabetes, celiac disease, and inflammatory bowel diseases. Reductions in risks of childhood leukemia and lymphoma have also been observed.

Sustainability aims to ensure vitality for generations to come. One way to ensure the well-being of future generations is to promote breastfeeding as optimal nutrition for infants and a positive health behavior for women. What if more women chose to breastfeed their infants? A recent analysis estimated the potential public health impact of near universal breastfeeding. By reducing the risk of a multitude of health problems, such as gastrointestinal and respiratory infections, researchers projected that 823,000 infant deaths and 20,000 cases of breast cancer could be prevented per year.[68] In fact, increasing the number of women who choose to breastfeed their infants would help to meet several of the United Nations' Sustainable Development Goals, including those focused on physical health, economic growth, education, and protecting the environment.[69]

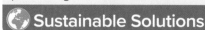 **Sustainable Solutions**

TABLE 14-6 ■ **Advantages of Breastfeeding**

| For Infant |
|---|
| Bacteriologically safe |
| Always fresh and ready to go |
| Provides antibodies and other substances that contribute to maturation of the immune system |
| Helps to establish a healthy gut microbiota and contributes to maturation of the gastrointestinal tract |
| Decreases risk of infections, such as diarrhea, respiratory disease, and ear infections |
| Reduces risk of atopic diseases, such as eczema and asthma |
| Reduces risk of celiac disease and inflammatory bowel diseases |
| Establishes habit of eating in moderation, linked to 15% to 30% lower risk of obesity and 40% lower risk of type 2 diabetes later in life |
| Contributes to proper development of jaws and teeth for better speech development |
| May enhance nervous system development and eventual learning ability |
| Decreases risk of childhood leukemia and lymphoma |
| **For Mother** |
| Contributes to earlier recovery from pregnancy due to the action of hormones that promote a quicker return of the uterus to its prepregnancy state |
| Decreases risk of several chronic diseases later in life, including hypertension, cardiovascular disease, and diabetes |
| Decreases the risk of ovarian and premenopausal breast cancer |
| Potential for quicker return to prepregnancy weight |
| Potential for delayed ovulation, thus reducing chances of pregnancy in the short term |

Some researchers propose that the passage of flavors from the mother's eating pattern into her milk affords an opportunity for the infant to learn about the flavor of the foods of its family long before solids are introduced. **What are the favorite flavors at your family meals?** Blend Images/SuperStock

**atopic disease** A condition involving an inappropriate immune response to environmental allergens; examples include asthma, eczema, and seasonal allergies.

**Lower Risk for Atopic Diseases.** Breastfeeding also reduces the chances of some **atopic diseases** (see Section 15.7). Currently, there is not enough evidence to draw conclusions about the duration of breastfeeding and the risk for food allergies. However, breastfeeding for at least 3 to 4 months lowers the risk for eczema and asthma.[66] Infants are also better able to tolerate human milk than formulas. Formulas sometimes must be switched several times until caregivers find the best one for the infant.

**Convenience and Cost.** Breastfeeding frees the mother from the time and expense involved in buying and preparing formula and washing bottles. Human milk is ready to go and safe to consume. This allows the mother to spend more time with her baby.

## POSSIBLE BARRIERS TO BREASTFEEDING

Widespread misinformation, the mother's need to return to a job, and social reticence are some possible barriers to breastfeeding.

**Misinformation.** The major barriers to breastfeeding stem from misinformation, such as the idea that one's breasts are too small, and the lack of role models. One positive note has been the widespread increase in the availability of lactation consultants over the past several years. First-time mothers who are interested in breastfeeding can find invaluable support from lactation consultants or by talking to women who have experienced it successfully. In almost every community, a group called La Leche League offers classes in breastfeeding and advises women who are facing challenges with breastfeeding (800-LALECHE or www.lalecheleague.org).

**Return to an Outside Job.** Certainly, continuing to breastfeed after the mother returns to work is beneficial for the infant and the mother. It is also in the employer's best interest to support continued breastfeeding because breastfed infants are usually sick less often, which will reduce absenteeism and medical costs.[70] However, working outside the home can complicate plans to breastfeed.

Employers can support breastfeeding mothers with flexible work hours or remote working arrangements. If the job necessitates on-site hours, the employer should provide appropriate time and space for breastfeeding mothers to express milk. In fact, companies with 50 employees or more are required by federal law to provide reasonable breaks and a private space (other than a bathroom) for breastfeeding mothers to pump milk. (See the earlier section on Expressing and Storing Human Milk.) State laws may provide additional accommodations.

Some women can juggle both a job and breastfeeding, but demanding schedules, lack of appropriate equipment or facilities, and unsupportive coworkers can make it difficult to keep up with breastfeeding. A compromise—balancing some breastfeedings, perhaps early morning and night, with infant formula feedings during the day—is possible. The use of supplemental feedings will decrease milk production, but some breast milk is better than no breast milk!

**Social Concerns.** Another barrier for some women is embarrassment about nursing a child in public. Historically, social mores in the U.S. have stressed modesty and discouraged public displays of breasts—even for as good a cause as nourishing babies. In the U.S., no state or territory has a law prohibiting breastfeeding. However, indecent exposure (including the exposure of women's breasts) has long been a common law or statutory offense. Now, all 50 states, the District of Columbia, and the Virgin Islands have specific laws that protect a woman's right to breastfeed in any location. Women who feel reluctant should be reassured that they do have social support and that breastfeeding can be done discreetly.

**galactosemia** Inborn error of metabolism in which the enzyme that converts galactose into glucose is missing or deficient.

**Medical Conditions Precluding Breastfeeding.** Breastfeeding may be ruled out by certain medical conditions in either the infant or the mother. For example, breastfeeding is contraindicated for infants with **galactosemia**, an inherited disorder in which the body cannot break down galactose. Lactose, the main carbohydrate in human breast milk, is made of glucose and galactose. When galactose is not broken down properly, its by-products can damage body organs. Infants with galactosemia cannot tolerate breast milk or cow's milk; they require lactose-free infant formula.

Certain medications can pass into the milk and adversely affect the infant. A breastfeeding mother should discuss all medications (prescription and over-the-counter) with a health care provider to make sure they are compatible with breastfeeding. In addition, a woman in North America or other developed region of the world who has a serious chronic disease (such as tuberculosis, AIDS, or HIV-positive status) or who is being treated with chemotherapy medications should not breastfeed.

**Cosmetic Alterations to the Breast.** Nipple piercings should have no impact on a woman's ability to breastfeed, but the jewelry should be removed before each feeding. Repeated removal and reinsertion of the jewelry may be inconvenient and irritating, so it may be best to leave the jewelry out for the entire period of breastfeeding. Breast tattoos will not impair breastfeeding, either. However, getting a new nipple piercing or breast tattoo while breastfeeding is not advised due to the pain of healing and possibility of infection. Past breast augmentation or reduction surgeries may impair a woman's ability to breastfeed if milk-producing tissue was damaged during the surgery.

**Environmental Contaminants in Human Milk.** There is some legitimate concern over the levels of various environmental contaminants in human milk. However, the benefits of human milk are well established, and the risks from environmental contaminants are still largely theoretical. A few measures a woman could take to counteract some known contaminants are to: (1) consume a variety of foods within each food group; (2) avoid freshwater fish from polluted waters; (3) carefully wash and peel fruits and vegetables (or choose organically raised produce, which has lower levels of pesticides than conventional produce); and (4) remove the fatty edges of meat, as pesticides can become concentrated in fat tissue. In addition, a woman should not try to lose weight rapidly while nursing (more than 0.75 to 1 pound per week) because contaminants stored in her fat tissue might be released into her bloodstream and then appear in her milk. If a woman questions whether her milk is safe, especially if she has lived in an area known to have a high concentration of toxic wastes or environmental pollutants, she should consult her local health department.

## DIETARY SUPPLEMENTS FOR BREASTFED INFANTS

There are some cases in which infant dietary supplements, used under a pediatrician's guidance, are recommended.

- The AAP recommends *all* infants, including exclusively breastfed infants, be given 10 micrograms of vitamin D per day, beginning shortly after birth and continuing until the infant consumes that much from food (e.g., at least 2 cups [0.5 liter] of infant formula per day).[71] Some sun exposure also helps in meeting vitamin D needs.
- Full-term infants are usually born with adequate iron stores to last for the first 4 to 6 months of life. However, to prevent anemia, the AAP recommends iron supplements for exclusively breastfed infants starting at 4 months of age. More aggressive iron supplementation may be needed for infants born preterm, LBW, or to mothers with iron deficiency.[72]
- The AAP does not advise fluoride supplements before 6 months of age. After 6 months, the pediatrician or dentist may recommend supplemental fluoride if the infant's exposure to fluoride from drinking water, foods, and oral hygiene products is insufficient.[73]
- Vitamin B-12 supplements are recommended for infants who are breastfed by mothers who follow a vegetarian dietary pattern.[74]

### ✓ CONCEPT CHECK 14.7

1. What do AND and AAP recommend for duration of breastfeeding?
2. Describe the physiological processes of milk production and let-down. Be sure to mention the hormones involved in these processes.
3. List three advantages of breastfeeding for the mother and three advantages for the infant.
4. Describe three potential barriers to breastfeeding, and suggest ways to overcome them.
5. Identify three micronutrients that may need to be supplemented in the dietary patterns of breastfed infants, and give the rationale for each.

**COVID CORNER**

The American Academy of Pediatrics recommends breastfeeding, even if a mother is positive for COVID-19. The mother may choose to breastfeed directly or pump her milk for the baby. The virus is not transmitted via breastmilk, but proper handwashing and use of face masks are encouraged for the mother to decrease the risk of other modes of transmission of the virus to the infant. Breastfeeding may offer immunological protection for the infant. In addition, during times when disruptions in the food supply may limit the availability of infant formula, breast milk is a fresh, safe source of infant nutrition.

For more information on successful breastfeeding, see *Your Guide to Breastfeeding* from the Office on Women's Health, available at www.womenshealth.gov/files/your-guide-to-breastfeeding.pdf.

## Reducing the Risk of Birth Defects

Floortje/Getty Images

Eating well for a healthy pregnancy not only supplies materials for fetal growth and development but also helps to direct the amazing process of building a new life. Considering the complexity of the human body and its more than 20,000 genes, it is not surprising that abnormalities of structure, function, or metabolism are sometimes present at birth. Birth defects impact 1 in every 33 babies born in the United States.[75] In some cases, they are so severe that a baby cannot survive or thrive. Defects in embryonic or fetal development are the presumed cause of many spontaneous abortions and are at the root of about 20% of infant deaths before 1 year of age. However, many babies with birth defects can go on to live healthy and productive lives.

A wide range of physical or mental disabilities result from birth defects. Heart defects are present in approximately 1 in every 100 to 200 newborn babies, accounting for a large proportion of infant deaths. Cleft lip and/or cleft palate are malformations of the lip or roof of the mouth and occur in approximately 1 in 700 to 1000 births. Neural tube defects are malformations of the brain or spinal cord that occur during embryonic development. Examples include **spina bifida,** in which all or part of the spinal cord is exposed, and **anencephaly,** in which some or all of the brain is missing (Fig. 14-11).

Babies born with spina bifida can survive to adulthood but in many cases have disabilities, such as paralysis, incontinence, and cognitive disorders. Babies born with anencephaly die soon after birth. Neural tube defects occur in 1 in 1000 births. Down syndrome, a condition in which an extra chromosome leads to intellectual disability and other physical alterations, occurs in about 1 in 800 births. Other common birth defects include musculoskeletal defects, gastrointestinal defects, and metabolic disorders.

What causes a birth defect? About 15% to 25% of birth defects are known to be genetic (i.e., inherited or spontaneous mutations of the genetic code). Another 10% are due to environmental influences (e.g., exposure to **teratogens**). The specific cause of the remaining 65% to 75% of birth defects is unknown. Although the etiology of birth defects is multifactorial and many elements are beyond human control, good nutrition practices can positively influence the outcome of pregnancy.

## Folic Acid

During the 1980s, researchers in the United Kingdom noticed a relationship between poor dietary habits and a high rate of neural tube defects among children of women living in poverty. Subsequent intervention studies demonstrated that administration of a multivitamin supplement during the periconceptional period—the months before conception and during early pregnancy—reduced the recurrence of these birth defects. The specific link between dietary folic acid and neural tube defects was tested and confirmed in several follow-up studies. Folate plays a leading role in the synthesis of DNA and the metabolism of amino acids. The rapid cell growth of pregnancy increases needs for folate during pregnancy to 600 micrograms DFE per day. Some women, for genetic reasons, may have an even higher requirement. Adequate folic acid in the periconceptional period decreases the risk of neural tube defects by about 70% and has also been associated with decreased risk of cleft lip/palate, heart defects, and Down syndrome.

In 1998, the U.S. Food and Drug Administration (FDA) mandated fortification of grain products to provide 140 micrograms of folic acid per 100 grams of grain consumed. In Canada, the level of fortification is 150 micrograms of folate per 100 grams of grain consumed. In general, this fortification program increases the average consumption of dietary folic acid by 100 micrograms per day.

Adequate folate status is crucial for all women of childbearing age because the neural tube closes within the first 28 days of pregnancy, a time when many women are unaware they are pregnant.

A surge in the popularity of low-carb diets may be undermining recent progress toward reducing rates of neural tube defects. By cutting out fortified grains (e.g., cereal, pasta, and bread), pregnant women who follow low-carb diets may miss out on important sources of folic acid.[76]

**teratogen** A compound (natural or synthetic) that may cause or increase the risk of a birth defect. Exposure to a teratogen does not always lead to a birth defect; its effects on the fetus depend on the dose, timing, and duration of exposure. Examples: alcohol, some drugs, some industrial or household chemicals, certain infections, and extreme heat.

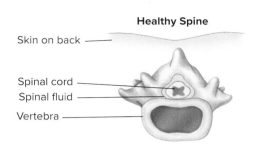

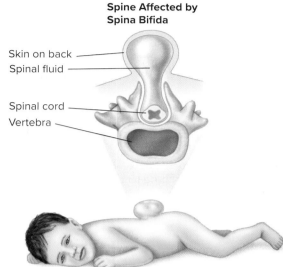

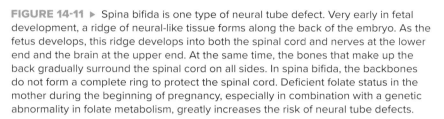

**FIGURE 14-11** ▶ Spina bifida is one type of neural tube defect. Very early in fetal development, a ridge of neural-like tissue forms along the back of the embryo. As the fetus develops, this ridge develops into both the spinal cord and nerves at the lower end and the brain at the upper end. At the same time, the bones that make up the back gradually surround the spinal cord on all sides. In spina bifida, the backbones do not form a complete ring to protect the spinal cord. Deficient folate status in the mother during the beginning of pregnancy, especially in combination with a genetic abnormality in folate metabolism, greatly increases the risk of neural tube defects.

A well-planned dietary pattern can meet the RDA for folate, but the U.S. Public Health Service, March of Dimes, and *Dietary Guidelines* recommend that all women of childbearing age take a daily multivitamin and mineral supplement that contains 400 micrograms of folic acid. Women who have already had a child with a neural tube defect are advised to consume megadoses of folic acid—4000 micrograms per day. They are to begin supplementation at least 1 month before any future pregnancy. This must be done under the supervision of a primary care provider.[77]

## Iodine

Low iodine status during the first trimester of pregnancy—a critical period of brain development—may lead to congenital hypothyroidism (formerly called *cretinism*).[78] If left untreated, consequences (which may vary in severity) include intellectual disability, stunting of growth, impaired hearing and speech, and infertility. Some other physical features are evident in Figure 14-12. When the defect is identified early (by newborn screening tests), these harmful effects can be prevented by treatment with thyroid hormones. Please note that iodine deficiency is just one possible cause of congenital hypothyroidism. The condition may also arise due to genetic abnormalities, autoimmune disorders, or exposure to some medications during pregnancy. Congenital hypothyroidism due to iodine deficiency is most common in developing nations without food fortification programs. With use of iodized salt, however, congenital hypothyroidism due to iodine deficiency is rare in developed nations.

## Antioxidants

A case can be made for antioxidants in the prevention of birth defects as well. Free radicals are constantly generated within the body as a result of normal metabolic processes. An abundance of free radicals results in damage of cells and their DNA, which can lead to gene mutations or tissue malformations. Some research points to free radicals as a source of damage during embryonic development and organogenesis. Antioxidant systems within the body act to minimize the damage caused by free radicals, and researchers hypothesize that dietary sources of antioxidants may aid in the prevention of birth defects. At this time, there is insufficient evidence to support supplementation of individual nutrients that participate in antioxidant

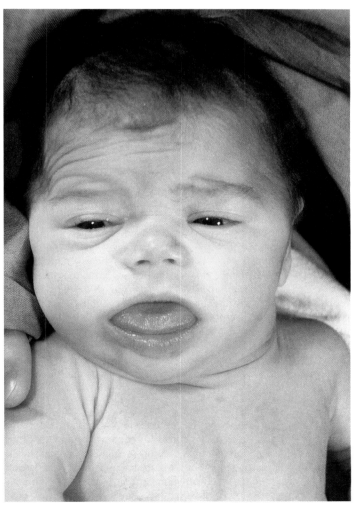

**FIGURE 14-12** ▲ Child with congenital hypothyroidism due to maternal iodine deficiency. This birth defect leads to intellectual disability, stunting of growth, and other physical defects, such as goiter, large tongue, enlarged head, and puffy eyes. Mediscan/Alamy Stock Photo

systems—vitamin E, vitamin C, selenium, zinc, and copper—for the prevention of birth defects. However, use of a balanced multivitamin and mineral supplement while consuming meals rich in whole grains, legumes, and a variety of fruits and vegetables will provide enough of these nutrients to meet current recommendations.

## Vitamin A

Although the requirements for most vitamins and minerals increase by about 30% during pregnancy, the RDA for vitamin A increases by only 10%. Studies have shown the teratogenic potential of vitamin A in doses as low as approximately 3000 micrograms RAE per day. This is just over three times the RDA of 770 micrograms RAE per day for pregnant adult women.

Fetal abnormalities resulting from vitamin A toxicity primarily include facial and cardiac defects, but a wide range of defects have been reported. It is rare that food sources of vitamin A would lead to toxicity. Preformed vitamin A is found in liver, fish, fish oils, fortified milk and yogurt, and eggs. Carotenoids, found in fruits and vegetables, are precursors of vitamin A that are converted into vitamin A in the small intestine, liver, and kidneys. However, the efficiency of absorption of carotenoids decreases as intake increases. Vitamin A excesses typically arise from high-dose dietary supplements rather than food sources.

Typical North American dietary patterns supply adequate vitamin A from foods, so supplemental use is not generally necessary. During pregnancy, supplemental preformed vitamin A should not exceed 3000 micrograms RAE per day. Most multivitamins and prenatal vitamins supply less than 1500 micrograms RAE per day. A balanced eating plan and prudent use of dietary supplements are actions that can sidestep potential problems with vitamin A toxicity.

## Caffeine

Caffeine has been scrutinized for its safety during pregnancy, especially for any link with the rate of birth defects. Caffeine decreases the mother's absorption of iron and may reduce blood flow through the placenta. In addition, the fetus is unable to detoxify caffeine. Research shows that as caffeine intake increases, so does the risk of miscarriage or delivering an LBW infant. Heavy caffeine use during pregnancy may also lead to caffeine withdrawal symptoms in the newborn. These risks are reported with caffeine intakes in excess of 500 milligrams, or the equivalent of about 5 cups of coffee per day. Moderate use of caffeine (up to 200 milligrams of caffeine, or the equivalent of 12 fluid ounces of regular coffee per day), however, is not associated with risk for birth defects.[24] Accounting for caffeine intake from tea, over-the-counter medicines containing caffeine, and chocolate is also important.

## Aspartame

Phenylalanine, a component of the artificial sweetener aspartame (NutraSweet® and Equal®), is a cause for concern for some women during pregnancy. High amounts of phenylalanine in maternal blood disrupt fetal brain development if the mother has a disease known as *phenylketonuria* (see "Obesity and Chronic Health Conditions" in this section). If the mother does not have this condition; however, it is unlikely that the baby will be affected by moderate aspartame use.

For most adults, diet soft drinks are the primary source of artificial sweeteners. Of greater concern than the safety of sweeteners during pregnancy is the quality of foods and beverages consumed. A high intake of diet soft drinks may crowd out healthier beverages, such as water and low-fat milk.

## Obesity and Chronic Health Conditions

Even before becoming pregnant, women of childbearing age should have regular medical checkups to keep an eye on any health conditions that already exist or to identify any developing health problems. In some cases, the condition itself increases risk for birth defects. Obesity, high blood pressure, and uncontrolled diabetes are common health problems known to increase the risk for birth defects, including neural tube defects. In other cases, medications used to control illnesses may pose a risk to the developing fetus. Other health issues, such as seizure disorders and metabolic disorders, could also affect fetal development. A preconception visit with a health professional can help to sort out and make plans to minimize such risks. Once a woman has become pregnant, early and regular prenatal care promote the success of a pregnancy.

Women with diabetes are two to three times more likely to give birth to a baby with birth defects compared to women with normal glucose metabolism. Examples of birth defects common in this group include malformations of the spine, legs, and blood vessels of the heart. Some experts speculate that the mechanism by which diabetes increases birth defects is via excessive free radicals, which lead to oxidative damage of DNA during early gestation. Careful control of blood glucose drastically lowers risk for women with diabetes. Optimal blood glucose control can be achieved through a combination of dietary modifications and medications. Given that diabetes is on the rise among women of childbearing age, this elevated rate of birth defects has become an area of heightened concern.

Another health condition for which maternal nutritional control is of utmost importance is PKU. Recall from Section 6.1 that PKU is an error of metabolism in which the liver lacks the ability to process phenylalanine, leading to an accumulation of this amino acid and its metabolites in body tissues. Babies born to women who have phenylketonuria that is not controlled by a PKU diet are at heightened risk for brain defects, such as microcephaly and intellectual disability.[79]

## Alcohol

Conclusive evidence shows that repeated consumption of four or more alcoholic drinks at one sitting harms the fetus.[80] Such binge drinking is especially perilous during the first 12 weeks of pregnancy, as this is when critical early developmental events take place in utero. Experts have not determined a safe level of alcohol intake during pregnancy. Until a safe level can be established, women are advised not to drink any alcohol—from beverages, foods, or medications (check the label)—during pregnancy or when there is a chance of conception. The embryo (and, at later stages, the fetus) has no means of detoxifying alcohol.

Women with alcohol use disorders give birth to children with a variety of physical and intellectual problems collectively called **fetal alcohol spectrum disorders (FASDs).** The most severe of these disorders is **fetal alcohol syndrome (FAS).** A diagnosis of FAS is based mainly on inadequate fetal and infant growth, physical deformities (especially of facial features), and intellectual disability (Fig. 14-13). Irritability, hyperactivity, short attention span, and

**fetal alcohol spectrum disorders (FASDs)** A group of irreversible physical and mental abnormalities in the infant that result from the mother's consumption of alcohol during pregnancy.

**fetal alcohol syndrome (FAS)** Severe form of FASD that involves abnormal facial features and problems with development of the nervous system and overall growth as a result of maternal alcohol consumption during pregnancy.

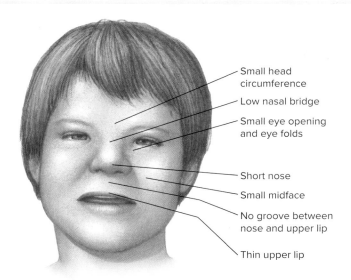

- Small head circumference
- Low nasal bridge
- Small eye opening and eye folds
- Short nose
- Small midface
- No groove between nose and upper lip
- Thin upper lip

**FIGURE 14-13** ▲ Fetal alcohol syndrome. The facial features shown are typical of affected children. Additional abnormalities in the brain and other internal organs accompany fetal alcohol syndrome but are not immediately apparent from simply looking at the child. Milder forms of alcohol-induced changes from a lower alcohol exposure to the fetus are known as alcohol-related neurodevelopmental disorders (ARNDs) and alcohol-related birth defects (ARBDs).

limited hand–eye coordination are other symptoms of FAS. Defects in vision, hearing, and mental processing may also develop over time. Other FASDs consist of some but not all of the defects of FAS. **Alcohol-related neurodevelopmental disorders (ARNDs)** include behavior and learning problems resulting from exposure to alcohol in utero. **Alcohol-related birth defects (ARBDs)** typically include malformations of the heart, kidneys, bones, and/or ears.

Exactly how alcohol causes these defects is not known. One line of research suggests that alcohol, or products of the metabolism of alcohol (e.g., acetaldehyde), cause faulty movement of cells in the brain during early stages of nerve cell development or block the action of certain neurotransmitters. In addition, inadequate nutrient intake, reduced nutrient and oxygen transfer across the placenta, concomitant tobacco and illicit drug use, and possibly other factors contribute to the overall result.

All major health authorities recommend that alcoholic beverages not be consumed by pregnant women. For more information about the effects of maternal alcohol use during pregnancy, visit the website www.cdc.gov/ncbddd/fasd.

## Environmental Contaminants

There is little evidence to link birth defects with the amounts of pesticides, herbicides, or other contaminants in foods or public water supplies in North America. However, evaluating such a link can be difficult, and many would argue that regulations concerning contaminants in the food and water supply are too permissive. Thus, it seems prudent to take measures to decrease intake of pesticides and other contaminants wherever possible. For fruits and vegetables, peeling, removing outer leaves, and/or thoroughly rinsing and scrubbing with a brush under running water will remove the majority of contaminants. In animal products, toxins are most likely to accumulate in fatty tissues. Therefore, removing skin, discarding drippings, and trimming visible fat will decrease exposure from meat, poultry, and fish.

For fish, mercury is of particular concern because it can harm the nervous system of the fetus. Thus, the FDA warns women to avoid swordfish, shark, king mackerel, and tilefish during pregnancy because of possible mercury contamination. Largemouth bass are also implicated. In general, intake of other fish and shellfish should not exceed 12 ounces per week. Canned albacore tuna is a potential mercury source, so it should not be consumed in amounts exceeding 6 ounces per week.[37] Most experts agree that the benefits of consuming fish far outweigh the potential risks of environmental contaminants. As a rule of thumb, consuming a *variety* of foods minimizes risk of exposure to any one contaminant from the food supply.

## In Summary

Although many risk factors for birth defects are beyond our control, parents-to-be can make some wise nutrition choices to improve the chances of having a healthy baby without birth defects. A varied and balanced dietary pattern, such as the food plan for pregnancy described in this chapter, along with a daily multivitamin and mineral supplement with 400 micrograms of folic acid, will ensure adequate nutrient status. It is estimated that daily use of a multivitamin and mineral supplement containing folic acid will decrease the rate of all birth defects by 50%. Discuss use of any other dietary supplements with a primary care provider or RDN to be sure that the fetus will not be exposed to toxic levels of vitamin A or other dangerous food components. Early and consistent prenatal care can help control obesity and any chronic health conditions that may complicate a pregnancy. Also, avoiding alcohol during pregnancy will eliminate any risk for fetal alcohol spectrum disorders.

Although it seems that advice for a healthy pregnancy is always directed at the mother, fathers-to-be are not off the hook! Health is a family affair, so encouraging healthy eating habits and avoiding smoking and alcohol are important for fathers, too. As you read in the beginning of this chapter, the genetics of the baby are certainly an outcome of both parents. Indeed, inadequate supplies of zinc, folate, antioxidants, and omega-3 fatty acids affect the quality of sperm. Overall, the periconceptional period is a time for good nutrition and careful lifestyle practices for mothers- and fathers-to-be.

**alcohol-related neurodevelopmental disorders (ARNDs)** One or more abnormalities of the central nervous system (e.g., small head size, impaired motor skills, hearing loss, or poor hand-eye coordination) related to confirmed alcohol exposure during gestation.

**alcohol-related birth defects (ARBDs)** One or more birth defects (e.g., malformations of the heart, bones, kidneys, eyes, or ears) related to confirmed alcohol exposure during gestation.

✓ **CONCEPT CHECK 14.8**

1. Why are women of childbearing age encouraged to obtain 400 micrograms per day of folic acid from supplements or fortified foods?

2. What is the recommendation for seafood intake during pregnancy? How can you balance the health benefits of regular seafood consumption with concerns about environmental contaminants in fish?

# Summary (Numbers refer to numbered sections in the chapter.)

**14.1** Energy imbalances can adversely affect fertility by altering hormone levels and promoting oxidative damage. Polycystic ovary syndrome, which tends to co-occur with upper-body obesity, is a condition of hormonal imbalance that causes infertility. Other than managing body weight, nutritional factors that may improve male and/or female fertility include low glycemic index carbohydrates, unsaturated fats, antioxidants, folate, iron, and zinc.

**14.2** Pregnancy is arbitrarily divided into three trimesters of 13 to 14 weeks each. The first trimester is characterized by a rapid increase in cell number as the zygote grows to be an embryo, then a fetus. During the first trimester, the growing organism is most susceptible to damage from exposure to toxic agents or nutrient deficiencies. By the start of the second trimester, the organs and limbs have formed and will continue to grow and develop. The third trimester is marked by rapid fetal growth and storage of nutrients in preparation for life outside the womb.

**14.3** A successful pregnancy results in optimal health for both the infant and the mother. Pregnancy success is defined as: (1) gestation longer than 37 weeks; and (2) birth weight greater than 5.5 pounds (2.5 kilograms). Factors that predict pregnancy success include early and regular prenatal care, maternal age within the range of 20 to 35 years, and adequate nutrition. Factors that contribute to poor pregnancy outcome include inadequate prenatal care, obesity, underweight, teenage pregnancy, smoking, alcohol consumption, use of certain prescription medications and all illicit drugs, inadequate nutrition, heavy caffeine use, and various infections, such as listeriosis.

**14.4** For women with a healthy prepregnancy BMI (18.5 to 24.9), total weight gain should be within the range of 25 to 35 pounds. Women who are underweight at conception and those carrying multiple fetuses should gain more; women who are overweight or obese at conception should gain less. During the first trimester, although she need not increase the quantity of food consumed, the woman should focus on the quality of the dietary pattern to meet increased requirements for protein, carbohydrate, essential fatty acids, fiber, water, and many vitamins and minerals.

A woman typically needs an additional 350 to 450 kcal per day during the second and third trimesters.

**14.5** Following a plan based on the *Dietary Guidelines* as exemplified by MyPlate is recommended during pregnancy and breastfeeding. The mother-to-be should especially emphasize good sources of vitamin B-6, folate, vitamin D, iron, zinc, and calcium. Vegetarian dietary patterns are safe during pregnancy, but mothers following a vegan dietary plan should specifically seek out good sources of iron, zinc, calcium, vitamin D, vitamin B-12, and omega-3 fatty acids. Prenatal multivitamin and mineral supplements are useful for meeting increased nutrient requirements during pregnancy.

**14.6** Gestational hypertension, gestational diabetes, heartburn, constipation, nausea, vomiting, edema, and anemia are all possible discomforts and complications of pregnancy. Nutrition therapy can help minimize some of these problems.

**14.7** Almost all women are physically able to breastfeed their infants. The nutritional composition of human milk is different from that of unaltered cow's milk; human milk is much more desirable for the infant. For the infant, the advantages of breastfeeding over formula feeding are numerous, including fewer infections, reduced risk of atopic diseases, and lower rates of obesity and type 2 diabetes throughout life. Benefits for the mother include reduced risk of certain cancers, earlier recovery from pregnancy, and faster return to prepregnancy weight. When medical conditions of the mother (e.g., AIDS) or infant (e.g., galactosemia) preclude breastfeeding or if the mother chooses not to breastfeed, infants can be adequately nourished with formula.

**14.8** During pregnancy, women can take several steps to reduce the risk of birth defects. They should achieve a healthy body weight before pregnancy and strive to gain weight within the ranges recommended by the National Academy of Medicine. Adequate intakes of folic acid, iodine, and antioxidant nutrients are essential for prevention of many types of birth defects. Excesses of vitamin A and caffeine should be avoided. There is no safe level of alcohol intake known during pregnancy. Dietary control of diseases (e.g., diabetes and PKU) will also protect the fetus.

# Check Your Knowledge (Answers are available at the end of this question set.)

1. Which of the following nutrition interventions is most likely to improve fertility?
   a. Taking a vitamin E supplement
   b. Losing excess body fat
   c. Consuming a low-carbohydrate diet
   d. Taking an iron supplement

2. Increased carbohydrate needs during pregnancy are set to
   a. prevent ketosis.
   b. alleviate nausea.
   c. prevent gestational hypertension.
   d. supply adequate folate.

3. An infant born at 38 weeks' gestation weighing 5 pounds can be described as
   a. preterm.
   b. LBW.
   c. SGA.
   d. LBW and SGA.

4. If a woman is 5 feet 2 inches tall and weighs 150 pounds before becoming pregnant, how much weight should she gain during pregnancy?
   a. 28 to 40 pounds (12.5 to 18 kilograms)
   b. 25 to 35 pounds (11.5 to 16 kilograms)
   c. 15 to 25 pounds (7 to 11.5 kilograms)
   d. As little as possible

5. Benefits of physical activity during pregnancy include
   a. preventing excessive gestational weight gain.
   b. improved sleep.
   c. lower risk for gestational diabetes.
   d. all of these.

6. Which of the following may help to alleviate nausea during pregnancy?
   a. Postponing meals until the afternoon
   b. Drinking large amounts of water
   c. Postponing use of iron supplements until the second trimester
   d. All of these

7. Physiologically, milk production requires _____ kcal per day.
   a. 300
   b. 500
   c. 800
   d. 1000

8. An eating pattern for a woman in the third trimester of pregnancy differs from her prepregnancy dietary plan in that
   a. fluid needs are higher.
   b. additional oils are allowed.
   c. there are more servings from the grains group.
   d. all of these apply.

9. Advantages of breastfeeding include
   a. decreased ear infections in the infant.
   b. decreased diarrheal diseases in the infant.
   c. decreased risk of breast cancer for the mother.
   d. all of these.

10. Consuming a single cup of coffee per day is associated with
    a. spontaneous abortion.
    b. LBW.
    c. birth defects.
    d. none of these.

Answer Key: 1. b (LO 14.1), 2. a (LO 14.2), 3. d (LO 14.3), 4. c (LO 14.4), 5. d (LO 14.5), 6. c (LO 14.6), 7. c (LO 14.7), 8. d (LO 14.8), 9. d (LO 14.9), 10. d (LO 14.10)

## Study Questions (Numbers refer to Learning Outcomes.)

1. Provide three key pieces of nutrition advice for parents seeking to maximize their chances of conceiving. Why did you identify those specific factors? **(LO 14.1)**

2. Identify four key nutrients for which intake should be significantly increased during pregnancy. **(LO 14.2)**

3. Describe some nutritional concerns related to teenage pregnancy. At what age do you think pregnancy is ideal? Why? **(LO 14.3)**

4. Outline current weight-gain recommendations for pregnancy. What is the basis for these recommendations? **(LO 14.4)**

5. Suggest several safe exercises for a pregnant woman. **(LO 14.5)**

6. What nutrition advice would you give to a friend who suffers from nausea during pregnancy? **(LO 14.6)**

7. Describe the physiological mechanisms that stimulate human milk production and release. How can knowing about these help mothers breastfeed successfully? **(LO 14.7)**

8. Starting in the second trimester of pregnancy, a woman typically needs 350 to 450 kcal more than her usual needs. Suggest a combination of nutrient-dense foods that would supply these extra calories. **(LO 14.8)**

9. Give three reasons a woman should give serious consideration to breastfeeding her infant. **(LO 14.9)**

10. Describe the importance of folic acid for conception and fetal development. **(LO 14.10)**

## Further Readings

1. Blau L and others: Women's experience and understanding of food cravings in pregnancy: A qualitative study in women receiving prenatal care at the University of North Carolina–Chapel Hill. *Journal of the Academy of Nutrition and Dietetics* 119:S2212, 2019. DOI:10.1016/j.jand.2019.09.020.

2. Fawcett EJ, Fawcett JM, and Mazmanian D: A meta-analysis of the worldwide prevalence of pica during pregnancy and the postpartum period. *International Journal of Gynecology and Obstetrics* 133:277, 2016. DOI:10.1016/j.ijgo.2015.10.012.

3. Gaskins AJ and Chavarro JE: Diet and fertility: A review. *American Journal of Obstetrics and Gynecology* 218:379, 2018. DOI:10.1016/j.ajog.2017.08.010.

4. Office on Women's Health: Polycystic ovary syndrome. May 27, 2016. Available at https://www.womenshealth.gov/files/documents/fact-sheet-pcos.pdf. Accessed January 31, 2020.

5. Teede H and others: International evidence-based guideline for the assessment and management of polycystic ovary syndrome. Melbourne, Australia, 2018. Available at www.monash.edu/medicine/sphpm/mchri/pcos. Accessed January 31, 2020.

6. Grassi A: Women's health: New data on polycystic ovary syndrome. *Today's Dietitian* 19:12, 2017.

7. Salas-Huetos A, Bullo M, and Salas-Salvado J: Dietary patterns, foods and nutrients in male fertility parameters and fecundability: A systematic review of observational studies. *Human Reproduction Update* 23:371, 2017. DOI:10.1093/humupd/dmx006.

8. Fallah A and others: Zinc is an essential element for male fertility: A review of Zn roles in men's health, germination, sperm quality, and fertilization. *Journal of Reproduction & Infertility* 19:69, 2018.

9. Schisterman EF and others: Effect of folic acid and zinc supplementation in men on semen quality and live birth among couples undergoing fertility treatment: A randomized controlled trial. *Journal of the American Medical Association* 323:35, 2020. DOI:10.1001/jama.2019.18714.

10. Chavarro JE and others: Diet and lifestyle in the prevention of ovulatory disorder infertility. *Obstetrics & Gynecology* 110:1050, 2007. DOI:10.1097/01.AOG.0000287293.25465.e1.

11. Chavarro JE and others: *The Fertility Diet: Groundbreaking Research Reveals Natural Ways to Boost Ovulation and Improve Your Chances of Getting Pregnant.* New York, NY: The McGraw-Hill Companies, 2008.

12. Koletzko B and others: Nutrition during pregnancy, lactation and early childhood and its implications for maternal and long-term child health: The Early Nutrition Project recommendations. *Annals of Nutrition and Metabolism* 74:93, 2019. DOI:10.1159/000496471.

13. March of Dimes: Premature babies cost employers $12.7 billion annually. Updated February 7, 2014. Available at https://www.marchofdimes .org/news /premature-babies-cost-employers-127-billion-annually.aspx. Accessed February 21, 2019.

14. U.S. Department of Health and Human Services, Office of Adolescent Health: Negative impacts of teen childbearing. March 28, 2019. Available at https://www.hhs.gov/ash/oah/adolescent-development/reproductive -health-and-teen-pregnancy/teen-pregnancy-and-childbearing/teen -childbearing/index.html. Accessed January 31, 2020.

15. Leftwich HK and Alves MV: Adolescent pregnancy. *Pediatric Clinics of North America* 64:381, 2017. DOI:10.1016/j.pcl.2016.11.007.

16. Sauer MV: Reproduction at an advanced maternal age and maternal health. *Fertility and Sterility* 103:1136, 2015. DOI:10.1016/j. fertnstert.2015.03.004.

17. American College of Obstetricians and Gynecologists: Interpregnancy care: Obstetric care consensus No. 8. *Obstetrics & Gynecology* 133:e51, 2019. DOI:10.1097/AOG.0000000000003025.

18. Centers for Disease Control and Prevention: Treating for two: Medicine and pregnancy. November 18, 2019. Available at https://www.cdc.gov /pregnancy/meds/treatingfortwo/index.html. Accessed January 31, 2020.

19. American College of Obstetricians and Gynecologists: Listeria and pregnancy. 2018. Available at https://www.acog.org/-/media/For-Patients /faq501.pdf?dmc=1&ts=20200131T1409217427. Accessed January 31, 2020.

20. Mohammadi M and others: The effect of prepregnancy body mass index on birth weight, preterm birth, Cesarean section, and preeclampsia in pregnant women. *Journal of Maternal-Fetal and Neonatal Medicine* 17:1, 2018. DOI:10.1080/14767058.2018.1473366.

21. Lisonkova S and others: Association between prepregnancy body mass index and severe maternal morbidity. *Journal of the American Medical Association* 318:1777, 2017. DOI:10.1001/jama.2017.16191.

22. Academy of Nutrition and Dietetics: Position of the Academy of Nutrition and Dietetics: Nutrition and lifestyle for a healthy pregnancy outcome. *Journal of the Academy of Nutrition and Dietetics* 114:1099, 2014. DOI:10.1016/j.jand.2014.05.005.

23. Schultz DJ and others: The impact of the 2009 Special Supplemental Nutrition Program for Women, Infants, and Children food package revisions on participants: A systematic review. *Journal of the Academy of Nutrition and Dietetics* 115:1832, 2015. DOI:10.1016/j.jand.2015.06.381.

24. Academy of Nutrition and Dietetics: Practice paper of the Academy of Nutrition and Dietetics: Nutrition and lifestyle for a healthy pregnancy outcome. *Journal of the Academy of Nutrition and Dietetics* 114:1447, 2014. DOI:10.1016/j.jand.2014.07.001.

25. Physical activity and exercise during pregnancy and the postpartum period: ACOG committee opinion summary, number 804. *Obstetrics and Gynecology* 135(4): 991, 2020. DOI:10.1097/AOG.0000000000003773.

26. Artal R: Exercise in pregnancy: Guidelines. *Clinical Obstetrics and Gynecology* 59:639, 2016. DOI:10.1097/GRF.0000000000000223.

27. U.S. Department of Health and Human Services: *Physical Activity Guidelines for Americans,* 2nd ed. 2018. Available at https://health.gov /paguidelines/second-edition/pdf/Physical_Activity_Guidelines_2nd _edition.pdf. Accessed January 31, 2020.

28. Institute of Medicine and National Research Council: *Weight Gain During Pregnancy: Reexamining the Guidelines.* Washington, DC: National Academies Press, 2009.

29. Goldstein RF and others: Association of gestational weight gain with maternal and infant outcomes: A systematic review and meta-analysis. *Journal of the American Medical Association* 317:2207, 2017. DOI:10.1001/ jama.2017.3635.

30. Academy of Nutrition and Dietetics: Position of the Academy of Nutrition and Dietetics: Obesity, reproduction, and pregnancy outcomes. *Journal of the Academy of Nutrition and Dietetics* 116:677, 2016. DOI:10.1016/j. jand.2016.01.008.

31. Berenson AB and others: Obesity risk knowledge, weight misperception, and diet and health-related attitudes among women intending to become pregnant. *Journal of the Academy of Nutrition and Dietetics* 116:69, 2016. DOI:10.1016/j.jand.2015.04.023.

32. Hamad R, Cohen AK, and Rehkopf DH: Changing national guidelines is not enough: The impact of 1990 IOM recommendations on gestational weight gain among US women. *International Journal of Obesity* 40:1529, 2016. DOI:10.1038/ijo.2016.97.

33. Getz L: Starving for two. *Today's Dietitian* 17:14, 2015.

34. Emmett PM and others: Pregnancy diet and associated outcomes in the Avon Longitudinal Study of Parents and Children. *Nutrition Reviews* 73:154, 2015. DOI:10.1093/nutrit/nuv053.

35. Sparling TM and others: The role of diet and nutritional supplementation in perinatal depression: A systematic review. *Maternal & Child Nutrition* 13:e12235, 2017. DOI:10.1111/mcn.12235.

36. Thompson M and others: Omega-3 fatty acid intake by age, gender, and pregnancy status in the United States: National Health and Nutrition Examination Survey 2003–2014. *Nutrients* 11:177, 2019. DOI:10.3390/ nu11010177.

37. U.S. Environmental Protection Agency and U.S. Food and Drug Administration: Advice about eating fish: What pregnant women and parents should know. January 2017. Available at https://www.fda.gov /media/102331/download. Accessed January 31, 2020.

38. Tone C: Omega-3 fats and pregnancy: Health benefits for both mom and baby. *Today's Dietitian* 18:14, 2016.

39. McKenzie AL and Armstrong LE: Monitoring body water balance in pregnant and nursing women: The validity of urine color. *Annals of Nutrition and Metabolism* 70(suppl 1):18, 2017. DOI:10.1159/000462999.

40. Gillman MW and others: Beverage intake during pregnancy and childhood adiposity. *Pediatrics* 140:e20170031, 2017. DOI:10.1542/ peds.2017- 0031.

41. Vincent J and others: Mothers' intake of sugar-containing beverages during pregnancy and body composition of their children during childhood: The Generation R Study. *American Journal of Clinical Nutrition* 105:834, 2017. DOI:10.3945/ajcn.116.147934.

42. Pilz S and others: The role of vitamin D in fertility and during pregnancy and lactation: A review of clinical data. *International Journal of Environmental Research and Public Health* 15:E2241, 2018. DOI:10.3390/ ijerph15102241.

43. U.S. Department of Agriculture and U.S. Department of Health and Human Services: *Dietary Guidelines for Americans, 2020–2025.* 9th Edition. December 2020. Available at DietaryGuidelines.gov. Accessed January 14, 2021.

44. American College of Obstetricians and Gynecologists: ACOG Committee Opinion No. 495: Vitamin D: Screening and supplementation during pregnancy. Reaffirmed 2019. Available at https://www.acog.org/-/media/Committee-Opinions/Committee-on-Obstetric-Practice/co495.pdf?dmc=1&ts=20200131T1519537666. Accessed January 31, 2020.

45. Grieger JA and Clifton VL: A review of the impact of dietary intakes in human pregnancy on infant birthweight. *Nutrients* 7:153, 2014. DOI:10.3390/nu7010153.

46. Banjari I: Iron deficiency anemia and pregnancy in *Current Topics in Anemia*. 2018. DOI:10.5772/intechopen.69114.

47. Blumfield ML and others: A systematic review and meta-analysis of micronutrient intakes during pregnancy in developed countries. *Nutrition Reviews* 71:118, 2013. DOI:10.1111/nure.12003.

48. Cantor AG and others: Routine iron supplementation and screening for iron deficiency anemia in pregnancy: A systematic review for the U.S. Preventive Services Task Force. *Annals of Internal Medicine* 162:566, 2015. DOI:10.7326/M14-2932.

49. Haider BA and Bhutta ZA: Multiple-micronutrient supplementation for women during pregnancy. *Cochrane Database of Systematic Reviews* 4:CD004905, 2017. DOI:10.1002/14651858.CD004905.pub5.

50. Lowensohn RI and others: Current concepts of maternal nutrition. *Obstetrical & Gynecological Survey* 71:413, 2016. DOI:10.1097/OGX.0000000000000329.

51. Academy of Nutrition and Dietetics: Position of the Academy of Nutrition and Dietetics: Vegetarian diets. *Journal of the Academy of Nutrition and Dietetics* 116:1970, 2016. DOI:10.1016/j.jand.2016.09.025.

52. American Academy of Family Physicians: Heartburn. December 2010. Available at https://www.aafp.org/afp/2010/1215/p1452.pdf. Accessed January 31, 2020.

53. American College of Obstetrics and Gynecology: ACOG Practice Bulletin No. 95: Anemia in pregnancy. *Obstetrics & Gynecology* 112:201, 2008.

54. American Diabetes Association: Management of diabetes in pregnancy: Standards of medical care in diabetes—2020. *Diabetes Care* 43(Suppl 1): S183, 2019. DOI:10.2337/dc20-S014.

55. Curry A: Special delivery: How diabetes affects the health of mother and child—and what you can do about it. *Diabetes Forecast* Sept/Oct 2016:51, 2016.

56. Centers for Disease Control and Prevention: High blood pressure during pregnancy. January 28, 2020. Available at https://www.cdc.gov/bloodpressure/pregnancy.htm. Accessed January 31, 2020.

57. Achamrah N and Ditisheim A: Nutritional approach to preeclampsia prevention. *Current Opinion in Clinical Nutrition & Metabolic Care* 21:168, 2018. DOI:10.1097/MCO.0000000000000462.

58. O'Callaghan KM and Kiely M: Systematic review of vitamin D and hypertensive disorders of pregnancy. *Nutrients* 10:294, 2018. DOI:10.3390/nu10030294.

59. Asayama K and Imai Y: The impact of salt intake during and after pregnancy. *Hypertension Research* 41:1, 2018. DOI:10.1038/hr.2017.90.

60. Sakuyama H and others: Influence of gestational salt restriction in fetal growth and in development of diseases in adulthood. *Journal of Biomedical Science* 23:12, 2016. DOI:10.1186/s12929-016-0233-8.

61. Academy of Nutrition and Dietetics: Position of the Academy of Nutrition and Dietetics: Promoting and supporting breastfeeding. *Journal of the Academy of Nutrition and Dietetics* 115:444, 2015. DOI:10.1016/j.jand.2014.12.014.

62. American Academy of Pediatrics: Policy Statement: Breastfeeding and the use of human milk. *Pediatrics* 129:e827, 2012. DOI:10.1089/bfm.2012.0067.

63. Centers for Disease Control and Prevention: Breastfeeding Report Card. 2018. Available at https://www.cdc.gov/breastfeeding/data/reportcard.htm. Accessed February 23, 2019.

64. Holt K and others: *Bright Futures Nutrition*, 3rd ed. Elk Grove Village, IL: American Academy of Pediatrics, 2011.

65. Plaza-Diaz J and others: Human milk oligosaccharides and immune system development. *Nutrients* 10:1038, 2018. DOI:10.3390/nu10081038.

66. Greer FR and others: The effects of early nutritional interventions on the development of atopic disease in infants and children: The role of maternal dietary restriction, breastfeeding, hydrolyzed formulas, and timing of introduction of allergenic complementary foods. *Pediatrics* 143:e20190281, 2019. DOI:10.1542/peds.2019-0281.

67. U.S. Department of Health and Human Services: Your Guide to Breastfeeding: Office on Women's Health. September 20, 2020. Available at https://www.womenshealth.gov/files/your-guide-to-breastfeeding.pdf. Accessed January 15, 2021.

68. Victora CG and others: Breastfeeding in the 21st century: Epidemiology, mechanisms, and lifelong effect. *The Lancet* 387:475, 2016. DOI:10.1016/S0140-6736(15)01024-7.

69. World Alliance for Breastfeeding Action: World Breastfeeding Week: Breastfeeding: A key to sustainable development. 2019. Available at http://waba.org.my/v3/wp-content/uploads/2019/12/SDG-WBW-Action-Folder-Insert.jpg. Accessed January 31, 2020.

70. Abdulwadud OA and Snow ME: Interventions in the workplace to support breastfeeding for women in employment. *Cochrane Database of Systematic Reviews* 2012:10, 2012. DOI: 10.1002/14651858.CD006177.pub3.

71. Wagner CL and others: Prevention of rickets and vitamin D deficiency in infants, children, and adolescents. *Pediatrics* 122(5):1142, 2008. DOI:10.1542/peds.2008-1862.

72. Baker RD and others: Clinical report—diagnosis and prevention of iron deficiency and iron-deficiency anemia in infants and young children (0–3 years of age). *Pediatrics* 126:1040, 2010. DOI:10.1542/peds.2010-2576.

73. Lewis CW: Fluoride and dental caries prevention in children. *Pediatrics in Review* 35:3, 2014. DOI:10.1542/pir.35-1-3.

74. Baroni L and others: Vegan nutrition for mothers and children: Practical tools for healthcare providers. *Nutrients* 11:5, 2019. DOI:10.3390/nu11010005.

75. Centers for Disease Control and Prevention: Data and statistics on birth defects. January 23, 2020. Available at https://www.cdc.gov/ncbddd/birthdefects/data.html. Accessed January 31, 2020.

76. Desrosiers TA and others: Low carbohydrate diets may increase risk of neural tube defects. *Birth Defects Research* 110:901, 2018. DOI:10.1002/bdr2.1198.

77. Van Gool JN and others: Folic acid and primary prevention of neural tube defects: A review. *Reproductive Toxicology* 80:73, 2018. DOI:10.1016/j.reprotox.2018.05.004.

78. Pearce EN and others: Consequences of iodine deficiency and excess in pregnant women: An overview of current knowns and unknowns. *American Journal of Clinical Nutrition* 104:918S, 2016. DOI:10.3945/ajcn.115.110429.

79. Committee on Genetics: Policy Statement: Maternal phenylketonuria. *Pediatrics* 122:445, 2008. Reaffirmed January 2013. DOI:10.1542/peds.2008-1485.

80. Viteri OA and others: Fetal anomalies and long-term effects associated with substance abuse in pregnancy: A literature review. *American Journal of Perinatology* 32:405, 2015. DOI:10.1055/s-0034-1393932.

# Chapter 15

# Nutrition from Infancy Through Adolescence

**Chapter 15 is designed to allow you to:**

**15.1** Describe the impact of nutrition on growth and physiological development from infancy through adolescence.

**15.2** List specific nutrients often found to be lacking in the dietary patterns of infants, toddlers, preschoolers, and teenagers, and provide recommendations for improvement.

**15.3** Identify appropriate dietary strategies to meet the basic nutritional needs for normal growth and development for an infant.

**15.4** Outline several challenges parents might face in promoting healthy eating behaviors during childhood and adolescence.

**15.5** Describe the long-term effects of childhood obesity, and suggest ways to prevent or treat the problem.

**15.6** Identify common food allergens, and suggest practices that may reduce the risk of food allergies.

## FAKE NEWS

Toddlers need special toddler drinks or "transition formulas" to meet their nutrient needs.

## THE FACTS

Packaged like infant formulas, toddler formulas may be perceived as a healthy step for child feeding after weaning from breast milk or infant formula. However, child nutrition experts do *not* advocate regular use of these products. Weaning is a critical time to expose children to a variety of tastes and textures. These sweet beverages may blunt the appetite and further decrease the child's acceptance of foods at meals; most provide added sugars and have less protein than cow's milk.[1] Offer water or milk to drink and provide a variety of nutrient-dense options at five or six meals or snacks each day. Let the child learn to rely on whole foods to meet nutrient needs!

Children born in North America have access to some of the best health care in the world. Even so, the numbers of children and teenagers with nutrition-related diseases are on the rise. Kids spend more time staring at electronic devices and less time in active play. Soft drinks and energy drinks have replaced much of the milk that children and teenagers previously consumed daily. Consumption of fruits, vegetables, and whole grains—although these are abundantly available—consistently falls short of recommendations. Evidently, there is room for improvement in the dietary patterns of most American children. How can we move children toward healthy eating behaviors?

Education designed to change nutrition behaviors should start early and involve the whole family. During infancy and early childhood, the family mainly controls food intake, informs food preferences, and models eating behaviors. Family mealtimes help children to establish healthy eating habits and aid in the prevention of both undernutrition and overnutrition. As children grow older, family meals are a time to teach children communication skills and improve self-esteem. Healthy dietary patterns and social skills learned early in life will equip children as they reach school age, when peer and media influences take on heightened importance. This chapter examines the dynamic nutritional needs of the growing years—from infancy through adolescence.

## 15.1 Assessing Growth

During infancy, future eating patterns and attitudes toward foods begin to take shape. If parents and other caregivers are flexible and model healthy nutrition behaviors, they can lead a child into lifelong healthful food habits. A family environment that encourages healthy eating will provide the nutrients needed to optimize physical growth and development. However, these advantages do not guarantee that a child will thrive.

Children also need specific attention focused on them; they need to grow in a stimulating environment, and they need a sense of security. For example, children hospitalized for growth failure gain weight more quickly when tactile stimulation, such as being held and rocked, accompanies proper nutrition.

### NORMAL GROWTH

It seems that all babies do is eat and sleep. There is a good reason for this! An infant's birth weight doubles in the first 4 to 6 months and triples within the first year. Never again is growth so rapid. Such swift growth requires a lot of nourishment and sleep. After the first year, growth is slower; it takes 5 more years to double the weight seen at 1 year. An infant also increases in length in the first year by 50% and then growth in height continues through the teen years. These gains are not necessarily continuous; spurts of rapid growth alternate with periods of relatively little change. Height is typically maximized by age 19, although growth may continue into the early twenties, especially for boys.

The human body needs a lot more food to support growth and development than it does to merely maintain its size once growth ceases. When nutrients are missing at critical phases of this process, growth and development may slow or even stop. Along with adequate energy, a few key nutrients are particularly important during these years: protein, calcium, iron, and zinc. In developing nations, about one-third of the children under 5 years of age are short and underweight for their ages. **Undernutrition** is at the heart of the problem.

### EFFECT OF UNDERNUTRITION ON GROWTH

Undernourished children are smaller versions of nutritionally fit children. In developing areas of the world, when breastfeeding ceases children are often fed a high-carbohydrate, low-protein diet (see Section 13.5). This eating pattern supports some growth but does not allow children to attain their full genetic potential. In developed regions of the world, such as North America, undernutrition exists, but overnutrition (i.e., childhood obesity) is a more prominent public health concern. As with the fetus in utero, nutritional problems in infancy and childhood may have enduring effects on health.

Growth is an excellent indicator of a child's nutritional status: gains in weight are a good reflection of nutritional status in the short term, whereas gains in height are an indicator of nutritional status over the long term.[2] An inadequate dietary pattern during a critical stage of infancy or childhood hampers the cell division that should occur at that stage. Mild zinc deficiencies among North American children, for example, have been linked to diminished growth and development. Unfortunately, improving dietary patterns after a period of nutrient deficiency will not completely compensate for losses in physical or mental development, because the hormonal and other conditions needed for growth will not likely be present. In addition, gains in height are no longer possible after the skeleton reaches full maturation. This happens as growth plates at the ends of the bones fuse, which begins around 14 years of age in girls and 15 years of age in boys. This process is usually complete by age 19 in girls and age 20 in boys. Furthermore, muscles can increase in diameter later in life, but their linear growth is limited by the length of the bone.

For these reasons, a 15-year-old female who has experienced undernutrition and is only 4 feet 8 inches tall cannot attain the adult height of a typical North American girl simply by changing her dietary intake. Girls experience their peak rate of growth right before the onset of menstrual periods. Once the time for growth ceases (for girls, this is about 5 years after they start menstruating), adequate nutrient intake helps maintain health and weight but cannot make up for missed growth in height.

## GROWTH CHARTS

Health professionals assess a child's pattern of growth by plotting measurements of height and weight on **growth charts.** The charts contain **percentile** divisions. A percentile represents the rank of the person among 100 peers matched for age and sex. If a young boy, for example, is at the 90th percentile of height-for-age, he is shorter than 10% and taller than 89% of children his age. A child at the 50th percentile is considered average. Fifty of 100 children will be taller than this child; 49 will be shorter.

A variety of growth charts for boys and girls are available from the Centers for Disease Control and Prevention (CDC) website (see Figures 15-1 and 15-2). For children from birth to age 2, growth charts developed by the World Health Organization (WHO) are used to assess length-for-age, weight-for-age, weight-for-length, and head-circumference-for-age measurements. WHO growth charts are based on data collected from children from various regions of the world who were raised *under conditions for optimal growth and development.* This means that they were breastfed as infants and had

**growth charts** Charts used by health professionals to compare the growth of an individual child to normal patterns of growth in weight, stature, and head circumference over time.

**percentile** With reference to a dataset, the value below which a given percentage of observations fall. For example, a male at the 75th percentile for BMI-for-age has a BMI higher than 74% of other males his age and lower than 25% of other males his age.

Children under 2 years of age are usually measured lying on their backs with legs fully extended, so the term *length* is used rather than *height* or *stature.*

Evaluate the child's growth relative to the percentile values shown at the ends of these growth curves. These show how the individual's size ranks among 100 peers.

Draw a vertical line from the child's age. Draw a horizontal line from the child's growth measurement. Plot the intersection of these two lines.

Measurements are given in both English and metric units.

For children from birth to age 2, use the growth charts developed by WHO in 2006.

Ella has been tracking around the 50th percentile for length throughout the first 2 years of her life.

Ella's weight, which started out near the 50th percentile, is now between the 75th and 90th percentiles.

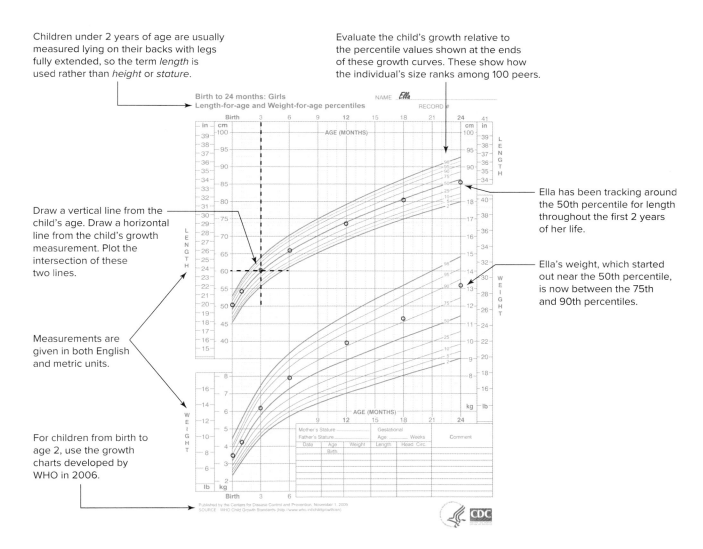

**FIGURE 15-1** ▲ Growth charts for assessment of children in the growing years. Length-for-age and weight-for-age plotted for a young girl (Ella).

Source: Centers for Disease Control and Prevention, based on WHO Child Growth Standards (2009).

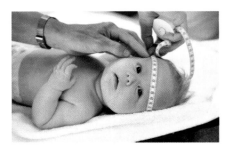

▲ Brain growth is faster in infancy than in any other stage of life. An infant's head needs to be large (about one-quarter of body length) to allow for such growth. By the time she reaches adulthood, this girl's head will only be about one-eighth of her height. Pixtal/age fotostock

caregivers who followed recommended infant and child feeding practices. In contrast, older growth charts for young children were based on data from primarily white children who were mostly formula-fed during infancy. The WHO growth standards emphasize that breastfeeding is the biological norm for infant nutrition.

For children from ages 2 to 20, growth charts developed by the CDC's National Center for Health Statistics are available to assess weight-for-age and stature-for-age. However, the preferred way to assess weight status for children and adolescents is body mass index (BMI)-for-age. For adults, BMI has fixed cutoff points (e.g., a BMI of 25 for an adult is considered overweight). Please note, as Figure 15-2 shows, these fixed, adult weight-status classifications do not apply for children. BMI reference ranges for children between 2 and 20 years of age are both sex and age specific.

In addition to taking measurements of weight and length, a health professional measures the infant's head circumference. Tracking head circumference-for-age on a growth chart (also available from CDC) is a means of assessing brain growth. The brain grows faster during the first year than at any other time of life. Unusual head circumference measurements can alert the health care team to various genetic disorders or growth problems. A small head circumference could be a result of malnutrition, infection, impaired mental development, or maternal substance abuse during pregnancy. An abnormally large measurement may be a sign of a tumor or fluid on the brain. Variations in head circumference may be due to harmless familial traits, but extreme

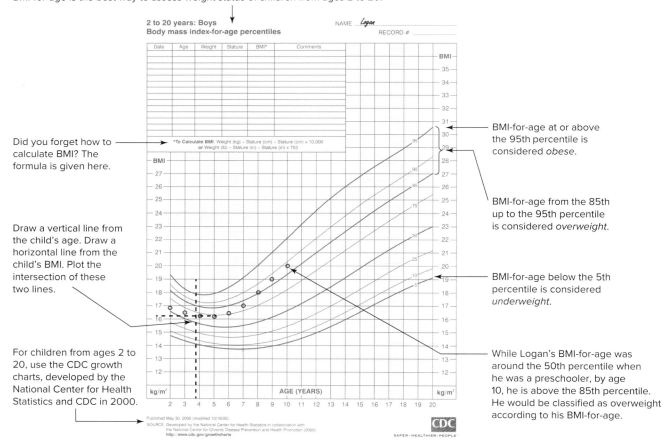

**FIGURE 15-2** ▲ BMI-for-age plotted for a young boy (Logan) up to age 10. See Table 15-1 for more information about interpreting BMI-for-age.

Source: Centers for Disease Control and Prevention, based on WHO Child Growth Standards (2009).

deviations from the norm or rapid changes in percentiles over time warrant further investigation.

Infants and children should have their growth assessed during regular health checkups. It takes 1 to 3 years for the genetic potential (in terms of percentile ranking on growth charts) of infant growth to be established. By 3 years of age, a child's measurements, such as length (height) for age, should track consistently along an established percentile. If the child's growth measurements deviate from his or her established percentile, the primary care provider needs to investigate whether a medical or nutritional problem is affecting growth. Likewise, when a child's BMI-for-age approaches the highest percentiles, caregivers should be concerned. A child from the 85th up to the 95th percentile for BMI-for-age is considered overweight. At or above the 95th percentile, a child is considered obese (Table 15-1).

The growth charts available from the CDC are intended for healthy children; they may not be appropriate for children born at **low birth weight** or those who have special health care needs. **Preterm** infants tend to be smaller than **full-term** infants in the early months but typically catch up in growth within 2 or 3 years. Catch-up growth requires that the child move up in percentiles. Specialized growth charts have been developed for children with special health care needs that affect growth and development, such as Down syndrome.

## GROWTH IN BODY FAT

Since 1970, researchers have speculated that overfeeding during infancy may increase the number of adipose tissue cells. Today, we know that the number of adipose cells can also increase with positive energy balance during adulthood. If energy intake is limited during infancy to minimize the number of adipose cells, the growth of other organ systems may also be severely restricted, especially brain and nervous system development. Furthermore, most overweight infants become normal-weight preschoolers without dietary restrictions. For these reasons, it is unwise for an infant's dietary pattern to be restrictive—especially related to fat intake. A healthy range of fat intake is 30% to 40% of total calories for ages 1 to 3 years and 25% to 35% of total calories for older children and teenagers.

## FAILURE TO THRIVE

About 5% to 10% of infants or children do not grow as expected. An infant may fail to reach important milestones, such as doubling birth weight by 6 months of age. On growth charts, weight-for-age may track below the 5th percentile. This condition of inadequate growth is termed **failure to thrive.**[3]

In some cases, failure to thrive has a specific medical cause. Physical problems that may reduce or limit energy intake include improper development of the mouth or digestive tract or problems with breastfeeding (e.g., improper latch-on). Even if a child consumes enough food and beverages, growth may still falter if there are digestive disorders, such as **celiac disease,** which, if left untreated, can compromise nutrient absorption. Last, some medical conditions, such as heart or lung disorders, lead to greatly increased energy requirements.

However, about 80% of infants or children who fail to thrive have no apparent medical condition; environmental or social problems are at the root of their undernutrition. Poverty is the most significant environmental risk factor for undernutrition leading to failure to thrive. Needing to stretch food dollars, caregivers may overdilute formula—a detrimental practice that fills the child's stomach without providing adequate calories. Sometimes the cause is unhealthy infant–parent interaction, including situations of abuse or neglect. Either too little or too much concern over child feedings can lead to feeding problems. Many cases of failure to thrive, however, arise from the parents' lack of education or inexperience with infant feeding rather than from intentional negligence.

**TABLE 15-1** ■ **Weight Status Classifications for Children, Ages 2 to 20**

| Weight Status Classification | BMI-for-Age Percentile |
| --- | --- |
| Underweight | < 5th percentile |
| Healthy weight | 5th up to 85th percentile |
| Overweight | 85th up to 95th percentile |
| Obese | ≥ 95th percentile |

Adapted from World Health Organization, 2018

**failure to thrive**  Condition of inadequate growth during infancy or early childhood caused by insufficient nutritional intake, inefficient nutrient absorption, or excessive energy expenditure; commonly defined as weight-for-age below the 5th percentile on multiple occasions or weight declining 2 or more major percentile lines on a standardized growth chart.

Children older than 2 years are less likely to experience failure to thrive because they can often get food for themselves. Younger children, for the most part, are limited to what caregivers provide. **What are some simple ways to monitor a child's nutritional status?** ONOKY-Photononstop/Alamy

Whatever the cause, the consequences of failure to thrive are serious and long-lasting. Possible outcomes include poor physical growth, impaired mental development, and behavioral problems. When health professionals encounter an infant or child failing to thrive, the true causes must be identified and then treated. If the problem stems from a lack of financial resources, appropriate referrals to social services should be made. Counseling about proper nutrition and the importance of healthy parent–child interactions can help to get a child's growth back on track.

### ✓ CONCEPT CHECK 15.1

1. How do health care providers assess the growth of infants and children?
2. Define childhood overweight and obesity in terms of BMI-for-age.
3. If a female adolescent is short for her age due to a brief period of undernutrition between ages 10 to 12, can she catch up in growth after proper nutrition is restored? Why or why not?
4. What is failure to thrive? List three possible causes.

## 15.2 Infant Nutritional Needs

Infants' nutritional needs vary as they grow. For the first 4 to 6 months of life, human milk or infant formula supplies all the required nutrients. Age-appropriate solid foods should be added to the infant's eating pattern around 6 months of age. (See the section about Expanding the Infant's Mealtime Choices for additional guidance on the timing of introduction of solid foods.) Even after solid foods are added, the foundation of an infant's dietary pattern for the first year is still human milk or infant formula. Because of the critical importance of adequate nutrition in infancy and the difficulties encountered in feeding some infants, there is more discussion in this chapter on infancy than on the later periods of childhood.

### ENERGY

Due to rapid growth and a high metabolic rate, the energy requirements per pound of body weight for an infant are the highest of any life stage (Table 15-2). For example, a 6-month-old infant requires two to four times more kilocalories per pound of body weight than an adult does.

Infants need a concentrated source of calories to meet these high demands. Exclusive feeding of either human milk or infant formula is ideal for the first 6 months of life; both are high in fat and supply about 640 kcal/quart (670 kcal/liter; Table 15-3). Beginning around 6 months of age, developmentally appropriate solid foods provide additional calories, nutrients, and variety for the developing infant.

**Healthy 6-month-old infant**

$$\frac{700 \text{ kcal}}{15 \text{ pounds}} = \frac{47 \text{ kcal}}{\text{pound}}$$

**Healthy 20-year-old woman**

$$\frac{2200 \text{ kcal}}{135 \text{ pounds}} = \frac{16 \text{ kcal}}{\text{pound}}$$

**TABLE 15-2 ■ Estimated Energy Requirements of Infants and Toddlers**

| Age | EER Equation |
|---|---|
| 0 to 3 months | (89 kcal × weight in kilograms) + 75 |
| 4 to 6 months | (89 kcal × weight in kilograms) − 44 |
| 7 to 12 months | (89 kcal × weight in kilograms) − 78 |
| 13 to 35 months | (89 kcal × weight in kilograms) − 80 |

**TABLE 15-3** ■ **Composition of Human and Cow's Milk and Infant Formulas (per Liter)[a]**

| | Energy (kcal) | Protein (grams) | Fat (grams) | Carbohydrate (grams) | Minerals[b] (grams) |
|---|---|---|---|---|---|
| **Milk** | | | | | |
| Human milk | 670[c] | 11 | 45 | 70 | 2 |
| Cow's milk, whole[d] | 670 | 36 | 36 | 49 | 7 |
| Cow's milk, fat-free[d] | 360 | 36 | 1 | 51 | 7 |
| **Casein/Whey-Based Formulas** | | | | | |
| Similac® | 680 | 14 | 36 | 71 | 3 |
| Enfamil® | 670 | 15 | 37 | 69 | 3 |
| Good Start® | 670 | 16 | 34 | 73 | 3 |
| **Soybean Protein-Based Formulas** | | | | | |
| ProSobee® | 670 | 20 | 35 | 67 | 4 |
| Isomil® | 680 | 16 | 36 | 68 | 4 |

[a] At 3 months of age, infants typically consume 0.75 to 1 liter of human milk or formula per day.
[b] Calcium, phosphorus, and other minerals.
[c] Rough estimate; ranges from 650 to 700 kcal per liter.
[d] Not appropriate for infant feeding, based primarily on high protein and mineral content.

## CARBOHYDRATES

Carbohydrate needs in infancy are 60 grams per day at 0 to 6 months and 95 grams per day at 7 to 12 months. These needs are based on the typical intakes of human milk by breastfed infants and their eventual intake of solid foods. These carbohydrate goals are easily satisfied by a developmentally appropriate eating pattern.

Do infants need fiber? There are no set AIs for fiber for infants and children younger than 1 year of age. For about the first 6 months of life, breast milk or formula, which contains no fiber, provides optimal nutrition. As solid foods are introduced, include some fruits, vegetables, and whole grains. Some experts recommend working up to about 5 grams of fiber per day by 1 year of age. Keep in mind that too much fiber can limit nutrient absorption because it binds to some minerals and speeds the passage of food through the GI tract. Let the child's bowel habits be your guide. If the child is constipated, try increasing fiber and fluid intakes. On the other hand, if the child is uncomfortably gassy or is having many soft bowel movements per day, decrease the amount of fiber in the dietary pattern.

## PROTEIN

Daily protein needs in infancy are about 9 grams per day for younger infants and about 11 grams per day for older infants. These recommendations are based on the typical consumption of human milk by infants who are breastfed for 0 to 6 months and then on the increased protein needs for growth for older infants. About half of total protein intake should come from **essential amino acids.** As with carbohydrate, protein needs are easily satisfied by either human milk or infant formula. However, protein metabolism generates waste products that must be excreted by the kidneys. Dietary patterns that are excessive in protein may stress the infant's immature kidneys, so protein intake should not greatly exceed these recommendations.

In North America, infant protein deficiency is unlikely, except in cases of inappropriate formula preparation, such as when an infant's formula is excessively diluted with water. Protein deficiency may also be induced by elimination diets used to identify **allergies** to certain foods. As foods are eliminated, infants may not be offered enough protein to compensate for that supplied by the suspected food **allergen** (see Section 15.7).

## FAT

Fat makes up the largest proportion (about 50%) of an infant's overall energy intake during the first year of life. Infants need about 30 grams of fat per day. **Essential fatty acids** should make up about 15% of total fat intake (about 5 grams per day). Both recommendations are again based on the typical consumption of human milk by breastfed infants and the eventual intake of solid foods. Fats are an important part of the infant's dietary pattern because they are vital to the development of the nervous system. Also, fats are a concentrated source of energy (9 kcal per gram). The stomach capacity of an infant is limited, so a concentrated source of calories is necessary to meet overall energy requirements. Thus, restriction of fat intake is not advised for infants or children under age 2.

**Arachidonic acid (ARA)** and **docosahexaenoic acid (DHA)** are two long-chain fatty acids that have important roles in infant development. These fatty acids can be made in the body from the essential fatty acids or they can be supplied by foods. The nervous system, especially the brain and eyes, depends on these fatty acids for proper development. During the last trimester, DHA and ARA provided by the mother accumulate in the brain and retinas of the eyes in the fetus. Breastfed infants continue to acquire these fatty acids from human milk, especially if their mothers are regularly eating fish. Since 2002, infant formula manufacturers have been adding ARA and DHA to their products to match the average fatty acid composition of human milk. Such infant formulas are particularly useful for feeding preterm infants.[4]

## VITAMINS OF SPECIAL INTEREST

All of the essential vitamins play important roles in infant growth and development, but three vitamins—K, D, and B-12—are of special interest for infants because their stores tend to be low and the consequences of deficiency are dire.

**Vitamin K.** Newborn infants have low levels of vitamin K because: (1) limited amounts of this vitamin are transferred from the mother to the fetus during gestation; (2) breast milk is not particularly high in vitamin K; and (3) newborn infants lack the intestinal bacteria that synthesize vitamin K. Infant vitamin K deficiency can lead to a rare but potentially fatal problem known as **vitamin K deficiency bleeding.** To prevent hemorrhage, vitamin K is routinely given by injection to all infants at birth.

**Vitamin D.** For bone health (to prevent **rickets**), immune function, and chronic disease prevention, the American Academy of Pediatrics (AAP) recommends that all infants and children consume 10 micrograms of vitamin D per day starting soon after birth. Supplemental vitamin D is necessary for all breastfed infants, as well as for formula-fed infants who consume less than 1 quart (approximately 1 liter) of formula per day.[5] However, no further benefits are seen beyond 10 micrograms per day, and toxicity is possible if intake exceeds 25 micrograms per day. Any use of supplemental vitamin D can be discontinued when dietary sources provide at least 10 micrograms of vitamin D per day.

**Vitamin B-12.** For infants who are breastfed by women who follow a vegan dietary pattern, ensuring adequate vitamin B-12 status is vitally important to prevent **anemia,** failure to thrive, and irreversible damage to the nervous system. Recall that vitamin B-12 is only found in foods of animal origin. Infant formula does contain vitamin B-12, but the breast milk of a mother who avoids all animal products may be deficient in vitamin B-12. A breastfeeding woman who follows a vegan dietary pattern should take care to obtain adequate vitamin B-12 from fortified foods or dietary supplements. If the mother's vitamin B-12 status is low, supplemental vitamin B-12 may be necessary for the infant.[6]

## MINERALS OF SPECIAL INTEREST

Two minerals of special interest in the dietary patterns of infants are iron and fluoride. Infants also need adequate amounts of zinc and iodine to support growth. However, when human milk and infant formula are provided in quantities to meet energy needs, zinc and iodine requirements are generally met.

**Iron.** Infants are born with some internal stores of iron. However, if food sources of iron are not part of the infant's dietary pattern, body iron stores will be depleted by about 6 months of age. If the mother was iron deficient during the pregnancy, the infant's iron stores will be exhausted even sooner. Iron-deficiency anemia can lead to poor cognitive development in infants. Several studies indicate that iron-deficiency anemia during infancy, even if corrected, has a lasting impact in terms of cognition, motor development, and behavior later in life.[7]

To maintain a desirable iron status, the AAP recommends that formula-fed infants should be given an iron-fortified formula from birth. Years ago, low-iron formulas were prescribed for infants with gastrointestinal distress; however, current evidence shows that these infant formulas do not improve GI symptoms; rather, low-iron infant formulas place an infant at risk for iron deficiency. Low-iron infant formulas are still available, but their use is strongly discouraged.

Breast milk is lower in iron than fortified infant formulas, but the form of iron in breast milk is much more bioavailable than the form in infant formula. Even so, by about 6 months of age, breastfed infants need solid foods to supply extra iron. This need for iron is a major consideration in the decision to introduce **complementary foods.** To prevent iron deficiency, the AAP recommends that exclusively breastfed infants should receive iron supplements starting at 4 months of age and continuing until dietary sources of iron are introduced. Preterm and low-birth-weight infants, those with blood disorders, or infants born to mothers who have iron-deficiency anemia or diabetes may need higher levels of supplemental iron.[8]

**complementary foods** Solid or semi-solid foods that are introduced to an infant's dietary pattern to complement (not replace) breast milk or infant formula during the latter part of the first year of life. These foods provide energy and essential nutrients to meet the infant's requirements for growth and development.

**Fluoride.** Breast milk is a poor source of fluoride, and formula manufacturers use fluoride-free water in formula preparation, so intake of this mineral during the first 6 months of life is low. However, fluoride supplementation is not advised before 6 months of age. After 6 months, the pediatrician or dentist may recommend fluoride supplements to aid in tooth development if fluoride supplied by tap water, foods, and toothpaste is inadequate.[9]

The American Dental Association does not recommend fluoridated bottled water for use by infants because it heightens the risk for enamel **fluorosis** during early tooth development.

## WATER

An infant needs about 3 cups (700 to 800 milliliters) of water per day to regulate body temperature and transport oxygen, nutrients, and wastes throughout the body. For the vast majority of infants, human milk or formula supply enough water to keep the infant well hydrated.

In infants, **dehydration** can occur rapidly and have devastating consequences. In the first few days after birth, improper feeding techniques can leave an infant deprived of water and nutrients. In addition, protracted episodes of vomiting or diarrhea can quickly deplete an infant of fluid and **electrolytes.**

To identify dehydration, look for these signs:[10]

- More than 6 hours without a wet diaper
- Dark-yellow or strong-smelling urine
- Unusually tired and fussy
- Dry mouth and lips
- Absence of tears when crying
- Eyes and soft spot on the head appear sunken
- Cold and splotchy hands and feet

Severe dehydration can result in rapid loss of kidney function and warrants medical intervention. In some cases, hospitalization and intravenous rehydration may be necessary. Most of the time, dehydration can be corrected with special fluid-replacement formulas containing electrolytes, such as sodium and potassium.[11] These oral rehydration solutions (e.g., Pedialyte®) are available in supermarkets and pharmacies to treat mild to moderate dehydration. A primary care provider should guide any use of these products.

Caregivers often wonder if breast milk or formula is sufficient to keep an infant hydrated, especially in hot weather. They may be tempted to give an infant supplemental water or fruit juice. In some stores, bottled water products marketed specifically for infants may be placed alongside infant formulas and oral rehydration solutions, giving the mistaken impression that bottled water products are an appropriate feeding beverage for infants. The AAP does not recommend supplemental water or juice during the first 6 months, even in hot weather. Excess water can cause electrolyte imbalances, such as hyponatremia, in infants. It is important to remember that excessive fluid can be harmful, especially to the brain.

Overall, it is best to rely exclusively on breast milk or infant formula to meet infant fluid needs up to 6 months of age, unless a health care provider suggests otherwise. In sum, extremes in fluid intake—either too little or too much—can lead to health problems.

**✓ CONCEPT CHECK 15.2**

1. Using the EER equations listed in Table 15-2, calculate the kilocalorie needs of a healthy 4-month-old infant who weighs 15 pounds (6.8 kilograms).

2. Review Table 15-3. Why is unaltered cow's milk *not* recommended for infant feeding?

3. Do infants require any supplemental vitamins or minerals? If so, which ones? What is the rationale for their use?

4. Antoinette is planning a family trip to the zoo on a hot summer day. How can she keep her 3-month-old baby from becoming dehydrated?

## **15.3** Guidelines for Infant Feeding

Oral nutrition for infants comes in two forms: human breast milk or infant formula. Breastfeeding is the preferred method of infant feeding (review Table 14-4). Besides its benefits for immune development, improved mother–infant bonding, and lower long-term risk for chronic diseases, breast milk offers optimal nutrition that is uniquely suited for human infants. For women who do not breastfeed, whether due to necessity or preference, infant formula is a workable substitute. In fact, formula manufacturers model their products on human milk. In areas of the world with safe water supplies, formula feeding is a safe and nutritionally adequate alternative to breastfeeding.

### BREAST MILK IS THE BEST MILK

Human milk is uniquely suited to meet the nutritional needs of human infants. Table 15-3 gives the composition of human milk, but these numbers are just estimates. Maternal eating patterns and nutritional status may impact the composition of breast milk, particularly for fatty acids and some micronutrients. In addition, the true composition of human milk changes over time as the infant matures and even within a feeding.

Breast milk provides up to 55% of total calories as fat. Fat is a concentrated source of calories, so it helps to meet the high energy needs of the growing infant even though the infant cannot consume a large volume of milk at one time. Perhaps most striking, the fat content of breast milk changes within each feeding. When the baby first latches on and begins suckling, the consistency of breast milk is thin and watery; the infant takes in necessary carbohydrates, protein, vitamins, and minerals. As the feeding goes on, the fat content of the milk increases to fulfill energy needs and satisfy the infant. The specific types of fat in human milk are ideal for infants, too. The short- and medium-chain fatty acids in breast milk are easily digested. Some of the fatty acids—arachidonic acid and docosahexaenoic acid—are essential for proper brain and eye development.

Breastfeeding takes some skill and patience on the part of the mother, especially in the first few weeks, but the physical and emotional benefits are worth the effort. **What are some of the benefits of breastfeeding you remember from Chapter 14?** Getty Images

If the mother's dietary pattern is rich in these fats, her milk will be a better source of them for her infant as well.[4]

Carbohydrates provide about 35% to 40% of the calories in human milk. The main carbohydrate in human milk is lactose, a disaccharide that tastes sweet and is easily digested in the human infant's digestive tract. Although **lactase** production tends to decline later in life, **lactose intolerance** is quite rare among infants. As described in Section 14.7, human milk also contains some oligosaccharides that have a prebiotic effect on the community of beneficial microorganisms in the infant's gut. A healthy microbial population in the GI tract influences development of the infant's immune system.

Protein supplies less than 10% of the total calories in human milk. The kidneys of the newborn infant are still immature, so they can be stressed by high protein intakes. The proteins that are present in human milk are easily digested and unlikely to trigger food allergies. These proteins do more than supply calories and building blocks for tissue synthesis; they also promote the proper development of the immune system and enhance nutrient absorption.

For the most part, the micronutrient needs of the infant can be met by human milk. A notable exception, described in Section 15.2, is vitamin D. Recall that the AAP recommends 10 micrograms per day of supplemental vitamin D for all infants (breastfed and formula-fed) until their dietary intake supplies this amount. The AAP also recommends routine iron supplementation for exclusively breastfed infants after 4 months of age, as well as for preterm or LBW infants. For infants who are breastfed by women who follow a vegan dietary pattern, have had bariatric surgery, or have pernicious anemia, vitamin B-12 supplements are recommended.

Breastfeeding during infancy impacts future feeding behaviors and weight management as well. Some studies show that exclusive breastfeeding for the first 6 months of life is associated with lower BMI later in childhood and into adulthood.[12] There are many possible explanations for a link between breastfeeding and healthy BMI. Breastfeeding may reinforce the natural ability of the infant or child to self-regulate food intake. Whereas infant formula flows easily from a bottle into the infant's mouth, an infant has to do some work to get milk from the mother's breast. Thus, it is unlikely that the infant will override satiety cues and overeat from the breast. Perhaps breastfed infants learn to eat based on internal cues of hunger and satiety, leading to better body weight management and lower risk for cardiovascular diseases and type 2 diabetes throughout life. Another explanation may lie in differences in the rate of growth early in life. Formula-fed infants tend to gain weight more rapidly than breastfed infants in the first months of life. This may impact future weight regulation.[13,14] Perhaps breastfeeding influences body weight regulation via manipulations of the microbiota.[15]

## FORMULA FEEDING FOR INFANTS

Young infants cannot tolerate unaltered cow's milk because of its high protein and mineral content. Cow's milk is perfect for the growth needs of calves but not for human infants! Thus, cow's milk must be altered by formula manufacturers to be safe for infant feeding. Altered forms of cow's milk, known as infant formulas, must conform to strict federal guidelines for nutrient composition and quality.

Formulas generally contain lactose and/or sucrose for carbohydrate, heat-treated proteins from cow's milk, and plant oils for fat (review Table 15-3). Soy protein–based formulas are available for infants whose parents want the infant to begin a vegan dietary pattern or those who cannot tolerate lactose or the types of proteins found in cow's milk. Infants with milk protein allergies are often sensitive to soy as well, so the best choice for infants with allergies is a **hydrolyzed protein formula.** In this type of formula, the proteins have been broken down into small polypeptides and individual amino acids. A variety of other specialized formulas are also available for specific medical conditions. In any case, it is important to use an iron-fortified formula unless a pediatrician recommends otherwise.

**hydrolyzed protein formula** Infant formula in which the proteins have been broken down into smaller peptides and amino acids to improve digestibility and reduce exposure to potential food allergens; sometimes called *predigested* or *hypoallergenic infant formula.*

▲ Bisphenol A (BPA) is a chemical used in the production of many plastics. Human exposure to BPA, mainly through leaching of the chemical from packaging into foods and beverages, is widespread. Concern about exposure stems from animal studies that link high doses of BPA with reproductive and developmental defects and cancer. The consensus among regulatory agencies in the United States and Canada is that current levels of BPA exposure are not harmful, even for infants. Nevertheless, in response to public concern, the U.S. Food and Drug Administration (FDA) has banned the use of BPA in the manufacture of baby bottles, sippy cups, and packaging for infant formulas. BSIP SA/Alamy Stock Photo

Some transition formulas/beverages have been introduced for older infants and toddlers (review Table 15-3). A few of these products are intended for use after 6 months of age if the infant is consuming solid foods, but most are intended for use only by toddlers (i.e., age 12 months and older). These transition products are lower in fat than human milk or standard infant formulas; their iron content is higher than that of cow's milk; and their overall mineral content is generally more like that of human milk than cow's milk. According to the manufacturers, the advantages of these transition formulas/beverages over standard formulas for older infants and toddlers include reduced cost and better flavor. Recent recommendations from experts in pediatric medicine, nutrition, and dentistry advise caregivers to skip these transition formulas.[16] Instead, offer milk or water as beverage choices after 1 year of age.

Infant formula comes in several different forms. Some infant formulas come in ready-to-feed, liquid form. These can be simply poured into a clean bottle and fed to the infant without further preparation. Powdered and concentrated fluid formula preparations are also commonly used. Powdered or concentrated formulas should be combined with clean, cold water, precisely following the directions on the formula label. The formula is then warmed, if desired, and fed immediately to the infant.

The following are some tips for safe preparation and storage of infant formula:

- All containers and utensils used to prepare formula should be washed with hot, soapy water and thoroughly rinsed with clean water before use. Household dishwashers are a good way to clean bottles and utensils, too. Unless the infant's immune system is compromised, it is not necessary to boil the containers and utensils prior to use.
- Use cold water to prepare infant formula. Hot water that sits in a hot water heater or runs through pipes made with lead is more likely to accumulate contaminants than cold water (see Section 15.6). For homes with older plumbing systems, it is prudent to let cold tap water run for 1 to 2 minutes before filling a bottle or cup.
- If well water will be used to make infant formula, it should be tested regularly for contaminants, such as naturally occurring nitrates, which can lead to a severe form of anemia (especially among babies younger than 1 year).
- If microbial contamination of well water or municipal tap water is a concern, water should be brought to a rolling boil (for 1 minute) and then cooled to room temperature (for up to 30 minutes) before use in formula preparation. Caregivers should pay close attention to local water advisories. Some pediatricians recommend boiling (then cooling) the water to be used in formula preparation for infants up to 6 months of age, regardless of reports about water safety.
- The American Dental Association does not recommend that formula be mixed with bottled "nursery water," which can be found alongside infant formula in most supermarkets. These bottled water products may contain high levels of fluoride, which can lead to tooth discoloration.
- To warm a bottle of formula, run hot water over it or place it briefly in a pan of simmering water. Infant formulas should not be heated in a microwave oven because hot spots may develop, which can burn the infant's mouth and esophagus.
- Refrigerating prepared formula for 1 day is safe. However, formula left over from a feeding should be discarded if not used within one hour because it will be contaminated by microbes and enzymes from the infant's saliva.

Dentists and pediatricians warn caregivers to avoid putting infants to bed (or in infant seats) with a bottle. This advice aims to prevent **early childhood caries** (Fig. 15-3). When carbohydrate-rich fluid continuously bathes the teeth, this provides an ideal growth medium for oral bacteria. These bacteria feast on the carbohydrates and produce acids, which dissolve tooth enamel and lead to decay. It is okay to feed an infant at bedtime, but remove the bottle from the infant's mouth when she falls asleep or before putting her in her crib so that formula (or expressed breast milk) does not pool in the infant's mouth.

**early childhood caries** Tooth decay that results from formula or juice (and even human milk) bathing the teeth as the child sleeps with a bottle in the mouth. The upper teeth are mostly affected as the lower teeth are protected by the tongue; formerly called *nursing bottle syndrome* or *baby bottle tooth decay.*

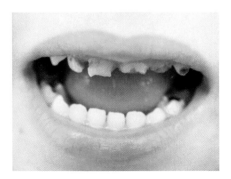

**FIGURE 15-3** ▲ Early childhood caries. This type of tooth decay may have resulted from frequently putting the child to bed with a bottle. Some of the upper teeth have decayed almost all the way to the gum line. Zoonar GmbH/Alamy Stock Photo

## FEEDING TECHNIQUE

Infants swallow a lot of air as they ingest either formula or human milk. To alleviate discomfort, it is important to burp an infant during feeding (every 1 to 2 ounces) and again at the end of the feeding. Spitting up a small amount of milk or formula is normal at this time.

Formula intake and feeding frequency may vary considerably from one infant to another and also from one day to the next. In general, a formula-fed infant should consume about 2½ fluid ounces (75 milliliters) of formula per pound of body weight each day (see Table 15-4), up to about 32 fluid ounces (960 milliliters) in one day. In most cases, however, the infant's appetite is a better guide than any standardized recommendations. When the infant begins acting full, bottle feeding should be stopped, even if some milk is left in the bottle. Common signals that an infant has had enough include turning the head away, being inattentive, falling asleep, or becoming playful. Breastfeeding infants usually have had enough to eat after about 20 minutes. Although it is difficult to tell how much milk breastfed infants are getting, they also give recognizable signs when they are full.

By carefully observing and responding appropriately to the infant's cues during feeding, caregivers can: (1) be assured that the infant's calorie needs are being met; (2) foster a climate of trust and responsiveness; and (3) help a child develop a habit of respecting internal cues of hunger and satiety.[17]

## EXPANDING THE INFANT'S MEALTIME CHOICES

By about 6 months of age, the infant is ready to start eating complementary foods. Complementary foods are age-appropriate solid foods that *complement* (rather than *replace*) breast milk or infant formula, at least initially.

In the first attempts to introduce solid foods, just getting the food into the infant's mouth may be a challenge. By the end of the first year, though, the infant should be eating a variety of protein sources, vegetables, fruits, and grains so that the the infant's daily menu begins to reflect a balanced pattern (Table 15-5). Throughout the process of expanding the infant's mealtime choices, the caregiver should respond to the infant's cues that he or she is hungry or has had enough to eat. This principle is called **responsive feeding** and should guide caregivers' feeding practices at all stages of infancy and childhood. Feeding behaviors developed through early exposures to food will set the stage for healthy eating to last a lifetime.[18]

**responsive feeding** A healthy feeding relationship between the caregiver and the child in which the caregiver pays attention to and respects the child's hunger and satiety.

**Determining the Infant's Readiness for Solid Foods.** How does the caregiver know it is time to introduce solid foods? Infant size can serve as a rough indicator of readiness: reaching a weight of at least 13 pounds (6 kilograms) is a preliminary sign of readiness for solid foods. Another physiological cue is frequency of feeding, such as consuming more than 32 ounces (1 liter) of formula daily or breastfeeding more than 8 to 10 times within 24 hours. Underlying these noticeable signals are several important developmental factors:[2]

1. *Nutritional need.* Before the infant is 6 months old, nutritional needs can generally be met with human milk and/or formula. After 6 months of age, however, many

#### TABLE 15-4 ■ Typical Formula Intake of Infants

| Age | Amount per Feeding | Frequency |
| --- | --- | --- |
| <1 month | 2 to 3 fluid ounces (60 to 90 ml) | Every 3 to 4 hours |
| 1 to 6 months | 4 to 6 fluid ounces (120 to 180 ml) | Every 4 to 6 hours |
| 6 to 12 months | 6 to 8 fluid ounces (180 to 240 ml)* | Every 4 to 6 hours |

Source: Data from American Academy of Pediatrics: *Caring for Your Baby and Young Child: Birth to Age 5*, 6th ed. New York: Bantam Books, 2014.
* With introduction of age-appropriate solid foods.

**TABLE 15-5 ■ Sample Daily Menu for a 1-Year-Old Child\***

| Breakfast | Snack |
|---|---|
| 1 tbsp unsweetened applesauce | 4 whole wheat crackers |
| ¼ cup toasted oat cereal | ½ cup water |
| ½ cup whole milk | |

| Snack | Dinner |
|---|---|
| ½ slice whole wheat toast | 1 ounce roast beef (finely diced) |
| ½ tsp smooth peanut butter | 2 tbsp mashed potatoes |
| ½ cup mandarin orange segments | 2 tbsp cooked carrots (diced) |
| ½ cup water | ½ cup whole milk |

| Lunch | Snack |
|---|---|
| 2 tbsp black beans | ½ banana |
| 2 tbsp rice with ½ tsp butter | 1 oatmeal cookie (no raisins) |
| 1 tbsp tomatoes | ½ cup whole milk |
| ½ cup whole milk | |

| Nutritional Analysis | |
|---|---|
| Total energy (kcal) | 775 |
| Approximate % energy from | |
| Carbohydrate | 50% |
| Protein | 20% |
| Fat | 30% |

*This menu is just a start. A 1-year-old may need more or less food. In those cases, serving sizes should be adjusted. The milk can be fed by cup; some can be put into a bottle if the child has not been fully weaned from the bottle.

infants need additional calories. In terms of individual nutrients, iron stores are exhausted by about 6 months of age. Developmentally appropriate, nutrient-dense foods will help to meet the infant's calorie and iron needs.

2. *Physiological capabilities.* Before about 3 months of age, an infant's digestive tract cannot readily digest starch. Also, kidney function is limited until about 4 to 6 weeks of age. Until then, waste products from excessive amounts of dietary protein or minerals are difficult to excrete. As the infant ages, the ability to digest and metabolize a wider range of food components improves.

3. *Physical ability.* Three observable markers indicate that a child is ready for solid foods: (1) the disappearance of the extrusion reflex (thrusting the tongue forward and pushing food out of the mouth); (2) head and neck control; and (3) the ability to sit up with support. These usually occur around 4 to 6 months of age, but they vary with each infant.

4. *Allergy prevention.* Because a newborn's digestive system is still immature, whole proteins can readily be absorbed from birth until 4 to 5 months of age. If the infant is exposed too early to some types of proteins, the infant may be predisposed to food allergies and autoimmune conditions (e.g., type 1 diabetes). However, there is no benefit to delaying introduction of solid foods beyond 6 months of age.

With these considerations in mind—nutritional need, physiological and physical readiness, and allergy prevention—the AAP and the *Dietary Guidelines* recommend introducing complementary foods to the infant's diet around 6 months of age.[19] Some infants may be developmentally ready for complementary foods earlier than 6 months of age, but caregivers are advised to wait until at least 4 months. Despite these recommendations, data from the National Health and Nutrition Examination Survey indicate that nearly 1 in 5 infants receives complementary foods before 4 months of age.[20]

One reason cited by caregivers for early introduction of solid foods is to satisfy infant hunger. Only occasionally does a rapidly growing infant need to add solid foods to meet calorie and nutrient needs before 6 months of age. In fact, most complementary foods are *less* energy dense than human milk or infant formula. Given the infant's small stomach capacity, providing complementary foods in place of human milk or infant formula before the infant is developmentally ready may actually lower the infant's energy and nutrient intake and hinder growth in the short term. Over the long term, some studies indicate that earlier introduction of solid foods is associated with overweight or obesity.[19]

Aside from hunger, many caregivers believe that early introduction of solid foods will help an infant sleep through the night (i.e., 6 to 8 hours of uninterrupted sleep). The age at which an infant begins to sleep through the night is of little consequence to the child's physical and mental outcomes,[21] yet sleep-deprived caregivers are eager for that blissful time of rest! Does early introduction of solid foods influence sleep duration? One recent analysis showed a modest but statistically significant relationship between early introduction of solid foods and longer sleep duration.[22] However, evidence has been mixed.[23] Overall, pediatric nutrition experts recommend waiting until at *least* 4 months—preferably closer to 6 months—before introducing complementary foods.

**Foods to Match Needs and Developmental Abilities During the First Year.** The primary goal of introducing complementary foods to the infant's dietary pattern is to meet nutrient needs, particularly for iron and zinc. Therefore, pediatric nutrition experts recommend iron-fortified infant cereals and lean ground (strained) meats as the first solid foods (Fig. 15-4).

When starting solid foods, begin with a teaspoon-size serving of a single-ingredient food item and increase the serving size gradually over the next few days. Once the new food has been fed for several days without adverse effects, another food can be added to the infant's eating pattern. There is no prescribed order for the introduction of specific types of foods or food groups, but food choices should be nutrient dense and developmentally appropriate.

Waiting about 3 to 5 days between the introduction of each new food is important because it can take that long for evidence of an allergy or intolerance to materialize. Also, it is best to avoid introducing mixed foods until each component of the combination dish has been given separately without an adverse reaction.[19] Signs of food allergies include diarrhea, vomiting, a rash, or wheezing. If one or more of these signs appears, the suspected problem food should be avoided for several weeks and then reintroduced in a small quantity. If the problem continues, consult a pediatrician. Fortunately, many babies outgrow food allergies later in childhood.

Until 2008, parents and caregivers were advised to avoid feeding children a wide range of potentially allergenic foods, including egg whites, chocolate, peanuts, tree nuts, fish, and other seafood. Now, pediatric nutrition experts acknowledge that there is no evidence that delaying introduction of solid foods—including these common food allergens—beyond 4 to 6 months of age is of any benefit for prevention of atopic diseases. Indeed, the most recent guidelines from the National Institute of Allergy and Infectious Diseases recommend early introduction of peanut protein—between 4 and 6 months of age—to infants at risk of food allergies.[24,25]

Many strained foods for infant feeding are available where groceries are sold. Single-food items are more desirable than mixed dinners and desserts, which are less nutrient dense. Most brands have no added salt, but some fruit desserts contain added sugar, which is not recommended for infant feeding.[26] Read food labels carefully to find the most nutrient-dense foods for infants.

As an alternative to store-bought baby foods, plain, unseasoned cooked foods—vegetables, fruits, and meats—can be ground up in an inexpensive baby food grinder at home. Another option is to purée a larger amount of food in a blender, freeze it in ice-cube portions, store in plastic bags, and defrost and warm as needed. Careful attention to food safety is necessary. Seasonings that may please the rest of the family should not be added to infant foods made at home. The infant does not notice the difference if salt, sugar, or spices are omitted. It is best to introduce infants to a variety of foods, so that by

Repeated exposure fosters acceptance of new tastes and textures. **Should you force an infant to eat a new food?** Corbis/PictureQuest

**FIGURE 15-4** ▶ Infant foods, like adult foods, are required to have a Nutrition Facts label; however, the information provided on infant food labels differs from that on adult food labels (see Fig. 2-19). There are separate Daily Values set for infants through 12 months of age and children 1–3 years of age.

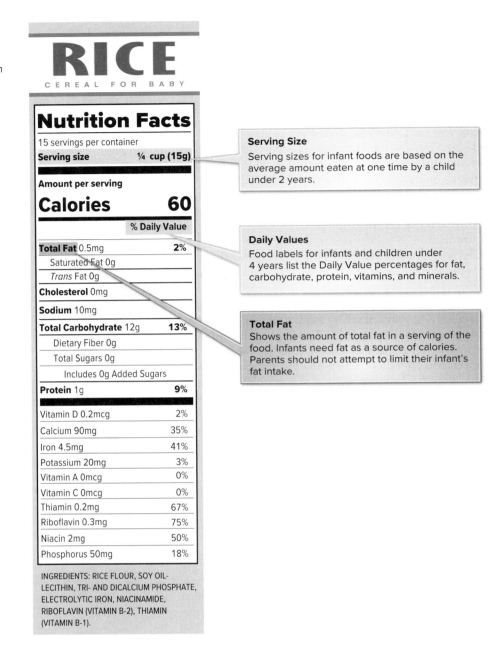

**Serving Size**
Serving sizes for infant foods are based on the average amount eaten at one time by a child under 2 years.

**Daily Values**
Food labels for infants and children under 4 years list the Daily Value percentages for fat, carbohydrate, protein, vitamins, and minerals.

**Total Fat**
Shows the amount of total fat in a serving of the food. Infants need fat as a source of calories. Parents should not attempt to limit their infant's fat intake.

the end of the first year, the infant is consuming many foods—human milk or formula, meats, fruits, vegetables, and grains.

To ease early attempts at feeding solid foods, consider the following tips:

- Use a baby-size spoon; a small spoon with a long handle is best.
- Hold the infant comfortably on the lap, as for breastfeeding or bottle feeding, but a little more upright to ease swallowing. When in this position, the infant expects food.
- Put a small dab of food on the spoon tip and gently place it on the infant's tongue.
- Convey a calm and casual approach to the infant, who needs time to get used to food.
- Expect the infant to take only two or three bites of the first meals.
- Present a new food on several consecutive days. Repeated exposures can promote acceptance of new food items. This advice holds true for toddlers and preschoolers, as well!

Self-feeding skills require coordination and can develop only if the infant is allowed to practice and experiment. By 6 to 7 months of age, most infants have learned to handle finger foods and transfer objects from one hand to the other with some dexterity. At about this time, teeth also begin to appear. By age 7 to 8 months, infants can push food around on a plate, handle a drinking cup, hold a bottle, and self-feed a cracker or a piece of toast. Through mastery of these manipulations, infants develop confidence and self-esteem. It is important that parents be patient and support these early self-feeding attempts, even though they appear inefficient.

At 9 to 10 months of age, the infant's desire to play with foods may hinder feeding. Food is used as a means to explore the environment, and therefore, feeding time is often very messy—a bowl of macaroni may end up in the child's hair! Caregivers should relax and take this phase of infant development in stride. By the end of the first year, finger-feeding becomes more efficient, and chewing is easier as more teeth erupt. Still, some sloppy experimentation is to be expected.

## BABY-LED INTRODUCTION TO SOLID FOODS

Traditionally, when we think about introducing complementary foods, we think about jars of baby food or homemade purées that can be fed to infants with a spoon. However, in recent years a growing number of caregivers and health professionals have embraced a different technique: **baby-led introduction to solid foods,** also known as baby-led weaning.

This technique has become quite popular in the United Kingdom and New Zealand. In the U.S., we use the term **weaning** to describe the gradual process of switching from breast milk or infant formula to other sources of nutrition. In the United Kingdom, *weaning* simply refers to adding solid foods to the infant's dietary pattern. For clarity, we will use the term *baby-led introduction to solid foods*.

With this technique, instead of caregivers being in control of the infant's feeding experience, the infant is allowed to self-feed from the first complementary food. At 6 months of age, most infants can pick up foods with a palmar grasp—the entire hand closes around the food and brings it to the mouth. Out of necessity, the first foods must be large, hand-held, and relatively soft. Within a few months, infants master the pincer grasp and can start to pick up smaller pieces of food between the thumb and forefinger. The process can be messy and inefficient at first. Also, some caregivers and clinicians are concerned about inadequate energy intake, nutrient deficiencies, and choking with the baby-led approach. However, advocates of baby-led introduction of solid foods assert that infants learn to regulate their food intake, develop important motor skills, and may end up accepting a wider variety of foods with this technique.

As with traditional spoon-feeding, caregivers are responsible for providing a variety of nutritious foods and a safe feeding environment. Registered dietitian nutritionists (RDNs) who work with families on the baby-led approach recommend offering one rich source of iron, one rich source of energy, and one palatable, easy-to-eat food at each feeding. If developmentally appropriate, nutrient-dense foods are offered and infant feeding is monitored, the growth and nutritional status of infants fed in this manner are no different than those of infants fed by traditional spoon-feeding.[27] Research comparing these feeding methods is in its infancy (pun intended), but some early results show less food fussiness (i.e., acceptance of a wider variety of foods) with the baby-led approach.[28] The influence of baby-led feeding on BMI throughout life is currently under investigation.

## WEANING FROM THE BREAST OR BOTTLE

Around the age of 6 months, expressed breast milk, formula, or water can be offered in a cup. The cup should have a wide, flat bottom and no lid (despite inevitable spills). Drinking from an open cup rather than from a bottle or sippy cup helps to prevent early childhood caries (review Fig. 15-3). In addition, learning to use a cup without a lid helps the infant to develop important oral and motor skills.

By about 10 months of age, infants are learning to self-feed and likewise to drink independently from a cup. As children drink from a cup more frequently, fewer bottle feedings and/or breastfeedings are necessary. Infants should at least begin drinking from a cup by 1 year of age and should be completely weaned from a bottle by 18 months of age. The added

**baby-led introduction to solid foods** A method of introducing developmentally appropriate complementary foods that allows the infant to have greater control over self-feeding; also called *baby-led weaning.*

**weaning** The gradual process of switching from breast milk or infant formula to other sources of nutrition for the infant.

Infants should begin drinking from a cup by 1 year of age. Cups with lids help to prevent spills, but caregivers should allow the child to practice without a lid as dexterity and coordination improve. **What is one benefit of weaning from bottle to cup?** tverdohlib/123RF

mobility of crawling and walking should naturally lead to gradual weaning from the bottle or breast. Even so, getting a baby out of the bedtime-bottle habit can be difficult. Determined caregivers can either endure a few nights of their baby's crying or slowly wean the baby away from the bottle with either a pacifier or water (for a week or so).

## WHAT TO FEED AN INFANT

It can be difficult for new parents to make sense of nutrition goals for infants in the face of changing dietary recommendations from health authorities, cultural preferences, and outdated advice from friends and family. In response to various controversies surrounding infant feeding, the AAP has issued a number of statements concerning infant dietary patterns.[19] The following guidelines are based on these statements:

- *Build to a variety of foods.* Dietary variety helps to ensure nutrient adequacy at all stages of the life cycle. When the infant is ready, start adding new foods, one at a time. During the first year, the goal is to teach an infant to enjoy a variety of nutritious foods. A lifetime of healthy eating habits begins during infancy!
- *Pay attention to your infant's appetite to avoid overfeeding or underfeeding.* Feed infants when they are hungry. Never force an infant to finish an unwanted serving of food. Watch for signs that indicate hunger or fullness. Responsive feeding will reinforce the infant's natural ability to self-regulate food intake.
- *Encourage adequate intake of dietary fat.* Although excessive intake of fat is implicated in many adult health problems, it is an essential source of calories for growing infants. Fat also helps the nervous system to develop properly. The infant's fat intake should not be restricted.
- *Offer a variety of fruits without added sugars.* Infants have a natural preference for sweet tastes, so fruits are an easy sell! Puréed or mashed fruits can be among the first foods offered to an infant with a spoon. As the infant's motor skills develop, bite-size pieces of soft, ripe fruits (e.g., bananas, peaches, pears) are perfect finger foods. Fresh, frozen, or canned fruits are all nutrient-dense options, but choose products without added sugars. In general, whole fruits are more nutrient dense than fruit juices, which are not recommended for infants. To reduce the chances of choking, avoid very firm fruits (e.g., raw apples) or whole, round shapes (e.g., grapes).
- *Vary the veggies.* During the second half of the first year, infants have a big appetite and are relatively open to new flavors. This is a perfect opportunity to introduce a variety of vegetables, *especially* those with stronger flavors, such as beets, broccoli, and spinach. If infants learn to eat a variety of vegetables early in life, they will be more likely to continue to eat them throughout childhood. However, studies show that by 1 year of age, vegetable choices are dominated by white potatoes. Continuing to offer many colorful options during late infancy and the toddler years will enhance intake of fiber, folate, vitamin A, vitamin C, vitamin E, magnesium, potassium, and phytochemicals.
- *Go for the grains.* Selecting a mix of whole grains and enriched or fortified grain products will provide several nutrients of concern. Whole grains provide fiber to promote healthy digestion. Enriched or fortified grain products are good sources of iron. Iron-fortified infant cereals made of oats, barley, rice, or a mixture of grains should be among the first foods offered to infants. As motor skills develop, fortified dry breakfast cereals are desirable finger foods.

## WHAT *NOT* TO FEED AN INFANT

The following are several foods and practices to avoid when feeding an infant:

- *Excessive infant formula or human milk.* After 6 months of age, solid foods should play an increasing role in satisfying an infant's growing appetite. Age-appropriate solid foods provide necessary calories and iron, plus they help the infant to develop motor skills. About 24 to 32 ounces (¾ to 1 liter) of human milk or formula daily is ideal after 6 months, with complementary foods supplying the rest of the infant's energy needs.
- *Foods that tend to cause choking.* Foods that are round or oval in shape, larger than ½ inch in diameter, or of a soft or sticky texture can easily get lodged in a child's

**Can infants have artificial sweeteners?** Although the FDA affirms the safety of all currently approved artificial sweeteners, foods that contain these additives (e.g., diet soft drinks, sugar-free candies) are typically not good sources of the nutrients infants need to support growth and development. Furthermore, artificial sweeteners given early in life may contribute to a preference for highly sweetened foods. A nutrient-dense snack with natural sugars, such as yogurt with fruit, would be a better option. FamVeld/Shutterstock

throat. These foods include hot dogs, hard or gummy candies, whole nuts, grapes, coarsely cut meats, raw carrots, popcorn, and large portions of nut butters. Caregivers should not allow younger children to gobble snack foods during playtime and should supervise all meals.

- *Potential food allergens before 4 months of age.* Cow's milk, egg whites, peanuts, tree nuts, soy, and wheat are responsible for 90% of food allergies in childhood. Until about 6 months of age, energy and nutrient needs of most infants can be met with exclusive breastfeeding or infant formula. If any solid foods are introduced before 4 months of age (not recommended), they should be iron-fortified infant cereals, puréed meats, vegetables, or fruits.

- *Cow's milk (as a replacement for human milk or infant formula).* Foods made with cow's milk, such as yogurt, can be safely introduced starting around 6 months of age. However, cow's milk is not recommended as a replacement for human milk or iron-fortified infant formula for infants until 1 year of age because it may displace good sources of iron in the infant's diet. In large amounts, cow's milk may also irritate the young infant's GI tract. In addition, the AAP strongly urges caregivers not to give reduced-fat or fat-free milk to children under age 2. Before age 2, the amount of reduced-fat or fat-free milk needed to meet energy needs would supply too many minerals, which could overwhelm the kidneys. Limiting fat intake at this age might also hinder nervous system development. After the second birthday, children can drink reduced-fat, 1%, or fat-free milk because by this age, they are consuming enough solid foods to meet calorie and fat requirements.

- *Other plant-based milk alternatives.* Plant-based milk alternatives (e.g., almond milk, rice milk) are not recommended for infants because of their low energy density, lower-quality protein, and lack of micronutrients that are critical for infant growth and development. Although many manufacturers do fortify plant-based milk alternatives with certain micronutrients (e.g., calcium and vitamin D), there is no federal mandate or standardization across the industry.

- *Goat's milk.* Although perceived by some to pose lower risk for food allergies, goat's milk is low in folate, iron, vitamin C, and vitamin D and should not be used as a source of nourishment for infants.

- *Do not overdo high-fiber foods.* The natural amounts of fiber in kid-size portions of fruits, vegetables, legumes, and grains are appropriate, but too much fiber can be harmful for infants. High-fiber dietary patterns are bulky and low in calories, so infants may feel full before calorie and nutrient needs are met. Furthermore, excessive fiber may limit the absorption of important minerals, such as iron, zinc, and calcium.

- *Skip the added sugars.* Natural sugars, such as those found in human milk, dairy products, and fruits, are excellent sources of energy for active, rapidly growing infants. However, the *Dietary Guidelines* state that added sugars should be avoided by infants and young children under age 2. Nutrient-dense foods are needed to support the infant's rapid growth and development. There is very little room in the infant's dietary pattern for sugar-sweetened beverages and sugary snacks—even those with fruit in their names. These choices supply calories without the benefits of fiber, vitamins, minerals, and phytochemicals.

- *Excessive fruit juice.* The fructose and sorbitol contained in some fruit juices, especially apple and pear juices, can lead to diarrhea because they are slowly absorbed. Also, if fruit juice or related drink products are displacing formula or milk in the dietary pattern, the infant may not be receiving adequate calories, calcium, or other nutrients essential for proper growth. Studies have shown a link between excessive amounts of fruit juice and failure to thrive, GI tract complications, obesity, short stature, and poor dental health. Fruit juice (even 100% fruit juice) is not recommended at all for infants. See Sections 15.4 and 15.5 for age-specific recommendations.

- *Heavily seasoned and processed foods.* Sodium is an essential mineral found naturally in almost all foods. As part of a healthy dietary pattern, infants need sodium for their bodies to work properly. However, average intakes of sodium among infants and toddlers are above the AI. Caregivers should refrain from offering heavily salted and highly processed foods.

### New Recommendations for Infant Feeding from the *Dietary Guidelines*

- **For about the first 6 months of life,** exclusively feed infants human milk. Continue to feed infants human milk through at least the first year of life, and longer if desired. Feed infants iron-fortified infant formula during the first year of life when human milk is unavailable.

- Provide infants with supplemental vitamin D beginning soon after birth.

- **At about 6 months,** introduce infants to nutrient-dense complementary foods.

- Introduce infants to potentially allergenic foods along with other complementary foods.

- Encourage infants and toddlers to consume a variety of foods from all food groups. Include foods rich in iron and zinc, particularly for infants fed human milk.

- Avoid foods and beverages with added sugars.

- Limit foods and beverages higher in sodium.

- As infants wean from human milk or infant formula, transition to a healthy dietary pattern.

- *Food safety hazards.* The immune system is still maturing during infancy and early childhood, so it is important to avoid potential sources of foodborne illness. For example, raw (unpasteurized) milk or soft cheeses (e.g., queso fresco) may be contaminated with bacteria or viruses. Meat, poultry, eggs, and seafood should be cooked to proper temperatures. In addition, honey may contain spores of *Clostridium botulinum,* which can lead to the potentially fatal foodborne illness known as *botulism.* Safe food handling starts with proper handwashing.
- *Excessive nutrient supplementation.* Supplemental vitamins or minerals that provide more than 100% of the RDA or AI for age can increase the risk for nutrient toxicities. Vitamin D supplementation is recommended for all infants until foods or beverages supply the RDA. For exclusively breastfed infants, iron supplements are recommended after 4 months of age until foods or beverages supply the RDA for iron. Except to correct a nutrient deficiency, other dietary supplements are not advised. Consult a pediatrician or RDN before giving dietary supplements to an infant.

### ✓ CONCEPT CHECK 15.3

1. List three similarities between human milk and infant formula. List three differences.
2. Describe four ways to assess an infant's readiness for solid foods.
3. Excessive intake of added sugars is common in late infancy. Describe several ways to limit intake of added sugars in an infant's eating pattern.
4. List three foods to avoid feeding infants during the first year of life.

## CASE STUDY  Undernutrition During Infancy

Damon is a 7-month-old boy who has been taken into a clinic for a routine checkup. On examination, he was found to be moderately underweight relative to his age and body length. His pediatrician scheduled a follow-up appointment in 3 months. At the 10-month visit, Damon appeared sluggish and was even more underweight for his age and length.

An RDN interviewed Damon's 16-year-old mother to collect information on Damon's dietary intake. His intake over the previous 24 hours consisted of two 8-ounce bottles of infant formula, three 8-ounce bottles of fruit punch, and a hot dog. Damon may have been fed some additional items on the nights that his mother left him with a neighbor so that she could go out with friends for a few hours. Thus, she was not aware of all that he ate.

1. Damon's mother did not specify what type of formula she gives to her child or how she prepares it. What questions would you ask about his formula?
2. Besides lagging growth, name three other potential consequences of inadequate calorie and nutrient intake during infancy.
3. What foods should Damon's caregivers offer that are appropriate for his age and nutritional needs?
4. What problems might arise from consumption of sugary drinks from a bottle?
5. Does Damon need any vitamin or mineral supplements?

*Complete the Case Study. Responses to these questions can be provided by your instructor.*

▲ At his well-baby check-up, Damon is underweight compared to other infants his age. How can Damon's caregivers help him eat well to achieve optimal growth? Kwame Zikomo/Purestock/SuperStock

## 15.4 Toddlers and Preschool Children: Nutrition Concerns

The rapid growth rate that characterizes infancy tapers off during the toddler and preschool years. The average annual weight gain is only 4.5 to 6.6 pounds (2 to 3 kilograms), and the average annual height gain is only 3 to 4 inches (7.5 to 10 centimeters) between the ages of 2 and 5. As the growth rate tapers off, energy needs decrease and eating behaviors change. For example, among toddlers, the decreased growth rate leads to a decreased appetite, which contributes to "picky eating."

*Relative* energy needs (i.e., kcal per kilogram of body weight) gradually decline from approximately 100 kcal per kilogram during infancy to about 80 or 90 kcal per kilogram for the preschooler. However, because body size is increasing, *absolute* calorie needs (i.e., kcal per day) gradually increase throughout childhood. For toddlers ages 12 to 23 months, typical energy needs are 700 to 1000 kcal per day (Table 15-6). From ages 2 through 5, energy requirements are 1000 to 1600 kcal per day (Table 15-7). Energy needs vary by age, biological sex, body size, physical activity level, and rate of growth.

Except in cases of poverty or homelessness, the dietary patterns of toddlers and preschoolers in the United States and Canada are adequate in calories and most nutrients. A few nutrients of particular concern among this age group are iron, calcium, and sodium.[29]

**Iron.** Childhood iron-deficiency anemia is most likely to appear in children between the ages of 6 and 24 months—a time when iron stores from gestation become depleted but intake of iron from food sources may be inadequate. This can lead to decreases in both stamina and learning ability, as well as lowered resistance to disease. The targeted efforts of the Special Supplemental Nutrition Program for Women, Infants, and Children (WIC) have helped to decrease the occurrence of iron deficiency among children, but it still remains a problem for almost 15% of toddlers and about 4% of preschoolers.[30]

The RDA for iron is 7 milligrams per day for children ages 1 to 3 and 10 milligrams per day for children ages 4 to 8. The best way to prevent iron-deficiency anemia in children is to provide foods that are rich sources of iron. Even though some animal products are high in saturated fat and cholesterol, the high proportion of heme iron in many animal foods allows the iron to be more readily absorbed than is iron from plant foods. Focus on lean cuts of meat, such as ground sirloin. Fortified breakfast cereals also contribute to meeting iron (and other nutrient) needs. Consuming a source of vitamin C will aid absorption of the less readily absorbed form of iron in plants, fortified foods, and supplements. While dietary changes can be effective for preventing iron-deficiency anemia, supplementation will be required to correct existing anemia (review Section 9.9).

**Calcium.** Childhood is a period of rapid bone growth and mineralization. The RDA for calcium for ages 1 to 3 is 700 milligrams per day. Between the ages of 4 and 8, calcium needs increase to 1000 milligrams per day. However, national surveys of food intake show that the dietary patterns of children—especially girls—fall short of the RDA for this important nutrient.[31]

---

### Quick Guide to Child Nutrition Needs

**Carbohydrates**
- 130 grams per day to supply energy for the central nervous system and prevent ketosis

**Protein**
- 13 to 19 grams per day (ages 1 to 3)
- 34 to 52 grams per day (older children)

**Fat**
- 30% to 40% of total kcal (ages 1 to 3)
- 25% to 35% of total kcal (older children)

---

**TABLE 15-6 ■ Healthy U.S.-Style Dietary Patterns for Toddlers Ages 12 Through 23 Months**[a]

Choose easy-to-chew whole **Fruits** rather than fruit juice. Choose options with little or no added sugars.

Choose a variety of easy-to-chew (i.e., cooked) **Vegetables** from each subgroup. Choose options that are prepared with less added fat and salt.

At least half of **Grain** choices should be whole grains.

Incorporate a variety of lean, unprocessed meats, plant sources of **Protein**, and low-mercury seafood choices. Avoid choking hazards (e.g., large chunks of tough meat, nuts, and large spoonfuls of nut butters).

Before age 2, choose whole milk. Strive to achieve the recommended daily amounts of **Dairy** foods to support bone health but remember that consuming too much milk can leave the dietary pattern short on iron.

**Oils** refer to nontropical plant oils, such as canola, corn, olive, peanut, safflower, soybean, and sunflower oils.

| Calorie Level | 700 | 800 | 900 | 1000 |
|---|---|---|---|---|
| Fruits (cups) | ½ | ¾ | 1 | 1 |
| Vegetables[b] (cups) | ⅔ | ¾ | 1 | 1 |
| Grains (oz-eq) | 1¾ | 2¼ | 2½ | 3 |
| Protein (oz-eq) | 2 | 2 | 2 | 2 |
| Dairy (cups) | 1⅔ | 1¾ | 2 | 2 |
| Oils (grams) | 9 | 9 | 8 | 13 |

[a]These dietary patterns are for toddlers who are no longer receiving human milk or infant formula.
[b]Beans, peas, and lentils can count either as vegetables or protein foods.

Source: U.S. Department of Agriculture and U.S. Department of Health and Human Services. *Dietary/Guidelines for Americans, 2020-2025.* 9th Edition. December 2020. Available at DietaryGuidelines.gov.

Cow's milk is a source of bioavailable calcium and vitamin D for toddlers and preschoolers, but overreliance on milk can crowd out other nutrient-dense foods. Children who drink more than 3 cups of milk per day are likely to consume inadequate amounts of iron and fiber. **What is the best way for toddlers and preschoolers to quench thirst?** Andrew Olney/age fotostock

**food jag** A period of time (usually a few days or weeks) during which a person will eat only a limited variety of foods.

---

**Choking is a preventable hazard for young children. Some suggestions for caregivers include:**

- Set a good example at the table by taking small bites and chewing foods thoroughly.
- During meals and snacks, limit distractions, such as television and electronic devices. Have children sit at the table, take their time, and focus on their food.
- Avoid giving children any foods that are round, firm, sticky, or cut into large chunks, especially before molars emerge (around age 4). For toddlers and preschoolers, some examples of foods to avoid are hot dogs, nuts, uncut grapes, raisins, popcorn, peanut butter (unless it is thinly spread on another food), caramel, marshmallows, and hard pieces of raw fruits or vegetables.

---

Milk and other dairy products are the primary source of calcium in the dietary patterns of children, but unfortunately milk consumption has declined as intake of sweetened beverages has increased. Consuming about 2 servings per day of milk or other foods from the dairy group will help toddlers and preschoolers meet their requirements for bone-building nutrients. Children up to 2 years of age should drink whole milk because they need the extra fat for energy, but after 2 years of age reduced-fat or fat-free milk are more nutrient-dense choices. For children who do not consume dairy products, whether due to choice or necessity, there are alternative sources of calcium and other bone-building nutrients. Fortified beverages (check the label to be sure), such as soy milk, almond milk, or orange juice, can supply as much calcium per serving as cow's milk. Some legumes and vegetables are sources of calcium, as well, but the mineral is not as bioavailable as it is from dairy products.

**Sodium.** While iron and calcium intakes fall short of needs in preschool children, excessive sodium intake is a concern.[32] For children 1 to 3 years of age, the CDRR for sodium is 1200 milligrams per day. For children 4 to 8 years of age, the CDRR is 1500 milligrams per day. High intakes of fast foods and highly processed foods elevate sodium intakes to about 1000 milligrams per day *more* than preschoolers need. Caregivers can lower sodium intake by preparing meals at home instead of relying on fast foods or frozen meals, by limiting salt added during cooking and at the table, by cutting back on use of highly processed foods (e.g., luncheon meats and hot dogs), by rinsing canned beans and vegetables before cooking, and by encouraging consumption of fruits, vegetables, and whole grains in place of prepackaged snacks.

Feeding skills are an important part of physical and cognitive development. Young children explore their environment through the tastes and textures of foods, develop dexterity using utensils and drinking from a cup, and begin to express their autonomy by refusing certain foods. At this time in life, children are also testing boundaries to find out what is acceptable in their little corner of the world. Messy mealtimes, food refusals, and **food jags** can be sources of tension in families. Creating a more harmonious family atmosphere at mealtime is an important way to keep these behaviors from becoming serious feeding problems (see "Understanding Picky Eating" later in this section). Caregivers must understand that these are normal phases of child development but should also be consistent about setting limits for behavior at the dinner table.

Because of the preschool child's reduced appetite, planning a dietary pattern that meets nutrient needs poses a special challenge. **Nutrient density** is an important consideration for this age group. Overall, caregivers should focus on offering a variety of healthy choices, allowing the child to exert some autonomy over the specific type of food and the amount eaten. To apply what you've learned about feeding toddlers and preschoolers, complete *Rate Your Plate: Getting Young Bill to Eat* in Connect.

Table 15-7 summarizes the healthy U.S.-style dietary patterns (from the *Dietary Guidelines*) at calorie levels appropriate for children ages 2 through 18. Keep in mind that Table 15-7 displays the quantity of food from each food group to be consumed throughout the course of a day. When it comes to planning individual meals, MyPlate is a useful, easy-to-understand tool for children. The *proportions* illustrated by MyPlate apply to all ages, even though the *portions* of foods at each meal will be smaller for children. Until a child is about 5 years of age, a good starting point for portion sizes in the vegetables group, fruits group, and protein group is about 1 tablespoon per year of life.

It is important to promote a healthy attitude about eating. While caregivers will want to focus on nutrient-dense foods, there is no reason to be overly restrictive about child food choices. In fact, when parents are extremely controlling about the family's food intake, children may be at risk for body dissatisfaction and disordered eating. There is room for occasional indulgences, a skipped meal or two, or once in a while "less than ideal" choices. It is eating and lifestyle patterns over the course of a month (and lifetime) that matter. Children develop healthy eating habits when adults set a good example, provide opportunities to learn, give support for exploration, and limit inappropriate behavior.

Next, we will consider some typical complaints and concerns of parents, explore the causes, and make suggestions for achieving optimal nutrition during the toddler and preschool years.

**TABLE 15-7 ■ Healthy U.S.-Style Dietary Patterns for Children Ages 2 Through 18**

Choose whole **Fruit** rather than fruit juice. Choose options with little or no added sugars. Avoid choking hazards (e.g., large pieces of firm fruit, dried fruit) until molars emerge (around age 4).

Choose a variety of **Vegetables** from each subgroup. Choose options that are prepared with less added fat and salt. Avoid choking hazards (e.g., firm, raw vegetables) until molars emerge.

At least half of **Grain** choices should be whole grains.

Incorporate a variety of lean, unprocessed meats, plant sources of **Protein**, and low-mercury seafood choices. Avoid choking hazards (e.g., large chunks of tough meat, nuts, and large spoonfuls of nut butters) until molars emerge.

Choose unsweetened and reduced-fat or fat-free **Dairy** foods. Strive to achieve the recommended daily amounts of dairy foods to support bone health but remember that consuming too much milk can leave the dietary pattern short on iron.

**Oils** refer to nontropical plant oils, such as canola, corn, olive, peanut, safflower, soybean, and sunflower oils.

**Calories for Other Uses** include added sugars and rich sources of saturated fat, such as butter, shortening, lard, or tropical plant oils (e.g., coconut oil).

|  | Toddlers & Preschoolers (2 to 5 years) | | | School-Age Children (6 to 12 years) | | | | | | Teenagers (13 to 18 years) | | |
|---|---|---|---|---|---|---|---|---|---|---|---|---|
| **Daily Amount of Food from Each Group Based on Calorie Level** | | | | | | | | | | | | |
| **Calorie Level** | 1000 | 1200 | 1400 | 1600 | 1800 | 2000 | 2200 | 2400 | 2600 | 2800 | 3000 | 3200 |
| Fruits (cups) | 1 | 1 | 1½ | 1½ | 1½ | 2 | 2 | 2 | 2 | 2½ | 2½ | 2½ |
| Vegetables[a] (cups) | 1 | 1½ | 1½ | 2 | 2½ | 2½ | 3 | 3 | 3½ | 3½ | 4 | 4 |
| Grains (oz-eq) | 3 | 4 | 5 | 5 | 6 | 6 | 7 | 8 | 9 | 10 | 10 | 10 |
| Protein (oz-eq) | 2 | 3 | 4 | 5 | 5 | 5½ | 6 | 6½ | 6½ | 7 | 7 | 7 |
| Dairy (cups) | 2 | 2½ | 2½[b] | 2½[b] | 2½[b] | 2½[b] | 3 | 3 | 3 | 3 | 3 | 3 |
| Oils (grams) | 15 | 17 | 17 | 22 | 22[c] | 24[c] | 29 | 31 | 34 | 36 | 44 | 51 |
| **Limit on Calories for Other Uses** | | | | | | | | | | | | |
| Calories for Other Uses (kcal) | 130 | 80 | 90[d] | 150[d] | 190[d] | 280[d] | 250 | 320 | 350 | 370 | 440 | 580 |

[a]Beans, peas, and lentils can count either as vegetables or protein foods.
[b]Older children at this Calorie level should have 3 cups per day from the Dairy group.
[c]Older children at this Calorie level need slightly more grams of Oils.
[d]Because they need more Dairy and Oils, older children at this Calorie level have a slightly lower allowance for Calories for Other Uses.

Source: U.S. Department of Agriculture and U.S. Department of Health and Human Services. *Dietary Guidelines for Americans, 2020-2025*. 9th Edition. December 2020. Available at DietaryGuidelines.gov.

## UNDERSTANDING PICKY EATING

Many parents are baffled by their toddler's erratic eating behaviors. Toddlers and pre-schoolers tend not to eat as much or as regularly as infants. One day, a young child may pick at his food and staunchly refuse to eat his green beans, but on the next day he might ask for a second helping. Parents often need reminding that toddlers and preschoolers cannot be expected to eat as voraciously as infants or to eat adult-size portions. Because the growth rate slows after infancy, a toddler's drive to eat is not so intense. In addition, children are sometimes more interested in playing and exploring than eating.

Youngsters also tend to be wary of new foods. One reason is that they have more taste buds, and their taste buds are more sensitive than those of adults. A general distrust of unfamiliar things is common in this age group. Thus, familiarity plays an important role in food acceptance. Adults can encourage young children to broaden their food repertoire by repeating exposure to new food choices. It may take 10 or more exposures to a new food before a child finds it acceptable, but if adults can be patient and persevere, children will build good food habits.[33]

Food preferences change rapidly in childhood and are influenced by food temperature, appearance, texture, and taste. The following are a few practical tips for improving acceptance of nutrient-dense foods.

- Build on what they know and accept. Pairing a new food item with a familiar one can help to foster acceptance of the new food.
- Enlist the child's aid in food selection and preparation. For example, let the child pick out the tomatoes and squash at the local farmers' market.
- Serve meals on a sectioned plate. Sometimes children object to having foods mixed, as in stews and casseroles, even if they normally like the ingredients separately.
- Keep it crunchy. Certain food characteristics, such as crisp textures and mild flavors, are appealing to children. Kids who reject mushy, cooked carrots may enjoy them raw or lightly steamed. (After about age 4, children can safely eat raw vegetables without fear of choking.)

▲ The *Physical Activity Guidelines for Americans* recommends that children should be physically active every day, starting as young as 3 years of age. Caregivers can encourage physical activity by setting aside time, providing a safe space, and engaging in active play with their children. Hero/Corbis/Glow Images

**avoidant/restrictive food intake disorder (ARFID)** Eating disorder characterized by failure to meet energy or nutrient needs, resulting in significant weight loss, nutritional deficiencies, or dependence on tube or intravenous feeding; the eating disturbance is not explained by lack of available food, a medical problem, or another eating disorder.

**neophobia** Fear of new things, such as new foods.

- Finger foods are fun. Preschoolers eventually develop skill with spoons and forks and can even use dull knives, but it is still a good idea to serve some finger foods, especially with healthy dips such as yogurt sauce or hummus.
- Save the best for last. If a child is prone to leave his or her chicken on the plate untouched, serve the chicken first. Hunger is the best means of getting a child to eat!

Although picky eating is usually not cause for alarm, a child's sudden loss of appetite may be a sign of underlying illness, such as an infection or gastrointestinal problem. Be alert for signs of eating disorders, as well. Extreme, self-imposed dietary restrictions could be an early sign of anorexia nervosa. **Avoidant/restrictive food intake disorder (ARFID)** is an eating disorder that is primarily diagnosed among children. With ARFID, a child lacks interest in eating specific foods (or all foods in general), which leads to weight loss or failure to grow as expected, as well as many nutrient deficiencies. It is most likely related to stress, anxiety, or depression.[34]

Be sure to read *Ask the RDN* in this section for a child nutrition expert's perspective on picky eating.

## ASK THE RDN  Picky Eating

*Dear RDN: I'm so frustrated by my toddler's picky eating! Our dining room table has become a battleground. It seems like all my son will eat is chicken nuggets and cheddar crackers. What can I do?!*

Food **neophobia**, fear of new food, tends to rear its head during the toddler years, and that can be frustrating if you're accustomed to a baby who gobbled up everything on her high-chair tray! Kids may not only start rejecting new foods but also refusing previously liked foods. And they may cycle through food jags, wanting the same food, meal after meal.

There are a couple of reasons why kids hit this finicky phase:

1. Children's appetites slow considerably after a growth explosion during infancy.
2. They're practicing their newfound independence—and realizing the power that one little word ("No!") can have on their caregivers.

How parents respond to these behaviors makes a difference. Resorting to bribing, cajoling, bargaining, or even punishing children at the dinner table can foster a negative feeding relationship that worsens picky eating.

It's also common for parents to give up on foods that children have rejected. Research shows that parents offer a food only three to five times before deciding their child doesn't like it. In reality, children need many more exposures to build comfort with an unfamiliar food.

Catering to a child's preferences by making separate meals may also prolong picky behaviors. Without incentive to try something new (or less preferred), kids can fall into a pattern of eating only a handful of accepted foods. It's better for long-term habits if *one* family meal is served, making sure that a favorite food is somewhere on the table, even if it's simply bread or fruit.

Finally, allowing children to snack throughout the day can also dull the appetite for meals, which can make children seem more difficult and picky than they actually are.

Parents may find relief by following the Division of Responsibility in Feeding, a highly regarded feeding practice created by dietitian Ellyn Satter.[35] In this approach, parents are only responsible for *what* is served, *when* it's served, and *where* it's served. Children are responsible for *whether* they eat it and *how much* they eat. Satter believes that children have a natural ability to eat what they need (and to learn to eat what their parents eat) and should be trusted to do so without pressure from caregivers. Instead of insisting on a certain number of bites or promising dessert for eating vegetables, parents should instead focus on providing regular meals and snacks and creating a pleasant mealtime environment—and let their children handle the rest.

Thankfully, for many kids, difficult feeding behaviors begin to wane during the school-age years. The growth and appetite surges that mark puberty can also cause picky eating to fade. Yet some behaviors go beyond garden-variety picky eating. If your child eats fewer than 20 foods, acts afraid or upset at mealtime, refuses to eat, or is experiencing weight loss or slowed growth, talk to your child's pediatrician about getting a referral to a therapist or a dietitian who specializes in feeding children.

Do away with dinner drama,

**Sally Kuzemchak, MS, RDN**

Author of *The 101 Healthiest Foods For Kids*, a contributor to magazines and websites, and founder of the blog Real Mom Nutrition

Sally Kuzemchak

## REDEFINE SNACKING

Parents may be concerned that frequent snacking will prevent children from eating well at mealtimes. However, children have small stomachs and need to eat every 3 to 4 hours. Sticking to three meals a day offers no special nutritional advantages; it is just a social custom. Instead, offering five or six small meals can help children meet their nutritional requirements more successfully than limiting them to three meals each day.[36] If the stretch between lunch and dinner is 6 hours, an afternoon snack about 2 hours before dinner could provide some needed nutrients and may preempt a cranky attitude at the evening meal. Try to plan so that snacks do not blunt the child's appetite for meals, but overall, let the child's appetite, rather than the clock, guide eating.

The location of snacking is important. Sitting calmly at the table instead of running around the house will decrease the risk for choking. Limiting distractions by turning off the television will help to prevent mindless eating. The caregiver could offer two or three nutrient-dense options and allow the child to choose one; responsibility for food choices by the child should start at an early age. Last, caregivers should promote hand-washing and good oral hygiene just as they would for a meal.

*When* we eat is not nearly as important as *what* we eat. Perhaps families simply need to redefine snacking altogether. A snack should not be synonymous with an indulgent dessert. Rather, a snack should be a small meal of nutrient-dense foods. Let hunger—not the clock—guide the timing of meals and snacks. It is important to plan ahead in order to have healthy foods available (Table 15-8). Fruits and vegetables (fresh, frozen, or canned) and whole grain breads and crackers are good snack choices (see *Farm to Fork* in this chapter).

> Helpful resources for planning nutritious, age-appropriate meals and snacks:
> www.MyPlate.gov
> www.fns.usda.gov/tn/team-nutrition
> brightfutures.aap.org

**TABLE 15-8 ■ Twenty Healthy Snack Ideas for Children**

| | Iron | Zinc | Calcium | Vitamin C | Fiber |
|---|---|---|---|---|---|
| Almonds (1 oz)* | | | ✓ | | ✓ |
| Unsweetened applesauce (½ cup) | | | | ✓ | ✓ |
| Bean and cheese burrito (1) | ✓ | ✓ | ✓ | | ✓ |
| Cheese (1 oz) and whole wheat crackers (6) | ✓ | ✓ | ✓ | | ✓ |
| Reduced-sugar dried cranberries (¼ cup) | | | | ✓ | ✓ |
| Frozen fruit pieces (1 cup) | | | | ✓ | ✓ |
| Fruit salad (1 cup) | | | | ✓ | ✓ |
| Fruit smoothie with bananas and strawberries (1 cup) | | | | ✓ | ✓ |
| Hard-boiled egg | ✓ | ✓ | | | |
| Hummus (2 tbsp) with bell pepper rings (1 cup) | | | | ✓ | ✓ |
| Low-fat microwave popcorn (3 tbsp unpopped)* | | | | | ✓ |
| Mini-pizzas on whole grain English muffins (2) | ✓ | ✓ | ✓ | ✓ | ✓ |
| Peanut butter (2 tbsp) and apple slices (1 cup)* | | ✓ | | | ✓ |
| Quick breads, such as banana bread, 1 slice | ✓ | | | | ✓ |
| String cheese (1 stick) | | ✓ | ✓ | | |
| Trail mix (¼ cup)* | ✓ | ✓ | | | ✓ |
| Tuna salad (½ cup) in whole wheat pita pocket | ✓ | ✓ | | | ✓ |
| Whole grain cereal (1 cup) | ✓ | ✓ | | ✓ | ✓ |
| Whole wheat pasta salad with veggies (1 cup) | ✓ | ✓ | | | ✓ |
| Yogurt (6 oz) with granola (2 tbsp) | | | ✓ | | ✓ |

Check marks indicate that the snack item is a good source of the nutrient.
* Snack items that are best suited for children older than age 4 due to potential for choking.

## FARM to FORK — Blueberries

U.S. Fish & Wildlife Service/Ryan Hagerty

From a health standpoint, it's tough to beat these AMAZING berries! The antioxidant and phytochemical activities of berries are four times greater than those of most other fruits, 10 times greater than those of most vegetables, and 40 times higher than those of most cereal grains. Berries may play a role in prevention of diabetes, cancer, high blood pressure, cardiovascular disease, and dementia!

### Grow

• Many urban landscapes are adding attractive berry patches as part of an edible environment. Although most blueberries thrive in cooler climates, some varieties do fine in warmer climates.
• Children love to pick berries! If you don't have your own berry patch, consider finding a local U-pick farm for building great family memories.

### Shop

• When shopping, examine the berries carefully and look for plump, firm, and colorful fruit.
• The most nutrient-dense frozen berries are flash-frozen to preserve phytochemicals and vitamin C. Flash-frozen wild berries are the best choice.
• If purchasing juice, read the ingredients to be sure it is 100% pure berry juice. Most berry juices contain more juice from apples and white grapes than from berries.

### Store

• Rinse fresh berries just prior to eating and eat them within several days to ensure the highest nutrient content.
• Before freezing your own berries, dust lightly with vitamin C powder or Fruit-Fresh® to retain nutrients.

### Prep

• Berries may be enjoyed fresh, frozen, stewed, and dried. They can be used to sweeten any dish.
• Frozen berries thawed quickly in the microwave actually retain double the nutrient content as those thawed at room temperature or in the refrigerator.
• Heating increases the bioavailability of phytochemicals and nutrients in berries. Thus, cooked or canned berries are healthy choices.

Source: Robinson J. Blueberries and blackberries: Extraordinarily nutritious, in *Eating on the Wild Side*. New York: Little, Brown and Company, 2013.

jenifoto/123RF

## CHOOSE DIETARY SUPPLEMENTS CAREFULLY

Major scientific groups, such as the Academy of Nutrition and Dietetics and the American Society for Nutrition, state that multivitamin and mineral supplements are generally unnecessary for healthy children; it is better to emphasize whole foods. In fact, consuming fortified foods and supplements may lead to intakes above the UL for some nutrients, such as vitamin A and zinc. Supplements for children that are made to look like candy may result in accidental overdose, particularly of iron. Rather than relying on supplements, choose fortified, low-sugar, ready-to-eat breakfast cereals with milk to close any gaps between current micronutrient intake and needs, such as for folate, vitamin D, vitamin E, iron, or zinc.[19,37]

For a child who is ill, has a very poor appetite, is extremely selective about foods, or adheres to dietary restrictions (e.g., food allergies, metabolic disorders), a children's multivitamin and mineral supplement not exceeding 100% of Daily Values for any nutrient may be beneficial. Still, as mentioned many times in this book, dietary supplements cannot substitute for an otherwise healthy dietary pattern. If current childhood feeding practices are to become more healthful, the focus should be on whole grain breads and cereals, fruits, vegetables, lean sources of protein, and low-fat milk and milk products.

## REDUCE LEAD POISONING

Humans may be exposed to lead from drinking contaminated water, consuming or inhaling lead dust (e.g., from cracked and peeling lead paint), contaminated dietary supplements (e.g., calcium supplements derived from bone meal), or foods stored or prepared in lead-containing vessels. In the United States, nearly half a million children between the ages of 1 and 5 have unacceptably high blood lead levels. Young children are particularly vulnerable to lead poisoning because they are small, absorb lead quickly, spend a lot of time on the floor, and are apt to put objects in their mouths. In the short term, symptoms of lead poisoning include gastrointestinal distress, lack of appetite, irritability, fatigue, and anemia. Over the long term, devastating effects include intellectual and behavioral impairments and increased risk for several chronic diseases in adulthood.

Although it does not address the source of exposure, proper nutrition can reduce the risks of lead poisoning for children. Consuming regular meals with moderate fat intake and ensuring adequate iron and calcium status are dietary practices known to reduce lead absorption. Adequate zinc, thiamin, and vitamin E intakes also reduce the harmful effects of absorbed lead. To minimize lead levels in drinking water, only use cold water for drinking and preparation of formula or food. Letting cold water run from the tap for 1 to 2 minutes after a long period of inactivity (e.g., overnight) will limit the amount of lead that has accumulated in tap water. If the public water supply contains a high concentration of lead, bottled water is a safer alternative, particularly for formula preparation. Overall, a balanced meal plan that offers a variety of whole grains, lean meats, and low-fat dairy products is especially useful for protecting children from lead poisoning.[38]

## ALLEVIATE CONSTIPATION WITH LIFESTYLE CHANGES

Constipation, a common problem among children, can be defined as hard, dry stools that are difficult to pass. Typically, a 4-year-old child has one bowel movement per day, but normal bowel habits vary widely. Therefore, the frequency of bowel movements is not as important as the consistency of stools. Pediatricians diagnose constipation after 2 or more weeks of delayed or difficult bowel movements. In rare situations, constipation can be a sign of a serious problem. If a child has a fever or vomiting along with constipation, if there is blood in the stool, or if the abdomen becomes swollen, caregivers should seek immediate medical attention.

What causes constipation? Although there could be a serious medical problem, most cases are related to lifestyle. Lack of physical activity contributes to constipation. Some cases may be due to inadequate fluid intake. Also, on average, children (and adults) in the United States barely obtain half of the AI for fiber. Altered bowel habits also may be a sign of a food allergy or intolerance to a food component such as cow's milk. The majority of the time, however, constipation results from the child withholding bowel movements. For children, a painful bowel movement can be so traumatic that they try to resist subsequent bowel movements. The longer they hold their stools, the harder and drier they get, leading to another painful movement. This cycle disrupts regular bowel habits, leading to distress and, if not treated, **fecal impaction**.[39]

When presented with a constipated child, a primary care provider first has to rule out a medical cause, such as an intestinal blockage. Treatment of fecal impaction may require evacuation of the bowels (e.g., with an enema). Once bowels have been evacuated, lifestyle changes are necessary to prevent future problems. Although various types of **laxatives** may be prescribed by the primary care provider in the short term, dietary and lifestyle strategies are safest over the long term. First, regular bowel habits must be established. For example, caregivers should set aside time for the child to use the toilet, without rushing, after each meal. Rewards, such as stickers on a chart, may be used to reinforce good habits. Increasing physical activity while cutting back on sedentary activities (e.g., watching television or using electronic devices) can help to promote regular bowel movements. The primary dietary interventions to alleviate constipation include eating more fiber and drinking more fluids. In the initial stages of treatment, providing certain fruit juices (e.g., prune, grape, and apple) and trying soy milk instead of cow's milk may relieve constipation.[40]

Ultimately, whole fruits (e.g., plums, peaches, and apricots) are better choices than juices because whole fruits are less concentrated sources of calories. Pediatric nutrition authorities recommend limiting fruit juice to just 4 fluid ounces per day for toddlers (ages 1 to 3).[1] Other foods to emphasize for fiber include vegetables, whole grain breads and cereals, and beans. The daily fiber goals for children set by the Food and Nutrition Board vary by age (see box). Few children meet these goals. It is important to increase fluid consumption along with fiber to avoid another fecal impaction. Accompanying fluid recommendations are 4 cups (900 milliliters) per day for toddlers and about 5 cups (1200 milliliters) per day for older children.

## VEGETARIAN DIETARY PATTERNS FOR YOUNG CHILDREN

Appropriately planned vegetarian dietary patterns can meet the young child's needs for growth and development. However, caregivers should be aware of a few potential nutrition risks. These include the possibility of developing iron-deficiency anemia, a

Between 2014 and 2015, the water supply in the city of Flint, Michigan, became contaminated with extremely high levels of lead. Read more about the far-reaching impact of this lead exposure in Section 12.2.

**fecal impaction** The presence of a mass of hard, dry feces that remains in the rectum as a result of chronic constipation.

**Fiber Recommendations for Children**

| Young Children | |
|---|---|
| 1–3 years | 19 grams/day |
| 4–8 years | 25 grams/day |
| **Boys** | |
| 9–13 years | 31 grams/day |
| 14–18 years | 38 grams/day |
| **Girls** | |
| 9–13 years | 26 grams/day |
| 14–18 years | 26 grams/day |

Source: Food and Nutrition Board.

The *Dietary Guidelines* provide a framework for healthy vegetarian dietary patterns for children, starting at 12 months of age. Notably, these patterns illustrate lactoovo-vegetarian eating patterns. Inclusion of dairy foods ensures nutrient-dense sources of protein, calcium, vitamin D, and vitamin B-12. See Appendix 3 of the *Dietary Guidelines* at DietaryGuidelines.gov.

deficiency of vitamin B-12, and rickets from a vitamin D deficiency. During the first few years of life, children also may not consume enough calories when following a bulky vegetarian eating pattern. These known pitfalls are easily avoided by informed meal planning (review Section 6.9). Dietary patterns for children who eat vegetarian fare should focus on the following:[41]

- A variety of plant sources of protein to provide a full complement of essential amino acids (e.g., beans, nuts, and grains)
- A synthetic source of vitamin B-12 (e.g., dietary supplement or fortified breakfast cereal)
- Plenty of plant sources of iron (e.g., beans, dried fruits, and fortified grain products)
- Good sources of zinc (e.g., whole grains, beans, nuts, and seeds)
- Foods that are fortified with vitamin D (e.g., fortified orange juice), along with regular sun exposure
- Rich plant sources of calcium (e.g., fortified milk or juice), almonds, some forms of tofu, and green, leafy vegetables

## PROMOTE GOOD ORAL HEALTH

A healthy dietary pattern goes a long way toward reducing the risk for **dental caries** in young children. In addition to beginning oral hygiene when teeth start to appear and seeking early pediatric dental care, the following nutrition-related tips can help reduce dental problems in children:[9]

- Drink water (fluoridated, if available) as opposed to carbohydrate-rich or acidic beverages (e.g., fruit juice, soft drinks, sports drinks, and energy drinks). If sugary or acidic beverages are consumed, it is better to drink them *with* meals rather than *between* meals. Sipping juice continuously between meals (e.g., from a sippy cup) exposes teeth to caries-promoting sugars and acids, which, over time, can erode enamel.
- Use small amounts of fluoridated toothpaste twice daily. In areas without fluoridated water, discuss fluoride needs with a dentist or pediatrician.
- Snack in moderation. Again, constant exposure of teeth to sugars and acids throughout the day (i.e., grazing) tends to promote caries.
- Make wise snack choices. We automatically think of sticky, sugary snacks as promoters of dental caries, but foods such as pretzels and popcorn provide a source of carbohydrates for oral bacteria as well. In contrast, crunchy fruits and vegetables, such as apples or celery, can help to brush away sticky food particles. Snacking on dairy products, such as cheese, can actually buffer the acids that lead to tooth decay.

**autism spectrum disorder** A disorder of neurological development characterized by problems with social interaction, verbal and nonverbal communication, and/or unusual, repetitive, or limited activities and interests.

*magnificent*

Researchers have observed differences between the gut microbiota of neurotypical children and those with autism spectrum disorder. Some researchers point to alterations in the gut microbiota as a causative factor for ASD. Perhaps some compounds produced by the microbiota are absorbed into the blood and can directly affect brain function. Scientists have seen that inoculating germ-free mice with the gut bacteria from a human with ASD led to ASD-like behaviors in the mice.[42] Scientists are interested in learning about the potential for manipulating the microbiota (e.g., with prebiotics or probiotics) to alter the course of ASD.

*microbiome*

## LINKS BETWEEN AUTISM AND NUTRITION

**Autism spectrum disorder (ASD)** is characterized by a range of problems with social interaction, verbal and nonverbal communication, and/or unusual, repetitive, or limited activities and interests. This disorder usually is diagnosed in early childhood and affects an estimated 1 in every 59 children, with higher prevalence in boys than girls. The causes for ASD are not well understood, but there is a definite genetic component.

ASD can both affect and be affected by nutritional status.[43] In addition to developmental and behavioral abnormalities, many children with ASD experience GI disorders, such as constipation, diarrhea, or reflux disease. Such disorders may impair nutrient intake or absorption. Medications used to treat behavioral problems may alter appetite.

Some children with ASD may have feeding problems related to developmental impairments. Also, selective eating behaviors may affect nutrient intake. Children with ASD can be very rigid with their food selections, rejecting foods or entire food groups based on sensory qualities such as texture, color, and temperature. Thus, careful attention to nutrient-dense food choices is of prime importance.

A new diagnosis of ASD can be bewildering, and the lack of treatment options leaves many families feeling helpless. A dietary intervention may seem like a reasonable, low-risk option. Thus, a variety of dietary restrictions and/or nutrient supplements are commonly employed by families affected by ASD.[44] Is there any evidence that nutritional interventions are effective?

A widely used nutritional intervention is the gluten-free, casein-free (GFCF) diet, which eliminates all wheat, barley, rye, and milk products. Proponents of the GFCF diet believe that some children have a "leaky gut," which allows food proteins to be absorbed intact from the GI tract, enter the bloodstream, and affect brain function. By eliminating the offending food proteins, could symptoms of autism be reduced? This sounds scientifically plausible, yet there is little science to support this theory. Certain proteins may cross the blood–brain barrier and exert druglike effects on the brain, but the levels of these druglike proteins in body fluids are no higher among children with autism than among children without autism.

As mentioned, many children on the autism spectrum display selective eating behaviors; they are more sensitive than other children to colors, tastes, temperatures, and textures. Also, children with autism prefer consistent routines and may refuse new foods. There is a strong possibility that a restrictive diet may exacerbate problems with social and emotional functioning.

In addition, restricting food choices may further limit an already marginal nutrient intake, leading to deficiencies of iron, calcium, vitamin D, and several B vitamins. Eliminating gluten will entail cutting out many types of breads, pastas, bakery products, crackers, and snack foods. Removing casein will require avoidance of milk, yogurt, cheese, butter, and many frozen desserts. Indeed, studies have shown that the dietary patterns of children with autism already may be deficient in key nutrients. Children who follow a casein-free diet may have low bone mass or delayed bone development related to inadequate intake of calcium and vitamin D. Children on a gluten-free diet have lower levels of folate and vitamin B-6. Nutrient shortfalls can have both short- and long-term implications for the child's health.

Clearly, the GFCF diet is not risk free. But would the risks be acceptable if the eating plan helps to improve symptoms? Unfortunately, there is very little evidence to support either the safety or the efficacy of the GFCF diet. The few studies that do show a positive outcome are of poor quality, meaning they were of short duration and had small numbers of subjects, poor study designs, and a high risk of bias.[45,46]

Knowing that children with autism are at risk for dietary inadequacies, the best dietary strategy is to offer a variety of nutrient-dense foods at each meal. A dietary restriction should only be used if there is evidence that a child has an allergy or intolerance to a food. Consulting with an RDN would be helpful to assess the child's dietary patterns and create a personalized plan to alleviate GI symptoms and ensure that nutritional needs are met. For a child with extremely limited dietary intake, multivitamin and mineral supplementation may be an option.

At this time, the AAP does not endorse any specific dietary plan or supplement as a treatment for autism. Despite the lack of evidence to support the efficacy of nutritional interventions for autism spectrum disorders, many parents will choose to try them anyway, hoping for positive results. There may, in fact, be a subset of children with autism who do respond to dietary treatments. Because of the rising incidence of ASD and the lack of curative treatments, nutritional interventions for ASD will continue to be an active area of research.

✓ **CONCEPT CHECK 15.4**

1.  Why is picky eating common among preschoolers? Provide three or more suggestions to help a preschooler choose nutritious foods.

2.  How often do preschoolers need to eat throughout the day? List three nutrient-dense snack ideas that would be appropriate for a 3-year-old child.

3.  Should toddlers and preschoolers take a multivitamin and mineral supplement? Why or why not?

4.  Explain the connections between nutrition and oral health. List three ways to reduce risk for dental caries with healthy eating habits.

5.  What are some nutrition concerns of children with autism spectrum disorder?

# **15.5** School-Age Children: Nutrition Concerns

The dietary patterns of many school-age children can be improved, particularly with regard to fruit, vegetable, whole grain, and beverage choices. In recent years, whole fruit consumption has gone up and fruit juice consumption has gone down, but overall fruit intake among school-age children remains below the targets of the *Dietary Guidelines*. Vegetable intake is low; white potatoes (including French fries) account for about 30% of vegetable intake.[47] Less than 1% of children meet the recommendation of the *Dietary Guidelines* to make half of grains whole.[48] Intakes of dairy foods, although they make a significant contribution to calcium, vitamin D, and potassium intakes among children, are below targets.[49] In general, the nutritional concerns and goals applicable to school-age children are the same as those discussed in relation to preschoolers. However, with the added pressures of peers, food advertisers, health messages from the media, and an increasing desire for independence, these goals may be harder to achieve as children grow older.

The healthy U.S.-style dietary patterns summarized in Table 15-7 are a good basis for meal planning. For school-age children from 6 to 12 years of age, daily calorie needs are usually within the range of 1200 to 2400 kcal per day, depending on age, biological sex, body size, activity level, and rate of growth. Continue to use MyPlate (Fig.15-5) as a guide for building healthy meals. Emphasize nutrient-dense sources of iron, zinc, calcium, and vitamin D while striving to limit intakes of saturated fat, added sugars, and sodium. Now let us look at several nutritional issues of particular concern during the school-age years.

## REVERSING TRENDS FOR OVERWEIGHT AND OBESITY

By far, the most troublesome nutritional problem facing children today is the rise in childhood obesity. Since the 1970s, the incidence of childhood overweight and obesity has more than tripled. In the past decade, the rate of childhood obesity has stabilized, but still, about one-third of U.S. school-age children are now classified as overweight or obese. The rates of childhood obesity are highest among children from low-income families and minority populations.

In the short term, the main consequences of childhood obesity are social concerns (e.g., bullying), psychiatric disorders (e.g., depression and anxiety), and short stature linked to early puberty. In the long term, significant health problems associated with obesity, such as cardiovascular disease, type 2 diabetes, hypertension, cancer, and osteoarthritis, may appear in adulthood or earlier. Childhood obesity is a serious health threat because about 40% of children with obesity (and about 80% of adolescents with obesity) remain obese as adults. To identify cases and reverse these trends, the United States Preventive Services Task Force recommends screening children for obesity starting at 6 years of age.[50]

Research points to many potential causes of childhood obesity. Recall the nature versus nurture discussion in Section 7.4. Some individuals are born with a genetic predisposition to obesity. They may experience lower metabolic rates, which means they use calories more efficiently and can store fat more easily. Despite any genetic predisposition, however, modifying behaviors can improve health outcomes.

Supersized portions of foods such as fast foods, snacks, and sugar-sweetened soft drinks are fueling a nation of supersized kids. **What strategies would you recommend to reduce calorie intake from fast-food restaurants?** image100/Corbis

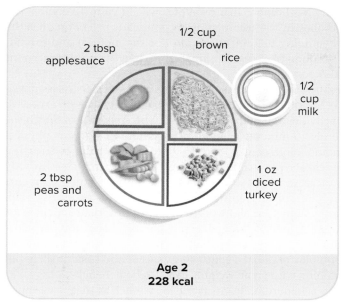

2 tbsp
applesauce

1/2 cup
brown
rice

1/2
cup
milk

2 tbsp
peas and
carrots

1 oz
diced
turkey

**Age 2**
**228 kcal**

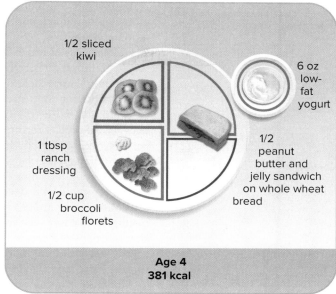

1/2 sliced
kiwi

6 oz
low-
fat
yogurt

1 tbsp
ranch
dressing

1/2
peanut
butter and
jelly sandwich
on whole wheat
bread

1/2 cup
broccoli
florets

**Age 4**
**381 kcal**

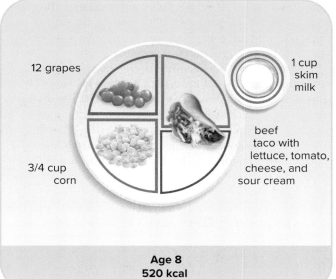

12 grapes

1 cup
skim
milk

beef
taco with
lettuce, tomato,
cheese, and
sour cream

3/4 cup
corn

**Age 8**
**520 kcal**

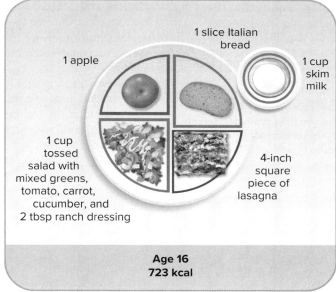

1 slice Italian
bread

1 apple

1 cup
skim
milk

1 cup
tossed
salad with
mixed greens,
tomato, carrot,
cucumber, and
2 tbsp ranch dressing

4-inch
square
piece of
lasagna

**Age 16**
**723 kcal**

**FIGURE 15-5** ▲ Using MyPlate to build a healthy meal for children. MyPlate is a useful tool for all Americans ages 2 and older. MyPlate proportions apply to children as well as adults, but portion sizes and food choices vary by age.

Researchers believe that although dietary patterns are an important factor, inactivity is also a major contributor to the increase in childhood obesity. Studies show that physical activity steadily declines and screen time increases as children age. Screen time includes time spent watching television, working at the computer, or using phones and other electronic devices. Only about 50% of children are getting the recommended 60 minutes of physical activity per day. It does not help that physical education classes are now elective in many high schools. Today's generation of children now engages in 7 hours per day of screen time, on average. For school-age children, parents should set appropriate limits on screen time, making sure to encourage plenty of physical activity.[51] In addition, excessive snacking, overreliance on fast-food restaurants, parental neglect, advertising and other messages in the media, lack of safe areas to play, and the abundant availability of high-calorie food choices contribute to childhood obesity. Soft drinks and other sugared beverages are especially implicated.

In 2016, the WHO released its Report of the Commission on Ending Childhood Obesity (ECHO). The ECHO Report quantifies global rates of childhood obesity and

To get kids involved in physical activity, new physical education classes have been introduced into some schools. Classes on rock climbing, kayaking, and martial arts help to promote activity because they take the focus away from teams and competition, which often discourage and embarrass kids who lack athletic talent. **What are the recommendations of the *Physical Activity Guidelines* for children?** Valerie Loiseleux/Getty Images

recognizes the problems of an environment that promotes weight gain. To combat the global epidemic of childhood obesity, the commission recommends promoting healthy foods and physical activity beginning before conception (for the mother) and extending all the way through childhood. To access the full report, see http://www.who.int /end-childhood-obesity/final-report/en.

The initial approach in treating a child who is overweight or obese is to assess physical activity. If a child spends too much time in sedentary activities (such as excessive screen time), more physical activities should be encouraged. The *Physical Activity Guidelines for Americans* recommends 60 minutes or more of moderate to vigorous physical activity per day for children and adolescents (Table 15-9). Making a habit of engaging in and enjoying regular physical activity will help children to maintain a healthy body weight throughout life. An increase in physical activity will not just happen; parents and other caregivers need to plan for it. Getting the family together for a brisk walk after dinner encourages healthy habits for all involved.[52] Age-appropriate activities for elementary school–age children include walking, dancing, jumping rope, and participation in organized sports that focus on fun rather than intense competition. For middle school–age children, more complex organized sports (e.g., soccer and basketball) are of interest, and some weight training with small weights can also be beneficial.

Moderation in calorie intake is important. High-calorie foods, such as sugar-sweetened soft drinks and whole milk, should be limited. The focus should be on more vitamin- and mineral-dense foods and healthy snacks. An emphasis on appropriate portion sizes may help youth learn to curb excessive food consumption. Making small changes, such as substituting low-fat for whole milk or fruit canned in its own juice instead of heavy syrup, can moderately cut calories without sacrificing taste or disrupting normal eating patterns.

Resorting to a weight-loss program is usually not necessary; it is best to change dietary and physical activity patterns to allow for weight maintenance. Children have an advantage over adults in dealing with obesity: their bodies can use stored energy for growth. An overweight child who maintains his or her body weight through a growth spurt will grow into a healthier BMI. This is one reason it is desirable to address obesity in childhood.

If weight loss is necessary in younger children, it should be gradual, about 0.5 to 1 pound per week. The child should be watched closely to ensure that the rate of growth continues to be normal; calorie intake should not be so low that gains in height diminish. In some cases, medications (e.g., orlistat [Xenical]) may be prescribed by a primary care provider. For the 1% to 2% of American children who are severely obese, bariatric surgery is an option for weight management.[53]

**TABLE 15-9 ■ Physical Activity Guidelines for Children**

| |
| --- |
| **Guidelines for Preschool-Age Children (age 3–5 years)** |
| Be physically active throughout the day to enhance growth and development. |
| Caregivers should encourage active play that includes a variety of types of activity. |
| **Guidelines for Children and Adolescents** |
| Provide opportunities and encouragement to participate in a variety of enjoyable, age-appropriate physical activities. |
| Engage in 60 minutes or more of moderate to vigorous physical activity daily, which should include: |
|     Mostly moderate- or vigorous-intensity aerobic physical activity; include vigorous-intensity physical activity on at least 3 days per week. |
|     Muscle-strengthening physical activity on at least 3 days per week. |
|     Bone-strengthening physical activity on at least 3 days per week. |

Source: *Physical Activity Guidelines for Americans*, 2nd ed. Available from https://health.gov/paguidelines /secondedition/pdf/Physical_Activity_Guidelines_2nd_edition.pdf.

## EARLY SIGNS OF CARDIOVASCULAR DISEASE

Parallel to the increase in childhood obesity, early signs of cardiovascular disease have become increasingly prevalent among children and adolescents. The CDC estimates that 7% of children have abnormal blood lipids.[54] Therefore, lifestyle modifications to delay the progression of the disease are important throughout the life span. The AAP recommends universal blood lipid screening for all children around the ages of 9 to 11. Even earlier screening is recommended for "at-risk" children who are overweight, have high blood pressure, smoke, or have diabetes; have a family history of cardiovascular disease; or whose family history is unknown.[55] For children whose cholesterol is elevated, lifestyle approaches such as weight management through dietary modification and increased physical activity are the first line of therapy.[56] An eating pattern that is consistent with the *Dietary Guidelines* would be appropriate for prevention of cardiovascular disease. To complement modifications to fat and sodium intakes, the American Heart Association recently released guidelines for children to limit added sugar intake to 25 grams per day.[57] This is quite a reduction from the average 80 grams of added sugars consumed by children and adolescents per day.

| Chronic Disease Risk Reduction Intake (CDRR) for Sodium for Children | |
| --- | --- |
| 1 to 3 years | 1200 milligrams per day |
| 4 to 8 years | 1500 milligrams per day |
| 9 to 13 years | 1800 milligrams per day |
| 14 to 18 years | 2300 milligrams per day |

## TYPE 2 DIABETES AMONG YOUTH

Type 2 diabetes was once regarded as an adult condition. However, an alarming increase in the frequency of the disease among children (and teenagers) has been documented. This is primarily due to the rise in obesity in this age group. Up to 85% of children with the disease are overweight at diagnosis.

Starting at age 10, children who are overweight or obese and who have risk factors for type 2 diabetes should be screened for type 2 diabetes every 2 years. Besides obesity and a sedentary lifestyle, examples of risk factors include having a close relative with the disease or belonging to a nonwhite population. In 2013, the AAP released the first-ever guidelines for management of type 2 diabetes in children.[58] These guidelines provide recommendations for monitoring of blood glucose, use of medications, weight management, and physical activity. Dietary management strategies include a regular schedule of meals and snacks; education on appropriate portion sizes; limiting sugar-sweetened beverages, high-fat foods, snacks, and fast foods; and focusing on incorporating more fruits, vegetables, and low-fat or fat-free dairy products. For physical activity, experts advise children to engage in moderate- or vigorous-intensity physical activity for at least 60 minutes each day.

## START THE DAY WITH BREAKFAST

You have heard it before: *Breakfast is the most important meal of the day.* Yet, as many as one-third of school-age children do not eat breakfast. The problem gets worse as children reach the teenage years. Children who skip breakfast are missing out on important nutrients that fuel the brain and the body. A fortified, low-sugar, ready-to-eat breakfast cereal offers lots of nutrition in a tasty and convenient package; it is typically the greatest source of iron, vitamin A, and folic acid for children ages 2 to 18. Although there is disagreement over the true benefit of breakfast for cognitive ability, children who eat breakfast are more likely to meet their daily needs for vitamins and minerals compared to children not eating breakfast.[59] Also, a growing body of research shows that starting the day with breakfast reduces risk for obesity.

Note that breakfast menus need not be limited to traditional fare. A little imagination can spark the interest of even the most reluctant eater. Instead of conventional breakfast foods, parents can offer leftovers from dinner, such as pizza, spaghetti, soups, yogurt topped with trail mix, chili with beans, or sandwiches. For lasting energy and satiety, combine traditional carbohydrate-rich breakfast foods with a source of protein, such as low-fat cheese, nuts, or eggs.

For kids who complain about waking up early to make time for breakfast, consider the convenience of healthy, grab-and-go options, such as leftover pancakes topped with peanut butter and bananas. **What quick and nutritious breakfast ideas would you recommend for a school-age child?** Alexis Joseph/McGraw-Hill Education

## CHOOSE HEALTHY FATS

Dietary patterns of school-age children should include a variety of foods from each major group, not necessarily excluding any specific food because of its fat content. Overemphasis on fat-reduced eating patterns during childhood has been linked to an increase in eating disorders and encourages an inappropriate "good food, bad food" attitude.

However, surveys of dietary intake among children show that they are consuming too much saturated fat, most of which comes from whole milk, other full-fat dairy products, and fatty meats.[49] Furthermore, few children (or adults) meet recommendations to include two servings of fish per week to ensure adequate intake of **omega-3 fatty acids**.[60] Emphasizing low-fat dairy products (from age 2 onward), offering broiled or baked fish, choosing leaner cuts of meat, trimming visible fat from meats, and removing the skin from poultry before serving foods will establish heart-healthy eating habits to last a lifetime. Snacks for children should emphasize fruits, vegetables, whole grains, and low-fat or fat-free dairy choices. Ideas for healthy snacks are found in Table 15-8.

## SELECT APPROPRIATE BEVERAGES

Maintaining proper hydration is important for children. The fluid needs of school-age children range from approximately 1½ to 2½ liters per day, depending on age and sex. However, over the past 30 years, beverage choices have shifted from water and milk to the empty calories provided by sugar-sweetened beverages. In fact, sugar-sweetened beverages (e.g, soft drinks, flavored fruit drinks, energy drinks, and sports drinks) and the sweeteners added to flavored milk account for about 200 empty kilocalories per day for school-age children.[61] The 135% increase in sugar-sweetened beverage consumption has paralleled the threefold rise in childhood obesity since the 1970s. Such high intakes of sugar-laden beverages are not only contributing excess calories but are also linked to increased levels of inflammation and worsened blood lipid profiles among children.[62] Even 100% fruit juices, which are perceived by many to be an important source of vitamin C and potassium for children, may contribute to excessive weight gain among children.[63]

Replacing sugar-sweetened beverages with water and choosing unflavored, low-fat, or fat-free milk instead of flavored milk would reduce sugar intake by about 10½ teaspoons per day and shrink overall calorie intake by about 10%. Furthermore, replacing 100% fruit juices with whole fruits would supply important nutrients in a lower-calorie package for children. Fruit juice should be limited to 6 fluid ounces per day for young children (ages 4 to 6) or 8 fluid ounces per day for older children (ages 7 to 18).[64] Overall, children should be given water and low-fat or fat-free milk as primary beverage choices.

## PROMOTE SOUND NUTRITION IN SCHOOLS

Children spend the majority of their waking hours in school, so it is a great place to learn about and practice healthy eating behaviors.[65] A strong emphasis on nutrition education in schools can help children understand why healthy dietary patterns will make them feel more energetic, look healthier, and work more efficiently. The USDA's Team Nutrition initiative supports child nutrition programs with education materials that promote healthy food choices and physical activity.

Most schools have included nutrition education in their health or science curricula, but until recently these healthy nutrition messages were not consistently backed up by the food offerings in school cafeterias. In 2010, the Healthy, Hunger-Free Kids Act extended funding for the National School Lunch Program, School Breakfast Program, and several other federal nutrition programs. The law also authorized the USDA to make significant changes to the nutritional quality of foods provided in schools.

The **Healthy Drinks. Healthy Kids.** campaign highlights appropriate beverage choices for children from birth to 5 years of age. All children up to 5 years of age should avoid beverages with added sugars, such as flavored milks, toddler formulas, sweetened soft drinks, and plant-based/nondairy milks.

**0 to 6 months**

Only provide breast milk or infant formula.

**6 to 12 months**

Continue to provide breast milk or infant formula as complementary foods are added to the infant's dietary pattern.

Introduce sips of plain water.

**12 to 24 months**

Introduce whole milk during weaning from breast milk or infant formula.

Continue to offer plain water.

If any fruit juice is offered, make sure it is 100% fruit juice.

**2 to 5 years**

Switch to low-fat or fat-free milk.

Milk and water should be the primary beverages.

If any fruit juice is offered, make sure it is 100% fruit juice.

Source: Academy of Nutrition and Dietetics, American Academy of Pediatric Dentistry, American Academy of Pediatrics, American Heart Association

Public school food-service programs now have to meet nutrition standards that stipulate the inclusion of fruits, vegetables, and whole grains, while limiting saturated fat and sodium content in meals.

Breakfasts and lunches prepared by school cafeterias are not the only targets of school nutrition reforms. In 2014, standards for the quality of competitive foods sold on school campuses (e.g., from snack bars and vending machines) went into effect. These guidelines set calorie limits on snacks and restrict the levels of saturated fat, added sugars, and sodium in foods that can be sold to students.

These school nutrition regulations are based on research studies that show how changing the quality of foods offered to students at school can stem the rise in children's BMI. Students' food choices, however, depend a lot on how the food choices taste. School food-service programs, although they receive some government reimbursement, often rely on cafeteria, snack bar, or vending machine sales to break even. It remains to be seen how these school nutrition reforms will influence children's eating behaviors. Research suggests that the new federal standards have increased fruit and vegetable consumption, improved nutrient density of school meals, and decreased energy density of school meals. So far, the data do not indicate any declines in students' participation in the school lunch program or increased food waste.[66,67]

Other strategies to engage children in positive nutrition behaviors include hands-on experience in cultivating school gardens and culinary education to prepare healthy foods.[65]

If we are to reduce childhood obesity, improving nutritional awareness at school is only part of the solution. Positive nutrition messages must extend beyond the classroom and into the home. Caregivers and other adult role models need to create safe opportunities for children to be active and must practice what they preach when it comes to healthy habits at home.[69]

## ✅ CONCEPT CHECK 15.5

1. Provide an example of a meal that resembles MyPlate and is appropriate for a 7-year-old child.

2. List three lifestyle changes to reduce childhood obesity. In what significant way do weight-management strategies for children differ from those for adults?

3. Describe appropriate beverage choices for school-age children. What are the implications of excessively consuming sugar-sweetened beverages?

4. Tim refuses to eat breakfast before school. He doesn't like cereal, toast, or any of the other usual breakfast foods. What can Tim's parents do to help him choose nutritious foods before leaving for school?

# 15.6 Teenage Years: Nutrition Concerns

Teenagers are on the cusp of adulthood; parents and schools may still be providing healthful food choices for them, yet they are capable of acquiring and preparing food for themselves. They pursue their independence, experience identity crises, seek peer acceptance, and worry about physical appearance. Advertisers push a vast array of products—candy, fast foods, soft drinks, and energy drinks—at the teenage market. Frequently, these foods crowd out nutrient-dense foods, thus limiting intake of calcium, iron, zinc, fat-soluble vitamins, and folate.

Teens often do not think about the long-term benefits of good health. Developmentally, they have a hard time relating today's actions to tomorrow's health outcomes. Still,

**COVID CORNER**

Households with children have been hardest hit by the economic fallout of the COVID-19 pandemic. Household budgets have been pinched by unemployment. Many families that rely on food pantries left empty-handed due to shortages of food, funds, and volunteers. School closures interrupted access to school breakfast and lunch for the 30 million children served by these programs each day. Many school systems were able to quickly mobilize their resources to supply meals to go for students. As a result of the Families First Coronavirus Response Act, SNAP benefits (and other nutrition assistance programs) have been increased to ensure access to adequate nutrition. The response from policy makers and communities has been rapid and effective, but many eligible children still fall through the cracks.[68]

Partnerships between local farms and schools can improve communities through better health and economic stability. Schools procure fresh, whole foods from farmers and ranchers in their community. Children benefit from access to a steady supply of seasonally fresh, nutrient-rich foods as well as opportunities to learn about nutrition and food production. Farmers benefit from job security and financial growth. Schools enjoy decreased food costs. Learn more about collaborations between schools and farmers from the National Farm to School Network: http://www.farmtoschool.org /Resources/BenefitsFactSheet.pdf.

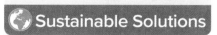
🌐 Sustainable Solutions

The physical changes of puberty cause body dissatisfaction for some adolescents. Late-blooming boys may be frustrated with slow gains in height and muscularity, whereas girls can be dissatisfied with gains in fat mass, which are a normal part of development. Be alert for signs of eating disorders (see Chapter 11).

positive dietary patterns do not require giving up favorite foods. In moderation, fast foods and sweet treats can occasionally fit into a healthy dietary pattern that is usually based on abundant fruits and vegetables, lean sources of protein, fat-free and reduced-fat dairy products, and whole grain products.

One of the most striking nutritional changes among adolescents is an increase in calorie intake. Most girls begin a rapid growth spurt between the ages of 10 and 13, and most boys experience rapid growth between the ages of 12 and 15. Early-maturing girls may begin their growth spurt as early as age 7 to 8, whereas early-maturing boys may begin growing by age 9 to 10. Nearly every organ and bone in the body grows during this adolescent growth spurt. Girls gain about 10 inches (25 centimeters) in height, and boys gain about 12 inches (30 centimeters). Girls tend to accumulate both lean and fat tissue, whereas boys tend to gain mostly lean tissue. This growth spurt provides about 50% of ultimate adult weight and about 15% of ultimate adult height (review Fig. 15-2).

As the growth spurt begins, teenagers begin to eat more. Physically active teenage boys, especially, seem driven to consume everything in sight! Teenage girls need 1600 to 2400 kcal per day, whereas teenage boys require 2000 to 3200 kcal per day. If teens choose nutrient-dense foods, they can take advantage of their ravenous appetites to easily satisfy their increased requirements for calcium, iron, and zinc. As discussed for younger age groups, the healthy U.S.-style dietary patterns and MyPlate can guide meal planning (see Table 15-7 and Fig. 15-5). Unfortunately, teens very often meet (or exceed) their increased energy needs with empty calories rather than nutrient-dense food options. About one-third of their calories come from solid fats and added sugars.[70]

Fruit and vegetable intake among teenagers is generally dismal; only about one-fourth of high school students regularly consume a minimum of five servings per day of fruits and vegetables. Sadly, potato chips and French fries make up more than one-third of the vegetable servings consumed by teens. Low consumption of fruits and vegetables correlates with inadequate intakes of vitamins A, C, and E; folate; magnesium; and fiber. Also, as teens (especially girls) trade their glasses of milk for bottles of soft drinks and other sugar-sweetened beverages, their intakes of calcium, phosphorus, and vitamin D fall short of recommendations. On the other hand, intakes of saturated fat, cholesterol, sodium, and sugars exceed the recommendations set by the American Heart Association, placing teens at risk for obesity and cardiovascular disease.

Childhood obesity, introduced in Section 15.5, continues to be a major nutritional problem into adolescence. Among teenagers, about 20% are obese.[71] There are sex and race disparities for adolescent obesity: black females and Hispanic males are at highest risk. Teens who are overweight or obese are likely to carry excess weight into adulthood and to develop comorbid conditions, such as type 2 diabetes, hypertension, cardiovascular diseases, sleep apnea, and joint problems. If a teen is still growing, he or she has an advantage in terms of weight management; by holding body weight steady while height gains are achieved, body mass index will decrease over time. However, if a teen attains ultimate adult height and is still obese, a weight-loss regimen may be necessary. Weight loss should be gradual, perhaps 1 pound per week, and generally follow the advice in Chapter 7.

**Calcium and Vitamin D.** Over the last 20 years, soft drinks and energy drinks have been replacing milk as the preferred beverage among children. Intake of milk is especially poor among adolescent females, who may view dairy products as a source of unwanted calories. This trend in milk consumption begins early in childhood, but we discuss it here because the gap between needs and actual intake of bone-building nutrients is greatest during the adolescent growth spurt. Less than 10% of girls and less than 25% of boys meet recommendations for calcium intake. To make matters worse, one in five children is deficient in vitamin D. Meanwhile, the adolescent growth spurt marks a critical time for bone development. About 50% of adult bone mass is accrued during adolescence. Calcium requirements for 14- to 18-year-old girls and boys are 1300 milligrams per day—higher than during any other time of life! Failure to maximize bone mineralization during childhood sets the stage for development of **osteoporosis** later in life.[19]

The *Dietary Guidelines* recommend 3 servings per day from the dairy group for all teenagers and young adults to meet calcium needs. If dairy products are not consumed, alternative calcium sources must be included. Nondairy sources of calcium include almonds, legumes, some green vegetables, and fortified foods (e.g., nondairy milks, fruit juices, cereal, and granola bars). However, it is important to note that these alternative sources of calcium may not provide other important nutrients supplied by dairy products, such as protein and vitamins A, D, and B-12.

**Iron.** About 10% of adolescent females have low iron stores or iron-deficiency anemia.[72] Iron-deficiency anemia is a highly undesirable condition for a teen. It can lead to fatigue and a decreased ability to concentrate and learn, such that academic and physical performance suffers. Iron-deficiency anemia sometimes appears in boys during their growth spurt, but adolescent females are at greatest risk of deficiency due to heavy menstrual flow and poor dietary intake. It is important that teenagers choose good food sources of iron, such as lean meats and fortified grain products. Teenage girls, in particular, need to eat good sources of iron. If dietary intake is insufficient to achieve optimal iron status, work with a primary care provider to determine whether dietary supplementation is necessary.

Many of the nutritional issues of adolescents—obesity, snacking, beverage choices, and skipping meals—have been adequately described with reference to younger children. Here, we present a few nutrition dilemmas that pertain especially to teenagers.

## BREAK THE FAST-FOOD HABIT

It is convenient, casual, inexpensive, and their friends work there. These are reasons why, on any given day, about 40% of the nation's youth eat food from a fast-food restaurant. Unfortunately, the average trip to a fast-food establishment yields about 300 extra calories, 14 additional grams of fat, and 400 milligrams of sodium *in excess* of typical home-prepared meals for teenagers.[32,73]

With some small changes, teens can still enjoy dining out with friends without detriment to their health. When building a sandwich, opt for one layer of meat instead of double or triple patties, and select grilled instead of fried meat. For deli sandwiches, choose moderate portions of lean meats, such as roasted turkey or chicken, rather than fatty slices of processed meats, such as bologna and salami. Skip the condiments or request them on the side; the mayonnaise on a typical fast-food sandwich supplies about 100 fat-laden kcal. Each slice of cheese supplies another 80 to 100 kcal. When it comes to choosing a side dish, a small baked potato or a garden salad with reduced-fat dressing will provide fewer calories and more nutrients than the typical 500-kcal large serving of fries. Calories from regular soft drinks—especially when free refills are available—can quickly add up. Teens should choose reduced-fat or fat-free milk as a nutrient-dense alternative or opt for water. Order a pizza with veggie toppings, low-fat cheese, and whole grain crust.

When burgers are measured in pounds instead of ounces, portion control is an issue. While already large, portion sizes at fast-food establishments continue to grow. Choosing items from the kids' menu can lessen the impact of dining out on adolescent wallets and waistlines. Supersized meals, while they may seem economical, should be avoided unless they are to be divided and shared among friends. All of us can benefit from the calorie information that many restaurants are now required to display on their menus.

The teenage years are noted for snacking. **Suggest some snack choices to fill common nutrient gaps among teens.** SW Productions/Getty Images

# Newsworthy Nutrition

## Glycemic load of food choices may influence acne

**INTRODUCTION:** About 80% to 90% of teens experience acne to some degree. Although it is popularly believed that nuts, chocolate, French fries, and pizza contribute to acne, scientific studies have failed to show a strong role for any of these dietary factors. Observational studies suggest a relationship between glycemic index or glycemic load of the dietary pattern and acne, but previous research has not adequately or accurately assessed dietary intake and biological factors relative to acne. **OBJECTIVE:** To examine differences in dietary intake (especially glycemic index and glycemic load of the dietary pattern) and biological markers of insulin resistance among young adults with or without acne. **METHODS:** Sixty-four adults between the ages of 18 and 40 with BMI within the range of 18.5 to 30 kg/m² completed 5-day diet records. The researchers collected blood samples to assess biological markers of insulin resistance. Digital photos were taken and assessed by trained dermatologists who scored acne severity. **RESULTS:** Compared to the dietary patterns of subjects with no acne, the dietary patterns of subjects with moderate or severe acne were higher in total carbohydrates and glycemic load. In addition, the subjects with acne had higher insulin resistance than subjects without acne. **CONCLUSION:** Glycemic load of the dietary pattern is associated with acne among young adults. Insulin resistance appears to play a role in development of acne. Choosing whole, unprocessed grains, vegetables, and fruits instead of refined grains and foods with added sugars may assist efforts to reduce acne.

Source: Burris J and others: Differences in dietary glycemic load and hormones in New York City adults with no and moderate/severe acne. *Journal of the Academy of Nutrition and Dietetics* 117:1375, 2017.

## CURB CAFFEINE INTAKE

The combined demands of school, work, extracurricular activities, social commitments, and late-night screen time leave many adolescents looking for a quick pick-me-up. Commonly, they are turning to caffeine, the most widely used stimulant on the planet. Soft drinks, a common choice among youth, provide about 25 milligrams of caffeine per serving. On average, 30% of adolescents report consuming energy beverages, which typically contain between 100 and 200 milligrams of caffeine per serving. Consumption of coffee and tea, which yield about 100 milligrams of caffeine per cup, is on the rise among teens. Various foods, including chocolate and some types of candies or sports nutrition products, contain caffeine as well. Average caffeine intake from all sources is just over 100 milligrams per day among teens. Many consumers are unaware of how much caffeine they are consuming; the exact amount of caffeine is not always listed on energy drink labels because: (1) it is not currently required by food labeling laws; and (2) some manufacturers (especially of energy drinks) consider it to be part of a "proprietary blend."

For children, the AAP advises limiting caffeine intake to 100 milligrams per day, if it is used at all. Some of the negative effects of caffeine at any age are gastrointestinal distress, sleep disturbances, anxiety, increased blood pressure, and irregular heartbeat. For children, in particular, there is concern that excessive caffeine intake could affect normal neurological and cardiovascular development. Furthermore, disturbances in normal sleep patterns could affect growth and learning ability. Alarmingly, there have been thousands of reports of caffeine poisoning—and even some deaths—as a result of excessive intake of energy drinks. Clearly, excessive caffeine intake has no place in eating patterns for children.[74]

## VEGETARIAN DIETARY PATTERNS DURING ADOLESCENCE

Teenagers, who strive to forge an identity by adopting dietary patterns different from those of their families, may choose to follow a plant-based dietary pattern. Vegetarians enjoy many health benefits, including lower body weight and better control of blood glucose and cholesterol. Indeed, an increased focus on plant foods is needed among adolescents, who often miss out on their recommended daily servings of fruits and vegetables. However, teens may not know enough about a vegetarian dietary pattern to keep from developing health problems, such as iron-deficiency anemia. The bulkiness of a plant-based eating pattern is not as much of a concern for teens as it is for younger children with smaller stomach capacity, but a strictly vegetarian eating pattern must be monitored for adequate energy, protein, iron, vitamin B-12, calcium, and vitamin D (the latter if sun exposure is not sufficient) at any age. These nutrients are particularly important in teenagers, as their dietary patterns are often already nutrient poor.

Teens often cite concern for the humane treatment of animals as their main reason for choosing vegetarian eating patterns, but be observant of teens who choose vegetarianism as a strategy to manage body weight. Vegetarian dietary patterns are sometimes used as a socially acceptable way to restrict food intake and, for some, can be an early sign of disordered eating.[75]

> The *Dietary Guidelines* outline healthy vegetarian dietary patterns at various calorie levels for ages 2 and older. For additional information, see Appendix 3 at DietaryGuidelines.gov.

## ALCOHOL ABUSE AMONG TEENS

In Section 15.5, we discussed how the beverage choices of school-age children are in need of improvement because they provide too much sugar and not enough micronutrients. The nutrient density of beverages continues to be a problem among teenagers, but a new problem arises: alcohol abuse. Developmentally, adolescents are prone to experimentation, rebellion, and risk taking, so use of this illegal and dangerous substance is common among teenagers. Results of the national Youth Risk Behavior Survey demonstrate that approximately 20% of teenagers have tried alcohol by the age of 13. At some point throughout the teenage years, about 70% of teens report drinking alcohol at least once, and 22% report binge drinking.

It is just harmless fun, right? Wrong! Alcohol use beginning in adolescence has severe consequences.[76] The adolescent's body and brain are still developing. Exposure to alcohol can decrease brain mass in the area of the brain involved in decision making, memory, and learning. This is evidenced by academic problems and poor decision making, which can lead to legal troubles, physical assault, and risky sexual behaviors. The most dangerous consequence of poor judgment is drinking and driving. About 1 in 10 teenagers admits to drinking and driving, a risky behavior that is implicated in about one-third of fatal motor vehicle accidents involving teens. Alcohol also contributes to other causes of accidental injuries and deaths, such as drowning, falls, and burns.

Adolescent alcohol abuse exacts a toll on long-term physical health as well. Studies show that alcohol abuse beginning during adolescence is a strong predictor of alcohol abuse during adulthood. Nutritional status can be affected because alcohol abuse is often accompanied by nutrient-poor dietary patterns. Also, weight gain from empty calories increases the risk for obesity-related diseases, such as hypertension and cardiovascular disease. These physical consequences may not surface until later in life, but it is certain that the effects of alcohol on the liver, brain, and cardiovascular system can start early.

Alcohol use by teenagers should not be viewed as a normal part of growing up. On the contrary, the physical, emotional, and intellectual consequences of underage drinking can be long-standing and devastating. Parents and other caregivers should talk to their children about the consequences of alcohol abuse, set clear rules, monitor their children's behavior, and be positive role models.

### ✔ CONCEPT CHECK 15.6

1. Which two minerals are most likely to be deficient in the dietary patterns of teens? Name two rich food sources of each of these minerals.

2. Design a meal for a teen that resembles MyPlate and can be purchased from a fast-food restaurant.

3. Are energy drinks safe for consumption by children of any age? Why or why not?

4. List three consequences of alcohol abuse that are specific to adolescents.

## Food Allergies and Intolerances

Corbis Super RF/Alamy

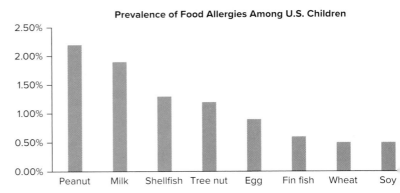

**Prevalence of Food Allergies Among U.S. Children**

FIGURE 15-6 ▲ Prevalence of food allergies among U.S. children. Source: Data from Gupta RS and others: The public health impact of parent-reported childhood food allergies in the United States. *Pediatrics* 142;e20181235, 2018.

Food allergies are on the rise. What used to be a rare medical incident is now the cause for 92,000 childhood emergency department visits and 84 child fatalities per year. Accounting for direct medical costs, special foods, and time lost from work, food allergies cost Americans $25 billion per year. Today, food allergies affect about 8% (5.9 million) of children in the United States. About 40% of children with food allergies are allergic to more than one type of food. The most commonly reported food allergies among children are peanuts, milk, shellfish, and tree nuts (see Fig. 15-6).[77]

Adverse reactions to foods—indicated by sneezing, coughing, nausea, vomiting, diarrhea, hives, and other rashes—are broadly classed as **food allergies** (also called *hypersensitivities*) or **food intolerances.** The term *food sensitivity* is ill defined but generally refers to any symptom that is perceived to be food related. In our discussion, we group adverse food reactions into two categories: those caused by an immune response are termed food allergies, and those not caused by an immune response are food intolerances.

## Food Allergies: Symptoms and Mechanisms

Symptoms of food allergies may affect the following:

- **Skin:** itching, tingling,* redness, hives, and swelling
- **GI tract:** nausea, vomiting, diarrhea, intestinal gas, bloating, pain, constipation, and indigestion

- **Respiratory tract:** runny nose, wheezing, congestion, and difficulty breathing*
- **Cardiovascular system:** low blood pressure* and rapid heart rate*

These symptoms usually set in shortly after consuming the offending food protein and may last for a few seconds or a few days. The symptoms marked with an asterisk (*) are signs of a rapid and potentially fatal type of allergic response called **anaphylaxis.** This severe allergic response results in low blood pressure and respiratory distress. A person who is extremely sensitive to a food may not be able to touch the food or even be in the same room where it is being cooked without reacting to it. Although any food can trigger anaphylaxis, the most common culprits are peanuts (a legume, not a nut), tree nuts (e.g., walnuts, pecans), shellfish, milk (also beware of an ingredient called casein), eggs (look for the ingredient albumin), soybeans, wheat, and fish. Other foods frequently identified with adverse reactions include meat and meat products, fruits, and cheese. For a small number of people, avoiding foods such as peanuts or shellfish is a matter of life and death.

Basically, allergies are an inappropriate response of the immune system. When immune cells identify a harmful foreign protein (**antigen**), they destroy it and produce antibodies to it, so that the next response to the harmful substance will be swift and effective. Almost all food allergies are caused by proteins in foods that act as antigens (also called **allergens**). In these cases, the immune system mistakes the food protein for a harmful substance and mounts an immune response, leading to symptoms such as hives, runny nose, and GI disturbances.

**food allergy** An adverse reaction to food that involves an immune response; also called *food hypersensitivity.*

**food intolerance** An adverse reaction to food that does not involve an allergic reaction.

**anaphylaxis** A severe allergic response that results in lowered blood pressure and respiratory distress. This can be fatal.

**allergen** A foreign protein, or antigen, that induces excess production of certain immune system antibodies; subsequent exposure to the same protein leads to allergic symptoms. Whereas all allergens are antigens, not all antigens are allergens.

No one is sure why the immune system sometimes overreacts to harmless proteins. The early introduction (e.g., before 4 months of age) of solid foods to infants may trigger food allergies. The reasoning is that the infant's GI tract is immature and "leaky," allowing some undigested proteins to be absorbed into the bloodstream. Gut permeability is beneficial for the absorption of immune proteins from breast milk; however, if some food proteins are introduced before the GI tract has matured, antigens may enter the bloodstream and stimulate an immune response.

The hygiene hypothesis offers another interesting explanation: in our "germophobic" society, with the protection of antibiotics, hand sanitizers, and antimicrobial soaps and cleaners, our immune systems are not vigorously challenged by antigens. As a result, the immune system may become sensitized to innocuous substances, such as food proteins. Current research supports the hygiene hypothesis. Children who grow up on farms or who have pets and are thereby exposed to many antigens have fewer allergies and a lower incidence of asthma than children who grow up in more sterile environments.

Researchers are currently interested in the connection between a healthy gut microbiota and risk for food allergies. Also, researchers have proposed a link between low levels of vitamin D and food allergies. The relationship between vitamin D and food allergies may be mediated by the vitamin's role in immune function.

## TESTING FOR A FOOD ALLERGY

The diagnosis of a food allergy can be a difficult task.[78] It requires the expertise of a skilled clinician. To determine whether a food allergy is present, the health professional will record a detailed history of symptoms, including the time from ingestion to onset of symptoms, duration of symptoms, most recent reaction, food suspected of causing a reaction, and quantity and nature of food needed to produce a reaction. A family history of allergic diseases can also help, as allergies tend to run in families. A physical examination may reveal evidence of an allergy, such as skin diseases and asthma. Various diagnostic tests can rule out other conditions (Table 15-10).

If the patient history and physical exam suggest a food allergy, the health professional then faces the task of identifying the source of the food allergy. The first step in diagnosing a food allergy is to eliminate from the eating pattern (for 1 to 2 weeks) all food components that appear to cause allergic symptoms. This is called an **elimination diet.** The person generally starts out eating foods to which almost no one reacts, such as rice, vegetables, noncitrus fruits, and fresh meats and poultry. If symptoms are still present, the person can more severely restrict the eating pattern or even use special formulas that are hypoallergenic.

Once a dietary pattern is found that causes no symptoms, foods can be added back one at a time. This type of food challenge is an option only when the culprit foods are known to pose no risk of anaphylaxis in the person. Doses of ½ to 1 teaspoon (2½ to 5 milliliters) are given at first. The amount is increased until the dose approximates usual intake. Any reintroduced food

**elimination diet** A restrictive diet that systematically tests foods that may cause an allergic response by first eliminating them for 1 to 2 weeks and then adding them back, one at a time.

**TABLE 15-10 ■ Diagnosing Food Allergies**

| | |
|---|---|
| History | Include description of symptoms, time between food ingestion and onset and severity of symptoms, duration of symptoms, most recent allergic episode, quantity of food required to produce reaction, suspected foods, and allergic diseases in other family members. |
| Physical examination | Look for signs of an allergic reaction (rash, itching, intestinal bloating, etc.). |
| Elimination diet | Remove the suspected food allergen for 1 to 2 weeks or until symptoms clear. |
| Food challenge | Add back small amounts of excluded foods, one at a time, as long as anaphylaxis is not a possible consequence. |
| Blood test | Determine the presence of antibodies in blood that bind to food antigens tested. |
| Skin test | Place a sample of the suspected allergen under the skin and watch for an inflammatory reaction. |

that causes significant symptoms to appear is identified as an allergen for the person.

Laboratory tests can also aid in diagnosis of food allergies. Skin testing involves pricking the skin with a small amount of purified food extract and observing any allergic response (e.g., a red eruption at the prick site). These types of tests are easy and safe, even for infants, but they may not clearly diagnose a food allergy. A positive skin-prick test merely indicates that a person has been sensitized to a food; it cannot clearly identify if that food is the cause for the symptoms in question. Newer types of blood testing, however, have more diagnostic value. Blood tests estimate the blood concentration of antibodies that bind certain foodborne antigens.

## LIVING WITH FOOD ALLERGIES

Once potential food allergens are identified, dietary modifications must be made.[79] In some cases, small amounts of the offending food can be consumed without an observable reaction. Also, some food allergens are destroyed by heating, so cooking may eliminate the allergic response. This is effective primarily for allergies to fruits or vegetables, not for the more common allergies to milk, peanuts, or seafood. For most cases, though, complete avoidance of allergy-causing food ingredients is the safest course of action. This makes careful reading of food labels essential. The Food Allergen Labeling and Consumer Protection Act requires manufacturers to clearly identify the presence of major food allergens (milk, eggs, fish, shellfish, peanuts, tree nuts, wheat, and soy) on food product labels. Sesame seed allergy is becoming more common and some advocacy groups have petitioned to see sesame seeds added to the list of food allergens that must be listed on food labels.

A major challenge when treating a person with a food allergy is to make sure that what remains in the dietary pattern can still provide essential nutrients. The small food intake of children permits less leeway in removing offending foods that may contain numerous

▶ **Most Common Allergenic Foods:**

| | |
|---|---|
| Peanuts | Fin fish |
| Tree nuts | Shellfish |
| Milk | Wheat |
| Eggs | Soy |

nutrients. An RDN can help guide the diet-planning process to ensure that the remaining food choices still meet nutrient needs. Dietary supplements may be necessary.[80]

Studies show that about 25% of young children with food allergies outgrow them. Food allergies diagnosed after 3 years of age are more likely to be lifelong. It is common for children to outgrow allergies to milk, soy, or eggs, but allergies to peanuts, tree nuts, and shellfish are likely to endure. Periodic reintroduction of offending foods can be tried every 6 to 12 months or so to see whether the allergic reaction has decreased. If no symptoms appear, tolerance to the food has developed.

Several strategies are being studied to ease the dietary restrictions imposed by food allergies. One possibility includes treatment with antibodies that will increase the threshold at which an allergic response occurs. For a person with an allergy to peanuts, for example, this would alleviate some anxiety about severe reactions to trace amounts of peanuts found in foods. Similarly, immunotherapy, which exposes allergic individuals to very small but progressively larger amounts of food allergens, may help some people build up a tolerance to certain food components. Vaccines are another area of research. Also, scientists are working on genetically engineered foods that do not contain common allergens.[81]

## PREVENTING FOOD ALLERGIES

With the rising number of cases of food allergies, many new parents wonder when and how to introduce new foods during infancy and early childhood. Over the past 25 years, expert recommendations

▲ In 2006, the Food Allergen Labeling and Consumer Protection Act mandated that food manufacturers make consumers aware of the presence of highly allergenic ingredients in their products. FoodIngredients/Alamy Stock Photo

for prevention of food allergies have undergone dramatic revisions. Out of an abundance of caution, experts once advised new parents to delay the introduction of potential food allergens—until after the third birthday for some foods. Allergy-prone women were advised to avoid highly allergenic foods during pregnancy and breastfeeding to limit the infant's exposure to allergens via the placenta or breast milk. Infants with a family history of food allergies were given extensively hydrolyzed infant formulas to limit their exposure to intact proteins. Based on observational studies comparing rates of food allergies among breastfed and formula-fed infants, experts promoted exclusive breastfeeding as a tactic to prevent food allergies.[19]

Did any of these recommendations reduce the prevalence of food allergies? No! As it turns out, there is not enough evidence to support any of this past guidance. The best evidence we have at this time centers around *early*—not late—introduction of peanut protein to prevent food allergies.[25] This protocol may prove useful for other food allergens, but the data are not yet clear.

So, in 2019, the AAP released updated recommendations for food allergy prevention through diet:[82]

- There is *no evidence* to recommend maternal dietary restrictions of potential food allergens during pregnancy or breastfeeding.
- There is *no evidence* to recommend any specific duration of breastfeeding.
- There is *very limited evidence* to recommend partially or extensively hydrolyzed infant formula for high-risk infants.
- There is *no evidence* to delay the introduction of any potentially allergenic foods beyond 4 to 6 months of age.
- There is evidence that early introduction (between 4 and 6 months of age) of peanut protein reduces the risk for peanut allergies.

The AAP, the American Academy of Allergy, Asthma, and Immunology, and the latest edition of the *Dietary Guidelines* advise waiting to introduce solid foods until at least 4 months of age, but preferably around 6 months of age for the lowest risk of food allergies. Delaying introduction of solid foods beyond 6 months of age is not advised. Even highly allergenic foods, such as peanuts, egg whites, and milk, can be introduced in forms that are safe for infants to eat (e.g., mixing a small amount of peanut butter into infant cereal or yogurt) when the family chooses to offer complementary foods.[19]

## FOOD INTOLERANCES

Food intolerances are adverse reactions to foods that do not involve immunologic mechanisms. Generally, larger amounts of an offending food are required to produce the symptoms of an intolerance than to trigger allergic symptoms. Common causes of food intolerances include the following:

- Constituents of certain foods (e.g., red wine, tomatoes, and pineapples) that have a drug-like activity, causing physiological effects such as changes in blood pressure
- Certain synthetic compounds added to foods, such as sulfites, food-coloring agents, and monosodium glutamate (MSG)

# Newsworthy Nutrition

### Early introduction of peanut protein reduces peanut allergy

**INTRODUCTION:** Based on a hypothesis that early introduction of food proteins to infants increased risk for food allergies, pediatricians once advised parents to delay introduction of potential food allergens (e.g., 2 years for eggs and 3 years for peanuts, tree nuts, and fish). However, in the 1990s, evidence started to accumulate that delaying introduction of a variety of foods provided no benefit for preventing food allergies. For instance, Jewish children raised in the United Kingdom, where peanuts were not introduced until after 1 year of age, were 10 times more likely to develop peanut allergies than Jewish children raised in Israel, where peanut-based foods are introduced within the first year of life. **OBJECTIVE:** The Learning Early About Peanut Allergy (LEAP) trial aimed to see if early introduction of peanut protein could prevent peanut allergies among at-risk children. **METHODS:** The randomized, controlled trial included 640 infants between 4 and 11 months of age who were at risk for food allergies (based on existing allergies, severe eczema, or both). Participants were divided into two groups based on previous sensitization to peanut protein (i.e., skin testing showed if the infants' immune systems had already reacted to peanut protein from dietary, skin, or respiratory exposure). Next, the infants were randomized to treatment or control groups. The treatment group received at least 6 grams of peanut protein per week in the form of a peanut-based snack food or smooth peanut butter, while the control group was advised to avoid dietary exposure to peanuts. **RESULTS:** At 5 years of age, the children were tested for peanut allergy using an oral food challenge. Among the children who were not sensitized to peanuts at baseline, peanut consumption reduced the risk of developing peanut allergy by 86.1% compared to controls. Among the children who were initially sensitized to peanut protein, treatment with peanut protein reduced the risk of developing peanut allergy by 70%. **CONCLUSION:** The researchers concluded that early (< 11 months), sustained peanut consumption reduced peanut allergy among children at risk of food allergies.

Source: Du Toit G and others: Randomized trial of peanut consumption in infants at risk for peanut allergy, *New England Journal of Medicine* 372:803, 2015.

- Food contaminants, including antibiotics and other chemicals used in the production of livestock and crops, as well as insect parts not removed during processing
- Toxic contaminants, which may be ingested with improperly handled and prepared foods containing *Clostridium botulinum*, *Salmonella* bacteria, or other foodborne microorganisms
- Deficiencies in digestive enzymes, such as lactase

Almost everyone is sensitive to one or more of these causes of food intolerance, many of which produce GI tract symptoms. Sulfites, added to foods and beverages as antioxidants, cause flushing, spasms of the airway, and a loss of blood pressure in susceptible people. Wine, dehydrated potatoes, dried fruits, gravy, soup mixes, and restaurant salad greens commonly contain sulfites. A reaction to MSG may include an increase in blood pressure, numbness, sweating, vomiting, headache, and facial pressure. MSG is commonly found in restaurant food and many processed foods (e.g., soups). A reaction to tartrazine, a yellow food-coloring additive, includes spasm of the airway, itching, and reddening skin. Tyramine, a derivative of the amino acid tyrosine, is commonly found in "aged" foods such as cheeses and red wines. This natural food constituent can cause high blood pressure in people taking monoamine oxidase (MAO) inhibitor medications, which may be prescribed for clinical depression.

The basic treatment for food intolerances is to avoid specific offending components. However, total elimination often is not required because people generally are not as sensitive to compounds causing food intolerances as they are to allergens.

### ✓ CONCEPT CHECK 15.7

1. Name the eight foods containing the common food allergens that must be listed on food labels in the United States.
2. What is the most common and most dangerous food allergen among children in the United States?
3. A new mom asks you how to prevent food allergies in her infant. What information can you provide?

## Summary (Numbers refer to numbered sections in the chapter.)

**15.1** Growth is rapid during infancy; birth weight doubles by 6 months of age, and length increases by 50% in the first year. An adequate dietary pattern, especially in terms of calories, protein, iron, and calcium, is essential to support normal growth. Growth charts can be used to assess changes in body weight, height (or length), head circumference, and body mass index over time.

**15.2** The energy needs of infants are highest per kilogram of body weight compared to any other life stage (about 100 kcal per kilogram). Fat should make up about 50% of total energy intake. DHA and ARA are important fatty acids for nervous system development. Carbohydrate needs range from 60 grams per day for younger infants to 95 grams per day for older infants. Protein needs are 9 grams per day for younger infants and 11 grams per day for older infants. Supplementation with vitamin D, iron, and fluoride may be appropriate for some infants. Adequate hydration can be maintained using only breast milk or formula; supplemental water is not recommended during the first 6 months of life.

**15.3** Infant nutrient needs can usually be met by human milk or iron-fortified infant formula for the first 6 months of life. Introduction of solid foods should begin no earlier than 4 months of age, but preferably around 6 months of age, based on an infant's nutritional needs, physical abilities, and developmental readiness. Solid foods should be introduced one at a time, starting with iron-fortified infant cereals or ground meats (sources of iron). Some foods to avoid giving infants in the first year include honey, unaltered cow's milk (especially fat-reduced varieties), foods with added salt or sugars, and foods that may cause choking.

**15.4** A slower growth rate results in decreased appetite among preschool children. Other common nutrition-related concerns include iron-deficiency anemia, constipation, and dental caries. With smaller portion sizes and picky eating behaviors, it is crucial to offer several small meals and snacks with a variety of nutrient-dense foods. Follow the example set forth by MyPlate, but use smaller portions (e.g., 1 tablespoon of food per year of life). For autism spectrum disorders, several nutritional interventions, including the gluten-free, casein-free diet, are under study, but none are endorsed by the AAP.

**15.5** Among school-age children, excessive energy intakes coupled with low levels of physical activity have led to an alarming increase in overweight, obesity, type 2 diabetes, and cardiovascular disease. Parents can provide healthful food choices and encourage at least 60 minutes of physical activity per day. When addressed early through dietary and physical activity interventions, BMI-for-age may deflect downward as the child continues to grow in height. Other important nutrition strategies for school-age children include starting the day with breakfast and selecting low-fat or fat-free milk or water instead of sugar-sweetened beverages. Recent changes in meal offerings through schools are aimed at curtailing the rise in childhood obesity.

**15.6** During the adolescent growth spurt, both boys and girls have increased needs for iron, calcium, and overall calories. Inadequate calcium intake by teenage girls is a major concern because it can set the stage for the development of osteoporosis later in life. Adolescents need to limit their intakes of high-fat and high-sugar fast foods and snacks and use caffeine in moderation (if at all). Alcohol abuse during adolescence has many severe consequences, including impaired brain development and increased risk for liver and cardiovascular diseases in adulthood.

**15.7** The most common food allergies are associated with peanuts, tree nuts, shellfish, milk, eggs, soybeans, wheat, and fish. Food allergies occur most often during infancy and young adulthood. Dietary treatment for food allergies involves complete avoidance of the food allergen.

## Check Your Knowledge (Answers are available at the end of this question set.)

1. Inadequate intake of which of the following results in poor growth?
   a. Calories
   b. Iron
   c. Zinc
   d. All of these

2. Cow's milk is a nutrient-dense source of all of the following except
   a. protein.
   b. iron.
   c. calcium.
   d. zinc.

3. To ensure adequate vitamin and mineral intake for a picky eater,
   a. provide a fortified breakfast cereal.
   b. promise dessert as a reward for eating meats and vegetables.
   c. use a multivitamin and mineral supplement.
   d. avoid all of these.

4. An 11-month-old female who weighs 19 pounds needs approximately ___ kcal per day.
   a. 690
   b. 810
   c. 845
   d. 930

5. Introduction of cow's milk should be delayed until 12 months of age because it
   a. contains too much fat.
   b. supplies too much lactose.
   c. contains too much protein.
   d. does all of these.

6. Your niece breaks out in hives and feels nauseous after eating a salad containing mango. She probably has a food
   a. sensitivity.
   b. allergy.
   c. intolerance.
   d. All of these are correct.

7. Which of the following is an outcome of consuming a fortified, ready-to-eat breakfast cereal instead of skipping breakfast?
   a. Increased intakes of saturated fat and cholesterol
   b. Improved intakes of iron and calcium
   c. Excessive weight gain
   d. All of these

8. Which of the following nutrition interventions is recommended by the AAP for treatment of autism spectrum disorder?
   a. Camel's milk
   b. Fish oil supplements
   c. Gluten-free, casein-free diet
   d. None of these

9. If moderate weight loss is needed, a school-age child should
   a. eat fewer meals.
   b. follow a low-carbohydrate eating plan.
   c. engage in physical activity for 60 minutes per day or more.
   d. avoid dairy products.

10. You are trying to introduce an apple and blueberry purée to a 7-month-old infant, but she rejects it. You should
    a. assume she doesn't like apples and blueberries.
    b. offer the food again on another day.
    c. force a spoonful into her mouth.
    d. do none of these.

Answer Key: 1. d (LO 15.1), 2. b (LO 15.2), 3. a (LO 15.2), 4. a (LO 15.3), 5. c (LO 15.3), 6. b (LO 15.6), 7. b (LO 15.4), 8. d (LO 15.4), 9. c (LO 15.5), 10. b (LO 15.3)

---

## Study Questions (Numbers refer to Learning Outcomes.)

1. List two factors that limit "catch-up" growth when a nutrient-deficient dietary pattern has been consumed throughout childhood. **(LO 15.1)**

2. What are two possible causes of failure to thrive? **(LO 15.1)**

3. Which two nutrients are of particular concern in planning dietary patterns for teenagers? Why does each deserve special attention? **(LO 15.2)**

4. List three nutrients of concern for a child who is following a vegetarian lifestyle. **(LO 15.2)**

5. Outline three key factors that help to determine when to introduce solid foods into an infant's dietary pattern. **(LO 15.3)**

6. List three foods or beverages to avoid feeding infants. Explain why these items should be avoided. **(LO 15.3)**

7. Describe the pros and cons of snacking. What is the basic advice for healthful snacking from childhood through the teenage years? **(LO 15.4)**

8. List three reasons why preschoolers are noted for picky eating. **(LO 15.4)**

9. What three factors are likely to contribute to obesity in a typical 10-year-old child? **(LO 15.5)**

10. What is the difference between a food allergy and a food intolerance? **(LO 15.6)**

---

## Further Readings

1. Lott M and others: Healthy beverage consumption in early childhood: Recommendations from key national health and nutrition organizations. Technical Scientific Report. Durham, NC: Healthy Eating Research, 2019. Available at http://healthyeatingresearch.org.

2. Holt K and others: *Bright Futures Nutrition,* 3rd ed. Elk Grove Village, IL: American Academy of Pediatrics, 2011.

3. Homan GJ: Failure to thrive: A practical guide. *American Family Physician* 94:295, 2016.

4. Delplanque D and others: Lipid quality in infant nutrition: Current knowledge and future opportunities. *Journal of Pediatric Gastroenterology and Nutrition* 61:8, 2015. DOI:10.1097/MPG.0000000000000818.

5. Wagner CL and others: Prevention of rickets and vitamin D deficiency in infants, children, and adolescents. *Pediatrics* 122(5):1142, 2008. DOI:10.1542/peds.2008-1862.

6. Baroni L and others: Vegan nutrition for mothers and children: Practical tools for healthcare providers. *Nutrients* 11:5, 2019. DOI:10.3390/nu11010005.

7. Lozoff B and others: Functional significance of early-life iron deficiency: Outcomes at 25 years. *Journal of Pediatrics* 163(5):1260, 2013. DOI:10.1016/j.jpeds.2013.05.015.

8. Baker RD and Greer FR: Clinical report—Diagnosis and prevention of iron deficiency and iron-deficiency anemia in infants and young children (0–3 years of age). *Pediatrics* 126:1040, 2010. DOI:10.1542/peds.2010-2576.

9. Lewis CW: Fluoride and dental caries prevention in children. *Pediatrics in Review* 35:3, 2014. DOI:10.1542/pir.35-1-3.

10. American Academy of Pediatrics: Signs of dehydration in infants. November 21, 2015. Available at https://www.healthychildren.org/English/health-issues/injuries-emergencies/Pages/dehydration.aspx. Accessed February 24, 2019.

11. American Academy of Pediatrics: Treating dehydration with electrolyte solution. November 21, 2015. Available at https://www.healthychildren.org/English/health-issues/conditions/abdominal/Pages/Treating-Dehydration-with-Electrolyte-Solution.aspx. Accessed February 24, 2019.

12. Gregory JW: Prevention of obesity and metabolic syndrome in children. *Frontiers in Endocrinology* 10:669, 2019. DOI:10.3389/fendo.2019.00669.

13. Rogers SL and Blissett J: Breastfeeding duration and its relation to weight gain, eating behaviours and positive maternal feeding practices in infancy. *Appetite* 108:399, 2017. DOI:10.1016/j.appet.2016.10.020.

14. Oddy WH and others: Early infant feeding and adiposity risk: From infancy to adulthood. *Annals of Nutrition and Metabolism* 64:262, 2014. DOI:10.1159/000365031.

15. Korpela K and others: Association of early-life antibiotic use and protective effects of breastfeeding. *Journal of the American Medical Association Pediatrics* 170(8):750, 2016. DOI:10.1001/jamapediatrics.2016.0585.

16. Pomeranz JL, Romo Palafox MJ, and Harris JL: Toddler drinks, formulas, and milks: Labeling practices and policy implications. *Preventive Medicine* 109:11, 2018. DOI:10.1016/j.ypmed.2018.01.009.

17. Paul IM and others: Effect of a responsive parenting educational intervention on childhood weight outcomes at 3 years of age: The INSIGHT randomized clinical trial. *Journal of the American Medical Association* 320:461, 2018. DOI:10.1001/jama.2018.9432.

18. Marshall Z and Delahunty C: The INSIGHT responsive parenting intervention reduced infant weight gain and overweight status. *Archives of Disease in Childhood: Education and Practice Edition* 103:57, 2018. DOI:10.1136/archdischild-2017-313112.

19. American Academy of Pediatrics Committee on Nutrition: *Pediatric Nutrition*, 8th ed. Itasca, IL: American Academy of Pediatrics, 2019.

20. Barrera CM and others: Timing of introduction of complementary foods to U.S. infants, National Health and Nutrition Examination Survey 2009-2014. *Journal of the Academy of Nutrition and Dietetics* 118:434, 2018. DOI:10.1016/j.jand.2017.10.020.

21. Pennestri M-H and others: Uninterrupted infant sleep, development, and maternal mood. *Pediatrics* 142(6):e20174330, 2018. DOI:10.1542/peds.2017-4330.

22. Perkin MR and others: Association of early introduction of solids with infant sleep: A secondary analysis of a randomized controlled trial. *Journal of the American Medical Association Pediatrics* 172(8):e180739, 2018. DOI:10.1001/jamapediatrics.2018.0739.

23. Nevarez MD and others: Associations of early life risk factors with infant sleep duration. *Academic Pediatrics* 10:187, 2010. DOI:10.1016/j.acap.2010.01.007.

24. Abrams EM: Introducing solid foods: Age of introduction and its effect on risk of food allergy and other atopic diseases. *Canadian Family Physician* 59:721, 2013.

25. Togias A and others: Addendum guidelines for the prevention of peanut allergy in the United States: Report of the National Institute of Allergy and Infectious Disease–sponsored expert panel. *Annals of Allergy, Asthma & Immunology* 118:166, 2017. DOI:10.1016/j.anai.2016.10.004.

26. American Heart Association: Added sugars and cardiovascular disease risk in children: A scientific statement from the American Heart Association. *Circulation* 134:1, 2016. DOI:10.1161/CIR.0000000000000439.

27. Williams Erickson L and others: Impact of a modified version of baby-led weaning on infant food and nutrient intakes: The BLISS randomized controlled trial. *Nutrients* 10:740, 2018. DOI:10.3390/nu10060740.

28. Taylor RW and others: Effect of a baby-led approach to complementary feeding on infant growth and overweight: A randomized clinical trial. *Journal of the American Medical Association Pediatrics* 171:838, 2017. DOI:10.1001/jamapediatrics.2017.1284.

29. Moag-Stahlberg A: The state of family nutrition and physical activity: Are we making progress? Report of the American Dietetic Association and American Dietetic Association Foundation, 2011. Available at http://eatrightfoundation.org/wp-content/uploads/2016/10/fnpa-report_2011.pdf. Accessed February 26, 2019.

30. Gupta PM and others: Iron, anemia, and iron deficiency anemia among young children in the United States. *Nutrients* 30:E330, 2016. DOI:10.3390/nu8060330.

31. Bailey RL and others: Estimation of total usual calcium and vitamin D intakes in the United States. *Journal of Nutrition* 140:817, 2010. DOI:10.3945/jn. 109.118539.

32. Quader ZS and others: Sodium intake among US school-aged children: National Health and Nutrition Examination Survey, 2011–2012. *Journal of the Academy of Nutrition and Dietetics* 117:39, 2017. DOI:10.1016/j.jand.2016.09.010.

33. Academy of Nutrition and Dietetics: Position of the Academy of Nutrition and Dietetics: Nutrition guidance for healthy children ages 2 to 11 years. *Journal of the Academy of Nutrition and Dietetics* 114:1257, 2014. DOI:10.1016/j.jand.2014.06.001.

34. Norris L, Spettigue WJ, and Katzman DK: Update on eating disorders: Current perspectives on avoidant/restrictive food intake disorder in children and youth. *Neuropsychiatric Disease and Treatment* 12:213, 2016. DOI:10.2147/NDT.S82538.

35. Satter E: Ellyn Satter's division of responsibility in feeding. 2016. Available at https://www.ellynsatterinstitute.org/wp-content/uploads/2016/11/handout-dor-tasks-cap-2016.pdf. Accessed February 11, 2020.

36. Palmer S: Snacking in young children. *Today's Dietitian* 18:8, 2016.

37. Ellis E: Does my child need a supplement? March 18, 2018. Available at https://www.eatright.org/food/vitamins-and-supplements/dietary-supplements/does-my-child-need-a-supplement. Accessed February 25, 2019.

38. United States Environmental Protection Agency: Lead poisoning and your children. December 2017. Available at https://www.epa.gov/sites/production/files/2018-02/documents/epa_lead_brochure-posterlayout_508.pdf. Accessed February 25, 2019.

39. National Institutes of Health National Institute of Diabetes and Digestive and Kidney Diseases: Constipation in children. 2018. Available at https://www.niddk.nih.gov/health-information/digestive-diseases/constipation-children. Accessed February 11, 2020.

40. Tabbers MM and others: Evaluation and treatment of functional constipation in infants and children: Evidence-based recommendations from ESPGHAN and NASPGHAN. *Journal of Pediatric Gastroenterology and Nutrition* 58:258, 2014. DOI:10.1097/MPG.0000000000000266.

41. Academy of Nutrition and Dietetics: Position of the Academy of Nutrition and Dietetics: Vegetarian diets. *Journal of the Academy of Nutrition and Dietetics* 116:1970, 2016. DOI:10.1016/j.jand.2016.09.025.

42. Sharon G and others: Human gut microbiota from autism spectrum disorder promote behavioral symptoms in mice. *Cell* 177:1600, 2019. DOI:10.1016/j.cell.2019.05.004.

43. Privett D: Autism spectrum disorder: Research suggests good nutrition may manage symptoms. *Today's Dietitian* 15:46, 2013.

44. Stewart PA and others: Dietary supplementation in children with autism spectrum disorders: Common, insufficient, and excessive. *Journal of the Academy of Nutrition and Dietetics* 115:1237, 2015. DOI:10.1016/j.jand.2015.03.026.

45. Piwowarczak A and others: Gluten-and casein-free diet and autism spectrum disorders in children: A systematic review. *European Journal of Nutrition* 27:433, 2018. DOI:10-1007/s00394-017-1483-2.

46. Sathe N and others: Nutritional and dietary interventions for autism spectrum disorder: A systematic review. *Pediatrics* 139:e20170346, 2017. DOI:10.1542/peds.2017-0346.

47. Kim SA and others: Vital signs: Fruit and vegetable intake among children—United States, 2003–2010. *Morbidity and Mortality Weekly Report* 63:671, 2014.

48. Albertson AM and others: Whole grain consumption trends and associations with body weight measures in the United States: Results from the cross sectional National Health and Nutrition Examination Survey 2001–2012. *Nutrition Journal* 15:8, 2016. DOI:10.1186/s12937-016-0126-4.

49. O'Neil CE and others: Food sources of energy and nutrients of public health concern and nutrients to limit with a focus on milk and other dairy foods in children 2 to 18 years of age: National Health and Nutrition Examination Survey, 2011–2014. *Nutrients* 10:1050, 2018. DOI:10.3390/nu10081050.

50. Grossman DC and others: Screening for obesity in children and adolescents: U.S. Preventive Services Task Force recommendation statement. *Journal of the American Medical Association* 317:2417, 2017. DOI:10.1001/jama.2017.6803.

51. Council on Communications and Media: Media use in school-aged children and adolescents. *Pediatrics* 138:e20162592, 2016. DOI:10.1542/peds.2016-2592.

52. Schaeffer J: Family-based weight loss. *Today's Dietitian* 16:26, 2014.

53. World Health Organization: *Report of the Commission on Ending Childhood Obesity*. Geneva: WHO Document Production Services, 2016.

54. Centers for Disease Control and Prevention: High cholesterol facts. February 6, 2019. Available at https://www.cdc.gov/cholesterol/facts.htm. Accessed February 25, 2019.

55. Bright Futures/American Academy of Pediatrics: 2019 recommendations for preventive pediatric health care. 2019. Available at https://www.aap.org/en-us/Documents/periodicity_schedule.pdf. Accessed February 26, 2019.

56. Kavey RW and others: Expert panel on integrated guidelines for cardiovascular health and risk reduction in children and adolescents: Summary report. *Pediatrics* 128:S213, 2011. DOI:10.1542/peds.2009-2107C.

57. American Heart Association: Added sugars and cardiovascular disease risk in children: A scientific statement from the American Heart Association. *Circulation* 134:1, 2016. DOI:10.1161/CIR.0000000000000439.

58. Copeland KC and others: Management of newly diagnosed type 2 diabetes mellitus (T2DM) in children and adolescents. *Pediatrics* 131:364, 2013. DOI:10.1542/peds.2012-3494.

59. Gibney MJ and others: Breakfast in human nutrition: The international breakfast research initiative. *Nutrients* 10:559, 2018. DOI:10.3390/nu10050559.

60. Sheppard KW and Cheatham CL: Omega-6/omega-3 fatty acid intake of children and older adults in the U.S.: Dietary intake in comparison to current dietary recommendations and the Healthy Eating Index. *Lipids in Health and Disease* 17:43, 2018. DOI:10.1186/s12944-018-0693-9.

61. Briefel RR and others: Reducing calories and added sugars by improving children's beverage choices. *Journal of the Academy of Nutrition and Dietetics* 113:269, 2013. DOI:10.1016/j.jand.2012.10.016.

62. Scharf RJ and DeBoer MD: Sugar-sweetened beverages and children's health. *Annual Review of Public Health* 37:273, 2016. DOI:10.1146/annurev-publhealth-032315-021528.

63. Auerbach BJ and others: Fruit juice and change in BMI: A meta-analysis. *Pediatrics* 139:e20162454, 2017. DOI:10.1542/peds.2016-2454.

64. Heyman MB and others: Fruit juice in infants, children, and adolescents: Current recommendations. *Pediatrics* 139(6):e20170967, 2017. DOI:10.1542/peds.2017-0967.

65. Academy of Nutrition and Dietetics: Position of the Academy of Nutrition and Dietetics, Society for Nutrition Education and Behavior, and School Nutrition Association: Comprehensive nutrition programs and services in schools. *Journal of the Academy of Nutrition and Dietetics* 118:913, 2018. DOI:10.1016/j.jand.2018.03.005.

66. Cullen KW and Dave JM: The new federal school nutrition standards and meal patterns: Early evidence examining the influence on student dietary behavior and the school food environment. *Journal of the Academy of Nutrition and Dietetics* 117:185, 2017. DOI:10.1016/j.jand.2016.10.031.

67. Mozer L and others: School lunch entrees before and after implementation of the Healthy, Hunger-Free Kids Act of 2010. *Journal of the Academy of Nutrition and Dietetics* 119(3):490, 2019. DOI:10.1016/j.jand.2018.09.009.

68. Dunn CG and others: Feeding low-income children during the COVID-19 pandemic. *The New England Journal of Medicine* 382(18):e40, 2020. DOI:10.1056/NEJMp2005638.

69. Perera T and others: Improving nutrition education in U.S. elementary schools: Challenges and opportunities. *Journal of Education and Practice* 6:41, 2015.

70. Das JK and others: Nutrition in adolescents: Physiology, metabolism, and nutritional needs. *Annals of the New York Academy of Sciences* 1393:21, 2017. DOI:10.1111/nyas.13330.

71. Centers for Disease Control and Prevention: Childhood obesity facts. August 13, 2018. Available at https://www.cdc.gov/obesity/data/childhood.html. Accessed February 26, 2019.

72. Looker AC and others: Prevalence of iron deficiency in the United States. *Journal of the American Medical Association* 277:973, 1997. DOI:10.1001/jama.1997.03540360041028.

73. Powell LM and Nguyen BT: Fast-food and full-service restaurant consumption among children and adolescents: Effect on energy, beverage, and nutrient intake. *Journal of the American Medical Association Pediatrics* 167:14, 2013. DOI:10.1001/jamapediatrics.2013.417.

74. Ruiz LD and Scherr RE: Risk of energy drink consumption to adolescent health. *American Journal of Lifestyle Medicine* 13:22, 2018. DOI:10.1177/1559827618803069.

75. Bardone-Cone AM and others: The inter-relationships between vegetarianism and eating disorders among females. *Journal of the Academy of Nutrition and Dietetics* 112:1247, 2012. DOI:10.1016/j.jand.2012.05.007.

76. Substance Abuse and Mental Health Services Administration. 2018. Underage drinking: Myths versus facts. Available at https://store.samhsa.gov/system/files/sma18-4299.pdf. Accessed February 26, 2019.

77. Gupta RS and others: The public health impact of parent-reported childhood food allergies in the United States. *Pediatrics* 142;e20181235, 2018. DOI:10.1542/peds.2018-1235.

78. National Institute of Allergy and Infectious Disease: Guidelines for the diagnosis and management of food allergy in the United States: Summary of the NIAID-sponsored expert panel report. *Nutrition* 27:253, 2011. DOI:10.1016/j.nut.2010.12.001.

79. Orenstein BW: Pediatric food allergies. *Today's Dietitian* 16:12, 2014.

80. Academy of Nutrition and Dietetics: Practice paper of the Academy of Nutrition and Dietetics: Role of the registered dietitian nutritionist in the diagnosis and management of food allergies. *Journal of the Academy of Nutrition and Dietetics* 116:1621, 2016. DOI:10.1016/j.jand.2016.07.018.

81. Coleman-Collins S: Food allergies/sensitivities: Will food allergies soon be eliminated? *Today's Dietitian* 21(11):12, 2019.

82. Greer FR and others: The effects of early nutritional interventions on the development of atopic disease in infants and children: The role of maternal dietary restriction, breastfeeding, hydrolyzed formulas, and timing of introduction of allergenic complementary foods. *Pediatrics* 143:e20190281, 2019. DOI:10.1542/peds.2019-0281.

# Chapter
## 16

# Nutrition During Adulthood

# Student Learning Outcomes

**Chapter 16 is designed to allow you to:**

**16.1** Discuss demographic trends among adults in North America and how they impact health care.

**16.2** List several hypotheses about the causes of aging.

**16.3** Describe how physiological changes of aging affect the nutritional status of adults.

**16.4** Compare the dietary patterns of adults with current recommendations.

**16.5** Identify nutrition-related health conditions common in the adult years, and describe the prevention and treatment options.

**16.6** List several nutritional programs available to help meet the nutritional needs of older adults.

**16.7** Summarize the role of nutrition in brain health and understand how dietary patterns and other modifiable behaviors may influence your risk of developing migraines, depression, and neurodegenerative diseases.

## FAKE NEWS

# Longevity is mostly in your genes.

## THE FACTS

Longevity—living a long life—*is* partly genetic. Researchers estimate that about 25% of the variation in life span between individuals is inherited from our parents.[1] There is not just one gene that predicts a long and healthy life. So far, a few genes have been identified, but there are likely many genes involved in regulating how long we can live, and we have no control over our DNA.

However, the good news is we *can* exert some control over the environmental and lifestyle factors that determine how long we will live. Exposures to environmental toxins, our physical activity level, and—you guessed it—our *food choices* play significant roles in not only the amount of years in our lives, but also the amount of life in our years!

Eating is one of our great pleasures. Guided by common sense, nutrition knowledge, and moderation, eating well is also a pathway to good health. Most of us want a long, productive life, free of illness. Unfortunately, many adults suffer from nutrition-related diseases such as cardiovascular disease, hypertension, stroke, type 2 diabetes, osteoporosis, and cancer. We can often prevent—or at least slow the progression of—these diseases by following a dietary pattern such as that exemplified by MyPlate or the Mediterranean diet. The cumulative effect of such a dietary pattern is most effective if we begin early and continue throughout adulthood.

Keep in mind that today's behaviors can significantly influence your health years from now. Many of the health problems that occur with age are not inevitable; they result from nutrition-related disease processes that influence physical and mental health. Much can be learned from healthy older people whose attention to healthy dietary and activity patterns—along with a little help from genetics—keeps them active and vibrant well beyond the average life span. Healthy aging is the goal.

# 16.1 Healthy Aging

Due to advances in health care and sanitation, the demographics of developed countries are shifting so that, as a population, we are getting older. In North America, the group constituting those aged 85+ years is the fastest-growing segment of the population. By 2040, the population aged 85+ years in the United States is expected to double its current size (Fig. 16-1).[2] Even more amazing is that 6 million or more people in the United States could be over 100 years old in 2050.

Although great news, the *graying* of North America poses some unique challenges. Individuals older than age 65 make up less than 15% of the U.S. population, but account for more than 35% of all prescription medications used, 35% of acute hospital stays, and almost 35% of the federal health budget.[3] Among older adults, 80% or more have chronic conditions, such as cardiovascular diseases, type 2 diabetes, hypertension, cancers, and osteoporosis.[4]

Preventing or postponing the onset of these chronic diseases for as long as possible can help control health care costs and improve quality of life. Health and independence contribute quality—not just quantity—to life and lessen the load on an already overburdened health care system. Keep in mind that aging is not a disease. Furthermore, diseases that commonly accompany old age—osteoporosis, cancers, and atherosclerosis, for example—are not an inevitable part of aging. Many can be prevented or managed by adhering to positive lifestyle behaviors (Fig. 16-2).

## CAUSES OF AGING

Adulthood, the longest stage of the normal life cycle, begins when an adolescent completes his or her physical growth. Unlike earlier stages of the life cycle, nutrients are used primarily to maintain the body rather than support physical growth. Recall that pregnancy is the only time during adulthood when substantial amounts of nutrients are used for growth. As adults get older, nutrient needs change. For example, vitamin D needs increase for older adults.

**aging** Time-dependent physical and physiological changes in body structure and function that occur normally and progressively throughout adulthood as humans mature and become older.

**Aging** can be defined as the physical and physiological changes in body structure and function that occur throughout adulthood as humans mature and become older. When we are young, aging is not apparent because the major metabolic activities are geared

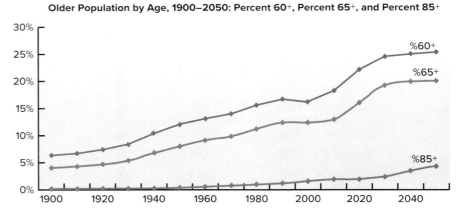

**Older Population by Age, 1900–2050: Percent 60+, Percent 65+, and Percent 85+**

**FIGURE 16-1** ▲ Growth of the U.S. population of older adults. This chart shows that the proportion of the total U.S. population composed of older adults has been steadily increasing over the past century and how these trends are expected to continue. The 85+ demographic group (green line), although still the smallest group in total numbers, is expected to experience the most rapid rate of growth. Conversely, most demographic groups under the age of 45 are shrinking as a percentage of overall population (not shown here). This means that fewer young people will be available to care for a growing population of older adults in years to come.

Source: U.S. Administration on Aging.

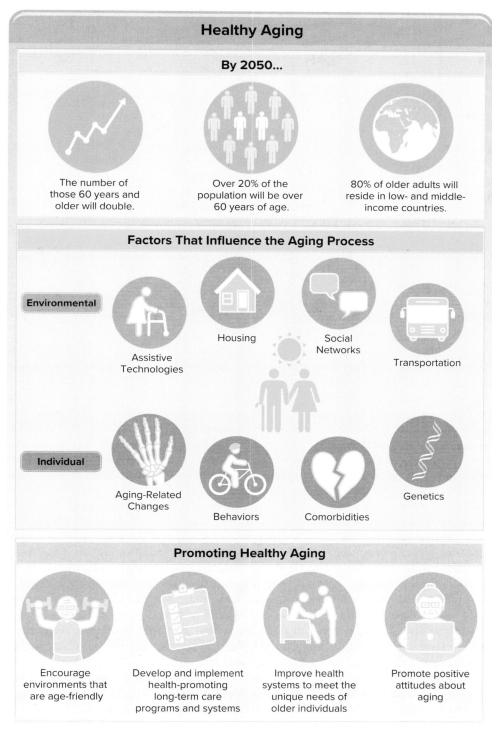

## Healthy Aging

### By 2050...

The number of those 60 years and older will double.

Over 20% of the population will be over 60 years of age.

80% of older adults will reside in low- and middle-income countries.

### Factors That Influence the Aging Process

**Environmental**

Assistive Technologies · Housing · Social Networks · Transportation

**Individual**

Aging-Related Changes · Behaviors · Comorbidities · Genetics

### Promoting Healthy Aging

Encourage environments that are age-friendly

Develop and implement health-promoting long-term care programs and systems

Improve health systems to meet the unique needs of older individuals

Promote positive attitudes about aging

**FIGURE 16-2** ▲ Healthy aging includes adopting a lifestyle with environments and opportunities that enable individuals to engage in meaningful and productive lives. This infographic emphasizes individual and environmental influences on aging and provides recommendations for healthy aging. Source: Adapted from World Health Organization.

The first baby boomers turned 70 in 2016, drawing increased attention to the health concerns of this aging population. **What are the main health concerns for baby boomers?** XiXinXing/Getty Images

toward growth and maturation. We produce plenty of active cells to meet physiological needs. During late adolescence and adulthood, the body's major task is to maintain cellular function. From the beginning of adulthood until age 30 or so, the body operates at peak performance: stature, stamina, strength, endurance, efficiency, and health. During this time, rates of cell synthesis are balanced with rates of cell breakdown in most tissues.

▲ The World Health Organization (WHO) defines *healthy aging* as the process of developing and maintaining the functional ability that enables well-being in older age. Healthy aging enables individuals to be and do what they value, including a person's ability to:

- Meet their basic needs
- Learn, grow, and make decisions
- Be mobile
- Build and maintain relationships
- Contribute to society

adamkaz/Getty Images

**reserve capacity** The extent to which an organ can preserve essentially normal function despite decreasing cell number or cell activity.

**nephrons** The functional units of kidney cells that filter wastes from the bloodstream and deposit them into the urine.

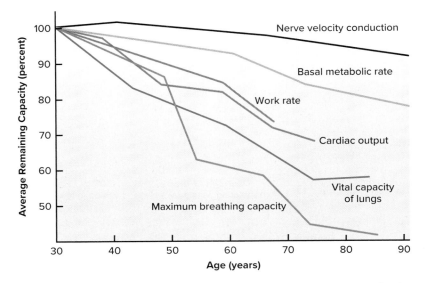

**FIGURE 16-3** ▲ Declines in physiological function often seen with aging. The decline in many body functions is especially evident in sedentary people.

Source: Shock NW: Some physiological aspects of aging in man. *Bulletin of the New York Academy of Medicine* 32:268, 1956.

Inevitably, though, cells begin to age and die. After ages 30 to 40, the rate of cell breakdown slowly begins to exceed the rate of cell renewal, leading to a gradual decline in organ size and efficiency.[5] Eventually, the body cannot adjust to meet all physiological demands, and body functioning begins to decline (Fig. 16-3). Still, body systems and organs usually retain enough **reserve capacity** to handle normal, everyday demands throughout one's life span. Problems caused by diminished capacity typically do not arise unless severe demands are placed on the aging body. For example, alcohol intake can overtax an aging liver. The stress of shoveling a snow-covered sidewalk can exceed the capacity of the heart and lungs. Coping with a serious illness can push an aged body beyond its capacity.

The specific mechanisms of aging remain a mystery. Most likely, the physiological changes of aging are the sum total of automatic cellular changes, lifestyle patterns, and environmental exposures, as listed in Table 16-1. Even with the most supportive environment and healthy lifestyle, cell structure and function decline over time. In some scenarios, programmed cell death is advantageous. For example, the eventual death of deteriorating cells can be beneficial in preventing diseases such as cancer. Unfortunately, as more cells in an organ die, organ function declines. For example, the number of **nephrons** in the kidneys declines as we age. In some people, this loss exhausts the kidneys' reserve capacity and ultimately leads to kidney dysfunction or disease.

Chronic diseases and degenerative processes have long been assumed to be unavoidable consequences of aging. Certainly, some of the declines are inevitable, such as gradual reductions in cellular function, graying hair, and reduced lung capacity. However, many age-related changes can, in fact, be minimized, prevented, and/or reversed by healthy lifestyle choices (e.g., maintaining a nutritious dietary pattern, engaging in regular physical activity, and getting adequate sleep) and avoiding adverse environmental factors (e.g., excessive exposure to sunlight and tobacco). These discoveries have led researchers to introduce the concepts of *usual aging* and *successful aging*.

## USUAL AND SUCCESSFUL AGING

All body cells eventually age. However, to a considerable extent, you can control how quickly you age throughout your adult years. *Usual aging* refers to those changes commonly thought to be a typical or expected part of aging, such as increasing body fatness, decreasing lean body mass, increasing blood pressure, declining bone mass, and the progression of chronic diseases. Researchers point out that many of these changes represent an acceleration of the aging process induced by modifiable lifestyle choices, adverse

**TABLE 16-1 ▪ Causes of Aging**

**Genetic mutations in DNA**
Copy errors during the cell cycle can lead to mutations that cause protein dysfunction, resulting in cell death.

**Oxidative free radicals damaging cellular components**
Electron-seeking free radicals can break down cell membranes and proteins.

**Hormonal changes**
Concentrations of many hormones, such as testosterone in men and estrogen in women, decline during the aging process.

**Glycosylation of proteins**
Chronically elevated blood glucose causes dysfunction and promotes an immune response. Parallel protein strands, found mostly in connective tissue, cross-link to each other, resulting in reduced flexibility in key body components.

**Immune system inefficiency**
With advancing age, the immune system is less able to recognize and counteract foreign substances, such as viruses, that enter the body.

**Autoimmune malfunctions**
The immune system malfunctions and begins to attack healthy cells, resulting in inflammation and cell damage or death.

**Programmed cell death**
Each human cell can divide about 50 times before it automatically succumbs.

**Excess calorie intake promoting more rapid aging**
Excess calorie intake leads to increased fat mass, hormonal alterations, and increased oxidative stress. Reducing calorie intake by about 30% has been shown to decrease markers of aging (in humans) and extend life span (in animals). Calorie restriction is currently the only known method to substantially slow the aging process.[6]

▲ Adopting eating patterns and lifestyle behaviors that minimize a decline in body function is an investment in your future health. Visit www.nia.nih.gov for many free resources on healthy aging.

**glycosylation** The process by which glucose attaches to other compounds, such as proteins.

environmental exposures, and/or chronic disease. For instance, blood pressure does not tend to rise with age among people who maintain a healthy body weight, engage in regular physical activity, and consume a primarily plant-based dietary pattern that is low in sodium. Lean body mass is maintained at much higher rates in older people who regularily engage in physical activity than in those who do not.

*Successful aging,* on the other hand, describes physiological declines that occur only because one grows older, not because lifestyle choices, environmental exposures, and chronic disease have accelerated the aging process and assaulted body tissues. Those who are successful agers experience age-related declines at a slower rate and the onset of chronic disease symptoms and disabilities at a later age than usual agers. Disability dramatically reduces quality of life for patients as well as their caregivers. Striving to have the greatest number of healthy years and the fewest disabled years is referred to as **compression of morbidity.** In other words, a person tries to delay the onset of disabilities and discomforts caused by chronic disease as long as possible and to compress, or reduce, age-related illness into the last few years—or months—of life.

**compression of morbidity** Delay of the onset of disabilities caused by chronic disease.

## FACTORS AFFECTING THE RATE OF AGING

**Life span** refers to the maximum number of years a human can live. The longest human life documented to date is 122 years for a woman and 116 years for a man. **Life expectancy,** alternatively, is the time an average person born in a specific year can expect to live. Life expectancy in North America is about 76.2 years for men and about 81.2 years for women, with a span of *healthy years* of about 64.[7]

The rate at which one ages is determined by genetics, lifestyle, and environment. With the exception of genetics, most of the factors that influence the rate of aging are directly linked to lifestyle behaviors that are largely under our control.

**life span** The potential oldest age a person can reach.

**life expectancy** The average length of life for a given group of people born in a specific year.

**thrifty metabolism** A genetic tendency toward efficient use of energy that results in below-average energy requirements and increased storage of calories as fat.

A plant-forward dietary pattern, exemplified by this bowl of chickpea salad with avocado and vegetables, is a common feature among populations who enjoy long, healthy, and productive lives. **What other lifestyle behaviors are associated with healthy aging?** Mizina/Getty Images

Besides having other long-lived family members, *centenarians*, or people who live to 100 years, generally:

- Do not use tobacco products
- Drink alcohol in moderation, if at all
- Maintain a healthy weight
- Eat many fruits and vegetables
- Perform daily physical activity
- Challenge their minds
- Maintain a positive outlook
- Have close friendships and social networks
- Are or were married (especially men)

**Genetics.** Living to an old age tends to run in families. If your parents and grandparents lived a long life, you are more likely to live to an old age, too. Studies of twins indicate that about 25% of longevity can be attributed to genetics.[1] For humans, as well as most other species, females tend to live longer than males. Another genetic characteristic that may influence longevity is metabolic efficiency. Some researchers hypothesize that individuals with a **thrifty metabolism** require fewer calories for metabolic processes and are able to store body fat more easily than those with faster metabolic rates. Throughout history, it was the individuals with thrifty metabolism who tended to live the longest because they efficiently stored fat during times of plenty and thus had the energy stores needed to survive frequent periods of food scarcity. In today's environment of labor-saving devices and abundant, energy-dense foods, however, a thrifty metabolism may actually reduce longevity. Accumulation of excessive body fat increases the risk of developing health problems (e.g., heart disease, hypertension, and many cancers) that reduce life expectancy.

Is DNA your destiny? Although genetics remain largely unchangeable, let us examine how you can enhance your lifestyle and environment to promote healthier aging. See *Newsworthy Nutrition* in this section for more support that lifestyle matters!

**Lifestyle.** Lifestyle includes one's pattern of living; it includes food choices, physical activity, and substance use (e.g., alcohol, drugs, and tobacco). Lifestyle choices can have a major impact on health and longevity, partly by regulating gene expression. If individuals have a family history of premature heart disease, they would be wise to adjust their dietary and activity patterns and to avoid tobacco products to prevent the onset or slow the progression of the disease. The converse is true, too—that is, lifestyle choices (e.g., high saturated-fat dietary pattern and sedentary lifestyle) can increase susceptibility to non-communicable diseases that hasten the rate of aging, ultimately shortening life expectancy, even if a person does not have a genetic predisposition to disease.

In an effort to unlock the secrets to living a long and healthy life, researchers have been very interested in studying lifestyle patterns of communities in which life expectancy is higher than average. Several communities in which people often live to see their 90th and 100th birthdays include Okinawa Island, Japan; some parts of the Mediterranean region; and areas of California that are home to members of the Seventh-day Adventist religious denomination.

The Okinawan dietary pattern, for example, is characterized by high consumption of vegetables, plant sources of protein (mostly soy), and healthy fats; moderate consumption of fish (mostly coastal) and alcohol; and low consumption of meats, dairy products, and sodium. The low energy density translates into a generally low calorie intake, and the average BMI of Okinawan adults is under 25.

Likewise, followers of the traditional Mediterranean diet also enjoy some of the lowest recorded rates of chronic disease in the world. As mentioned many times throughout this text, the Mediterranean diet features abundant daily intake of fruits, vegetables, whole grains, beans, nuts, and seeds. Olive oil, a source of heart-healthy monounsaturated fat, is the main dietary fat. Beans and fish are emphasized as sources of protein, whereas dairy products, eggs, poultry, and meats are consumed less frequently. Daily physical activity is a way of life. In addition, many Mediterraneans consume wine in moderation at mealtimes.

Note the similarities in lifestyle behaviors of healthy agers. All tend to focus on unprocessed, fiber-rich foods, healthy sources of fat (e.g., vegetable oils and fish), and plant-based or lean sources of protein. Besides dietary patterns, physical activity is a major component of their daily routines. See the sidebar for a summary of lifestyle behaviors that are associated with longevity.

**Environment.** Income, education, health care, shelter, and other socioecological factors exert a powerful influence on the rate of aging. For instance, being able to access and purchase nutritious foods, obtain optimal health care, and reside in safe housing all decrease the rate of aging. Having the education to earn sufficient living wages, as well as the knowledge to make wise lifestyle choices, also can slow the aging process. In addition, the

# Newsworthy Nutrition

## Impact of dietary patterns on biomarkers of aging

**INTRODUCTION:** Telomeres are sections of DNA at the ends of chromosomes (i.e., genetic material) that protect the genetic code. Each time a cell divides, some DNA is lost and the telomeres shorten. Exposure to excessive free radicals may also shorten telomeres. When telomeres have been reduced to a critical length, cell division stops and the cell will die. In research studies, cellular telomere length (TL) is a biomarker of aging, with shorter TL being associated with a shorter life span. Assessing the impact of dietary patterns and specific food groups on TL contributes additional insights into the effect of dietary patterns on health and longevity. **OBJECTIVE:** To evaluate the relationship between dietary patterns, food groups, and TL in human cohorts. **METHODS:** This *systematic review* searched open access databases (PubMed, Science Direct, The Cochrane Library, and Google Scholar) for relevant studies through late 2015. **RESULTS:** A Mediterranean dietary pattern was related to longer TL in three studies. A beneficial effect on TL length was noted in five studies with higher fruits or vegetables intake. In seven studies, an inverse relationship was reported between TL and intake of cereals, processed meat, and fats and oils. **CONCLUSION:** This systematic review supports the health benefits of adherence to the Mediterranean diet on TL. Food categories such as processed meat, cereals, and sugar-sweetened beverages may shorten TL length.

Source: Rafie N and others: Dietary patterns, food groups and telomere length: A systematic review of current studies. *European Journal of Clinical Nutrition* 71(2):151–158, 2017.

ability and willingness to seek health care promptly when it is needed, the health literacy to understand a health care provider, and the ability to accept responsibility for one's own health can slow the rate of aging. Likewise, safe shelter that protects individuals from physical danger, environmental toxins, climate extremes, and sun exposure slows the aging process. Allowing people to make at least some decisions for themselves and control their own activities (autonomy), as well as providing psychosocial support (informational and emotional resources), promote successful aging and psychological well-being. In contrast, aging is likely to accelerate if any or all of the converse are true—that is, insufficient income, low education level, lack of health care, inadequate shelter, and/or lack of autonomy and psychosocial support. Do you think you are aging in a healthy and successful manner? Visit the *Rate Your Plate: Am I Aging Healthfully?* activity in Connect to find out!

## ✔ CONCEPT CHECK 16.1

1. Describe three causes of aging.
2. What is the difference between *usual* and *successful* aging?
3. Provide one example each to describe how genetics, lifestyle, and environment influence aging.

# 16.2 Nutrient Needs During Adulthood

The challenge of the adult years is to maintain the body, preserve optimal function, and avoid chronic disease—that is, to age successfully. A healthy dietary pattern can help achieve this goal. One blueprint for a healthy eating pattern comes from the *Dietary Guidelines*. The advice from those guidelines includes the following goals:

- Follow a healthy eating pattern across the life span with special attention to potassium, calcium, vitamin D, vitamin B-12, and dietary fiber.
- Enjoy nutrient-dense foods and beverages that reflect personal preferences, cultural traditions, and budgetary considerations.
- Focus on meeting food group needs with nutrient-dense foods and beverages while staying within calorie limits.
- Limit foods and beverages higher in added sugars, saturated fat, and sodium, and limit alcoholic beverages.

Meeting one's nutrient needs delays the onset of certain diseases; improves the management of existing diseases; speeds recovery from acute illnesses; and increases mental, physical, and social well-being.[8] As you will recall, overweight and obesity increase the risk of noncommunicable chronic diseases. Common dietary excesses are calories, saturated fat, sodium, and, for some, alcohol. Yet the dietary patterns of adult women tend to fall short of the recommended amounts of vitamins D and E, folate, magnesium, calcium, zinc, and fiber. The dietary patterns of adult men tend to be low in the same nutrients, except vitamin D, which becomes more problematic after age 50.[9]

People age 65 and older, particularly those in long-term care facilities and hospitals, are at heightened risk for malnutrition.[10] This group is often underweight and shows signs of numerous micronutrient deficiencies (e.g., vitamins B-6, B-12, folate). There are many nutrition screening tools to help pinpoint older adults who are at risk for malnutrition. One tool that has been frequently used in the community setting is the Nutrition Screening Initiative's *DETERMINE Your Nutritional Health* checklist. (Fig. 16-4). Notice the *DETERMINE* mnemonic outlines a variety of factors that can impact nutritional status. Of these factors, two truly stand out as the strongest indicators of risk for malnutrition: *recent involuntary weight loss* and *lack of appetite.* In fact, the Academy of Nutrition and Dietetics recently recommended use of a simplified tool, called the *Malnutrition Screening Tool,* which relies on these two criteria to quickly identify individuals

**FIGURE 16-4** ▶ *DETERMINE Your Nutritional Health* Checklist.

Source: Nutrition Screening Initiative, a project of the American Academy of Family Physicians, Academy of Nutrition and Dietetics and the National Council on Aging, Inc., and funded in part by a grant from Abbott Nutrition.

## DETERMINE Your Nutritional Health

Circle the number of points for each statement that applies. Then compute the total and check it against the nutritional score.

| Possible Problem | Points | Description |
|---|---|---|
| Disease | 2 | The person has a chronic illness or current condition that has changed the kind or amount of food eaten. |
| Eating Poorly | 3 | The person eats fewer than two full meals per day. |
| | 2 | The person eats few fruits, vegetables, or milk products. |
| | 2 | The person drinks three or more servings of beer, liquor, or wine almost every day. |
| Tooth Loss/ Mouth Pain | 2 | The person has tooth or mouth problems that make eating difficult. |
| Economic Hardship | 4 | The person does not have enough money for food. |
| Reduced Social Contact | 1 | The person eats alone most of the time. |
| Multiple Medications | 1 | The person takes three or more different prescription or over-the-counter drugs each day. |
| Involuntary Weight Loss/Gain | 2 | The person has unintentionally lost or gained 10 pounds within the last 6 months. |
| Needs Assistance In Self-Care | 2 | The person cannot always shop, cook, or feed himself or herself. |
| Elder Above Age 80 | - | The person is above the age of 80. |
| **Total** | | |

**Nutritional Score:**

**0–2: Good.** Recheck in 6 months.

**3–5: Marginal.** A local agency on aging has information about nutrition programs for the elderly. The National Association of Area Agencies on Aging can assist in finding help; call (800) 677-1116. Recheck in 6 months.

**6 or more: High risk.** A doctor should review this test and suggest how to improve nutritional health.

at nutritional risk.[11,12] Such screening tools are helpful because they are fast, noninvasive, and require minimal training to administer. Once an individual is identified as "at risk," a referral to an RDN will help to ensure that the individual receives professional, personalized advice to optimize nutritional status.

The DRIs for adults (Appendix G) are organized by sex and age. These changes in nutrient needs take into consideration aging-related physiological alterations in body composition, metabolism, and organ function.

## CALORIES

After the age of 30, total calorie needs of physically inactive adults tend to steadily decline.[13] There are several explanations for the lower calorie requirements of older adults. Basal metabolic rate declines by about 2% per decade after age 30, such that overall energy needs of a 70-year-old man are reduced by 100 to 150 kcal per day compared to those of a 30-year-old man. Losses of lean body mass and decreases in physical activity also tend to accompany aging. To a considerable extent, adults can exert control over this reduction in calorie need by exercising. Physical activity can halt, slow, and even reverse reductions in lean body mass and subsequent declines in energy needs. Being able to consume more calories makes it much easier to meet micronutrient needs without dietary supplements.

## PROTEIN

The protein intake of younger adults in North America typically exceeds the current RDA (0.8 gram per kilogram of body weight) and falls within the recommended range of 10% to 35% of total calories. Among older adults, however, several studies indicate that consuming protein in amounts slightly higher than the RDA (in the range of 1.0 to 1.2 grams per kilogram of body weight) may help preserve muscle and bone mass.[14,15] As with calorie needs, protein requirements should be determined in relation to routine physical activity. Furthermore, evenly distributing protein intake throughout the day (e.g., 25 to 30 grams of high-quality protein at each meal) appears to be best for preserving lean mass for aging adults.[16] Breakfast is a great opportunity for increased high-quality protein intake because many individuals, including older adults, typically start their day with a meal that is low in protein and dominated by carbohydrate.

Inclusion of animal proteins can be helpful because these foods generally contain a higher proportion of the amino acid leucine, which plays a key role in stimulating muscle protein synthesis. Plant-based proteins, such as soy products (tofu, soy milk, and soy yogurt), lentils, beans, nuts, and seeds are excellent choices for vegetarians. Adults who have limited food budgets, have difficulty chewing meat, or are lactose intolerant may not get enough protein, especially animal protein. Keep in mind that any protein consumed in excess of that needed for the maintenance of body tissue will be broken down and used as energy or stored as fat. The waste products of metabolism of protein must be removed by the kidneys; excessive protein intake may accelerate kidney function decline.

## FAT

The typical fat intake of adults of all ages is near the upper end of the 20% to 35% of total calories recommended by the Food and Nutrition Board. Looking specifically at saturated fat, usual intakes are slightly above the 10% of total calories limit suggested by the *Dietary Guidelines*.[18] It is a good idea for almost all adults to reduce their saturated fat intake because of the strong link between high saturated fat and obesity, heart disease, and certain cancers. A few strategies that would help most American adults to align with current recommendations for fat intake would be: (1) choose seafood, skinless poultry, lean meats, beans, peas, or lentils as protein sources for most meals; (2) prepare foods by steaming, grilling, broiling, or sautéing in a small amount of plant oil instead of deep-frying; and (3) choose skim or low-fat dairy products instead of full-fat varieties.

**COVID CORNER**

Older adults, especially those with existing chronic illnesses, are most at risk for COVID-19. Among this population, nutritional status is a crucial consideration. Malnutrition impairs immune function, complicates recovery, and increases mortality from infection. The Academy of Nutrition and Dietetics recommends use of the *Malnutrition Screening Tool* to identify COVID-19 patients at risk for malnutrition. Those who are identified as "at risk" should undergo comprehensive nutritional assessment by an RDN. Patients with malnutrition benefit from small but frequent meals and/or oral nutritional supplements to meet their needs for protein and energy. Beyond correcting specific nutrient deficiencies, there is no evidence to support vitamin or mineral supplementation for prevention or treatment of COVID-19.[17]

## CARBOHYDRATES

Although total carbohydrate intake is adequate, fewer than 1 in 10 Americans follow the advice to *make half your grains whole grains.* Many adults need to shift their carbohydrate choices to emphasize complex carbohydrates and whole grains while minimizing the intake of added sugars and refined grains. Remember that the healthy eating pattern described in the *Dietary Guidelines* limits added sugars to 10% of calories per day and emphasizes the following carbohydrates:

• A variety of vegetables from all of the subgroups: dark green; red and orange; beans, peas, and lentils; starchy vegetables; and other vegetables
• Fruits, especially whole fruits
• Grains, at least half of which are whole grains
• Low-fat dairy, including milk, yogurt, cheese, and/or fortified soy beverages

A dietary pattern rich in complex carbohydrates helps us meet our nutrient needs without excess calories. Replacing sweets and refined grains with fiber-rich whole grains also improves blood glucose control. This is particularly helpful because inactivity and increasing body fatness that often accompany aging are connected to insulin resistance. Carbohydrate metabolism dysfunction is so common that more than 25% of those age 65 years or older have diabetes. Beyond glucose control, a dietary pattern rich in fiber helps adults reduce their risk for heart disease and some forms of cancer. It also minimizes constipation. The typical American adult gets slightly more than half the recommended amount of dietary fiber. Most Americans could benefit from incorporating more whole grains into their dietary patterns.

## WATER

Many adults, especially as they approach their later years, fail to consume adequate quantities of water. In fact, many may be in a constant state of mild dehydration and at risk of electrolyte imbalances. Low fluid intakes in older adults may be caused by blunted thirst mechanisms, chronic diseases, and/or conscious reductions in fluid intake in order to reduce urination. Some may have alterations in fluid output due to certain medications (i.e., diuretics and laxatives), and/or experience an age-related decline in the kidneys' ability to concentrate urine. Dehydration is very dangerous and, among other symptoms, can cause disorientation and mental confusion, constipation, fecal impaction, and even death. Initiatives to improve fluid intake, especially in older adults, should include assessments of barriers to drinking, hydration education, close intake monitoring, frequent prompting, offering a choice of healthy drinks, and addressing continence issues.

## MINERALS AND VITAMINS

Dietary requirements for many nutrients change throughout the adult years (Figs. 16-5 and 16-6). Most adults can get all the nutrients they need from foods. Adults who have impaired nutrient absorption, who have low overall food intake, or who are unable to follow a nutritious dietary pattern may benefit from mineral or vitamin supplements matched with their needs. For example, supplements or fortified foods can be especially helpful when it comes to meeting the RDAs for vitamin D and vitamin B-12. As always, it is best to discuss supplements with a doctor or an RDN. If older adults need to supplement their eating pattern, they should look for a supplement that supplies the needed nutrient(s) without other unnecessary ingredients. Megadoses of nutrients can be harmful.

In general, adults would benefit from lowering their intakes of sodium and consuming more food sources of calcium, vitamin D, iron, zinc, magnesium, vitamin B-6, folate, vitamin B-12, vitamin E, and carotenoids. Let's explore these dietary components in further detail in the next few pages.

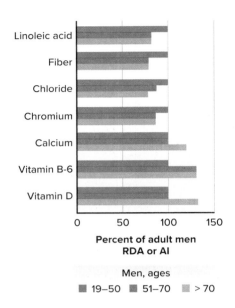

**FIGURE 16-5** ▲ Relative nutrient requirements for aging adult men. Only nutrients that vary by age are shown.

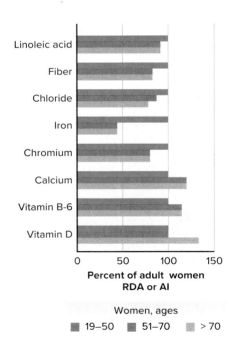

**FIGURE 16-6** ▲ Relative nutrient requirements for aging adult women. Only nutrients that vary by age are shown.

**Calcium and Vitamin D.** These bone-building nutrients tend to be low in the dietary patterns of all adults. They become particularly problematic after age 50. Inadequate intake of these nutrients, combined with their decreased absorption, the reduced synthesis of vitamin D in the skin, and the kidneys' decreased ability to put vitamin D in its active form, greatly contributes to the development of osteoporosis. Getting enough of these nutrients is a challenge for many older adults because food sources of vitamin D are limited and the major sources—fatty fish and fortified milk—are not widely consumed by older adults. Plus, with increasing age, lactase production frequently decreases. As you will recall, one of the richest and most absorbable sources of these nutrients—milk—contains lactose. To get the vitamin D and calcium they need, individuals with lactose intolerance may be able to consume small amounts of milk at mealtime with no ill effects. Calcium-fortified foods, cheese, yogurt, fish eaten with bones (e.g., canned sardines or salmon), and dark-green leafy vegetables can help those with lactose intolerance meet their calcium needs—but these sources often do not provide vitamin D. Just 10 to 15 minutes per day of sunlight can make a large difference in vitamin D status.

**Iron.** Iron deficiency is the most common nutrient deficiency during all stages of the life cycle. Recall that iron deficiency can impair red blood cell synthesis, which leads to weakness, fatigue, shortness of breath, confusion, and disorientation. During their reproductive years, women are at risk for iron deficiency because their dietary patterns do not provide enough iron to compensate for the iron lost monthly during menstruation. Other common causes of iron deficiency in adults of all ages include digestive tract injuries that cause bleeding (i.e., bleeding ulcers or hemorrhoids) and the use of medicines, such as aspirin, that cause blood loss. Impaired iron absorption due to age-related declines in stomach-acid production may also contribute to iron deficiency in older adults. Remember that iron deficiency can be present before any observable signs of anemia. Dietary sources of iron include fortified grains; meat, fish, and poultry; seafood; beans; dark-green leafy vegetables; and peas.

**Zinc.** Not only is dietary zinc intake less than optimal, diminished stomach acid production impairs zinc absorption as adults age. Poor zinc status may contribute to impairments in the sense of taste, mental lethargy, declines in immune function, and delayed wound healing that many older adults experience. Some dietary sources of zinc include oysters, meats, seafood, fortified grains, yogurt, and chickpeas.

**Magnesium.** This mineral tends to be low in the adult eating pattern. Inadequate magnesium intakes may contribute to the bone loss, muscular weakness, and mental

## FARM to FORK    Grapes and Raisins

Steve Hyde/Moment/Getty Images

Light green Thompson seedless grapes are the nation's most popular variety of grapes, outselling others by 1000 to 1. Yet the red, purple, and black varieties have 50% to 75% more phytochemicals than the pale Thompson. Sun-dried raisins, from Thompson grapes, are also the most popular dried fruit in the United States. Both are fantastic sources of nutrients and fiber for all ages.

### Grow
- Home gardeners can successfully grow grapes if they select the correct site and cultivar, and an effective training and trellis system.
- Fertility and pest management programs are also necessary along with pruning the grapevines annually.
- A well-maintained grapevine can produce up to 20 pounds of grapes per year for over 50 years!

### Shop
- As with many fruits, grapes are often harvested well before maximum ripeness to ensure stability during shipping.
- When shopping, look for vine-ripe grapes for maximum flavor and texture. Look for grapes that are firm and plump. Stems should be bright green and flexible, not dry and brittle. To check freshness, give the stem a gentle shake, and the grapes should remain on the vine.

### Store
- Once harvested, cool grapes quickly in the coldest part of the refrigerator to extend shelf life and nutrient content.
- Store grapes in the plastic grape bag from the grocery store.
- Rinse grapes just prior to eating to reduce decay.

### Prep
- The beauty of grapes is that they are a no-prep healthy snack for any time of the day!
- To add variety, try frozen grapes, grapes or raisins in salads, or pair with low-fat cheeses.
- Home drying can be accomplished in a dehydrator or oven or by baking in the sun.

Source: Robinson J. Grapes and raisins: From muscadines to Thompson seedless. In *Eating on the Wild Side*, New York: Little, Brown and Company, 2013.

Alexis Joseph/McGraw-Hill Education

▲ Many seniors exceed sodium recommendations. Reading labels is a good way to begin controlling sodium intake at home. Blend Images/Getty Images

confusion seen in some older adults. It also can lead to sudden death from heart rhythm dysfunction and is linked to the development of cardiovascular disease, osteoporosis, and diabetes. The best source of magnesium comes from food because supplements may cause side effects such as diarrhea. Dietary sources of magnesium include nuts, spinach, soy milk, beans, and peanut butter.

**Sodium.** The declining sense of taste that typically accompanies aging often contributes to a preference for highly salted foods. The *Dietary Guidelines* advise adults to consume less than 2300 milligrams of sodium per day (about 1 teaspoon of table salt). Yet, the average sodium intakes of American adults are over 3500 milligrams per day. A heavy reliance on highly processed foods and restaurant meals is mostly to blame for the high sodium intakes of Americans. The most widely recognized consequence of high sodium intake is hypertension, but high sodium intake has also been linked to osteoporosis secondary to increased calcium excretion in the urine. Excessive dietary sodium may also overtax poorly functioning kidneys of older adults.

Even though excessive sodium receives most of the attention, low blood sodium (*hyponatremia*) is also a concern for older adults. Adults older than age 70, especially those who take diuretic medications or who have poor kidney function, are at increased risk for hyponatremia. The consequences of mild hyponatremia include lightheadedness, confusion, and unsteady gait, which can certainly increase the risk for falls among older adults. Other problems include fatigue, muscle cramps, and lack of appetite. There is no reason to severely restrict sodium among older adults who do not have kidney disease, but lowering sodium intakes closer to the AI would improve overall health for most adults.

**Folate and Vitamins B-6 and B-12.** Sufficient folate, because of its role in prevention of neural tube defects, is very important to women during the childbearing years. In later years, folate and vitamins B-6 and B-12 are especially important because they are required to clear the amino acid homocysteine from the bloodstream. Elevated blood concentrations of homocysteine are associated with the increased risk of cardiovascular disease, stroke, bone fracture, and neurological decline seen in some older people. Vitamin B-12 is a particular problem for the older population because a deficiency may exist even when intake appears to be adequate. As people age, the stomach slows its production of acid and intrinsic factor, which leads to poor absorption of vitamin B-12. If vitamin B-12 is depleted, anemia and nerve damage could result. Adults age 51 years and older often must meet vitamin B-12 needs with supplements or fortified foods because synthetic vitamin B-12 is more readily absorbed than natural forms of B-12.

**Vitamin E.** The dietary intake of most of the population falls short of recommendations for vitamin E. Low vitamin E intake means that the body has a reduced supply of antioxidants, which may increase the degree of cell damage caused by free radicals, promote the progression of chronic diseases and cataracts, and accelerate the aging process. In addition, low vitamin E levels can lead to declines in physical abilities.

**Carotenoids.** Dietary intakes of certain carotenoids have been shown to have a variety of important antiaging, anticancer, and other protective effects. Specifically, lutein and zeaxanthin have been linked with the prevention of cataracts and age-related macular degeneration.[19] Dietary patterns high in fruits and vegetables, the major sources of carotenoids and other beneficial phytochemicals, are consistently shown to be protective against a wide variety of age-related conditions.

## ARE ADULTS FOLLOWING DIETARY RECOMMENDATIONS?

In general, adults in North America are trying to follow many of the dietary recommendations described in this chapter. Since the mid-1950s, they have consumed less saturated fat as more people substitute fat-free and low-fat milk for cream and whole milk. However, they eat more cheese, usually a concentrated form of saturated fat. Since 1963, they have eaten less butter, fewer eggs, less animal fat, and more plant oils and fish. Animal breeders are raising animals leaner than those produced in 1950, which also helps reduce saturated fat intake.

Other aspects of average adult eating patterns are still in need of improvement. The latest nutrition survey data show that the major contributors of calories to the adult dietary pattern are still white bread, beef, doughnuts, cakes and cookies, soft drinks, milk, poultry, cheese, alcoholic beverages, salad dressing, mayonnaise, potatoes, and sugars/syrups/jams. If Americans were truly lowering their intakes of sugar, saturated fat, and sodium while increasing their intakes of fiber, many of these foods would not appear at the top of the list.

### ✓ CONCEPT CHECK 16.2

1. What factors contribute to the declining basal metabolic rate that accompanies the aging process?

2. Identify one nutrient that should be limited in the dietary patterns of most American adults. Suggest three specific dietary changes that would help to reduce intake of this nutrient. (Note: Calories are not nutrients.)

3. Name three nutrients that are commonly lacking in dietary patterns of adults. Suggest one rich food source of each of these nutrients.

## 16.3 Factors Related to Nutritional Status of Adults

Dietary adequacy is influenced by physiological, psychosocial, and economic factors. Figure 16-7 summarizes the nutritional implications of many of the physiological changes that occur during adulthood. Some of the changes listed (e.g., tooth loss and changes in taste and smell) can directly influence dietary intake. Other changes (e.g., loss of lean body tissue) can alter nutrient and/or calorie needs. Some body changes (e.g., reduced stomach acidity, diminished kidney function) can affect nutrient utilization. Furthermore, chronic diseases and the medications used to manage them may influence food intake and nutrient needs. In this section, explore the influences of several of these factors on nutritional status during adulthood.

### BODY COMPOSITION

The primary changes in body composition that occur with aging are diminished lean body mass, increased fat stores, and decreased body water. A focus on physical activity, as detailed below, can attenuate many of these unwanted changes.

The loss of lean body mass is termed **sarcopenia.** Some muscle cells shrink, others are lost as muscles age, and some muscles lose their elasticity as they accumulate fat and collagen. Loss of muscle mass leads to a decrease in basal metabolism, muscle strength, and energy needs. Less muscle mass also leads to lower fitness performance, which makes the prognosis for maintaining muscle even worse. Clearly, it is best to prevent this downward spiral.

As lean tissue declines with age, body fat often increases, a condition called **sarcopenic obesity.**[20] Much of this increase in body fat results from overconsumption of calories

**sarcopenia** Loss of muscle tissue. Among older adults, this loss of lean mass greatly increases their risk of illness and death.

**sarcopenic obesity** Advanced muscle loss accompanied by gains in fat mass.

## Physiological Changes of Aging

**↓ Appetite**
- Monitor changes in weight and report unintentional weight loss to your primary care provider.
- Eat small, frequent, calorie- and nutrient-dense meals throughout the day.
- Incorporate nutrient-rich drinks or smoothies between meals.

**↓ Bone mass**
- Meet nutrient needs, especially protein, calcium, and vitamin D (including regular sun exposure).
- Perform regular physical activity, especially weight-bearing activities.
- Aim to maintain a healthy body weight (i.e., BMI between 18.5 and 24.9).

**↓ Bowel function**
- Emphasize a dietary pattern rich in fruits, vegetables, and whole grains.
- Meet fluid needs and monitor hydration.
- Increase physical activity.

**↓ Cardiovascular function**
- Achieve and maintain a healthy body weight.
- Choose a dietary pattern rich in fruits, vegetables, whole grains, plant-based protein, and fatty fish.
- Stay physically active.

**↓ Chewing or swallowing ability**
- Work with a dentist or trained therapist to maximize chewing and swallowing ability.
- Modify food consistency as necessary.

**↓ Cognitive function**
- Choose a dietary pattern rich in fruits, vegetables, whole grains, and plant sources of protein.
- Consume seafood, a source of omega-3 fatty acids, twice per week.
- Perform regular physical activity.
- Obtain adequate sleep, rest, and relaxation.

**↑ Fat stores**
- Avoid overconsumption of calorie-dense foods and beverages.
- Perform regular physical activity.

**↓ Immune function**
- Meet nutrient needs, especially protein, vitamin A, vitamin C, vitamin E, and zinc.
- Perform regular physical activity.

**↓ Insulin function**
- Maintain a healthy body weight.
- Choose whole grains; limit refined grains and added sugars.
- Perform regular physical activity.

**↓ Kidney function**
- Consume adequate fluids.
- Maintain normal blood pressure and weight.
- Be cautious about using medications and dietary supplements.

**↓ Lactase production**
- Reduce serving sizes of milk.
- Substitute yogurt or cheese for milk.
- Use reduced-lactose or lactose-free products and seek nondairy calcium sources.

**↓ Liver function**
- Consume alcohol in moderation, if at all.
- Avoid consuming dietary supplements that contain more than 100% of the Daily Value of nutrients, especially vitamin A.

**↓ Lung function**
- Avoid tobacco.
- Perform regular physical activity.

**↓ Muscle mass**
- Meet nutrient needs, especially protein.
- Perform regular physical activity, including strength training.

**↓ Sense of taste and smell**
- Vary the colors and textures of foods.
- Experiment with sodium-free herbs and spices.

**↓ Sense of thirst**
- Monitor fluid intake.
- Drink water throughout the day.
- Stay alert for evidence of dehydration (e.g., dark-colored urine).

**↓ Stomach acidity**
- Include lean meats and iron-fortified foods in the dietary pattern.
- Consume vitamin C–rich foods to enhance absorption of nonheme iron from foods.
- Choose foods fortified with vitamin B-12.

**↓ Vision**
- Choose a dietary pattern rich in fruits, vegetables, and whole grains.
- Regularly consume fatty fish, a source of omega-3 fatty acids.
- Wear sunglasses in sunny conditions.
- Avoid tobacco.

**FIGURE 16-7** ▲ Physiological Changes of Aging.
Source: Adapted from NIH, National Institute on Aging.

and inadequate physical activity, although even physically fit men and women typically gain some additional body fatness after age 50. A small fat gain in adulthood may not compromise health, but large gains are problematic. Recall that obesity increases risks for hypertension, cardiovascular disease, type 2 diabetes, osteoarthritis, cancers, and many other serious chronic conditions. Ultimately, these changes in body composition may diminish a person's ability to perform daily tasks, such as getting up from a chair and climbing a flight of stairs.

On the other hand, decreases in body weight can also be a problem for adults age 70 and older. About 2% of older adults are underweight (BMI less than 18.5). Unintended weight loss increases the risk of malnutrition, which alters an individual's ability to cope with illnesses and injuries and could ultimately lead to death. Potential causes for unintended weight loss among older adults (many of which can be detected using the DETERMINE nutrition screening tool in Figure 16-4) include the following:

- Mental or physical illness
- Depression/social isolation
- Side effects of medications
- Changes in taste or smell
- Reduced chewing ability
- Limited financial resources
- Decreased dexterity or strength
- Transportation barriers

## BONES AND JOINTS

Recall that some bone loss is an expected consequence of aging. In women, bone loss rapidly occurs after menopause. For men, bone loss is slow and steady from middle age throughout later life. Many older adults may suffer from undiagnosed osteomalacia, a condition mainly caused by insufficient vitamin D. Osteoporosis can limit the ability of older people to shop, prepare food, and engage in physical activity. Consuming adequate vitamin D, calcium, and protein; not smoking; avoiding excessive alcohol; and engaging in weight-bearing physical activity can help preserve bone mass.

There are over 100 forms of arthritis, a disease that causes the degeneration of the cartilage that covers and cushions the joints. Such changes in the joints cause them to ache and become inflamed and painful to move. Severe arthritis can cause permanent joint changes, and some types of arthritis affect the heart, eyes, lungs, kidneys, and skin in addition to the joints. Over 54 million U.S. adults suffer from some type of arthritis.[21] **Osteoarthritis,** which affects about 30 million U.S. adults, is the leading cause of disability among older persons. **Rheumatoid arthritis,** which affects about 1.5 million U.S. adults, is more prevalent in younger adults. **Gout,** a very painful form of arthritis that comes on suddenly and can be related to changes in eating patterns, affects about 6 million men and 2 million women in the United States. The estimated annual medical costs related to these conditions are over $140 billion.

Although precise causes or cures are unknown, many unproven arthritis remedies have been publicized. Fad diets, food restrictions, and nutrient supplementation are some of the more popular treatments. Maintaining a healthy weight, which reduces stress on painful arthritic joints, is the best strategy to offer relief from arthritis. Although no special diet, food, or nutrient has been proven to reliably prevent, relieve, or cure arthritis, some research shows that a dietary pattern (e.g., Mediterranean diet) that is rich in antioxidant nutrients, anti-inflammatory phytochemicals, and omega-3 fatty acids can help to reduce the inflammation that underlies these conditions.[22] As far as supplements go, results remain controversial. Discuss the pros and cons of supplement use for these conditions with your primary care provider or dietitian.

LWA/Dann Tardif/Blend Images LLC

Researchers believe that maintaining lean muscle mass may be the most important strategy for successful aging because doing so:

- Maintains basal metabolic rate, which helps to decrease the risk of obesity
- Keeps body fat within a healthy range, which helps in the management of blood lipids and blood glucose
- Maintains body water, which decreases the risk of dehydration and improves body temperature regulation
- Promotes healthy bones
- Helps to prevent falls
- Helps a person to preserve independence because functional abilities are maintained over time.

**osteoarthritis** A degenerative joint condition caused by a breakdown of cartilage in joints. It often results from wear and tear due to repetitive motions or the pressure of excess body weight.

**rheumatoid arthritis** A degenerative joint condition resulting from an autoimmune disease that causes inflammation in the joints and other sites of the body.

**gout** A form of arthritis caused by the buildup of uric acid crystals in the joints.

**ASK THE RDN** Anti-Inflammatory Diet

**Dear RDN:** *My mother has been battling osteoarthritis for several years and wonders if Dr. Weil's Anti-Inflammatory Diet can help her cope with her disease. How is Dr. Weil's plan different from the Mediterranean diet?*

When you are sick or injured, chronic inflammation wreaks havoc on the body and is now thought to be at the root of many chronic diseases, including heart disease, Alzheimer's disease, and arthritis.

There is good evidence that an anti-inflammatory eating pattern may reduce inflammation, the risk of chronic diseases, and some symptoms of arthritis. The basic principles of an anti-inflammatory eating plan include the following:

- Eating plenty of fruits, vegetables, and whole grains
- Limiting or avoiding saturated fats
- Including good sources of omega-3 fatty acids
- Choosing lean sources of protein
- Avoiding refined carbohydrates and highly processed foods
- Including herbs and spices with known anti-inflammatory effects

To ease arthritis, the #1 thing you can do is to keep your weight within a healthy range. Although Dr. Weil's plan is not aimed at weight loss, it is not surprising that when individuals follow a primarily plant-based dietary pattern and reduce their intakes of solid fats and highly processed foods, they often drop some pounds.

Carbohydrates make up 40% to 50% of daily calories on Dr. Weil's plan. Whole or cracked grains, beans, and abundant fruits and vegetables are recommended. When it comes to selecting produce, eat a wide variety of colorful fruits and vegetables. Many of the phytochemicals in colorful plant foods have anti-inflammatory effects.

Dr. Weil also emphasizes healthy fats, which should account for about 30% of your daily calories. He recommends seeking out sources of omega-3 fatty acids, which are found in foods such as avocados; nuts and nut butters; fortified eggs; flaxseeds; hemp seeds; and fish such as salmon, sardines, black cod, and herring. Extra-virgin olive oil is the preferred oil for cooking and flavoring foods.

Protein should make up 20% to 30% of your daily calories on Dr. Weil's plan, but he stresses the importance of plant sources of protein. In fact, he recommends limiting most animal proteins to two servings per week. Instead, choose beans and products made from soybeans, along with fish.

If the Anti-Inflammatory Diet sounds a lot like the Mediterranean diet you read about earlier, that's because Dr. Weil based his plan on this healthy dietary pattern. He also adds specific advice about anti-inflammatory herbs and spices, such as garlic, ginger, and turmeric, and touts the health benefits of tea and small amounts of dark chocolate.

As RDNs, we fully support this food-based strategy to fight chronic diseases. However, we would be cautious when it comes to purchasing dietary supplements, including fish oil, ginger, turmeric, and several antioxidant compounds (see Table 16-3). There is little evidence that these supplements will provide benefits beyond what whole foods supply.

Wholesome thoughts,

**Colleen Spees, PhD, RDN, LD, FAND**

Associate Professor, Researcher, The Ohio State University, and Author of *Contemporary Nutrition*

Ralphoto Studio

## PHYSICAL ACTIVITY

Many physical changes of aging can be traced back to a sedentary lifestyle. As you might predict, an active lifestyle helps preserve muscle mass and decrease body fat. Physical activity increases muscle strength and mobility, improves balance which decreases the risk of falling, eases daily tasks that require strength, improves sleep, slows bone loss,

**TABLE 16-2** ■ **Physical Activity Guidelines for Older Adults**

| Guidelines for all adults |
| --- |
| Adults should move more and sit less throughout each day. Any physical activity is better than none and results in health benefits. |
| For substantial health benefits, adults should engage in at least 150 minutes (2 hours and 30 minutes) to 300 minutes (5 hours) a week of moderate-intensity, or 75 minutes (1 hour and 15 minutes) to 150 minutes (2 hours and 30 minutes) a week of vigorous-intensity, aerobic physical activity, or an equivalent combination of moderate- and vigorous-intensity aerobic activity. Preferably, aerobic activity should be spread throughout the week. |
| Additional health benefits are gained by engaging in physical activity beyond the equivalent of 300 minutes (5 hours) of moderate-intensity physical activity a week. |
| Adults should also complete muscle-strengthening activities of moderate or greater intensity that involve all major muscle groups on 2 or more days a week, as these activities provide additional health benefits. |

| Guidelines for older adults |
| --- |
| As part of their weekly physical activity, older adults should engage in multicomponent physical activity that includes balance training as well as aerobic and muscle-strengthening activities. |
| Older adults should determine their level of effort for physical activity relative to their level of fitness. |
| Older adults with chronic conditions should understand whether and how their conditions affect their ability to do regular physical activity safely. |
| When older adults cannot do 150 minutes of moderate-intensity aerobic activity a week because of chronic conditions, they should be as physically active as their abilities and conditions allow. |

Source: *Physical Activity Guidelines for Americans*, 2nd ed. Available from https://health.gov/paguidelines/secondedition/pdf/Physical_Activity_Guidelines_2nd_edition.pdf.

and increases joint movement, thus reducing injuries. It also has a positive impact on a person's mental outlook. Ideally, an active lifestyle should be maintained throughout life and include activities to build endurance, strength, balance, and flexibility. The *Physical Activity Guidelines for Americans* provide recommendations specifically for older adults (Table 16-2).

All older adults should avoid inactivity. Having a comprehensive physical activity plan, developed with a health professional to accommodate individual health risks and needs, will enhance success for older adults. Men older than age 40; women older than age 50; those with heart conditions, diabetes, or joint problems; and anyone who has been sedentary should consult a primary care provider before beginning a physical activity program.

**Aerobic Activity.** All adults should engage in either moderate-intensity aerobic physical activity for at least 150 to 300 minutes per week, or vigorous-intensity aerobic activity for 75 to 150 minutes per week, or an equivalent combination of the two. Moderate-intensity activity requires medium effort. Thinking back to the Rating of Perceived Exertion (Section 10.2), moderate activity would be equal to a 4 to 6 (out of 10) and produce increases in breathing rate and heart rate. Vigorous activity might begin at a 7 or 8 (out of 10) and produces larger increases in breathing and heart rate. This amount of aerobic activity improves endurance and aids in prevention of chronic diseases. A longer duration of daily physical activity may be required for weight loss or weight maintenance. Weight-bearing exercises are particularly helpful for preservation of bone mass. For older adults who have not been physically active, it is important to increase the pace gradually. Remember, any amount of physical activity is better than none at all.

**Muscle-Strengthening Activities.** To maintain lean tissue and basal metabolic rate, muscle-strengthening activities should be performed 2 to 3 days per week and involve all major muscle groups (legs, hips, abdomen, chest, back, shoulders, and arms).

To keep active, adults age 50 and older are eligible to participate in local and national Senior Games (https://nsga.com/), including walking events as shown here. **What activities could you begin now that could be enjoyed throughout adulthood?** ©Anne Smith

Specific recommendations for muscle strengthening include performing at least 1 set of 8 to 12 repetitions, although 2 or 3 sets may be more effective. Start slowly, concentrate on breathing and technique, rest between circuits and sets, and stop an activity if it becomes painful.

**Balance Activities.** For those at risk for falling, activities that improve balance are recommended. Specific guidelines include performing balance activities about 3 times per week. Examples of safe activities include heel-to-toe walking, repetitive standing from a sitting position, and using a wobble board. Muscle strengthening of the back, abdomen, and legs also improves balance.

**Flexibility, Warm-Up, and Cool-Down.** Stretching each major muscle group should accompany aerobic or muscle strengthening. Improving flexibility can make it easier to perform many simple tasks, such as tying shoes. All fitness programs should include warm-up and cool-down activities. A safe warm-up before aerobic activities allows a gradual increase in heart rate and breathing. An adequate cool-down after physical activity allows a gradual heart rate decrease at the end of the session.

**Multicomponent Physical Activity.** These types of activities include incorporating more than one type of physical activity into your physical fitness program. Dancing, tai chi, gardening, or sports are considered multicomponent since they often incorporate several different types of physical activity.

## DIGESTIVE SYSTEM

As you will recall from previous chapters, digestion begins in the mouth. Over 25% of older adults have no natural teeth, and many more are missing some teeth.[23] The problem of tooth loss is worse among low-income populations. Even with properly fitting dentures, chewing ability may be limited. Older adults with poor dentition often avoid meats or crunchy fruits and vegetables, thereby missing out on key nutrients such as protein, iron, and zinc (from meat) as well as potassium and fiber (from fruits and vegetables). Ground meats, beans, peas, lentils, and cooked or finely chopped vegetables are easier options for older adults with chewing problems.[24]

Further along the GI tract, the production of hydrochloric acid, intrinsic factor, and some digestive enzymes (e.g., lactase) declines with advancing age.[25] In addition, some medications affect acid production. As a result of low acid production, absorption of some minerals, such as iron, zinc, and calcium, can be impaired. Low levels of acid and intrinsic factor reduce the digestion and absorption of vitamin B-12. Thus, even with adequate intakes of iron and vitamin B-12, older adults may become anemic. Symptoms of lactose intolerance can lead to avoidance of dairy products, which can limit the availability of bone-building nutrients. Again, fortified foods or supplements can help older adults overcome these problems with digestion and absorption of nutrients.

Constipation is also a common problem for older people. To prevent constipation, older people should have a primarily plant-based dietary pattern to meet fiber needs, drink plenty of fluids, and engage in regular physical activity. Fiber supplements may be useful when overall food consumption does not allow for adequate fiber intake. Because some medications can be habit forming, a primary care provider should be consulted to determine if a laxative or stool softener is necessary.

In addition to changes in the GI tract, the functions of the accessory organs decline as we age. For instance, the liver functions less efficiently. A history of significant alcohol consumption or liver disease will intensify existing problems with liver function. As liver efficiency declines, its ability to detoxify substances, including medications, alcohol, and vitamin and mineral supplements, drops. This increases the possibility for vitamin toxicity.

Poor dentition contributes to decreased food intake and digestive problems. **What strategies might be effective in improving the nutritional status of adults who are missing some or all of their natural teeth?**
Stockbyte/PunchStock

The gallbladder also functions less efficiently in later years. Gallstones can block the flow of bile out of the gallbladder into the small intestine, thereby interfering with fat digestion. Obesity is a major risk factor for gallbladder disease, especially in older women. Gradual (as opposed to rapid) loss of excess body weight and a low-fat dietary pattern may reduce symptoms.

Although pancreatic function may decline with age, this organ has a large reserve capacity. One sign of a failing pancreas is high blood glucose, although this can occur as the result of several conditions. The pancreas may be secreting less insulin, or cells may be resisting insulin action (as is commonly seen in people with android obesity). Where appropriate, improved nutrient intake, regular physical activity, and loss of excess body weight can improve insulin action and blood glucose regulation.

## NERVOUS SYSTEM

A gradual loss of nerve cells may decrease perceptions of taste and smell and impair neuro-muscular coordination, reasoning, and memory. Both hearing and vision typically decline with age. Hearing impairment is the greatest in those who have been exposed consistently to loud noises, such as urban traffic, lawnmowers, and loud music.

Declining eyesight, frequently caused by retina degeneration and cataracts, can affect a person's abilities to grocery shop, read labels for nutritional content, and prepare foods safely at home. Vision losses also may cause people to curtail social interactions, reduce physical activity, and have inadequate daily personal health and grooming routines. Age-related macular degeneration, a common cause of failing eyesight, affects about 2 million U.S. adults.[26] A major risk factor is tobacco use. Dietary patterns rich in carotenoids and omega-3 fatty acids help to reduce the risk of certain eye diseases among older adults.[27]

Loss of neuromuscular coordination may also impact food access and preparation for older adults. Physical tasks as simple as opening food packages can become so difficult that individuals restrict dietary intake to foods that require little preparation and depend on others to provide food that is ready to eat. Eating may become difficult, too. Loss of coordination may make it a challenge to grasp cup handles and manipulate eating utensils. As a result, older adults often avoid foods that can be easily spilled (e.g., soups and juices) or that need to be cut (e.g., meats, large vegetables). Some may even withdraw from social activites and eat alone, which often means eating less.

## IMMUNE SYSTEM

With age, the immune system often operates less efficiently. Recurrent illnesses and delayed wound healing are warning signs that nutrient deficiencies (especially protein and zinc) may be hindering immune function. On the other hand, overnutrition appears to be equally harmful to the immune system. For example, obesity and excessive fat, iron, and zinc intakes can suppress immune function. Meeting daily requirements for protein, vitamins (especially folate and vitamins A, C, D, and E), iron, and zinc will help to maximize immune function. Increased attention to food safety is crucial to prevent foodborne illness.

## ENDOCRINE SYSTEM

As adulthood progresses, the rate of hormone synthesis and release can diminish. Declining thyroid hormone production, for example, can decrease basal metabolic rate, leading to unexpected weight gain. A decrease in insulin release or sensitivity to insulin, for instance, means that it takes longer for blood glucose levels to return to normal after a meal. Maintaining a healthy weight, engaging in regular physical activity, adhering to a dietary pattern that is low in saturated fat and high in fiber, and avoiding highly processed foods can enhance the body's ability to use insulin and restore elevated blood glucose levels to normal after a meal.

▲ Immune function declines with age, so food safety becomes increasingly important for older adults. **List three simple consumer food safety practices that would help to stave off foodborne illness.**
CDC/Cade Martin

## CHRONIC DISEASE

The prevalence of obesity, heart disease, osteoporosis, cancers, hypertension, and diabetes rises with age. More than half of older adults have one of these chronic and potentially debilitating diseases. Four out of 10 older adults have at least two chronic conditions contributing to $3.5 trillion in annual health care costs.[28] One small change can trigger a chain of events that results in poor health. Chronic diseases may have a strong impact on dietary patterns. For instance, obesity, heart disease, and osteoporosis may impair physical mobility to the extent that victims are unable to shop for and prepare food. Chronic disease also can influence nutrient requirements. Cancer, for example, boosts nutrient and calorie needs. Hypertension may indicate a need to lower sodium intake. Nutrient utilization can be affected by chronic disease, too. For instance, diabetes alters the body's ability to utilize glucose. Chronic kidney diseases may impair the kidneys' ability to reabsorb glucose, amino acids, and vitamin C.

## MEDICATIONS

Older adults are major consumers of medications (both prescription and over-the-counter) and nutritional supplements. The CDC reports that over 90% of older adults take at least one prescription medication daily, and 40% of all people over age 65 take five or more medicines each day.[29] The rate of supplement use increases throughout adulthood such that by the time they reach retirement age, approximately 70% of all adults are using supplements.[30] Physiological declines that occur during aging (e.g., reduced body water and reduced liver and kidney function) may exaggerate and prolong the effects of medications and dietary supplements in older adults.

Medications can eradicate infections and control chronic diseases, but some also adversely affect nutritional status, particularly of those who are older and/or take many different medications. For instance, some medications depress taste and smell acuity or cause anorexia or nausea that can blunt interest in eating and lead to reduced dietary intake. Some medications alter nutrient needs. Aspirin, for example, increases the likelihood of stomach bleeding, so long-term use may elevate the need for iron, as well as other nutrients. Antibiotics kill beneficial bacteria along with pathogens, so they can limit the amount of vitamin K that is synthesized by bacteria in the large intestine. Some medications may impair nutrient utilization; diuretics and laxatives may cause excessive excretion of water and minerals. Even vitamin and mineral supplements may have unanticipated effects on nutritional status. Iron supplements taken in large doses can interfere with the functioning of zinc and copper. Folate supplements can mask vitamin B-12 deficiencies.

People who must take medications should eat nutrient-dense foods and avoid any specific food or supplement that interferes with the function of their medications. For example, vitamin K can reduce the action of oral anticoagulants, aged cheese can interfere with certain drugs used to treat hypertension and depression, and grapefruit can interfere with medications such as tranquilizers and those that lower cholesterol levels. An RDN can help to plan a dietary pattern that meets nutritional needs while avoiding harmful food/medication interactions.

The CDC reports that over 40% of seniors (over 65 years) take five or more medications per day. In some cases, drugs can affect nutrient status. **What are two examples of medications that can directly alter one's nutrient status?** Hill Street Studios/Blend Images LLC

## COMPLEMENTARY AND ALTERNATIVE MEDICINE

Approximately 70% of U.S. adults report using some kind of dietary, botanical, or herbal supplement.[30] Recall that the safety, purity, and effectiveness of dietary supplements are not tightly monitored by the FDA. Table 16-3 reviews some popular herbal products used by older adults. Note from the table that these products can pose health risks in certain people. In addition, they may be expensive—in some cases, more than $100 per month—and are not covered by health insurance plans. In recent years, the use of many herbal products has declined because of expense and documented negative effects.

TABLE 16-3 ■ **Popular Herbal Products**

| Product | Purported Effects[a] | Side Effects | Use Caution |
|---|---|---|---|
| Black cohosh | Reduce symptoms of menopause (possibly effective) | Nausea<br>Liver damage | Women who have had breast cancer<br>Pregnant women<br>Anyone taking estrogen, or hypertension or blood-thinning medications[b]<br>Anyone with abnormal liver function |
| Cranberry | Prevent and treat urinary tract infections (possibly effective)<br>Improve blood sugar (possibly ineffective) | GI upset and diarrhea<br>Kidney stones | People susceptible to kidney stones<br>Anyone taking antidepressants, prescription painkillers, or blood thinners |
| Echinacea | Prevention or treatment of colds or other infections (possibly effective) | Nausea<br>Skin irritation<br>Allergic reactions<br>GI tract upset<br>Increased urination | Anyone with an autoimmune disease<br>Pre- or postsurgical patients<br>Anyone with allergies to daisies |
| Garlic | Improve blood sugar, cholesterol, blood pressure (possibly effective)<br>Decrease risk of cancer (e.g., breast, gastric, lung) (possibly ineffective) | GI tract upset<br>Unpleasant odor<br>Allergic reactions | Pre- or postsurgical patients<br>Perinatal women<br>Anyone taking blood-thinning medications or AIDS medications |
| Ginger | Decrease nausea and vomiting (possibly effective)<br>Reduce symptoms of osteoarthritis (possibly effective)<br>Decreased muscle soreness (possibly ineffective) | Heartburn<br>Diarrhea<br>Increased menstrual bleeding | Pregnant women<br>People with bleeding disorders<br>Anyone with a heart condition<br>People taking blood glucose-lowering or blood-thinning medications |
| Ginkgo biloba | Reduce symptoms of anxiety and dementia (possibly effective)<br>Improve glaucoma damage (possibly effective)<br>Improve cognition (without dementia) and blood pressure (possibly ineffective) | Mild headache<br>GI tract upset<br>Allergic reactions<br>Irritability<br>Reduced blood clotting<br>Seizures | People with bleeding disorders<br>Pre- or postsurgical patients<br>Concurrent use of feverfew, garlic, ginseng, dong quai, or red clover<br>Anyone taking diabetes medications, blood-thinning medications, vitamin E supplements, antidepressants, or diuretics |
| Ginseng | Improve symptoms of Alzheimer's and cognition (possibly effective)<br>Prevent influenza (possibly effective)<br>Improve lung disease symptoms (possibly effective) | Hypertension<br>Asthma<br>Irregular heartbeat<br>Hypoglycemia<br>Insomnia<br>Headache<br>Nervousness<br>GI tract upset<br>Reduced blood clotting<br>Menstrual irregularities and breast tenderness | Anyone who takes a prescription drug<br>Women who have had breast cancer<br>Anyone with chronic GI tract disease<br>Anyone with uncontrolled hypertension |
| Glucosamine sulfate | Improve symptoms of osteoarthritis (likely effective) | GI tract upset | People with asthma or shellfish allergies |
| Milk thistle | Improve blood sugar and indigestion (possibly effective) | GI distress<br>Pain<br>Anorexia | Pregnant women<br>People taking blood glucose-lowering medications<br>People with hormone-sensitive cancers<br>People with allergies to plants in the Asteraceae family |

*(continued)*

**TABLE 16-3** ■ *(continued)*

| Product | Purported Effects[a] | Side Effects | Use Caution |
|---|---|---|---|
| Saw palmetto | Improve prostate health (possibly ineffective) | Dizziness<br>Headache<br>GI distress | Pregnant women<br>People with bleeding disorders who take blood-thinning medications |
| St. John's wort | Reduce symptoms of depression (likely effective)<br>Improve wound healing (possibly effective)<br>Reduce menopause symptoms (possibly effective) | GI tract upset<br>Rash<br>Fatigue<br>Restlessness<br>Increased sensitivity to sunlight | Anyone who takes a prescription drug<br>People with UV sensitivity[c]<br>People with bipolar disorder, major depression, schizophrenia, and Alzheimer's disease<br>Anyone recovering from a graft or organ transplant |
| Turmeric | Reduce symptoms of depression and osteoarthritis (possibly effective)<br>Improve cholesterol (possibly effective) | GI tract upset<br>Dizziness | People with gallbladder problems<br>People with GERD<br>Pre- or postsurgical patients |

[a]Ratings of effectiveness from Natural Medicines Comprehensive Database; [b]Coumadin®, aspirin, Heparin®, Lovenox®, or Fragmin®; [c]UV sensitivity may be induced by some drugs, such as sulfa medications, anti-inflammatory medications, or acid-reflux medications.

For additional information about herbal products, access the following websites:

National Institutes of Health
National Center for Complementary and Integrative Health
www.nccih.nih.gov

National Institutes of Health
Office of Dietary Supplements
ods.od.nih.gov

Natural Medicines
naturalmedicines.therapeuticresearch.com

When it comes to herbal products, proceed with caution. Significant health risks—including death—have been associated with the use of many herbal products. The FDA advises anyone who experiences adverse side effects from an herbal product to contact a primary care provider. Clinicians are then encouraged to report adverse events to the FDA, state and local health departments, and consumer protection agencies.

Pregnant or breastfeeding women, children under 2 years of age, anyone over the age of 65 years, and anyone with a chronic disease should not take supplements unless under the guidance of a primary care provider. A concern has been raised with regard to patients who abruptly end any alternative medicine at the start of hospital treatments or deny that they are using alternative therapies. Interactions between alternative therapies and pharmaceutical drugs can be drastic and include complications such as delirium, clotting abnormalities, and rapid heartbeat, resulting in the need for intensive care. Full disclosure of all prescription and nonprescription treatments aids in prevention of such complications. Experts recommend that, if time permits, patients stop taking herbal products for about a week before a scheduled surgery or otherwise take all original supplement containers to the hospital, so that the anesthesiologist can evaluate what was taken.

### ✓ CONCEPT CHECK 16.3

1. What is sarcopenia? What two dietary and lifestyle changes would you suggest to avoid this condition?
2. Describe three ways aging affects the processes of digestion, absorption, and utilization of nutrients.
3. List three risks associated with consuming dietary supplements, herbals, or botanicals.

## 16.4 Healthful Dietary Patterns for the Adult Years

Recommended dietary practices for later years would be to increase nutrient density and ensure fiber and fluid intakes are adequate. In addition, although plant proteins are always recommended, lean meats can be especially helpful for meeting vitamin B-6, vitamin B-12, iron, and zinc needs. Table 16-4 outlines healthy U.S.-style dietary patterns at various calorie levels commonly needed by older adults.

**Home-Delivered Meal Kits**

Participation in fresh-food, home-delivered meals is a fast-growing industry, predicted to double by 2022, reaching over $11 billion in sales.[31] With more than a hundred companies occupying this growing space, these kits are attractive for busy consumers who are short on time and cooking skills. Benefits include convenience, minimal food waste, portion control, and meal variety—often at a lower cost than the same meal served at a restaurant.

**TABLE 16-4 ■ Healthy U.S.-Style Dietary Patterns for Adults Ages 60 and Older**

Choose whole **Fruit** rather than fruit juice. Consider soft, cooked, or pureed fruits if chewing ability is impaired. Canned and frozen fruits with no added sugars are nutrient-dense, yet affordable options.

Choose a variety of **Vegetables** from each subgroup. Choose options that are prepared with less added fat and salt. Select softer, cooked vegetables if chewing ability is impaired. No-salt or low-salt canned or frozen vegetables with are nutrient-dense, yet affordable options.

At least half of **Grain** choices should be whole grains. Use fortified breakfast cereals as a source of many essential vitamins and minerals.

Incorporate a variety of lean, unprocessed meats, plant sources of **Protein**, eggs, and seafood choices. Choose tender cuts of meat if chewing ability is impaired.

Choose unsweetened and reduced-fat or fat-free **Dairy** foods. Strive to achieve the recommended daily amounts of dairy foods to support bone health.

**Oils** refer to nontropical plant oils, such as canola, corn, olive, peanut, safflower, soybean, and sunflower oils.

**Calories for Other Uses** include added sugars and rich sources of saturated fat, such as butter, shortening, lard, or tropical plant oils (e.g., coconut oil).

| Calorie Level | Daily Amount of Food from Each Group Based on Calorie Level | | | | | |
| --- | --- | --- | --- | --- | --- | --- |
| | 1600 | 1800 | 2000 | 2200 | 2400 | 2600 |
| Fruits (cups) | 1½ | 1½ | 2 | 2 | 2 | 2 |
| Vegetables (cups) | 2 | 2½ | 2½ | 3 | 3 | 3½ |
| Grains (oz-eq) | 5 | 6 | 6 | 7 | 8 | 9 |
| Protein (oz-eq) | 5 | 5 | 5½ | 6 | 6½ | 6½ |
| Dairy (cups) | 3 | 3 | 3 | 3 | 3 | 3 |
| Oils (grams) | 22 | 24 | 27 | 29 | 31 | 34 |
| **Limit on Calories for Other Uses** | | | | | | |
| Calories for Other Uses (kcal) | 100 | 140 | 240 | 250 | 320 | 350 |

Note: Beans, peas, and lentils can count either as vegetables or protein foods. Source: U.S. Department of Agriculture and U.S. Department of Health and Human Services. *Dietary Guidelines for Americans, 2020–2025.* 9th Edition. December 2020. Available at DietaryGuidelines.gov.

Singles of all ages face the following logistical barriers obtaining adequate nutrients: purchasing, preparing, storing, and using food with minimal waste. Value-priced packages of meats and vegetables are normally too large to be useful for a single person. Many singles live in small dwellings, some without kitchens and freezers. Creating an adequate dietary pattern to accommodate a limited budget and a single appetite requires special considerations. Table 16-5 provides some practical suggestions for eating healthfully in later years.

Nutritional deficiencies and protein-calorie malnutrition have been identified among some aging populations, particularly those in hospitals, nursing homes, or long-term care facilities.[32] Friends, relatives, and health care professionals should be alert for unintentional changes in weight among older people. If there are signs of inadequate dietary intake, an RDN can assess nutrition concerns and offer professional and personalized advice.

As you have learned throughout these chapters, optimizing nutritional status is important throughout the life cycle. Consider the benefits for older adults, specifically the following:

• Delays the onset of some diseases
• Improves the management of existing diseases
• Hastens recovery from many illnesses
• Increases mental, physical, and social well-being
• Improves quality of life

MyPlate is a useful guide for planning healthy meals, but older adults must be sure to emphasize certain nutrients: potassium, calcium, vitamin D, vitamin B-12, vitamin C, and fiber. MyPlate for Older Adults illustrates nutrient-dense food choices that are accessible and appealing for this population group (Fig. 16-8). This figure also affirms the importance of adequate fluid intake and modified physical activity goals. In some cases, dietary supplementation may be appropriate to address the unique needs of older adults. Discuss this with your primary care provider or dietitian.

Obtaining enough food may be difficult for some older persons, especially if they have a low income, are unable to drive, or do not have social networks that can assist with cooking or shopping. For an older person, a request for help may be equated

After spending much of their adulthood cooking for a large family, empty-nesters may find it difficult to adjust their grocery shopping and food preparation habits to suit just one or two adults. Here are some ideas to save money and reduce food waste.

- Build a collection of your favorite single-serving recipes.
- Before shopping, make a plan for the week. If you buy a large package of one item, find several recipes that will use the food ingredients in a variety of ways. Consider how you will use (or share) the leftovers.
- Make wise use of your freezer. If you purchase and cook large amounts of food, divide it into single-serve portions, then freeze them for later use.
- For perishable items that are not suitable for freezing, buy only what you will use. Small containers may be more expensive, but letting food spoil is also costly.
- Consider buying one ripe, one medium-ripe, and one unripe piece of fruit to ensure the best quality over several days.
- Fresh produce is delicious and nutrient dense, but canned and frozen fruits and vegetables are also excellent options. Select products with less sodium and no added sugar.

**TABLE 16-5 ■ Practical Guidance for Healthful Eating in the Later Years**

Eat regularly and attempt to eat larger meals early in the day. Use plant-based, nutrient-dense foods as a basis for all meals and snacks.

Use labor-saving cooking methods and healthier convenience foods, but try to incorporate fresh foods into daily menus.

Try new foods, new seasonings, and new ways of preparing foods. Use low-sodium canned goods or frozen produce in cooking.

Keep easy-to-prepare nutritious foods on hand for times when you feel tired.

Allow yourself to enjoy a treat occasionally, perhaps an expensive cut of fish or small piece of dark chocolate.

Eat in a well-lit or sunny area; serve meals attractively; and experiment with different flavors, colors, shapes, textures, and smells.

Arrange kitchen and eating areas so that food preparation and cleanup are simple.

Eat with friends, with relatives, or at a community center when possible.

Share cooking responsibilities with a neighbor or close friend.

Use community resources for help with shopping and other daily care needs.

Stay physically active and well hydrated.

If possible, take a walk before eating to stimulate the appetite.

When necessary, chop, grind, or blend hard-to-chew foods. Softer, protein-rich foods (e.g., beans, shredded meats, eggs) can be substituted for whole pieces of meat when poor dentition prevents normal food intake. Prepare soups, stews, cooked whole grains, and casseroles.

Dry milk powder is an inexpensive way to add protein to baked goods (e.g., muffins), smoothies, soups, and mixed dishes (e.g., mashed potatoes). Keep a box of dry milk powder in the pantry to boost the protein content of your meals.

Source: NIH National Institute for Aging and NIDDK

**hospice care** A program offering care that emphasizes comfort and dignity at the end of life.

to a loss of independence. In these cases, family and friends can assist. Special transportation arrangements may also be available through community agencies, local transit companies, or lift services. Indeed, many eligible older people are missing meals and are poorly nourished because they do not realize that programs are available to help them. Irregular meal patterns and weight loss are warning signs that malnutrition may be developing. An effort should be made to identify poorly nourished seniors and inform them of the services available in their communities.

## COMMUNITY NUTRITION SERVICES FOR OLDER ADULTS

Health care advice and services for older people can come from clinics, private practitioners, hospitals, and health maintenance organizations. Home health care agencies, adult day care programs, 24-hour-care programs, and **hospice care** (for the terminally ill) also provide daily care for those who qualify.

One in five older adults (about 11 million) are served by the Older Americans Act (OAA).[34] Originally enacted in 1965, the OAA supports a range of home- and community-based services, including Meals on Wheels and other nutrition programs. The OAA nutrition programs serve over 230 million meals each year to approximately 2.5 million U.S. adults over the age of 60. Federal standards mandate that these meals supply at least one-third of adult energy and nutrient requirements.

Some meals, such as Meals on Wheels, are delivered directly to older adults in their homes. Although home-delivered meals can make a valuable contribution to the positive nutritional status of homebound older adults, services are usually limited

**FIGURE 16-8** ▲ **MyPlate for Older Adults,** Copyright © 2016 Tufts University, All rights reserved. Reprinted with permission. MyPlate for Older Adults graphic and accompanying website were developed with support from AARP Foundation. "Tufts University" and "AARP Foundation" are registered trademarks and may not be reproduced apart from their inclusion in the "MyPlate for Older Adults" graphic without express permission from their respective owners.

The World Health Organization recommends eating 5 portions of fruits and vegetables every day, but a new meta-analysis[33] found that boosting intake to 10 portions of fruits and vegetables a day could:

- reduce cardiovascular disease by 28%.
- reduce stroke by 33%.
- reduce cancer by 13%.
- prevent 7.5 million premature deaths globally.

to one or two meals per day. If a recipient has a poor appetite, the food may end up stored for later or simply thrown away. If foods are not eaten immediately upon delivery or not stored properly, risk for foodborne illness could be a concern. Other meals are provided by congregate meal programs, which usually serve lunch at a central location. With congregate meals, the social aspect of eating tends to improve nutritional intake. However, programs generally provide just one meal per day on 5 days per week.

In addition to congregate and home-delivered meals, federal commodity distribution is available in some areas of the United States to low-income older people. Individuals whose incomes are below the poverty level can benefit from the SNAP program. The Senior Farmers Market Nutrition Program provides low-income seniors with coupons that can be exchanged for eligible foods (fruits, vegetables, honey, and fresh-cut herbs) at farmers' markets, roadside stands, and community-supported agriculture programs. Food cooperatives and a variety of clubs and religious and social organizations provide additional aid.

To learn more about resources available for older adults in your area, check out the following websites:

Food Assistance Programs for Seniors: www.nutrition.gov/topics/food-assistance-programs

Elder Care Locator: eldercare.acl.gov/Public/Index.aspx

National Institute on Aging: www.nia.nih.gov

American Geriatrics Society: www.americangeriatrics.org

Administration for Community Living: acl.gov

## ✓ CONCEPT CHECK 16.4

1. Gerald is a 76-year-old man whose wife recently passed away. He now lives alone for the first time in his life and is not accustomed to preparing meals for himself. What three recommendations would you give Gerald about eating well?

2. List three nutrition resources for an older adult with limited financial resources.

# CASE STUDY Dietary Assistance for an Older Adult

Frances is an 82-year-old woman who suffers from macular degeneration, osteoporosis, and arthritis. Since her husband died a year ago, she has moved from their family home to a small one-bedroom apartment. Her eyesight is progressively getting worse, making it difficult to go to the grocery store or even to cook (for fear of burning herself). She is often lonely; her only son lives an hour away and works two jobs, but he visits her as often as he can. Frances has lost her appetite and, as a result, often skips meals during the week. She has resorted to eating mostly cold foods. These are simple to prepare but seriously limit the variety and palatability of her overall intake. Also, she wears dentures and has trouble chewing tough meats and foods with crisp textures. She is slowly losing weight as a result of her eating patterns and loss of appetite.

Her typical dietary intake usually consists of a breakfast that may include 1 slice of wheat toast with margarine, honey, and cinnamon, and 1 cup of hot tea. If she has lunch, she normally has a can of peaches, half of a turkey sandwich, and a glass of water. For dinner, she might have half of a tuna fish sandwich made with mayonnaise and 1 cup of iced tea. She usually includes one or two soft cookies at bedtime.

1. What nutrients are likely to be inadequate in Frances's current dietary pattern?
2. What potential effects might Frances's limited dietary pattern have on her health status?
3. Which physiological changes of aging will add to the effects of her inadequate dietary intake (review Figure 16-7)?
4. Use the DETERMINE checklist (Figure 16-4) to estimate Frances's Nutritional Risk score.
5. What services are available in the community that could help Frances improve her dietary pattern?
6. What other convenience foods could be included in her eating pattern to make it more healthful and varied?

*Complete the Case Study. Responses to these questions can be provided by your instructor.*

▲ Many older adults, like Frances, would benefit from nutritional assessment and counseling. RDNs are the nutrition experts specifically trained to tailor individual recommendations to meet the specific needs of individuals like Frances. Keith Brofsky/Getty Images

## Brain Health: Food for Thought

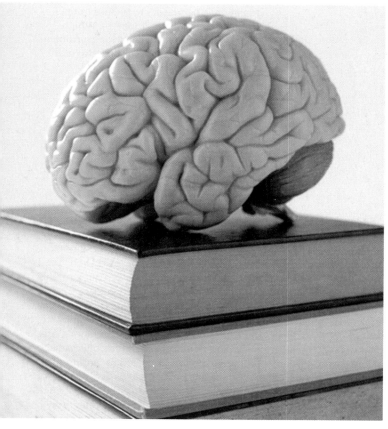

The human brain is a metabolically active organ that requires a continual supply of energy—preferably glucose—to meet its energy demands. Even though the average adult brain weighs only 3 pounds (about 2% of body weight), 20% of your blood supply is directed to the brain to supply enough oxygen, glucose, and other nutrients necessary to support proper cognitive function. Indeed, your abilities to speak and understand others, focus on tasks, learn new information, and make decisions, as well as your overall mood, may be significantly impacted by your dietary choices.

So, what does it take to keep your brain's 100 billion neurons alive and functioning optimally? First, let's understand the roles of key nutrients—essential fatty acids, the B vitamins, choline, antioxidant nutrients, and some minerals—in brain function. Then, we'll examine how dietary patterns and other lifestyle choices may influence your risk of developing migraines, depression, and neurodegenerative diseases.

## NUTRIENTS THAT FUNCTION IN BRAIN HEALTH

**Formation of Brain Tissue.** Other than adipose tissue, nervous tissue is the fattiest tissue in the body—the brain is 60% fat by dry weight. Omega-3 and omega-6 fatty acids (including the two essential fatty acids, alpha-linolenic acid and linoleic acid) are used to form phospholipids, which are key players in the formation and maintenance of healthy cell membranes in the brain and nerve cells. Deficiencies of omega-3 or omega-6 fatty acids in the prenatal period or during infancy are detrimental to optimal brain health. Recall that choline is vital for the synthesis of two phospholipids, phosphatidylcholine and sphingomyelin, that are highly concentrated in nervous tissue.

Iodine is a vital nutrient during brain growth and development. Iodine is involved in the synthesis of thyroid hormones, which regulate growth, development, and metabolism. In utero and shortly after birth, iodine participates in the myelination of nerves. Iodine deficiency during early development can result in intellectual disabilities.

Iron is another trace mineral that is vital during brain formation. Iron participates in the pathways that yield energy to fuel the brain, the myelination of nerve tissue, and the formation of neurotransmitters. A deficiency of iron during brain growth and development may result in impaired learning ability and behavioral problems.

**Fueling the Brain.** The preferred source of fuel for the brain and nervous tissue is glucose, but the brain does not have any way to store appreciable amounts of this carbohydrate. Most of the time, the brain relies on the glucose that is circulating in the blood from your last meal. During times of fasting (e.g., overnight, very-low-carbohydrate diets), glucose can be derived from the breakdown of liver glycogen or the conversion of amino acids to glucose in the liver and kidneys. When the body is in a state of starvation, the brain is able to utilize ketone bodies (by-products of fatty-acid metabolism) as a source of fuel. However, ketones are acidic, so high levels of ketones can cause acidosis, which has negative and potentially harmful implications for every body system.

Recall that the metabolism of glucose for fuel requires many micronutrient cofactors. Thus, deficiencies of thiamin, riboflavin, niacin, pantothenic acid, biotin, iron, magnesium, manganese, and vitamin B-12 can adversely affect brain function.

**Nervous System: Communication Superhighway.** Communication between the brain and other body tissues occurs via nerve cells (also known as neurons), using a combination of electrical and chemical signals. Recall how sodium and potassium are involved in nerve impulse transmission. Shifts in the concentrations of these electrolyte nutrients across the nerve cell membrane allow for the transmission of an electrical signal along the length of the nerve cell. The speed at which electrical signals move depends on the myelin sheath that surrounds the nerve cell. Thiamin, folate, vitamin B-6, vitamin B-12, and iron are involved in the formation and function of the myelin sheath.

When an electrical signal reaches the end of a nerve cell, the message must be converted into a chemical signal. Neurotransmitters are chemical messengers that are released from nerve cells to transmit a signal to a target cell, such as another nerve cell, an organ, or a gland. Acetylcholine, dopamine, norepinephrine, and serotonin are some examples of neurotransmitters. Amino acids are the building blocks of neurotransmitters. The synthesis and function of these chemical messengers also depends on several B vitamins, including thiamin, riboflavin, niacin, vitamin B-6, folate, and vitamin B-12.

**Protecting Nervous Tissue.** As described, the brain is a metabolically active organ. Energy metabolism naturally generates some damaging free radicals. Free radicals can cause oxidative damage to cell membranes and DNA. With its high concentration of polyunsaturated fatty acids, brain tissue is highly susceptible to the damaging effects of free radicals. Fortunately, the body's antioxidant systems are able to limit this type of cellular damage. An adequate supply of the antioxidant nutrients vitamin C, vitamin E, selenium, and zinc, as well as the phytochemical beta-carotene, help to protect the brain from oxidative damage.

One compound that has been linked to declines in brain health is homocysteine. As you learned earlier, high levels of homocysteine have been identified as a risk factor for poor health outcomes, including heart disease, cancer, stroke, and Alzheimer's disease. Metabolic pathways to convert homocysteine into less harmful amino acids require adequate levels of folate and vitamins B-6 and B-12. Choline may also participate in these metabolic processes.

Evidence clearly shows that the accumulation of excess body fat (starting in childhood) negatively impacts cognitive function.[35] Insulin resistance appears to be a mediating factor in the relationship between excess body fat and brain health. Recall that body weight management requires calorie control and physical activity. Independent of body weight, physical activity also improves brain development, cognition, attention, and memory.[36]

## PROMOTING BRAIN HEALTH THROUGHOUT THE LIFE CYCLE

The brain is the body's command center. From brain development in utero until old age, food and lifestyle choices affect brain health. Now that you recognize the importance of nutrients for the growth, development, and function of the nervous system, let us examine how dietary patterns can affect brain health at various stages of the life cycle.

**Gestation and Infancy.** Gestation and early infancy are critical periods for brain growth and development. Nutrient deficiencies at this stage of the life cycle can profoundly affect neurological health throughout life. At this early phase, it is the biological mother's responsibility to ensure that the developing fetus has optimal nutrients for brain growth and development. Insufficient maternal intakes of folate, vitamin B-12, omega-3 fatty acids, and iron have all been related to reduced cognitive function in offspring.[37] As you learned in Chapter 14, the RDAs for pregnant women for these nutrients increase to as much as 150% of the requirements for nonpregnant women. A dietary pattern that includes lean meats, plant sources of protein, fatty fish, and fortified breakfast cereals will support proper fetal brain growth and development. While good food sources of these nutrients are important, a pregnant woman is likely to require a dietary supplement, especially to meet her needs for iron and folic acid.

After birth, breastfeeding is the preferred method to nourish an infant. For brain health, the presence of long-chain polyunsaturated fatty acids (e.g., EPA and DHA) in human milk promotes optimal cognitive development. For a breastfed infant, the adequacy of the mother's dietary pattern will influence the nutrient content of the human milk and impact infant health. However, if breastfeeding is not possible or preferred, iron-fortified infant formula is a safe and healthy alternative. Infant formulas are designed to mimic the nutritional composition of human milk, which is the ideal food for infants.

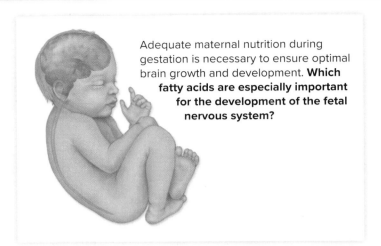

Adequate maternal nutrition during gestation is necessary to ensure optimal brain growth and development. **Which fatty acids are especially important for the development of the fetal nervous system?**

**Childhood and Adolescence.** Throughout childhood and adolescence, iron status is correlated with learning ability and behavior. Children who are deficient in iron experience decreased attention, poor memory, and may display disruptive behavior. In addition, good iron status can help to prevent lead poisoning, which is extremely detrimental to the nervous system. As you learned in Section 15.4, young children are at heightened risk for iron deficiency. Overall intake of iron-rich foods can be inadequate because of low appetite, limited chewing ability, and food selectivity (i.e., picky eating). Caregivers should be sure to offer age-appropriate sources of iron, such as bite-size pieces of lean meats, beans, peas, lentils, and fortified cereals. In addition, they should ensure that excessive intakes of other foods (even nutrient-rich foods, like milk) do not crowd out rich food sources of iron. As children reach puberty, they enter a growth spurt, which increases the demand for iron to build lean tissue and expand blood volume. Adolescent girls also require additional iron to compensate for monthly blood loss after menstrual periods begin.

### Adulthood
***Eating Patterns and Migraine Headaches.*** A migraine is a severe, intense, and recurring throbbing pain affecting one or both sides of the head. Migraines affect 38 million adults and children in the United States and can severely impact quality of life.[38] Depression, anxiety, and sleep disturbances are common for those with chronic migraines. Other potentially debilitating neurological symptoms may include visual disturbances; nausea and vomiting; dizziness; sensitivity to sound, light, touch, and smell; and tingling or numbness in the extremities or face.

Both biologic and lifestyle factors contribute to migraines. Some nonmodifiable risk factors include family history, age, sex, and hormonal changes.[39] Individuals with obesity are at increased risk of migraines, possibly related to higher levels of inflammation.

Dehydration and meal skipping are two often overlooked culprits for migraines. Prolonged fasting can deprive the brain of its go-to fuel source: glucose. Meeting the AI for fluid intake and eating small, frequent meals throughout the day to prevent dehydration and hypoglycemia are simple dietary strategies that may help some people avoid migraine headaches.

There is also a wide range of potential dietary triggers for migraine headaches. The most commonly cited food triggers are aged cheeses, fermented foods, cured meats, alcohol, chocolate, and citrus fruits. These foods contain certain amino acids that may lead to changes in blood flow and neurotransmitter release in the brain.

When it comes to migraine headaches, caffeine is a touchy subject! Caffeine causes blood vessels to constrict. Since migraines are linked to dilation of the blood vessels that supply the brain, caffeine-induced vasoconstriction can help to treat symptoms. Indeed, some medications for migraine relief combine pain relievers with a low dose of caffeine. For some people who are habitual caffeine users, caffeine withdrawal can precipitate headaches. Most experts recommend that people who suffer from frequent headaches avoid or at least limit caffeine consumption to 200 milligrams (about 2 cups of coffee) per day.

At this time, there is no evidence-based *migraine diet*—individuals vary in their sensitivities to food triggers and, for many people who suffer from migraines, there may be some underlying pathology that does not involve food at all.[40] Overall, a dietary pattern that complies with the *Dietary Guidelines*—plentiful in fruits, vegetables, whole grains, nuts, seeds, beans, peas, and lentils but low in sodium, solid fats, alcohol, and added sugars—has been linked to fewer complaints of migraine headaches.[41]

***Nutrition and Depression.*** Major depressive disorders affect about 7% of adults in the United States.[42] Clinical depression is more than just a case of the blues. It is characterized by feelings of sadness, hopelessness, or despair that last for more than 2 weeks and interfere with daily living. Some symptoms of depression include loss of interest in activities that were once enjoyable (e.g., work, hobbies), changes in sleep habits (sleeping either too much or too little), and changes in appetite, which can lead to either weight loss or weight gain.

The causes of depression are varied. Stress, grief, illness, substance abuse, and some medications may certainly lead to depression, but biological changes in the brain are also implicated. For example, people who experience depression may have altered synthesis or activity of hormones (e.g., cortisol) or neurotransmitters (e.g., serotonin).

Does nutrition play a role in the biological basis for depression? Considering the macronutrients, carbohydrates certainly affect mood. Glucose is the brain's primary source of fuel. Nevertheless, studies indicate that dietary patterns high in added sugars may be a risk factor for depression.[43] This may be due to the rapid rises and falls in blood glucose levels that accompany diets high in added sugars. In addition, high sugar intake tends to promote inflammation, which is associated with development of depression. Ensuring an adequate and steady supply of carbohydrates to the brain throughout the day may help to prevent mood swings.

Evidence supports a role of omega-3 fatty acids in the prevention or treatment of depression. The importance of omega-3 fatty acids to brain health is threefold. First, long-chain polyunsaturated fatty acids, particularly DHA, are incorporated into cell membranes throughout the brain. Changes in membrane fluidity may affect how well chemical signals are transmitted by neurotransmitters from nerve cells to target cells. Second, omega-3 fatty acids improve vascular health by reducing the risk of blood clots and atherosclerosis. These changes ensure an adequate blood supply to the brain. Third, omega-3 fatty acids are involved in biochemical pathways that decrease inflammation and are associated with a reduction in depression.[44]

Regarding micronutrients, folic acid, vitamin B-6, and vitamin B-12 are involved in the synthesis and activity of neurotransmitters and have been studied most extensively in relation to depression.

## magnificent microbiome

For several decades, researchers have been interested in the *gut-brain axis*. This refers to the communication that occurs along the vagus nerve, transmitting signals back and forth between the central nervous system and the GI tract. More recently, scientists have observed that changes in the gut microbiota may influence brain health. For example, some toxic compounds produced by pathogenic bacteria can be transported via blood and cross the blood–brain barrier. In the brain, these compounds may be promoting inflammation, depression, and neurodegenerative diseases (e.g., Alzheimer's disease and Parkinson's disease). Conversely, metabolic by-products of beneficial gut microbes may be responsible for decreasing inflammation and reducing the risk for these mental health problems.[45,46,47] As we learn about the benefits of plant-forward dietary patterns, such as the MIND diet, let us look beyond how the nutrients nourish our cells and consider how they fuel the microbial cells living within us!

Research also suggests that vitamin D and zinc may be particularly important in the prevention or treatment of depressive disorders.

Overall, following a balanced dietary pattern with a slant toward plants, such as the *Dietary Guidelines* or the DASH diet, can benefit the mind as well as the body. Including plenty of fruits, vegetables, and whole grains will supply B vitamins and antioxidant nutrients. Emphasizing healthy sources of fat, such as walnuts, canola oil, flaxseed oil, and fatty fish, will boost levels of omega-3 fatty acids in the body.[44]

***Preventing Neurodegenerative Diseases.*** *Neurodegenerative disease* is a term that encompasses a variety of conditions characterized by gradual, progressive deterioration of neurons. Alterations of these cells result in abnormal functioning, plaque formation, and eventual cell death. The most common neurodegenerative diseases in the United States include Alzheimer's disease, **Parkinson's disease,** and **multiple sclerosis.** In the past, research focused on genetics and aging as the primary risk factors for diseases of the brain. Now, researchers are turning their focus to lifestyle choices, with energy balance and optimal body fatness

**Parkinson's disease** Disease that belongs to a group of conditions called motor system disorders, which are the result of the loss of dopamine-producing brain cells. The four primary symptoms of PD are tremor, or trembling in hands, arms, legs, jaw, and face; rigidity, or stiffness of the limbs and trunk; bradykinesia, or slowness of movement; and postural instability, or impaired balance and coordination.

**multiple sclerosis** An unpredictable disease of the central nervous system that can range from relatively benign to somewhat disabling to devastating, as communication between the brain and other parts of the body is disrupted.

viewed as very important for optimizing brain health. Research actually links both undernutrition and overnutrition to reduced cognitive function, via mechanisms that involve changes in hormone levels in the body.[37] Compared to adults with a healthy BMI, adults with a BMI in the obese range are three times more likely to suffer impaired cognitive function and dementia.

The most common neurodegenerative disease is Alzheimer's disease, which affects about 5.5 million adults in the United States.[48] It is an irreversible, progressive deterioration of the brain that causes victims to steadily lose the ability to remember, reason, and comprehend. Alzheimer's disease takes a terrible toll on the mental and eventual physical health of people who are older. Scientists have proposed various causes, including alterations in cell development or protein production in the brain, strokes, altered blood lipoprotein composition, poor blood glucose regulation (e.g., diabetes), high blood pressure, viral infections, and high free radical levels. The 10 warning signs of Alzheimer's disease are listed below.

**Ten Warning Signs of Alzheimer's Disease**

1. Recent memory loss that affects job performance
2. Difficulty performing familiar tasks
3. Problems with language
4. Faulty or decreased judgment
5. Problems with abstract thinking
6. Tendency to misplace things
7. Changes in mood or behavior
8. Changes in personality
9. Loss of initiative
10. Withdrawal from work or social activities

Source: National Institute on Aging.

▲ Patients with advanced dementia may be easily distracted during meals, forget to eat, be unable to prepare food safely, and have trouble feeding themselves or swallowing. Caregivers and health care providers need to monitor the patient's weight to ensure maintenance of a healthy weight and nutritional state.
Ingram Publishing

Early efforts at prevention of Alzheimer's disease should receive the most attention, because the process of cognitive decline begins 10 to 20 years before warning signs appear. Preventive measures for Alzheimer's disease focus on engaging in regular physical activity, adhering to a dietary pattern rich in fruits and vegetables, controlling blood pressure, and maintaining brain activity through lifelong learning. Fruits and vegetables provide antioxidant nutrients, such as vitamin C, vitamin E, and selenium, that help to protect the body from the damaging effects of free radicals. In addition, various phytochemicals with antioxidant and anti-inflammatory properties, including polyphenols and carotenoids (e.g., lutein and zeaxanthin), have been shown to protect the brain throughout life.[49] Good food sources of lutein and zeaxanthin include green leafy vegetables, yellow-orange fruits and vegetables, and egg yolks. Polyphenols are found in foods such as berries, coffee, green or black tea, and chocolate.

Adequate intakes of folate, vitamin B-6, vitamin B-12, and choline are important to decrease blood homocysteine, which is a risk factor for neurodegenerative and cardiovascular disease (see *Newsworthy Nutrition* in this section). Dietary fats, too, may play a role in keeping Alzheimer's disease at bay. Individuals with dietary patterns rich in omega-3 fatty acids and low in saturated and *trans* fatty acids have a reduced risk of Alzheimer's disease. However, the effects of supplements are not clear.[50] The role of nutrition in preventing or minimizing the risk of this disease continues to be studied.[37,51,52]

# Newsworthy Nutrition

### Mediterranean-style diet linked to reduced Alzheimer's disease

**INTRODUCTION:** Epidemiological studies estimate that more than 30% of Alzheimer's disease may be attributed to modifiable risk factors and, thus, potentially preventable. A growing body of evidence has linked higher adherence to a Mediterranean-style diet to a lower risk of cognitive decline and dementia. **OBJECTIVE:** To examine the effects of adherence to a Mediterranean diet on biomarkers of Alzheimer's disease in a *cohort study* of adults during their midlife. **METHODS:** Seventy cognitively normal adults, ages 30 to 60 years, were followed over a span of more than 2 years. Based upon food frequency questionnaires, participants were categorized as high or low adherers to a Mediterranean diet. **RESULTS:** At baseline, > 90% of the participants reported stability of their eating patterns for ≥ 5 years. Of the 70 participants at baseline, 51% were categorized as having low Mediterranean diet adherence and 49% with high adherence. Results showed that higher adherence to the Mediterranean diet was associated with a reduced emergence and progression of Alzheimer's disease. **CONCLUSION:** Higher adherence to a Mediterranean dietary pattern was inversely associated with markers of Alzheimer's disease among middle-aged adults. This study provides support for potential protective effects of a Mediterranean eating pattern in delaying Alzheimer's disease.

Source: Berti V and others: Mediterranean diet and 3-year Alzheimer brain biomarker changes in middle-aged adults. *Neurology* 90:1789, 2018.

The MIND Diet is a dietary intervention that shows promise for protecting the brain against cognitive decline. MIND stands for Mediterranean-DASH Intervention for Neurodegenerative Delay, a hybrid of the nutrient-rich Mediterranean eating pattern with the low-calorie, low-sodium DASH diet. The MIND Diet's 10 brain-healthy foods include green leafy vegetables, other vegetables, nuts, berries, beans, whole grains, fish, poultry, olive oil, and wine. The five foods to limit include red meats, butter and stick margarine, cheese, pastries and sweets, and fried or fast food. Each day, followers of the MIND Diet should consume a minimum of three servings of whole grains, a salad plus one other vegetable, and one glass of wine. Nuts, beans, poultry, and berries are recommended regularly, but butter, cheese, and fast food should be limited.[53,54] In population studies, adults who adhered to the MIND Diet had 54% lower rates of Alzheimer's disease. Even more surprising, those that followed the MIND pattern, even periodically, also reduced their risk of the disease by 35%.[55] Recently, the MIND Diet was ranked among the top 5 best diets overall in *U.S. News & World Report's* Best Diets Rankings.

▲ The MIND Diet includes three servings of whole grains, a salad plus one other vegetable, and a glass of wine each day. Alexis Joseph/McGraw-Hill Education

## IN SUMMARY: FOOD FOR THOUGHT

When it comes to studies of individual nutrients and brain health, the results are not that impressive. Recognize, however, that we seldom eat just one nutrient. It is most relevant to examine how overall dietary patterns influence health. Time and time again, we see that dietary patterns rich in fruits, vegetables, beans, peas, lentils, nuts, seeds, whole grains, and fish are beneficial for brain health.

Now that you have studied the nutrients and phytochemicals, you can understand that folate, vitamin B-6, and vitamin B-12 work together to keep homocysteine levels in check. You can describe how vitamins E and C from fruits and vegetables work as antioxidants to prevent damage and decrease inflammation. You understand the importance of polyunsaturated fatty acids in fish oils for brain function. Although a few studies have documented that micronutrient supplementation improves mood and psychological well-being, the overall evidence to support use of dietary supplements for brain health remains equivocal. Dietary supplements may seem like a quick fix to fill nutrient gaps to promote optimal brain health, but when we focus on isolated nutrients, we miss out on the synergistic effects of the full complement of brain-healthy nutrients and phytochemicals in whole foods.[56,57]

---

### ✓ CONCEPT CHECK 16.5

1. List four potential food triggers for migraines.
2. Name three brain-healthy foods promoted by the MIND Diet. Name three foods that should be limited on the MIND Diet.
3. List five early warning signs of Alzheimer's disease.

## Summary (Numbers refer to numbered sections in the chapter.)

**16.1** Although maximum life span has not changed, life expectancy has increased dramatically over the past century. For many societies, this means that an increasing proportion of the population is over 65 years of age. As health care costs rise, chronic diseases are a leading concern.

The physiologic changes of aging are the sum of cellular changes, lifestyle behaviors, and environmental influences. Many of these changes can be minimized, prevented, and/or reversed by healthy lifestyles. Usual aging refers to the age-related physical and physiological changes that are commonly thought to be typical of aging. Successful aging describes the declines in physical and physiological function that occur because one grows older. Striving to have the greatest number of healthy years and the fewest years of illness is referred to as compression of morbidity.

**16.2** A dietary pattern based on MyPlate and the *Dietary Guidelines* can help one to preserve body function, avoid chronic disease, and age successfully. American adults are fairly well nourished, although common dietary excesses are calories, saturated fat, sodium, and, for some, alcohol. Common dietary inadequacies include vitamins D and E, folate, magnesium, calcium, zinc, and fiber. The DRIs for adults are divided by sex and age to reflect how nutrient needs change as adults grow older. These changes in nutrient needs take into consideration the aging-related physiological alterations in body composition, metabolism, and organ function. People ages 65 and older, particularly those in

long-term care facilities and hospitals, are at risk for malnutrition. The Nutrition Screening Initiative's *DETERMINE* checklist can help to identify older adults at risk of nutrient deficiencies.

**16.3** Chronic diseases, changes in body composition, and declining function of body systems can influence nutritional status. Of particular concern is sarcopenia, the loss of muscle mass that frequently accompanies aging. Changes in GI tract function can affect nutrient intake and utilization. Some medications and dietary supplements can negatively impact nutritional status. Use of medications and supplements should be guided by a primary care provider.

**16.4** Dietary patterns for adults should be based on nutrient-dense foods and need to be individualized for existing health problems, physical abilities, the presence of drug–nutrient interactions, possible depression, and economic constraints. Community nutrition assistance programs, including congregate or home-delivered meal systems, SNAP, and commodity distribution, make wholesome, nutritious foods more accessible for low-income and older adults.

**16.5** Good nutritional status is essential for proper brain development and function. The brain requires a steady supply of glucose; micronutrient deficiencies adversely affect brain development and function. The key nutrients involved in brain function are the essential fatty acids, B vitamins, choline, antioxidant nutrients, and iron.

## Check Your Knowledge (Answers are available at the end of this question set.)

1. The reason the incidence of obesity increases with age is that
   a. the basal metabolic rate decreases with age.
   b. physical activity often decreases with age.
   c. energy intake exceeds energy expenditure.
   d. any of these can occur.

2. Congregate meal programs provide
   a. 100% of nutrient needs.
   b. a social atmosphere for eating.
   c. food stamps.
   d. all of these.

3. Among the older population of the United States, the age of the fastest-growing segment is _____ years.
   a. 65
   b. 74
   c. 79
   d. 85+

4. Which of the following accurately portrays a theory about the causes of aging?
   a. Increases in testosterone and estrogen affect cell processes.
   b. Blood sugar decreases, failing to supply adequate energy to brain cells.
   c. Inadequate calorie intake speeds body breakdown.
   d. Excess free radicals damage cell components.

5. The immune system becomes less efficient with age, so it is especially important to consume adequate _____ and _____, nutrients that contribute to immune function.
   a. vitamin A, potassium
   b. protein, zinc
   c. zinc, iodine
   d. vitamin A, vitamin K

6. Which of the following is a useful strategy to prevent or delay the onset of Alzheimer's disease?
   a. Avoid stressing the brain with challenging mental tasks.
   b. Limit intake of dairy products.
   c. Consume a Mediterranean-type dietary pattern.
   d. Increase the ratio of omega-6 to omega-3 fatty acids.

7. To maintain optimal nutritional status and healthy weight, the dietary pattern of an older person should have a _____ nutrient density and be _____ in energy content.
   a. low, high
   b. low, low
   c. high, moderate
   d. high, high

8. Choline is important for brain growth and development through its role in the synthesis of
   a. phospholipids.
   b. cholesterol.
   c. omega-3 fatty acids.
   d. amino acids.

9. The preferred energy source for the brain's metabolic activity is
   a. fatty acids.
   b. glucose.
   c. amino acids.
   d. fructose.

10. The MIND Diet is a hybrid of the Mediterranean diet and the
    a. Blood-Type Diet.
    b. Paleo Diet.
    c. DASH Diet.
    d. Diabetes Diet.

Answer Key: 1. d (LO 16.5), 2. b (LO 16.6), 3. d (LO 16.1), 4. d (LO 16.2), 5. b (LO 16.3), 6. c (LO 16.7), 7. c (LO 16.4), 8. a (LO 16.7), 9. b (LO 16.7), 10. c (LO 16.7).

## Study Questions (Numbers refer to Learning Outcomes.)

1. What is the difference between life span and life expectancy? **(LO 16.1)**

2. Describe two hypotheses proposed to explain the causes of aging, and note evidence for each in your daily life experiences. **(LO 16.2)**

3. List four warning signs of undernutrition in older people that are part of the acronym DETERMINE. Briefly justify the inclusion of each. **(LO 16.3)**

4. List four organ systems that can decline in function in later years. Describe the dietary strategies that can help an individual cope with each problem. **(LO 16.3)**

5. List three important points made by the *Dietary Guidelines for Americans* for the general population, and give an example of why each one may be difficult for older adults to implement. What are some suggestions for overcoming these barriers? **(LO 16.4)**

6. Describe two ways the nutritional needs of older people differ from those of younger people. How are their needs similar? Be specific. **(LO 16.4)**

7. Why is it important for older adults to engage in physical activity (especially resistance training)? **(LO 16.5)**

8. List three common herbal remedies. What are the possible benefits and risks of each one? If your grandmother were considering using any of these herbal remedies, what advice would you give her? **(LO 16.5)**

9. What three resources in a community are widely available to aid older adults in maintaining nutritional health? **(LO 16.6)**

10. Identify three key elements of a dietary pattern to promote brain health. **(LO 16.7)**

## Further Readings

1. U.S. National Library of Medicine: Is longevity determined by genetics? 2020. Available at https://ghr.nlm.nih.gov/primer/traits/longevity. Accessed February 7, 2020.

2. Administration for Community Living: 2018 Profile of Older Americans. 2018. Available at https://acl.gov/sites/default/files/Aging%20and%20Disability%20in%20America/2018OlderAmericansProfile.pdf. Accessed February 7, 2020.

3. Centers for Medicare and Medicaid Services: NHE Fact Sheet. 2019. Available at https://www.cms.gov/Research-Statistics-Data-and-Systems/Statistics-Trends-and-Reports/NationalHealthExpendData/NHE-Fact-Sheet. Accessed February 7, 2020.

4. National Council on Aging: Facts about healthy aging. 2018. Available at https://www.ncoa.org/resources/fact-sheet-healthy-aging. Accessed February 7, 2020.

5. U.S. National Library of Medicine. Aging changes in organs, tissues, and cells. 2019. Available at https://medlineplus.gov/ency/article/004012.htm. Accessed February 7, 2020.

6. Fontana L and others: Promoting health and longevity through diet: From model organisms to humans. *Cell* 161:1, 2015. DOI:10.1016/j.cell.2015.02.020.

7. Kochanek KD, Anderson RN, and Arias E: Health E-Stats: Changes in life expectancy at birth, 2010–2018. 2020. Available at https://www.cdc.gov/nchs/data/hestat/life-expectancy/life-expectancy-2018.htm. Accessed February 7, 2020.

8. Bernstein M and others: Position of the Academy of Nutrition and Dietetics: Food and nutrition for older adults: Promoting health and wellness. *Journal of the Academy of Nutrition and Dietetics* 112:1255, 2012. DOI:10.1016/j.jand.2012.06.015.

9. Thalheimer JC: Nutrition and healthy aging for men. *Today's Dietitian* 17(6):44, 2015.

10. Fávaro-Moreira NC and others: Risk factors for malnutrition in older adults: A systematic review of the literature based on longitudinal data. *Advances in Nutrition* 7:507, 2016. DOI:10.3945/an.115.011254.

11. Skipper A and others: Position of the Academy of Nutrition and Dietetics: Malnutrition (undernutrition) screening tools for all adults. *Journal of the Academy of Nutrition and Dietetics* 120:709, 2020. DOI:10.1016/j.jand.2019.09.011.

12. Abbott Laboratories: Malnutrition Screening Tool (MST). 2013. Available at https://abbottnutrition.com/tools-for-patient-care/rd-toolkit. Accessed July 1, 2020.

13. Manini TM: Energy expenditure and aging. *Ageing Research Review* 9:1, 2010. DOI: 10.1016/j.arr.2009.08.002.

14. Deer RR and Volpi E: Protein intake and muscle function in older adults. *Current Opinion in Clinical Nutrition & Metabolic Care* 18(3):248–253, May 2015. DOI:10.1097/MCO.0000000000000162.

15. Genaro P and others: Dietary protein intake in elderly women: Association with muscle and bone mass. *Nutrition in Clinical Practice* 30:283, 2015. DOI:10.1177/0884533614545404.

16. Paddon-Jones D and others: Protein and healthy aging. *American Journal of Clinical Nutrition* 101:s1339, 2015. DOI:10.3945/ajcn.114.084061.

17. Handu D and others: Malnutrition care during the COVID-19 pandemic: Considerations for registered dietitian nutritionists. *Journal of the Academy of Nutrition and Dietetics* May 14 (Epub ahead of print; DOI:10.1016/j.jand.2020.05.012), 2020.

18. USDA, Agricultural Research Service: Usual Nutrient Intake from Food and Beverages, by Gender and Age, What We Eat in America, NHANES 2013–2016. 2019. Available at http://www.ars.usda.gov/nea/bhnrc. Accessed February 7, 2020.

19. Higdon J and others: Carotenoids. 2016. Available at https://lpi.oregonstate.edu/mic/dietary-factors/phytochemicals/carotenoids. Accessed February 8, 2020.

20. Buch A and others: Muscle function and fat content in relation to sarcopenia, obesity and frailty of old age—An overview. *Experimental Gerontology* 16(10), 2016. DOI:10.1016/j.exger.2016.01.008.

21. Centers for Disease Control and Prevention, National Center for Chronic Disease Prevention and Health Promotion: Arthritis. 2019. Available at

https://www.cdc.gov/chronicdisease/resources/publications/factsheets/arthritis.htm. Accessed February 8, 2020.

22. Oliviero F and others: How the Mediterranean diet and some of its components modulate inflammatory pathways in arthritis. *Swiss Medical Weekly* 145:w14190, 2015. DOI:10.4414/smw.2015.14190.

23. National Institutes of Health, National Institute of Dental and Craniofacial Research: Tooth loss in seniors. 2018. Available at https://www.nidcr.nih.gov/research/data-statistics/tooth-loss/seniors. Accessed February 8, 2020.

24. Touger-Decker R and Mobley C: Position of the American Academy of Nutrition and Dietetics: Oral health and nutrition. *Journal of the Academy of Nutrition and Dietetics* 113:693, 2013. DOI:10.1016/j.jand.2013.03.001.

25. Scarlata K: Digestive wellness: The link between aging and digestive disorders. *Today's Dietitian* 17(7):12, 2015.

26. National Institutes of Health, National Eye Institute: Eye health data and statistics. 2019. Available at https://www.nei.nih.gov/learn-about-ee-health/resources-for-health-educators/eye-health-data-and-statistics. Accessed February 8, 2020.

27. National Institutes of Health: NIH Research Matters. How diet may affect age-related macular degeneration. Updated 2017. Available at https://www.nih.gov/news-events/nih-research-matters/how-diet-may-affect-age-related-macular-degeneration. Accessed February 8, 2020.

28. Centers for Disease Control and Prevention, National Center for Chronic Disease Prevention and Health Promotion: Chronic diseases in America. 2019. Available at https://www.cdc.gov/chronicdisease/resources/infographic/chronic-diseases.htm. Accessed February 8, 2020.

29. Centers for Disease Control and Prevention, National Center for Health Statistics: Data Finder—Health, United States, 2018. 2019. Available at https://www.cdc.gov/nchs/hus/contents2018.htm#Table_038. Accessed February 8, 2020.

30. Gahche J and others: Dietary supplement use was very high among older adults in the United States in 2011–2014. *Journal of Nutrition* 147:1968, 2017. DOI:10.3945/jn.117.255984.

31. Statista: Dossier: Online meal kit delivery services in the U.S. 2019. Available at https://www.statista.com/study/41244/online-meal-kit-delivery-services-in-the-us-statista-dossier. Accessed February 8, 2020.

32. Favaro-Moreira NC and others: Risk factors for malnutrition in older adults: A systematic review of the literature based on longitudinal data. *Advances in Nutrition* 7:507, 2016. DOI:10.3945/an.115.011254.

33. Aune D and others: Fruit and vegetable intake and the risk of cardiovascular disease, total cancer and all-cause mortality—A systematic review and dose-response meta-analysis of prospective studies. *International Journal of Epidemiology* 43(3):1029, 2017. DOI:10.1093/ije/dyw319.

34. Bunis D: House approves update of landmark Older Americans Act. 2019. Available at https://www.aarp.org/politics-society/government-elections/info-2019/house-older-americans-act-funding.html. Accessed February 8, 2020.

35. Khan NA and others: IV. The cognitive implications of obesity and nutrition in childhood. *Monographs of the Society for Research in Child Development* 79:51, 2014. DOI:10.1111/mono.12130.

36. Meeusin R: Exercise, nutrition and the brain. *Sports Medicine* 44:S47, 2014. DOI:10.1007/s40279-014-0150-5.

37. Dauncey MJ: Nutrition, the brain and cognitive decline: Insights from epigenetics. *European Journal of Clinical Nutrition* 68:1179, 2014. DOI:10.1038/ejcn.2014.173.

38. Hribar C: Migraine statistics. 2019. Available at https://migraine.com/migraine-statistics. Accessed February 8, 2020.

39. Appold K: Migraine headaches—Here's how to identify food triggers and reduce debilitating symptoms. *Today's Dietitian* 14:14, 2012.

40. Slavin M and Ailani J: A clinical approach to addressing diet with migraine patients. *Current Neurology and Neuroscience Reports* 17:17, 2017. DOI:10.1007/s11910-017-0721-6.

41. Evans EW and others: Dietary intake patterns and diet quality in a nationally representative sample of women with and without severe headache or migraine. *Headache* 55:550, 2015. DOI:10.1111/head.12527.

42. National Institutes of Health, National Institute of Mental Health: Major Depression. 2019. Available at https://www.nimh.nih.gov/health/statistics/major-depression.shtml. Accessed February 8, 2020.

43. Gangwisch JE and others: High glycemic index diet as a risk factor for depression: Analyses from the Women's Health Initiative. *American Journal of Clinical Nutrition* 102:454, 2015. DOI:10.3945/ajcn.114.103846.

44. Lassale C and others: Healthy dietary indices and risk of depressive outcomes: A systematic review and meta-analysis of observational studies. *Molecular Psychiatry* 24(7):965–986. 2018. DOI:10.1038/s41380-018-0237-8.

45. Swift KM: Integrative nutrition therapy for mood disorders. *Today's Dietitian* 19:36, 2017.

46. Angelucci F and others: Antibiotics, gut microbiota, and Alzheimer's disease. *Journal of Neuroinflammation* 16:108, 2019. DOI:10.1186/s12974-019-1494-4.

47. Jackson A and others: Diet in Parkinson's disease: Critical role for the microbiome. *Frontiers in Neurology* 10:1245, 2019. DOI:10.3389/fneur.2019.01245.

48. National Institutes of Health, National Institute on Aging: Alzheimer's disease fact sheet. 2019. Available at https://www.nia.nih.gov/health/alzheimers-disease-fact-sheet. Accessed February 8, 2020.

49. Johnson EJ: Role of lutein and zeaxanthin in visual and cognitive function throughout the lifespan. *Nutrition Reviews* 72:605, 2014. DOI:10.1111/nure.12133.

50. Rutjes AWS and others: Vitamin and mineral supplementation for maintaining cognitive function in cognitively healthy people in mid and late life. *Cochrane Database of Systematic Reviews*, 2018. DOI:10.1002/14651858.CD011906.pub2.

51. Danthir V and others: An 18-mo randomized, double-blind, placebo-controlled trial of DHA-rich fish oil to prevent age-related cognitive decline in cognitively normal adults. *American Journal of Clinical Nutrition* 107:754, 2018. DOI:10.1093/ajcn/nqx077.

52. Radd-Vagenas S and others: Effect of the Mediterranean diet on cognition and brain morphology and function: A systematic review of randomized controlled trials. *American Journal of Clinical Nutrition* 107:389, 2018. DOI:10.1093/ajcn/nqx070.

53. Marcason W: What are the components to the MIND diet? *Journal of the Academy of Nutrition and Dietetics* 115:1744, 2015. DOI:10.1016/j.jand.2015.08.002.

54. Thalheimer JC: Nutrients for a sharp memory. *Today's Dietitian* 19:24, 2017.

55. Morris MC and others: MIND diet associated with reduced incidence of Alzheimer's disease. *Alzheimer's and Dementia* 11(9):1007, 2015. DOI:10.1016/j.jalz.2014.11.009.

56. Barberger-Gateau P: Nutrition and brain aging: How can we move ahead? *European Journal of Clinical Nutrition* 68:1245, 2014. DOI:10.1038/ejcn.2014.177.

57. Ruscigno M: Brain food for older adults. *Today's Dietitian* 18:22, 2016.

# Appendix A
## Daily Values Used on Food Labels

**TABLE A-1** ■ Daily Values Used on Food Labels in the United States, with a Comparison to the Latest RDAs and Other Nutrient Standards*

| Dietary Constituent | Unit of Measure | Current Daily Values for People Over 4 Years of Age | RDA or Other Current Dietary Standard Males 19–30 Years Old | RDA or Other Current Dietary Standard Females 19–30 Years Old |
|---|---|---|---|---|
| Total fat† | grams (g) | 78 | — | — |
| Saturated fatty acids† | grams (g) | 20 | — | — |
| Protein† | grams (g) | 50 | 56 | 46 |
| Cholesterol§ | milligrams (mg) | 300 | — | — |
| Carbohydrate† | grams (g) | 275 | 130 | 130 |
| Added sugars | grams (g) | 50 | — | — |
| Dietary fiber | grams (g) | 28 | 38 | 25 |
| Vitamin A | micrograms (mcg) Retinol Activity Equivalents (RAE) | 900 | 900 | 700 |
| Vitamin D | micrograms (mcg) | 20 | 15 | 15 |
| Vitamin E | milligrams (mg) alpha tocopherol | 15 | 15 | 15 |
| Vitamin K | micrograms (mcg) | 120 | 120 | 90 |
| Vitamin C | milligrams (mg) | 90 | 90 | 75 |
| Folate | micrograms (mcg) Dietary Folate Equivalents (DFE) | 400 | 400 | 400 |
| Thiamin | milligrams (mg) | 1.2 | 1.2 | 1.1 |
| Riboflavin | milligrams (mg) | 1.3 | 1.3 | 1.1 |
| Niacin | milligrams (mg) | 16 | 16 | 14 |
| Vitamin B-6 | milligrams (mg) | 1.7 | 1.3 | 1.3 |
| Vitamin B-12 | micrograms (mcg) | 2.4 | 2.4 | 2.4 |
| Biotin | micrograms (mcg) | 30 | 30 | 30 |
| Pantothenic acid | milligrams (mg) | 5 | 5 | 5 |
| Calcium | milligrams (mg) | 1300 | 1000 | 1000 |
| Phosphorus | milligrams (mg) | 1250 | 700 | 700 |
| Iodine | micrograms (mcg) | 150 | 150 | 150 |
| Iron | milligrams (mg) | 18 | 8 | 18 |
| Magnesium | milligrams (mg) | 420 | 400 | 310 |
| Copper | milligrams (mg) | 0.9 | 0.9 | 0.9 |
| Zinc | milligrams (mg) | 11 | 11 | 8 |
| Sodium‡ | milligrams (mg) | 2300 | 1500 | 1500 |
| Potassium‡ | milligrams (mg) | 4700 | 3400 | 2600 |
| Chloride‡ | milligrams (mg) | 2300 | 2300 | 2300 |
| Manganese | milligrams (mg) | 2.3 | 2.3 | 1.8 |
| Selenium | micrograms (mcg) | 55 | 55 | 55 |
| Chromium | micrograms (mcg) | 35 | 35 | 25 |
| Molybdenum | micrograms (mcg) | 45 | 45 | 45 |

* Daily Values are generally set at the highest nutrient recommendation in a specific age and gender category. Some changes were made in 2016 to the Daily Values and their units as part of the new Nutrition Facts label. Most changes occurred because of the dietary-related diseases that are common in the United States and to more closely match the RDA or AI values for the nutrient.

† These Daily Values are based on a 2000-kcal diet, instead of RDAs, with a caloric distribution of 30% from fat (and one-third of this total from saturated fat), 60% from carbohydrate, and 10% from protein.

§ Based on recommendations of U.S. federal agencies.

‡ The considerably higher Daily Value for sodium is there to allow for more diet flexibility, but the extra amount is not needed to maintain health.

# Appendix B
## Diabetes Menu-Planning Tools

## Food Lists, Eating Plans, and Healthy Lifestyles: Dietary Recommendations for Individuals Living with Diabetes

Registered dietitian nutritionists and other diabetes educators work closely with their clients to help them understand how the foods they eat directly impact their day-to-day quality of life. Providing **food lists** that promote individual choices is one way diabetes educators can help people to plan their dietary patterns to better manage their blood sugar. The first food lists for people with diabetes were developed by the American Dietetic Association (now the Academy of Nutrition and Dietetics), American Diabetes Association, and U.S. Public Health Service 60 years ago. Since then, they have been revised to reflect advances in nutrition recommendations and the ever-expanding variety of foods in the marketplace. The most recent version, *Choose Your Foods: Food Lists for Diabetes,* was published in 2019.

Food lists organize the many details of the nutrient composition of foods into a manageable framework based on calorie and macronutrient content. In the *Food Lists for Diabetes,* individual foods are placed into three broad groups: carbohydrates, proteins, and fats. Within these groups are lists that contain foods of similar macronutrient composition: various types of milk and milk substitutes, fruits, vegetables, starches, other carbohydrates, proteins, and fats. There are even lists that show how to account for alcohol, combination foods (e.g., casseroles), and a wide variety of fast foods. These lists are designed so that when the given serving size is observed, each food on a list provides roughly the same amount of carbohydrate, protein, fat, and calories. The patient and a registered dietitian nutritionist first tailor a healthy eating pattern to meet the client's energy and specific macronutrient needs. Then, the client can select **choices** from each of the various lists that fit into the plan without having to look up or memorize the nutrient values of numerous foods.

Table B-1 summarizes the basic nutrient composition of foods in each food list. The serving sizes of individual foods in a list may vary, but general estimates are given. The protein and milk and milk substitutes lists are divided into subclasses, which vary in fat content and, thus, in the amount of calories they provide. You can see that each food list is unique in the calories and macronutrients it supplies. A healthy meal plan should include foods from each of the lists to ensure nutrient adequacy. Study Table B-1 to become familiar with the food groupings, the approximate size of choices on each food list, and the amounts of carbohydrate, protein, fat, and calories per choice.

Because the *Food Lists for Diabetes* offer a quick way to estimate the calorie, carbohydrate, protein, and fat content in any food or meal, they are a valuable menu-planning tool for individuals without diabetes, as well. In fact, the Academy of Nutrition and Dietetics and the American Diabetes Association have published a related guide, *Choose Your Foods: Food Lists for Weight Management.*

Recognize that these food lists group foods somewhat differently than MyPlate. For the *Food Lists,* the focus is more on nutrient composition and the food's eventual physiological effect on blood sugar than on its plant or animal origin. For example, the starch list includes not only breads, cereals, grains and pasta, starchy vegetables, crackers and snacks, but also beans, peas, and lentils. Although beans and peas are vegetables, their macronutrient composition resembles that of bread more than that of broccoli. In addition, many foods that would traditionally be categorized as dairy products do not appear with the milk and milk substitutes list. Instead, cheeses are grouped as proteins, whereas cream and cream cheese show up on the fats lists.

**food lists** A system for classifying foods into numerous lists based on the foods' macronutrient composition, and establishing serving sizes, so that one serving of each food on a list contains the same amount of carbohydrate, protein, fat, and calories.

**choice** The serving size of a food on a specific food list.

## TABLE B-1 ■ Approximate Nutrient Composition of Food Choices from *Choose Your Foods: Food Lists for Diabetes, 2019*

| Groups/Lists | Household Measures* | Carbohydrate (g) | Protein (g) | Fat (g) | Energy (kcal) |
|---|---|---|---|---|---|
| **Carbohydrates** | | | | | |
| Starch (e.g., breads, cereals, pasta, starchy vegetables, crackers, snacks, beans, peas, and lentils) | 1 slice, ¾ cup raw, or ½ cup cooked | 15 | 3 | 1 or less† | 80 |
| Fruits | 1 small/medium piece | 15 | — | — | 60 |
| Milk and milk substitutes | 1 cup | | | | |
| Fat-free, low-fat, 1% | | 12 | 8 | 0–3† | 100 |
| Reduced-fat, 2% | | 12 | 8 | 5 | 120 |
| Whole | | 12 | 8 | 8 | 160 |
| Nonstarchy vegetables | 1 cup raw or ½ cup cooked | 5 | 2 | — | 25 |
| Sweets, desserts, and other carbohydrates | Varies | 15 | Varies | Varies | Varies |
| **Proteins** | 1 ounce | | | | |
| Lean | | — | 7 | 2 | 45 |
| Medium-fat | | — | 7 | 5 | 75 |
| High-fat | | — | 7 | 8 | 100 |
| Plant-based | | Varies | 7 | Varies | Varies |
| **Fats** | 1 teaspoon | — | — | 5 | 45 |
| **Alcohol** | Varies | Varies | — | — | 100 |

* An estimate; see food lists for actual amounts.
† Calculated as 1 gram for purposes of calorie contribution.
Source: *Choose Your Foods: Food Lists for Diabetes*, 2019, which is the basis of a meal-planning system designed by a committee of the American Diabetes Association and the Academy of Nutrition and Dietetics.

In some instances, a food counts for more than one choice at a time. In the category of sweets, desserts, and other carbohydrates, you will find a variety of snack items and condiments that count as carbohydrates and fats. The food lists also provide the user with some guidance on accounting for a wide variety of combination foods, such as pizza, casseroles, and soups. A list of free foods includes choices such as reduced-fat or fat-free foods, condiments, seasonings, and sugar-free drinks that, when consumed in moderation, have little or no impact on energy intake and blood sugar.

## USING THE *FOOD LISTS* TO DEVELOP DAILY MENUS

Now let us use the *Food Lists* to plan a 1-day menu. Our calorie target will be 2000 kcal, with 55% derived from carbohydrates (1100 kcal), 15% from protein (300 kcal), and 30% from fat (600 kcal). This can be achieved with 2 reduced-fat milk choices, 3 nonstarchy vegetable choices, 5 fruit choices, 11 starch choices, 4 lean protein choices, and 6 fat choices (Table B-2). This is just one of many possible combinations; the *Food Lists* offer great flexibility. For example, more protein choices could be included if less milk were used.

Table B-3 arbitrarily distributes these choices into breakfast, lunch, dinner, and a snack. Breakfast includes 1 reduced-fat milk choice, 2 fruit choices, 2 starch choices, and 1 fat choice. Drawing from the food lists, this plan could be achieved with ¾ cup of a ready-to-eat breakfast cereal, 1 cup of reduced-fat milk, 1 slice of bread with 1 teaspoon margarine, and 1 cup of orange juice.

## TABLE B-2 ■ Possible Food Choice Patterns That Yield Approximately 55% of Calories as Carbohydrate, 30% as Fat, and 15% as Protein

| Food List | 1200* | 1600* | 2000 | 2400 | 2800 | 3200 | 3600 |
|---|---|---|---|---|---|---|---|
| | | | | | Daily Kcal Intakes | | |
| | | | **Choices Per Day** | | | | |
| Milk (2% reduced-fat) | 2 | 2 | 2 | 2 | 2 | 2 | 2 |
| Nonstarchy vegetable | 3 | 3 | 3 | 4 | 4 | 4 | 4 |
| Fruit | 3 | 4 | 5 | 6 | 8 | 9 | 9 |
| Starch | 5 | 8 | 11 | 13 | 15 | 18 | 21 |
| Protein (lean) | 4 | 4 | 4 | 5 | 6 | 7 | 8 |
| Fats | 2 | 4 | 6 | 8 | 10 | 11 | 13 |

*Calorie intakes of 1200 and 1600 kcal contain 20% of calories as protein and 50% of calories as carbohydrate to allow for greater flexibility in meal and snack planning.

**TABLE B-3** ▪ **Sample 1-Day 2000-kcal Menu Based on the Food Lists for Diabetes Plan***

**Breakfast**

| | |
|---|---|
| 1 reduced-fat milk choice | 1 cup 2% reduced-fat milk (some on cereal) |
| 2 fruit choices | 1 cup (8 oz) orange juice |
| 2 starch choices | ¾ cup ready-to-eat breakfast cereal and 1 slice (1 oz) whole wheat toast |
| 1 fat choice | 2 tablespoons (1 oz) avocado on toast |

**Lunch**

| | |
|---|---|
| 4 starch choices | 2 slices (2 oz) whole wheat bread and 6 graham crackers (2½ inches by 2½ inches) |
| 2 fat choices | 1 slice bacon, 1 teaspoon mayonnaise |
| 1 nonstarchy vegetable choice | 1 sliced tomato |
| 2 fruit choices | 1 banana (8 inches or 8 oz) |
| 1 reduced-fat milk choice | 1 cup 2% reduced-fat milk |

**Snack**

| | |
|---|---|
| 1 starch choice | ¾ ounce pretzels |

**Dinner**

| | |
|---|---|
| 4 lean protein choices | 4 ounces lean steak (well trimmed) |
| 2 starch choices | 1 medium baked potato (6 oz) |
| 1 fat choice | 1 teaspoon soft tub margarine |
| 2 nonstarchy vegetable choices | 1 cup cooked broccoli |
| 1 fruit choice | ½ cup sliced kiwi |

**Snack**

| | |
|---|---|
| 2 starch choices | ½ large bagel (2 oz) |
| 2 fat choices | 2 tablespoons (1 oz) regular cream cheese |

*The target plan was a 2000-kcal intake, with 55% of calories from carbohydrate, 15% from protein, and 30% from fat. Computer analysis indicates that this menu yielded approximately 2040 kcal, with 53% of calories from carbohydrate, 16% from protein, and 31% from fat—in close agreement with the targeted goals.

This 1-day menu is only one of endless possibilities with the *Food Lists for Diabetes.* Apple juice could replace the orange juice; two apples could be exchanged for the banana. For simplicity, we have used a variety of individual foods to achieve the total number of choices for this healthy eating plan. However, the *Food Lists* also include some commonly used combination foods. For instance, a 1-cup serving of lasagna typically provides 2 medium-fat meat choices plus 2 carbohydrate choices. With practice, you will be able to estimate the choices from complex foods on your own (Fig. B-1). For now, using individual foods makes learning the *Food Lists* much easier. Finally, you might want to prove to yourself that the food choices listed in Table B-3 really fulfill the plan set forth in Table B-2.

| Food List | Total Food Choices to Be Consumed Daily | Food Choices Consumed at Each Meal | | | |
|---|---|---|---|---|---|
| | | Breakfast | Lunch | Dinner | Snacks |
| Milk and milk substitutes | | | | | |
| Nonstarchy vegetables | | | | | |
| Fruits | | | | | |
| Starch | | | | | |
| Proteins | | | | | |
| Fats | | | | | |

**FIGURE B-1** ◄ Record the *Food Lists for Diabetes* pattern you have chosen in the left-hand column. Then distribute the food choices throughout the day, noting the food to be used and the serving size.

# Examples of Food Choices from
# *Food Lists for Diabetes*

In this section, you will find just a few examples of the many food choices that are included in the most recent edition of *Choose Your Foods: Food Lists for Diabetes*.

## Starches

Starches provide 15 grams of carbohydrate, 0 to 3 grams of protein, 0 to 1 gram of fat, and about 80 kcal per serving. Keep in mind that the serving sizes for starch choices on these food lists are usually smaller than those recommended by MyPlate. Typically, 1 starch choice equals ½ cup of cooked cereal, grain, or starchy vegetable; ⅓ cup of cooked rice or pasta; 1 slice (1 oz) of bread; or ¾ to 1 ounce of crackers or grain-based snack foods. Also, some foods with high fat content, such as biscuits or hash browns, may be counted as 1 starch plus 1 or 2 fats. Beans, peas, and lentils count as 1 starch plus 1 lean protein choice.

### BREAD

| Serving Size | Food | |
| --- | --- | --- |
| ¼ large (1 oz) | Bagel | |
| 1 slice (1 oz) | Bread | |
| ½ | English muffin | |
| ½ (¾ oz) | Hamburger bun | |
| 1 oz | Naan (3¼-inch square) | |
| 1 | Pancake (4-inch diameter) | |
| 1 | Tortilla, flour (6-inch diameter) | |

### CEREALS

| Serving Size | Food | |
| --- | --- | --- |
| ½ cup | Cooked cereal (e.g., oatmeal) | |
| ¼ cup | Granola cereal | |
| 1½ cups | Puffed cereal (e.g., puffed rice) | |
| ½ cup | Sugar-coated cereal (e.g., Frosted Flakes®) | |
| ¾ cup | Unsweetened, ready-to-eat cereal (e.g., Cheerios®) | |

### GRAINS

| Serving Size | Food | |
| --- | --- | --- |
| ⅓ cup | Pasta, cooked | |
| ⅓ cup | Quinoa, cooked | |
| ⅓ cup | Rice, cooked (e.g., white and brown) | |
| ½ cup | Wild rice, cooked | |

### STARCHY VEGETABLES

| Serving Size | Food |
| --- | --- |
| ½ cup | Corn, kernel |
| 1 cup | Mixed vegetables (e.g., corn, peas, and carrots) |
| ¼ (3 oz) | Potato, baked with skin |
| ½ cup | Potatoes, mashed with milk and fat (1 starch + 1 fat) |
| ½ cup | Spaghetti sauce |
| ½ cup | Sweet potato |
| 1 cup | Winter squash (e.g., acorn and butternut) |

### CRACKERS AND SNACKS

| Serving Size | Food |
| --- | --- |
| 8 | Animal crackers |
| 6 | Butter crackers (e.g., Ritz®; 1 starch + 1 fat) |
| 3 | Graham crackers (2½-inch squares) |
| 3 cups | Popcorn (no fat added) |
| ¾ ounce | Pretzels |
| 13 | Tortilla chips (1 starch + 2 fats) |

### BEANS, PEAS, AND LENTILS
### (count as 1 starch and 1 lean protein)

| Serving Size | Food |
| --- | --- |
| ⅓ cup | Baked beans |
| ½ cup | Beans, cooked or canned (e.g., black, garbanzo, and kidney) |
| ½ cup | Lentils, cooked |
| ½ cup | Peas, cooked (e.g., black-eyed and split) |

# Fruits

One choice from the Fruits list provides 15 grams of carbohydrate, 0 grams of protein, 0 grams of fat, and 60 kcal. Typically, 1 fruit choice is equal to ½ cup of unsweetened canned or frozen fruit, 1 small fresh fruit (about 2½ inch diameter), ½ cup (4 oz) of unsweetened 100% fruit juice, or 2 tablespoons of dried fruit. Recognize that the fruit you buy at the grocery store may amount to more than one fruit choice; a large banana, for example, counts for 2 fruit choices. The serving sizes of fruit juices and dried fruit are small because these are more concentrated sources of carbohydrates and energy.

## FRUITS

| Serving Size | Food |
| --- | --- |
| 1 small (4 oz) | Apple, unpeeled |
| ½ cup | Applesauce, unsweetened |
| 1 (4 oz) | Banana |
| ¾ cup | Blueberries |
| 12 (3½ oz) | Cherries, fresh |
| 17 (3 oz) | Grapes |
| 1 cup | Honeydew melon, diced |

| Serving Size | Food |
| --- | --- |
| ½ cup | Kiwi, sliced |
| 1 (6½ oz) | Orange |
| ½ cup | Pineapple, canned |
| ½ cup | Pomegranate seeds (arils) |
| 3 | Prunes |
| 1¼ cup | Strawberries, whole |
| 1¼ cup | Watermelon, diced |

## FRUIT JUICE

| Serving Size | Food |
| --- | --- |
| ½ cup (4 oz) | Apple juice or cider |
| ⅓ cup (2.7 oz) | Grape juice |
| ½ cup (4 oz) | Orange juice |
| ⅓ cup (2.7 oz) | Prune juice |

# Milk and Milk Substitutes

Milk and milk substitutes are divided into subcategories based on their fat content. All milk and yogurt products provide 12 grams of carbohydrate and 8 grams of protein but may vary in fat content from 0 to 8 grams per choice. The subcategory of other milk foods and milk substitutes includes some products that may be used in place of milk in the eating pattern (e.g., soy milk) but have a slightly different nutrient profile than traditional milk and yogurt products. Those foods, as indicated below, are counted as a combination of carbohydrate (15 grams of carbohydrate, 60 kcal) and fat (5 grams of fat, 45 kcal) choices. Please note that other products used as dairy alternatives (e.g., almond milk) are listed with fat choices.

## FAT-FREE (SKIM) AND LOW-FAT MILK AND YOGURT
### (12 grams of carbohydrate, 8 grams of protein, 0–3 grams of fat, and 100 kcal)

| Serving Size | Food |
| --- | --- |
| ½ cup (4 oz) | Canned, evaporated, fat-free milk |
| 1 cup (8 oz) | Fat-free (skim) milk, low-fat (1%) milk, and buttermilk |

| Serving Size | Food |
| --- | --- |
| ⅔ cup (6 oz) | Yogurt (fat-free plain or fat-free Greek, unsweetened) (1 milk + 1 carb) |
| 1 cup (8 oz) | Chocolate milk (1 milk + 1 carb) |

## REDUCED-FAT (2%) MILK AND YOGURT
**(12 grams of carbohydrate, 8 grams of protein, 5 grams of fat, and 120 kcal)**

| Serving Size | Food |
|---|---|
| 1 cup (8 oz) | Reduced-fat (2%) milk, acidophilus milk, and kefir |
| ⅔ cup (6 oz) | Yogurt, 2% reduced-fat, plain |

## WHOLE MILK AND YOGURT
**(12 grams of carbohydrate, 8 grams of protein, 8 grams of fat, and 160 kcal)**

| Serving Size | Food |
|---|---|
| ½ cup (4 oz) | Evaporated whole milk |
| 1 cup (8 oz) | Whole milk, buttermilk, and goat's milk |
| 1 cup (8 oz) | Yogurt, whole milk, plain |

### OTHER MILK FOODS AND MILK SUBSTITUTES

| Serving Size | Food | Choices |
|---|---|---|
| 1 cup (8 oz) | Almond milk, plain | ½ carbohydrate + ½ fat |
| 1 cup (8 oz) | Coconut milk, flavored | 1 carbohydrate + 1 fat |
| ⅓ cup (2.7 oz) | Eggnog, whole milk | 1 carbohydrate + 1 fat |
| 1 cup (8 oz) | Rice milk, flavored, low-fat | 2 carbohydrates |
| 1 cup (8 oz) | Soy milk, regular, plain | 1 carbohydrate + 1 fat |
| ⅔ cup (6 oz) | Yogurt with fruit, low-fat | 1 fat-free milk + 1 carbohydrate |

# Nonstarchy Vegetables

Nonstarchy vegetables still provide carbohydrates, but not as much as their starchy counterparts. One nonstarchy vegetable provides 5 grams of carbohydrate, 2 grams of protein, 0 grams of fat, and 25 kcal. Typically, a choice is equal to ½ cup of cooked vegetables, 1 cup of raw vegetables, 3 cups of salad or leafy greens, or ½ cup (4 fl oz) vegetable juice. Large servings of nonstarchy vegetables (i.e., three choices) should be counted as one carbohydrate choice (15 grams of carbohydrate, 60 kcal) rather than multiple nonstarchy vegetables. Because of their low carbohydrate content, salad greens (e.g., iceberg, romaine, and endive) actually count as free foods. To comply with advice from the *Dietary Guidelines for Americans*, it is important to select a variety of starchy and nonstarchy vegetables each day because each has a distinct micronutrient and phytochemical profile. Take extra care to select 2 or 3 nonstarchy vegetables with deep colors, such as spinach, carrots, and beets, per day.

| Serving Size | Food |
|---|---|
| ½ cup | Asparagus, cooked |
| ½ cup | Beets, cooked |
| ½ cup | Broccoli, cooked |
| 1 cup | Carrots, sliced, raw |
| ½ cup | Collard greens, cooked |
| 1 cup | Cucumber, raw |

| Serving Size | Food |
|---|---|
| ½ cup | Green beans, cooked |
| 3 cups | Salad greens, raw |
| ½ cup | Summer squash, cooked |
| ½ cup | Tomatoes, stewed |
| ½ cup | Vegetable juice |

# Sweets, Desserts, and Other Carbohydrates

Foods on this list may not match the nutrient profiles of other starches, but they are commonly consumed and must be accounted for in the eating pattern. Sweetened beverages, desserts, and sweeteners and condiments that we add to foods can be counted as a combination of carbohydrate (15 grams of carbohydrates, 60 kcal) and fat (5 grams of fat, 45 kcal) choices.

## BEVERAGES, SODA, AND SPORTS DRINKS

| Serving Size | Food | Choices |
|---|---|---|
| 1 can (8 oz) | Energy drink | 2 carbohydrates |
| 1 cup (8 oz) | Fruit drink or lemonade | 2 carbohydrates |
| 1 can (12 oz) | Soft drink, regular | 2½ carbohydrates |
| 1 cup (8 oz) | Sports drink | 1 carbohydrate |

## BROWNIES, CAKE, COOKIES, GELATIN, PIE, AND PUDDING

| Serving Size | Food | Choices |
|---|---|---|
| 1/12 cake (2 oz) | Angel food cake, unfrosted | 2 carbohydrates |
| 1¼-inch square (1 oz) | Brownie, unfrosted | 1 carbohydrate + 1 fat |
| 2-inch square (2 oz) | Cake, frosted | 2 carbohydrates + 1 fat |
| 2 small | Chocolate chip cookies | 1 carbohydrate + 2 fats |
| ½ cup | Gelatin, regular | 1 carbohydrate |
| ½ cup | Pudding, regular, 2% milk | 2 carbohydrates |
| ⅛ pie | Pumpkin pie | 1½ carbohydrates + 1½ fats |
| 5 pieces | Vanilla wafers | 1 carbohydrate + 1 fat |

## CANDY, SPREADS, SWEETS, SWEETENERS, SYRUPS, AND TOPPINGS

| Serving Size | Food | Choices |
|---|---|---|
| 1 tbsp | Agave, syrup | 1 carbohydrate |
| 5 | Chocolate kisses | 1 carbohydrate + 1 fat |
| 1 tbsp | Honey | 1 carbohydrate |
| 1 tbsp | Jam or jelly, regular | 1 carbohydrate |
| 2 tbsp | Liquid nondairy coffee creamer | 1 carbohydrate |
| 1 tbsp | Pancake syrup, regular | 1 carbohydrate |

## CONDIMENTS AND SAUCES

| Serving Size | Food | Choices |
|---|---|---|
| 3 tbsp | Barbecue sauce | 1 carbohydrate |
| ½ cup | Gravy | ½ carbohydrate + ½ fat |
| 3 tbsp | Ketchup | 1 carbohydrate |
| 3 tbsp | Salad dressing, fat-free, cream-based | 1 carbohydrate |

## DOUGHNUTS, MUFFINS, PASTRIES, AND SWEET BREADS

| Serving Size | Food | Choices |
|---|---|---|
| 1 (2½ oz) | Danish | 2½ carbohydrates + 2 fats |
| 1 (2 oz) | Glazed doughnut | 2 carbohydrates + 2 fats |
| 1 (4 oz) | Muffin, regular | 4 carbohydrates + 2½ fats |

## FROZEN BARS, FROZEN DESSERTS, FROZEN YOGURT, AND ICE CREAM

| Serving Size | Food | Choices |
|---|---|---|
| 1 (3 oz) | Frozen 100% fruit juice bar | 1 carbohydrate |
| ½ cup | Greek frozen yogurt, low-fat | 1½ carbohydrates |
| ½ cup | Ice cream, no sugar added | 1 carbohydrate + 1 fat |
| ½ cup | Ice cream, regular | 1 carbohydrate + 2 fats |
| ½ cup | Sherbet, sorbet | 2 carbohydrates |

# Protein

Similar to the choices on the Milk and Milk Substitutes list, protein choices vary in fat and calorie content. Lean protein choices, such as egg whites and skinless poultry, provide 0 grams of carbohydrate, 7 grams of protein, 2 grams of fat, and 45 kcal. Medium-fat protein choices, such as whole eggs and poultry with skin, provide 0 grams of carbohydrate, 7 grams of protein, 5 grams of fat, and 75 kcal. High-fat protein choices, including many types of sausage and bacon, provide 0 grams of carbohydrate, 7 grams of protein, 8 grams of fat, and 100 kcal. Plant-based protein choices usually contain some carbohydrates, so they count as a combination of carbohydrate or starch and protein choices. Note that choices are very small (1-oz portions); a typical hamburger would count as 3 or 4 protein choices.

## LEAN PROTEIN (0 grams of carbohydrate, 7 grams of protein, 2 grams of fat, and 45 kcal)

| Serving Size | Food |
|---|---|
| 1 oz | Beef with 10% or lower fat (e.g., round and sirloin) |
| 1 oz | Cheese with 3 grams of fat or less (e.g., fat-free mozzarella) |
| 1 oz | Deli meats with 3 grams of fat or less per serving (e.g., turkey and ham) |
| 2 | Egg whites |
| 1 oz | Fish, not fried (e.g., catfish, cod, and tuna canned in water) |
| 1 oz | Lean pork (e.g., ham and tenderloin) |
| 1 oz | Poultry, without skin |
| 1 oz | Shellfish (e.g., shrimp and crab) |
| 1 oz | Wild game (e.g., buffalo and venison) |

## MEDIUM-FAT PROTEIN (0 grams of carbohydrate, 7 grams of protein, 5 grams of fat, and 75 kcal)

| Serving Size | Food |
|---|---|
| 1 oz | Beef with 15% or higher (e.g., rib roast and ground beef) |
| 1 oz | Cheese with 4 to 7 grams of fat per ounce (e.g., feta and mozzarella) |
| 1 | Egg |
| 1 oz | Fish, fried |
| 1 oz | Pork (e.g., cutlet and shoulder roast) |
| 1 oz | Poultry, with skin |

## HIGH-FAT PROTEIN (0 grams of carbohydrate, 7 grams of protein, 8 grams of fat, and 100 kcal)

| Serving Size | Food |
|---|---|
| 2 slices | Bacon, pork |
| 1 oz | Cheese (e.g., American, Cheddar, Parmesan, and Swiss) |
| 1 oz | Deli meat with 8 grams of fat or more per serving (e.g., bologna and salami) |
| 1 | Hot dog |
| 1 oz | Sausage (e.g., bratwurst and summer sausage) |

## PLANT-BASED PROTEINS

| Serving Size | Food | Choices |
|---|---|---|
| ½ cup | Beans, cooked or canned (e.g., black, kidney, and pinto) | 1 starch + 1 lean protein |
| ½ cup | Edamame, shelled | ½ carbohydrate + 1 lean protein |
| ⅓ cup | Hummus | 1 carbohydrate + 1 medium-fat protein |
| 1 (2½ oz) | Meatless burger, soy-based | ½ carbohydrate + 2 lean proteins |
| 1 tbsp | Peanut butter | 1 high-fat protein |
| ½ cup | Refried beans, canned | 1 carbohydrate + 1 lean protein |
| ½ cup (4 oz) | Tofu | 1 medium-fat protein |

# Fats

One fat choice is 5 grams of fat and 45 kcal. Fats are subdivided into unsaturated fats, which come mainly from plant sources, and saturated fats, which come mainly from animal sources. In line with recommendations from other major health authorities, the *Food Lists for Diabetes* advises people to choose monounsaturated and polyunsaturated fats in place of saturated fats.

## UNSATURATED FATS—MONOUNSATURATED FATS (5 grams of fat and 45 kcal)

| Serving Size | Food | Serving Size | Food |
|---|---|---|---|
| 1 cup | Almond milk, unsweetened | 1 tsp | Oil (e.g., canola and olive) |
| 6 | Almonds | 8 | Olives, black |
| 2 tbsp (1 oz) | Avocado | 10 | Peanuts |
| 1½ tsp | Nut butter (e.g., almond and peanut) | 16 | Pistachios |

## UNSATURATED FATS—POLYUNSATURATED FATS (5 grams of fat and 45 kcal)

| Serving Size | Food | Serving Size | Food |
|---|---|---|---|
| 1½ tbsp | Flaxseed, ground | 1 tsp | Oil (e.g., corn, safflower, and sunflower) |
| 1 tbsp | Low-fat vegetable oil spread | 1 tbsp | Reduced-fat mayonnaise |
| 1 tsp | Margarine, stick and tub | 2 tbsp | Salad dressing, reduced-fat (may contain carbohydrate) |
| 1 tsp | Mayonnaise, regular | 1 tbsp | Salad dressing, regular |

## SATURATED FATS (5 grams of fat and 45 kcal)

| Serving Size | Food |
|---|---|
| 1 slice | Bacon, cooked |
| 1 tbsp | Butter, reduced-fat |
| 1½ tsp | Butter, regular |

| Serving Size | Food |
|---|---|
| 1 tbsp (½ oz) | Cream cheese, regular |
| 2 tbsp | Coconut, shredded |
| 1 tsp | Coconut oil |
| 2 tbsp | Sour cream, regular |

# Free Foods

A *free food* is any food or drink choice that contains less than 20 kcal or less than 5 grams of carbohydrate per serving. When eaten in small amounts throughout the day, these foods have little impact on blood sugar. Foods with a serving size listed should be limited to 3 servings per day. Foods listed without a serving size (indicated with a *) can be eaten as often as you like. However, many free foods are high in sodium, so moderation is important.

## LOW-CARBOHYDRATE FOODS

| Serving Size | Food |
|---|---|
| 1 piece | Candy, hard or sugar-free |
| ¼ cup | Cooked nonstarchy vegetables (e.g., carrots, cauliflower, and green beans) |
| * | Gelatin, sugar-free |
| 2 tsp | Jam or jelly, light or no-sugar-added type |

| Serving Size | Food |
|---|---|
| ½ cup | Raw nonstarchy vegetables (e.g., broccoli, carrots, cucumber, and tomato) |
| * | Salad greens (no dressing) |
| * | Sugar substitutes |

## REDUCED-FAT OR FAT-FREE FOODS

| Serving Size | Food |
|---|---|
| 1 tbsp | Cream cheese, fat-free |
| 4 tsp | Coffee creamer, liquid, sugar-free, flavored |
| 1 tsp | Margarine spread, reduced-fat |

| Serving Size | Food |
|---|---|
| 1 tbsp | Mayonnaise, fat-free |
| 1 tbsp | Salad dressing, fat-free |
| 2 tbsp | Whipped topping, light or fat-free |

## CONDIMENTS

| Serving Size | Food |
|---|---|
| 2 tsp | Barbecue sauce |
| 1½ | Dill pickles (medium) |
| * | Hot pepper sauce |
| 1 tbsp | Ketchup |

| Serving Size | Food |
|---|---|
| * | Mustard (e.g., brown, Dijon, or yellow) |
| 1 tbsp | Parmesan cheese, grated |
| 1 tbsp | Soy sauce |

## DRINKS/MIXES

| Serving Size | Food |
|---|---|
| * | Bouillon or broth |
| * | Club soda |
| * | Coffee, unsweetened or artificially sweetened |

| Serving Size | Food |
|---|---|
| * | Diet soft drinks, sugar-free |
| * | Tea, unsweetened or with sugar substitute |
| * | Water |
| * | Water, flavored, sugar-free |

## SEASONINGS

| Serving Size | Food |
|---|---|
| * | Garlic, fresh or powder |
| * | Herbs, fresh or dried |
| * | Spices |

# Combination Foods

These foods contain a mixture of ingredients and cannot be grouped into one food list. Typical examples of combination foods include casseroles, sandwiches, frozen meals, and fast foods containing multiple ingredients. These foods may be consumed at home, dining out, or after home delivery. Many of these foods are high in sodium.

## MAIN DISHES/ENTREES

| Serving Size | Food | Choices |
| --- | --- | --- |
| 1 cup (8 oz) | Casserole-type entrees (e.g., tuna noodle, lasagna, and spaghetti with meatballs) | 2 carbohydrates + 2 medium-fat proteins |
| 1 cup (8 oz) | Stews (meat and vegetables) | 1 carbohydrate + 1 medium-fat protein + 0 to 3 fats |
| 8 to 10 oz | Vegetarian bowl | 3 carbohydrates + 1 lean protein + 1 fat |

## FROZEN MEALS/ENTREES

| Serving Size | Food | Choices |
| --- | --- | --- |
| 1 (5 oz) | Burrito (beef and bean) | 3 carbohydrates + 1 lean protein + 2 fats |
| 9 to 12 oz | Dinner-type healthy meal (< 400 kcal) | 2 to 3 carbohydrates + 1 to 2 lean proteins + 1 fat |
| ¼ of a 12-inch (5 oz) | Pizza with thin crust and meat toppings | 2 carbohydrates + 2 medium-fat proteins + 1½ fats |
| 1 (4½ oz) | Pocket sandwich | 3 carbohydrates + 1 lean protein + 1 to 2 fats |

## SALADS (DELI-STYLE)

| Serving Size | Food | Choices |
| --- | --- | --- |
| ½ cup | Coleslaw, creamy | 1 carbohydrate + 1½ fats |
| ½ cup | Macaroni salad | 2 carbohydrates + 3 fats |
| ½ cup (3½ oz) | Tuna salad or chicken salad | ½ carbohydrate + 2 lean proteins + 1 fat |

## SOUPS

| Serving Size | Food | Choices |
| --- | --- | --- |
| 1 cup (8 oz) | Bean, lentil, or split pea soup | 2 carbohydrates + 1 lean protein |
| 1 cup (8 oz) | Broth-based soups with vegetables and meat | 1 carbohydrate + 1 lean protein |
| 1 cup (8 oz) | Chowder (made with milk) | 1 carbohydrate + 1 lean protein + 1½ fats |
| 1 cup (8 oz) | Cream soup (made with water) | 1 carbohydrate + 1 fat |
| 1 cup (8 oz) | Ramen noodle soup | 2 carbohydrates + 2 fats |
| 1 cup (8 oz) | Tomato soup (made with water) | 1 carbohydrate |

# Fast Foods

Fast foods are high in sodium and fat. These should be consumed in moderation, if at all. It is much easier to control carbohydrate, fat, sodium, and calorie intake when you prepare your own foods at home rather than relying on restaurants.

## MAIN DISHES/ENTREES

| Serving Size | Food | Choices |
| --- | --- | --- |
| 1 (7 oz) | Chicken breast, breaded and fried | 1 carbohydrate + 6 medium-fat proteins |
| 6 pieces | Chicken nuggets or tenders | 1 carbohydrate + 2 medium-fat proteins + 1 fat |
| 1 (2 oz) | Chicken wing, breaded and fried | ½ carbohydrate + 2 medium-fat proteins |
| ⅛ of 14-inch | Pizza, thick crust with or without meat toppings | 2½ carbohydrates + 1 high-fat protein + 1 fat |

## ASIAN

| Serving Size | Food | Choices |
|---|---|---|
| 1 (3 oz) | Egg roll with meat filling | 1½ carbohydrates + 1 lean protein + 1½ fats |
| 1 cup | Fried rice, meatless | 2½ carbohydrates + 2 fats |
| 1 cup (6 oz) | Meat with vegetables in sauce | 1 carbohydrate + 2 lean proteins + 1 fat |
| 1 cup | Pad Thai with chicken | 3 carbohydrates + 2 lean proteins + 2 fat |
| 1 cup | Sushi, California rolls | 1 carbohydrates + 1 fat |

## MEXICAN

| Serving Size | Food | Choices |
|---|---|---|
| 1 small (6 oz) | Burrito with beans and cheese | 3½ carbohydrates + 1 medium-fat protein + 1 fat |
| 1 small (3 oz) | Crisp taco with meat and cheese | 1 carbohydrate + 1 medium-fat protein + 1½ fat |
| 8 chips | Nachos with cheese | 2½ carbohydrates + 1 high-fat protein + 2 fats |
| 1 salad (1 lb) | Taco salad with chicken and tortilla bowl | 3½ carbohydrates + 4 medium-fat proteins + 3 fats |

## SANDWICHES

| Serving Size | Food | Choices |
|---|---|---|
| 1 small (4 oz) | Breakfast burrito with sausage, egg, and cheese | 1½ carbohydrates + 2 high-fat proteins |
| 1 (8½ oz) | Cheeseburger (4 oz) with condiments | 3 carbohydrates + 4 medium-fat proteins + 2½ fats |
| 1 (5 oz) | Fried fish fillet sandwich with cheese and tartar sauce | 2½ carbohydrates + 2 medium-fat proteins + 1½ fats |
| 1 (7½ oz) | Grilled chicken sandwich with lettuce, tomato, and spread | 3 carbohydrates + 4 lean proteins |
| 1 6-inch | Submarine sandwich (no cheese or sauce) | 3 carbohydrates + 2 lean proteins + 1 fat |

## SIDE DISHES

| Serving Size | Food | Choices |
|---|---|---|
| 1 medium (5 oz) | French fries | 3½ carbohydrates + 3 fats |
| 8 (4 oz) | Onion rings | 3½ carbohydrates + 4 fats |
| 1 small | Side salad (no cheese, croutons, or dressing) | 1 nonstarchy vegetable |

## BEVERAGES AND DESSERTS

| Serving Size | Food | Choices |
|---|---|---|
| 12 fl oz | Coffee, latte, with fat-free milk | 1 fat-free milk |
| 1 small | Ice cream cone | 2 carbohydrates + ½ fat |
| 16 fl oz | Milk shake | 7 carbohydrates + 4 fats |

# Alcohol

For people with diabetes, up to 1 or 2 drinks per day for women and men, respectively, can fit into a healthy eating plan. Alcohol itself does not raise blood glucose, but alcoholic drinks often contain carbohydrates that must be counted. One alcohol equivalent provides 100 kcal. One carbohydrate choice provides 15 grams of carbohydrate and 60 kcal. Alcohol should be consumed with a meal to lower the risk of hypoglycemia.

| Serving Size | Drink | Choices |
|---|---|---|
| 12 fl oz | Beer, regular | 1 alcohol equivalent + 1 carbohydrate |
| 4 fl oz | Champagne | 1 alcohol equivalent |
| 3½ fl oz | Dessert wine | 1 alcohol equivalent + 1 carbohydrate |
| 1½ fl oz | Distilled spirits (e.g., rum and vodka) | 1 alcohol equivalent |
| 5 fl oz | Wine | 1 alcohol equivalent |

# Appendix C
## Dietary Assessment

Although it may seem overwhelming at first, it is easy to track the foods you eat. One tip is to record all foods and beverages as soon as possible after consumption.

I.  **Fill in the food record form that follows.** This appendix contains a blank copy, Table C-2 (see the completed example in Table C-1). Your instructor may ask you to record 1 day or several days. To get a good estimate of your usual nutrient intake, it is best to record several days (e.g., 2 weekdays and 1 weekend day). As you record your intake for use on the nutrient analysis form that follows, consider the following tips:
    - Measure and record the amounts of foods eaten in portion sizes of cups, teaspoons, tablespoons, ounces, slices, or inches (or convert metric units to these units).
    - Record brand names of all food products, such as "Quick Quaker Oats®."
    - Measure and record all those little extras, such as gravies, salad dressings, taco sauces, pickles, jelly, sugar, ketchup, and butter.
    - For beverages:
        - List the type of milk, such as whole, fat-free, 1%, evaporated, chocolate, reconstituted dry, or specific plant-based milk (e.g., soy milk, almond milk).
        - Indicate whether fruit juice is fresh, frozen, or canned.
        - Indicate type for other beverages, such as fruit drink, fruit-flavored drink, Kool-Aid®, and hot chocolate made with water or milk.
    - For fruits:
        - Indicate whether fresh, frozen, dried, or canned.
        - If whole, record number eaten and size with approximate measurements (such as 1 apple—3 inches in diameter).
        - Indicate whether processed in water, light syrup, or heavy syrup.
    - For vegetables:
        - Indicate whether fresh, frozen, dried, or canned.
        - Record as portion of cup, teaspoon, or tablespoon, or as pieces (such as carrot sticks—4 inches long, ½ inch thick).
        - Record preparation method.
    - For cereals:
        - Record cooked cereals in portions of tablespoon or cup (a level measurement after cooking).
        - Record dry cereal in level portions of tablespoon or cup.
        - If butter, milk, sugar, fruit, or something else is added, measure and record amount and type.
    - For breads:
        - Indicate whether whole wheat, rye, white, and so on.
        - Measure and record number and size of portion (biscuit—2 inches across, 1 inch thick; slice of homemade rye bread—3 inches by 4 inches, ¼ inch thick).
        - Sandwiches: list all ingredients (lettuce, mayonnaise, tomato, and so on).
    - For meat, fish, poultry, and cheese:
        - Give size (length, width, and thickness) in inches or weight in ounces after cooking for meat, fish, and poultry (such as cooked hamburger patty— 3 inches across, ½ inch thick).
        - Record measurements only for the cooked, edible part—without bone or fat left on the plate.
        - Describe how meat, poultry, or fish was prepared.
        - Give size (length, width, and thickness) in inches or weight in ounces for cheese.

- For eggs:
  - Record as soft- or hard-cooked, fried, scrambled, poached, or omelet.
  - If milk, butter, or drippings are used, specify types and amount.
- For desserts:
  - List commercial brand or "homemade" or "bakery" under brand.
  - Purchased candies, cookies, and cakes: specify kind and size.
  - Measure and record portion size of cakes, pies, and cookies by specifying thickness, diameter, and width or length, depending on the item.

**TABLE C-1 ■ Example of a 1-Day Food Record**

| Time | Minutes Spent Eating | M or S* | H† (0–3) | Activity While Eating | Place of Eating | Food and Quantity | Others Present | Reason for Choice |
|---|---|---|---|---|---|---|---|---|
| 7:10 A.M. | 15 | M | 2 | Standing, fixing lunch | Kitchen | Orange juice, 1 cup | — | Health |
| | | | | | | Crispix cereal, 1 cup | | Habit |
| | | | | | | Nonfat milk, ½ cup | | Health |
| | | | | | | Sugar, 2 tsp | | Taste |
| | | | | | | Black coffee, 1 cup | | Habit |
| 10:00 A.M. | 4 | S | 1 | Sitting, taking notes | Classroom | Diet cola, 12 oz | Class | Weight control |
| 12:15 P.M. | 40 | M | 2 | Sitting, talking | Student union | Chicken sandwich with lettuce and mayonnaise (3 oz grilled chicken, 2 slices whole wheat bread, 1 leaf iceberg lettuce, and 2 tsp mayonnaise) | Friends | Taste |
| | | | | | | Pear, 1 medium | | Health |
| | | | | | | Nonfat milk, 1 cup | | Health |
| 2:30 P.M. | 10 | S | 1 | Sitting, studying | Library | Regular cola, 12 oz | Friend | Hunger |
| 6:30 P.M. | 35 | M | 3 | Sitting, talking | Kitchen | Pork chop, broiled, 1 | Boyfriend | Convenience |
| | | | | | | Baked potato, 1 | | Health |
| | | | | | | Butter, 2 tbsp | | Taste |
| | | | | | | Lettuce and tomato salad, 1 cup | | Health |
| | | | | | | Ranch dressing, 2 tbsp | | Taste |
| | | | | | | Peas, steamed, ½ cup | | Health |
| | | | | | | Whole milk, 1 cup | | Habit |
| | | | | | | Cherry pie, 1 piece | | Taste |
| | | | | | | Iced tea, 12 oz | | Health |
| 9:10 P.M. | 10 | S | 2 | Sitting, studying | Living room | Apple, 1 medium | — | Weight control |
| | | | | | | Water, 1 cup | | Weight control |

*M or S: Meal or snack.

†H: Degree of hunger (0 none; 3 maximum).

**TABLE C-2 ■ 1-Day Food Record**

| Time | Minutes Spent Eating | M or S* | H† (0–3) | Activity While Eating | Place of Eating | Food and Quantity | Others Present | Reason for Choice |
|------|----------------------|---------|----------|-----------------------|-----------------|-------------------|----------------|-------------------|
|      |                      |         |          |                       |                 |                   |                |                   |
|      |                      |         |          |                       |                 |                   |                |                   |
|      |                      |         |          |                       |                 |                   |                |                   |
|      |                      |         |          |                       |                 |                   |                |                   |
|      |                      |         |          |                       |                 |                   |                |                   |
|      |                      |         |          |                       |                 |                   |                |                   |
|      |                      |         |          |                       |                 |                   |                |                   |
|      |                      |         |          |                       |                 |                   |                |                   |
|      |                      |         |          |                       |                 |                   |                |                   |
|      |                      |         |          |                       |                 |                   |                |                   |
|      |                      |         |          |                       |                 |                   |                |                   |
|      |                      |         |          |                       |                 |                   |                |                   |
|      |                      |         |          |                       |                 |                   |                |                   |

*M or S: Meal or snack.

†H: Degree of hunger (0 none; 3 maximum).

II. **Enter all the foods and beverages you consumed into NutritionCalc Plus** (see Fig. C-1). As you get started, you will find many useful tutorial videos on the NutritionCalc Plus site. If you have not already done so, you will need to create a profile in NutritionCalc Plus. For each food or drink, select the meal or snack and the appropriate serving size. If you have data for more than 1 day, be sure to enter your foods and beverages on separate days. If the NutritionCalc Plus database does not contain the exact food or drink you consumed, you may need to enter a reasonably close substitute. For example, the chicken sandwich from the student union in the food record example in Table C-1 could be entered as a Subway® 6" chicken sandwich on white bread.

**FIGURE C-1** ▶ Entering food and beverage intake data using NutritionCalc Plus.

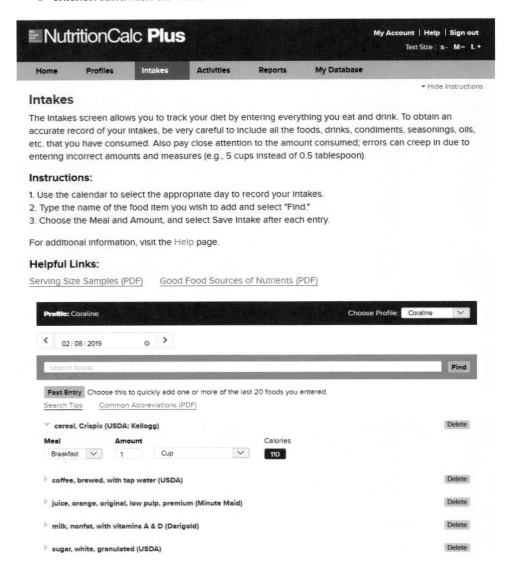

III. **Use NutritionCalc Plus to generate a report of your nutrient intake** (see Fig. C-2). After you have finished entering all the foods and beverages you consumed, save your intake data and click on the **Reports** tab. Select **Bar Graph** to see a report of your nutrient intake for a particular day or several days compared to your nutrient needs. Be sure to enter your student ID information (name, instructor, and course), choose the correct profile for comparison, then select the day(s) and meals(s) you would like to analyze. Then you may save your report as a PDF or Excel spreadsheet or e-mail it as an attachment to yourself or your instructor.

# Bar Graph Report

The Bar Graph Report displays graphically the amount of the nutrient consumed and compares that to the dietary intake recommendations.

**FIGURE C-2** ◄ Sample Bar Graph Report from NutritionCalc Plus.

### Profile Info

Personal: Coraline   Female   23 yrs   5 ft 7 in   140 lb

Student Info:   HUMN NTR 2210

Day(s): 2019 Feb 8 (All)

Activity Level: Low Active   (Strive for an Active activity level.)

BMI: 21.9   Normal is 18.5 to 25.

Weight Change: None   Best not to exceed 2 lbs per week.

| Nutrient | Value | DRI Goal | Percent | 0 | 50 | 100 | 150 |
|---|---|---|---|---|---|---|---|
| **Basic Components** | | | | | | | |
| Calories | 2,114.10 | 2,251.46 | 94 % | | | | |
| Calories from Fat | 651.63 | 630.41 | 103 % | | | | |
| Calories from SatFat | 175.39 | 202.63 | 87 % | | | | |
| Protein (g) | 90.40 | 50.80 | * 178 % | | | | |
| Protein (% Calories) | 17.10 | 9.03 | * 190 % | | | | |
| Carbohydrates (g) | 283.39 | 309.58 | 92 % | | | | |
| Carbohydrates (% Calories) | 53.62 | 55.00 | 97 % | | | | |
| Total Sugars (g) | 165.68 | ^ | | | | | |
| Dietary Fiber (g) | 17.05 | 31.52 | 54 % | | | | |
| Soluble Fiber (g) | 1.83 | | | | | | |
| InSoluble Fiber (g) | 9.92 | | | | | | |
| Fat (g) | 72.40 | 70.05 | 103 % | | | | |
| Fat (% Calories) | 30.82 | 28.00 | 110 % | | | | |
| Saturated Fat (g) | 19.49 | 22.51 | ~87 % | | | | |
| Trans Fat (g) | 4.44 | | | | | | |
| Mono Fat (g) | 22.74 | 25.02 | 91 % | | | | |
| Poly Fat (g) | 16.77 | 22.51 | 74 % | | | | |
| Cholesterol (mg) | 173.09 | 300.00 | ~58 % | | | | |
| Water (g) | 1,681.50 | 2,700.00 | 62 % | | | | |
| **Vitamins** | | | | | | | |
| Vitamin A - RAE (mcg) | 659.74 | 700.00 | 94 % | | | | |
| Vitamin B1 - Thiamin (mg) | 2.20 | 1.10 | 200 % | | | | |
| Vitamin B2 - Riboflavin (mg) | 1.83 | 1.10 | 166 % | | | | |
| Vitamin B3 - Niacin (mg) | 18.29 | 14.00 | 131 % | | | | |
| Vitamin B6 (mg) | 2.17 | 1.30 | 167 % | | | | |
| Vitamin B12 (mcg) | 3.74 | 2.40 | 156 % | | | | |
| Vitamin C (mg) | 163.50 | 75.00 | 218 % | | | | |
| Vitamin D - mcg (mcg) | 9.43 | 15.00 | 63 % | | | | |
| Vitamin E - a-Toco (mg) | 4.11 | 15.00 | 27 % | | | | |
| Folate (mcg) | 516.38 | 400.00 | 129 % | | | | |
| **Minerals** | | | | | | | |
| Calcium (mg) | 1,094.00 | 1,000.00 | 109 % | | | | |
| Iron (mg) | 16.25 | 18.00 | 90 % | | | | |
| Magnesium (mg) | 187.37 | 310.00 | 60 % | | | | |
| Phosphorus (mg) | 912.31 | 700.00 | 130 % | | | | |
| Potassium (mg) | 2,889.11 | 4,700.00 | 61 % | | | | |
| Sodium (mg) | 1,879.29 | 2,300.00 | ~82 % | | | | |
| Zinc (mg) | 6.73 | 8.00 | 84 % | | | | |
| **Other** | | | | | | | |
| Omega-3 (g) | 1.85 | + | | | | | |
| Omega-6 (g) | 14.77 | + | | | | | |
| Alcohol (g) | 0.00 | | | | | | |
| Caffeine (mg) | 212.22 | | | | | | |

\* Protein is not adjusted for endurance/strength athletes at an Active or Very Active activity level.

^ Total Sugars includes those naturally occuring in food and added sugars.

~ This value is a recommended consumption limit, not a goal.

+ There is no established recommendation for Omega-3 and Omega-6.

IV. **Use NutritionCalc Plus to compare your dietary pattern to the dietary pattern recommended by the** *Dietary Guidelines for Americans* (see Fig. C-3). On the **Reports** tab, choose **MyPlate.** Enter your student information, choose the correct profile for comparison, and select the day(s) and meal(s) you would like to analyze. You may save your report, print it, or e-mail it as an attachment to yourself or your instructor.

**FIGURE C-3** ▶ Sample MyPlate Report from NutritionCalc Plus.

## MyPlate

The MyPlate Food Guide report displays graphically how close the foodlist compares to the lastest USDA Dietary Guidelines (see ChooseMyPlate.gov for more info).

### Profile Info

Personal: Coraline    Female    23 yrs    5 ft 7 in    140 lb

Student Info:          HUMN NTR 2210

Day(s): 2019 Feb 8 (All)

| | |
|---|---|
| Activity Level: Low Active | (Strive for an Active activity level.) |
| BMI: 21.9 | Normal is 18.5 to 25. |
| Weight Change: None | Best not to exceed 2 lbs per week. |

## Intake vs. Recommendation

### 2200 Calorie Pattern

| Group | Percent | Comparison | Amount | * |
|---|---|---|---|---|
| Grains Intake | 60 | % | 4.2 | oz equivalent |
| Grains Recommendation | | | 7.0 | oz equivalent |
| Vegetables Intake | 80 | % | 2.4 | cup equivalent |
| Vegetables Recommendation | | | 3.0 | cup equivalent |
| Fruits Intake | 131 | % | 2.6 | cup equivalent |
| Fruits Recommendation | | | 2.0 | cup equivalent |
| Dairy Intake | 73 | % | 2.2 | cup equivalent |
| Dairy Recommendation | | | 3.0 | cup equivalent |
| Protein Foods Intake | 103 | % | 6.2 | oz equivalent |
| Protein Foods Recommendation | | | 6.0 | oz equivalent |

### Make Half Your Grains Whole

Aim for at least 3.5 oz equivalents whole grains a day

### Oils & Empty Calories

Aim for 6.0 teaspoons of oils a day

Limit your extra fats & sugars to 290 Calories a

### Vary Your Vegetables

| | | |
|---|---|---|
| Dark Green Vegetables | 3.0 | cups weekly |
| Orange Vegetables | 2.0 | cups weekly |
| Dry Beans & Peas | 3.0 | cups weekly |
| Starchy Vegetables | 6.0 | cups weekly |
| Other Vegetables | 7.0 | cups weekly |

\* oz equivalent is a 1 ounce estimate, rounded to consumer friendly units. For example, an oz equivalent of Grains is 1 slice of bread, or 1/2 cup of rice. An oz equivalent of Protein Foods 1 oz of meat, 1 egg, or 1/4 cup cooked beans.

V. **Evaluate your dietary pattern.** Answer the following questions about the results of your own dietary assessment and suggest ways that you could improve your dietary pattern.

1. How did your kilocalorie intake compare to the goal recommended by NutritionCalc Plus? If it was much higher or lower than your goal, what specific changes could you make to adjust your energy intake?

2. Do your meals usually resemble MyPlate? If not, what specific changes could you make to improve your meal planning?

3. Use the following worksheet to calculate your fat intake as a percentage of total kilocalories. How does this compare to the Acceptable Macronutrient Distribution Range (AMDR) of 20% to 35% of total kilocalories from fat? If your fat intake does not fall within the AMDR, what specific changes could you make?

> Calculating Percent of Kilocalories from Fat:
>
> _____ grams of fat × 9 kcal per gram = _____ kcal from fat
>
> _____ kcal from fat / _____ total kcal = _____% of kcal from fat

4. Use the following worksheet to calculate your saturated fat intake as a percentage of total kilocalories. How does this compare to the *Dietary Guidelines* recommendation to limit saturated fat to 10% of total kilocalories? If your intake of saturated fat was higher than 10% of total kcal, what specific dietary changes could you make to lower your intake of saturated fat?

> Calculating Percent of Kilocalories from Saturated Fat:
>
> _____ grams of saturated fat × 9 kcal per gram = _____ kcal from saturated fat
>
> _____ kcal from saturated fat / _____ total kcal = _____% of kcal from saturated fat

5. The *Dietary Guidelines* advises Americans to limit their consumption of added sugars. What were some sources of added sugars in your diet on the days you recorded? List several nutrient-dense food choices you could use to replace sources of added sugars in your dietary pattern.

6. What was your average fiber intake for the 3 days? How close did you come to meeting the AI for fiber? If your intake of fiber was lower than the AI, what specific dietary changes could you make to improve your intake of fiber?

7. How did your average intake of sodium compare to your AI and UL for sodium? If your sodium intake was higher than the UL, what specific dietary changes could you make to lower your intake of sodium?

8. Are your intakes of any vitamins or minerals less than 75% of the RDA or AI? Choose one of these nutrients and discuss how you could change your *dietary* habits to increase your intake of this nutrient.

9. Besides sodium, do your intakes of any vitamins or minerals exceed the UL? (Also consider any micronutrients consumed in supplement form.) If so, what negative consequences could this have for your health?

10. If you consume alcohol, are your drinking habits consistent with the *Dietary Guidelines* recommendations to limit alcohol to 1 drink per day (women) or 2 drinks per day (men)? List one potential benefit of moderate alcohol consumption. List one negative consequence of excessive alcohol consumption.

# Appendix D

## Chemical Structures Important in Nutrition

### Amino Acids

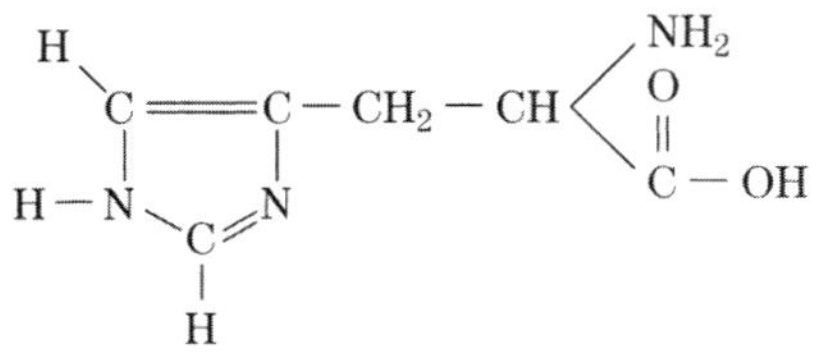

Histidine (His)
(essential)

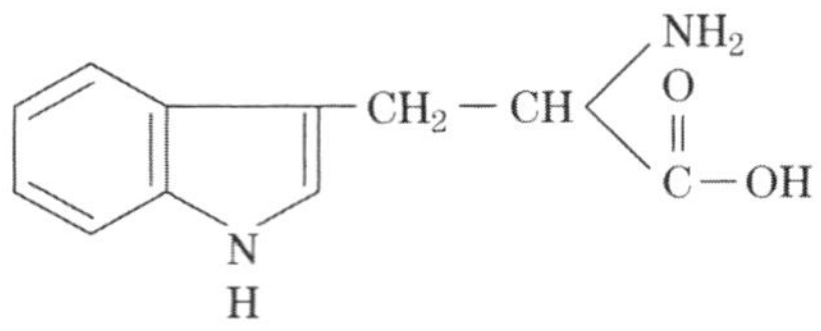

Tryptophan (Trp)
(essential)

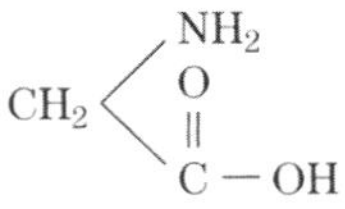

Glycine (Gly)

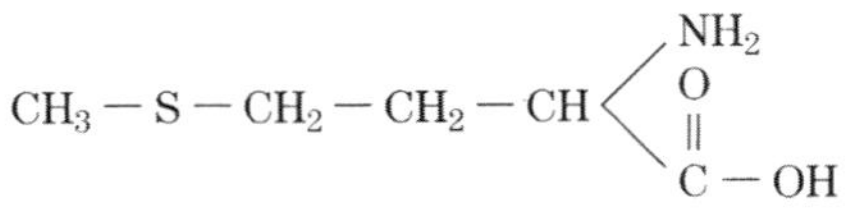

Methionine (Met)
(essential)

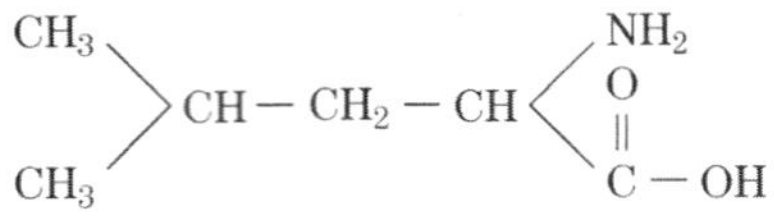

Leucine (Leu)
(essential)

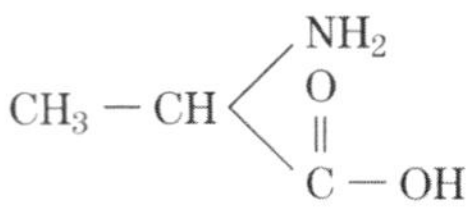

Alanine (Ala)

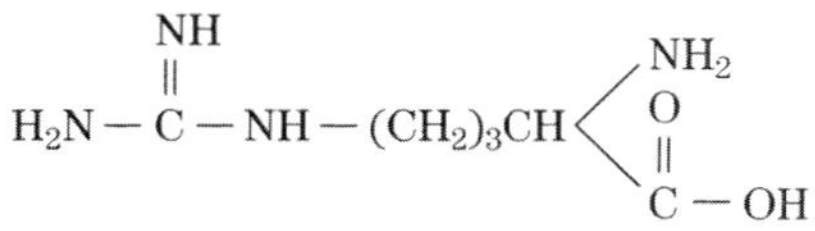

Arginine (Arg)
(essential in infancy)

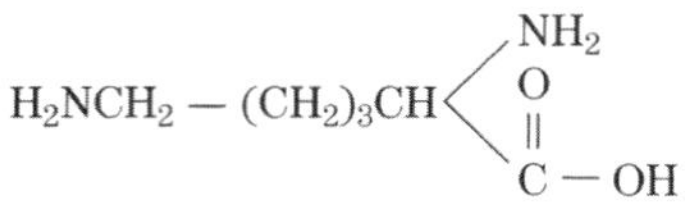

Lysine (Lys)
(essential)

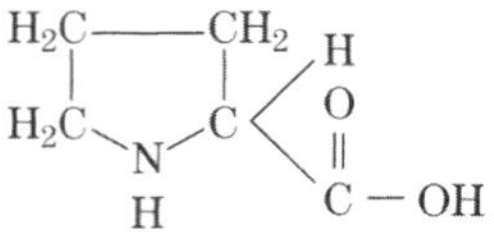

Proline (Pro)

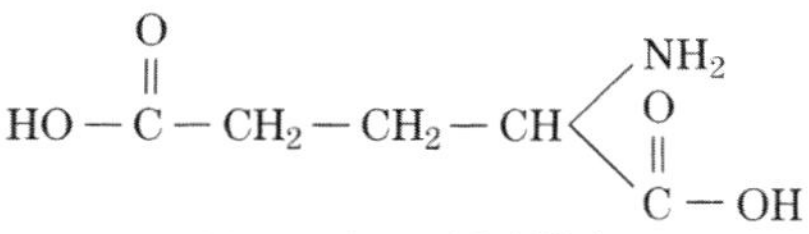

Glutamic Acid (Glu)

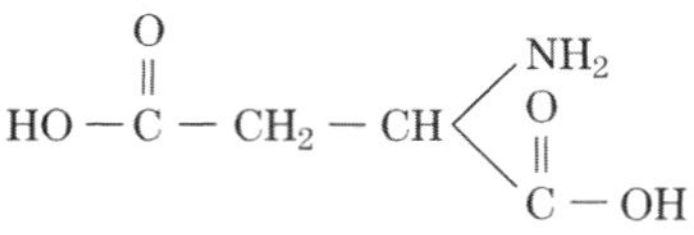

Aspartic Acid (Asp)

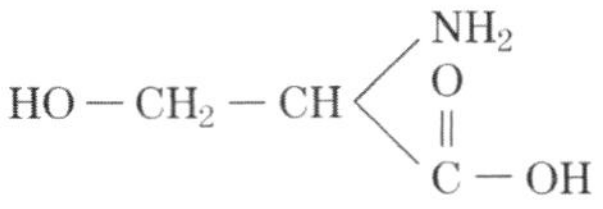

Serine (Ser)

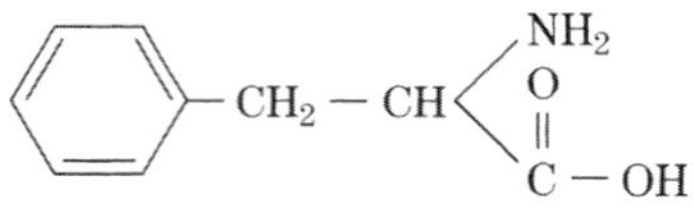

Phenylalanine (Phe)
(essential)

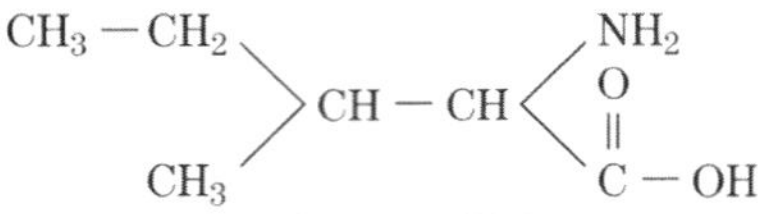

Isoleucine (Ile)
(essential)

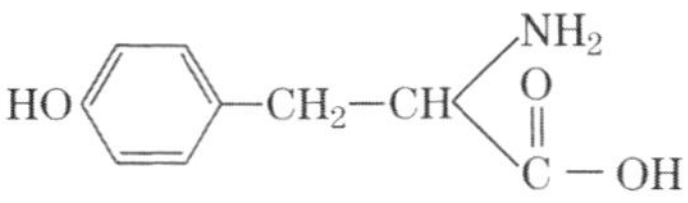

Tyrosine (Tyr)

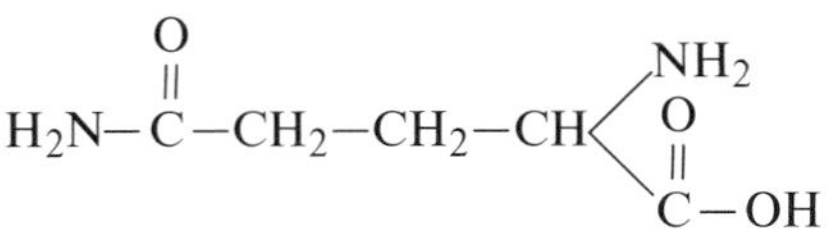

Glutamine (Gln)

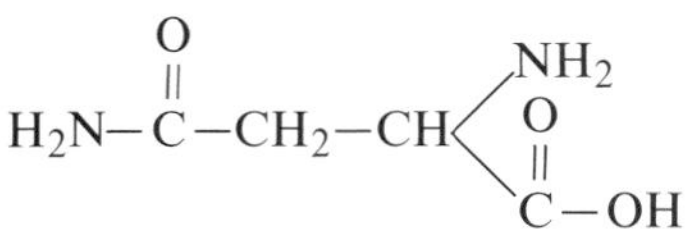

Asparagine (Asn)

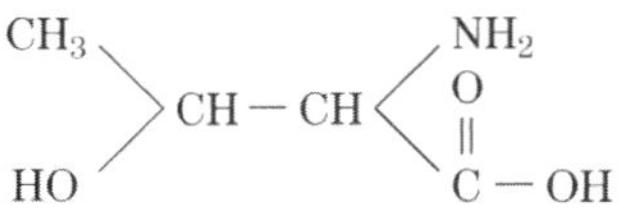

Threonine (Thr)
(essential)

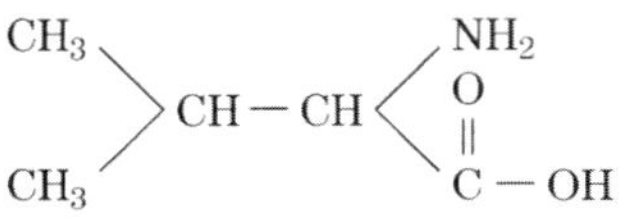

Valine (Val)
(essential)

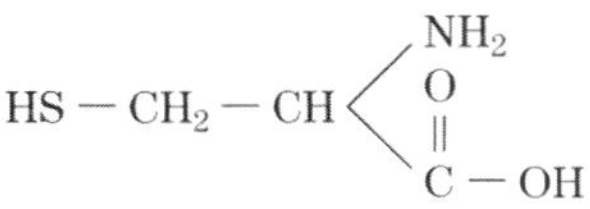

Cysteine (Cys)

# Vitamins

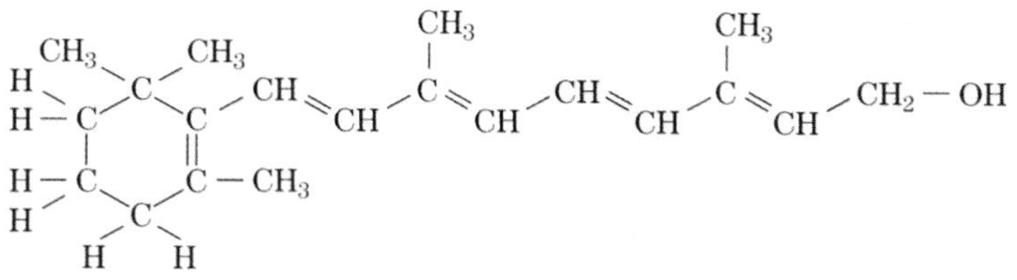

**Vitamin A (retinol)**

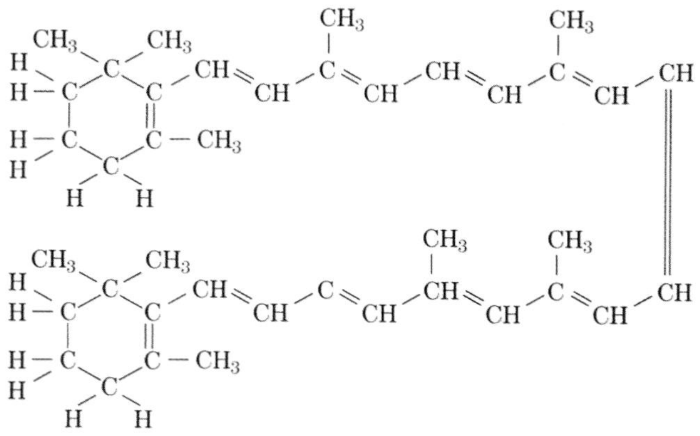

**Beta-carotene**

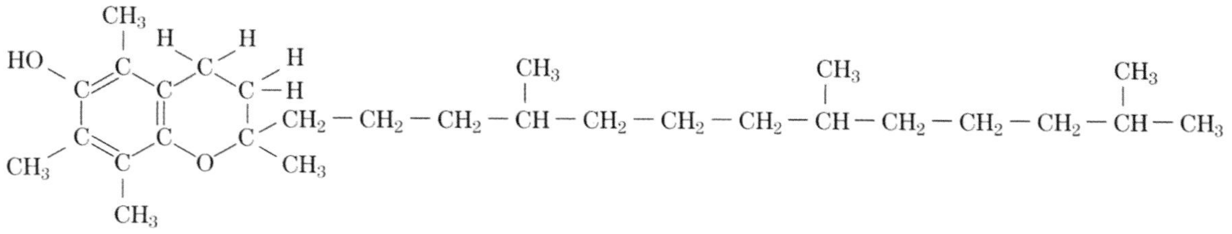

**Vitamin E**

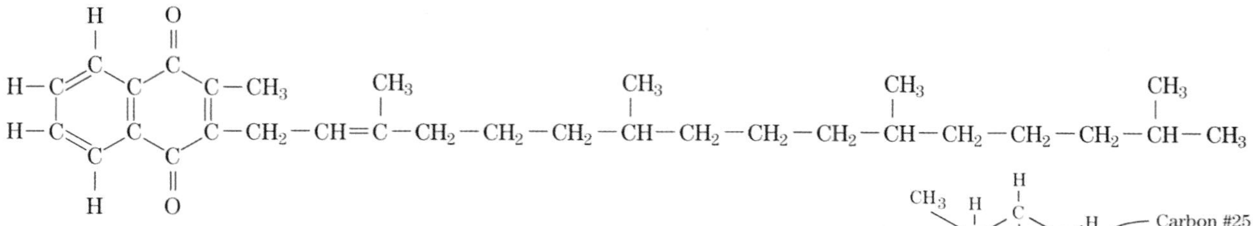

**Vitamin K**

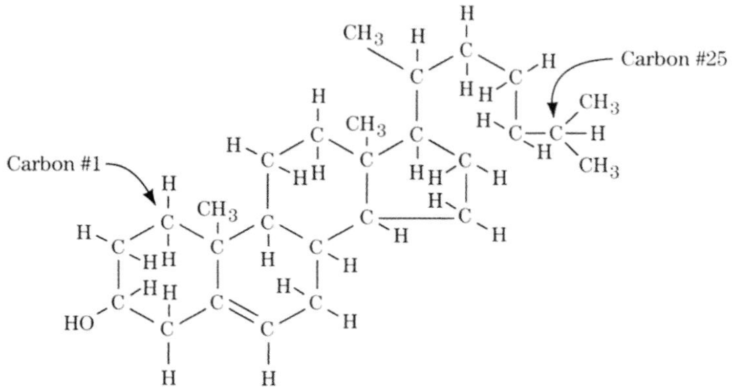

**7-Dehydrocholesterol**

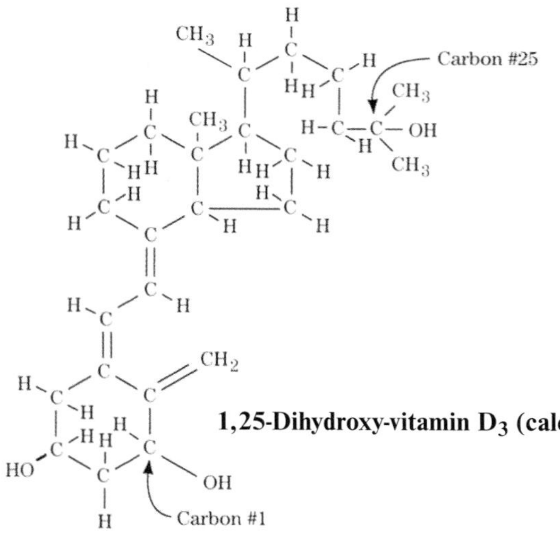

**1,25-Dihydroxy-vitamin D$_3$ (calcitriol)**

**Active vitamin D (calcitriol) and its precursor 7-dehydrocholesterol**

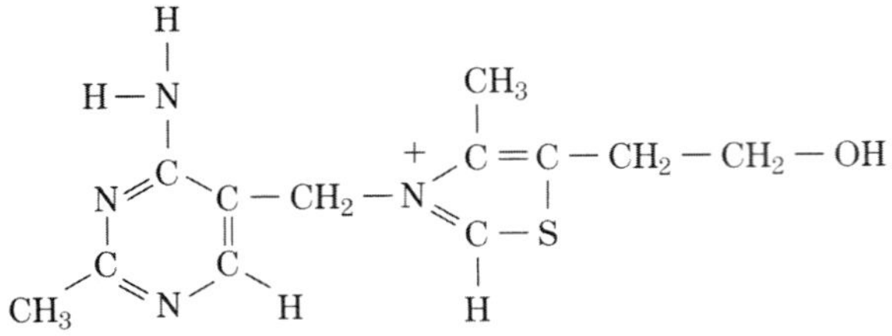

**Thiamin**

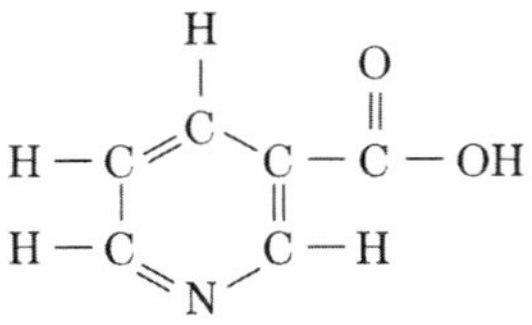

**Niacin (nicotinic acid and nicotinamide)**

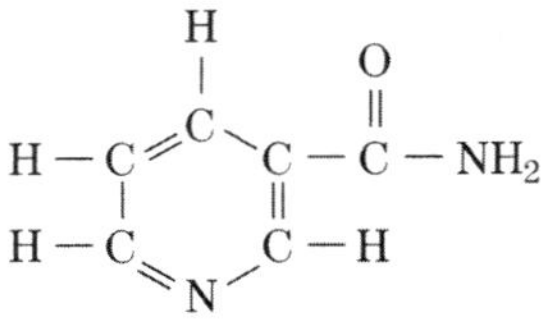

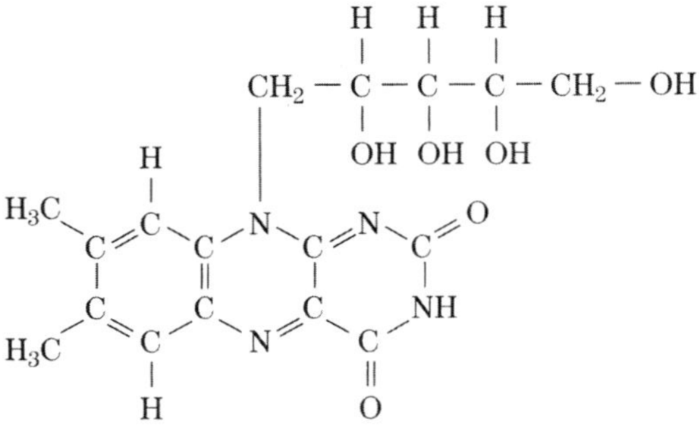

Nicotinic acid

Nicotinamide

**Riboflavin**

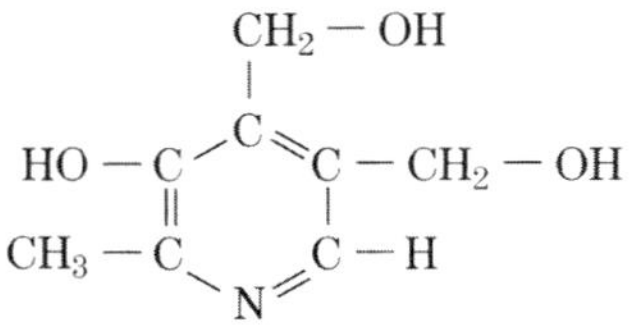

Pyridoxine

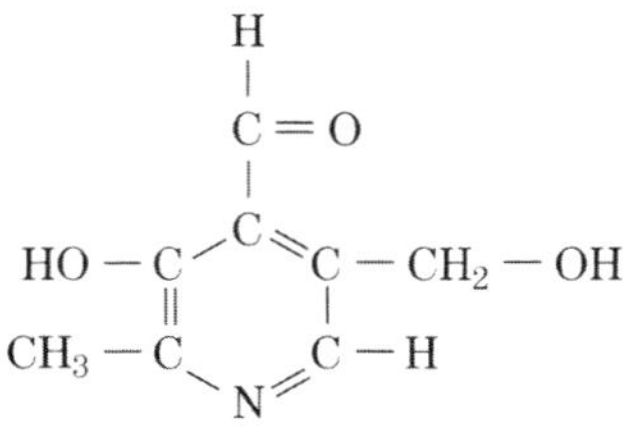

Pyridoxal

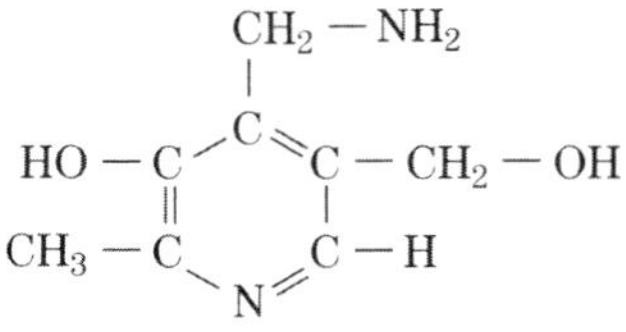

Pyridoxamine

**Vitamin B-6 (a general name for three compounds—
pyridoxine, pyridoxal, and pyridoxamine)**

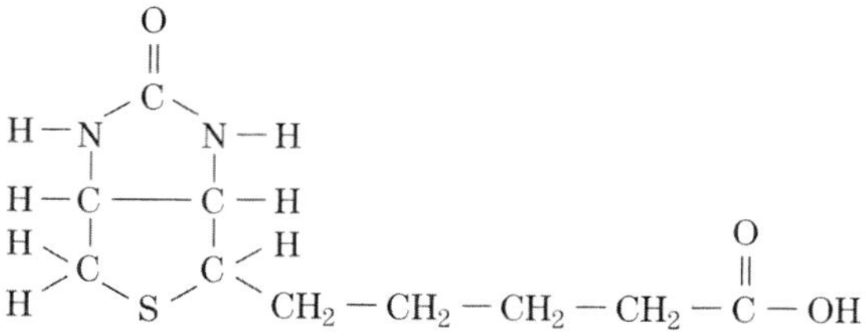

**Biotin**

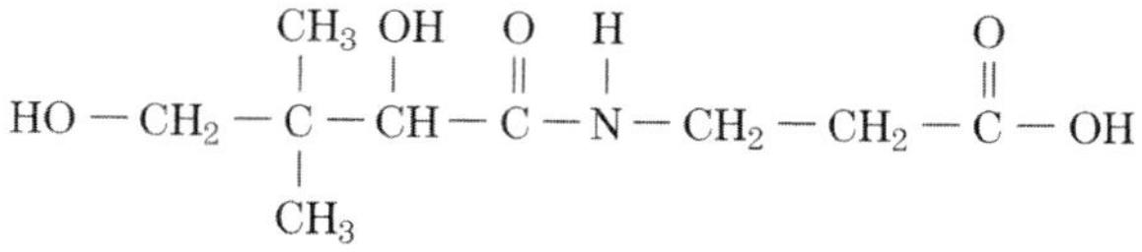

**Pantothenic acid**

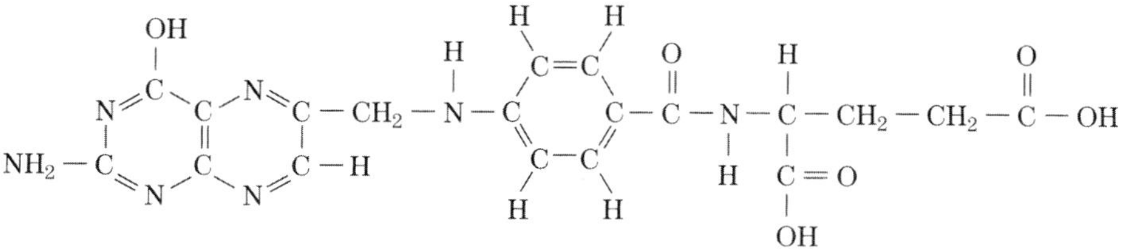

**Folate (folic acid form)**

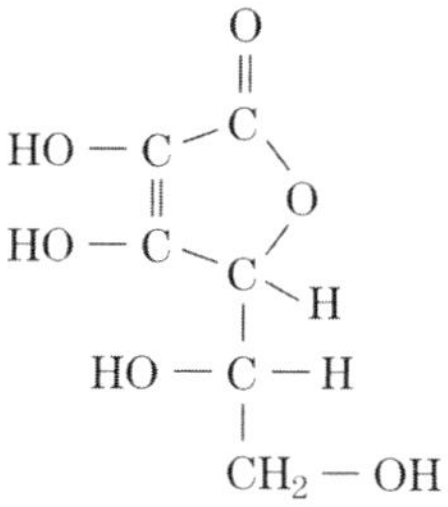

**Vitamin C (ascorbic acid)**

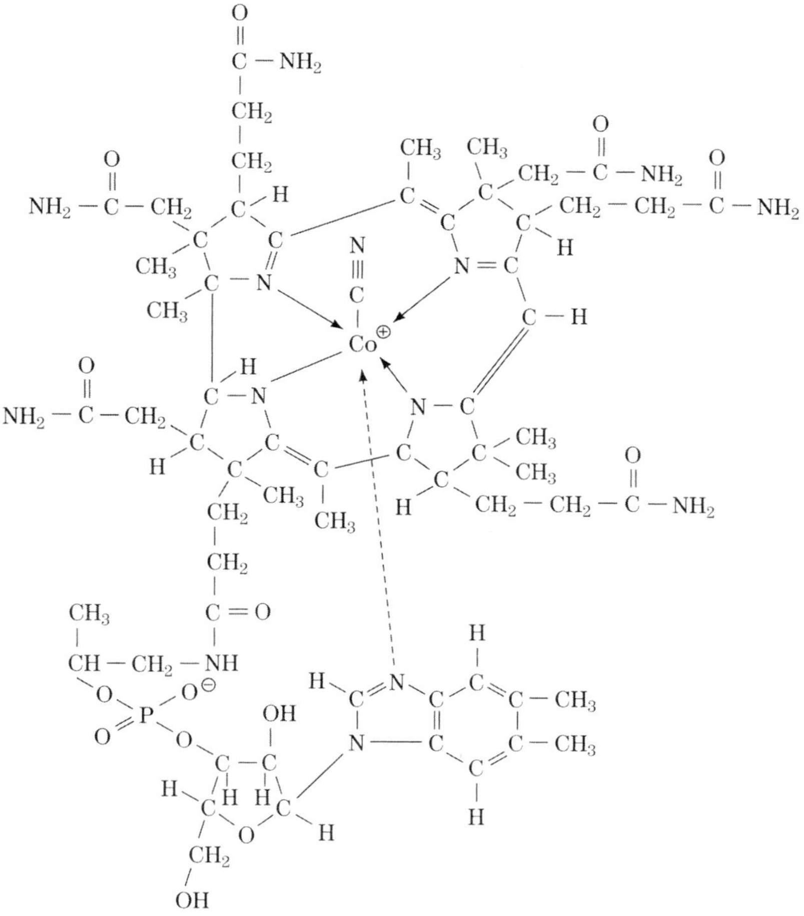

**Vitamin B-12 (cyanocobalamin)** The arrows in this diagram indicate that the spare electrons on the nitrogens are attracted to the cobalt atom.

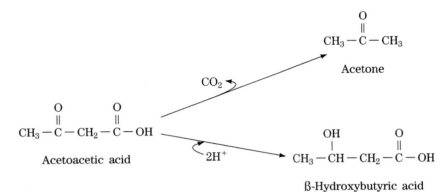

**Ketone bodies**

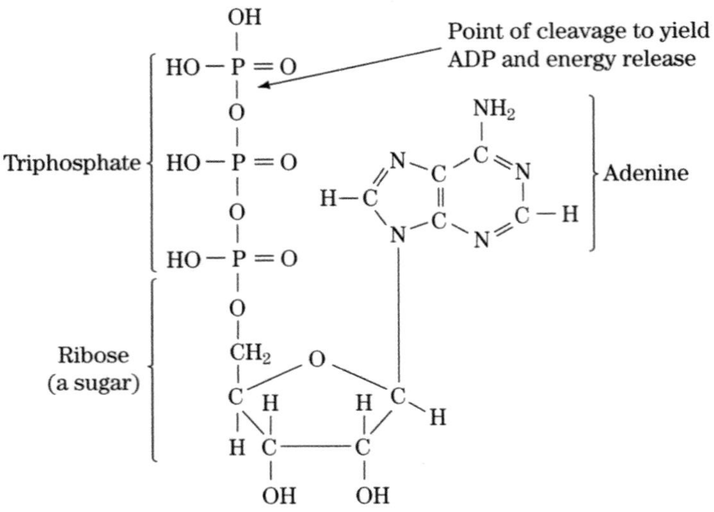

$$CH_3-\overset{\overset{\displaystyle O}{\|}}{C}-CH_3$$

Acetone

$CO_2$

$$CH_3-\overset{\overset{\displaystyle O}{\|}}{C}-CH_2-\overset{\overset{\displaystyle O}{\|}}{C}-OH$$

Acetoacetic acid

$2H^+$

$$CH_3-\overset{\overset{\displaystyle OH}{|}}{CH}-CH_2-\overset{\overset{\displaystyle O}{\|}}{C}-OH$$

ß-Hydroxybutyric acid

Point of cleavage to yield ADP and energy release

Triphosphate

Adenine

Ribose (a sugar)

**Adenosine triphosphate (ATP)**

# Appendix E

## Sources of Nutrition Information

Consider the following reliable sources of food and nutrition information:

### Journals That Regularly Cover Nutrition Topics

*American Family Physician**
*American Journal of Clinical Nutrition*
*American Journal of Epidemiology*
*American Journal of Medicine*
*American Journal of Obstetrics and Gynecology*
*American Journal of Public Health*
*American Scientist*
*Annals of Internal Medicine*
*Annual Review of Medicine*
*Annual Review of Nutrition*
*Appetite*
*Archives of Disease in Childhood*
*BMJ (British Medical Journal)*
*British Journal of Nutrition*
*Canadian Journal of Dietetic Practice and Research*
*Cancer Research*
*Clinical Nutrition*
*Critical Reviews in Food Science and Nutrition*
*Current Opinion in Clinical Nutrition and Metabolic Care*
*Current Opinion in Endocrinology, Diabetes and Obesity*

*Diabetes*
*Diabetes Care*
*Food and Chemical Toxicology*
*Food & Nutrition Magazine**
*Gastroenterology*
*International Journal of Behavioral Nutrition and Physical Activity*
*International Journal of Obesity*
*Journal of the Academy of Nutrition and Dietetics**
*Journal of the American College of Nutrition**
*Journal of the American Geriatrics Society*
*JAMA (Journal of the American Medical Association)*
*JNCI (Journal of the National Cancer Institute)*
*Journal of Clinical Investigation*
*Journal of Food Science*
*Journal of Functional Foods*
*Journal of Human Nutrition and Dietetics*
*Journal of Nutrition*
*Journal of Nutrition Education and Behavior**
*Journal of Nutrition in Gerontology and Geriatrics*

*Journal of Pediatrics*
*The Lancet*
*Maternal and Child Health*
*Medicine & Science in Sports & Exercise*
*Nature*
*The New England Journal of Medicine*
*Nutrition*
*Nutrition & Dietetics*
*Nutrition in Clinical Practice**
*Nutrition Journal*
*Nutrition, Metabolism and Cardiovascular Diseases*
*Nutrition Research Reviews*
*Nutrition Reviews*
*Nutrition Today**
*Obesity*
*Pediatric Obesity*
*Pediatrics*
*Proceedings of the Nutrition Society*
*Public Health Nutrition*
*Science*
*Science News**
*Scientific American*
*Today's Dietitian**

The majority of these journals are available in college and university libraries or in a specialty library on campus, such as one designated for health sciences. Most of them are now available online. As indicated, a few journals will be filed under their abbreviations, rather than the first word in their full name. A reference librarian can help you locate any of these sources. The journals with an asterisk (*) are ones you may find especially interesting and useful because of the number of nutrition articles presented each month or the less technical nature of the presentation.

### Textbooks and Other Sources for Advanced Study of Nutrition Topics

Gropper SS and others: *Advanced nutrition and human metabolism,* 7th ed. Boston, MA: Cengage Learning, 2018.
Marriott B and others: *Present knowledge in nutrition,* 11th ed. Cambridge, MA: Academic Press, 2020.
Raymond JL and Morrow K: *Krause's food and the nutrition care process,*

15th ed. St. Louis, MO: Elsevier Saunders, 2020.
Rodwell VW and others: *Harper's illustrated biochemistry,* 31st ed. New York, NY: McGraw-Hill Education, 2018.
Stipanuk MH and Caudill MA: *Biochemical, physiological, and molecular aspects of human nutrition,* 4th ed. St. Louis, MO: Elsevier Saunders, 2019.

### Newsletters That Cover Nutrition Issues on a Regular Basis

*Berkeley Wellness*
University of California at Berkeley
www.berkeleywellness.com

*Consumer Health Digest*
www.consumerhealthdigest.com

*Environmental Nutrition*
https://universityhealthnews
 .com/publication/
 environmental-nutrition

*Harvard Health Letter* (and others)
Harvard Medical School
www.health.harvard.edu/newsletters

*Health & Nutrition Letter*
Tufts University
www.nutritionletter.tufts.edu

*Mayo Clinic Health Letter*
Mayo Clinic
https://healthletter.mayoclinic.com

*Nutrition Action Healthletter*
Center for Science in the Public Interest
www.cspinet.org

*Soy Connection*
United Soybean Board
www.soyconnection.com

*Women's Nutrition Connection
 Newsletter*
Weill Cornell Medical College
www.womensnutritionconnection.com

### Professional Organizations

Academy of Nutrition and Dietetics
www.eatright.org

American Academy of Pediatrics
www.aap.org

American Cancer Society
www.cancer.org

American College of Sports Medicine
www.acsm.org

American Dental Association
www.ada.org

American Diabetes Association
www.diabetes.org

American Geriatrics Society
www.americangeriatrics.org

American Heart Association
www.heart.org

American Institute for Cancer Research
www.aicr.org

American Medical Association
www.ama-assn.org

American Public Health Association
www.apha.org

American Society for Nutrition
www.nutrition.org

Canadian Nutrition Society
www.cns-scn.ca

Diabetes Canada
www.diabetes.ca

Dietitians of Canada
www.dietitians.ca

Institute of Food Technologists
www.ift.org

National Academy of Medicine
https://nam.edu

National Council on Aging
www.ncoa.org

National Osteoporosis Foundation
www.nof.org

Society for Nutrition Education and
 Behavior
www.sneb.org

### Professional Organizations with a Commitment to Nutrition Issues

Bread for the World Institute
www.bread.org

Food Research & Action Center
frac.org

Institute for Food and
 Development Policy
foodfirst.org

La Leche League International
www.llli.org

March of Dimes
www.marchofdimes.org

National Council Against
 Health Fraud
www.ncahf.org

National WIC Association
www.nwica.org

Overeaters Anonymous
www.oa.org

Oxfam America
www.oxfamamerica.org

### Local Resources for Advice on Nutrition Issues

Cooperative extension agents in county
 extension offices
Nutrition faculty affiliated with
 departments of food and nutrition,
 and dietetics
Registered dietitian nutritionists
 (RDNs) in health care, city, county,
 or state agencies, as well as in
 private practice

### Government Agencies Concerned with Nutrition Issues or That Distribute Nutrition Information

*United States*
Agricultural Research Service
U.S. Department of Agriculture
www.ars.usda.gov

Food and Drug Administration
www.fda.gov

Food Safety and Inspection Service
U.S. Department of Agriculture
www.fsis.usda.gov

Health Information from the U.S.
 government
www.usa.gov/health-resources

MyPlate
www.choosemyplate.gov

National Agricultural Library
www.nal.usda.gov

National Cancer Institute
www.cancer.gov

National Center for Health Statistics
www.cdc.gov/nchs

National Heart, Lung, and Blood Institute
www.nhlbi.nih.gov

National Institute on Aging
www.nia.nih.gov

U.S. Government Publishing Office
www.gpo.gov

*Canada*
Canadian Food Inspection Agency
www.inspection.gc.ca

Health Canada
www.hc-sc.gc.ca

*United Nations*
Food and Agriculture Organization
www.fao.org

World Health Organization
www.who.int

### Trade Organizations and Companies That Distribute Nutrition Information

Abbott Nutrition
abbottnutrition.com

American Peanut Council
https://www.peanutsusa.com

Beech-Nut Nutrition
www.beechnut.com

California Avocado Commission
www.californiaavocado.com

Dannon Company
www.dannon.com

Del Monte Foods
www.delmonte.com

Gatorade Sports Science Institute
https://www.gssiweb.org/en

General Mills/Pillsbury
www.generalmills.com

Gerber
www.gerber.com

Idaho Potato Commission
www.idahopotato.com

International Tree Nut Council
www.nuthealth.org

Kellogg Company
www.kelloggs.com

Kraft Foods Group, Inc.
www.kraftrecipes.com

Kraft Heinz
www.heinzbaby.com/en-ca

Mead Johnson Nutrition
www.meadjohnson.com

National Dairy Council
www.nationaldairycouncil.org

Nature's Path Organic
https://www.naturespath.com/en-us

North American Meat Institute
www.meatinstitute.org

Oldways Whole Grains Council
https://wholegrainscouncil.org

Sunkist Growers
www.sunkist.com

U.S. Dry Bean Council
www.usdrybeans.com

# Appendix F

## English-Metric Conversions and Metric Units

## Metric-English Conversions

### LENGTH

| English (USA) | Metric |
|---|---|
| inch (in) | = 2.54 cm, 25.4 mm |
| foot (ft) | = 0.30 m, 30.48 cm |
| yard (yd) | = 0.91 m, 91.4 cm |
| mile (statute) (5280 ft) | = 1.61 km, 1609 m |
| mile (nautical) (6077 ft, 1.15 statute mi) | = 1.85 km, 1850 m |

| Metric | English (USA) |
|---|---|
| millimeter (mm) | = 0.039 in (thickness of a dime) |
| centimeter (cm) | = 0.39 in |
| meter (m) | = 3.28 ft, 39.37 in |
| kilometer (km) | = 0.62 mi, 1091 yd, 3273 ft |

### WEIGHT

| English (USA) | Metric |
|---|---|
| grain | = 64.80 mg |
| ounce (oz) | = 28.35 g |
| pound (lb) | = 453.60 g, 0.45 kg |
| ton (short—2000 lb) | = 0.91 metric ton (907 kg) |

| Metric | English (USA) |
|---|---|
| milligram (mg) | = 0.002 grain (0.000035 oz) |
| gram (g) | = 0.04 oz (1/28 of an oz) |
| kilogram (kg) | = 35.27 oz, 2.20 lb |
| metric ton (1000 kg) | = 1.10 tons |

### VOLUME

| English (USA) | Metric |
|---|---|
| cubic inch | = 16.39 cc |
| cubic foot | = 0.03 m$^3$ |
| cubic yard | = 0.765 m$^3$ |
| teaspoon (tsp) | = 5 ml |
| tablespoon (tbsp) | = 15 ml |
| fluid ounce | = 0.03 liter (30 ml)* |
| cup (c) | = 237 ml |
| pint (pt) | = 0.47 liter |
| quart (qt) | = 0.95 liter |
| gallon (gal) | = 3.79 liters |

| Metric | English (USA) |
|---|---|
| milliliter (ml) | = 0.03 oz |
| liter (L) | = 2.12 pt |
| liter | = 1.06 qt |
| liter | = 0.27 gal |

1 liter ÷ 1000 = 1 milliliter or 1 cubic centimeter ($10^{-3}$ liter)
1 liter ÷ 1,000,000 = 1 microliter ($10^{-6}$ liter)

\* Note: 1 ml = 1 cc.

## Metric and Other Common Units

| Unit/Abbreviation | Other Equivalent Measure |
|---|---|
| milligram/mg | 1/1000 of a gram |
| microgram/μg | 1/1,000,000 of a gram |
| deciliter/dl | 1/10 of a liter (about ½ cup) |
| milliliter/ml | 1/1000 of a liter (5 ml is about 1 tsp) |
| International Unit/IU | Crude measure of vitamin activity generally based on growth rate seen in animals |

## Fahrenheit-Celsius Conversion Scale

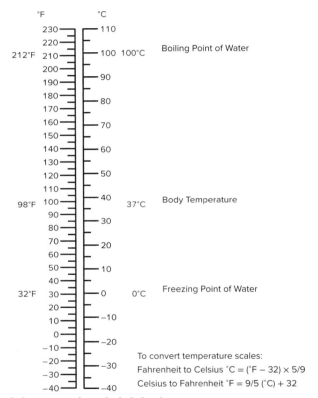

To convert temperature scales:
Fahrenheit to Celsius °C = (°F − 32) × 5/9
Celsius to Fahrenheit °F = 9/5 (°C) + 32

## Household Units

| | | |
|---|---|---|
| 3 teaspoons | = 1 tablespoon | = 15 grams |
| 4 tablespoons | = ¼ cup | = 60 grams |
| 5⅓ tablespoons | = ⅓ cup | = 80 grams |
| 8 tablespoons | = ½ cup | = 120 grams |
| 10⅔ tablespoons | = ⅔ cup | = 160 grams |
| 16 tablespoons | = 1 cup | = 240 grams |
| 1 tablespoon | = ½ fluid ounce | = 15 milliliters |
| 1 cup | = 8 fluid ounces | = 237 milliliters |
| 1 cup | = ½ pint | = 240 grams |
| 2 cups | = 1 pint | = 480 grams |
| 4 cups | = 1 quart | = 960 grams = 1 liter |
| 2 pints | = 1 quart | = 960 grams = 1 liter |
| 4 quarts | = 1 gallon | = 3840 grams = 4 liters |

# DIETARY REFERENCE INTAKES (DRIs): RECOMMENDED INTAKES FOR INDIVIDUALS, VITAMINS
Food and Nutrition Board, Institute of Medicine, National Academies

| Life Stage Group | Vitamin A (µg/d)[a] | Vitamin C (mg/d) | Vitamin D (µg/d)[b,c] | Vitamin E (mg/d)[d] | Vitamin K (µg/d) | Thiamin (mg/d) | Riboflavin (mg/d) | Niacin (mg/d)[e] | Vitamin B-6 (mg/d) | Folate (µg/d)[f] | Vitamin B-12 (µg/d) | Pantothenic Acid (mg/d) | Biotin (µg/d) | Choline (mg/d)[g] |
|---|---|---|---|---|---|---|---|---|---|---|---|---|---|---|
| **Infants** | | | | | | | | | | | | | | |
| 0–6 mo | 400* | 40* | 10 | 4* | 2.0* | 0.2* | 0.3* | 2* | 0.1* | 65* | 0.4* | 1.7* | 5* | 125* |
| 7–12 mo | 500* | 50* | 10 | 5* | 2.5* | 0.3* | 0.4* | 4* | 0.3* | 80* | 0.5* | 1.8* | 6* | 150* |
| **Children** | | | | | | | | | | | | | | |
| 1–3 y | 300 | 15 | 15 | 6 | 30* | 0.5 | 0.5 | 6 | 0.5 | 150 | 0.9 | 2* | 8* | 200* |
| 4–8 y | 400 | 25 | 15 | 7 | 55* | 0.6 | 0.6 | 8 | 0.6 | 200 | 1.2 | 3* | 12* | 250* |
| **Males** | | | | | | | | | | | | | | |
| 9–13 y | 600 | 45 | 15 | 11 | 60* | 0.9 | 0.9 | 12 | 1.0 | 300 | 1.8 | 4* | 20* | 375* |
| 14–18 y | 900 | 75 | 15 | 15 | 75* | 1.2 | 1.3 | 16 | 1.3 | 400 | 2.4 | 5* | 25* | 550* |
| 19–30 y | 900 | 90 | 15 | 15 | 120* | 1.2 | 1.3 | 16 | 1.3 | 400 | 2.4 | 5* | 30* | 550* |
| 31–50 y | 900 | 90 | 15 | 15 | 120* | 1.2 | 1.3 | 16 | 1.3 | 400 | 2.4 | 5* | 30* | 550* |
| 51–70 y | 900 | 90 | 15 | 15 | 120* | 1.2 | 1.3 | 16 | 1.7 | 400 | 2.4[h] | 5* | 30* | 550* |
| >70 y | 900 | 90 | 20 | 15 | 120* | 1.2 | 1.3 | 16 | 1.7 | 400 | 2.4[h] | 5* | 30* | 550* |
| **Females** | | | | | | | | | | | | | | |
| 9–13 y | 600 | 45 | 15 | 11 | 60* | 0.9 | 0.9 | 12 | 1.0 | 300 | 1.8 | 4* | 20* | 375* |
| 14–18 y | 700 | 65 | 15 | 15 | 75* | 1.0 | 1.0 | 14 | 1.2 | 400[i] | 2.4 | 5* | 25* | 400* |
| 19–30 y | 700 | 75 | 15 | 15 | 90* | 1.1 | 1.1 | 14 | 1.3 | 400[i] | 2.4 | 5* | 30* | 425* |
| 31–50 y | 700 | 75 | 15 | 15 | 90* | 1.1 | 1.1 | 14 | 1.3 | 400[i] | 2.4 | 5* | 30* | 425* |
| 51–70 y | 700 | 75 | 15 | 15 | 90* | 1.1 | 1.1 | 14 | 1.5 | 400 | 2.4[h] | 5* | 30* | 425* |
| >70 y | 700 | 75 | 20 | 15 | 90* | 1.1 | 1.1 | 14 | 1.5 | 400 | 2.4[h] | 5* | 30* | 425* |
| **Pregnancy** | | | | | | | | | | | | | | |
| 14–18 y | 750 | 80 | 15 | 15 | 75* | 1.4 | 1.4 | 18 | 1.9 | 600[j] | 2.6 | 6* | 30* | 450* |
| 19–30 y | 770 | 85 | 15 | 15 | 90* | 1.4 | 1.4 | 18 | 1.9 | 600[j] | 2.6 | 6* | 30* | 450* |
| 31–50 y | 770 | 85 | 15 | 15 | 90* | 1.4 | 1.4 | 18 | 1.9 | 600[j] | 2.6 | 6* | 30* | 450* |
| **Lactation** | | | | | | | | | | | | | | |
| 14–18 y | 1200 | 115 | 15 | 19 | 75* | 1.4 | 1.6 | 17 | 2.0 | 500 | 2.8 | 7* | 35* | 550* |
| 19–30 y | 1300 | 120 | 15 | 19 | 90* | 1.4 | 1.6 | 17 | 2.0 | 500 | 2.8 | 7* | 35* | 550* |
| 31–50 y | 1300 | 120 | 15 | 19 | 90* | 1.4 | 1.6 | 17 | 2.0 | 500 | 2.8 | 7* | 35* | 550* |

mg = milligram, µg = microgram

NOTE: This table (taken from the *DRI reports*; see www.nap.edu) presents Recommended Dietary Allowances (RDAs) in **bold type** and Adequate Intakes (AIs) in ordinary type followed by an asterisk (*). RDAs and AIs may both be used as goals for individual intake. RDAs are set to meet the needs of almost all (97 to 98%) individuals in a group. For healthy breastfed infants, the AI is the mean intake. The AI for other life stage and gender groups is believed to cover needs of all individuals in the group, but lack of data or uncertainty in the data prevents being able to specify with confidence the percentage of individuals covered by this intake.

[a] As retinol activity equivalents (RAEs). 1 RAE = 1 µg retinol, 12 µg β-carotene, 24 µg α-carotene, or 24 µg β-cryptoxanthin. To calculate RAEs from REs of provitamin A carotenoids in foods, divide the REs by 2. For preformed vitamin A in foods or supplements and for provitamin A carotenoids in supplements, 1 RE = 1 RAE.

[b] Cholecalciferol. 1 µg cholecalciferol = 40 IU vitamin D.

[c] In the absence of adequate exposure to sunlight.

[d] As α-tocopherol. α-Tocopherol includes RRR-α-tocopherol, the only form of α-tocopherol that occurs naturally in foods, and the 2R-stereoisomeric forms of α-tocopherol (RRR-, RSR-, RRS-, and RSS-α-tocopherol) that occur in fortified foods and supplements. It does not include the 2S-stereoisomeric forms of α-tocopherol (SRR-, SSR-, SRS-, and SSS-α-tocopherol), also found in fortified foods and supplements.

[e] As niacin equivalents (NE). 1 mg of niacin = 60 mg of tryptophan; 0–6 months = preformed niacin (not NE).

[f] As dietary folate equivalents (DFE). 1 DFE = 1 µg food folate = 0.6 µg of folic acid from fortified food or as a supplement consumed with food = 0.5 µg of a supplement taken on an empty stomach.

[g] Although AIs have been set for choline, there are few data to assess whether a dietary supply of choline is needed at all stages of the life cycle, and it may be that the choline requirement can be met by endogenous synthesis at some of these stages.

[h] Because 10% to 30% of older people may malabsorb food-bound B-12, it is advisable for those older than 50 years to meet their RDA mainly by consuming foods fortified with B-12 or a supplement containing B-12.

[i] In view of evidence linking folate intake with neural tube defects in the fetus, it is recommended that all women capable of becoming pregnant consume 400 µg from supplements or fortified foods in addition to intake of food folate from a varied diet.

[j] It is assumed that women will continue consuming 400 µg from supplements or fortified food until their pregnancy is confirmed and they enter prenatal care, which ordinarily occurs after the end of the periconceptional period—the critical time for formation of the neural tube.

Source: Adapted from the *Dietary Reference Intakes* series, National Academies Press. Copyright 1997, 1998, 2000, 2001, 2011, by the National Academy of Sciences. The full reports are available from the National Academies Press at www.nap.edu.

# DIETARY REFERENCE INTAKES (DRIs): RECOMMENDED INTAKES FOR INDIVIDUALS, ELEMENTS
Food and Nutrition Board, Institute of Medicine, National Academies

| Life Stage Group | Calcium (mg/d) | Chromium (μg/d) | Copper (μg/d) | Fluoride (mg/d) | Iodine (μg/d) | Iron (mg/d) | Magnesium (mg/d) | Manganese (mg/d) | Molybdenum (μg/d) | Phosphorus (mg/d) | Selenium (μg/d) | Zinc (mg/d) |
|---|---|---|---|---|---|---|---|---|---|---|---|---|
| **Infants** | | | | | | | | | | | | |
| 0–6 mo | 200* | 0.2* | 200* | 0.01* | 110* | 0.27* | 30* | 0.003* | 2* | 100* | 15* | 2* |
| 7–12 mo | 260* | 5.5* | 220* | 0.5* | 130* | 11 | 75* | 0.6* | 3* | 275* | 20* | 3 |
| **Children** | | | | | | | | | | | | |
| 1–3 y | 700 | 11* | 340 | 0.7* | 90 | 7 | 80 | 1.2* | 17 | 460 | 20 | 3 |
| 4–8 y | 1000 | 15* | 440 | 1* | 90 | 10 | 130 | 1.5* | 22 | 500 | 30 | 5 |
| **Males** | | | | | | | | | | | | |
| 9–13 y | 1300 | 25* | 700 | 2* | 120 | 8 | 240 | 1.9* | 34 | 1250 | 40 | 8 |
| 14–18 y | 1300 | 35* | 890 | 3* | 150 | 11 | 410 | 2.2* | 43 | 1250 | 55 | 11 |
| 19–30 y | 1000 | 35* | 900 | 4* | 150 | 8 | 400 | 2.3* | 45 | 700 | 55 | 11 |
| 31–50 y | 1000 | 35* | 900 | 4* | 150 | 8 | 420 | 2.3* | 45 | 700 | 55 | 11 |
| 51–70 y | 1000 | 30* | 900 | 4* | 150 | 8 | 420 | 2.3* | 45 | 700 | 55 | 11 |
| >70 y | 1200 | 30* | 900 | 4* | 150 | 8 | 420 | 2.3* | 45 | 700 | 55 | 11 |
| **Females** | | | | | | | | | | | | |
| 9–13 y | 1300 | 21* | 700 | 2* | 120 | 8 | 240 | 1.6* | 34 | 1250 | 40 | 8 |
| 14–18 y | 1300 | 24* | 890 | 3* | 150 | 15 | 360 | 1.6* | 43 | 1250 | 55 | 9 |
| 19–30 y | 1000 | 25* | 900 | 3* | 150 | 18 | 310 | 1.8* | 45 | 700 | 55 | 8 |
| 31–50 y | 1000 | 25* | 900 | 3* | 150 | 18 | 320 | 1.8* | 45 | 700 | 55 | 8 |
| 51–70 y | 1200 | 20* | 900 | 3* | 150 | 8 | 320 | 1.8* | 45 | 700 | 55 | 8 |
| >70 y | 1200 | 20* | 900 | 3* | 150 | 8 | 320 | 1.8* | 45 | 700 | 55 | 8 |
| **Pregnancy** | | | | | | | | | | | | |
| 14–18 y | 1300 | 29* | 1000 | 3* | 220 | 27 | 400 | 2.0* | 50 | 1250 | 60 | 12 |
| 19–30 y | 1000 | 30* | 1000 | 3* | 220 | 27 | 350 | 2.0* | 50 | 700 | 60 | 11 |
| 31–50 y | 1000 | 30* | 1000 | 3* | 220 | 27 | 360 | 2.0* | 50 | 700 | 60 | 11 |
| **Lactation** | | | | | | | | | | | | |
| 14–18 y | 1300 | 44* | 1300 | 3* | 290 | 10 | 360 | 2.6* | 50 | 1250 | 70 | 13 |
| 19–30 y | 1000 | 45* | 1300 | 3* | 290 | 9 | 310 | 2.6* | 50 | 700 | 70 | 12 |
| 31–50 y | 1000 | 45* | 1300 | 3* | 290 | 9 | 320 | 2.6* | 50 | 700 | 70 | 12 |

NOTE: This table presents Recommended Dietary Allowances (RDAs) in **bold type** and Adequate Intakes (AIs) in ordinary type followed by an asterisk (*). RDAs and AIs may both be used as goals for individual intake. RDAs are set to meet the needs of almost all (97% to 98%) individuals in a group. For healthy breastfed infants, the AI is the mean intake. The AI for other life stage and gender groups is believed to cover needs of all individuals in the group, but lack of data or uncertainty in the data prevents being able to specify with confidence the percentage of individuals covered by this intake.

Sources: *Dietary Reference Intakes for Calcium, Phosphorus, Magnesium, Vitamin D, and Fluoride* (1997); *Dietary Reference Intakes for Thiamin, Riboflavin, Niacin, Vitamin B-6, Folate, Vitamin B-12, Pantothenic Acid, Biotin, and Choline* (1998); *Dietary Reference Intakes for Vitamin C, Vitamin E, Selenium, and Carotenoids* (2000); *Dietary Reference Intakes for Vitamin A, Vitamin K, Arsenic, Boron, Chromium, Copper, Iodine, Iron, Manganese, Molybdenum, Nickel, Silicon, Vanadium, and Zinc* (2001); and *Dietary Reference Intakes for Calcium and Vitamin D* (2011). These reports may be accessed via www.nap.edu. Adapted from the *Dietary Reference Intake* series, National Academies Press. Copyright 1997, 1998, 2000, 2001, and 2011 by the National Academy of Sciences. The full reports are available from the National Academies Press at www.nap.edu.

# DIETARY REFERENCE INTAKES (DRIs): RECOMMENDED INTAKES FOR INDIVIDUALS, MACRONUTRIENTS
Food and Nutrition Board, Institute of Medicine, National Academies

| Life Stage Group | Carbohydrate (g/d) | Total Fiber (g/d) | Fat (g/d) | Linoleic Acid (g/d) | α-Linolenic Acid (g/d) | Protein[a] (g/d) |
|---|---|---|---|---|---|---|
| Infants | | | | | | |
| 0–6 mo | 60* | ND | 31* | 4.4* | 0.5* | 9.1* |
| 7–12 mo | 95* | ND | 30* | 4.6* | 0.5* | **11.0** |
| Children | | | | | | |
| 1–3 y | **130** | 19* | ND[b] | 7* | 0.7* | **13** |
| 4–8 y | **130** | 25* | ND | 10* | 0.9* | **19** |
| Males | | | | | | |
| 9–13 y | **130** | 31* | ND | 12* | 1.2* | **34** |
| 14–18 y | **130** | 38* | ND | 16* | 1.6* | **52** |
| 19–30 y | **130** | 38* | ND | 17* | 1.6* | **56** |
| 31–50 y | **130** | 38* | ND | 17* | 1.6* | **56** |
| 51–70 y | **130** | 30* | ND | 14* | 1.6* | **56** |
| >70 y | **130** | 30* | ND | 14* | 1.6* | **56** |
| Females | | | | | | |
| 9–13 y | **130** | 26* | ND | 10* | 1.0* | **34** |
| 14–18 y | **130** | 26* | ND | 11* | 1.1* | **46** |
| 19–30 y | **130** | 25* | ND | 12* | 1.1* | **46** |
| 31–50 y | **130** | 25* | ND | 12* | 1.1* | **46** |
| 51–70 y | **130** | 21* | ND | 11* | 1.1* | **46** |
| >70 y | **130** | 21* | ND | 11* | 1.1* | **46** |
| Pregnancy | | | | | | |
| 14–18 y | **175** | 28* | ND | 13* | 1.4* | **71** |
| 19–30 y | **175** | 28* | ND | 13* | 1.4* | **71** |
| 31–50 y | **175** | 28* | ND | 13* | 1.4* | **71** |
| Lactation | | | | | | |
| 14–18 y | **210** | 29* | ND | 13* | 1.3* | **71** |
| 19–30 y | **210** | 29* | ND | 13* | 1.3* | **71** |
| 31–50 y | **210** | 29* | ND | 13* | 1.3* | **71** |

NOTE: This table presents Recommended Dietary Allowances (RDAs) in **bold type** and Adequate Intakes (AIs) in ordinary type followed by an asterisk (*). RDAs and AIs may both be used as goals for individual intake. RDAs are set to meet the needs of almost all (97% to 98%) individuals in a group. For healthy breastfed infants, the AI is the mean intake. The AI for other life stage and gender groups is believed to cover needs of all individuals in the group, but lack of data or uncertainty in the data prevents being able to specify with confidence the percentage of individuals covered by this intake.

[a]Based on 0.8 g protein/kg body weight for reference body weight.

[b]ND = not determinable at this time.

Sources: *Dietary Reference Intakes for Energy, Carbohydrate, Fiber, Fat, Fatty Acids, Cholesterol, Protein, and Amino Acids* (2002). This report may be accessed via www.nap.edu.

Adapted from the *Dietary Reference Intake* series, National Academies Press. Copyright 1997, 1998, 2000, 2001, by the National Academy of Sciences. The full reports are available from the National Academies Press at www.nap.edu.

## DIETARY REFERENCE INTAKES (DRIs): RECOMMENDED INTAKES FOR INDIVIDUALS, ELECTROLYTES AND WATER
Food and Nutrition Board, Health and Medicine Division, National Academies

| Life Stage Group | Sodium (mg/d) | Sodium CDRR[a] | Potassium (mg/d) | Chloride (mg/d) | Water (L/d) |
|---|---|---|---|---|---|
| **Infants** | | | | | |
| 0–6 mo | 110* | ND[b] | 400* | 180* | 0.7* |
| 7–12 mo | 370* | ND[b] | 860* | 570* | 0.8* |
| **Children** | | | | | |
| 1–3 y | 800* | Reduce intakes if above 1200 mg/day[c] | 2000* | 1500* | 1.3* |
| 4–8 y | 1000* | Reduce intakes if above 1500 mg/day[c] | 2300* | 1900* | 1.7* |
| **Males** | | | | | |
| 9–13 y | 1200* | Reduce intakes if above 1800 mg/day[c] | 2500* | 2300* | 2.4* |
| 14–18 y | 1500* | Reduce intakes if above 2300 mg/day[c] | 3000* | 2300* | 3.3* |
| 19–30 y | 1500* | Reduce intakes if above 2300 mg/day | 3400* | 2300* | 3.7* |
| 31–50 y | 1500* | Reduce intakes if above 2300 mg/day | 3400* | 2300* | 3.7* |
| 51–70 y | 1500* | Reduce intakes if above 2300 mg/day | 3400* | 2000* | 3.7* |
| > 70 y | 1500* | Reduce intakes if above 2300 mg/day | 3400* | 1800* | 3.7* |
| **Females** | | | | | |
| 9–13 y | 1200* | Reduce intakes if above 1800 mg/day[c] | 2300* | 2300* | 2.1* |
| 14–18 y | 1500* | Reduce intakes if above 2300 mg/day[c] | 2300* | 2300* | 2.3* |
| 19–30 y | 1500* | Reduce intakes if above 2300 mg/day | 2600* | 2300* | 2.7* |
| 31–50 y | 1500* | Reduce intakes if above 2300 mg/day | 2600* | 2300* | 2.7* |
| 51–70 y | 1500* | Reduce intakes if above 2300 mg/day | 2600* | 2000* | 2.7* |
| > 70 y | 1500* | Reduce intakes if above 2300 mg/day | 2600* | 1800* | 2.7* |
| **Pregnancy** | | | | | |
| 14–18 y | 1500* | Reduce intakes if above 2300 mg/day[c] | 2600* | 2300* | 3.0* |
| 19–30 y | 1500* | Reduce intakes if above 2300 mg/day | 2900* | 2300* | 3.0* |
| 31–50 y | 1500* | Reduce intakes if above 2300 mg/day | 2900* | 2300* | 3.0* |
| **Lactation** | | | | | |
| 14–18 y | 1500* | Reduce intakes if above 2300 mg/day[c] | 2500* | 2300* | 3.8* |
| 19–30 y | 1500* | Reduce intakes if above 2300 mg/day | 2800* | 2300* | 3.8* |
| 31–50 y | 1500* | Reduce intakes if above 2300 mg/day | 2800* | 2300* | 3.8* |

NOTE: This table is adapted from the *DRI reports*. See www.nap.edu. Adequate Intakes (AIs) are followed by an asterisk (*). These may be used as a goal for individual intake. For healthy breastfed infants, the AI is the average intake. The AI for other life stage and gender groups is believed to cover the needs of all individuals in the group, but lack of data prevents being able to specify with confidence the percentage of individuals covered by this intake; therefore, no Recommended Dietary Allowance (RDA) was set.

[a]CDRR = Chronic Disease Risk Reduction Intakes; [b]Not determined owing to insufficient strength of evidence for causality and intake-response; [c]Extrapolated from the adult CDRR based on sedentary Estimated Energy Requirements.

Sources: National Academies of Sciences, Engineering, and Medicine. *Dietary Reference Intakes for Sodium and Potassium*. The National Academies Press, Washington, DC, 2019; and *Dietary Reference Intakes for Water, Potassium, Sodium, Chloride, and Sulfate* (2005). These reports may be accessed via www.nap.edu.

## ACCEPTABLE MACRONUTRIENT DISTRIBUTION RANGES

| Macronutrient | Range (percent of energy) | | |
|---|---|---|---|
| | Children, 1–3 y | Children, 4–18 y | Adults |
| Fat | 30–40 | 25–35 | 20–35 |
| omega-6 polyunsaturated fats (linoleic acid) | 5–10 | 5–10 | 5–10 |
| omega-3 polyunsaturated fats[a] (α-linolenic acid) | 0.6–1.2 | 0.6–1.2 | 0.6–1.2 |
| Carbohydrate | 45–65 | 45–65 | 45–65 |
| Protein | 5–20 | 10–30 | 10–35 |

[a]Approximately 10% of the total can come from longer-chain n-3 fatty acids.

Source: *Dietary Reference Intakes for Energy, Carbohydrate, Fiber, Fat, Fatty Acids, Cholesterol, Protein, and Amino Acids* (2002). The report may be accessed via www.nap.edu.

Adapted from the *Dietary Reference Intakes* series, National Academies Press. Copyright 1997, 1998, 2000, 2001, 2011, by the National Academy of Sciences. The full reports are available from the National Academies Press at www.nap.edu.

# DIETARY REFERENCE INTAKES (DRIs): TOLERABLE UPPER INTAKE LEVELS (UL[a]), VITAMINS
## Food and Nutrition Board, Institute of Medicine, National Academies

| Life Stage Group | Vitamin A (µg/d)[b] | Vitamin C (mg/d) | Vitamin D (µg/d) | Vitamin E (mg/d)[c,d] | Vitamin K | Thiamin | Riboflavin | Niacin (mg/d)[d] | Vitamin B-6 (mg/d) | Folate (µg/d)[d] | Vitamin B-12 | Pantothenic Acid | Biotin | Choline (g/d) | Carotenoids[e] |
|---|---|---|---|---|---|---|---|---|---|---|---|---|---|---|---|
| **Infants** | | | | | | | | | | | | | | | |
| 0–6 mo | 600 | ND[f] | 25 | ND | ND | ND | ND | ND | ND | ND | ND | ND | ND | ND | ND |
| 7–12 mo | 600 | ND | 38 | ND | ND | ND | ND | ND | ND | ND | ND | ND | ND | ND | ND |
| **Children** | | | | | | | | | | | | | | | |
| 1–3 y | 600 | 400 | 63 | 200 | ND | ND | ND | 10 | 30 | 300 | ND | ND | ND | 1.0 | ND |
| 4–8 y | 900 | 650 | 75 | 300 | ND | ND | ND | 15 | 40 | 400 | ND | ND | ND | 1.0 | ND |
| **Males, Females** | | | | | | | | | | | | | | | |
| 9–13 y | 1700 | 1200 | 100 | 600 | ND | ND | ND | 20 | 60 | 600 | ND | ND | ND | 2.0 | ND |
| 14–18 y | 2800 | 1800 | 100 | 800 | ND | ND | ND | 30 | 80 | 800 | ND | ND | ND | 3.0 | ND |
| 19–70 y | 3000 | 2000 | 100 | 1000 | ND | ND | ND | 35 | 100 | 1000 | ND | ND | ND | 3.5 | ND |
| >70 y | 3000 | 2000 | 100 | 1000 | ND | ND | ND | 35 | 100 | 1000 | ND | ND | ND | 3.5 | ND |
| **Pregnancy** | | | | | | | | | | | | | | | |
| 14–18 y | 2800 | 1800 | 100 | 800 | ND | ND | ND | 30 | 80 | 800 | ND | ND | ND | 3.0 | ND |
| 19–50 y | 3000 | 2000 | 100 | 1000 | ND | ND | ND | 35 | 100 | 1000 | ND | ND | ND | 3.5 | ND |
| **Lactation** | | | | | | | | | | | | | | | |
| 14–18 y | 2800 | 1800 | 100 | 800 | ND | ND | ND | 30 | 80 | 800 | ND | ND | ND | 3.0 | ND |
| 19–50 y | 3000 | 2000 | 100 | 1000 | ND | ND | ND | 35 | 100 | 1000 | ND | ND | ND | 3.5 | ND |

[a] UL = The maximum level of daily nutrient intake likely to pose no risk of adverse effects. Unless otherwise specified, the UL represents total intake from food, water, and supplements. Due to lack of suitable data, ULs could not be established for vitamin K, thiamin, riboflavin, vitamin B-12, pantothenic acid, biotin, or carotenoids. In the absence of ULs, extra caution may be warranted in consuming levels above recommended intakes.

[b] As preformed vitamin A only.

[c] As α-tocopherol; applies to any form of supplemental α-tocopherol.

[d] The ULs for vitamin E, niacin, and folate apply to synthetic forms obtained from supplements, fortified foods, or a combination of the two.

[e] β-Carotene supplements are advised only to serve as a provitamin A source for individuals at risk of vitamin A deficiency.

[f] ND = Not determinable due to lack of data of adverse effects in this age group and concern with regard to lack of ability to handle excess amounts. Source of intake should be from food only to prevent high levels of intake.

Sources: Dietary Reference Intakes for Calcium and Vitamin D (2011); Dietary Reference Intakes for Calcium, Phosphorus, Magnesium, Vitamin D, and Fluoride (1997); Dietary Reference Intakes for Thiamin, Riboflavin, Niacin, Vitamin B-6, Folate, Vitamin B-12, Pantothenic Acid, Biotin, and Choline (1998); Dietary Reference Intakes for Vitamin C, Vitamin E, Selenium, and Carotenoids (2000); and Dietary Reference Intakes for Vitamin A, Vitamin K, Arsenic, Boron, Chromium, Copper, Iodine, Iron, Manganese, Molybdenum, Nickel, Silicon, Vanadium, and Zinc (2001). These reports may be accessed via www.nap.edu.

Adapted from the Dietary Reference Intakes series, National Academies Press. Copyright 1997, 1998, 2000, 2001, 2011, by the National Academy of Sciences. The full reports are available from the National Academies Press at www.nap.edu.

# DIETARY REFERENCE INTAKES (DRIs): TOLERABLE UPPER INTAKE LEVELS (UL[a]), ELEMENTS AND ELECTROLYTES[b]
## Food and Nutrition Board, Institute of Medicine, National Academies

| Life Stage Group | Boron (mg/d) | Calcium (g/d) | Copper (µg/d) | Fluoride (mg/d) | Iodine (µg/d) | Iron (mg/d) | Magnesium (mg/d)[c] | Manganese (mg/d) | Molybdenum (µg/d) | Nickel (mg/d) | Phosphorus (g/d) | Selenium (µg/d) | Vanadium (mg/d)[d] | Zinc (mg/d) | Sodium | Potassium | Chloride (mg/d) |
|---|---|---|---|---|---|---|---|---|---|---|---|---|---|---|---|---|---|
| Infants | | | | | | | | | | | | | | | | | |
| 0–6 mo | ND[e] | 1 | ND | 0.7 | ND | 40 | ND | ND | ND | ND | ND | 45 | ND | 4 | ND | ND | ND |
| 7–12 mo | ND | 1.5 | ND | 0.9 | ND | 40 | ND | ND | ND | ND | ND | 60 | ND | 5 | ND | ND | ND |
| Children | | | | | | | | | | | | | | | | | |
| 1–3 y | 3 | 2.5 | 1000 | 1.3 | 200 | 40 | 65 | 2 | 300 | 0.2 | 3 | 90 | ND | 7 | ND | ND | 2300 |
| 4–8 y | 6 | 2.5 | 3000 | 2.2 | 300 | 40 | 110 | 3 | 600 | 0.3 | 3 | 150 | ND | 12 | ND | ND | 2900 |
| Males, Females | | | | | | | | | | | | | | | | | |
| 9–13 y | 11 | 3 | 5000 | 10 | 600 | 40 | 350 | 6 | 1100 | 0.6 | 4 | 280 | ND | 23 | ND | ND | 3400 |
| 14–18 y | 17 | 3 | 8000 | 10 | 900 | 45 | 350 | 9 | 1700 | 1.0 | 4 | 400 | ND | 34 | ND | ND | 3600 |
| 19–70 y | 20 | 2.5[f] | 10000 | 10 | 1100 | 45 | 350 | 11 | 2000 | 1.0 | 4 | 400 | 1.8 | 40 | ND | ND | 3600 |
| >70 y | 20 | 2 | 10000 | 10 | 1100 | 45 | 350 | 11 | 2000 | 1.0 | 3 | 400 | 1.8 | 40 | ND | ND | 3600 |
| Pregnancy | | | | | | | | | | | | | | | | | |
| 14–18 y | 17 | 3 | 8000 | 10 | 900 | 45 | 350 | 9 | 1700 | 1.0 | 3.5 | 400 | ND | 34 | ND | ND | 3600 |
| 19–50 y | 20 | 2.5 | 10000 | 10 | 1100 | 45 | 350 | 11 | 2000 | 1.0 | 3.5 | 400 | ND | 40 | ND | ND | 3600 |
| Lactation | | | | | | | | | | | | | | | | | |
| 14–18 y | 17 | 3 | 8000 | 10 | 900 | 45 | 350 | 9 | 1700 | 1.0 | 4 | 400 | ND | 34 | ND | ND | 3600 |
| 19–50 y | 20 | 2.5 | 10000 | 10 | 1100 | 45 | 350 | 11 | 2000 | 1.0 | 4 | 400 | ND | 40 | ND | ND | 3600 |

[a] UL = The maximum level of daily nutrient intake that is likely to pose no risk of adverse effects. Unless otherwise specified, the UL represents total intake from food, water, and supplements. Due to lack of suitable data, ULs could not be established for arsenic, chromium, and silicon. In the absence of ULs, extra caution may be warranted in consuming levels above recommended intakes.

[b] Although silicon has not been shown to cause adverse effects in humans, there is no justification for adding silicon to supplements.

[c] The ULs for magnesium represent intake from a pharmacological agent only and do not include intake from food and water.

[d] Although vanadium in food has not been shown to cause adverse effects in humans, there is no justification for adding vanadium to food, and vanadium supplements should be used with caution. The UL is based on adverse effects in laboratory animals, and this data could be used to set a UL for adults but not children and adolescents.

[e] ND = Not determinable due to lack of data of adverse effects in this age group and concern with regard to lack of ability to handle excess amounts.

[f] Upper Limit declines to 2 after age 50.

Sources: Dietary Reference Intakes for Calcium and Vitamin D (2011); Dietary Reference Intakes for Calcium, Phosphorus, Magnesium, Vitamin D, and Fluoride (1997); Dietary Reference Intakes for Thiamin, Riboflavin, Niacin, Vitamin B-6, Folate, Vitamin B-12, Pantothenic Acid, Biotin, and Choline (1998); Dietary Reference Intakes for Vitamin C, Vitamin E, Selenium, and Carotenoids (2000); Dietary Reference Intakes for Vitamin A, Vitamin K, Arsenic, Boron, Chromium, Copper, Iodine, Iron, Manganese, Molybdenum, Nickel, Silicon, Vanadium, and Zinc (2001); Dietary Reference Intakes for Water, Potassium, Sodium, Chloride, and Sulfate (2004); and Dietary Reference Intakes for Sodium and Potassium (2019). These reports may be accessed via www.nap.edu.

Adapted from the Dietary Reference Intakes series, National Academies Press. Copyright 1997, 1998, 2000, 2001, 2011, and 2019, by the National Academy of Sciences. The full reports are available from the National Academies Press at www.nap.edu.

# Glossary

**1,25-dihydroxyvitamin D$_3$** Biologically active form of vitamin D; also called *calcitriol* or abbreviated *1,25(OH)D$_3$*.

**25-hydroxyvitamin D$_3$** Intermediate form of vitamin D found in blood; also called *calcidiol* or *calcifediol* or abbreviated *25(OH)D$_3$*.

**7-dehydrocholesterol** Precursor of vitamin D found in the skin.

**absorption** The process by which substances are taken up from the GI tract and enter the bloodstream or the lymph.

**absorptive cells** The intestinal cells that line the villi and participate in nutrient absorption; also known as *enterocytes*.

**Acceptable Daily Intake (ADI)** Estimate of the amount of a sweetener that an individual can safely consume daily over a lifetime. ADIs are given as milligrams per kilogram of body weight per day.

**acesulfame-K** Alternative sweetener that yields no energy to the body; 200 times sweeter than sucrose.

**acetaldehyde dehydrogenase** An enzyme used in ethanol metabolism that eventually converts acetaldehyde into carbon dioxide and water.

**acid group** In chemistry, a functional group that consists of a carbon atom that shares bonds with two oxygen atoms. This is the site where fatty acids are linked to glycerol to form triglycerides.

**acquired immunodeficiency syndrome (AIDS)** Late stage HIV when the body's immune system is badly damaged because of the virus.

**active absorption** Movement of a substance across a semipermeable membrane from an area of lower solute concentration to an area of higher solute concentration. This type of transport requires energy and a carrier.

**adaptive thermogenesis** The ability of humans to regulate body temperature within narrow limits (thermoregulation) in response to changes in dietary patterns or environmental temperatures.

**added sugars** Nutritive sweeteners (e.g., sugars and syrups) that are not naturally present in foods, but are added during processing for the purpose of flavoring and/or preserving foods.

**additives** Substances added to foods, either intentionally or incidentally.

**adenosine diphosphate (ADP)** A breakdown product of ATP. ADP is synthesized into ATP using energy from foodstuffs and a phosphate group (abbreviated $P_i$).

**adenosine triphosphate (ATP)** The main energy currency used by cells. ATP energy is used to promote ion pumping, enzyme activity, and muscular contraction.

**Adequate Intake (AI)** Nutrient intake amount set for any nutrient for which insufficient research is available to establish an RDA. AIs are based on estimates of intakes that appear to maintain a defined nutritional state in a specific life stage.

**adipose tissue** Connective tissue made up of cells that store fat; also cushions and insulates the body.

**advantame** Similar in structure to aspartame, this sweetener is 20,000 times sweeter than sucrose.

**aerobic** Requiring oxygen; with reference to physical activity, all forms of activity that are intense enough and performed long enough to maintain or improve an individual's cardiorespiratory fitness.

**aging** Time-dependent physical and physiological changes in body structure and function that occur normally and progressively throughout adulthood as humans mature and become older.

**air displacement** A method for estimating body composition that makes use of the volume of space taken up by a body inside a small chamber (Bod Pod®). This tool is also known as *air displacement plethysmography*.

**alcohol** Ethyl alcohol or ethanol (CH$_3$CH$_2$OH) is the compound in alcoholic beverages.

**alcohol dehydrogenase** An enzyme used in alcohol (ethanol) metabolism that converts alcohol into acetaldehyde.

**alcohol use disorder** Problem drinking characterized by a compulsive pattern of alcohol use that leads to significant impairment or distress.

**alcohol-related birth defects (ARBDs)** One or more birth defects (e.g., malformations of the heart, bones, kidneys, eyes, or ears) related to confirmed alcohol exposure during gestation.

**alcohol-related neurodevelopmental disorders (ARNDs)** One or more abnormalities of the central nervous system (e.g., small head size, impaired motor skills, hearing loss, or poor hand-eye coordination) related to confirmed alcohol exposure during gestation.

**aldosterone** A hormone produced by the adrenal glands when blood volume is low. It acts on the kidneys to conserve sodium (and therefore water) to increase blood volume.

**allergen** A foreign protein, or antigen, that induces excess production of certain immune system antibodies; subsequent exposure to the same protein leads to allergic symptoms. Whereas all allergens are antigens, not all antigens are allergens.

**allergy** A hypersensitive immune response that occurs when the immune system identifies a harmless protein (e.g., a food protein) as a harmful pathogen (i.e., antigen) and attempts to destroy it.

**allulose** Naturally occurring sugar in some food sources that is about 70% as sweet as sugar.

**alpha-linolenic acid** An essential omega-3 fatty acid with 18 carbons and three double bonds.

**amenorrhea** Absence of menstrual periods in a woman of reproductive age.

**amino acid** The building block for proteins containing a central carbon atom with nitrogen and other atoms attached.

**amylase** Starch-digesting enzyme produced by the salivary glands and the pancreas.

**amylopectin** A digestible branched-chain type of starch composed of glucose units.

**amylose** A digestible straight-chain type of starch composed of glucose units.

**anabolic** Relating to pathways that use small, simple compounds to build larger, more complex compounds.

**anaerobic** Not requiring oxygen; with reference to physical activity, high intensity activity that exceeds the capacity of the cardiovascular system to provide oxygen to muscle cells for the usual oxygen-consuming metabolic pathways.

**anal sphincters** A group of two sphincters (inner and outer) that help control expulsion of feces from the body.

**anaphylaxis** A severe allergic response that results in lowered blood pressure and respiratory distress. This can be fatal.

**anemia** A decreased oxygen-carrying capacity of the blood. This can be caused by many factors, such as iron deficiency or blood loss.

**anencephaly** Type of neural tube defect characterized by the absence of some or all of the brain and skull.

**angiotensin** A hormone produced by the liver and activated by enzymes from the kidneys. It signals the adrenal glands to produce aldosterone and also directs the kidneys to conserve sodium (and therefore water). Both of these actions have the effect of increasing blood volume.

**angular cheilitis** Inflammation of the corners of the mouth with painful cracking; also called *cheilosis* or *angular stomatitis*.

**animal experiments** Use of animals to study disease to understand more about human disease.

**anorexia nervosa** An eating disorder characterized by extreme restriction of energy intake relative to requirements, leading to significantly low body weight.

**anthropometric assessment** Measurement of body weight and the lengths and proportions of parts of the body.

**antibody** Blood protein that binds foreign proteins found in the body; also called *immunoglobulin*.

**antidiuretic hormone (ADH)** A hormone secreted by the pituitary gland when blood concentration of solutes is high. It causes the kidneys to decrease water excretion, which increases blood volume.

**antigen** Any substance that induces a state of sensitivity and/or resistance to microorganisms or toxic substances after a lag period; a foreign substance that stimulates a specific aspect of the immune system.

**antioxidant** A substance that has the ability to prevent or repair the damage caused by oxidation.

**anus** Last portion of the GI tract; serves as an outlet for the digestive system.

**appetite** The primarily psychological (external) influences that encourage us to find and eat food, often in the absence of obvious hunger.

**arachidonic acid (AA)** An omega-6 fatty acid made from linoleic acid with 20 carbon atoms and four double bonds.

**ariboflavinosis** A deficiency disease resulting from a riboflavin deficiency and often characterized by mouth sores, dermatitis, glossitis, and/or angular cheilitis.

**artery** A blood vessel that carries blood away from the heart.

**ascending colon** Segment of the large intestine that carries feces from the cecum, up the right side of the abdomen, to the transverse colon.

**aseptic processing** A method by which food and container are separately and simultaneously sterilized; it allows manufacturers to produce boxes of milk that can be stored at room temperature.

**aspartame** Alternative sweetener made of two amino acids and methanol; about 200 times sweeter than sucrose.

**atherosclerosis** A buildup of fatty material (plaque) in the arteries, including the arteries that supply blood to the heart.

**atom** Smallest combining unit of an element, such as iron or calcium. Atoms consist of protons, neutrons, and electrons.

**atopic disease** A condition involving an inappropriate immune response to environmental allergens; examples include asthma, eczema, and seasonal allergies.

**atypical anorexia nervosa** A subthreshold eating disorder in which a person meets most of the criteria for diagnosis of anorexia nervosa, except BMI is within a normal range.

**autism spectrum disorder** A disorder of neurological development characterized by problems with social interaction, verbal and nonverbal communication, and/or unusual, repetitive, or limited activities and interests.

**avoidant/restrictive food intake disorder (ARFID)** Eating disorder characterized by failure to meet energy or nutrient needs, resulting in significant weight loss, nutritional deficiencies, or dependence on tube or intravenous feeding; the eating disturbance is not explained by lack of available food, a medical problem, or another eating disorder.

**β-glucan** Oats and barley are rich sources of these glucose polymers.

**baby-led introduction to solid foods** A method of introducing developmentally appropriate complementary foods that allows the infant to have greater control over self-feeding; also called *baby-led weaning*.

**bacteria** Single-cell microorganisms; some produce poisonous toxins, which cause illness in humans. Bacteria can be carried by water, animals, and people. They survive on skin, clothes, and hair and thrive in foods at room temperature. Some can live without oxygen and survive by means of spore formation.

**bariatrics** The medical specialty focusing on the treatment of obesity.

**basal metabolism** The minimal amount of calories the body uses to support itself in a fasting state when resting and awake in a warm, quiet environment. It amounts to roughly 1 kcal per kilogram per hour for men and 0.9 kcal per kilogram per hour for women; these values are often referred to as *basal metabolic rate (BMR)*.

**benign** Noncancerous; tumors that do not spread.

**beriberi** The thiamin-deficiency disorder characterized by muscle weakness, loss of appetite, nerve degeneration, and sometimes edema.

**beta-carotene** The orange-yellow pigment in carrots; beta-carotene is the only carotenoid that can be sufficiently absorbed and converted into retinol in the body.

**BHA, BHT** Butylated hydroxyanisole and butylated hydroxytoluene: two common synthetic antioxidants added to foods.

**bicarbonate** Alkaline compound produced as part of the body's buffer systems. For example, the pancreas secretes bicarbonate to neutralize the hydrochloric acid in chyme in the small intestine.

**bile** A liver secretion stored in the gallbladder and released through the common bile duct into the first segment of the small intestine. It is essential for the digestion and absorption of fat.

**bile acid** A compound produced by the liver. Bile acids are the main component of bile, which aids in emulsification of fat during digestion in the small intestine.

**binge drinking** Drinking sufficient alcohol within a 2-hour period to increase blood alcohol content to 0.08% or higher; for men, consuming 5 or more drinks in a row; for women, consuming 4 or more drinks in a row.

**binge eating** Consuming an abnormally large amount of food within a short time period (e.g., 2 hours).

**binge eating disorder** An eating disorder characterized by recurrent episodes of binge eating that are associated with marked distress and lack of control over behavior, but not followed by inappropriate compensatory behaviors to prevent weight gain.

**binge eating disorder of limited duration** A subthreshold eating disorder in which a person meets all of the criteria for diagnosis of binge eating disorder, except the duration of the disordered eating behavior is less than 3 months.

**binge eating disorder of low frequency** A subthreshold eating disorder in which a person meets all of the criteria for diagnosis of binge eating disorder, except the frequency of binges is less than once per week.

**bioavailability** The degree to which an ingested nutrient is digested and absorbed and thus is available to the body.

**biochemical assessment** Measurement of biochemical functions (e.g., concentrations of nutrient by-products or biologic activities in the blood, feces, or urine) related to a nutrient's function.

**bioelectrical impedance analysis (BIA)** The method to estimate total body fat that uses a low-energy electrical current. The more fat storage a person has, the more impedance (resistance) to electrical flow will be exhibited.

**biofortification** Use of selective breeding or other biotechnology to enhance the nutrient content of crops.

**biological pest management** Control of agricultural pests by using natural predators, parasites, or pathogens. For example, ladybugs can be used to control an aphid infestation.

**biotechnology** A collection of processes that involves the use of biological systems for altering and, ideally, improving the characteristics of plants, animals, and other forms of life.

**bisphosphonates** Drugs that bind minerals and prevent osteoclast breakdown of bone. Examples are alendronate (Fosamax) and risedronate (Actonel).

**Bitot's spots** Dry, foamy spots made from an accumulation of keratin (a protein) on the surface of the eye; caused by vitamin A deficiency.

**body mass index (BMI)** Weight (in kilograms) divided by height (in meters) squared; a value of 25 and above indicates overweight, and a value of 30 and above indicates obesity.

**bolus** A moistened mass of food swallowed from the oral cavity into the pharynx.

**bomb calorimeter** An instrument used to determine the calorie content of a food.

**bond** A linkage between two atoms formed by the sharing of electrons, or attractions.

**branched-chain amino acids** Amino acids with a branching carbon backbone; these are leucine, isoleucine, and valine. All are essential amino acids.

**brown adipose tissue** A specialized form of adipose (fat) tissue that produces large amounts of heat by metabolizing energy-yielding nutrients without synthesizing much useful energy for the body. The unused energy is released as heat.

**buffer** Compounds that cause a solution to resist changes in acid–base conditions.

**bulimia nervosa** An eating disorder characterized by recurrent episodes of binge eating followed by inappropriate compensatory behaviors to prevent weight gain.

**bulimia nervosa of limited duration** A subthreshold eating disorder in which a person meets all of the criteria for diagnosis of bulimia nervosa, except the duration of the disordered eating behavior is less than 3 months.

**bulimia nervosa of low frequency** A subthreshold eating disorder in which a person meets all of the criteria for diagnosis of bulimia nervosa, except the frequency of binge-compensate cycles is less than once per week.

**cancer** A condition characterized by uncontrolled growth of abnormal cells.

**capillary** A microscopic blood vessel that connects the smallest arteries and veins; site of nutrient, oxygen, and waste exchange between body cells and the blood.

**capillary bed** Network of one-cell-thick vessels that create a junction between arterial and venous circulation. It is here that gas and nutrient exchange occurs between body cells and the blood.

**carbohydrate** A compound containing carbon, hydrogen, and oxygen atoms. Most are known as *sugars, starches,* and *fibers.*

**carbohydrate loading** A process in which a high-carbohydrate diet is consumed for several days before an athletic event while tapering exercise duration in an attempt to increase muscle glycogen stores.

**carbon footprint** The greenhouse gas emissions caused by an organization, event, product, or individual.

**carbon skeleton** Amino acid structure that remains after the amino group ($-NH_2$) has been removed.

**cardiovascular disease** A general term that refers to any disease of the heart and circulatory system. This disease is generally characterized by the deposition of fatty material in the blood vessels (hardening of the arteries), which in turn can lead to organ damage and death. Also termed *coronary heart disease (CHD)* or simply, *heart disease,* as the vessels of the heart are the primary sites of the disease.

**cardiovascular system** The body system consisting of the heart, blood vessels, and blood. This system transports nutrients, waste products, gases, and hormones throughout the body and plays an important role in immune responses and regulation of body temperature.

**case reports** Descriptive studies based on uncontrolled observations of individual patients.

**case-control study** A study in which individuals who have a disease or condition, such as lung cancer, are compared with individuals who do not have the condition.

**catabolic** Relating to pathways that break down large compounds into smaller compounds.

**catalase** Enzyme that catalyzes the decomposition of hydrogen peroxide into water and oxygen.

**cecum** A pouch at the first part of the large intestine that houses many bacteria.

**celiac disease** Chronic, immune-mediated disease precipitated by exposure to dietary gluten in genetically predisposed people.

**cell** The structural basis of plant and animal organization. In animals it is bounded by a cell membrane. Cells have the ability to take up compounds from and excrete compounds into their surroundings.

**cellular differentiation** The process of a less-specialized cell becoming a more specialized type. An example is when stem cells in the bone marrow become red and white blood cells.

**cellulose** An indigestible polysaccharide made of glucose molecules and found in plant cell walls.

**ceruloplasmin** Copper-containing protein in the blood; functions in the transport of iron.

**chain-breaking** Breaking the link between two or more behaviors that encourage overeating, such as snacking while watching television.

**chemical reaction** An interaction between two chemicals that changes both chemicals.

**choice** The serving size of a food on a specific food list.

**cholecystokinin** A hormone produced by the small intestinal cells that stimulates enzyme release from the pancreas and bile release from the gallbladder.

**cholesterol** A waxy lipid found in all body cells. It has a structure containing multiple chemical rings. Cholesterol is found only in food ingredients of animal origin.

**chromosome** A single, large DNA molecule and its associated proteins; contains many genes to store and transmit genetic information.

**chronic** Long-standing, developing over time. When referring to disease, this term indicates that the disease process, once developed, is slow and lasting. A good example is cardiovascular disease.

**Chronic Disease Risk Reduction Intake (CDRR)** Category of DRIs based upon chronic disease risk.

**chylomicron** Lipoprotein that carries dietary fats (mainly triglycerides) from the small intestine to cells throughout the body. Chylomicrons are first absorbed into the lymphatic system, and then transported via the bloodstream.

**chylomicron remnant** Lipoprotein that remains after triglycerides have been removed from a chylomicron; composed of protein, phospholipids, and cholesterol.

**chyme** A mixture of stomach secretions and partially digested food.

**cirrhosis** A loss of functioning liver cells, which are replaced by nonfunctioning connective tissue. Any substance that poisons liver cells can lead to cirrhosis. The most common cause is chronic, excessive alcohol intake. Exposure to certain industrial chemicals also can lead to cirrhosis.

*cis* **fatty acid** A form of an unsaturated fatty acid that has the hydrogens lying on the same side of the carbon-carbon double bond.

**clinical assessment** Examination of general appearance of skin, eyes, and tongue; sense of touch; ability to cough and walk; and evidence of rapid hair loss.

**coagulation** Formation of a blood clot.

**coenzyme** An organic compound that combines with an inactive enzyme to form a catalytically active form. In this manner, coenzymes aid in enzyme function.

**cognitive behavioral therapy** Psychological therapy in which the person's assumptions about dieting, body weight, and related issues are confronted. New ways of thinking are explored and then practiced by the person. In this way, an individual can learn new ways to control disordered eating behaviors and related life stress.

**cognitive restructuring** Changing one's frame of mind regarding eating; for example, instead of using a difficult day as an excuse to overeat, a person would substitute other pleasures for rewards, such as a relaxing walk with a friend.

**cohort studies** Observational studies that look at large groups of people, prospectively or retrospectively, studying their exposure to certain risk factors for disease.

**colostrum** The first fluid secreted by the breast during late pregnancy and the first few days after birth. This thick fluid is rich in immune factors and protein.

**community-supported agriculture (CSA)** Farms that are supported by a community of growers and consumers who provide mutual support and share the risks and benefits of food production, usually including a system of weekly delivery or pickup of vegetables and fruit, and sometimes dairy products and meat.

**compensatory behaviors** Actions taken to rid the body of excess calories and/or to alleviate guilt or anxiety associated with a binge; examples include vomiting, misuse of laxatives, or excessive exercise.

**complementary foods** Solid or semi-solid foods that are introduced to an infant's dietary pattern to complement (not replace) breast milk or infant formula during the latter part of the first year of life. These foods provide energy and essential nutrients to meet the infant's requirements for growth and development.

**complementary proteins** Two food protein sources that make up for each other's inadequate supply of specific essential amino acids; together, they yield a sufficient amount of all nine and so provide high-quality (complete) protein for the diet.

**complex carbohydrate** Carbohydrate composed of many monosaccharide molecules. Examples include glycogen, starch, and fiber.

**compression of morbidity** Delay of the onset of disabilities caused by chronic disease.

**conditionally essential amino acids** Nonessential amino acids that cannot be made in adequate amounts to support the body's increased requirements during conditions of rapid growth, disease, or metabolic stress, and therefore become essential (i.e., must be obtained from food).

**congenital hypothyroidism** The stunting of body growth and poor development in the offspring that result from inadequate maternal intake of iodine during pregnancy that impairs thyroid hormone synthesis (formerly called *cretinism*).

**congenital lactase deficiency** A rare birth defect resulting in the inability to produce lactase, such that a lactose-free diet is required from birth.

**connective tissue** Protein tissue that holds different structures in the body together. Some body structures are made up of connective tissue—notably, tendons and cartilage. Connective tissue also forms part of bone and the nonmuscular structures of arteries and veins.

**constipation** A condition characterized by difficult and/or infrequent bowel movements (i.e., fewer than three bowel movements per week).

**contingency management** Forming a plan of action to respond to a situation in which overeating is likely, such as when snacks are within arm's reach at a party.

**control group** Participants in an experiment who are not given the treatment being tested.

**creatine** An organic (i.e., carbon-containing) molecule in muscle cells that serves as part of a high-energy compound (termed *creatine phosphate* or *phosphocreatine*) capable of synthesizing ATP from ADP.

**cross-contamination** Process by which bacteria or other microorganisms are unintentionally transferred from one substance or object to another, with harmful effect.

**cross-sectional study** Type of observational study that analyzes data from a population group at one specific point in time and based on particular variables of interest.

**cytoplasm** The fluid and organelles (except the nucleus) in a cell; also called *cytosol*.

**Daily Value (DV)** Quantity (expressed in percentage) of a specific nutrient that corresponds to the total percentage of the daily requirements for a particular nutrient based on a 2000 kcal diet.

**danger zone** Temperature range (40°F to 140°F) where bacteria grow most rapidly.

**dehydration** A harmful condition in which water intake is inadequate to replace losses.

**Delaney Clause** A clause to the 1958 Food Additives Amendment of the Pure Food and Drug Act in the United States that prevents the intentional (direct) addition to foods of a compound shown to cause cancer in laboratory animals or humans.

**dementia** A general loss or decrease in mental function.

**denaturation** Alteration of a protein's three-dimensional structure, usually because of treatment by heat, enzymes, acid or alkaline solutions, or agitation.

**dental caries** Erosions in the surface of a tooth caused by acids made by bacteria as they metabolize sugars.

**deoxyribonucleic acid (DNA)** Double strand of nucleic acids that carries hereditary information in cells; DNA directs the synthesis of cell proteins.

**dermatitis** Condition that involves itchy, dry skin or a rash on swollen, reddened skin.

**descending colon** Segment of the large intestine that carries feces from the transverse colon, down the left side of the abdomen, to the sigmoid colon.

**diabetes** A group of diseases characterized by high blood glucose. Type 1 diabetes involves insufficient or no release of the hormone insulin by the pancreas and therefore requires daily insulin therapy. Type 2 diabetes results from either insufficient release of insulin or general inability of insulin to act on certain body cells, such as muscle cells. Persons with type 2 diabetes may or may not require insulin therapy.

**diarrhea** Increased fluidity, frequency, or amount of bowel movements (i.e., three or more loose stools per day).

**diastolic blood pressure** The pressure in the arteries when the heart rests between beats and fills with blood and receives oxygen.

**dietary assessment** Estimation of typical food choices relying mostly on the recounting of one's usual intake or a record of one's previous days' intake.

**dietary fiber** Indigestible fiber found in food.

**Dietary Reference Intakes (DRIs)** Term used to encompass nutrient recommendations made by the Food and Nutrition Board of the National Academies of Sciences, Engineering, and Medicine. These include RDAs, AIs, EERs, CDRRs, and ULs.

**digestion** Process by which large ingested molecules are mechanically and chemically broken down to produce basic nutrients that can be absorbed across the wall of the GI tract.

**digestive system** System consisting of the gastrointestinal tract and accessory structures (liver, gallbladder, and pancreas). This system performs the mechanical and chemical processes of digestion, absorption of nutrients, and elimination of wastes.

**diglyceride** A breakdown product of a triglyceride consisting of two fatty acids attached to a glycerol backbone.

**direct calorimetry** A method of determining a body's energy use by measuring heat released from the body. An insulated metabolic chamber is typically used.

**direct food additives** Additives knowingly (intentionally) incorporated into food products by manufacturers.

**disaccharide** Class of sugars formed by the chemical bonding of two monosaccharides.

**disordered eating** Mild and short-term changes in eating patterns that occur in relation to a stressful event, an illness, or a desire to modify one's dietary pattern for a variety of health and personal appearance reasons.

**diuretic** A substance that increases urinary fluid excretion.

**diverticula** Pouches that protrude through the exterior wall of the large intestine.

**diverticulitis** Inflammation of the diverticula, which may be related to bacterial activity inside the diverticula.

**diverticulosis** The condition of having many diverticula in the large intestine.

**docosahexaenoic acid (DHA)** An omega-3 fatty acid with 22 carbons and six carbon-carbon double bonds. It is present in large amounts in fatty fish and is slowly synthesized in the body from alpha-linolenic acid. In the human body, high levels of DHA are found in the retina and brain.

**double-blind** A study or trial in which any information which may influence the behavior of the tester or the subject is withheld until after the test.

**dual energy x-ray absorptiometry (DXA)** A scientific tool used to measure bone mineral density and body composition.

**duodenum** First segment of the small intestine that receives chyme from the stomach and digestive juices from the pancreas and gallbladder. This is the site of most chemical digestion of nutrients; approximately 10 inches in length.

**dysbiosis** A disturbance in the balance of beneficial and pathogenic microorganisms in the microbiota.

**early childhood caries** Tooth decay that results from formula or juice (and even human milk) bathing the teeth as the child sleeps with a bottle in the mouth. The upper teeth are mostly affected as the lower teeth are protected by the tongue; formerly called *nursing bottle syndrome* or *baby bottle tooth decay.*

**eating disorder** Severe alterations in eating patterns linked to physiological changes. The alterations are associated with food restriction, binge eating, inappropriate compensatory behaviors, and fluctuations in weight. They also involve a number of emotional and cognitive changes that affect the way a person perceives and experiences his or her body.

**eclampsia** A severe form of gestational hypertension characterized by protein in the urine and seizures; formerly called *toxemia.*

**edema** The buildup of excess fluid in extracellular spaces of tissues.

**eicosanoids** A class of hormone compounds, including the prostaglandins, derived from the essential fatty acids. These signaling compounds are involved in cellular activity that affects practically all important functions in the body.

**eicosapentaenoic acid (EPA)** An omega-3 fatty acid with 20 carbons and five carbon-carbon double bonds. It is present in large amounts in fatty fish and is slowly synthesized in the body from alpha-linolenic acid.

**electrolyte** A mineral that separates into positively or negatively charged ions in water. Electrolytes are able to transmit an electrical current.

**elimination diet** A restrictive diet that systematically tests foods that may cause an allergic response by first eliminating them for 1 to 2 weeks and then adding them back, one at a time.

**embryo** In humans, the developing offspring in utero from about the beginning of the third week to the end of the eighth week after conception.

**emulsifier** A compound that can suspend fat in water by isolating individual fat droplets, using a shell of water molecules or other substances to prevent the fat from coalescing.

**endocrine gland** A hormone-producing gland.

**endocrine system** The body system consisting of the various glands and the hormones these glands secrete. This system has major regulatory functions in the body, such as reproduction and cell metabolism.

**endoplasmic reticulum (ER)** An organelle composed of a network of canals running through the cytoplasm. Part of the endoplasmic reticulum contains ribosomes.

**endorphins** Natural body tranquilizers that function in pain reduction and may be involved in the feeding response.

**energy balance** The state in which energy (calorie) intake, in the form of food and beverages, matches the energy expended, primarily through basal metabolism and physical activity.

**energy density** A comparison of the calorie (kcal) content of a food with the weight of the food. An energy-dense food is high in calories but weighs very little (e.g., potato chips), whereas a food low in energy density has few calories but weighs a lot (e.g., an orange).

**enterohepatic circulation** A continual recycling of compounds such as bile acids between the small intestine and the liver.

**environmental assessment** Includes details about living conditions, education level, and the ability of the person to purchase, transport, and prepare food. The person's weekly budget for food purchases is also a key factor to consider.

**enzyme** A compound that speeds up the rate of a chemical reaction but is not altered by the reaction. Almost all enzymes are proteins (some are made of genetic material).

**epidemiology** The study of how disease rates vary among different population groups.

**epigenetics** Study of heritable changes in gene function that are independent of DNA sequence. For example, malnutrition during pregnancy may modify gene expression in the fetus and affect long-term body weight regulation in the offspring.

**epigenome** A network of chemical compounds surrounding DNA that modify the genome without altering the DNA sequences and have a role in determining which genes are active (expressed) or inactive (silenced) in a particular cell.

**epiglottis** The flap that folds down over the trachea during swallowing.

**epinephrine** A hormone that is released by the adrenal glands at times of stress. It acts to increase glycogen breakdown in the liver, among other functions. Also known as *adrenaline.*

**epithelial tissue** The surface cells that line the outside of the body and all external passages within it.

**ergogenic** A mechanical, nutritional, psychological, pharmacological, or physiological substance or treatment intended to directly improve exercise performance.

**esophagus** A tube in the GI tract that connects the pharynx with the stomach.

**essential amino acids** The amino acids that cannot be synthesized by humans in sufficient amounts or at all and therefore must be included in the dietary pattern; there are nine essential amino acids. These are also called *indispensable amino acids.*

**essential fatty acids** Fatty acids that must be supplied by the diet to maintain health. Currently, only linoleic acid and alpha-linolenic acid are classified as essential.

**essential nutrient** In nutritional terms, a substance that, when left out of a dietary pattern, leads to signs of poor health. The body either cannot produce this nutrient or cannot produce enough of it to meet its needs. If added back to a dietary pattern before permanent damage occurs, the affected aspects of health are restored.

**Estimated Energy Requirement (EER)** Estimate of the energy (kcal) intake needed to match the energy use of an average person in a specific life stage.

**ethanol** Chemical term for the form of alcohol found in alcoholic beverages.

**exercise** Physical activities that are planned, repetitive, and intended to improve physical fitness.

**extracellular fluid** Fluid found outside the cells; it represents about one-third of body fluid.

**extracellular space** The space outside cells; represents one-third of body fluid.

**facilitated diffusion** Movement of a substance across a semipermeable membrane from an area of higher solute concentration to an area of lower solute concentration. This type of transport does not require energy, but it does require a carrier.

**failure to thrive** Condition of inadequate growth during infancy or early childhood caused by inadequate nutritional intake, inefficient nutrient absorption, or excessive energy expenditure; commonly defined as weight-for-age below the 5th percentile on multiple occasions or weight declining two or more major percentile lines on a standardized growth chart.

**famine** An extreme shortage of food, which leads to massive starvation in a population; often associated with crop failures, war, and political unrest.

**fat adaptation** Manipulating the diet and physical training regimen so that muscles become more efficient at metabolizing fat as fuel during aerobic activity. Also known as *ketoadaptation*.

**fat-soluble** Soluble in fats, oils, or fat solvents.

**fat-soluble vitamins** Vitamins that dissolve in fat and some chemical compounds but not readily in water. These vitamins are A, D, E, and K.

**fecal impaction** The presence of a mass of hard, dry feces that remains in the rectum as a result of chronic constipation.

**feces** Mass of water, fiber, tough connective tissues, bacterial cells, and sloughed intestinal cells that passes through the large intestine and is excreted through the anus; also called *stool*.

**federal poverty level (FPL)** An economic measure used to determine if an individual or family qualifies for specific federal benefits and programs; also called *poverty line*.

**fermentation** The conversion of carbohydrates to alcohols, acids, and carbon dioxide without the use of oxygen.

**ferritin** A protein that stores iron and releases it in a controlled manner; acts as a buffer against iron deficiency and iron overload.

**fetal alcohol spectrum disorders (FASDs)** A group of irreversible physical and mental abnormalities in the infant that result from the mother's consumption of alcohol during pregnancy.

**fetal alcohol syndrome (FAS)** Severe form of FASD that involves abnormal facial features and problems with development of the nervous system and overall growth as a result of maternal alcohol consumption during pregnancy.

**fetal macrosomia** A condition in which an infant grows excessively large (e.g., birth weight > 4000 grams) in utero, usually as a consequence of maternal hyperglycemia.

**fetal origins hypothesis** A theory that links nutritional and other environmental factors during gestation to the future health of the offspring.

**fetus** The developing organism from about the beginning of the ninth week after conception until birth.

**fiber** Substances in plant foods not digested in the human stomach or small intestine. These add bulk to feces. Fiber naturally found in foods is also called *dietary fiber*.

**fluorosis** Discoloration of tooth enamel sometimes accompanied with pitting due to consuming a large amount of fluoride for an extended period.

**foam cells** Lipid-loaded white blood cells that have surrounded large amounts of a fatty substance, usually cholesterol, on the blood vessel walls.

**FODMAPs** **F**ermentable **o**ligosaccharides, **d**isaccharides, **m**onosaccharides, **a**nd **p**olyols. These carbohydrates may be poorly digested and lead to GI symptoms such as bloating, gas, and diarrhea in some people.

**food allergy** An adverse reaction to food that involves an immune response; also called *food hypersensitivity*.

**food desert** Areas where residents do not live near supermarkets or other food retailers that carry affordable and nutritious food.

**food insecure** The state of being without reliable access to a sufficient quantity of affordable, nutritious food.

**food intolerance** An adverse reaction to food that does not involve an allergic reaction.

**food jag** A period of time (usually a few days or weeks) during which a person will eat only a limited variety of foods.

**food lists** A system for classifying foods into numerous lists based on the foods' macronutrient composition, and establishing serving sizes, so that one serving of each food on a list contains approximately the same amount of carbohydrate, protein, fat, and calories.

**food waste** Food that is edible or fit for consumption which is being discarded as plate waste by consumers and by retailers due to color or appearance.

**foodborne illness** Sickness caused by the ingestion of food containing harmful substances.

**foodborne infection** Occurs when a person eats food containing harmful microorganisms, which then grow in the intestinal tract and cause illness.

**foodborne intoxication** Results when a person eats food containing toxins that cause illness.

**fructose** A six-carbon monosaccharide that usually exists in a ring form; found in fruits and honey; also known as *fruit sugar*.

**fruitarian** Referring to a dietary pattern that primarily includes fruits, nuts, honey, and vegetable oils.

**full term** Referring to an infant born from 37 up to 42 weeks of gestation; also called *term*.

**functional fiber** Indigestible carbohydrates that have beneficial physiological effects in humans.

**fungi** Simple parasitic life forms, including molds, mildews, yeasts, and mushrooms. They live on dead or decaying organic matter. Fungi can grow as single cells, like yeast, or as a multicellular colony, as seen with molds.

**galactose** A six-carbon monosaccharide that usually exists in a ring form; closely related to glucose.

**galactosemia** Inborn error of metabolism in which the enzyme that converts galactose into glucose is missing or deficient.

**gallbladder** An organ attached to the underside of the liver; site of bile storage, concentration, and eventual secretion.

**gastroesophageal reflux disease (GERD)** Disease that results from stomach acid backing up into the esophagus. The acid irritates the lining of the esophagus, causing pain.

**gastrointestinal (GI) tract** The main sites in the body used for digestion and absorption of nutrients. It consists of the mouth, esophagus, stomach, small intestine, large intestine, rectum, and anus. Also called the *digestive tract*.

**gene** A specific segment on a chromosome. Genes provide the blueprint for the production of cell proteins.

**gene expression** Use of DNA information on a gene to produce a protein.

**generally recognized as safe (GRAS)** A list of food additives that in 1958 were considered safe for consumption. The FDA continues to bear responsibility for proving they are not safe and can remove unsafe products from the list.

**generational poverty** Chronic state of poverty lasting for two generations or longer.

**genetic engineering** Manipulation of the genetic makeup of any organism with recombinant DNA technology. This includes DNA insertion, deletion, modification, or replacement. Also referred to as gene *editing* or *genetic editing*.

**genetically modified organism (GMO)** Any organism created by genetic engineering.

**gestation** The period of intrauterine development of offspring, from conception to birth; in humans, normal gestation is 38 to 42 weeks.

**gestational diabetes** A high blood glucose concentration that develops during pregnancy and returns to normal after birth;

one cause is the placental production of hormones that antagonize the regulation of blood glucose by insulin.

**gestational hypertension** Blood pressure of 140/90 mmHg or higher that is first diagnosed after 20 weeks of gestation. This may evolve into preeclampsia or eclampsia. Note that the current cutoff for gestational hypertension is slightly higher than the recently updated threshold for diagnosis of chronic hypertension among the general population.

**ghrelin** A hormone produced by stomach cells and the brain that stimulates appetite.

**Global Hunger Index (GHI)** A tool that tracks the state of hunger worldwide and calls attention to locations where immediate hunger relief is urgently needed.

**global maternal mortality ratio** The number of maternal deaths per 100,000 live births.

**glossitis** Inflammation and swelling of the tongue.

**glucagon** A hormone made by the pancreas that stimulates the breakdown of glycogen in the liver into glucose; this ends up increasing blood glucose. Glucagon also increases the generation of glucose from noncarbohydrate substances.

**glucose** A six-carbon sugar that exists in a ring form; found as such in blood and in table sugar bound to fructose; also known as *dextrose,* it is one of the simple sugars.

**gluten** Poorly digested protein found in wheat, barley, and rye.

**glycemic index (GI)** The blood glucose response of a given food, compared to a standard (typically, glucose or white bread). Glycemic index is influenced by starch structure, fiber content, food processing, physical structure, and macronutrients.

**glycemic load (GL)** A measure of both the quality (GI value) and quantity (grams per serving) of a carbohydrate in a meal.

**glycerol** A three-carbon alcohol used to form triglycerides.

**glycogen** A carbohydrate made of multiple units of glucose with a highly branched structure. It is the storage form of glucose in humans and is synthesized (and stored) in the liver and muscles.

**glycosylation** The enzymatic process by which glucose attaches to other compounds, such as proteins.

**goiter** An enlargement of the thyroid gland; this is often caused by insufficient iodine in the dietary pattern.

**Golgi complex** The cell organelle near the nucleus that packages proteins and lipids for secretion or distribution to other organelles.

**gout** A form of arthritis caused by the buildup of uric acid crystals in the joints.

**green revolution** Refers to increases in crop yields that accompanied the introduction of new agricultural technologies in less-developed countries. The key technologies were high-yielding, disease-resistant strains of rice, wheat, and corn; greater use of fertilizer and water; and improved cultivation practices.

**growth charts** Charts used by health professionals to compare the growth of an individual child to normal patterns of growth in weight, stature, and head circumference over time.

**gruel** A thin mixture of grains or legumes in milk or water.

**gums** Viscous polysaccharides often found in seeds and often used in the food industry for its thickening and stabilizing properties.

**gut-associated lymphoid tissues (GALT)** Clusters of lymphoid cells located throughout the gastrointestinal tract that destroy pathogens.

**hard water** Water that contains high levels of calcium and magnesium.

**hazard** Chance that injury will result from use of a substance.

**Healthy Eating Index (HEI)** A measure of diet quality that can be used to assess compliance with the *Dietary Guidelines.*

**heart attack** Rapid fall in heart function caused by reduced blood flow through the heart's blood vessels. Often part of the heart dies in the process. Technically called a *myocardial infarction.*

**heavy drinking** Any pattern of alcohol consumption defined as consuming 15 drinks or more per week for men and 8 drinks or more per week for women.

**helminth** Parasitic worm that can contaminate food, water, feces, animals, and other substances.

**hematocrit** The percentage of blood made up of red blood cells.

**heme iron** Iron provided from animal tissues in the form of hemoglobin and myoglobin. Approximately 40% of the iron in meat, fish, and poultry is heme iron; it is readily absorbed.

**hemicellulose** A group of polysaccharides that are found in plant cell walls.

**hemochromatosis** A disorder of iron metabolism characterized by increased iron absorption and deposition in the liver and heart. This eventually poisons the cells in those organs.

**hemoconcentration** Decrease in plasma volume, causing an increase in the concentration of red blood cells and other constituents of the blood.

**hemoglobin** The iron-containing part of the red blood cell that carries oxygen to the cells and carbon dioxide away from the cells. The heme iron portion is also responsible for the red color of blood.

**hemolytic uremic syndrome (HUS)** Disease characterized by anemia caused by destruction of red blood cells (hemolytic), acute kidney failure (uremic), and a low platelet count.

**hemorrhagic stroke** Damage to part of the brain resulting from rupture of a blood vessel and subsequent bleeding within or over the internal surface of the brain.

**hemorrhoid** A swollen vein in the rectum or anus.

**hemostasis** The process of stopping blood loss.

**hepatic portal circulation** The portion of the circulatory system that uses a large vein (portal vein) to carry nutrient-rich blood from capillaries in the intestines and portions of the stomach to the liver.

**hepatic portal vein** Large vein that carries absorbed nutrients from the gastrointestinal tract to the liver.

**hidden hunger** A lack of vitamins and minerals that occurs when the quality of foods people eat does not meet their nutrient requirements.

**high-density lipoprotein (HDL)** Lipoprotein that picks up cholesterol from dying cells and other sources and transfers it to the other lipoproteins in the bloodstream or directly to the liver. Higher HDL levels are associated with decreased risk for cardiovascular disease, so it is sometimes called *good cholesterol.*

**high-fructose corn syrup (HFCS)** Corn syrup that has been manufactured to contain from 42% to 55% fructose.

**high-intensity interval training (HIIT)** Short, intense aerobic workout that alternates short periods of intense anaerobic exercise with less-intense recovery periods to provide improved athletic capacity and condition, improved glucose metabolism, and improved fat burning.

**high-quality proteins** Dietary proteins that contain ample amounts of all nine essential amino acids; also called *complete proteins.*

**homocysteine** An amino acid that arises from the metabolism of methionine. Vitamin B-6, folate, vitamin B-12, and choline are required for its metabolism. Elevated levels are associated with an increased risk of cardiovascular disease.

**hospice care** A program offering care that emphasizes comfort and dignity at the end of life.

**human immunodeficiency virus (HIV)** A virus that attacks the body's immune cells, which normally fight infection. As HIV progresses, immune function may be severely compromised, leading to acquired immune deficiency syndrome (AIDS).

**human milk oligosaccharides** Small, indigestible carbohydrates made in the human breast from lactose and other simple sugars.

**hunger** The primarily physiological (internal) drive to find and eat food.

**hydrogenation** The addition of hydrogen to a carbon-carbon double bond, producing a single carbon-carbon bond with two hydrogens attached to each carbon.

**hydrolyzed protein formula** Infant formula in which the proteins have been broken down into smaller peptides and amino acids to improve digestibility and reduce exposure to potential food allergens; sometimes called *predigested* or *hypoallergenic infant formula.*

**hydroponics** This type of agriculture involves growing plants in a nutrient solution root medium in a controlled, soilless environment.

**hydroxyapatite** Crystalline compound containing calcium, phosphorus, and sometimes fluoride, also known as *bone mineral.*

**hyperglycemia** High blood glucose, typically defined as above 125 mg/dl while in a fasted state.

**hyperkeratosis** A condition in which patches of skin become thicker, rougher, or drier than usual; a possible consequence of vitamin A deficiency.

**hypertension** A condition in which blood pressure remains persistently elevated. Obesity, inactivity, alcohol intake, excess salt intake, and genetics may each contribute to the problem.

**hypertonic** Having high concentration of solutes.

**hypoglycemia** A condition caused by low levels of blood sugar that is often related to the treatment of diabetes and often defined by 70 mg/dl or less.

**hyponatremia** Dangerously low blood sodium level.

**hypothalamus** A region of the forebrain that controls body temperature, thirst, and hunger.

**hypotheses** Tentative explanations by a scientist to explain a phenomenon.

**hypotonic** Having low concentration of solutes.

**identical twins** Two offspring that develop from a single ovum and sperm and, consequently, are born with the same genetic makeup.

**ileocecal sphincter** The ring of smooth muscle between the end of the small intestine and the beginning of the large intestine.

**ileum** Last segment of the small intestine; approximately 5 feet in length.

**indirect calorimetry** A method to measure energy use by the body by measuring oxygen uptake and carbon dioxide output. Formulas are then used to convert this gas exchange value into energy use, estimating the proportion of energy nutrients that are being oxidized for energy in the fuel mix.

**indirect food additives** Additives that appear in food products incidentally, from environmental contamination of food ingredients or during the manufacturing process.

**infertility** Inability of a couple to conceive after 1 year of unprotected intercourse.

**inorganic** Any substance lacking carbon atoms bonded to hydrogen atoms in the chemical structure.

**insoluble fiber** A fiber that does not dissolve in water and may pass through the GI tract intact so is not a source of calories; also known as *roughage.*

**insulin** A hormone produced by the pancreas. Insulin allows for the movement of glucose from the blood into body cells and signals the synthesis of glycogen.

**inulin** A mixture of fructose chains that vary in length and occur naturally in plants.

**intracellular fluid** Fluid contained within a cell; it represents about two-thirds of body fluid.

**intrinsic factor** A protein-like compound produced by the stomach that enhances vitamin B-12 absorption in the ileum.

**ion** A positively or negatively charged atom.

**irradiation** A process in which radiation energy is applied to foods, creating compounds (free radicals) within the food that destroy cell membranes, break down DNA, link proteins, limit enzyme activity, and alter a variety of other proteins and cell functions of microorganisms that can lead to food spoilage. This process does not make the food radioactive.

**ischemic stroke** Damage to part of the brain resulting from lack of blood flow to the brain.

**isotonic** Having equal concentration of solutes.

**jejunum** Middle segment of the small intestine; approximately 4 feet in length.

**ketone bodies** Partial breakdown products of fat that contain three or four carbons.

**ketosis** The condition of having a high concentration of ketone bodies and related breakdown products in the bloodstream and tissues.

**kilocalorie (kcal)** Heat energy needed to raise the temperature of 1000 grams (1 L) of water 1 degree Celsius.

**kwashiorkor** A form of protein-calorie malnutrition occurring primarily in young children who have an existing disease and consume a marginal amount of calories and insufficient protein in relation to needs. The child generally suffers from infections and exhibits edema, poor growth, weakness, and an increased susceptibility to further illness.

**kyphosis** Abnormally increased bending of the spine; commonly known as *dowager's hump.*

**lactase** An enzyme made by absorptive cells of the small intestine; this enzyme digests lactose to glucose and galactose.

**lactate** A three-carbon acid formed during anaerobic cell metabolism; a partial breakdown product of glucose; also called *lactic acid.*

**lactation** The period of milk secretion following pregnancy; typically called *breastfeeding.*

**lactation consultant** Health care professional (often a registered nurse or RDN) with special training to provide education and support for breastfeeding mothers and their infants.

**lacteal** Lymphatic vessel that absorbs fats from the small intestine.

**lactoovovegetarian** Referring to a dietary pattern that is primarily plant-based but also includes dairy products and eggs.

**lactose** A disaccharide consisting of glucose bonded to galactose; also known as *milk sugar.*

**lactose intolerance** A condition in which symptoms such as abdominal gas and bloating appear as a result of severe lactose maldigestion.

**lactovegetarian** Referring to a dietary pattern that is primarily plant-based but also includes dairy products.

**lanugo** Downlike hair that appears after a person has lost much body fat through semistarvation. The hair stands erect and traps air, acting as insulation for the body to compensate for the relative lack of body fat, which usually functions as insulation.

**laxative** A medication or other substance that stimulates evacuation of the intestinal tract.

**lean body mass** Body weight minus fat storage weight equals lean body mass. This includes organs such as the brain, muscles, and liver, as well as bone and blood and other body fluids.

**lecithin** A group of phospholipid compounds that are major components of cell membranes.

**leptin** A hormone made by adipose tissue in proportion to total fat stores in the body that influences long-term regulation of fat mass. Leptin also influences appetite and the release of insulin.

**let-down reflex** A reflex stimulated by infant suckling that causes the release (ejection) of milk from milk ducts in the mother's breasts; also called *milk ejection reflex.*

**life expectancy** The average length of life for a given group of people born in a specific year.

**life span** The potential oldest age a person can reach.

**lignin** A noncarbohydrate dietary fiber that is found in cell walls of woody plants and seeds.

**limiting amino acid** The essential amino acid in lowest concentration in a food or diet relative to body needs.

**linoleic acid** An essential omega-6 fatty acid with 18 carbons and two double bonds.

**lipase** Fat-digesting enzyme produced by the salivary glands, stomach, and pancreas.

**lipid** A compound containing much carbon and hydrogen, little oxygen, and sometimes other atoms. Lipids do not dissolve in water and include fats, oils, and cholesterol.

**lipoprotein** A compound found in the bloodstream containing a core of triglycerides and cholesterol surrounded by a shell of protein and phospholipids.

**lipoprotein lipase** An enzyme attached to the cells that form the inner lining of blood vessels; it breaks down triglycerides into free fatty acids and glycerol.

**lobules**  Saclike structures in the breast that store milk; also called *alveoli.*

**locavore**  Someone who eats food grown or produced locally or within a certain radius such as 50, 100, or 150 miles.

**long-chain fatty acid**  A fatty acid that contains 12 or more carbons.

**low birth weight (LBW)**  Referring to any infant weighing less than 2.5 kilograms (5.5 pounds) at birth; most commonly results from preterm birth.

**low-density lipoprotein (LDL)**  Lipoprotein that remains after most of the triglycerides have been removed from a VLDL; transports cholesterol from the liver to cells throughout the body. Elevated LDL is strongly linked to cardiovascular disease risk, so it is sometimes called *bad cholesterol.*

**lower esophageal sphincter**  A circular muscle that constricts the opening of the esophagus to the stomach. Also called the *gastroesophageal sphincter* or the *cardiac sphincter.*

**lower-body obesity**  The type of obesity in which fat storage is primarily located in the buttocks and thigh area. Also known as *gynoid* or *gynecoid obesity.*

**lower-quality proteins**  Dietary proteins that are low in or lack one or more essential amino acids; also called *incomplete proteins.*

**lumen**  The hollow opening inside a tube, such as the GI tract.

**luo han guo**  Extract of the monk fruit, this alternative sweetener is 100 to 250 times sweeter than sucrose.

**lymph**  A clear fluid that flows through lymph vessels; carries most forms of fat after their absorption by the small intestine.

**lymph nodes**  Clusters of lymphoid tissue, situated along the lymph vessels, that trap and destroy pathogens.

**lymphatic system**  A system of vessels and lymph that accepts fluid surrounding cells and large particles, such as products of fat absorption. Lymph eventually passes into the bloodstream from the lymphatic system.

**lymphoid tissue**  Specialized cells that participate in the immune response; includes the thymus, spleen, lymph nodes, and white blood cells.

**lysosome**  A cellular organelle that contains digestive enzymes for use inside the cell for turnover of cell parts.

**macrocyte**  A large, immature red blood cell that results from the inability of the cell to divide normally; also called *megaloblast.*

**macrocytic anemia**  Anemia characterized by the presence of abnormally large red blood cells; also called *megaloblastic anemia.*

**macronutrient**  A nutrient needed in gram quantities in a dietary pattern.

**major mineral**  Vital to health, a mineral that is required in the dietary pattern in amounts greater than 100 milligrams per day.

**malignant**  Malicious; in reference to a tumor, the property of invading surrounding tissues and spreading to distant sites.

**malnutrition**  Failing health that results from chronic eating practices that do not coincide with nutritional needs.

**maltase**  An enzyme made by absorptive cells of the small intestine; this enzyme digests maltose to two glucose molecules.

**maltose**  A disaccharide consisting of glucose bonded to glucose; also known as *malt sugar.*

**marasmus**  A form of protein-calorie malnutrition resulting from consuming a grossly insufficient amount of protein and calories. Victims have little or no fat stores, little muscle mass, and poor strength. Death from infections is common.

**market basket**  Food the FDA buys, prepares, and analyzes as part of the ongoing Total Diet Study, which monitors levels of about 800 contaminants and nutrients in the average U.S. dietary pattern; the number varies slightly from year to year. About 280 kinds of foods and beverages from representative areas of the country are included four times a year.

**megadose**  Large intake of a nutrient well beyond estimates of needs or what would be found in a balanced diet; 2 to 10 times above human needs is typically a starting point.

**Menkes syndrome**  An inherited X-linked recessive pattern disorder that affects copper levels in the body.

**menopause**  The cessation of the menstrual cycle in women, usually beginning at about 50 years of age.

**meta-analysis**  A statistical examination of data from multiple scientific studies of the same subject in order to determine overall trends.

**metabolic syndrome**  A condition in which a person has poor blood glucose regulation, hypertension, increased blood triglycerides, and other health problems. This condition is usually accompanied by obesity, lack of physical activity, and a dietary pattern high in refined carbohydrates. Also called *Syndrome X.*

**metabolic water**  Water formed as a by-product of carbohydrate, lipid, and protein metabolism.

**metabolism**  Chemical processes in the body by which energy is provided in useful forms and vital activities are sustained.

**metastasize**  The spreading of disease from one part of the body to another, even to parts of the body that are remote from the site of the original tumor. Cancer cells can spread via blood vessels, the lymphatic system, or direct growth of the tumor.

**methyl group**  In chemistry, a carbon atom that shares bonds with three hydrogen atoms. The methyl group is the omega end of a fatty acid.

**microbiome**  Entire collection of microorganisms, their genes, and their environment.

**microbiota**  Community of microorganisms living in a particular region; with regard to our discussion of probiotics, the community of microorganisms coexisting on and within the human body.

**micronutrient**  A nutrient needed in milligram or microgram quantities in a dietary pattern.

**microvilli**  Extensive folds on the mucosal surface of the absorptive cells.

**mineral**  Element used in the body to promote chemical reactions and to form body structures.

**mitochondria**  (singular, mitochondrion) Organelles that are the main sites of energy-yielding chemical reactions in a cell.

**moderate drinking**  For men, consuming no more than two drinks per day, and for women, consuming no more than one drink per day.

**moderate-intensity aerobic physical activity**  Aerobic activity that increases a person's heart rate and breathing to some extent. Examples include brisk walking, dancing, swimming, or bicycling.

**monoglyceride**  A breakdown product of a triglyceride consisting of one fatty acid attached to a glycerol backbone.

**monosaccharide**  Simple sugar, such as glucose, that is not broken down further during digestion.

**monounsaturated fatty acid**  A fatty acid containing one carbon-carbon double bond.

**motility**  Generally, the ability to move spontaneously. In this context, it refers to movement of food through the GI tract.

**mucilage**  A gelatinous substance of plants that contains protein and polysaccharides and is similar to plant gums.

**mucus**  A thick fluid secreted by many cells throughout the body. It contains a compound that has both carbohydrate and protein parts. It acts as a lubricant and means of protection for cells.

**multiple sclerosis**  An unpredictable disease of the central nervous system that can range from relatively benign to somewhat disabling to devastating, as communication between the brain and other parts of the body is disrupted.

**muscle dysmorphia**  A psychological disorder in which an individual perceives himself or herself as too thin, desires a highly muscular physique, and engages in obsessive behaviors to gain muscle mass.

**muscle tissue**  A type of tissue adapted to contract to cause movement.

**muscle-strengthening activity** Physical activity that increases skeletal muscle strength, power, endurance, and mass. Examples include lifting free weights, using weight machines, and calisthenics (e.g., push-ups).

**myelin** A combination of lipids and proteins that covers nerve fibers.

**myoglobin** Iron-containing protein that binds oxygen in muscle tissue.

**negative energy balance** The state in which energy intake is less than energy expended, resulting in weight loss.

**negative protein balance** A state in which protein intake is less than related protein losses, such as often seen during acute illness.

**neophobia** Fear of new things, such as new foods.

**neotame** General-purpose, nonnutritive sweetener that is approximately 7000 to 13,000 times sweeter than table sugar. It has a chemical structure similar to aspartame.

**nephrons** The functional units of kidney cells that filter wastes from the bloodstream and deposit them into the urine.

**nervous system** The body system consisting of the brain, spinal cord, nerves, and sensory receptors. This system detects sensations, directs movements, and controls physiological and intellectual functions.

**nervous tissue** Tissue composed of highly branched, elongated cells that transport nerve impulses from one part of the body to another.

**neural tube defect** A defect in the formation of the neural tube occurring during early fetal development. This type of defect results in various nervous system disorders, such as spina bifida. Folate deficiency in the pregnant woman increases the risk that the fetus will develop this disorder.

**neuron** The structural and functional unit of the nervous system. Consists of a cell body, dendrites, and an axon.

**neurotransmitter** A compound made by a nerve cell that allows for communication between it and other cells.

**night blindness** Vitamin A deficiency disorder that results in loss of the ability to see under low-light conditions.

**night eating syndrome** An eating disorder characterized by consumption of a large volume of food in the late evening and nocturnal awakenings with ingestion of food.

**nitrosamine** A carcinogen formed from nitrates and breakdown products of amino acids; associated with cancer risk.

**nonceliac wheat sensitivity (NCWS)** One or more of a variety of immune-related conditions with symptoms similar to celiac disease that are precipitated by the ingestion of gluten in people who do not have celiac disease.

**nonessential amino acids** Amino acids that can be synthesized by a healthy body in sufficient amounts; there are 11 nonessential amino acids. These are also called *dispensable amino acids.*

**nonheme iron** Iron provided from plant sources, supplements, and animal tissues other than in the forms of hemoglobin and myoglobin. Nonheme iron is less efficiently absorbed than heme iron; absorption is closely dependent on body needs.

**nonspecific immunity** Defenses that stop the invasion of pathogens; requires no previous encounter with a pathogen; also called *innate immunity.*

**norepinephrine** A neurotransmitter from nerve endings and a hormone from the adrenal gland. It is released in times of stress and is involved in hunger regulation, blood glucose regulation, and other body processes.

**NSAIDs** Nonsteroidal anti-inflammatory drugs; include aspirin, ibuprofen (Advil®), and naproxen (Aleve®).

**nucleus** Membrane-bound organelle that contains genetic information (DNA) for protein synthesis and cell replication.

**nutrient density** The ratio derived by dividing a food's nutrient content by its calorie content. When the food's overall nutrient contribution exceeds its contribution to our calorie needs, the food is considered to have a favorable nutrient density.

**nutrients** Chemical substances in food that contribute to health, many of which are essential parts of a dietary pattern. Nutrients nourish us by providing calories to fulfill energy needs, materials for building body parts, and factors to regulate necessary chemical processes in the body.

**nutrigenetics** A branch of nutritional genomics that studies the effects of genes on nutritional health, such as variations in nutrient requirements and responsiveness to dietary modifications.

**nutrigenomics** A branch of nutritional genomics that studies how food impacts health through its interaction with our genes and its subsequent effect on gene expression.

**nutrition security** Secure access to a nutritious diet coupled with a sanitary environment and adequate health services and care.

**nutritional genomics** Study of interactions between nutrition and genetics; includes nutrigenetics and nutrigenomics.

**nutritional status** The nutritional health of a person as determined by anthropometric measurements (height, weight, circumferences, and so on), biochemical measurements of nutrients or their by-products in blood and urine, a clinical (physical) examination, a dietary analysis, and economic evaluation; also called *nutritional state.*

**obesity** Ratio of weight to height that is significantly higher than what is associated with optimal health, usually due to excessive body fat. For adults, this is defined as BMI of 30 or higher. For children, this is defined as BMI-for-age at the 95th percentile or higher.

**oleogustus** A taste for fat. The presence of fatty acids in foods stimulates taste receptors in the mouth; this sensation is unpleasant.

**omega-3 (ω-3) fatty acid** An unsaturated fatty acid with the first double bond on the third carbon from the methyl end ($-CH_3$).

**omega-6 (ω-6) fatty acid** An unsaturated fatty acid with the first double bond on the sixth carbon from the methyl end ($-CH_3$).

**organ** A group of tissues designed to perform a specific function; for example, the heart, which contains muscle tissue, nervous tissue, and so on.

**organ system** A collection of organs that work together to perform an overall function.

**organelles** Compartments, particles, or filaments that perform specialized functions within a cell.

**organic compounds** In chemistry, any chemical compounds that contain carbon.

**organic food** Food grown without use of pesticides, synthetic fertilizers, sewage sludge, genetically modified organisms, antibiotics, hormones, or ionizing radiation.

**orthorexia** A proposed psychological disorder characterized by an obsession with proper or healthful eating.

**osmosis** The passage of water through a membrane from a less concentrated compartment to a more concentrated compartment.

**osteoarthritis** A degenerative joint condition caused by a breakdown of cartilage in joints. It often results from wear and tear due to repetitive motions or the pressure of excess body weight.

**osteoblast** Bone cells that initiate the synthesis of new bone.

**osteoclast** Bone cells that break down bone and subsequently release bone minerals into the blood.

**osteomalacia** Adult form of rickets. The bones have low mineral density and consequently are at risk for fracture.

**osteopenia** A bone disease defined by low mineral density.

**osteoporosis** The presence of a stress-induced fracture or a T-score of −2.5 or lower. The bones are porous and fragile due to low mineral density.

**overnutrition** A state in which nutritional intake greatly exceeds the body's needs.

**overweight** A ratio of weight-to-height that is slightly higher than what is associated with optimal health. For adults, this is defined as BMI within the range of 25.0 up to 30. For children, this is defined as BMI-for-age from the 85th up to the 95th percentile.

**ovovegetarian** Referring to a dietary pattern that is primarily plant-based but also includes egg products.

**ovum** The egg cell from which a fetus eventually develops if the egg is fertilized by a sperm cell.

**oxalic acid** An organic acid found in spinach, rhubarb, and sweet potatoes that can depress the absorption of certain minerals present in the food, such as calcium; also called *oxalate.*

**oxidize** In the most basic sense, the loss of an electron or gain of an oxygen by a chemical substance. This change typically alters the shape and/or function of the substance.

**oxytocin** A hormone secreted by the pituitary gland. It causes contraction of the musclelike cells surrounding the ducts of the breasts and the smooth muscle of the uterus.

**parasite** An organism that lives in or on another organism and derives nourishment from it.

**parathyroid hormone (PTH)** A hormone made by the parathyroid glands that increases synthesis of the vitamin D hormone and aids calcium release from bone and calcium conservation by the kidneys, among other functions.

**Parkinson's disease** Disease that belongs to a group of conditions called motor system disorders, which are the result of the loss of dopamine-producing brain cells. The four primary symptoms of PD are tremor, or trembling in hands, arms, legs, jaw, and face; rigidity, or stiffness of the limbs and trunk; bradykinesia, or slowness of movement; and postural instability, or impaired balance and coordination.

**passive diffusion** Movement of a substance across a semipermeable membrane from an area of higher solute concentration to an area of lower solute concentration. This type of transport does not require a carrier and does not require energy.

**pathogen** A microorganism that can cause disease.

**pasteurizing** The process of heating food products to kill pathogenic microorganisms and reduce the total number of bacteria.

**pectin** Polysaccharides made of 300 to 1000 monosaccharides that are soluble viscous fibers and abundant in berries and other fruit.

**pellagra** Niacin-deficiency disease characterized by dementia, diarrhea, and dermatitis, and possibly leading to death.

**pepsin** A protein-digesting enzyme produced by the stomach.

**peptic ulcer** Erosion of the tissue lining, usually in the stomach or the upper small intestine.

**peptide bond** A chemical bond formed between amino acids in a protein.

**peer review** Evaluation of work by professionals of similar competence (peers) to the producers of the work to maintain standards of quality and credibility. Scholarly peer review is used to determine if a scientific study is suitable for publication.

**percentile** With reference to a dataset, the value below which a given percentage of observations fall. For example, a male at the 75th percentile for BMI-for-age has a BMI higher than 74% of other males his age and lower than 25% of other males his age.

**perforation** A hole made by boring or piercing. With reference to the gastrointestinal tract, the hole is in the wall of the esophagus, stomach, intestine, rectum, or gallbladder. Complications include bleeding and infection.

**periodization** Cycling the volume, intensity, and activities of workouts throughout the training season.

**peristalsis** A coordinated muscular contraction used to propel food down the gastrointestinal tract.

**pernicious anemia** The anemia that results from a lack of vitamin B-12 absorption; it is *pernicious* because of associated nerve degeneration that can result in eventual paralysis and death.

**peroxisome** A cell organelle that destroys toxic products within the cell.

**pescovegetarian** Referring to a dietary pattern that is primarily plant-based but also includes fish and other aquatic animal protein. Also called *pescatarian.*

**pH** A measure of relative acidity or alkalinity of a solution. The pH scale is 0 to 14. A pH of 7 is neutral; a pH below 7 is acidic; a pH above 7 is alkaline.

**phagocytosis** A process in which a cell forms an indentation, and solid particles enter the indentation and are engulfed by the cell.

**pharynx** A cavity located at the back of the oral and nasal cavities, commonly known as the throat. It is part of the digestive tract and the respiratory tract.

**phenylketonuria (PKU)** Disease caused by a genetic defect in the liver's ability to metabolize the amino acid phenylalanine; untreated, toxic by-products of phenylalanine build up in the body and lead to brain damage and severe health issues.

**phosphocreatine (PCr)** A high-energy compound that can be used to reform ATP. It is used primarily during bursts of activity, such as lifting and jumping.

**phospholipid** Any of a class of fat-related substances that contain phosphorus, fatty acids, and a nitrogen-containing component. Phospholipids are an essential part of every cell.

**photosynthesis** Process by which plants use energy from the sun to synthesize energy-yielding compounds, such as glucose.

**physical activity** Any movement of skeletal muscles that requires energy.

**physical fitness** The ability to perform moderate to vigorous activity without undue fatigue.

**physiological anemia** The normal decrease in red blood cell concentration in the blood due to increased blood volume during pregnancy; also called *hemodilution.*

**phytic acid** A constituent of plant fibers that binds positive ions to its multiple phosphate groups; also called *phytate.*

**phytochemical** A chemical found in plants. Some phytochemicals may contribute to a reduced risk of cancer or cardiovascular disease in people who consume them regularly.

**pica** The practice of eating nonfood items, such as dirt, laundry starch, or clay.

**pinocytosis** A process in which a cell forms an indentation, and fluid enters the indentation and is engulfed by the cell.

**placebo** Generally, an inactive medicine or treatment used to disguise the treatments given to the participants in an experiment.

**placenta** An organ that forms in the uterus in pregnant women. Through this organ, oxygen and nutrients from the mother's blood are transferred to the fetus, and fetal wastes are removed. The placenta also releases hormones that maintain the state of pregnancy.

**plaque** A cholesterol-rich substance deposited in the blood vessels; it contains various white blood cells, smooth muscle cells, various proteins, cholesterol and other lipids, and eventually calcium.

**plasma** The fluid, extracellular portion of blood.

**platelets** Protoplasmic discs in the blood that promote coagulation; also called *thrombocytes.*

**pollovegetarian** Referring to a dietary pattern that is primarily plant-based but also includes chicken, turkey, and other poultry.

**polycystic ovary syndrome (PCOS)** A condition of hormonal imbalance (e.g., elevated testosterone and insulin) in a woman that can lead to infertility, weight gain in the abdominal region, excessive growth of body hair, and acne.

**polypeptide** A group of 10 to 2000 or more amino acids bonded together to form proteins.

**polysaccharides** Carbohydrates containing many glucose units, from 10 to 1000 or more. Also called *complex carbohydrates.*

**polyunsaturated fatty acid** A fatty acid containing two or more carbon-carbon double bonds.

**pool** The amount of a nutrient stored within the body that can be mobilized when needed.

**positive energy balance** The state in which energy intake is greater than energy expended, generally resulting in weight gain.

**positive protein balance** A state in which protein intake exceeds related protein losses, as is needed during times of growth.

**postbiotics** Metabolic by-products of the microorganisms that colonize the human body.

**prebiotic** Selectively fermented ingredient that results in specific changes in the composition and/or activity of the gastrointestinal microbiota, thus conferring benefits upon the host.

**preeclampsia** A form of gestational hypertension characterized by protein in the urine.

**preservatives** Compounds that extend the shelf life of foods by inhibiting microbial growth or minimizing the destructive effect of oxygen and metals.

**preterm** Referring to an infant born before 37 weeks of gestation; also referred to as *premature.*

**primary hypertension** Blood pressure of 130/80 mmHg or higher with no identified cause; also called *essential hypertension.*

**primary lactose maldigestion** Develops at about age 3 to 5 years when the production of the enzyme lactase decreases. When significant symptoms develop after lactose intake, it is then called *lactose intolerance.*

**probiotics** Live microorganisms that, when administered in adequate amounts, confer health benefits on the host.

**progression** Incremental increase in frequency, intensity, and time spent in each type of physical activity over several weeks or months.

**prolactin** A hormone secreted by the pituitary gland that stimulates the synthesis of milk in the breast.

**protein** Food and body compounds made of more than 100 amino acids; proteins contain carbon, hydrogen, oxygen, nitrogen, and sometimes other atoms in a specific configuration. Proteins contain the form of nitrogen most easily used by the human body.

**protein equilibrium** A state in which protein intake is equal to related protein losses; the person is said to be in *protein balance.*

**protein isolate** Protein powder that has been processed more than a protein concentrate to remove lower protein portions and collect pure protein fractions.

**protein turnover** The process by which cells break down old proteins and resynthesize new proteins. In this way, the cell will have the proteins it needs to function at that time.

**protein–calorie malnutrition (PCM)** A condition resulting from regularly consuming insufficient amounts of calories and protein. The deficiency eventually results in body wasting, primarily of lean tissue, and an increased susceptibility to infections. Also known as *protein-energy malnutrition (PEM).*

**proton pump inhibitor (PPI)** A medication that inhibits the ability of gastric cells to secrete hydrogen ions.

**protozoa** (singular, protozoan) One-celled animals that are more complex than bacteria. Disease-causing protozoa can be spread through food and water.

**provitamin A** A substance that can be converted into vitamin A.

**psyllium** Mostly soluble type of dietary fiber found in the seeds of the plantago plant; common ingredient in bulk-forming laxatives, such as Metamucil®.

**purging disorder** An eating disorder characterized by repeated purging (e.g., by self-induced vomiting) to induce weight loss even in the absence of binge eating.

**pyloric sphincter** Ring of smooth muscle between the stomach and the small intestine.

**pyruvate** A three-carbon compound formed during glucose metabolism; also called *pyruvic acid.*

**R-proteins** Proteins produced by the salivary glands that bind to free vitamin B-12 in the stomach and protect it from stomach acid.

**radiation** Literally, energy that is emitted from a center in all directions. Various forms of radiation energy include X rays and ultraviolet rays from the sun.

**rancidity** Production of decomposed fatty acids that have an unpleasant flavor and odor.

**randomized controlled trial** An experimental design that is double-blind and placebo controlled.

**receptor** A site in a cell at which compounds (such as hormones) bind. Cells that contain receptors for a specific compound are partially controlled by that compound.

**Recommended Dietary Allowance (RDA)** Nutrient intake amount sufficient to meet the needs of 97% to 98% of the individuals in a specific life stage.

**rectum** Terminal section of the large intestine where feces are held prior to expulsion.

**red blood cells** Cells that transport oxygen and carbon dioxide through the blood; also called *erythrocytes.*

**regenerative agriculture** A system of farming and grazing practices that aims to reverse climate change by rebuilding soil organic matter and restoring degraded soil biodiversity, resulting in both carbon drawdown and improving the water cycle.

**registered dietitian (RD)** A person who has completed a baccalaureate degree program approved by the Accreditation Council for Education in Nutrition and Dietetics (ACEND), performed at least 1200 hours of supervised professional practice, passed a registration examination, and complies with continuing education requirements.

**registered dietitian nutritionist (RDN)** The RDN is the updated credential formerly abbreviated RD. The credential was updated to better reflect the scope of practice of the dietitian and to align with the new name of the professional organization for dietitians, the Academy of Nutrition and Dietetics.

**relapse prevention** A series of strategies used to help prevent and cope with weight-control lapses, such as recognizing high-risk situations and deciding beforehand on appropriate responses.

**relative energy deficiency in sport (RED-S)** A syndrome of altered metabolism, immune function, and mental health caused by low energy availability in athletes, which may be due to unintentional failure to meet the high energy demands of sports or intentional restriction of energy intake to control weight.

**responsive feeding** A healthy feeding relationship between the caregiver and the child in which the caregiver pays attention to and respects the child's hunger and satiety.

**reserve capacity** The extent to which an organ can preserve essentially normal function despite decreasing cell number or cell activity.

**resistant starch** Indigestible dietary fiber sequestered in plant walls. Bananas and legumes are rich sources.

**resting metabolism** The amount of calories the body uses when the person has not eaten in 4 hours and is resting (e.g., 15 to 30 minutes) and awake in a warm, quiet environment. It is usually slightly higher (~10%) than basal metabolism due to the more flexible testing criteria; often referred to as *resting metabolic rate (RMR).*

**retina** A light-sensitive lining in the back of the eye. It contains retinal.

**retinal** Aldehyde form of vitamin A.

**retinoic acid** Acid form of vitamin A.

**retinoids** Chemical forms of preformed vitamin A found in animal foods.

**retinol** Alcohol form of vitamin A.

**retinyl** Storage form of vitamin A.

**reverse cholesterol transport** Process by which HDL picks up cholesterol from the tissues and blood vessels and takes it to the liver for metabolism or excretion.

**rheumatoid arthritis** A degenerative joint condition resulting from an autoimmune disease that causes inflammation in the joints and other sites of the body.

**ribonucleic acid (RNA)** The single-stranded nucleic acid involved in the transcription of genetic information and translation of that information into protein structure.

**ribosomes** Cytoplasmic particles that mediate the linking together of amino acids to form proteins; may exist freely in the cytoplasm or attached to endoplasmic reticulum.

**rickets** A disease characterized by poor mineralization of newly synthesized bones because of low calcium content. Arising in infants and children, this deficiency is caused by insufficient amounts of vitamin D in the body.

**risk factors** A term used frequently when discussing the factors contributing to the development of a disease. A risk factor is an aspect of our lives, such as heredity, lifestyle choices (e.g., use of tobacco products), or nutritional habits.

**Russell's sign** Evidence of abrasion that appears on the knuckles of a person who repeatedly induces vomiting by using the fingers to trigger the gag reflex in the back of the throat; named after the psychiatrist who first identified bulimia nervosa.

**saccharin** Alternative sweetener that yields no energy to the body; 200 to 700 times sweeter than sucrose.

**safety** Relative certainty that a substance will not cause injury.

**saliva** Watery fluid, produced by the salivary glands in the mouth, which contains lubricants, enzymes, and other substances.

**sarcopenia** Loss of muscle tissue. Among older adults, this loss of lean mass greatly increases their risk of illness and death.

**sarcopenic obesity** Advanced muscle loss accompanied by gains in fat mass.

**satiety** A state in which there is no longer a desire to eat; a feeling of satisfaction.

**saturated fatty acid** A fatty acid containing no carbon-carbon double bonds.

**scavenger cells** Specific form of white blood cells that can bury themselves in the artery wall and accumulate LDL. As these cells take up LDL, they contribute to the development of atherosclerosis.

**scurvy** The vitamin C deficiency disease characterized by weakness, fatigue, slow wound healing, bone pain, fractures, sore and bleeding gums, diarrhea, and pinpoint hemorrhages on the skin.

**secondary hypertension** Blood pressure of 130/80 mmHg or higher as a result of disease (e.g., kidney dysfunction or sleep apnea) or drug use.

**secondary lactose maldigestion** Temporary condition in which lactase production is decreased in response to illness or surgery.

**secretory vesicles** Membrane-bound vesicles produced by the Golgi complex; contain protein and other compounds to be secreted by the cell.

**self-monitoring** Tracking foods eaten and conditions affecting eating; actions are usually recorded in a diary, along with location, time, and state of mind. This is a tool to help people understand more about their eating behaviors.

**sequestrants** Compounds that bind free metal ions. By so doing, they reduce the ability of ions to cause rancidity in foods containing fat.

**serotonin** A neurotransmitter involved in the regulation of mood, sleep, and appetite.

**set point theory** The theory that changes in energy metabolism and appetite work to maintain a steady body weight throughout adulthood.

**sickle cell disease** An illness that results from a malformation of the red blood cell because of an incorrect structure in part of its hemoglobin protein chains; also called *sickle cell anemia.*

**sigmoid colon** Last segment of the large intestine that carries feces from the descending colon to the rectum.

**simple sugar** Monosaccharide or disaccharide in the diet.

**situational poverty** State of poverty caused by specific circumstances such as death, illness, divorce, or catastrophe.

**skinfold measurements** Skinfold or caliper testing is a common method to determine body fat percentage. This utilizes prediction equations that are population specific to estimate fat.

**small for gestational age (SGA)** Referring to infants who weigh less than the expected weight for their length of gestation. This corresponds to less than 2.5 kilograms (5.5 pounds) in a full-term newborn.

**soft water** Water that contains a high level of sodium.

**soluble fiber** A fiber that dissolves in water to form a thick gel-like substance and is broken down by bacteria in the large intestine, providing some calories.

**solvent** A liquid substance in which other substances dissolve.

**sorbitol** Alcohol derivative of glucose that yields about 3 kcal/g but is slowly absorbed from the small intestine; used in some sugarless gums and dietetic foods.

**specific immunity** Function of white blood cells directed at specific antigens; also called *adaptive immunity.*

**spina bifida** Type of neural tube defect resulting from improper closure of the neural tube during embryonic development. The

spinal cord or fluid may bulge outside the spinal column.

**spontaneous abortion** Cessation of pregnancy and expulsion of the embryo or non-viable fetus prior to 20 weeks of gestation. This is the result of natural causes, such as a genetic defect or developmental problem; also called *miscarriage.*

**sports anemia** Exercise-induced iron deficiency anemia caused by plasma volume expansion, low hemoglobin synthesis, or increased destruction of red blood cells.

**starch** A carbohydrate made of multiple units of glucose attached together in a form the body can digest; also known as *complex carbohydrate.*

**sterol** A compound containing a multi-ring (steroid) structure and a hydroxyl group (–OH). Cholesterol is a typical example.

**stevia** Alternative sweetener derived from South American shrub; 200 to 400 times sweeter than sucrose.

**stimulus control** Altering the environment to minimize the stimuli for eating, such as removing foods from sight by storing them in kitchen cabinets.

**stress fracture** A fracture that occurs from repeated jarring of a bone. Common sites include bones of the foot and shins.

**stroke** A decrease or loss in blood flow to the brain that results from a blood clot or other change in arteries in the brain. This in turn causes the death of brain tissue. Also called a *cerebrovascular accident.*

**stunting** Impaired growth and development characterized by low height-for-age.

**subclinical** Stage of a disease or disorder not severe enough to produce symptoms that can be detected or diagnosed.

**subthreshold eating disorder** A clinically recognized eating disorder that meets some, but not all of the criteria for diagnosis of anorexia nervosa, bulimia nervosa, or binge eating disorder.

**sucralose** Alternative sweetener that has chlorines in place of 3 hydroxyl (–OH) groups on sucrose; 600 times sweeter than sucrose.

**sucrase** An enzyme made by absorptive cells of the small intestine; this enzyme digests sucrose to glucose and fructose.

**sucrose** Disaccharide composed of fructose bonded to glucose; also known as *table sugar.*

**sugar** A simple carbohydrate with the chemical composition $(CH_2O)_n$. The basic unit of all sugars is *glucose.*

**sustainable agriculture** Agricultural system that provides a secure living for farm families; maintains the natural environment and resources; supports the rural community; and offers respect and fair treatment to all involved, from farm workers to consumers to the animals raised for food.

**sustainable development** Economic growth that will simultaneously reduce poverty, protect the environment, and preserve natural capital.

**sustainable intensification** Agricultural practices that consider whole landscapes, territories, and ecosystems to optimize resource utilization and management.

**symptom** A change in health status noted by the person with the problem, such as stomach pain.

**synapse** The space between one neuron and another neuron (or cell).

**synbiotic** Combination of pro- and prebiotics taken to confer health benefits on the host.

**systematic review** A thorough summary of the results of available carefully designed health care studies (controlled trials) in a particular area.

**systolic blood pressure** The pressure on blood vessels when the heart beats, squeezing and pushing blood through the arteries to the rest of the body.

**teratogen** A compound (natural or synthetic) that may cause or increase the risk of a birth defect. Exposure to a teratogen does not always lead to a birth defect; its effects on the fetus depend on the dose, timing, and duration of exposure. Examples: alcohol, some drugs, some industrial or household chemicals, certain infections, and extreme heat.

**tetany** A body condition marked by sharp contraction of muscles and failure to relax afterward; usually caused by abnormal calcium metabolism.

**theory** An explanation for a phenomenon that has numerous lines of evidence to support it.

**therapeutic phlebotomy** Periodic blood removal, as a blood donation, for the purpose of ridding the body of excess iron.

**thermic effect of food (TEF)** The increase in metabolism that occurs during the digestion, absorption, and metabolism of energy-yielding nutrients. This typically represents 8% to 15% of calories consumed. Also called *diet-induced thermogenesis.*

**thrifty metabolism** A genetic tendency toward efficient use of energy that results in below-average energy requirements and increased storage of calories as fat.

**thyroid hormones** Hormones produced by the thyroid gland that regulate growth and metabolic rate.

**tissue saturation** The limited storage capacity of water-soluble vitamins in the tissues.

**tissues** Collections of cells adapted to perform a specific function.

**Tolerable Upper Intake Level (UL)** Maximum chronic daily intake level of a nutrient that is unlikely to cause adverse health effects in almost all people in a specific life stage.

**tolerance** Needing more of a substance to achieve the desired effect (e.g., intoxication) or experiencing diminished effects of a given amount of a substance after repeated use.

**toxicity** Capacity of a substance to produce injury or illness at some dosage.

**toxicology** Scientific study of harmful substances.

**toxin-mediated infection** Occurs when a person eats food containing harmful bacteria. While in the intestinal tract, the bacteria produce toxins that cause illness.

**toxins** Poisonous compounds produced by an organism that can cause disease.

**trace mineral** Vital to health, a mineral that is required in the dietary pattern in amounts less than 100 milligrams per day.

**trachea** The airway that extends from the throat, down the neck, to the lungs; also called the *windpipe.*

**trans fatty acid** A form of an unsaturated fatty acid, usually a monounsaturated one when found in food, in which the hydrogens lie on opposite sides of the carbon-carbon double bond.

**transcription** The process by which the code or gene for a protein on a DNA sequence is copied into a single-stranded mRNA molecule that is ready to leave the nucleus.

**transferrin** Iron-binding protein; controls the level of free iron in blood.

**translation** The process of adding amino acids one at a time to a growing polypeptide chain, according to the instructions on the mRNA.

**transverse colon** Segment of the large intestine that carries feces from the ascending colon, from right to left across the top of the abdomen, to the descending colon.

**triglyceride** The major form of lipid in the body and in food. It is composed of three fatty acids attached to glycerol.

**trimesters** Three 13- to 14-week periods into which the normal pregnancy (on average, 40 weeks) is divided somewhat arbitrarily for purposes of discussion and analysis. Development of the offspring, however, is continuous throughout pregnancy, with no specific physiological markers demarcating the transition from one trimester to the next.

**trypsin** A protein-digesting enzyme secreted by the pancreas to act in the small intestine.

**tumor** Mass of cells; may be cancerous (malignant) or noncancerous (benign).

**type 1 diabetes** A form of diabetes characterized by total insulin deficiency due to destruction of insulin-producing cells of the pancreas. Insulin therapy is required.

**type 1 osteoporosis** Porous trabecular bone characterized by rapid bone demineralization following menopause.

**type 2 diabetes** A form of diabetes characterized by insulin resistance and often associated with obesity. Insulin therapy may be required in advanced stages of the disease.

**type 2 osteoporosis** Porous trabecular and cortical bone observed in men and women after the age of 70.

**ultra-high temperature (UHT) processing** Method of sterilizing food by heating it above 275°F (135°C) for 2 to 5 seconds. Also called *ultra-heat treatment.*

**ultratrace mineral** A mineral present in the human diet in trace amounts but that has not been shown to be essential to human health.

**umami** A brothy, meaty, savory flavor in some foods. Monosodium glutamate enhances this flavor when added to foods.

**undernutrition** Failing health that results from a long-standing dietary intake that is suboptimal and does not meet nutritional needs.

**underwater weighing** A method of estimating total body fat by weighing the individual on a standard scale, then weighing him or her again once submerged in water. The difference between the two weights is used to estimate total body volume. Also known as *hydrostatic weighing* or *hydrodensitometry.*

**underweight** Ratio of weight to height that is lower than what is associated with optimal health. For adults, this is defined as BMI less than 18.5. For children, this is defined as BMI-for-age below the 5th percentile.

**upper-body obesity** The type of obesity in which fat is stored primarily in the abdominal area; defined as a waist circumference more than 40 inches (102 centimeters) in men and more than 35 inches (88 centimeters) in women; closely associated with a high risk for cardiovascular disease, hypertension, and type 2 diabetes. Also known as *android obesity, visceral obesity,* or *central obesity.*

**urea** Nitrogenous waste product of protein metabolism; major source of nitrogen in the urine.

**ureter** Tube that transports urine from the kidney to the urinary bladder.

**urethra** Tube that transports urine from the urinary bladder to the outside of the body.

**urinary system** The body system consisting of the kidneys, urinary bladder, and the ducts that carry urine. This system removes waste products from the circulatory system and regulates blood acid–base balance, overall chemical balance, and water balance in the body.

**vasoconstriction** The narrowing of blood vessels. As part of hemostasis, this temporarily reduces the flow of blood to the site of injury while a clot is being formed.

**vegan** Referring to a dietary pattern that only includes foods of plant origin.

**vegetarian** Referring to a dietary pattern that includes primarily foods of plant origin.

**vein** A blood vessel that carries blood to the heart.

**very-low-calorie diet (VLCD)** This diet allows a person fewer than 800 kcal per day, often in liquid form. Of this, 120 to 480 kcal are typically from carbohydrate, and the rest are mostly from high-quality protein.

**very-low-density lipoprotein (VLDL)** Lipoprotein that carries triglycerides (and cholesterol) from the liver to cells thoughout the body.

**vigorous-intensity aerobic physical activity** Aerobic activity that greatly increases a person's heart rate and breathing. Examples include jogging, tennis, swimming, or bicycling uphill.

**villi** (singular, villus) The fingerlike protrusions into the small intestine that participate in digestion and absorption of food.

**virus** One of the smallest known types of infectious agents, many of which cause disease in humans. A virus is essentially a piece of genetic material surrounded by a coat of protein. Viruses do not metabolize, grow, or move by themselves. They reproduce only with the aid of a living cellular host.

**vitamin** An essential organic (carbon-containing) compound needed in small amounts in the dietary pattern to help regulate and support chemical reactions and processes in the body.

**vitamin D$_2$** Form found in nonanimal sources, such as in some mushrooms. Also synthetically produced and included in many supplements. Also called *ergocalciferol.*

**vitamin D$_3$** Previtamin form synthesized in the skin and found naturally in some animal sources, including fish and egg yolks; also called *cholecalciferol.*

**vitamin K deficiency bleeding** Hemorrhage caused by a lack of vitamin K, especially among infants from birth to 6 months of age; also called *hemorrhagic disease of the newborn.*

**wasting** Low weight for height (thinness) that typically indicates a recent and severe process of weight loss, often associated with acute starvation or severe disease.

**water** The universal solvent; chemically, $H_2O$. The body is composed of about 60% water. Water (fluid) needs are about 9 (women) or 13 (men) cups per day; needs are greater if one exercises heavily.

**water intoxication** Potentially fatal condition that occurs with a high intake of water, which results in a severe dilution of the blood and other fluid compartments.

**water-soluble** Capable of dissolving in water.

**water-soluble vitamins** Vitamins that dissolve in water. These vitamins are the B vitamins and vitamin C.

**weight bias** Negative attitudes toward, and beliefs about, others because of their weight that are often manifested by stereotypes or prejudice toward people with overweight and obesity.

**weaning** The gradual process of switching from breast milk or infant formula to other sources of nutrition for the infant.

**white blood cells** Variety of immune cells that circulate in the lymph and blood and work to neutralize, detoxify, and/or destroy pathogens and other foreign proteins; also called *leukocytes.*

**whole grains** Grains containing the entire seed of the plant, including the bran, germ, and endosperm (starchy interior). Examples are whole wheat bread and brown rice.

**Wilson's disease** A genetic disorder that results in accumulation of copper in the tissues; characterized by damage to the liver, nervous system, and other organs.

**withdrawal** Physical symptoms related to cessation of substance use, such as sweating, rapid pulse, shakiness, insomnia, nausea and vomiting, anxiety, and even seizures.

**xerophthalmia** Hardening of the cornea and drying of the surface of the eye, which can result in blindness as a result of vitamin A deficiency.

**xylitol** Alcohol derivative of the five-carbon monosaccharide xylose. Absorbed more slowly than sucrose, xylitol supplies 40% fewer calories than table sugar.

**zoochemicals** Chemicals found in animal products that have health-protective actions.

**zygote** The fertilized ovum; the cell resulting from the union of an egg cell (ovum) and sperm until it divides.

# Index

Page numbers followed by *f* and *t* indicate figures and tables, respectively.

## A

Gastrointestinal (GI) tract
  absorption along, 99*t*
  alcohol consumption and, 30*f*
  defined, 90
  health, 103–104
  lymphatic circulation in, 84
  motility, 107
  peristalsis, 93, 93*f*, 109
  physiology of, 90*f*
  portal circulation in, 83
Gelatin, 203
Gels/chews, carbohydrates, 435, 435*t*
Gender differences. *See* Men; Women
Gender equality, 544–545
Gender expression, 59
Gender identity, 59
Gene expression, 78
Generally recognized as safe (GRAS) list,
  510–511
Generally recognized as safe (GRAS)
  substance, 379
Generational poverty, 534, 535
Genes, 13, 78, 200, 224
Genetic Alliance, website, 224
Genetically modified organisms (GMOs), 484,
  485, 486, 547
Genetic engineering, 484, 547
Genetic Information Nondiscrimination Act,
  224
Genetics
  aging and, 652
  eating disorders and, 451, 452
  insulin resistance and, 154
  medical family tree, 224, 225*f*
  nutrigenetics, 222, 222*f*
  nutrigenomics, 222–223, 222*f*
  nutrition and, 202, 222–225
  profile, 224, 225*f*
  type 2 diabetes and, 152
  on weight management, 250–252
Genetic Science Learning Center, website,
  224
Genetic testing, 224, 225*f*
Genogram, 224, 225*f*
Gestation, 564
  brain growth and, 674
Gestational diabetes, 150, 581
Gestational hypertension, 581–582
Ghrelin, 88
Global Hunger Index (GHI), 538
Global maternal mortality ratio, 538
Global nutrition, 525–553
Glossitis, 308
Glucagon, 88, 141–142, 142*f*
β-glucans, 124
Glucose. *See also* Blood glucose
  active absorption, 139
  from amino acids, 211
  anaerobic breakdown, 417, 417*f*
  defined, 9
  in disaccharides, 121
  formation of, 121

honey and, 133
  as simple sugar, 10, 11
Glucose tolerance test, 151, 151*f*
Glutathione peroxidase, 390
Gluten, 112, 207
Gluten-free casein-free (GFCF) diet, 627
Gluten-free diet, 114
Gluten sensitivity/intolerance, 113, 207
Glycemic index (GI), 143, 152, 636
Glycemic load (GL), 143, 152
Glycerol, 160
Glycogen, 120, 123*f*, 124, 417, 432–434, 436
Glycosylation, 651
Goat's milk, 617
Goiter, 392, 393, 528
Golgi complex, 76*f*, 78
Good cholesterol. *See* High-density
  lipoprotein (HDL)
Gooseflesh appearance, 289
Gout, 661
Grab-and-go food options, 632
Graham crackers, 184
Grains. *See also* Whole grains
  as B vitamin source, 304–305
  fortification, 318, 547, 590
Grapes, 657
Grazing, 466
Great Lakes Region, goiter and, 392
Green revolution, 539
Greens, 575
Growth and development
  assessment of, 600–604, 601*f*
  of infants and children, 601–603, 601*f*, 602*f*,
    603*t*
  overview, 599
  prenatal, 561–564
  undernutrition, impact on, 600–601
Growth charts
  CDC, 601, 601*f*, 602*f*, 603
  defined, 601
  WHO, 601–602
Growth spurts, 634
Gruel, 220
Gums, 124, 125
Gut-associated lymphoid tissues (GALT), 89,
  104
Gut-brain axis, 675
Gut microbiota, 102, 102*f*, 240
  anorexia nervosa and, 458
  brain health and, 675
  components of, 102–103, 103*f*
  heart health and, 192
  iron and, 384
  protein sources on, 209
  vitamins and, 306
Gut transit time, 334
Gynoid obesity, 249–250, 249*f*

# H

H₂ blockers, 107, 109
Habits, as factor in food choice, 5

Hair loss, 456
Hard water, 350, 351
Hazards, defined, 511
HDLs, 175, 175*t*, 176–177
Headaches, 674–675
Head circumference, brain growth and, 602
Health claims on food labels, 65–66
Health Food Financing Initiative of 2010
  (HFFI), 536
Health outcomes, diet quality on, 47
Healthy Eating Index (HEI), 43
Healthy eating patterns
  balancing calories within, 44
  customized food components, 44
  food components to limit, 44–45
Healthy, Hunger-Free Kids Act, 632
Healthy aging, 648–653, 649*f*
Healthy Drinks. Healthy Kids campaign, 632
Healthy eating patterns, 43, 48
Healthy Food Financing Initiative (HFFI), 536
Healthy food priority areas, 535
Healthy Mediterranean-Style Dietary Pattern,
  43
*Healthy People 2030* (HHS)
  features of, 23–24
  COVID-19 pandemic and, 23
  Nutrition and Healthy Eating objectives,
    23–24
  on physical activity for adults, 411
Healthy U.S.-Style Dietary Pattern, 43, 49*t*,
  619*t*, 621*t*, 628, 669*t*
Healthy Vegetarian Dietary Pattern, 43
Healthy weight, 24–25
Healthy weight gain, 271
Heart attacks, 56, 187
  calcium supplements and, 376, 377
  common triggers for, 189
  warning signs of, 187
Heartburn, 106–107, 106*f*, 579
  nutrition and lifestyle recommendations,
    107*t*
Heart defects, 590
Heart disease. *See* Cardiovascular disease
Heart rate, 413, 413*f*, 456
Heat cramps, 431*t*
Heat exhaustion, 431*t*
Heat-related illnesses, 431, 431*t*
Heatstroke, 350, 431*t*
Heavy drinking, 27, 29–32, 30*f*
*Helicobacter pylori,* 75, 109
Helminths, 505–507
Heme iron, 384
Hemicelluloses, 124, 124*f*
Hemochromatosis, 386–387
Hemoconcentration, 348
Hemoglobin, 150, 383
Hemoglobin A1c (HbA1c) test, 150
Hemolytic uremic syndrome (HUS), 500
Hemorrhage, 302
Hemorrhagic stroke, 182
Hemorrhoids, 580

Sterols, 163f, 164, 191
    defined, 160
    ergosterol, 294, 296
    sources of, 168–169, 169t
    in synthesis of hormones, 179
Stevia, 135t, 136
Stimulus control, 261
Stomach, 93–95, 94f
    absorption in, 94, 99t
    carbohydrate digestion and absorption, 137, 137f
    fat digestion, 172, 173f
    hunger sensations and, 256-257
    microbiota, 102f
    protein digestion/absorption, 208, 209f
Stone fruits, 253
Stoneground wheat, 129
Stool, 99
Storage, infant formula, 610
Strength training, 258, 271, 436–439, 663–664
Stress fractures, 429, 456
Stroke, 9, 187
Subclinical deficiency, 54
Submucosa, 96f
Subthreshold eating disorders, 451
Successful aging, 651
Suckling, 583, 584f. See also Breastfeeding
Sucralose, 135t, 136
Sucrase, 138
Sucrose, 10, 122f, 122, 132, 133
Sugar alcohols, 134
Sugarless gum, 136
Sugars. See also Added sugars
    defined, 121
    dietary quality and, 146–147
    digestion, 137–138
    in energy drinks, 430, 431t
    hyperactivity and, 148
    intake recommendations, 132, 146–149
    oral health and, 148–149
    overconsumption of, 146–149
    raw, 134
    types, 134
Sugar-sweetened beverages (SSBs), 22, 127, 133, 136
    decreasing consumption of, 149
Sulfites, 640
Sulfur, 354
Sunflower oil, 298
Sunlight, vitamin D and, 292, 292f, 294, 295, 374
Superfoods, 13
Supplemental Nutrition Assistance Program (SNAP), 531, 532t, 536, 568, 633, 671
Supplements. See Dietary supplements
The Surgeon General's Call to Action to Control Hypertension (Call to Action), 403
Sustainability, 7
Sustainable agriculture, 482, 483, 487–488, 539, 545–547, 547f
Sustainable development, 545

Sustainable Development Goals (United Nations), 525, 541–542, 543, 543f
Sustainable diets, 243
Sustainable intensification, 545
Sustainable seafood, 488
Sustainable world food output, 537–538
Swallowing process, 92–93, 92f
Sweets, in diabetes menu planning, A–7–A–9
Symptoms, defined, 54
Synapse, 86
Synbiotics, 102
Syndrome X, 153–154, 153f
Synthetic chemicals, 511
Synthetic vitamins, 281
Systematic review, 18–19, 19f
Systolic blood pressure, 359

## T

Table salt, 357, 358. See also Sodium
Table sugar (sucrose), 10, 121, 122f, 122
Talk test, 414
Tapering, 432
Tapeworms, 506t, 507
Tap water, 351
Target heart rate (THR), 413
Taurine, 327
Tears, 346
Teenagers. See also Adolescents
    sugar overconsumption in, 145
Temperature
    foodborne illnesses and, 503, 504, 517–519
Tequila, 134
Teratogen, 590
Terrorism, 495
Tetany, 364
Tetrodotoxin, 513t
Texture of foods, 5, 170
"The Clean 15," 494
"The Dirty Dozen," 494
Theory, defined, 18
Therapeutic phlebotomy, 387
Thermic effect of food (TEF), 242
Thermogenesis, 240, 242
    adaptive, 240
    non-shivering, 243
    shivering, 243
    thermic effect of food (TEF), 242
Thiamin. See Vitamin B-1 (thiamin)
Thiaminase, 512, 513t
Third trimesters, 564
Thirst, 349, 350, 429
3500-kcal rule, 253
Thrifty metabolism, 250, 652
Thyroid hormones, 89, 252, 390, 392
Time, as factor in food choice, 6–7
Tissues, 78
Tissue saturation, 282
Toadskin appearance, 289
Tobacco use
    chronic diseases and, 25
    vitamin C and, 323

Tocopherols. See Vitamin E
Toddlers, nutrition concerns for, 618–627. See also Children
Healthy U.S.-Style Dietary Pattern, 619t, 621t, 628Tolerable Upper Intake Levels (ULs)
    calcium, 376
    chloride, 363
    choline, 327
    chromium, 398
    defined, 57
    Dietary Reference Intakes and, A–34, A–35
    fluoride, 396
    folate, 318
    iodine, 393
    iron, 386
    magnesium, 382
    manganese, 398
    niacin, 310
    phosphorus, 380
    selenium, 391
    use of, 58, 59
    vitamin A, 290, 624
    vitamin B-6, 314
    vitamin C, 323
    vitamin D, 297
    vitamin E, 300
    zinc, 389, 624
Tolerance, to alcohol use, 30
Tomatoes, 13, 14, 282
Tooth loss, 456, 463. See also Dental caries
Toothpaste, dental caries and, 149
Topamirate (Topamax®), 467
Total calorie intake, calculation of, 16
Total diet, 25
Total fat, 180–181
Total vegetarians. See Vegans
Toxicity, defined, 12, 511
Toxicology, defined, 511
Toxin-mediated infections, 500
Toxins, defined, 496
Toxoplasma gondii, 506t
Trace minerals, 354
    defined, 12
Trachea, 92, 92f
Training diet case study, 440
Training effect, 420–421
Transcription, 201
trans fatty acids, 161, 162f, 169, 171, 181
    fertility and, 560
    tips for cutting back on, 185t
Transferrin, 356, 383
Transfer RNA (tRNA), 201–202
Transgender individuals, 59
Translation, 201
Transverse colon, 98, 98f
Treadmills, 259
Trichinella spiralis, 507, 507t
Triglycerides, 162–163
    chemical forms, 163f
    defined, 160
    as energy source, 419